Probability on Trees and Networks

Starting around the late 1950s, several research communities began relating the geometry of graphs to stochastic processes on these graphs. This book, twenty years in the making, ties together research in the field, encompassing work on percolation, isoperimetric inequalities, eigenvalues, transition probabilities, and random walks. Written by two leading researchers, the text emphasizes intuition, while giving complete proofs and more than 850 exercises. Many recent developments, in which the authors have played a leading role, are discussed, including percolation on trees and Cayley graphs, uniform spanning forests, the mass-transport technique, and connections on random walks on graphs to embedding in Hilbert space.

   This state-of-the-art account of probability on networks will be indispensable for graduate students and researchers alike.

RUSSELL LYONS is James H. Rudy Professor of Mathematics at Indiana University, Bloomington. He obtained his PhD at the University of Michigan in 1983. He has written seminal papers concerning probability on trees and random spanning trees in networks, and written foundational papers on unimodular random networks and determinantal probability measures. Lyons was a Sloan Foundation Fellow and has been an Invited Speaker at the International Congress of Mathematicians and the Joint Mathematics Meetings. He is a Fellow of the American Mathematical Society.

YUVAL PERES obtained his PhD at the Hebrew University, Jerusalem in 1990. Later, he was a faculty member there and at the University of California, Berkeley, and a Principal Researcher at Microsoft Research. He has written more than 300 research papers in probability, ergodic theory, analysis, and theoretical computer science. Peres has coauthored books on Brownian motion and Markov chain mixing times. He was awarded the Rollo Davidson Prize in 1995, the Loève Prize in 2001, and the David P. Robbins Prize in 2011. He was elected to the National Academy of Sciences in 2016.

A love and respect of trees has been characteristic of mankind since the beginning of human evolution. Instinctively, we understood the importance of trees to our lives before we were able to ascribe reasons for our dependence on them.

– James and Louise Bush-Brown, *America's Garden Book*

This series of high-quality upper-division textbooks and expository monographs covers all aspects of stochastic applicable mathematics. The topics range from pure and applied statistics to probability theory, operations research, optimization, and mathematical programming. The books contain clear presentations of new developments in the field and also of the state of the art in classical methods. While emphasizing rigorous treatment of theoretical methods, the books also contain applications and discussions of new techniques made possible by advances in computational practice.

A complete list of books in the series can be found at www.cambridge.org/statistics.
Recent titles include the following:

Probability on Trees and Networks

Russell Lyons
Indiana University, Bloomington

Yuval Peres
Kent State University

CAMBRIDGE
UNIVERSITY PRESS

CAMBRIDGE
UNIVERSITY PRESS

One Liberty Plaza, 20th Floor, New York, NY 10006, USA

Cambridge University Press is part of the University of Cambridge.

It furthers the University's mission by disseminating knowledge in the pursuit of education, learning, and research at the highest international levels of excellence.

www.cambridge.org
Information on this title: www.cambridge.org/9781107160156

First published 2016
Paperback edition with corrections 2021

Printed in the United Kingdom by TJ Books Limited, Padstow Cornwall

A catalogue record for this publication is available from the British Library.

Library of Congress Cataloging-in-Publication Data

Names: Lyons, Russell. | Peres, Y. (Yuval)
Title: Probability on trees and networks / Russell Lyons, Indiana University, Bloomington, Yuval Peres, Microsoft Research, Redmond, Washington.
Description: New York NY : Cambridge University Press, 2016. | Series: Cambridge series in statistical and probabilistic mathematics | Includes bibliographical references and index.
Identifiers:
LCCN 2016021811 | ISBN 9781107160156 (hardback : alk. paper)
Subjects: LCSH: Stochastic processes. | Trees (Graph theory)
Classification: LCC QA274 .L956 2016 | DDC 511/.52–dc23
LC record available at https://lccn.loc.gov/2016021811

ISBN 978-1-107-16015-6 Hardback
ISBN 978-1-108-73272-7 Paperback

Contents

Color plate section follows page xvi

Preface

This book is concerned with certain aspects of discrete probability on infinite graphs that are currently in vigorous development. Of course, finite graphs are analyzed as well, but usually with the aim of understanding infinite graphs and networks. These areas of discrete probability are full of interesting, beautiful, and surprising results, many of which connect to other areas of mathematics and theoretical computer science. Numerous fascinating questions are still open.

Our major topics include random walks and their intimate connection to electrical networks; uniform spanning trees, their limiting forests, and their marvelous relationships with random walks and electrical networks; branching processes; percolation and the powerful, elegant mass-transport technique; isoperimetric inequalities and how they relate to both random walks and percolation; minimal spanning trees and forests and their connections to percolation; Hausdorff dimension, capacity, and how to understand them via trees; and random walks on Galton–Watson trees. Connections among our topics are pervasive and rich, making for surprising and enjoyable proofs.

There are three main classes of graphs on which discrete probability is most interesting, namely, trees, Cayley graphs of groups (or, more generally, transitive, or even quasi-transitive, graphs), and planar graphs. More classical discrete probability has tended to focus on the special and important case of the Euclidean lattices, $\mathbb{Z}^d$, which are prototypical Cayley graphs. This book develops the general theory of various probabilistic processes on graphs and then specializes to the three broad classes listed, always seeing what we can say in the case of $\mathbb{Z}^d$.

Besides their intrinsic interest, there are several reasons for a special study of trees. Since in most cases, analysis is easier on trees, analysis can be carried further. Then one can often either apply the results from trees to other situations or transfer to other situations the techniques developed by working on trees. Trees also occur naturally in many situations, either combinatorially or as descriptions of compact sets in Euclidean space, $\mathbb{R}^d$.

In choosing our topics, we have been swayed by those results we find most striking as well as by those that do not require extensive background. Thus, the only prerequisite is basic knowledge of Markov chains and conditional expectation with respect to a σ-algebra. For Chapter 17, basic knowledge of ergodic theory is also required, though we review it there. Of course, we are highly biased by our own research interests and knowledge. We include the best proofs available of recent as well as classic results.

Most exercises that appear in the text, as opposed to those at the ends of the chapters, are ones that will be particularly helpful to do when they are reached. They either facilitate one's understanding or will be used later in the text. These in-text exercises are also collected at the end of each chapter for easy reference, just before additional exercises are presented. In each chapter, the additional exercises appear in the order that the corresponding material appears in the text.

Some general notation we use is $\langle \cdots \rangle$ for a sequence (or, sometimes, more general function), $\upharpoonright$ for the restriction of a function or measure to a set, $\mathbf{E}[X \, ; \, A]$ for the expectation of X on the event A, and $|\cdot|$ for the cardinality of a set. Also, "decreasing" will mean "nonincreasing" unless we say "strictly decreasing," and likewise "increasing" will mean "nondecreasing." Defined terms are in ***bold italics***. Some definitions are repeated in different chapters to enable more selective reading.

A question labeled as **Question** $m.n$ is one to which the answer is unknown (as of the time of original writing), where m and n are numbers. Unattributed results are usually not due to us. Items such as theorems are numbered in this book as $C.n$, where C is the chapter number and n is the item number in that chapter.

Major chapter dependencies are indicated in the first color plate. The plate section, which is between the preface and the first chapter, contains color figures that appear in text as grayscale. Such a figure in text is indicated by the words $\genfrac{}{}{0pt}{}{\text{color}}{\text{plate}}$ preceding its caption.

It is possible to choose only small parts of various chapters to make a coherent course on specific subjects. For example, a judicious choice of material from the following sections can be used for a one-semester course on relationships of probability to geometric group theory: 3.4, 7.1, 6.1–3, 6.6, 6.7, 13.1–2, 14.1–4, 5.1, 7.2–7, 8.1, 8.3, 8.4, 11.1–4, 11.6, 2.1–5, 6.9, 4.1, 4.2, 9.1, 9.3, 9.4, 10.1, 10.2, 10.9.

In the electronic version of this book, most symbols that are used with a fixed meaning are hyperlinked to their definitions, although the fact that such hyperlinks exist is not made visible.

Many exercises at varying levels of difficulty are included, with many comments, hints, or solutions in the back of the book.

This book began as lecture notes for an advanced graduate course called "Probability on Trees" that Lyons gave in Spring 1993, but its emphasis has been transformed over the intervening years. We are grateful to Rabi Bhattacharya for having suggested that he teach such a course. We have attempted to preserve the informal flavor of lectures.

After Peres joined as a coauthor, writing and research became intertwined, and many delays ensued. Over the course of many months together in Jerusalem, Berkeley, and Redmond, the authors planned the content of most chapters, but the great majority of the actual writing was done by Lyons. Exceptions include especially Chapters 13 and 14 as well as a few sections of other chapters that were mostly written by Peres. Several chapters are based on joint works with Itai Benjamini, Robin Pemantle, and Oded Schramm. A few of the authors' new results appear here for the first time; they are due to both authors in about equal measure. Lyons was responsible for all other aspects of authorship of the book, such as drawing figures, preparing the index, ensuring consistent notation, and typography; most remaining errors can be attributed to him.

Lyons is grateful to the Institute for Advanced Studies and the Institute of Mathematics, both at the Hebrew University of Jerusalem, and to Microsoft Research for support during some of the writing. We are grateful to Brian Barker, Jochen Geiger, Janko Gravner, Yiping (Kenneth) Hu, Svante Janson, Tri Minh Lai, Steve Morrow, Peter Mörters, Minwoo Park, Perla Sousi, Jason Schweinsberg, Jeff Steif, Pengfei Tang, and Ádám Timár for noting several corrections to the manuscript. Section 6.6 and much of Chapter 13 are based on lectures that Peres gave in Berkeley, which were scribed by Asaf Nachmias. In addition, Gábor Pete helped with editing a few sections and provided a careful reading and thoughtful comments throughout. Special thanks are due to Jacob Magnusson for his very thorough and careful reading, which uncovered many small mistakes and possible improvements.

The paperback edition of this book incorporates minor corrections, improvements, and updates.

Russell Lyons
Indiana University
rdlyons@indiana.edu
https://rdlyons.pages.iu.edu/

Yuval Peres
Kent State University
yuval@yuvalperes.com
http://yuvalperes.com

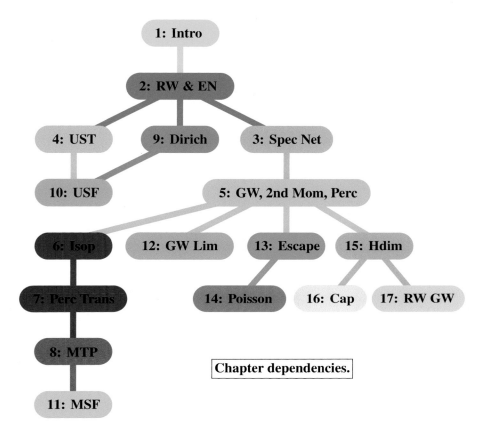

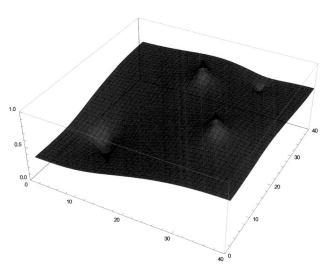

Figure 2.1. A harmonic function on a 40×40 square grid with four specified values where it is not harmonic.

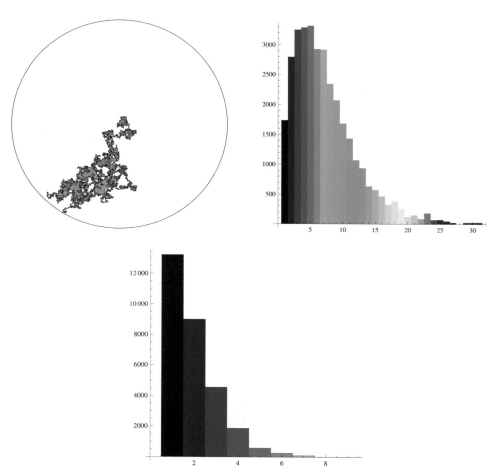

Figure 2.3. Random walk until it goes distance 200 from its starting point, colored according to the number of visits at a vertex. The histogram shows the time spent at vertices that were visited n times for each $n \geq 1$, with the same color coding. For three dimensions, only the histogram is shown, which is approximately a geometric distribution.

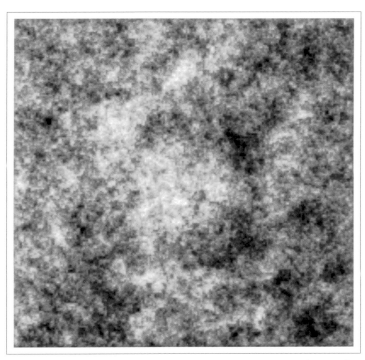

Figure 2.5. The net turns of the paths from a fixed vertex in a uniform spanning tree in a 200×200 grid on the left, with a key on the right showing the correspondence of visual colors to numbers.

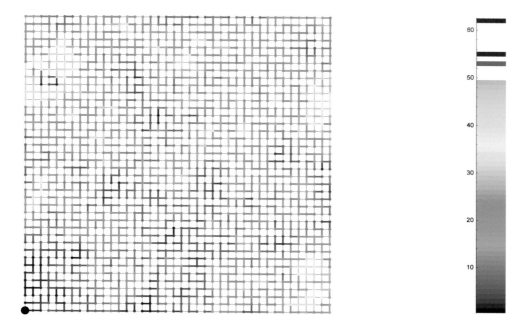

Figure 4.3. A colored uniform spanning tree in a 40×40 grid on the left, with a key on the right showing the correspondence of visual colors to numbered colors.

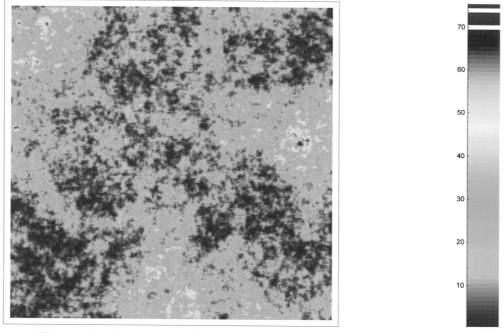

Figure 4.4. The colors of a uniform spanning tree in a 200×200 grid on the left, with a key on the right showing the correspondence of visual colors to numbered colors.

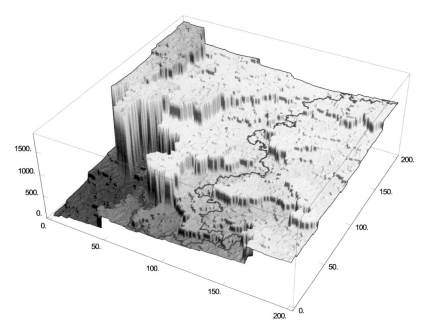

Figure 4.5. The distances to the root in a uniform spanning tree in a 200×200 grid, together with the path from the opposite corner.

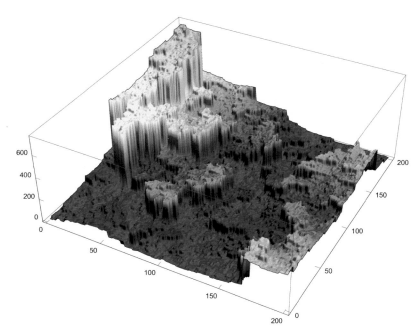

Figure 4.6. The distances in the tree to the path between opposite corners in a uniform spanning tree in a 200×200 grid.

Figure 4.12.   The Peano-like curve surrounding a uniform spanning tree on a 99×99 grid, with hue showing progress along the curve.

Figure 7.1.   Bernoulli bond percolation on a 40×40 square grid graph at levels $p = 0.4, 0.5, 0.6$. Each cluster is given a different color.

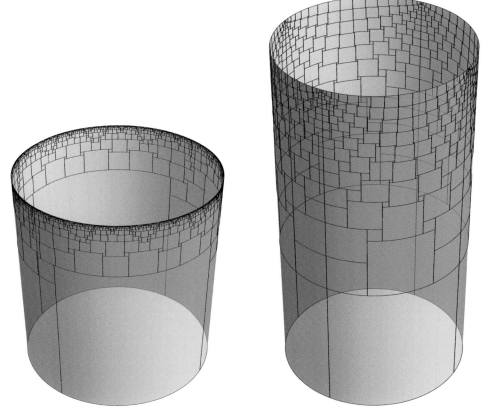

Figure 9.4. Tilings of cylinders by squares corresponding to the (2,3,7)-triangle tessellation of the hyperbolic plane on the left and the 21×21 grid with current from its center vertex to its boundary vertices on the right.

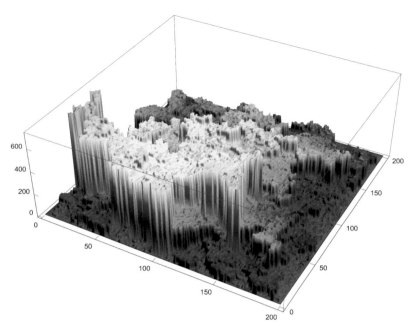

Figure 10.2. The distances to the outer boundary in a uniform spanning tree of a wired 200×200 square grid.

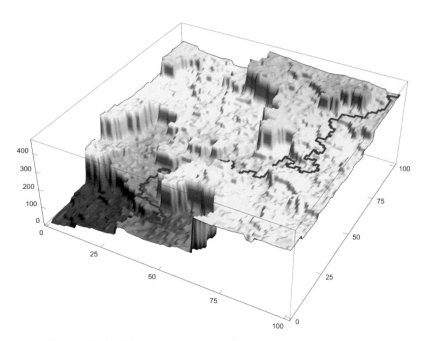

Figure 11.1. The distances to a vertex in a minimal spanning tree in a 100×100 grid and the path between the opposite corners.

1 | Some Highlights

This chapter surveys a few of the highlights to be encountered in this book, mainly, Chapters 2, 3, 4, 5, 15, and 16. Several of the topics in the book do not appear at all here since they are not as suitable to a quick overview. Also, we concentrate in this overview on trees, since it is easiest to use them to illustrate many of our themes.

1.1 Graph Terminology

For later reference, we introduce in this section the basic notation and terminology for graphs. A *graph* is a pair $G = (\mathsf{V}, \mathsf{E})$, where V is a set of *vertices* and E is a symmetric irreflexive subset of $\mathsf{V} \times \mathsf{V}$, called the *edge* set. *Irreflexive* means that E contains no element of the form (x, x). The word *symmetric* means that $(x, y) \in \mathsf{E}$ iff $(y, x) \in \mathsf{E}$; here, x and y are called the *endpoints* of (x, y). The symmetry assumption is usually phrased by saying that the graph is *undirected* or that its edges are *unoriented*. Without this symmetry assumption, the graph is called *directed*. If we need to distinguish the two, we write an unoriented edge as $[x, y]$, whereas an oriented edge is written as $\langle x, y \rangle$. An unoriented edge can be thought of as the pair of oriented edges with the same endpoints. If $(x, y) \in \mathsf{E}$, then we call x and y *adjacent* or *neighbors*, and we write $x \sim y$. The *degree* of a vertex is the number of its neighbors. If this is finite for each vertex, we call the graph *locally finite*. If the degree of every vertex is the same number d, then the graph is called *regular* or *d-regular*. If x is an endpoint of an edge e, then we also say that x and e are *incident*, whereas if two edges share an endpoint, then we call those edges *adjacent*. If we have more than one graph under consideration, we distinguish the vertex and edge sets by writing $\mathsf{V}(G)$ and $\mathsf{E}(G)$. A *subgraph* of a graph G is a graph whose vertex set is a subset of $\mathsf{V}(G)$ and whose edge set is a subset of $\mathsf{E}(G)$. One can define the product of two graphs $G_i = (\mathsf{V}_i, \mathsf{E}_i)$ $(i = 1, 2)$ in various ways. The one we use almost exclusively is the *Cartesian product* $G = (\mathsf{V}, \mathsf{E})$ with $\mathsf{V} := \mathsf{V}_1 \times \mathsf{V}_2$ and

$$\mathsf{E} := \left\{ \big((x_1, x_2), (y_1, y_2) \big) ; \ \big(x_1 = y_1, \ (x_2, y_2) \in \mathsf{E}_2 \big) \text{ or } \big((x_1, y_1) \in \mathsf{E}_1, \ x_2 = y_2 \big) \right\};$$

this product graph is denoted $G = G_1 \,\square\, G_2$.

A *path** in a graph is a sequence of vertices where each successive pair of vertices is an edge in the graph; it is said to *join* its first and last vertices. When a path does not pass

---

* In graph theory, a path is necessarily self-avoiding. What we call a path is called in graph theory a *walk*. However, to avoid confusion with random walks, we do not adopt that terminology.

through any vertex (resp., edge) more than once, we will call it **vertex simple** (resp., **edge simple**). We'll just say **simple** also to mean vertex simple, which implies edge simple. A finite path with at least one edge and whose first and last vertices are the same is called a **cycle**. A cycle is called **simple** if no pair of vertices are the same except for its first and last ones. A graph is **connected** if, for each pair $x \neq y$ of its vertices, there is a path joining x to y. The **distance** between x and y is the minimum number of edges among all paths joining x and y, denoted either $d(x, y)$ or $\mathrm{dist}(x, y)$. The **distance** between a vertex x and an edge e is the minimum number of edges on a path that includes both x and e. A graph with no cycles is called a **forest**; a connected forest is a **tree**.

If there are numbers (weights) $c(e)$ assigned to the edges e of a graph, the resulting object is called a **network**. Given a network $G = (V, E)$ with weights $c(\cdot)$ and a subset K of its vertices, the **induced subnetwork** $G \restriction K$ is the subnetwork with vertex set K, edge set $(K \times K) \cap E$, and weights $c \restriction ((K \times K) \cap E)$.

Sometimes we work with objects more general than graphs, called multigraphs. A **multigraph** is a pair of sets, V and E, together with a pair of maps $E \to V$, denoted $e \mapsto e^-$ and $e \mapsto e^+$. The images of e are called the **endpoints** of e, the former being its **tail** and the latter its **head**. If $e^- = e^+ = x$, then e is a **loop** at x. Edges with the same set of endpoints are called **parallel** or **multiple**. If the multigraph is undirected, then for every edge $e \in E$, there is an edge $-e \in E$ such that $(-e)^- = e^+$ and $(-e)^+ = e^-$. For a vertex x of an undirected multigraph, its **degree** is $|\{e \,;\, e^- = x\}|$. Sometimes we use **paths** of edges rather than of vertices; in this case, the head of each edge must equal the tail of the next edge. Given a subset $K \subseteq V$, the multigraph G/K obtained by **identifying** K to a single vertex $z \notin V$ is the multigraph whose vertex set is $(V \setminus K) \cup \{z\}$ and whose edge set is obtained from E by replacing the tail and head maps so that every tail or head that took a value in K now takes the value z. A similar operation is **contraction** of an edge e, which is the result of first deleting e and then identifying e^- and e^+; we denote this graph by G/e. A multigraph that is a graph is called a **simple graph**, and a network on a simple graph is called a **simple network**.

Let $G_1 = (V_1, E_1)$ and $G_2 = (V_2, E_2)$ be two (multi)graphs. A **homomorphism** of G_1 to G_2 is a map $\phi\colon G_1 \to G_2$ such that whenever x and e are incident in G_1, then so are $\phi(x)$ and $\phi(e)$ in G_2. When the graph is directed, then ϕ must also preserve orientation of edges, that is, if the head and tail of e are x and y, respectively, then the head and tail of $\phi(e)$ must be $\phi(x)$ and $\phi(y)$, respectively. If in addition, these graphs come with weight functions c_1 and c_2, so that they are networks, then a **network homomorphism** is a graph homomorphism ϕ that satisfies $c_1(e) = c_2(\phi(e))$ for all edges $e \in E_1$. If ϕ induces bijections of V_1 to V_2 and of E_1 to E_2, then ϕ is called an **isomorphism**. When $G_1 = G_2$, an isomorphism is called an **automorphism**. A homomorphism $\phi\colon G_1 \to G_2$ extends to map each subset A of G_1 to a subset $\phi(A)$ of G_2 by mapping all elements of A by ϕ. We also extend ϕ to collections $\mathcal{A}$ of subsets of G_1 by applying ϕ to all elements of $\mathcal{A}$.

1.2 Branching Number

Our trees will usually be **rooted**, meaning that some vertex is designated as the root, denoted o. We imagine the tree as growing (upward) away from its root. Each vertex then has branches leading to its children, which are its neighbors that are farther from the root. For the purposes of this chapter, we do not allow the possibility of vertices without children.

How do we assign an average branching number to an arbitrary infinite, locally finite tree? If the tree is a binary tree, as in Figure 1.1, then clearly the answer will be 2. But in the general case, since the tree is infinite, no straight average is available. We must take some kind of limit or use some other procedure, but we will be amply rewarded for our efforts.

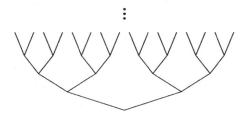

Figure 1.1. The binary tree.

One simple idea is as follows. Let T_n be the set of vertices at distance n from the root, o, called the *nth level* of T. Define the *lower (exponential) growth rate* of the tree to be

$$\underline{\mathrm{gr}}\, T := \liminf_{n \to \infty} |T_n|^{1/n} \,.$$

This certainly will give the number 2 to the binary tree. One can also define the *upper (exponential) growth rate*

$$\overline{\mathrm{gr}}\, T := \limsup_{n \to \infty} |T_n|^{1/n}$$

and the *(exponential) growth rate*

$$\mathrm{gr}\, T := \lim_{n \to \infty} |T_n|^{1/n}$$

when the limit exists. However, notice that these notions of growth barely account for the structure of the tree: only $|T_n|$ matters, not how the vertices at different levels are connected to each other. Of course, if T is *spherically symmetric*, meaning that for each n, every vertex at distance n from the root has the same number of children (which may depend on n), then there is really no more information in the tree than that contained in the sequence $\langle |T_n| \,;\, n \geq 0 \rangle$. For more general trees, however, we will use a different approach.

Consider the tree as a network of pipes and imagine water entering the network at the root. However much water enters a pipe leaves at the other end and splits up among the outgoing pipes (edges). Formally, this means that we consider a nonnegative function θ on the edges of T, called a *flow*, with the property that for every vertex x other than the root, if x has parent z and children $y_1, \ldots, y_d$, then $\theta((z, x)) = \sum_{i=1}^d \theta((x, y_i))$. We say that $\theta(e)$ is the amount of water flowing along e and that the total amount of water flowing from the root to infinity is $\sum_{j=1}^k \theta((o, x_j))$, where the children of the root o are $x_1, \ldots, x_k$.

Consider the following sort of restriction on a flow: given $\lambda \geq 1$, suppose that the amount of water that can flow through an edge at distance n from o is only λ^{-n}. In other words, if $x \in T_n$ has parent z, then the restriction is that $\theta((z, x)) \leq \lambda^{-n}$. If λ is too big, then perhaps no positive amount of water can flow from the root to infinity. Indeed, consider the binary tree. Then the *equally splitting flow* that sends an amount 2^{-n} through each edge at distance n from the root will satisfy the restriction imposed when $\lambda \leq 2$ but not for any $\lambda > 2$. In fact, it is intuitively clear that there is no way to get any water to flow when $\lambda > 2$. Obviously, this critical value of 2 for λ is the same as the branching number of the binary tree – if the tree were ternary, then the critical value would be 3. So let us make a general definition: the **branching number** of a tree T is the supremum of those λ that admit a positive total amount of water to flow through T; denote this critical value of λ by $\operatorname{br} T$.

Let's spend some time on this new concept. For a vertex x other than the root, let $e(x)$ denote the edge that joins x to its parent. The total amount of water flowing is, by definition, $\sum_{x \in T_1} \theta(e(x))$. If we apply the flow condition to each x in T_1, then we see that this sum also equals $\sum_{x \in T_2} \theta(e(x))$. Induction shows, in fact, that it equals $\sum_{x \in T_n} \theta(e(x))$ for every $n \geq 1$. When the flow is constrained in the way we have specified, then this sum is at most $\sum_{x \in T_n} \lambda^{-n} = |T_n| \lambda^{-n}$. Now if we choose $\lambda > \operatorname{\underline{gr}} T$, then $\liminf_{n \to \infty} |T_n| \lambda^{-n} = 0$, whence for such λ, no water can flow. Conclusion:

$$\operatorname{br} T \leq \operatorname{\underline{gr}} T . \tag{1.1}$$

Often, as in the case of the binary tree, equality holds here. However, there are many examples of strict inequality.

Before we give an example of strict inequality, here is another case of equality in (1.1).

Example 1.1. If T is a tree such that vertices at even distances from o have two children whereas the rest have three children, then $\operatorname{br} T = \operatorname{gr} T = \sqrt{6}$. Why? It is easy to see that $\operatorname{gr} T = \sqrt{6}$, whence by (1.1), it remains to show that $\operatorname{br} T \geq \sqrt{6}$. In other words, it remains to show that, given $\lambda < \sqrt{6}$, a positive amount of water can flow to infinity under the constraints described. Indeed, we can use the water flow with amount $6^{-n/2}$ flowing on those edges at distance n from the root when n is odd and with amount $6^{-(n-1)/2}/3$ flowing on those edges at distance n from the root when n is even.

More generally, one can show (Exercise 1.2) that equality holds in (1.1) whenever T is spherically symmetric.

Now we give an example where strict inequality holds in (1.1).

Example 1.2. (The 1–3 Tree) We will construct a tree T embedded in the upper half-plane with o at the origin. We'll have $|T_n| = 2^n$, but we'll connect them in a funny way. List T_n in clockwise order as $\langle x_1^n, \ldots, x_{2^n}^n \rangle$. Let x_k^n have one child if $k \leq 2^{n-1}$ and three children otherwise; see Figure 1.2. Define a **ray** in a tree to be an infinite path from the root that doesn't backtrack. If x is a vertex of T that does not have the form $x_{2^n}^n$, then there are only finitely many rays that pass through x. This means that water cannot flow to infinity through such a vertex x when $\lambda > 1$. That leaves only the possibility of water flowing along the single ray consisting of the vertices $x_{2^n}^n$, but that's impossible too. Hence $\operatorname{br} T = 1$, yet $\operatorname{gr} T = 2$.

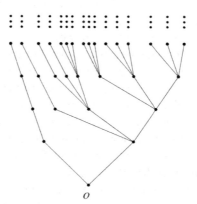

o

Figure 1.2. A tree with branching number 1 and growth rate 2.

Example 1.3. If $T^{(1)}$ and $T^{(2)}$ are trees, form a new tree $T^{(1)} \vee T^{(2)}$ from disjoint copies of $T^{(1)}$ and $T^{(2)}$ by joining their roots to a new point taken as the root of $T^{(1)} \vee T^{(2)}$ (Figure 1.3). Then

$$\mathrm{br}\!\left(T^{(1)} \vee T^{(2)}\right) = \mathrm{br}\, T^{(1)} \vee \mathrm{br}\, T^{(2)}$$

since water can flow in the join $T^{(1)} \vee T^{(2)}$ iff water can flow in one of the trees. Here, as usual in probability, we use $a \vee b$ to mean $\max\{a, b\}$ when a and b are real numbers.

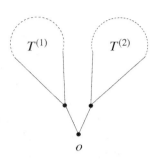

Figure 1.3. Joining two trees.

Although $\mathrm{gr}\, T$ is easy to compute, $\mathrm{br}\, T$ may not be. Nevertheless, it is the branching number that is much more important. Theorems to be described shortly will bear out this assertion. We will develop tools to compute $\mathrm{br}\, T$ in many common situations.

1.3 Electric Current

We can ask another flow question on trees, this one concerning electrical current. All electrical terms are given precise mathematical definitions in Chapter 2, but for now, we give some bare definitions to sketch the arc of some of the fascinating and surprising connections that we develop later. If positive numbers $c(e)$ are assigned to the edges e of a tree, we may call these numbers **conductances**, and in that case, the **energy** of a flow θ is defined to be $\sum_e \theta(e)^2/c(e)$. We say that electrical **current flows** from the root to infinity if there is a nonzero flow with finite energy.

Here's our new flow question: if λ^{-n} is the conductance of edges at distance n from the root of T, will current flow?

Example 1.4. Consider the binary tree. The equally splitting flow has finite energy for every $\lambda < 2$, so in those cases, electrical current does flow. One can show that when $\lambda \geq 2$, not only does the equally splitting flow have infinite energy, but so does every nonzero flow (Exercise 1.4). Thus, current flows in the infinite binary tree iff $\lambda < 2$. Note the slight difference to water flow: when $\lambda = 2$, water can still flow on the binary tree.

In general, there will be a critical value of λ below which current flows and above which it does not. In the example of the binary tree that we just analyzed, this critical value was the same as that for water flow. Is this equality special to nice trees, or does it hold for all trees? We have seen an example of a strange tree (another is in Exercise 1.3), so we might doubt its generality. However, it is indeed a general fact (Lyons, 1990):

Theorem 1.5.* *If $\lambda < \mathrm{br}\, T$, then electrical current flows, but if $\lambda > \mathrm{br}\, T$, then it does not.*

---

* This will follow from Theorem 3.5 and the discussion in Section 2.2.

1.4 Random Walks

There is a striking, but easily established, correspondence between electrical networks and random walks on graphs (or on networks). Namely, given a finite, connected graph G with conductances (that is, positive numbers) assigned to the edges, we consider the random walk that can go from a vertex only to an adjacent vertex and whose transition probabilities from a vertex are proportional to the conductances along the edges to be taken. That is, if x is a vertex with neighbors $y_1, \ldots, y_d$ and the conductance of the edge (x, y_i) is c_i, then the transition probability from x to y_j is $p(x, y_j) := c_j / \sum_{i=1}^d c_i$. Now consider two fixed vertices a_0 and a_1 of G. A **voltage function** on the vertices is then a function v such that $v(a_i) = i$ for $i = 0, 1$ and for every other vertex $x \neq a_0, a_1$, the equation $v(x) \sum_{i=1}^d c_i = \sum_{i=1}^d c_i v(y_i)$ holds, where the neighbors of x are $y_1, \ldots, y_d$. In other words, $v(x)$ is a weighted average of the values at the neighbors of x. We will see in Section 2.1 that voltage functions exist and are unique. The following proposition provides the basic connection between random walks and electrical networks:

Proposition 1.6. (Voltage as Probability) *For every vertex x, the voltage at x equals the probability that when the corresponding random walk starts at x, it will visit a_1 before it visits a_0.*

The proof of this proposition will be simple: In outline, there is a discrete Laplacian (a difference operator) that will define a notion of harmonic function. Both the voltage and the probability mentioned are harmonic functions of x. The two functions clearly have the same values at a_i (the "boundary" points), and the uniqueness principle holds for this Laplacian, whence the functions agree at all vertices x. This is developed in detail in Section 2.1.

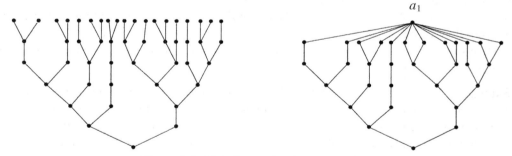

Figure 1.4. Identifying a level to a vertex, a_1.

What does this say about our trees? Given N, identify all the vertices of level N, that is, T_N, to one vertex, a_1 (see Figure 1.4). Use the root as a_0. Then, according to Proposition 1.6, the voltage at x is the probability that the random walk visits level N before it visits the root when it starts from x. When $N \to \infty$, the limiting voltages are all 0 iff the limiting probabilities are all 0, which is the same thing as saying that on the infinite tree, the probability of visiting the root from any vertex is 1, in other words, the random walk is recurrent. Although we have not yet defined "current," we'll see that no current flows across edges whose endpoints have the same voltage. This will imply, then, that no electrical current flows iff the random walk is recurrent. Contrapositively, electrical current flows iff the random walk is transient. In this way, electrical networks will be a powerful tool to help us decide whether a random walk is recurrent or transient. These ideas are detailed in Section 2.2.

Earlier we considered conductances λ^{-n} on edges at distance n from the root. In this case, the conductances decrease by a factor of λ as the distance increases by 1, so the relative weights at a vertex other than the root are as shown in Figure 1.5. That is, the edge leading back toward the root is λ times as likely to be taken as each edge leading away from the root. Denoting the dependence of the random walk on the parameter λ by RW_λ, we may translate Theorem 1.5 into a probabilistic form (Lyons, 1990):

Figure 1.5. The relative weights at a vertex. The tree is growing upwards.

Theorem 1.7.* *If $\lambda < \operatorname{br} T$, then RW_λ is transient, whereas if $\lambda > \operatorname{br} T$, then RW_λ is recurrent.*

Is this form intuitive? Consider a vertex other than the root with, say, d children. If we consider only the distance from o, which increases or decreases at each step of the random walk, a balance at this vertex between increasing and decreasing occurs when $\lambda = d$. If d is constant, then the distance from the root undergoes a random walk with a constant bias (for a fixed λ), so it is easy to see that indeed d is the critical value separating transience from recurrence. What Theorem 1.7 says is that this same heuristic can be used in the general case, provided we substitute the "average" $\operatorname{br} T$ for d.

We will also see how to use electrical networks to prove Pólya's wonderful, seminal theorem that simple random walk on the hypercubic lattice $\mathbb{Z}^d$ is recurrent for $d \le 2$ and transient for $d \ge 3$.

1.5 Percolation

Suppose that we remove edges at random from a tree, T. To be specific, we keep each edge with some fixed probability p and make these decisions independently for different edges. This random process is called **percolation**. As we'll see, by Kolmogorov's zero-one law, the probability that an infinite connected component remains in the tree is either 0 or 1. On the other hand, we'll see that this probability is monotonic in p, whence there is a critical value $p_c(T)$ where it changes from 0 to 1. It is also intuitively clear that the "bigger" the tree, the more likely it is that there will be an infinite component for a given p. That is, the "bigger" the tree, the smaller is the critical value p_c. Thus, p_c is vaguely inversely related to a notion of average size or maybe average branching number. Surprisingly, this vague heuristic is precise and general (Lyons, 1990):

---

* This will be proved as Theorem 3.5.

Theorem 1.8.[*] *For any tree, $p_c(T) = 1/\operatorname{br} T$.*

What is this telling us? If a vertex x has d children, then the expected number of children remaining after percolation is dp. If dp is "usually" less than 1, then one would not expect that an infinite component would remain, whereas if dp is "usually" greater than 1, then one might guess that an infinite component would be present somewhere. Theorem 1.8 says that this intuition becomes correct when one replaces the "usual" d by $\operatorname{br} T$. Both Theorems 1.5 and 1.8 say that the branching number of a tree is a single number that captures enough of the complexity of a general tree to give the critical value for a stochastic process on the tree. There are other examples as well of this striking phenomenon. Altogether, they make a convincing case that the branching number is indeed the most important single number to attach to an infinite tree.

1.6 Branching Processes

In the preceding section, we looked at existence of infinite components after percolation on a tree. Although this event has probability 0 or 1, if we restrict attention to the connected component of the root, its probability of being infinite is between 0 and 1. These are equivalent ways to approach the issue, since, as we'll see, there is an infinite component somewhere with probability 1 iff the component of the root is infinite with positive probability. But looking at the component of the root also suggests a different stochastic process.

Percolation on a fixed tree produces random trees by random pruning, but there is a way to *grow* trees randomly that was invented by Bienaymé in 1845. Given probabilities p_k adding to 1 ($k = 0, 1, 2, \ldots$), we begin with one individual, and let it reproduce according to these probabilities, that is, it has k children with probability p_k. Each of these children (if there are any) then reproduce independently with the same law, and so on forever or until some generation goes extinct. The family trees produced by such a process are called **(Bienaymé–)Galton–Watson trees**. A fundamental theorem in the subject (Proposition 5.4) is that extinction is a.s. iff $m \leq 1$ and $p_1 < 1$, where $m := \sum_k k p_k$ is the mean number of offspring per individual. This provides further justification for our intuitive understanding of Theorem 1.8. It also raises a natural question: Given that a Galton–Watson family tree is nonextinct (infinite), what is its branching number? All the intuition suggests that it is m a.s., and indeed it is. This was first proved by Hawkes (1981). But here is the idea of a very simple proof (Lyons, 1990).

According to Theorem 1.8, to determine $\operatorname{br} T$, we may determine $p_c(T)$. Thus, let T grow according to a Galton–Watson process, then perform percolation on T, that is, keep edges with probability p. Focus on the component of the root. Looked at as a random tree in itself, this component appears simply as some other Galton–Watson tree; its mean is mp by independence of the growing and the "pruning" (percolation). Hence, the component of the root is infinite with positive probability iff $mp > 1$. This implies that $p_c = 1/m$ a.s. on nonextinction, thus $\operatorname{br} T = m$ a.s. on nonextinction. We'll flesh out the details when we prove Proposition 5.9.

---

[*] This will be proved as Theorem 5.15.

Now let's consider another way to measure the size of Galton–Watson trees. Let Z_n be the size of the nth generation in a Galton–Watson process. How quickly does Z_n grow? It will be easy to calculate that $\mathbf{E}[Z_n] = m^n$. Moreover, a martingale argument will show that the limit $W := \lim_{n\to\infty} Z_n/m^n$ always exists (and is finite). When $1 < m < \infty$, do we have that $W > 0$ a.s. on the event of nonextinction? When $W > 0$, the growth rate of the tree is asymptotically Wm^n; this implies the cruder asymptotic $\mathrm{gr}\, T = m$. It turns out that indeed $W > 0$ a.s. on the event of nonextinction, under a very mild hypothesis:

The Kesten–Stigum Theorem (1966). *When $1 < m < \infty$, the following are equivalent:*
(i) *$W > 0$ a.s. on the event of nonextinction;*
(ii) *$\sum_{k=1}^{\infty} p_k k \log k < \infty$.*

This will be shown in Section 12.2. Although condition (ii) appears technical and suggests some possibly unpleasant analysis, we will enjoy a conceptual proof of the theorem that uses only extremely simple estimates.

1.7 Random Spanning Trees

The fertile and fascinating field of random spanning trees is one of the oldest areas to be studied in this book but one of the newest to be explored in depth. A ***spanning tree*** of a (connected) graph is a subgraph that is connected, contains every vertex of the whole graph, and contains no cycle: see Figure 1.6 for an example. These trees are usually not rooted. The subject of random spanning trees of a graph goes back to Kirchhoff (1847), who showed its

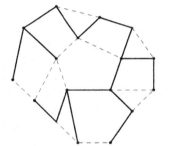

Figure 1.6. A spanning tree in a graph, where the edges of the graph not in the tree are dashed.

surprising relation to electrical networks. (Actually, Kirchhoff did not think probabilistically; rather, he considered quotients of the number of spanning trees with a certain property divided by the total number of spanning trees. See Kirchhoff's effective-resistance formula in Section 4.2 and Exercise 4.30.) One of Kirchhoff's results expresses the probability that a uniformly chosen spanning tree will contain a given edge in terms of electrical current in the graph.

To get our feet wet, let's begin with a very simple finite graph. Namely, consider the ladder graph of Figure 1.7. Among all spanning trees of this graph, what proportion contain the bottom rung (edge)? In other words, if we were

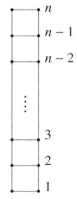

Figure 1.7. A ladder graph.

to choose uniformly at random a spanning tree, what is the chance that it would contain the bottom rung? We have illustrated in Figure 1.8 the entire probability spaces for the smallest ladder graphs.

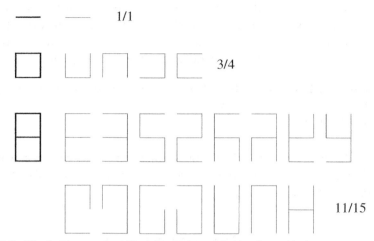

Figure 1.8. The ladder graphs of heights 0, 1, and 2, together with their spanning trees.

As shown, the probabilities in these cases are $1/1$, $3/4$, and $11/15$. The next one is $41/56$. Do you see any pattern? One thing that is fairly evident is that these numbers are decreasing but hardly changing. Amusingly, these numbers are every other term of the continued fraction expansion of $\sqrt{3} - 1 = 0.73^+$ and, in particular, converge to $\sqrt{3} - 1$. In the limit, then, the probability of using the bottom rung is $\sqrt{3} - 1$, and even before taking the limit, this gives an excellent approximation to the probability. How can we easily calculate such numbers? In this case, there is a rather easy recursion to set up and solve, but we will use this example to illustrate the more general theorem of Kirchhoff that we mentioned earlier. In fact, Kirchhoff's theorem will show us why these probabilities are decreasing even before we calculate them.

For the next two paragraphs, we will assume some familiarity with electrical networks; those who do not know these terms will find precise mathematical definitions in Sections 2.1 and 2.2. Suppose that each edge of our graph (any graph – say, the ladder graph) is an electric conductor of unit conductance. Hook up a battery between the endpoints of any edge e – say, the bottom rung (Figure 1.9). Kirchhoff (1847) showed that the proportion of current that flows directly along e is then equal to the probability that e belongs to a randomly chosen spanning tree!

Now current flows in two ways: some flows directly across e and some flows through the rest of the network. It is intuitively clear (and justified by Rayleigh's

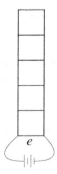

Figure 1.9. A battery is hooked up between the endpoints of e.

monotonicity principle in Section 2.4) that the higher the ladder, the greater the effective conductance of the ladder minus the bottom rung, hence the less current proportionally will flow along e, whence by Kirchhoff's theorem, the less the chance that a random spanning tree contains the bottom rung. This confirms our observations.

It turns out that generating spanning trees at random according to the uniform measure is of interest to computer scientists, who have developed various algorithms over the years for random generation of spanning trees. In particular, this is closely connected to generating a random state from any Markov chain. See Propp and Wilson (1998) for more on this issue.

Early algorithms for generating a random spanning tree used the matrix-tree theorem, which counts the number of spanning trees in a graph via a determinant (Section 4.4). A better algorithm than these early ones, especially for probabilists, was introduced by Aldous (1990) and Broder (1989). It says that if you start a simple random walk at *any* vertex of a finite (connected) graph G and draw every edge it traverses except when it would complete a cycle (that is, except when it arrives at a previously visited vertex), then when no more edges can be added without creating a cycle, what will be drawn – amazingly – is a uniformly chosen spanning tree of G. (To be precise: if X_n ($n \geq 0$) is the path of the random walk, then the associated spanning tree is the set of edges $\{[X_n, X_{n+1}]; \ X_{n+1} \notin \{X_0, X_1, \ldots, X_n\}\}$.) This beautiful algorithm is quite efficient and useful for theoretical analysis, yet as a graduate student, Wilson (1996) found an even better one that we'll describe in Section 4.1.

Return for a moment to the ladder graphs. We saw that as the height of the ladder tends to infinity, there is a limiting probability that the bottom rung of the ladder graph belongs to a uniform spanning tree. What about uniform spanning trees in other sequences of growing finite graphs? Suppose that G is an infinite graph. Let G_n be finite (connected) subgraphs with $G_1 \subset G_2 \subset G_3 \subset \cdots$ and $\bigcup G_n = G$. Take the uniform spanning tree probability measure on each G_n. This gives a sequence of probability measures on subsets of edges of G. Does this sequence converge in a reasonable sense? Lyons conjectured that it does, and Pemantle (1991) verified that the weak limit exists. (In other words, if μ_n denotes the uniform spanning tree measure on G_n and B, B' are finite sets of edges in G, then $\lim_n \mu_n[B \subset T_n, \ B' \cap T_n = \varnothing]$ exists, where T_n denotes a random spanning tree in G_n.) This limit is now called the ***free uniform spanning forest**** on G, denoted FUSF or just FSF. Considerations of electrical networks play the dominant role in Pemantle's proof. Pemantle (1991) discovered the astounding fact that on $\mathbb{Z}^d$, the uniform spanning forest is a single tree a.s. if $d \leq 4$, but when $d \geq 5$, there are infinitely many trees a.s.! We'll prove this as Theorem 10.30.

---

* In graph theory, "spanning forest" usually means a maximal subgraph without cycles, that is, a spanning tree in each connected component. We mean, instead, a subgraph without cycles that contains every vertex.

1.8 **Hausdorff Dimension**

We've used water flow on trees to define the branching number, where the amount of water that can flow through an edge at distance n from the root of a tree is constrained to be at most λ^{-n}. There is a useful way to reformulate this via what's known as the max-flow min-cut theorem, proved in Section 3.1. Namely, consider a set Π of edges whose removal leaves the root o in a finite component. We call such a set a ***cutset*** (separating o from infinity). If θ is a flow from o to infinity, then all the water must flow through Π, so one expects that an upper bound on the total that can flow is $\sum_{e(x)\in\Pi}\lambda^{-|x|}$, where $e(x)$ denotes the edge that joins x to its parent, as before, and $|x|$ denotes the distance of a vertex x to the root. This expectation turns out to be correct, so that the most that can flow is

$$\inf\left\{\sum_{e(x)\in\Pi}\lambda^{-|x|}\;;\;\Pi\text{ is a cutset}\right\}. \tag{1.2}$$

Remarkably, this upper bound is always achievable, that is, there is a flow with this amount in total flowing from the root to infinity; this is the content (in a special case) of the max-flow min-cut theorem. We are going to use this now to understand Hausdorff dimension, but a much more detailed and varied motivation of Hausdorff dimension is given in Chapter 15.

A vertex of degree 1 in a tree is called a ***leaf***. By analogy with the leaves of a finite tree, we call the set of rays of T the ***boundary*** (at infinity) of T, denoted ∂T. (Recall that a ray is an infinite, simple path from the root, so ∂T does not include any leaves of T.) Now there is a natural metric on ∂T: if $\xi, \eta \in \partial T$ have exactly n edges in common, define their distance to be $d(\xi, \eta) := e^{-n}$. Thus, if $x \in T$ has more than one child with infinitely many descendants, then the set of rays going through x,

$$B_x := \{\xi \in \partial T\;;\;\xi_{|x|} = x\}, \tag{1.3}$$

has diameter $\operatorname{diam} B_x = e^{-|x|}$. We call a collection $\mathscr{C}$ of subsets of ∂T a ***cover*** if

$$\bigcup_{B\in\mathscr{C}} B = \partial T.$$

▷ **Exercise 1.1.**
　　Let T be an infinite, locally finite tree.
　　　　(a) **(Kőnig's Lemma)** Show that $\partial T \neq \varnothing$.
　　　　(b) Show that ∂T is compact.

　　Note that

$$\Pi\text{ is a cutset (separating }o\text{ from }\infty\text{) iff }\{B_x\;;\;e(x)\in\Pi\}\text{ is a cover.} \tag{1.4}$$

The ***Hausdorff dimension*** of ∂T is defined to be

$$\dim\partial T := \sup\left\{\alpha\;;\;\inf_{\mathscr{C}\text{ a countable cover}}\sum_{B\in\mathscr{C}}(\operatorname{diam} B)^\alpha > 0\right\}.$$

This number is just a disguised version of the branching number.* Indeed,

$$\operatorname{br} T = \sup\left\{\lambda\,;\ \text{water can flow through pipe capacities } \lambda^{-|x|}\right\}.$$

Now use the condition (1.2) to write this as

$$\sup\left\{\lambda\,;\ \inf_{\Pi \text{ a cutset}} \sum_{e(x)\in\Pi} \lambda^{-|x|} > 0\right\}.$$

Replace λ by e^α to rewrite it as

$$\exp\sup\left\{\alpha\,;\ \inf_{\Pi \text{ a cutset}} \sum_{e(x)\in\Pi} e^{-\alpha|x|} > 0\right\},$$

and then use the correspondence (1.4) between cutsets and covers to write this as

$$\exp\sup\left\{\alpha\,;\ \inf_{\mathscr{C} \text{ a cover}} \sum_{B\in\mathscr{C}} (\operatorname{diam} B)^\alpha > 0\right\}.$$

Now we see the disguise revealed as

$$\operatorname{br} T = \exp\dim\partial T.$$

Soon we'll see how this helps us to analyze Hausdorff dimension in Euclidean space.

1.9 Capacity

In Section 1.3, we made the definition

electrical current flows from the root of an infinite tree

$$\Longleftrightarrow \qquad\qquad (1.5)$$

there is a flow with finite energy.

A unit flow on a tree T from the root to infinity is a flow where a total of 1 unit flows from the root. By identifying vertices x with edges $e(x)$, we may identify a unit flow with a function θ on the vertices of T that is 1 at the root and that has the property that for all vertices x,

$$\theta(x) = \sum_i \theta(y_i),$$

where y_i are the children of x. The energy of a flow for the conductances that we've been using as our basic example is then

$$\sum_{x\in T} \theta(x)^2 \lambda^{|x|},$$

---

* Historically, the branching number was defined by Lyons (1990) only after Furstenberg (1970) considered the Hausdorff dimension of the boundaries of trees, which served as the former's inspiration.

whence we may write Theorem 1.5 as

$$\operatorname{br} T = \sup\left\{\lambda \, ; \text{ there exists a unit flow } \theta \quad \sum_{x \in T} \theta(x)^2 \lambda^{|x|} < \infty\right\}. \tag{1.6}$$

We can also identify unit flows θ on T with Borel probability measures μ on ∂T via

$$\mu(B_x) = \theta(x)$$

(see Section 15.4). A bit of algebra (Proposition 16.1) will show that (1.6) is equivalent to

$$\operatorname{br} T = \exp \sup\left\{\alpha \, ; \, \exists \text{ a probability measure } \mu \text{ on } \partial T \quad \iint \frac{d\mu(\xi)\, d\mu(\eta)}{d(\xi, \eta)^\alpha} < \infty\right\}.$$

For $\alpha > 0$, define the α-**capacity** of ∂T to be the reciprocal of the minimum energy of a unit flow for $\lambda = e^\alpha$. When we express this purely in terms of probability measures on the boundary, this will turn out to be the same as the following definition:

$$\operatorname{cap}_\alpha(\partial T)^{-1} := \inf\left\{\iint \frac{d\mu(\xi)\, d\mu(\eta)}{d(\xi, \eta)^\alpha} \, ; \, \mu \text{ a probability measure on } \partial T\right\}.$$

Then statement (1.5) says that for $\alpha > 0$,

$$\text{random walk with parameter } \lambda = e^\alpha \text{ is transient} \iff \operatorname{cap}_\alpha(\partial T) > 0. \tag{1.7}$$

It follows from Theorem 1.7 that

$$\text{the critical value of } \alpha \text{ for positivity of } \operatorname{cap}_\alpha(\partial T) \text{ is } \dim \partial T. \tag{1.8}$$

Theorem 1.8 told us that these same critical values for random walk, electrical networks, Hausdorff dimension, or capacity are also critical for percolation. But it did not tell us what happens at the critical value, unlike, say, (1.7) does for random walk. This is more subtle to analyze for percolation but is also known (Lyons, 1992):

Theorem 1.9.[*] **(Tree Percolation and Capacity)** *For $\alpha > 0$, percolation with parameter $p = e^{-\alpha}$ yields an infinite component a.s. iff $\operatorname{cap}_\alpha(\partial T) > 0$. Moreover,*

$$\operatorname{cap}_\alpha(\partial T) \le \mathbf{P}[\textit{the component of the root is infinite}] \le 2\operatorname{cap}_\alpha(\partial T).$$

Although this appears rather abstract, it is very useful. First of all, when T is spherically symmetric and $p = e^{-\alpha}$, we can calculate the capacities easily (Exercise 16.1):

$$\operatorname{cap}_\alpha(\partial T) = \left(1 + (1 - p)\sum_{n=1}^{\infty} \frac{1}{p^n |T_n|}\right)^{-1}.$$

Second, one can use this theorem in combination with (1.7); this allows us to translate problems freely between the domains of random walks and percolation (Lyons, 1992). Third, we describe how it can be used to analyze Brownian motion in the next section.

---

[*] This will be proved as Theorem 16.3. The case of the first part of this theorem where all the degrees are uniformly bounded was shown earlier by Fan (1989, 1990).

1.10 **Embedding Trees into Euclidean Space**

The results described previously, especially those concerning percolation, can be translated to give interesting results on closed sets in Euclidean space. We describe only the simplest such correspondence here.*

Let $b \geq 2$ be an integer. An interval of the form $[k/b^n, (k + 1)/b^n]$ for integers k and n is called *b-adic of order n*. For a closed nonempty set $E \subseteq [0, 1]$, consider the system of b-adic subintervals of $[0, 1]$. We'll associate a tree to E as follows: Those intervals whose intersection with E is nonempty will form the vertices of the tree. Two such intervals are connected by an edge iff one contains the other and the ratio of their lengths is b. The root of this tree is $[0, 1]$. Denote this tree by $T_{[b]}(E)$. An example is illustrated in Figure 1.10 with $b = 4$. Were it not for the fact that certain numbers have two representations in base b, we could identify $\partial T_{[b]}(E)$ with E. Because of this multiplicity of representation, there are other trees whose boundary we could identify with E. That is, given a tree T, suppose that we associate to each $x \in T_n$ a b-adic interval $I_x \subseteq [0, 1]$ of order n in such a way that $|I_x \cap I_y| \leq 1$ for $|x| = |y|$, $x \neq y$, and that I_x is contained in I_z when z is the parent of x. Then the tree T *codes* the closed set $E := \bigcap_{n \geq 0} \bigcup_{x \in T_n} I_x$. The difference between ∂T

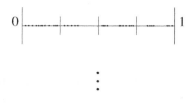

Figure 1.10. Coding by trees.

and $\partial T_{[b]}(E)$ is at most countable. As we will see, this implies, for example, that these two boundaries have the same Hausdorff dimension.

Hausdorff dimension is defined for subsets of $[0, 1]$ just as we defined it for ∂T: A *cover* of E is a collection $\mathscr{C}$ of sets whose union contains E, and

$$\dim E := \sup\left\{\alpha \; ; \; \inf_{\mathscr{C} \text{ a cover of } E} \sum_{B \in \mathscr{C}} (\operatorname{diam} B)^\alpha > 0\right\},$$

where diam B denotes the (Euclidean) diameter of E. When T codes E, covers of ∂T by sets of the form B_x (as in (1.3)) correspond to covers of E by b-adic intervals, but of diameter $b^{-|x|}$, rather than $e^{-|x|}$. One can show that restricting to covers of E by b-adic intervals does not change the computation of Hausdorff dimension, whence we may conclude (compare the calculation at the end of Section 1.8) that

$$\dim E = \frac{\dim \partial T}{\log b} = \log_b(\operatorname{br} T). \tag{1.9}$$

Example 1.10. Let E be the Cantor middle-thirds set. If $b = 3$, then the binary tree codes E (when the obvious 3-adic intervals are associated to the binary tree), whence (1.9) tells us

---

* This correspondence was part of Furstenberg's motivation in 1970 for looking at the dimension of the boundary of a tree.

that the Hausdorff dimension of E is $\log_3 2 = \log 2 / \log 3$. If we use a different base, b, to code E by a tree T, we will have $\operatorname{br} T = b^{\log_3 2}$.

Capacity in Euclidean space is also defined as we defined it on the boundary of a tree:

$$(\operatorname{cap}_\alpha E)^{-1} := \inf \left\{ \iint \frac{d\mu(x)\,d\mu(y)}{|x - y|^\alpha} \; ; \; \mu \text{ a probability measure on } E \right\}.$$

It was shown by Benjamini and Peres (1992) and Pemantle and Peres (1995b) (see Section 16.3) that when T codes E,

$$\frac{1}{2} \operatorname{cap}_\alpha E \le \frac{1}{1 - b^{-\alpha}} \operatorname{cap}_{\alpha \log b} \partial T \le 3b \operatorname{cap}_\alpha E. \tag{1.10}$$

This means that the percolation criterion Theorem 1.9 can be used in Euclidean space. This, and similar extensions, will allow us in Section 16.4 to analyze Brownian motion in $\mathbb{R}^d$ by replacing the path of Brownian motion by an "intersection-equivalent" random fractal that is much easier to analyze, being an embedding of a Galton–Watson tree. This will allow us to determine whether Brownian motion has double points, triple points, etc., in a very easy fashion.

1.11 Notes

The product of two graphs, G_1 and G_2, with $\mathsf{V} := \mathsf{V}_1 \times \mathsf{V}_2$ and the choice

$$\mathsf{E} := \left\{ ((x_1, x_2), (y_1, y_2)) \; ; \; (x_1, y_1) \in \mathsf{E}_1 \text{ and } (x_2, y_2) \in \mathsf{E}_2 \right\}$$

is called the ***tensor product***, since its adjacency matrix is the tensor product of the adjacency matrices corresponding to E_1 and E_2. It is denoted $G = G_1 \times G_2$. The union of $G_1 \mathbin{\square} G_2$ and $G_1 \times G_2$ is denoted $G_1 \boxtimes G_2$. Terminology for graph products is not universal; other terms include "sum" for what we called the Cartesian product and "product" for the tensor product.

Some other ways that $\operatorname{br} T$ serves as a critical value on trees arise in the Ising model (Lyons, 1989), the Heisenberg model (Pemantle and Steif, 1999), information broadcasting (Evans, Kenyon, Peres, and Schulman, 2000), and the firefighting game (Lehner, 2019).

Recent books that cover material related to the topics of this book include *Probability on Graphs* by Geoffrey Grimmett, *Reversible Markov Chains and Random Walks on Graphs* by David Aldous and Jim Fill (preliminary version online), *Coarse Geometry and Randomness* by Itai Benjamini, *Markov Chains and Mixing Times* by David A. Levin, Yuval Peres, and Elizabeth L. Wilmer, *Probability: The Classical Limit Theorems* by Henry McKean, *Random Trees: An Interplay between Combinatorics and Probability* by Michael Drmota, *A Course on the Web Graph* by Anthony Bonato, *Random Graph Dynamics* by Rick Durrett, *Complex Graphs and Networks* by Fan Chung and Linyuan Lu, *The Random-Cluster Model* by Geoffrey Grimmett, *Superfractals* by Michael Fielding Barnsley, *Introduction to Mathematical Methods in Bioinformatics* by Alexander Isaev, *Gaussian Markov Random Fields* by Håvard Rue and Leonhard Held, *Conformally Invariant Processes in the Plane* by Gregory F. Lawler, *Random Networks in Communication* by Massimo Franceschetti and Ronald Meester, *Percolation* by Béla Bollobás and Oliver Riordan, *Probability and Real Trees* by Steven Evans, *Random Trees, Lévy Processes and Spatial Branching Processes* by Thomas Duquesne and Jean-François Le Gall, *Combinatorial Stochastic Processes* by Jim Pitman, *Random Geometric Graphs* by Mathew Penrose, *Random Graphs* by Béla

Bollobás, *Random Graphs* by Svante Janson, Tomasz Łuczak, and Andrzej Ruciński, *Phylogenetics* by Charles Semple and Mike Steel, *Stochastic Networks and Queues* by Philippe Robert, *Random Walks on Infinite Graphs and Groups* by Wolfgang Woess, *Random Walk: A Modern Introduction* by Gregory F. Lawler and Vlada Limic, *Percolation* by Geoffrey Grimmett, *Noise Sensitivity of Boolean Functions and Percolation* by Christophe Garban and Jeffrey E. Steif, *Stochastic Interacting Systems: Contact, Voter and Exclusion Processes* by Thomas M. Liggett, and *Discrete Groups, Expanding Graphs and Invariant Measures* by Alexander Lubotzky.

1.12 Collected In-Text Exercises

1.1. Let T be an infinite, locally finite tree.

(a) (Kőnig's Lemma) Show that $\partial T \neq \varnothing$.

(b) Show that ∂T is compact.

1.13 Additional Exercises

1.2. Show that $\operatorname{br} T = \underline{\operatorname{gr}} T$ when T is a spherically symmetric tree.

1.3. Here we'll look more closely at the joining construction of Example 1.3. We will put together two spherically symmetric trees $T^{(1)}$ and $T^{(2)}$ such that $\operatorname{br}(T^{(1)} \vee T^{(2)}) = 1$, yet at the same time, $\underline{\operatorname{gr}}(T^{(1)} \vee T^{(2)}) > 1$. Let $n_k \uparrow \infty$. Let $T^{(1)}$ (resp., $T^{(2)}$) be a tree such that x has one child (resp., two children) for $n_{2k} \leq |x| \leq n_{2k+1}$ and two (resp., one) otherwise; this is shown schematically in Figure 1.11. If n_k increases sufficiently rapidly, then $\operatorname{br} T^{(1)} = \operatorname{br} T^{(2)} = 1$, so $\operatorname{br}(T^{(1)} \vee T^{(2)}) = 1$. Prove that if $\langle n_k \rangle$ increases sufficiently rapidly, then $\underline{\operatorname{gr}}(T^{(1)} \vee T^{(2)}) = \sqrt{2}$. Furthermore, show that the set of possible values of $\underline{\operatorname{gr}}(T^{(1)} \vee T^{(2)})$ over *all* sequences $\langle n_k \rangle$ is $[\sqrt{2}, 2]$.

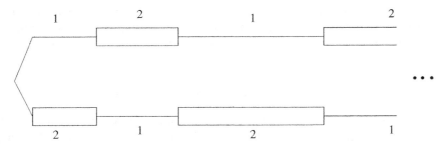

Figure 1.11. A schematic of a tree with branching number 1 and growth rate $\sqrt{2}$.

1.4. Complete Example 1.4 by showing that when λ^{-n} is the conductance of edges at distance n from the root of a binary tree T, current does not flow for $\lambda \geq 2$.

2 | Random Walks and Electric Networks

The two topics of the title of this chapter do not sound related to each other, but in fact, they are intimately connected in several extraordinarily useful ways. This is a discrete version of profound and detailed connections between continuous potential theory and Brownian motion, which we describe briefly in Section 2.9. The next chapter applies most of our work here to particularly interesting classes of networks. For example, we'll prove that the critical parameter λ separating transience from recurrence for the biased random walk RW_λ on a general tree is equal to the branching number of the tree (as mentioned in Theorem 1.7). Then Chapter 4 explains a marvelous link to a third topic, uniform spanning trees. Later, in Chapter 9, we examine some of the subtleties inherent in infinite electrical networks. Those ideas are then combined in Chapter 10 with the web of ideas in Chapter 4 to study the analogues of uniform spanning trees in infinite networks.

Our principal interest in this chapter centers around transience and recurrence of irreducible, reversible Markov chains, otherwise known as network-based random walks. Although we develop mathematically rigorous tools from electrical network theory for this purpose, these tools will have the added benefit of allowing us to estimate hitting and cover times in finite networks. They also give variances for a field of Gaussian random variables that is connected to the network; this field is known variously as the canonical Gaussian field or the discrete Gaussian free field. Techniques from the linear algebra of inner-product spaces give electrical network theory a rich structure, which will be extended in Chapter 9. Many supplementary results, often not requiring reversibility, are in the exercises at the end of the chapter.

2.1 Circuit Basics and Harmonic Functions

If a Markov chain starts at a state x, how can we determine whether it is bound to visit another given state a, that is, whether the chance that it ever visits a is 1 or is less than 1?

Our theory will apply only to reversible Markov chains, where we call a Markov chain *reversible* if there is a positive function $x \mapsto \pi(x)$ on the state space such that the transition probabilities satisfy $\pi(x)p(x, y) = \pi(y)p(y, x)$ for all pairs of states x, y. (Such a function $\pi(\cdot)$ will then provide a stationary measure: see Exercise 2.1 for this and for why we use the name "reversible." *Note that $\pi(\cdot)$ is not generally a probability measure.*) In this case, make a graph G (possibly with loops) by taking the states of the Markov chain for the vertices of G and joining two vertices x, y by an edge when $p(x, y) > 0$. Assign weight

$$c(x, y) := \pi(x)p(x, y) \tag{2.1}$$

to that edge; note that the condition of reversibility ensures that this weight is the same no matter in what order we take the endpoints of the edge. With this network in hand, the Markov chain may be described as a random walk on G: when the walk is at a vertex x, it chooses randomly among the vertices adjacent to x with transition probabilities proportional to the weights of the edges incident to x. Conversely, every connected graph with (positive) weights on the edges such that the sum of the weights incident to every vertex is finite gives rise to a random walk with transition probabilities proportional to the weights. Such a random walk is an irreducible reversible Markov chain: define $\pi(x)$ to be the sum of the weights incident to x.* Often, all the edge weights are equal; we call the random walk in this case **simple random walk**. When a graph is given without weights, we take the weights to be identically 1 as a default.

The most well known example of a reversible Markov chain is gambler's ruin. A gambler needs \$$n$ but has only \$$k$ ($1 \le k \le n-1$). He plays games that give him chance p of winning \$1 and chance $q := 1 - p$ of losing \$1 each time. When his fortune is either \$$n$ or 0, he stops. What is his chance of ruin (that is, of reaching 0 before n)? We will answer this in Example 2.4 by using the following weighted graph. The vertices are $\{0, 1, 2, \ldots, n\}$, the edges are between consecutive integers, and the weights are $c(i, i+1) = c(i+1, i) = (p/q)^i$.

Although we are interested ultimately in recurrence or transience of infinite networks, we begin by studying random walks on finite networks. In fact, our first results will be about finite Markov chains that need not be reversible. Suppose that A is a subset of states. Write τ_A for the first time that the Markov chain visits ("hits") some vertex in A; if the chain happens to start in A, then $\tau_A = 0$. Occasionally, we will use τ_A^+, which is the first time after 0 that the chain visits A; this is different from τ_A only when the chain starts in A. Usually A will be a singleton.

Consider the probability that the Markov chain visits A before it visits a disjoint subset Z as a function of its starting point x:

$$F(x) := \mathbf{P}_x[\tau_A < \tau_Z]. \tag{2.2}$$

The key idea here is to let x vary, even if we are interested in a particular x. Recall that $\restriction$ indicates the restriction of a function to a set. Clearly $F \restriction A \equiv 1$, $F \restriction Z \equiv 0$, and for $x \notin A \cup Z$,

$$F(x) = \sum_y \mathbf{P}_x[\text{first step is to } y] \, \mathbf{P}_x[\tau_A < \tau_Z \mid \text{first step is to } y] = \sum_{x \sim y} p(x, y) F(y).$$

In the reversible case, we can write further that

$$F(x) = \frac{1}{\pi(x)} \sum_{x \sim y} c(x, y) F(y),$$

---

* Suppose that we consider an edge e of G to have length $c(e)^{-1}$. Run a Brownian motion on G and observe it only when it reaches a vertex different from the previous one. Then we see the random walk on G just described (if we ignore the fact that the times between observations vary). There are several equivalent ways to define rigorously Brownian motion on G; one way is described in Section 2.9. See Georgakopoulos and Winkler (2014) for an interesting analysis of Brownian motion on finite networks.

where $x \sim y$ indicates that x and y are adjacent in G. In the special case of simple random walk, this equation becomes

$$F(x) = \frac{1}{\deg x} \sum_{x \sim y} F(y),$$

where $\deg x$ is the degree of x, that is, the number of edges incident to x. In words, $F(x)$ is the average of the values of F at the neighbors of x. In general, this is still true, but the average is taken with weights.

This averaging property is so important that it has a name: a function f is **harmonic** at x when

$$f(x) = \sum_{x \sim y} p(x, y) f(y).$$

If f is harmonic at each point of a set W, then we say that f is harmonic on W. Harmonic functions satisfy a maximum principle. To state it, we use the following concept: the chain **absorbed** off of W is the Markov chain whose transition probabilities are modified only so that $p(x, x) = 1$ for $x \notin W$.

Maximum Principle. *Let W be a set of states of a Markov chain on a finite or countable state space V. If $f: \mathsf{V} \to \mathbb{R}$ is a function that is harmonic on W and the supremum of f on V is achieved at some element $x_0 \in W$, then f is constant on all states accessible from x_0 in the chain absorbed off of W.*

Proof. Let $K := \{y \in \mathsf{V} ; \ f(y) = \sup f\}$. Note that if $x \in W \cap K$ and $p(x, y) > 0$, then $y \in K$ because f is harmonic at x. Hence the conclusion follows.   ◄

This leads to the

Uniqueness Principle. *Let W be a finite, proper subset of states of a Markov chain on a finite or countable state space V. Suppose that $\mathsf{V} \setminus W$ is accessible from every state in W for the chain absorbed off of W. If $f, g: \mathsf{V} \to \mathbb{R}$ are two functions that are both harmonic on W and agree off W (that is, $f(x) = g(x)$ for all $x \notin W$), then $f = g$.*

Proof. Let $h := f - g$. We claim that $h \leq 0$. This suffices to establish the corollary, since then $h \geq 0$ by symmetry, whence $h \equiv 0$.

Now $h = 0$ off W. Since W is finite, h achieves its overall supremum at some point $x_0 \in \mathsf{V}$. If $x_0 \notin W$, then $h \leq 0$, as desired. On the other hand, if $x_0 \in W$, then by the maximum principle, $h(x_0) \leq \sup h \upharpoonright (\mathsf{V} \setminus W) = 0$, which again shows that $h \leq 0$.   ◄

Here are two consequences of the uniqueness principle: (1) The harmonicity of the function $x \mapsto \mathbf{P}_x[\tau_A < \tau_Z]$ on a finite, irreducible Markov chain (together with its values where it is not harmonic) characterizes this function. (2) If f, f_1, and f_2 are harmonic on some finite, proper subset $W \subset \mathsf{V}$ and $a_1, a_2 \in \mathbb{R}$ with $f = a_1 f_1 + a_2 f_2$ on $\mathsf{V} \setminus W$, then $f = a_1 f_1 + a_2 f_2$ everywhere. This is one form of the **superposition principle**.

Given a function defined on a subset of states, the **Dirichlet problem** asks whether the given function can be extended to all states of the Markov chain so as to be harmonic wherever it was not originally defined. The answer is often yes:

Existence Principle. *Let W be a proper subset of states of a Markov chain on a finite or countable state space V. If $f_0 \colon \mathsf{V} \setminus W \to \mathbb{R}$ is bounded, then $\exists f \colon \mathsf{V} \to \mathbb{R}$ such that $f {\upharpoonright} (\mathsf{V} \setminus W) = f_0$ and f is harmonic on W.*

Proof. For any starting point x of the Markov chain, let X be the first vertex in $\mathsf{V} \setminus W$ visited by the Markov chain if $\mathsf{V} \setminus W$ is indeed visited. Let $Y := f_0(X)$ when $\mathsf{V} \setminus W$ is visited and $Y := 0$ otherwise. It is easily checked that $f(x) := \mathbf{E}_x[Y]$ works by using first-step analysis, that is, the same method as we used to see that the function F of (2.2) is harmonic. ◀

An example for simple random walk is shown in Figure 2.1, where the function was specified to be 1 at two vertices, 0.5 at another, and 0 at a fourth; the function is harmonic elsewhere.

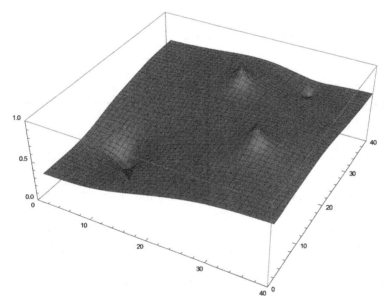

color plate **Figure 2.1.** A harmonic function on a 40×40 square grid with four specified values where it is not harmonic.

The function F of (2.2) is the particular case of the existence principle where $W = \mathsf{V} \setminus (A \cup Z)$, $f_0 {\upharpoonright} A \equiv 1$, and $f_0 {\upharpoonright} Z \equiv 0$.

For finite Markov chains, we could have immediately deduced the existence principle from the uniqueness principle: The Dirichlet problem on a finite state space consists of a finite number of linear equations, one for each state in W. Since the number of unknowns is equal to the number of equations, the uniqueness principle implies the existence principle.

To make further progress toward our goal of determining whether certain states are bound to be visited, we will need to assume reversibility, which we do from now on, unless stated otherwise. The following exercise contains some background information and facts that we will use about reversible Markov chains. Additional background on Markov chains, not necessarily reversible, is in Exercises 2.42 and 2.43.

▷ **Exercise 2.1.**
(Reversible Markov Chains) All Markov chains in this exercise are denumerable.

(a) Let P be the transition probability matrix of a Markov chain. Show that if P is π-reversible, that is, $\pi(x)p(x, y) = \pi(y)p(y, x)$ for all states x, y, then π is P-stationary, that is, $\sum_x \pi(x)p(x, y) = \pi(y)$ for all states y.

(b) Let P be the transition probability matrix of a Markov chain and π be P-stationary. Define the **reversed Markov chain** to have transition probabilities

$$\widehat{p}(x, y) := \pi(y)p(y, x)/\pi(x)$$

and write $\widehat{P}$ for the associated transition matrix. Show that $\widehat{P}$ is indeed a transition matrix and that $\forall x_1, x_2, \ldots, x_n$,

$$\pi(x_1) \prod_{i=1}^{n-1} p(x_i, x_{i+1}) = \pi(x_n) \prod_{i=1}^{n-1} \widehat{p}(x_{n+1-i}, x_{n-i}) \,.$$

Of course, P is π-reversible iff $P = \widehat{P}$.

(c) Show that if a Markov chain is reversible, then $\forall x_1, x_2, \ldots, x_n$ with $x_1 = x_n$,

$$\prod_{i=1}^{n-1} p(x_i, x_{i+1}) = \prod_{i=1}^{n-1} p(x_{n+1-i}, x_{n-i}) :$$

the chance of traversing a cycle is the same in either direction. Show too that this equation implies reversibility for irreducible chains.

(d) Let $\langle X_n \rangle$ be a random walk on a network G, and let x and y be two vertices in G. Let $\mathcal{P}$ be a path from x to y and $\mathcal{P}'$ be its reversal, a path from y to x. Show that

$$\mathbf{P}_x\big[\langle X_n ; \ n \le \tau_y \rangle = \mathcal{P} \mid \tau_y < \tau_x^+\big] = \mathbf{P}_y\big[\langle X_n ; \ n \le \tau_x \rangle = \mathcal{P}' \mid \tau_x < \tau_y^+\big],$$

where τ_w denotes the first time the random walk visits w, τ_w^+ denotes the first time after 0 that the random walk visits w, and $\mathbf{P}_u$ denotes the law of random walk started at u.

(e) Consider a random walk on a network G that is either transient or is stopped on the first visit to a set of vertices Z. Let $\mathscr{G}(x, y)$ be the expected number of visits to y for a random walk started at x; if the walk is stopped at Z, we count only those visits that occur strictly before visiting Z. Show that for every pair of vertices x and y,

$$\pi(x)\,\mathscr{G}(x, y) = \pi(y)\,\mathscr{G}(y, x) \,.$$

(f) Show that random walk on a connected network G is positive recurrent (that is, has a stationary probability distribution, which is therefore unique) iff $\sum_{x,y} c(x, y) < \infty$, in which case the stationary probability distribution is proportional to $\pi(\cdot)$. Show that if the random walk is not positive recurrent, then $\pi(\cdot)$ is a stationary infinite measure.

To study the solution to the Dirichlet problem, especially for a sequence of subgraphs of an infinite graph, we will discover that electrical networks are wonderfully suited. Electrical networks, of course, have a physical meaning whose intuition is useful to us, but also they can be used as a rigorous mathematical tool. We now spend the rest of the chapter developing and exploiting this tool.

Mathematically, an electrical network is just a weighted graph.* But now we call the weights of the edges ***conductances***; their reciprocals are called ***resistances***. (Note that, later, we will encounter effective conductances and resistances; these are not the same.) We denote $c(x, y)^{-1}$ by $r(x, y)$. The reason for this and more new terminology is that not only does it match physics, but the physics can aid our intuition. We will carefully define everything in pure mathematical terms but also give a little of the physical background. Whenever we speak of physics, we will use the symbol 🖈 🖈 in the margin to indicate that it is merely for intuition and is not used in any proofs whatsoever. Given two subsets A and Z of vertices of a network, a ***voltage*** function is a function on the vertices of the network that is harmonic at all $x \notin A \cup Z$. Usually, the voltage will be specified to be 1 on A and 0 on Z. For example, our hitting probability function F, defined in (2.2), is such a voltage function. Given a voltage function v, we define the associated ***current*** function i on the edges by

$$i(x, y) := c(x, y)[v(x) - v(y)] \,.$$

Notice that $i(x, y) = -i(y, x)$ and that current flows in the direction of decreasing voltage, by which we mean that $i(x, y) > 0$ iff $v(x) > v(y)$. Notice also that whenever v is harmonic at a vertex x, we have

$$0 = v(x) \sum_{x \sim y} c(x, y) - \sum_{x \sim y} c(x, y)v(y) = \sum_{x \sim y} i(x, y) \,.$$

This property is sufficiently important that it gets a name in a broader context. Namely, a function θ on ordered pairs of neighboring vertices is called a ***flow*** between A and Z if $\theta(x, y) = -\theta(y, x)$ for all neighbors x, y and $\sum_{y \sim x} \theta(x, y) = 0$ for all $x \notin A \cup Z$.

This definition of current and this property of current are usually called "laws" as follows.

Ohm's Law: If $x \sim y$, the current $i(x, y)$ from x to y satisfies

$$v(x) - v(y) = i(x, y)r(x, y) \,.$$

Kirchhoff's Node Law: The current is a flow between A and Z.

Now if we sum the differences of a function, such as the voltage v, on the edges of a cycle, we get 0. Thus, by Ohm's law, we deduce

Kirchhoff's Cycle Law: If $x_1 \sim x_2 \sim \cdots \sim x_n \sim x_{n+1} = x_1$ is a cycle, then

$$\sum_{k=1}^{n} i(x_k, x_{k+1}) \, r(x_k, x_{k+1}) = 0 \,.$$

_____

* We are ignoring capacitors and inductors, whose usefulness to probability theory is not clear.

One can also deduce Ohm's law from Kirchhoff's two laws, in other words, a flow that satisfies Kirchhoff's cycle law is a current. A somewhat more general statement is in the following exercise.

▷ **Exercise 2.2.**
Suppose that an antisymmetric function j (meaning that $j(x, y) = -j(y, x)$) on the edges of a finite, connected network satisfies Kirchhoff's cycle law and satisfies Kirchhoff's node law in the form $\sum_{y \sim x} j(x, y) = 0$ for all $x \in W$. Show that j is the current corresponding to some voltage function whose values are specified off W and that the voltage function is unique up to an additive constant.

✔ ✔   The remainder of this section gives some physical background for this terminology. Edges of the network are made of conducting wires. We hook up a battery or batteries between A and Z so that the voltage at every vertex in A is 1 and in Z is 0 (or more generally, so that the voltages on $V \setminus W$ are given by some function f_0). Sometimes, voltages are called ***potentials*** or potential differences. Voltages v are then established at every vertex and current i runs through the edges. These functions are observed in experiments to satisfy Ohm's law and Kirchhoff's law. Physically, Ohm's law, which is usually stated as $v = ir$ in engineering, is an empirical statement about linear response to voltage differences – certain components obey this law over a wide range of voltage differences. Kirchhoff's node law expresses the fact that charge does not build up at a node (current being the passage rate of charge per unit time). If we count the currents in the wires corresponding to the batteries, then the sum of the currents at every vertex is 0, not merely at $x \notin A \cup Z$.

2.2 More Probabilistic Interpretations

Let G be a finite network. Suppose that $A = \{a\}$ is a singleton. What is the chance that a random walk starting at a will hit Z before it returns to a? Write this as

$$\mathbf{P}[a \to Z] := \mathbf{P}_a[\tau_Z < \tau_a^+].$$

Impose a voltage of $v(a)$ at a and 0 on Z. Since $v(\cdot)$ is linear in $v(a)$ by the superposition principle, we have that $\mathbf{P}_x[\tau_a < \tau_Z] = v(x)/v(a)$, whence

$$\mathbf{P}[a \to Z] = \sum_x p(a, x)\left(1 - \mathbf{P}_x[\tau_a < \tau_Z]\right) = \sum_x \frac{c(a, x)}{\pi(a)}\left[1 - \frac{v(x)}{v(a)}\right]$$

$$= \frac{1}{v(a)\pi(a)} \sum_x c(a, x)[v(a) - v(x)] = \frac{1}{v(a)\pi(a)} \sum_x i(a, x).$$

In other words,

$$v(a) = \frac{\sum_x i(a, x)}{\pi(a)\,\mathbf{P}[a \to Z]}. \tag{2.3}$$

Since $\sum_x i(a, x)$ is the ***total amount of current flowing into the circuit at*** a (by definition), we may regard the entire circuit between a and Z as a single conductor (between two new vertices) of ***effective conductance***

$$C_{\mathrm{eff}} := \pi(a)\,\mathbf{P}[a \to Z] =: \mathscr{C}(a \leftrightarrow Z), \tag{2.4}$$

where the last notation indicates the dependence on a and Z. If we need to indicate the dependence on G, we will write $\mathscr{C}(a \leftrightarrow Z; G)$. (Recall that $\pi(\cdot)$ is not generally a probability measure.) The similarity to (2.1) can provide a good mnemonic, but the analogy should not be pushed too far. We define the ***effective resistance*** $\mathscr{R}(a \leftrightarrow Z)$ to be the reciprocal of the effective conductance; in case $a \in Z$, then we also define $\mathscr{R}(a \leftrightarrow Z) := 0$. One answer to our earlier question is thus $\mathbf{P}[a \to Z] = \mathscr{C}(a \leftrightarrow Z)/\pi(a)$. In Sections 2.3 and 2.4, we will see some ways to compute effective conductances.

Now the number of visits to a before hitting Z is a geometric random variable with mean $\mathbf{P}[a \to Z]^{-1} = \pi(a)\mathscr{R}(a \leftrightarrow Z)$. According to (2.3), this can also be expressed as $\pi(a)v(a)$ when there is unit current flowing from a to Z and the voltage is 0 on Z. This generalizes as follows. Let $\mathscr{G}_Z(a, x)$ be the expected number of visits to x strictly before hitting Z by a random walk started at a. Thus, $\mathscr{G}_Z(a, x) = 0$ for $x \in Z$ and

$$\mathscr{G}_Z(a, a) = \mathbf{P}[a \to Z]^{-1} = \pi(a)\mathscr{R}(a \leftrightarrow Z). \tag{2.5}$$

The function $\mathscr{G}_Z(\cdot, \cdot)$ is the ***Green*** function for the random walk absorbed (or "killed") on Z. By definition of current, scaling the voltage by multiplying all values by a constant also scales the current by the same factor. Whereas in the preceding section, it was useful to take the voltage to have values 0 and 1 at special vertices, here it will be useful to scale so that the total current flow is 1, in other words, the current is a ***unit flow*** or ***unit current flow***.

Proposition 2.1. (Green Function as Voltage) *Let G be a finite, connected network. When a voltage is imposed on $\{a\} \cup Z$ so that a unit current flows from a to Z and the voltage is 0 on Z, then the voltage function satisfies $v(x) = \mathscr{G}_Z(a, x)/\pi(x)$ for all x.*

Proof. We have just shown that this is true for $x \in \{a\} \cup Z$, so it suffices to establish that $\mathscr{G}_Z(a, x)/\pi(x)$ is harmonic elsewhere. But by Exercise 2.1, we have that $\mathscr{G}_Z(a, x)/\pi(x) = \mathscr{G}_Z(x, a)/\pi(a)$, and the harmonicity of $\mathscr{G}_Z(x, a)$ is established just as for the function of (2.2). ◄

Given that we now have two probabilistic interpretations of voltage, we naturally wonder whether current has a probabilistic interpretation. We might guess one by the following ✎ ✎ unrealistic but simple model of electricity: positive particles enter the circuit at a, they do Brownian motion on G (being less likely to pass through small conductors), and, when they hit Z, they are removed. The net flow rate of particles across an edge would then be the current on that edge. It turns out that in our mathematical model, this is basically correct:

Proposition 2.2. (Current as Edge Crossings) *Let G be a finite, connected network. Start a random walk at a and absorb it when it first visits Z. For $x \sim y$, let S_{xy} be the number of transitions from x to y. Then $\mathbf{E}[S_{xy}] = \mathscr{G}_Z(a, x)p(x, y)$ and $\mathbf{E}[S_{xy} - S_{yx}] = i(x, y)$, where i is the current in G when a potential is applied between a and Z in such an amount that unit current flows in at a.*

Note that we count a transition from y to x when $y \notin Z$ but $x \in Z$, although we do not count this as a visit to x in computing $\mathscr{G}_Z(a, x)$.

Proof. We have

$$
\mathbf{E}[S_{xy}] = \mathbf{E}\left[\sum_{k=0}^{\infty} \mathbf{1}_{[X_k=x, X_{k+1}=y]}\right] = \sum_{k=0}^{\infty} \mathbf{P}[X_k = x, X_{k+1} = y]
$$

$$
= \sum_{k=0}^{\infty} \mathbf{P}[X_k = x]\, p(x, y) = \mathbf{E}\left[\sum_{k=0}^{\infty} \mathbf{1}_{[X_k=x]}\right] p(x, y) = \mathscr{G}_Z(a, x) p(x, y).
$$

Hence, by Proposition 2.1, we have

$$
\forall x, y \qquad \mathbf{E}[S_{xy} - S_{yx}] = \mathscr{G}_Z(a, x) p(x, y) - \mathscr{G}_Z(a, y) p(y, x)
$$
$$
= v(x)\pi(x) p(x, y) - v(y)\pi(y) p(y, x)
$$
$$
= [v(x) - v(y)] c(x, y) = i(x, y). \qquad \blacktriangleleft
$$

Effective conductance is a key quantity because of the following relationship to the question of transience and recurrence when G is infinite. For an infinite network G, we assume that

$$
\forall x \qquad \sum_{x \sim y} c(x, y) < \infty, \tag{2.6}
$$

so that the associated random walk is well defined. (Of course, this is true when G is *locally finite*, that is, the number of edges incident to every given vertex is finite.) It will be convenient to allow more than one edge between a given pair of vertices: each such edge has its own conductance. We'll also allow loops (edges with only one endpoint), but these may be ignored for our present purposes since they only delay the random walk. Strictly speaking, then, G may be a *multigraph*, not a graph. When a random walk moves from x to y in a multigraph that has several edges connecting x to y, then we think of the walk as moving along one of those edges, chosen with probability proportional to its conductance. Thus, the multigraph form of Proposition 2.2 is $\mathbf{E}[S_e] = \mathscr{G}_Z(a, e^-) p(e)$ and $\mathbf{E}[S_e - S_{-e}] = i(e)$. However, we will usually ignore the extra notational complications that arise for multigraphs. In fact, we have not yet used anywhere that G has only finitely many edges:

▷ **Exercise 2.3.**
Verify that Propositions 2.1 and 2.2 are valid when the number of edges is infinite but the number of vertices is finite.

The way we approach infinite networks in this chapter is by taking large finite subgraphs. More precisely, for an infinite network G, let $\langle G_n \rangle$ be any sequence of finite subgraphs of G that *exhaust* G, that is, $G_n \subseteq G_{n+1}$ and $G = \bigcup G_n$. Each edge in G_n is an edge in G, so we simply give it the same conductance it has in G. We also assume that G_n is the graph induced in G by $\mathsf{V}(G_n)$. Let Z_n be the set of vertices in $G \setminus G_n$. Let G_n^{W} be the graph obtained from G by identifying Z_n to a single vertex, z_n, and then removing loops (but keeping multiple edges). This graph will have finitely many vertices but may have infinitely many edges even when loops are deleted if some vertex of G_n has infinite degree. Given a network random walk on G, if we stop it the first time it reaches Z_n, then we obtain a network random walk on G_n^W until it reaches z_n. Now for every $a \in G$, the events $[a \to Z_n]$ are decreasing in n, so the limit $\lim_n \mathbf{P}[a \to Z_n]$ is the probability of never returning to a in G, which we call the **escape** probability from a. This is positive iff the random walk on G is transient. Hence, by (2.4), $\lim_{n \to \infty} \mathscr{C}(a \leftrightarrow z_n; G_n^W) > 0$ iff the random walk on G is transient. We call $\lim_{n \to \infty} \mathscr{C}(a \leftrightarrow z_n)$ the **effective conductance** from a to ∞ in G and denote it by $\mathscr{C}(a \leftrightarrow \infty)$ or, if a is understood, by C_{eff}. Its reciprocal, **effective resistance**, is denoted R_{eff}. We have shown:

Theorem 2.3. (Transience and Effective Conductance) *Random walk on an infinite, connected network is transient iff the effective conductance from any of its vertices to infinity is positive.*

▷ **Exercise 2.4.**
For a fixed vertex a in G, show that $\lim_n \mathscr{C}(a \leftrightarrow Z_n)$ is the same for every sequence $\langle G_n \rangle$ of induced subgraphs that exhausts G.

▷ **Exercise 2.5.**
When G is finite but A is not a singleton, define $\mathscr{C}(A \leftrightarrow Z)$ to be $\mathscr{C}(a \leftrightarrow Z)$ if all the vertices in A were to be identified to a single vertex, a. Show that if voltages are applied at the vertices of $A \cup Z$ so that $v{\restriction}A$ and $v{\restriction}Z$ are constants, then $v{\restriction}A - v{\restriction}Z = I_{AZ}\mathscr{R}(A \leftrightarrow Z)$, where $I_{AZ} := \sum_{x \in A} \sum_y i(x, y)$ is the total amount of current flowing from A to Z.

2.3 Network Reduction

How do we calculate effective conductance of a network between, say, two vertices a and z? Since we want to replace a network by an equivalent single conductor, it is natural to attempt this by replacing more and more of G through simple transformations, leaving a and z but possibly removing other vertices. There are, in fact, three such simple transformations: series, parallel, and star-triangle. Remarkably, these three transformations suffice to reduce all finite planar networks according to a theorem of Epifanov; see Truemper (1989).

I. Series Law. Two resistors* r_1 and r_2 in series are equivalent to a single resistor $r_1 + r_2$. In other words, if $w \in \mathsf{V}(G) \setminus (A \cup Z)$ is a node of degree 2 with neighbors u_1, u_2 and we replace the edges (u_i, w) by a single edge (u_1, u_2) having resistance $r(u_1, w) + r(w, u_2)$, then all potentials and currents in $G \setminus \{w\}$ are unchanged and the current that flows from u_1 to u_2 equals $i(u_1, w)$.

Proof. It suffices to check that Ohm's and Kirchhoff's laws are satisfied on the new network for the voltages and currents given. This is easy. ◀

---

* A resistor r is an edge with resistance r. We have drawn such edges using the squiggly notation common to physics, but this only indicates that they have weights.

▷ **Exercise 2.6.**
Give two harder but instructive proofs of the series equivalence as follows. Since voltages determine currents, it suffices to check that the voltages are as claimed on the new network G'. (1) Show that $v \upharpoonright (V(G) \setminus \{w\})$ is harmonic on $V(G') \setminus (A \cup Z)$. (2) Use the "craps principle" (Pitman (1993), p. 210) to show that $\mathbf{P}_x[\tau_A < \tau_Z]$ is unchanged for $x \in V(G) \setminus \{w\}$.

Example 2.4. Consider simple random walk on $\mathbb{Z}$. Let $0 \le k \le n$. What is $\mathbf{P}_k[\tau_0 < \tau_n]$? It is the voltage at k when there is a unit voltage imposed at 0 and zero voltage at n. (Here, we ignore vertices outside $[0, n]$ and thus work on a finite network.) If we replace the resistors in series from 0 to k by a single resistor with resistance k and the resistors from k to n by a single resistor of resistance $n - k$, then the voltage at k does not change. But now this voltage is simply the probability of taking a step to 0, which is thus $(n - k)/n$.

For the more general gambler's ruin, rather than simple random walk, we have the conductances $c(i, i + 1) = (p/q)^i$. The replacement of edges in series by single edges now gives one edge from 0 to k of resistance $\sum_{i=0}^{k-1}(q/p)^i$ and one edge from k to n of resistance $\sum_{i=k}^{n-1}(q/p)^i$. The probability of ruin is therefore $\sum_{i=k}^{n-1}(q/p)^i / \sum_{i=0}^{n-1}(q/p)^i = [(p/q)^{n-k} - 1]/[(p/q)^n - 1]$.

II. Parallel Law. Two conductors* c_1 and c_2 in parallel are equivalent to one conductor $c_1 + c_2$. In other words, if two edges e_1 and e_2 that both join vertices $w_1, w_2 \in V(G)$ are replaced by a single edge e joining w_1, w_2 of conductance $c(e) := c(e_1) + c(e_2)$, then all voltages and currents in $G \setminus \{e_1, e_2\}$ are unchanged and the current $i(e)$ equals $i(e_1) + i(e_2)$ (if e, e_1 and e_2 have the same orientations, that is, same tail and head). This transformation is valid even for an infinite number of edges in parallel.

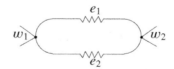

Proof. Check Ohm's and Kirchhoff's laws with $i(e) := i(e_1) + i(e_2)$.   ◀

▷ **Exercise 2.7.**
Give two more proofs of the parallel equivalence as in Exercise 2.6.

Before explaining the star-triangle transformation, we give two amusing examples of the series and parallel transformations, as well as a useful general consequence.

Example 2.5. Suppose that each edge in the following network has equal conductance. What is $\mathbf{P}[a \to z]$? We may assume that the edge conductances are all 1, since the probability is not affected by a change in scale of the conductances. Following the transformations indicated in the figure, we obtain $\mathscr{C}(a \leftrightarrow z) = 7/12$, so that

$$\mathbf{P}[a \to z] = \frac{\mathscr{C}(a \leftrightarrow z)}{\pi(a)} = \frac{7/12}{3} = \frac{7}{36}.$$

---

* A conductor c is an edge with conductance c.

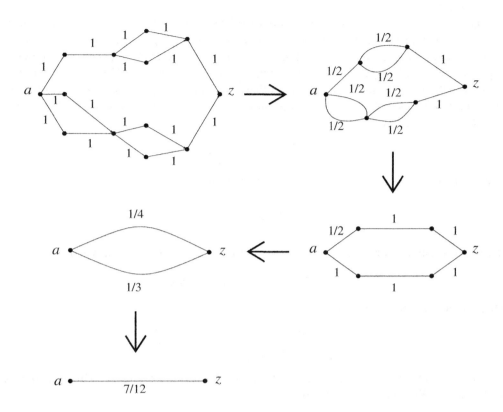

Note that in any network G with voltage applied from a to z, if it happens that $v(x) = v(y)$, then we may identify x and y to a single vertex, obtaining a new network $G/\{x, y\}$ in which the voltages at all vertices are the same as in G.

Example 2.6. What is $\mathbf{P}[a \to z]$ in the following network?

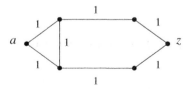

There are two ways to deal with the vertical edge:

(1) Remove it: by symmetry, the voltages at its endpoints are equal, whence no current flows on it.

(2) Contract it, that is, remove it but combine its endpoints into one vertex (we could also combine the other two unlabeled vertices with each other): the voltages are the same, so they may be combined.

In either case, we get $\mathscr{C}(a \leftrightarrow z) = 2/3$, whence $\mathbf{P}[a \to z] = 1/3$.

▷ **Exercise 2.8.**

Let (G, c) be a network. A ***network automorphism*** of (G, c) is a map $\phi \colon G \to G$ that is a bijection of the vertex set with itself and a bijection of the edge set with itself such that if x and e are incident, then so are $\phi(x)$ and $\phi(e)$ and such that $c(e) = c(\phi(e))$ for all edges e. Suppose that (G, c) is ***spherically symmetric*** about o, meaning that if x and y are any two vertices at the same distance from o, then there is an automorphism of (G, c) that leaves o fixed and that takes x to y. Let C_n be the sum of $c(e)$ over all edges e with $d(e^-, o) = n - 1$ and $d(e^+, o) = n$. Show that

$$\mathscr{R}(o \leftrightarrow \infty) = \sum_{n \geq 1} \frac{1}{C_n},$$

whence the network random walk on G is transient iff

$$\sum_{n \geq 1} \frac{1}{C_n} < \infty.$$

III. Star-Triangle Law. The configurations in Figure 2.2 are equivalent when

$$\forall i \in \{1, 2, 3\} \qquad c(w, u_i) c(u_{i-1}, u_{i+1}) = \gamma,$$

where indices are taken mod 3 and

$$\gamma := \frac{\prod_i c(w, u_i)}{\sum_i c(w, u_i)} = \frac{\sum_i r(u_{i-1}, u_{i+1})}{\prod_i r(u_{i-1}, u_{i+1})}.$$

We won't use this equivalence, except in Example 2.7 and the exercises. This is also called the "Y-Δ" or "Wye-Delta" transformation.

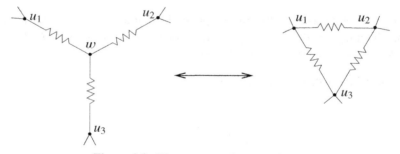

Figure 2.2. The star-triangle equivalence.

▷ **Exercise 2.9.**

Give at least one proof of the star-triangle equivalence.

Actually, there is a fourth trivial transformation: we may prune (or add) vertices of degree 1 (and attendant edges) as well as loops.

▷ **Exercise 2.10.**

Find a (finite) graph with two vertices a and z that can't be reduced to a single edge between a and z by the four transformations pruning, series, parallel, and star-triangle.

One can also use the star-triangle transformation in either direction to reduce the network in Example 2.6.

Example 2.7. What is $\mathbf{P}_x[\tau_a < \tau_z]$ in the following network? Following the transformations indicated in the figure, we obtain

$$\mathbf{P}_x[\tau_a < \tau_z] = \frac{20/33}{20/33 + 15/22} = \frac{8}{17}.$$

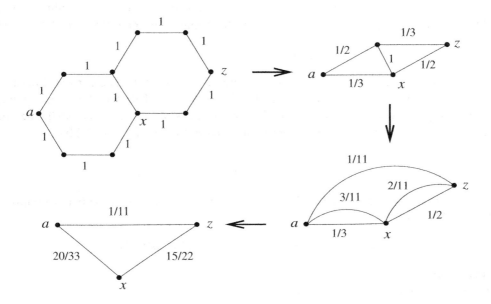

2.4 Energy

We now come to another extremely useful concept, energy.* We will begin with some convenient notation and some facts about this notation. Unfortunately, there is actually a fair bit of notation. But once we have it all in place, we will be able to quickly reap some valuable consequences. In particular, we will prove a powerful monotonicity principle due to Rayleigh: a transient network whose edge conductances are increased remains transient. This should be contrasted with the lack of monotonicity of return probabilities $p_n(a, a)$, for example, whose summability determines transience.

We will often contract some vertices in a graph, which may produce a multigraph. When we say that a graph is *finite*, we mean that V and E are finite. In this section, we consider only finite networks. Define $\ell^2(\mathsf{V})$ to be the usual real Hilbert space of functions on V with inner product

$$(f, g) := \sum_{x \in \mathsf{V}} f(x)g(x).$$

---

* Although the term energy is used for mathematical reasons, the physical concept is actually power dissipation.

Since we are interested in flows on E, it is natural to consider that what flows one way is the negative of what flows the other. ***From now on, each edge occurs with both orientations.*** Thus, define $\ell_-^2(\mathsf{E})$ to be the space of ***antisymmetric*** functions θ on E (that is, $\theta(-e) = -\theta(e)$ for each edge e) with the inner product

$$(\theta, \theta') := \frac{1}{2} \sum_{e \in \mathsf{E}} \theta(e)\theta'(e) = \sum_{e \in \mathsf{E}_{1/2}} \theta(e)\theta'(e),$$

where $\mathsf{E}_{1/2} \subset \mathsf{E}$ is a set of edges containing exactly one of each pair $e, -e$. Since voltage differences across edges lead to currents, define the ***coboundary operator*** $d \colon \ell^2(\mathsf{V}) \to \ell_-^2(\mathsf{E})$ by

$$(df)(e) := f(e^-) - f(e^+).$$

(Note that this is the negative of the more natural definition; but since current flows from greater to lesser voltage, it is the more useful definition for us.) This operator is clearly linear. Conversely, given an antisymmetric function on the edges, we are interested in the net flow out of a vertex, whence we define the ***boundary operator*** $d^* \colon \ell_-^2(\mathsf{E}) \to \ell^2(\mathsf{V})$ by

$$(d^*\theta)(x) := \sum_{e^-=x} \theta(e).$$

This operator is also clearly linear. We use the superscript * because these two operators are adjoints of each other:

$$\forall f \in \ell^2(\mathsf{V}) \ \forall \theta \in \ell_-^2(\mathsf{E}) \qquad (\theta, df) = (d^*\theta, f).$$

▷ **Exercise 2.11.**
Prove that d and d^* are adjoints of each other.

One use of this notation is that the calculation left here for Exercise 2.11 need not be repeated every time it arises – and it arises a lot. Another use is the following compact forms of the network laws. Let i be a current.

> **Ohm's Law:** $dv = ir$, that is, $\forall e \in \mathsf{E} \ \ dv(e) = i(e)r(e)$.
> **Kirchhoff's Node Law:** $d^*i(x) = 0$ if $x \notin A \cup Z$.

It will be useful to study flows other than current to discover a special property of the current flow. We can imagine water flowing through a network of pipes. Let $\theta \in \ell_-^2(\mathsf{E})$ be a function, which we think of as a flow. The amount of water flowing into the network at a vertex a is $d^*\theta(a)$. Thus, we call $\theta \in \ell_-^2(\mathsf{E})$ a ***flow between*** A to Z if $d^*\theta$ is 0 off of A and Z; if $d^*\theta$ is nonnegative on A and nonpositive on Z, then we say that θ is a flow ***from*** A to Z. The ***total amount flowing into the network*** is then $\sum_{a \in A} d^*\theta(a)$; not surprisingly, this is also the total amount flowing out of the network, as the next lemma shows. We call

$$\text{Strength}(\theta) := \sum_{a \in A} d^*\theta(a)$$

the ***strength*** of the flow θ. A flow of strength 1 is called a ***unit flow***.

Lemma 2.8. (Flow Conservation) *Let G be a finite graph and A and Z be two disjoint subsets of its vertices. If θ is a flow between A and Z, then*

$$\sum_{a \in A} d^* \theta(a) = -\sum_{z \in Z} d^* \theta(z) \,.$$

Proof. We have

$$\sum_{x \in A} d^* \theta(x) + \sum_{x \in Z} d^* \theta(x) = \sum_{x \in A \cup Z} d^* \theta(x) = (d^* \theta, \mathbf{1}) = (\theta, d\mathbf{1}) = (\theta, \mathbf{0}) = 0$$

since $d^* \theta(x) = 0$ for $x \notin A \cup Z$. ◀

The following consequence will be useful in a moment.

Lemma 2.9. *Let G be a finite graph and A and Z be two disjoint subsets of its vertices. If θ is a flow from A to Z and $f \!\restriction\! A$, $f \!\restriction\! Z$ are constants α and ζ, respectively, then*

$$(\theta, df) = \mathsf{Strength}(\theta)(\alpha - \zeta) \,.$$

Proof. We have $(\theta, df) = (d^* \theta, f) = \sum_{a \in A} d^* \theta(a)\alpha + \sum_{z \in Z} d^* \theta(z)\zeta$. Now apply Lemma 2.8. ◀

When a current i flows through a resistor of resistance r and voltage difference v, energy is dissipated at rate $P = iv = i^2 r = i^2/c = v^2 c = v^2/r$. We are interested in the total power (= energy per unit time) dissipated.

Notation. Write

$$(f, g)_h := (fh, g) = (f, gh)$$

and

$$\|f\|_h := \sqrt{(f, f)_h} \,.$$

Definition. For an antisymmetric function θ, define its **energy** to be

$$\mathscr{E}(\theta) := \|\theta\|_r^2 \,,$$

where r is the collection of resistances.

Thus $\mathscr{E}(i) = (i, i)_r = (i, dv)$. If i is a unit current flow from A to Z with voltages v_A and v_Z that are constant on A and on Z, respectively, then by Lemma 2.9 and Exercise 2.5,

$$\mathscr{E}(i) = v_A - v_Z = \mathscr{R}(A \leftrightarrow Z) \,. \tag{2.7}$$

This will be an important tool to estimate effective resistances.

The inner product $(\cdot, \cdot)_r$ is important not only for its squared norm $\mathscr{E}(\cdot)$. For example, we may express Kirchhoff's laws as follows. Let $\chi^e := \mathbf{1}_{\{e\}} - \mathbf{1}_{\{-e\}}$ denote the unit flow along e represented as an antisymmetric function in $\ell^2_-(\mathsf{E})$. Note that for every antisymmetric function θ and every e, we have

$$(\chi^e, \theta)_r = \theta(e)r(e),$$

so that

$$\left(\sum_{e^-=x} c(e)\chi^e, \theta \right)_r = d^*\theta(x). \tag{2.8}$$

Let i be any current.

Kirchhoff's Node Law: For every vertex $x \notin A \cup Z$, we have

$$\left(\sum_{e^-=x} c(e)\chi^e, i \right)_r = 0.$$

Kirchhoff's Cycle Law: If $e_1, e_2, \ldots, e_n$ is an oriented cycle in G, then

$$\left(\sum_{k=1}^{n} \chi^{e_k}, i \right)_r = 0.$$

Now for our last bit of notation before everything comes into focus, let $\sum_{e^-=x} c(e)\chi^e$ be the *star* at x, and let $\bigstar$ denote the subspace in $\ell^2_-(\mathsf{E})$ spanned by all the stars. Let $\Diamond$ denote the subspace spanned by the *cycles* $\sum_{k=1}^{n} \chi^{e_k}$, where $e_1, e_2, \ldots, e_n$ forms an oriented cycle. We call these subspaces the *star space* and the *cycle space* of G. These subspaces are clearly orthogonal to each other; ***here and subsequently, orthogonality refers to the inner product*** $(\cdot, \cdot)_r$. To indicate that we use the inner product $(\cdot, \cdot)_r$ on $\ell^2_-(\mathsf{E})$, we write the space as $\ell^2_-(\mathsf{E}, r)$. Moreover, the sum of $\bigstar$ and $\Diamond$ is all of $\ell^2_-(\mathsf{E}, r)$, which is the same as saying that only the zero vector is orthogonal to both $\bigstar$ and $\Diamond$. To see that this is the case, suppose that $\theta \in \ell^2_-(\mathsf{E}, r)$ is orthogonal to both $\bigstar$ and $\Diamond$. Since θ is orthogonal to $\Diamond$, there is a function F such that $\theta = c\,dF$ by Exercise 2.2 (use $W := \varnothing$ there). Since θ is orthogonal to $\bigstar$, the function F is harmonic. Since G is finite, the uniqueness principle implies that F is constant on each component of G, whence $\theta = \mathbf{0}$, as desired.

Thus, Kirchhoff's cycle law says that i, being orthogonal to $\Diamond$, is in $\bigstar$. Furthermore, any $i \in \bigstar$ is a current by Exercise 2.2 (let $W := \{x \,;\, (d^*i)(x) = 0\}$). Now if θ is any flow with the same sources and sinks as i, more precisely, if θ is any antisymmetric function such that $d^*\theta = d^*i$, then $\theta - i$ is a sourceless flow; in other words, by (2.8), it is orthogonal to $\bigstar$ and thus is an element of $\Diamond$. Therefore, the expression

$$\theta = i + (\theta - i)$$

is the orthogonal decomposition of θ relative to $\ell^2_-(\mathsf{E}, r) = \bigstar \oplus \Diamond$. This hints that the orthogonal projection $P_{\bigstar} : \ell^2_-(\mathsf{E}, r) \to \bigstar$ plays a crucial role in network theory. In particular,

$$i = P_{\bigstar}\theta \tag{2.9}$$

and

$$\|\theta\|_r^2 = \|i\|_r^2 + \|\theta - i\|_r^2. \tag{2.10}$$

This leads to the following all-important principle:

Thomson's Principle. *Let G be a finite network and A and Z be two disjoint subsets of its vertices. Let θ be a flow from A to Z and i be the current flow from A to Z with $d^*i = d^*\theta$. Then $\mathscr{E}(\theta) > \mathscr{E}(i)$ unless $\theta = i$.*

Proof. The result is an immediate consequence of (2.10). ◀

Note that given θ, the corresponding current i such that $d^*i = d^*\theta$ is unique (and given by (2.9)).

Recall that $E_{1/2} \subset E$ is a set of edges containing exactly one of each pair $e, -e$. This gives a convenient orthogonal basis $\{\chi^e \; ; \; e \in E_{1/2}\}$ of $\ell^2_-(E, r)$; these vectors are not necessarily unit vectors, as $\|\chi^e\|^2_r = r(e)$. What is the matrix of $P_\bigstar$ in the orthogonal basis $\{\chi^e \; ; \; e \in E_{1/2}\}$? We have

$$(P_\bigstar \chi^e, \chi^{e'})_r = (i^e, \chi^{e'})_r = i^e(e')r(e'), \qquad (2.11)$$

where i^e is the unit current flow from e^- to e^+. Therefore, the matrix coefficient at (e', e) equals $(P_\bigstar \chi^e, \chi^{e'})_r / (\chi^{e'}, \chi^{e'})_r = i^e(e') =: Y(e, e')$, the current that flows across e' when a unit current is imposed between the endpoints of e. This matrix is called the ***transfer-current matrix***. This matrix will be extremely useful for our study of random spanning trees and forests in Chapters 4 and 10. Since $P_\bigstar$, being an orthogonal projection, is self-adjoint, we have $(P_\bigstar \chi^e, \chi^{e'})_r = (\chi^e, P_\bigstar \chi^{e'})_r$, whence

$$Y(e, e')r(e') = Y(e', e)r(e). \qquad (2.12)$$

This is called the ***reciprocity law***.

Recall that the escape probabilities $\mathbf{P}[a \to Z]$ were important to determining whether a network was recurrent or transient. Let's use our new concepts to analyze these probabilities. For example, how do the escape probabilities change when an edge is removed from G? When an edge is added? When the conductance of an edge is changed? These questions are not easy to answer probabilistically but yield to the ideas we have developed. Since $\mathbf{P}[a \to Z] = \mathscr{C}(a \leftrightarrow Z)/\pi(a)$, if no edge incident to a is affected, then we need analyze only the change in effective conductance.

▷ **Exercise 2.12.**
Show that $\mathbf{P}[a \to Z]$ can increase in some situations and decrease in others when an edge incident to a is removed.

The following powerful principle tells us how effective conductance changes. We use subscripts to indicate the edge conductances used.

Rayleigh's Monotonicity Principle. *Let G be a connected graph with two assignments, c and c', of conductances on G with $c \le c'$ (everywhere).*
  (i) *If G is finite and A and Z two disjoint subsets of its vertices, then $\mathscr{C}_c(A \leftrightarrow Z) \le \mathscr{C}_{c'}(A \leftrightarrow Z)$.*
  (ii) *If G is infinite and a is one of its vertices, then $\mathscr{C}_c(a \leftrightarrow \infty) \le \mathscr{C}_{c'}(a \leftrightarrow \infty)$. In particular, if (G, c) is transient, then so is (G, c').*

Proof. Part (ii) is immediate from part (i), so we concentrate on part (i). By (2.7), we have $\mathscr{C}(A \leftrightarrow Z) = 1/\mathscr{E}(i)$ for a unit current flow i from A to Z. Now

$$\mathscr{E}_c(i_c) \geq \mathscr{E}_{c'}(i_c) \geq \mathscr{E}_{c'}(i_{c'}),$$

where the first inequality follows from the definition of energy and the second from Thomson's principle (after identifying Z to a singleton). Taking reciprocals gives the result. ◄

In particular, removing an edge decreases effective conductance, so if the edge is not incident to a, then its removal decreases $\mathbf{P}[a \to Z]$. In addition, contracting an edge (called "shorting" in electrical network theory), that is, identifying its two endpoints and removing the resulting loop, increases the effective conductance between any sets of vertices. This is intuitive from thinking of increasing to infinity the conductance on the edge to be contracted, so we will still refer to it as part of Rayleigh's monotonicity principle. To prove it rigorously, let i be the unit current flow from A to Z. If the graph G with the edge e contracted is denoted G/e, then the edge set of G/e may be identified with $\mathsf{E}(G) \setminus \{e\}$. If e does not connect A to Z, then the restriction θ of i to the edges of G/e is a unit flow from A to Z, whence the effective resistance between A and Z in G/e is at most $\mathscr{E}(\theta)$, which is at most $\mathscr{E}(i)$, which equals the effective resistance in G.

▷ **Exercise 2.13.**

Given disjoint vertex sets, A and Z, in a finite network, we may express the effective resistance between A and Z by Thomson's principle as

$$\mathscr{R}(A \leftrightarrow Z) = \min \left\{ \sum_{e \in \mathsf{E}_{1/2}} r(e)\theta(e)^2 \ ; \ \theta \text{ is a unit flow from } A \text{ to } Z \right\}.$$

Prove the following dual expression for the effective conductance, known as ***Dirichlet's principle***:

$$\mathscr{C}(A \leftrightarrow Z) = \min \left\{ \sum_{e \in \mathsf{E}_{1/2}} c(e)dF(e)^2 \ ; \ F{\upharpoonright}A \equiv 1, \ F{\upharpoonright}Z \equiv 0 \right\}.$$

2.5 Transience and Recurrence

We have seen that effective conductance from any vertex to ∞ is positive iff the random walk is transient. Thus, a lower bound on the effective resistance between vertices in a network can be useful to show recurrence. To use the energy formulation of effective resistance, (2.7), we use the following notion. Let A and Z be two disjoint sets of vertices. A set Π of edges ***separates A and Z*** if every path with one endpoint in A and the other endpoint in Z must include an edge in Π; we also call Π a ***cutset***.

The Nash-Williams Inequality. *If a and z are distinct vertices in a finite network that are separated by pairwise disjoint cutsets $\Pi_1, \ldots, \Pi_n$, then*

$$\mathscr{R}(a \leftrightarrow z) \geq \sum_{k=1}^{n} \left(\sum_{e \in \Pi_k} c(e) \right)^{-1} . \tag{2.13}$$

Proof. By (2.7), it suffices to show that the unit current flow i from a to z has energy at least the right-hand side. Now given a finite cutset Π that separates a from z, let Z be the set of endpoints of Π that are separated by Π from a. Let K denote the set of vertices that are *not* separated from a by Π. Let $H := G {\upharpoonright} (K \cup Z)$ be the subnetwork of G induced by $K \cup Z$. Then i induces a unit flow i_H from a to Z, whence Lemma 2.8 applied to H gives $1 = -\sum_{x \in Z} d^* i_H(x) = -\sum_{e^- \in Z, e \in H} i(e)$. If both the head and tail of e happen to lie in Z, then $i(e)$ occurs together with $i(-e)$ in that sum, so they cancel. Also, all edges in H with only one endpoint in Z must lie in Π. Therefore, $\sum_{e \in \Pi} |i(e)| \geq 1$, and so the Cauchy–Schwarz inequality gives

$$\sum_{e \in \Pi} i(e)^2 r(e) \sum_{e \in \Pi} c(e) \geq \left(\sum_{e \in \Pi} |i(e)| \right)^2 \geq 1 .$$

In other words,

$$\sum_{e \in \Pi} i(e)^2 r(e) \geq \left(\sum_{e \in \Pi} c(e) \right)^{-1} .$$

Substitute $\Pi = \Pi_k$ and add for $k = 1, \ldots, n$. ◀

To apply this to infinite networks, say that a set Π of edges *separates a and ∞* if every infinite, simple path from a must include an edge in Π; we also call Π a *cutset*.

The Nash-Williams Criterion. *If $\langle \Pi_n \rangle$ is a sequence of pairwise disjoint, finite cutsets in a locally finite network G, each of which separates a from ∞, then*

$$\mathscr{R}(a \leftrightarrow \infty) \geq \sum_{n} \left(\sum_{e \in \Pi_n} c(e) \right)^{-1} . \tag{2.14}$$

In particular, if the right-hand side is infinite, then G is recurrent.

Proof. For $n \geq 1$, choose a finite subnetwork G_n that contains $\bigcup_{k=1}^{n} \Pi_k$ and identify its complementary set of vertices to a single vertex, z_n, as usual, to form the finite network G_n^{W}. Then $\mathscr{R}(a \leftrightarrow \infty) = \lim_{n \to \infty} \mathscr{R}(a \leftrightarrow z_n)$, so (2.13) gives the conclusion. ◀

Remark 2.10. If the cutsets can be ordered so that Π_1 separates a from Π_2 and for $n > 1$, Π_n separates Π_{n-1} from Π_{n+1}, then the sum appearing in the statement of this criterion has a natural interpretation: Short together (that is, join by edges of infinite conductance, or, in other words, identify) all the vertices between Π_n and Π_{n+1} into one vertex U_n. Short all the vertices that Π_1 separates from ∞ into one vertex U_0. Then only parallel edges of Π_n join U_{n-1} to U_n. Replace these edges by a single edge of resistance $\left(\sum_{e \in \Pi_n} c(e) \right)^{-1}$. This new network is a series network with effective resistance from U_0 to ∞ equal to the right-hand side of (2.14). Thus, Rayleigh's monotonicity principle shows that the effective resistance from a to ∞ in G is at least the right-hand side of (2.14).

The Nash-Williams criterion allows us to prove the first part of Pólya's (1921) famous and beautiful theorem concerning random walk on the integer lattices.

Pólya's Theorem (first part). *Simple random walk on the nearest-neighbor graph of $\mathbb{Z}^d$ is recurrent for $d = 1, 2$.*

Proof. For $d = 1, 2$, we can use the Nash-Williams criterion with cutsets

$$\Pi_n := \left\{ e \, ; \, d(\mathbf{0}, e^-) = n - 1, \, d(\mathbf{0}, e^+) = n \right\},$$

where $\mathbf{0}$ is the origin and $d(\cdot, \cdot)$ is the graph distance.   ◄

To show that simple random walk on $\mathbb{Z}^d$ is transient for $d \geq 3$, we need another technique. It is more involved than the very simple technique we just used to prove recurrence, but it is also very powerful. In fact, it involves a condition that is both necessary and sufficient for transience.

If $G = (\mathsf{V}, \mathsf{E})$ is a denumerable network, let

$$\ell^2(\mathsf{V}) := \left\{ f : \mathsf{V} \to \mathbb{R} \, ; \, \sum_{x \in \mathsf{V}} f(x)^2 < \infty \right\}$$

with the inner product $(f, g) := \sum_{x \in \mathsf{V}} f(x) g(x)$. Define the Hilbert space

$$\ell_-^2(\mathsf{E}, r) := \left\{ \theta : \mathsf{E} \to \mathbb{R} \, ; \, \forall e \quad \theta(-e) = -\theta(e) \text{ and } \sum_{e \in \mathsf{E}_{1/2}} \theta(e)^2 r(e) < \infty \right\}$$

with the inner product $(\theta, \theta')_r := \sum_{e \in \mathsf{E}_{1/2}} \theta(e) \theta'(e) r(e)$ and $\mathscr{E}(\theta) := (\theta, \theta)_r$. Define $df(e) := f(e^-) - f(e^+)$ as before. If $\sum_{e^- = x} |\theta(e)| < \infty$, then we also define $(d^*\theta)(x) := \sum_{e^- = x} \theta(e)$.

Suppose now that V is finite and $\sum_e |\theta(e)| < \infty$. Then the calculation of Exercise 2.11 shows that we still have $(\theta, df) = (d^*\theta, f)$ for all f. (Here, we use (ϕ, ψ) to mean the sum $\sum_\alpha \phi(\alpha) \psi(\alpha)$ for functions on any space where the sum converges absolutely.) Likewise, under these hypotheses, we have Lemmas 2.8 and 2.9 still holding. The remainder of Section 2.4 also then holds because of the following consequence of the Cauchy–Schwarz inequality:

$$\forall x \in \mathsf{V} \quad \sum_{e^- = x} |\theta(e)| \leq \sqrt{\sum_{e^- = x} \theta(e)^2 / c(e) \cdot \sum_{e^- = x} c(e)} \leq \sqrt{\mathscr{E}(\theta) \pi(x)}. \tag{2.15}$$

In particular, if $\mathscr{E}(\theta) < \infty$, then $d^*\theta$ is defined.

▷ **Exercise 2.14.**

Let $G = (\mathsf{V}, \mathsf{E})$ be denumerable and $\theta_n \in \ell_-^2(\mathsf{E}, r)$ be such that $\mathscr{E}(\theta_n) \leq M < \infty$ and $\theta_n \to \theta$ edgewise, that is, $\theta_n(e) \to \theta(e)$ for each $e \in \mathsf{E}$. Show that θ is antisymmetric, $\mathscr{E}(\theta) \leq \liminf_n \mathscr{E}(\theta_n) \leq M$, and $\forall x \in \mathsf{V} \ d^*\theta_n(x) \to d^*\theta(x)$.

Call an antisymmetric function θ on the edges E of a possibly infinite graph a *flow* if

$$\forall x \in \mathsf{V} \quad \sum_{e^- = x} |\theta(e)| < \infty.$$

If, in addition, θ satisfies $(d^*\theta)(x) = \mathbf{1}_{\{a\}}(x)$, then θ is a **unit flow** from $a \in \mathsf{V}$ to ∞.

Our main theorem is the following criterion for transience due to T. Lyons (1983). It is adapted from a theorem of Royden (1952).

Theorem 2.11. (Energy and Transience) *Let G be a denumerable, connected network. Random walk on G is transient iff there is a unit flow on G of finite energy from some (every) vertex to ∞.*

Proof. Let G_n be finite, induced subgraphs that exhaust G. Recall that G_n^W is the graph obtained from G by identifying the vertices outside G_n to a single vertex, z_n, and then removing loops (but keeping multiple edges). Fix any vertex $a \in G$, which, without loss of generality, belongs to each G_n. We have, by definition, $\mathscr{R}(a \leftrightarrow \infty) = \lim \mathscr{R}(a \leftrightarrow z_n)$. Let i_n be the unit current flow in G_n^W from a to z_n and v_n be the corresponding voltage. Then $\mathscr{E}(i_n) = \mathscr{R}(a \leftrightarrow z_n)$, so $\mathscr{R}(a \leftrightarrow \infty) < \infty \Leftrightarrow \lim \mathscr{E}(i_n) < \infty$.

Note that each edge of G_n^W comes from an edge in G and may be identified with it, even though one endpoint may be different.

If θ is a unit flow on G from a to ∞ that has finite energy, then the restriction $\theta \restriction G_n^W$ of θ to G_n^W is a unit flow from a to z_n, whence Thomson's principle gives

$$\mathscr{E}(i_n) \le \mathscr{E}(\theta \restriction G_n^W) \le \mathscr{E}(\theta) < \infty .$$

In particular, $\lim \mathscr{E}(i_n) < \infty$, and so the random walk is transient.

Conversely, suppose that G is transient. Then there is some $M < \infty$ such that $\mathscr{E}(i_n) \le M$ for all n. Start a random walk at a. Let $Y_n(x)$ be the number of visits to x before hitting $G \setminus G_n$ and $Y(x)$ be the total number of visits to x. Then $Y_n(x)$ increases to $Y(x)$, whence the monotone convergence theorem and Proposition 2.1 imply that $\mathbf{E}[Y(x)] = \lim_{n \to \infty} \mathbf{E}[Y_n(x)] = \lim_{n \to \infty} \pi(x)v_n(x) =: \pi(x)v(x)$. By transience, we know that $\mathbf{E}[Y(x)] < \infty$, whence $v(x) < \infty$. Hence $i := c \cdot dv = \lim_{n \to \infty} c \cdot dv_n = \lim_{n \to \infty} i_n$ exists and is a unit flow from a to infinity of energy at most M by Exercise 2.14. ◀

This allows us to carry over the remainder of the electrical apparatus to infinite networks:

Proposition 2.12. *Let G be a transient, connected network and G_n be finite, induced subnetworks that contain a vertex a and that exhaust G. Identify the vertices outside G_n to z_n, forming G_n^W. Let i_n be the unit current flow in G_n^W from a to z_n. Then $\langle i_n \rangle$ has a pointwise limit i on G, which is the unique unit flow on G from a to ∞ of minimum energy. Let v_n be the voltages on G_n^W corresponding to i_n and with $v_n(z_n) := 0$. Then $v := \lim v_n$ exists on G and has the following properties:*

$$dv = ir ,$$

$$v(a) = \mathscr{E}(i) = \mathscr{R}(a \leftrightarrow \infty) ,$$

$$\forall x \qquad v(x)/v(a) = \mathbf{P}_x[\tau_a < \infty] .$$

Start a random walk at a. For all vertices x, the expected number of visits to x is $\mathscr{G}(a, x) = \pi(x)v(x)$. For all edges e, the expected signed number of crossings of e is $i(e)$.

Proof. We saw in the proof of Theorem 2.11 that v and i exist, that $dv = ir$, and that $\mathscr{G}(a, x) = \pi(x)v(x)$. The proof of Proposition 2.2 now applies as written for the last claim of the proposition. Since the events $[\tau_a < \tau_{G \setminus G_n}]$ are increasing in n with union $[\tau_a < \infty]$, we have (with superscript indicating on which network the random walk takes place)

$$v(x)/v(a) = \lim_n v_n(x)/v_n(a) = \lim_n \mathbf{P}_x^{G_n^W}[\tau_a < \tau_{z_n}] = \lim_n \mathbf{P}_x^G[\tau_a < \tau_{G \setminus G_n}] = \mathbf{P}_x^G[\tau_a < \infty] .$$

Now $v(a) = \lim v_n(a) = \lim \mathscr{E}(i_n) = \lim \mathscr{R}(a \leftrightarrow z_n) = \mathscr{R}(a \leftrightarrow \infty)$. By Exercise 2.14, $\mathscr{E}(i) \leq \liminf \mathscr{E}(i_n)$. Since $\mathscr{E}(i_n) \leq \mathscr{E}(i)$ as in the proof of Theorem 2.11, we have $\mathscr{E}(i) = \lim \mathscr{E}(i_n) = v(a)$. Likewise, $\mathscr{E}(i_n) \leq \mathscr{E}(\theta)$ for every unit flow from a to infinity, whence i has minimum energy.

Finally, we establish uniqueness of a unit flow (from a to ∞) with minimum energy. Note that $\forall \theta, \theta'$

$$\frac{\mathscr{E}(\theta) + \mathscr{E}(\theta')}{2} = \mathscr{E}\left(\frac{\theta + \theta'}{2}\right) + \mathscr{E}\left(\frac{\theta - \theta'}{2}\right). \tag{2.16}$$

Therefore, if θ and θ' both have minimum energy, so does $(\theta + \theta')/2$, and hence $\mathscr{E}((\theta - \theta')/2) = 0$, which gives $\theta = \theta'$.   ◄

Thus, we may call i the unit current flow and v the voltage on G. We may think of G as grounded (that is, has 0 voltage) at infinity.

By Theorem 2.11 and Rayleigh's monotonicity principle, the **type** of a random walk, that is, transient or recurrent, does not change when the conductances are changed by bounded factors. This fact is by no means clear probabilistically. An extensive generalization of this is given in Theorem 2.17.

The question now arises: how do we determine whether there is a flow from a to ∞ of finite energy? There is no recipe, but a very useful technique involves flows created from random paths. Suppose that $\mathbf{P}$ is a probability measure on paths $\langle e_n ; n \geq 0 \rangle$ from a to z on a finite graph or from a to ∞ on an infinite graph. (An infinite path is said to **go to** ∞ when no vertex is visited infinitely many times.) Define

$$\theta(e) := \sum_{n \geq 0} \left(\mathbf{P}[e_n = e] - \mathbf{P}[e_n = -e]\right), \tag{2.17}$$

provided

$$\sum_{n \geq 0} \left(\mathbf{P}[e_n = e] + \mathbf{P}[e_n = -e]\right) < \infty. \tag{2.18}$$

For example, the summability condition (2.18) holds when the paths are edge-simple, since the sum on the left in (2.18) equals the expected number of times that e is traversed in either direction. Each path $\langle e_n ; n \geq 0 \rangle$ determines a unit flow ψ from a to z (or to ∞) by sending 1 along each edge in the path:

$$\psi := \sum_{n \geq 0} \chi^{e_n}.$$

If (2.18) holds for all e, then θ is defined everywhere. Now θ is an expectation of a random unit flow, so that θ is a unit flow itself. We saw in Propositions 2.2 and 2.12 that this is precisely how network random walks and unit electric current are related (where the walk $\langle X_n ; n \geq 0 \rangle$ gives rise to the path $\langle e_n ; n \geq 0 \rangle$ with $e_n := \langle X_n, X_{n+1} \rangle$). However, there are other useful pairs of random paths and their expected flows as well.

We now illustrate the preceding techniques. First, we complete Pólya's theorem by the random-path method. The resulting flow is essentially the same as the one used by T. Lyons in his 1983 proof of Pólya's theorem (which also occurs on p. 173 of Mori (1954)).

Pólya's Theorem (second part). *Simple random walk on the nearest-neighbor graph of $\mathbb{Z}^d$ is transient for all $d \geq 3$.*

Proof. By Rayleigh's monotonicity principle, it suffices to do $d = 3$. Let L be a random uniformly distributed ray from the origin $\mathbf{0}$ of $\mathbb{R}^3$ to ∞ (that is, a straight half-line with uniform intersection on the unit sphere). Let $\mathcal{P}(L)$ be a simple path in $\mathbb{Z}^3$ from $\mathbf{0}$ to ∞ that stays within distance 4 of L; choose $\mathcal{P}(L)$ measurably, such as recursively growing a path by successively adding a vertex closest to L. Define the flow θ from the law of $\mathcal{P}(L)$ via (2.17). Then θ is a unit flow from $\mathbf{0}$ to ∞; we claim it has finite energy. There is some constant A such that if e is an edge whose midpoint is at Euclidean distance R from $\mathbf{0}$, then $\mathbf{P}[e \in \mathcal{P}(L)] \leq A/R^2$. Since all edge centers are separated from each other by Euclidean distance at least $1/\sqrt{2}$, there is also a constant B such that there are at most Bn^2 edge centers whose distance from the origin is between n and $n + 1$. It follows that the energy of θ is at most $\sum_n A^2 B n^2 n^{-4}$, which is finite. Now transience follows from Theorem 2.11. ◀

Remark 2.13. The continuous case, that is, Brownian motion in $\mathbb{R}^3$, is easier to handle (after establishing a similar relationship to an electrical framework) because of its spherical symmetry; see Section 2.9, the notes to this chapter. Here, we are approximating this continuous case in our solution. One can in fact use the transience of the continuous case to deduce that of the discrete case (or vice versa); see Theorem 2.26 in the notes.

The difference between two and three dimensions is illustrated in Figure 2.3. For information on the asymptotic behavior of these figures in dimension 2, see Dembo, Peres, Rosen, and Zeitouni (2001).

Because the harmonic series, which arises in the recurrence of $\mathbb{Z}^2$, just barely diverges, it seems that the change from recurrence to transience occurs "just after" dimension 2, rather than somewhere else in $[2, 3]$. One way to explore this is to ask about the type of spaces intermediate between $\mathbb{Z}^2$ and $\mathbb{Z}^3$. For example, consider the wedge

$$W_f := \left\{ (x, y, z) \, ; \, |z| \leq f(|x|) \right\},$$

where $f \colon \mathbb{N} \to \mathbb{N}$ is an increasing function. The number of edges that leave the portion $W_f \cap \{(x, y, z) \, ; \, |x| \vee |y| \leq n\}$ is of the order $n(f(n) + 1)$, so that according to the Nash-Williams criterion,

$$\sum_{n \geq 1} \frac{1}{n(f(n) + 1)} = \infty \tag{2.19}$$

is sufficient for recurrence.

▷ **Exercise 2.15.**
Show that (2.19) is also necessary for recurrence if $f(n + 1) \leq f(n) + 1$ for all n.

Because simple random walk on $\mathbb{Z}^2$ is recurrent, the effective resistance from the origin to distance n tends to infinity – but how quickly? Our techniques are good enough to answer this within a constant factor.

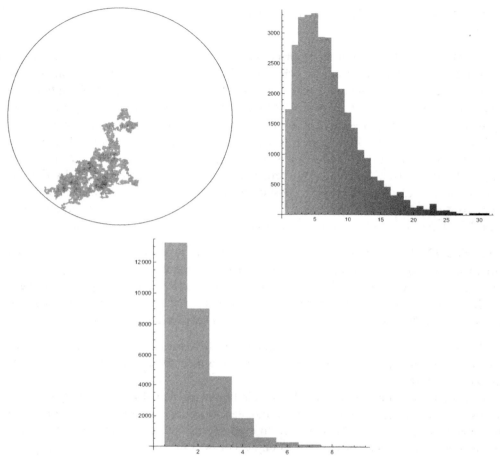

^{color}_{plate} **Figure 2.3.** Random walk until it goes distance 200 from its starting point, colored
according to the number of visits at a vertex. The histogram shows the time spent at vertices
that were visited n times for each $n \geq 1$, with the same color coding. For three dimensions,
only the histogram is shown, which is approximately a geometric distribution.

Proposition 2.14. *There are positive constants C_1, C_2 such that if one identifies to a single
vertex z_n all vertices of $\mathbb{Z}^2$ that are at distance more than n from $\mathbf{0}$, then*

$$C_1 \log n \leq \mathscr{R}(\mathbf{0} \leftrightarrow z_n) \leq C_2 \log n .$$

Proof. The lower bound is an immediate consequence of (2.13) applied to the cutsets Π_k
used in our proof of Pólya's theorem. The upper bound follows from the estimate of the
energy of the unit flow analogous to that used for the transience of $\mathbb{Z}^3$. That is, $\theta(e)$ is defined
via (2.17) from a uniform ray emanating from the origin. Then θ defines a unit flow from $\mathbf{0}$
to z_n and its energy is bounded by $C_2 \log n$.   ◀

We can extend Proposition 2.14 as follows.

Proposition 2.15. *For $d \geq 2$, there is a positive constant C_d such that if G_n is the subnetwork of $\mathbb{Z}^d$ induced on the vertices in a box of side length n, then for any pair of vertices x, y in G_n at mutual distance k,*

$$\mathscr{R}(x \leftrightarrow y; G_n) \in \begin{cases} (C_d^{-1} \log k, C_d \log k) & \text{if } d = 2 \\ (C_d^{-1}, C_d) & \text{if } d \geq 3. \end{cases}$$

Proof. The lower bounds follow from (2.13). For the upper bounds, we give the details for $d = 2$ only. There is a straight-line segment L of length k inside the portion of $\mathbb{R}^2$ that corresponds to G_n such that L meets the straight line M joining x and y at the midpoint of M in a right angle, as in the figure. Let Q be a random uniform point on L. Write $L(Q)$ for the union of two straight-line segments, one from x to Q and the other from Q to y. Let $\mathcal{P}(Q)$ be a path in G_n from x to y that is closest to $L(Q)$. Use the law $\mathbf{P}$ of $\mathcal{P}(Q)$ to define the unit flow θ as in (2.17). Then $\mathscr{E}(\theta) \leq C_2 \log k$ for some C_2, as in the proof of Proposition 2.14. ◀

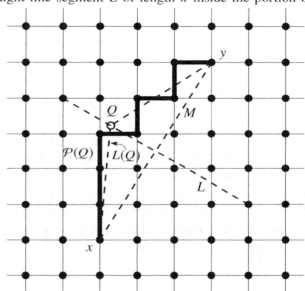

2.6 Rough Isometries and Hyperbolic Graphs

The most direct proof of Pólya's theorem goes by calculation of the Green function and is not hard; see Exercise 2.100. However, that calculation depends on the precise structure of the graph. The proof in the preceding section begins to show that the type doesn't change when fairly drastic changes are made in the lattice graph. Suppose, for example, that diagonal edges are added to the square lattice in the plane. Then clearly we can still use the Nash-Williams criterion to show recurrence. Of course, a similar addition of edges in higher dimensions preserves transience simply by Rayleigh's monotonicity principle. But suppose that in $\mathbb{Z}^3$, we remove each edge $[(x, y, z), (x, y, z + 1)]$ with $x + y$ odd. Is the resulting graph H still transient? If so, by how much has the effective resistance to infinity changed?

Notice that graph distances haven't changed much after these edges are removed. In fact, if we define the *k-fuzz* of a graph G by adjoining to the edges of G a new edge between every pair of vertices whose distance in G lies between 2 and k, then the graph H above has a 3-fuzz that includes the original graph on $\mathbb{Z}^3$. Thus, we can solve problems like the preceding by the following theorem:

Theorem 2.16. *Let G be a connected graph of bounded degree and k a positive integer. Then G and the k-fuzz of G have the same type, that is, both are transient or both are recurrent.*

We will establish an even more powerful result. To motivate an extension of the preceding from graphs to networks, think of the resistance $r(e)$ as the length of the edge e.

Given two networks G and G' with resistances r and r', we say that a map ϕ from the vertices of G to the vertices of G' is a ***rough embedding*** if there are constants $\alpha, \beta < \infty$ and a map Φ defined on the edges of G such that

(i) for every edge $\langle x, y \rangle \in G$, $\Phi(\langle x, y \rangle)$ is a nonempty, simple, oriented path of edges in G' from $\phi(x)$ to $\phi(y)$ with

$$\sum_{e' \in \Phi(\langle x,y \rangle)} r'(e') \le \alpha r(x, y)$$

and $\Phi(\langle y, x \rangle)$ is the reverse of $\Phi(\langle x, y \rangle)$;

(ii) for every edge $e' \in G'$, there are no more than β edges in G whose image under Φ contains e'.

If we need to refer to the constants, we call such a map (α, β)-***rough***. We call two networks ***roughly equivalent*** if there are rough embeddings in both directions. For example, every two Euclidean lattices of the same dimension are roughly equivalent. Also, for every graph G of bounded degree and every k, the graph G and its k-fuzz are roughly equivalent. Kanai (1986) showed that rough embeddings preserve transience:

Theorem 2.17. (Rough Embeddings and Transience) *If G and G' are roughly equivalent connected networks, then G is transient iff G' is transient. In fact, if there is a rough embedding from G to G' and G is transient, then G' is transient.*

Proof. Suppose that G is transient and ϕ is an (α, β)-rough embedding from G to G'. Let θ be a unit flow on G of finite energy from a to infinity. We will use Φ to carry the flow θ to a unit flow θ' on G' that will have finite energy. Namely, define

$$\theta'(e') := \sum_{e' \in \Phi(e)} \theta(e).$$

(The sum goes over all edges, not merely those in $\mathsf{E}_{1/2}$.) It is easy to see that θ' is antisymmetric and $d^*\theta'(x') = \sum_{x \in \phi^{-1}(\{x'\})} d^*\theta(x)$ for all $x' \in G'$. Thus, θ' is a unit flow from $\phi(a)$ to infinity.

Now

$$\theta'(e')^2 \le \beta \sum_{e' \in \Phi(e)} \theta(e)^2$$

by the Cauchy–Schwarz inequality and the condition (ii). Therefore,

$$\sum_{e' \in \mathsf{E}'} \theta'(e')^2 r'(e') \le \beta \sum_{e' \in \mathsf{E}'} \sum_{e' \in \Phi(e)} \theta(e)^2 r'(e') = \beta \sum_{e \in \mathsf{E}} \sum_{e' \in \Phi(e)} \theta(e)^2 r'(e')$$

$$\le \alpha\beta \sum_{e \in \mathsf{E}} \theta(e)^2 r(e) < \infty. \qquad \blacktriangleleft$$

▷ **Exercise 2.16.**
Show that if we remove each edge $[(x, y, z), (x, y, z + 1)]$ in $\mathbb{Z}^3$ with $x + y$ odd, then we obtain a transient graph with effective resistance to infinity at most 6 times what it was before removal.

A closely related notion is that of rough isometry, also called quasi-isometry. Given two graphs $G = (\mathsf{V}, \mathsf{E})$ and $G' = (\mathsf{V}', \mathsf{E}')$, call a function $\phi : \mathsf{V} \to \mathsf{V}'$ a **_rough isometry_** if there are positive constants α and β such that for all $x, y \in \mathsf{V}$,

$$\alpha^{-1} \operatorname{dist}(x, y) - \beta \le \operatorname{dist}'\big(\phi(x), \phi(y)\big) \le \alpha \operatorname{dist}(x, y) + \beta \tag{2.20}$$

and such that every vertex in G' is within distance β of the image of V. Here, dist and dist$'$ denote the usual graph distances on G and G'. The function ϕ need not be a bijection. In fact, the same definition applies to metric spaces, with "vertex" replaced by "point." Thus, $\mathbb{Z}^d$ is roughly isometric to $\mathbb{R}^d$.

▷ **Exercise 2.17.**
Show that being roughly isometric is an equivalence relation.

For example, if G and G' are both the usual graph on $\mathbb{Z}$ and $\phi(x) := 4x$, then ϕ is a rough isometry; similarly if $\phi(x) := \lfloor x/4 \rfloor$. Also, if G is any graph and H is any finite graph, then G and $G \times H$ are roughly isometric for most reasonable notions of product graph.

▷ **Exercise 2.18.**
Show that $\mathbb{Z}$ and $\mathbb{Z}^2$ are not roughly isometric graphs.

Proposition 2.18. (Rough Isometry and Rough Equivalence) *Let G and G' be two infinite, roughly isometric graphs with conductances c and c'. If c, c', c^{-1}, c'^{-1} are all bounded and the degrees in G and G' are all bounded, then G is roughly equivalent to G'.*

▷ **Exercise 2.19.**
Prove Proposition 2.18.

Not only can we also use rough isometries and Theorem 2.17 to understand lots of perturbations of the regular graph $\mathbb{Z}^d$, but we can also use them to give a very simple proof of Pólya's theorem itself. First, consider simple random walk in one dimension. The probability of return to the origin after $2n$ steps is exactly $\binom{2n}{n} 2^{-2n}$. Stirling's formula shows that this is asymptotic to $1/\sqrt{\pi n}$. Since this series is not summable, the random walk is recurrent. If we consider random walk in d dimensions where each coordinate is independent of the other coordinates and does simple random walk in one dimension, then the return probability after $2n$ steps is $\left(\binom{2n}{n} 2^{-2n}\right)^d \sim (\pi n)^{-d/2}$. This is summable precisely when $d \ge 3$. On the other hand, this independent-coordinate walk is simple random walk on another graph whose

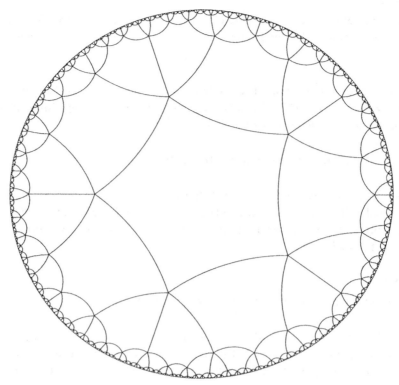

Figure 2.4. A graph in the hyperbolic disc formed from congruent regular hyperbolic pentagons of interior angle $2\pi/5$. This was drawn by a program created by Don Hatch.

vertices are a subset of $\mathbb{Z}^d$, and this other graph is clearly roughly isometric to the usual graph on $\mathbb{Z}^d$. Thus, we deduce Pólya's theorem.

We now go beyond Euclidean space to examine another very nice family of graphs that will serve as useful examples throughout the rest of the text. These graphs are roughly isometric to hyperbolic spaces, whose geometry we explain briefly. Let $\mathbb{H}^d$ denote the standard hyperbolic space of dimension $d \geq 2$; it has scalar curvature -1 everywhere. See Figure 2.4 for one graph roughly isometric to $\mathbb{H}^2$. This drawing uses the Poincaré disc model of $\mathbb{H}^2$, in which the unit disc $\{z \in \mathbb{C} \,;\, |z| < 1\}$ is given the arc-length metric $2\,|dz|/(1 - |z|^2)$. The corresponding ball model of $\mathbb{H}^d$ uses the unit ball $\{x \in \mathbb{R}^d \,;\, |x| < 1\}$ with the arc-length metric $2\,|dx|/(1 - |x|^2)$. Here, we write $|x|$ for the Euclidean norm usually written as $\|x\|$. The length of a smooth curve $t \mapsto x(t)$ parametrized by $t \in [0, 1]$ is

$$\int_0^1 \frac{2\,|dx(t)/dt|}{1 - |x(t)|^2}\, dt \,.$$

The minimum of such lengths among curves joining $x_1, x_2 \in \mathbb{H}^d$ is the hyperbolic distance between x_1 and x_2. A curve that achieves the minimum is called a ***geodesic***. For example, if x_1 is the origin and $|x_2| = R \in (0, 1)$, then a geodesic between x_1 and x_2 is a Euclidean straight-line segment. To see this, note that $|dx(t)| \geq |d\rho(t)|$, where $\rho(t) := |x(t)|$, whence the preceding integral is at least

$$\int_0^1 \frac{2\,|d\rho(t)/dt|}{1 - \rho(t)^2}\,dt \geq \int_0^R \frac{2\,ds}{1 - s^2} = \log\frac{1+R}{1-R},$$

and this distance is achieved by the Euclidean line segment.

For each point a in the unit ball, there is a hyperbolic isometry that takes a to the origin, namely,

$$x \mapsto a^* + \frac{|a^*|^2 - 1}{|x - a^*|^2}(x - a^*),$$

where $a^* := a/|a|^2$; see, for example, Matsuzaki and Taniguchi (1998) for the calculation. Such a map preserves Euclidean angles and the class of Euclidean circular arcs, including Euclidean straight lines as a special case. Therefore, infinite geodesics, which are curves minimizing hyperbolic distance locally, are images of Euclidean diameters of the ball and hence are Euclidean circular arcs perpendicular to the Euclidean unit sphere.

A key difference to Euclidean space is that for each point $o \in \mathbb{H}^d$, the sphere of hyperbolic radius r about o has hyperbolic surface area asymptotic to $\alpha e^{r(d-1)}$ for some positive constant α depending on d. (Therefore, the hyperbolic volume of the ball of hyperbolic radius r is asymptotic to $(d-1)^{-1}\alpha e^{r(d-1)}$.) Indeed, if $|x| = R$, then the hyperbolic distance between the origin and x was seen earlier to be

$$r = \log\frac{1+R}{1-R},$$

so that

$$R = \frac{e^r - 1}{e^r + 1}.$$

The hyperbolic surface area of this sphere centered at the origin is therefore

$$\int_{|x|=R} \frac{2^{d-1}\,dS}{\left(1 - |x|^2\right)^{d-1}},$$

where dS is the element of Euclidean surface area in $\mathbb{R}^d$. Integrating gives the value

$$C\left(\frac{R}{1-R^2}\right)^{d-1} = \alpha(e^r - e^{-r})^{d-1}$$

for some constants C and α, as we claimed. Therefore, there is a positive constant A such that the following hold for any fixed point $o \in \mathbb{H}^d$:

(1) the hyperbolic volume of the shell of points whose distance from o is between r and $r+1$ is at most $Ae^{r(d-1)}$;

(2) the solid angle subtended at o by a spherical cap of hyperbolic area δ on the sphere centered at o of radius r is at most $A\delta e^{-r(d-1)}$.

For more background on hyperbolic space, see, for example, Cannon, Floyd, Kenyon, and Parry (1997), Ratcliffe (2006), or Benedetti and Petronio (1992). Graphs that are roughly isometric to $\mathbb{H}^d$ often arise as Cayley graphs of groups (see Section 3.4) or, more generally, as nets. Here, a graph G is called an ϵ-**net** of a metric space M if the vertices of G form a maximal ϵ-separated subset of M and edges join distinct vertices iff their distance in M is at most 3ϵ. (A set is ϵ-**separated** if all nonzero distances between points are at least ϵ.)

Theorem 2.19. (Transience of Hyperbolic Space) *If G is roughly isometric to a hyperbolic space $\mathbb{H}^d$ with $d \geq 2$, then simple random walk on G is transient.*

Proof. By Theorem 2.17, given $d \geq 2$, it suffices to show transience for one such G. Our proof is quite like our first proof of Pólya's theorem. Let G be a 1-net of $\mathbb{H}^d$. We take the edges of G to be geodesic segments. Let L be a random uniformly distributed geodesic ray from some point $o \in G$ to ∞. (In the ball model of $\mathbb{H}^d$, if o is at the origin, then a geodesic ray is simply a Euclidean ray of Euclidean length 1 starting at the origin.) Let $\mathcal{P}(L)$ be a simple path in G from o to ∞ whose vertices stay within distance 1 of L; choose $\mathcal{P}(L)$ measurably. (By choice of G, for all $p \in L$, there is a vertex $x \in G$ within distance 1 of p.) Define the flow θ from the law of $\mathcal{P}(L)$ via (2.17). Then θ is a unit flow from o to ∞; we claim it has finite energy. There is some constant C such that if e is an edge whose midpoint is at hyperbolic distance r from o, then $\mathbf{P}[e \in \mathcal{P}(L)] \leq Ce^{-r(d-1)}$. Given an edge center s, there is a bound on the number of edge centers whose hyperbolic distance from s is at most 1. Therefore, there is also a constant D such that there are at most $De^{n(d-1)}$ edge centers whose hyperbolic distance from the origin is between n and $n+1$. It follows that the energy of θ is at most $\sum_n C^2 D e^{-2n(d-1)} e^{n(d-1)}$, which is finite. Now transience follows from Theorem 2.11. ◀

2.7 Hitting, Commute, and Cover Times

The remaining two (main) sections of the chapter concern topics other than recurrence and transience, but they use some of the tools we have developed. How can we calculate the expected time it takes for a random walk to hit some set of vertices? (This is also referred to as the ***mean first passage time***.) The following answer is due to Tetali (1991). Recall that $\pi(\cdot)$ is not generally a probability measure.

Proposition 2.20. (Hitting-Time Identity) *Given a finite network with a vertex a and a disjoint subset of vertices Z, let $v(\cdot)$ be the voltage when a unit current flows from a to Z. We have $\mathbf{E}_a[\tau_Z] = \sum_{x \in \mathsf{V}} \pi(x)v(x)$.*

Proof. By Proposition 2.1, we have $\mathbf{E}_a[\tau_Z] = \sum_x \mathscr{G}_Z(a, x) = \sum_x \pi(x)v(x)$. ◀

The expected time for a random walk started at a to visit z and then return to a, that is, $\mathbf{E}_a[\tau_z] + \mathbf{E}_z[\tau_a]$, is called the ***commute time*** between a and z. This turns out to have a particularly pleasant expression, as shown by Chandra, Raghavan, Ruzzo, Smolensky, and Tiwari (1996/1997):

Corollary 2.21. (Commute-Time Identity) *Let G be a finite network and $\gamma := \sum_{e \in \mathsf{E}} c(e)$, where in the sum, each edge with two endpoints occurs twice and each loop occurs once. Let a and z be two vertices of G. The commute time between a and z is $\gamma \mathscr{R}(a \leftrightarrow z)$.*

Proof. The time $\mathbf{E}_a[\tau_z]$ is expressed via Proposition 2.20 using voltages $v(x)$. Now the voltage at x for a unit-current flow from z to a is equal to $v(a) - v(x)$. Thus, $\mathbf{E}_z[\tau_a] = \sum_{x \in \mathsf{V}} \pi(x)[v(a) - v(x)]$. Adding these two hitting times, we get that the commute time is $v(a) \sum_x \pi(x) = \gamma v(a)$. Finally, we use that $v(a) = \mathscr{R}(a \leftrightarrow z)$. ◀

Another interesting quantity is the **cover time** Cov of a finite-state Markov chain $\langle X_j ; j \geq 0 \rangle$, which is the first (random) time the process visits all states V, that is,

$$\mathsf{Cov} := \min\{t ; \forall x \in \mathsf{V} \quad \exists j \leq t \quad X_j = x\}.$$

For the complete graph, the cover time is studied in the coupon-collector problem; in particular, its expectation in the case of n vertices is exactly $(n-1)\sum_{k=1}^{n-1} 1/k$. (It takes no time to visit the starting vertex, which is why $n-1$ appears in place of the usual n.) What is its expectation for general networks? It turns out that this is hard to calculate, but it can be estimated by hitting times (even without reversibility), as shown with a beautiful argument by Matthews (1988). Remarkably, his upper bound is sharp in view of the case of the complete graph.

Theorem 2.22. (Cover-Time Upper Bound) *Given an irreducible, finite Markov chain whose state space V has size n and starting state o, we have*

$$\mathbf{E}_o[\mathsf{Cov}] \leq \left(\max_{a,b \in \mathsf{V}} \mathbf{E}_a \tau_b\right)\left(1 + \frac{1}{2} + \cdots + \frac{1}{n-1}\right).$$

Proof. It takes no time to visit the starting state, so order all of V except for the starting state according to a random permutation, $\langle J_1, \ldots, J_{n-1}\rangle$. Let T_k be the first time by which all $\{J_1, \ldots, J_k\}$ were visited, and let $L_k := X_{T_k}$ be the state the chain is in at time T_k. In other words, L_k is the last state visited among the states $\{J_1, \ldots, J_k\}$. In particular, $\mathbf{P}_o[L_k = J_k] = 1/k$ because of the random permutation. Considering the two cases $L_k = J_k$ and $L_k \neq J_k$, we can make the following somewhat unusual calculation:

$$\mathbf{E}_o[T_k - T_{k-1} \mid X_{T_{k-1}}, X_{T_k}, J_1, \ldots, J_k] = \mathbf{E}_{L_{k-1}}[\tau_{J_k} \mid J_1, \ldots, J_k]\mathbf{1}_{\{L_k = J_k\}}.$$

Taking unconditional expectations, we conclude that

$$\mathbf{E}_o[T_k - T_{k-1}] \leq \left(\max_{a,b \in \mathsf{V}} \mathbf{E}_a \tau_b\right)\frac{1}{k},$$

and summing over k yields the result. ◄

The same technique leads to lower bounds as well. For $A \subseteq \{1, \ldots, n\}$, consider the cover time of A, denoted by Cov_A. Clearly $\mathbf{E}_o[\mathsf{Cov}] \geq \mathbf{E}_o[\mathsf{Cov}_A]$. Let $t_{\min}^A := \min_{a,b \in A, a \neq b} \mathbf{E}_a \tau_b$. Then, similarly to the last proof, we have

$$\mathbf{E}_o[\mathsf{Cov}_A] \geq t_{\min}^A\left(1 + \frac{1}{2} + \cdots + \frac{1}{|A|-1}\right),$$

which gives the following result of Matthews (1988). Again, the case of the complete graph shows that this is sharp.

Theorem 2.23. (Cover-Time Lower Bound) *For any irreducible, finite Markov chain on a state space V with starting state o,*

$$\mathbf{E}_o[\mathsf{Cov}] \geq \max_{A \subseteq \mathsf{V}} t_{\min}^A\left(1 + \frac{1}{2} + \cdots + \frac{1}{|A|-1}\right).$$

▷ **Exercise 2.20.**
Prove Theorem 2.23.

2.8 The Canonical Gaussian Field

This section concerns a model that represents an electrical network via Gaussian random variables. It is known variously as the ***canonical Gaussian field***, the ***Gaussian network***, the ***discrete Gaussian free (or massless) field***, ***random network potentials***, or in some contexts, the ***harmonic crystal***.* More information on it is in the exercises at the end of this chapter and in Exercises 4.17, 4.35, and 10.29–10.32. Some examples of its use are given in Section 2.9.

The canonical Gaussian field arose first in a statistics problem. We explain this origin to motivate the model. Suppose we want to measure the altitudes at a finite set of locations V on land. Assume we know the altitude at some location $o \in$ V. We find the other altitudes by measuring the differences in altitudes between certain pairs E of them. However, each measurement Y has an error that is normally distributed. To be precise, if $Y(e)$ is a measurement of the difference in the altitudes from $x \in$ V to $y \in$ V, then $Y(e) \sim$ $N(\alpha(x) - \alpha(y), \sigma_e^2)$, where $\alpha(x)$ is the true altitude at x and the variances σ_e^2 are assumed known. The numbers $\alpha(x)$ are simply constants; the case where all $\alpha(x) = 0$ is already interesting. We assume all measurements are independent. Let $G = (V, E)$ be the multigraph associated to the measurements. (There are multiple edges between vertices when multiple measurements are made of the same difference, but there are no loops.) Assume G is connected. Make this into a network by assigning the resistances $r(e) := \sigma_e^2$. The ***maximum likelihood estimate*** of the altitudes given these measurements is the function $\widehat{\alpha} \colon$ V $\to \mathbb{R}$ with $\widehat{\alpha}(o) = \alpha(o)$ that maximizes the ***likelihood***

$$\frac{1}{\prod_{e \in E_{1/2}} \sqrt{2\pi r(e)}} \exp\left\{ -\frac{1}{2} \sum_{e \in E_{1/2}} \left[Y(e) - (d\widehat{\alpha})(e) \right]^2 / r(e) \right\},$$

which is what the joint density would be at the observed values Y if the true altitudes were $\widehat{\alpha}$. The random variables $\widehat{\alpha}$ form the ***canonical Gaussian field***; they are functions of the random variables Y. Of course, maximizing the likelihood is the same as minimizing the sum of squares in the exponent. Since $Y(e) - (d\widehat{\alpha})(e) = \left[Y(e) - d\alpha(e) \right] - (d\widehat{\alpha} - d\alpha)(e)$, and α is a nonrandom field, $Z := \widehat{\alpha} - \alpha$ minimizes $\sum_{e \in E_{1/2}} \left[X(e) - (dZ)(e) \right]^2 / r(e)$, where $X(e) := Y(e) - d\alpha(e)$. Therefore, we will henceforth work with X and Z in place of Y and $\widehat{\alpha}$; equivalently, we take $\alpha \equiv 0$.

We restate the definition now of the canonical Gaussian field, shorn of the statistical motivation: Given a network (G, c) and a fixed vertex $o \in$ V, let $X(e)$ ($e \in E_{1/2}$) be independent, normal random variables with mean 0 and variance $r(e)$. Define the random variables $Z(x)$ ($x \in$ V) by the condition that they minimize $\sum_{e \in E_{1/2}} \left[X(e) - (dZ)(e) \right]^2 / r(e)$ and that $Z(o) = 0$. The joint distribution of $Z(x)$ ($x \in$ V) is the ***canonical Gaussian field***.

We consider two simple examples where we can easily understand this field. For the first example, suppose that G is the usual nearest-neighbor graph on the integers $\{0, 1, \ldots, n\}$ and that all resistances are 1. Also, take $o = 0$. Then $X(e) \sim N(0, 1)$ for all edges e and $dZ = X$, whence Z is just n steps of random walk with each step a standard normal random variable.

---

* The term discrete Gaussian free field is the one most commonly used in mathematics. However, the variant where there are masses (Exercise 2.138) is no longer free, and on infinite networks, it would be awkward to refer to the wired and free versions (Exercise 10.29) if we adopted that terminology.

More generally, for the second example, if G is a tree rooted at o, then Z is a random walk indexed by the tree in the sense that when two paths starting at the root branch off from each other, then the random walks along those paths, which were identical, have independent (normal) increments thereafter.

One of the striking properties of the canonical Gaussian field is that

$$\text{Var}\big(Z(x) - Z(y)\big) = \mathscr{R}(x \leftrightarrow y) \tag{2.21}$$

for all $x, y \in \mathsf{V}$. Proving this will be relatively easy once we calculate the joint distribution of dZ, which we proceed to do.

Given a network, define the **gradient** of a function f on V to be the antisymmetric function

$$(\nabla f)(e) := \frac{(df)(e)}{r(e)}, \text{ that is, } \nabla f := c\, df$$

on E. Thinking of resistance of an edge as its length makes this a natural name.

In this notation, Z is the function β with $\beta(o) = 0$ that minimizes $\|X/r - \nabla\beta\|_r$. Since $\nabla\mathbf{1}_{\{x\}}$ is the star at $x \in \mathsf{V}$, it follows that the set of all functions of the form $\nabla\beta$ for some function β equals $\bigstar$, whence we are looking for the element of $\bigstar$ closest to X/r. Such a minimization is achieved through orthogonal projection, so that $\nabla Z = P_{\bigstar}(X/r)$. Since $X/r = \sum_{e \in \mathsf{E}_{1/2}} \chi^e X(e)/r(e)$, applying $P_{\bigstar}$ to both sides yields

$$\nabla Z = \sum_{e \in \mathsf{E}_{1/2}} i^e X(e)/r(e). \tag{2.22}$$

In particular, the random variables ∇Z are linear combinations of independent, normal random variables, so themselves are jointly normal. This explains the name "Gaussian." Since all $X(e)$ have mean 0, so do all $\nabla Z(e)$.

Another way to look at this orthogonal projection is as follows. An orthonormal basis for the space $\ell^2_-(\mathsf{E}, r)$ is $\langle \chi^e/\sqrt{r(e)} ; e \in \mathsf{E}_{1/2}\rangle$. If $(\Omega, \mathbf{P})$ is a probability space on which the random variables $X(e)$ are defined, then $\langle X(e)/\sqrt{r(e)} ; e \in \mathsf{E}_{1/2}\rangle$ are orthonormal in $L^2(\Omega, \mathbf{P})$. Thus, if $\mathscr{H}$ denotes the linear span of the random variables $X(e)$, then $\Phi \colon \chi^e \mapsto X(e)$ is an isometric isomorphism from $\ell^2_-(\mathsf{E}, r)$ to $\mathscr{H}$. From (2.22), we have

$$dZ(e) = r(e)\nabla Z(e) = \sum_{e' \in \mathsf{E}_{1/2}} r(e)Y(e', e)X(e')/r(e') = \sum_{e' \in \mathsf{E}_{1/2}} Y(e, e')X(e') \tag{2.23}$$

using the definition $Y(e, e') := i^e(e')$ from Section 2.4 and then the reciprocity law (2.12). On the other hand, we have trivially that

$$i^e = \sum_{e' \in \mathsf{E}_{1/2}} Y(e, e')\chi^{e'}. \tag{2.24}$$

Comparing (2.23) with (2.24) shows that Φ takes i^e to $dZ(e)$. In particular,

$$\text{Cov}\big(dZ(e), dZ(e')\big) = \mathbf{E}\big[dZ(e)\, dZ(e')\big] = (i^e, i^{e'})_r = (i^e, \chi^{e'})_r = Y(e, e')r(e').$$

This is the same as the voltage difference across e' when unit current flows from e^- to e^+. (The matrix of such (e, e') entries is called the **transfer impedance matrix**.) Since the means and covariances determine the distribution uniquely for jointly normal random variables, we could regard this as a definition of dZ; since dZ determines Z because $Z(o) = 0$, this could also be used as a definition of Z. Other properties that could be used as definitions follow.

The isomorphism Φ takes the subspace $\bigstar$ to a subspace that we will denote $\bigstar(\mathcal{H})$. This latter subspace is simply the linear span of $\sum_{e^-=x} X(e)/r(e)$ over $x \in \mathsf{V}$. Furthermore, since $i^e = P_{\bigstar}\chi^e$, it follows from the isomorphism that $dZ(e) = P_{\bigstar(\mathcal{H})}X(e)$. Since $Z(o) = \mathbf{0}$, we may write $Z(x)$ for $x \in \mathsf{V}$ by summing $-dZ$ along a path ψ from o to x; this gives $Z(x) = -\sum_{e \in \psi} dZ(e) = \Phi\left(-\sum_{e \in \psi} i^e\right) = \Phi(i_{x,o})$, where $i_{x,o}$ is the unit current flow from x to o. Therefore,

$$Z(x) = \sum_{e \in \mathsf{E}_{1/2}} i_{x,o}(e)X(e). \tag{2.25}$$

Proposition 2.24. *Let Z be the preceding canonical Gaussian field.*
(i) *The random variables Z are jointly normal with*

$$Z(x) - Z(y) \sim \mathrm{N}\big(0, \mathscr{R}(x \leftrightarrow y)\big)$$

for $x \neq y \in \mathsf{V}$.
(ii) *The covariance of $Z(x) - Z(y)$ and $Z(z) - Z(w)$ equals $v(z) - v(w)$ when v is the voltage associated to a unit current flow from x to y (with $x \neq y$).*
(iii) *We have $\mathrm{Cov}\big(Z(x), Z(z)\big) = \mathscr{G}_o(x, z)/\pi(z)$, where $\mathscr{G}_o(x, z)$ is the expected number of visits to z of the network random walk started at x, counting only visits that occur before visiting o.*

Proof. Part (iii) follows from (ii) by putting $y := w := o$ and using Proposition 2.1. Part (ii) extends (i), so we prove only (ii). Let $v_{x,y}$ be the voltage function corresponding to a unit current flow from x to y. Let $\psi \in \ell^2(\mathsf{E}, r)$ represent a path from w to z, that is, $\psi = \sum_e \chi^e$, where the sum ranges over the edges in a path from w to z, oriented in the direction of this path. Then as above, we have that $Z(x) - Z(y) = \Phi(i_{x,y})$, whence

$$\begin{aligned}
\mathrm{Cov}\big(Z(x) - Z(y), Z(z) - Z(w)\big) &= (i_{x,y}, i_{z,w})_r = (\nabla v_{x,y}, -P_{\bigstar}\psi)_r \\
&= (\nabla v_{x,y}, -\psi)_r = v_{x,y}(z) - v_{x,y}(w),
\end{aligned}$$

as desired. ◀

Since the random vector X/r is a standard normal vector in $\ell^2_-(\mathsf{E}, r)$, that is, in (any) orthonormal coordinates, its components are independent, standard normal random variables, its image ∇Z under the orthogonal projection $P_{\bigstar}$ is a standard normal vector in $\bigstar$. With the change of notation $\|\nabla Z\|_r^2 = \|dZ\|_c^2$, it follows that the joint density of the random variables $\langle Z(x); x \neq o \rangle$ is

$$C \exp\left\{-\frac{1}{2}\|dZ\|_c^2\right\} \tag{2.26}$$

for some constant C. The constant C is evaluated in Exercise 4.17.

This result could also be used as the definition of the field Z. Finally, one can also define Z by the requirement that dZ has the distribution of X conditioned to sum to zero along every cycle; see Janson (1997), Section 9.4.

2.9 **Notes**

A superb elementary introduction to the ideas of the first five sections of this chapter is given by Doyle and Snell (1984). For detailed study of simple random walk in $\mathbb{Z}^d$, see Révész (2005) and Dembo (2005).

One way to define Brownian motion on a graph is the following. If x is a vertex and the lengths of the d edges incident to x are $\ell_1 \leq \cdots \leq \ell_d$ with $\ell_1 = \cdots = \ell_m < \ell_{m+1}$ (or $m = d$), then define Brownian motion starting at x as follows: Let $B(t)$ be standard Brownian motion on $\mathbb{R}$ for $t \geq 0$ with $B(0) = 0$. Let $\tau := \min\{t\,;\, |B(t)| = \ell_1\}$ and $\sigma := \max\{t < \tau\,;\, B(t) = 0\}$. Consider the excursions up to time τ, that is, the open intervals $I \subset [0, \tau)$ on which $B(t) \neq 0$ for $t \in I$ but for which $B(t) = 0$ at the endpoints of I. Assign each excursion independently and uniformly to one of the d edges incident to x by letting $|B(t)|$ be the distance from x along that edge for $t \in I$. Also, assign the bridge $\{|B(t)|\,;\, \sigma \leq t \leq \tau\}$ independently and uniformly to one of the d edges in the same way. If it happens that the bridge is assigned to one of the m shortest edges, then the motion is at the other endpoint of that edge at time τ. Otherwise, it is in the middle of the ith edge for some $i > m$, and we continue Brownian motion on $\mathbb{R}$ until the first time after τ that $B(t) = 0$ or $|B(t)| = \ell_i$, at which time we are either back at x to start again without having reached a new vertex, or we are at the other endpoint of the ith edge.

The continuous classical analogue of harmonic functions, the Dirichlet problem, and its solution via Brownian motion are as follows. Let D be an open subset of $\mathbb{R}^d$. If $f : D \to \mathbb{R}$ is Lebesgue integrable on each ball contained in D and for all x in D, $f(x)$ is equal to the average value of f on each ball in D centered at x, then f is called **harmonic** in D. If f is locally bounded, then this is equivalent to $f(x)$ being equal to its average on each sphere in D centered at x. Harmonic functions are always infinitely differentiable and satisfy $\Delta f = 0$. Conversely, if f has two continuous partial deriviatives and $\Delta f = 0$ in D, then f is harmonic in D. If D is bounded and connected and f is harmonic on D and continuous on its closure, $\overline{D}$, then $\max_{\overline{D}} f = \max_{\partial D} f$. Let $S(x, r)$ denote the sphere of radius r about x. Let ν denote surface area measure and $|S(x, r)| := \nu(S(x, r))$. If f has two continuous partial derivatives in a neighborhood of x, then

$$(\Delta f)(x) = \lim_{r \to 0} \frac{2d}{|S(x, r)|} \int_{S(x, r)} \big[f(x) - f(z) \big]\, d\nu(z) \,.$$

The Dirichlet problem is the following. Given a bounded connected open set D and a continuous function f on ∂D, is there a continuous extension of f to $\overline{D}$ that is harmonic in D? The answer is yes when D satisfies certain regularity conditions. In this case, the solution can be given via Brownian motion X_t in D as $f(x) := \mathbf{E}_x\big[f(X_\tau) \big]$, where $\tau := \inf\{t \geq 0\,;\, X_t \notin D\}$. See, for example, Bass (1995), pp. 83–90, for details or Doob (1984) for a comprehensive treatment.

Brownian motion in $\mathbb{R}^d$ is analogous to simple random walk in $\mathbb{Z}^d$. The electrical analogue to a discrete graph is a uniformly conducting material. The analogue of a flow is a vector field whose divergence is 0 off of some specified part, whereas the analogue of a current is such a vector field whose curl is 0. There is a similar relationship to an electrical framework for reversible diffusions, even on Riemannian manifolds: Let M be a complete Riemannian manifold. Given a function $\sigma(x)$ which is Borel-measurable, locally bounded, and locally bounded below, called the **(scalar) conductivity**, we associate the diffusion whose generator is $\big(2\sigma(x)\sqrt{g(x)} \big)^{-1} \sum \partial_i \sigma(x) \sqrt{g(x)} g^{ij}(x) \partial_j$ in coordinates, where the metric is g_{ij} with inverse g^{ij} and determinant g. In coordinate-free notation, this is $(1/2)\Delta + (1/2)\nabla \log \sigma$. In other words, the diffusion is Brownian motion with drift equal to half the gradient of the log of the conductivity. The main result of Ichihara (1978) [see also the exposition by Durrett (1986), p. 75; Fukushima (1980), Theorem 1.5.1, and Fukushima (1985); or Grigor'yan (1985)] gives the following test for transience, an analogue to Exercise 2.93.

Theorem 2.25. *On a complete, connected Riemannian manifold M, the diffusion corresponding to the* *scalar conductivity $\sigma(x)$ is transient iff*

$$\inf\left\{\int_M |\nabla u(x)|^2 \sigma(x)\,dx\,;\ u \in C_0^\infty(M),\ u{\restriction}B_1(o) \equiv 1\right\} > 0,$$

where dx is the volume form and $o \in M$ is any fixed point.

One can use networks to decide the recurrence or transience of a Riemannian manifold, and vice versa. Recall that a graph G is called an ϵ-net of M if the vertices of G form a maximal ϵ-separated subset of M and edges join distinct vertices iff their distance in M is at most 3ϵ. When a conductivity σ is given on M, we assign conductances c to the edges of G by

$$c(u, w) := \int_{B_\epsilon(u)} \sigma(x)\,dx + \int_{B_\epsilon(w)} \sigma(x)\,dx.$$

An evident modification of the proof of Theorem 2 of Kanai (1986) shows the following analogue to Theorem 2.17. A manifold M is said to have **bounded geometry** if its Ricci curvature is bounded below and the injectivity radius is positive. If the Ricci curvature is bounded below, then nets have bounded degree (Kanai (1985), Lemma 2.3). We say that σ is ϵ-***slowly varying*** if

$$\sup\left\{\sigma(x)/\sigma(y)\,;\ \mathrm{dist}(x, y) \leq \epsilon\right\} < \infty.$$

Theorem 2.26. *Suppose that M is a complete, connected Riemannian manifold of bounded geometry, that ϵ is at most half the injectivity radius of M, that σ is an ϵ-slowly varying Borel-measurable conductivity on M, and that G is an ϵ-net in M. Then the associated diffusion on M is transient iff the associated random walk on G is transient.*

The transformations of a network described in Section 2.3 can be used for several other purposes as well. As we will see (Chapter 4), spanning trees are intimately connected to electrical networks, so it will not be surprising that such network reductions can be used to count the number of spanning trees of a graph. See Colbourn, Provan, and Vertigan (1995) for this, as well as for applications to the Ising model and perfect matchings (also known as domino tilings). For a connection to knot theory, see Goldman and Kauffman (1993).

The Nash-Williams criterion was proved by Nash-Williams (1959).

The precise asymptotics for the effective resistance in Proposition 2.14 are given in Exercise 2.99, whereas similar asymptotics for Proposition 2.15 are given in Exercise 4.52.

In 1994, Aldous and Fill (Open Problem 6.35 in Section 6.8.3 of Aldous and Fill (2002)) asked for a deterministic estimate of cover times $\mathbf{E}_x[\mathrm{Cov}]$, for reversible Markov chains, up to a constant factor that can be computed in polynomial time. Let $C_* := \max_{x \in \mathsf{V}} \mathbf{E}_x[\mathrm{Cov}]$ denote the cover time, maximized over the starting state. Kahn, Kim, Lovász, and Vu (2000) focused on the problem of estimating C_*. They noted that for a single edge, with n loops added to one endpoint, the lower bound in Theorem 2.23 is quite far from C_*. Denote by M the larger of the right-hand side in Theorem 2.23 and the maximal expected hitting time between two states in V. By Theorem 2.23, C_* is at least M. Kahn, Kim, Lovász, and Vu (2000) showed that $C_* = O\big(M(\log \log n)^2\big)$ (in the reversible case).

The statistical model of Section 2.8 and its connection to random walks (which is equivalent to electrical networks) is due to Borre and Meissl (1974); see Tjur (1991). The maximum likelihood estimate is also the best linear unbiased estimate, a general fact about linear Gaussian models. See Constantine (2003) for some additional information on unbiased estimates in this model. This model is similar to Dynkin's isomorphism, which is for continuous-time Markov processes; see Dynkin (1980). A different connection of networks to Gaussian fields, obtained by using the network Laplacian (defined

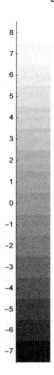

_{color plate} **Figure 2.5.** The net turns of the paths from a fixed vertex in a
uniform spanning tree in a 200×200 grid on the left, with a key on
the right showing the correspondence of visual colors to numbers.

in Exercise 2.62) as covariance matrix, is due to Diaconis and Evans (2002). A relationship of the
canonical Gaussian field to the expected cover time was shown by Ding, Lee, and Peres (2012), which
resolved several open questions on cover times. Scaling limits of canonical Gaussian fields and other
models sometimes give what's known as the (continuum) Gaussian free field; for example, see Kenyon
(2001, 2008), Rider and Virág (2007), and Sheffield (2007). For one such example, consider a spanning
tree T in an $n \times n$ square grid. Associate to the path from the lower left corner to a vertex x its net number
$h(x)$ of turns (also called its winding number), that is, the number of times it turns left minus the number
of times it turns right. Let h_T be the mean value of $h(x)$ over all $x \in V(T)$. When T is chosen uniformly
at random (the topic of Chapter 4), the scaling limit of the distribution of $x \mapsto (\sqrt{\pi}/4)\big(h(x) - h_T\big)$ is
the Gaussian free field, as shown by Kenyon (2001) (and conjectured in looser form by Itai Benjamini);
a picture for $n = 200$ appears in Figure 2.5, where in this sample, the mean winding number h_T is about
0.917. A vast extension of this result on the turns in the uniform spanning tree was proved by Berestycki,
Laslier, and Ray (2020); beware, though, that slightly different normalizations of turns and of the
Gaussian free field are used. The canonical Gaussian field was also introduced in molecular biology
by Bahar, Atilgan, and Erman (1997), where it facilitates very useful computation. Here, the edges are
regarded as springs. Because the correlations are positive, the random variables $Z(x)$ of Exercise 2.138
are positively associated (see Section 5.8 for the definition) by the main result of Pitt (1982); see
Joag-Dev, Perlman, and Pitt (1983) for a simpler proof of this implication for normal random variables.

Problems about random walks that sound similar to those we have analyzed, yet to which the tools of
electrical network theory do not apply, can be very vexing and can often behave contrary to what our
intuition tells us. For example, Rayleigh's monotonicity principle tells us that subgraphs of recurrent
graphs are themselves recurrent. This would suggest that if we do simple random walk in $\mathbb{Z}^2$, except

that at certain times determined in advance, we step either to the right or left only, with equal probability, then the resulting (time-inhomogeneous) Markov chain would still be recurrent. However, this is false!

▷ **Exercise 2.21.**
Find a sequence of times $S \subset \mathbb{N}$ such that if $\langle X_n \, ; \, n \in \mathbb{N} \rangle$ is the Markov chain on $\mathbb{Z}^2$ with increments that are uniformly distributed on $\{(1,0), (-1,0), (0,1), (0,-1)\}$ at times $n \notin S$ and uniformly distributed on $\{(1,0), (-1,0)\}$ at times $n \in S$, then the chain is transient. *Hint:* Let S consist of very long intervals separated by long intervals.

The following variant is still open:

Question 2.27. Define the non-Markovian process $\langle X_n \rangle$ taking values in $\mathbb{Z}^2$ in the following manner. Put $X_0 := (0,0)$. For $n \geq 1$, let $X_{n+1} - X_n$ be either $(1,0)$ or $(-1,0)$ with equal probability if $X_n \notin \{X_0, \ldots, X_{n-1}\}$, whereas $X_{n+1} - X_n$ is either $(0,1)$ or $(0,-1)$ with equal probability if $X_n \in \{X_0, \ldots, X_{n-1}\}$. Does $\langle X_n \rangle$ visit any site infinitely often with positive probability? This question is due to Benjamini, Kozma, and Schapira (2011), who analyze a version of this in four dimensions. A similar question was answered in three dimensions by Peres, Schapira, and Sousi (2016).

A notoriously difficult, though innocent-sounding, variation on random walk on graphs, due to Coppersmith and Diaconis, is the following. Let G be a graph. To model the idea that a random walker may prefer to traverse edges previously visited, we will allow the edge weights to change with time. Thus, define a non-Markovian process $\langle X_n \, ; \, n \geq 0 \rangle$ as follows. Start with a fixed vertex X_0 and some constant $a > 0$. Let $L_n(e)$ be the number of traversals of e by $\langle X_k \, ; \, 0 \leq k \leq n \rangle$. When choosing X_{n+1}, use edge weights $c_n(e) := a + L_n(e)$. This is called ***edge-reinforced random walk***. It is not hard to show that if $G = \mathbb{Z}$, then $X_n \not\to \infty$ a.s.

More generally, when G is a tree, there is an alternative representation of edge-reinforced random walk that allows it to be analyzed, as done first by Pemantle (1988). He proved that this model exhibits a phase transition: when a is small, the random walk is recurrent (that is, it visits every vertex infinitely often a.s.), whereas when a is large, it is transient. The precise critical value of a is known, even for general trees: see Lyons and Pemantle (1992) (and Lyons and Pemantle (2003)). To see this alternative representation, consider first a directed graph, G, and only increase the weights of directed edges when traversed. Let $C_n(x)$ be the vector of weights on the outgoing edges from x at the nth visit to x. We could define this for all n regardless of how many times x is visited. In fact, these sequences $\langle C_n(x) \, ; \, n \geq 0 \rangle$ are independent for different $x \in \mathsf{V}(G)$ and are the same as choosing from a Pólya urn that starts with weight a on each outgoing edge from x. Since the draws from a Pólya urn are exchangeable, they could equally well be represented as a mixture of i.i.d. choices by de Finetti's theorem; it turns out that the mixture is a Dirichlet distribution. Thus, one can alternatively describe the directed edge-reinforced random walk as follows: First pick a random transition probability distribution for each vertex in an i.i.d. way from a certain Dirichlet distribution. Then do a Markovian random walk according to these transition probabilities. This is known as a ***random walk in a random environment*** (RWRE). When G is a tree, we do not need to use directed edges for this representation, since each edge is traversed twice when returning to a vertex before any other incident edge is traversed. It turns out that for all levels of reinforcement of directed edges in $\mathbb{Z}^d$ and all $d \geq 3$, the directed edge-reinforced random walk is transient: see Sabot (2011). Nothing is known for $d = 2$, however.

When $G = \mathbb{Z}^d$ for $d \geq 2$, (undirected) edge-reinforced random walk is harder to analyze and there was little progress for many years. Merkl and Rolles (2009) showed recurrence for small a on graphs obtained by subdividing the edges of $\mathbb{Z}^2$. Finally, Angel, Crawford, and Kozma (2014) and Sabot and Tarrès (2015) showed independently that on every graph G of bounded degree, if $a > 0$ is sufficiently small, then edge-reinforced random walk on G is recurrent. Sabot and Zeng (2019) showed that edge-reinforced random walk on $\mathbb{Z}^2$ is recurrent for all $a > 0$. Conversely, Disertori, Sabot, and Tarrès

(2015) proved that if $d \geq 3$ and a is sufficiently large, then the edge-reinforced random walk in $\mathbb{Z}^d$ is transient.

It turns out that a similar-sounding process, where one uses vertex weights instead of edge weights, but again, the weight of a vertex is increased every time a vertex is visited, behaves quite differently and is considerably easier to understand: see Volkov (2001) and Sabot, Tarrès, and Zeng (2017). This process is called **vertex-reinforced random walk**.

The topic of RWRE, with any i.i.d. transition probabilities, is quite natural and extensive but only partially understood, except on trees.

▷ **Exercise 2.22.**
Let $\langle A_k \, ; \ k \in \mathbb{Z} \rangle$ be i.i.d. random variables with values in $(0, 1)$. Let $\langle X_n \rangle$ be the RWRE on $\mathbb{Z}$ with transition probability at k given by stepping to $k + 1$ with probability A_k and to $k - 1$ with probability $1 - A_k$. Show that this walk tends to ∞ a.s. when $\mathbf{E}[\log A_0 - \log(1 - A_0)] > 0$, tends to $-\infty$ a.s. when $\mathbf{E}[\log A_0 - \log(1 - A_0)] < 0$, and is a.s. recurrent when $\mathbf{E}[\log A_0 - \log(1 - A_0)] = 0$ (when these expectations are defined). *Hint:* Use a random electrical network.

To see how difficult the topic of RWRE in $\mathbb{Z}^d$ is when $d \geq 2$, consider symmetric random environments. Only in very special cases is it known whether the RWRE is recurrent or transient. Even when the environment is not symmetric but has a nonzero average drift in some direction, transience is not always established. See Zeitouni (2004) and Sznitman (2004) for surveys of RWRE.

2.10 Collected In-Text Exercises

2.1. (Reversible Markov Chains) All Markov chains in this exercise are denumerable.

(a) Let P be the transition probability matrix of a Markov chain. Show that if P is π-reversible, that is, $\pi(x)p(x, y) = \pi(y)p(y, x)$ for all states x, y, then π is P-stationary, that is, $\sum_x \pi(x)p(x, y) = \pi(y)$ for all states y.

(b) Let P be the transition probability matrix of a Markov chain and π be P-stationary. Define the **reversed Markov chain** to have transition probabilities

$$\widehat{p}(x, y) := \pi(y)p(y, x)/\pi(x)$$

and write $\widehat{P}$ for the associated transition matrix. Show that $\widehat{P}$ is indeed a transition matrix and that $\forall x_1, x_2, \ldots, x_n$,

$$\pi(x_1) \prod_{i=1}^{n-1} p(x_i, x_{i+1}) = \pi(x_n) \prod_{i=1}^{n-1} \widehat{p}(x_{n+1-i}, x_{n-i}).$$

Of course, P is π-reversible iff $P = \widehat{P}$.

(c) Show that if a Markov chain is reversible, then $\forall x_1, x_2, \ldots, x_n$ with $x_1 = x_n$,

$$\prod_{i=1}^{n-1} p(x_i, x_{i+1}) = \prod_{i=1}^{n-1} p(x_{n+1-i}, x_{n-i}) :$$

the chance of traversing a cycle is the same in either direction. Show too that this equation implies reversibility for irreducible chains.

(d) Let $\langle X_n \rangle$ be a random walk on a network G, and let x and y be two vertices in G. Let $\mathcal{P}$ be a path from x to y and $\mathcal{P}'$ be its reversal, a path from y to x. Show that

$$\mathbf{P}_x\big[\langle X_n \, ; \ n \leq \tau_y \rangle = \mathcal{P} \mid \tau_y < \tau_x^+\big] = \mathbf{P}_y\big[\langle X_n \, ; \ n \leq \tau_x \rangle = \mathcal{P}' \mid \tau_x < \tau_y^+\big],$$

where τ_w denotes the first time the random walk visits w, τ_w^+ denotes the first time after 0 that the random walk visits w, and $\mathbf{P}_u$ denotes the law of random walk started at u.

(e) Consider a random walk on a network G that is either transient or is stopped on the first visit to a set of vertices Z. Let $\mathscr{G}(x, y)$ be the expected number of visits to y for a random walk started at x; if the walk is stopped at Z, we count only those visits that occur strictly before visiting Z. Show that for every pair of vertices x and y,

$$\pi(x)\,\mathscr{G}(x, y) = \pi(y)\,\mathscr{G}(y, x)\,.$$

(f) Show that random walk on a connected network G is positive recurrent (that is, has a stationary probability distribution, which is therefore unique) iff $\sum_{x,y} c(x, y) < \infty$, in which case the stationary probability distribution is proportional to $\pi(\cdot)$. Show that if the random walk is not positive recurrent, then $\pi(\cdot)$ is a stationary infinite measure.

2.2. Suppose that an antisymmetric function j (meaning that $j(x, y) = -j(y, x)$) on the edges of a finite, connected network satisfies Kirchhoff's cycle law and satisfies Kirchhoff's node law in the form $\sum_{y \sim x} j(x, y) = 0$ for all $x \in W$. Show that j is the current corresponding to some voltage function whose values are specified off W and that the voltage function is unique up to an additive constant.

2.3. Verify that Propositions 2.1 and 2.2 are valid when the number of edges is infinite but the number of vertices is finite.

2.4. For a fixed vertex a in G, show that $\lim_n \mathscr{C}(a \leftrightarrow Z_n)$ is the same for every sequence $\langle G_n \rangle$ of induced subgraphs that exhausts G.

2.5. When G is finite but A is not a singleton, define $\mathscr{C}(A \leftrightarrow Z)$ to be $\mathscr{C}(a \leftrightarrow Z)$ if all the vertices in A were to be identified to a single vertex, a. Show that if voltages are applied at the vertices of $A \cup Z$ so that $v{\restriction}A$ and $v{\restriction}Z$ are constants, then $v{\restriction}A - v{\restriction}Z = I_{AZ}\mathscr{R}(A \leftrightarrow Z)$, where $I_{AZ} := \sum_{x \in A} \sum_y i(x, y)$ is the total amount of current flowing from A to Z.

2.6. Give two harder but instructive proofs of the series equivalence as follows. Since voltages determine currents, it suffices to check that the voltages are as claimed on the new network G'. (1) Show that $v{\restriction}(\mathsf{V}(G) \setminus \{w\})$ is harmonic on $\mathsf{V}(G') \setminus (A \cup Z)$. (2) Use the "craps principle" (Pitman (1993), p. 210) to show that $\mathbf{P}_x[\tau_A < \tau_Z]$ is unchanged for $x \in \mathsf{V}(G) \setminus \{w\}$.

2.7. Give two more proofs of the parallel equivalence as in Exercise 2.6.

2.8. Let (G, c) be a network. A ***network automorphism*** of (G, c) is a map $\phi: G \to G$ that is a bijection of the vertex set with itself and a bijection of the edge set with itself such that if x and e are incident, then so are $\phi(x)$ and $\phi(e)$ and such that $c(e) = c(\phi(e))$ for all edges e. Suppose that (G, c) is ***spherically symmetric*** about o, meaning that if x and y are any two vertices at the same distance from o, then there is an automorphism of (G, c) that leaves o fixed and that takes x to y. Let C_n be the sum of $c(e)$ over all edges e with $d(e^-, o) = n - 1$ and $d(e^+, o) = n$. Show that

$$\mathscr{R}(o \leftrightarrow \infty) = \sum_{n \geq 1} \frac{1}{C_n}\,,$$

whence the network random walk on G is transient iff

$$\sum_{n \geq 1} \frac{1}{C_n} < \infty\,.$$

2.9. Give at least one proof of the star-triangle equivalence.

2.10. Find a (finite) graph with two vertices a and z that can't be reduced to a single edge between a and z by the four transformations pruning, series, parallel, and star-triangle.

2.11. Prove that d and d^* are adjoints of each other.

2.12. Show that $\mathbf{P}[a \to Z]$ can increase in some situations and decrease in others when an edge incident to a is removed.

2.13. Given disjoint vertex sets, A and Z, in a finite network, we may express the effective resistance between A and Z by Thomson's principle as

$$\mathscr{R}(A \leftrightarrow Z) = \min \left\{ \sum_{e \in E_{1/2}} r(e)\theta(e)^2 \; ; \; \theta \text{ is a unit flow from } A \text{ to } Z \right\}.$$

Prove the following dual expression for the effective conductance, known as **_Dirichlet's principle_**:

$$\mathscr{C}(A \leftrightarrow Z) = \min \left\{ \sum_{e \in E_{1/2}} c(e)dF(e)^2 \; ; \; F{\restriction}A \equiv 1, \; F{\restriction}Z \equiv 0 \right\}.$$

2.14. Let $G = (\mathsf{V}, \mathsf{E})$ be denumerable and $\theta_n \in \ell^2_-(\mathsf{E}, r)$ be such that $\mathscr{E}(\theta_n) \le M < \infty$ and $\theta_n \to \theta$ edgewise, that is, $\theta_n(e) \to \theta(e)$ for each $e \in \mathsf{E}$. Show that θ is antisymmetric, $\mathscr{E}(\theta) \le \liminf_n \mathscr{E}(\theta_n) \le M$, and $\forall x \in \mathsf{V} \; d^*\theta_n(x) \to d^*\theta(x)$.

2.15. Show that (2.19) is also necessary for recurrence if $f(n + 1) \le f(n) + 1$ for all n.

2.16. Show that if we remove each edge $[(x, y, z), (x, y, z + 1)]$ in $\mathbb{Z}^3$ with $x + y$ odd, then we obtain a transient graph with effective resistance to infinity at most 6 times what it was before removal.

2.17. Show that being roughly isometric is an equivalence relation.

2.18. Show that $\mathbb{Z}$ and $\mathbb{Z}^2$ are not roughly isometric graphs.

2.19. Prove Proposition 2.18.

2.20. Prove Theorem 2.23.

2.21. Find a sequence of times $S \subset \mathbb{N}$ such that if $\langle X_n \; ; \; n \in \mathbb{N}\rangle$ is the Markov chain on $\mathbb{Z}^2$ with increments that are uniformly distributed on $\{(1, 0), (-1, 0), (0, 1), (0, -1)\}$ at times $n \notin S$ and uniformly distributed on $\{(1, 0), (-1, 0)\}$ at times $n \in S$, then the chain is transient. *Hint:* Let S consist of very long intervals separated by long intervals.

2.22. Let $\langle A_k \; ; \; k \in \mathbb{Z}\rangle$ be i.i.d. random variables with values in $(0, 1)$. Let $\langle X_n \rangle$ be the RWRE on $\mathbb{Z}$ with transition probability at k given by stepping to $k + 1$ with probability A_k and to $k - 1$ with probability $1 - A_k$. Show that this walk tends to ∞ a.s. when $\mathbf{E}\big[\log A_0 - \log(1 - A_0)\big] > 0$, tends to $-\infty$ a.s. when $\mathbf{E}\big[\log A_0 - \log(1 - A_0)\big] < 0$, and is a.s. recurrent when $\mathbf{E}\big[\log A_0 - \log(1 - A_0)\big] = 0$ (when these expectations are defined). *Hint:* Use a random electrical network.

2.11 Additional Exercises

In all the exercises, assume the networks are connected. Recall that $\pi(\cdot)$ is not necessarily a probability measure.

2.23. A function f on the states of a Markov chain is called *subharmonic* at x if

$$f(x) \le \sum p(x, y) f(y)$$

and *superharmonic* if the opposite inequality holds. Show that the maximum principle extends to subharmonic functions and that there is a corresponding minimum principle for superharmonic functions.

2.24. Let $A \subset V$ be a subset of states of a Markov chain such that $\mathbf{P}_x[\tau_A < \infty] = 1$ for all $x \in V$. Suppose that $f: V \to \mathbb{R}$ is subharmonic at all $x \notin A$ (as defined in Exercise 2.23).
  (a) Show that $f(x) \le \sup_{y \in A} f(y)$ for all $x \in V$ or else $\sup_{x \notin A} f(x) = \infty$.
  (b) Find an example where $f \restriction A$ is bounded but f is not.

2.25. Give another proof of the existence principle along the following lines. Given f_0 off W, let $f(x) := \inf g(x)$ over all functions g that are superharmonic on W, that agree with f_0 off W, and such that $g \ge \inf f_0$. (See Exercise 2.23 for the definition of "superharmonic.") Then f is harmonic and agrees with f_0 off W.

2.26. Let G be a transient network and $x, y \in V$. Show that

$$\pi(x) \mathbf{P}_x[\tau_y < \infty] \mathscr{G}(y, y) = \pi(y) \mathbf{P}_y[\tau_x < \infty] \mathscr{G}(x, x).$$

2.27. Let G be a transient network and $f: V \to \mathbb{R}$ satisfy $\sum_y \mathscr{G}(x, y) |f(y)| < \infty$ for every x. Define $(\mathscr{G}f)(x) := \sum_y \mathscr{G}(x, y) f(y)$. Let I be the identity operator and P be the *transition operator* (that is, $(Pg)(x) := \sum_y p(x, y) g(y)$). Show that $(I - P)(\mathscr{G}f) = f$.

2.28. Let G be transient and u be a nonnegative, superharmonic function. Show that there exist unique functions f and h such that $f \ge 0$, h is harmonic, and $u = \mathscr{G}f + h$, where $\mathscr{G}f$ is defined in Exercise 2.27. Show that also $f = (I - P)u$, $h \ge 0$, and $h \ge g$ whenever $g \le u$ is harmonic, where I and P are defined in Exercise 2.27. *Hint:* Define $h := \lim_n P^n u$ and $f := (I - P)u$.

2.29. (Starr's Maximal Inequality) Let G be a positive recurrent network with stationary probability distribution π. Let P be the transition operator as in Exercise 2.27. Use the following steps to prove that for every $p \in (1, \infty)$ and $f \in \ell^p(V, \pi)$,

$$\left\| \sup_{n \ge 0} P^{2n} f \right\|_p \le \left(\frac{p}{p-1} \right) \|f\|_p.$$

(This exercise assumes familiarity with martingale theory.) Let $X_0 \sim \pi$ and $f \ge 0$. Write $R_n := \mathbf{E}[f(X_{2n}) \mid X_n]$ and $h_N := \max_{0 \le n \le N} P^{2n} f$.
  (a) Show that $(P^{2n} f)(X_0) = \mathbf{E}[R_n \mid X_0]$.
  (b) Show that $(X_n, X_{n+1}, \ldots, X_{2n}) \overset{\mathscr{D}}{=} (X_n, X_{n-1}, \ldots, X_0)$.
  (c) Show that $R_n = \mathbf{E}[f(X_0) \mid X_n, X_{n+1}, \ldots]$, whence $\langle R_N, R_{N-1}, \ldots, R_0 \rangle$ is a martingale.
  (d) Show that $\|\max_{0 \le n \le N} R_n\|_p \le \frac{p}{p-1} \|f\|_p$.
  (e) Show that $h_N(X_0) \le \mathbf{E}[\max_{0 \le n \le N} R_n \mid X_0]$.
  (f) Show that $\|h_N\|_p \le \frac{p}{p-1} \|f\|_p$.
  (g) Take $N \to \infty$ to deduce Starr's inequality.

2.30. Suppose that $\langle X_n \rangle$ is a stationary sequence with values in some measurable space. Let the distribution of X_0 be μ. Fix a measurable set A of possible values and let $\tau_A^+ := \inf\{n \geq 1 \,;\, X_n \in A\}$. Write μ_A for the distribution of X_0 given that $X_0 \in A$.

(a) **(Poincaré Recurrence Theorem)** Show that $\mathbf{P}[\tau_A^+ < \infty \mid X_0 \in A] = 1$.

(b) Show that the conditional distribution of $X_{\tau_A^+}$ given that $X_0 \in A$ is also μ_A.

(c) **(Kac Lemma)** Assume that $\mathbf{P}[\tau_A^+ < \infty] = 1$. Show that $\mathbf{E}[\tau_A^+ \mid X_0 \in A] = 1/\mu(A)$.

2.31. Given a finite graph G and two of its vertices a and z, let $i_c(\cdot)$ be the unit current flow from a to z when conductances $c(\cdot)$ are assigned to the edges. Show that i_c is a continuous function of $c(\cdot)$.

2.32. Let G be a network. If $h: \mathsf{V} \to [0, \infty)$ is harmonic at every vertex of $W \subseteq \mathsf{V}$, while $h(x) > 0$ iff $x \in W$, there is a Markov chain on W associated to h called ***Doob's h-transform***; its transition probabilities are defined to be $p^h(x, y) := p(x, y)h(y)/h(x)$ for $x, y \in W$. Check that these are indeed transition probabilities for a reversible Markov chain. Find corresponding conductances.

2.33. Let G be a finite network and $W \subsetneq \mathsf{V}$. At every visit to a vertex $x \in W$, a random walker collects a payment of $g(x)$. When reaching a vertex $y \notin W$, the walker receives a final retirement package of $h(y)$ and stops moving. Let $f(x)$ denote the expected total payment the walker receives starting from x.

(a) Show that f is finite.

(b) Write a set of linear equations that the values $f(x)$ for $x \in W$ must satisfy (one equation for each such vertex x).

(c) Uniqueness: Show that these equations specify f.

(d) Existence: Without using the probabilistic interpretation, prove there is a solution to this set of equations.

(e) Let i be the current associated to the voltage function f, that is, $i(x, y) := c(x, y)[f(x) - f(y)]$. Show that the amount of current flowing into the network at x, that is, $\sum_y i(x, y)$, equals $\pi(x)g(x)$ for $x \in W$. Thus, currents can be specified by giving voltages h on one set of vertices and giving flow amounts $\pi(x)g(x)$ on the complementary set of vertices. (Recall that $\pi(\cdot)$ is not generally a probability measure.)

2.34. If voltages are given at vertices a and z of a finite network and thus are harmonic elsewhere, must the voltages of the vertices be monotone along every shortest path between a and z?

2.35. Let x, y, z be vertices in a finite network. Show that $\mathscr{C}(x \leftrightarrow \{y, z\}) \leq \mathscr{C}(x \leftrightarrow y) + \mathscr{C}(x \leftrightarrow z)$.

2.36. Let A and Z be two sets of vertices in a finite network. Show that for any vertex $x \notin A \cup Z$, we have

$$\mathbf{P}_x[\tau_A < \tau_Z] \leq \frac{\mathscr{C}(x \leftrightarrow A)}{\mathscr{C}(x \leftrightarrow A \cup Z)} .$$

2.37. Show that on every finite network, $|\mathbf{E}[S_{xy}] - \mathbf{E}[S_{yx}]| \leq 1$ for all x, y, where S_{xy} is defined as in Proposition 2.2.

2.38. When a voltage is imposed so that a unit current flows from a to Z in a finite network and $v{\restriction}Z \equiv 0$, show that the expected total number of times an edge $[x, y]$ is crossed by a random walk starting at a and absorbed at Z equals $c(x, y)[v(x) + v(y)]$.

2.39. Define S_{xy} as in Proposition 2.2. Show that $\mathbf{E}[S_{xy}]$ is monotone increasing in $c(x, y)$.

2.40. Show that every transient network contains a locally finite, transient subnetwork.

2.41. Let G be a network such that $\gamma := \sum_{e \in \mathsf{E}} c(e) < \infty$ (for example, G could be finite). Here, the sum counts each edge with two endpoints twice and each loop once. For every vertex $a \in G$, show that the expected time for a random walk started at a to return to a is $\gamma/\pi(a)$.

2.42. Let $\langle X_n \rangle$ be a recurrent, irreducible Markov chain, not necessarily reversible. Let $\tau > 0$ be a stopping time and a a state such that $\mathbf{P}_a[X_\tau = a] = 1$. Let $\mathcal{G}_\tau(x, y) := \mathbf{E}_x\big[\sum_{0 \le n < \tau} \mathbf{1}_{[X_n = y]}\big]$. Assume that $\mathcal{G}_\tau(a, a) < \infty$; for example, we could use $\tau = \tau_a^+$.

(a) Show $x \mapsto \mathcal{G}_\tau(a, x)$ is finite for each x and gives a stationary measure.

(b) Assume in addition that $\langle X_n \rangle$ is positive recurrent, that is, it has a stationary probability measure $\pi(\cdot)$, not necessarily reversible. Show that $\mathbf{E}_a[\tau] < \infty$ and for all states x, we have $\pi(x) = \mathcal{G}_\tau(a, x)/\mathbf{E}_a[\tau]$. In particular, $\mathbf{E}_a[\tau_a^+] = 1/\pi(a)$. Give another proof of the formula of Exercise 2.41 from this.

(c) Show that if $\mathbf{E}_a[\tau_a^+] < \infty$, then $\langle X_n \rangle$ is positive recurrent.

(d) Show that if $\langle X_n \rangle$ is positive recurrent, then $\mathbf{E}_\pi[\tau] = \mathbf{E}_a[\tau^2]/(2\,\mathbf{E}_a[\tau]) + 1/2$, where $\mathbf{P}_\pi$ is the law of the chain when the distribution of X_0 is a stationary probability measure, π.

2.43. Let $\langle X_n \rangle$ be an irreducible Markov chain. Suppose that the Markov chain is recurrent.

(a) Show that there are no bounded harmonic functions other than the constants.

(b) Show that there are no nonnegative harmonic functions other than the constants.

(c) Existence of a stationary measure was shown in Exercise 2.42. Show that the stationary measure is unique up to a multiplicative constant.

2.44. Let $\langle X_n \rangle$ be an irreducible Markov chain. Show that the Markov chain is recurrent iff every nonnegative, superharmonic function is constant. (See Exercise 2.23 for the definition of "superharmonic.")

2.45. Let $\langle X_n \rangle$ be a positive recurrent, irreducible, aperiodic Markov chain with stationary probability measure $\pi(\cdot)$, not necessarily reversible. Write $\mathbf{P}_\pi$ for the law of the chain when the distribution of X_0 is π. Show that for all states x, we have

$$\pi(x)\,\mathbf{E}_\pi[\tau_x] = \sum_{n \ge 0} [p_n(x, x) - \pi(x)].$$

The right-hand side can be thought of as the expected excess number of visits to x when starting at x compared to starting according to π.

2.46. Let G be a network such that $\sum_{e \in \mathsf{E}} c(e) = \infty$. For every vertex $a \in G$, show that the expected time for a random walk started at a to return to a is ∞.

2.47. Let G be a finite network and A and Z be two disjoint subsets of vertices in G. Show that

$$\mathscr{C}(A \leftrightarrow Z) = \sum_{x \in A} \pi(x)\,\mathbf{P}_x[\tau_Z < \tau_A^+].$$

(Recall that $\pi(\cdot)$ is not generally a probability measure.)

2.48. Let $\langle X_n \rangle$ be a positive recurrent, irreducible Markov chain with stationary probability measure $\pi(\cdot)$, not necessarily reversible. Show that the expected hitting time of a random π-distributed target does not depend on the starting state. That is, show that if

$$f(x) := \sum_y \pi(y)\,\mathbf{E}_x[\tau_y],$$

then $f(x)$ is the same for all x. In case the state space is finite, show that

$$f(x) = \sum_i \frac{1}{1 - \lambda_i},$$

where the sum is over the eigenvalues λ_i of the transition matrix P (with multiplicity) other than 1.

2.49. Suppose that a tree T is transient for simple random walk $\langle X_n \rangle$. If we iteratively erase backtracking from the path of the walk, then we obtain a.s. a ray $\xi \in \partial T$ that intersects the path infinitely often. We say that $\langle X_n \rangle$ **converges** to ξ. Prove that if $\langle X_n \rangle$ and $\langle X_n' \rangle$ are independent simple random walks on T, then a.s. they converge to distinct rays.

2.50. Let $\langle X_n \rangle$ be a Markov chain and A be a set of states such that $\tau_A < \infty$ a.s. The distribution of X_{τ_A} is called **harmonic measure** on A. In the case that $\langle X_n \rangle$ is a random walk on a network $G = (V, E)$ starting from a vertex z and $A \subseteq V$ is finite, let μ be harmonic measure and define

$$v(x) := \mathbf{P}_x[\tau_z < \tau_A^+] \quad \text{for } x \in A.$$

(a) Show that

$$\mu(x) = \mathbf{P}_z[\tau_x < \tau_{A \setminus \{x\}} \mid \tau_A < \tau_z^+] \quad \text{for all } x \in A.$$

(b) Show that

$$\mu(x) = \mathscr{R}(A \leftrightarrow z)\pi(x)v(x) \quad \text{for all } x \in A.$$

(Recall that $\pi(\cdot)$ is not generally a probability measure.)

(c) Let G be a transient network and $\langle G_n \rangle$ be an exhaustion of G by finite, induced subnetworks. Let G_n^{W} be the network obtained from G by identifying the vertices outside G_n to a single vertex, z_n, and removing loops at z_n. Fix a finite set $A \subset V$. Let μ_n be harmonic measure on A for the network G_n^{W} from z_n. Show that **wired harmonic measure from infinity** $\mu := \lim_{n \to \infty} \mu_n$ exists and satisfies

$$\mu(x) = \mathscr{R}(A \leftrightarrow \infty)\pi(x) \, \mathbf{P}_x[\tau_A^+ = \infty] \quad \text{for all } x \in A.$$

2.51. Let G be a finite network with a fixed vertex, a. Fix $s \in (0, 1)$. Add a new vertex, Δ, which is joined to each vertex x with an edge of conductance $w(x)$ chosen so that at x, the probability of taking a step to Δ is equal to $1 - s$. Call the new network G'. Prove that

$$\sum_{k \geq 0} p_k s^k = \pi(a)\mathscr{R}(a \leftrightarrow \Delta; G')/s ,$$

where p_k is the probability that the network random walk on G (not on G') starting from a is back at a at time k. The preceding series is the generating function for the return probabilities and is sometimes called the "Green function," despite the other notion of Green function defined in this chapter.

2.52. Let G be a transient network with a fixed vertex, a. Fix $s \in (0, 1)$. Add a new vertex, Δ, which is joined to each vertex x with an edge of conductance $w(x)$ chosen so that at x, the probability of taking a step to Δ is equal to $1 - s$. Call the new network G'. Define the effective resistance between a and $\{\Delta, \infty\}$ to be the limit of the effective resistance between a and $V(G') \setminus V(G_n)$, where G_n is an exhaustion of G (not of G') by induced subnetworks. Prove that the limit defining the effective resistance between a and $\{\Delta, \infty\}$ exists and that

$$\sum_{k \geq 0} p_k s^k = \pi(a)\mathscr{R}(a \leftrightarrow \Delta, \infty; G')/s ,$$

where p_k is the probability that the network random walk on G (not on G') starting from a is back at a at time k.

2.53. Give an example of two graphs $G_i = (V, E_i)$ on the same vertex set ($i = 1, 2$) such that both graphs are connected and recurrent, yet their union $(V, E_1 \cup E_2)$ is transient.

2.54. Consider nearest-neighbor random walk on $\mathbb{N}$ that steps $+1$ with probability $3/4$ and -1 with probability $1/4$ unless the walker is at a multiple of 3, in which case the transition probabilities are $1/10$ and $9/10$, respectively. (Of course, at 0, the walker always moves to 1.) Show that the walk is recurrent. On the other hand, show that if before taking each step, a fair coin is tossed and one uses the transition probabilities of this biased walk when the coin shows heads and moves right or left with equal probability when the coin shows tails, then the walk is transient. In this latter case, show that the walk tends to infinity at a positive linear rate.

2.55. In the following networks, each edge has unit conductance.

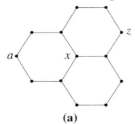
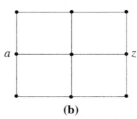
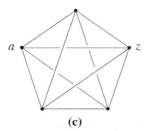

(a)   (b)   (c)

(a) What are $\mathbf{P}_x[\tau_a < \tau_z]$, $\mathbf{P}_a[\tau_x < \tau_z]$, and $\mathbf{P}_z[\tau_x < \tau_a]$?

(b) What is $\mathscr{C}(a \leftrightarrow z)$? (Or: show a sequence of transformations that could be used to calculate $\mathscr{C}(a \leftrightarrow z)$.)

(c) What is $\mathscr{C}(a \leftrightarrow z)$? (Or: show a sequence of transformations that could be used to calculate $\mathscr{C}(a \leftrightarrow z)$.)

2.56. Suppose that G is a finite network and voltages are given to be 1 at a vertex a and 0 at a vertex z. Let x and y be two other vertices of G, and let $G' := G/\{x, y\}$ be the graph obtained by shorting x and y, that is, identifying them. Show that the voltage at the shorted vertex in G' lies between the original voltages at x and y in G.

2.57. Let W be a set of vertices in a finite network G. Let $j \in \ell_-^2(\mathsf{E})$ satisfy $\sum_{i=1}^n j(e_i) r(e_i) = 0$ whenever $\langle e_1, e_2, \ldots, e_n \rangle$ is a cycle; and $d^* j \restriction (\mathsf{V} \setminus W) = 0$. According to Exercise 2.2, the values of $d^* j \restriction W$ determine j uniquely. Show that the map $d^* j \restriction W \mapsto j$ is linear. This is another form of the superposition principle.

2.58. Let (G, c) be a finite network, $z \in \mathsf{V}(G)$, and $A \subseteq \mathsf{V}(G) \setminus \{z\}$. Consider two voltage functions v, v' on G specified to be 0 at z and arbitrary on A, with the property that $v(a) \le v'(a)$ for each $a \in A$. Let i, i' be the corresponding currents. Show that $d^* i(z) \ge d^* i'(z)$.

2.59. Let A and Z be subsets of vertices in a finite network. Show that

$$\frac{2}{|A||Z|} \sum_{a \in A, z \in Z} \mathscr{R}(a \leftrightarrow z) \ge \frac{1}{|A|^2} \sum_{a, b \in A} \mathscr{R}(a \leftrightarrow b) + \frac{1}{|Z|^2} \sum_{y, z \in Z} \mathscr{R}(y \leftrightarrow z),$$

with equality iff $A = Z$.

2.60. Let G be a finite network and $f: \mathsf{V} \to \mathbb{R}$ satisfy $\sum_x f(x) = 0$. Pick $z \in \mathsf{V}$. Let $\mathscr{G}_z(\cdot, \cdot)$ be the Green function for the random walk on G absorbed at z. Consider the voltage function $v(x) := \sum_y \mathscr{G}_z(x, y) f(y)/\pi(y)$. Show that the current $i = c \cdot dv$ satisfies $d^* i = f$ and $\mathscr{E}(i) = \sum_{x,y} \mathscr{G}_z(x, y) f(x) f(y)/\pi(y)$.

2.61. A *cut* in a graph G is a set of edges of the form $\{(x, y); x \in A, y \notin A\}$ for some proper nonempty vertex set A of G. Show that for every finite network G, the linear span of

$$\left\{ \sum_{e \in \Pi} c(e) \chi^e \; ; \; \Pi \text{ is a cut of } G \right\}$$

equals the star space.

2.62. Let G be a finite, connected, simple network. The **network Laplacian** is the $\mathsf{V} \times \mathsf{V}$ matrix Δ_G whose (x, y) entry is $-c(x, y)$ if $x \neq y$ and is $\pi(x)$ if $x = y$. Thus, Δ_G is symmetric and all its row sums are 0. Write $\Delta_G[a]$ for the matrix obtained from Δ_G by deleting the row and column indexed by a. Let $v_a(x, y)$ be the voltage at x when a unit current $i_{y,a}$ flows from y to a (so that the voltage at a is 0) if $y \neq a$ and be 0 otherwise. (By Proposition 2.1, $v_a(x, y) = \mathscr{G}_a(x, y)/\pi(y)$ for the random walk killed at a.) Prove the following statements:

(a) if $x, y \neq a$, then $v_a(x, y)$ is the (x, y)-entry of $\Delta_G[a]^{-1}$;

(b) $v_a(x, y) = v_a(y, x)$;

(c) $v_a(x, x) = v_x(a, a)$;

(d) for all $a, x, y \in \mathsf{V}$, we have $\mathscr{R}(x \leftrightarrow y) = v_a(x, x) - 2v_a(x, y) + v_a(y, y)$;

(e) if $x, y \neq a$, then $v_a(x, y) = (i_{x,a}, i_{y,a})_r$;

(f) for all $f \in \ell^2(\mathsf{V})$, we have $(f, \Delta_G f) = \|df\|_c^2$;

(g) for all $x, y \in \mathsf{V}$, we have $\mathscr{R}(x \leftrightarrow y) = \big(\mathbf{1}_{\{x\}} - \mathbf{1}_{\{y\}}, (\Delta_G + J)^{-1}(\mathbf{1}_{\{x\}} - \mathbf{1}_{\{y\}})\big)$ for any nonzero matrix J with all entries equal.

2.63. Let G be a finite, connected, simple network. Show that $\big\langle \mathscr{R}(x \leftrightarrow y) \,;\, x, y \in \mathsf{V}(G) \big\rangle$ determines $\big\langle c(x, y) \,;\, x, y \in \mathsf{V}(G) \big\rangle$, even if one does not know $\mathsf{E}(G)$.

2.64. Let G be a finite network. Show that if a voltage is 1 at a and 0 at z, then the corresponding current flow from a to z is the projection of the star at a on the orthocomplement of the span of all the other stars except that at z.

2.65. (Foster's Theorem) Show that if G has n vertices, then $\sum_{e \in \mathsf{E}_{1/2}} i^e(e) = n - 1$, where i^e denotes the unit current flow from e^- to e^+.

2.66. Let G be a graph with unit conductances and $e, e' \in \mathsf{E}(G)$. Show that $i^e(e) \geq i^e(e')$.

2.67. Show that in every finite network, for every three vertices u, x, and w, we have

$$\mathscr{R}(u \leftrightarrow x) + \mathscr{R}(x \leftrightarrow w) \geq \mathscr{R}(u \leftrightarrow w).$$

2.68. Show that in every finite network, for every three vertices a, x, and z, we have

$$\mathbf{P}_x[\tau_z < \tau_a] = \frac{\mathscr{R}(a \leftrightarrow x) - \mathscr{R}(x \leftrightarrow z) + \mathscr{R}(a \leftrightarrow z)}{2\mathscr{R}(a \leftrightarrow z)}.$$

2.69. The star-triangle equivalence can be extended as follows. Suppose that (G, c) and (G', c') are two finite networks with a common subset W of vertices that has the property that for all $x, y \in W$, the effective resistance between x and y is the same in each network. Then say that G and G' are **W-equivalent**.

(a) Let G and G' be W-equivalent. Show that specifying voltages on W leads to the same current flows out of W in each of the two networks. More precisely, let $f_0 \colon W \to \mathbb{R}$, and let f and f' be the extensions of f_0 to G and G', respectively, that are harmonic off W. Show that $d^* c d f = d^* c' d f'$ on W.

(b) Let G and G' be W-equivalent. Suppose that H is another network with subset W of vertices, but H is otherwise disjoint from G and G'. Form two new networks $G \cup H$ and $G' \cup H$ by identifying the copies of W. Show that if the same voltages are established at some vertices of H in each of these two networks, then the same voltages and currents will be present in each of these two copies of H.

(c) Given G and a vertex subset W with $|W| = 3$, show that there is a 4-vertex network G' with underlying graph a tree that is W-equivalent to G.

(d) For $x, y \in W$, let $p_W(x, y)$ be the probability that the network random walk on G starting at x is at y when it first returns to W; possibly $x = y$. Define the network $G' := (W, c')$ with

$c'(x, y) := \pi(x)p_W(x, y)$ for all $x \neq y \in W$. Show that $c'(x, y) = c'(y, x)$ for all $x \neq y \in W$ and that G and G' are W-equivalent. *Hint:* Consider adding loops to G'.

  (e) **(The Star-Clique Transformation)** Let $z \in V(G)$ and N be the set of the neighbors of z. Form the network G' from G by deleting z and adding an edge between each pair of distinct vertices $x, y \in N$ of conductance $c(z, x)c(z, y)/\pi(z)$. Show that G and G' are $V(G')$-equivalent. Note that when $|N| = 1$, this is the same as pruning vertices of degree 1; when $|N| = 2$, this is the same as the series transformation; and when $|N| = 3$, this is the same as the star-triangle transformation.

  2.70. Show the following quantitative forms of Rayleigh's monotonicity principle in every finite network:

  (a) If $r(e)$ denotes the resistance of the edge e and i is the unit current flow from a to z, then

$$\frac{\partial}{\partial r(e)} \mathscr{R}(a \leftrightarrow z) = i(e)^2 .$$

  (b) If $c(e)$ denotes the conductance of the edge e and v is the unit voltage from a to z, then

$$\frac{\partial}{\partial c(e)} \mathscr{C}(a \leftrightarrow z) = \left(dv(e)\right)^2 .$$

  2.71. Let G be a recurrent network with an exhaustion by induced subnetworks $\langle G_n \rangle$. Suppose that $a, z \in V(G_n)$ for all n. Let v_n be the voltage function on G_n that arises from a unit voltage at a and 0 voltage at z. Let i_n be the unit current flow on G_n from a to z. Let $\mathscr{G}_z(\cdot, \cdot; G_n)$ be the Green function on G_n for random walk absorbed at z.

  (a) Show that $v := \lim_n v_n$ exists pointwise and that $v(x) = \mathbf{P}_x[\tau_a < \tau_z]$ for all $x \in V$.

  (b) Show that $i := \lim_n i_n$ exists pointwise and that if θ is a finitely supported unit flow from a to z, then $i = P_{\diamondsuit^\perp}\theta$. Here, $\diamondsuit$ is the closure of the linear span of the cycles in G.

  (c) Show that $\mathscr{E}(i)\,dv = ir$.

  (d) Show that the effective resistance between a and z in G_n is monotone decreasing with limit $\mathscr{E}(i)$. We define $\mathscr{R}(a \leftrightarrow z; G) := \mathscr{E}(i)$.

  (e) Show that $\mathscr{G}_z(a, x; G_n) \to \mathscr{G}_z(a, x; G) = \mathscr{E}(i)\pi(x)v(x)$ for all $x \in V$.

  (f) Show that $i(e)$ is the expected number of signed crossings of e.

  2.72. With all the notation of Exercise 2.71, also let G_n^W be the graph obtained from G by identifying the vertices outside G_n to a single vertex, z_n. Let v_n^W and i_n^W be the associated unit voltage and unit current from a to z.

  (a) Show that v_n^W and i_n^W have the same limits v and i as v_n and i_n.

  (b) Show that the effective resistance between a and z in G_n^W is monotone increasing with limit $\mathscr{E}(i)$.

  2.73. Let G be a recurrent network. Define $\bigstar$ to be the closure of the linear span of the stars and $\diamondsuit$ to be the closure of the linear span of the cycles. Show that $\ell_-^2(E, r) = \bigstar \oplus \diamondsuit$.

  2.74. Show that R_{eff} is a concave function of the collection of resistances $\langle r(e) \rangle$.

  2.75. Show that C_{eff} is a concave function of the collection of conductances $\langle c(e) \rangle$.

  2.76. Give another proof of Rayleigh's monotonicity principle by using Exercise 2.13.

  2.77. Show that if a unit voltage is imposed between two vertices of a finite network, then for each fixed edge e, we have that $|dv(e)|$ is a decreasing function of $c(e)$.

2.78. (Extremal Length) Given disjoint vertex sets A, Z in a finite network, prove that

$$\mathscr{C}(A \leftrightarrow Z) = \min \left\{ \sum_{e \in \mathsf{E}_{1/2}} c(e)\ell(e)^2 \right\},$$

where ℓ is an assignment of nonnegative lengths so that the minimum distance from every point in A to every point in Z is 1.

2.79. Extend Exercise 2.13 to the full form of Dirichlet's principle in the finite setting: Let $A \subset \mathsf{V}$ and let $F_0 \colon A \to \mathbb{R}$ be given. Let $F \colon \mathsf{V} \to \mathbb{R}$ be the extension of F_0 that is harmonic at each vertex not in A. Then F is the unique extension of F_0 that minimizes $\mathscr{E}(c\, dF)$.

2.80. Let G be a finite graph and $W \subsetneq \mathsf{V}$. Suppose that $f \colon \mathsf{V} \to \mathbb{R}$ satisfies $df(e) \neq 0$ for every $e \in \mathsf{E}$ and

$$\forall x \notin W \ \exists y \sim x \ \exists z \sim x \ \ f(y) < f(x) < f(z). \tag{2.27}$$

Let $g \colon \mathsf{E} \to (0, \infty)$ be symmetric.

  (a) Show that there is an assignment of conductances $c(\cdot)$ to G such that if the voltages imposed on W are $f \upharpoonright W$, then the energies are g with current flowing in the direction of df, that is, the voltages satisfy $c(e)\, dv(e)^2 = g(e)$ for all $e \in \mathsf{E}$ and $\operatorname{sgn} dv = \operatorname{sgn} df$. *Hint:* Maximize $\sum_{e \in \mathsf{E}_{1/2}} g(e) \log |dh(e)|$ over $h \colon \mathsf{V} \to \mathbb{R}$ with $h \upharpoonright W = f \upharpoonright W$ and $\operatorname{sgn} dh = \operatorname{sgn} df$. Call the maximizer $v(\cdot)$ and define $c := g/dv^2$.

  (b) Show that the maximizer for the hint in part (a) is unique. Therefore, given $f_0 \colon W \to \mathbb{R}$, the number of assignments of $c(\cdot)$ whose energies are a given g is the same for all g, namely, it is the number of choices of $\mathsf{E}_{1/2}$ such that for some extension f of f_0, we have $df(e) > 0$ for all $e \in \mathsf{E}_{1/2}$ and (2.27) holds.

  (c) Show that there exists $f \colon \mathsf{V} \to \mathbb{R}$ such that $df(e) \neq 0$ for all e and (2.27) holds iff there is a choice of $\mathsf{E}_{1/2}$, that is, an orientation of E, such that the only sources and sinks are in W and there are no (oriented) cycles. (An orientation has a source (resp., sink) if there is a vertex which is the tail (resp., head) of all its incident edges.)

2.81. Let H be a proper subnetwork of G. Let $\langle X_n \rangle$ be the network random walk on G. Show that if H is recurrent, then $\mathbf{P}[\forall n \ X_n \in H] = 0$.

2.82. Prove that (2.14) holds even when the network is not locally finite and the cutsets Π_n may be infinite.

2.83. Find a counterexample to the converse of the Nash-Williams criterion. More specifically, find a tree of bounded degree on which simple random walk is recurrent, yet every sequence $\langle \Pi_n \rangle$ of pairwise disjoint cutsets separating the root and ∞ satisfies $\sum_n |\Pi_n|^{-1} \leq C$ for some constant $C < \infty$.

2.84. Show that if $\langle \Pi_n \rangle$ is a sequence of pairwise disjoint cutsets separating the root and ∞ of a tree T without leaves, then $\sum_n |\Pi_n|^{-1} \leq \sum_n |T_n|^{-1}$.

2.85. Give a probabilistic proof as follows of the Nash-Williams criterion in case the cutsets are nested as in Remark 2.10.

  (a) Show that it suffices to prove the criterion for all networks but only for cutsets that consist of all edges incident to some set of vertices.

  (b) Let Π_n be the set of edges incident to the set of vertices W_n. Let A_n be the event that a random walk starting at a visits W_n exactly once before returning to a. Let μ_n be the distribution of the first vertex of W_n visited by a random walk started at a. Show that

$$\mathbf{P}(A_n) = \mathscr{C}(a \leftrightarrow W_n) \sum_{x \in W_n} \pi(x)^{-1} \mu_n(x)^2$$

$$\geq \mathscr{C}(a \leftrightarrow W_n) \left(\sum_{x \in W_n} \pi(x) \right)^{-1} = \mathscr{C}(a \leftrightarrow W_n) \left(\sum_{e \in \Pi_n} c(e) \right)^{-1}.$$

(c) Conclude.

2.86. Show that if G is a graph such that for some $o \in \mathsf{V}(G)$ and some constant $C < \infty$, the ball of radius n about o contains at most Cn^2 edges for infinitely many n, then G is recurrent. More generally, show that if (G, o) is a random rooted graph such that for some constant $C < \infty$, the ball of radius n about o contains at most Cn^2 edges in expectation for infinitely many n, then G is a.s. recurrent.

2.87. Show that if $\theta \in \ell^2_-(\mathsf{E}, r)$, then $\sum_x \pi(x)^{-1} d^* \theta(x)^2 \leq 2 \mathscr{E}(\theta)$.

2.88. It follows from (2.16) that the function $\theta \mapsto \mathscr{E}(\theta)$ is convex. Show the stronger statement that $\theta \mapsto \mathscr{E}(\theta)^{1/2}$ is a convex function. Why is this a stronger inequality?

2.89. Let i be the unit current flow from a to z on a finite network or from a to ∞ on a transient infinite network. Consider the random walk started at a and, if the network is finite, stopped at z. Let S_e be the number of times the edge e is traversed (in the same direction as e).
     (a) Show that $i(e) = \mathbf{E}[S_e - S_{-e} \mid \tau_a^+ = \infty]$.
     (b) Show that if $e^- = a$, then $i(e)$ is the probability that e is traversed following the last visit to a.

2.90. Show that the current i of Exercise 2.71 is the unique unit flow from a to z of minimum energy.

2.91. Let G be a transient network and $\langle X_n \rangle$ the corresponding random walk. Show that if v is the unit voltage between a and ∞ (with $v(a) = 1$), then $v(X_n) \to 0$ a.s.

2.92. Let G be a transient network with $\sup_x \pi(x) < \infty$. Show that $\mathbf{P}_a[\tau_x < \infty] \to 0$ as $d(a, x) \to \infty$. (To avoid an incorrect solution, think about the case of $\mathbb{N}$ attached to $\mathbb{Z}^3$.)

2.93. Show that if (G, c) is an infinite network and A is a finite subset of vertices, then

$$\inf \left\{ \sum_{e \in \mathsf{E}_{1/2}} dF(e)^2 c(e) \,;\, F {\restriction} A \equiv 1 \text{ and } F \text{ has finite support} \right\} = \mathscr{C}(A \leftrightarrow \infty).$$

2.94. Let (G, c) be an infinite network and $o \in \mathsf{V}(G)$. For $f: \mathsf{V}(G) \to \mathbb{R}$, we say that $\lim_{x \to \infty} f(x) = \infty$ if, for all $s \in \mathbb{R}$, there are only finitely many vertices x where $f(x) < s$. Show that (G, c) is recurrent iff there is some $f: \mathsf{V}(G) \to \mathbb{R}$ with $\lim_{x \to \infty} f(x) = \infty$ and $\sum_{e \in \mathsf{E}} df(e)^2 c(e) < \infty$.

2.95. Let (G, c) be an infinite network and $o \in \mathsf{V}(G)$. Show that (G, c) is recurrent iff there is an assignment ℓ of positive lengths to the edges so that the corresponding distance function d_ℓ satisfies $\lim_{x \to \infty} d_\ell(o, x) = \infty$ (in the sense of Exercise 2.94) and $\sum_{e \in \mathsf{E}} c(e) \ell(e)^2 < \infty$.

2.96. Suppose that G is a graph with random resistances $R(e)$ having finite means $r(e) := \mathbf{E}[R(e)]$. Show that if (G, r) is transient, then (G, R) is a.s. transient.

2.97. Suppose that G is a graph with random conductances $C(e)$ having finite means $c(e) := \mathbf{E}[C(e)]$. Show that if (G, c) is recurrent, then (G, C) is a.s. recurrent.

2.98. Let (G, c) be a transient network and $o \in \mathsf{V}(G)$. Consider the voltage function v when the voltage is 1 at o and 0 at infinity. For $t \in (0, 1)$, let $A_t := \{x \in \mathsf{V} \,;\, v(x) > t\}$. Normally, A_t is finite. Show that even if A_t is infinite, the subnetwork it induces is recurrent.

2.99. Sharpen Proposition 2.14 to show that if one identifies to a single vertex z_n all vertices of $\mathbb{Z}^2$ that are at distance more than n from $\mathbf{0}$, then as $n \to \infty$,

$$\mathscr{R}(\mathbf{0} \leftrightarrow z_n) \sim \frac{1}{2\pi} \log n \,.$$

2.100. Consider random walk on $\mathbb{Z}^d$.

(a) Complete the following outline of a Fourier proof of Pólya's theorem. Define the function $\psi(\alpha) := d^{-1} \sum_{k=1}^d \cos 2\pi \alpha_k$ for $\alpha = (\alpha_1, \ldots, \alpha_d) \in \mathbb{T}^d := (\mathbb{R}/\mathbb{Z})^d$. For all $n \in \mathbb{N}$, we have $p_n(\mathbf{0}, \mathbf{0}) = \int_{\mathbb{T}^d} \psi(\alpha)^n \, d\alpha$. Therefore $\sum_n p_n(\mathbf{0}, \mathbf{0}) = \int_{\mathbb{T}^d} 1/\big(1 - \psi(\alpha)\big) \, d\alpha < \infty$ iff $d \geq 3$.

(b) Show that $p_n(\mathbf{0}, \mathbf{0}) \leq c_d n^{-d/2}$ for some constant c_d by using the estimate $\cos 2\pi t \leq 1 - 4\pi t^2$ for $|t| \leq 1/4$ to get $\psi(\alpha) \leq e^{-4\pi|\alpha|^2/d}$ for $\alpha \in [-1/4, 1/4]^d$. Similarly, show that when n is even, $p_n(\mathbf{0}, \mathbf{0}) \geq c'_d n^{-d/2}$ for some constant c'_d by using reverse estimates.

(c) Use a similar method for a random walk on $\mathbb{Z}^d$ that has mean-0, nonconstant, bounded jumps to prove transience and recurrence as in (a) and bounds of the order $n^{-d/2}$ for $p_n(\mathbf{0}, \mathbf{0})$ as in (b).

2.101. Consider an urn with three balls of different colors. Pick a ball uniformly at random, and return it to the urn together with another ball of the same color. Repeating this indefinitely yields a process known as **Pólya's urn**.

(a) Model this process as a random path $\langle S_n \,;\, n \geq 0 \rangle$ in the lattice $\mathbb{Z}^3$ starting from $S_0 := (1, 1, 1)$. Show that S_n is uniformly distributed over $\{(x, y, z) \in \mathbb{Z}^3 \,;\, x + y + z = n + 3, \ x, y, z > 0\}$ for every $n \geq 0$.

(b) Deduce that $\theta(e) := \mathbf{P}\big[\langle S_n, S_{n+1} \rangle = e\big]$ defines a unit flow of finite energy in $\mathbb{Z}^3$.

(c) Use the same method to show that for every $n \geq 1$, the effective resistance from $(0, 0)$ to (n, n) in $\mathbb{Z}^2$ is at most $C \log n$ for some universal constant C, as in Proposition 2.15.

2.102. Give another proof of Theorem 2.17 by using Exercise 2.93.

2.103. Let (G, r) and (G', r') be two finite networks. Let $\phi \colon (G, r) \to (G', r')$ be an (α, β)-rough embedding. Show that for all vertices $x, y \in G$, we have $\mathscr{R}\big(\phi(x) \leftrightarrow \phi(y)\big) \leq \alpha \beta \, \mathscr{R}(x \leftrightarrow y)$.

2.104. Show that all regular trees of degree at least 3 are roughly isometric.

2.105. Let $B_r(G, o)$ denote the set of vertices within distance r of o in the graph G. Suppose that G and G' are roughly isometric, infinite, connected graphs of bounded degree.

(a) Show that G and G' have the same polynomial growth rates, that is, if $\lim_{r \to \infty} \dfrac{\log |B_r(G, o)|}{\log r} = \alpha$, then $\lim_{r \to \infty} \log |B_r(G', o')|/\log r = \alpha$.

(b) Show that if G has exponential growth, then so does G', that is, if $\liminf_{r \to \infty} r^{-1} \log |B_r(G, o)| > 0$, then $\liminf_{r \to \infty} r^{-1} \log |B_r(G', o')| > 0$.

2.106. Let G be an infinite graph and $x, y \in \mathsf{V}$. Show that simple random walk on G satisfies $\sum_{n=0}^{\infty} p_n(x, y)^3 < \infty$. In fact, show that an upper bound for this sum is $c \cdot \deg_G(y)^3$ for some absolute constant c.

2.107. Show that if G is a graph that is roughly isometric to hyperbolic space, then the number of vertices within distance n of a fixed vertex of G grows exponentially fast in n.

2.108. Show that in every finite network,

$$\mathbf{E}_a[\tau_z] = \frac{1}{2} \sum_{x \in \mathsf{V}} \pi(x)\big[\mathscr{R}(a \leftrightarrow z) + \mathscr{R}(z \leftrightarrow x) - \mathscr{R}(x \leftrightarrow a)\big] \,.$$

(Recall that $\pi(\cdot)$ is not generally a probability measure.)

2.109. Consider a positive recurrent Markov chain, not necessarily reversible. Write $H(x, y) := \mathbf{E}_x[\tau_y]$ for the matrix of hitting times. Let A be a subset of the state space, V. Write $\mu_z^+(x) := \mathbf{P}_z[X_{\tau_A^+} = x]$.

(a) Show that for every $z \in \mathsf{V}$ and $y \in A$, we have

$$H(z, y) = \begin{cases} \mathbf{E}_z[\tau_A^+] + \sum_{x \in A} \mu_z^+(x) H(x, y) & \text{if } y \neq z \\ 0 & \text{otherwise.} \end{cases}$$

(b) Assume now that $2 \leq |A| < \infty$. Show that the $(A \times A)$-submatrix H_A of H is invertible.

(c) Let $K_{z,A}$ denote the row vector $\mathbf{E}_z[\tau_y^+] - \mathbf{E}_z[\tau_A^+]$ ($y \in A$). Show that

$$\mu_z^+ = K_{z,A} H_A^{-1}.$$

(d) Show that if there is a group acting transitively on the state space that preserves the transition probabilities, then

$$\frac{1}{|\mathsf{V}|} \sum_{z \in \mathsf{V}} \mu_z^+ = c\mathbf{1}^T H_A^{-1},$$

where c is a constant so that the right-hand side adds to 1, in other words, $c^{-1} = \mathbf{1}^T H_A^{-1} \mathbf{1}$.

2.110. Let G be a network. Suppose that x and y are two fixed vertices such that there is an automorphism of G that takes x to y (though it might not take y to x). Show that for every k, we have $\mathbf{P}_x[\tau_y = k] = \mathbf{P}_y[\tau_x = k]$. *Hint:* Show the equality with "$\leq k$" in place of "$= k$."

2.111. Let G be a finite network whose automorphism group acts transitively. Let $A \subseteq \mathsf{V}$. Write R_A for the $(A \times A)$-matrix whose (x, y)-entry is $\mathscr{R}(x \leftrightarrow y)$. Define μ_z^+ as in Exercise 2.109. Show that

$$\frac{1}{|\mathsf{V}|} \sum_{z \in \mathsf{V}} \mu_z^+ = c\mathbf{1}^T R_A^{-1},$$

where c is a constant so that the right-hand side adds to 1.

2.112. Let G be a network such that $\gamma := \sum_{e \in \mathsf{E}} c(e) < \infty$ (for example, G could be finite). Here, the sum counts each edge with two endpoints twice and each loop once. Let x, y, z be three distinct vertices in G. Write $\tau_{x,y,z}$ for the first time that the network random-walk trajectory contains the vertices x, y, z in that order. Show that

$$\mathbf{P}_x[\tau_{y,z,x} \leq \tau_x^+] = \mathbf{P}_x[\tau_{z,y,x} \leq \tau_x^+] \tag{2.28}$$

and

$$\mathbf{E}_x[\tau_{y,z,x}] = \mathbf{E}_x[\tau_{z,y,x}] = \frac{\gamma}{2}\big[\mathscr{R}(x \leftrightarrow y) + \mathscr{R}(y \leftrightarrow z) + \mathscr{R}(z \leftrightarrow x)\big]. \tag{2.29}$$

2.113. Let G be a finite network and x, y be two distinct vertices in G. Write $\tau_{x,y}$ for the first time that the network random-walk trajectory contains the vertices x, y in that order. Let π be the stationary probability measure. Write $\mathbf{P}_\pi(A) := \sum_x \pi(x) \mathbf{P}(A)$ for events A.

(a) Show that

$$\mathbf{P}_\pi[\tau_{x,y} = k] = \mathbf{P}_\pi[\tau_{y,x} = k]$$

for all k.

(b) Define $f(w, z) := \mathbf{E}_w[\tau_z] - \mathbf{E}_\pi[\tau_z]$. Show that $f(w, z) = f(z, w)$ for all vertices w, z.

(c) Define the binary relation $w \preccurlyeq z$ to mean $\mathbf{E}_w[\tau_z] \leq \mathbf{E}_z[\tau_w]$. Show that this relation is a total order.

(d) Suppose that independent random walks X and Y start from x and y, respectively. However, you are allowed to decide which of the walks to move at each time step. At time n, you are allowed to look at all the steps that have been made up to time n. Thus, if by time n, the walk X has made s_n steps and Y has made t_n steps, where $s_n + t_n = n$, then the choice whether $s_{n+1} = s_n + 1$ or $t_{n+1} = t_n + 1$ can depend on $\langle X_k ; k \leq s_n \rangle$ and $\langle Y_k ; k \leq t_n \rangle$. Let $\tau := \inf\{n ; X_{s_n} = Y_{t_n}\}$ be the first time the controlled random walks collide. Define $M_n := f(X_{s_n}, Y_{t_n}) + n$. Show that $\langle M_{n \wedge \tau} ; n \geq 0 \rangle$ is a martingale.

(e) Show that $\mathbf{E}[\tau] \leq 2 \max_{w,z} \mathbf{E}_w[\tau_z]$.

2.114. Let G be a network and x, y, z be three distinct vertices in G. Write $\tau_{x,y,z}$ for the first time that the network random-walk trajectory contains the vertices x, y, z in that order. Strengthen and generalize the first equality of (2.29) by showing that

$$\mathbf{P}_x[\tau_{y,z,x} = k] = \mathbf{P}_x[\tau_{z,y,x} = k]$$

for all k.

2.115. Consider a Markov chain that is not necessarily reversible. Let a, x, and z be three of its states. Show that

$$\mathbf{P}_x[\tau_z < \tau_a] = \frac{\mathbf{E}_x[\tau_a] + \mathbf{E}_a[\tau_z] - \mathbf{E}_x[\tau_z]}{\mathbf{E}_z[\tau_a] + \mathbf{E}_a[\tau_z]} . \tag{2.30}$$

Use this in combination with (2.29) and Corollary 2.21 to give another solution to Exercise 2.68. *Hint:* Consider whether the chain visits z on the way to a.

2.116. Consider a Markov chain that is not necessarily reversible. Let a, x, and z be three of its states. Show that

$$\mathbf{E}_x[\tau_a \wedge \tau_z] = \frac{\mathbf{E}_x[\tau_a]\,\mathbf{E}_a[\tau_z] + \mathbf{E}_x[\tau_z]\,\mathbf{E}_z[\tau_a] - \mathbf{E}_a[\tau_z]\,\mathbf{E}_z[\tau_a]}{\mathbf{E}_z[\tau_a] + \mathbf{E}_a[\tau_z]} .$$

2.117. Let T be a tree and x and y be two of its vertices. For a vertex z on the path from x to y, including x but not y, let $A_y(z)$ be the number of edges that are separated from y by z. Show that $\mathbf{E}_x[\tau_y] = 2 \sum_z A_y(z) + \operatorname{dist}(x, y)$.

2.118. Let G be a finite graph and x and y be two of its vertices such that there is a unique simple path in G that joins these particular vertices; for example, G could be a tree (possibly with loops), and then x and y could be any of its vertices. Show that $\mathbf{E}_x[\tau_y] \in \mathbb{N}$. In the case that G is a tree possibly with loops, show that $\mathbf{E}_x\!\left[\binom{\tau_y}{k}\right] \in \mathbb{N}$ for all $k \in \mathbb{N}$.

2.119. Let G be a transient network and $R_n := |\{X_0, X_1, \dots, X_n\}|$ be the number of vertices visited by time n. Show that for all n and all $o \in \mathsf{V}(G)$,

$$\frac{\mathbf{E}_o[R_n]}{n + 1} \geq \inf_{x \in \mathsf{V}(G)} \frac{\mathscr{C}(x \leftrightarrow \infty)}{\pi(x)} .$$

2.120. Consider a finite, irreducible Markov chain that is not necessarily reversible. Let a and z be two of its states. Let π be the stationary probability distribution. Show that $\pi(a)\,\mathbf{P}_a[\tau_z < \tau_a^+]$ is the reciprocal of the commute time between a and z. Deduce that $\pi(a)\,\mathbf{P}_a[\tau_z < \tau_a^+] = \pi(z)\,\mathbf{P}_z[\tau_a < \tau_z^+]$.

2.121. Consider a positive recurrent, irreducible Markov chain that is not necessarily reversible. Let W be a subset of its states.

(a) Show that there is a Markov chain on W all whose commute times are the same as the corresponding commute times of the original chain.

(b) Show that (a) fails if "commute" is replaced by "hitting."

2.122. Consider a finite, irreducible Markov chain that is not necessarily reversible. Let a and z be two of its states. Let π be the stationary probability distribution. Using the reversed Markov chain $\widehat{P}$, define the **symmetrized Markov chain** to have transition probabilities $\bar{p}(x, y) := [p(x, y) + \widehat{p}(x, y)]/2$.

(a) Show that $\bar{P}$ is reversible with stationary measure π.

(b) Write $L := I - P$ and $\widehat{L} := I - \widehat{P}$. Show that $\widehat{L}$ is the adjoint of L with respect to the inner product $(\cdot, \cdot)_\pi$.

(c) Show that for all real-valued functions f on the state space, $(Lf, f)_\pi \geq 0$, with equality iff f is constant.

(d) Write $r(a, z) := \left(\mathbf{E}_a[\tau_z] + \mathbf{E}_z[\tau_a]\right)^{-1}$ for the reciprocal of the commute time between a and z. Write $h(x) := \mathbf{P}_x[\tau_a < \tau_z]$ and $\widehat{h}(x) := \widehat{\mathbf{P}}_x[\tau_a < \tau_z]$. Show that

$$\pi(x)(Lh)(x) = r(a, z)\big(\mathbf{1}_{\{a\}}(x) - \mathbf{1}_{\{z\}}(x)\big) = \pi(x)(\widehat{L}\widehat{h})(x).$$

(e) Show that for all real-valued f, we have $(Lh, f)_\pi = (Lf, \widehat{h})_\pi = r(a, z)\big[f(a) - f(z)\big]$.

(f) Show that if $f(a) = 1$ and $f(z) = 0$, then $(Lh, 2f - h)_\pi = r(a, z)$.

(g) Put $\bar{h} := (h + \widehat{h})/2$. Suppose that $f + g = 2\bar{h}$, where $f(a) = g(a) = 1$ and $f(z) = g(z) = 0$. Show that $(Lf, g)_\pi \leq r(a, z)$, with equality iff $f = h$ and $g = \widehat{h}$.

(h) Show that

$$r(a, z) = \min_\varphi \ \max_{f+g=2\varphi} (Lf, g)_\pi\,,$$

where φ, f, g are real-valued functions that all take the value 1 at a and 0 at z.

(i) Show that the commute time between a and z for the symmetrized chain is at least that for the original chain, with equality iff $h = \widehat{h}$.

2.123. Show that if two finite, irreducible (not necessarily reversible) Markov chains on the same state space determine the same functions $(a, z, x) \mapsto \mathbf{P}_x[\tau_a < \tau_z]$ and each of the chains have only 0 on the diagonal of their transition matrices, then the chains have the same transition probabilities.

2.124. Consider an irreducible Markov chain on a state space V of size n, not necessarily reversible. Let π be the stationary probability distribution. Show that $\sum_{x,y\in\mathsf{V}} \pi(x)p(x, y)\,\mathbf{E}_y[\tau_x] = n - 1$. Give another proof of Foster's theorem (Exercise 2.65) from this.

2.125. Consider a finite, irreducible Markov chain that is not necessarily reversible. Let π be the stationary probability distribution.

(a) Prove that $\sum_{x,y\in\mathsf{V}} \pi(x)p(x, y)\,\mathbf{E}_x[\tau_y] \leq n - 1$.

(b) Show that equality holds in (a) iff the Markov chain is reversible.

2.126. Let G be a finite network and let a and z be two vertices of G. Let $x \sim y$ in G. Show that the expected number of transitions from x to y for a random walk started at a and stopped at the first return to a that occurs after visiting z is $c(x, y)\mathscr{R}(a \leftrightarrow z)$. Give another proof of Corollary 2.21 by using this formula.

2.127. Show that Corollary 2.21 and Exercises 2.126, 2.68, and 2.108 hold for all recurrent networks.

2.128. Let G be a network with n vertices and consider two of its vertices, a and z. Consider a random walk $\langle X_k\,;\ 0 \leq k \leq \tau\rangle$ that starts at a, visits z, and is then stopped at its first return τ to a after visiting z. Show that $\mathbf{E}\big[\sum_{k=0}^{\tau-1} \mathscr{R}(X_k \leftrightarrow X_{k+1})\big] = 2(n - 1)\mathscr{R}(a \leftrightarrow z)$.

2.129. Given two vertices a and z of a finite network (G, c), show that the commute time between a and z is at least twice the square of the graph distance between a and z. *Hint:* Consider the cutsets between a and z that are determined by spherical shells.

2.130. Show that the expected cover time of a graph $G = (\mathsf{V}, \mathsf{E})$ is at most $2|\mathsf{E}| \cdot |\mathsf{V}|$.

2.131. A graph is called **edge transitive** if, for each pair e, e' of (unoriented) edges, there is an automorphism of the graph that takes e to e'. Show that the expected cover time of an edge-transitive graph $G = (\mathsf{V}, \mathsf{E})$ is at most $2|\mathsf{V}|^2$.

2.132. Let G be a connected, simple graph on n vertices. Define

$$\delta(e) := \mathscr{R}(e^- \leftrightarrow e^+) - (\deg e^- + 1)^{-1} - (\deg e^+ + 1)^{-1}.$$

For a spanning tree T, define $R(T) := \sum_{e \in T} \mathscr{R}(e^- \leftrightarrow e^+)$.
  (a) Show that $\delta(e) \geq 0$ for all $e \in \mathsf{E}$.
  (b) Show that $\sum_{e \in \mathsf{E}_{1/2}} \delta(e) = \sum_{x \in \mathsf{V}} (\deg x + 1)^{-1} - 1$.
  (c) Show that if G is d-regular, then

$$\frac{2(n-1)}{d+1} \leq R(T) \leq \frac{3(n-1)}{d+1}$$

for all spanning trees T of G.
  (d) Show that if G is regular, then the expected cover time of G is at most $3n^2$.
  (e) Show that the commute time between any pair of vertices in G is less than the quantity $6|\mathsf{E}| \sum_{x \in \mathsf{V}} (\deg x + 1)^{-1}$.
  (f) Show that there exist $\epsilon_n \to 0$ such that for all connected, simple graphs G on n vertices, the commute time between any pair of vertices in G is less than $(4/27 + \epsilon_n)n^3$.

2.133. The **hypercube** of dimension n is the subgraph induced on the set $\{0, 1\}^n$ in the usual nearest-neighbor graph on $\mathbb{Z}^n$.
  (a) Find the first-order asymptotics of the effective resistance between the hypercube's opposite corners $(0, 0, \ldots, 0)$ and $(1, 1, \ldots, 1)$ and the first-order asymptotics of the commute time between them. (To find first-order asymptotics for some function $f(n)$ means to find a simple expression $g(n)$ such that $\lim_{n \to \infty} f(n)/g(n) = 1$.)
  (b) Find the first-order asymptotics of $\mathbf{E}_x[\tau_y]$ for every $x \neq y$ in the hypercube.
  (c) Find the first-order asymptotics of the cover time of the hypercube.

2.134. Let (G, c) be a finite network and $o \in \mathsf{V}(G)$. Let $X(e)$ be independent, normal random variables with variance $r(e)$ for $e \in \mathsf{E}_{1/2}$; put $X(-e) := -X(e)$ for $e \notin \mathsf{E}_{1/2}$. Given random walks starting at each $x \in \mathsf{V}(G)$ with each one stopped when it reaches o, define $S(x)$ to be the sum of $X(e)$ over the edges e traversed by the random walk starting at x. Show that $x \mapsto \mathbf{E}[S(x) \mid X]$ has the law of the canonical Gaussian field.

2.135. Let (G, c) be a finite network with $o \in \mathsf{V}(G)$; let Z be the associated canonical Gaussian field. Let W be a nonempty, proper subset of vertices that includes o. Show that the squared distance in $L^2(\mathbf{P})$ from $Z(x)$ to the linear span of $Z(w)$ ($w \in W$) equals $\mathscr{R}(x \leftrightarrow W)$.

2.136. Let (G, c) be a finite network with $o \in \mathsf{V}(G)$; let Z be the associated canonical Gaussian field. Show that if b_x are constants such that $\sum_{x \in \mathsf{V}} b_x Z(x)$ is constant a.s., then $b_x = 0$ for all $x \neq o$.

2.137. Let (G, c) be a finite network. Fix a nonempty, proper subset $W \subsetneq \mathsf{V}$ and a function $u: W \to \mathbb{R}$. Let $Z: \mathsf{V} \to \mathbb{R}$ be the jointly normal random variables such that $Z = u$ on W and the joint density of $Z{\upharpoonright}(\mathsf{V} \setminus W)$ is

$$C \exp\left\{-\frac{1}{2}\|dZ\|_c^2\right\}$$

for some constant C. These random variables are called the **canonical Gaussian field pinned on** W. Let v be the harmonic extension of u to V.
  (a) Show that $\|dZ\|_c^2 = \|d(Z - v)\|_c^2 + \|dv\|_c^2$. Deduce that the law of $Z - v$ on G/W is that of the usual canonical Gaussian field on G/W, where G/W is the network obtained from G by identifying W to a single vertex, where the field is 0.
  (b) Show that $\mathbf{E}[Z(x)] = v(x)$ for all x.

(c) Show that

$$Z(x) - Z(y) \sim \mathrm{N}\big(v(x) - v(y), \mathscr{R}(x \leftrightarrow y; G/W)\big)$$

for $x \neq y \in \mathsf{V}$, as long as x and y are not both in W.

(d) Show that the covariance of $Z(x) - Z(y)$ and $Z(z) - Z(w)$ equals $v'(z) - v'(w)$ when v' is the voltage associated to a unit current flow from x to y in the network G/W (with $x \neq y$).

2.138. Let (G, c) be a finite network and $m > 0$. Fix a nonempty, proper subset $W \subsetneq \mathsf{V}$ and a function $u\colon W \to \mathbb{R}$. Let $Z\colon \mathsf{V} \to \mathbb{R}$ be the jointly normal random variables such that $Z = u$ on W and the joint density is

$$C \exp\left\{ -\frac{1}{2}\left(\|dZ\|_c^2 + m \sum_{x \in \mathsf{V}} Z(x)^2 \right) \right\}$$

for some constant C. These random variables are called the ***canonical Gaussian field pinned on*** W ***with mass*** m. Calculate $\mathbf{E}\big[Z(x) - Z(y)\big]$ and $\mathrm{Cov}\big(Z(x) - Z(y), Z(z) - Z(w)\big)$ for $x, y, z, w \in \mathsf{V}$.

2.139. Use results on canonical Gaussian fields to solve Exercises 2.59, 2.62(b)–(e), and 2.67.

3 | Special Networks

In this chapter, we use our tools from Chapter 2 to analyze transience and recurrence of networks on trees and on Cayley graphs of groups. First, we study flows that are not necessarily current flows. This involves some tools that are very general and useful. In particular, the surprising max-flow min-cut theorem has a wealth of specializations and applications, some of which are in the exercises to this chapter and some of which will be in other chapters. After applying these to trees, we will define Cayley graphs and give several examples. When we analyze the type of certain biased random walks on Cayley graphs, trees will actually be a key tool to show that in a certain sense, Cayley graphs are like spherically symmetric graphs, even though they are rarely truly spherically symmetric and can even be quite far from spherically symmetric. Other special networks, planar and hyperbolic, are studied in Section 9.4.

3.1 Flows, Cutsets, and Random Paths

Notice that if there is a flow from a to ∞ of finite energy on some network with conductances $c(\cdot)$ and if i is the unit current flow with corresponding voltage function v, then $|i(e)| = |c(e) \cdot (dv)(e)| \leq v(a)c(e) = \mathscr{R}(a \leftrightarrow \infty)c(e)$ for all edges e. In particular, there is a nonzero flow bounded on each edge by $c(\cdot)$ (namely, $i/v(a)$).* The existence of flows that are bounded by some given numbers on the edges is an interesting and valuable property in itself. We call such flows ***admissible***. To determine whether there is a nonzero, admissible flow, we turn to the powerful max-flow min-cut theorem of Ford and Fulkerson (1962). For finite networks, the theorem reads as follows. We call a set Π of edges a ***cutset*** separating A and Z if every path that starts in A and ends in Z must include an edge in Π. We call $c(e)$ the ***capacity*** of e in this context. Think of water flowing through pipes. A flow between A and Z is an antisymmetric function θ such that $d^*\theta$ equals 0 off $A \cup Z$. Recall from Section 2.4 that the strength of a flow between A and Z, that is, the total amount flowing into the network at vertices in A (and out at Z) is $\mathsf{Strength}(\theta) := \sum_{a \in A} d^*\theta(a)$. Since all the water must flow through every cutset Π, it is intuitively clear that the strength of every admissible flow is at most $\inf_\Pi \sum_{e \in \Pi} c(e)$. Remarkably, this upper bound is always achieved. This is the content of the max-flow min-cut theorem.

It will be useful for the proof of this theorem, as well as later, to establish a more general statement about flows in *directed* networks. In a directed network, the capacity function c on

---

* This also follows from the Nash-Williams criterion and the upcoming max-flow min-cut theorem.

edges is not necessarily symmetric, even if both orientations of an edge occur. Define the vertex-edge incidence function

$$\phi(x, e) := \mathbf{1}_{\{e^- = x\}} - \mathbf{1}_{\{e^+ = x\}} \, .$$

A *nonnegative* function θ on the edges is now called a ***flow between A and Z*** if it satisfies the condition $\sum_e \phi(x, e)\theta(e) = 0$ for all $x \notin A \cup Z$. In particular, flows are *not* necessarily antisymmetric (nor symmetric) functions. We have that

$$\sum_x \sum_e \phi(x, e)\theta(e) = \sum_e \sum_x \phi(x, e)\theta(e) = 0 \, ,$$

whence, similarly to Lemma 2.8, we have

$$\sum_{x \in A} \sum_e \phi(x, e)\theta(e) = - \sum_{x \in Z} \sum_e \phi(x, e)\theta(e) \, .$$

The common value here is called the **strength** of a flow from A to Z. A flow θ is called **admissible** if $\theta(e) \le c(e)$ for every edge e. **Cutsets** separating A from Z are required to intersect every directed path from A to Z. To reduce the study of undirected networks to that of directed ones, we simply replace each undirected edge by a pair of parallel directed edges with opposite orientations and the same capacity. A flow on the undirected network is replaced by a flow on the directed network that is nonzero on only one edge of each new parallel pair, whereas a flow on the resulting directed network yields a flow on the undirected network by subtracting the values on each parallel oppositely oriented pair of edges.

The Max-Flow Min-Cut Theorem. *Let A and Z be disjoint sets of vertices in a (directed or undirected) finite network G. The maximum strength of an admissible flow from A and Z equals the minimum cutset sum of the capacities. In other words, in the directed case,*

$$\max \left\{ \text{Strength}(\theta) \, ; \, \theta \text{ flows from } A \text{ to } Z \text{ satisfying } \forall e \ \ 0 \le \theta(e) \le c(e) \right\}$$

$$= \min \left\{ \sum_{e \in \Pi} c(e) \, ; \, \Pi \text{ separates } A \text{ and } Z \right\} ,$$

whereas in the undirected case,

$$\max \left\{ \text{Strength}(\theta) \, ; \, \theta \text{ flows from } A \text{ to } Z \text{ satisfying } \forall e \ |\theta(e)| \le c(e) \right\}$$

$$= \min \left\{ \sum_{e \in \Pi} c(e) \, ; \, \Pi \text{ separates } A \text{ and } Z \right\} .$$

Proof. It suffices to establish the case of directed networks. Because the network is finite, the set of flows from A to Z bounded by c on each edge is a compact set in $\mathbb{R}^E$, whence there is a flow of maximum strength. Let θ be a flow of maximum strength. If Π is any cutset separating A from Z, let A' denote the set of vertices that are not separated from A by Π. Since $A \subseteq A'$ and $A' \cap Z = \varnothing$, we have

$$\text{Strength}(\theta) = \sum_{x \in A} \sum_{e \in E} \phi(x, e)\theta(e) = \sum_{x \in A'} \sum_{e \in E} \phi(x, e)\theta(e)$$

$$= \sum_{e \in E} \theta(e) \sum_{x \in A'} \phi(x, e) \le \sum_{e \in \Pi} \theta(e) \le \sum_{e \in \Pi} c(e), \tag{3.1}$$

since $\sum_{x \in A'} \phi(x, e)$ is 0 when e joins two vertices in A', is 1 when e leads out of A', and is -1 when e leads into A'. This proves the intuitive half of the desired equality.

For the more surprising reverse inequality, given an admissible flow θ, call a sequence of vertices $x_0, x_1, \ldots, x_k$ an ***augmentable path*** if $x_0 \in A$ and for all $i = 1, \ldots, k$, either there is an edge e from x_{i-1} to x_i with $\theta(e) < c(e)$ or there is an edge e' from x_i to x_{i-1} with $\theta(e') > 0$. Let B denote the set of vertices x such that there exists an augmentable path (possibly just one vertex) from a vertex of A to x. If there were an augmentable path $x_0, x_1, \ldots, x_k$ with $x_k \in Z$, then we could obtain from θ a stronger flow bounded by c as follows: For each $i = 1, \ldots, k$ where there is an edge e from x_{i-1} to x_i, let $\theta^*(e) := \theta(e) + \epsilon$, whereas if there is an edge e' from x_i to x_{i-1}, let $\theta^*(e') := \theta(e') - \epsilon$. By taking $\epsilon > 0$ sufficiently small, we would contradict maximality of θ. Therefore $Z \subset B^c$. Let Π be the set of edges connecting B to B^c. Then Π is a cutset separating A from Z. For every edge e leading from B to B^c, necessarily $\theta(e) = c(e)$, whereas θ must vanish on every edge from B^c to B. Therefore a calculation as in (3.1) shows that

$$\text{Strength}(\theta) = \sum_{e \in E} \theta(e) \sum_{x \in B} \phi(x, e) = \sum_{e \in \Pi} \theta(e) = \sum_{e \in \Pi} c(e).$$

In conjunction with (3.1), this completes the proof. ◄

Let θ be a nonnegative function on the edges of a directed network. We call a vertex x a ***source*** of θ if $\sum_e \phi(x, e)\theta(e) > 0$ and a ***sink*** is this sum is negative. If A and Z are disjoint, then there is an admissible flow from A to Z of maximum strength such that no vertex in A is a sink and no vertex in Z is a source. In fact, there is such a flow θ with the property that if $e^+ \in A$, then $\theta(e) = 0$ and if $e^- \in Z$, then $\theta(e) = 0$. Indeed, let θ be an admissible flow of maximum strength that minimizes $\sum_e |\theta(e)|$. If there is some $e^+ \in A$ with $\theta(e) > 0$, then there must be a directed path from a vertex in A to e^+ along which $\theta \ge c > 0$ (because there is no augmenting path). We may then subtract c from θ for every edge in this path, which does not change the strength of the flow but diminishes the sum that was supposed to be minimum, a contradiction. The argument is similar for edges leading out of Z.

Suppose now that $G = (\mathsf{V}, \mathsf{E})$ is a countable, directed or undirected network and a is one of its vertices. As usual, we assume that $\forall x \; \sum_{e^- = x} c(e) < \infty$. We want to extend the max-flow min-cut theorem to G for flows from a to ∞. Recall that a cutset Π separates a and ∞ if every infinite, simple path from a must include an edge in Π. A flow of maximum strength exists, since a maximizing sequence of flows has an edgewise limit point, which is a maximizing flow bounded by c in light of the dominated convergence theorem. We claim that this maximum strength is equal to the infimum of the cutset sums:

Theorem 3.1. *If a is a vertex in a countable, directed network G, then*

$$\max \left\{ \text{Strength}(\theta) \,;\; \theta \text{ flows from } a \text{ to } \infty \text{ satisfying } \forall e \;\; 0 \le \theta(e) \le c(e) \right\}$$

$$= \inf \left\{ \sum_{e \in \Pi} c(e) \,;\; \Pi \text{ separates } a \text{ and } \infty \right\}.$$

Proof. In the proof, "cutset" will always mean "cutset separating a and ∞." Let θ be a flow of maximum strength among flows from a to ∞ that are bounded by $c(\cdot)$. Given $\epsilon > 0$, let D be a possibly empty set of edges such that $\sum_{e \in D} c(e) < \epsilon$ and $G' := (V, E \setminus D)$ is locally finite. Let $\mathscr{P}$ be the set of simple paths in G' from a to ∞. Define the distance between two elements $\langle e_n \rangle$, $\langle e_n' \rangle$ of $\mathscr{P}$ to be $\inf\{1/(n+1)\,;\; e_k = e_k' \text{ for } 1 \le k \le n\}$. The important aspect of this is that the set $\mathscr{P}$ is compact: Given paths $\mathcal{P}_m = \langle e_{m,n} \rangle$, we may choose a subsequence $\langle m_k \rangle$ such that $e_{m_k,1}$ is the same for all k because G' is locally finite. Then we may choose a further subsequence where the second edge is constant. Continuing this way, a diagonal argument provides a limit path of the original sequence of paths.

If we associate to an edge e the set of paths in $\mathscr{P}$ that pass through e, then a cutset becomes associated to a cover of $\mathscr{P}$. Compactness of $\mathscr{P}$ therefore means that for any cutset Π in G, there is a finite cutset $\Pi' \subseteq \Pi$ in G'. Also, Π' separates only finitely many vertices A' from ∞ in G'. Therefore,

$$\text{Strength}(\theta) = \sum_{e \in E} \phi(a, e)\theta(e) = \sum_{x \in A'} \sum_{e \in E} \phi(x, e)\theta(e)$$

$$= \sum_{x \in A'} \sum_{e \in E \setminus D} \phi(x, e)\theta(e) + \sum_{x \in A'} \sum_{e \in D} \phi(x, e)\theta(e)$$

$$= \sum_{e \in E \setminus D} \theta(e) \sum_{x \in A'} \phi(x, e) + \sum_{x \in A'} \sum_{e \in D} \phi(x, e)\theta(e)$$

[since the first sum is finite]

$$\le \sum_{e \in \Pi'} \theta(e) + \epsilon \le \sum_{e \in \Pi'} c(e) + \epsilon \le \sum_{e \in \Pi} c(e) + \epsilon.$$

Since this holds for all $\epsilon > 0$, we obtain one inequality of the desired equality.

For the inequality in the other direction, let $C(H)$ denote the infimum cutset sum in any network H. Given $\epsilon > 0$, let D and G' be as before. Then $C(G) \le C(G') + \epsilon$, since we may adjoin D to any cutset of G' to obtain a cutset of G. Let $\langle G_n' \rangle$ be an exhaustion of G' by finite, connected, induced networks with $a \in G_n'$ for all n. Identify the vertices outside G_n' to a single vertex z_n and remove loops at z_n to form the finite network G_n^W from G. Then $C(G') = \inf_n C(G_n^W)$, since every minimal cutset of G' is finite (it separates only finitely many vertices from ∞), where minimal means with respect to inclusion. Let θ_n be a flow on G_n^W of maximum strength among flows from a to z_n that are bounded by $c \restriction G_n^W$. Then $\text{Strength}(\theta_n) = C(G_n^W) \ge C(G')$. Let θ be a limit point of $\langle \theta_n \rangle$. Then θ is a flow on G' with $\text{Strength}(\theta) \ge C(G') \ge C(G) - \epsilon$. ◄

In Section 2.5, we constructed a unit flow from a random path. The reverse is also useful; we show how to do this now. We return to undirected graphs for this, so that a flow θ satisfies $\theta(-e) = -\theta(e)$ for all edges e. Suppose that

$$
\left.\begin{array}{l}
\theta \text{ is a unit flow from } a \text{ to } z \text{ on a finite graph or from } a \text{ to } \infty \\
\text{on an infinite graph such that if } e^- = a, \text{ then } \theta(e) \geq 0 \text{ and,} \\
\text{in the finite case, if } e^+ = z, \text{ then } \theta(e) \geq 0.
\end{array}\right\} \tag{3.2}
$$

Use θ to define a random path as the trajectory of a Markov chain $\langle Y_n \rangle$ as follows. The initial state is $Y_0 := a$ and z is an absorbing state. For a vertex $x \neq z$, set

$$
\theta_{\text{out}}(x) := \sum_{\substack{e^- = x, \\ \theta(e) > 0}} \theta(e),
$$

which is the amount flowing out of x. The transition probability from x to w is then $(\theta(x, w) \vee 0)/\theta_{\text{out}}(x)$. This gives us our random path. Now go back and construct a unit flow from this random path as on p. 40 in Section 2.5, that is, define

$$
\theta'(e) := \sum_{n \geq 0} \left\{ \mathbf{P}[\langle Y_n, Y_{n+1} \rangle = e] - \mathbf{P}[\langle Y_{n+1}, Y_n \rangle = e] \right\}.
$$

How is θ' related to θ? We call θ **acyclic** if there is no cycle of oriented edges on each of which $\theta > 0$. For example, current flows are acyclic because they minimize energy (or because they equal $c \, dv$).

Proposition 3.2. *Suppose that θ is an acyclic unit flow satisfying the preceding conditions (3.2). With the preceding notation, for every edge e with $\theta(e) > 0$, we have*

$$
0 \leq \theta'(e) \leq \theta(e) \tag{3.3}
$$

with equality on the right if G is finite or if θ is the unit current flow from a to ∞.

Proof. Since the Markov chain travels only in the direction of θ, it clear that $\theta'(e) \geq 0$ when $\theta(e) > 0$.

For edges e with $\theta(e) > 0$, set

$$
p_N(e) := \mathbf{P}[\exists n \leq N \ \langle Y_n, Y_{n+1} \rangle = e].
$$

Because θ is acyclic, $p_N(e) \to \theta'(e)$. Thus, to show that $\theta'(e) \leq \theta(e)$, it suffices to show that $p_N(e) \leq \theta(e)$ for all N. We proceed by induction on N. This is clear for $N = 0$. For vertices x, define $p_N(x) := \mathbf{P}[\exists n \leq N \ Y_n = x]$. Suppose that $p_N(e) \leq \theta(e)$ for all edges e with $\theta(e) > 0$. Then also $p_{N+1}(x) \leq \sum_{e^+ = x, \theta(e) > 0} \theta(e) = \theta_{\text{out}}(x)$ for all vertices $x \neq a, z$. Hence $p_{N+1}(x) \leq \theta_{\text{out}}(x)$ for all vertices $x \neq z$. Therefore, for every edge e with $\theta(e) > 0$, if we put $x := e^-$, then we get $p_{N+1}(e) = p_{N+1}(x)\theta(e)/\theta_{\text{out}}(x) \leq \theta(e)$. This completes the induction and proves (3.3).

If G is finite, then $\theta'' := \theta - \theta'$ is a sourceless, acyclic flow, since it is positive only where θ is positive. If there were an edge e_1 where $\theta''(e_1) > 0$, then there would exist an edge e_2 whose head is the tail of e_1 where also $\theta''(e_2) > 0$, and so on. Eventually, this would close a cycle, which is impossible. Thus, $\theta' = \theta$.

If θ is the unit current flow from a to ∞, then θ has minimum energy among all unit flows from a to ∞. Thus, (3.3) implies that $\theta' = \theta$. ◀

Remark. Of course, other random paths or other rules for transporting mass through the network according to the flow θ will lead to the same result.

▷ **Exercise 3.1.**
Show that if equality holds on the right-hand side of (3.3), then for all x, we have

$$\sum_{\substack{e^+=x, \\ \theta(e)>0}} \theta(e) \le 1 \,.$$

▷ **Exercise 3.2.**
Suppose that simple random walk is transient on G and $a \in \mathsf{V}$. Show that there is a random, edge-simple path from a to ∞ such that the expected number of edges common to two such independent paths is equal to $\mathscr{R}(a \leftrightarrow \infty)$ (for unit conductances on G).

Here is one use of these random paths:

Corollary 3.3. (Monotone-Voltage Paths) *Let G be a transient, connected network and v be the voltage function from the unit current flow i from a vertex a to ∞ with $v(\infty) = 0$. For every vertex x, there is a path of vertices from a to x along which v is monotone. If x is incident to some edge e with $i(e) > 0$, then there is such a path along which v is strictly monotone.*

Proof. Let W be the set of vertices incident to some edge e with $i(e) > 0$. By Proposition 3.2, if $i(e) > 0$, then a path $\langle Y_n \rangle$ chosen at random (as defined earlier) will cross e with positive probability. Thus, each vertex in W has a positive chance of being visited by $\langle Y_n \rangle$. Clearly v is strictly monotone along every path $\langle Y_n \rangle$. Thus, there is a path from a to any vertex in W along which v is strictly monotone. Now any vertex x not in W can be connected to some $w \in W$ by edges along which $i = 0$. Extending the path from a to w by such a path from w to x gives the required path from a to x. ◀

Curiously, we do not know a deterministic construction of a path satisfying the conclusion of Corollary 3.3.

3.2 Trees

Flows and electrical networks on trees can be analyzed with greater precision than on general graphs. One easy reason for this is that we know which direction the flow goes, by which we mean the following. Fix a root o in a tree T and denote by $|e|$ the distance from an edge $e \in T$ to o, that is, the number of edges on the shortest path that includes both o and e. Choose a unique orientation for each edge, namely, the one leading away from o. Given any network on T, we claim that there is an admissible flow of maximal strength from the root to infinity that does not have negative flow on any edge (with this orientation). Indeed, it suffices to prove this for trees of finite height and flows from the root to the leaves (when the leaves are identified to a single vertex), since a maximal flow to infinity is the edgewise limit of maximal flows on trees of finite height (replace each subtree T^x for x in level n with a single edge $[x, x']$ of capacity equal to the maximal strength of a flow in T^x from x to infinity). For a tree of finite height, our claim follows from our discussion on p. 77 after the proof of the max-flow min-cut theorem. Likewise, if the tree network is transient, then the unit current flow does not have negative flow on any edge. This is immediate from Proposition 2.12 that equates current flow on an edge with the expected number of signed crossings of that edge. For these reasons, in our considerations, we may restrict to flows that are nonnegative.

▷ **Exercise 3.3.**
Let T be a locally finite tree and Π be a minimal finite cutset separating o from ∞. Let θ be a flow from o to ∞. Show that
$$\text{Strength}(\theta) = \sum_{e \in \Pi} \theta(e).$$

The Nash-Williams criterion gave a condition sufficient for recurrence, but it was not necessary for recurrence. However, a useful partial converse to the Nash-Williams criterion for *trees* can be stated as follows.

Proposition 3.4. *Let c be conductances on a locally finite, infinite tree T and w_n be positive numbers with $\sum_{n \geq 1} w_n < \infty$. Every flow θ on T satisfying $0 \leq \theta(e) \leq w_{|e|}c(e)$ for all edges e has finite energy.*

Proof. Apply Exercise 3.3 to the cutset formed by the edges at distance n from o to obtain
$$\sum_{e \in T} \theta(e)^2 r(e) = \sum_{n \geq 1} \sum_{|e|=n} \theta(e)[\theta(e) r(e)] \leq \sum_{n \geq 1} w_n \sum_{|e|=n} \theta(e) = \sum_{n \geq 1} w_n \text{ Strength}(\theta) < \infty.$$

(This special case of Exercise 3.3 was also shown by induction in Section 1.2.) ◄

Let's consider some particular conductances. Since trees tend to grow exponentially, let the conductance of an edge decrease exponentially with distance from o, say, $c(e) := \lambda^{-|e|}$, where $\lambda > 1$. Let $\lambda_c = \lambda_c(T)$ denote the critical λ for nonzero, admissible flow from o to ∞, in other words, "water" can flow for $\lambda < \lambda_c$ but not for $\lambda > \lambda_c$. What is the critical λ for current flow? We saw at the start of Section 3.1 that if current flows for a certain value of λ (that is, the associated random walk is transient), then so does water, whence $\lambda \leq \lambda_c$. Conversely, for $\lambda < \lambda_c$, we claim that current flows: choose $\lambda' \in (\lambda, \lambda_c)$ and set $w_n := (\lambda/\lambda')^n$. Of course, $\sum_n w_n < \infty$; by definition of λ_c, there is a nonzero flow θ satisfying $0 \leq \theta(e) \leq (\lambda')^{-|e|} = w_{|e|}\lambda^{-|e|}$, whence Proposition 3.4 shows that this flow has finite energy and so current indeed flows. Thus, the same λ_c is the critical value for current flow. Since λ_c balances the growth of T while taking into account the structure of the tree, we call it the ***branching number*** of T:
$$\text{br } T := \sup \left\{ \lambda ; \, \exists \text{ a nonzero flow } \theta \text{ on } T \text{ with } \forall e \in T \quad 0 \leq \theta(e) \leq \lambda^{-|e|} \right\}.$$

Of course, the max-flow min-cut theorem gives an equivalent formulation as
$$\text{br } T = \sup \left\{ \lambda ; \, \inf_{\Pi} \sum_{e \in \Pi} \lambda^{-|e|} > 0 \right\}, \tag{3.4}$$

where the inf is over cutsets Π separating o from ∞. Denote by RW_λ the random walk associated to the conductances $e \mapsto \lambda^{-|e|}$. We may summarize some of our work in the following theorem of Lyons (1990).

Theorem 3.5. (Branching Number and Random Walk) *If T is a locally finite, infinite tree, then $\lambda < \mathrm{br}\,T \Rightarrow \mathrm{RW}_\lambda$ is transient and $\lambda > \mathrm{br}\,T \Rightarrow \mathrm{RW}_\lambda$ is recurrent.*

In particular, we see that for simple random walk to be transient, it suffices that $\mathrm{br}\,T > 1$.

▷ **Exercise 3.4.**
For simple random walk on T to be transient, is it necessary that $\mathrm{br}\,T > 1$?

We might call RW_λ *homesick random walk* for $\lambda > 1$ since the random walker has a tendency to walk toward its starting place, the root.

▷ **Exercise 3.5.**
Find an example where $\mathrm{RW}_{\mathrm{br}\,T}$ is transient and an example where it is recurrent.

▷ **Exercise 3.6.**
Show that $\mathrm{br}\,T$ is independent of which vertex in T is the root.

Let's try to understand the significance of $\mathrm{br}\,T$, which will turn out to be a very important number. If T is an n-ary tree (that is, every vertex has n children), then the distance of RW_λ from o is simply a biased random walk on $\mathbb{N}$. It follows that $\mathrm{br}\,T = n$.

▷ **Exercise 3.7.**
Show directly from the definition that $\mathrm{br}\,T = n$ for an n-ary tree T. Deduce that if every vertex of T has between n_1 and n_2 children, then $\mathrm{br}\,T$ is between n_1 and n_2.

Since λ_c balances the number of edges leading away from a vertex over all of T, it is reasonable to think of $\mathrm{br}\,T$ as an average number of branches per vertex.

3.3 Growth of Trees

In this section, we again consider only locally finite, infinite trees. To understand better their branching number, we will look at the simpler notion of growth. For a vertex $x \in T$, let $|x|$ denote its distance from o. Define the *lower (exponential) growth rate* of a tree T by

$$\underline{\mathrm{gr}}\,T := \liminf_{n \to \infty} |T_n|^{1/n},$$

where $T_n := \{x \in T \,;\, |x| = n\}$ is *level n* of T. Similarly, the *upper (exponential) growth rate* of T is $\overline{\mathrm{gr}}\,T := \limsup |T_n|^{1/n}$. If $\underline{\mathrm{gr}}\,T = \overline{\mathrm{gr}}\,T$, then the common value is called the *(exponential) growth rate* of T and is denoted $\mathrm{gr}\,T$.

In most of the examples of trees so far, the branching number was equal to the lower growth rate. In general, we have the inequality

$$\mathrm{br}\,T \le \underline{\mathrm{gr}}\,T,$$

as we showed in Section 1.2. There are various ways to construct a tree whose branching number is different from its growth: see Section 1.2 and Exercise 1.3.

▷ **Exercise 3.8.**
We have seen that if $\operatorname{br} T > 1$, then simple random walk on T is transient. Is $\underline{\operatorname{gr}} T > 1$ sufficient for transience?

We call T **spherically symmetric** if $\deg x$ depends only on $|x|$. Recall from Exercise 1.2 that $\operatorname{br} T = \underline{\operatorname{gr}} T$ when T is spherically symmetric.

Notation. Write $x \le y$ if x is on the shortest path from o to y; $x < y$ if $x \le y$ and $x \ne y$; $x \to y$ if $x \le y$ and $|y| = |x| + 1$ (that is, y is a child of x); and T^x for the subtree of T containing the vertices $y \ge x$.

There is an important class of trees whose structure is "periodic" in a certain sense. To exhibit these trees, we review some elementary notions from combinatorial topology. Let G be a finite, connected graph and x_0 be any vertex in G. Define a tree T in the following way: its vertices are the finite paths $\langle x_0, x_1, x_2, \ldots, x_n \rangle$ that never backtrack, that is, $x_i \ne x_{i+2}$ for $0 \le i \le n - 2$. Join two vertices in T by an edge when one path is an extension by one vertex of the other. The tree T is called the **universal cover** (based at x_0) of G. See Figure 3.1 for an example.

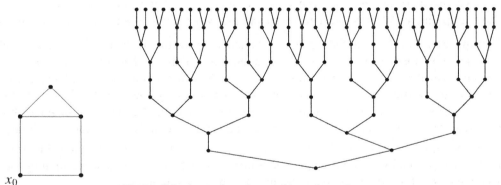

Figure 3.1. A graph and part of its universal cover.

This idea can be extended. Suppose that G is a finite, directed multigraph and x_0 is any vertex in G. That is, edges are not required to appear with both orientations, and two vertices can have many edges joining them. Loops are also allowed. Define the **directed cover** (based at x_0) of G to be the tree T whose vertices are the finite paths of edges $\langle e_1, e_2, \ldots, e_n \rangle$ in G that start at x_0 (we use paths of edges rather than of vertices in case there are multiple edges). The root is the empty path. We join two vertices in T as we did before. See Figure 3.2 for an example.

The periodic aspect of these trees can be formalized as follows.

Definition. Let $N \ge 0$. An infinite tree T is called **N-periodic** (resp., **N-subperiodic**) if $\forall x \in T$ there exists an adjacency-preserving bijection (resp., injection) $f : T^x \to T^{f(x)}$ with $|f(x)| \le N$. A tree is **periodic** (resp., **subperiodic**) if there is some N for which it is N-periodic (resp., N-subperiodic).

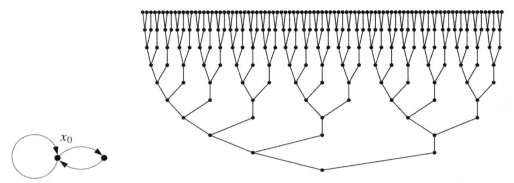

Figure 3.2. A graph and part of its directed cover. This tree is also called the ***Fibonacci tree***.

All universal and directed covers are periodic. Conversely, every periodic tree is a directed cover of a finite graph, G: If T is an N-periodic tree, take $\{x \in T ; |x| \leq N\}$ for the vertex set of G. For $|x| \leq N$, let f_x be the identity map and for $x \in T_{N+1}$, let f_x be a bijection as guaranteed by the definition. Let the edges of G be $\{\langle x, f_y(y)\rangle; |x| \leq N$ and y is a child of $x\}$. Then T is the directed cover of G based at the root.

If two (sub)periodic trees are joined as in Example 1.3, then clearly the resulting tree is also (sub)periodic. We now present some examples where subperiodic trees arise naturally.

Example 3.6. Consider the finite paths in the lattice $\mathbb{Z}^2$ starting at the origin that go through no vertex more than once. These paths are called self-avoiding and are of substantial interest to mathematical physicists. Form a tree T whose vertices are the finite self-avoiding paths and with two such vertices joined when one path is an extension by one step of the other. Then T is 0-subperiodic and has infinitely many leaves. Its growth rate has been estimated at about 2.64; see Madras and Slade (1993). For the hexagonal lattice in the plane, the growth rate is exactly $\sqrt{2 + \sqrt{2}} = 2\cos(\pi/8)$. This was conjectured by Nienhuis (1982) based on ideas from physics and proved by Duminil-Copin and Smirnov (2012), who were also inspired by ideas from physics to find some subtle and beautiful symmetries in the problem.

Example 3.7. Suppose that E is a closed set in $[0, 1]$ and $T_{[b]}(E)$ is the associated tree that represents E in base b, as in Section 1.10. Then $T_{[b]}(E)$ is 0-subperiodic iff E is invariant under the map $x \mapsto bx \pmod 1$, that is, iff the fractional part of bx lies in E for every $x \in E$.

How can we calculate the growth rate of a periodic tree? If T is a periodic tree, let G be a finite, directed graph of which T is the directed cover based at x_0. We may assume that G contains only vertices that can be reached from x_0, since the others do not contribute to T. The key to analysis is the ***directed adjacency matrix*** A of G, that is, the square matrix indexed by the vertices of G with the (x, y)-entry equal to the number of edges from x to y. Recall that the ***norm*** of a square matrix M is $\|M\| := \sup\{\|Mv\|; \|v\| = 1\}$, where $\|v\| = \sqrt{(v, v)}$ is the usual ℓ^2-norm of the vector v. Also, the ***spectral radius*** of M is the maximum of $|\lambda|$ over all complex eigenvalues λ of M. Gelfand's formula says that the spectral radius of M equals $\lim_{n \to \infty} \|M^n\|^{1/n}$; see, for example, Corollary 5.6.14 of Horn and Johnson (2013). In our case, all entries of A are nonnegative, so the Perron–Frobenius theorem (Minc (1988),

Theorem 4.2) says that the spectral radius of A is equal to its largest positive eigenvalue, λ_*, and that there is a λ_*-eigenvector v_* all of whose entries are nonnegative. We call λ_* the **Perron eigenvalue** of A and v_* a **Perron eigenvector** of A. To see how these are useful, let **1** denote a column vector all of whose entries are 1, $\mathbf{1}_x$ denote the column vector that is 1 at x and 0 elsewhere, and $\mathbf{1}_x'$ denote its transpose. Then the number of paths in G with n edges, which is $|T_n|$, is $\mathbf{1}_{x_0}' A^n \mathbf{1}$. Now $\mathbf{1}_{x_0}' A^n \mathbf{1} = (A^n \mathbf{1}, \mathbf{1}_{x_0}) \le \|A^n \mathbf{1}\| \cdot \|\mathbf{1}_{x_0}\| \le \|A^n\| \cdot \|\mathbf{1}\| \cdot \|\mathbf{1}_{x_0}\|$, whence Gelfand's formula yields $\limsup_{n\to\infty} |T_n|^{1/n} \le \lambda_*$. On the other hand, let x be any vertex such that $v_*(x) > 0$, let j be such that $A^j(x_0, x) > 0$, and let $c > 0$ be such that $\mathbf{1} \ge c v_*$. Then

$$|T_{j+n}| \ge \mathbf{1}_x' A^n \mathbf{1} \ge c \mathbf{1}_x' A^n v_* = c\lambda_*^n \mathbf{1}_x' v_* = c v_*(x) \lambda_*^n \,.$$

Therefore $\liminf_{n\to\infty} |T_n|^{1/n} \ge \lambda_*$. We conclude that $\operatorname{gr} T = \lambda_*$.

The "regularity" of periodic trees leads us to expect that their growth rate equals their branching number. One very nice feature of subperiodic trees is that this equality also holds for them! This fact, as well as its proof, is analogous to a classical fact about sequences of reals, known as Fekete's lemma (Exercise 3.9). A sequence $\langle a_n \rangle$ of real numbers is called **subadditive** if

$$\forall m, n \ge 1 \qquad a_{m+n} \le a_m + a_n \,.$$

A simple example is $a_n := \lceil \beta n \rceil$ for some real $\beta > 0$.

▷ **Exercise 3.9.**

    (a) (Fekete's Lemma) Show that for every subadditive sequence $\langle a_n \rangle$, the sequence $\langle a_n/n \rangle$ converges to its infimum:

$$\lim_{n\to\infty} \frac{a_n}{n} = \inf \frac{a_n}{n} \,.$$

    (b) Show that Fekete's lemma holds even if a finite number of the a_n are infinite.
    (c) Show that for every 0-subperiodic tree T, the limit $\lim_{n\to\infty} |T_n|^{1/n}$ exists.

The equality $\operatorname{br} T = \operatorname{gr} T$ for subperiodic trees T is due to Furstenberg (1967). To prove it, we use the following notation: given $\lambda > 0$ and a cutset Π in a tree T, denote

$$\|\Pi\|_\lambda := \sum_{e(x) \in \Pi} \lambda^{-|x|} \,,$$

where $e(x)$ is the edge from the parent of x to x. Note $|e(x)| = |x|$. Of course, the **parent** of a vertex $x \ne o$ is the neighbor of x that is closer to o.

Theorem 3.8. (Subperiodicity and Branching Number) *For every subperiodic, infinite tree T, the growth rate of T exists and* $\operatorname{br} T = \operatorname{gr} T$. *Moreover,*

$$\inf_{\Pi} \|\Pi\|_{\operatorname{br} T} > 0 \,;$$

in particular,

$$\inf_n |T_n| (\operatorname{br} T)^{-n} > 0 \,.$$

Proof. First, suppose that T has no leaves and is 0-subperiodic. We will show that if, for some cutset Π and some $\lambda_1 > 0$, we have

$$\|\Pi\|_{\lambda_1} < 1 , \tag{3.5}$$

then

$$\limsup_{n \to \infty} |T_n|^{1/n} < \lambda_1 . \tag{3.6}$$

Since $\inf_\Pi \|\Pi\|_{\lambda_1} = 0$ for all $\lambda_1 > \mathrm{br}\, T$, this implies that

$$\mathrm{br}\, T = \mathrm{gr}\, T \quad \text{and} \quad \inf_\Pi \|\Pi\|_{\mathrm{br}\, T} \geq 1 . \tag{3.7}$$

So suppose that (3.5) holds. We may assume that Π is finite and minimal with respect to inclusion by Exercise 3.23. To get strict inequality in (3.6), we'll need a little wiggle room, so choose $\lambda \in (0, \lambda_1)$ such that

$$\|\Pi\|_\lambda < 1 . \tag{3.8}$$

Let $d := \max_{e(x) \in \Pi} |x|$ denote the maximal level of Π. By 0-subperiodicity, for each $e(x)$ in Π, there is a cutset $\Pi(x)$ of T^x such that $\sum_{e(w) \in \Pi(x)} \lambda^{-|w-x|} \leq \|\Pi\|_\lambda < 1$ and $\max_{e(w) \in \Pi(x)} |w - x| \leq d$. Thus, $\|\Pi(x)\|_\lambda = \sum_{e(w) \in \Pi(x)} \lambda^{-|w|} < \lambda^{-|x|}$. (Note that $|w|$ always denotes the distance from w to the root of T.) This allows us to replace the cutset Π by another in several ways while preserving (3.5): for any given $A \subseteq \Pi$, if we replace those edges $e(x)$ in A by the edges of the corresponding $\Pi(x)$, then we obtain a cutset $\widetilde{\Pi} := (\Pi \setminus A) \cup \bigcup_{e(x) \in A} \Pi(x)$ that satisfies

$$\|\widetilde{\Pi}\|_\lambda = \sum_{e(x) \in \Pi \setminus A} \lambda^{-|x|} + \sum_{e(x) \in A} \|\Pi(x)\|_\lambda \leq \|\Pi\|_\lambda < 1 .$$

Given $n > d$, we may iterate this process for all edges $e(x)$ in the cutset with $|x| < n$ until we obtain a cutset Π^* lying between levels n and $n + d$ with $\|\Pi^*\|_\lambda < 1$. Therefore $|T_n| \lambda^{-(n+d)} \leq \|\Pi^*\|_\lambda < 1$, so that $\limsup |T_n|^{1/n} \leq \limsup \lambda^{1+d/n} = \lambda < \lambda_1$. This establishes (3.6).

Now let T be N-subperiodic but still without leaves. Let $\widehat{T}$ be the union of disjoint copies of the descendant trees $\{T^x ; |x| \leq N\}$ with their roots identified (which is not exactly the same as Example 1.3). It is easy to check that $\widehat{T}$ is 0-subperiodic and $\mathrm{gr}\, \widehat{T} \geq \overline{\mathrm{gr}}\, T$. Moreover, for every cutset Π of T with $\min_{e(x) \in \Pi} |x| \geq N$, there is a corresponding cutset Π' of $\widehat{T}$ such that

$$\forall \lambda > 0 \quad \|\Pi'\|_\lambda \leq \sum_{k=0}^{N} |T_k| \lambda^k \|\Pi\|_\lambda ,$$

whence $\mathrm{br}\, \widehat{T} = \mathrm{br}\, T$. In conjunction with (3.7) for $\widehat{T}$, this completes the proof.

Finally, if T has leaves, consider the tree T' obtained from T by adding to each leaf of T an infinite ray. Then T' is subperiodic, so

$$\limsup |T_n|^{1/n} \geq \mathrm{br}\, T = \mathrm{br}\, T' = \limsup |T'_n|^{1/n} \geq \limsup |T_n|^{1/n} .$$

Also, except in the trivial case that $\mathrm{br}\, T = 1$, every cutset Π of T can be extended to a cutset Π' of T' with $\|\Pi'\|_{\mathrm{br}\, T}$ arbitrarily close to $\|\Pi\|_{\mathrm{br}\, T}$.   ◀

For another proof of Theorem 3.8, see Section 15.5.

Next, we consider a notion dual to subperiodicity. Although it sounds just as natural, it actually does not arise very often. However, such trees behave similarly to subperiodic trees, which will be easy to prove.

Definition. Let $N \geq 0$. A tree T is called N-**superperiodic** if $\forall x \in T$ there is an adjacency-preserving injection $f : T \to T^{f(o)}$ with $f(o) \in T^x$ and $|f(o)| - |x| \leq N$.

For example, every 0-periodic tree is 0-superperiodic, although 1-periodic trees are not necessarily 1-superperiodic. For another example, consider the finite paths in the lattice $\mathbb{Z}^2$ starting at the origin that stay in the right half-plane. Form the tree T whose vertices are the finite paths of this type and with two such vertices joined when one path is an extension by one step of the other. Then T is 0-superperiodic. For more examples, see Exercise 3.33.

Theorem 3.9. *Let $N \geq 0$. Any N-superperiodic tree T with $\overline{\mathrm{gr}}\, T < \infty$ satisfies $\mathrm{br}\, T = \mathrm{gr}\, T$ and $|T_n| \leq (\mathrm{gr}\, T)^{n+N}$ for all n.*

Proof. Consider the case $N = 0$. In this case, $|T_{n+m}| \geq |T_n| \cdot |T_m|$. By Exercise 3.9, $\mathrm{gr}\, T$ exists and $|T_n| \leq (\mathrm{gr}\, T)^n$ for all n. Fix any positive integer k. Let θ be the unit flow from o to T_k that is uniform on T_k. By 0-superperiodicity, we can extend θ in a periodic fashion to a flow from o to infinity that satisfies $\theta(e(x)) \leq |T_k|^{-\lfloor |x|/k \rfloor}$ for all vertices x. Consequently, $\mathrm{br}\, T \geq |T_k|^{1/k}$. Letting $k \to \infty$ completes the proof for $N = 0$. ◄

▷ **Exercise 3.10.**
Prove the case $N > 0$ of Theorem 3.9.

3.4 Cayley Graphs

Suppose we investigate RW_λ on graphs other than trees. What does RW_λ mean in this context? Fix a vertex o in a graph G. If e is an edge at distance n from o, let the conductance of e be λ^{-n}. Again, by Rayleigh's monotonicity principle, there is a critical value of λ, denoted $\lambda_c(G)$, that separates the transient regime from the recurrent regime. To understand what $\lambda_c(G)$ measures, consider the class of spherically symmetric graphs, where we call G **spherically symmetric** about o if, for all pairs of vertices x and y at the same distance from o, there is an automorphism of G fixing o that takes x to y. Let $\widetilde{M}_n$ be the number of edges that lead from a vertex at distance $n - 1$ from o to a vertex at distance n. Then the critical value of λ is the growth rate of G:

$$\lambda_c(G) = \liminf_{n \to \infty} \widetilde{M}_n^{1/n}\,.$$

In fact, we have the following more precise criterion for transience:

▷ **Exercise 3.11.**
Show that if G is spherically symmetric about o, then RW_λ is transient iff $\sum_n \lambda^n / \widetilde{M}_n < \infty$.

Next, consider the Cayley graphs of finitely generated groups: We say that a group Γ is **generated** by a subset S of its elements if the smallest subgroup containing S is all of Γ. In other words, every element of Γ can be written as a product of elements of the form s or s^{-1} with $s \in S$. If Γ is generated by S, then we form the associated **Cayley graph** G with vertices Γ and (unoriented) edges $\{[x, xs];\ x \in \Gamma,\ s \in S\} = \{(x, y) \in \Gamma^2;\ x^{-1}y \in S \cup S^{-1}\}$. Because S generates Γ, the graph is connected. Cayley graphs are highly symmetric: they look the same from every vertex, since left multiplication by yx^{-1} is an automorphism of G that carries x to y. These automorphisms, left multiplication by a group element, are called **translations** of the Cayley graph.

▷ **Exercise 3.12.**
Show that different Cayley graphs of the same finitely generated group are roughly isometric.

The Euclidean lattices are the most well-known Cayley graphs. It is useful to keep in mind other Cayley graphs as well, so we will look at some constructions of groups. First recall the **Cartesian** or **direct product** of two groups Γ and Γ', where the multiplication on $\Gamma \times \Gamma'$ is defined coordinatewise: $(\gamma_1, \gamma_1')(\gamma_2, \gamma_2') := (\gamma_1\gamma_2, \gamma_1'\gamma_2')$. A similar definition is made for the direct product of any sequence of groups. It is convenient to rephrase this definition in terms of presentations.

First, recall that the **free group** generated by a set S is the set of all finite words in s and s^{-1} for $s \in S$ with the empty word as the identity and concatenation as multiplication, with the further stipulation that if a word contains either ss^{-1} or $s^{-1}s$, then the pair is eliminated from the word. The group defined by the **presentation** $\langle S \mid R \rangle$ is the quotient of the free group generated by the set S by the normal subgroup generated by the set R, where R consists of finite words, called **relators**, in the elements of S. (We think of R as giving a list of products that must equal the identity; other identities are consequences of these ones and of the definition of a group.) For example, the free group on two letters $\mathbb{F}_2$ is $\langle \{a, b\} \mid \varnothing \rangle$, usually written $\langle a, b \mid \rangle$, whereas $\mathbb{Z}^2$ is (isomorphic to) $\langle \{a, b\} \mid \{aba^{-1}b^{-1}\} \rangle$, usually written $\langle a, b \mid aba^{-1}b^{-1} \rangle$, also known as the **free abelian group** on two letters (or of **rank** 2). In this notation, if $\Gamma = \langle S \mid R \rangle$ and $\Gamma' = \langle S' \mid R' \rangle$ with $S \cap S' = \varnothing$, then $\Gamma \times \Gamma' = \langle S \cup S' \mid R \cup R' \cup [S, S'] \rangle$, where $[S, S'] := \{ss's^{-1}s'^{-1};\ s \in S, s' \in S'\}$. On the other hand, the **free product** of Γ and Γ' is $\Gamma * \Gamma' := \langle S \cup S' \mid R \cup R' \rangle$. A similar definition is made for the free product of any sequence of groups. Interesting free products to keep in mind as we examine various phenomena are $\mathbb{Z} * \mathbb{Z}$ (the free group on two letters), $\mathbb{Z}_2 * \mathbb{Z}_2 * \mathbb{Z}_2$, $\mathbb{Z} * \mathbb{Z}_2$ (whose Cayley graph is isomorphic to that of $\mathbb{Z}_2 * \mathbb{Z}_2 * \mathbb{Z}_2$, that is, a 3-regular tree, when the usual generators are used), $\mathbb{Z}_2 * \mathbb{Z}_3$, $\mathbb{Z}^2 * \mathbb{Z}$, and $\mathbb{Z}^2 * \mathbb{Z}_2$. (Here, we write $\mathbb{Z}_n$ for the cyclic group $\mathbb{Z}/n\mathbb{Z}$ on n elements.) Write $\mathbb{T}_{b+1}$ for the regular tree of degree $b + 1$ (so it has branching number b). It is a Cayley graph of the free product of $b + 1$ copies of $\mathbb{Z}_2$. Its Cartesian product with $\mathbb{Z}^d$ is another interesting graph. Some examples of Cayley graphs with respect to natural generating sets appear in Figure 3.3.

A presentation is called **finite** when it uses only finite sets of generators and relators. For example, $\mathbb{Z}_2 * \mathbb{Z}_3$ has the presentation $\langle a, b \mid a^2, b^3 \rangle$. Another Cayley graph was shown in Figure 2.4, which corresponds to the presentation $\langle a, b, c, d, e \mid a^2, b^2, c^2, d^2, e^2, abcde \rangle$ (see Chaboud and Kenyon (1996)).

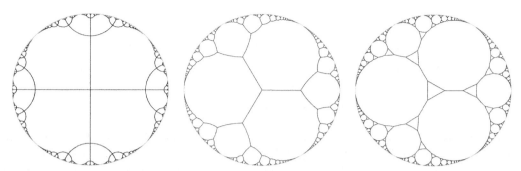

Figure 3.3. The Cayley graphs of the free group on two letters; the free product of $\mathbb{Z}_2$ with itself three times, $\mathbb{Z}_2 * \mathbb{Z}_2 * \mathbb{Z}_2$; and the free product $\mathbb{Z}_2 * \mathbb{Z}_3$. These are drawn, without vertices, in the hyperbolic disc by a program created by Don Hatch. The infinite faces have infinite area, but there is a fundamental domain of finite measure for the group of isometries acting on the hyperbolic disc. Consequently, it is possible to consider an invariant random embedding of any of these Cayley graphs.

Finitely presentable groups arise often "in practice": for example, fundamental groups of compact topological manifolds are finitely presentable, and each finitely presentable group is the fundamental group of a compact 4-manifold (see, for example, Massey (1991), pp. 114–115). The fundamental group of a compact manifold is roughly isometric to the universal cover of the manifold.

The following infinitely presented group will serve as a useful example in several chapters. This is the group $\mathbb{Z}^{\circ}$, also known as the ***lamplighter group*** over $\mathbb{Z}$. It is defined as a (restricted) wreath product, which is a special kind of semidirect product. First, the group $\sum_{x \in \mathbb{Z}} \mathbb{Z}_2$, the direct sum of copies of $\mathbb{Z}_2$ indexed by $\mathbb{Z}$, is the group of maps $\psi \colon \mathbb{Z} \to \mathbb{Z}_2$ with $\psi^{-1}(\{1\})$ finite and with componentwise addition mod 2, which we denote $\oplus$; that is, $(\psi \oplus \psi')(j) := \psi(j) + \psi'(j) \pmod{2}$. Let $\mathcal{S}$ be the left shift, $\mathcal{S}(\psi)(j) := \psi(j + 1)$. Now define $\mathbb{Z}^{\circ} := \left(\sum_{x \in \mathbb{Z}} \mathbb{Z}_2 \right) \rtimes \mathbb{Z}$, which is the set $\left(\sum_{x \in \mathbb{Z}} \mathbb{Z}_2 \right) \times \mathbb{Z}$ with the following group operation: for $\psi, \psi' \in \sum_{x \in \mathbb{Z}} \mathbb{Z}_2$ and $m, m' \in \mathbb{Z}$, we multiply by the rule

$$(\psi, m)(\psi', m') := (\psi \oplus \mathcal{S}^{-m}\psi', m + m') \, .$$

We call an element $\psi \in \sum_{m \in \mathbb{Z}} \mathbb{Z}_2$ a ***configuration*** and call $\psi(k)$ the ***bit*** at k. We identify $\mathbb{Z}_2$ with $\{0, 1\}$. The second component of an element $x = (\psi, m) \in \mathbb{Z}^{\circ}$ is called the ***position of the marker in the state*** x. (Another notation for $\mathbb{Z}^{\circ}$ is the ***restricted wreath product*** $\mathbb{Z}_2 \wr \mathbb{Z}$.) One nice set of generators of $\mathbb{Z}^{\circ}$ is $\{(\mathbf{0}, 1), (\mathbf{0}, -1), (\mathbf{1}_{\{0\}}, 0)\}$. The reason for the name of this group is that we may think of a streetlamp at each integer with the configuration ψ representing which lights are on, namely, those where $\psi = 1$. We also may imagine a lamplighter at the position of the marker. The first two generators of $\mathbb{Z}^{\circ}$ correspond (for right multiplication) to the lamplighter taking a step either to the right or to the left (leaving the lights unchanged); the third generator corresponds to flipping the light at the position of the lamplighter. See Figure 3.4.

The lamplighter group $\mathbb{Z}^{\circ}$ has exponential growth: Consider the subset T_n of group elements at distance n from the identity that can be arrived at from the identity by using only the generators $(\mathbf{0}, 1)$ and $(\mathbf{1}_{\{0\}}, 0)$, in other words, by never allowing the lamplighter

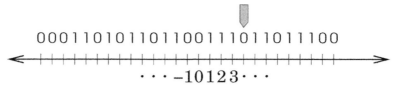

Figure 3.4. A typical element of $\mathbb{Z}^\circ$.

to move leftward. Since T_n is a disjoint union of $(\mathbf{0}, 1)T_{n-1}$ and $(\mathbf{1}_{\{0\}}, 0)(\mathbf{0}, 1)T_{n-2}$, we have $|T_n| = |T_{n-1}| + |T_{n-2}|$. Thus, $\langle |T_n| \, ; \, n \geq 0 \rangle$ is the sequence of Fibonacci numbers. Therefore the exponential growth rate of $|T_n|$ equals the golden mean, $(1 + \sqrt{5})/2$. In fact, it is easy to check that this is the growth rate of balls in $\mathbb{Z}^\circ$.

Despite the beautiful symmetry of Cayley graphs, they are rarely spherically symmetric. Still, if M_n denotes the number of vertices at distance n from the identity, then $\lim M_n^{1/n}$ exists since $M_{m+n} \leq M_m M_n$; thus, we may apply Fekete's lemma (Exercise 3.9) to $\langle \log M_n \rangle$. Note that this also implies that the exponential growth rate of the balls in G equals $\lim M_n^{1/n}$; we refer to this common number as the *(exponential) growth rate of G*. When the growth rate is 1, we say that the Cayley graph has **subexponential growth**, and otherwise that it has **exponential growth**. Our analysis of spherically symmetric graphs, though it does not apply to Cayley graphs, may still suggest the question, Is $\lambda_c(G) = \lim M_n^{1/n}$?

First of all, if $\lambda > \lim M_n^{1/n}$, then RW_λ is positive recurrent, since for such λ,

$$\sum_{e \in \mathsf{E}_{1/2}} c(e) < d \sum_{x \in G} \lambda^{-|x|} = d \sum_{n \geq 0} M_n \lambda^{-n} < \infty ,$$

where d is the degree of G and $|x|$ denotes the distance of x to the identity. (One could also use the Nash-Williams criterion to get merely recurrence.) Second, to prove transience for a given λ, it suffices, by Rayleigh's monotonicity principle, to prove that a subgraph is transient. The easiest subgraph to analyze would be a subtree of G, while to have the greatest likelihood of being transient, it should be as big as possible, that is, a **spanning tree** (one that includes every vertex). Here is one: Assume that the inverse of each generator is also in the generating set. Order the generating set $S = \langle s_1, s_2, \ldots, s_d \rangle$. For each $x \in G$, there is a unique word $(s_{i_1}, s_{i_2}, \ldots, s_{i_n})$ in the generators such that $x = s_{i_1} s_{i_2} \cdots s_{i_n}$, $n = |x|$, and $(s_{i_1}, \ldots, s_{i_n})$ is **lexicographically minimal** with these properties, that is, if $(s_{i'_1}, \ldots, s_{i'_n})$ is another word whose product is x and m is the first j such that $i_j \neq i'_j$, then $i_m < i'_m$. Call this lexicographically minimal word w_x. Let T be the subgraph of G containing all vertices and with y adjacent to x when either $|y| + 1 = |x|$ and w_y is an initial segment of w_x, or vice versa.

▷ **Exercise 3.13.**
Show that T is a subperiodic tree when rooted at the identity.

Because T is spanning and distances to the identity in T are the same as in G, we have $\mathrm{gr}\, T = \lim M_n^{1/n}$. Since T is subperiodic, we have $\mathrm{br}\, T = \lim M_n^{1/n}$ by Theorem 3.8. Hence RW_λ is transient on T for $\lambda < \lim M_n^{1/n}$ by Theorem 3.5, whence on G as well. We have proved the following theorem of Lyons (1995):

Theorem 3.10. (Group Growth and Random Walk) RW_λ *on an infinite Cayley graph has critical value λ_c equal to the exponential growth rate of the graph.*

Of course, this theorem makes Cayley graphs look spherically symmetric from a probabilistic point of view. Such a conclusion, however, should not be pushed too far, for there are Cayley graphs with the following very surprising properties; the lamplighter group is one such. Define the *speed* (or rate of escape) of RW_λ as the limit of the distance from the identity at time n divided by n as $n \to \infty$, if the limit exists. The speed is monotonically decreasing in λ on spherically symmetric graphs and is positive for any positive λ less than the growth rate. However, there are Cayley graphs of exponential growth for which the speed of simple random walk is 0 (a topic studied further in Chapter 14). This already shows that such a Cayley graph is far from spherically symmetric. Even more surprisingly, on the lamplighter group, which has growth rate $(1 + \sqrt{5})/2$, the speed is 0 at $\lambda = 1$[*] yet is strictly positive when $1 < \lambda < (1 + \sqrt{5})/2$ (Lyons, Pemantle, and Peres, 1996b). Perhaps this surprising example is actually part of a general phenomenon in one aspect:

Question 3.11. If G is a Cayley graph of growth rate b and $1 < \lambda < b$, must the speed of RW_λ exist and be positive?

Another manifestation of the lack of spherical symmetry in the Cayley graph of $\mathbb{Z}^\circ$ is the presence of *dead ends*, that is, elements all of whose neighbors are closer to o. Indeed, the element $(1_{[-n,n]}, 0)$ is such a dead end.

Question 3.12. If T is a spanning tree of a graph G rooted at some vertex o, we call T a *geodesic* subtree if $\text{dist}(o, x)$ is the same in T as in G for all vertices x. If T is a geodesic spanning tree of the Cayley graph of a finitely generated group G of growth rate b, is $\text{br}\, T = b$? We saw that this is the case for the lexicographically minimal spanning tree constructed earlier.

Denote the growth rate of a group Γ with respect to a finite generating set S by $\text{gr}_S \Gamma$. By Exercise 3.12, every pair of Cayley graphs of the same group are roughly isometric to each other, whence if $\text{gr}_S \Gamma > 1$ for some generating set, then $\text{gr}_S \Gamma > 1$ for every generating set. In this case, is $\inf_S \text{gr}_S \Gamma > 1$? This question was posed by Gromov (1981b) (see also Gromov (1999) for a revised version in English), and for a long time, it remained open. It is known to hold for certain classes of groups, but finally a counterexample was found by Wilson (2004b); see also Bartholdi (2003) and Wilson (2004c). In Theorem 10.13, we will use random spanning forests to give examples of groups with uniform exponential growth. See Mann (2012) for more on the growth of groups.

We have begun to see how the behavior of some probabilistic processes on Cayley graphs is related to geometric properties of the underlying groups. This is a fascinating theme in contemporary research. We will see some more examples in Chapters 6, 7, 8, 10, 11, and 14. In particular, we will see in Theorem 6.40 that simple random walk is transient on Cayley graphs whose volume growth is at least cubic.

---

[*] This part is easy to see: after n steps, the marker has visited only locations within distance roughly $\sqrt{n}$ from 0.

3.5 Notes

A Cayley graph is spherically symmetric iff it is 2-point homogeneous, that is, given vertices u, v, w, x such that $d(u, v) = d(w, x)$, there is an automorphism taking u to w and v to x. These graphs are characterized by Macpherson (1982).

The lamplighter group $\mathbb{Z}^\circ$ was denoted G_1 by Kaimanovich and Vershik (1983).

3.6 Collected In-Text Exercises

3.1. Show that if equality holds on the right-hand side of (3.3), then for all x, we have

$$\sum_{\substack{e^+=x, \\ \theta(e)>0}} \theta(e) \le 1 \,.$$

3.2. Suppose that simple random walk is transient on G and $a \in \mathsf{V}$. Show that there is a random, edge-simple path from a to ∞ such that the expected number of edges common to two such independent paths is equal to $\mathscr{R}(a \leftrightarrow \infty)$ (for unit conductances on G).

3.3. Let T be a locally finite tree and Π be a minimal finite cutset separating o from ∞. Let θ be a flow from o to ∞. Show that

$$\mathsf{Strength}(\theta) = \sum_{e \in \Pi} \theta(e) \,.$$

3.4. For simple random walk on T to be transient, is it necessary that $\operatorname{br} T > 1$?

3.5. Find an example where $\mathsf{RW}_{\operatorname{br} T}$ is transient and an example where it is recurrent.

3.6. Show that $\operatorname{br} T$ is independent of which vertex in T is the root.

3.7. Show directly from the definition that $\operatorname{br} T = n$ for an n-ary tree T. Deduce that if every vertex of T has between n_1 and n_2 children, then $\operatorname{br} T$ is between n_1 and n_2.

3.8. We have seen that if $\operatorname{br} T > 1$, then simple random walk on T is transient. Is $\operatorname{gr} T > 1$ sufficient for transience?

3.9. (a) (Fekete's Lemma) Show that for every subadditive sequence $\langle a_n \rangle$, the sequence $\langle a_n/n \rangle$ converges to its infimum:

$$\lim_{n \to \infty} \frac{a_n}{n} = \inf \frac{a_n}{n} \,.$$

(b) Show that Fekete's lemma holds even if a finite number of the a_n are infinite.

(c) Show that for every 0-subperiodic tree T, the limit $\lim_{n \to \infty} |T_n|^{1/n}$ exists.

3.10. Prove the case $N > 0$ of Theorem 3.9.

3.11. Show that if G is spherically symmetric about o, then RW_λ is transient iff $\sum_n \lambda^n/\widetilde{M}_n < \infty$.

3.12. Show that different Cayley graphs of the same finitely generated group are roughly isometric.

3.13. Show that the lexicographically minimal spanning tree T of a Cayley graph is a subperiodic tree when rooted at the identity.

3.7 **Additional Exercises**

3.14. There are two other versions of the max-flow min-cut theorem that are useful. We state them for directed, finite networks using the notation of our proof of the max-flow min-cut theorem.

(a) Suppose that each *vertex* x is given a capacity $c(x)$, meaning that an admissible flow θ is required to satisfy (i) $\theta(e) \geq 0$ for all edges e and (ii) for all x other than the sources A and sinks Z, $\sum_{e \in E} \phi(x, e)\theta(e) = 0$ and $\sum_{e^+ = x} \theta(e) \leq c(x)$, where ϕ is the vertex-edge incidence function defined in Section 3.1. A cutset now consists of *vertices* that intersect every directed path from A to Z. Show that the maximum strength of an admissible flow from A to Z equals the minimum cutset sum of the capacities.

(b) Suppose that each edge *and* each vertex has a capacity, with the restrictions that each of these imply. A cutset now may consist of both vertices and edges. Again, show that the maximum strength of an admissible flow from A to Z equals the minimum cutset sum of the capacities.

3.15. Show that if all the edge capacities $c(e)$ in a directed, finite network are integers, then among the admissible flows θ of maximal strength, there is one such that all $\theta(e)$ are also integers. Show the same for networks with capacities assigned to the vertices or to both edges and vertices, as in Exercise 3.14.

3.16. (Menger's Theorem)

(a) Let a and z be vertices in a graph that are not adjacent. Show that the maximum number of paths from a to z that are pairwise disjoint (except at a and z) is equal to the minimum cardinality of a set W of vertices such that W is disjoint from a and z, but such that every path from a to z passes through W.

(b) Let a be a vertex in an infinite graph. Show that the maximum number of paths from a to ∞ that are pairwise disjoint (except at a) is equal to the minimum cardinality of a set W of vertices such that W is disjoint from a, but such that every path from a to ∞ passes through W.

3.17. A ***perfect matching*** of a graph G is a subset M of its edges such that each vertex of G belongs to exactly one edge in M. Let G be a finite ***bipartite*** graph, that is, its vertex set can be partitioned into two parts, A and Z, such that all edges of G have one endpoint in A and one in Z.

(a) (Kőnig's Theorem) Show that if G is regular, then it has a perfect matching.

(b) (Hall's Theorem) Show that if $|A| = |Z|$ and for each $K \subseteq A$, the number of vertices adjacent to some vertex of K is at least $|K|$, then G has a perfect matching.

3.18. Show that the maximum strength of an admissible flow from A to Z (in a finite, undirected network) also equals

$$\min \left\{ \sum_{e \in E_{1/2}} c(e)\ell(e) \right\},$$

where ℓ is an assignment of nonnegative lengths to the edges so that the minimum distance from every point in A to every point in Z is 1.

3.19. Suppose that θ is a flow from A to Z in a finite, undirected network. Show that if Π is a cutset separating A from Z that is minimal with respect to inclusion, then $\mathsf{Strength}(\theta) = \sum_{e \in \Pi} \theta(e)$.

3.20. Let G be a finite network and a and z be two of its vertices. Show that $\mathscr{R}(a \leftrightarrow z)$ is the minimum of $\sum_{e \in E_{1/2}} r(e) \mathbf{P}[e \in \mathcal{P} \text{ or } -e \in \mathcal{P}]^2$ over all probability measures $\mathbf{P}$ on paths $\mathcal{P}$ from a to z.

3.21. Let G be an undirected graph and $o \in \mathsf{V}$. Recall that E consists of both orientations of each edge. Suppose that $q: \mathsf{E} \to [0, \infty)$ satisfies the following three conditions:

(i) For every vertex $x \neq o$, we have

$$\sum_{\{u \,;\, (u,x) \in \mathsf{E}\}} q(u, x) \leq \sum_{\{w \,;\, (x,w) \in \mathsf{E}\}} q(x, w) ;$$

(ii) $$\sum_{\{u\,;\,(u,o)\in \mathsf{E}\}} q(u,o) = 0 \quad \text{and} \quad \sum_{\{w\,;\,(o,w)\in \mathsf{E}\}} q(o,w) > 0\,;\text{ and}$$

(iii) there exists $K < \infty$ such that for every directed path $(u_0, u_1, \ldots)$ in G starting at $u_0 = o$, we have

$$\sum_{i=0}^{\infty} q(u_i, u_{i+1}) \le K\,.$$

Show that simple random walk on (the undirected graph) G starting at o is transient.

3.22. Let G be a finite graph and a, z be two vertices of G. Let the edges be labeled by positive resistances $r(\,\cdot\,)$. Two players, a passenger and a troll, simultaneously pick edge-simple paths from a to z. The passenger then pays the troll the sum of $\pm r(e)$ for all the edges e common to both paths; if e is traversed in the same direction by the two paths, then the $+$ sign is used, otherwise the $-$ sign is used. Show that the troll has a strategy of picking a random path in such a way that no matter what path is picked by the passenger, the troll's gain has expectation equal to the effective resistance between a to z. Show further that the passenger has a similar strategy that has expected loss equal to $\mathscr{R}(a \leftrightarrow z)$ no matter what the troll does.

3.23. Let T be a locally finite tree and Π be a cutset separating o from ∞. Show that there is a finite cutset $\Pi' \subseteq \Pi$ separating o from ∞ that does not properly contain any other cutset.

3.24. Show that a network (T, c) on a tree is transient iff there exists a function F on the vertices of T such that $F \ge 0$, $\forall e\ dF(e) \ge 0$, and $\inf_\Pi \sum_{e \in \Pi} dF(e)c(e) > 0$. Here, edges are oriented away from the root; the infimum is over cutsets separating the root from infinity.

3.25. Given a tree T and $k \ge 1$, form the tree $T^{[k]}$ by taking the vertices x of T for which $|x|$ is a multiple of k and joining x and y by an edge in $T^{[k]}$ when their distance is k in T. Show that $\operatorname{br} T^{[k]} = (\operatorname{br} T)^k$.

3.26. Show that RW_λ is positive recurrent on a tree T if $\lambda > \overline{\operatorname{gr}} T$ but not if $\lambda < \overline{\operatorname{gr}} T$.

3.27. Let $U(T)$ be the set of unit flows on a tree T (from o to ∞). For $\theta \in U(T)$, define its **Frostman exponent** to be

$$\operatorname{Frost}(\theta) := \liminf_{|x| \to \infty} \theta(x)^{-1/|x|}\,.$$

Show that

$$\operatorname{br} T = \sup_{\theta \in U(T)} \operatorname{Frost}(\theta)\,.$$

3.28. Let $k \ge 1$. Show that if T is a 0-periodic (resp., 0-subperiodic) tree, then for all vertices x with $|x| \ge k$, there is an adjacency-preserving bijection (resp., injection) $f : T^x \to T^{f(x)}$ with $|f(x)| = k$.

3.29. Given a finite, *directed* multigraph G, one can also define another covering tree by using as vertices all directed paths of the form $\langle x_0, x_1, \ldots, x_n \rangle$ or $\langle x_{-n}, \ldots, x_{-1}, x_0 \rangle$, with the former a child of $\langle x_0, x_1, \ldots, x_{n-1} \rangle$ and the latter a child of $\langle x_{-(n-1)}, \ldots, x_{-1}, x_0 \rangle$. Show that this tree is also periodic.

3.30. Given an integer $k \ge 0$, construct a periodic tree T with $|T_n|$ approximately equal to n^k for all n.

3.31. Show that critical homesick random walk (that is, $\mathsf{RW}_{\operatorname{br} T}$) is recurrent on each periodic tree.

3.32. Construct a subperiodic tree for which critical homesick random walk (that is, $\mathsf{RW}_{\operatorname{br} T}$) is transient.

3.33. Let $N \geq 0$ and $0 < \alpha < 1$. Identify the binary tree with the set of all finite sequences of 0s and 1s. Let T be the subtree of the binary tree that contains the vertex corresponding to $(x_1, \ldots, x_n)$ iff $\forall k \leq n \; \sum_{i=1}^{k} x_i \leq \alpha(k + N)$. Show that T is N-superperiodic but not $(N - 1)$-superperiodic. Also, determine $\operatorname{br} T$.

3.34. Roth's theorem says that a subset of $\mathbb{N}$ that contains no three-term arithmetic progression must have density 0. Identify the binary tree with the set of all finite sequences of 0s and 1s. Let T be the subtree of the binary tree that contains the vertex corresponding to $(x_1, \ldots, x_n)$ iff $x_i x_{i+j} x_{i+2j} = 0$ whenever $1 \leq i < i + 2j \leq n$. Show that $\operatorname{gr} T = 1$.

3.35. Let $T(1)$ and $T(2)$ be two trees rooted at o_1 and o_2, respectively. Define their ***product tree*** $T(1) \cdot T(2)$ to be the tree with vertex set $\{(x_1, x_2) \, ; \; x_1 \in T(1), \; x_2 \in T(2), \; |x_1| = |x_2|\}$, rooted at (o_1, o_2), and such that $(x_1, x_2) \to (y_1, y_2)$ iff $x_1 \to y_1$ and $x_2 \to y_2$. For example, if $T(i)$ is a b_i-ary tree, then their product tree is a $b_1 b_2$-ary tree.
  (a) Show that $\operatorname{br} T(1) \cdot \operatorname{br} T(2) \leq \operatorname{br}(T(1) \cdot T(2))$.
  (b) Show that if $T(i)$ are 0-subperiodic, then so is their product tree.
  (c) Identify the binary tree with the set of all finite sequences of 0s and 1s. For a subset $S \subseteq \mathbb{N}$, define the spherically symmetric tree $T(S)$ be the subtree of the binary tree that contains the vertex corresponding to $(x_1, \ldots, x_n)$ iff $x_k \leq \mathbf{1}_S(k)$ for all $k \leq n$. That is, S is the set of heights where there is branching. Show that if both S and $\mathbb{N} \setminus S$ have lower density 0, then $\operatorname{br} T(S) = \operatorname{br} T(\mathbb{N} \setminus S) = 1$ and $\operatorname{br}(T(S) \cdot T(\mathbb{N} \setminus S)) = 2$.

3.36. Suppose that S generates the group Γ and that Γ' is a subgroup of Γ. The ***Schreier graph*** of the coset space $\Gamma' \backslash \Gamma$ with respect to S has as vertices the right cosets $\Gamma' \gamma$ for $\gamma \in \Gamma$ and as edges $[\Gamma' \gamma, \Gamma' \gamma s]$ for $s \in S$. When Γ' is normal in Γ, this is a Cayley graph of the quotient group. Show that different finite generating sets of Γ give roughly isometric Schreier graphs of the same coset space.

3.37. We have defined the right Cayley graph of a finitely generated group and noted that left multiplication is a graph automorphism. The left Cayley graph is defined similarly. Show that the right and left Cayley graphs are isomorphic.

3.38. Show that there are Cayley graphs G_n of $\mathbb{Z}$ such that for each r, the balls of radius r in G_n are isomorphic to the ball of radius r in the usual Cayley graph of $\mathbb{Z}^2$ for all large n. Similarly, show that there are Cayley graphs of $\mathbb{Z}$ where the balls are eventually isomorphic to those in the usual Cayley graph of $\mathbb{Z}^3$.

3.39. Extend Theorem 3.10 to all infinite, (locally finite) transitive graphs.

4 | Uniform Spanning Trees

One lesson of Chapter 2 is that in many ways, electrical networks and random walks are two faces of the same underlying object. Here we discover an appealing third face, which will appear at first to be completely unrelated.

Every connected graph has a ***spanning tree***, that is, a subgraph that is a tree and that includes every vertex. Special spanning trees of Cayley graphs were used in Section 3.4. Here, we consider finite and, more generally, recurrent graphs and properties of their spanning trees when such trees are chosen randomly. We will exhibit an amazing way to generate spanning trees uniformly at random. In Chapter 10, we will look at how to extend these notions to transient graphs, where the connections to random walks and geometric group theory flourish. Other natural ways of choosing random spanning trees and forests will be studied in Chapter 11, but those ways will be connected to percolation rather than to random walks.

Notation. In an undirected graph, a spanning tree is also composed of undirected edges. However, we will be using the ideas and notations of Chapter 2 concerning random walks and electrical networks, so that we will be making use of directed edges as well. Sometimes, e will even denote an undirected edge on one side of an equation and a directed edge on the other side; see, for example, Kirchhoff's effective-resistance formula. This abuse of notation, we hope, will be easier for the reader than would the use of different notations for directed and undirected edges.

4.1 Generating Uniform Spanning Trees

A graph typically has an enormous number of spanning trees. Because of this, it is not obvious how to choose one uniformly at random in a reasonable amount of time. We are going to present an algorithm that works quickly by exploiting some hidden independence in Markov chains. This algorithm is of enormous *theoretical* importance for us. Although we are interested in spanning trees of undirected graphs, it turns out that for this algorithm, it is just as easy, and somewhat more clear, to work with directed graphs coming from Markov chains.

Let $p(\cdot, \cdot)$ be the transition probability function of a finite-state, irreducible Markov chain. The directed graph associated to this chain has for vertices the states and for edges all $\langle x, y \rangle$ for which $p(x, y) > 0$. Edges e are oriented from tail e^- to head e^+. We call a connected subgraph a ***spanning tree*** if it includes every vertex, there is no cycle, and there is one vertex, the ***root***, such that every vertex other than the root is the tail of exactly one edge in the tree. Thus, the edges in a spanning tree point toward its root. For any vertex r, there is at least one spanning tree rooted at r: Pick some vertex other than r, and draw a path from it to r that does not contain any cycles. Such a path exists by irreducibility. This starts the tree. Then continue with another vertex not on the part of the tree already drawn, draw a cycle-free path from it to the partial tree, and so on. Remarkably, with a little care, this naive method of drawing spanning trees leads to a very powerful algorithm.

We are going to choose spanning trees at random according not only to uniform measure but, in general, proportional to their weights, where, for a spanning tree T, we define its ***weight*** to be

$$\Psi(T) := \prod_{e \in T} p(e).$$

In case the original Markov chain is reversible, let us see what the weights $\Psi(T)$ are. Given conductances $c(e)$ with $\pi(x) = \sum_{e^- = x} c(e)$, the transition probabilities are $p(e) := c(e)/\pi(e^-)$, so the weight of a spanning tree T is

$$\Psi(T) = \prod_{e \in T} p(e) = \prod_{e \in T} c(e) \Big/ \prod_{\substack{x \in G, \\ x \neq \mathrm{root}(T)}} \pi(x).$$

Since the root is fixed, a tree T is picked with probability proportional to $\Psi(T)/\pi(\mathrm{root}(T))$, which is proportional to

$$\Xi(T) := \prod_{e \in T} c(e).$$

Note that $\Xi(T)$ is independent of the root of T. This new expression, $\Xi(T)$, has a nice interpretation. If $c = \mathbf{1}$, then all spanning trees are equally likely. If all the weights $c(e)$ are positive integers, then we could replace each edge e by $c(e)$ parallel copies of e and interpret the uniform spanning tree measure in the resulting multigraph as the probability measure above with the probability of T proportional to $\Xi(T)$. If all weights are divided by the same constant, then the probability measure does not change, so the case of rational weights can still be thought of as corresponding to a uniform spanning tree. Since the case of general weights is a limit of rational weights, we use the term ***weighted uniform spanning tree*** for such a probability measure. Similar comments apply to the nonreversible case.

Now suppose we have some method of choosing a rooted spanning tree at random proportionally to the weights $\Psi(\cdot)$ for a reversible Markov chain. Consider any vertex u on a weighted undirected graph. If we choose a random spanning tree rooted at u proportionally to the weights $\Psi(\cdot)$ and forget about the orientation of its edges and also about the root, then we obtain an unrooted spanning tree of the undirected graph, chosen proportionally to the weights $\Xi(\cdot)$. In particular, if the conductances are all equal, which corresponds to the Markov chain being simple random walk, then we get a uniformly chosen spanning tree.

---

* For directed graphs, these are usually called ***spanning arborescences***. In this case, "connected" means "weakly connected", in other words, the underlying undirected graph of the arborescence is connected.

The method we now describe for generating random spanning trees is the fastest method known. It is due to Wilson (1996) (see also Propp and Wilson (1998)).

To describe Wilson's method, we define the important idea of loop erasure* of a path, due to Lawler (1980). If $\mathcal{P}$ is any finite path $\langle x_0, x_1, \ldots, x_l \rangle$ in a directed or undirected graph G, we define the **loop erasure** of $\mathcal{P}$, denoted $\mathsf{LE}(\mathcal{P}) = \langle u_0, u_1, \ldots, u_m \rangle$, by erasing cycles in $\mathcal{P}$ in the order they appear. More precisely, set $u_0 := x_0$. If $x_l = x_0$, we set $m = 0$ and terminate; otherwise, let u_1 be the first vertex in $\mathcal{P}$ after the last visit to x_0, that is, $u_1 := x_{i+1}$, where $i := \max\{j \, ; \, x_j = x_0\}$. If $x_l = u_1$, then we set $m = 1$ and terminate; otherwise, let u_2 be the first vertex in $\mathcal{P}$ after the last visit to u_1, and so on. For example, the loop erasure of the planar path shown in Figure 2.3 appears in Figure 4.1. In the case of a multigraph, one cannot notate a path merely by the vertices it visits. However, the notion of loop erasure should still be clear.

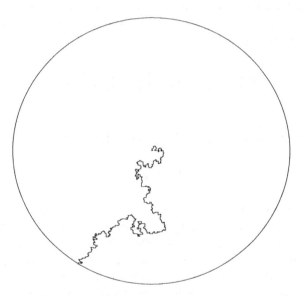

Figure 4.1. A loop-erased simple random walk in $\mathbb{Z}^2$
until it reaches distance 200 from its starting point.

Now to generate a random spanning tree with a given root r with probability proportional to the weights $\Psi(\cdot)$ for a given Markov chain, create a growing sequence of trees $T(i)$ $(i \geq 0)$ as follows. Choose any ordering of the vertices $\mathsf{V} \setminus \{r\}$. Let $T(0) := \{r\}$. Suppose that $T(i)$ is known. If $T(i)$ spans G, we are done. Otherwise, pick the first vertex x in our ordering of V that is not in $T(i)$ and take an independent sample from the Markov chain beginning at x until it hits $T(i)$. Now create $T(i + 1)$ by adding to $T(i)$ the loop erasure of this path from x to $T(i)$. Marvellously, the final tree in this growing sequence has the desired distribution. We call this **Wilson's method** of generating random spanning trees.

---

* This refers to loops in the topological sense, which are cycles in the graph-theoretical sense.

Theorem 4.1. *Given any finite-state, irreducible Markov chain and any state r, Wilson's method yields a random spanning tree rooted at r with distribution proportional to $\Psi(\cdot)$. Therefore, for any finite, connected, undirected graph, Wilson's method yields a random spanning tree that, when the orientation and root are forgotten, has distribution proportional to $\Xi(\cdot)$.*

In particular, this says that the distribution of the spanning tree resulting from Wilson's method does not depend on the choice made in ordering V. Actually, you need not order V in advance; you can choose where to start the next loop-erased path depending on what you have already constructed, but you still cannot change the distribution of the spanning tree! In fact, we'll see that in some sense, the tree itself cannot be changed.

To state this precisely, we construct the Markov chain in a special way. Every time we are at a state x, the next state will have a given probability distribution; and the choices of which states follow the visits to x are independent of each other. This, of course, is just part of the definition of a Markov chain. Constructively, however, we implement this as follows. Let $\langle S_i^x \ ; \ x$ is a state and $i \geq 1 \rangle$ be independent with each S_i^x being a state chosen according to the transition probability distribution from x. When we make the ith visit to x (if ever), then the Markov chain will move next to S_i^x.

The following image is useful. For each x, think of $\langle S_i^x \ ; \ i \geq 1 \rangle$ as a stack lying under the state x with S_1^x being on top, then S_2^x, and so forth. To run the Markov chain starting from state x_0, we simply "pop off" (that is, remove) the top state of the stack lying under x_0 and move there, then repeat the same procedure from the new state for as long as we want to run the chain. In other words, from the current state at any time, the next state is the first state in the stack under the current state. This state is then removed from that stack and we repeat with the next state as the current state.

Now, our aim is not to generate the Markov chain but a random spanning tree rooted at r. Thus, we make one small variation: give r an empty stack. We use the stacks as follows. Observe that at any time, the top items of the stacks determine a directed graph, namely, the directed graph whose vertices are the states and whose edges are the pairs (x, y) where y is the top item of the stack under x. Call this the ***visible graph*** at that time. If it happens that the visible graph contains no (directed) cycles, then it is a spanning tree rooted at r. In that case, we do nothing more. Otherwise, we ***pop*** a cycle, meaning that we remove the top items of the stacks under the vertices of a cycle. Then we pop a remaining cycle, if any, and so on. We claim that this process will stop with probability 1 at a spanning tree and that this spanning tree has the desired distribution. Note that we do not pop the top of a stack unless it belongs to a cycle. We will also show that this way of generating a random spanning tree is the same as Wilson's method.

To prove these statements, we will keep track of the locations in the stacks of edges that are popped, which we will call colors. That is, say that an edge (x, S_i^x) has ***color*** i. A ***colored cycle*** is simply a cycle all of whose edges are colored like this (the colors of the edges in a cycle do not have to be the same as each other). Thus, the initial visible graph has all edges colored 1, whereas later visible graphs will not generally have all their edges the same color. While a cycle of vertices might be popped many times, a colored cycle can be popped at most once. See Figure 4.2.

We begin with a deterministic lemma, which is the heart of this algorithm.

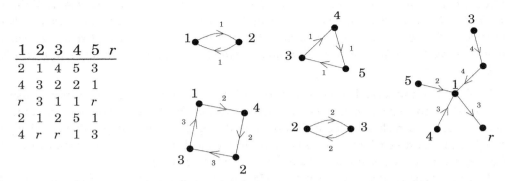

1	2	3	4	5	r
2	1	4	5	3	
4	3	2	2	1	
r	3	1	1	r	
2	1	2	5	1	
4	r	r	1	3	

Figure 4.2. This Markov chain has six states, one called r, which is the root. The first five elements of each stack are listed under the corresponding states. Colored cycles are popped as shown clockwise, leaving the colored spanning tree shown.

Lemma 4.2. *Given any stacks under the states, the order in which cycles are popped is irrelevant in the sense that every order pops an infinite number of cycles or every order pops the same (finite set of) colored cycles, thus leaving the same colored spanning tree on top in the latter case.*

Proof. We will show that if C is any colored cycle that can be popped, that is, there is some sequence $C_1, C_2, \ldots, C_n = C$ that may be popped in that order, but some colored cycle $C' \neq C_1$ happens to be the first colored cycle popped, then (1) $C = C'$, or else (2) C can still be popped from the stacks after C' is popped. Once we show this, we are done, since if there are an infinite number of colored cycles that can be popped, the popping can never stop; whereas in the alternative case, every colored cycle that can be popped will be popped.

Now if all the vertices of C' are disjoint from those of $C_1, C_2, \ldots, C_n$, then of course C can still be popped. Otherwise, let C_k be the first cycle that has a vertex in common with C'. Now, all the edges of C' have color 1. Consider any vertex x in $C' \cap C_k$. Since $x \notin C_1 \cup C_2 \cup \cdots \cup C_{k-1}$, the edge in C_k leading out of x also has color 1, so it leads to the same vertex as it does in C'. We can repeat the argument for this successor vertex of x, then for its successor, and so on, until we arrive at the conclusion that $C' = C_k$. Thus, $C' = C$ or we can pop C in the order $C', C_1, C_2, \ldots, C_{k-1}, C_{k+1}, \ldots, C_n$. ◄

Proof of Theorem 4.1. Wilson's method (using loop-erased parts of a Markov chain) certainly stops with probability 1 at a spanning tree. Using stacks to run the Markov chain and noting that loop erasure in order of cycle creation is one way of popping cycles, we see that Wilson's method pops all the cycles lying over a spanning tree. Because of Lemma 4.2, popping cycles in any other manner also stops with probability 1 and with the same distribution. Furthermore, if we think of the stacks as given in advance, then we see that all our choices inherent in Wilson's method have no effect whatsoever on the resulting spanning tree.

Now to show that the distribution is the desired one, think of a given set of stacks as defining a finite set O of colored cycles lying over a noncolored spanning tree T. We don't need to keep track of the colors in the spanning tree, since they are easily recovered from the colors in the cycles over it. Let X be the set of all pairs (O, T) that can arise from stacks corresponding to our given Markov chain. If $(O, T) \in X$, then also $(O, T') \in X$ for any other spanning tree T': indeed, anything at all can be in the stacks under any finite set O of colored cycles. That is, $X = X_1 \times X_2$, where X_1 is a certain collection of sets of colored cycles and X_2 is the set of all noncolored spanning trees. Extend our definition of $\Psi(\cdot)$ to colored cycles C by $\Psi(C) := \prod_{e \in C} p(e)$ and to sets O of colored cycles by $\Psi(O) := \prod_{C \in O} \Psi(C)$. What is the chance of seeing a given set O of colored cycles lying over a given spanning tree T? It is simply the probability of seeing all the arrows in $\bigcup O \cup T$ in their respective places, which is simply the product of $p(e)$ for all $e \in \bigcup O \cup T$, in other words, $\Psi(O)\Psi(T)$. Letting $\mathbf{P}$ be the law of (O, T), we get $\mathbf{P} = \mu_1 \times \mu_2$, where μ_i are probability measures proportional to $\Psi(\cdot)$ on X_i. Therefore, the set of colored cycles seen is independent of the colored spanning tree and the probability of seeing the tree T is proportional to $\Psi(T)$. This shows that Wilson's method does what we claimed it does. ◄

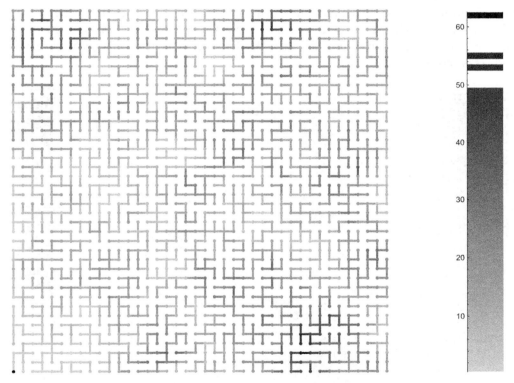

color plate **Figure 4.3.** A colored uniform spanning tree in a 40×40 grid on the left, with a key on the right showing the correspondence of visual colors to numbered colors.

An actual example of Wilson's algorithm showing the colored uniform spanning tree on a 40×40 grid is shown in Figure 4.3. Since the colors are determined by the popped cycles, which are independent of the spanning tree, it follows that if we attach the colors to the vertices instead of to the edges, then the colors are independent of the spanning tree (the colors are naturally attached to the vertices, since the stacks correspond to vertices). Just the colors are shown for a uniform spanning tree on a 200×200 grid in Figure 4.4. The colors have a multivariate negative binomial distribution, as shown in Corollary 6 (p. 81) and (4.3) with $\alpha = 1$ of Le Jan (2011) (but note that the 1 in the exponent on the left-hand side of (4.3) should be α in general and $G_{x,y}$ should be $G^{x,y}$). In Figure 4.5, we show the distances in the tree to the lower left vertex, together with the path from the upper right vertex. This seems to be the best way of viewing a large spanning tree. Although the distances do not determine the tree, all spanning trees consistent with the given distances are, of course, equally likely. Furthermore, given the distances, one can easily sample from the consistent spanning trees by working one's way out from the root: the vertices at distance 1 from the root must be attached to the root, while the vertices at distance 2 can be attached uniformly at random to their neighbors at distance 1, and so on. The distance in the tree from the root to the opposite corner, say, grows like $n^{5/4}$ in an $n \times n$ square: This was first conjectured by Guttmann and Bursill (1990) from numerical simulations, then calculated by Duplantier (1992) and Majumdar (1992) using nonrigorous conformal field theory. Kenyon (2000a) proved a form of this using domino tilings associated to spanning trees. It was extended to other planar lattices by Masson (2009) and strengthened by Barlow and Masson (2010). An alternative view is given in Figure 4.6, where the distances in the tree to the path between the corners is shown.

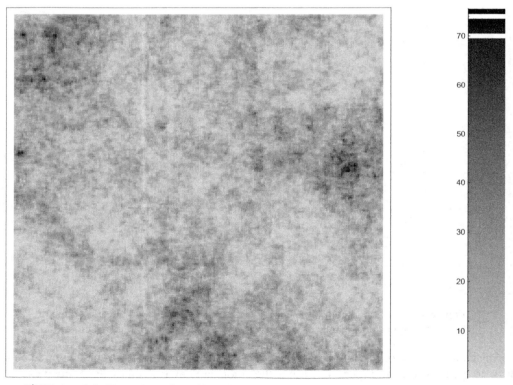

color plate **Figure 4.4.** The colors of a uniform spanning tree in a 200×200 grid on the left, with a key on the right showing the correspondence of visual colors to numbered colors.

We will see here and in Chapter 10 some of the far-reaching consequences of Wilson's method. First, we record the following obvious consequence of Wilson's algorithm:

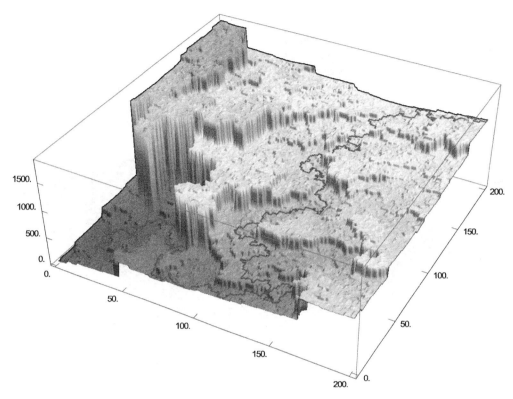

color plate **Figure 4.5.** The distances to the root in a uniform spanning tree in a 200×200 grid, together with the path from the opposite corner.

Corollary 4.3. *Given vertices x and y in a finite network, the distribution of the path in the weighted uniform spanning tree from x to y equals the distribution of loop-erased random walk from x to y.* ◄

This corollary was first proved by Pemantle (1991) using a different method that relies on an algorithm of Aldous and Broder (see Corollary 4.9). By conditioning on the path in the tree and then contracting it, one can immediately deduce Wilson's algorithm for the whole tree. This was observed by Wilson (1996) and Propp and Wilson (1998).

Of course, another immediate corollary is the invariance of loop-erased random walk under time reversal, which was known already to Lawler (1983):

Corollary 4.4. *Given vertices x and y in a finite network, the distribution of loop-erased random walk from x to y equals the distribution of the reversal of loop-erased random walk from y to x.* ◄

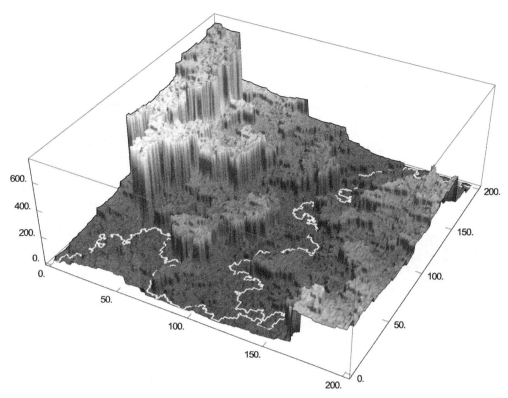

color plate **Figure 4.6.**　　The distances in the tree to the path between
opposite corners in a uniform spanning tree in a 200×200 grid.

Next, we use Wilson's algorithm to prove Cayley's formula for the number of spanning trees on a ***complete graph***, that is, a graph in which every pair of distinct vertices is joined by an edge. A number of proofs are known of this result; for a collection of them, see Moon (1967). The following proof is inspired by the one of Aldous (1990), Proposition 19.

Corollary 4.5. (Cayley, 1889) *The number of labeled, unrooted trees with n vertices is n^{n-2}. Here, a labeled tree is one whose vertices are labeled 1 through n.*

To prove this, we use the result of the following exercise.

▷ **Exercise 4.1.**
Suppose that Z is a set of states in a Markov chain and that x_0 is a state not in Z. Assume that when the Markov chain is started in x_0, then it visits Z with probability 1. Define the random path $Y_0, Y_1, \ldots$ by $Y_0 := x_0$ and then recursively by letting Y_{n+1} have the distribution of one step of the Markov chain starting from Y_n given that the chain will visit Z before visiting any of $Y_0, Y_1, \ldots, Y_n$ again. However, if $Y_n \in Z$, then the path is stopped and Y_{n+1} is not defined. Show that $\langle Y_n \rangle$ has the same distribution as loop-erasing a sample of the Markov chain started

from x_0 and stopped when it reaches Z. In the case of a random walk, the conditioned path $\langle Y_n \rangle$ is called the **Laplacian random walk** from x_0 to Z.

Proof of Corollary 4.5. We show that the uniform probability of a specific spanning tree of the complete graph on $\{1, 2, \ldots, n\}$ is $1/n^{n-2}$. Take the tree to be the path $\langle 1, 2, 3, \ldots, n \rangle$. We will calculate the probability of this tree by using Wilson's algorithm started at 1 and rooted at n. Since the root is n and the tree is a path from 1 to n, this tree probability is just the chance that loop-erased random walk from 1 to n is this particular path. By Exercise 4.1, we must show that the Laplacian random walk $\langle Y_n \rangle$ from 1 to n is precisely this path with probability $1/n^{n-2}$. Recall the following notation from Chapter 2: $\mathbf{P}_i$ denotes simple random walk started at state i; the first time ≥ 0 that the walk visits state k is denoted τ_k; and the first time ≥ 1 that the walk visits state k is denoted τ_k^+. Let $\langle X_n \rangle$ be the usual simple random walk.

Consider first the distribution of Y_1. The definition of Y_1 gives that for all $i \in [2, n]$,

$$\mathbf{P}[Y_1 = i] = \mathbf{P}_1[X_1 = i \mid \tau_n < \tau_1^+] = \frac{\mathbf{P}_1[X_1 = i, \tau_n < \tau_1^+]}{\mathbf{P}_1[\tau_n < \tau_1^+]}$$

$$= \frac{\mathbf{P}_1[X_1 = i] \, \mathbf{P}_i[\tau_n < \tau_1]}{\mathbf{P}_1[\tau_n < \tau_1^+]} = \frac{\mathbf{P}_i[\tau_n < \tau_1]}{(n-1) \, \mathbf{P}_1[\tau_n < \tau_1^+]} . \qquad (4.1)$$

Now

$$\mathbf{P}_i[\tau_n < \tau_1] = \begin{cases} 1/2 & \text{if } i \neq n \\ 1 & \text{if } i = n. \end{cases} \qquad (4.2)$$

Since the probabilities for Y_1 add to 1, combining (4.1) and (4.2) yields that $\mathbf{P}_1[\tau_n < \tau_1^+] = n/[2(n-1)]$, whence by (4.1) again, $\mathbf{P}[Y_1 = i] = 1/n$ for $1 < i < n$. Similarly, for $j \in [1, n-2]$ and $i \in [j+1, n]$, we have

$$\mathbf{P}[Y_j = i \mid Y_1 = 2, \ldots, Y_{j-1} = j] = \mathbf{P}_j[X_1 = i \mid \tau_n < \tau_1 \wedge \tau_2 \wedge \cdots \wedge \tau_{j-1} \wedge \tau_j^+]$$

$$= \frac{\mathbf{P}_j[X_1 = i, \tau_n < \tau_1 \wedge \tau_2 \wedge \cdots \wedge \tau_{j-1} \wedge \tau_j^+]}{\mathbf{P}_j[\tau_n < \tau_1 \wedge \tau_2 \wedge \cdots \wedge \tau_{j-1} \wedge \tau_j^+]}$$

$$= \frac{\mathbf{P}_j[X_1 = i] \, \mathbf{P}_i[\tau_n < \tau_1 \wedge \tau_2 \wedge \cdots \wedge \tau_{j-1} \wedge \tau_j]}{\mathbf{P}_j[\tau_n < \tau_1 \wedge \tau_2 \wedge \cdots \wedge \tau_{j-1} \wedge \tau_j^+]} .$$

Now the minimum of $\tau_1, \ldots, \tau_j, \tau_n$ for simple random walk starting at $i \in (j, n)$ is equally likely to be any one of these. Therefore,

$$\mathbf{P}_i[\tau_n < \tau_1 \wedge \tau_2 \wedge \cdots \wedge \tau_{j-1} \wedge \tau_j] = \begin{cases} 1/(j+1) & \text{if } j < i < n \\ 1 & \text{if } i = n. \end{cases}$$

Since $\mathbf{P}_j[X_1 = i] = 1/(n-1)$, we obtain $\mathbf{P}_j[\tau_n < \tau_1 \wedge \tau_2 \wedge \cdots \wedge \tau_{j-1} \wedge \tau_j^+] = n/[(j+1)(n-1)]$, and thus

$$\mathbf{P}[Y_j = j+1 \mid Y_1 = 2, \ldots, Y_{j-1} = j] = 1/n$$

for all $j \in [1, n-2]$. Of course,

$$\mathbf{P}[Y_{n-1} = n \mid Y_1 = 2, \ldots, Y_{n-2} = n-1] = 1 .$$

Multiplying together these conditional probabilities gives the result. ◀

4.2 Electrical Interpretations

We return now to undirected graphs and networks, except for occasional parenthetical remarks about more general Markov chains. We don't have to restrict ourselves to finite networks: Wilson's method for generating spanning trees will also give a random spanning tree (a.s.) on any recurrent network (or for any recurrent irreducible Markov chain), provided we start the new loop-erased walks in such a way as to guarantee that every vertex belongs to the final tree. The distribution of the resulting spanning tree again does not depend on the choice of root nor on the ordering of vertices: The corresponding version of Lemma 4.2 is that given any stacks under the states, we pop cycles in any order, provided that the stack under each vertex is considered for popping infinitely often. In that case, the order in which cycles are popped is irrelevant in the sense that every order pops the same colored cycles. The proof is similar to that for finite networks because cycles are finite. Can we interpret such random spanning trees in terms of uniform spanning trees when the network is infinite? Suppose that G' is a finite, connected subnetwork of G and consider T_G and $T_{G'}$, random spanning trees generated by Wilson's method on G and G', respectively. After describing a connection to electrical networks, we will show that for any event $\mathscr{B}$ depending on only finitely many edges, we can make $|\mathbf{P}[T_G \in \mathscr{B}] - \mathbf{P}[T_{G'} \in \mathscr{B}]|$ arbitrarily small by choosing G' sufficiently large.* Thus, the random spanning tree of G looks locally like a tree chosen with probability proportional to $\Xi(\cdot)$. Also, these local probabilities determine the distribution of T_G uniquely. In particular, when simple random walk is recurrent, such as on $\mathbb{Z}^2$, we may regard T_G as a "uniform" random spanning tree of G.

We will study uniform spanning trees on recurrent networks further and also "uniform" spanning forests on transient networks in Chapter 10. For now, though, we will deduce some important theoretical consequences of the connection between random walk and spanning trees on finite and recurrent networks. For recurrent networks, the definitions and relations among random walks and electrical networks appear in Exercises 2.71, 2.72, and 2.73; some are also covered in Section 9.1 and Corollary 9.6, but we won't need any material from Chapter 9 here.

Kirchhoff's Effective Resistance Formula. *Let T be an unrooted, weighted uniform spanning tree of a recurrent network G and e be an edge of G. Then*

$$\mathbf{P}[e \in T] = \mathbf{P}_{e^-}[\textit{first hit } e^+ \textit{ via traveling along } e] = i(e) = c(e)\mathscr{R}(e^- \leftrightarrow e^+),$$

where i is the unit current from e^- to e^+.

Remark. That $\mathbf{P}[e \in T] = i(e)$ in finite networks is due to Kirchhoff (1847); he didn't say anything about random walks.

Proof. The first equality follows by taking the vertex e^+ as the root of T and then starting the construction of Wilson's method at e^-. The second equality then follows from the probabilistic interpretation (Proposition 2.2 and its extension Exercise 2.71 to infinite, recurrent networks) of i as the expected number of crossings of e minus the expected number of crossings of the reversed edge $-e$ for a random walk started at e^- and stopped at e^+: e is crossed once or not at all and $-e$ is never crossed. The third equality comes from the definition of effective resistance. ◀

---

* This can also be proved by coupling the constructions using Wilson's method.

▷ **Exercise 4.2.**
Consider the ladder graph L_n of height n shown in Figure 1.7. Choose a spanning tree $T(n)$ of L_n uniformly. Use Kirchhoff's effective-resistance formula to determine the chance $\mathbf{P}[\text{rung 1 is in } T(n)]$ and its limiting behavior as $n \to \infty$.

Kirchhoff's fundamental result tells us the single-edge marginals of uniform spanning trees. What about the marginals for several edges? Suppose e and f are two distinct edges of a finite graph G. If T is a random spanning tree chosen uniformly from all spanning trees of G, then we might expect that the events $[e \in T]$ and $[f \in T]$ are negatively correlated, that is, the probability that both happen is at most the product of the probabilities that each happens: Intuitively, the presence of f would make e less needed for connecting everything *and* more likely to create a cycle. Furthermore, since the number of edges in a spanning tree is constant, namely, one fewer than the number of vertices in the graph, the negative correlation is certainly true "on average." We will now prove this negative correlation by using Kirchhoff's effective-resistance formula. Interestingly, no direct proof is known, although a combinatorial version of the electrical proof was given by Feder and Mihail (1992).

In fact, we can use Kirchhoff's effective-resistance formula to compute the chance that certain edges are in T and certain others are not. To see this, denote the dependence of T on G by T_G. The **contraction** G/e of a graph G along an edge e is obtained by removing the edge e and identifying its endpoints. Note that this may give a multigraph even if G is a simple graph. Deleting e without identifying its endpoints gives the graph denoted $G \backslash e$. In both cases, we may identify the edges of G other than e with the edges of G/e and of $G \backslash e$. We think of a spanning tree primarily as a set of edges. Now the distribution of T_G/e (the contraction of T_G along e) given that $e \in T_G$ is the same as that of $T_{G/e}$, and the distribution of T_G given $e \notin T_G$ is the same as that of $T_{G \backslash e}$. This gives a recursive method to compute $\mathbf{P}[e_1, \ldots, e_k \in T_G, e_{k+1}, \ldots, e_l \notin T_G]$: for example, if $e \neq f$, then

$$\mathbf{P}[e, f \in T_G] = \mathbf{P}[e \in T_G]\,\mathbf{P}[f \in T_G \mid e \in T_G] = \mathbf{P}[e \in T_G]\,\mathbf{P}[f \in T_{G/e}]$$

and

$$\mathbf{P}[e \notin T_G, f \in T_G] = \mathbf{P}[e \notin T_G]\,\mathbf{P}[f \in T_{G \backslash e}]\,.$$

Thus, we may deduce that the events $e \in T$ and $f \in T$ are negatively correlated:

▷ **Exercise 4.3.**
By using Kirchhoff's effective-resistance formula, show that if $e \neq f$, then the events $e \in T$ and $f \in T$ are negatively correlated.

We can now also establish our claim at the beginning of this section that on a recurrent graph, the random spanning tree looks locally like that of large finite, connected subnetworks. For example, given an edge e and a subnetwork G' of G, the current in G' flowing (directly) along e arising from a unit current (in total) between the endpoints of e will be very close to the corresponding current along e in G, provided G' is sufficiently large, by Exercise 2.71. That means that $\mathbf{P}[e \in T_{G'}]$ will be very close to $\mathbf{P}[e \in T_G]$.

▷ **Exercise 4.4.**
Write out the rest of the proof that for any event $\mathscr{B}$ depending on only finitely many edges, $\big|\mathbf{P}[T_{G'} \in \mathscr{B}] - \mathbf{P}[T_G \in \mathscr{B}]\big|$ is arbitrarily small for sufficiently large G'.

What does contraction of some edges do to a current in the setting of the inner-product space $\ell_-^2(\mathsf{E}, r)$? Let i^e denote the unit current from the tail of e to the head of e in a finite network G. Contract the edges $f \in F$ to obtain the graph G/F, and let $\widehat{i^e}$ be the unit current flowing in G/F, where $e \notin F$ and, moreover, e does not form any undirected cycle with the edges of F, so that e does not become a loop when the edges of F are contracted. Note that the restriction of any $\theta \in \ell_-^2(\mathsf{E}, r)$ to $\mathsf{E} \setminus F$ yields an antisymmetric function on the edges of the contracted graph G/F. Let Z be the linear span of $\{i^f \; ; \; f \in F\}$. We claim that

$$\widehat{i^e} = (P_Z^\perp i^e){\restriction}(\mathsf{E} \setminus F) \tag{4.3}$$

and

$$(P_Z^\perp i^e){\restriction}F = \mathbf{0}, \tag{4.4}$$

where $P_Z^\perp$ denotes the orthogonal projection onto the orthocomplement of Z in $\ell_-^2(\mathsf{E}, r)$. These equations may look a little forbidding at first, but a second or third look ought to reveal their inner simplicity. In fact, if instead of removing the edges in F when we contract them, we leave them as loops, then these equations say that $\widehat{i^e} = P_Z^\perp i^e$.

To prove these equations, write $[F]$ for the linear span of $\{\chi^f \; ; \; f \in F\}$. We will identify the edges of G/F with $\mathsf{E} \setminus F$ and $\ell_-^2(\mathsf{E} \setminus F, r)$ with $[F]^\perp$ in $\ell_-^2(\mathsf{E}, r)$. Then the equations (4.3) and (4.4) again are equivalent to the one equation $\widehat{i^e} = P_Z^\perp i^e$. Note that every star in G/F is a sum θ of stars in G such that $\theta(f) = 0$ for all $f \in F$, that is, $\bigstar(G/F) \subseteq \bigstar \cap [F]^\perp$. In fact, equality holds here: since $\Diamond(G/F) \subseteq \Diamond + [F]$, we have $\bigstar(G/F) \supseteq \Diamond^\perp \cap [F]^\perp = \bigstar \cap [F]^\perp$, as claimed. Finally, we may write $\bigstar(G/F) = \bigstar \cap Z^\perp$, because the elements of $\bigstar$ that are orthogonal to $[F]$ are precisely those which are orthogonal to $P_{\bigstar}[F]$, which equals Z. It follows that $P_{\bigstar(G/F)} = P_Z^\perp P_{\bigstar}$. Recall from (2.9) or (2.11) that $P_{\bigstar}\chi^e = i^e$ and likewise $P_{\bigstar(G/F)}\chi^e = \widehat{i^e}$. Thus, $\widehat{i^e} = P_{\bigstar(G/F)}\chi^e = P_Z^\perp i^e$, as claimed.

Although we have indicated that successive contractions can be used for computing $\mathbf{P}[e_1, \ldots, e_k \in T]$, this requires computations of effective resistance on k different graphs. The formula we just gave allows us to replace the different graphs with computations of orthogonal projections for the original graph, but this is not necessarily pleasant, either. However, it turns out that these computations can be organized in a marvellous way, as shown by the following wonderful theorem of Burton and Pemantle (1993) and its extension, Exercise 4.41:

The Transfer-Current Theorem. *For any distinct edges* $e_1, \ldots, e_k \in G$,

$$\mathbf{P}[e_1, \ldots, e_k \in T] = \det\big[Y(e_i, e_j)\big]_{1 \le i, j \le k}. \tag{4.5}$$

Recall that $Y(e, f) = i^e(f)$. Note that, in particular, we get a quantitative version of the negative correlation between $\{e \in T\}$ and $\{f \in T\}$: for distinct edges e, f, we have

$$\mathbf{P}[e, f \in T] - \mathbf{P}[e \in T]\,\mathbf{P}[f \in T] = -Y(e, f)Y(f, e) = -c(e)r(f)Y(e, f)^2$$

by the reciprocity law (2.12).

Proof. It suffices to show the result for finite G, because taking limits of this result implies it holds for infinite, recurrent G by Exercise 2.71.

We first show that if some cycle can be formed from the edges $e_1, \ldots, e_k$, then a linear combination of the corresponding columns of $[Y(e_i, e_j)]$ is zero: Suppose that such a cycle is $\sum_j a_j \chi^{e_j}$, where $a_j \in \{-1, 0, 1\}$. Then, for $1 \le m \le k$, we have

$$\sum_j a_j r(e_j) Y(e_m, e_j) = \sum_j a_j r(e_j) i^{e_m}(e_j) = 0$$

by the cycle law applied to the current i^{e_m}. Therefore, both sides of (4.5) are 0. For the remainder of the proof, then, we may assume that there are no such cycles.

We next proceed by induction. When $k = 1$, (4.5) is the same as Kirchhoff's effective-resistance formula. For $1 \le m \le k$, let

$$Y_m := \left[Y(e_i, e_j) \right]_{1 \le i,j \le m}. \tag{4.6}$$

To carry the induction from $m = k - 1$ to $m = k$, we must show that

$$\det Y_k = \mathbf{P}[e_k \in T \mid e_1, \ldots, e_{k-1} \in T] \det Y_{k-1}. \tag{4.7}$$

Now we know that

$$\mathbf{P}[e_k \in T \mid e_1, \ldots, e_{k-1} \in T] = \widehat{i^{e_k}}(e_k) \tag{4.8}$$

for the current $\widehat{i^{e_k}}$ in the graph $G/\{e_1, \ldots, e_{k-1}\}$. In addition,

$$P_Z^\perp i^{e_k} = i^{e_k} - \sum_{m=1}^{k-1} a_m i^{e_m}$$

for some constants a_m, where Z is the linear span of $\{i^{e_1}, \ldots, i^{e_{k-1}}\}$. Subtracting these same multiples of the first $k - 1$ rows from the last row of Y_k leads to a matrix $\widehat{Y}$ whose (m, j)-entry is that of Y_{k-1} for $m, j < k$ and whose (k, j)-entry is

$$i^{e_k}(e_j) - \sum_{m=1}^{k-1} a_m i^{e_m}(e_j) = (P_Z^\perp i^{e_k})(e_j) = \begin{cases} 0 & \text{if } j < k \\ \widehat{i^{e_k}}(e_k) & \text{if } j = k \end{cases}$$

by (4.4) and (4.3). Therefore expansion of $\det \widehat{Y}$ along the kth row is very simple and gives that

$$\det Y_k = \det \widehat{Y} = \widehat{i^{e_k}}(e_k) \det Y_{k-1}.$$

Combining this with (4.8), we obtain (4.7). (At bottom, we are using, or proving, the fact that the determinant of a Gram matrix is the square of the volume of the parallelepiped determined by the vectors whose inner products give the entries.)   ◀

It turns out that there is a more general negative correlation than that between the presence of two given edges. Regard a spanning tree as simply a set of edges. We may extend our probability measure **P** on the set of spanning trees to the product σ-field on $2^{E(G)}$ by defining the probability to be 0 of the event that the set of edges do not form a spanning tree. This may sound unhelpful, but surprisingly, it is useful. Call an event $\mathscr{A} \subseteq 2^{E(G)}$ ***increasing*** (also called ***upwardly closed***) if the addition of any edge to any set in $\mathscr{A}$ results in another set in $\mathscr{A}$, that is, $A \cup \{e\} \in \mathscr{A}$ for all $A \in \mathscr{A}$ and all $e \in \mathsf{E}$. For example, $\mathscr{A}$ could be the collection of all subsets of $E(G)$ that contain at least two of the edges $\{e_1, e_2, e_3\}$. We say that an event $\mathscr{A}$ ***ignores an edge*** e if $A \cup \{e\} \in \mathscr{A}$ and $A \setminus \{e\} \in \mathscr{A}$ for all $A \in \mathscr{A}$. In the prior example, e is ignored provided $e \notin \{e_1, e_2, e_3\}$. We also say that $\mathscr{A}$ ***depends (only) on a set*** $F \subseteq \mathsf{E}(G)$ if, for every pair $\omega_1, \omega_2 \in 2^{\mathsf{E}}$ that agree on F, we have either both ω_1, ω_2 are in $\mathscr{A}$ or neither are in $\mathscr{A}$.

▷ **Exercise 4.5.**
Suppose that $\mathscr{A}$ is an increasing event on a graph G and $e \in \mathsf{E}$. Note that $\mathsf{E}(G/e) = \mathsf{E}(G \backslash e) = \mathsf{E}(G) \backslash \{e\}$. Define $\mathscr{A}/e := \{F \subseteq \mathsf{E}(G/e) ; F \cup \{e\} \in \mathscr{A}\}$ and $\mathscr{A} \backslash e := \{F \subseteq \mathsf{E}(G \backslash e) ; F \in \mathscr{A}\}$. Show that these are increasing events on G/e and $G \backslash e$, respectively.

Pemantle conjectured (personal communication, 1990) that $\mathscr{A}$ and $[e \in T]$ are negatively correlated when $\mathscr{A}$ is an increasing event that ignores e. Though unaware that Pemantle had conjectured this, Feder and Mihail (1992) proved it:

Theorem 4.6. *Let G be a finite network. If $\mathscr{A}$ is an increasing event that ignores some edge e, then $\mathbf{P}[\mathscr{A} \mid e \in T] \le \mathbf{P}[\mathscr{A}]$.*

As we will see in Chapter 10, this result is quite useful.

Proof. We induct on the number of edges in G. The case of exactly one (undirected) edge is trivial. Now assume that the number of (undirected) edges is $m \ge 2$ and that we know the result for graphs with $m - 1$ edges. Let G have m edges. If $\mathbf{P}[f \in T] = 1$ for some $f \in \mathsf{E}$, then we could contract f and reduce to the case of $m - 1$ edges by Exercise 4.5, so assume this is not the case. If $|\mathsf{V}| = 2$ and G has parallel edges, then the result also follows without using induction, so assume that $|\mathsf{V}| \ge 3$. Fix an increasing event $\mathscr{A}$ and an edge e ignored by $\mathscr{A}$. We may assume that $\mathbf{P}[\mathscr{A} \mid e \in T] > 0$ to prove our inequality. The graph G/e has only $m - 1$ edges, and every spanning tree of G/e has $|\mathsf{V}| - 2$ edges. This latter simple fact leads to the key equation

$$\sum_{f \in \mathsf{E} \backslash \{e\}} \mathbf{P}[\mathscr{A}, f \in T \mid e \in T] = (|\mathsf{V}| - 2) \mathbf{P}[\mathscr{A} \mid e \in T] = \mathbf{P}[\mathscr{A} \mid e \in T] \sum_{f \in \mathsf{E} \backslash \{e\}} \mathbf{P}[f \in T \mid e \in T].$$

Therefore there is some $f \in \mathsf{E} \setminus \{e\}$ with $\mathbf{P}[f \in T \mid e \in T] > 0$ such that

$$\mathbf{P}[\mathscr{A}, f \in T \mid e \in T] \ge \mathbf{P}[\mathscr{A} \mid e \in T] \mathbf{P}[f \in T \mid e \in T],$$

which is the same as $\mathbf{P}[\mathscr{A} \mid f, e \in T] \ge \mathbf{P}[\mathscr{A} \mid e \in T]$. This also means that

$$\mathbf{P}[\mathscr{A} \mid f, e \in T] \ge \mathbf{P}[\mathscr{A} \mid f \notin T, e \in T]; \tag{4.9}$$

in case this is not evident, one can deduce it from

$$\mathbf{P}[\mathscr{A} \mid e \in T] = \mathbf{P}[f \in T \mid e \in T]\,\mathbf{P}[\mathscr{A} \mid f, e \in T] + \mathbf{P}[f \notin T \mid e \in T]\,\mathbf{P}[\mathscr{A} \mid f \notin T, e \in T].$$
(4.10)

Now we also have

$$\mathbf{P}[f \in T \mid e \in T] \le \mathbf{P}[f \in T]$$

by Exercise 4.3. Because of (4.9), it follows that

$$\mathbf{P}[\mathscr{A} \mid e \in T] \le \mathbf{P}[f \in T]\,\mathbf{P}[\mathscr{A} \mid f, e \in T] + \mathbf{P}[f \notin T]\,\mathbf{P}[\mathscr{A} \mid f \notin T, e \in T]:$$
(4.11)

we have replaced a convex combination in (4.10) by another in (4.11) that puts more weight on the larger term. We also have

$$\mathbf{P}[\mathscr{A} \mid f, e \in T] \le \mathbf{P}[\mathscr{A} \mid f \in T]$$
(4.12)

by the induction hypothesis applied to the event $\mathscr{A}/f$ on the network G/f (see Exercise 4.5), and

$$\mathbf{P}[\mathscr{A} \mid f \notin T, e \in T] \le \mathbf{P}[\mathscr{A} \mid f \notin T]$$
(4.13)

by the induction hypothesis applied to the event $\mathscr{A} \backslash f$ on the network $G \backslash f$. By (4.12) and (4.13), we have that the right-hand side of (4.11) is

$$\le \mathbf{P}[f \in T]\,\mathbf{P}[\mathscr{A} \mid f \in T] + \mathbf{P}[f \notin T]\,\mathbf{P}[\mathscr{A} \mid f \notin T] = \mathbf{P}[\mathscr{A}]. \qquad \blacktriangleleft$$

▷ **Exercise 4.6.**

(Negative Association) Let G be a finite network. Extend Theorem 4.6 to show that if $\mathscr{A}$ and $\mathscr{B}$ are both increasing events and they depend on disjoint sets of edges, then they are negatively correlated. Still more generally, show the following. Say that a random variable $X: 2^{\mathsf{E}(G)} \to \mathbb{R}$ *depends on a set* $F \subseteq \mathsf{E}(G)$ if X is measurable with respect to the σ-field consisting of events that depend on F. Say also that X is *increasing* if $X(H) \le X(H')$ whenever $H \subset H'$. If X and Y are increasing random variables that depend on disjoint sets of edges, then $\mathbf{E}[XY] \le \mathbf{E}[X]\,\mathbf{E}[Y]$. This property of $\mathbf{P}$ is called *negative association*.

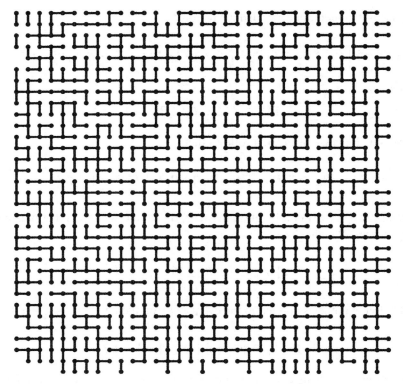

Figure 4.7. A portion of a uniformly chosen spanning tree on $\mathbb{Z}^2$, drawn by David Wilson.

4.3 The Square Lattice $\mathbb{Z}^2$

Uniform spanning trees on the nearest-neighbor graph on the square lattice $\mathbb{Z}^2$ are particularly appealing. A portion of one is shown in Figure 4.7. This can be thought of as an infinite maze. In Section 10.6, we will show that there is exactly one way to get from any square to any other square without backtracking and exactly one way to get from any square to infinity without backtracking. Thus, if escaping the maze means finding a path to infinity from a given starting square, then there is exactly one way to do it without backtracking, and it is also possible to get lost anywhere else. For now, though, we will consider only the "walls" of the maze, that is, the actual spanning tree.

What is the distribution of the degree of a vertex with respect to a uniform spanning tree in $\mathbb{Z}^2$? It turns out that although the distribution is not so easy to calculate, the expected degree is easy to calculate and is part of a quite general result. This uses the amenability of $\mathbb{Z}^2$. What's that? If G is a graph and $K \subset V$, the *edge boundary* of K is the set $\partial_E K$ of (unoriented) edges that connect K to its complement. We say that G is *edge amenable* if there are finite $V_n \subset V$ with

$$\lim_{n\to\infty} |\partial_E V_n|/|V_n| = 0\,.$$

▷ **Exercise 4.7.**
Let G be an edge-amenable, infinite graph as witnessed by the sequence $\langle V_n \rangle$. Show that the average degree of vertices in any spanning tree of G is 2. That is, if $\deg_T(x)$ denotes the

degree of x in a spanning tree T of G, then

$$\lim_{n\to\infty} |V_n|^{-1} \sum_{x\in V_n} \deg_T(x) = 2 \,.$$

Every infinite, recurrent graph can be shown to be edge-amenable by various results from Chapter 6 that we'll look at later, such as Theorems 6.5, 6.7, or 6.42. (The origin of the term "amenable" will be discussed in Section 6.1.) Deduce that for the uniform spanning tree measure on a recurrent graph,

$$\lim_{n\to\infty} |V_n|^{-1} \sum_{x\in V_n} \mathbf{E}\big[\deg_T(x)\big] = 2 \,.$$

In particular, if G is also transitive, such as $\mathbb{Z}^2$, meaning that for every pair of vertices x and y, there is a bijection of V with itself that preserves adjacency and takes x to y, then every vertex has expected degree 2.

By symmetry, each edge of $\mathbb{Z}^2$ has the same probability to be in a uniform spanning tree of $\mathbb{Z}^2$. Since the expected degree of a vertex is 2 by Exercise 4.7, it follows that

$$\mathbf{P}[e \in T] = 1/2 \tag{4.14}$$

for each $e \in \mathbb{Z}^2$. By Kirchhoff's effective-resistance formula, this means that if unit current flows from the tail to the head of e, then $1/2$ of the current flows directly across e and that the effective resistance between two adjacent vertices is $1/2$. These electrical facts are classic engineering puzzles.

To calculate the distribution of the degree, we will use the transfer-current theorem. The result is rather surprising, namely, the degree has the following distribution:

Degree	Probability	
1	$\dfrac{8}{\pi^2}\left(1 - \dfrac{2}{\pi}\right)$	$= .294^+$
2	$\dfrac{4}{\pi}\left(2 - \dfrac{9}{\pi} + \dfrac{12}{\pi^2}\right)$	$= .447^-$
3	$2\left(1 - \dfrac{2}{\pi}\right)\left(1 - \dfrac{6}{\pi} + \dfrac{12}{\pi^2}\right)$	$= .222^+$
4	$\left(\dfrac{4}{\pi} - 1\right)\left(1 - \dfrac{2}{\pi}\right)^2$	$= .036^+$

$$(4.15)$$

To find the transfer currents $Y(e_1, e_2)$, we will first find voltages, then use $i = dv$. (We assume unit conductances on the edges.) When i is a unit flow from x to y, we have

$d^*i = \mathbf{1}_{\{x\}} - \mathbf{1}_{\{y\}}$. Hence the voltages satisfy $\Delta v := d^*dv = \mathbf{1}_{\{x\}} - \mathbf{1}_{\{y\}}$; here, Δ is called the **graph Laplacian**. We are interested in solving this equation when $x := e_1^-$, $y := e_1^+$; then we compute $v(e_2^-) - v(e_2^+)$. Our method is to use Fourier analysis. We begin with a formal (that is, heuristic) derivation of the solution, then prove that the formula we get is correct.

Let $\mathbb{T}^2 := (\mathbb{R}/\mathbb{Z})^2$ be the two-dimensional torus. For $(x_1, x_2) \in \mathbb{Z}^2$ and $(\alpha_1, \alpha_2) \in \mathbb{T}^2$, write $(x_1, x_2) \cdot (\alpha_1, \alpha_2) := x_1\alpha_1 + x_2\alpha_2 \in \mathbb{R}/\mathbb{Z}$. For a function f on $\mathbb{Z}^2$, define the function $\widehat{f}$ on $\mathbb{T}^2$ by

$$\widehat{f}(\alpha) := \sum_{x \in \mathbb{Z}^2} f(x)e^{-2\pi i x \cdot \alpha}.$$

We are not worrying here about whether this converges in any sense, but certainly $\widehat{\mathbf{1}_{\{x\}}}(\alpha) = e^{-2\pi i x \cdot \alpha}$. Now a formal calculation shows that for a function f on $\mathbb{Z}^2$, we have

$$\widehat{\Delta f}(\alpha) = \varphi(\alpha)\widehat{f}(\alpha),$$

where

$$\varphi((\alpha_1, \alpha_2)) := 4 - \left(e^{2\pi i \alpha_1} + e^{-2\pi i \alpha_1} + e^{2\pi i \alpha_2} + e^{-2\pi i \alpha_2}\right) = 4 - 2\left(\cos 2\pi\alpha_1 + \cos 2\pi\alpha_2\right).$$

Hence, to solve $\Delta f = g$, we may try to solve $\widehat{\Delta f} = \widehat{g}$ by using $\widehat{f} := \widehat{g}/\varphi$ and then finding f. In fact, a formal calculation shows that we may recover f from $\widehat{f}$ by the formula

$$f(x) = \int_{\mathbb{T}^2} \widehat{f}(\alpha)e^{2\pi i x \cdot \alpha}\, d\alpha,$$

where the integration is with respect to Lebesgue measure. This is the approach we will follow. Note that we need to be careful about the nonuniqueness of solutions to $\Delta f = g$, since there are nonzero functions f with $\Delta f = 0$.

▷ **Exercise 4.8.**
Show that $(\widehat{\mathbf{1}_{\{x\}}} - \widehat{\mathbf{1}_{\{y\}}})/\varphi \in L^1(\mathbb{T}^2)$ for all $x, y \in \mathbb{Z}^2$.

▷ **Exercise 4.9.**
Show that if $F \in L^1(\mathbb{T}^2)$ and $f(x) = \int_{\mathbb{T}^2} F(\alpha)e^{2\pi i x \cdot \alpha}\, d\alpha$, then

$$(\Delta f)(x) = \int_{\mathbb{T}^2} F(\alpha)e^{2\pi i x \cdot \alpha}\varphi(\alpha)\, d\alpha.$$

Proposition 4.7. (Voltage on $\mathbb{Z}^2$) *The voltage at u when a unit current flows from x to y in $\mathbb{Z}^2$ and when $v(y) = 0$ is*

$$v(u) = v'(u) - v'(y),$$

where

$$v'(z) := \int_{\mathbb{T}^2} \frac{e^{-2\pi i x \cdot \alpha} - e^{-2\pi i y \cdot \alpha}}{\varphi(\alpha)} e^{2\pi i z \cdot \alpha}\, d\alpha.$$

Proof. By Exercises 4.8 and 4.9, we have

$$\Delta v'(z) = \int_{\mathbb{T}^2} \left(e^{-2\pi i x \cdot \alpha} - e^{-2\pi i y \cdot \alpha} \right) e^{2\pi i z \cdot \alpha} \, d\alpha = \mathbf{1}_{\{x\}}(z) - \mathbf{1}_{\{y\}}(z) \, ;$$

that is, $\Delta v' = \mathbf{1}_{\{x\}} - \mathbf{1}_{\{y\}}$. Since v satisfies the same equation, we have $\Delta(v' - v) = 0$. In other words, $v' - v$ is harmonic at every point in $\mathbb{Z}^2$. Furthermore, v' is bounded in absolute value by the L^1 norm of $(\widehat{\mathbf{1}_{\{x\}}} - \widehat{\mathbf{1}_{\{y\}}})/\varphi$. Since v is also bounded (by $v(x)$), it follows that $v' - v$ is bounded. Since the only bounded harmonic functions on $\mathbb{Z}^2$ are the constants (by, say, Exercise 2.43), this means that $v' - v$ is constant. Since $v(y) = 0$, we obtain $v = v' - v'(y)$, as desired. ◀

Let v and v' be as defined in Proposition 4.7. We now need to find a good method to compute the integral v'. Set

$$H(u) := 4 \int_{\mathbb{T}^2} \frac{1 - e^{2\pi i u \cdot \alpha}}{\varphi(\alpha)} \, d\alpha \, . \tag{4.16}$$

Note that the integrand is integrable by Exercise 4.8 applied to $x := (0,0)$ and $y := -u$. (Also, $H(u)/2$ is the effective resistance between $(0,0)$ and u by Proposition 4.7.) The integral H is useful because $v'(u) = [H(u - y) - H(u - x)]/4$. (The factor of 4 is introduced in H only to conform to the usage of other authors.) For given edges e_1 and e_2, put $x := e_1^-$ and $y := e_1^+$ to get

$$Y(e_1, e_2) = v(e_2^-) - v(e_2^+)$$
$$= \frac{1}{4} \left[H(e_2^- - e_1^+) - H(e_2^- - e_1^-) - H(e_2^+ - e_1^+) + H(e_2^+ - e_1^-) \right]. \tag{4.17}$$

Thus, we concentrate on calculating H. Now $H(0,0) = 0$, and a direct calculation as in Exercise 4.9 shows that $\Delta H = -4 \cdot \mathbf{1}_{\{(0,0)\}}$. Furthermore, the symmetries of φ show that H is invariant under reflection in the axes and in the 45° line. Therefore, all the values of H can be computed from those on the 45° line by computing values at gradually increasing distance from the origin and from the 45° line. (For example, we first compute $H(1, 0) = 1$ from the equations $H(0,0) = 0$ and $(\Delta H)(0, 0) = 4H(0,0) - H(0, 1) - H(1, 0) - H(0, -1) - H(-1, 0) = -4$, then $H(2, 1)$ from the value of $H(1, 1)$ and the equation $(\Delta H)(1, 1) = 4H(1, 1) - H(1, 0) - H(0, 1) - H(1, 2) - H(2, 1) = 0$, then $H(2, 0)$ from $(\Delta H)(1, 0) = 0$, then $H(3, 2)$, and so on.)

The reflection symmetries we observed for H imply that $H(u) = H(-u)$, whence H is real. Thus, we can write the diagonal values as

$$H(n, n) = 4 \int_{\mathbb{T}^2} \frac{1 - \cos 2\pi n(\alpha_1 + \alpha_2)}{\varphi(\alpha)} \, d\alpha$$
$$= \int_0^1 \int_0^1 \frac{1 - \cos 2\pi n(\alpha_1 + \alpha_2)}{1 - \cos(\pi(\alpha_1 + \alpha_2)) \cos(\pi(\alpha_1 - \alpha_2))} \, d\alpha_1 \, d\alpha_2$$

for $n \geq 1$. This new integrand has various symmetries shown in Figure 4.8. These symmetries imply that if we change variables to $\theta_1 := \pi(\alpha_1 + \alpha_2)$ and $\theta_2 := \pi(\alpha_1 - \alpha_2)$, then

$$H(n, n) = \frac{1}{\pi^2} \int_0^\pi \int_0^\pi \frac{1 - \cos 2n\theta_1}{1 - \cos \theta_1 \cos \theta_2} \, d\theta_1 \, d\theta_2 \, . \tag{4.18}$$

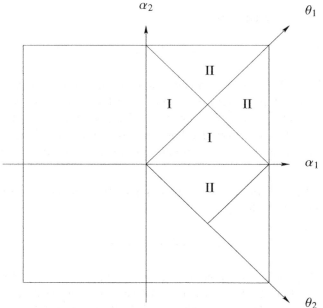

Figure 4.8. The integrals over the regions labeled I
are all equal, as are those labeled II.

Only a little bit of calculus remains before we have our answer. Now for $0 < a < 1$, we have

$$\int_0^\pi \frac{d\theta}{1 - a\cos\theta} = \frac{2}{\sqrt{1 - a^2}} \tan^{-1} \frac{\sqrt{1 - a^2}\,\tan(\theta/2)}{1 - a} \Bigg|_0^\pi = \frac{\pi}{\sqrt{1 - a^2}}.$$

Therefore integration on θ_2 in (4.18) gives

$$H(n, n) = \frac{1}{\pi} \int_0^\pi \frac{1 - \cos 2n\theta_1}{\sin\theta_1}\, d\theta_1.$$

Note that $(1 - \cos 2n\theta_1)/\sin\theta_1 = 2\sum_{k=1}^n \sin(2k - 1)\theta_1$, as can be seen by using complex exponentials. Therefore

$$H(n, n) = \frac{2}{\pi} \int_0^\pi \sum_{k=1}^n \sin(2k - 1)\theta_1\, d\theta_1 = \frac{4}{\pi} \sum_{k=1}^n \frac{1}{2k - 1}. \tag{4.19}$$

▷ **Exercise 4.10.**
Deduce the distribution of the degree of a vertex in the uniform spanning tree of $\mathbb{Z}^2$, that is, the table (4.15).

We may also make use of the preceding work, without the actual values being needed, to prove the following remarkable fact. Edges of the uniform spanning tree in $\mathbb{Z}^2$ along diagonals are like fair coin flips! We have seen in (4.14) that each edge has 50 percent chance to be in the tree. The independence we are now asserting is the following theorem.

Theorem 4.8. (Independence on Diagonals) *Let e be any edge of $\mathbb{Z}^2$. For $n \in \mathbb{Z}$, let X_n be the indicator that $e + (n, n)$ lies in the spanning tree. Then X_n are i.i.d. Likewise for $e + (n, -n)$.*

Proof. By symmetry, it suffices to prove the first part. We may also assume that e is the edge from the origin to $(1, 0)$. By the transfer-current theorem, it suffices to show that $Y(e, e + (n, n)) = 0$ for all $n \neq 0$. The formula (4.17) shows that

$$4Y(e, e + (n, n)) = H(n + 1, n) - H(n, n) - H(n, n) + H(n - 1, n).$$

Now the symmetries we have noted already of the function $H(\cdot, \cdot)$ show that $H(n, n + 1) = H(n + 1, n)$ and $H(n, n - 1) = H(n - 1, n)$. Since $H(n, n)$ is the average of these four numbers for $n \neq 0$, it follows that $H(n + 1, n) - H(n, n) = H(n, n) - H(n - 1, n)$. This proves the result. ◀

4.4 Notes

There is another important connection of spanning trees to Markov chains:

The Markov Chain Tree Theorem. *The stationary distribution of a finite-state, irreducible Markov chain is proportional to the measure that assigns the state x the measure*

$$\sum_{\text{root}(T)=x} \Psi(T).$$

It is for this reason that generating spanning trees at random is very closely tied to generating a state of a Markov chain at random according to its stationary distribution. This latter topic is especially interesting in computer science. See Propp and Wilson (1998) for more details. Some of the history of the Markov chain tree theorem can be found in Anantharam and Tsoucas (1989).

To prove the Markov chain tree theorem, we associate to the original Markov chain a new Markov chain on spanning trees. Given a spanning tree T and an edge e with $e^- = \text{root}(T)$, define two new spanning trees:

forward procedure: This creates a new spanning tree denoted $F(T, e)$. First, add e to T. This creates a cycle. Delete the edge $f \in T$ out of e^+ that breaks the cycle. See Figure 4.9.

backward procedure: This creates a new spanning tree denoted $B(T, e)$. Again, first add e to T. This creates a cycle. Break it by removing the appropriate edge $g \in T$ that leads into e^-. See Figure 4.9.

Note that in both procedures, it is possible that $f = -e$ or $g = -e$. Also, note that

$$B(F(T, e), f) = F(B(T, e), g) = T,$$

where f and g are as specified in the definitions of the forward and backward procedures.

Now define transition probabilities on the set of spanning trees by

$$p(T, F(T, e)) := p(e) = p(\text{root}(T), \text{root}(F(T, e))). \tag{4.20}$$

Thus $p(T, \widetilde{T}) > 0 \iff \exists e \ \widetilde{T} = F(T, e) \iff \exists g \ T = B(\widetilde{T}, g)$.

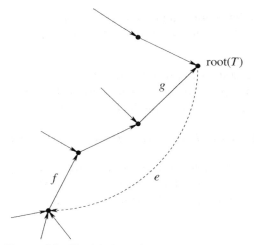

Figure 4.9. The Markov chain on spanning trees.

▷ **Exercise 4.11.**
Prove that the Markov chain on trees given by (4.20) is irreducible.

▷ **Exercise 4.12.**
   (a) Show that the weight $\Psi(\,\cdot\,)$ is a stationary measure for the Markov chain on trees given by (4.20).
   (b) Prove the Markov chain tree theorem.

Another way to express the relationship between the original Markov chain and this associated one on trees is as follows. Recall that one can create a stationary Markov chain $\langle X_n \rangle_{-\infty}^{\infty}$ indexed by $\mathbb{Z}$ with the original transition probabilities $p(\cdot, \cdot)$ by, say, Kolmogorov's existence theorem. Set

$$L_n(w) := \max\{m < n \,;\ X_m = w\}\,.$$

This is well defined a.s. by recurrence. Let Y_n be the tree formed by the edges

$$\{\langle w, X_{L_n(w)+1}\rangle \,;\ w \in \mathsf{V} \setminus \{X_n\}\}\,.$$

Then Y_n is rooted at X_n and $\langle Y_n \rangle$ is a stationary Markov chain with the transition probabilities (4.20).
   A method due to Aldous (1990) and Broder (1989) of generating weighted uniform spanning trees comes from reversing these Markov chains; related ideas were in the air at that time, and both these authors thank Persi Diaconis for discussions. Let $\langle X_n \rangle_{-\infty}^{\infty}$ be a stationary Markov chain on a finite state space. Then so is the ***reversed*** process $\langle X_{-n} \rangle$: the definition of the Markov property via independence of the past and the future given the present shows this immediately. We can also find the transition probabilities $\widehat{p}$ for the reversed chain: Let π be the stationary probability $\pi(a) := \mathbf{P}[X_0 = a]$. Then clearly π is also the stationary probability for the reversed chain. Comparing the chance of seeing state a followed by state b for the forward chain with the equal probability of seeing state b followed by state a for the reversed chain, we see that

$$\pi(a)p(a, b) = \pi(b)\widehat{p}(b, a)\,,$$

whence

$$\widehat{p}(b, a) = \frac{\pi(a)}{\pi(b)} p(a, b),$$

just as in Exercise 2.1.

If we reverse these chains $\langle X_n \rangle$ and $\langle Y_n \rangle$, then we find that Y_{-n} can be expressed in terms of X_{-n} as follows: Let

$$H_n(w) := \min\{m > n;\ X_{-m} = w\}.$$

Then Y_{-n} has edges $\{\langle w, X_{-[H_n(w)-1]} \rangle;\ w \in V \setminus \{X_{-n}\}\}$.

▷ **Exercise 4.13.**

Prove that the transition probabilities of Y_{-n} are

$$\widehat{p}(T, B(T, e)) = p(e).$$

Since the stationary measure of $\langle Y_n \rangle$ is still proportional to $\Psi(\cdot)$, we get the following algorithm for generating a random spanning tree with distribution proportional to $\Psi(\cdot)$: Find the stationary probability π of the chain giving rise to the weights $\Psi(\cdot)$, run the reversed chain starting at a state chosen according to π, and construct the tree via H_0. That is, we draw an edge from u to w the first time ≥ 1 that the reversed chain hits u, where w is the state preceding the visit to u.

In case the chain is reversible, this construction simplifies. From the discussion in Section 4.1, we have the following:

Corollary 4.9. (Aldous–Broder Algorithm) *Let $\langle X_n \rangle_0^\infty$ be a random walk on a finite, connected network G with X_0 arbitrary (not necessarily random). Let $H(u) := \min\{m > 0;\ X_m = u\}$ and let T be the unrooted tree with edges $\{(u, X_{H(u)-1});\ X_0 \neq u \in G\}$. Then the distribution of T is proportional to $\Xi(\cdot)$. In particular, simple random walk on a finite, connected graph gives a uniform unrooted random spanning tree.*

Proof. We have seen that if X_0 has the stationary distribution, then as a rooted spanning tree with edges oriented toward the root, T has probability proportional to $\Psi(T)$. We also know that as an unoriented, unrooted tree, the conditional probability of T is proportional to $\Xi(T)$ given the root. Hence the same holds when X_0 is fixed, as desired. ◀

This method of generating uniform spanning trees can be and was used in place of Wilson's method for the purposes of this chapter. However, Wilson's method is much better suited to the study of uniform spanning forests, the topic of Chapter 10.

▷ **Exercise 4.14.**

Let G be a cycle and $x \in V$. Start simple random walk at x and stop when all edges but one have been traversed at least once. Show that the edge that has not been traversed is equally likely to be any edge.

▷ **Exercise 4.15.**

Suppose that the graph G has a Hamiltonian path, $\langle x_k;\ 1 \leq k \leq n \rangle$, that is, a simple path that is a spanning tree. Let q_k be $\mathbf{P}_{x_k}[\tau_{x_k}^+ > \tau_{x_{k+1},\ldots,x_n}]$ for simple random walk on G. Show that the number of spanning trees of G equals $\prod_{k<n} q_k \deg_G x_k$.

The first use of Wilson's method for infinite, recurrent networks was by Benjamini, Lyons, Peres, and Schramm (2001), hereinafter referred to as BLPS (2001). The transfer-current theorem was shown for the case of two edges by Brooks, Smith, Stone, and Tutte (1940). The proof here is due to BLPS (2001). The interest of Brooks, Smith, Stone, and Tutte (1940) was elicited by their discovery of the connection of electrical networks to square tilings: see Section 9.6. An extension of the transfer-current theorem to nonreversible Markov chains is given by Avena and Gaudillière (2018) in terms of continuous-time random walks.

We are grateful to David Wilson for permission to include Figure 4.7. It was created using the linear algebraic techniques of Wilson (1997) for generating domino tilings; the needed matrix inversion was accomplished using the formulas of Kenyon (1997). The resulting tiling gives dual spanning trees by the bijection of Temperley: see the problem section, pp. 202–204 of McDonough and Mavron (1974). One of the trees is Figure 4.7. (This bijection was extended to the infinite setting by Burton and Pemantle (1993) and to directed and weighted graphs by Kenyon, Propp, and Wilson (2000).) As mentioned, we will see in Section 10.6 that there is a unique path to infinity in the spanning tree a.s. The domino tiling also gives the orientation of each edge toward infinity; this is shown in Figure 4.10.

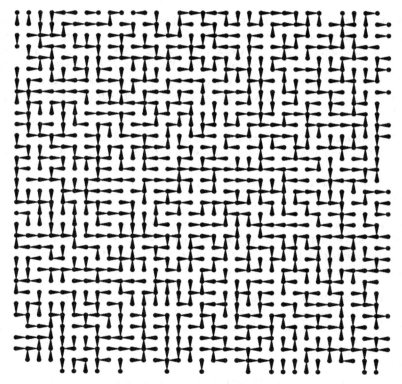

Figure 4.10.   A portion of a uniformly chosen spanning tree on $\mathbb{Z}^2$ showing the orientation toward infinity.

The function $H(u)$ from (4.16) is equal to the ***potential kernel*** of simple random walk on $\mathbb{Z}^2$, that is, $\lim_{n\to\infty} \sum_{k=0}^{n} [p_k(\mathbf{0}, \mathbf{0}) - p_k(\mathbf{0}, u)]$; see Exercise 4.49 or Section 12 of Spitzer (1976).

Theorem 4.8 is due to R. Lyons and is published here for the first time. In 2002, Ciucu and Lyons observed that this independence result and even stronger ones can be proved by Lemma 1.1 or its proof in Ciucu (1997). Namely, let $A \subset \mathbb{Z}$. As alluded to in Section 4.3, Theorem 10.36 shows that a.s., from each vertex, there begins only one infinite, simple path contained in the uniform spanning tree of $\mathbb{Z}^2$. Thus, we may orient each edge in the tree toward that "end". One extension of Theorem 4.8 is that, in the notation of that theorem, given which of the edges $e + (n, n)$ belong to the tree for $n \in A$ and given their orientations, the random variables X_n for $n \notin A$ are i.i.d. Oriented edges are not independent; for example, the chance that e and $e + (1, 1)$ both belong to the spanning tree and point in the same direction (that is, both toward or both away from the end) is $1/8 + 2(1/\pi - 1/4)^2$, whereas the chance they belong and point in opposite directions is $1/8 - 2(1/\pi - 1/4)^2$. (For this calculation, see Fig. 5 of Kenyon (1997).) Another result states that if Y_n is the indicator that the path from (n, n) to infinity in the uniform spanning tree has its first step either to $(n + 1, n)$ or to $(n, n - 1)$, then Y_n are i.i.d. More information on the path from $(0, 0)$ to infinity in the uniform spanning tree is provided by Kenyon and Wilson (2015). For example, the expected number of the four $\mathbb{Z}^2$-neighbors of $(0, 0)$ that lie on that path is $5/4$, as also shown by Poghosyan, Priezzhev, and Ruelle (2011).

Another way to state Theorem 4.8 is that if independent, fair coin flips are used to decide which of the edges $\{e + (n, n) ; n \in \mathbb{Z}\}$ will be present, for some fixed edge e, then there exists a percolation on the remaining edges of $\mathbb{Z}^2$ that will lead in the end to a percolation on all of $\mathbb{Z}^2$ with the distribution of the uniform spanning tree. A related surprising result of Lyons and Steif (2003) says that we can independently determine some of the horizontal edges and then decide the remaining edges to get a uniform spanning tree. To be precise, fix a horizontal edge, e. Suppose that $\langle U(e + x) ; x \in \mathbb{Z}^2 \rangle$ are i.i.d. uniform $[0, 1]$ random variables. Let $K_0 := \{e + x ; U(e + x) \leq e^{-4G/\pi}\}$ and $K_1 := \{e + x ; U(e + x) \geq 1 - e^{-4G/\pi}\}$, where

$$G := \sum_{k=0}^{\infty} \frac{(-1)^k}{(2k + 1)^2}$$

is *Catalan's constant*. (We have that $e^{-4G/\pi} = 0.3115^+$.) Then there exists a percolation ω on $\mathbb{Z}^2$ such that $(\omega \cup K_1) \setminus K_0$ has the distribution of the uniform spanning tree.

Additional information on loop-erased random walk and another proof of Wilson's algorithm can be found in Marchal (2000).

Enumeration of spanning trees in graphs is an old topic. There are many proofs of Cayley's formula, Corollary 4.5, which was proved earlier by Borchardt (1860). The shortest proof is due to Joyal (1981) and goes as follows. Denote $[n] := \{1, \ldots, n\}$. First note that because every permutation can be represented as a product of disjoint directed cycles, it follows that for any finite set S, the number of sets of cycles of elements of S (each element appearing exactly once in some cycle) is equal to the number of linear arrangements of the elements of S. The number of functions from $[n]$ to $[n]$ is clearly n^n. To each such function f, we may associate its functional digraph, which has a directed edge from i to $f(i)$ for each i in $[n]$. Every weakly connected component of the functional digraph can be represented by a cycle of rooted trees. So n^n is also the number of linear arrangements of rooted trees on $[n]$. We claim now that $n^n = n^2 t_n$, where t_n is the number of trees on $[n]$.

It is clear that $n^2 t_n$ is the number of triples (x, y, T), where $x, y \in [n]$ and T is a tree on $[n]$. Given such a triple, we obtain a linear arrangement of rooted trees by removing all directed edges on the unique path from x to y and taking the nodes on this path to be the roots of the trees that remain. This correspondence is bijective, and thus $t_n = n^{n-2}$. ◀

The matrix-tree theorem of graph theory gives a way to calculate the number of spanning trees on any graph. Namely, it says that the number of spanning trees of a graph G equals $\det \Delta_G[x]$ for each $x \in \mathsf{V}$, where Δ_G is the graph Laplacian defined in Exercise 2.62 and $[x]$ indicates striking the row and column indexed by x. More generally, the sum $\Xi(G)$ of the weights $\Xi(T) = \prod_{e \in T} c(e)$ over spanning trees T in a simple network equals $\det \Delta_G[x]$. There is also a version for directed graphs. A proof that uses techniques from this book is given in the following exercise.

▷ **Exercise 4.16.**
Prove the matrix-tree theorem by using Kirchhoff's effective-resistance formula and Exercise 2.62. *Hint:* Pick a spanning tree t and successively contract the edges to calculate $\Xi(t)/\Xi(G)$.

▷ **Exercise 4.17.**
Show that the constant C in (2.26) equals $\sqrt{\Xi(G)/(2\pi)^{|V|-1}}$.

Asymptotics of the number of spanning trees are connected to mathematical physics. For example, if one combines the entropy result for domino tilings proved by Montroll (1964) with the Temperley (1974) bijection, then one gets that

$$\lim_{n \to \infty} \frac{1}{n^2} \log\left(\text{number of spanning trees of } [1,n]^2 \text{ in } \mathbb{Z}^2\right)$$

$$= \int_0^1 \int_0^1 \log(4 - 2\cos 2\pi x - 2\cos 2\pi y)\, dx\, dy$$

$$= \frac{4G}{\pi} = 1.166^+,$$

where G is Catalan's constant, as earlier (see Kasteleyn (1961) or Montroll (1964) for the evaluation of the integral). This result first appeared explicitly in Burton and Pemantle (1993). Thus, $e^{1.166^+} = 3.21^-$ can be thought of as the average number of independent choices per vertex to make a spanning tree of $\mathbb{Z}^2$. See, for example, Burton and Pemantle (1993) and Shrock and Wu (2000) and the references therein for this and several other such examples. Very general methods of calculating and comparing asymptotics were given by Lyons (2005, 2010), where a key tool is the rate of convergence of random walks on finite graphs to their stationary distribution. Much more precise asymptotics for the number of spanning trees in portions of the square lattice graph inside a region were obtained by Kenyon (2000a); the results depend on the area as well as the perimeter of the enclosing region. The simplest case was done by Duplantier and David (1988), whose equation (4.23) shows that for an $m \times n$ rectangular grid in the square lattice graph $\mathbb{Z}^2$, the logarithm of the number of spanning trees is

$$\frac{4Gmn}{\pi} - (m+n)\log(\sqrt{2}+1) - \frac{1}{2}\log m + \frac{5}{4}\log 2 + \log \eta(ni/m) + O\left(\frac{1}{mn}\right),$$

where $\eta(z) := e^{2\pi i z/24} \prod_{k=1}^{\infty}\left(1 - e^{2k\pi i z}\right)$ is Dedekind's eta function for $\Im z > 0$. For example, $\eta(i) = \Gamma(1/4)/(2\pi^{3/4})$.

▷ **Exercise 4.18.**
Consider simple random walk on $\mathbb{Z}^2$. Let $A := \{(x,y) \in \mathbb{Z}^2 \;;\; y < 0 \text{ or } (y = 0 \text{ and } x < 0)\}$. Show that $\mathbf{P}_{(0,0)}[\tau_{(0,0)}^+ > \tau_A] = e^{4G/\pi}/4$, where G is Catalan's constant.

One may consider the uniform spanning tree on $\mathbb{Z}^2$ embedded in $\mathbb{R}^2$. In fact, consider it on $\epsilon\mathbb{Z}^2$ in $\mathbb{R}^2$ and let $\epsilon \to 0$. In appropriate senses, one can describe the limit and show that it has a conformal invariance property. Partial results for this were first proved by Kenyon (2000b); for example, he calculated the limiting distribution of the "meeting point" of the subtree determined by three vertices on the boundary of a domain. The full result was proved by Lawler, Schramm, and Werner (2004a). The stochastic Loewner evolution*, SLE, introduced by Schramm (2000) initially for this very purpose, plays the central role. For the uniform spanning tree, there are two ways SLE enters the analysis: One is

---

* After Schramm's death, this became known as the "Schramm–Loewner evolution".

the scaling limit of loop-erased random walk, which is the path between two vertices and fundamental to this chapter. The second is less obvious. If we draw a curve around the spanning tree in a bounded region, as in Figure 4.11, we obtain a cycle in another graph. That cycle visits every vertex and is very reminiscent of Peano's space-filling curve. It is called the UST Peano curve, and it, too, has a scaling limit described by SLE. Figure 4.12 shows the curve from a uniform spanning tree in a 99×99 square grid, where the hue represents progress along the curve. SLE is also central to the study of scaling limits of other planar processes, including percolation.

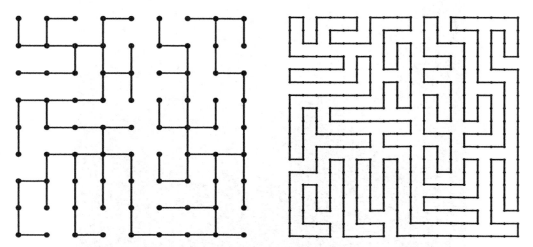

Figure 4.11. A uniform spanning tree in a 9×9 grid on the left, with its surrounding Peano-like curve in an 18×18 grid on the right.

Lawler (1980) introduced loop-erased random walk originally as a model that was similar to self-avoiding walk but easier to understand because of its Markov property. Many basic aspects of self-avoiding walk remain beyond proof, despite precise conjectures from physics. For example, Nienhuis (1982) conjectured that for all two-dimensional lattices, the number of self-avoiding paths of length n starting at the origin is asymptotic to $A\mu^n n^{\gamma-1}$, where $\gamma = 43/32$, and that the average squared distance between the origin and the other endpoint of such a path is asymptotic to $Dn^{2\nu}$, where $\nu = 3/4$; the other constants, A, μ, and D, depend on the lattice. If the scaling limit of self-avoiding walk in the plane exists and is conformally invariant, as is believed to be the case, then Lawler, Schramm, and Werner (2004b) proved that it must be $\text{SLE}_{8/3}$. This would in turn likely imply the preceding values of γ and ν. For lattices in dimensions 5 and higher, corresponding statements have been proved by Hara and Slade (1990, 1992), with $\gamma = 1$ and $\nu = 1/2$. See Slade (2011) for a survey of current knowledge of self-avoiding walk.

It turns out that uniform spanning trees arise as a limiting case within a wide class of probability measures on graphs. To say what this is, we first define ***Bernoulli percolation*** with parameter $p \in [0, 1]$ as the product measure $\mathbf{P}_p$ on subsets of edges where each edge is retained with probability p. A more general two-parameter model of random subgraphs, known as the random-cluster model, was introduced by Fortuin and Kasteleyn (1972) and Fortuin (1972a, 1972b). The two parameters of random-cluster measures are $p \in [0, 1]$ and $q > 0$. Given a finite, connected graph G and $\omega \in 2^E$, write $\|\omega\|$ for the number of components of ω. The ***random-cluster measure*** with parameters (p, q) on G, denoted $\text{FRC}(p, q) = \text{FRC}^G(p, q)$, is the probability measure on E proportional to $q^{\|\omega\|}\mathbf{P}_p(\omega)$, that is, the Bernoulli(p) percolation measure $\mathbf{P}_p$ biased by $q^{\|\omega\|}$ (and renormalized). Thus, when $q = 1$, this is merely $\mathbf{P}_p$. The limit $\text{FRC}(p, 0)$ of $\text{FRC}(p, q)$ as $q \to 0$ exists and is concentrated on connected

color
plate **Figure 4.12.**   The Peano-like curve surrounding a uniform spanning
tree on a 99×99 grid, with hue showing progress along the curve.

subgraphs of G. For example, $\mathrm{FRC}(1/2, 0)$ is the uniform random connected subgraph. The limit $\lim_{p \to 0} \mathrm{FRC}(p, 0)$ is the uniform spanning tree. The limit $\lim_{p \to 0} \mathrm{FRC}(p, p)$ is the uniform forest.

On infinite graphs G, there are several ways to define random-cluster measures. We restrict ourselves to $q \geq 1$, since the measures with $q < 1$ behave rather differently and are poorly understood. Indeed, it is a major open problem to understand the case $q < 1$; for example, it is unknown whether they have negative associations, as they do in the limiting case of uniform spanning trees. This is unknown even for the special cases of the uniform random connected subgraph and the uniform forest. The advantage of $q \geq 1$ is that then the measures have *positive* associations (defined in Section 5.8). Let $\langle G_n \rangle$ be an exhaustion of an infinite, connected graph G by connected, finite subgraphs. Define $\mathrm{FRC}^G(p, q)$ to be the weak* limit of $\mathrm{FRC}^{G_n}(p, q)$; this is called the *free random-cluster measure* on G. Define the *wired random-cluster measure* $\mathrm{WRC}^G(p, q)$ to be the weak* limit of $\mathrm{FRC}^{G_n^*}(p, q)$. These limits always exist (see, for example, Aizenman, Chayes, Chayes, and Newman (1988)). Furthermore, they have positive associations, and so the free random-cluster measure is stochastically dominated by the wired random-cluster measure (Aizenman, Chayes, Chayes, and Newman, 1988). When q is an integer, the random-cluster measure can be used to construct the Potts model; when $q = 2$, the Potts model is called the Ising model. See Grimmett (2006) for more details on random-cluster measures, especially on $\mathbb{Z}^d$. In the plane, scaling limits of Ising measures are known to exist and to have conformal invariance properties: see Smirnov (2010).

The transfer-current theorem shows that weighted uniform spanning tree measures have their marginals given by simple determinants. This property leads to what are called determinantal probability measures; see Lyons (2003) for their properties. The degree distribution at a vertex, exhibited in (4.15) for the uniform spanning tree on $\mathbb{Z}^2$, is the mixed binomial distribution with success probabilities equal to the eigenvalues of the submatrix of the transfer-current matrix given by the edges incident to the vertex in question. In fact, for the weighted uniform spanning tree T on any graph G and every $F \subset E(G)$, the distribution of $|T \cap F|$ is the mixed binomial distribution corresponding to the eigenvalues of $[Y(e, f)]_{e, f \in F}$, as shown by Bapat (1992). A similar property holds for all determinantal probability measures: see Lemma 3.4 of Shirai and Takahashi (2003a), (2.38) of Shirai and Takahashi (2003b), or Hough, Krishnapur, Peres, and Virág (2009). In addition, the negative association property holds for all determinantal probability measures.

A property that is even stronger than negative association, and which again holds for all determinantal probability measures, is called **strongly Rayleigh**. Namely, if $\mathbf{P}$ is a probability measure on subsets of a finite set E, then define the multivariate complex polynomial $f(\mathbf{z}) := \sum_{F \subseteq E} \mathbf{P}(F) \prod_{e \in F} z_e$, where $\mathbf{z} := \langle z_e \; ; \; e \in E \rangle$. Borcea, Brändén, and Liggett (2009) call $\mathbf{P}$ strongly Rayleigh if $f(\mathbf{z}) \neq 0$ when all z_e have strictly positive imaginary parts. Borcea, Brändén, and Liggett (2009) show that the strongly Rayleigh property is preserved under symmetrization and under symmetric exclusion processes. The strongly Rayleigh property is also preserved under conditioning on events such as $|F \cap A| = k$, where $A \subseteq E$ and k are fixed, as long as we restrict attention to A: see Lemma 4.16 of Borcea, Brändén, and Liggett (2009). Thus, if we consider the uniform spanning tree T on a graph G and condition on $|T \cap A| = k$ for some fixed $A \subset E(G)$, then $T \cap A$ has negative associations.

Another type of negative correlation is as follows. It was conjectured to hold by BLPS (2001) and is still open. We say that $\mathscr{A}, \mathscr{B} \subset 2^E$ **occur disjointly** for $F \subset E$ if there are disjoint sets $F_1, F_2 \subset E$ such that $F' \in \mathscr{A}$ for every F' with $F' \cap F_1 = F \cap F_1$ and $F' \in \mathscr{B}$ for every F' with $F' \cap F_2 = F \cap F_2$. For example, $\mathscr{A}$ may be the event that x and y are joined by a path of length at most 5, while $\mathscr{B}$ may be the event that z and w are joined by a path of length at most 6. If there are disjoint paths of lengths at most 5 and 6 joining the first and second pair of vertices, respectively, then $\mathscr{A}$ and $\mathscr{B}$ occur disjointly.

Conjecture 4.10. *Let $\mathscr{A}, \mathscr{B} \subset 2^E$ be increasing. Then the probability that $\mathscr{A}$ and $\mathscr{B}$ occur disjointly for the weighted uniform spanning tree T is at most $\mathbf{P}[T \in \mathscr{A}] \, \mathbf{P}[T \in \mathscr{B}]$.*

The BK inequality of van den Berg and Kesten (1985) says that this inequality holds when T is a random subset of E chosen according to any product measure on 2^E; it was extended by Reimer (2000) to allow $\mathscr{A}$ and $\mathscr{B}$ to be any events, confirming a conjecture of van den Berg and Kesten (1985). However, we cannot allow arbitrary events for uniform spanning trees: consider the case where $\mathscr{A} := \{e \in T\}$ and $\mathscr{B} := \{f \notin T\}$, where $e \neq f$.

There is a basic connection of uniform spanning trees to the sandpile model on finite graphs; see Holroyd, Levine, Mészáros, Peres, Propp, and Wilson (2008) for a survey. See also Kassel and Wilson (2016) for more, such as a calculation that the probability that $(1, 0)$ lies on the path from $(0, 0)$ to ∞ equals $5/16$.

4.5 Collected In-Text Exercises

4.1. Suppose that Z is a set of states in a Markov chain and that x_0 is a state not in Z. Assume that when the Markov chain is started in x_0, then it visits Z with probability 1. Define the random path $Y_0, Y_1, \ldots$ by $Y_0 := x_0$ and then recursively by letting Y_{n+1} have the distribution of one step of the Markov chain starting from Y_n given that the chain will visit Z before visiting any of $Y_0, Y_1, \ldots, Y_n$ again. However, if $Y_n \in Z$, then the path is stopped and Y_{n+1} is not defined. Show that $\langle Y_n \rangle$ has the same distribution as loop-erasing a sample of the Markov chain started from x_0 and stopped when it reaches Z. In the case of a random walk, the conditioned path $\langle Y_n \rangle$ is called the **Laplacian random walk** from x_0 to Z.

4.2. Consider the ladder graph L_n of height n shown in Figure 1.7. Choose a spanning tree $T(n)$ of L_n uniformly. Use Kirchhoff's effective-resistance formula to determine the chance $\mathbf{P}[\text{rung 1 is in } T(n)]$ and its limiting behavior as $n \to \infty$.

4.3. By using Kirchhoff's effective-resistance formula, show that if $e \neq f$, then the events $e \in T$ and $f \in T$ are negatively correlated.

4.4. Write out the rest of the proof that for any event $\mathscr{B}$ depending on only finitely many edges, $\left|\mathbf{P}[T_{G'} \in \mathscr{B}] - \mathbf{P}[T_G \in \mathscr{B}]\right|$ is arbitrarily small for sufficiently large G'.

4.5. Suppose that $\mathscr{A}$ is an increasing event on a graph G and $e \in \mathsf{E}$. Note that $\mathsf{E}(G/e) = \mathsf{E}(G\backslash e) = \mathsf{E}(G) \setminus \{e\}$. Define $\mathscr{A}/e := \{F \subseteq \mathsf{E}(G/e) ;\ F \cup \{e\} \in \mathscr{A}\}$ and $\mathscr{A}\backslash e := \{F \subseteq \mathsf{E}(G\backslash e) ;\ F \in \mathscr{A}\}$. Show that these are increasing events on G/e and $G\backslash e$, respectively.

4.6. (Negative Association) Let G be a finite network. Extend Theorem 4.6 to show that if $\mathscr{A}$ and $\mathscr{B}$ are both increasing events and they depend on disjoint sets of edges, then they are negatively correlated. Still more generally, show the following. Say that a random variable $X : 2^{\mathsf{E}(G)} \to \mathbb{R}$ *depends on a set* $F \subseteq \mathsf{E}(G)$ if X is measurable with respect to the σ-field consisting of events that depend on F. Say also that X is *increasing* if $X(H) \leq X(H')$ whenever $H \subset H'$. If X and Y are increasing random variables that depend on disjoint sets of edges, then $\mathbf{E}[XY] \leq \mathbf{E}[X]\mathbf{E}[Y]$. This property of $\mathbf{P}$ is called *negative association*.

4.7. Let G be an edge-amenable, infinite graph as witnessed by the sequence $\langle \mathsf{V}_n \rangle$. Show that the average degree of vertices in any spanning tree of G is 2. That is, if $\deg_T(x)$ denotes the degree of x in a spanning tree T of G, then

$$\lim_{n \to \infty} |\mathsf{V}_n|^{-1} \sum_{x \in \mathsf{V}_n} \deg_T(x) = 2 \,.$$

Every infinite, recurrent graph can be shown to be edge-amenable by various results from Chapter 6 that we'll look at later, such as Theorems 6.5, 6.7, or 6.42. (The origin of the term "amenable" will be discussed in Section 6.1.) Deduce that for the uniform spanning tree measure on a recurrent graph,

$$\lim_{n \to \infty} |\mathsf{V}_n|^{-1} \sum_{x \in \mathsf{V}_n} \mathbf{E}[\deg_T(x)] = 2 \,.$$

In particular, if G is also transitive, such as $\mathbb{Z}^2$, meaning that for every pair of vertices x and y, there is a bijection of V with itself that preserves adjacency and takes x to y, then every vertex has expected degree 2.

4.8. Show that $(\widehat{\mathbf{1}_{\{x\}}} - \widehat{\mathbf{1}_{\{y\}}})/\varphi \in L^1(\mathbb{T}^2)$ for all $x, y \in \mathbb{Z}^2$.

4.9. Show that if $F \in L^1(\mathbb{T}^2)$ and $f(x) = \int_{\mathbb{T}^2} F(\alpha)e^{2\pi i x \cdot \alpha}\, d\alpha$, then

$$(\Delta f)(x) = \int_{\mathbb{T}^2} F(\alpha)e^{2\pi i x \cdot \alpha} \varphi(\alpha)\, d\alpha \,.$$

4.10. Deduce the distribution of the degree of a vertex in the uniform spanning tree of $\mathbb{Z}^2$, that is, the table (4.15).

4.11. Prove that the Markov chain on trees given by (4.20) is irreducible.

4.12. (a) Show that the weight $\Psi(\cdot)$ is a stationary measure for the Markov chain on trees given by (4.20).

(b) Prove the Markov chain tree theorem.

4.13. Let $\langle Y_n \rangle$ be a stationary Markov chain with transition probabilities (4.20) and consider its reversal. Prove that the transition probabilities of Y_{-n} are

$$\widehat{p}(T, B(T, e)) = p(e).$$

4.14. Let G be a cycle and $x \in \mathsf{V}$. Start simple random walk at x and stop when all edges but one have been traversed at least once. Show that the edge that has not been traversed is equally likely to be any edge.

4.15. Suppose that the graph G has a Hamiltonian path, $\langle x_k ; 1 \leq k \leq n \rangle$, that is, a simple path that is a spanning tree. Let q_k be $\mathbf{P}_{x_k}[\tau_{x_k}^+ > \tau_{x_{k+1}, \ldots, x_n}]$ for simple random walk on G. Show that the number of spanning trees of G equals $\prod_{k<n} q_k \deg_G x_k$.

4.16. Prove the matrix-tree theorem by using Kirchhoff's effective-resistance formula and Exercise 2.62. *Hint:* Pick a spanning tree t and successively contract the edges to calculate $\Xi(t)/\Xi(G)$.

4.17. Show that the constant C in (2.26) equals $\sqrt{\Xi(G)/(2\pi)^{|\mathsf{V}|-1}}$.

4.18. Consider simple random walk on $\mathbb{Z}^2$. Let $A := \{(x, y) \in \mathbb{Z}^2 ; y < 0 \text{ or } (y = 0 \text{ and } x < 0)\}$. Show that $\mathbf{P}_{(0,0)}[\tau_{(0,0)}^+ > \tau_A] = e^{4G/\pi}/4$, where G is Catalan's constant.

4.6 Additional Exercises

In all the exercises, assume the networks are connected.

4.19. Show that not every probability distribution on spanning trees of an undirected graph is proportional to a weight distribution, where the weight of a tree equals the product of the weights of its edges.

4.20. Given a probability measure $\mathbf{P}$ on spanning trees of a finite graph G, there is the vector of marginal edge probabilities, $\mu(e) := \mathbf{P}[e \in T]$ for $e \in \mathsf{E}$. The set of such vectors for all possible $\mathbf{P}$ forms a polytope, called the ***spanning tree polytope***. Show that this polytope consists precisely of those vectors μ that satisfy
 (i) $\mu(e) \geq 0$ for all $e \in \mathsf{E}(G)$,
 (ii) $\sum_{e \in \mathsf{E}(G)} \mu(e) = |\mathsf{V}(G)| - 1$, and
 (iii) $\sum_{e \in \mathsf{E}(G \upharpoonright K)} \mu(e) \leq |K| - 1$ for all $\emptyset \neq K \subsetneq \mathsf{V}(G)$.
Show in addition that if G has no ***cut-vertices*** (vertices whose removal disconnects G), then the relative interior of this polytope (that is, the interior as a subset of the affine span of the polytope) equals the set of such μ that satisfy strict inequality in every instance of (i) and (iii) with $|K| > 1$ in (iii).

4.21. Given a probability measure $\mathbf{P}$ on spanning trees of a finite graph G, there is the vector of marginal edge probabilities, $\mu_{\mathbf{P}}(e) := \mathbf{P}[e \in T]$ for $e \in \mathsf{E}$. The ***entropy*** of $\mathbf{P}$ is defined to be $H(\mathbf{P}) := -\sum_T \mathbf{P}(T) \log \mathbf{P}(T)$, where $0 \log 0 := 0$.
 (a) Show that if $\mathbf{P}$ is a weighted uniform spanning tree measure and Q is any probability measure on spanning trees with the same edge marginals $\mu_{\mathbf{P}} = \mu_Q$, then $H(\mathbf{P}) > H(Q)$ unless $Q = \mathbf{P}$.
 (b) Suppose that G has no cut-vertices. Show that if μ lies in the relative interior of the spanning tree polytope (see Exercise 4.20), then there is a unique weighted spanning tree measure whose edge marginal equals μ.

4.22. Let G be a finite or recurrent network and $a \neq z$ be two of its vertices. Let i be the unit current flow from a to z. Show that for every edge e, the probability that loop-erased random walk from a to z crosses e minus the probability that it crosses $-e$ equals $i(e)$.

4.23. Let G be a finite or recurrent network and $a \neq z$ be two of its vertices. Let i be the unit current flow from a to z. Let T be the uniform spanning tree in G and i_T be the associated unit current flow from a to z. Show that $i = \mathbf{E}[i_T]$.

4.24. Show that the following procedure also gives a.s. a random spanning tree rooted at r with distribution proportional to $\Psi(\cdot)$. Let $G_0 := \{r\}$. Given G_i, if G_i spans G, stop. Otherwise, choose any vertex $x \neq r$ that does not have an edge in G_i that leads out of x and add a (directed) edge from x picked according to the transition probability $p(x, \cdot)$ independently of the past. Add this edge to G_i and remove any cycle it creates to make G_{i+1}.

4.25. How efficient is Wilson's method? What takes time is to generate a random successor state of a given state. Call this a step of the algorithm. Show that the expected number of steps to generate a random spanning tree rooted at r for a finite-state, irreducible Markov chain is

$$\sum_{x \text{ a state}} \pi(x)\big(\mathbf{E}_x[\tau_r] + \mathbf{E}_r[\tau_x]\big),$$

where π is the stationary probability distribution for the Markov chain. Show that another expression for this expected time is the trace of $(I - P_r)^{-1}$, where I is the identity matrix and P_r is the transition matrix with the row and column corresponding to r deleted. In the case of a random walk on a network (V, E), this is

$$\sum_{e \in \mathsf{E}_{1/2}} c(e)\big(\mathscr{R}(e^- \leftrightarrow r) + \mathscr{R}(e^+ \leftrightarrow r)\big),$$

where edge e has conductance $c(e)$ and endpoints e^- and e^+, and $\mathscr{R}$ denotes effective resistance.

4.26. Let $\langle X_n \rangle$ be a transient Markov chain. Then its loop erasure $\langle Y_n \rangle$ is well defined a.s. Show that $\mathbf{P}_{x_0}[Y_1 = x_1] = p(x_0, x_1)\,\mathbf{P}_{x_1}[\tau_{x_0} = \infty]/\mathbf{P}_{x_0}[\tau_{x_0}^+ = \infty]$.

4.27. Suppose that x and y are two vertices in the complete graph K_n. Show that the probability that the distance between x and y is k in a uniform spanning tree of K_n is

$$\frac{k+1}{n^k}\prod_{i=1}^{k-1}(n-i-1).$$

4.28. Prove Cayley's formula another way as follows: Let t_{n-1} be a spanning tree of the complete graph on n vertices and $t_1 \subset t_2 \subset \cdots \subset t_{n-2} \subset t_{n-1}$ be subtrees such that t_i has i edges. Then

$$\mathbf{P}[T = t_{n-1}] = \mathbf{P}[t_1 \subseteq T] \cdot \prod_{i=1}^{n-2}\mathbf{P}[t_{i+1} \subseteq T \mid t_i \subseteq T].$$

Show that $\mathbf{P}[t_1 \subseteq T] = 2/n$ and

$$\mathbf{P}[t_{i+1} \subseteq T \mid t_i \subseteq T] = \frac{i+2}{n(i+1)}.$$

4.29. (Foster's Theorem) Exercise 2.65 showed (in slightly different notation) that if G has n vertices, then $\sum_{e \in \mathsf{E}_{1/2}} c(e)\mathscr{R}(e^- \leftrightarrow e^+) = n - 1$. Give another proof using spanning trees.

4.30. Kirchhoff (1847) generalized his effective-resistance formula in two ways. One of them is in Exercise 4.23. To express the other, let G be a finite network and $a \neq z \in G$ be two of its vertices. Denote the sum of $\Xi(T)$ over all spanning trees of G by $\Xi(G)$. Show that the effective conductance between a and z is given by

$$\mathscr{C}(a \leftrightarrow z) = \frac{\Xi(G)}{\Xi(G/\{a, z\})}, \tag{4.21}$$

where $G/\{a, z\}$ indicates the network G with a and z identified.

4.31. Let (G, c) be a finite network. Denote the sum of $\Xi(T)$ over all spanning trees of G by $\Xi(G)$. Show that $\mathbf{P}[e \in T] = d \log \Xi(G)/dc(e)$.

4.32. Jacobi's determinant identity says that for a square invertible matrix M with a block decomposition

$$M = \begin{bmatrix} A & B \\ C & D \end{bmatrix}$$

and corresponding block decomposition

$$M^{-1} = \begin{bmatrix} X & Y \\ Z & W \end{bmatrix},$$

where A and X are square and have the same size, we have

$$\det W = \frac{\det A}{\det M}.$$

Proof: equate determinants in

$$\begin{bmatrix} A & B \\ C & D \end{bmatrix} \begin{bmatrix} I & Y \\ 0 & W \end{bmatrix} = \begin{bmatrix} A & 0 \\ C & I \end{bmatrix}.$$

    (a) Use this and the matrix-tree theorem to give another proof of Kirchhoff's effective-resistance formula (other than the part about random walks).

    (b) Let G be a finite network and $H = G \restriction K$ be a connected subnetwork induced by $\varnothing \neq K \subsetneq \mathsf{V}(G)$. Let $v_a(x, y)$ be the voltage at x when a unit current flows from y to a (so that the voltage at a is 0) if $y \neq a$ and be 0 otherwise. Fix $o \in K$. Let T be a weighted uniform spanning tree of G and t be a fixed spanning tree of H. Use the matrix-tree theorem and Exercise 2.62 to show that

$$\mathbf{P}[T \restriction H = t] = \Xi(t) \det \left[v_o(x, y) \right]_{x, y \in K \setminus \{o\}}.$$

    (c) Show that $\det \left[v_o(x, y) \right]_{x, y \in K \setminus \{o\}}$ takes the same value no matter which vertex $o \in K$ is chosen.

4.33. Suppose that G is a graph with two sets of positive conductances, c and c', and no cut-vertices. Show that if, for every edge e, we have $c(e)\mathscr{R}(e^- \leftrightarrow e^+; c) = c'(e)\mathscr{R}(e^- \leftrightarrow e^+; c')$, then c/c' is constant.

4.34. Let (G, c) be a finite network. Recall from Exercise 2.67 that $(x, y) \mapsto \mathscr{R}(x \leftrightarrow y)$ is a metric on V.

    (a) Show that V with this effective-resistance metric can be embedded isometrically into some ℓ^1 space.

    (b) Show that if $f \colon \mathsf{V} \to \mathbb{Z}$ satisfies $\sum_{x \in \mathsf{V}} f(x) = 1$, then $\sum_{x, y \in \mathsf{V}} f(x)f(y)\mathscr{R}(x \leftrightarrow y) \leq 0$.

4.35. Let (G, c) be a finite network with $o \in \mathsf{V}(G)$. Let Z be the canonical Gaussian field on G, defined via independent normal random variables $X(e)$ with variance $r(e)$ for $e \in \mathsf{E}_{1/2}$. Let T be the uniform spanning tree on G and Z_T be the associated canonical Gaussian field, where the conductances c from G are used on T and the same $X(e)$ are used for $e \in \mathsf{E}(T)$. Show that $Z = \mathbf{E}[Z_T \mid X]$. Since Z_T is easily constructed via summing X along the edges of T starting from o, this identity shows how to construct Z in a probabilistic way.

4.36. Let G be a finite network. Let $a, z \in \mathsf{V}(G)$ and $e \in \mathsf{E}(G)$. Let T_G denote a uniform spanning tree of G.

(a) Show that if a and z are in the same component of $G \backslash e$, then

$$\mathscr{R}(a \leftrightarrow z; G) = \mathbf{P}[e \in T_G]\,\mathscr{R}(a \leftrightarrow z; G/e) + \mathbf{P}[e \notin T_G]\,\mathscr{R}(a \leftrightarrow z; G \backslash e),$$

and otherwise

$$\mathscr{R}(a \leftrightarrow z; G) = \mathscr{R}(a \leftrightarrow z; G/e) + r(e).$$

(b) Show that if a and z are not the endpoints of e, then

$$\mathscr{C}(a \leftrightarrow z; G) = \mathbf{P}[e \in T_{G/\{a,z\}}]\,\mathscr{C}(a \leftrightarrow z; G/e) + \mathbf{P}[e \notin T_{G/\{a,z\}}]\,\mathscr{C}(a \leftrightarrow z; G \backslash e),$$

and otherwise

$$\mathscr{C}(a \leftrightarrow z; G) = \mathscr{C}(a \leftrightarrow z; G \backslash e) + c(e).$$

4.37. Consider the doubly infinite ladder graph, the Cayley graph G of $\mathbb{Z} \times \mathbb{Z}_2$ with respect to its natural generators. Show that the uniform spanning tree T on G has the following description: The "rungs" $[(n, 0), (n, 1)]$ in T form a stationary renewal process with inter-rung distance being k with probability $2k(2 - \sqrt{3})^k$ ($k \geq 1$). Given two successive rungs in T, all the other edges between the rungs of the form $[(n, 0), (n + 1, 0)]$ and $[(n, 1), (n + 1, 1)]$ lie in T, with one exception chosen uniformly and independently for different pairs of successive rungs.

4.38. Let G be a finite network. Let e and f be two edges that are not parallel and such that $G \backslash f$ is still connected. Let $\widehat{i}^e$ be the unit current in G/f between the endpoints of e and let $\widetilde{i}^e$ be the unit current in $G \backslash f$ between the endpoints of e. Let

$$i_{\mathrm{c}}^e(g) := \begin{cases} \widehat{i}^e(g) & \text{if } g \neq f \\ 0 & \text{if } g = f, \end{cases}$$

and

$$i_{\mathrm{d}}^e(g) := \begin{cases} \widetilde{i}^e(g) & \text{if } g \neq f \\ 0 & \text{if } g = f. \end{cases}$$

(a) From (4.3), we have $i_{\mathrm{c}}^e = P_{\star f}^\perp i^e$. Show that

$$i^e = i_{\mathrm{c}}^e + \frac{Y(e, f)}{Y(f, f)} i^f.$$

(b) Show that $\chi^e - i_{\mathrm{d}}^e = P_{\chi^f - i^f}^\perp (\chi^e - i^e)$ and that

$$i^e = i_{\mathrm{d}}^e + \frac{Y(e, f)}{1 - Y(f, f)} (\chi^f - i^f).$$

(c) Show that

$$i^e = Y(f, f) i_{\mathrm{c}}^e + [1 - Y(f, f)] i_{\mathrm{d}}^e + Y(e, f) \chi^f.$$

4.39. Let G be a finite network and i be a current on G. If the conductance on the edge f is changed to $c'(f)$, then let i' be the current with the same sources and sinks as i, that is, so that $d^* i' = d^* i$. Show that

$$i = i' + \frac{[c(f) - c'(f)] i(f)}{c(f)[1 - Y(f, f)] + c'(f) Y(f, f)} (\chi^f - i^f),$$

where i^f is the unit current with the original conductances from f^- to f^+ as defined after (2.11), and deduce that

$$\frac{di}{dc(f)} = i(f) r(f) (\chi^f - i^f).$$

4.40. Let G be a finite, transitive graph of degree 3 and n vertices such that the automorphism group of G induces all six permutations on the three neighbors of any vertex. For example, G could be the 1-skeleton of the tetrahedron, the cube, or the dodecahedron. Show that the probability that the degree of a vertex in the uniform spanning tree on G is 1, 2, or 3 is, respectively, $(1 + 2/n)^2/4$, $(1 - 4/n^2)/2$, and $(1 - 2/n)^2/4$.

4.41. Given any numbers x_i $(i = 1, \ldots, k)$, let X be the diagonal matrix with entries $x_1, \ldots, x_k$. Show that $\det(Y_k + X) = \mathbf{E}\big[\prod_i (\mathbf{1}_{\{e_i \in T\}} + x_i)\big]$, where Y_k is as in (4.6). Deduce that

$$\mathbf{P}[e_1, \ldots, e_m \notin T, \, e_{m+1}, \ldots, e_k \in T] = \det Z_m \,,$$

where

$$Z_m(i, j) := \begin{cases} 1 - Y(e_i, e_j) & \text{if } j = i \leq m, \\ -Y(e_i, e_j) & \text{if } j \neq i \text{ and } i \leq m, \\ Y(e_i, e_j) & \text{if } i > m. \end{cases}$$

4.42. Give another proof of Cayley's formula (Corollary 4.5) by using the transfer-current theorem.

4.43. Consider the weighted uniform spanning tree measure on an infinite, recurrent network G. Let X and Y be increasing random variables with finite second moments that depend on disjoint sets of edges. Show that $\mathbf{E}[XY] \leq \mathbf{E}[X]\,\mathbf{E}[Y]$.

4.44. Let (G, c) be a finite network. Let e a fixed edge in G and $\mathscr{A}$ be an increasing event that ignores e. Suppose that a new network is formed from (G, c) by increasing the conductance on e while leaving unchanged all other conductances. Show that in the new network, the chance of $\mathscr{A}$ under the weighted spanning tree measure is no larger than it was in the original network.

4.45. Let E be a finite set and $k < |E|$. Let $\mathbf{P}$ be a weighted uniform measure on subsets of E of size k, that is, for some set of weights $w_e > 0$ $(e \in E)$, we have $\mathbf{P}(B) = \prod_{e \in B} w_e$ for $B \subseteq E$ of size k. Show that if X and Y are increasing random variables that depend on disjoint subsets of E, as defined in Exercise 4.6, then $\mathbf{E}[XY] \leq \mathbf{E}[X]\,\mathbf{E}[Y]$.

4.46. Given two probability measures μ_1 and μ_2 on $\mathbb{R}$, we say that μ_1 **stochastically dominates** μ_2 if, for all $r \in \mathbb{R}$, we have $\mu_1(r, \infty) \geq \mu_2(r, \infty)$.

(a) Show that μ_1 stochastic dominates μ_2 iff there exist random variables $X_1 \sim \mu_1$ and $X_2 \sim \mu_2$ on a common probability space such that $X_1 \geq X_2$ a.s.

(b) Let E be a finite set and $k < |E|$. Let X be a uniform random subset of E of size k. Show that if $\mathscr{A}$ is an increasing event that depends only on $F \subset E$, then the conditional distribution of $|X \cap F|$ given $\mathscr{A}$ stochastically dominates the unconditional distribution of $|X \cap F|$.

4.47. Let G be the hexagonal lattice. Show that the probability that the degree of a vertex in the uniform spanning tree on G is 1, 2, or 3 is, respectively, $1/4$, $1/2$, and $1/4$.

4.48. Let T be the uniform spanning tree in $\mathbb{Z}^2$. Let x and y be neighbors in $\mathbb{Z}^2$, and let L be the length of the path in T that joins x to y. Since $L < \infty$ a.s., we have $\lim_{n \to \infty} \mathbf{P}[L \geq n] = 0$. How quickly do these probabilities decay? This is hard to answer; here we give a soft bound that holds for every automorphism-invariant random spanning tree.

(a) Show that $\mathbf{P}[L \geq n] \geq 1/(8n)$.

(b) Show that if the law of T is only assumed to be invariant under translations of $\mathbb{Z}^2$, then $\mathbf{E}[L] = \infty$.

4.49. Let H be as in (4.16) and $u \in \mathbb{Z}^2$. Consider simple random walk on $\mathbb{Z}^2$. Show that $H(u) = \lim_{n \to \infty} \sum_{k=0}^{n} [p_k(\mathbf{0}, \mathbf{0}) - p_k(\mathbf{0}, u)]$.

4.50. Find the effective resistance between the origin and the vertex $(2, 1)$ in $\mathbb{Z}^2$.

4.51. Show that for $n \geq 1$, the probability that simple random walk on $\mathbb{Z}^2$ starting at $(0,0)$ visits (n,n) before returning to $(0,0)$ equals

$$\frac{\pi}{8} \left(\sum_{k=1}^{n} \frac{1}{2k-1} \right)^{-1}.$$

4.52. Show that in $\mathbb{Z}^2$, we have that as $x \to \infty$,

$$\mathscr{R}(\mathbf{0} \leftrightarrow x) \sim \frac{1}{\pi} \log \|x\|.$$

4.53. Let $A \subset \mathbb{Z}^2$ be finite. Write $\mu_z(x) := \mathbf{P}_z[X_{\tau_A} = x]$ for the harmonic measure on A with respect to z, as in Exercise 2.50.

  (a) Show that $\lim_{z \to \infty} \mu_z$ exists. We call the limit, μ, **harmonic measure from infinity**. *Hint:* It suffices to show that μ_z and μ_w are close when z and w are on the boundary of a large square with corners $(\pm N, \pm N)$. Consider $z = (-N, N)$ and $w = (N, N)$. Couple the walks from z and w by having them do the same vertical steps but opposite horizontal steps until they meet, and then keep them together. With high probability, they meet before either one visits A.

  (b) Write R_A for the $(A \times A)$-matrix whose (x,y)-entry is $\mathscr{R}(x \leftrightarrow y)$. Show that $\mu = c\mathbf{1}^T R_A^{-1}$, where c is a constant so that the right-hand side adds to 1.

  (c) Show that if $A = \{(-1,-1), (0,0), (1,1)\}$, then $\mu = (3/8, 1/4, 3/8)$.

  (d) Show that if $A = \{(0,1), (0,0), (1,0)\}$, then

$$\mu = \left(\frac{\pi}{4(\pi-1)}, \frac{\pi-2}{2(\pi-1)}, \frac{\pi}{4(\pi-1)} \right).$$

  (e) Show that if $A = \{(-1,0), (0,0), (1,0)\}$, then $\mu = (\pi/8, 1 - \pi/4, \pi/8)$.

4.54. For a function $f \in L^1(\mathbb{T}^2)$ and integers x, y, define

$$\widehat{f}(x, y) := \int_{\mathbb{T}^2} f(\alpha) e^{-2\pi i(x\alpha_1 + y\alpha_2)} \, d\alpha.$$

Let Y be the transfer-current matrix for the square lattice $\mathbb{Z}^2$. Let $e_{x,y}^{\mathrm{h}} := [(x,y), (x+1,y)]$ and $e_{x,y}^{\mathrm{v}} := [(x,y), (x,y+1)]$. Show that $Y(e_{0,0}^{\mathrm{h}}, e_{x,y}^{\mathrm{h}}) = \widehat{f}(x,y)$ and $Y(e_{0,0}^{\mathrm{h}}, e_{x,y}^{\mathrm{v}}) = \widehat{g}(x,y)$, where

$$f(\alpha_1, \alpha_2) := \frac{\sin^2 \pi\alpha_1}{\sin^2 \pi\alpha_1 + \sin^2 \pi\alpha_2}$$

and

$$g(\alpha_1, \alpha_2) := \frac{(1 - e^{2\pi i \alpha_1})(1 - e^{-2\pi i \alpha_2})}{4(\sin^2 \pi\alpha_1 + \sin^2 \pi\alpha_2)}.$$

4.55. Consider the ladder graph on $\mathbb{Z} \times \mathbb{Z}_2$ that is the doubly infinite limit of the ladder graphs shown in Figure 1.7. Calculate its transfer-current matrix.

5 | Branching Processes, Second Moments, and Percolation

Consider groundwater percolating down through soil and rock. How can we model the effects of the irregularities of the medium through which the water percolates? One common approach is to use a model in which the medium is random. More specifically, the pathways by which the water can travel are randomly chosen out of some regular set of possible pathways. For example, one may treat the ground as a half-space in which possible pathways are the rectangular lattice lines. Thus, we consider the nearest-neighbor graph on the vertices $\mathbb{Z} \times \mathbb{Z} \times \mathbb{Z}^-$, and each edge is independently chosen to be open (allowing water to flow) or closed. Commonly, the marginal probability that an edge is open, p, is the same for all edges. In this case, the only parameter in the model is p, and one studies how p affects large-scale behavior of possible water flow.

In fact, this model of percolation is used in many other contexts to have a simple model that nevertheless captures some important aspects of an irregular situation. In particular, it has an interesting phase transition. Some information about percolation on $\mathbb{Z}^d$ and other transitive graphs is given in this chapter and in Section 6.9, but a thorough study of percolation on transitive graphs, especially on nonamenable graphs, is deferred to Chapters 7 and 8.

In this chapter, we consider percolation mostly on trees rather than on lattices. This turns out to be interesting and also useful for other seemingly unrelated probabilistic processes and questions. For example, we'll find another fundamental interpretation of the branching number of a tree.

We begin by studying a beautiful way of growing trees at random known as Galton–Watson branching processes. We then move to general trees and develop some basic analytic methods of probability known as the first- and second-moment methods. These will fit remarkably well with our study of random walks on the connected components of percolation in $\mathbb{Z}^d$. After that interlude on $\mathbb{Z}^d$, we return to Galton–Watson processes to understand better how they behave and what flows are possible on random networks based on Galton–Watson trees. Deeper results on Galton–Watson branching processes are proved in Chapter 12. The last chapter of this book, Chapter 17, is devoted primarily to random walks on Galton–Watson trees.

5.1 Galton–Watson Branching Processes

Percolation on a tree breaks up the tree into random subtrees. Historically, the first random trees to be considered were a model of genealogical (family) trees. Since such trees will be an important source of examples and an important tool in later work, we too will consider their basic theory before turning to percolation. They are also beautiful processes in themselves.

Galton–Watson branching processes are most often defined as Markov chains $\langle Z_n ; \; n \geq 0 \rangle$ on the nonnegative integers, where Z_n represents the size of the nth generation of a family, but we will be interested as well in the underlying family trees. Given numbers $p_k \in [0, 1]$ with $\sum_{k \geq 0} p_k = 1$, the process is defined as follows. We start with one particle, $Z_0 \equiv 1$, unless specified otherwise. It has k children with probability p_k. Then each of these children (should there be any) also has children with the same progeny (or "offspring") distribution $\langle p_k ; \; k \geq 0 \rangle$, independently of the others and of its parent. This continues forever or until there are no more children. To be formal, let L be a random variable with $\mathbf{P}[L = k] = p_k$, and let $\langle L_i^{(n)} ; \; n, i \geq 1 \rangle$ be independent copies of L. The generation sizes of the branching process are then defined inductively by

$$Z_{n+1} := \sum_{i=1}^{Z_n} L_i^{(n+1)} . \tag{5.1}$$

The ***probability generating function (p.g.f.)*** of L is very useful and is denoted

$$f(s) := \mathbf{E}[s^L] = \sum_{k \geq 0} p_k s^k .$$

This is defined for $0 \leq s \leq 1$, and possibly for other s as well. Note that we interpret $0^0 = 1$, so that $f(0) = \mathbf{P}[L = 0] = p_0$. We call the event $[\exists n \; Z_n = 0]$ ***extinction***; this, of course, is the same as the event $[Z_n \to 0]$. We will often omit the superscripts on L when not needed. The family (or genealogical) tree associated to a branching process is obtained simply by having one vertex for each particle ever produced and joining two by an edge if one is the parent of the other. See Figure 5.1 for an example. We will give a formal definition later of trees and the associated probability measures on them.

The first basic result on Galton–Watson processes is that on the event of nonextinction, the population size explodes, except in the trivial case that $p_1 = 1$:

Proposition 5.1. *On the event of nonextinction, $Z_n \to \infty$ a.s. provided $p_1 \neq 1$.*

Proof. We want to see that 0 is the only nontransient state of the Markov chain $\langle Z_n \rangle$. If $p_0 = 0$, this is clear, whereas if $p_0 > 0$, then from any state $k \geq 1$, eventually returning to k requires not immediately becoming extinct, whence it has probability $\leq 1 - p_0^k < 1$.   ◄

What is $q := \mathbf{P}[\text{extinction}]$? To find out, we use the following very handy property of the p.g.f.:

Proposition 5.2. $\mathbf{E}[s^{Z_n}] = \underbrace{f \circ \cdots \circ f}_{n \text{ times}}(s) =: f^{(n)}(s) \, \text{for } 0 \leq s \leq 1.$

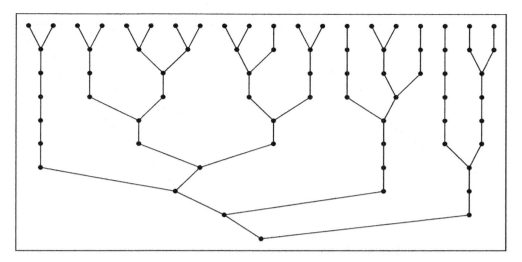

Figure 5.1. Generations 0 to 9 of a typical Galton–Watson tree for $f(s) = (s + s^2)/2$.

Proof. We have

$$\mathbf{E}[s^{Z_n}] = \mathbf{E}\left[\mathbf{E}\left[s^{\sum_{i=1}^{Z_{n-1}} L_i} \mid Z_{n-1}\right]\right] = \mathbf{E}\left[\mathbf{E}\left[\prod_{i=1}^{Z_{n-1}} s^{L_i} \mid Z_{n-1}\right]\right]$$

$$= \mathbf{E}\left[\prod_{i=1}^{Z_{n-1}} \mathbf{E}[s^{L_i}]\right] = \mathbf{E}\left[\mathbf{E}[s^L]^{Z_{n-1}}\right] = \mathbf{E}\left[f(s)^{Z_{n-1}}\right],$$

where the random variables $L_i := L_i^{(n)}$ are independent of each other and of Z_{n-1} and have the same distribution as L. Iterate this equation n times. ◀

Note that within this proof is the identity

$$\mathbf{E}[s^{Z_n} \mid Z_0, Z_1, \ldots, Z_{n-1}] = f(s)^{Z_{n-1}}. \tag{5.2}$$

Corollary 5.3. (Extinction Probability) *The extinction probability is $q = \lim_n f^{(n)}(0)$.*

Proof. Since extinction is the increasing union of the events $[Z_n = 0]$, it follows that $q = \lim_n \mathbf{P}[Z_n = 0] = \lim_n f^{(n)}(0)$. ◀

Looking at a graph of the increasing convex function f (Figure 5.2), we discover the most-used result in the field and value of q:

Proposition 5.4. (Extinction Criterion) *Provided $p_1 \neq 1$, we have*
 (i) *$q = 1 \Leftrightarrow f'(1) \leq 1$;*
 (ii) *q is the smallest root of $f(s) = s$ in $[0, 1]$ – the only other possible root being 1.* ◀

When we differentiate f at 1, we mean the left-hand derivative. Note that

$$f'(1) = \mathbf{E}[L] =: m = \sum k p_k, \tag{5.3}$$

the mean number of offspring. We call m simply the ***mean*** of the branching process.

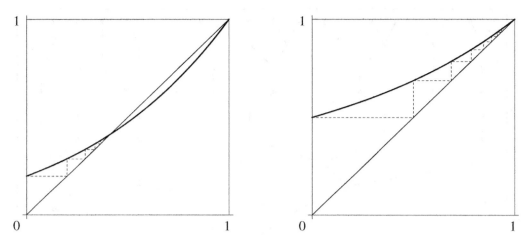

Figure 5.2. Typical graphs of f when $m > 1$ and $m \leq 1$.

▷ **Exercise 5.1.**
Justify the differentiation in (5.3). Show too that $\lim_{s\uparrow 1} f'(s) = m$.

Because of Proposition 5.4, a branching process is called **subcritical** if $m < 1$, **critical** if $m = 1$, and **supercritical** if $m > 1$.

How quickly does $Z_n \to \infty$ on the event of nonextinction? The most naive guess would be that it grows approximately like m^n. This is essentially correct. Our first result is that a martingale appears when we divide Z_n by m^n:

Proposition 5.5. *The sequence $\langle Z_n/m^n \rangle$ is a martingale when $0 < m < \infty$.*

In particular, $\mathbf{E}[Z_n] = m^n$ since $\mathbf{E}[Z_0] = \mathbf{E}[\mathbf{1}] = 1$.

Proof. We have

$$
\mathbf{E}\!\left[\frac{Z_{n+1}}{m^{n+1}} \,\middle|\, Z_n\right] = \mathbf{E}\!\left[\frac{1}{m^{n+1}} \sum_{i=1}^{Z_n} L_i \,\middle|\, Z_n\right] = \frac{1}{m^{n+1}} \sum_{i=1}^{Z_n} \mathbf{E}[L_i \mid Z_n] = \frac{1}{m^{n+1}} \sum_{i=1}^{Z_n} m = \frac{Z_n}{m^n} \,.
$$

Actually, we have not verified that we are computing conditional expectations of integrable random variables. One way to avoid calculating (in a similar manner) the unconditional expectation first is to note that all random variables are nonnegative.* Another way is to use the fact that Z_n takes only countably many values, so that we may work with expectations conditioned on events, rather than on a random variable.   ◀

Since this martingale $\langle Z_n/m^n \rangle$ is nonnegative, it has a finite limit a.s., denoted W. Thus, when $W > 0$, the generation sizes Z_n grow as expected, that is, like m^n up to a random factor. Otherwise, they grow more slowly. Our attention is thus focused on the following two questions.

---

* See the end of Section 12.1 for a review of conditional expectation for nonnegative random variables that may not be integrable.

Question 1. When is $W > 0$?

Question 2. When $W = 0$ and the process does not become extinct, what is the rate at which $Z_n \to \infty$?

To answer these questions, we first note a general zero-one property of Galton–Watson branching processes. Call a property of trees **inherited** if every finite tree has this property and if whenever a tree has this property, so do *all* the descendant trees of the children of the root.

Proposition 5.6. *Every inherited property has conditional probability either 0 or 1 given nonextinction.*

Proof. Let A be the set of trees possessing a given inherited property. For a tree T with k children of the root, let $T^{(1)}, \dots, T^{(k)}$ be the descendant trees of these children. Then

$$\mathbf{P}(A) = \mathbf{E}\Big[\mathbf{P}[T \in A \mid Z_1]\Big] \leq \mathbf{E}\Big[\mathbf{P}[T^{(1)} \in A, \dots, T^{(Z_1)} \in A \mid Z_1]\Big]$$

by definition of inherited. Since $T^{(1)}, \dots, T^{(Z_1)}$ are i.i.d. given Z_1, the last quantity in the display is equal to $\mathbf{E}\big[\mathbf{P}(A)^{Z_1}\big] = f(\mathbf{P}(A))$. Thus, $\mathbf{P}(A) \leq f(\mathbf{P}(A))$. On the other hand, $\mathbf{P}(A) \geq q$, since every finite tree is in A. It follows upon inspection of a graph of f that $\mathbf{P}(A) \in \{q, 1\}$, from which the desired conclusion follows. ◄

Corollary 5.7. *Suppose that $0 < m < \infty$. Either $W = 0$ a.s. or $W > 0$ a.s. on nonextinction. In other words, $\mathbf{P}[W = 0] \in \{q, 1\}$.*

Proof. The property that $W = 0$ is clearly inherited, whence this is an immediate consequence of Proposition 5.6. ◄

In answer to the preceding two questions, we have the following two theorems.

The Kesten–Stigum Theorem (1966). *The following are equivalent when $1 < m < \infty$:*
  (i) $\mathbf{P}[W = 0] = q$;
  (ii) $\mathbf{E}[W] = 1$;
  (iii) $\mathbf{E}[L \log^+ L] < \infty$.

This will be shown in Section 12.2. The sufficiency of $\mathbf{E}[L^2] < \infty$ for (i) and (ii) is much easier and follows from Exercise 5.27. Since (iii) requires barely more than the existence of a mean, generation sizes "typically" do grow as expected. When (iii) fails, however, the means m^n overestimate the rate of growth. Yet there is still an essentially deterministic rate of growth, as shown by Seneta (1968) and Heyde (1970), which is only slightly less than m^n:

The Seneta–Heyde Theorem. *If $1 < m < \infty$, then there exist constants c_n such that*
  (i) $\lim Z_n/c_n$ *exists a.s. in* $[0, \infty)$;
  (ii) $\mathbf{P}[\lim Z_n/c_n = 0] = q$;
  (iii) $c_{n+1}/c_n \to m$.

Proof. We will find another martingale to do our work. Choose $s_0 \in (q, 1)$ and set $s_{n+1} := f^{-1}(s_n)$ for $n \geq 0$. Then $s_n \uparrow 1$. By (5.2), we have that $\langle s_n^{Z_n} \rangle$ is a martingale. Being positive and bounded, it converges a.s. and in L^1 to a limit $Y \in [0, 1]$ such that $\mathbf{E}[Y] = \mathbf{E}[s_0^{Z_0}] = s_0$. Now we reformulate these exponentials. Set $c_n := -1/\log s_n$. Then $s_n^{Z_n} = e^{-Z_n/c_n}$, so that $\lim Z_n/c_n$ exists a.s. in $[0, \infty]$. By l'Hôpital's rule and Exercise 5.1,

$$\lim_{s \uparrow 1} \frac{-\log f(s)}{-\log s} = \lim_{s \uparrow 1} \frac{f'(s)s}{f(s)} = m \,.$$

Considering this limit along the sequence $\langle s_n \rangle$, we get (iii). It follows from (iii) that the property that $\lim Z_n/c_n = 0$ is inherited, whence by Proposition 5.6 and the fact that $\mathbf{E}[Y] = s_0 < 1$, we deduce (ii). Likewise, the property that $\lim Z_n/c_n < \infty$ is inherited and has probability 1 since $\mathbf{E}[Y] > q$. This implies (i). ◄

The proof of the Seneta–Heyde theorem gives a prescription for calculating the constants c_n but does not immediately provide estimates for them. Another approach gives a different prescription that leads sometimes to an explicit estimate: see Asmussen and Hering (1983), pp. 45–49.

We will often want to consider random trees produced by a Galton–Watson branching process. Up to now, we have avoided that by giving theorems just about the random variables Z_n (except for Proposition 5.6, but that was used so far only for studying the limiting behavior of Z_n). One approach to formalize tree-valued random variables is as follows. A ***rooted, labeled tree*** T is a nonempty collection of finite sequences of positive integers such that if $\langle i_1, i_2, \dots, i_n \rangle \in T$, then

   (i) for every $k \in [0, n]$, also the initial segment $\langle i_1, i_2, \dots, i_k \rangle \in T$, where the case $k = 0$ means the empty sequence, and

   (ii) for every $j \in [1, i_n]$, also the sequence $\langle i_1, i_2, \dots, i_{n-1}, j \rangle \in T$.

The root of the tree is the empty sequence, $\varnothing$. Thus, $\langle i_1, \dots, i_n \rangle$ is the i_nth child of the i_{n-1}th child of ... of the i_1th child of the root. If $x = \langle i_1, i_2, \dots, i_n \rangle \in T$, then we define $T^x := \{ \langle j_1, j_2, \dots, j_k \rangle ; \ \langle i_1, i_2, \dots, i_n, j_1, j_2, \dots, j_k \rangle \in T \}$ to be the ***descendant tree*** of the vertex x in T. The ***height*** of a tree is the supremum of the lengths of the sequences in the tree. If T is a tree and $n \in \mathbb{N}$, write the ***truncation*** of T to its first n levels as $T \!\restriction\! n := \{ \langle i_1, i_2, \dots, i_k \rangle \in T ; \ k \le n \}$. This is a tree of height at most n. A tree is called ***locally finite*** if its truncation to every finite level is finite. Let $\mathcal{T}$ be the space of rooted, labeled, locally finite trees. We define a metric on $\mathcal{T}$ by setting $d(T, T') := \left(1 + \sup\{ n ; \ T \!\restriction\! n = T' \!\restriction\! n \} \right)^{-1}$.

▷ **Exercise 5.2.**
Verify that d is a metric and that $(\mathcal{T}, d)$ is complete and separable.

▷ **Exercise 5.3.**
Define the measure **GW** formally on the space $\mathcal{T}$ of Exercise 5.2; your measure should be the law of a random tree produced by a Galton–Watson process with arbitrary given offspring distribution.

We can now use this formalism to give meaning to statements such as that in the following exercise.

▷ **Exercise 5.4.**
Show that for any Galton–Watson process with mean $m > 1$, the family tree T has growth rate $\operatorname{gr} T = m$ a.s. given nonextinction. (Don't use the Kesten–Stigum theorem to show this, as we have not yet proved that theorem.)

5.2 The First-Moment Method

Let G be a countable, possibly unconnected, graph. The most common percolation on G is ***Bernoulli bond percolation with constant survival parameter*** p, or ***Bernoulli(p)*** ***percolation*** for short; here, for fixed $p \in [0, 1]$, each edge is kept with probability p and removed otherwise, independently of the other edges. Often, the edges kept are called ***open***, whereas the edges removed are called ***closed***. Denote the random subgraph of G that remains by ω. The connected components of ω are called ***clusters***. Given a vertex x of G, we are often interested in the cluster of x in ω, written $K(x)$, and we'd especially like to know whether the diameter of $K(x)$ is infinite with positive probability. The first-moment method explained in this section gives a simple upper bound on this probability. In fact, this method is so simple that it works in complete generality: Suppose that ω is *any* random subgraph of G. The only measurability needed is that for each vertex x and each edge e, the sets $\{\omega ; x \in \omega\}$ and $\{\omega ; e \in \omega\}$ are measurable. We will call such a random subgraph a ***general percolation*** on G. We will say that a set Π of edges of G ***separates*** x ***from infinity*** if the removal of Π leaves x in a component of finite diameter. Denote by $[x \leftrightarrow a]$ the event that a is in the cluster of x for a vertex or edge, a, and by $[x \leftrightarrow \infty]$ the event that x is in a cluster of infinite diameter.

▷ **Exercise 5.5.**
Show that for a general percolation, the events $[x \leftrightarrow a]$ and $[x \leftrightarrow \infty]$ are indeed measurable.

Proposition 5.8. *Given a general percolation on* G,

$$\mathbf{P}[x \leftrightarrow \infty] \le \inf \left\{ \sum_{e \in \Pi} \mathbf{P}[x \leftrightarrow e] \, ; \, \Pi \text{ separates } x \text{ from infinity} \right\}. \tag{5.4}$$

Proof. For any Π separating x from infinity, we have

$$[x \leftrightarrow \infty] \subseteq \bigcup_{e \in \Pi} [x \leftrightarrow e]$$

by definition. Therefore $\mathbf{P}[x \leftrightarrow \infty] \le \sum_{e \in \Pi} \mathbf{P}[x \leftrightarrow e]$.   ◀

The method used in the proof of Proposition 5.8 is also called the "union bound." The reason for the name "first-moment method" is that another way to get the same bound is to write

$$\mathbf{1}_{[x \leftrightarrow \infty]} \le \sum_{e \in \Pi} \mathbf{1}_{[x \leftrightarrow e]}$$

and then

$$\mathbf{P}[x \leftrightarrow \infty] = \mathbf{E}\big[\mathbf{1}_{[x \leftrightarrow \infty]}\big] \leq \sum_{e \in \Pi} \mathbf{E}\big[\mathbf{1}_{[x \leftrightarrow e]}\big] = \sum_{e \in \Pi} \mathbf{P}[x \leftrightarrow e].$$

That is, we are using the first moment of a random variable, namely, the sum of certain indicator random variables. Of course, the first-moment method is extremely popular and surprisingly powerful.

Returning to Bernoulli percolation with constant survival parameter p, denote the law of ω by $\mathbf{P}_p$. By Kolmogorov's zero-one law,

$$\mathbf{P}_p[\omega \text{ has a cluster of infinite diameter}] \in \{0, 1\}.$$

It is intuitively clear that this probability is increasing in p. For a rigorous proof of this, we couple all the percolation processes at once as follows. Let $U(e)$ be i.i.d. uniform $[0, 1]$ random variables indexed by the edges of G. If ω_p is the graph containing all the vertices of G and exactly those edges e with $U(e) < p$, then the law of ω_p is precisely $\mathbf{P}_p$. This coupling is referred to as the **standard coupling** of Bernoulli percolation. But now when $p \leq q$, the event that ω_p has a cluster of infinite diameter is contained in the event that ω_q has a cluster of infinite diameter. Hence the probability of the first event is at most the probability of the second. This leads us to define the **critical probability**

$$p_c(G) := \sup\{p \,;\; \mathbf{P}_p[\exists \text{ infinite-diameter cluster}] = 0\}.$$

If G is connected and x is any given vertex of G, then

$$p_c(G) = \sup\{p \,;\; \mathbf{P}_p[x \leftrightarrow \infty] = 0\}. \tag{5.5}$$

▷ **Exercise 5.6.**
Prove this.

Again, the standard coupling provides a rigorous proof that $\mathbf{P}_p[x \leftrightarrow \infty]$ is increasing in p.

Generally, $p_c(G)$ is extremely difficult to calculate. Clearly $p_c(\mathbb{Z}) = 1$. After long efforts, it was shown that $p_c(\mathbb{Z}^2) = 1/2$ (Kesten, 1980). There is not even a conjecture for the value of $p_c(\mathbb{Z}^d)$ for any $d \geq 3$. Now Proposition 5.8 provides a lower bound for p_c provided we can estimate $\mathbf{P}[o \leftrightarrow e]$. If G is a tree T, such an estimate is easy to find, since we actually know the exact value: $\mathbf{P}[o \leftrightarrow e] = p^{|e|}$. Hence, the definition of branching number gives immediately that

$$p_c(T) \geq 1/\operatorname{br} T. \tag{5.6}$$

In fact, we have equality here (Theorem 5.15), but this requires the second-moment method. Nevertheless, there are some cases among nonlattice graphs where it is easy to determine p_c exactly, even without the first-moment method. One example is given in the next exercise:

▷ **Exercise 5.7.**
Show that for $p \geq p_c(G)$, we have $p_c(\omega) = p_c(G)/p$ for $\mathbf{P}_p$-a.e. ω. Physicists often refer to p as the "density" of edges in ω, and this helps the intuition.

For another example, if T is an n-ary tree, then the cluster of the root under percolation is a Galton–Watson tree with progeny distribution Bin(n, p). Thus, this cluster is infinite with positive probability iff $np > 1$, whence $p_c(T) = 1/n$. This reasoning may be extended to all Galton–Watson trees (in which case, $p_c(T)$ is a random variable):

Proposition 5.9. (Lyons, 1990) *Let T be the family tree of a Galton–Watson process with mean $m > 1$. Then $p_c(T) = 1/m$ a.s. given nonextinction.*

Proof. Let T be a given tree and write K for the cluster of the root of T after percolation on T with survival parameter p. When T has the law of a Galton–Watson tree with mean m, we claim that K has the law of another Galton–Watson tree having mean mp: if Y_i represent i.i.d. Bin($1, p$) random variables that are also independent of L, then

$$\mathbf{E}\left[\sum_{i=1}^{L} Y_i\right] = \mathbf{E}\left[\mathbf{E}\left[\sum_{i=1}^{L} Y_i \mid L\right]\right] = \mathbf{E}\left[\sum_{i=1}^{L} \mathbf{E}[Y_i]\right] = \mathbf{E}\left[\sum_{i=1}^{L} p\right] = pm .$$

Hence K is finite a.s. iff $mp \le 1$. Since

$$\mathbf{E}\left[\mathbf{P}\big[|K| < \infty \mid T\big]\right] = \mathbf{P}\big[|K| < \infty\big], \tag{5.7}$$

this means that for almost every Galton–Watson tree* T, the cluster of its root is finite a.s. if $p \le 1/m$. On the other hand, for fixed p, the property $\{T ; \mathbf{P}_p[|K| < \infty] = 1\}$ is inherited, and so has probability q or 1. If it has probability 1, then (5.7) shows that $mp \le 1$. That is, if $mp > 1$, this property has probability q, so that the cluster of the root of T will be infinite with positive probability a.s. on the event of nonextinction. Considering a sequence $p_n \downarrow 1/m$, we see that this holds a.s. on the event of nonextinction for all $p > 1/m$ at once, not just for a fixed p. We conclude that $p_c(T) = 1/m$ a.s. on nonextinction. ◄

We may easily deduce the branching number of Galton–Watson trees, as shown by Lyons (1990):

Corollary 5.10. (Branching Number of Galton–Watson Trees) *If T is a Galton–Watson tree with mean $m > 1$, then $\mathrm{br}\, T = m$ a.s. given nonextinction.*

Proof. By Proposition 5.9 and (5.6), we have $\mathrm{br}\, T \ge m$ a.s. given nonextinction. On the other hand, $\mathrm{br}\, T \le \mathrm{gr}\, T = m$ a.s. given nonextinction by Exercise 5.4. ◄

This corollary was shown in the language of Hausdorff dimension (which will be explained in Chapter 15) by Hawkes (1981) under the assumption that $\mathbf{E}[L(\log L)^2] < \infty$.

---

* This typical abuse of language means "for almost every tree with respect to Galton–Watson measure."

5.3 The Weighted Second-Moment Method

We obtained a simple upper bound on $\mathbf{P}[o \leftrightarrow \infty]$ for a general percolation on a graph G by the first-moment method. A lower bound can be obtained by the second-moment method, which is a powerful method of wide applicability. Fix $o \in G$ and let Π be a minimal set of edges that separates o from ∞. The second-moment method consists in calculating the second moment of the number of edges in Π that are connected to o; this is then compared to the first moment. However, we will see that it can be much better to use a weighted count, rather than a pure count, of the number of connected edges. We will use $\mu \in \mathcal{P}(\Pi)$ to make such weights, where $\mathcal{P}(\Pi)$ is the set of probability measures on Π. We will assume for simplicity that $\mathbf{P}[o \leftrightarrow e] > 0$ for each $e \in \mathsf{E}$.

If, as before, we write $K(o)$ for the cluster of o in the percolation graph ω, and if we set

$$X(\mu) := \sum_{e \in \Pi} \mu(e)\mathbf{1}_{[e \in K(o)]} / \mathbf{P}\big[e \in K(o)\big], \tag{5.8}$$

then

$$\mathbf{E}\big[X(\mu)\big] = 1 .$$

Write $o \leftrightarrow \Pi$ for the event that $o \leftrightarrow e$ for some $e \in \Pi$. This event is implied by the event that $X(\mu) > 0$. We are looking for a lower bound on the probability of $o \leftrightarrow \infty$. Since

$$\mathbf{P}[o \leftrightarrow \infty] = \inf\big\{\mathbf{P}[o \leftrightarrow \Pi] ; \ \Pi \text{ separates } o \text{ from } \infty\big\} , \tag{5.9}$$

we seek a lower bound on $\mathbf{P}[X(\mu) > 0]$. This will be a consequence of an upper bound on the second moment of $X(\mu)$ as follows. (A slightly more general inequality due to Paley and Zygmund (1932) is given in Section 5.5.)

Proposition 5.11. *Given a general percolation on G,*

$$\mathbf{P}\big[o \leftrightarrow \Pi\big] \geq 1/\mathbf{E}\big[X(\mu)^2\big]$$

for every $\mu \in \mathcal{P}(\Pi)$.

Proof. Given $\mu \in \mathcal{P}(\Pi)$, the Cauchy–Schwarz inequality yields

$$1 = \mathbf{E}\big[X(\mu)\big]^2 = \mathbf{E}\big[X(\mu)\mathbf{1}_{[X(\mu)>0]}\big]^2 \leq \mathbf{E}\big[X(\mu)^2\big]\mathbf{E}\big[\mathbf{1}_{[X(\mu)>0]}^2\big]$$
$$= \mathbf{E}\big[X(\mu)^2\big]\mathbf{P}\big[X(\mu) > 0\big] \leq \mathbf{E}\big[X(\mu)^2\big]\mathbf{P}\big[o \leftrightarrow \Pi\big],$$

because $X(\mu) > 0$ implies $o \leftrightarrow \Pi$. Therefore, $\mathbf{P}[o \leftrightarrow \Pi] \geq 1/\mathbf{E}[X(\mu)^2]$.   ◀

Clearly we want to choose the weight function $\mu \in \mathcal{P}(\Pi)$ that optimizes this lower bound. Now

$$\mathbf{E}\big[X(\mu)^2\big] = \sum_{e_1, e_2 \in \Pi} \mu(e_1)\mu(e_2)\frac{\mathbf{P}\big[e_1, e_2 \in K(o)\big]}{\mathbf{P}\big[e_1 \in K(o)\big]\mathbf{P}\big[e_2 \in K(o)\big]} . \tag{5.10}$$

We denote this quantity by $\mathscr{E}(\mu)$ and call it the ***energy*** of μ.

Why is this called an "energy"? In general, like the energy of a flow in Chapter 2, energy is a quadratic form, usually positive definite. In electrostatics, if μ is a charge distribution confined to a conducting region Ω in space, then μ will minimize the energy

$$\int_\Omega \int_\Omega \frac{d\mu(x)\,d\mu(y)}{|x - y|^2} .$$

One could also write (5.10) as a double integral to put it in a form closer to this. We too are interested in minimizing the energy. Thus, from Proposition 5.11, we obtain the following:

Proposition 5.12. *Given a general percolation on G,*

$$\mathbf{P}[o \leftrightarrow \infty] \geq \inf \left\{ \frac{1}{\inf_{\mu \in \mathcal{P}(\Pi)} \mathcal{E}(\mu)} ; \ \Pi \ separates \ o \ from \ \infty \right\}.$$

Proof. We have shown that $\mathbf{P}[o \leftrightarrow \Pi] \geq 1/\mathcal{E}(\mu)$ for every $\mu \in \mathcal{P}(\Pi)$. Hence, the same holds when we take the sup of the right-hand side over μ. Then the result follows from (5.9). ◄

Of course, for Proposition 5.12 to be useful, one has to find a way to estimate such energies. As for the first-moment method, the case of trees is most conducive to such analysis.

Consider first the case of ***independent percolation***, that is, with $[e \in \omega]$ mutually independent events for all edges e. If $\mu \in \mathcal{P}(\Pi)$, write $\mu(x)$ for $\mu(e(x))$. Then we have

$$\mathcal{E}(\mu) = \sum_{e(x),e(y)\in\Pi} \mu(x)\mu(y) \frac{\mathbf{P}[o \leftrightarrow x, o \leftrightarrow y]}{\mathbf{P}[o \leftrightarrow x]\,\mathbf{P}[o \leftrightarrow y]} = \sum_{e(x),e(y)\in\Pi} \frac{\mu(x)\mu(y)}{\mathbf{P}[o \leftrightarrow x \wedge y]}, \qquad (5.11)$$

where $x \wedge y$ denotes the farthest vertex from o that is a common ancestor of both x and y (we say that z is an ***ancestor*** of w if z lies on the shortest path between o and w, so that $z = w$ is not excluded). This looks suspiciously similar to the result of the following calculation: Consider a network of conductances on a finite tree T and some flow θ from o to the leaves, which we'll write as $\partial_L T$, of T. (We do not include o in $\partial_L T$, even if o happens to be a leaf.) Write $\theta(x)$ for $\theta(e(x))$.

Lemma 5.13. *Let θ be a flow on a finite tree T from o to $\partial_L T$. Then*

$$\mathcal{E}(\theta) = \sum_{x,y\in\partial_L T} \theta(x)\theta(y)\mathcal{R}(o \leftrightarrow x \wedge y).$$

Proof. We use the fact that $\sum_{\{x\in\partial_L T \,;\, e\leq x\}} \theta(x) = \theta(e)$ for any edge e (see Exercise 3.3). Thus, we have

$$\sum_{x,y\in\partial_L T} \theta(x)\theta(y)\mathcal{R}(o \leftrightarrow x \wedge y) = \sum_{x,y\in\partial_L T} \theta(x)\theta(y) \sum_{e\leq x\wedge y} r(e)$$

$$= \sum_{e\in T} r(e) \sum_{\substack{x,y\in\partial_L T \\ x,y\geq e}} \theta(x)\theta(y) = \sum_{e\in T} r(e)\theta(e)^2 = \mathcal{E}(\theta). \quad ◄$$

We now want to relate these two calculations and thereby show that percolation on trees is related to electrical networks. Write $\overline{\Pi} := \{x ;\ e(x) \in \Pi\}$. Let Π be a minimal set of edges that separates o from ∞. If we happen to have $\mathbf{P}[o \leftrightarrow x] = \mathcal{C}(o \leftrightarrow x)$ and if θ is the flow induced by μ from o to $\overline{\Pi}$, that is,

$$\theta(e) := \sum_{e\leq x\in\overline{\Pi}} \mu(x),$$

then we see from (5.11) and Lemma 5.13 that $\mathcal{E}(\mu) = \mathcal{E}(\theta)$, and we can hope to profit from our understanding of electrical networks and random walks. However, if $x = o$, then the desired equation cannot hold, since $\mathbf{P}[o \leftrightarrow o] = 1$ and $\mathscr{C}(o \leftrightarrow o) = \infty$. But suppose we have, instead, that

$$1/\mathbf{P}[o \leftrightarrow x] = 1 + \mathscr{R}(o \leftrightarrow x) \tag{5.12}$$

for all vertices x. Then we find that

$$\mathscr{E}(\mu) = 1 + \mathscr{E}(\theta),$$

which is hardly worse. This suggests that given a percolation problem, we choose conductances so that (5.12) holds in order to use our knowledge of electrical networks. Let us say that conductances are **adapted** to a percolation, and vice versa, if (5.12) holds.

Let the survival probability of $e(x)$ be p_x. To solve (5.12) explicitly for the conductances in terms of the survival parameters, write $\overleftarrow{x}$ for the parent of x. Note that the right-hand side of (5.12) is a resistance of a series, whence

$$r(e(x)) = \left[1 + \mathscr{R}(o \leftrightarrow x)\right] - \left[1 + \mathscr{R}(o \leftrightarrow \overleftarrow{x})\right] = 1/\mathbf{P}[o \leftrightarrow x] - 1/\mathbf{P}[o \leftrightarrow \overleftarrow{x}]$$
$$= (1 - p_x)/\mathbf{P}[o \leftrightarrow x],$$

or, in other words,

$$c(e(x)) = \frac{\mathbf{P}[o \leftrightarrow x]}{1 - p_x} = \frac{1}{1 - p_x} \prod_{o < u \leq x} p_u. \tag{5.13}$$

In particular, for a given $p \in (0, 1)$, we have that $p_x \equiv p$ iff $c(e(x)) = (1 - p)^{-1} p^{|x|}$; these conductances correspond to the random walk $\mathsf{RW}_{1/p}$ defined in Section 3.2.

▷ **Exercise 5.8.**
Show that, conversely, the survival parameters adapted to given edge resistances are

$$p_x = \frac{1 + \sum_{o < u < x} r(e(u))}{1 + \sum_{o < u \leq x} r(e(u))}.$$

For example, simple random walk ($c \equiv 1$) is adapted to $p_x = |x|/(|x| + 1)$.

These notions lead us to the following conclusion:

Theorem 5.14. (Lyons, 1992) *For an independent percolation and adapted conductances on the same tree, we have*

$$\frac{\mathscr{C}(o \leftrightarrow \infty)}{1 + \mathscr{C}(o \leftrightarrow \infty)} \leq \mathbf{P}[o \leftrightarrow \infty].$$

Proof. We first estimate the infimum of energies in Proposition 5.12: Given a minimal set Π of edges that separates o from ∞, let $\mu \in \mathcal{P}(\Pi)$ be the measure in $\mathcal{P}(\Pi)$ that has minimum energy and let θ be the flow induced by μ. We have

$$\mathscr{E}(\mu) = 1 + \mathscr{E}(\theta) = 1 + \mathscr{R}(o \leftrightarrow \overline{\overline{\Pi}})$$

by Thomson's principle. Therefore,

$$\mathbf{P}[o \leftrightarrow \infty] \geq \inf_{\Pi} 1/\left[1 + \mathscr{R}(o \leftrightarrow \overline{\overline{\Pi}})\right] = 1/\left[1 + \mathscr{R}(o \leftrightarrow \infty)\right],$$

as desired.   ◀

An immediate corollary of this combined with (5.6) and Theorems 2.3 and 3.5 is:

Theorem 5.15. (Lyons, 1990) *For every locally finite, infinite tree T,*

$$p_c(T) = \frac{1}{\operatorname{br} T} .$$ ◀

This reinforces the idea of $\operatorname{br} T$ as an average number of branches per vertex.

Question 5.16. This result shows that the first-moment method correctly identifies the critical value for Bernoulli percolation on trees. Does it in general? In other words, if the right-hand side of (5.4) is strictly positive for Bernoulli(p) percolation on a connected graph G, then must it be the case that $p \geq p_c(G)$? This is known to hold on $\mathbb{Z}^d$ and on "tree-like" graphs; see Lyons (1989). However, Kahn (2003) gave a counterexample and suggested the following modification of the question: Write $A(x, e, \Pi)$ for the event that there is an open path from x to e that is disjoint from $\Pi \setminus \{e\}$. If

$$\inf \left\{ \sum_{e \in \Pi} \mathbf{P}_p[A(x, e, \Pi)] \,;\; \Pi \text{ separates } x \text{ from infinity} \right\} > 0 ,$$

then is $p \geq p_c(G)$? In 2019, Pengfei Tang (in preparation) gave a positive answer.

It turns out that the inequality in Theorem 5.14 can be reversed up to a factor of 2. We show this by a stopping-time method in Section 5.6.

5.4 Quasi-independent Percolation

We have now achieved a fairly good understanding of independent percolation on trees. In the next section, we apply the second-moment method to Bernoulli percolation in $\mathbb{Z}^d$. Here, however, we remain with trees, but we weaken the assumption of independence.

We begin with an interesting example that this will allow us to analyze. Suppose that we label the edges e of a tree T by independent random variables $Z(e)$ that take the values ± 1 with probability 1/2 each. Fix an integer $N > 0$. Define $S(x) := \sum_{e \leq x} Z(e)$. Consider the percolation

$$\omega_N := \{e \,;\; S(e^-) \in [0, N],\; S(e^+) \in [0, N]\} . \tag{5.14}$$

Obviously the component of the root in ω_1 is the same as the component of the root in Bernoulli(1/2) percolation on T. In particular, the root belongs to an infinite cluster with positive probability for $\operatorname{br} T > 2$ but not for $\operatorname{br} T < 2$. The next case, ω_2, is almost as simple:

$$\mathbf{P}[e(x) \in \omega_2 \mid e(\bar{x}) \in \omega_2] = \begin{cases} 1/2 & \text{if } |x| \text{ is odd} \\ 1 & \text{if } |x| \text{ is even,} \end{cases}$$

so by Exercise 3.25, the root belongs to an infinite cluster with positive probability for $\operatorname{br} T > \sqrt{2}$ but not for $\operatorname{br} T < \sqrt{2}$. However, the succeeding cases ω_N for $N \geq 3$ are more complicated, as there is dependency in the percolation that was not there before. Luckily, the dependency is not very large; we will show that it is an example of the following kind of percolation, from which we will deduce that the critical branching number for infinite clusters in ω_N is $1\Big/\cos\dfrac{\pi}{N+2}$.

We call a percolation **quasi-independent*** if $\exists M < \infty$ $\forall x, y$ with $\mathbf{P}[o \leftrightarrow x \wedge y] > 0$,

$$\mathbf{P}[o \leftrightarrow x,\ o \leftrightarrow y \mid o \leftrightarrow x \wedge y] \le M\,\mathbf{P}[o \leftrightarrow x \mid o \leftrightarrow x \wedge y]\,\mathbf{P}[o \leftrightarrow y \mid o \leftrightarrow x \wedge y],\quad (5.15)$$

or, what is the same, if $\mathbf{P}[o \leftrightarrow x]\,\mathbf{P}[o \leftrightarrow y] > 0$, then

$$\frac{\mathbf{P}[o \leftrightarrow x,\, o \leftrightarrow y]}{\mathbf{P}[o \leftrightarrow x]\,\mathbf{P}[o \leftrightarrow y]} \le \frac{M}{\mathbf{P}[o \leftrightarrow x \wedge y]}\,.$$

Example 5.17. Sometimes this condition holds for easy reasons: if

$$\inf_{x \ne o} \mathbf{P}[o \leftrightarrow x \mid o \leftrightarrow \widetilde{x}] > 0$$

and

$$\mathbf{P}[o \leftrightarrow x,\ o \leftrightarrow y \mid o \leftrightarrow x \wedge y] = \mathbf{P}[o \leftrightarrow x \mid o \leftrightarrow \widetilde{x}]\,\mathbf{P}[o \leftrightarrow \widetilde{x},\ o \leftrightarrow y \mid o \leftrightarrow \widetilde{x} \wedge y]$$

whenever $\widetilde{x} \ne x \wedge y$, then the percolation is quasi-independent.

▷ **Exercise 5.9.**
Verify the assertion of Example 5.17.

Example 5.18. Now we verify that the percolation ω_N of (5.14) is quasi-independent. Indeed, fix N and write $q_k(n)$ for the probability that simple random walk on $\mathbb{Z}$ stays in the interval $[0, N]$ for n steps when it starts at k. There is clearly a constant M such that for all $n \ge 0$ and all $k, k' \in [0, N]$, we have $q_k(n) \le M q_{k'}(n)$. We claim that this M works in (5.15). To see this, fix x and y and put $r := |x \wedge y|$, $m := |x| - r$, and $n := |y| - r$. Also, write p_k for the probability that simple random walk at time r is at location k given that it stays in $[0, N]$ for r steps when it starts at 0. We have

$$\mathbf{P}[o \leftrightarrow x,\ o \leftrightarrow y \mid o \leftrightarrow x \wedge y] = \sum_{k=0}^{N} q_k(m) q_k(n) p_k \le M \min_k q_k(n) \cdot \sum_{k=0}^{N} q_k(m) p_k$$

$$\le M \sum_{k=0}^{N} q_k(n) p_k \sum_{k=0}^{N} q_k(m) p_k$$

$$= M\,\mathbf{P}[o \leftrightarrow x \mid o \leftrightarrow x \wedge y]\,\mathbf{P}[o \leftrightarrow y \mid o \leftrightarrow x \wedge y].$$

This shows that our percolation is indeed quasi-independent.

The virtue of a quasi-independent percolation is that it obeys essentially the same lower bound of Theorem 5.14 as Bernoulli percolation:

---

* This was called "quasi-Bernoulli" in Lyons (1989, 1992).

Theorem 5.19. (Lyons, 1989) *For a quasi-independent percolation with constant M and adapted conductances, we have*

$$\frac{1}{M} \frac{\mathscr{C}(o \leftrightarrow \infty)}{1 + \mathscr{C}(o \leftrightarrow \infty)} \le \mathbf{P}[o \leftrightarrow \infty].$$

Proof. For $\mu \in \mathcal{P}(\Pi)$, write

$$\mathscr{E}'(\mu) := \sum_{e(x), e(y) \in \Pi} \frac{\mu(x)\mu(y)}{\mathbf{P}[o \leftrightarrow x \wedge y]}.$$

Then the definition of quasi-independent gives $\mathscr{E}(\mu) \le M\mathscr{E}'(\mu)$, where $\mathscr{E}(\mu)$ is still defined as in (5.10). Also, if θ is the flow induced by μ, then $\mathscr{E}'(\mu) = 1 + \mathscr{E}(\theta)$. Hence

$$\mathbf{E}[X(\mu)^2] = \mathscr{E}(\mu) \le M\mathscr{E}'(\mu) = M[1 + \mathscr{E}(\theta)],$$

and the rest of the proof of Theorem 5.14 can be followed to the desired conclusion. ◄

Example 5.20. Let's apply this to Example 5.18. If we consider simple random walk on $[0, N]$ killed on exiting the interval, the corresponding substochastic transition matrix P is symmetric and so real diagonalizable. Let λ_k be its eigenvalues and v_k be a corresponding system of orthonormal eigenvectors. Thus,

$$P^n(i, j) = \sum_k \lambda_k^n v_k(i)v_k(j).$$

By the Perron–Frobenius theorem, $|\lambda_k| \le \rho$, where ρ is the largest positive eigenvalue and the corresponding eigenvector, say, v_1, has positive entries. Since this Markov chain has period 2, it follows that $P^n(i, j) \sim 2v_1(i)v_1(j)\rho^n$ when n and $i - j$ have the same parity; otherwise $P^n(i, j) = 0$. Now in our case, the top eigenvalue equals $\cos \frac{\pi}{N+2}$ (for example, see Spitzer (1976), Chapter 21, Proposition 1), whence $\mathbf{P}[o \leftrightarrow x] \sim a_{|x|} \left(\cos \frac{\pi}{N+2} \right)^{|x|}$ as $|x| \to \infty$ for some constants $a_m > 0$, where a_m depends only on the parity of m. This means that for the conductances $c(e)$ adapted to this percolation, there are constants a_1' and a_2' such that $a_1' \left(\cos \frac{\pi}{N+2} \right)^{|e|} \le c(e) \le a_2' \left(\cos \frac{\pi}{N+2} \right)^{|e|}$. Thus, Theorem 5.19 yields that $\mathbf{P}[o \leftrightarrow \infty] > 0$ if RW_λ is transient on T for $\lambda := 1 / \cos \frac{\pi}{N+2}$, which holds, in particular, if $\mathrm{br}\, T > 1 / \cos \frac{\pi}{N+2}$. This result is due to Benjamini and Peres (1994b).

▷ **Exercise 5.10.**
Show that if $\mathrm{br}\, T < 1 / \cos \frac{\pi}{N+2}$, then the root belongs to an infinite cluster in ω_N with probability zero.

5.5 Transience of Percolation Clusters in $\mathbb{Z}^d$

If T is a tree with $p_c(T) < 1$, then for $p_c(T) < p \le 1$, consider Bernoulli(p) percolation on T and its open subgraph ω_p. By Exercise 5.7, we have $p_c(\omega_p) < 1$ a.s., whence by (5.6), some component of ω_p has branching number larger than 1. By Theorem 3.5 in turn, this means that some component of ω_p is transient for simple random walk (among other random walks). In this section, we look at this same property (transience of percolation clusters for simple random walk), but for percolation on graphs that are not trees, such as $\mathbb{Z}^d$. Of course, we can hope for transience of a percolation cluster in $\mathbb{Z}^d$ only when $d \ge 3$ by Pólya's theorem and Rayleigh's monotonicity principle. The technique we use is quite similar to the methods of the previous sections on the second-moment method, and it also uses random paths and their connection to flows, as discussed in Section 3.1. It will use a slight improvement of Proposition 5.11, which was really the special case of the following where $t = 0$:

The Paley–Zygmund Inequality (1932). *If X is a random variable with mean 1 and $t < 1$, then*

$$\mathbf{P}[X > t] \ge \frac{(1-t)^2}{\mathbf{E}[X^2]}.$$

Proof. Let A be the event that $X > t$. The Cauchy–Schwarz inequality gives

$$\mathbf{E}[X^2]\,\mathbf{P}(A) = \mathbf{E}[X^2]\,\mathbf{E}[\mathbf{1}_A^2] \ge \mathbf{E}[X\mathbf{1}_A]^2 = \left(1 - \mathbf{E}[X\mathbf{1}_{A^c}]\right)^2 \ge (1-t)^2. \quad \blacktriangleleft$$

Whereas before we considered probability measures on cutsets, now we consider probability measures on paths. These latter probability measures induce the former probability measures by looking at the first intersection of a path with a cutset. But random paths also induce flows, as we saw in Section 3.1, and in order to show transience, we want to find a flow on the percolation cluster that has finite energy.

We'll start with finite paths, which we think of as sets of edges. Finite paths will arise from infinite paths by considering the initial segments of the paths from a starting point to a given finite distance. If μ is a probability measure on paths ξ from a to z and $\mathbf{P}$ is Bernoulli(p) percolation on the graph, then we combine μ and $\mathbf{P}$ as follows: if ω is the set of open edges for a percolation, then assign the positive measure $Y_\mu(\omega)$ to the open paths from a to z by letting the measure of ξ be $\mu(\xi)/\mathbf{P}[\xi \text{ open}]$ when $\xi \subseteq \omega$ and 0 otherwise. The measure $Y_\mu(\omega)$ induces a flow $\theta_\mu(\omega)$ from a to z by letting the amount of flow along an edge e be the measure that $e \in \xi$ minus the measure that $-e \in \xi$. Because $Y_\mu(\omega)$ is not necessarily a probability measure, $\theta_\mu(\omega)$ is not necessarily a unit flow. Instead, the strength of $\theta_\mu(\omega)$ is $X_\mu(\omega) := \sum_{\xi \subseteq \omega} \mu(\xi)/\mathbf{P}[\xi \text{ open}]$. Thus, $\mathbf{E}[X_\mu] = 1$ and

$$\mathbf{E}[X_\mu^2] = \sum_{\xi,\xi'} \mu(\xi)\mu(\xi')\frac{\mathbf{P}[\xi \cup \xi' \text{ open}]}{\mathbf{P}[\xi \text{ open}]\,\mathbf{P}[\xi' \text{ open}]} = \sum_{\xi,\xi'} \mu(\xi)\mu(\xi')p^{-|\xi \cap \xi'|}. \tag{5.16}$$

This is pleasingly analogous to (5.10). On the other hand,

$$\mathcal{E}\big(\theta_\mu(\omega)\big) \le \sum_e \left(\sum_{\xi \ni e} \mu(\xi)\mathbf{1}_{[\xi \subseteq \omega]}/\mathbf{P}[\xi \text{ open}] \right)^2,$$

whence

$$\mathbf{E}\big[\mathscr{E}(\theta_\mu)\big] \leq \sum_e \sum_{\xi,\xi' \ni e} \mu(\xi)\mu(\xi') \frac{\mathbf{P}[\xi \cup \xi' \text{ open}]}{\mathbf{P}[\xi \text{ open}]\,\mathbf{P}[\xi' \text{ open}]}$$

$$= \sum_{\xi,\xi'} |\xi \cap \xi'| \mu(\xi)\mu(\xi') p^{-|\xi \cap \xi'|}$$

$$= \sum_{n \geq 1} n p^{-n} \cdot (\mu \times \mu)\big[|\xi \cap \xi'| = n\big] . \tag{5.17}$$

Here, the intersection $\xi \cap \xi'$ is counted without regard to orientation of edges. Thus, our attention is focused particularly on the μ-probability that two independent paths have n edges in common.

We say that a probability measure μ on infinite paths that start at o has **exponential intersection tails with parameter** ζ, or EIT(ζ) for short, if there is some constant C such that for all n,

$$(\mu \times \mu)\big[|\xi \cap \xi'| = n\big] \leq C\zeta^n . \tag{5.18}$$

This definition and the following application of it are due to Benjamini, Pemantle, and Peres (1998).

Proposition 5.21. *If there is a probability measure on infinite paths in a graph G that has* EIT(ζ), *then for* $\zeta < p < 1$, *Bernoulli(p) percolation on G has a.s. a transient open cluster.*

Proof. The existence of a transient open cluster does not depend on the status of any finite set of edges, whence it is a tail event and has probability either 0 or 1 by Kolmogorov's zero-one law. Thus, it suffices to prove that this event has positive probability. Let μ be a measure on infinite paths starting at o that satisfies (5.18). If we identify the complement of the ball about o of radius r to a vertex z_r, then μ induces a probability measure μ_r on paths from o to z_r. Write θ_r and X_r for the random flows θ_{μ_r} and their strengths X_{μ_r} that we associated earlier to μ_r. By Thomson's principle,

$$\mathscr{R}(o \leftrightarrow z_r; \omega) \leq \mathscr{E}(\theta_r(\omega))/X_r(\omega)^2 .$$

We need to get an upper bound on the numerator and a lower bound on the denominator, with probability bounded from below. By our assumption (5.18), we have

$$\mathbf{E}[X_r^2] = \sum_{n \geq 1} p^{-n}(\mu \times \mu)\big[|\xi \cap \xi'| = n\big] \leq \sum_{n \geq 1} C(\zeta/p)^n = C\zeta/(p - \zeta) .$$

Since $\mathbf{E}[X_r] = 1$, we may deduce that

$$\mathbf{P}[X_r > 1/2] \geq \frac{p - \zeta}{4C\zeta} =: \delta > 0 \tag{5.19}$$

by the Paley–Zygmund inequality. Now our calculation (5.17) gives

$$\mathbf{E}\big[\mathscr{E}(\theta_r)\big] \leq \sum_{n \geq 1} Cn(\zeta/p)^n = Cp\zeta/(p - \zeta)^2 =: M ,$$

whence

$$\mathbf{P}\big[\mathscr{E}(\theta_r) > \beta\big] < M/\beta$$

for $\beta > 0$. Putting these bounds together, we find that

$$\begin{aligned}
\mathbf{P}\big[\mathscr{R}(o \leftrightarrow z_r; \omega) < 4\beta\big] &\geq \mathbf{P}\big[X_r > 1/2, \mathscr{E}(\theta_r) \leq \beta\big] \\
&\geq \mathbf{P}\big[X_r > 1/2\big] - \mathbf{P}\big[\mathscr{E}(\theta_r) > \beta\big] \\
&> \delta - M/\beta = \delta/2
\end{aligned}$$

if we choose $\beta := 2M/\delta$. The events $[\mathscr{R}(o \leftrightarrow z_r; \omega) < 4\beta]$ are decreasing in r, since the effective resistance is increasing, whence their intersection $[\mathscr{R}(o \leftrightarrow \infty; \omega) \leq 4\beta]$ has positive probability. But this means that the cluster of o is transient with positive probability. (In particular, the cluster is infinite with positive probability, but this already follows from (5.19), which arises from essentially the same calculations as in Section 5.3.)   ◀

▷ **Exercise 5.11.**
Show that Proposition 5.21 is sharp on trees in the sense that when T is a tree, for all $\zeta > p_c(T)$, there is a probability measure with EIT(ζ). Of course, there is no $p < p_c(T)$ such that Bernoulli(p) percolation on T has a.s. a transient open cluster.

To apply this criterion to $\mathbb{Z}^d$, we need random paths with exponential intersection tails in $\mathbb{Z}^d$. Such random paths were constructed first by Kesten for $d \geq 4$ (published by Cox and Durrett (1983)):

Proposition 5.22. *When $d \geq 4$, there is a probability measure on paths in $\mathbb{Z}^d$ with EIT(ζ) for some $\zeta < 1$.*

Proof. It clearly suffices to prove this for $d = 4$. Consider the random walk starting at $\mathbf{0}$ that takes a step in a *positive* coordinate direction with probability 1/4 for each direction. For two independent such random walks, ξ and ξ', their difference $\langle \xi(n) - \xi'(n); n \geq 0 \rangle$ is a reversible random walk on the subset $V_3 := \{(x_1, x_2, x_3, x_4); \sum_{i=1}^4 x_i = 0\}$. In addition, $|\xi \cap \xi'| \leq |\{n; \xi(n) = \xi'(n)\}|$ since the random walks on $\mathbb{Z}^4$ move only in the positive directions. The associated network on V_3 is clearly roughly isometric to the usual graph on $\mathbb{Z}^3$, whence it is transient and returns of $\xi - \xi'$ to $\mathbf{0}$ have a geometric distribution. Thus, (5.18) holds with $C = 1$ and $\zeta = \mathbf{P}\big[\exists n \geq 1\ \xi(n) - \xi'(n) = \mathbf{0}\big]$.   ◀

This result also holds in $\mathbb{Z}^3$, but obviously the same method does not work to prove it. The reader may be interested in the challenge of finding such random paths in $\mathbb{Z}^3$; this was solved by Benjamini, Pemantle, and Peres (1998).

Corollary 5.23. *There is a constant $p_0 < 1$ such that for $d \geq 4$ and $p > p_0$, Bernoulli(p) percolation in $\mathbb{Z}^d$ has a.s. a transient open cluster.*   ◀

To put this in another context, we state a few basic facts about percolation on $\mathbb{Z}^d$, but they will be proved only later. The case $d = 1$ is not of interest here, but for other d, we have that $0 < p_c(\mathbb{Z}^d) < 1$ (Proposition 7.12 and Theorem 7.16). Furthermore, when there is an infinite cluster, it is a.s. unique (Theorem 7.6).

Grimmett, Kesten, and Zhang (1993) proved Corollary 5.23 for all $p > p_c(\mathbb{Z}^d)$ and all $d \geq 3$. This was a difficult result, but a much simpler proof was found by Pete (2008), relying on ideas that we discuss in Section 6.9, among others. In any case, the upshot is that in Euclidean lattices, transience is preserved when the whole lattice is replaced by an infinite percolation cluster. We return to this issue for other graphs in Section 6.9.

5.6 Reversing the Second-Moment Inequality

The first- and second-moment methods give inequalities in very general situations. These two inequalities usually give fairly close estimates of a probability, but not so close as to agree up to a constant factor. Thus, one must search for additional information if one wants finer estimates. Usually, the estimate that the second-moment method gives is sharper than the one provided by the first moment. A method for showing the sharpness of the estimate given by the second-moment method is described here in the context of percolation; it depends on a Markov-like structure (see also Exercise 16.10).

This method seems to be due to Hawkes (1970/71) and Shepp (1972). It was applied to trees by Lyons (1992) and to Markov chains (with a slight improvement) by Benjamini, Pemantle, and Peres (1995) (see Exercise 16.10). Consider independent percolation on a tree. Embed the tree in the upper half-plane with its root at the origin. Given a minimal cutset Π separating o from ∞, order Π clockwise. Call this linear ordering $\preccurlyeq$. This has the property that for each $e \in \Pi$, the events $[o \leftrightarrow e']$ for $e' \preccurlyeq e$ are conditionally independent of the events $[o \leftrightarrow e'']$ for $e'' \succcurlyeq e$ given that $o \leftrightarrow e$. On the event $o \leftrightarrow \Pi$, define e^* to be the least edge in Π that is in the cluster of o; on $o = \Pi$, define e^* to take some value not in Π. Note that e^* is a random variable. Let σ be the (possibly defective) hitting measure

$$\sigma(e) := \mathbf{P}[e^* = e] \qquad (e \in \Pi),$$

so that

$$\sigma(\Pi) = \mathbf{P}[o \leftrightarrow \Pi].$$

Provided $\mathbf{P}[o \leftrightarrow \Pi] > 0$, we may define the probability measure

$$\mu := \sigma / \mathbf{P}[o \leftrightarrow \Pi] \in \mathcal{P}(\Pi).$$

For all $e \in \Pi$, we have

$$\sum_{e' \preccurlyeq e} \sigma(e') \frac{\mathbf{P}[o \leftrightarrow e', o \leftrightarrow e]}{\mathbf{P}[o \leftrightarrow e']} = \sum_{e' \preccurlyeq e} \mathbf{P}[e^* = e'] \mathbf{P}[o \leftrightarrow e \mid o \leftrightarrow e']$$

$$= \sum_{e' \preccurlyeq e} \mathbf{P}[e^* = e'] \mathbf{P}[o \leftrightarrow e \mid e^* = e']$$

$$= \sum_{e' \in \Pi} \mathbf{P}[e^* = e'] \mathbf{P}[o \leftrightarrow e \mid e^* = e'] = \mathbf{P}[o \leftrightarrow e].$$

Thus

$$\sum_{e' \preccurlyeq e} \mu(e') \frac{\mathbf{P}[o \leftrightarrow e', o \leftrightarrow e]}{\mathbf{P}[o \leftrightarrow e'] \mathbf{P}[o \leftrightarrow e]} = \frac{1}{\mathbf{P}[o \leftrightarrow \Pi]}.$$

By symmetry, it follows that

$$\mathscr{E}(\mu) \le 2 \sum_{e \in \Pi} \sum_{e' \le e} \mu(e)\mu(e') \frac{\mathbf{P}[o \leftrightarrow e', o \leftrightarrow e]}{\mathbf{P}[o \leftrightarrow e']\mathbf{P}[o \leftrightarrow e]} = 2 \sum_{e \in \Pi} \frac{\mu(e)}{\mathbf{P}[o \leftrightarrow \Pi]} = \frac{2}{\mathbf{P}[o \leftrightarrow \Pi]} .$$

Therefore,

$$\mathbf{P}[o \leftrightarrow \Pi] \le \frac{2}{\mathscr{E}(\mu)} \le \frac{2}{\inf_{v \in \mathcal{P}(\Pi)} \mathscr{E}(v)} .$$

To sum up, provided such orderings on cutsets Π exist, we are able to reverse the inequality of Proposition 5.12 up to a factor of 2 (Lyons, 1992). In particular, for independent percolation on trees, we get the following inequalities:

Theorem 5.24. (Tree Percolation and Conductance)   *For an independent percolation* $\mathbf{P}$ *on a tree with adapted conductances (that is, such that (5.12) holds), we have*

$$\frac{\mathscr{C}(o \leftrightarrow \infty)}{1 + \mathscr{C}(o \leftrightarrow \infty)} \le \mathbf{P}[o \leftrightarrow \infty] \le 2 \frac{\mathscr{C}(o \leftrightarrow \infty)}{1 + \mathscr{C}(o \leftrightarrow \infty)} , \tag{5.20}$$

which is the same as

$$\frac{\mathbf{P}[o \leftrightarrow \infty]}{2 - \mathbf{P}[o \leftrightarrow \infty]} \le \mathscr{C}(o \leftrightarrow \infty) \le \frac{\mathbf{P}[o \leftrightarrow \infty]}{1 - \mathbf{P}[o \leftrightarrow \infty]} . \qquad ◀$$

Consequently, we obtain a sharp refinement of Theorem 5.15:

Corollary 5.25. *We have* $\mathbf{P}[o \leftrightarrow \infty] > 0$ *in percolation on* T *iff random walk on* T *is transient for corresponding adapted conductances (satisfying (5.12)).*   ◀

This shows that the connection between percolation and random walks hinted at by Theorems 5.15 and 3.5 goes much deeper than just sharing critical parameters.

Sometimes it is useful to consider percolation on a finite portion of T (which is, in fact, how our proofs have proceeded). Recall that $\partial_L T$ denotes the set of leaves of T (other than possibly its root). We have shown that for finite trees,

$$\frac{\mathscr{C}(o \leftrightarrow \partial_L T)}{1 + \mathscr{C}(o \leftrightarrow \partial_L T)} \le \mathbf{P}[o \leftrightarrow \partial_L T] \le 2 \frac{\mathscr{C}(o \leftrightarrow \partial_L T)}{1 + \mathscr{C}(o \leftrightarrow \partial_L T)} , \tag{5.21}$$

which is the same as

$$\frac{\mathbf{P}[o \leftrightarrow \partial_L T]}{2 - \mathbf{P}[o \leftrightarrow \partial_L T]} \le \mathscr{C}(o \leftrightarrow \partial_L T) \le \frac{\mathbf{P}[o \leftrightarrow \partial_L T]}{1 - \mathbf{P}[o \leftrightarrow \partial_L T]} .$$

Remark 5.26. These inequalities take a nicer form if we add a new vertex Δ to T by joining it to o with an edge of conductance 1. Then doing random walk on this new tree $T \cup \{\Delta\}$, we have

$$\frac{\mathscr{C}(o \leftrightarrow \partial_L T)}{1 + \mathscr{C}(o \leftrightarrow \partial_L T)} = \mathbf{P}_o[\tau_{\partial_L T} \le \tau_\Delta] = \mathscr{C}(\Delta \leftrightarrow \partial_L T),$$

so (5.21) is equivalent to

$$\mathscr{C}(\Delta \leftrightarrow \partial_L T) = \mathbf{P}_o[\tau_{\partial_L T} \le \tau_\Delta] \le \mathbf{P}[o \leftrightarrow \partial_L T] \le 2\,\mathbf{P}_o[\tau_{\partial_L T} \le \tau_\Delta] = 2\,\mathscr{C}(\Delta \leftrightarrow \partial_L T).$$

Likewise, on an infinite tree, (5.20) is equivalent to

$$\mathscr{C}(\Delta \leftrightarrow \infty) = \mathbf{P}_o[\tau_\Delta = \infty] \le \mathbf{P}[o \leftrightarrow \infty] \le 2\,\mathbf{P}_o[\tau_\Delta = \infty] = 2\,\mathscr{C}(\Delta \leftrightarrow \infty). \tag{5.22}$$

Note also that the condition of being adapted, (5.12), also becomes nicer:

$$\mathbf{P}[o \leftrightarrow x] = \mathscr{C}(\Delta \leftrightarrow x). \tag{5.23}$$

▷ **Exercise 5.12.**
Give a tree for which percolation does and a tree for which percolation does not occur at criticality.

▷ **Exercise 5.13.**
Show that critical homesick random walk on supercritical Galton–Watson trees is a.s. recurrent in two ways: (1) by using Corollary 5.25; (2) by using the Nash-Williams criterion.

As in Exercise 5.13, Theorem 5.24 can be used to solve problems about random walk on deterministic or random trees. Indeed, sometimes percolation is crucial to such solutions (see, for example, Lyons (1992)).

We saw in Proposition 5.1 that when a Galton–Watson process survives, the number of survivors tends to infinity a.s. We can strengthen this by Proposition 5.9 to say that T' has generation sizes tending to infinity a.s. when the Galton–Watson tree T is infinite. In other words, when T is infinite, it contains infinitely many rays a.s. Does the same hold for Bernoulli(p) percolation on a general tree, T? It turns out that it does. Moreover, we can prove this by adapting the idea of the proof of Proposition 5.9, that is, by doing a further percolation after the Bernoulli percolation. This technique uses the inequality (5.20) that the percolation probability is captured up to a factor of 2 by effective conductance. In other words, we will show how a uniform quantitative bound on percolation probability implies a qualitative property of percolation.

To prove this, we will use the following exercise that gives an alternative expression for the energy of a flow.

▷ **Exercise 5.14.**
Let (T, c) be a transient network on a tree. Suppose that each ray of T is recurrent. Let θ be a flow from the root to infinity of finite energy.
  (a) Show that $\inf\{\mathscr{R}(o \leftrightarrow x)\,;\ |x| = n\} \to \infty$ as $n \to \infty$.
  (b) Show that $\sum_{|x|=n} \theta(e(x))^2 \downarrow 0$ as $n \to \infty$.
  (c) Show that $\mathscr{E}(\theta) = \sum_x \mathscr{R}(o \leftrightarrow x)\big[\theta(e(x))^2 - \sum_{\overleftarrow{y}=x} \theta(e(y))^2\big]$.

Proposition 5.27. (Surviving Rays in Independent Percolation) *For* $0 < p < 1$ *and every tree T, the number of surviving rays from the root under Bernoulli(p) percolation on T a.s. either is 0 or has the cardinality of the continuum. More generally, the same holds for every independent percolation on T such that each ray in T individually has probability 0 to survive.*

Proof. Let R be the event that there is exactly one surviving ray. We begin our proof by showing that the probability of R is 0, which we do for the more general hypothesis. For $n \in \mathbb{N}$, let $\mathscr{F}_n$ denote the σ-field generated by the events $[o \leftrightarrow x]$ for $|x| = n$. If $\mathbf{P}(R) > 0$, then by the Lévy zero-one law, $\mathbf{P}(R \mid \mathscr{F}_n) \to \mathbf{1}_R$ a.s. as $n \to \infty$. Thus, we may choose m and an event $B \in \mathscr{F}_m$ of positive probability so that $\mathbf{P}(R \mid B) > 3/4$. Moreover, we may choose B so that the set of x with $|x| = m$ and $o \leftrightarrow x$ is a constant on B. Denote those x by $x_1, \ldots, x_k$. Whether R occurs given B now depends only on the descendant trees T^{x_i} for $1 \leq i \leq k$, so we may consider the tree formed from T^{x_i} obtained by identifying all x_i to one vertex, which we take to be the root of a new tree, T'. Let R' be the event that the induced percolation on T' has exactly one surviving ray. Thus, $\mathbf{P}(R') = \mathbf{P}(R \mid B) > 3/4$. If R_0 denotes the event that percolation on T' has no surviving rays, then it follows that $\mathbf{P}(R_0) < 1/4$.

We are now going to define an additional percolation η on T', independent of the given percolation ω, such that $\mathbf{P}[\omega \cap \eta \notin R_0] > 1/4$ and $\mathbf{P}[\omega \cap \eta \in R_0 \mid \omega \in R'] = 1$. The second of these implies that $\mathbf{P}[\omega \cap \eta \in R_0] \geq \mathbf{P}[\omega \cap \eta \in R_0, \omega \in R'] = \mathbf{P}[\omega \in R'] > 3/4$, which contradicts the first of these.

Consider the adapted edge conductances on T' given by (5.13). As in Remark 5.26, add a new vertex Δ to T' by joining it to the root with an edge of conductance 1. Let i be the unit current flow from Δ to ∞, so that $\mathscr{E}(i) = \mathscr{R}(\Delta \leftrightarrow \infty)$ on T'. Now use Exercise 5.14 and (5.23) to write

$$\mathscr{E}(i) = \sum_x \mathbf{P}[o \leftrightarrow_\omega x]^{-1} \left[i(x)^2 - \sum_{\overset{\leftarrow}{y} = x} i(y)^2 \right], \tag{5.24}$$

where the subscript ω denotes the percolation and we write $i(x) := i(e(x))$. This expression allows us to define η to have the properties we wish. Namely, note first that all terms in (5.24) are nonnegative since $i(x) = \sum_{\overset{\leftarrow}{y} = x} i(y)$. Since $\mathscr{E}(i) < \infty$, we may choose a sequence $\langle n_k \rangle$ increasing quickly enough that for each $k \geq 1$, we have

$$\sum_{|x| \geq n_k} \mathbf{P}[o \leftrightarrow_\omega x]^{-1} \left[i(x)^2 - \sum_{\overset{\leftarrow}{y} = x} i(y)^2 \right] < \frac{\mathscr{E}(i)}{6^k}. \tag{5.25}$$

Define η to be the independent percolation where every edge is kept with probability 1 except for those $e(x)$ with $|x| = n_k$ for some k. In the latter case, we set the probability that $e(x) \in \eta$ to be $1/2$. Thus, we have that

$$\mathbf{P}[o \leftrightarrow_{\omega \cap \eta} x] = \mathbf{P}[o \leftrightarrow_\omega x] \prod_{n_k \leq |x|} \frac{1}{2}.$$

It follows from this, (5.24), and (5.25) that

$$\sum_x \mathbf{P}[o \leftrightarrow_{\omega \cap \eta} x]^{-1} \left[i(x)^2 - \sum_{\overset{\leftarrow}{y} = x} i(y)^2 \right] < \frac{3\mathscr{E}(i)}{2}. \tag{5.26}$$

To see this, break the sum over x into blocks where $|x| < n_1$, $n_1 \leq |x| < n_2$, and so on. Then apply (5.25) to the block where $n_k \leq |x| < n_{k+1}$. Therefore, the effective resistance for the conductances adapted to the percolation $\omega \cap \eta$ is less than $3\mathscr{E}(i)/2$, whence by (5.22), it follows that

$$\mathbf{P}[\omega \cap \eta \notin R_0] > \frac{2}{3\mathscr{E}(i)} \geq \frac{1}{3}\,\mathbf{P}[\omega \notin R_0] > \frac{1}{3} \cdot \frac{3}{4} = \frac{1}{4}.$$

This establishes the first property we desired for η. The second property is easy, since every given ray of T' a.s. has an edge not in η, and on the event R', there is only one ray in ω.

Thus, we have shown that $\mathbf{P}(R) = 0$. By applying this result to every descendant tree T^x, we may deduce that a.s. the set of surviving rays has no isolated point (as a subset of ∂T).

Now note that the set of surviving rays forms a closed subset of ∂T. For the culmination of the proof, assume first that T is locally finite. Then ∂T is compact by Exercise 1.1, so the surviving rays a.s. form a perfect set, which, when it is nonempty, must have the cardinality of the continuum. This completes the proof in this first case. Now let T be general. Enumerate the children of each vertex x as $x(1), x(2), \ldots$ in any order. Let R^x be the event that x is on at least one surviving ray. Fix $\epsilon > 0$. Let $N(x)$ be minimal so that $\mathbf{P}\big[\bigcup_{n=1}^{N(x)} R^{x(k)}\big] \geq (1 - \epsilon/2^{|x|+1})\,\mathbf{P}[R^x]$. Define V' to be the set of vertices $x(n)$ such that $n \leq N(x)$ for some x, as well as the root, o, of T, and let T' be the largest tree with vertices from V' that contains o. Then T' is locally finite, and the probability that T' contains a surviving ray is at least $(1 - \epsilon)\,\mathbf{P}[R^o]$, whence the probability that T' contains a continuum of surviving rays is at least $(1 - \epsilon)\,\mathbf{P}[R^o]$. Since this holds for all $\epsilon > 0$, the proof is done. ◄

This proof can be modified to handle various other random processes, as in Exercise 16.8. A different proof is given in Section 5.8, which allows other sorts of modifications, such as in Exercise 5.68. Another useful variant is in the notes, Proposition 5.36, with an application in Exercise 5.19.

5.7 Surviving Galton–Watson Trees

What does the cluster of a vertex look like in Bernoulli percolation given that the cluster is infinite? This question is easy to answer on regular trees. In fact, the analogous question for Galton–Watson trees is also easy to answer. We actually give two kinds of answers. The first answer describes the Galton–Watson tree given survival in terms of other Galton–Watson processes. The second answer tells us how large d is so that we can find d-ary subtrees of a Galton–Watson tree. The first answer will suggest some similar properties for other types of percolation on trees, which we discuss briefly.

We begin with a Galton–Watson process $\langle Z_n \rangle$ and notation as in Section 5.1. Let Z_n^* be the number of particles in generation n that have an infinite line of descent. A little thought reveals that $\langle Z_n^* \rangle$ is a Galton–Watson process where each individual has $k \geq 1$ children with probability $\sum_{j \geq k} p_j \binom{j}{k}(1 - q)^{k-1} q^{j-k}$. Now, a little thought also reveals that given extinction, $\langle Z_n \rangle$ is a Galton–Watson process. Finally, a little more thought reveals that given nonextinction, the family tree of $\langle Z_n \rangle$ has the same law as a tree grown first by $\langle Z_n^* \rangle$, then adding "bushes" independently and in the appropriate number to each node. We calculate the details and prove all this as follows.

Let T denote the genealogical tree of a Galton–Watson process with p.g.f. f and T' denote the **reduced** subtree of particles with an infinite line of descent. (Thus, $T' = \varnothing$ iff T is finite.) Let $Y_{i,j}^{(n)}$ be the indicator that the jth child of the ith particle in generation n has an infinite line of descent. Write

$$\bar{q} := 1 - q,$$

$$L_i^{*(n+1)} := \sum_{j=1}^{L_i^{(n+1)}} Y_{i,j}^{(n)},$$

$$Z_{n+1}^* := \sum_{i=1}^{Z_n} L_i^{*(n+1)},$$

and $Z_0^* = \mathbf{1}_{\text{nonextinction}}$. Note that all the random variables $Y_{i,j}^{(n)}$ for $i \le Z_n$ and $j \le L_i^{(n+1)}$ have the same distribution. In addition, for fixed n, they are independent. Likewise, all $\left(L_i^{(n+1)}, L_i^{*(n+1)}\right)$ have the same distribution for $i \le Z_n$ and, for fixed n, are independent. Thus, we will write $(L, L^*, Y_1, Y_2, \ldots, Y_L)$ for random variables having the common distribution $\left(L_i^{(n+1)}, L_i^{*(n+1)}, Y_{i,1}^{(n)}, Y_{i,2}^{(n)}, \ldots, Y_{i,L_i^{(n+1)}}^{(n)}\right)$. Parts (iii) and (iv) of the following proposition are due to Lyons (1992). Parts (i) and (ii) are illustrated in Figure 5.3.

Proposition 5.28. (Decomposition) *Suppose that* $0 < q < 1$.

(i) *The law of* T' *given nonextinction is the same as that of a Galton–Watson process with p.g.f.*

$$f^*(s) := [f(q + \bar{q}s) - q]/\bar{q}.$$

(ii) *The law of* T *given extinction is the same as that of a Galton–Watson process with p.g.f.*

$$\widetilde{f}(s) := f(qs)/q.$$

(iii) *The joint p.g.f. of* $L - L^*$ *and* L^* *is*

$$\mathbf{E}[s^{L-L^*} t^{L^*}] = f(qs + \bar{q}t).$$

More generally, the joint p.g.f. of $Z_n - Z_n^*$ *and* Z_n^* *is*

$$\mathbf{E}[s^{Z_n - Z_n^*} t^{Z_n^*}] = f^{(n)}(qs + \bar{q}t).$$

(iv) *The law of* T *given nonextinction is the same as that of a tree* $\overline{T}$ *generated as follows: Let* T^* *be the tree of a Galton–Watson process with p.g.f.* f^* *as in (i). To each vertex* x *of* T^* *having* d_x *children, attach* U_x *independent copies of a Galton–Watson tree with p.g.f.* $\widetilde{f}$ *as in (ii), where* U_x *has the p.g.f.*

$$\mathbf{E}[s^{U_x}] = \frac{(D^{d_x} f)(qs)}{(D^{d_x} f)(q)},$$

where d_x *derivatives of* f *are indicated; all* U_x *and all trees added are mutually independent given* T^*. *The resultant tree is* $\overline{T}$.

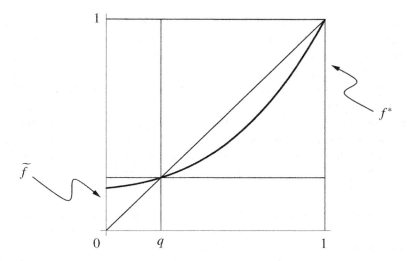

Figure 5.3. The graph of f embeds scaled versions of the graphs of $\widetilde{f}$ and f^*.

Proof. We begin with (iii). We have

$$\mathbf{E}[s^{L-L^*}t^{L^*}] = \mathbf{E}\Big[\mathbf{E}[s^{L-L^*}t^{L^*} \mid L]\Big] = \mathbf{E}\Big[\mathbf{E}\big[s^{\sum_{j=1}^{L}(1-Y_j)}t^{\sum_{j=1}^{L}Y_j} \mid L\big]\Big]$$

$$= \mathbf{E}\Big[\mathbf{E}[s^{1-Y_1}t^{Y_1}]^L\Big] = \mathbf{E}\big[(qs + \bar{q}t)^L\big] = f(qs + \bar{q}t)\,.$$

The other part of (iii) follows by a precisely parallel calculation.

To show (i), we need to show that on the event $L_i^{*(n)} \neq 0$ given the σ-field $\mathscr{F}_{n,i}$ generated by $L_k^{*(n)}$ ($k \neq i$) and $L_k^{*(m)}$ ($m < n$, $k \geq 1$), the p.g.f. of $L_i^{*(n)}$ is f^*. Indeed, on the event $L_i^{*(n)} \neq 0$, we have

$$\mathbf{E}[s^{L_i^{*(n)}} \mid \mathscr{F}_{n,i}] = \mathbf{E}[s^{L_i^{*(n)}} \mid L_i^{*(n)} \neq 0] = \mathbf{E}[s^{L^*}\mathbf{1}_{[L^* \neq 0]}]/\bar{q} = \mathbf{E}[s^{L^*}(1 - \mathbf{1}_{[L^*=0]})]/\bar{q}$$

$$= \big\{\mathbf{E}[1^{L-L^*}s^{L^*}] - \mathbf{P}[L^* = 0]\big\}/\bar{q} = [f(q1 + \bar{q}s) - q]/\bar{q} = f^*(s)$$

by (iii).

Similarly, (ii) comes from the fact that on the event of extinction,

$$\mathbf{E}[s^{L_i^{(n)}} \mid \mathscr{F}_{n,i}] = \mathbf{E}[s^{L_i^{(n)}} \mid L_i^{*(n)} = 0] = \mathbf{E}[s^L\mathbf{1}_{[L^*=0]}]/q$$

$$= \mathbf{E}[s^{L-L^*}0^{L^*}]/q = f(qs + \bar{q}0)/q = \widetilde{f}(s)\,,$$

since $0^x = \mathbf{1}_{\{0\}}(x)$.

Finally, (iv) follows once we show that the function claimed to be the p.g.f. of U_x is the same as the p.g.f. of $L - L^*$ given $L^* = d_x$. Once again, by (iii), we have that for some constants c and c',

$$\mathbf{E}[s^{L-L^*} \mid L^* = d] = c\left(\frac{\partial}{\partial t}\right)^d f(qs + \bar{q}t)\bigg|_{t=0} = c'(D^d f)(qs)\,.$$

Substitution of $s = 1$ yields $c' = 1/(D^d f)(q)$.   ◀

Remark. Part (iii), which led to all the other parts of the proposition, is an instance of the following general calculation. Suppose that we are given a nonnegative random number X of particles with X having p.g.f. F. One interpretation of F is that if we color each particle red independently with probability r, then the probability that all particles are colored red is $F(r)$. Now suppose that we first categorize each particle independently as small or big, with the probability of a particle being categorized small being q. Then independently color each small particle red with probability s and color each big particle red with probability t. Since the chance that a given particle is colored red is then $qs + \bar{q}t$, we have (by our interpretation) that the probability that all particles are colored red is $F(qs + \bar{q}t)$. On the other hand, if Y is the number of big particles, then by conditioning on X, we see that this is also equal to the joint p.g.f. of $X - Y$ and Y.

Another interesting question concerning the surviving Galton–Watson trees is whether binary subtrees can be found among the survivors. In fact, this issue first arose in various applications: see, for example, the proof of Theorem 5.33, due to Chayes, Chayes, and Durrett (1988), or Lemma 5 of Pemantle (1988). More generally, let $\tau(d)$ be the probability that a Galton–Watson tree contains a d-ary subtree beginning at the initial individual. Thus, $\tau(1)$ is the survival probability, $1 - q$. Pakes and Dekking (1991) found a method to determine $\tau(d)$, after special cases were solved by Chayes, Chayes, and Durrett (1988) and Dekking and Meester (1990):

Theorem 5.29. (*d*-ary Subtrees) *Let f be the p.g.f. of a supercritical Galton–Watson process. Set*

$$G_d(s) := \sum_{j=0}^{d-1} (1 - s)^j \frac{(D^j f)(s)}{j!} .$$

Then $1 - \tau(d)$ is the smallest fixed point of G_d in $[0, 1]$.

Note that $G_1 = f$. Thus, this answer is a nice extension of Proposition 5.4.

Proof. We first reinterpret G_d. Let $g_d(s)$ be the probability that the root has at most $d - 1$ marked children when each child is marked independently with probability $1 - s$. This function is clearly monotonic increasing in s. By considering how many children the root has in total, we see that

$$g_d(s) = \sum_{k=0}^{\infty} p_k \sum_{j=0}^{d-1} \binom{k}{j}(1 - s)^j s^{k-j} . \tag{5.27}$$

After changing the order of summation in (5.27), we obtain

$$g_d(s) = \sum_{j=0}^{d-1} \frac{(1 - s)^j}{j!} \sum_{k \geq j} p_k \, k(k - 1) \cdots (k - j + 1) s^{k-j}$$

$$= \sum_{j=0}^{d-1} \frac{(1 - s)^j (D^j f)(s)}{j!} = G_d(s) ,$$

that is, $g_d = G_d$.

Now we are ready to conclude in a similar fashion to the proof of Proposition 5.4. Let q_n be the chance that the Galton–Watson tree does not contain a d-ary subtree of height n at the initial individual; notice that $1 - q_n \downarrow \tau(d)$. By marking a child of the root when it has a d-ary subtree of height $n - 1$, we see that $q_n = G_d(q_{n-1})$. Since $q_0 = 0$ and G_d is nonnegative, increasing, and satisfies $G_d(1) = 1$, letting $n \to \infty$ shows that $\lim_n q_n = 1 - \tau(d)$ is the smallest fixed point of G_d. ◀

▷ **Exercise 5.15.**

Show that for a Galton–Watson tree with offspring distribution $\mathrm{Bin}(d + 1, p)$, we have that $\tau(d) > 0$ for p sufficiently close to 1.

An interesting feature of the phase transitions from $\tau(d) = 0$ to $\tau(d) > 0$ as the parameter is varied in, for example, binomial, geometric or Poisson offspring distributions is that unlike the case of usual percolation $d = 1$, for $d \geq 2$ the probability of having the d-ary subtree is positive at criticality. For integers $n \geq 2$ and $1 \leq d \leq n$, define $\pi(n, d)$ to be the infimum of probabilities p such that $\tau(d) > 0$ in a Galton–Watson tree with offspring distribution $\mathrm{Bin}(n, p)$. Of course, such offspring distributions also describe the law of the clusters in Bernoulli(p) percolation on an n-ary tree.

•

Proposition 5.30. (Nontrivial Phase Transition) *Let T be an n-ary tree and $2 \leq d < n$. Then $\pi(n, d) \in [1/n, 1)$. For Bernoulli(p) percolation on T, the probability that there is an open d-ary subtree is 0 for $p < \pi(n, d)$ and is 1 for $p \geq \pi(n, d)$.*

Proof. The fact that $\pi(n, d) < 1$ follows from Exercise 5.15, and the fact that $\pi(n, d) \geq 1/n$ follows from the fact that $p_c(T) = 1/n$.

Let $\theta(p)$ denote the probability (called $\tau(d)$ earlier) that the root of T belongs to an open d-ary subtree. Similarly, let $\theta_k(p)$ denote the probability that the root of T belongs to an open d-ary subtree of height at least k. Then $\theta_k(p) \downarrow \theta(p)$ as $k \to \infty$. Choose $\alpha_0 \in \left(0, \binom{n}{d}^{-1/(d-1)}\right)$; thus, there exists $c_0 < 1$ such that for all $\alpha \in [0, \alpha_0]$, we have $\binom{n}{d}\alpha^d \leq c_0\alpha$. We claim that if p and k_0 are such that $\theta_{k_0}(p) \in [0, \alpha_0]$, then $\theta(p) = 0$. To see this, note that since $\theta_{k+1}(p)$ is the probability that there exist at least d open edges incident to the root that lead to children who begin open d-ary subtrees of height at least k, it follows from a union bound that $\theta_{k+1}(p) \leq \binom{n}{d}\theta_k(p)^d$, and this, in turn, is at most $c_0\theta_k(p)$ if $\theta_k(p) \leq \alpha_0$. Thus, $\theta_{k+1}(p) \leq c_0\theta_k(p)$ for all $k \geq k_0$, and so, letting $k \to \infty$, we find that $\theta(p) = 0$.

Now suppose that $\theta(\pi(n, d)) = 0$. Then, for some k, we have $\theta_k(\pi(n, d)) < \alpha_0$, whence also $\theta_k(p) < \alpha_0$ for some $p > \pi(n, d)$. But this implies by the above claim that $\theta(p) = 0$, contradicting the definition of $\pi(n, d)$. Kolmogorov's 0-1 law completes the proof. ◀

For a binomial offspring distribution, the critical mean value is asymptotically d, as shown by the following result. This was proved by Balogh, Peres, and Pete (2006), where it was applied to bootstrap percolation, a model we discuss in Section 7.8.

We now show how to calculate $\pi(n, d)$ and that for large n and d, we have $\pi(n, d) \sim d/n$, which is as small as it could be since $\mathrm{Bin}(n, d/n)$ has mean d.

Proposition 5.31. (Asymptotics of the Critical Probability) *Consider Bernoulli offspring distributions. The critical probability $\pi(n, d)$ is the infimum of all p for which the equation*

$$\mathbf{P}\big[\mathrm{Bin}(n, (1 - s)p) \leq d - 1\big] = s \tag{5.28}$$

has a real root $s \in [0, 1)$. For any constant $\gamma \in [0, 1]$ and sequence of integers d_n with $\lim_{n \to \infty} d_n/n = \gamma$, we have

$$\lim_{n \to \infty} \pi(n, d_n) = \gamma. \tag{5.29}$$

Proof. In the proof of Theorem 5.29, we saw that $G_d = g_d$; in the present case, it is clear that $g_d(s) = \mathbf{P}\big[\mathrm{Bin}(n, (1 - s)p) \leq d - 1\big] =: B_{n,d,p}(s)$. If the probability of not having the required subtree is denoted by $y = y(p)$, then by Theorem 5.29, y is the smallest fixed point of the function $B_{n,d,p}(s)$ in $s \in [0, 1]$. One fixed point is $s = 1$, and $\pi(n, d)$ is the infimum of the p values for which there is a fixed point $s \in [0, 1)$. It is easy to see that

$$\frac{\partial}{\partial s} B_{n,d,p}(s) = np\, \mathbf{P}\big[\mathrm{Bin}(n - 1, (1 - s)p) = d - 1\big],$$

which is positive for $s \in [0, 1)$, with at most one extremal point (a maximum) in $(0, 1)$. Thus $B_{n,d,p}(s)$ is a monotone strictly increasing function, with $B_{n,d,p}(0) > 0$ when $p < 1$, and with at most one inflection point in $(0, 1)$. When $(n - 1)p < d - 1$, there is no inflection point, and $B_{n,d,p}(s)$ is concave in $[0, 1]$.

If $\lim_{n \to \infty} d_n/n = \gamma$, then for any fixed p and s, by the weak law of large numbers,

$$B_{n,d_n,p}(s) = \mathbf{P}\left[\frac{\mathrm{Bin}(n, (1 - s)p)}{n} \leq \frac{d_n - 1}{n}\right] \to \begin{cases} 1 & \text{if } (1 - s)p < \gamma \\ 0 & \text{if } (1 - s)p > \gamma \end{cases}$$

as $n \to \infty$. Solving the equation $(1 - s)p = \gamma$ for s gives a critical value $s_c = 1 - \gamma/p$. Thus for $p < \gamma$ we have $\lim_{n \to \infty} B_{n,d_n,p}(s) \to 1$ for all $s \in [0, 1]$; since for large enough n, $B_{n,d_n,p}(s)$ is concave in $[0, 1]$, there is no positive root $s < 1$ of $B_{n,d_n,p}(s) = s$. On the other hand, for $p > \gamma$ there must be a root $s = s(n)$ for large enough n. These prove (5.29). ◀

Pakes and Dekking (1991) used Theorem 5.29 to show that the critical mean value for a geometric offspring distribution to produce with positive probability a d-ary subtree in the Galton–Watson tree is asymptotic to ed as $d \to \infty$ (where e is the base of the natural logarithm), and it is asymptotically d for a Poisson offspring distribution.

We now give an application of Proposition 5.30 to fractal percolation. Given an integer $b \geq 2$ and $p \in [0, 1]$, consider the natural tiling of the unit square $[0, 1]^2$ by b^2 closed squares of side $1/b$. Let $\mathcal{K}_1$ be a random subcollection of these squares, where each square has probability p of belonging to $\mathcal{K}_1$, and these events are mutually independent. (Thus the cardinality $|\mathcal{K}_1|$ of $\mathcal{K}_1$ is a binomial random variable.) In general, if $\mathcal{K}_n$ is a collection of squares of side b^{-n}, tile each square $Q \in \mathcal{K}_n$ by b^2 closed subsquares of side b^{-n-1} (with disjoint interiors) and include each of these subsquares in $\mathcal{K}_{n+1}$ with probability p (independently). Finally, define

$$A_n := A_{n,b}(p, \mathcal{K}_{n-1}) := \bigcup \mathcal{K}_n \quad \text{and} \quad Q_b(p) := \bigcap_{n=1}^{\infty} A_n.$$

In the construction of $Q_b(p)$, the cardinalities $|\mathcal{K}_n|$ of $\mathcal{K}_n$ form a Galton–Watson branching process where the offspring distribution is $\text{Bin}(b^2, p)$. The process $\langle A_n ; n \geq 1 \rangle$ is called *fractal percolation*, whereas $Q_b(p)$ is the *limit set*.

We prove the following theorem of Chayes, Chayes, and Durrett (1988): if the probability p is close enough to 1, then with positive probability, the limit set of the planar fractal percolation $Q_2(p)$ contains a *left-to-right crossing* of the unit square, that is, a continuous path $(x(t), y(t)): [0, 1] \to [0, 1]^2$ such that $x(0) = 0$, $x(1) = 1$.

We begin with a discrete analogue. A *left-to-right crossing of squares at level n* is a sequence of distinct squares $a_1, \ldots, a_r$ contained in $\mathcal{K}_n$ so that a_1 shares a side with the left side of $[0, 1]^2$, all pairs of successive squares a_j, a_{j+1} share a side, and a_r shares a side with the right side of $[0, 1]^2$.

Lemma 5.32. *Consider fractal percolation with $b \geq 3$. If each square retained at level n always contains at least $b^2 - 1$ surviving subsquares at level $n + 1$, then there is a left-to-right crossing of squares at all levels n.*

Proof. Clearly there is a left-to-right crossing of squares at level 1. More generally, all the squares in $\mathcal{K}_1$ can be connected via paths of squares in $\mathcal{K}_1$ with consecutive pairs sharing a side; let us say that $\mathcal{K}_1$ is *side connected* for the purposes of this proof.

We proceed by induction. Suppose there is a left-to-right crossing of squares $a_1, \ldots, a_r$ at level n. For notational convenience, let a_0 and a_{r+1} both be the unit square. Let S_i be the common side of a_i with a_{i+1} for $0 \leq i \leq r$. Since $b \geq 3$, for each i, there is a pair of squares in $\mathcal{K}_{n+1}$, each having a side on S_i, call them $c_i \subset a_i$ and $d_i \subset a_{i+1}$, such that c_i and d_i share a side (of length b^{-n-1}); see Figure 5.4. Since the subset of $\mathcal{K}_{n+1}$ contained in a_{i+1} is side connected, each d_i is connected to c_{i+1} by squares in $\mathcal{K}_{n+1}$ that are contained in a_{i+1}. It follows that there is a left-to-right crossing in $\mathcal{K}_{n+1}$ from d_0 to c_r. This proves the induction step. ◀

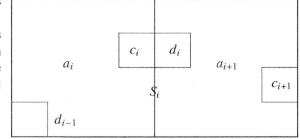

Figure 5.4. Notation for the proof of Lemma 5.32.

▷ **Exercise 5.16.**
Is Lemma 5.32 true for $b = 2$?

Define $\theta_n(p)$ as the probability of a left-to-right crossing of squares at level n. The sequence θ_n is decreasing in n, hence its limit $\theta_\infty(p)$, the chance of a left-to-right crossing in $Q_2(p)$, exists.

Theorem 5.29 and Lemma 5.32 can be combined to give an easy proof that there is a nontrivial phase where there exist left-to-right crossings in the limit set.

Theorem 5.33. (Left-to-Right Crossing in Fractal Percolation) *For p close enough to 1, the left-to-right crossing probability $\theta_\infty(p)$ is positive.*

Proof. We do the case $b \geq 3$, leaving $b = 2$ for an exercise. By Lemma 5.32, it suffices to show that with positive probability there exist $\mathcal{K}'_n \subseteq \mathcal{K}_n$ with the properties that for all $n \geq 1$, we have $\bigcup \mathcal{K}'_n \supseteq \bigcup \mathcal{K}'_{n+1}$ and each square of $\mathcal{K}'_n$ contains at least $b^2 - 1$ subsquares in

$\mathcal{K}'_{n+1}$. This event occurs if and only if the tree associated with $\langle \mathcal{K}_n \rangle$ contains a $(b^2 - 1)$-ary descendant subtree from the root. Exercise 5.15 shows that such subtrees exist with positive probability provided p is large enough.   ◄

Exercise 5.67 shows that Theorem 5.33 holds for $b = 2$ as well.

More information on fractal percolation can be found in Exercises 5.70, 5.71, and 15.22 as well as in Example 15.9.

5.8 Harris's Inequality

A property that is valid for Bernoulli percolation on any graph is an inequality due to Harris (1960), though it is nowadays usually called the FKG inequality due to an extension by Fortuin, Kasteleyn, and Ginibre (1971). In fact, Harris's inequality holds for independent percolation more generally.

This inequality permits us to conclude such things as the positive correlation of the events $\{x \leftrightarrow y\}$ and $\{u \leftrightarrow v\}$ for any vertices x, y, u, and v. The special property that these two events have is that they are increasing, where an event $A \subseteq 2^{\mathsf{E}}$ is called ***increasing*** if whenever $\omega \in A$ and $\omega \subseteq \omega'$, then $\omega' \in A$. As a natural extension, we call a random variable X on 2^{E} ***increasing*** if $X(\omega) \leq X(\omega')$ whenever $\omega \subseteq \omega'$. Thus, $\mathbf{1}_A$ is increasing iff A is increasing. Similar definitions apply for site processes, that is, random subsets of vertices.

Harris's Inequality. *Let* $\mathbf{P}$ *be an independent percolation on a graph.*

(i) *If* A *and* B *are both increasing events, then* A *and* B *are positively correlated:* $\mathbf{P}(AB) \geq \mathbf{P}(A)\,\mathbf{P}(B)$.

(ii) *If* X *and* Y *are both increasing random variables with finite second moments, then* $\mathbf{E}[XY] \geq \mathbf{E}[X]\,\mathbf{E}[Y]$. *Equality holds iff* X *and* Y *are independent.*

Property (i) of Harris's inequality says, by definition, that the measure $\mathbf{P}$ has ***positive associations*** (in contrast to the negative associations of the uniform spanning tree measure in Section 4.2).

▷ **Exercise 5.17.**

Let $\mathbf{P}$ be an independent percolation. Suppose that X is an increasing random variable with finite expectation, $F \subset \mathsf{E}$ is finite, and $\mathscr{F}$ is the (finite) σ-field generated by the coordinate functions $\omega \mapsto \omega(e)$ ($\omega \in 2^{\mathsf{E}}$) for the edges $e \in F$. Show that $\mathbf{E}[X \mid \mathscr{F}]$ is an increasing random variable.

Harris's inequality is essentially the following lemma, where the case $d = 1$ is due to Chebyshev.

Lemma 5.34. *Suppose that* $\mu_1, \dots, \mu_d$ *are probability measures on* $\mathbb{R}$ *and* $\mu := \mu_1 \times \cdots \times \mu_d$. *Let* $f, g \in L^2(\mathbb{R}^d, \mu)$ *be functions that are increasing in each coordinate separately. Then*

$$\int fg \, d\mu \geq \int f \, d\mu \int g \, d\mu.$$

Proof. We proceed by induction on d. Suppose first that $d = 1$. Note that

$$[f(x) - f(y)][g(x) - g(y)] \geq 0$$

for all $x, y \in \mathbb{R}$. Therefore

$$0 \leq \iint [f(x) - f(y)][g(x) - g(y)] \, d\mu(x) \, d\mu(y) = 2 \int fg \, d\mu - 2 \int f \, d\mu \int g \, d\mu,$$

which gives the desired inequality.

Now suppose that the inequality holds for $d = k$, and let us prove the case $d = k + 1$. Define

$$f_1(x_1) := \int_{\mathbb{R}^k} f(x_1, x_2, \ldots, x_{k+1}) \, d\mu_2(x_2) \cdots d\mu_{k+1}(x_{k+1})$$

and similarly define g_1. Clearly f_1 and g_1 are increasing functions, whence

$$\int f_1 g_1 \, d\mu_1 \geq \int f_1 \, d\mu_1 \int g_1 \, d\mu_1 = \int f \, d\mu \int g \, d\mu. \tag{5.30}$$

On the other hand, the inductive hypothesis tells us that for each fixed x_1,

$$f_1(x_1) g_1(x_1) \leq \int_{\mathbb{R}^k} f(x_1, x_2, \ldots, x_{k+1}) \, g(x_1, x_2, \ldots, x_{k+1}) \, d\mu_2(x_2) \cdots d\mu_{k+1}(x_{k+1}),$$

whence $\int f_1 g_1 \, d\mu_1 \leq \iint fg \, d\mu$. In combination with (5.30), this proves the inequality for $d = k + 1$ and completes the proof. ◄

Proof of Harris's inequality. The proofs for bond and site processes are identical; our notation will be for bond processes.

Since (i) derives from (ii) by using indicator random variables, it suffices to prove (ii). If X and Y depend on only finitely many edges, then the inequality is a consequence of Lemma 5.34, since $\omega(e)$ are mutually independent random variables for all e. To prove it in general, write $\mathsf{E} = \{e_1, e_2, \ldots\}$. Let X_n and Y_n be the expectations of X and Y conditioned on $\omega(e_1), \ldots, \omega(e_n)$. According to Exercise 5.17, the random variables X_n and Y_n are increasing, whence $\mathbf{E}[X_n Y_n] \geq \mathbf{E}[X_n] \mathbf{E}[Y_n]$. Since $X_n \to X$ and $Y_n \to Y$ in L^2 by the martingale convergence theorem, which implies that $X_n Y_n \to XY$ in L^1, we may deduce the desired inequality for X and Y.

For the equality condition, decomposing X and Y into their positive and negative parts reduces it to the case that $X, Y \geq 0$. Then write $X = \int_0^\infty \mathbf{1}_{[X \geq s]} \, ds$ and $Y = \int_0^\infty \mathbf{1}_{[Y \geq t]} \, dt$. Thus, $\mathbf{E}[XY] - \mathbf{E}[X] \mathbf{E}[Y] = \iint \big(\mathbf{P}[X \geq s, Y \geq t] - \mathbf{P}[X \geq s] \mathbf{P}[Y \geq t]\big) \, ds \, dt \geq 0$, with equality iff for a.e. s and t, the events $[X \geq s]$ and $[Y \geq t]$ are independent. ◄

▷ **Exercise 5.18.**
Let $\mathbf{P}$ be an independent percolation. Show that if $A_1, \ldots, A_n$ are increasing events, then for all p, $\mathbf{P}\big(\bigcap A_i\big) \geq \prod \mathbf{P}(A_i)$ and $\mathbf{P}\big(\bigcap A_i^c\big) \geq \prod \mathbf{P}(A_i^c)$.

We can use Harris's inequality to give a short proof of Proposition 5.27:

Proposition 5.27. (Surviving Rays in Independent Percolation) *For $0 < p < 1$ and every tree T, the number of surviving rays from the root under Bernoulli(p) percolation on T a.s. either is 0 or has the cardinality of the continuum. More generally, the same holds for every independent percolation on T such that each ray in T individually has probability 0 to survive.*

Proof. Let R_k be the event that there are exactly k open rays. We must show that for every finite $k \geq 1$, the probability of R_k is 0, which we do for the more general hypothesis. As in our first proof of this proposition, it then follows that the number of open rays is a.s. either 0 or $2^{\aleph_0}$. For $x \in V(T)$, write Rays_x for the set of surviving rays (starting at o) that pass through x. If $\mathbf{P}(R_k) > 0$, then by the Lévy zero-one law, $\mathbf{P}(R_k \mid \mathscr{F}_n) \to \mathbf{1}_{R_k}$ a.s. Thus, we may choose m so that $\mathbf{P}(R_k \mid \mathscr{F}_m) \geq 8/9$ on an event $B \in \mathscr{F}_m$ of positive probability. We may also choose $n \geq m$ so large that for all $x \in T_n$, we have $\mathbf{P}[|\mathsf{Rays}_x| \geq 1 \mid B] < 1/3$, as otherwise there would be a fixed ray that is open with probability at least $\mathbf{P}(B)/3$. Thus, we may partition $T_n = A_1 \cup A_2$ in such a way that $\mathbf{P}(\mathscr{A}_1 \mid B) \in (1/3, 2/3)$, where $\mathscr{A}_1 := \left[\sum_{x \in A_1} |\mathsf{Rays}_x| \geq k\right]$. Define $\mathscr{A}_2 := \left[\sum_{x \in A_2} |\mathsf{Rays}_x| \geq 1\right]$. Then $\mathbf{P}(\mathscr{A}_2 \mid B) \geq 8/9 - \mathbf{P}(\mathscr{A}_1 \mid B)$ and $\mathscr{A}_i$ are positively correlated given B by Harris's inequality, whence $\mathbf{P}(\mathscr{A}_1 \cap \mathscr{A}_2 \mid B) > 4/27 > 1/9$. Since occurrence of $\mathscr{A}_1 \cap \mathscr{A}_2$ precludes occurrence of R_k, this contradicts the choice of B. ◀

5.9 Galton–Watson Networks

We have seen examples in Corollary 5.10 and Exercise 5.13 of the application of percolation to questions that do not appear to involve percolation; other examples will appear in Exercise 5.55 and Proposition 13.3. We will give yet another example in this section. Consider the following ***Galton–Watson network*** generated by a random variable $\mathscr{L} := (L, A_1, \ldots, A_L)$, where $L \in \mathbb{N}$, $A_i \in (0, 1]$. First, use L as an offspring random variable to generate a Galton–Watson tree. Let the number of children of particle x be L_x. Complete these random variables to i.i.d. random variables $\mathscr{L}_x$. Thus, $\mathscr{L}_x = (L_x, A_{y_1}, \ldots, A_{y_{L_x}})$, where $y_1, \ldots, y_{L_x}$ are the children of x.[*] Use these random variables to assign edge capacities (or weights or conductances)

$$c(e(x)) := \prod_{w \leq x} A_w.$$

We will see the usefulness of such networks for the study of random fractals in Section 15.3. When can water flow to ∞ in such a random network? Note that if $A_i \equiv 1$, then water can flow to ∞ iff the tree is infinite. Let

$$\gamma := \mathbf{E}\left[\sum_{i=1}^{L} A_i\right].$$

This is the expected value of the constraint on the total that can flow out of a vertex compared to what can flow into that vertex. The following theorem of Falconer (1986) shows that

---

[*] A formal definition using the framework of Exercise 5.2 makes independent random variables $\mathscr{L}_x$ for each finite string $x = \langle i_1, i_2, \ldots, i_n \rangle$ of positive integers.

the condition for positive flow to infinity with positive probability on these Galton–Watson networks is $\gamma > 1$. Of course, when $A_i \equiv 1$, this reduces to the usual survival criterion for Galton–Watson processes. In general, this theorem confirms the intuition that for flow to infinity to be possible, more water must be able to flow from parent to children, on average, than from grandparent to parent.

Theorem 5.35. (Flow in Galton–Watson Networks) *If $\gamma \leq 1$, then a.s. no flow is possible unless $\sum_1^L A_i = 1$ a.s. If $\gamma > 1$, then flow is possible a.s. on nonextinction.*

Proof. As usual, let T_1 denote the set of individuals (or vertices) of the first generation. Let F be the maximum strength of an admissible flow, that is, of a flow that satisfies the capacity constraints on the edges.

For $x \in T_1$, let F_x be the maximum strength of an admissible flow from x to infinity through the subtree T^x with renormalized capacities $e \mapsto c(e)/A_x$. Thus, F_x has the same law as F has. It is easily seen that $F = \sum_{x \in T_1} \big(A_x \wedge (A_x F_x) \big) = \sum_{x \in T_1} A_x (1 \wedge F_x)$. Now suppose that $\gamma \leq 1$. Taking expectations in the preceding equation yields

$$
\mathbf{E}[F] = \mathbf{E}\big[\mathbf{E}[F \mid \mathscr{L}_o]\big] = \mathbf{E}\left[\mathbf{E}\left[\sum_{x \in T_1} A_x(1 \wedge F_x) \,\Big|\, \mathscr{L}_o\right]\right] = \mathbf{E}\left[\sum_{x \in T_1} \mathbf{E}[A_x(1 \wedge F_x) \mid \mathscr{L}_o]\right]
$$

$$
= \mathbf{E}\left[\sum_{x \in T_1} A_x\, \mathbf{E}[1 \wedge F_x \mid \mathscr{L}_o]\right] = \mathbf{E}\left[\sum_{x \in T_1} A_x\, \mathbf{E}[1 \wedge F]\right] = \gamma\, \mathbf{E}[1 \wedge F] \leq \mathbf{E}[1 \wedge F].
$$

Therefore $F \leq 1$ almost surely and $\mathbf{P}[F > 0] > 0$ only if $\gamma = 1$. In addition, we have, by independence,

$$
\|F\|_\infty = \left\| \sum_{x \in T_1} A_x \right\|_\infty \|F\|_\infty .
$$

If $\|F\|_\infty > 0$, it follows that $\left\| \sum_{x \in T_1} A_x \right\|_\infty = 1$. In combination with $\gamma = 1$, this implies that $\sum_{x \in T_1} A_x = 1$ a.s.

For the second part, we introduce percolation as in the proof of Corollary 5.10 via Proposition 5.9. Namely, augment the probability space so that for each vertex $u \neq o$, there is a random variable X_u with the following properties. Denote by $\mathscr{F}$ the σ-field generated by the random variables $\mathscr{L}_u$. Then given $\mathscr{F}$, all X_u are independent and each X_u takes the value 1 with probability A_u and 0 otherwise.* Consider the subtree of the Galton–Watson tree consisting of the initial individual o together with those individuals u such that $\prod_{w \leq u} X_w = 1$. This subtree has, unconditionally, the law of a Galton–Watson branching process with progeny distribution the unconditional law of $\sum_{u \in T_1} X_u$. Let Q be the probability that this subtree is infinite conditional on $\mathscr{F}$. Now the unconditional mean of the new process is

$$
\mathbf{E}\left[\sum_{u \in T_1} X_u\right] = \mathbf{E}\left[\mathbf{E}\left[\sum_{u \in T_1} X_u \,\Big|\, \mathscr{L}_o\right]\right] = \mathbf{E}\left[\sum_{u \in T_1} A_u\right] = \gamma,
$$

---

* A formal definition uses independent, uniform $[0, 1]$ random variables M_u and defines X_u to be the indicator that $M_u \leq A_u$.

whence if $\gamma > 1$, then this new Galton–Watson branching process survives with positive probability, say, Q. Of course, $Q = \mathbf{E}[Q]$. On the other hand, for any cutset Π of the original Galton–Watson tree, $\sum_{e \in \Pi} c(e)$ is the expected number, given $\mathscr{F}$, of edges in Π that are also in the subtree. This expectation is at least the probability that the number of such edges is at least one, which, in turn, is at least Q:

$$F = \inf_{\Pi} \sum_{e \in \Pi} c(e) \geq Q \,.$$

Hence $F > 0$ on the event $Q > 0$. The event $Q > 0$ has positive probability since $Q > 0$, whence $\mathbf{P}[F > 0] > 0$. Since the event that $F = 0$ is clearly inherited, it follows that $\mathbf{P}[F = 0] = q$. ◀

We will return to percolation on trees in Chapter 16.

5.10 Notes

Galton–Watson processes are sometimes called Bienaymé–Galton–Watson processes, because Bienaymé, in 1845, was the first to give the fundamental theorem, Proposition 5.4. However, while he stated the correct result, he gave barely a hint of his proof. See Heyde and Seneta (1977), pp. 116–120, and Guttorp (1991), pp. 1–3, for some of the history. The first published proof of Bienaymé's theorem appears on pp. 83–86 of Cournot (1847), where the context is gambling: An urn contains tickets marked with nonnegative integers, a proportion p_k of them being marked k. Pierre begins with 1 écu, which he gives to Paul for the right of drawing a ticket from the urn. If the ticket is marked k, then Paul gives Pierre k écus. The ticket is returned to the urn. This is the end of the first round. For the second and succeeding rounds, if Pierre is not broke, then for each of the écus he has, he repeats the procedure of the first round. The problem was to determine the probability, for each n, that Pierre is broke at the end of the nth round. If one keeps track of Pierre's fortune at all times, not merely at the ends of rounds, then one obtains a coding of the associated tree by a random walk as in Exercise 5.35.

For other codings of the trees as random walks, as well as various uses, see Geiger (1995), Pitman (1998), Bennies and Kersting (2000), Dwass (1969), Harris (1952), Le Gall and Le Jan (1998), Duquesne and Le Gall (2002), Marckert and Mokkadem (2003), Marckert (2008), Lamperti (1967), and Kendall (1951).

For additional material on branching processes, see Chapter 12, the books by Athreya and Ney (1972) and Asmussen and Hering (1983), and the review articles by Vatutin and Zubkov (1985, 1993).

The study of percolation was initiated by Broadbent and Hammersley (1957). Exact values of $p_c(G)$ are rarely known. For some classes of Cayley graphs that are not too far from trees, exact values have been calculated by Kozáková (2008) and Špakulová (2009).

Tree-indexed random walks are used in Section 5.4 to create examples of quasi-independent percolation. They are studied further in Section 13.8.

A more subtle example of a quasi-independent percolation than the one of Example 5.18 is obtained by replacing the requirement $S(x) \in [0, N]$ by $S(x) \geq 0$. That this is quasi-independent is proved by Benjamini and Peres (1994b) in the course of proving their Theorem 1.1. For more on this particular case, see Pemantle and Peres (1995a).

Raoufi and Yadin (2017) proved that there is a probability measures on paths with exponential intersection tails on each transient Cayley graph of a group that has a finite-index subgroup admitting a nontrivial homomorphism to $\mathbb{R}$. In particular, this includes all groups of polynomial growth with order at least cubic, due to the proof of a theorem of Gromov (1981a).

Proposition 5.27 is due to Pemantle and Peres (1996), Lemma 4.2(i). The following variant is new. For $x \in \mathsf{V}(T)$, let $\mathscr{F}^x$ denote the σ-field generated by the events $[o \leftrightarrow y]$ for all y except $y > x$. For $x \in \mathsf{V}(T)$, write Rays_x for the set of surviving rays (starting at o) that pass through x.

Proposition 5.36. *Let* **P** *be a general percolation on a general tree T with the following property: there is some $\epsilon > 0$ such that for all $x \in V(T)$,*

$$\mathbf{P}\big[|\mathsf{Rays}_x| \geq 2 \mid \mathscr{F}^x\big] \geq \epsilon \mathbf{1}_{[o \leftrightarrow x]} . \tag{5.31}$$

Then the number of open rays is a.s. 0 or $2^{\aleph_0}$.

Proof. As in the second proof of Proposition 5.27 on p. 164, let R_k be the event that there are exactly k open rays. We must show that for every finite $k \geq 1$, the probability of R_k is 0. As in the proof of Proposition 5.27, it then follows that the number of open rays is a.s. either 0 or $2^{\aleph_0}$. Suppose that $\mathbf{P}(R_k) > 0$ for some $k \in [1, \infty)$. For $n \in \mathbb{N}$, let $\mathscr{F}_n$ denote the σ-field generated by the events $[o \leftrightarrow x]$ for $|x| = n$. By the Lévy zero-one law, $\mathbf{P}(R_k \mid \mathscr{F}_n) \to \mathbf{1}_{R_k}$ a.s. Write V_n for the random set $\{|x| = n \,;\, x \leftrightarrow o\}$. Then $\mathbf{P}\big[|V_n| \geq k \mid R_k\big] \to 1$ as $n \to \infty$. Thus, we may choose an integer n and an event $A \in \mathscr{F}_n$ of positive probability such that

$$\mathbf{P}(R_k \mid A) > 1 - \epsilon^k \tag{5.32}$$

and $A \subseteq \big[|V_n| \geq k\big]$. Let $X_1, \ldots, X_k$ be an $\mathscr{F}_n$-measurable choice of k distinct vertices in V_n, or $X_i := o$ if $|V_n| < k$; for example, if we embed T in the upper half-plane with its root at the origin, we can choose the k left-most vertices in V_n. Consider nonrandom distinct vertices $x_1, \ldots, x_k \in T_n$ such that $\mathbf{P}(C) > 0$, where C is the event $[X_1 = x_1, \ldots, X_k = x_k]$. Define B_i to be the event that $|\mathsf{Rays}_{x_i}| \geq 2$. Since $A, C \in \mathscr{F}_n \subseteq \mathscr{F}^{x_i}$ and so $B_1 \cap \cdots \cap B_{i-1} \cap A \cap C \in \mathscr{F}^{x_i}$ for $1 \leq i \leq k$, we may apply (5.31) to get

$$\mathbf{P}(B_i \mid B_1 \cap \cdots \cap B_{i-1} \cap A \cap C) \geq \epsilon,$$

whence

$$\mathbf{P}(B_1 \cap \cdots \cap B_k \mid A \cap C) \geq \epsilon^k .$$

Therefore, $\mathbf{P}\big[|\mathsf{Rays}_o| \geq 2k \mid A \cap C\big] \geq \epsilon^k$. Since this holds for all choices of C, it follows that $\mathbf{P}\big[|\mathsf{Rays}_o| \geq 2k \mid A\big] \geq \epsilon^k$. This contradicts (5.32), whence $\mathbf{P}(R_k) = 0$, as desired. ◄

▷ **Exercise 5.19.**
Let T be a binary tree. Give random labels $A(x)$ to its vertices as follows: Begin with $A(o) := 0$. The other labels are defined recursively by dividing $A(x)$ at random (as integers) between the children of x and adding 1 to each child. That is, if the two children of x are x_1 and x_2, then $A(x_1)$ is uniform in $\{1, 2, \ldots, A(x) + 1\}$ and $A(x_2) = A(x) + 2 - A(x_1)$. Given $A(x)$, all labels $A(y)$ for $y \notin T^x$ are conditionally independent of all $A(z)$ for $z > x$.

   (a) Let $\xi = \langle \xi_n \,;\, n \geq 0 \rangle$ be a ray in T starting at some vertex $\xi_0 = x$. Show that $\big\langle 2^n(A(\xi_n) - 2) \big\rangle$ is a martingale.

   (b) Show that for all $n \geq 1$, we have $\mathbf{P}\big[A(\xi_k) \geq 4 \text{ for } 1 \leq k \leq n \mid A(\xi_0)\big] \leq 2^{-n}\big(A(\xi_0) - 2\big)/2$ on the event $[A(\xi_0) \geq 4]$.

   (c) Show that a.s. there is no infinite path in T all of whose labels are at least 4.

   (d) If the labels are real numbers instead of integers with the sole change that $A(x_1)$ is uniform in $[1, A(x) + 1]$, then show that again a.s. there is no infinite path in T all of whose labels are at least 4.

   The proof of Proposition 5.30 is based on a similar argument in Chayes, Chayes, and Durrett (1988). The result itself was first given by Pakes and Dekking (1991).

   Theorem 5.33 was suggested by Mandelbrot (1982), Chapter 23. One might wonder whether for large p, there is also positive probability that a directed (horizontally monotonic) left-to-right crossing exists. However, this is not so, as was proved by Chayes (1995b). An extension was given by Chayes, Pemantle, and Peres (1997).

The proof presented here of Theorem 5.35 is new. The proof of the first part of the theorem uses the same idea as the proof of Lemma 4.4(b) of Falconer (1987). The result of Theorem 5.35 was extended in a few ways to general fixed trees by Lyons and Pemantle (1992) and to Galton–Watson networks by Faraud (2011).

Recursions on trees are a very powerful and general tool. Exercise 5.60 gives some simple examples. Additional ones along these lines for more general processes are analyzed by Pemantle and Peres (2010).

5.11 Collected In-Text Exercises

5.1. Justify the differentiation in (5.3). Show too that $\lim_{s \uparrow 1} f'(s) = m$.

5.2. Let $\mathscr{T}$ be the space of rooted, labeled, locally finite trees. Verify that the function $d(T, T') := (1 + \sup\{n\,;\ T{\upharpoonright}n = T'{\upharpoonright}n\})^{-1}$ is a metric on $\mathscr{T}$ and that $(\mathscr{T}, d)$ is complete and separable.

5.3. Define the measure **GW** formally on the space $\mathscr{T}$ of Exercise 5.2; your measure should be the law of a random tree produced by a Galton–Watson process with arbitrary given offspring distribution.

5.4. Show that for any Galton–Watson process with mean $m > 1$, the family tree T has growth rate $\operatorname{gr} T = m$ a.s. given nonextinction. (Don't use the Kesten–Stigum theorem to show this, as we have not yet proved that theorem.)

5.5. Show that for a general percolation, the events $[x \leftrightarrow a]$ and $[x \leftrightarrow \infty]$ are indeed measurable.

5.6. Prove (5.5).

5.7. Show that for $p \geq p_c(G)$, we have $p_c(\omega) = p_c(G)/p$ for $\mathbf{P}_p$-a.e. ω. Physicists often refer to p as the "density" of edges in ω, and this helps the intuition.

5.8. Show that, conversely to (5.13), the survival parameters adapted to given edge resistances are

$$p_x = \frac{1 + \sum_{o < u < x} r(e(u))}{1 + \sum_{o < u \leq x} r(e(u))}\,.$$

For example, simple random walk ($c \equiv 1$) is adapted to $p_x = |x|/(|x| + 1)$.

5.9. Verify the assertion of Example 5.17.

5.10. Show that if $\operatorname{br} T < 1\big/\cos\dfrac{\pi}{N+2}$, then the root belongs to an infinite cluster in ω_N of (5.14) with probability zero.

5.11. Show that Proposition 5.21 is sharp on trees in the sense that when T is a tree, for all $\zeta > p_c(T)$, there is a probability measure with EIT(ζ). Of course, there is no $p < p_c(T)$ such that Bernoulli(p) percolation on T has a.s. a transient open cluster.

5.12. Give a tree for which percolation does and a tree for which percolation does not occur at criticality.

5.13. Show that critical homesick random walk on supercritical Galton–Watson trees is a.s. recurrent in two ways: (1) by using Corollary 5.25; (2) by using the Nash-Williams criterion.

5.14. Let (T, c) be a transient network on a tree. Suppose that each ray of T is recurrent. Let θ be a flow from the root to infinity of finite energy.
   (a) Show that $\inf\{\mathscr{R}(o \leftrightarrow x)\,;\ |x| = n\} \to \infty$ as $n \to \infty$.
   (b) Show that $\sum_{|x|=n} \theta(e(x))^2 \downarrow 0$ as $n \to \infty$.
   (c) Show that $\mathscr{E}(\theta) = \sum_x \mathscr{R}(o \leftrightarrow x)\big[\theta(e(x))^2 - \sum_{\overleftarrow{y}=x} \theta(e(y))^2\big]$.

5.15. Show that for a Galton–Watson tree with offspring distribution $\text{Bin}(d+1, p)$, we have that $\tau(d) > 0$ for p sufficiently close to 1, where $\tau(d)$ is the probability that the tree contains a d-ary subtree beginning at the initial individual.

5.16. Is Lemma 5.32 true for $b = 2$?

5.17. Let $\mathbf{P}$ be an independent percolation. Suppose that X is an increasing random variable with finite expectation, $F \subset \mathsf{E}$ is finite, and $\mathscr{F}$ is the (finite) σ-field generated by the coordinate functions $\omega \mapsto \omega(e)$ ($\omega \in 2^{\mathsf{E}}$) for the edges $e \in F$. Show that $\mathbf{E}[X \mid \mathscr{F}]$ is an increasing random variable.

5.18. Let $\mathbf{P}$ be an independent percolation. Show that if $A_1, \ldots, A_n$ are increasing events, then for all p, $\mathbf{P}(\bigcap A_i) \geq \prod \mathbf{P}(A_i)$ and $\mathbf{P}(\bigcap A_i^c) \geq \prod \mathbf{P}(A_i^c)$.

5.19. Let T be a binary tree. Give random labels $A(x)$ to its vertices as follows: Begin with $A(o) := 0$. The other labels are defined recursively by dividing $A(x)$ at random (as integers) between the children of x and adding 1 to each child. That is, if the two children of x are x_1 and x_2, then $A(x_1)$ is uniform in $\{1, 2, \ldots, A(x) + 1\}$ and $A(x_2) = A(x) + 2 - A(x_1)$. Given $A(x)$, all labels $A(y)$ for $y \notin T^x$ are conditionally independent of all $A(z)$ for $z > x$.

  (a) Let $\xi = \langle \xi_n ; n \geq 0 \rangle$ be a ray in T starting at some vertex $\xi_0 = x$. Show that $\langle 2^n (A(\xi_n) - 2) \rangle$ is a martingale.

  (b) Show that for all $n \geq 1$, we have $\mathbf{P}[A(\xi_k) \geq 4 \text{ for } 1 \leq k \leq n \mid A(\xi_0)] \leq 2^{-n}(A(\xi_0) - 2)/2$ on the event $[A(\xi_0) \geq 4]$.

  (c) Show that a.s. there is no infinite path in T all of whose labels are at least 4.

  (d) If the labels are real numbers instead of integers with the sole change that $A(x_1)$ is uniform in $[1, A(x) + 1]$, then show that again a.s. there is no infinite path in T all of whose labels are at least 4.

5.12 Additional Exercises

5.20. Consider a rooted Galton–Watson tree (T, o) whose offspring distribution is $\text{Poisson}(c)$ for some $c > 0$. This is sometimes called a ***Poisson–Galton–Watson(c) tree***. If the total number of vertices of T is $k < \infty$, then label the vertices of T uniformly with the integers $1, \ldots, k$. (These labels are different from the sequence labels used in Section 5.1, where the purpose of the sequence labels was merely to make a formal definition.) Show that every rooted, labeled tree on k vertices arises with probability $e^{-ck} c^{k-1}/k!$. Consequently, if we condition that $|\mathsf{V}(T)| = k$, then the rooted, labeled tree is uniformly distributed among all rooted, labeled trees on k vertices, that is, it has the distribution of a uniformly rooted uniform spanning tree on the complete graph on $\{1, 2, \ldots, k\}$.

5.21. Consider Bernoulli(c/n) percolation on the complete graph K_n with fixed $c > 0$ and any $n \geq c$. Fix a vertex o of K_n. Let $C(o)$ be the cluster of o "rooted" at o.

  (a) Show that as $n \to \infty$, the distribution of $C(o)$ tends to that of a rooted Galton–Watson tree (T, o) whose offspring distribution is $\text{Poisson}(c)$ in the sense that for every r, one can couple the ball of radius r about o in $C(o)$ with the ball of radius r about the root in a Galton–Watson tree in such a way that they are equal with probability tending to 1 as $n \to \infty$.

  (b) Let C_{L} be the result of labeling the vertices of $C(o)$ uniformly by the integers $1, \ldots, |C(o)|$. Show that for $k < \infty$, if we condition that $|C(o)| = k$, then the distribution of the labeled cluster C_{L} tends to that of a uniformly labeled Poisson–Galton–Watson(c) tree conditioned to have size k.

  (c) Show that for a rooted, labeled tree T of size $k < \infty$,

$$\lim_{n \to \infty} \mathbf{P}[C_{\mathrm{L}} = T] = \frac{e^{-ck} c^{k-1}}{k!}.$$

This gives another solution to Exercise 5.20.

5.22. Suppose that L_n are offspring random variables that converge in law to an offspring random variable L with $\mathbf{P}[L = 1] < 1$. Let the corresponding extinction probabilities be q_n and q. Show that $q_n \to q$ as $n \to \infty$.

5.23. Give another proof that $m \leq 1 \Rightarrow$ a.s. extinction unless $p_1 = 1$ as follows. Let Z_1^* be the number of particles of the first generation with an infinite line of descent. Let $Z_1^{(i)*}$ be the number of children of the ith particle of the first generation with an infinite line of descent. Show that $Z_1^* = \sum_{i=1}^{Z_1} 1 \wedge Z_1^{(i)*}$. Take the L^1 and L^∞ norms of this equation to get the desired conclusion.

5.24. Give another proof of Propositions 5.1 and 5.4(i) as follows. Write

$$Z_{n+1} = Z_n + \sum_{i=1}^{Z_n} \left(L_i^{(n+1)} - 1 \right).$$

This is a randomly sampled random walk, that is, a random subsequence of the locations of a random walk on $\mathbb{Z}$ whose steps have the same distribution as $L - 1$. Apply the strong law of large numbers or the Chung–Fuchs theorem (Durrett (2019), Theorem 5.4.8), as appropriate.

5.25. Let $\langle X_n \rangle$ be a Markov chain on $\mathbb{N}$ such that the law of X_{n+1} given X_n is Poisson with parameter X_n. Show that a.s. for all large n, we have $X_n = 0$.

5.26. Show that $\langle Z_n \rangle$ is a nonnegative supermartingale when $m \leq 1$ to give another proof of a.s. extinction when $m \leq 1$ and $p_1 \neq 1$.

5.27. Suppose that the offspring random variable L in a Galton–Watson process has first moment $m > 1$ and a finite second moment. Show that the second moments of Z_n/m^n are bounded. Deduce that $Z_n/m^n \to W$ in L^2.

5.28. Show directly that if $0 < r < 1$ satisfies $f(r) = r$, then $\langle r^{Z_n} \rangle$ is a martingale. Use this to give another proof of Proposition 5.1 and Proposition 5.4 in the case $m > 1$.

5.29. **(Grey, 1980)** Let $\langle Z_n \rangle$, $\langle Z_n' \rangle$ be independent Galton–Watson processes with identical offspring distribution and arbitrary, possibly random, Z_0, Z_0' with $Z_0 + Z_0' \geq 1$ a.s., and set

$$Y_n := \begin{cases} Z_n/(Z_n + Z_n') & \text{if } Z_n + Z_n' \neq 0 \\ Y_{n-1} & \text{if } Z_n + Z_n' = 0. \end{cases}$$

(a) Fix n. Let A be the event

$$A := \left[Z_{n+1} + Z_{n+1}' \neq 0 \right]$$

and $\mathscr{F}_n$ be the σ-field generated by $Z_0, \ldots, Z_n, Z_0', \ldots, Z_n'$. Use symmetry to show that

$$\mathbf{E}\left[L_i^{(n+1)}/(Z_{n+1} + Z_{n+1}') \mid \mathscr{F}_n \right] = 1/(Z_n + Z_n') \quad \text{a.s. on the event } A.$$

(b) Show that $\langle Y_n \rangle$ is a martingale with a.s. limit Y.

(c) If $1 < m < \infty$, show that $0 < Y < 1$ a.s. on the event $Z_n \not\to 0$ and $Z_n' \not\to 0$. *Hint:* Let $Y^{(k)} := \lim_{n\to\infty} Z_n/(Z_n + Z_{k+n}')$. Then $\mathbf{E}[Y^{(k)} \mid Z_0, Z_k'] = Z_0/(Z_0 + Z_k')$ and $\mathbf{P}[Y = 1, Z_n \not\to 0, Z_n' \not\to 0] = \mathbf{P}[Y^{(k)} = 1, Z_n \not\to 0, Z_n' \not\to 0] \leq \mathbf{E}[Y^{(k)} \mathbf{1}_{[Z_k' > 0]}] \to 0$ as $k \to \infty$.

5.30. In the notation of Exercise 5.29, show that if $Z_0 \equiv Z_0' \equiv 1$ and $1 < m < \infty$, then

$$\mathbf{P}\left[Y \in (0, 1) \right] = (1 - q)^2 + p_0^2 + \sum_{n \geq 1} \left[f^{(n+1)}(0) - f^{(n)}(0) \right]^2.$$

5.31. Deduce the Seneta–Heyde theorem from Grey's theorem (Exercise 5.29).

5.32. Show that for any Galton–Watson process with $m < \infty$, we have $Z_{n+1}/Z_n \to m$ a.s. on nonextinction.

5.33. Let f be the p.g.f. of a Galton–Watson process. Show that

$$\mathbf{E}\big[s^{\sum_{n=0}^{N} Z_n}\big] = g_N(s),$$

where $g_0(s) := s$ and $g_{n+1}(s) := sf\big(g_n(s)\big)$ for $n \geq 1$. Define $s_0 := \sup\{t/f(t) \,;\, t \geq 1\} \geq 1$. Show that

$$\mathbf{E}\big[s^{\sum_{n \geq 0} Z_n}\big] = g_\infty(s),$$

where $g_\infty(s) := \lim_{N \to \infty} g_N(s) = sf\big(g_\infty(s)\big)$ is finite for $s < s_0$. Show that if the process is subcritical and $f(s) < \infty$ for some $s > 1$, then $s_0 > 1$.

5.34. Let $0 < p < 1$ and consider a branching process such that each individual has two children with probability p and 0 children otherwise. Show that the probability generating function for the total number of individuals is

$$s \mapsto \frac{1 - \sqrt{1 - 4p(1-p)s^2}}{2ps}.$$

5.35. Let L be the offspring random variable of a Galton–Watson process. There are various useful ways to encode by a random walk the family tree when it is finite. We consider one such way here. Suppose that the process starts with k individuals, that is, $Z_0 = k$. Let $Z_{\mathrm{tot}} := \sum_{n \geq 0} Z_n$ be the total size of the process. Let $S_n := \sum_{j=1}^{n}(L_j - 1)$, where L_j are independent copies of L.

    (a) Show that $\mathbf{P}[Z_{\mathrm{tot}} = n] = \mathbf{P}[S_n = -k, \ \forall i < n \ S_i > -k]$.

    (b) (Otter–Dwass Formula) Show that $\mathbf{P}[Z_{\mathrm{tot}} = n] = \frac{k}{n}\,\mathbf{P}[S_n = -k]$.

    (c) Show that in the noncritical case, the expected number of visits to $-k$ of the random walk $\langle S_n \,;\, n \geq 0 \rangle$ is $q^k/(1 - f'(q))$, where q is the extinction probability of the Galton–Watson process and f is the probability generating function of L. Show that in the critical case, this expectation is infinite.

    (d) Let τ_n be the time of the nth visit to $-k$ of the random walk $\langle S_n \,;\, n \geq 0 \rangle$, where $\tau_n := \infty$ if fewer than n visits are made to $-k$. Show that $\mathbf{P}[\tau_1 < \infty] = q^k$ and $\mathbf{E}\big[\sum_n 1/\tau_n\big] = q^k/k$.

5.36. Consider a Galton–Watson process with offspring distribution equal to Poisson(1) and total size $Z_{\mathrm{tot}} := \sum_{n \geq 0} Z_n$.

    (a) Show that $\mathbf{P}[Z_{\mathrm{tot}} = n] = n^{n-1}e^{-n}/n!$. *Hint:* Use the Otter–Dwass formula from Exercise 5.35(b).

    (b) By comparing with Exercise 5.20, derive Cayley's formula that the number of trees on n vertices is n^{n-2}.

5.37. Consider a Galton–Watson process with offspring distribution equal to Poisson(1). Let o be the root and X be a random uniform vertex of the tree. Show that $\mathbf{P}[X = o] = 1/2$.

5.38. Let T be a tree. Show that for $p < 1/\overline{\mathrm{gr}}\,T$, the expected size of the cluster of a vertex is finite for Bernoulli(p) percolation, whereas it is infinite for $p > 1/\overline{\mathrm{gr}}\,T$.

5.39. Let T be an infinite, locally finite tree. Define the random tree $T(p)$ by contracting $e(x)$ for each $x \neq o$ independently with probability p. Show that $\sup\{p \,;\, T(p) \text{ is locally finite a.s.}\} = p_{\mathrm{c}}(T)$ and calculate the distribution of $\mathrm{br}\,T(p)$ for $p < p_{\mathrm{c}}(T)$.

5.40. Deduce Corollary 5.10 from Hawkes's earlier result (that is, from the special case where it is assumed that $\mathbf{E}[L(\log L)^2] < \infty$) by considering truncation of the progeny random variable, L.

5.41. Let $\theta_n(p)$ denote the probability that the root of an n-ary tree has an infinite cluster under Bernoulli(p) percolation. Thus, $\theta_n(p) = 0$ iff $p \leq 1/n$. Calculate and graph $\theta_2(p)$ and $\theta_3(p)$. Show that for all $n \geq 2$, the left-hand derivative of θ_n at 1 is 0 and the right-hand derivative of θ_n at $1/n$ is $2n^2/(n-1)$.

5.42. Let Π be a cutset in a locally finite graph on which is defined some percolation measure. Show that for $\mu_1, \mu_2 \in \mathcal{P}(\Pi)$,

$$\mathscr{E}\left(\frac{\mu_1 + \mu_2}{2}\right) + \mathscr{E}\left(\frac{\mu_1 - \mu_2}{2}\right) = \frac{\mathscr{E}(\mu_1) + \mathscr{E}(\mu_2)}{2},$$

where $\mathscr{E}(\,\cdot\,)$ is extended in the obvious way to nonprobability measures on Π from the definition (5.10).

5.43. Let Π be a cutset in a locally finite graph on which is defined some percolation measure, **P**. Show that if $\mathbf{P}\big[e_1, e_2 \subseteq K(o)\big] \neq 0$ for every pair $e_1, e_2 \in \Pi$, then $\mathscr{E}(\,\cdot\,)$ has a unique minimum on $\mathcal{P}(\Pi)$.

5.44. Let T be the family tree of a supercritical Galton–Watson process. Show that a.s. on the event of nonextinction, simple random walk on T is transient.

5.45. Let T be the family tree of a supercritical Galton–Watson process without extinction and with mean m. For $0 < \lambda < m$, consider RW_λ with the conductances $c(e) = \lambda^{-|e|}$.
  (a) Show that $\mathbf{E}\big[\mathscr{C}(o \leftrightarrow \infty; T)\big] \le m - \lambda$ and $\mathbf{E}\big[\mathscr{R}(o \leftrightarrow \infty; T)\big] \ge 1/(m - \lambda)$.
  (b) Show that $\mathbf{E}\big[\mathbf{P}[\tau_o^+ = \infty]\big] \le 1 - \lambda/m$.

5.46. Given a quasi-independent percolation on a locally finite tree T with $\mathbf{P}[o \leftrightarrow u \mid o \leftrightarrow \overleftarrow{u}] \equiv p$, show that if $p < (\mathrm{br}\,T)^{-1}$, then $\mathbf{P}[o \leftrightarrow \infty] = 0$ whereas if $p > (\mathrm{br}\,T)^{-1}$, then $\mathbf{P}[o \leftrightarrow \infty] > 0$.

5.47. Let T be any tree on which simple random walk is transient. Let U be a random variable uniform on $[0, 1]$. Define a percolation on T by taking the subtree of all edges at distance at most $1/U$. Show that the inequality of Theorem 5.19 fails if conductances are adapted to this percolation.

5.48. Let T be any infinite tree. Let $U(x)$ be i.i.d. random variables uniform on $[0, 1]$ indexed by the vertices of T. Define a percolation on T by taking the subgraph spanned by all vertices x such that $U(x) \le U(o)$. Show that the root belongs to an infinite cluster with positive probability iff $\mathrm{br}\,T > 1$. Show that this percolation is not quasi-independent.

5.49. Improve the Paley–Zygmund inequality to Cantelli's inequality: if X is a random variable with mean 1 and $t < 1$, then

$$\mathbf{P}[X > t] \ge \frac{(1 - t)^2}{(1 - t)^2 + \mathrm{Var}(X)}.$$

5.50. Consider the usual graph on $\mathbb{Z}^d$ and orient each edge in its positive coordinate direction. For Bernoulli percolation on $\mathbb{Z}^d$, the set of oriented, open paths is called ***oriented percolation***. The critical value $p_c(d)$ for oriented percolation is the supremum of those p such that in Bernoulli(p) percolation, there is no infinite, oriented, open path. Show that $\lim_{d \to \infty} d\,p_c(d) = 1$.

5.51. Consider independent percolation on a tree T with the survival probability of each edge bounded away from 1. Let m_n denote the expected number of vertices of T at level n that are connected to the root, o.
  (a) Show that if $\sum_n m_n^{-1} = \infty$, then $\mathbf{P}[o \leftrightarrow \infty] = 0$.
  (b) Show that if T is spherically symmetric, then the converse of (a) holds.
  (c) Give an example of a nonspherically symmetric T of bounded degree where the converse of (a) fails.

5.52. Consider Bernoulli(1/2) percolation on a tree T. Let $p_n := \mathbf{P}_{1/2}[o \leftrightarrow T_n]$.
  (a) Show that if there is a constant c such that for every n we have $|T_n| \le c 2^n$, then $p_n \le 4c/(n+2c)$ for all n.
  (b) Show that if there are constants c_1 and c_2 such that for every n we have $|T_n| \ge c_1 2^n$ and also for every vertex x we have $|T_n^x| \le c_2 2^n$, then $p_n \ge c_3/(n + c_3)$ for all n, where $c_3 := 2c_1^2/c_2^3$.

5.53. Let $T(n)$ be a binary tree of height n and consider Bernoulli(1/2) percolation. For large n, which inequalities in (5.21) are closest to equalities?

5.54. Let T be a binary tree and consider Bernoulli(p) percolation with $p \in (1/2, 1)$. Which inequalities in Theorem 5.24 have the ratios of the two sides closest to 1?

5.55. Randomly stretch a tree T by adding vertices to ("subdividing") the edges to make the edge $e(x)$ a path of length $L_x(\omega)$ with L_x i.i.d. Call the resulting tree $T(\omega)$. Calculate the distribution of $\operatorname{br} T(\omega)$ in terms of $\operatorname{br} T$ and of the distribution of L_x. Show that if $\operatorname{br} T > 1$, then $\operatorname{br} T(\omega) > 1$ a.s.

5.56. Suppose that $\operatorname{br} T = \overline{\operatorname{gr}} T$. Consider the stretched tree $T(\omega)$ from Exercise 5.55. Show that $\operatorname{br} T(\omega) = \overline{\operatorname{gr}} T(\omega)$ a.s.

5.57. Consider the stretched tree $T(\omega)$ from Exercise 5.55. If we assume only that simple random walk on T is transient, must simple random walk on $T(\omega)$ be transient a.s.?

5.58. Consider a percolation on T for which there is some $M' > 0$ such that, for all $x, y \in T$ and $A \subseteq T$ with the property that the removal of $x \wedge y$ would disconnect y from every vertex in A,

$$\mathbf{P}[o \leftrightarrow y \mid o \leftrightarrow x, \ o = A] \geq M' \, \mathbf{P}[o \leftrightarrow x \mid o \leftrightarrow x \wedge y] \,.$$

Show that the adapted conductances satisfy

$$\mathbf{P}[o \leftrightarrow \infty] \leq \frac{2}{M'} \frac{\mathscr{C}(o \leftrightarrow \infty)}{1 + \mathscr{C}(o \leftrightarrow \infty)} \,.$$

5.59. Consider the percolation of Example 5.20. Sharpen Exercise 5.10 by showing that if the root belongs to an infinite cluster with positive probability, then RW_λ is transient on T for $\lambda := 1 \big/ \cos \dfrac{\pi}{N+2}$.

5.60. The inequalities (5.21) can also be proved by entirely elementary means.
  (a) Prove that if $0 < x_n \leq 1$, then

$$\sum \frac{1 - x_n}{x_n} \leq \frac{1 - \prod x_n}{\prod x_n} \quad \text{and} \quad \sum \frac{1 - x_n}{1 + x_n} \geq \frac{1 - \prod x_n}{1 + \prod x_n} \,.$$

  (b) Use induction to deduce (5.21) from the inequalities of part (a).
  (c) Prove that for $C \geq 0$ and $C/(1 + C) < p \leq 1$,

$$p \left(1 - \exp \left(\frac{-2C}{p(1 + C) - C} \right) \right) \leq 1 - e^{-2C} \,.$$

  (d) Use induction to deduce from part (c) the following sharper form of the right-hand inequality of (5.21):

$$\mathbf{P}[o \leftrightarrow \partial_L T] \leq 1 - \exp\left(-2\mathscr{C}(o \leftrightarrow \partial_L T)\right) \,.$$

5.61. Show that if f is the p.g.f. of a supercritical Galton–Watson process, then the p.g.f. of Z_n given survival is

$$[f^{(n)}(s) - f^{(n)}(qs)]/\overline{q} \,.$$

5.62. A *multitype Galton–Watson branching process* has J types of individuals, with an individual of type i generating k_j individuals of type j for $j = 1, \ldots, J$ with probability $p_{\mathbf{k}}^{(i)}$, where $\mathbf{k} := (k_1, \ldots, k_J)$. As in the single-type case, all individuals generate their children independently of each other. Show that a supercritical single-type Galton–Watson tree given survival has the following alternative description as a two-type Galton–Watson tree. Let $\langle p_k \rangle$ be the offspring distribution and q be the extinction probability. Let type 1 have offspring distribution obtained as follows: Begin with k children with probability $p_k(1 - q^k)/(1 - q)$ for $k \geq 1$. Then make each child type 1 with probability $1 - q$ and type 2 with probability q, independently but conditional on having at least one type-1 child. The type-2 offspring distribution is simpler: it has k children of type 2 (only) with probability $p_k q^{k-1}$. Let the root be type 1.

5.63. Let $1 \le b < d < n$. Recall the critical probabilities $\pi(\cdot, \cdot)$ of Proposition 5.30. Show that $\pi(n, d)\pi(d, b) \ge \pi(n, b)$.

5.64. Let $d \ge 2$. Consider a Galton–Watson process with offspring distribution being geometric with parameter p. Write $\tau_p(d)$ for the probability that the Galton–Watson tree contains a d-ary subtree beginning with the initial individual. Show that there exists $p_0 \in (0, 1)$ such that for $p < p_0$, we have $\tau_p(d) = 0$ and for $p \ge \tau_p(d)$, we have $\tau_p(d) > 0$. Prove a similar statement when the offspring distribution is Poisson with parameter λ.

5.65. Show that the extinction probability $y(p)$ introduced in the proof of Proposition 5.31 satisfies $y(p) \to 0$ as $p \to 1$.

5.66. Show that

$$\pi(n, n-1) = \frac{(n-1)^{2n-3}}{n^{n-1}(n-2)^{n-2}}$$

for the critical probability of Section 5.7. What is the probability of having an $(n-1)$-ary subtree exactly at this value of p?

5.67. This exercise uses the notion of stochastic domination, as explained in Exercise 4.46 for real-valued random variables and in Sections 7.4 and 10.2 with respect to inclusion.

　(a) Show that for $p \in (0, 1)$, there exists $q \in (0, 1)$ such that the first stage $A_{1,b^2}(p)$ of one fractal percolation is stochastically dominated by the second stage $A_{1,b}(q) \cap A_{2,b}(q)$ of another fractal percolation. *Hint:* Let Bern(q) denote a Bernoulli random variable with parameter q, that is, a random variable that takes the value 1 with probability q and 0 otherwise. Take q so that Bern(q) dominates the maximum of b^2 independent copies of a Bern$(\sqrt{p})$ random variable.

　(b) Deduce that Theorem 5.33 holds for $b = 2$ as well.

5.68. Fix an infinite tree T. Let $\varphi\colon \mathsf{V}(T) \to \mathbb{R}$ be given, as well as real-valued independent random variables $A(x)$ for $x \in \mathsf{V}(T)$. Define the associated *T-indexed random walk* $S(x) := \sum_{y \le x} A(y)$. Let Φ be the set of rays $\xi \in \partial T$ such that $S(x) > \varphi(x)$ for all $x \in \xi$. Suppose that for all $\xi \in \partial T$, the probability that $\xi \in \Phi$ is 0. Show that a.s. $|\Phi| \in \{0, 2^{\aleph_0}\}$. *Hint:* Modify the proof in Section 5.8 of Proposition 5.27.

5.69. Let T be a tree such that $\sup_n |T_n|/\sqrt{n} < \infty$. Assign ± 1 labels to each vertex independently with probability 1/2 each. Show that a.s. there is no ray along which the sum of the labels stays positive.

5.70. Consider fractal percolation in the rectangle $[0, 1] \times [0, 2]$ with parameters b and p. That is, take a union of two independent copies of fractal percolation $Q_b(p)$ in the unit square as in Theorem 5.33, one translated vertically by 1. Let $\lambda_k(p)$ denote the probability of a left-to-right crossing with retained squares of side length b^{-k} and write $\lambda_\infty(p) := \lim_k \lambda_k(p)$.

　(a) Show that there exists $p_{1\times2} \in [1/b^2, 1)$ such that $\lambda_\infty(p) = 0$ for $p < p_{1\times2}$ and $\lambda_\infty(p) > 0$ for $p > p_{1\times2}$.

　(b) Show that $p_{1\times2} \ge 1/b$.

　(c) Prove that $\lambda_{k+1}(p) \le \big((4b-3)\lambda_k(p)\big)^2$. *Hint:* Consider first the case $b = 2$. Show that crossing $[0, 1/2] \times [0, 2]$ in level $k+1$ requires a horizontal crossing of at least one of three specific rectangles of shape $1/2 \times 1$, or a vertical crossing of at least one of two specific squares of side $1/2$.

　(d) Deduce that $\lambda_\infty(p_{1\times2}) > 0$.

5.71. Consider fractal percolation $Q_b(p)$ in the unit square $[0, 1] \times [0, 1]$ with parameters b and p. Recall that $\theta_k(p)$ denotes the probability of a left-to-right crossing with retained squares of side length b^{-k} and

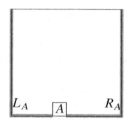

Figure 5.5. See Exercise 5.71.

$\theta_\infty(p) := \lim_k \theta_k(p)$. Let $p_{1\times1} := \inf\{p \in [0,1]; \; \theta_\infty(p) > 0\}$. For a level m square A on the bottom of the unit square, let L_A be the left side of the unit square union the bottom side to the left of A, and define R_A analogously (see Figure 5.5 on the preceding page). For $k \geq m$, write $\theta_k(p, A)$ for the probability there exists a path of retained squares at level k that connects L_A to R_A and that does not use A. We also adopt the notation of Exercise 5.70.

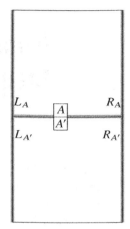

(a) Show that $\lambda_k(p) \leq 2p^m + 2\theta_k(p, A)$ for all $m \leq k$ and every A. *Hint:* See Figure 5.6 at right.

(b) Show that there exists A such that $\theta_k(p, A)^{b^m} \leq \theta_k(p)$. *Hint:* Use Harris's inequality.

(c) Deduce that $\theta_\infty(p) = 0$ implies $\lambda_\infty(p) = 0$.

(d) Conclude that $p_{1\times1} = p_{1\times2}$ and that $\theta_\infty(p_{1\times1}) > 0$.

Figure 5.6. See Exercise 5.71.

*The remaining exercises use the following notions. Let X and Y be real-valued random variables. Say that X is at least Y in the **increasing convex order** if* $\mathbf{E}[h(X)] \geq \mathbf{E}[h(Y)]$ *for all nonnegative, increasing convex functions* $h: \mathbb{R} \to \mathbb{R}$.

5.72. Let X and Y be real-valued random variables.

(a) Show that X is at least Y in the increasing convex order iff $\int_a^\infty \mathbf{P}[X > t]\, dt \geq \int_a^\infty \mathbf{P}[Y > t]\, dt$ for all $a \in \mathbb{R}$.

(b) Suppose that X is at least Y in the increasing convex order. Show that $\mathbf{E}[h(X)] \geq \mathbf{E}[h(Y)]$ for every convex function $h: \mathbb{R} \to \mathbb{R}$ if and only if X and Y have the same mean. When the means of X and Y are the same, one also says that X is **stochastically more variable** than Y.

5.73. Suppose that X_i are nonnegative, independent identically distributed random variables, that Y_i are nonnegative, independent identically distributed random variables, and that X_i is stochastically more variable than Y_i for each $i \geq 1$.

(a) Show that $\sum_{i=1}^n X_i$ is at least $\sum_{i=1}^n Y_i$ in the increasing convex order for each $n \geq 1$.

(b) Let M and N be nonnegative integer-valued random variables independent of all X_i and Y_i. Suppose that M is at least N in the increasing convex order. Show that $\sum_{i=1}^M X_i$ is at least $\sum_{i=1}^N X_i$ in the increasing convex order, which, in turn, is at least $\sum_{i=1}^N Y_i$ in the increasing convex order.

5.74. Suppose that $L^{(1)}$ is an offspring random variable that is at least $L^{(2)}$ in the increasing convex order and that $Z_n^{(i)}$ are the corresponding generation sizes of Galton–Watson processes beginning with one individual each.

(a) Show that $Z_n^{(1)}$ is at least $Z_n^{(2)}$ in the increasing convex order for each n.

(b) Show that $\mathbf{P}[Z_n^{(1)} = 0] \geq \mathbf{P}[Z_n^{(2)} = 0]$ for each n if the means of $L^{(1)}$ and $L^{(2)}$ are the same.

(c) Show that the conditional distribution of $Z_n^{(1)}$ given $Z_n^{(1)} > 0$ is at least the conditional distribution of $Z_n^{(2)}$ given $Z_n^{(2)} > 0$ in the increasing convex order for each n.

(d) Show that the following is an example where $L^{(1)}$ is at least $L^{(2)}$ in the increasing convex order. Write $p_k^{(i)}$ for $\mathbf{P}[L^{(i)} = k]$. Suppose that $p_0^{(1)} > 0$, that $a := \min\{k \geq 1; \; p_k^{(1)} > 0\}$, that $p_k^{(2)} = 0$ for $k > K$, that $p_k^{(2)} = p_k^{(1)}$ for $a < k \leq K$, and that $\mathbf{E}[L^{(1)}] = \mathbf{E}[L^{(2)}]$.

6 | Isoperimetric Inequalities

Just as the branching number of a tree is for most purposes more important than the growth rate, there is a number for a general graph that is more important for many purposes than its growth rate. In the present chapter, we consider this number, or, rather, several variants of it, called isoperimetric or expansion constants. This is not an extension of the branching number, however; for that, the reader can see Section 13.7.

Our main interest in expansion constants is to apply them to random walks and percolation on infinite graphs. In particular, whether the expansion is zero or positive plays a crucial role in determining qualitative behavior of these probabilistic processes. This will be seen here as well as in the later Chapters 7, 8, 10, and 11. A similar role is played on finite graphs, but we touch on finite graphs only briefly in Section 6.4.

The second half of this chapter concerns isoperimetric and expansion profiles, functions that measure the size of the boundary of a set compared to the size of the set itself, when such a ratio is *not* necessarily bounded away from 0. Again, we present applications to random walks and to percolation.

6.1 Flows and Submodularity

A common illegal scheme for making money, known as a pyramid scheme or Ponzi scheme, goes essentially as follows. You convince 10 people to send you $100 each and to ask 10 others in turn to send them $100. Everyone who manages to do this will profit $900 (and you will profit $1000). Of course, some people will lose $100 in the end. But suppose that we had an infinite number of people. Then no one need lose money and indeed everyone can profit $900. But what if people can ask only people they know? Suppose that people are at the vertices of the square lattice and know only their four nearest neighbors. Is it now possible for everyone to profit $900? If the amount of money allowed to change hands (that is, the amount crossing any edge) is unbounded, then certainly this is possible. But what if the amount crossing any edge is bounded by, say, $1,000,000? The answer now is no. In fact, although it is still possible for everyone to profit, the profit cannot be bounded away from 0.

Why is this? Consider first the case that there are only a finite number of people. If we simply add up the total net gains, we obtain 0, whence it is impossible for everyone to gain a strictly positive amount. For the lattice case, consider all the people lying within distance n of the origin. What is the *average* net gain of these people? The only reason this average might not be 0 is that money can cross the boundary. However, because of our assumption that the money crossing any edge is bounded, it follows that for the average net gain, this boundary crossing is negligible in the limit as $n \to \infty$. Hence the average net gain of everyone is 0 and it cannot be that everyone profits $900 (or even just one cent).

But what if the neighbor graph was that of the hyperbolic tessellation of Figure 2.4? The figure suggests that the preceding argument, which depended on finding finite subsets of vertices with relatively few edges leading out of them, will not work. Does that mean everyone could profit $900? What is the maximum profit everyone could make? The picture also suggests that the graph is somewhat like a tree, which suggests that significant profit is possible. Indeed, this is true, as we now show. We will generalize this problem a little by restricting the amount that can flow over edge e to be at most $c(e)$ and by supposing that the person at location x has as a goal to profit $D(x)$. We ask, What is the maximum α such that each person x can profit $\alpha D(x)$? Thus, on the Euclidean lattice, we saw that $\alpha = 0$ when c is bounded above and D is bounded below.

To answer this question, we introduce some terminology. Let G be a connected graph. Weight the edges by positive numbers $c(e)$ and the vertices by positive numbers $D(x)$. We call (G, c, D) a **network**, which is a little more general than the networks we have used before where we had weights only for the edges. We assume that G is locally finite, or, more generally, that for each x,

$$\sum_{e^- = x} c(e) < \infty . \tag{6.1}$$

For $K \subset V$, we define the **edge boundary** $\partial_E K$ to be the set of (unoriented) edges that connect K to its complement. Write

$$|K|_D := \sum_{x \in K} D(x)$$

and

$$|\partial_E K|_c := \sum_{e \in \partial_E K} c(e) .$$

Define the **edge-expansion constant** or **edge-isoperimetric constant** of (G, c, D) by

$$\Phi_E(G) := \Phi_E(G, c, D) := \inf \left\{ \frac{|\partial_E K|_c}{|K|_D} ; \; \varnothing \neq K \subset V \text{ is finite} \right\} .$$

This is a measure of how small we can make the boundary effects in our preceding argument. The most common choices for (c, D) are $(\mathbf{1}, \mathbf{1})$ and $(\mathbf{1}, \deg)$. For a network with conductances c, one often chooses (c, π) (as in Section 6.2). When $\Phi_E(G) = 0$, we say the network is **edge amenable** (we'll explain the origin of the name after we define "vertex amenable" later). Thus, the square lattice is edge amenable; but note that there are large sets in $\mathbb{Z}^2$ that also have large boundary. The point is that there *exist* sets that have relatively small boundary. This is impossible for a regular tree of degree at least 3.

▷ **Exercise 6.1.**
Show that $\Phi_E(\mathbb{T}_{b+1}, \mathbf{1}, \mathbf{1}) = b - 1$ for all $b \geq 1$, where $\mathbb{T}_{b+1}$ is the regular tree of degree $b + 1$.

The opposite of "amenable" is **nonamenable**; one also says that a nonamenable network satisfies a **strong isoperimetric inequality**.

The following theorem (and Exercise 6.2) shows that $\Phi_E(G)$ is precisely the maximum proportional profit α that we seek (actually, because of our choice of d^*, we need to take the negative of the function θ guaranteed by Theorem 6.1). The theorem thus dualizes the infimum in the definition of $\Phi_E(G)$ not only to a supremum, but to a maximum. It is due to Benjamini, Lyons, Peres, and Schramm (1999b), hereinafter referred to as BLPS (1999b).

Theorem 6.1. (Duality for Edge Expansion) *For any network (G, c, D), we have*

$$\Phi_E(G, c, D) = \max\{\alpha \geq 0 \,; \, \exists \theta \; \forall e \; |\theta(e)| \leq c(e) \text{ and } \forall x \; d^*\theta(x) = \alpha D(x)\},$$

where θ runs over the antisymmetric functions on E.

Proof. Denote by A the set of α of which the maximum is taken on the right-hand side of the desired equation. Thus, we want to show that $\max A = \Phi_E(G)$. In fact, we will show that $A = [0, \Phi_E(G)]$ as a consequence of the max-flow min-cut theorem.

Given $\alpha \geq 0$ and any finite, nonempty $K \subset$ V, define the network $G(K) = G(K, \alpha)$ with vertices K and two extra vertices, a and z. The edges are those of G with both endpoints in K, those in $\partial_E K$ where the endpoint not in K is replaced by z, and an edge between a and each point in K. Give each edge e with both endpoints in K the capacity $c(e)$. Give each edge $[x, z]$ the capacity of the corresponding edge in $\partial_E K$. Give the other edges capacity $c(a, x) := \alpha D(x)$.

In view of the cutset consisting of all edges incident to z, it is clear that any admissible flow from a to z in $G(K)$ has strength at most $|\partial_E K|_c$. Suppose that $\alpha \in A$, in other words, that there is a function θ satisfying $\forall e \; |\theta(e)| \leq c(e)$ and $\forall x \; d^*\theta(x) = \alpha D(x)$. This function θ induces an admissible flow on $G(K)$ from a to z of strength $\alpha |K|_D$ by putting the flow $\alpha D(x)$ on each edge (a, x) for $x \in K$, whence $\alpha \leq |\partial_E K|_c / |K|_D$. Since this holds for all K, we get $\alpha \leq \Phi_E(G)$.

In the other direction, if $\alpha \leq \Phi_E(G)$, then for all finite K, we claim that there is an admissible flow from a to z in $G(K)$ of strength $\alpha |K|_D$. Consider any cutset Π separating a and z. Let K' be the vertices in K that Π separates from z. Then $\partial_E K' \subseteq \Pi$ and $[a, x] \in \Pi$ for all $x \in K \setminus K'$. Therefore,

$$\sum_{e \in \Pi} c(e) \geq |\partial_E K'|_c + \alpha |K \setminus K'|_D \geq \Phi_E(G) |K'|_D + \alpha |K \setminus K'|_D$$

$$\geq \alpha |K'|_D + \alpha |K \setminus K'|_D = \alpha |K|_D \,.$$

Thus, our claim follows from the max-flow min-cut theorem. Note that the flow along $[a, x]$ is $\alpha D(x)$ for every $x \in K$, since the flow's strength is $\alpha |K|_D$.

Now let K_n be finite sets increasing to V, and let θ_n be admissible flows on $G(K_n)$ with strength $\alpha |K_n|_D$. There is a subsequence $\langle n_i \rangle$ such that for all edges $e \in$ E, the limit $\theta(e) := \theta_{n_i}(e)$ exists; clearly $\forall e \; |\theta(e)| \leq c(e)$ and, by (6.1) and the dominated convergence theorem, $\forall x \; d^*\theta(x) = \alpha D(x)$. Thus, $\alpha \in A$. ◀

▷ **Exercise 6.2.**
Show that for any network (G, c, D), we have

$$\Phi_\mathsf{E}(G, c, D) = \max\{\alpha \geq 0 \,;\, \exists \theta \ \forall e \ |\theta(e)| \leq c(e) \text{ and } \forall x \ d^* \theta(x) \geq \alpha D(x)\},$$

where θ runs over the antisymmetric functions on E.

Is the infimum in the definition of $\Phi_\mathsf{E}(G)$ a minimum? Certainly not in the edge-amenable case. The reader should find an example where it is a minimum, however. It turns out that in the transitive case, it is never a minimum, as shown by BLPS (1999b). Here, we say that a network (G, c, D) is **transitive** if, for every pair $x, y \in V(G)$, there is an automorphism of the graph G that takes x to y and that preserves the edge weights c and vertex weights D. To prove that transitive networks do not have minimizing sets, we will use the following concept.

A function b on finite subsets of V is called **submodular** if

$$\forall K, K' \quad b(K \cup K') + b(K \cap K') \leq b(K) + b(K'). \tag{6.2}$$

For example, $K \mapsto |K|_D$ is obviously submodular with equality holding in (6.2):

$$\forall K, K' \quad |K \cup K'|_D + |K \cap K'|_D = |K|_D + |K'|_D. \tag{6.3}$$

The identity

$$|\partial_\mathsf{E}(K \cup K')|_c + |\partial_\mathsf{E}(K \cap K')|_c + 2\,|\partial_\mathsf{E}(K \setminus K') \cap \partial_\mathsf{E}(K' \setminus K)|_c = |\partial_\mathsf{E}K|_c + |\partial_\mathsf{E}K'|_c \tag{6.4}$$

is easy, though tedious, to check. It shows that $b \colon K \mapsto |\partial_\mathsf{E}K|_c$ is submodular, with equality in (6.2) iff $\partial_\mathsf{E}(K \setminus K') \cap \partial_\mathsf{E}(K' \setminus K) = \varnothing$, which is the same as $K \setminus K'$ not adjacent to $K' \setminus K$.

Theorem 6.2. *If (G, c, D) is an infinite, transitive network, then for all finite, nonempty $K \subset \mathsf{V}$, we have $|\partial_\mathsf{E}K|_c/|K|_D > \Phi_\mathsf{E}(G)$.*

Proof. At first, we do not need to suppose that G is transitive. By the submodularity of $b(K) := |\partial_\mathsf{E}K|_c$ and (6.3), we have for any finite K and K',

$$\frac{b(K \cup K') + b(K \cap K')}{|K \cup K'|_D + |K \cap K'|_D} \leq \frac{b(K) + b(K')}{|K|_D + |K'|_D},$$

with equality iff $K \setminus K'$ is not adjacent to $K' \setminus K$. Now when a, b, c, d are positive numbers, we have

$$\min\{a/b, c/d\} \leq (a + c)/(b + d) \leq \max\{a/b, c/d\}.$$

Therefore

$$\min\left\{\frac{b(K \cup K')}{|K \cup K'|_D}, \frac{b(K \cap K')}{|K \cap K'|_D}\right\} \leq \max\left\{\frac{b(K)}{|K|_D}, \frac{b(K')}{|K'|_D}\right\}, \tag{6.5}$$

with equality iff $K \setminus K'$ is not adjacent to $K' \setminus K$ *and* all four quotients appearing in (6.5) are equal. (In case $K \cap K'$ is empty, omit it on the left-hand side.)

Now we use the hypothesis that G is transitive. Suppose for a contradiction that there is some finite set K with $b(K)/|K|_D = \Phi_\mathsf{E}(G)$. Choose some such set K of minimal cardinality. Let $o \in K$ and choose an automorphism γ of G such that γo is outside K but adjacent to some vertex in K. Define $K' := \gamma K$. Note that $b(K')/|K'|_D = b(K)/|K|_D$. If $K \cap K' \neq \varnothing$, then our choice of K implies that equality cannot hold in (6.5), whence $K \cup K'$ has a strictly smaller quotient, a contradiction. But if $K \cap K' = \varnothing$, then $K \setminus K'$ and $K' \setminus K$ are adjacent, whence (6.5) shows again that $K \cup K'$ has a strictly smaller quotient, a contradiction. ◀

Sometimes it is useful to look at boundary vertices rather than boundary edges. Thus, given positive numbers $D(x)$ on the vertices of a graph G, define the **(outer) vertex boundary** $\partial_V K$

$$\partial_V K := \{x \notin K \; ; \; \exists y \in K \;\; y \sim x\}$$

and

$$\Phi_V(G) := \Phi_V(G, D) := \inf\left\{\frac{|\partial_V K|_D}{|K|_D} \; ; \; \varnothing \neq K \subset V \text{ is finite}\right\},$$

called the **vertex-expansion constant** or **vertex-isoperimetric constant** of G. The most common choice is $D = 1$. We call G **vertex amenable** if its vertex-expansion constant is 0. For two positive functions f_1 and f_2 on the same domain, write $f_1 \asymp f_2$ to mean that $\inf f_1/f_2 > 0$ and $\sup f_1/f_2 < \infty$. Note that if (G, c, D) satisfies $c \asymp 1 \asymp D \asymp \deg$, then (G, c, D) is edge amenable iff (G, D) is vertex amenable. We'll call (G, c, D) simply **amenable** if it is both edge amenable and vertex amenable.

▷ **Exercise 6.3.**
Suppose that G is a graph such that for some $o \in V$, we have subexponential growth of balls: $\liminf_{n\to\infty} |\{x \in V \; ; \; d(o, x) \le n\}|^{1/n} = 1$, where $d(\cdot, \cdot)$ denotes the graph distance in G. Show that $(G, 1)$ is vertex amenable.

▷ **Exercise 6.4.**
Show that every Cayley graph of a finitely generated, abelian group is amenable.

▷ **Exercise 6.5.**
Suppose that G_1 and G_2 are roughly isometric graphs with bounded degrees and having both edge and vertex weights $\asymp 1$. Show that G_1 is amenable iff G_2 is.

Because of Exercises 6.5 and 3.12, either all Cayley graphs of a group are amenable or none are; that is, amenability is a property of the group. In fact, the concept of amenability comes from groups, not graphs. This origin also explains the name "amenable" in the following way. Let Γ be any countable group, and let $\ell^\infty(\Gamma)$ be the Banach space of bounded, real-valued functions on Γ. A linear functional on $\ell^\infty(\Gamma)$ is called a **mean** if it maps the constant function **1** to 1 and nonnegative functions to nonnegative numbers. If $f \in \ell^\infty(\Gamma)$ and $\gamma \in \Gamma$, we write $(R_\gamma f)(\gamma') := f(\gamma'\gamma)$. We call a mean μ **invariant** if $\mu(R_\gamma f) = \mu(f)$ for all $f \in \ell^\infty(\Gamma)$ and all $\gamma \in \Gamma$. Finally, we say that Γ is **amenable** if there is an invariant mean on $\ell^\infty(\Gamma)$. Thus, "amenable" was introduced as a play on words that evoked the word "mean." How is this related to the definitions we have given? Suppose that Γ is finitely generated and that G is one of its Cayley graphs. If $(G, 1)$ is amenable, then there is a sequence of finite sets K_n with

$$|\partial_V K_n|/|K_n| \to 0. \tag{6.6}$$

Now consider the sequence of means

$$f \mapsto \mu_n(f) := \frac{1}{|K_n|} \sum_{x \in K_n} f(x).$$

Then given $f \in \ell^\infty(\Gamma)$, for every generator or its inverse γ of Γ, we see that $|\mu_n(f) - \mu_n(R_\gamma f)| \to 0$ as $n \to \infty$, whence the same holds for all $\gamma \in \Gamma$. One can use a weak* limit point of the means μ_n to obtain an invariant mean and therefore show that Γ is amenable. The converse was established by Følner (1955) and is usually stated in the form that for every nonempty finite $B \subset \Gamma$ and all $\epsilon > 0$, there is a nonempty finite set $A \subset \Gamma$ such that $|BA \triangle A| \le \epsilon|A|$; see Paterson (1988), Theorem 4.13, for a proof of this converse. In this case, one often refers informally to A as a Følner set. More properly, a sequence $\langle K_n \rangle$ satisfying (6.6) is called a ***Følner sequence***. In conclusion, a finitely generated group Γ is amenable iff any of its Cayley graphs is.

We claim that for any graph G, the balls $B_G(x, n)$ in G about every point x of radius n satisfy

$$|B_G(x, n)|^{1/n} \ge 1 + \Phi_V(G, \mathbf{1}). \tag{6.7}$$

Indeed, with $c = \Phi_V(G, \mathbf{1})$, we have that $|\partial_V B(x, n)| \ge c|B(x, n)|$, which is the same as $|B(x, n+1)| \ge (1+c)|B(x, n)|$. This implies (6.7). In particular, if a group Γ is nonamenable, then all its Cayley graphs G have exponential growth.

Surprisingly, the converse fails, and various classes of counterexamples are known. One counterexample is the lamplighter group $\mathbb{Z}^\circ$ defined in Section 3.4, where we saw that it has growth rate $(1 + \sqrt{5})/2$. On the other hand, to see that $\mathbb{Z}^\circ$ is amenable, consider the "boxes"

$$K_n := \left\{ (\psi, m) ; \ m \in [-n, n], \ \psi^{-1}(\{1\}) \subseteq [-n, n] \right\}. \tag{6.8}$$

Then $|K_n| = (2n + 1)2^{2n+1}$, while $|\partial_E K_n| = 2^{2n+2}$.

The analogue for vertex amenability of Theorem 6.1 is due to Benjamini and Schramm (1997). It involves the amount flowing along edges into vertices,

$$\mathsf{flow}_+(\theta, x) := \sum_{e^+ = x} \left(\theta(e) \vee 0 \right).$$

Theorem 6.3. (Duality for Vertex Expansion) *For any graph G with vertex weights D, we have*

$$\Phi_V(G, D) = \max\left\{ \alpha \ge 0 ; \ \exists \theta \ \forall x \ \mathsf{flow}_+(\theta, x) \le D(x) \ \text{and} \ d^*\theta(x) = \alpha D(x) \right\},$$

where θ runs over the antisymmetric functions on E.

Proof. This time we use the max-flow min-cut theorem in the version of Exercise 3.14, where capacity constraints are imposed on vertices as well as on edges. Given $\alpha \ge 0$ and any finite, nonempty $K \subset$ V, define the network $G(K)$ with vertices $K \cup \partial_V K$ and two extra vertices, a and z. The edges of $G(K)$ are those of G that have both endpoints in $K \cup \partial_V K$, an edge between a and each point in K, and an edge between z and each point in $\partial_V K$. Give all edges incident to a capacity $c(a, x) := \alpha D(x)$. Let the capacity of the vertices in K be $c(x) := (\alpha + 1)D(x)$, and the capacity of the vertices in $\partial_V K$ be $D(x)$. The remaining edges and vertices are given infinite capacity.

In view of the cutset consisting of all vertices in $\partial_V K$, it is clear that any admissible flow from a to z in $G(K)$ has strength at most $|\partial_V K|_D$. Now a function θ satisfying

$\forall x$ $\mathsf{flow}_+(\theta, x) \leq D(x)$ and $d^*\theta(x) = \alpha D(x)$ induces an admissible flow on $G(K)$ from a to z of strength $\alpha|K|_D$, whence $\alpha \leq |\partial_V K|_D/|K|_D$. Since this holds for all K, we get $\alpha \leq \Phi_V(G)$.

In the other direction, if $\alpha \leq \Phi_V(G)$, then for all nonempty K, we claim that there is an admissible flow from a to z in $G(K)$ of strength $\alpha|K|_D$. Consider any cutset Π of edges and vertices separating a and z. We will show that $|\Pi|_c \geq \alpha|K|_D$, whence our claim will follow from the max-flow min-cut theorem. If Π contains an edge of infinite capacity, then this is trivial, so assume it does not. Let K' be the vertices in K that Π separates from z. Then $\partial_V K' \subset \Pi$ and $[a, x] \in \Pi$ for all $x \in K \setminus K'$. Therefore,

$$\sum_{e \in \Pi \cap E} c(e) + \sum_{x \in \Pi \cap V} c(x) \geq \alpha|K \setminus K'|_D + |\partial_V K'|_D \geq \alpha|K \setminus K'|_D + \Phi_V(G)|K'|_D \geq \alpha|K|_D.$$

Thus, our claim follows. Note that the flow along $[a, x]$ is $\alpha D(x)$ for every $x \in K$.

Now let K_n be finite, nonempty sets increasing to V, and let θ_n be the corresponding admissible flows on $G(K_n)$. There is a subsequence $\langle n_i \rangle$ such that for all edges $e \in E$, the limit $\theta(e) := \theta_{n_i}(e)$ exists; clearly $\forall x$ $\mathsf{flow}_+(\theta, x) \leq D(x)$ and $d^*\theta(x) = \alpha D(x)$. ◀

Not surprisingly, an analogue of Theorem 6.2 holds, that the infimum in the definition of $\Phi_V(G)$ is not attained for transitive G; again, this is due to BLPS (1999b). To show it, we first check that the function $K \mapsto |\partial_V K|_D$ is submodular. In fact, the following identity holds, where $\overline{K} := K \cup \partial_V K$:

$$|\partial_V(K \cup K')|_D + |\partial_V(K \cap K')|_D + |(\overline{K} \cap \overline{K'}) \setminus \overline{K \cap K'}|_D = |\partial_V K|_D + |\partial_V K'|_D. \quad (6.9)$$

Of course, in the transitive case, we must have D is constant.

Theorem 6.4. *If G is an infinite, transitive graph, then for all finite, nonempty $K \subset V$, we have $|\partial_V K|/|K| > \Phi_V(G)$.*

Proof. By the submodularity of $b(K) := |\partial_V K|$, we have for any finite K and K', as in the proof of Theorem 6.2, that

$$\min\left\{ \frac{b(K \cup K')}{|K \cup K'|}, \frac{b(K \cap K')}{|K \cap K'|} \right\} \leq \frac{b(K) + b(K')}{|K| + |K'|}, \quad (6.10)$$

with equality iff both terms on the left-hand side are equal to the right-hand side. (In case $K \cap K'$ is empty, omit it on the left-hand side.)

Now suppose for a contradiction that G is transitive and that K is a finite set minimizing $|K|$ among those K with $b(K)/|K| = \Phi_V(G)$. Let γ be any automorphism of G such that $\gamma K \cap K \neq \varnothing$. Define $K' := \gamma K$. Then (6.10) shows that $K' = K$. Thus, if, instead, we choose γ so that $\gamma K \cap \partial_V K \neq \varnothing$, then $\gamma K \cap K = \varnothing$, whence (6.9) shows that $K'' := K \cup \gamma K$ satisfies $b(K'')/|K''| < b(K)/|K|$, a contradiction. ◀

An interesting aspect of Theorem 6.3 is that it enables us to find (virtually) regular subtrees in many nonamenable graphs, as shown by Benjamini and Schramm (1997):

Theorem 6.5. (Regular Subtrees in Nonamenable Graphs) *Let G be any graph with* $n := \lfloor \Phi_V(G, \mathbf{1}) \rfloor \geq 1$. *Then G has a spanning forest in which every tree has one vertex of degree n and all others of degree n + 2.*

Proof. The proof of Theorem 6.3 in combination with Exercise 3.15 shows that there is an integer-valued θ satisfying

$$\forall x \quad \mathsf{flow}_+(\theta, x) \leq 1 \text{ and } d^*\theta(x) = n \,. \tag{6.11}$$

If there is an oriented cycle along which $\theta = 1$, then we may change the values of θ to be 0 on the edges of this cycle without changing the validity of (6.11). Thus, we may assume that there is no such oriented cycle. After this, there is no (unoriented) cycle in the support of θ because this would force a flow ≥ 2 into some vertex x on the cycle, and hence $\mathsf{flow}_+(\theta, x) \geq n + 2 > 1$. We may similarly assume that there is no oriented, bi-infinite path along which $\theta = 1$. Thus, (6.11) shows that for every x with $\mathsf{flow}_+(\theta, x) = 1$, there are exactly $n + 1$ edges leaving x with flow out of x equal to 1. Furthermore, the lack of an oriented, bi-infinite path where $\theta = 1$ shows that each component of the support of θ contains a vertex into which there is no flow, and (6.11) implies that such a vertex has n edges leading out with flow 1 each. This immediately implies also that no component of the support of θ contains two vertices into which there is no flow. Thus, the support of θ is the desired spanning forest.
◀

This result can be extended to graphs with $\Phi_V > 0$, but it is more complicated; see Benjamini and Schramm (1997). It follows from Theorem 6.5 that if $\Phi_V(G, \mathbf{1}) \geq 1$, then G contains a spanning tree T with $\Phi_V(T, \mathbf{1}) \geq 1$: just add edges to the spanning forest from Theorem 6.5. However, it is unknown whether "≥ 1" can be replaced here by "> 0," a question asked by Benjamini and Schramm (1997). It is also unknown whether every Cayley graph of exponential growth has a nonamenable subtree, another question asked by Benjamini and Schramm (1997).

6.2 Spectral Radius

In most infinite networks of interest, the probability of return of random walk to its starting point decays to 0 as the number of steps tends to infinity. Intuitively, the more the network spreads out, the more quickly the return probabilities will decay. This section makes precise this connection; network spread will be measured by the expansion constant, while decay of return probabilities will be measured by the spectral radius, to be defined.

As in Chapter 2, we use the inner-product notation

$$(f, g)_h := (fh, g) = (f, gh)$$

and

$$\|f\|_h := \sqrt{(f, f)_h} \,.$$

Also, let $\mathbf{D}_{00}$ denote the collection of functions on V with finite support.

Suppose that $\langle X_n \rangle$ is a Markov chain on a countable state space V with a stationary measure π. We define the ***transition operator***

$$(Pf)(x) := \mathbf{E}_x[f(X_1)] = \sum_{y \in \mathsf{V}} p(x, y) f(y) \,.$$

Then P maps $\ell^2(\mathsf{V}, \pi)$ to itself with norm

$$\|P\|_\pi := \|P\|_{\ell^2(\mathsf{V}, \pi)} := \sup\left\{ \frac{\|Pf\|_\pi}{\|f\|_\pi} \; ; \; f \neq \mathbf{0} \right\}$$

at most 1.

▷ **Exercise 6.6.**
Prove that $\|P\|_\pi \leq 1$.

As the reader should recall from the theory of Markov chains, we have that

$$(P^n f)(x) = \sum_{y \in \mathsf{V}} p_n(x, y) f(y)$$

when $p_n(x, y) := \mathbf{P}_x[X_n = y]$.

Let G be a (connected) graph with conductances $c(e) > 0$ on the edges and $\pi(x)$ be the sum of the conductances incident to a vertex x. The operator P that we defined earlier is self-adjoint: For functions $f, g \in \mathbf{D}_{00}$, we have

$$(Pf, g)_\pi = \sum_{x \in \mathsf{V}} \pi(x)(Pf)(x)g(x) = \sum_{x \in \mathsf{V}} \pi(x)\left[\sum_{y \in \mathsf{V}} p(x, y)f(y) \right] g(x)$$

$$= \sum_{x \in \mathsf{V}} \sum_{y \in \mathsf{V}} c(x, y)f(y)g(x) \,.$$

Since this is symmetric in f and g, it follows that $(Pf, g)_\pi = (f, Pg)_\pi$. Since the functions with finite support are dense in $\ell^2(\mathsf{V}, \pi)$, we get this identity for all $f, g \in \ell^2(\mathsf{V}, \pi)$.

There are two other expressions for the norm $\|P\|_\pi$ that will be useful to us. One is in the following exercise, while the other is in our next proposition.

▷ **Exercise 6.7.**
(**Rayleigh Quotient**) Show that

$$\|P\|_\pi = \sup\left\{ \frac{|(Pf, f)_\pi|}{(f, f)_\pi} \; ; \; f \in \mathbf{D}_{00} \setminus \{\mathbf{0}\} \right\} = \sup\left\{ \frac{(Pf, f)_\pi}{(f, f)_\pi} \; ; \; f \in \mathbf{D}_{00} \setminus \{\mathbf{0}\} \right\}.$$

Proposition 6.6. *For any two vertices* $x, y \in \mathsf{V}$, *we have*

$$\|P\|_\pi = \limsup_{n \to \infty} \sup_z \left(p_n(x, z)/\sqrt{\pi(z)} \right)^{1/n} = \limsup_{n \to \infty} p_n(x, y)^{1/n} \,. \tag{6.12}$$

Moreover,

$$\forall n \quad p_n(x, y) \leq \sqrt{\pi(y)/\pi(x)} \, \|P\|_\pi^n \,. \tag{6.13}$$

Proof. Irreducibility of the Markov chain implies that the right-most quantity in (6.12) does not depend on the choice of x, y. Thus, define

$$\rho(G) := \limsup_{n \to \infty} p_n(x, y)^{1/n}. \qquad (6.14)$$

It is clear that the middle term in (6.12) is at least $\rho(G)$. To see that this middle term is at most $\|P\|_\pi$, use $p_n(x, z) = (\mathbf{1}_{\{x\}}, P^n \mathbf{1}_{\{z\}})_\pi / \pi(x)$ and the Cauchy–Schwarz inequality to deduce that $p_n(x, z) \le \sqrt{\pi(z)/\pi(x)} \, \|P^n\|_\pi \le \sqrt{\pi(z)/\pi(x)} \, \|P\|_\pi^n$. This also implies (6.13).

Finally, to show the converse inequality that $\|P\|_\pi \le \rho(G)$, suppose that $0 \le f \in \mathbf{D}_{00} \setminus \{\mathbf{0}\}$. Then self-adjointness of P and the Cauchy–Schwarz inequality yield

$$\|P^{n+1} f\|_\pi^4 = (P^n f, P^{n+2} f)^2 \le \|P^n f\|_\pi^2 \, \|P^{n+2} f\|_\pi^2.$$

This means that $\|P^{n+1} f\|_\pi / \|P^n f\|_\pi$ is increasing, whence it has a limit, L. Since the product of these quotients for $n = 0, \dots, N - 1$ is $\|P^N f\|_\pi / \|f\|_\pi$, we have that

$$L = \lim_{N \to \infty} \left(\|P^N f\|_\pi / \|f\|_\pi \right)^{1/N} = \lim_{N \to \infty} \|P^N f\|_\pi^{1/N}. \qquad (6.15)$$

Now $p_{2n}(x, x) \ge p_n(x, x)^2$, since one way to return after $2n$ steps is to return after n steps and then again after n steps. Therefore,

$$\rho(G) = \limsup_{n \to \infty} p_{2n}(x, x)^{1/2n}.$$

Thus, for $0 \le f \in \mathbf{D}_{00} \setminus \{\mathbf{0}\}$, we have, by self-adjointness of P,

$$\limsup_{n \to \infty} \|P^n f\|_\pi^{1/n} = \limsup_{n \to \infty} (P^{2n} f, f)_\pi^{1/2n}$$

$$= \limsup_{n \to \infty} \left[\sum_{x,y} \pi(x) f(x) f(y) p_{2n}(x, y) \right]^{1/2n}$$

$$= \rho(G).$$

Combining this with (6.15), we deduce that $L = \rho(G)$. In particular, $\|Pf\|_\pi / \|f\|_\pi \le \rho(G)$. When $f \in \mathbf{D}_{00}$ is not assumed to be nonnegative, we may use the inequality $\|Pf\|_\pi \le \|P|f|\|_\pi$ to deduce that $\|Pf\|_\pi \le \rho(G)\|f\|_\pi$ in general, in other words, $\|P\|_\pi \le \rho(G)$. ◄

Thus, we have proved that the norm $\|P\|_\pi$ equals the ***spectral radius***,* denoted (as in the proof) $\rho(G)$.

For example, the spectral radius of simple random walk in $\mathbb{Z}^d$ is 1, since the return probabilities decay only polynomially fast. If the walk were biased, this might no longer hold. Likewise, simple random walk on a regular tree of degree $d \ge 3$ has spectral radius less than 1; we will calculate it exactly in several ways (Exercise 6.9, (6.25), Exercise 6.62, and Proposition 7.35).

Our main result in this section is the following comparison between the expansion constant and the spectral radius. Recall that π denotes the sum of the conductances $c(\cdot)$ at a vertex.

---

* For a general operator on a Banach space, its spectral radius is defined to be $\max |z|$ for z in the spectrum of the operator. For reversible Markov operators acting on ℓ^2, this agrees with the definition we gave; see, for example, Taylor and Lay (1980), Theorem VI.3.3. We will not need this general representation explicitly. For a nonreversible Markov chain, the norm of the transition operator is not generally equal to its spectral radius.

Theorem 6.7. (Expansion and Spectral Radius) *Let (G, c, π) be a connected, infinite network and $\Phi_E(G) := \Phi_E(G, c, \pi)$ be its edge-expansion constant. The spectral radius $\rho(G)$ of the associated network random walk satisfies*

$$\Phi_E(G)^2/2 \le 1 - \sqrt{1 - \Phi_E(G)^2} \le 1 - \rho(G) \le \Phi_E(G) . \tag{6.16}$$

The most important consequence of this is the qualitative statement that the spectral radius $\rho(G)$ is less than 1 iff the network (G, c, π) is nonamenable.

To prove Theorem 6.7, we will use the following easy calculation:

▷ **Exercise 6.8.**
Show that for $f \in \mathbf{D}_{00}$, we have $d^*(c\, df) = \pi(f - Pf)$.

The meaning of the equation in this exercise is this: Recall from Section 2.8 that the ***gradient*** of a function f on V is the antisymmetric function

$$\nabla f := c\, df$$

on E. If we define the ***divergence*** by $\operatorname{div} \theta := \pi^{-1} d^* \theta$, then Exercise 6.8 says that $\operatorname{div} \nabla = I - P$, where I is the identity. This is the discrete probabilistic Laplace operator. In particular, $\operatorname{div} \nabla f = 0$ iff $Pf = f$, in other words, f is harmonic according to our definition of "harmonic" in Section 2.1.

Now by Exercise 6.8, we have that for $f \in \mathbf{D}_{00}$,

$$(df, df)_c = (c\, df, df) = \big(d^*(c\, df), f\big) = \big(\pi(f - Pf), f\big) = (f, f)_\pi - (Pf, f)_\pi . \tag{6.17}$$

Combine this with Exercise 6.7 and Proposition 6.6 to write

$$\rho(G) = \sup\left\{\frac{(Pf, f)_\pi}{(f, f)_\pi} \, ; \ f \in \mathbf{D}_{00} \setminus \{\mathbf{0}\}\right\} = \sup\left\{\frac{(f, f)_\pi - (df, df)_c}{(f, f)_\pi} \, ; \ f \in \mathbf{D}_{00} \setminus \{\mathbf{0}\}\right\}$$

$$= 1 - \inf\left\{\frac{(df, df)_c}{(f, f)_\pi} \, ; \ f \in \mathbf{D}_{00} \setminus \{\mathbf{0}\}\right\} . \tag{6.18}$$

Choosing $f := \mathbf{1}_K \in \mathbf{D}_{00}$ shows the last inequality in (6.16). This also illustrates the primary difference between $\rho(G)$ and $\Phi_E(G)$: the former is related to all $f \in \mathbf{D}_{00}$, whereas the latter is related only to indicator functions.

We will also use the following lemma for the proof of Theorem 6.7.

Lemma 6.8. *For any nonnegative $f \in \mathbf{D}_{00}$, we have*

$$\Phi_E(G, c, \pi) \sum_{x \in V} f(x)\, \pi(x) \le \sum_{e \in E_{1/2}} |df(e)|\, c(e) .$$

Proof. For $t > 0$, we may use $K := \{x \, ; \ f(x) > t\}$ in the definition of Φ_E to see that

$$\Phi_E \cdot |\{x \, ; \ f(x) > t\}|_\pi \le \sum_{x, y \in V} c(x, y)\, \mathbf{1}_{\{f(x) > t \ge f(y)\}} . \tag{6.19}$$

Now

$$\int_0^\infty |\{x \, ; \ f(x) > t\}|_\pi \, dt = \sum_{x \in V} f(x)\, \pi(x)$$

and

$$\int_0^\infty \mathbf{1}_{\{f(x) > t \ge f(y)\}} \, dt = f(x) - f(y)$$

if $f(x) \ge f(y)$. Therefore, integrating (6.19) on $t \in (0, \infty)$ gives the desired result. ◀

Proof of Theorem 6.7. We have already seen the last inequality in (6.16). The first inequality in (6.16) comes from the elementary inequality $1 - x/2 \geq \sqrt{1-x}$. To prove the crucial middle inequality, let $f \in \mathbf{D}_{00}$. Apply Lemma 6.8 to f^2 to get

$$(f,f)_\pi^2 \leq \Phi_E(G)^{-2} \left(\sum_{e \in E_{1/2}} c(e) \left| f(e^+)^2 - f(e^-)^2 \right| \right)^2 \tag{6.20}$$

$$= \Phi_E(G)^{-2} \left(\sum_e c(e) |f(e^+) - f(e^-)| \cdot |f(e^+) + f(e^-)| \right)^2$$

$$\leq \Phi_E(G)^{-2} \left(\sum_e c(e)\, df(e)^2 \right) \left(\sum_e c(e)(f(e^+) + f(e^-))^2 \right)$$

by the Cauchy–Schwarz inequality. The last factor here is

$$\sum_e c(e) \left[f(e^+)^2 + f(e^-)^2 + 2f(e^+)f(e^-) \right] = \sum_x \pi(x)f(x)^2 + \sum_{x \sim y} c(x,y)f(x)f(y)$$

$$= (f,f)_\pi + \sum_x f(x)\pi(x) \sum_{y \sim x} p(x,y)f(y)$$

$$= (f,f)_\pi + (f, Pf)_\pi = 2(f,f)_\pi - (df, df)_c$$

by (6.17). Therefore,

$$(f,f)_\pi^2 \leq \Phi_E(G)^{-2} \left[2(f,f)_\pi - (df,df)_c \right] (df,df)_c .$$

A little algebra transforms this to

$$\left(1 - \frac{(df,df)_c}{(f,f)_\pi} \right)^2 \leq 1 - \Phi_E(G)^2 .$$

This gives the middle inequality of (6.16) when combined with (6.18). ◀

▷ **Exercise 6.9.**
Show that for simple random walk on $\mathbb{T}_{b+1}$, we have $\rho(\mathbb{T}_{b+1}) = 2\sqrt{b}/(b+1)$.

As one application of knowing the spectral radius, we give a sufficient condition for random walk on a network to have positive speed. Later, in Section 13.7, we will give some necessary conditions for positive speed. Define the ***upper (exponential) growth rate*** of a graph to be $\limsup_{n \to \infty} |B(o,n)|^{1/n}$, where $B(o,n)$ is the ***ball*** of radius n centered at o, that is, the set of vertices whose graph distance from o is at most n.

Proposition 6.9. (Speed and Spectral Radius) *Let G be a graph with upper exponential growth rate $b \in (1, \infty)$. Suppose that the edges are weighted so that the spectral radius $\rho(G) < 1$ and π is bounded. Then the network random walk $\langle X_n \rangle$ on G has positive liminf speed, that is,*

$$\liminf_{n \to \infty} \mathrm{dist}_G(X_0, X_n)/n \geq \log \rho(G)^{-1} / \log b \qquad a.s.$$

Note that $\mathrm{dist}_G(\cdot, \cdot)$ denotes the graph distance, so that the quotient whose limit we are taking is distance by time. This is the reason we call it "speed."

Proof. Without loss of generality, assume X_0 is a fixed vertex, o. Let $\alpha < -\log \rho(G)/\log b$, so that $\rho(G)b^\alpha < 1$. Choose $\lambda > b$ so that $\rho(G)\lambda^\alpha < 1$. Equation (6.13), our assumption that π is bounded, and the definition of b ensure that there is some constant $M < \infty$ so that for all $n \geq 0$,

$$\forall y \in V \quad p_n(o, y) \leq M\rho(G)^n$$

and

$$\left|\{y;\ \mathrm{dist}_G(o, y) \leq n\}\right| \leq M\lambda^n .$$

Therefore,

$$\mathbf{P}_o\left[\mathrm{dist}_G(o, X_n) \leq \alpha n\right] = \sum_{\mathrm{dist}_G(o,y)\leq\alpha n} p_n(o, y) \leq M^2\rho(G)^n\lambda^{\alpha n} = M^2\left[\rho(G)\lambda^\alpha\right]^n .$$

Since this is summable in n, it follows by the Borel–Cantelli lemma that $\mathrm{dist}_G(o, X_n) \leq \alpha n$ only finitely often a.s. ◄

In Section 7.7, we apply Theorem 6.7 to percolation on nonamenable graphs.

6.3 Nonbacktracking Paths and Cogrowth

A path in a multigraph is called ***nonbacktracking*** if no edge is immediately followed by its reversal. Note that a loop is its own reversal. Nonbacktracking random walks are almost as natural as ordinary random walks, though more difficult to analyze in most situations. Moreover, they can be more useful than ordinary random walks when random walks are used to search for something, as they explore more quickly, not wasting time immediately backtracking; see Exercise 6.64. In this section, however, we use them to analyze the spectral radius of ordinary random walks on regular graphs.

We begin with some calculations that count paths and nonbacktracking paths for general d-regular graphs, G. Write $b := d - 1$. Let A be the ***adjacency matrix*** of G, where $A(x, y)$ is the number of edges from x to y.

Let A_k be the matrix whose (x, y)-entry is the number of nonbacktracking paths of length k from x to y for $x, y \in V(G)$. Define $A_{-1} := 0$ and note that $A_0 = I$, $A_1 = A$. We claim that

$$A_{k+1} = \begin{cases} AA_k - bA_{k-1} & \text{for } k \in \mathbb{N} \setminus \{1\} \\ AA_1 - dI & \text{for } k = 1. \end{cases} \tag{6.21}$$

Indeed, the (x, y)-entry of AA_k equals the number of paths $x = x_0, x_1, \ldots, x_{k+1} = y$ such that the path from x_1 to x_{k+1} does not backtrack. This includes once each nonbacktracking path of length $k + 1$ from x to y, but also includes backtracking paths when $x_2 = x$. For each nonbacktracking path $x_2, \ldots, x_{k+1}$ of length $k - 1$ from x to y, there are b choices of edges from x to some x_1 that make a path counted by AA_k, namely, all the edges incident to x other than the one used from x_2 to x_3. (We have to speak of edges in case of multiple edges or loops.) This exception (to x_3) does not occur, however, when $k = 1$.

Now define $F(z) := \sum_{k\geq 0} A_k z^k$. Multiplying the recursion (6.21) by z^k and summing over $k \geq 0$, we get

$$\frac{F(z) - I}{z} = AF(z) - bzF(z) - zI$$

whenever all the series involved converge. Since $\|A\| \leq d$ and thus $\|A_k\| \leq \|A\|^k \leq d^k$, all the series do converge when $|z| < 1/d$. Solving this equation for $F(z)$ gives

$$F(z) = (1 - z^2)\big((1 + bz^2)I - zA\big)^{-1} \tag{6.22}$$

for $|z| < 1/d$.

Next, let the number of cycles of length n starting from some fixed $o \in V(G)$ be $c_n(G)$, whereas the number of those that are nonbacktracking is $b_n(G)$. Factor out $1 + bz^2$ from (6.22) and expand the inverse as a geometric series. Then look at the (o, o)-entry. Since $|z| < 1/(d + 1)$ implies that $|z/(1 + bz^2)| < 1/d$, we get that for $|z| < 1/(d + 1)$,

$$\sum_{n \geq 0} b_n(G)z^n = \frac{1 - z^2}{1 + bz^2} \sum_{n \geq 0} c_n(G)\Big(\frac{z}{1 + bz^2}\Big)^n . \tag{6.23}$$

Now consider the special case where $G = \mathbb{T}_{b+1}$, the d-regular tree. In that case, $b_n(G) = 0$ for all $n \geq 1$ and $b_0(G) = 1$. Put $w := z/(1+bz^2)$. Then $z/w = 1+bz^2$, whence $1 = w/z+bwz$. Therefore, $\big(\sum_{n \geq 0} c_n(\mathbb{T}_{b+1})w^n\big)^{-1} = (1-z^2)w/z = 1-dwz$. Since $z = \big(1-\sqrt{1 - 4bw^2}\big)/(2bw)$, we get that

$$\sum_{n \geq 0} c_n(\mathbb{T}_{b+1})w^n = \frac{2b}{b - 1 + (b + 1)\sqrt{1 - 4bw^2}} \tag{6.24}$$

for small w. The radius of convergence of the series on the left is the reciprocal of the exponential growth rate of c_{2n}, in other words, is equal to $1/(d\rho(\mathbb{T}_{b+1}))$. The right-hand side of (6.24) shows that the radius of convergence is equal to $1/(2\sqrt{b})$ – the function cannot be analytically continued from a neighborhood of 0 to $w = 1/(2\sqrt{b})$ – whence

$$\rho(\mathbb{T}_{b+1}) = \frac{2\sqrt{b}}{b + 1} , \tag{6.25}$$

as in Exercise 6.9. (Recall that a function is analytic at a point if it is equal to a convergent power series in a (complex) neighborhood of that point. If $f_1 : \Omega_1 \to \mathbb{C}$ is an analytic function and $\Omega' \subseteq \mathbb{C}$ (possibly a singleton), then we say that f_1 can be analytically continued to Ω' if there exist an open connected set $\Omega_2 \supseteq \Omega_1 \cup \Omega'$ and an analytic function $f_2 : \Omega_2 \to \mathbb{C}$ such that $f_2 \restriction \Omega_1 = f_1$. If Ω_1 is open, then f_2 is unique when it exists.)

To compare $\rho(G)$ to $\rho(\mathbb{T}_d)$ for other d-regular G, we continue analyzing (6.23) for general d-regular G. Let $\mathrm{cogr}(G) := \limsup_{n \to \infty} b_n(G)^{1/n}$, the exponential growth rate of the number of nonbacktracking cycles containing o. This number is called the **cogrowth** of G. The reason for this name is the following: The universal cover of G in the sense of p. 83 of Section 3.3 is $\mathbb{T}_d$. Vertices in $\mathbb{T}_d$ are finite, nonbacktracking paths in G starting at o. The **covering map** $\varphi : \mathbb{T}_d \to G$ maps such a path in G to its final endpoint. Then the cogrowth of G equals the exponential growth rate of $\varphi^{-1}(o)$ inside $\mathbb{T}_d$. One can see by using this covering map that $\mathrm{cogr}(G)$ does not depend on o. In particular, if G is the Cayley graph of Γ with respect to S and Γ is (isomorphic to) the quotient of the free group on S by the normal subgroup N, then $\mathrm{cogr}(G)$ is the growth rate of N inside the free group.

The central result about cogrowth is the following formula (6.27), due to Grigorchuk (1980) for Cayley graphs and extended by Northshield (1992) to all regular graphs:

Theorem 6.10. (Cogrowth Formula) *If G is a d-regular connected multigraph, then*

$$\mathrm{cogr}(G) > \sqrt{d-1} \quad \textit{iff} \quad \rho(G) > \frac{2\sqrt{d-1}}{d}, \tag{6.26}$$

in which case

$$d\rho(G) = \frac{d-1}{\mathrm{cogr}(G)} + \mathrm{cogr}(G). \tag{6.27}$$

If the conditions of (6.26) fail, then $\rho(G) = 2\sqrt{d-1}/d$ and $\mathrm{cogr}(G) \le \sqrt{d-1}$.

Proof. Suppose that G is d-regular. We may assume that G is not a tree. We follow the notation previously established. Let $g(z)$ be an analytic continuation of $\sum_{n \ge 0} b_n(G) z^n$ and $h(z)$ be an analytic continuation of $\sum_{n \ge 0} c_n(G) z^n$. Write $\psi(z) := z/(1 + bz^2)$. Thus, (6.23) becomes in this notation

$$g(z) = \frac{1 - z^2}{1 + bz^2} h(\psi(z)).$$

The idea now is to equate radii of convergence, but because of the composition with $\psi(z)$ on the right, we have to do this carefully. In fact, we are going to look only at analytic continuation along the real line.

The radius of convergence for $g(z)$ about $z = 0$ is $z_1 := 1/\mathrm{cogr}(G) \le 1$, whereas that for $h(z)$ is $z_2 := 1/(d\rho(G))$. Equation (6.27) is equivalent to $\psi(z_1) = z_2$. Because both series have nonnegative coefficients, Pringsheim's theorem (see Exercise 6.58) tells us that g can be analytically continued to $[0, z_1)$ but not to $[0, z_1]$, and likewise h can be analytically continued to $[0, z_2)$ but not to $[0, z_2]$. Note that $\psi(0) = 0$, $\psi(z)$ is strictly monotone increasing for $0 < z < 1/\sqrt{b}$, and $\psi(z)$ is strictly monotone decreasing for $z > 1/\sqrt{b}$.

The fact that g is analytic on $[0, z_1)$ implies that $h(\psi(z))$ is analytic on that same interval, whence $\psi(z_1) \le z_2$.

Now suppose that $\mathrm{cogr}(G) > \sqrt{d-1}$, that is, $z_1 < 1/\sqrt{b}$. Since z_1 is a singularity of g and ψ is strictly increasing on $[0, z_1)$, it must be that $\psi(z_1)$ is a singularity of h, whence $\psi(z_1) \ge z_2$. Together with the result of the preceding paragraph, this gives $\psi(z_1) = z_2$, that is, (6.27).

Next, suppose that $\rho(G) > 2\sqrt{d-1}/d$, that is, $z_2 < \psi(1/\sqrt{b})$. Then the smaller value of $\psi^{-1}(z_2)$ is the first singularity of $h(\psi(z))$ on the positive real line, whence it equals z_1. This is (6.27) again.

Since each condition of (6.26) implies (6.27), it is clear that the two conditions of (6.26) are equivalent.

Finally, since $\mathbb{T}_{b+1}$ covers G, we may couple simple random walk on G with simple random walk on $\mathbb{T}_{b+1}$ in such a way that it returns to $o \in \mathsf{V}(G)$ if it returns to its starting point in $\mathbb{T}_{b+1}$. Thus, $\rho(G) \ge \rho(\mathbb{T}_{b+1})$.   ◀

A great advantage of the cogrowth formula is that it allows us to show that $\rho(G) > \rho(\mathbb{T}_{b+1})$ in many common situations by showing that $\mathrm{cogr}(G) > \sqrt{b}$. Kesten (1959b) proved the following result and various extensions for Cayley graphs:

Theorem 6.11. *If $d \geq 3$ and G is a d-regular, transitive multigraph that is not a tree, then $\rho(G) > \rho(\mathbb{T}_d)$.*

It follows that (6.27) holds for all transitive multigraphs, G, other than trees.

Proof. Let L be the length of the shortest cycle in G (which is 1 if there is a loop). Consider a nonbacktracking random walk $\langle Y_n ; n \geq 1 \rangle$, where each edge Y_{n+1} is chosen uniformly among the edges incident to the head Y_n^+ of Y_n, other than the reversal of Y_n. We are going to handle loops differently than other cycles, so it will be convenient to let

$$L' := \begin{cases} L & \text{if } L > 1 \\ 3 & \text{if } L = 1. \end{cases}$$

Let A_n be the event that $Y_{n+1}, \ldots, Y_{n+L'}$ is a nonbacktracking cycle. For $n \geq 0$,

$$\mathbf{P}(A_n \mid Y_1, \ldots, Y_n) \geq \frac{1}{db^{L'-1}}, \tag{6.28}$$

since if $L > 1$, then there is a way to traverse a simple cycle starting at Y_n^+ and not using the reversal of Y_n, whereas if $L = 1$, then the walk can first choose an edge other than the reversal of Y_n, then traverse a loop, and then return by the reversal of Y_{n+1}. The inequality (6.28) implies that we may couple the events $A_{kL'}$ to Bernoulli trials with probability $1/(db^{L'-1})$ each so that the kth successful trial implies $A_{kL'}$. Therefore if we choose $\epsilon < 1/(db^{L'-1})$, then in nL' steps, at least ϵn events $A_{kL'}$ will occur for $0 \leq k < n$ with probability tending to 1 as $n \to \infty$.

Consider the following transformation of a path $\mathcal{P} = (Y_1, \ldots, Y_{nL'})$ to a "reduced" path $\mathcal{P}'$: For each k such that $A_{kL'}$ occurs, remove the edges $Y_{k+1}, \ldots, Y_{k+L'}$. Next, combine $\mathcal{P}$ and $\mathcal{P}'$ to form a nonbacktracking cycle $\mathcal{P}''$ by following $\mathcal{P}$ with a nonbacktracking cycle of length L' that does not begin with the reversal of $Y_{nL'}$, and then by returning to the tail of Y_1 by $\mathcal{P}'$ in reverse order. Note that the map $\mathcal{P} \mapsto \mathcal{P}''$ is 1-1.

When at least ϵn events $A_{kL'}$ occur, the length of $\mathcal{P}''$ is at most $(2n+1-\epsilon n)L'$. The number of nonbacktracking paths $Y_1, \ldots, Y_n$ equals db^{n-1}, whence $\sum_{k \leq (2n+1-\epsilon n)L'} b_k(G) \geq db^{nL'-1}/2$ for large n. This gives that $\mathrm{cogr}(G) > \sqrt{b}$, which implies the result by Theorem 6.10. ◄

Remark 6.12. An alternative way of handling loops in the preceding proof is to use Exercise 6.41. On the other hand, the proof can be modified to go beyond transitive multigraphs, provided that there is a nonbacktracking cycle of length at most L at every vertex.

▷ **Exercise 6.10.**
Let G be a d-regular multigraph. Show that $\rho(G) = 1$ iff $\mathrm{cogr}(G) = d - 1$.

▷ **Exercise 6.11.**
Give an example of a d-regular graph G where $1 < \mathrm{cogr}(G) < \sqrt{d-1}$.

6.4 Relative Mixing Rate, Spectral Gap, and Expansion in Finite Networks

In this section we investigate the rate at which random walk on a finite network converges to the stationary distribution. This is an analogue to spectral radius on infinite networks, and there is an inequality analogous to Theorem 6.7 that relates this rate to an expansion constant.

There are many ways to measure convergence to the stationary distribution. We'll consider one here and another in Section 13.3. Namely, at any time t, we'll consider here the relative distance to the stationary distribution, maximized over the initial and current state:

$$\max_{x, y \in V} \left| \frac{p_t(x, y) - \pi(y)}{\pi(y)} \right|.$$

In this section, all inner products are with respect to the stationary *probability* measure π, that is,

$$\langle f, g \rangle := (f, g)_\pi = \sum_{x \in V} \pi(x) f(x) g(x).$$

As we have seen in the preceding section, the transition operator P is self-adjoint with respect to this inner product. Now the norm of P is at most 1 by Exercise 6.6. Because we are working now on a finite network, P has $\mathbf{1}$ as an eigenvector with eigenvalue 1. Thus, we may write the eigenvalues of P as $-1 \leq \lambda_n \leq \cdots \leq \lambda_1 = 1$, where $n := |V|$. Since we still want to use the notation that $\pi(x) = \sum_{e^- = x} c(e)$, we now assume that the conductances are normalized so that $\sum_{e \in E_{1/2}} c(e) = 1/2$.

We will show that when the Markov chain is aperiodic, the chain converges to its stationary distribution at an exponential rate in t, with the exponent given by the gap between 1 and the absolute values of the other eigenvalues. The idea is that any function can be expanded in a basis of eigenfunctions. This expansion shows clearly how P^t acts on the given function. Those parts of the function multiplied by small eigenvalues go quickly to 0 as $t \to \infty$.

▷ **Exercise 6.12.**
Show that $\lambda_2 < 1$ iff the network is connected and that $\lambda_n > -1$ iff the random walk is aperiodic.

Theorem 6.13. (Relative Mixing Rate and Absolute Spectral Gap) *Consider an aperiodic random walk on a finite, connected network. Let $\pi_{\min} := \min_x \pi(x)$ and $\lambda_* := \max_{i \geq 2} |\lambda_i|$. Write $g_* := 1 - \lambda_*$. Then, for all vertices x and y,*

$$\left| \frac{p_t(x, y) - \pi(y)}{\pi(y)} \right| \leq \frac{e^{-g_* t}}{\pi_{\min}}.$$

Proof. Consider the functions $\psi_x := \mathbf{1}_{\{x\}}/\pi(x)$ on the vertices, indexed by vertices x. Since for all x and y, we have $(P^t \psi_y)(x) = p_t(x, y)/\pi(y)$, we get that

$$\frac{p_t(x, y) - \pi(y)}{\pi(y)} = \langle \psi_x, P^t \psi_y - \mathbf{1} \rangle. \tag{6.29}$$

Since $P\mathbf{1} = \mathbf{1}$, we have

$$\langle \psi_x, P^t \psi_y - \mathbf{1} \rangle = \langle \psi_x, P^t(\psi_y - \mathbf{1}) \rangle \le \|\psi_x\|_\pi \|P^t(\psi_y - \mathbf{1})\|_\pi . \tag{6.30}$$

Let $\langle f_i \rangle_{i=1}^n$ be a basis of orthogonal eigenvectors of P, where f_i is an eigenvector of eigenvalue λ_i. Since $\psi_y - \mathbf{1}$ is orthogonal to $\mathbf{1}$, there exist constants $\langle a_i \rangle_{i=2}^n$ such that $\psi_y - \mathbf{1} = \sum_{i=2}^n a_i f_i$, whence

$$\begin{aligned}
\|P^t(\psi_y - \mathbf{1})\|_\pi^2 &= \left\| \sum_{i=2}^n \lambda_i^t a_i f_i \right\|_\pi^2 = \sum_{i=2}^n |\lambda_i^t a_i|^2 \|f_i\|_\pi^2 \\
&\le \sum_{i=2}^n \lambda_*^{2t} |a_i|^2 \|f_i\|_\pi^2 = \lambda_*^{2t} \left\| \sum_{i=2}^n a_i f_i \right\|_\pi^2 \\
&= \lambda_*^{2t} \|\psi_y - \mathbf{1}\|_\pi^2 .
\end{aligned} \tag{6.31}$$

Since $(\psi_y - \mathbf{1}) \perp \mathbf{1}$, an orthogonal decomposition of ψ_y is $(\psi_y - \mathbf{1}) + \mathbf{1}$, which gives $\|\psi_y - \mathbf{1}\|_\pi \le \|\psi_y\|_\pi$, and thus by (6.29), (6.30), and (6.31),

$$\left| \frac{p_t(x, y) - \pi(y)}{\pi(y)} \right| \le \lambda_*^t \|\psi_x\|_\pi \|\psi_y\|_\pi = \frac{\lambda_*^t}{\sqrt{\pi(x)\pi(y)}} \le \frac{e^{-g_* t}}{\pi_{\min}} . \qquad \blacktriangleleft$$

▷ **Exercise 6.13.**
Show that the rate of exponential convergence in Theorem 6.13 cannot be faster than λ_*. More precisely, show that there is an x for which

$$\lim_{t \to \infty} -\frac{\log |p_t(x, x) - \pi(x)|}{t} = -\log \lambda_* .$$

Because of the bound in Theorem 6.13, we'd like to know how we can estimate the ***absolute spectral gap*** $g_* := 1 - \max_{i \ge 2} |\lambda_i|$ defined in that theorem. Intuitively, if a Markov chain has a "bottleneck," that is, a large set of states with large complement that is difficult to transition into or out of, then it will take it more time to mix, that is, to become close to the stationary distribution. To formulate this intuition and relate it to the absolute spectral gap, we define what is known as the expansion constant, the analogue for finite networks of the expansion constant we defined in Section 6.1 for infinite networks.

For any two subsets of vertices A and B, let $c(A, B) := \sum_{a \in A, b \in B} c(a, b)$ and $\pi(A) := \sum_{a \in A} \pi(a)$ (which we have also denoted $|A|_\pi$).

Definition 6.14. The ***expansion constant*** of a finite network is

$$\Phi_* := \min\{\Phi_S ; \ 0 < \pi(S) \le 1/2\},$$

where for $S \subseteq \mathsf{V}$,

$$\Phi_S := \frac{c(S, S^c)}{\pi(S)} .$$

Note that $0 \le \Phi_S \le 1$ and Φ_S is the probability that one step from a π-random state in S will lead to S^c. (The standard notation Φ may suggest a graph cut in two.) A more symmetrical form of Φ_* is

$$\Phi_* = \min\left\{\frac{c(S, S^c)}{\min\{\pi(S), \pi(S^c)\}} \, ; \, 0 < \pi(S) < 1\right\}.$$

Some people prefer to define Φ_* as the slightly different quantity

$$\min\left\{\frac{c(S, S^c)}{\pi(S)\pi(S^c)} \, ; \, 0 < \pi(S) < 1\right\},$$

which is equal to our definition up to a factor of at most 2.

The following theorem is the analogue of Theorem 6.7. Combined with the previous theorem, it connects the expansion properties of a network with its mixing time via its **spectral gap** $g := 1 - \lambda_2$. We assume that the chain is **lazy**, that is, for any state x we have that $p(x, x) \ge 1/2$. In that case, $P = (I + \widetilde{P})/2$ where $\widetilde{P}$ is the transition operator of the random walk on another network, and hence all the eigenvalues of P are in $[0, 1]$, so $\lambda_* = \lambda_2$. Note that laziness implies aperiodicity. If the chain is not lazy, then we can always consider the new chain with transition matrix $(I + P)/2$, which is lazy.

Theorem 6.15. (Expansion and Spectral Gap) *Let λ_2 be the second eigenvalue of a reversible and lazy Markov chain and $g := 1 - \lambda_2$. Then*

$$\frac{\Phi_*^2}{2} \le g \le 2\Phi_* \, .$$

We will use the following lemma in the proof of the lower bound. The proof is the same as that for Lemma 6.8.

Lemma 6.16. *Let $\psi \ge 0$ be a function on the vertices of a network with stationary probability distribution π. If $\pi[\psi > 0] \le 1/2$, then*

$$\Phi_* \sum_x \psi(x)\pi(x) \le \frac{1}{2} \sum_{x,y} |\psi(x) - \psi(y)| \, c(x, y) \, .$$

We also need the following analogue of Exercise 6.7:

$\triangleright$ **Exercise 6.14.**
Show that

$$\lambda_2 = \max_{f \perp \mathbf{1}} \frac{\langle Pf, f \rangle}{\langle f, f \rangle} \, .$$

Proof of Theorem 6.15. The upper bound is easier. By Exercise 6.14, we have

$$g = \min_{f \perp 1} \frac{\langle (I - P)f, f \rangle}{\langle f, f \rangle} \, . \tag{6.32}$$

As we have seen before in (6.17), expanding the numerator gives

$$\langle (I - P)f, f \rangle = \frac{1}{2} \sum_{x,y} \pi(x) p(x, y) \big[f(y) - f(x) \big]^2 \, .$$

To obtain $g \leq 2\Phi_*$, consider any S with $\pi(S) \leq 1/2$. Define a function f by $f(x) := \pi(S^c)$ for $x \in S$ and $f(x) := -\pi(S)$ for $x \notin S$. Then $\sum_x f(x)\pi(x) = 0$, so $f \perp 1$. Using this f in (6.32), we get that

$$g \leq \frac{2c(S, S^c)}{2\pi(S)\pi(S^c)} \leq \frac{2c(S, S^c)}{\pi(S)} = 2\Phi_S \, ,$$

and so $g \leq 2\Phi_*$.

To prove the lower bound, take an eigenfunction f_2 such that $P f_2 = \lambda_2 f_2$ and $\pi[f_2 > 0] \leq 1/2$ (if this inequality does not hold, take $-f_2$). Define a new function $f := \max\{f_2, 0\}$. We claim that

$$\forall x \quad \big[(I - P)f \big](x) \leq g f(x) \, .$$

This is because if $f(x) = 0$, this inequality translates to $-(Pf)(x) \leq 0$, which is true since $f \geq 0$, whereas if $f(x) > 0$, then $\big[(I - P)f \big](x) \leq \big[(I - P)f_2 \big](x) = g f_2(x) = g f(x)$. Since $f \geq 0$, we get

$$\langle (I - P)f, f \rangle \leq g \langle f, f \rangle \, ,$$

or equivalently,

$$g \geq \frac{\langle (I - P)f, f \rangle}{\langle f, f \rangle} =: R \, .$$

(This looks like a contradiction to (6.32), but it is not, since f is not orthogonal to $\mathbf{1}$.) Then just as in the proof of Theorem 6.7, we obtain

$$1 - \frac{\Phi_*^2}{2} \geq \sqrt{1 - \Phi_*^2} \geq 1 - R \geq 1 - g \, . \qquad \blacktriangleleft$$

Families of graphs with rapid mixing are useful in a variety of applications in theoretical computer science; see, for example, Hoory, Linial, and Wigderson (2006). To ensure rapid mixing, one usually bounds the expansion constants from below. A family of d-regular graphs $\{G_n\}$ is said to be a (d, c)-***expander*** family if the expansion constant of the simple random walk on G_n satisfies $\Phi_*(G_n) \geq c$ for all n. Note that for a d-regular graph G, we have

$$\Phi_*(G) = \min \left\{ \frac{|\partial_\mathsf{E} S|}{d|S|} \, ; \; S \subset \mathsf{V}, \, 0 < |S| \leq |\mathsf{V}|/2 \right\} \, .$$

Although for applications, explicit families of expanders are needed, they are more difficult to construct than random families. We now construct a 3-regular family $\langle G_n \rangle$ of expander multigraphs. This was the first construction of an expander family of bounded degree, and it is due to Pinsker (1973).* We will construct $G_n = (\mathsf{V}, \mathsf{E})$ as a bipartite graph with parts A and B, each with n vertices. Although A and B are distinct, we will denote them both by $\{1, \ldots, n\}$. Draw uniformly at random two permutations σ_1, σ_2 of $\{1, \ldots, n\}$ and take the edge set to be $\mathsf{E} = \big\{ (i, i), (i, \sigma_1(i)), (i, \sigma_2(i)) ;\ 1 \le i \le n \big\}$. We'll call this **Pinsker's model** on $2n$ vertices.

Theorem 6.17. *There exists $\delta > 0$ such that with probability tending to 1 as $n \to \infty$, Pinsker's model on $2n$ vertices satisfies $\forall S \subset \mathsf{V}$ with $0 < |S| \le n$,*

$$\frac{|\partial_\mathsf{E} S|}{|S|} > \delta \,.$$

Proof. We claim that it is enough to prove that for some $\delta > 0$, with probability tending to 1, every nonempty $S \subset A$ of size $k \le \lfloor n/2 \rfloor$ has at least $\lceil (1 + \delta)k \rceil$ neighbors $\partial_\mathsf{V} S$ in B. To see this, consider any $S \subset \mathsf{V}$ with $0 < |S| \le n$. Write $S = S_1 \cup S_2$ with $S_1 \subseteq A$ and $S_2 \subseteq B$. We may assume that $|S_1| \ge |S_2|$. If $|S_1| > n/2$, then let S' be a subset of S_1 of cardinality $\lfloor n/2 \rfloor$; otherwise, let $S' := S_1$. In either case, we have $|S'| \ge |S_2|$, and so

$$|\partial_\mathsf{E} S| \ge |\partial_\mathsf{V} S'| - |S_2| \ge \big\lceil (1 + \delta)|S'| \big\rceil - |S'| \ge \delta |S'| \ge \delta |S|/2$$

if our condition holds.

So let $S \subset A$ be a set of size $0 < k \le \lfloor n/2 \rfloor$. We wish to bound from above the probability that $|\partial_\mathsf{V} S| \le \lfloor (1 + \delta)k \rfloor$. Since (i, i) is an edge for every $1 \le i \le n$, we always have that $|\partial_\mathsf{V} S| \ge k$. Consider therefore the possible vertex sets of size $\lfloor (1 + \delta)k \rfloor$ that could contain $\partial_\mathsf{V} S$, and calculate the probability that both $\sigma_1(S)$ and $\sigma_2(S)$ fall within that set. A first-moment argument (union bound) then gives

$$\mathbf{P}\Big[|\partial_\mathsf{V} S| \le \lfloor (1 + \delta)k \rfloor \Big] \le \frac{\binom{n}{\lfloor \delta k \rfloor} \binom{\lfloor (1+\delta)k \rfloor}{k}^2}{\binom{n}{k}^2} \,.$$

Considering now all possible S, we obtain

$$\mathbf{P}\Big[\exists S \subset A\ \ 0 < |S| \le \lfloor \tfrac{n}{2} \rfloor,\ |\partial_\mathsf{V} S| \le \lfloor (1 + \delta)k \rfloor \Big] \le \sum_{k=1}^{\lfloor n/2 \rfloor} \binom{n}{k} \frac{\binom{n}{\lfloor \delta k \rfloor} \binom{\lfloor (1+\delta)k \rfloor}{\lfloor \delta k \rfloor}^2}{\binom{n}{k}^2} \,,$$

which tends to 0 for $\delta > 0$ small enough by the following calculation.

Since for any integer ℓ, we have $\log \ell! = \sum_{i=2}^{\ell} \log i \ge \int_1^\ell \log x \, dx = \ell \log \ell - \ell + 1$, in other words, $\ell! \ge e(\ell/e)^\ell$, it follows that

$$\binom{n}{\lfloor \delta k \rfloor} \le \frac{n^{\lfloor \delta k \rfloor}}{\lfloor \delta k \rfloor!} < \left(\frac{en}{\lfloor \delta k \rfloor} \right)^{\lfloor \delta k \rfloor} \le \left(\frac{en}{\delta k} \right)^{\delta k} \,,$$

---

* An earlier random construction with bounded mean degree was due to Kolmogorov and Barzdin' (1967).

where the last inequality holds because $t \mapsto (en/t)^t$ is increasing for $t \in (0, n)$. Similarly bound $\binom{\lfloor (1+\delta)k \rfloor}{\lfloor \delta k \rfloor}$, while $\binom{n}{k} \geq \frac{n^k}{k^k}$. This gives

$$\sum_{k=1}^{\lfloor n/2 \rfloor} \frac{\binom{n}{\lfloor \delta k \rfloor} \binom{\lfloor (1+\delta)k \rfloor}{\lfloor \delta k \rfloor}^2}{\binom{n}{k}} < \sum_{k=1}^{\lfloor n/2 \rfloor} \left(\frac{k}{n} \right)^{(1-\delta)k} \left[\frac{e^3(1+\delta)^2}{\delta^3} \right]^{\delta k}.$$

Each term clearly tends to 0 as n tends to ∞ given any $\delta \in (0, 1)$, and since $\frac{k}{n} \leq \frac{1}{2}$ and $\left(\frac{1}{2} \right)^{(1-\delta)} \left[\frac{e^3(1+\delta)^2}{\delta^3} \right]^{\delta} < 1$ for $\delta \leq 0.05$, for any such δ the whole sum tends to 0 as n tends to ∞ by the dominated convergence theorem.   ◀

6.5 Planar Graphs

With rare exceptions, planar graphs are the only graphs we can draw in a nice way and gaze at. See Figures 6.1 and 6.2 for some examples drawn by a program created by Don Hatch. They also make great art; see Figure 6.3 for a transformation by Doug Dunham of a print by Escher.

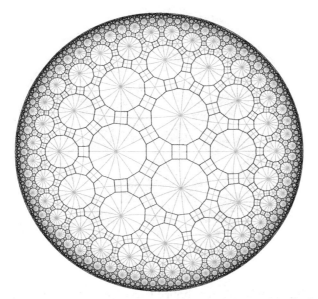

Figure 6.1. A Cayley graph in the hyperbolic disc and its dual in light gray, with triangles of interior angles $\pi/2$, $\pi/3$, and $\pi/7$.

These are good reasons for studying planar graphs separately. But a mathematical reason is that they often exhibit special behavior, as we will see several times in this book, and there are special techniques available as well. Planar duality is a prototype for more general types of duality, which is yet another reason to study it.

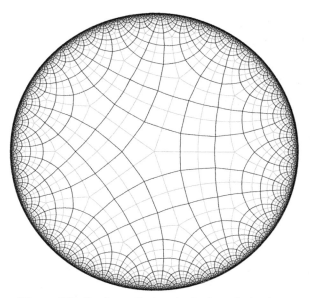

Figure 6.2. Dual tessellations in the hyperbolic disc.

Figure 6.3. A transformed Escher print, *Circle Limit I*, based on a (6,4)-tessellation of the hyperbolic disc.

A *planar* graph is one that can be drawn in the plane in such a way that edges do not cross; an actual such embedding is called a *plane* graph. If *G* is a plane graph such that each bounded set in the plane intersects only finitely many vertices and edges of *G*, then *G* is said to be *properly embedded* in the plane. *We will always assume without further mention that plane graphs are properly embedded.* A *face* of a plane graph is a connected component of the complement of the graph in the plane. If G is a plane (multi)graph, then the *plane dual* $G^\dagger$ of G is the (multi)graph formed as follows: The vertices of $G^\dagger$ are the faces formed by G. Two faces of G are joined by an edge in $G^\dagger$ precisely when they share an edge in G. A face of G also gets a loop for each edge in G on both sides of which lie that same face. Thus, $\mathsf{E}(G)$ and $\mathsf{E}(G^\dagger)$ are in a natural one-to-one correspondence. Furthermore, if one draws each vertex of $G^\dagger$ in the interior of the corresponding face of G and each edge of $G^\dagger$ so that it crosses only the corresponding edge of G, then the dual of $G^\dagger$ is G.

In this section, drawn from Häggström, Jonasson, and Lyons (2002), we will calculate the edge-expansion constants of certain regular planar graphs that arise from tessellations of the hyperbolic plane. The same results were found independently (with different proofs) by Higuchi and Shirai (2003). Of course, edge graphs of Euclidean tessellations are amenable, whence their expansion constants are 0. During this section, we assume that G and its plane dual $G^\dagger$ are locally finite, whence each graph has one end, that is, the deletion of any finite set of vertices leaves exactly one infinite component. We first examine the combinatorial difference between Euclidean and hyperbolic tessellations. See Section 2.6 for background on the hyperbolic plane.

A regular Euclidean polygon with $d^\dagger$ sides has interior angles $\pi(1 - 2/d^\dagger)$. For such polygons to form a tessellation of the plane with d polygons meeting at each vertex, we must have $\pi(1 - 2/d^\dagger) = 2\pi/d$, in other words, $1/d + 1/d^\dagger = 1/2$, or, equivalently, $(d-2)(d^\dagger - 2) = 4$. There are three such cases, and in all three, tessellations have been well known since antiquity. On the other hand, in the hyperbolic plane, the interior angles of a regular geodesic polygon with $d^\dagger$ sides can take any value in $\left(0, \pi(1 - 2/d^\dagger)\right)$, whence a tessellation of degree d exists only if $1/d + 1/d^\dagger < 1/2$, or, equivalently, $(d - 2)(d^\dagger - 2) > 4$; again, this condition is also sufficient for the existence of a hyperbolic tessellation, as has been known since the 19th century. Furthermore, in the hyperbolic plane, any two (regular geodesic) d-gons with interior angles all equal to some number α are congruent. (There are no homotheties of the hyperbolic plane.) The edges of a tessellation form the associated *edge graph*. Clearly, when the tessellation is any of these under discussion, the associated edge graph is regular and its dual is regular as well. An example is drawn in Figure 2.4. (We remark that the cases $(d - 2)(d^\dagger - 2) < 4$ correspond to the spherical tessellations that arise from the five regular solids.)

Moreover, we claim that if G is a plane, regular graph with regular dual, then G is transitive, as is $G^\dagger$, and that G is the edge graph of a tessellation by congruent regular polygons. The proof of this claim is a nice application of geometry to graph theory.

First, suppose we are given the edge graphs of any two tessellations by congruent regular polygons (in the Euclidean or hyperbolic plane, as necessary) of the same type $(d, d^\dagger)$ and one fixed vertex in each edge graph. Then there is an isomorphism of the two edge graphs that takes one fixed vertex to the other. This is easy to see by going out ring by ring around the fixed vertices, since the polygons are all congruent. In particular, taking the two edge graphs the same but with different fixed vertices, we see that such edge graphs are transitive.

Now to prove the general statement, it suffices to prove that any (proper) tessellation of a plane with degree d and codegree $d^\dagger$ has an edge graph that is isomorphic to the edge graph of the corresponding tessellation by congruent polygons. In case $(d - 2)(d^\dagger - 2) = 4$, replace each face by a congruent copy of a flat polygon; in case $(d-2)(d^\dagger-2) > 4$, replace it by a congruent copy of a regular hyperbolic polygon (with curvature -1) of $d^\dagger$ sides and interior angles $2\pi/d$; whereas if $(d-2)(d^\dagger-2) < 4$, replace it by a congruent copy of a regular spherical polygon (with curvature $+1$) of $d^\dagger$ sides and interior angles $2\pi/d$. Glue these polygons together along their edges. We get a metrically complete Riemannian surface of curvature 0, -1, or $+1$, correspondingly, that is homeomorphic to the plane, since our assumption is that the plane is the union of the faces, edges, and vertices of the tessellation, without needing any limit points. A theorem of Riemann says that the surface is isometric to either the Euclidean plane or the hyperbolic plane (the spherical case is impossible). That is, we now have a tessellation by congruent polygons, as desired. (One could also prove the existence of tessellations by congruent polygons in a similar manner. We also remark that either the graph or its dual is a Cayley graph; see Chaboud and Kenyon (1996).)

We write d_G for the degree of vertices in G when G is regular. Our main result will be the following calculation of the edge-expansion constant $\Phi_E(G, \mathbf{1}, \mathbf{1})$:

Theorem 6.18. *If G is an infinite, plane, regular graph with regular dual $G^\dagger$, then*

$$\Phi_E(G, \mathbf{1}, \mathbf{1}) = (d_G - 2)\sqrt{1 - \frac{4}{(d_G - 2)(d_{G^\dagger} - 2)}}.$$

Compare this to the regular tree of degree d, where the left-hand side is equal to $d - 2$ (Exercise 6.1). (Note that in Section 6.2, $\Phi_E(G)$ denoted $\Phi_E(G, \mathbf{1}, \deg)$ for an unweighted graph G, which differs in the regular case by a factor of d_G from $\Phi_E(G, \mathbf{1}, \mathbf{1})$ used here.)

To prove Theorem 6.18, we unfortunately need to introduce a few more ways to measure expansion. For $K \subseteq V$, define

$$E(K) := \{[x, y] \in E ; \ x, y \in K\} \tag{6.33}$$

and

$$E^*(K) := \{[x, y] \in E ; \ x \in K \text{ or } y \in K\}.$$

Thus, $\partial_E K = E^*(K) \setminus E(K)$ and the graph induced by K is $G{\upharpoonright}K = (K, E(K))$. Write

$$\Phi'_E(G) := \lim_{N \to \infty} \inf\left\{\frac{|\partial_E K|}{|K|} ; \ K \subset V, \ G{\upharpoonright}K \text{ connected}, \ N \le |K| < \infty\right\},$$

$$\beta(G) := \lim_{N \to \infty} \inf\left\{\frac{|K|}{|E(K)|} ; \ K \subset V, \ G{\upharpoonright}K \text{ connected}, \ N \le |K| < \infty\right\},$$

$$\delta(G) := \lim_{N \to \infty} \sup\left\{\frac{|K|}{|E^*(K)|} ; \ K \subset V, \ G{\upharpoonright}K \text{ connected}, \ N \le |K| < \infty\right\}.$$

When G is regular, we have for all finite K that $2|E(K)| = d_G|K| - |\partial_E K|$ and, similarly, $2|E^*(K)| = d_G|K| + |\partial_E K|$, whence

$$\beta(G) = \frac{2}{d_G - \Phi'_E(G)} \tag{6.34}$$

and

$$\delta(G) = \frac{2}{d_G + \Phi'_{\mathsf{E}}(G)}. \tag{6.35}$$

Combining this with Theorem 6.2, we see that when G is transitive, $\Phi'_{\mathsf{E}}(G) = \Phi_{\mathsf{E}}(G)$ and

$$\delta(G) = \sup\left\{\frac{|K|}{|E^*(K)|} \; ; \; K \subset \mathsf{V} \text{ finite and nonempty}\right\}. \tag{6.36}$$

A short bit of algebra shows that Theorem 6.18 follows from applying the following identity to G, as well as to $G^\dagger$, then solving the resulting two equations using (6.34) and (6.35):

Theorem 6.19. *For any plane, regular graph G with regular dual $G^\dagger$, we have*

$$\beta(G) + \delta(G^\dagger) = 1.$$

Proof. We begin with a sketch of the proof in the simplest case. Suppose that $K \subset \mathsf{V}(G)$ and that the graph $G \restriction K$ looks like Figure 6.4 in the following sense: if K^f denotes the vertices of $G^\dagger$ that are faces of $G \restriction K$, then K^f consists of all the faces, other than the outer face, of $G \restriction K$; and $|E(K)| = |E^*(K^f)|$. In this nice case, Euler's formula applied to the graph $G \restriction K$ gives

$$|K| - |E(K)| + \left[|K^f| + 1\right] = 2,$$

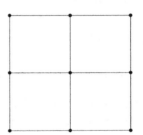

Figure 6.4. The simplest case.

which is equivalent to

$$|K|/|E(K)| + |K^f|/|E^*(K^f)| = 1 + 1/|E(K)|.$$

Now if we also assume that K can be chosen so that the first term above is arbitrarily close to $\beta(G)$ and the second term is arbitrarily close to $\delta(G^\dagger)$ with $|E(K)|$ arbitrarily large, then the desired formula follows at once. Thus, our work will consist in reducing or comparing things to such a nice situation.

Let $\epsilon > 0$, and let K be a finite set in $\mathsf{V}(G^\dagger)$ such that $G \restriction K$ is connected, $|K|/|E^*(K)| \geq \delta(G^\dagger) - \epsilon$, and $|E^*(K)| > 1/\epsilon$. Regarding each element of K as a face of G, let $K' \subset \mathsf{V}(G)$ be the set of vertices bounding these faces, and let $E' \subset \mathsf{E}(G)$ be the set of edges bounding these faces. Then $|E(K')| \geq |E'| = |E^*(K)|$. Since the number of faces F of the graph (K', E') is at least $|K| + 1$, we have

$$|K'|/|E(K')| + |K|/|E^*(K)| \leq |K'|/|E'| + |K|/|E'|$$
$$\leq |K'|/|E'| + (F - 1)/|E'|$$
$$= 1 + 1/|E'| < 1 + \epsilon, \tag{6.37}$$

where the identity comes from Euler's formula applied to the graph (K', E'). Our choice of K then implies that

$$|K'|/|E(K')| + \delta(G^\dagger) \leq 1 + 2\epsilon.$$

Since $G{\upharpoonright}K'$ is connected and $|K'| \to \infty$ when $\epsilon \to 0$, it follows that $\beta(G) + \delta(G^\dagger) \leq 1$.

To prove that $\beta(G) + \delta(G^\dagger) \geq 1$, note that the constant $\Phi'_{\mathsf{E}}(G)$ is unchanged if, in its definition, we require K to be connected and simply connected when K is regarded as a union of closed faces of $G^\dagger$ in the plane. This is because filling in holes increases $|K|$ and decreases $|\partial_{\mathsf{E}} K|$. Since G is regular, the same holds for $\beta(G)$ by (6.34). Now let $\epsilon > 0$. Let $K \subset V(G)$ be connected and simply connected (when regarded as a union of closed faces of $G^\dagger$ in the plane) such that $|K|/|E(K)| \leq \beta(G) + \epsilon$. Let K^f be the set of vertices in $G^\dagger$ that correspond to faces of $G{\upharpoonright}K$. Since $|E^*(K^f)| \leq |E(K)|$ and the number of faces of the graph $G{\upharpoonright}K$ is precisely $|K^f| + 1$, we have

$$|K|/|E(K)| + |K^f|/|E^*(K^f)| \geq |K|/|E(K)| + |K^f|/|E(K)| = 1 + 1/|E(K)| \geq 1$$

by Euler's formula applied to the graph $G{\upharpoonright}K$. (In case K^f is empty, a comparable calculation shows that $|K|/|E(K)| \geq 1$.) Because $G^\dagger$ is transitive, (6.36) allows us to conclude that

$$\beta(G) + \delta(G^\dagger) + \epsilon \geq \beta(G) + \epsilon + |K^f|/|E^*(K^f)| \geq |K|/|E(K)| + |K^f|/|E^*(K^f)| \geq 1.$$

Since ϵ is arbitrary, the desired inequality follows.   ◄

Question 6.20. What is $\Phi_{\mathsf{V}}(G)$ for regular, co-regular plane graphs G? What are $\Phi_{\mathsf{E}}(G)$ and $\Phi_{\mathsf{V}}(G)$ for more general transitive, plane graphs G? Let $\alpha_d := \left(d - 6 - \sqrt{(d-2)(d-6)}\right)/2$. Haslegrave and Panagiotis (2021) have shown that if G is regular of degree d and co-regular of codegree $d^\dagger$, then $\Phi_{\mathsf{V}}(G) = \alpha_d$ for $d^\dagger = 3$ and $\Phi_{\mathsf{V}}(G) = \alpha_{d+2}$ for $d^\dagger = 4$. They have also shown that for every plane G with minimal degree $d \geq 7$, we have $\Phi_{\mathsf{V}}(G) \geq \alpha_d$, and for every G with minimal degree $d \geq 5$ and minimal codegree at least 4, we have $\Phi_{\mathsf{V}}(G) \geq \alpha_{d+2}$. See Oh (2020) for extensions.

Question 6.21. Suppose that G is a planar graph with all degrees in $[d_1, d_2]$ and all co-degrees in $[d_1^\dagger, d_2^\dagger]$. Do we have

$$(d_1 - 2)\sqrt{1 - \frac{4}{(d_1 - 2)(d_1^\dagger - 2)}} \leq \Phi_{\mathsf{E}}(G, \mathbf{1}, \mathbf{1}) \leq (d_2 - 2)\sqrt{1 - \frac{4}{(d_2 - 2)(d_2^\dagger - 2)}} \quad ?$$

This has been proved by Oh (2020).

6.6 **Euclidean Lattices and Entropy**

Euclidean space is amenable and so does not satisfy the kind of strong isoperimetric inequality that we have studied in the earlier sections of this chapter. However, it is the origin of isoperimetric inequalities that continue to be useful today. The main result of this section is the following discrete analogue of the classical isoperimetric inequality for balls in space.

Theorem 6.22. (Discrete Isoperimetric Inequality) *If $A \subset \mathbb{Z}^d$ is a finite set, then*

$$|\partial_E A| \geq 2d|A|^{\frac{d-1}{d}} \ .$$

Observe that the $2d$ constant in the inequality is the best possible, as the example of the d-dimensional cube shows: if $A = [0, n)^d \cap \mathbb{Z}^d$, then $|A| = n^d$ and $|\partial_E A| = 2dn^{d-1}$. The same inequality without the sharp constant follows from Theorem 6.29.

To prove this inequality, we will develop other very useful tools concerning entropy.

For every $1 \leq i \leq d$, define the projection $\mathcal{P}_i \colon \mathbb{Z}^d \to \mathbb{Z}^{d-1}$ simply as the function dropping the ith coordinate, that is, $\mathcal{P}_i(x_1, \ldots, x_d) = (x_1, \ldots, x_{i-1}, x_{i+1}, \ldots, x_d)$. Theorem 6.22 will follow easily from the following beautiful inequality of Loomis and Whitney (1949).

Lemma 6.23. (Discrete Loomis and Whitney Inequality) *For any finite $A \subset \mathbb{Z}^d$,*

$$|A|^{d-1} \leq \prod_{i=1}^{d} |\mathcal{P}_i(A)| \ .$$

Before proving Lemma 6.23, we show how it gives our isoperimetric inequality.

Proof of Theorem 6.22. We claim that

$$|\partial_E A| \geq 2 \sum_{i=1}^{d} |\mathcal{P}_i(A)| \ . \tag{6.38}$$

To see this, observe that any vertex in $\mathcal{P}_i(A)$ matches to a straight line in the ith coordinate direction which intersects A. Thus, since A is finite, to any vertex in $\mathcal{P}_i(A)$, we can always match two distinct edges in $\partial_E A$: the first and last edges on the straight line that intersects A.

Using Lemma 6.23, the arithmetic mean–geometric mean inequality, and (6.38), we get

$$|A|^{d-1} \leq \prod_{i=1}^{d} |\mathcal{P}_i(A)| \leq \left(\frac{1}{d} \sum_{i=1}^{d} |\mathcal{P}_i(A)| \right)^d \leq \left(\frac{|\partial_E A|}{2d} \right)^d ,$$

as desired. ◀

To prove Lemma 6.23, we introduce the powerful notion of entropy. Let X be a random variable taking values $x_1, \ldots, x_n$. Denote $p(x) := \mathbf{P}[X = x]$, and define the **entropy** of X to be

$$H(X) := \sum_{i=1}^{n} p(x_i) \log \frac{1}{p(x_i)} = -\sum_{i=1}^{n} p(x_i) \log p(x_i) \ .$$

Clearly $H(X)$ depends only on the law $\mu_X = p(\cdot)$ of X, and it will be convenient to write this functional as $H[\mu_X]$. The entropy is the same as the expectation of $\log(1/p(X))$. This logarithm can be thought of as the "surprise" (or information) of seeing the actual value of X, measured in bits when the logarithm is to base 2. For example, if one of the values of X has probability $1/2^5$, then seeing that value gives 5 bits of surprise. (Note, though, that we use log to mean natural logarithm.) Suppose, however, that one has an incorrect idea of the distribution of X, thinking that it is the function $q(x)$. In this case, one's surprise will be $\log(1/q(X))$, whose expectation is at least $H(X)$, because

$$
\sum_{i=1}^{n} p(x_i) \log \frac{1}{q(x_i)} - \sum_{i=1}^{n} p(x_i) \log \frac{1}{p(x_i)} = -\sum_{i=1}^{n} p(x_i) \log \frac{q(x_i)}{p(x_i)}
$$

$$
\geq -\sum_{i=1}^{n} p(x_i) \left(\frac{q(x_i)}{p(x_i)} - 1 \right) = 0 ,
$$

(6.39)

due to the inequality $\log t \leq t - 1$ for $t > 0$, with equality iff $t = 1$. The difference, or excess surprise from using the wrong distribution, is

$$
\sum_{i=1}^{n} p(x_i) \log \frac{1}{q(x_i)} - \sum_{i=1}^{n} p(x_i) \log \frac{1}{p(x_i)} = \sum_{i=1}^{n} p(x_i) \log \frac{p(x_i)}{q(x_i)} =: D_{\mathrm{KL}}(p \,\|\, q) ,
$$

which is called the **Kullback–Leibler divergence of q from p** or the **relative entropy for using q instead of p**. Inequality (6.39) is due to Gibbs. Note that $D_{\mathrm{KL}}(p \,\|\, q) = 0$ iff $p = q$. As an example, if we use the uniform measure $q(x_i) := 1/n$, then we see that

$$
H(X) \leq \log n ,
$$

(6.40)

with equality iff X is uniform on n values.

The fact that $D_{\mathrm{KL}}(p \,\|\, q) = 0$ iff $p = q$ turns out to be useful to incentivize accurate predictions. Suppose that we want to predict the value of X and that someone else knows the distribution p of X. We ask that person for p, but the answer might be q instead. To promote truthfulness, we offer to pay that person some fixed amount minus a penalty of $\log 1/q(x_i)$ when the outcome is x_i. Since the expected penalty is minimized when $q = p$, we hope that p will indeed be reported. This is part of the field called proper scoring rules; see Winkler (1969) and Gneiting and Raftery (2007).

Now suppose we are given a pair (X, Y) of discrete random variables with joint distribution $\mu_{(X,Y)}$ and marginal distributions μ_X and μ_Y. We define the **mutual information** $I(X, Y)$ of X and Y to be

$$
I(X, Y) := D_{\mathrm{KL}}\big(\mu_{(X,Y)} \,\|\, \mu_X \otimes \mu_Y\big) = H(X) + H(Y) - H(X, Y) ;
$$

in this calculation, we used the fact that

$$
\log \frac{1}{\mu_X(x)\mu_Y(y)} = \log \frac{1}{\mu_X(x)} + \log \frac{1}{\mu_Y(y)} .
$$

Thus, $I(X, Y) \geq 0$ with equality iff X and Y are independent. We also define the ***conditional entropy*** $H(X \mid Y)$ ***of*** X ***given*** Y as

$$H(X \mid Y) := H(X, Y) - H(Y) = H(X) - I(X, Y). \tag{6.41}$$

The reason this is called "conditional entropy" is that it can be written in a way using conditional distributions as follows. Write $\mu_{X|Y}$ for the conditional distribution of X given Y; this is a (vector-valued) random variable, a function of Y, with $\mathbf{E}[\mu_{X|Y}] = \mu_X$.

Proposition 6.24. *Given two discrete random variables X and Y on the same space, we have* $H(X \mid Y) = \mathbf{E}[H[\mu_{X|Y}]]$.

Proof. We have

$$\mathbf{E}[H[\mu_{X|Y}]] = -\sum_y \mathbf{P}[Y = y] \sum_x \mathbf{P}[X = x \mid Y = y] \log \mathbf{P}[X = x \mid Y = y]$$

$$= -\sum_{x,y} \mathbf{P}[X = x, Y = y](\log \mathbf{P}[X = x, Y = y] - \log \mathbf{P}[Y = y])$$

$$= H(X, Y) - H(Y). \qquad \blacktriangleleft$$

Corollary 6.25. (Shannon's Inequalities) *For any discrete random variables X, Y, and Z, we have*

$$0 \leq H(X \mid Y) \leq H(X), \tag{6.42}$$

$$H(X, Y) \leq H(X) + H(Y), \tag{6.43}$$

and

$$H(X, Y \mid Z) \leq H(X \mid Z) + H(Y \mid Z). \tag{6.44}$$

Proof. Since entropy is nonnegative, so is conditional entropy by Proposition 6.24. This proves the first part of (6.42). Since Kullback–Leibler divergence is nonnegative, we have (6.43). Combined with the definition of $H(X \mid Y)$, (6.43) gives the second part of (6.42). Because of Proposition 6.24, (6.43) also gives (6.44). $\blacktriangleleft$

Our last step before proving Lemma 6.23 is the following inequality of Han (1978):

Theorem 6.26. (Han's Inequality) *For any discrete random variables $X_1, \ldots, X_k$, we have*

$$(k - 1)H(X_1, ..., X_k) \leq \sum_{i=1}^k H(X_1, ..., X_{i-1}, X_{i+1}, ..., X_k).$$

Proof. Write X_i^* for the vector $(X_1, ..., X_{i-1}, X_{i+1}, ..., X_k)$. Then a telescoping sum and (6.42) give

$$H(X_1, ..., X_k) = \sum_{i=1}^k H(X_i \mid X_1, ..., X_{i-1})$$

$$\geq \sum_{i=1}^k H(X_i \mid X_i^*) = \sum_{i=1}^k [H(X_1, ..., X_k) - H(X_i^*)].$$

Rearranging the terms gives Han's inequality. $\blacktriangleleft$

We now prove the Loomis–Whitney inequality.

Proof of Lemma 6.23. Take random variables $X_1, \ldots, X_d$ such that $(X_1, \ldots, X_d)$ is distributed uniformly on A. Clearly $H(X_1, \ldots, X_d) = \log |A|$, and by (6.40),

$$H(X_1, \ldots X_{i-1}, X_{i+1}, \ldots, X_d) \le \log |\mathcal{P}_i(A)|.$$

Now use Theorem 6.26 on $X_1, \ldots, X_d$ to find that

$$(d-1) \log |A| \le \sum_{i=1}^{d} \log |\mathcal{P}_i(A)|,$$

as desired. ◀

The proof of Han's inequality, while short, leaves some mystery as to why the inequality is true. In fact, a more general and very beautiful inequality due to Shearer, discovered in the same year but not published until later by Chung, Graham, Frankl, and Shearer (1986), has a proof that shows not only why the inequality holds but also why it must hold. We present this now. Shearer's inequality has many applications in combinatorics.

Lemma 6.27. *Given random variables $X_1, \ldots, X_k$ and $S \subseteq [1, k]$, write X_S for the random variable $\langle X_i ; i \in S \rangle$. The function $S \mapsto H(X_S)$ is submodular, where $H(X_\varnothing) := 0$.*

Proof. Given $S, T \subseteq [1, k]$, we wish to prove that

$$H(X_{S \cup T}) + H(X_{S \cap T}) \le H(X_S) + H(X_T).$$

Subtracting $2H(X_{S \cap T})$ from both sides, we see that this is equivalent to

$$H(X_{S \cup T} \mid X_{S \cap T}) \le H(X_S \mid X_{S \cap T}) + H(X_T \mid X_{S \cap T}),$$

which is (6.44). ◀

Theorem 6.28. (Shearer's Inequality) *In the notation of Lemma 6.27, let $\mathcal{S}$ be a collection of subsets of $[1, k]$ such that each integer in $[1, k]$ appears in exactly r of the sets in $\mathcal{S}$. Then*

$$rH(X_1, \ldots, X_k) \le \sum_{S \in \mathcal{S}} H(X_S).$$

Proof. Apply submodularity to the right-hand side by combining summands in pairs as much as possible and iteratively: When we apply submodularity to the right-hand side, we take a pair of index sets and replace them by their union and their intersection. We get a new sum that is smaller. It does not change the number of times that any element appears in the collection of sets. We repeat on any pair we wish. This won't change anything if one of the pair is a subset of the other, but it will otherwise. So we keep going until we can't change anything, that is, until *every* remaining pair has the property that one index set is contained in the other. (This procedure must terminate because in every step that changes the sets, the sum of squares of set sizes increases.) Since each element appears exactly r times in $\mathcal{S}$, it follows that we are left with r copies of $[1, k]$ (and some copies of $\varnothing$, which may be ignored). ◀

▷ **Exercise 6.15.**

Prove the following generalization of the Loomis–Whitney inequality. Let $A \subset \mathbb{Z}^d$ be finite and $\mathscr{S}$ be a collection of subsets of $\{1, \ldots, d\}$ such that each integer in $[1, d]$ appears in exactly r of the sets in $\mathscr{S}$. Write $\mathcal{P}_S$ for the projection of $\mathbb{Z}^d \to \mathbb{Z}^S$ onto the coordinates in S. Then

$$|A|^r \leq \prod_{S \in \mathscr{S}} |\mathcal{P}_S A|.$$

6.7 Expansion Profiles and Decay of Transition Probabilities

The next two sections look at the expansion of sets as a function of their size, without assuming that the network is nonamenable. Recall that nonamenability is equivalent to the spectral radius being less than 1, in other words, that transition probabilities decay exponentially fast. Here we will look at transition probabilities that decay more slowly than exponentially.

This section examines what expansion tells us about long-term transition probabilities, whereas the next section looks at transience. Both sections extend well-known results for $\mathbb{Z}^d$.

Kesten (1967) asked whether the recurrent Cayley graphs are precisely those whose growth is at most quadratic. The results here will allow us to answer Kesten's question positively. The reason is that the growth rate of a Cayley graph implies a bound on its expansion, as we see from the following theorem due to Coulhon and Saloff-Coste (1993). Define the ***inner vertex boundary*** of a set K as $\partial_V^{\mathrm{in}} K := \{x \in K \,;\, \exists y \notin K \ \ y \sim x\}$.

Theorem 6.29. (Expansion of Cayley Graphs) *Let G be a Cayley graph. Let $R(m)$ be the smallest radius of a ball in G that contains at least m vertices. Then, for all finite $K \subset \mathsf{V}$, we have*

$$\frac{|\partial_V^{\mathrm{in}} K|}{|K|} \geq \frac{1}{2R(2|K|)}.$$

Proof. Let s be a generator or its inverse of the group Γ used for the right Cayley graph G. The bijection $x \mapsto xs$ moves x to a neighbor of x. Thus, it moves at most $|\partial_V^{\mathrm{in}} K|$ vertices of K to the complement of K. If γ is the product of r generators or their inverses, then the map $x \mapsto x\gamma$ is a composition of r steps of distance 1, each of which moves at most $|\partial_V^{\mathrm{in}} K|$ points of K out of K, whence the composition moves at most $r|\partial_V^{\mathrm{in}} K|$ points of K out of K. Let $R := R(2|K|)$. Now by choice of R, a random γ in the ball of radius R about the identity has chance at least $1/2$ of moving any given $x \in K$ out of K, so that, in expectation, a random γ moves at least $|K|/2$ points of K out of K. Hence there is some γ that moves this many points, whence $R|\partial_V^{\mathrm{in}} K| \geq |K|/2$. This is the desired inequality. ◀

The same proof yields the same lower bound for the outer vertex boundary, $\partial_V K$. An extension to transitive graphs is given in Lemma 10.46.

Example 6.30. One might expect that the outer boundary of a finite set in an infinite Cayley graph is at least as big as its inner boundary, especially if the set is a ball. However, this is not true. Consider the usual Cayley graph of $\mathbb{Z} \times \mathbb{Z}_{2n}$, and $K := \{(x, y) \,;\, |x| \leq n - 1, y \neq n\}$.

Then K is a "square" whose inner boundary has size $8n - 8$ but whose outer boundary has size $6n - 1$, which is smaller when $n \geq 4$. To make K a ball, add the generators $(1, 1)$ and $(1, -1)$.

Note that both sides of Theorem 6.29 are of order $1/r$ for balls of radius r in $\mathbb{Z}^d$ ($d \geq 1$) and also of order r for "boxes" (6.8) of "radius" r in the lamplighter group $\mathbb{Z}^\circ$, so, up to constant factors, this bound is sharp for such groups. Of course, it is not sharp for nonamenable groups.

Let $p(\cdot, \cdot)$ be the transition probabilities of an irreducible Markov chain on a countable state space V. In this section, ***we assume that the chain has an infinite stationary measure*** π and obtain upper bounds on n-step transition probabilities $p_n(\cdot, \cdot)$ using information on expansion of finite sets of varying size.

For $x, y \in \mathsf{V}$, let $Q(x, y) := \pi(x)p(x, y)$. For $S, A \subset \mathsf{V}$, define $Q(S, A) := \sum_{s \in S, a \in A} Q(s, a)$ and $|\partial_E S|_Q := Q(S, S^c)$. The edge expansion of a finite set $S \subset \mathsf{V}$ is $\Phi_S := |\partial_E S|_Q / \pi(S)$. Write $\pi_{\min} := \inf_{x \in \mathsf{V}} \pi(x)$. The ***expansion profile*** of the chain is defined for $u \in (\pi_{\min}, \infty)$ by

$$\Phi(u) := \inf\{\Phi_S \; ; \; 0 < \pi(S) \leq u\} \, . \tag{6.45}$$

For $u \in (0, \pi_{\min}]$, we set $\Phi(u) := \infty$. Recall that a Markov chain on V is called ***lazy*** if $p(x, x) \geq 1/2$ for all $x \in \mathsf{V}$. The following bound is the main result of this section.

Theorem 6.31. *Suppose that the Markov chain* (V, P) *is either lazy or reversible. If*

$$n \geq 1 + \int_{\pi(x) \wedge \pi(y)}^{4/\epsilon} \frac{16 \, du}{u \Phi^2(u)} \, , \tag{6.46}$$

then

$$\frac{p_n(x, y)}{\pi(y)} \leq \epsilon \, . \tag{6.47}$$

See Exercise 6.77 for a version of the theorem where the laziness assumption is relaxed. Before we prove the theorem, we describe some applications.

Corollary 6.32. *Suppose that the Markov chain* (V, P) *is lazy or reversible.*
  (i) *If* $\Phi(u) \geq \varphi_0$ *for some* $\varphi_0 > 0$ *and all* $u > 0$, *then*

$$\frac{p_n(x, y)}{\pi(y)} \leq \frac{4}{\pi_{\min}} \exp\left(-\frac{\varphi_0^2(n - 1)}{16}\right) \quad \textit{for all } x, y \in \mathsf{V} \textit{ and } n \geq 1 \, .$$

  (ii) *Let* $d > 0$. *If* $\Phi(u) \geq cu^{-1/d}$ *for some* $c > 0$ *and all* $u > 0$, *then*

$$\frac{p_n(x, y)}{\pi(y)} \leq C'n^{-d/2} \quad \textit{for all } x, y \in \mathsf{V} \textit{ and } n \geq 1 \, ,$$

where $C' = C'(d, c)$. *In particular, by Theorem 6.29, this applies if the chain is simple random walk on a Cayley graph where the balls satisfy* $|B(o, r)| \geq c_0 r^d$ *for some* $c_0 > 0$, *so* $R(k) = O(k^{1/d})$.

(iii) *If $\Phi(u) \geq c/\log(bu)$ for some $b, c > 0$ and all $u > 0$, then*

$$\frac{p_n(x, y)}{\pi(y)} \leq C_1 \exp(-C_2 n^{1/3}) \quad \text{for all } x, y \in V \text{ and } n \geq 1,$$

where C_1, C_2 depend on b, c. In particular, this applies if the chain is simple random walk on a Cayley graph of exponential growth because of Theorem 6.29.

Proof. (i) Given $n > 1$, let ϵ satisfy

$$n = 1 + \int_{\pi_{\min}}^{4/\epsilon} 16\varphi_0^{-2} u^{-1} \, du = 1 + 16\varphi_0^{-2} \log\left(\frac{4}{\epsilon \pi_{\min}}\right).$$

By Theorem 6.31,

$$p_n(x, y)/\pi(y) \leq \epsilon = \frac{4}{\pi_{\min}} e^{-\varphi_0^2(n-1)/16}.$$

(For lazy chains, 16 in the exponent can be replaced by 2; see Exercise 6.76.)

(ii) Given $\epsilon \in (0, 1)$, let $n \geq 1 + \int_0^{4/\epsilon} 16c^{-2} u^{2/d-1} \, du = 1 + c_1 \epsilon^{-2/d}$, where c_1 is a constant. By Theorem 6.31, $p_n(x, y) \leq \epsilon \pi(y)$. Choosing the minimum ϵ in terms of n proves the claim.

(iii) This part is proved similarly, using $\int_{\pi_{\min}}^{4/\epsilon} \frac{\log^2(bu)}{u} \, du \leq c_2 \log^3(4b/\epsilon)$. ◀

The bound of Corollary 6.32(ii) in terms of the expansion profile is more general than the Fourier method discussed in Exercise 2.100: if, say, the edges of $\mathbb{Z}^d$ get weights that are bounded and bounded away from 0, then the same upper bound on $p_n(x, y)$ holds, up to a bounded factor.

The proof of Theorem 6.31 will use a set-valued Markov chain closely tied to the original walk. This set-valued Markov chain is often referred to as **evolving sets**. Given V, π, and Q as earlier, consider the Markov chain $\langle S_n ; n \geq 0 \rangle$ on subsets of V with the following transition rule. If the current state S_n is $S \subset V$, choose U uniformly in $[0, 1]$ (independently of $\langle S_j \rangle_{j \leq n}$), and let the next state S_{n+1} be

$$S' := \{y ; \ Q(S, y) \geq U\pi(y)\}.$$

Since $Q(V, y) = \pi(y)$ by stationarity, we know that $Q(S, y) \leq \pi(y)$. Consequently,

$$\mathbf{P}[y \in S' \mid S] = \mathbf{P}[Q(S, y) \geq U\pi(y) \mid S] = \frac{Q(S, y)}{\pi(y)}. \tag{6.48}$$

Write $\mathbf{P}_S(\cdot) := \mathbf{P}(\cdot \mid S_0 = S)$ and similarly for $\mathbf{E}_S[\cdot]$.

Lemma 6.33. *The sequence $\langle \pi(S_n) \rangle_{n \geq 0}$ forms a martingale.*

Proof. By (6.48), we have

$$\mathbf{E}[\pi(S_{n+1}) \mid S_n] = \sum_{y \in V} \pi(y) \mathbf{P}[y \in S_{n+1} \mid S_n] = \sum_{y \in V} Q(S_n, y) = \pi(S_n). \quad ◀$$

The following two propositions relate the nth order transition probabilities of the original chain to the evolving-set process.

Proposition 6.34. *For all $n \geq 0$ and $x, y \in V$, we have*

$$p_n(x, y) = \frac{\pi(y)}{\pi(x)} \mathbf{P}_{\{x\}}[y \in S_n].$$

Proof. We use induction on n. The case $n = 0$ is trivial. Fix $n > 0$ and suppose the claim holds for $n - 1$. Let U be the uniform variable used to generate S_n from S_{n-1}. Then

$$\pi(x)p_n(x, y) = \pi(x) \sum_{z \in V} p_{n-1}(x, z)p(z, y) = \sum_{z \in V} \mathbf{P}_{\{x\}}[z \in S_{n-1}]\pi(z)\, p(z, y)$$

$$= \mathbf{E}_{\{x\}}\left[\sum_{z \in V} \mathbf{1}_{[z \in S_{n-1}]} Q(z, y)\right] = \pi(y)\, \mathbf{E}_{\{x\}}\left[\frac{1}{\pi(y)} Q(S_{n-1}, y)\right]$$

$$= \pi(y)\, \mathbf{P}_{\{x\}}[y \in S_n]$$

by (6.48).   ◀

Proposition 6.35. *For all $n \geq 0$ and $x \in V$, we have*

$$\left\|\frac{p_n(x, \cdot)}{\pi(\cdot)}\right\|_\pi \leq \frac{\mathbf{E}\left[\sqrt{\pi(S_n)}\right]}{\pi(x)}. \tag{6.49}$$

Proof. Let $\langle S_n \rangle_{n \geq 0}$ and $\langle \Lambda_n \rangle_{n \geq 0}$ be two independent replicas of the evolving-set process, with $S_0 = \Lambda_0 = \{x\}$. Then by Proposition 6.34,

$$\left\|\frac{p_n(x, \cdot)}{\pi(\cdot)}\right\|_\pi^2 = \sum_{y \in V} \frac{p_n(x, y)^2}{\pi(y)} = \sum_{y \in V} \pi(y) \frac{\mathbf{P}[y \in S_n]^2}{\pi(x)^2}$$

$$= \frac{1}{\pi(x)^2} \sum_{y \in V} \pi(y)\, \mathbf{P}[\{y \in S_n\} \cap \{y \in \Lambda_n\}]$$

$$= \frac{\mathbf{E}[\pi(S_n \cap \Lambda_n)]}{\pi(x)^2} \leq \frac{\mathbf{E}[\pi(S_n) \wedge \pi(\Lambda_n)]}{\pi(x)^2} \leq \frac{\mathbf{E}\left[\sqrt{\pi(S_n)\pi(\Lambda_n)}\right]}{\pi(x)^2}. \tag{6.50}$$

◀

By Jensen's inequality and Lemma 6.33, $\left\langle \sqrt{\pi(S_n)} \right\rangle$ is a supermartingale, so the right-hand side of (6.49) is decreasing. To quantify the rate of decrease, we will need the following four lemmas.

Lemma 6.36. *For every real number $\varphi \in [0, \frac{1}{2}]$, we have*

$$\frac{\sqrt{1 + 2\varphi} + \sqrt{1 - 2\varphi}}{2} \leq \sqrt{1 - \varphi^2} \leq 1 - \varphi^2/2.$$

Proof. Squaring gives the second inequality and converts the first inequality into

$$1 + \sqrt{1 - 4\varphi^2} \leq 2(1 - \varphi^2).$$

This last inequality is easily verified by subtracting 1 and squaring again.   ◀

The next lemma relates evolving sets to expansion.

Lemma 6.37. *Let* $\varnothing \ne S \subset \mathsf{V}$. *If* $p(y, y) \ge 1/2$ *for all* $y \in \mathsf{V}$, *then*

$$\mathbf{E}_S\left[\sqrt{\pi(S')}\right] \le (1 - \Phi_S^2/2)\sqrt{\pi(S)}.$$

Proof. Let U be the uniform variable used to generate S' from S. The laziness assumption yields

$$\pi(y)\,\mathbf{P}_S\left[y \in S' \mid U < \tfrac{1}{2}\right] = \begin{cases} \pi(y) & \text{if } y \in S \\ 2Q(S, y) & \text{if } y \in S^c. \end{cases}$$

Summing over $y \in \mathsf{V}$, we infer that

$$\mathbf{E}_S\left[\pi(S') \mid U < \tfrac{1}{2}\right] = \pi(S) + 2Q(S, S^c) = \pi(S) + 2|\partial_E S|_Q.$$

Thus

$$\mathbf{E}_S\left[\pi(S') \mid U < \tfrac{1}{2}\right] = \pi(S)\left(1 + 2\Phi_S\right). \tag{6.51}$$

By Lemma 6.33, $\mathbf{E}_S\left[\pi(S')\right] = \pi(S)$, and comparing this to (6.51) gives

$$\mathbf{E}_S\left[\pi(S') \mid U \ge \tfrac{1}{2}\right] = \pi(S)\left(1 - 2\Phi_S\right).$$

Therefore, by Jensen's inequality and Lemma 6.36, we have

$$\mathbf{E}_S\left[\sqrt{\pi(S')}\right] \le \tfrac{1}{2}\sqrt{\mathbf{E}_S\left[\pi(S') \mid U \le \tfrac{1}{2}\right]} + \tfrac{1}{2}\sqrt{\mathbf{E}_S\left[\pi(S') \mid U > \tfrac{1}{2}\right]}$$

$$= \left[\tfrac{1}{2}\sqrt{1 + 2\Phi_S} + \tfrac{1}{2}\sqrt{1 - 2\Phi_S}\right]\sqrt{\pi(S)} \le \left(1 - \frac{\Phi_S^2}{2}\right)\sqrt{\pi(S)}. \qquad \blacktriangleleft$$

Lemma 6.38. *Let* $f: [0, \infty) \to [0, 1]$ *be an increasing function. If* $\langle L_n \rangle_{n \ge 0}$ *satisfy* $L_n \ge 0$ *and* $L_n - L_{n+1} \ge L_n f(L_n)$ *for all* n, *then for every* $n \ge \int_\delta^{L_0} \frac{du}{u f(u)}$, *we have* $L_n \le \delta$.

Proof. It suffices to show that for every n, we have

$$\int_{L_n}^{L_0} \frac{du}{u\, f(u)} \ge n. \tag{6.52}$$

For all $k \ge 0$, the inequality $L_{k+1} \le L_k\left[1 - f(L_k)\right] \le L_k e^{-f(L_k)}$ holds, whence

$$\int_{L_{k+1}}^{L_k} \frac{du}{u f(u)} \ge \frac{1}{f(L_k)} \int_{L_{k+1}}^{L_k} \frac{du}{u} = \frac{1}{f(L_k)} \log \frac{L_k}{L_{k+1}} \ge 1.$$

Summing this over $k \in \{0, 1, \ldots, n - 1\}$ gives (6.52). $\qquad \blacktriangleleft$

Lemma 6.39. *Suppose that $Z \geq 0$ is a random variable and $f_0 : [0, \infty) \to [0, \infty)$ is an increasing function. Define $f(u) := f_0(u/2)/2$. Then $\mathbf{E}[Z \cdot f_0(Z)] \geq \mathbf{E}Z \cdot f(\mathbf{E}Z)$.*

Proof. Let $A := [Z \geq \mathbf{E}Z/2]$. Then $\mathbf{E}[Z\mathbf{1}_{A^c}] \leq \mathbf{E}Z/2$, so $\mathbf{E}[Z\mathbf{1}_A] \geq \mathbf{E}Z/2$. Therefore,

$$\mathbf{E}[Z \cdot f_0(Z)] \geq \mathbf{E}[Z\mathbf{1}_A \cdot f_0(\mathbf{E}Z/2)] \geq \mathbf{E}Z \cdot f(\mathbf{E}Z). \qquad \blacktriangleleft$$

Proof of Theorem 6.31. First, suppose the chain is lazy. Our starting point is the result of Proposition 6.35:

$$\|p_n(x, \cdot)/\pi\|_\pi \leq L_n := \frac{1}{\pi(x)} \mathbf{E}\big[\sqrt{\pi(S_n)}\big]. \tag{6.53}$$

Let $S_n^\bullet$ have the π-biased law of the evolving set S_n, that is,

$$\mathbf{P}[S_n^\bullet = A] := \frac{\pi(A)}{\pi(x)} \mathbf{P}_{\{x\}}[S_n = A] \quad \text{for every finite } A \subset \mathsf{V}.$$

For every nonnegative function F defined on finite subsets of V,

$$\mathbf{E}F(S_n^\bullet) = \mathbf{E}_x\left[\frac{\pi(S_n)}{\pi(x)} F(S_n)\right]; \tag{6.54}$$

in particular, $\mathbf{E}\big[\pi(S_n^\bullet)^{-1/2}\big] = L_n$. By Lemma 6.37,

$$\frac{1}{\pi(x)} \mathbf{E}\left[\sqrt{\pi(S_{n+1})} \,\Big|\, S_n\right] \leq \frac{\sqrt{\pi(S_n)}}{\pi(x)}\left(1 - \frac{\Phi^2(\pi(S_n))}{2}\right).$$

Taking expectations with respect to $\mathbf{P}_{\{x\}}$, using (6.53) and (6.54), gives

$$L_{n+1} \leq \mathbf{E}\left[\pi(S_n^\bullet)^{-1/2}\left(1 - \frac{\Phi^2(\pi(S_n^\bullet))}{2}\right)\right] = \mathbf{E}[Z_n(1 - f_0(Z_n))], \tag{6.55}$$

where $Z_n := \pi(S_n^\bullet)^{-1/2}$ and $f_0(z) := \Phi^2(z^{-2})/2$. Since f_0 is an increasing function, (6.55) and Lemma 6.39 imply that $L_{n+1} \leq L_n(1 - f(L_n))$, where $f(z) := f_0(z/2)/2$. Note that $L_0 = \pi(x)^{-1/2}$. By Lemma 6.38, for all

$$n \geq \int_{\sqrt{\epsilon}}^{L_0} \frac{dz}{zf(z)} = \int_{\sqrt{\epsilon}}^{L_0} \frac{4\,dz}{z\Phi^2(4/z^2)} = \int_{4\pi(x)}^{4/\epsilon} \frac{2\,du}{u\Phi^2(u)} \leq \int_{\pi(x)}^{4/\epsilon} \frac{2\,du}{u\Phi^2(u)}, \tag{6.56}$$

we have, by Proposition 6.35, $\|p_n(x, \cdot)/\pi\|_\pi \leq L_n \leq \sqrt{\epsilon}$. (We used the change of variable $u = 4/z^2$.)

For the reversible case, the next remark is not needed: Recall from Exercise 2.1 that the reversed chain has transition probabilities $\widehat{p}(x, y) = \pi(y)p(y, x)/\pi(x)$ and stationary measure π. By the upcoming Exercise 6.16, the reversed chain $\widehat{p}(\cdot, \cdot)$ has the same expansion profile as $p(\cdot, \cdot)$.

Thus, for all

$$m, n \geq \int_{\pi(x) \wedge \pi(y)}^{4/\epsilon} \frac{2 \, du}{u \Phi^2(u)} \, ,$$

we have $\|p_n(x, \cdot)/\pi\|_\pi \leq \sqrt{\epsilon}$ and $\|\widehat{p}_m(x, \cdot)/\pi\|_\pi \leq \sqrt{\epsilon}$. Observe that

$$\left| \frac{p_{n+m}(x, z)}{\pi(z)} \right| = \left| \sum_{y \in V} \frac{p_n(x, y) p_m(y, z)}{\pi(z)} \right| = \left| \left\langle \frac{p_n(x, \cdot)}{\pi(\cdot)}, \frac{\widehat{p}_m(z, \cdot)}{\pi(\cdot)} \right\rangle_\pi \right|.$$

By Cauchy–Schwarz, this is at most $\|p_n(x, \cdot)/\pi\|_\pi \cdot \|\widehat{p}_m(x, \cdot)/\pi\|_\pi \leq \epsilon$. This establishes Theorem 6.31 for lazy chains, with 4 instead of 16 in the numerator. If the chain is reversible but not lazy, then we can reduce to the lazy case using the upcoming Exercise 6.17, since for reversible chains, $\|p_n(x, \cdot)/\pi\|_\pi^2 = p_{2n}(x, x)/\pi(x)$. ◄

For certain cases, the preceding proof used the next two exercises.

▷ **Exercise 6.16.**
For the reversed chain, define $\widehat{Q}(x, y) := \pi(x) \widehat{p}(x, y) = Q(y, x)$ for all x, y. Prove that for any finite $S \subset V$, we have $|\partial_E S|_Q = |\partial_E S|_{\widehat{Q}}$.

▷ **Exercise 6.17.**
The lazy version of a transition matrix P is given by $\widetilde{P} = (P + I)/2$. Show that the corresponding expansion profile satisfies $\widetilde{\Phi}(r) = \Phi(r)/2$ for all $r > \pi_{\min}$. If the original matrix P is reversible, show that $p_{2n}(x, x)$ is decreasing in n for all $x \in V$, and deduce that $p_{2n}(x, x) \leq 2 \widetilde{p}_{2n}(x, x)$.

We are now able to fulfill our promise to answer Kesten's question whether the recurrent Cayley graphs are precisely those whose growth is at most quadratic. Exercise 2.86 showed that all graphs of at most quadratic growth are recurrent. Gromov (1981a) showed that Cayley graphs whose growth is not at most quadratic have at least cubic growth. That Cayley graphs of at least cubic growth are indeed transient was shown finally by Varopoulos (1986). We can now show this as a consequence of our preceding work:

Theorem 6.40. (Transience of Cayley Graphs) *If G is a Cayley graph of at least cubic growth, then simple random walk on G is transient.*

Proof. Because of Theorem 6.29, the hypotheses of Corollary 6.32(ii) are satisfied with $d = 3$. Since $\sum_n n^{-3/2} < \infty$, transience results. ◄

6.8 Anchored Isoperimetric Profiles and Transience

Write

$$\psi(G, t) := \inf\{|\partial_E K|_c ; t \leq |K|_\pi < \infty\}.$$

This is a functional elaboration of the expansion constant: the constant $\Phi_E(G, c, \pi)$ measures only whether $\psi(G, t)$ grows linearly in t. We saw that such linear growth is equivalent to exponential decay of return probabilities of the network random walk. The function ψ is similar to the profile function Φ of (6.45), but the former imposes a lower bound on $|K|_\pi$, whereas the latter imposes an upper bound on $|K|_\pi$; the latter also uses a quotient. Another connection to random walk arises from this isoperimetric "profile" via a relationship to effective resistance, which will allow us to give another proof that Cayley graphs of at least cubic growth are transient; it will also prove useful to us in Section 10.6. Since we will consider effective resistance from a set to infinity, we will use the still more refined function

$$\psi(G, A, t) := \inf\{|\partial_E K|_c ; A \subseteq K, K/A \text{ is connected}, t \leq |K|_\pi < \infty\} \qquad (6.57)$$

for $A \subset V(G)$. Here, we say that K/A is connected when the graph induced by K in G/A is connected, where we have identified all of A to a single vertex. If $A = \{a\}$, then we'll write more simply $\psi(G, a, t)$ for $\psi(G, \{a\}, t)$. The requirement that K contain A and K/A be connected is what leads to the adjective "anchored." When this anchored isoperimetric profile grows linearly, we will not need to use the profile function; applications of this to percolation will be in the following section.

It is commonly the case that $\psi(t) = \psi(G, A, t) \geq f(t)$ for some increasing function f on $[|A|_\pi, \infty)$ that satisfies $0 < f(t) \leq t$ and $f(2t) \leq \alpha f(t)$ for some α. For example, f has the latter two properties when f has the form $f(t) = \beta t^a$ for some $\beta \in (0, |A|_\pi^{-1} \wedge 1]$ and $a \in [0, 1]$. In this case, we obtain the following upper bound on effective resistance:

Theorem 6.41. *Let A be a finite set of vertices in a network G with $|V(G)|_\pi = \infty$. Suppose that $\psi(t) := \psi(G, A, t)$ has the property that $\psi(t) \geq f(t)$ for some increasing function f on $[|A|_\pi, \infty)$ that satisfies $0 < f(t) \leq t$ and $f(2t) \leq \alpha f(t)$ for some α. Then*

$$\mathcal{R}(A \leftrightarrow \infty) \leq \int_{|A|_\pi}^\infty \frac{4\alpha^2}{f(t)^2} \, dt.$$

Theorem 6.41 is an easy consequence of a result that does not assume any regularity on $\psi(G, A, t)$, namely:

Theorem 6.42. *Let A be a finite set of vertices in a network G with $|V(G)|_\pi = \infty$. Let $\psi(t) := \psi(G, A, t)$. Define $s_0 := |A|_\pi$ and $s_{k+1} := s_k + \psi(s_k)/2$ recursively for $k \geq 0$. Then*

$$\mathcal{R}(A \leftrightarrow \infty) \leq \sum_{k \geq 0} \frac{2}{\psi(s_k)}.$$

Before proving Theorem 6.42, we show how it implies Theorem 6.41.

Proof of Theorem 6.41. Define $t_0 := |A|_\pi$ and $t_{k+1} := t_k + f(t_k)/2$ recursively. We have that $s_k \geq t_k$ and $t_k \leq t_{k+1} \leq 2t_k$, whence for $t_k \leq t \leq t_{k+1}$, we have $f(t) \leq f(2t_k) \leq \alpha f(t_k)$, so that

$$\int_{|A|_\pi}^\infty \frac{4\alpha^2}{f(t)^2}\, dt \geq \sum_{k \geq 0} \int_{t_k}^{t_{k+1}} \frac{4\alpha^2}{f(t)^2}\, dt \geq \sum_{k \geq 0} \int_{t_k}^{t_{k+1}} \frac{4}{f(t_k)^2}\, dt = \sum_{k \geq 0} \frac{4(t_{k+1} - t_k)}{f(t_k)^2}$$

$$= \sum_{k \geq 0} \frac{2f(t_k)}{f(t_k)^2} \geq \sum_{k \geq 0} \frac{2}{\psi(t_k)} \geq \sum_{k \geq 0} \frac{2}{\psi(s_k)} \geq \mathscr{R}(A \leftrightarrow \infty). \qquad \blacktriangleleft$$

To show Theorem 6.42, we'll prove an analogue for finite networks, Lemma 6.43, which gives Theorem 6.42 immediately by means of the following exercise:

▷ **Exercise 6.18.**
Let A be a finite set of vertices in a connected network G with $|V(G)|_\pi = \infty$. Let H be a finite, connected subnetwork containing A. In H^W, identify A to a single vertex a, and let z be the (wired) boundary vertex; call the new graph H'. Define $\psi(G, A, t)$ and $\langle s_k \rangle$ as in Theorem 6.42 and $\phi(H', t)$ with its associated sequence $\langle s_k' \rangle$ as in Lemma 6.43. Show that for all $t \geq 0$,

$$\phi(H', \pi(a) + t) \geq \psi(G, A, |A|_\pi + t),$$

and that for all $k \geq 0$, we have $\phi(H', s_k') \geq \psi(G, A, s_k)$.

Lemma 6.43. *Let a and z be two distinct vertices in a finite, connected network G. Define*

$$\phi(t) := \phi(G, t) := \min\{|\partial_E W|_c \; ; \; a \in W, \; z \notin W, \; W \text{ is connected}, \; t \leq |W|_\pi\}$$

for $t \leq |V(G) \setminus \{z\}|_\pi$ and $\phi(t) := \infty$ for $t > |V(G) \setminus \{z\}|_\pi$. Define $s_0' := \pi(a)$ and $s_{k+1}' := s_k' + \phi(s_k')/2$ recursively for $k \geq 0$. Then

$$\mathscr{R}(a \leftrightarrow z) \leq \sum_{k=0}^\infty \frac{2}{\phi(s_k')}.$$

Proof. Let $v(\cdot)$ be the voltage corresponding to the unit current flow i from z to a, with $v(a) = 0$.

For $t \geq 0$, let $W(t) := \{x \in V \; ; \; v(x) \leq t\}$, and for $0 \leq t < t'$, let $E(t, t')$ be the set of directed edges from $W(t)$ to $\{x \in V \; ; \; v(x) \geq t'\}$. Define $t_0 := 0$ and recursively,

$$t_{k+1} := \inf\{t > t_k \; ; \; |E(t_k, t)|_c \leq |\partial_E W(t_k)|_c/2\}.$$

Set $\bar{k} := \min\{j \; ; \; z \in W(t_j)\} = \min\{j \; ; \; t_{j+1} = \infty\}$. Fix some $k < \bar{k}$. Note that $i(e) \leq 0$ for every $e \in \partial_E W(t_k)$ (where edges in $\partial_E W(t_k)$ are oriented away from $W(t_k)$). Since $E(t_k, t_{k+1}) \subseteq \partial_E W(t_k)$,

$$1 = \sum_{e \in \partial_E W(t_k)} |i(e)| \geq \sum_{e \in E(t_k, t_{k+1})} c(e)\big(v(e^+) - v(e^-)\big)$$

$$\geq \sum_{e \in E(t_k, t_{k+1})} c(e)(t_{k+1} - t_k) \geq (t_{k+1} - t_k) \frac{|\partial_E W(t_k)|_c}{2},$$

where the last inequality follows from the definition of t_{k+1}.

Thus

$$t_{k+1} - t_k \leq 2/\phi(|W_k|_\pi), \tag{6.58}$$

where we abbreviate $W_k := W(t_k)$. Clearly,

$$|W_{k+1}|_\pi = |W_k|_\pi + |W_{k+1} \setminus W_k|_\pi \geq |W_k|_\pi + \frac{1}{2}|\partial_E W_k|_c \geq |W_k|_\pi + \frac{1}{2}\phi(|W_k|_\pi).$$

Since ϕ is an increasing function, it follows by induction that $|W_k|_\pi \geq s_k'$ for $k < \bar{k}$, and so (6.58) gives

$$\mathscr{R}(a \leftrightarrow z) = v(z) = t_{\bar{k}} - t_0 \leq \sum_{k=0}^{\bar{k}-1} \frac{2}{\phi(|W_k|_\pi)} \leq \sum_{k=0}^{\bar{k}-1} \frac{2}{\phi(s_k')}. \qquad \blacktriangleleft$$

We can now give a second proof of transience for Cayley graphs of at least cubic growth:

Proof of Theorem 6.40. In the notation of Theorem 6.29, we have $R(m) \leq cm^{1/3}$ for some positive constant c. By that theorem, therefore, we have $\psi(G, o, m) \geq c'm^{2/3}$ for some other positive constant c'. Hence, transience is a consequence of Theorem 6.41. $\qquad \blacktriangleleft$

6.9 Anchored Expansion and Percolation

Recall the probability measure $\mathbf{P}_p$ defining Bernoulli percolation from Section 5.2 and the critical probability p_c. There, we took a random subset of edges, but an alternative is to take the subgraph induced by a random subset of vertices. This alternative is called ***site percolation***. The adjective Bernoulli applies when each vertex is present independently with the same probability. When we need different notation for these two processes, we use $\mathbf{P}_p^{\text{site}}$ and $\mathbf{P}_p^{\text{bond}}$ for the two product measures on 2^V and 2^E, respectively, and p_c^{site} and p_c^{bond} for the two critical probabilities. If we don't indicate whether the percolation is bond or site and both make sense in context, then results we state should be taken to apply to both types of percolation.

Grimmett, Kesten, and Zhang (1993) showed that simple random walk on the infinite cluster of Bernoulli percolation in $\mathbb{Z}^d$ (when $p > p_c$) is transient for $d \geq 3$; in other words, in Euclidean lattices, transience is preserved when the whole lattice is replaced by an infinite percolation cluster. (We gave a proof of part of this result in Section 5.5, where we also noted that another proof based on ideas in the present section was given by Pete (2008).) In general, what aspects of Cayley graphs are preserved under percolation?

Conjecture 6.44. (Percolation and Transience) *If G is a transient Cayley graph, then a.s. every infinite cluster of Bernoulli percolation on G is transient.*

Conjecture 6.45. (Percolation and Speed) *Let G be a Cayley graph. Then simple random walk on G has positive speed iff simple random walk on infinite clusters of Bernoulli percolation has positive speed a.s.*

These conjectures were made by Benjamini, Lyons, and Schramm (1999), who proved that simple random walk on an infinite cluster of any nonamenable Cayley graph has positive speed. One might hope to use Proposition 6.9 to establish this result, but, in fact, the infinite clusters are amenable:

▷ **Exercise 6.19.**

For any $p < 1$, every infinite cluster K of Bernoulli(p) percolation on any graph G of bounded degree has $\Phi_E(K) = 0$ a.s.

On the other hand, the infinite clusters might satisfy the following weaker "anchored expansion" property, which is known to imply positive speed (Theorem 6.53).

Fix $o \in V(G)$. The *anchored expansion constants* of G are

$$\Phi_E^*(G) := \lim_{n \to \infty} \inf\left\{ \frac{|\partial_E K|}{|K|} \; ; \; o \in K \subset V, \; G\!\restriction\!K \text{ is connected}, \; n \le |K| < \infty \right\}$$

and

$$\Phi_V^*(G) := \lim_{n \to \infty} \inf\left\{ \frac{|\partial_V K|}{|K|} \; ; \; o \in K \subset V, \; G\!\restriction\!K \text{ is connected}, \; n \le |K| < \infty \right\}.$$

These are closely related to the number $\psi(G, o, 0)$ of (6.57), but have the advantage that $\Phi_E^*(G)$ and $\Phi_V^*(G)$ do not depend on the choice of the basepoint o. We say that a graph G has *anchored expansion* if $\Phi_E^*(G) > 0$.

▷ **Exercise 6.20.**

Show that if G is a transitive graph, then $\Phi_E(G) = \Phi_E^*(G)$ and $\Phi_V(G) = \Phi_V^*(G)$.

▷ **Exercise 6.21.**

Show that for every graph G, the balls $B_G(x, n)$ in G about every point x of radius n satisfy

$$\liminf_{n \to \infty} |B_G(x, n)|^{1/n} \ge 1 + \Phi_V^*(G). \tag{6.59}$$

An important feature of anchored expansion is that several probabilistic implications of nonamenability remain true with this weaker assumption. Furthermore, anchored expansion is quite stable under percolation, as we will see.

First, we give a simple relationship between anchored expansion and percolation via the following upper bound on p_c due to Benjamini and Schramm (1996b):

Theorem 6.46. (Percolation and Anchored Expansion) *For any graph G, we have* $p_c^{\text{bond}}(G) \le 1/\bigl(1 + \Phi_E^*(G)\bigr)$ *and* $p_c^{\text{site}}(G) \le 1/\bigl(1 + \Phi_V^*(G)\bigr)$.

Note that equality holds in both inequalities when G is a regular tree. For the proof, as well as later, we will identify a subset $\omega \subseteq E$ with its indicator function, so that $\omega(e)$ takes the value 0 or 1 depending on whether e lies in the subset.

Proof. The proofs of both inequalities are completely analogous, so we prove only the first. In fact, we prove it with $\Phi_E(G)$ in place of $\Phi_E^*(G)$, leaving the improvement to Exercise 6.22.

Choose any ordering $\langle e_1, e_2, \ldots \rangle$ of E so that o is an endpoint of e_1. Fix $p > 1/\bigl(1 + \Phi_E(G)\bigr)$ and let $\langle Y_k \rangle$ and $\langle Y_k' \rangle$ be independent $\{0, 1\}$-valued Bernoulli(p) random variables. If A is the event that

$$\frac{1}{n} \sum_{k=1}^{n} Y_k > \frac{1}{1 + \Phi_E(G)}$$

for all $n \geq 1$, then A has positive probability by the strong law of large numbers.

Define $E_0 := \varnothing$. We will look at a finite or infinite subsequence of edges $\langle e_{n_j} \rangle$ via a recursive procedure and define a percolation ω as we go. Suppose that the edges $E_k := \langle e_{n_1}, \ldots, e_{n_k} \rangle$ have been selected and that $\omega(e_{n_j}) = Y_j$ for $j \leq k$. Let V_k be the union of $\{o\}$ and the endpoints of the open edges of E_k. Let n_{k+1} be the smallest index of an edge in $\mathsf{E} \setminus E_k$ that has exactly one endpoint in V_k, if any. If there are none, then stop; the cluster $K(o)$ of o is finite and we set $\omega(e_j) := Y_j'$ for the remaining edges $e_j \in \mathsf{E} \setminus E_k$. Otherwise, let $\omega(e_{n_{k+1}}) := Y_{k+1}$.

If this procedure never ends, then $K(o)$ is infinite; assign $\omega(e_j) := Y_j'$ for any remaining edges $e_j \in \mathsf{E} \setminus E_k$.

In both cases (whether $K(o)$ is finite or infinite), ω is a fair sample of Bernoulli(p) percolation on G.

We claim that $K(o)$ is infinite on the event A. This would imply that $p \geq p_c^{\mathrm{bond}}(G)$ and would complete the proof.

For suppose that $K(o)$ is finite and contains m vertices. Let E_n be the final set of selected edges. Note that E_n contains $\partial_{\mathsf{E}} K(o)$ (all edges of which are closed) and a spanning tree of $K(o)$ (all edges of which are open). This implies that $n \geq |\partial_{\mathsf{E}} K(o)| + m - 1$ and $\sum_{k=1}^n Y_k = m - 1$. Since $|\partial_{\mathsf{E}} K(o)|/m \geq \Phi_{\mathsf{E}}(G)$, we have

$$
\frac{1}{n} \sum_{k=1}^{n} Y_k \leq \frac{m-1}{|\partial_{\mathsf{E}} K(o)| + m - 1} = \frac{1}{1 + |\partial_{\mathsf{E}} K(o)|/(m-1)} < \frac{1}{1 + |\partial_{\mathsf{E}} K(o)|/m} \leq \frac{1}{1 + \Phi_{\mathsf{E}}(G)}
$$

and the event A does not occur.   ◀

▷ **Exercise 6.22.**
Prove the first inequality of Theorem 6.46 as written with anchored expansion.

Similar ideas show the following general property. Given two multigraphs G and G', a homomorphism $\phi: G \to G'$ is a **weak covering map** if, for every vertex $x \in \mathsf{V}(G)$ and every edge e' incident to $\phi(x)$, there is some edge e incident to x such that $\phi(e) = e'$. For example, if $\phi: \Gamma \to \Gamma'$ is a group homomorphism that maps a generating set S for Γ onto a generating set S' for Γ', and if G, G' are the corresponding Cayley graphs, then ϕ is also a weak covering map. In the case $|S| = |S'|$ and G, G' are simple, we get a stronger notion than weak covering map, one which is closer to the topological notion of covering map; we now define it for networks. Given two graphs $G = (\mathsf{V}, \mathsf{E})$ and $G' = (\mathsf{V}', \mathsf{E}')$ with edges weighted by c, c', respectively, and vertices weighted by D, D', respectively, call a surjection $\phi: \mathsf{V} \to \mathsf{V}'$ a **covering map** if for every vertex $x \in \mathsf{V}$, the restriction $\phi: T(x) \to T(\phi(x))$ is a network isomorphism, where $T(x)$ denotes the star at x, that is, the network induced on the edges incident to x. If there is such a covering map, then we call G a **covering network** of G'. To illustrate the distinction, map the nearest-neighbor graph on $\{-1, 0, 1\}$ with unit weights to the edge between 0 and 1 by mapping -1 to 1. This provides a weak covering map but not a covering map. The following result is due to Campanino (1985), but our proof is modeled on that of Benjamini and Schramm (1996b).

Theorem 6.47. (Covering and Percolation) *Suppose that $\phi: G \to G'$ is a weak covering map of multigraphs. Then, for any $x \in V(G)$ and $p \in (0, 1)$, we have*

$$\mathbf{P}_p[x \leftrightarrow \infty] \geq \mathbf{P}_p[\phi(x) \leftrightarrow \infty].$$

Therefore $p_c(G) \leq p_c(G')$.

Proof. Write $G = (V, E)$ and $G' = (V', E')$. We prove the theorem for bond percolation, the proof for site percolation being almost identical. We will construct a coupling of the percolation measures on the two graphs. That is, given $\omega' \in 2^{E'}$, we will define $\omega \in 2^E$ in such a way that, first, if ω' has distribution $\mathbf{P}_p$ on G', then ω has distribution $\mathbf{P}_p$ on G; and, second, if $K(\phi(x))$ is infinite, then so is $K(x)$.

Choose any ordering $\langle e_1', e_2', \ldots \rangle$ of E' so that $\phi(x)$ is an endpoint of e_1' and so that for each $k > 1$, one endpoint of e_k' is also an endpoint of some e_j' with $j < k$.

Let e_1 be any edge that ϕ maps to e_1'. Define $\omega(e_1) := \omega'(e_1')$ and set $n_1 := 1$. We will select a subsequence of edges $\langle e_{n_j}' \rangle$ via a recursive procedure. Suppose that $E_k' := \{e_{n_1}', \ldots, e_{n_k}'\}$ have been selected and edges e_j that ϕ maps onto e_{n_j}' ($j \leq k$) have been chosen. Let n_{k+1} be the smallest index of an edge in $E' \setminus E_k'$ that shares an endpoint with at least one of the open edges in E_k', if any. If there are none, then stop; $K(\phi(x))$ is finite and $\omega(e)$ for the remaining edges $e \in E$ may be assigned independently in any order. Otherwise, if $e_{n_{k+1}}'$ is incident with $e_{n_j}' \in E_k'$, then let e_{k+1} be any edge that ϕ maps to $e_{n_{k+1}}'$ and that is incident with e_j; such an edge exists because ϕ is a weak covering map. Set $\omega(e_{k+1}) := \omega'(e_{n_{k+1}}')$.

If this procedure never ends, then $K(\phi(x))$ is infinite; assigning $\omega(e)$ for any remaining edges $e \in E$ independently in any order gives a fair sample of Bernoulli percolation on G that has $K(x)$ infinite. This proves the theorem. ◀

In the appendix to Chen and Peres (2004), G. Pete strengthened the conclusion of Theorem 6.46 as follows:

Theorem 6.48. (Anchored Expansion of Clusters) *Consider Bernoulli(p) bond percolation on a graph G with $\Phi_E^*(G) > 0$. If $p > 1/(1 + \Phi_E^*(G))$, then almost surely on the event that the open cluster K containing o is infinite, it satisfies $\Phi_E^*(K) > 0$. Likewise, for $p > 1/(1 + \Phi_V^*(G))$, we have $\mathbf{P}_p[\Phi_V^*(K) > 0 \mid |K| = \infty] = 1$.*

Proof. We prove only the first assertion, as the second is similar. Define

$$\mathscr{A}_n := \left\{K \subset V(G); \; o \in K, \; G{\restriction}K \text{ is connected and finite}, \; |\partial_E K| = n\right\}. \tag{6.60}$$

We will consider edge boundaries with respect to both $G{\restriction}K$ and G, so we denote them by ∂_E^K and ∂_E^G, respectively. Note that in Bernoulli(p) bond percolation, for any $0 < \alpha < p$ and $S \in \mathscr{A}_n$, we can estimate the conditional probability

$$\mathbf{P}\left[\frac{|\partial_E^K S|}{|\partial_E^G S|} \leq \alpha \;\middle|\; S \subseteq K\right] = \mathbf{P}[\text{Bin}(n, p) \leq \alpha n] \leq e^{-nI_p(\alpha)}, \tag{6.61}$$

where the large deviation rate function

$$I_p(\alpha) := \alpha \log \frac{\alpha}{p} + (1 - \alpha) \log \frac{1 - \alpha}{1 - p} \tag{6.62}$$

is continuous in α and $-\log(1-p) = I_p(0) > I_p(\alpha) > 0$ for $0 < \alpha < p$ (see Billingsley (1995), p. 151, or Dembo and Zeitouni (1998), Exercise 2.2.23(b)). Therefore,

$$\mathbf{P}\left[\exists S \in \mathcal{A}_n, S \subseteq K \; ; \; \frac{|\partial_E^K S|}{|\partial_E^G S|} \leq \alpha\right] \leq \sum_{S \in \mathcal{A}_n} \mathbf{P}\left[S \subseteq K, \; \frac{|\partial_E^K S|}{|\partial_E^G S|} \leq \alpha\right]$$

$$\leq \sum_{S \in \mathcal{A}_n} e^{-n I_p(\alpha)} \, \mathbf{P}[S \subseteq K]$$

$$= e^{n[I_p(0)-I_p(\alpha)]} \sum_{S \in \mathcal{A}_n} (1-p)^n \, \mathbf{P}[S \subseteq K]$$

$$= e^{n[I_p(0)-I_p(\alpha)]} \sum_{S \in \mathcal{A}_n} \mathbf{P}[K = S]$$

$$= e^{n[I_p(0)-I_p(\alpha)]} \, \mathbf{P}\big[|K| < \infty, \, |\partial_E^G K| = n\big]. \quad (6.63)$$

To estimate $\mathbf{P}\big[|K| < \infty, \, |\partial_E^G(K)| = n\big]$ for $p > 1/(1 + \Phi_E^*(G))$, recall the argument of Theorem 6.46. Choose $h < \Phi_E^*(G)$ such that $p > 1/(1+h)$. Then there exists $n_h < \infty$ such that $|\partial_E^G S|/|S| > h$ for all $S \in \mathcal{A}_n$ with $n > n_h$. We showed that

$$\big\{|K| < \infty, \, |\partial_E^G(K)| = n\big\} \subset \bigcup_{N=n}^{\infty} B_N \,,$$

where

$$B_N := \left[\sum_{j=1}^{N} Y_j \leq \frac{N}{1+h}\right]$$

and $\langle Y_j \rangle$ is an i.i.d. sequence of Bernoulli(p) random variables.

As earlier, $\mathbf{P}[B_N] \leq e^{-N\delta_p}$ where $\delta_p := I_p\left(\frac{1}{1+h}\right) > 0$, since $p > 1/(1+h)$. Thus for some constant $C_p < \infty$,

$$\mathbf{P}\big[|K| < \infty, \, |\partial_E^G(K)| = n\big] \leq \sum_{N=n}^{\infty} e^{-N\delta_p} \leq C_p e^{-n\delta_p} \,. \quad (6.64)$$

Taking $\alpha > 0$ in (6.63) so small that $I_p(0) - I_p(\alpha) < \delta_p$, we deduce that (6.63) is summable in n. By the Borel–Cantelli lemma,

$$\lim_{n \to \infty} \inf\left\{\frac{|\partial_E^K S|}{|\partial_E^G S|} \; ; \; o \in S \subset \mathsf{V}(K), \, S \text{ is connected}, \, n \leq |\partial_E^G S|\right\} \geq \alpha \quad \text{a.s.,}$$

whence

$$\Phi_E^*(K) \geq \alpha \Phi_E^*(G) > 0$$

almost surely on the event that K is infinite.   ◀

The following possible extension was open:

Question 6.49. If $\Phi_E^*(G) > 0$, does every infinite cluster K in a Bernoulli percolation satisfy $\Phi_E^*(K) > 0$ a.s.?

A partial result was due to Pete (2008). This was extended by Hermon and Hutchcroft (2019b) to give a positive answer to Question 6.49 for every $p > p_c(G)$ when G is a nonamenable, transitive graph. They also gave a counterexample when G is not assumed transitive. A converse is known, namely, that if G is a transitive, amenable graph, then for every invariant percolation on G, a.s. each cluster has 0 anchored expansion constant; see Corollary 8.38.

Percolation is one way of randomly thinning a graph. Another way is to replace an edge by a random path of edges. What happens to expansion then? We will use the following notation.

Let G be an infinite graph of bounded degree, and pick a probability distribution ν on the positive integers. Replace each edge $e \in \mathsf{E}(G)$ by a path of $L_e \geq 1$ edges, where the random variables $\langle L_e \rangle_{e \in \mathsf{E}(G)}$ are independent with law ν. Let G^ν denote the random graph obtained from G in this way. We call G^ν a ***random subdivision*** of G. Say that ν has an ***exponential tail*** if, for some $\epsilon > 0$ and all sufficiently large ℓ, we have $\nu[\ell, \infty) < e^{-\epsilon\ell}$. This is equivalent to the condition that if $X \sim \nu$, then $\mathbf{E}[s^X] < \infty$ for some $s > 1$.

▷ **Exercise 6.23.**
Show that if the support of ν is unbounded, then $\Phi_E(G^\nu) = 0$ a.s.

Define another anchored expansion constant,

$$\Phi_E^{**}(G) := \lim_{n \to \infty} \inf\left\{ \frac{|\partial_E K|}{|E(K)|} \; ; \; o \in K \subset \mathsf{V}, \, G{\restriction}K \text{ is connected}, \, n \leq |K| < \infty \right\},$$

with notation as in (6.33). Since $|E(K)| \geq |K| - 1$, we have

$$\Phi_E^{**}(G) \leq \Phi_E^*(G)$$

for every graph G. Conversely, if the maximum degree of G is D, then

$$\Phi_E^{**}(G) \geq \Phi_E^*(G)/D$$

since $|E(K)| \leq D|K|$. On the other hand, for trees G, we have $|E(K)| = |K| - 1$, so $\Phi_E^{**}(G) = \Phi_E^*(G)$ for trees.

Theorem 6.50. (Anchored Expansion and Subdivision) *Suppose that $\Phi_E^{**}(G) > 0$. If ν has an exponential tail, then the random subdivision satisfies $\Phi_E^{**}(G^\nu) > 0$ a.s. In particular, if G has bounded degree and $\Phi_E^*(G) > 0$, then $\Phi_E^*(G^\nu) > 0$ a.s.*

To prove this, we use the following combinatorial aspect of anchored expansion.

Proposition 6.51. *As in (6.60), let*

$$\mathcal{A}_n := \{K \subset V(G) \,;\; o \in K, \; G{\upharpoonright}K \text{ is connected and finite, } |\partial_E K| = n\} \tag{6.65}$$

and

$$h_n := \inf\left\{\frac{|\partial_E K|}{|K|} \,;\; K \in \mathcal{A}_n\right\}. \tag{6.66}$$

Then

$$|\mathcal{A}_n| \le \Psi(h_n)^n \,, \tag{6.67}$$

where $\Psi(\,\cdot\,)$ is the monotone decreasing function

$$\Psi(h) := (1+h)^{1+\frac{1}{h}}/h, \qquad \Psi(0) := \infty \,.$$

Proof. Consider Bernoulli(p) bond percolation in G. Let $K(o)$ be the open cluster containing o. For any $K \in \mathcal{A}_n$, we have $|E(K)| \ge |K| - 1$, since a spanning tree on K has $|K| - 1$ edges; also, $|\partial_E K| \ge h_n|K|$. Therefore,

$$\mathbf{P}\big[\mathsf{V}(K(o)) = K\big] \ge p^{|K|-1}(1-p)^{|\partial_E K|} \ge p^{n/h_n}(1-p)^n \,,$$

whence

$$1 \ge \mathbf{P}\big[\mathsf{V}(K(o)) \in \mathcal{A}_n\big] = \sum_{K \in \mathcal{A}_n} \mathbf{P}\big[\mathsf{V}(K(o)) = K\big] \ge |\mathcal{A}_n| p^{n/h_n}(1-p)^n \,.$$

Thus,

$$|\mathcal{A}_n| \le \left(\frac{1}{p}\right)^{n/h_n}\left(\frac{1}{1-p}\right)^n$$

for every $p \in (0, 1)$. Letting $p := 1/(1 + h_n)$ concludes the proof. ◄

Proof of Theorem 6.50. Since v has an exponential tail, there is an increasing convex rate function $I(\,\cdot\,)$ such that $I(c) > 0$ for $c > EL_i$ and $\mathbf{P}\big[\sum_{i=1}^n L_i > cn\big] \le \exp(-nI(c))$ for all n (see Dembo and Zeitouni (1998), Theorem 2.2.3) when $L_i \sim v$ are independent. Fix $h < \Phi_E^*(G)$. Choose c large enough that $I(c) > \log \Psi(h)$. For $S \in \mathcal{A}_n$,

$$\mathbf{P}\left[\frac{\sum_{e \in E^*(S)} L_e}{|E^*(S)|} > c\right] \le \exp(-|E^*(S)|I(c)) \le \exp(-|\partial_E S|I(c))$$

since $\partial_E S \subseteq E^*(S)$. Therefore for all n,

$$\mathbf{P}\left[\exists S \in \mathcal{A}_n \,;\; \frac{\sum_{e \in E^*(S)} L_e}{|E^*(S)|} > c\right] \le |\mathcal{A}_n| e^{-I(c)n} \,,$$

which is summable by (6.67) since h_n (defined in (6.66)) is strictly larger than h for all large n. By the Borel–Cantelli lemma, with probability one, we have

$$\limsup_{n \to \infty} \sup_{S \in \mathcal{A}_n} \frac{\sum_{e \in E^*(S)} L_e}{|E^*(S)|} \le c \,.$$

Therefore

$$\liminf_{n\to\infty}\left\{\frac{|\partial_E S|}{\sum_{e\in E^*(S)} L_e}\;;\; o\in S\subset \mathsf{V}(G),\; G{\upharpoonright}S \text{ is connected},\; n\le|\partial_E S|\right\} \ge \frac{\Phi_E^{**}(G)}{c\big(1+\Phi_E^{**}(G)\big)}$$

a.s., since $E^*(S) = E(S)\cup\partial_E S$.

Since G^v is obtained from G by adding new vertices, $\mathsf{V}(G)$ can be embedded into $\mathsf{V}(G^v)$ as a subset. In particular, we can choose the same basepoint o in G^v and in G. For S connected in G such that $o\in S\subset \mathsf{V}(G)$, there is a unique *maximal* connected $\widetilde{S}\subset \mathsf{V}(G^v)$ such that $\widetilde{S}\cap\mathsf{V}(G) = S$; it satisfies $|E(\widetilde{S})| \le \sum_{e\in E^*(S)} L_e$. In computing $\Phi_E^*(G^v)$, it suffices to consider only such maximal sets $\widetilde{S}$, so we conclude that $\Phi_E^{**}(G^v) \ge \Phi_E^{**}(G)/\big(c\big(1+\Phi_E^{**}(G)\big)\big) > 0$. ◄

The exponential tail condition is necessary to ensure the positivity of $\Phi_E^*(G^v)$; see Exercise 6.88.

Do Galton–Watson trees have anchored expansion? Clearly they do when the offspring distribution $\langle p_k\rangle$ satisfies $p_0 = p_1 = 0$. On the other hand, when $p_1 \in (0, 1)$, the tree can be obtained from a different Galton–Watson tree with $p_1 = 0$ by randomly subdividing the edges. This will allow us to use Theorem 6.50 to establish that Galton–Watson trees do indeed have anchored expansion when $p_0 = 0$, and another argument will cover the case $p_0 > 0$.

Theorem 6.52. (Anchored Expansion of Galton–Watson Trees) *For a supercritical Galton–Watson tree T, given nonextinction we have $\Phi_E^*(T) > 0$ a.s.*

Proof. Case (i): $p_0 = p_1 = 0$. For every finite $S\subset \mathsf{V}(T)$, we have

$$|S| \le |\partial_E S|\left(\frac{1}{2} + \frac{1}{2^2} + \cdots\right) \le |\partial_E S|\,.$$

So $\Phi_E^*(T) \ge \Phi_E(T) \ge 1$.

Case (ii): $p_0 = 0$, $p_1 > 0$. In this case, let x be the vertex closest to the root that has at least two children. Then T^x has the law of a random subdivision G^v of another Galton–Watson tree G, and T differs from T^x by a finite path. Here, G is generated according to the offspring distribution $\langle p'_k\;;\; k\ge 0\rangle$, where $p'_k := p_k/(1 - p_1)$ for $k = 2, 3, \ldots$ and $p'_0 := p'_1 := 0$, and v is the geometric distribution with parameter $1 - p_1$. By Theorem 6.50 and the fact that $\Phi_E^{**} = \Phi_E^*$ for trees, $\Phi_E^*(T) = \Phi_E^*(T^x) = \Phi_E^*(G^v) > 0$ a.s.

Case (iii): $p_0 > 0$. Let $A(n, h)$ be the event that there is a subtree $S\subset T$ having n vertices, including the root of T, and satisfying $|\partial_E S| \le hn$. We claim that

$$\mathbf{P}\big[A(n, h)\big] \le e^{nf(h)}\,\mathbf{P}\big[n\le|\mathsf{V}(T)| < \infty\big] \tag{6.68}$$

for some function $f:\mathbb{R}^+ \to \mathbb{R}^+$ that satisfies $\lim_{h\downarrow 0} f(h) = 0$. The idea is that the event $A(n, h)$ is "close" to the event that $\mathsf{V}(T)$ is finite but at least n. That is, it could have happened that the hn leaves of the growing Galton–Watson tree had no children after it already had n vertices, and for $T\in A(n, h)$, this alternative scenario isn't too unlikely compared to what actually happened.

For the proof, we can map any tree T in $A(n, h)$ to a finite tree $\phi(T)$ with at least n vertices as follows: Given $x\in\mathsf{V}(T)$, label its children from 1 to the number of children of x. Use this to place a canonical total order on all finite subtrees of T that include the root. (This can be done in a manner similar to the lexicographic order of finite strings.) Choose the first n-vertex S in this order such that the edge boundary of S in T has at most hn edges. Define $\phi(T)$ from T by retaining all edges in S and its edge boundary in T, while deleting all other edges. Note that for each vertex $x \in T$, the tree $\phi(T)$ contains either all children of x or none.

Any finite tree t with m vertices arises as $\phi(T)$ from at most $\sum_{k \leq hm} \binom{m}{k}$ choices of S, because for $n \leq m$, there are at most $hn \leq hm$ edges in $t \setminus S$, whereas for $n > m$, there are no choices of S. Now $\sum_{k \leq hm} \binom{m}{k} \leq \exp(m f_1(h))$, where

$$f_1(\alpha) := -\alpha \log \alpha - (1 - \alpha) \log (1 - \alpha),$$

by (6.61). Given S and a possible tree t in the image of ϕ on $A(n, h)$, we have

$$\mathbf{P}\big[\phi(T) = t\big] \leq p_0^{-hn} \, \mathbf{P}[T = t].$$

Indeed, let $L(t)$ denote the leaves of t and $J(t) := \mathsf{V}(t) \setminus L(t)$. Let $d(x)$ be the number of children in t of $x \in J(t)$. Then

$$\mathbf{P}\big[\phi(T) = t\big] = \prod_{x \in J(t)} p_{d(x)} = p_0^{-|L(t)|} \, \mathbf{P}[T = t] \leq p_0^{-hn} \, \mathbf{P}[T = t].$$

Thus, if we let $f(\alpha) := f_1(\alpha) - \alpha \log p_0$, we obtain (6.68).

Now a supercritical Galton–Watson process conditioned on extinction is the same as a subcritical process with p.g.f. $\widetilde{f}(s) := f(qs)/q$ by Proposition 5.28(ii). Since $\widetilde{f}(1/q) < \infty$ and $1/q > 1$, the total size of this subcritical Galton–Watson process decays exponentially by Exercise 5.33. Therefore, the last term of (6.68) decays exponentially in n. By choosing h small enough, we can ensure that also the left-hand side $\mathbf{P}\big[A(n, h)\big]$ also decays exponentially in n. ◄

6.10 Notes

Kesten (1959a, 1959b) proved the qualitative statement that a countable group Γ is nonamenable iff some (or every) symmetric random walk with support generating Γ has spectral radius less than 1. Making this quantitative, as in Theorem 6.7, was accomplished by Cheeger (1970) in the continuous setting; he dealt with the bottom of the spectrum of the Laplacian, rather than any spectral radius, but this is equivalent: in the discrete case, the Laplacian is $I - P$, so the bottom of the spectrum of $I - P$ equals 1 minus the spectral radius of P. Cheeger's inequality states the following: Let M be a closed n-dimensional Riemannian manifold. Let $\lambda_1(M)$ denote the smallest positive eigenvalue of the Laplace–Beltrami operator on M. Let $h(M)$ be the infimum of $V_{n-1}(E)/\min\{V_n(A), V_n(B)\}$ when M is divided into two pieces A and B by an $(n - 1)$-submanifold E and V_k denotes k-dimensional volume. Then

$$\lambda_1(M) \geq h(M)^2/4.$$

An inequality in the opposite direction was proved later by Buser (1982), who showed that in this context, if the Ricci curvature of M is always at least $-(n - 1)a^2$, then

$$\lambda_1(M) \leq 2a(n - 1)h(M) + 10h(M)^2.$$

In the discrete case, the direction of Buser's inequality is the easy one. Cheeger's result was transferred to the discrete setting in various contexts of infinite graphs by Dodziuk (1984), Dodziuk and Kendall (1986), Varopoulos (1985a), Ancona (1988), Gerl (1988), Biggs, Mohar, and Shawe-Taylor (1988), and Kaimanovich (1992). Cheeger's method of proof is used in all of these. We have incorporated an improvement due to Mohar (1988). Similar inequalities were proved independently for finite graphs, again inspired by Cheeger (1970). The first results were by Alon and Milman (1985) and Alon (1986), and the final form was given by Jerrum and Sinclair (1989) and Lawler and Sokal (1988) independently. Analogous inequalities for measure-preserving actions of groups are due to Lyons and Nazarov (2011).

Theorem 6.7 is used mostly to deduce whether $\rho(G)$ is less than 1, depending on whether G is nonamenable or amenable. However, it has also been used to deduce amenability of some groups that had not been known to be amenable, by analyzing simple random walks thereon. This proved that the class of amenable groups is not the closure (under natural operations) of the class of groups of subexponential growth. See Bartholdi and Virág (2005), Kaimanovich (2005), Brieussel (2009), and Bartholdi, Kaimanovich, and Nekrashevych (2010).

Lubotzky, Phillips, and Sarnak (1988) defined a connected d-regular graph G with $d \geq 3$ to be **Ramanujan** if every eigenvalue λ of its adjacency matrix is either $\pm d$ or satisfies $|\lambda| \leq 2\sqrt{d-1}$. Such expanders (which, in light of the Alon–Boppana theorem discussed in Exercise 6.47, have an asymptotically optimal spectral gap) were first constructed by Lubotzky, Phillips, and Sarnak (1988) and independently by Margulis (1988). The mixing time (see Section 13.3) of simple random walk on these graphs was determined by Lubetzky and Peres (2016): on a d-regular non-bipartite Ramanujan graph with n vertices,

$$\left| t_{\mathrm{mix}}(\epsilon) - \frac{d}{d-2} \log_{d-1} n \right| \leq C\sqrt{\log n}$$

for every $\epsilon \in (0, 1)$, where $C = C(d, \epsilon)$. The key tool in the proof is the analysis of nonbacktracking walks following Exercise 6.64.

See Bartholdi (1999) for a relationship between the numbers of cycles of different lengths at a vertex and the numbers of cycles of different lengths with given numbers of backtracking edge pairs. His analysis extends that of the generating functions used in Section 6.3. Our proof of Theorem 6.11 is taken from Lyons and Peres (2015). This method of proof can be shown to work with even weaker hypotheses: see Lyons and Peres (2015), inspired by Theorem 5 of Abért, Glasner, and Virág (2016). The latter authors prove quantitative versions and other extensions, with conclusions about spectral radii also derived from the analysis of nonbacktracking walk.

Much more is known about spectral gap for random regular graphs than what is implied by Theorem 6.17; see Puder (2015) for some history and new results.

The fact proved in Section 6.5 that proper tessellations of the same type are isomorphic and transitive is folklore. The sizes of the spheres in the tessellations analyzed in Section 6.5 are given by explicit rational generating functions: see Paul and Pippenger (2011).

Loomis and Whitney (1949) proved an inequality analogous to Lemma 6.23 for bodies in $\mathbb{R}^d$. It implies the inequality we stated by taking a cube in $\mathbb{R}^d$ centered at each point of A.

Our proof of Theorem 6.29 is modeled on the one presented by Gromov (1999), p. 348. Results similar to Theorem 6.29 were proved earlier by Aldous (1987) and Babai (1991).

Upper bounds for transition probabilities using expansion profiles, as in Theorem 6.31, were first discovered by Varopoulos (1985a). These were later refined by Coulhon (1996) and Coulhon, Grigor'yan, and Pittet (2001). For surveys of the analytic approach to these bounds, see Pittet and Saloff-Coste (2001) and Section 14 of Woess (2000). The probabilistic approach to Theorem 6.31 is from Morris and Peres (2005), where similar bounds on mixing times of finite Markov chains are also obtained using the evolving set process. A related process was investigated earlier by Diaconis and Fill (1990), who discovered a powerful coupling of the evolving set process with the underlying Markov chain; see also

Levin and Peres (2017) for an exposition of this coupling. Mixing-time bounds involving the same integral as in Theorem 6.31 were first obtained by Lovász and Kannan (1999).

Fix a real $d > 0$, a graph $G = (\mathsf{V}, \mathsf{E})$ and a vertex $x \in \mathsf{V}$. For simple random walk on G, an upper bound on transition probabilities of the form $p_n(x, x) = O(n^{-d/2})$ as obtained in Corollary 6.32(ii), in conjunction with an upper bound $|B(x, r)| = O(r^d)$ on the growth of balls, imply a matching lower bound $p_n(x, x) \geq cn^{-d/2}$ for some $c > 0$. See Pittet and Saloff-Coste (2001) or Theorems 14.12 and 14.19 in Woess (2000). By Gromov's theorem, discussed in Section 7.9, and its extension by Trofimov (1984), the hypotheses hold for some integer d if G is a transitive graph of polynomial growth.

The notion of anchored expansion was implicit in Thomassen (1992) and made explicit by Benjamini, Lyons, and Schramm (1999). Theorem 6.42 is due to Lyons, Morris, and Schramm (2008); it refines Thomassen (1992) and is adapted from a similar result of He and Schramm (1995). It is very similar to an independent result of Benjamini and Kozma (2005). The relevance of anchored expansion to random walks, beyond the issue of transience, is exhibited by the following theorem of Virág (2000a), the first part of which was conjectured by Benjamini, Lyons, and Schramm (1999).

Theorem 6.53. *Let G be a bounded-degree graph with $\Phi_\mathsf{E}^*(G) > 0$. For a vertex x, denote by $|x|$ the distance from x to the basepoint o in G. Then simple random walk $\langle X_n \rangle$ in G, started at o, satisfies $\liminf_{n\to\infty} |X_n|/n > 0$ a.s., and there exists $C > 0$ such that $\mathbf{P}[X_n = o] \leq \exp(-Cn^{1/3})$ for all $n \geq 1$.*

Note that this theorem, combined with Theorem 6.48, implies positive speed on the infinite clusters of Bernoulli(p) percolation on any G with $\Phi_\mathsf{E}^*(G) > 0$, provided $p > 1/(1 + \Phi_\mathsf{E}^*(G))$. This partially answers Question 6.49. Furthermore, in conjunction with Theorem 6.52, we get that the speed of simple random walk on supercritical Galton–Watson trees is positive, a result first proved in Lyons, Pemantle, and Peres (1995b); see Theorem 17.13 and Exercise 17.6, where a formula for the speed is given.

▷ **Exercise 6.24.**
Show that the bound of Theorem 6.53 on the return probabilities is sharp by giving an example of a graph with anchored expansion that has $\mathbf{P}[X_n = o] \geq \exp(-Cn^{1/3})$ for some $C < \infty$. *Hint:* Take a Galton–Watson tree T with offspring distribution $p_1 = p_2 = 1/2$, rooted at o. Look for a long pipe (of length $n^{1/3}$) starting at level $n^{1/3}$ of T.

Conditions for a strict inequality in Theorem 6.47 are given by Martineau and Severo (2019).

Benjamini, Lyons, and Schramm (1999) initiated a systematic study of the properties of a transitive graph G that are preserved for infinite percolation clusters. Both Theorem 6.50 and Proposition 6.51 are from Chen and Peres (2004). The idea of the proof of the latter originates in Kesten (1982). Theorem 6.52 is due to Chen and Peres (2004), but the original proof was incomplete.

A combination of ideas from Sections 6.8, 6.9, and 6.6 was used by Pete (2008) to show that transient wedges in $\mathbb{Z}^3$ (see (2.19)) also have transient percolation clusters for supercritical percolation, provided a mild technical condition on the function f of (2.19) is satisfied. The result is true without this technical condition; see Angel, Benjamini, Berger, and Peres (2006).

6.11 Collected In-Text Exercises

6.1. Show that $\Phi_E(\mathbb{T}_{b+1}, \mathbf{1}, \mathbf{1}) = b - 1$ for all $b \geq 1$, where $\mathbb{T}_{b+1}$ is the regular tree of degree $b + 1$.

6.2. Show that for any network (G, c, D), we have

$$\Phi_E(G, c, D) = \max\{\alpha \geq 0 \, ; \, \exists \theta \, \forall e \, |\theta(e)| \leq c(e) \text{ and } \forall x \, d^*\theta(x) \geq \alpha D(x)\},$$

where θ runs over the antisymmetric functions on E.

6.3. Suppose that G is a graph such that for some $o \in V$, we have subexponential growth of balls: $\liminf_{n \to \infty} |\{x \in V \, ; \, d(o, x) \leq n\}|^{1/n} = 1$, where $d(\cdot, \cdot)$ denotes the graph distance in G. Show that $(G, \mathbf{1})$ is vertex amenable.

6.4. Show that every Cayley graph of a finitely generated, abelian group is amenable.

6.5. Suppose that G_1 and G_2 are roughly isometric graphs with bounded degrees and having both edge and vertex weights $\asymp \mathbf{1}$. Show that G_1 is amenable iff G_2 is.

6.6. Prove that $\|P\|_\pi \leq 1$.

6.7. (Rayleigh Quotient) Show that

$$\|P\|_\pi = \sup\left\{\frac{|(Pf, f)_\pi|}{(f, f)_\pi} \, ; \, f \in \mathbf{D}_{00} \setminus \{0\}\right\} = \sup\left\{\frac{(Pf, f)_\pi}{(f, f)_\pi} \, ; \, f \in \mathbf{D}_{00} \setminus \{0\}\right\}.$$

6.8. Show that for $f \in \mathbf{D}_{00}$, we have $d^*(c \, df) = \pi(f - Pf)$.

6.9. Show that for simple random walk on $\mathbb{T}_{b+1}$, we have $\rho(\mathbb{T}_{b+1}) = 2\sqrt{b}/(b + 1)$.

6.10. Let G be a d-regular multigraph. Show that $\rho(G) = 1$ iff $\mathrm{cogr}(G) = d - 1$.

6.11. Give an example of a d-regular graph G where $1 < \mathrm{cogr}(G) < \sqrt{d - 1}$.

6.12. Show that $\lambda_2 < 1$ iff the network is connected and that $\lambda_n > -1$ iff the random walk is aperiodic.

6.13. Show that the rate of exponential convergence in Theorem 6.13 cannot be faster than λ_*. More precisely, show that there is an x for which

$$\lim_{t \to \infty} -\frac{\log |p_t(x, x) - \pi(x)|}{t} = -\log \lambda_*.$$

6.14. Show that

$$\lambda_2 = \max_{f \perp 1} \frac{\langle Pf, f \rangle}{\langle f, f \rangle}.$$

6.15. Prove the following generalization of the Loomis–Whitney inequality. Let $A \subset \mathbb{Z}^d$ be finite and $\mathcal{S}$ be a collection of subsets of $\{1, \dots, d\}$ such that each integer in $[1, d]$ appears in exactly r of the sets in $\mathcal{S}$. Write $\mathcal{P}_S$ for the projection of $\mathbb{Z}^d \to \mathbb{Z}^S$ onto the coordinates in S. Then

$$|A|^r \leq \prod_{S \in \mathcal{S}} |\mathcal{P}_S A|.$$

6.16. For the reversed chain, define $\widehat{Q}(x, y) := \pi(x)\widehat{p}(x, y) = Q(y, x)$ for all x, y. Prove that for any finite $S \subset V$, we have $|\partial_E S|_Q = |\partial_E S|_{\widehat{Q}}$.

6.17. The lazy version of a transition matrix P is given by $\widetilde{P} = (P+I)/2$. Show that the corresponding expansion profile satisfies $\widetilde{\Phi}(r) = \Phi(r)/2$ for all $r > \pi_{\min}$. If the original matrix P is reversible, show that $p_{2n}(x, x)$ is decreasing in n for all $x \in \mathsf{V}$, and deduce that $p_{2n}(x, x) \le 2\,\widetilde{p}_{2n}(x, x)$.

6.18. Let A be a finite set of vertices in a connected network G with $|\mathsf{V}(G)|_\pi = \infty$. Let H be a finite, connected subnetwork containing A. In H^{W}, identify A to a single vertex a, and let z be the (wired) boundary vertex; call the new graph H'. Define $\psi(G, A, t)$ and $\langle s_k \rangle$ as in Theorem 6.42 and $\phi(H', t)$ with its associated sequence $\langle s'_k \rangle$ as in Lemma 6.43. Show that for all $t \ge 0$,

$$\phi\big(H', \pi(a) + t\big) \ge \psi\big(G, A, |A|_\pi + t\big),$$

and that for all $k \ge 0$, we have $\phi(H', s'_k) \ge \psi(G, A, s_k)$.

6.19. For any $p < 1$, every infinite cluster K of Bernoulli(p) percolation on any graph G of bounded degree has $\Phi_{\mathsf{E}}(K) = 0$ a.s.

6.20. Show that if G is a transitive graph, then $\Phi_{\mathsf{E}}(G) = \Phi_{\mathsf{E}}^*(G)$ and $\Phi_{\mathsf{V}}(G) = \Phi_{\mathsf{V}}^*(G)$.

6.21. Show that for every graph G, the balls $B_G(x, n)$ in G about every point x of radius n satisfy

$$\liminf_{n\to\infty} |B_G(x, n)|^{1/n} \ge 1 + \Phi_{\mathsf{V}}^*(G). \tag{6.59}$$

6.22. Prove the first inequality of Theorem 6.46 as written with anchored expansion.

6.23. Show that if the support of ν is unbounded, then $\Phi_{\mathsf{E}}(G^\nu) = 0$ a.s., where G^ν is the random subdivision defined in Section 6.9.

6.24. Show that the bound of Theorem 6.53 on the return probabilities is sharp by giving an example of a graph with anchored expansion that has $\mathbf{P}[X_n = o] \ge \exp(-Cn^{1/3})$ for some $C < \infty$. *Hint:* Take a Galton–Watson tree T with offspring distribution $p_1 = p_2 = 1/2$, rooted at o. Look for a long pipe (of length $n^{1/3}$) starting at level $n^{1/3}$ of T.

6.12 Additional Exercises

6.25. If (G_1, c_1, D_1) and (G_2, c_2, D_2) are networks, consider the ***Cartesian product*** graph $(\mathsf{V}, \mathsf{E}) = G_1 \,\square\, G_2$ defined by $\mathsf{V} := \mathsf{V}_1 \times \mathsf{V}_2$,

$$\mathsf{E} := \Big\{\big((x_1, x_2), (y_1, y_2)\big);\; \big(x_1 = y_1,\; (x_2, y_2) \in \mathsf{E}_2\big) \text{ or } \big((x_1, y_1) \in \mathsf{E}_1,\; x_2 = y_2\big)\Big\}$$

with the weights $D\big((x_1, x_2)\big) := D_1(x_1)D_2(x_2)$ on the vertices and

$$c\big([(x_1, x_2), (x_1, y_2)]\big) := D_1(x_1)c_2([x_2, y_2]) \quad \text{and} \quad c\big([(x_1, x_2), (y_1, x_2)]\big) := D_2(x_2)c_1([x_1, y_1])$$

on the edges. Show that with these weights, $\Phi_{\mathsf{E}}(G_1 \,\square\, G_2) = \Phi_{\mathsf{E}}(G_1) + \Phi_{\mathsf{E}}(G_2)$.

6.26. Let $G = (\mathsf{V}, \mathsf{E})$ be an infinite graph and M a ***matching*** in G, that is, a set of edges that are pairwise nonadjacent. Show that $\Phi_{\mathsf{E}}(G \backslash M, \mathbf{1}, \mathbf{1}) \ge \Phi_{\mathsf{E}}(G, \mathbf{1}, \mathbf{1}) - 1$.

6.27. Use Theorem 6.1 and its proof to give another proof of Theorem 6.2.

6.28. Refine Theorem 6.2 to show that if K is any finite vertex set in a transitive, infinite graph G, then $|\partial_{\mathsf{E}} K|/|K| \ge \Phi_{\mathsf{E}}(G) + 1/|K|$.

6.29. Show that if G is a transitive graph of degree d and the edge-expansion constant $\Phi_E(G, 1, 1) = d - 2$, then G is a tree.

6.30. Show that if G is a finite, transitive network, then the minimum of $|\partial_E K|_c / |K|$ over all K of size at most $|V|/2$ occurs only for $|K| > |V|/4$.

6.31. Suppose that we had used the **inner vertex boundary** $\partial_V^{in} K := \{x \in K ;\ \exists y \notin K\ y \sim x\}$ of sets K in place of the outer vertex boundary to define vertex amenability. Show that this would not change the class of networks that are vertex amenable. Moreover, show that

$$\inf\left\{\frac{|\partial_V^{in} K|_D}{|K|_D} ;\ \varnothing \ne K \subset V \text{ is finite}\right\} = \frac{\Phi_V(G, D)}{1 + \Phi_V(G, D)} .$$

6.32. Let G and $G^\dagger$ be plane dual graphs such that $G^\dagger$ has bounded degrees. Show that if G is amenable, then so is $G^\dagger$.

6.33. Show that every finitely generated subgroup of an amenable, finitely generated group is itself amenable.

6.34. Use Theorem 6.3 and its proof to give another proof of Theorem 6.4.

6.35. Refine Theorem 6.4 to show that if K is any finite vertex set in a transitive, infinite graph G, then $|\partial_V K|/|K| \ge \Phi_V(G) + 1/|K|$.

6.36. Recall from Exercise 6.31 the inner vertex boundary of sets K. Let $\Phi_V^{int}(G)$ be the corresponding expansion constant. Show that if G is a transitive network, then for all finite K, we have $|\partial_V^{in} K|/|K| > \Phi_V^{int}(G)$.

6.37. Let G be a transitive graph and b be a submodular function that is invariant under the automorphisms of G and is such that if K and K' are disjoint but adjacent, then strict inequality holds in (6.2). Show that there is no finite set K that minimizes $b(K)/|K|$.

6.38. Show that a transitive graph G is nonamenable iff there exists a function $f : V(G) \to V(G)$ having the two properties that (i) $\sup_{x \in V(G)} \text{dist}_G(x, f(x)) < \infty$, and (ii) for all $x \in V(G)$, the cardinality of $f^{-1}(x)$ is at least 2.

6.39. Does every subperiodic tree with exponential growth have a nonamenable subtree?

6.40. Consider any network random walk and $x, y \in V$ and $n \ge 0$.
  (a) Show that

$$p_{2n}(x, y) \le \sqrt{\frac{\pi(y)}{\pi(x)}} \sqrt{p_{2n}(x, x) p_{2n}(y, y)}$$

and

$$p_{2n+1}(x, y) \le \sqrt{\frac{\pi(y)}{\pi(x)}} \sqrt{p_{2n}(x, x) p_{2n+2}(y, y)} .$$

Hint: Look at the proof of Proposition 6.6.
  (b) Show that if the walk is lazy, then

$$p_n(x, y) \le \sqrt{\frac{\pi(y)}{\pi(x)}} \sqrt{p_n(x, x) p_n(y, y)} .$$

  (c) Find a 3-regular infinite graph G and vertices $y_n \in V(G)$ for all n such that lazy simple random walk on G satisfies $p_n(y_0, y_n)/p_n(y_0, y_0) \to \infty$ as $n \to \infty$.

6.41. Consider a random walk on a graph with spectral radius ρ. Suppose that we introduce a delay so that each step goes nowhere with probability p_{delay}, and otherwise chooses a neighbor with the same distribution as before. Show that the new spectral radius equals $p_{\text{delay}} + (1 - p_{\text{delay}})\rho$.

6.42. Show that if G is a covering network of G', then $\Phi_E(G) \geq \Phi_E(G')$, $\Phi_V(G) \geq \Phi_V(G')$, and $\rho(G) \leq \rho(G')$.

6.43. Show that if G is a graph of maximum degree d, then its edge-expansion constant satisfies $\Phi_E(G, \mathbf{1}, \mathbf{1}) \leq d - 2$.

6.44. Let T be a tree and $\mathbb{T}_{b+1}$ be the regular tree of degree $b + 1$.
  (a) Show that if the degree of each vertex in T is at least $b + 1$, then $\rho(T) \leq \rho(\mathbb{T}_{b+1})$.
  (b) Show that if, for every r, the ball of radius r in $\mathbb{T}_{b+1}$ is isomorphic to some ball in T, then $\rho(T) \geq \rho(\mathbb{T}_{b+1})$.

6.45. Let G be a graph and H be a transitive graph. Which of the following extensions of Exercise 6.44 are valid?
  (a) If each vertex in G is contained in a subgraph of G that is isomorphic to H, then $\rho(G) \leq \rho(H)$.
  (b) If for every r, the ball of radius r in H is isomorphic to some ball in G, then $\rho(G) \geq \rho(H)$.

6.46. Give another proof of (6.20) by using Theorem 6.1.

6.47. Let G be a d-regular finite, connected graph of diameter at least $2k$ for some integer $k \geq 2$. Let λ_2 be the second-largest eigenvalue of its adjacency matrix. Show that

$$\lambda_2 > 2\sqrt{d-1} - \frac{2\sqrt{d-1} - 1}{k - 1} .$$

Hint: Choose $a, z \in V$ so that their distance is at least $2k$. Write $b := d - 1$. Consider the function f on V defined by

$$f(x) := \begin{cases} 1 & \text{if } x = a, \\ b^{-(i-1)/2} & \text{if } \text{dist}(x, a) = i \in [1, k-1], \\ cb^{-(i-1)/2} & \text{if } \text{dist}(x, z) = i \in [1, k-1], \\ c & \text{if } x = z, \\ 0 & \text{otherwise,} \end{cases}$$

where c is chosen so that $\sum_x f(x) = 0$. Show that $(\Delta_G f, f) < 1 + b - 2\sqrt{b} + (2\sqrt{b} - 1)/(k - 1)$, where Δ_G is the graph Laplacian of Exercise 2.62.

6.48. Show that for a network (G, c, D), we have

$$\Phi_E(G, c, D) = \inf\left\{ \frac{\|df\|_{\ell_1(c)}}{\|f\|_{\ell_1(D)}} \; ; \; 0 < \|f\|_{\ell_1(D)} < \infty \right\} .$$

6.49. For a network (G, c, D) with $0 < |V(G)|_D < \infty$, one of the alternative definitions of the *expansion constant* (also known, unfortunately, as the *conductance*) is

$$\Phi_{c,D}(G) := \inf\left\{ \frac{|\partial_E K|_c}{\min\{|K|_D, |V \setminus K|_D\}} \; ; \; K \subset V, \, 0 < |K|_D < |V|_D \right\} .$$

Show that

$$\Phi_{c,D}(G) = \inf\left\{ \frac{\|df\|_{\ell_1(c)}}{\inf_{a \in \mathbb{R}} \|f - a\|_{\ell_1(D)}} \; ; \; 0 < \|f\|_{\ell_1(D)} < \infty \right\} .$$

6.50. Let G be a network with spectral radius $\rho(G)$, and let A be a set of vertices in G. Show that for any $x \in V$ and $n \in \mathbb{N}$, we have $\mathbf{P}_x[X_n \in A] \leq \rho(G)^n \sqrt{|A|_\pi / \pi(x)}$.

6.51. Let (G, c) be a network. For a finite, nonempty set of vertices A, let $\pi_A(\cdot) = \pi(\cdot)/\pi(A)$ be the normalized restriction of π to A. Write $\mathbf{P}_{\pi_A}$ for the network random walk $\langle X_n \rangle$ started at a point $x \in A$ with probability $\pi(x)/\pi(A)$. Show that (G, c) is nonamenable iff there is some function $f \colon \mathbb{N} \to [0, 1]$ that tends to 0 at infinity and that has the following property: for all finite A and all $n \in \mathbb{N}$, we have $\mathbf{P}_{\pi_A}[X_n \in A] \leq f(n)$.

6.52. Let (G, c) be a network with spectral radius $\rho < 1$. Let $v(\cdot)$ be the voltage function from a fixed vertex o to infinity. Show that $\sum_{x \in V} \pi(x) v(x)^2 < \infty$.

6.53. Let (G, c) be a network with spectral radius $\rho < 1$. Let $A \subset V$ be a nonempty set of states with $\pi(A) < \infty$ and let $\pi_A(\cdot) = \pi(\cdot)/\pi(A)$ be the normalized restriction of π to A. Show that when the chain is started according to π_A, the chance that it never returns to A is at least $1 - \rho$:

$$\mathbf{P}_{\pi_A}[X_n \text{ never returns to } A] \geq 1 - \rho.$$

Hint: Consider the function $f(x)$ defined as the chance that starting from x, the set A will ever be visited. Use Exercise 6.7.

6.54. Let G be the Cayley graph of a group Γ with respect to a finite generating set S. Without assuming that S is closed under inverses, let A be the associated averaging operator that includes the identity, that is, $(Af)(x) := \left(f(x) + \sum_{s \in S} f(xs)\right)/(|S| + 1)$ for $f \in \ell^2(\Gamma)$. Show that $\|A\| < 1$ iff Γ is nonamenable.

6.55. Suppose that G is a network with bounded π. Improve Proposition 6.9 to show that

$$\liminf_{n \to \infty} \operatorname{dist}_G(X_0, X_n)/n \geq 2 \log \rho(G)^{-1}/\log b \qquad \text{a.s.}$$

6.56. Give an example of a graph on which simple random walk has speed 0 a.s. and spectral radius less than 1.

6.57. Nonbacktracking random walk can be thought of as a Markov chain on directed edges. Show that if G is a regular graph of degree at least 3, then simple random walk on G is transient iff nonbacktracking random walk is transient.

6.58. (Pringsheim's Theorem) Let $\langle a_n ; \ n \geq 0 \rangle$ be a sequence of nonnegative real numbers and $f(z) := \sum_{n \geq 0} a_n z^n$ be convergent for some positive z. Suppose that $\delta := \limsup_{n \to \infty} a_n^{1/n} > 0$. Show that there is no analytic function whose domain includes $[0, 1/\delta]$ and that agrees with $f(z)$ in a neighborhood of 0. *Hint:* Assume the contrary. Write $R := 1/\delta$. Then there is a $z_0 \in (0, R)$ and an $\epsilon > 0$ so that f has a power series in a disc about z_0 that includes $R + \epsilon$. Calculate its coefficients. Note they are nonnegative. Rearrange terms to show it converges at $R + \epsilon$, a contradiction.

6.59. Let G be a connected graph.
  (a) Show that if G has no simple, nonloop cycle and at most one loop, then $\operatorname{cogr}(G) = 0$.
  (b) Show that if G has one simple cycle and no loop or no simple, nonloop cycle and two loops, then $\operatorname{cogr}(G) = 1$ and, if G is d-regular, $\rho(G) = 2\sqrt{d-1}/d$.
  (c) Show that in all other cases, $\operatorname{cogr}(G) > 1$.

6.60. Let G be a d-regular multigraph. Suppose that there are some $L, M < \infty$ such that for every vertex $x \in V(G)$, there is a simple cycle of length at most L that is at distance at most M from x. Show that $\rho(G) > 2\sqrt{d-1}/d$.

6.61. Let G be a graph. Let the number of nonbacktracking cycles of length n starting from $x \in V(G)$ be $b_n(x)$. Among those, let $b_n^*(x)$ count the ones whose first edge is not the reverse of its last edge, or is a loop. Write $S(x) := \{n ;\ b_n(x) \neq 0\}$ and $S^*(x) := \{n ;\ b_n^*(x) \neq 0\}$.

   (a) Show that $b_m^*(x) b_n^*(x)/2 \leq b_{m+n}^*(x)$.

   (b) Show that $\lim_{S^*(x) \ni n \to \infty} b_n^*(x)^{1/n}$ exists and $b_n^*(x) \leq 2 \operatorname{cogr}(G)^n$.

   (c) Show that $\lim_{S(x) \ni n \to \infty} b_n(x)^{1/n}$ exists and does not depend on x.

6.62. Here we give another approach to proving (6.23) and (6.24). Let G be a d-regular graph with root o. Write $b := d - 1$. Let the number of cycles of length n starting from o be $c_n(G)$, whereas the number of those that are nonbacktracking is $b_n(G)$. Write $H_0(z) := \sum_{n \geq 0} c_n(\mathbb{T}_{b+1}) z^n$ and $H(z) := \sum_{n \geq 0} c_n(T) z^n$, where T is a b-ary tree.

   (a) Show that

$$\sum_{n \geq 0} c_n(G) z^n = \sum_{n \geq 0} b_n(G) z^n H(z)^n H_0(z).$$

   (b) Show that

$$H(z) = \sum_{n \geq 0} (bz^2)^n H(z)^n$$

and that

$$H(z) = \frac{2}{1 + \sqrt{1 - 4bz^2}}.$$

   (c) Show that $H_0(z) = H(z)/(1 - z^2 H(z)^2)$ and that

$$H_0(z) = \frac{2b}{b - 1 + (b + 1)\sqrt{1 - 4bz^2}}.$$

6.63. Let G be a finite, connected graph. Recall that every edge of G comes with both orientations and $\ell_-^2(E)$ is the Hilbert space of square-summable, antisymmetric functions on E. Let $\ell_+^2(E)$ denote the Hilbert space of square-summable, symmetric functions on E. Thus, $\ell^2(E) = \ell_-^2(E) \oplus \ell_+^2(E)$. Given $f \in \ell^2(V)$, let $\theta_-(e) := f(e^+) - f(e^-)$ and $\theta_+(e) := f(e^+) + f(e^-)$. Note that $\bigstar(G)$ is the set of θ_- of this form; denote by $\bigstar_+(G)$ the set of θ_+ of this form. Show that $\dim \bigstar_+(G)$ is $|V| - 1$ if G is bipartite and is $|V|$ otherwise.

6.64. Let G be a d-regular graph. Write $b := d - 1$. Let A be the adjacency matrix of G. Write B for the matrix indexed by the oriented edges of G such that $B((x, y), (y, z)) = 1$ when $(x, y), (y, z) \in E(G)$ and $x \neq z$, with all other entries of B equal to 0. We will use the notation from Exercise 6.63.

   (a) Show that B is invertible and that bB^{-1} has integer entries.

   (b) Let $C := bB^{-1} + B$. Show that the subspaces $\ell_-^2(E)$ and $\ell_+^2(E)$ are invariant under C.

   (c) Show that C is self-adjoint.

   (d) Given $f \in \ell^2(V)$, let $\theta_-(e) := f(e^+) - f(e^-)$ and $\theta_+(e) := f(e^+) + f(e^-)$. Show that if $Af = \lambda f$, then $C\theta_- = \lambda \theta_-$ and $C\theta_+ = \lambda \theta_+$.

   (e) More generally, show that if λ belongs to the spectrum $\sigma(A)$ of A, then it belongs to the spectrum of C.

   (f) Show that if $\psi \perp \bigstar(G)$ and $\psi \in \ell_-^2(E)$, then $C\psi = d\psi$, whereas if $\psi \perp \bigstar_+(G)$ and $\psi \in \ell_+^2(E)$, then $C\psi = -d\psi$.

   (g) Show that if $\lambda \neq \pm d$ belongs to the spectrum of C, then λ belongs to the spectrum of A.

   (h) Show that the set $\{b/\kappa + \kappa ;\ \kappa \in \sigma(B)\}$ equals $\sigma(A) \cup \{\pm d\}$ and that $|\kappa| = \sqrt{b}$ for all $\kappa \in \sigma(B) \setminus \mathbb{R}$.

   (i) Show that when $d \geq 3$ and G is a finite, non-bipartite graph of diameter at least 6, nonbacktracking random walk mixes faster on G than does simple random walk (in a spectral sense). In other words, let M_- and M_+ be the incidence matrices whose rows are indexed by vertices and columns by oriented edges, where $M_-(x, e)$ is the indicator that x is the tail of e and $M_+(x, e)$ is the indicator that x is the head of e. The (x, y)-entry of $Q_k := M_- B^k M_+^T/(db^k)$ is the probability that a nonbacktracking random walk from x is at y at time $k + 1$, where the superscript T indicates transpose. Show that nonbacktracking random walk is aperiodic and for all x, y, we have

$$\limsup_{k \to \infty} \left| Q_k(x, y) - 1/|\mathsf{V}| \right|^{1/k} < \lim_{k \to \infty} \left| P^k(x, y) - 1/|\mathsf{V}| \right|^{1/k},$$

where P is the transition matrix for simple random walk.

6.65. For a general irreducible, positive recurrent Markov chain with stationary probability measure π, prove that for any set S of states, we have

$$\sum_{x \in S} \pi(x) \mathbf{E}_x[\tau_S^+] = 1$$

and

$$\sum_{x \in S} \sum_{y \in S^c} \pi(x) p(x, y) \mathbf{E}_y[\tau_S] = \pi(S^c).$$

This shows that starting at the stationary measure conditioned on having just made a transition from S to S^c, the expected time to hit S again is $1/\Phi_{S^c}$. *Hint:* Write $\mathbf{E}_x[\tau_S^+] = 1 + \sum_y p(x, y) \mathbf{E}_y[\tau_S]$ and observe that in the last sum, only $y \in S^c$ contribute.

6.66. Let $\langle X_n \rangle$ be a reversible Markov chain with a stationary probability distribution π on the state space V. Let A be a set of states with $|A| \geq 2$. Consider the chain $\langle Y_n \rangle$ induced on A, that is, $\mathbf{P}[Y_0 = x] = \pi(x)/\pi(A)$ and $\mathbf{P}[Y_{n+1} = y \mid Y_n = x] = \mathbf{P}[X_{\tau_A^+} = y \mid X_0 = x]$. Show that the spectral gap for the chain $\langle Y_n \rangle$ is at least that for the chain $\langle X_n \rangle$.

6.67. Show that if G is a plane, regular graph with regular dual, then $\Phi_{\mathsf{E}}(G)$ is either 0 or irrational.

6.68. Show that if G is a plane, regular graph with nonamenable, regular dual $G^\dagger$, then

$$\beta(G) + \beta(G^\dagger) > 1.$$

6.69. Let G be a plane, regular graph with regular dual $G^\dagger$. Write K' for the set of vertices incident to the faces corresponding to K, for both $K \subset \mathsf{V}$ and for $K \subset \mathsf{V}^\dagger$. Likewise, let $\widehat{K}$ denote the faces inside the outermost cycle of $E(K')$. Let $K_0 \subset \mathsf{V}$ be an arbitrary finite, connected set and recursively define $L_n := (\widehat{K}_n)' \subset \mathsf{V}^\dagger$ and $K_{n+1} := (\widehat{L}_n)' \subset \mathsf{V}$. Show that $|\partial_{\mathsf{E}} K_n|/|K_n| \to \Phi'_{\mathsf{E}}(G)$ and $|\partial_{\mathsf{E}} L_n|/|L_n| \to \Phi'_{\mathsf{E}}(G^\dagger)$.

6.70. Let G be a plane graph in the hyperbolic plane whose dual $G^\dagger$ has geodesic edges.

(a) Show that $\Phi_{\mathsf{E}}(G, \mathbf{1}, \mathbf{1})$ is at least π^{-1} times the infimum of the hyperbolic areas of the faces of $G^\dagger$. *Hint:* Use Theorem 6.1 and the fact that geodesic triangles have area at most π.

(b) Show that if all degrees in G are at least d and all degrees in $G^\dagger$ are at least $d^\dagger$, then $\Phi_{\mathsf{E}}(G, \mathbf{1}, \mathbf{1}) \geq [(d - 2)(d^\dagger - 2) - 4]/d^\dagger$. *Hint:* The area of a geodesic polygon of n sides equals $(n - 2)\pi$ minus the sum of the interior angles.

6.71. Show that equality holds on the left in (6.42) iff X is a function of Y, whereas equality holds on the right iff X and Y are independent. Show that equality holds in (6.44) iff X and Y are independent given Z.

6.72. (Data Processing Inequality) Suppose that X, Y, and Z are discrete random variables with X and Z being conditionally independent given Y (in other words, (X, Y, Z) is a Markov chain).

(a) Show that $H(X \mid Y) \leq H(X \mid Z)$.

(b) Show that $I(X, Y) \geq I(X, Z)$.

6.73. Show that the functional $\mu \mapsto H[\mu]$ is concave. Use this concavity to prove (6.40) and the second part of (6.42).

6.74. Show that there exists a constant $C_d > 0$ such that if A is a subgraph of the box $\{0, \ldots, n-1\}^d$ and $|A| < n^d/2$, then

$$|\partial_E A| \geq C_d |A|^{\frac{d-1}{d}} .$$

Here, $\partial_E A$ refers to the edge boundary within the box, so this is not a special case of Theorem 6.22.

6.75. Consider a stationary Markov chain $\langle X_n \rangle$ on a finite state space V with stationary measure $\pi(\cdot)$ and transition probabilities $p(\cdot, \cdot)$. Show that for $n \geq 0$, we have

$$H(X_0, \ldots, X_n) = H[\pi] - n \sum_{x, y \in V} \pi(x) p(x, y) \log p(x, y) .$$

6.76. Suppose that the chain (V, P) is lazy and $\Phi \geq \varphi$, where the function $z \mapsto z\varphi^2(z^{-2})/2$ is convex on $(0, \infty)$. Show that if

$$n \geq 1 + \int_{\pi(x) \wedge \pi(y)}^{1/\epsilon} \frac{2 \, du}{u\varphi^2(u)} ,$$

then $p_n(x, y) \leq \epsilon\pi(y)$.

6.77. Suppose that the chain (V, P) is not lazy, but there exist $j \geq 1$ and $0 < \eta < 1/2$ so that $p_j(x, x) \geq \eta$ for all $x \in V$. Adapt the proof of Theorem 6.31 to show that, under this assumption, if

$$\lfloor n/j \rfloor \geq 1 + \frac{(1 - \eta)^2}{\eta^2} \int_{\pi_{\min}}^{4/\epsilon} \frac{4 \, du}{u\Phi^2(u)} , \tag{6.69}$$

then

$$\frac{p_n(x, y)}{\pi(y)} \leq \epsilon . \tag{6.70}$$

6.78. Show that simple random walk $\langle (\Psi_j, X_j) \rangle$ on the Cayley graph of $\mathbb{Z}^\circ$ (with the standard generators) satisfies $p_{2n}(o, o) \geq \exp(-Cn^{1/3})$ for some C and all n. (The matching upper bound was proved in Corollary 6.32(iii).)

6.79. The upper bound on effective resistance given in Lemma 6.43, while useful for applications to resistance to infinity as in Theorem 6.42, is very weak for finite networks. Prove the following better version and some consequences.

(a) Let a and z be two distinct vertices in a finite, connected network G. Define

$$\psi(t) := \min\left\{ |\partial_E W|_c \; ; \; a \in W, \; z \notin W, \; G{\restriction}W \text{ is connected, } t \leq \min\{|W|_\pi, |V(G) \setminus W|_\pi\}\right\}$$

when this set is nonempty and $\psi(t) := \infty$ otherwise. Define $s_0 := \pi(a)$ and $s_{k+1} := s_k + \psi(s_k)/2$ recursively for $k \geq 0$. Then

$$\mathscr{R}(a \leftrightarrow z) \leq \sum_{k=0}^{\infty} \frac{4}{\psi(s_k)} .$$

(b) Let a and z be two distinct vertices in a finite, connected network G with $c(e) \geq 1$ for all edges e. Show that

$$\mathscr{R}(a \leftrightarrow z) \leq \frac{12}{\Phi_*} + 4 ,$$

where Φ_* is the expansion constant of Definition 6.14.

(c) Give another proof of the upper bound of Proposition 2.14 that uses (a).

6.80. Show that there is a function $C: (1, \infty) \times (1, \infty) \to (0, 1)$ such that every graph G with the property that the cardinality of each of its balls of radius r lies in $[b^r/a, ab^r]$ satisfies

$$\frac{|\partial_V K|}{|K|} \geq \frac{C(a, b)}{\log (1 + |K|)}$$

for each finite, nonempty $K \subset V(G)$. An outline of its proof is in the following parts (a)–(f). Define $f(x, y) := b^{-d(x, y)}$, where the distance is measured in G. Fix K. Set $Z := \sum_{x \in K} \sum_{y \in \partial_V K} f(x, y)$. Estimate Z in two ways, depending on the order of summation.

(a) Fix $x \in K$. Choose R so that $|B(x, R)| \geq 2|K|$ and let $W := B(x, R) \setminus K$. For $w \in W$, fix a geodesic (that is, shortest) path from x to w, and let w' be the first vertex in $\partial_V K$ on this path. Let $B := |\{(w, w'); \ w \in W\}|$. Show that $B \geq C b^R$.

(b) Show that $B \leq C' b^R \sum_{y \in \partial_V K} f(x, y)$.

(c) Deduce that $Z \geq C|K|$.

(d) Fix $y \in \partial_V K$. Show that $\sum_{x \in K} f(x, y) \leq C' \log (1 + |K|)$.

(e) Deduce that $Z \leq C'|\partial_V K| \log (1 + |K|)$.

(f) Deduce the result.

(g) Find a tree with bounded degree and such that every ball of radius r has cardinality in $[2^{\lfloor r/2 \rfloor}, 3 \cdot 2^r]$, yet there are arbitrarily large finite subsets with only one boundary vertex.

(h) Show that if a tree satisfies $|\partial_V K| \geq 3$ for every vertex set K of size at least m, where m is fixed, then the tree is nonamenable.

6.81. Show that there is a function $C: (1, \infty) \times (1, \infty) \to (0, \infty)$ such that every graph G with the property that the cardinality of each of its balls of radius r lies in $[b^r/a, ab^r]$ satisfies $\mathscr{R}(x \leftrightarrow \infty) \leq C(a, b)$ for all $x \in V(G)$.

6.82. Find a transient graph of bounded degree that does not contain any transient tree as a subgraph.

6.83. Let G be a Cayley graph of growth rate b. Show that when $p > 1/b$, a.s. some infinite open cluster of Bernoulli(p) percolation on G is transient.

6.84. Write out the proof of the second inequality of Theorem 6.46.

6.85. Write out the proof of the second inequality of Theorem 6.48.

6.86. Show that $\Phi_E^*(G) > 0$ implies that for p sufficiently close to 1, in Bernoulli(p) percolation the open cluster $K(o)$ of any vertex $o \in V(G)$ satisfies $\mathbf{P}_p \big[|V(K(o))| < \infty, \ |\partial_E V(K(o))| = n \big] < Cq^n$ for some $q < 1$ and $C < \infty$.

6.87. Let $\mathscr{A}_n$ and h_n be as in (6.65) and (6.66). Suppose that G satisfies an ***anchored (at-least-) two-dimensional isoperimetric inequality***, that is, $h_n > c/n$ for a fixed $c > 0$ and all n. Show that $|\mathscr{A}_n| \leq e^{Cn \log n}$ for some $C < \infty$. Give an example of a graph G that satisfies an anchored two-dimensional isoperimetric inequality, has $p_c(G) = 1$, and $|\mathscr{A}_n| \geq e^{c_1 n \log n}$ for some $c_1 > 0$.

6.88. Let G be a binary tree. Show that if v has a tail that decays slower than exponentially, then $\Phi_E^*(G) > 0$ yet $\Phi_E^*(G^v) = 0$ a.s., where G^v is the random subdivision defined in Section 6.9.

7 | Percolation on Transitive Graphs

How many infinite clusters does Bernoulli(p) percolation have on a given graph? How does this change as p changes? How do the infinite clusters themselves change as p increases? As we saw in Section 5.2, once p is large enough that there is an infinite cluster a.s., then the same holds for all larger p. For general graphs, the number of infinite clusters can be any nonnegative integer or infinity, and can change in an irregular fashion as p increases. However, many nice properties of infinite clusters ensue when the underlying graph is itself nice. What do we mean?

Many natural graphs look the same from every vertex. To make this notion precise, recall that an ***automorphism*** of a graph $G = (\mathsf{V}, \mathsf{E})$ is a bijection $\phi\colon \mathsf{V} \to \mathsf{V}$ such that $[\phi(x), \phi(y)] \in \mathsf{E}$ iff $[x, y] \in \mathsf{E}$. We write $\mathrm{Aut}(G)$ for the group of automorphisms of G. If G has the property that for every pair of vertices x, y, there is an automorphism of G that takes x to y, then G is called ***(vertex-) transitive***. The most prominent examples of transitive graphs are Cayley graphs, defined in Section 3.4. Of course, these include the usual Euclidean lattices $\mathbb{Z}^d$, on which the classical theory of percolation has been built.

Our purpose is not to develop the classical theory of percolation, for which Grimmett (1999) is an excellent source, but we will now briefly state some of the important facts from that theory that motivate some of the questions that we will treat. Recall from Section 6.9 the probability measures $\mathbf{P}_p^{\mathrm{site}}$ and $\mathbf{P}_p^{\mathrm{bond}}$ for the two product measures on 2^{V} and 2^{E} defining Bernoulli percolation and the associated critical probabilities $p_{\mathrm{c}}^{\mathrm{bond}}$ and $p_{\mathrm{c}}^{\mathrm{site}}$. As we said there, if we don't indicate whether the percolation is bond or site and both make sense in context, then results we state should be taken to apply to both types of percolation. As we will prove, we have $0 < p_{\mathrm{c}}(\mathbb{Z}^d) < 1$ for all $d \geq 2$. It was conjectured about 1955 that $p_{\mathrm{c}}^{\mathrm{bond}}(\mathbb{Z}^2) = 1/2$, but this was not proved until Kesten (1980); see Figure 7.1 for an illustration. How many

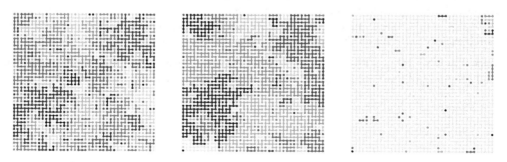

Figure 7.1. Bernoulli bond percolation on a 40×40 square grid graph at levels $p = 0.4, 0.5, 0.6$. Each cluster is given a different color.

infinite clusters are there when $p \geq p_c$? We will see in Theorem 7.5 that for each p, this number is a random variable that is constant a.s. More precisely, Aizenman, Kesten, and Newman (1987) showed that there is a.s. only one infinite cluster when $p > p_c(\mathbb{Z}^d)$, and one of the central conjectures in the field is that there is a.s. no infinite cluster when $p = p_c(\mathbb{Z}^d)$ ($d \geq 2$). This was proved for $d = 2$ partially by Harris (1960) and fully by Kesten (1980), and for $d \geq 19$ by Hara and Slade (1990, 1994) (and for $d \geq 7$ when bonds between all pairs of vertices within distance L of each other are added, for some L). After considerable time, the result of Hara and Slade was extended to all $d \geq 11$ by Fitzner and van der Hofstad (2015).

The conventional notation for $\mathbf{P}_p[x$ belongs to an infinite cluster$]$ is $\theta_x(p)$, not to be confused with the notation for a flow used in other chapters. For a transitive graph, it is clear that this probability does not depend on x, so the subscript is usually omitted. For all $d \geq 2$, van den Berg and Keane (1984) showed that $\theta(p)$ is continuous for all $p \neq p_c$ and is continuous at p_c iff $\theta(p_c) = 0$; see Exercise 7.33 for a more general result. Thus, θ is a continuous function on all of $[0, 1]$ iff the conjecture above [that $\theta(p_c) = 0$] holds. More results that lend support to this conjecture are that for all $d \geq 2$,

$$\lim_{k \to \infty} p_c(\mathbb{Z}^2 \,\square\, [0, k]^{d-2}) = p_c(\mathbb{Z}^d) = p_c(\mathbb{Z}^{d-1} \,\square\, \mathbb{Z}^+) \tag{7.1}$$

(Grimmett and Marstrand, 1990) and $\theta(p_c) = 0$ on the graph $\mathbb{Z}^{d-1} \,\square\, \mathbb{Z}^+$ (Barsky, Grimmett, and Newman, 1991). Also, $\theta(p_c) = 0$ on every graph of the form $\mathbb{Z}^2 \,\square\, H$ for H a finite, connected graph (Duminil-Copin, Sidoravicius, and Tassion, 2016). We will not prove any results regarding $\theta(p_c)$ on $\mathbb{Z}^d$ (except that $\theta(1/2) = 0$ for bond percolation on $\mathbb{Z}^2$), but in Section 8.4, we will prove that $\theta(p_c) = 0$ on nonamenable Cayley graphs.

This chapter is devoted to the basics of percolation theory and especially to the possible existence of a double phase transition on nonamenable graphs: one phase transition is at p_c, where the number of infinite clusters changes from 0 to positive, while another may occur at a point denoted p_u, where the number of infinite clusters changes from one positive number to another. This latter phase transition is considerably subtler than the former. Further progress on these issues requires (at present) another tool, the mass-transport technique, which is explained in the succeeding chapter. That tool works wonders on Cayley graphs but is not very effective on all transitive graphs.

In this chapter, all graphs are assumed to be locally finite without explicit mention.

A slight generalization of transitive graphs is the class of *quasi-transitive* graphs, which are those that have only finitely many orbits for the action of the automorphism group on the vertex set. Note that an orbit is an equivalence class of vertices if two vertices are equivalent when there is an automorphism that takes one vertex to the other. Most results concerning quasi-transitive graphs can be deduced from corresponding results for transitive graphs or can be deduced in a similar fashion but with some additional attention to details. In this context, the following construction is useful: Suppose that $\Gamma \subseteq \mathrm{Aut}(G)$ acts *quasi-transitively* on V, that is, the orbit space V/Γ is finite. Let o be a vertex in G. Let r be such that every vertex in G is within distance r of some vertex in the orbit Γo. Form the graph G' from the vertex set Γo by joining two vertices by an edge if their distance in G is at most $2r + 1$. It is easy to see that G' is connected: if $f : \mathsf{V} \to \Gamma o$ is a map such that $\mathrm{dist}_G(x, f(x)) \leq r$ for all x, then any path $x_0, x_1, \dots, x_n$ in G between two vertices of Γo maps to a path

$x_0, f(x_1), f(x_2), \ldots, f(x_{n-1}), x_n$ in G'. Also, restriction of the elements of Γ to G' yields a subgroup $\Gamma' \subseteq \operatorname{Aut}(G')$ that acts transitively on G', that is, V' is a single orbit. We call G' a *transitive representation* of G.

We begin with a section giving additional background on Cayley graphs before we turn to percolation theory.

7.1 Groups and Amenability

In Section 3.4, we looked at some basic constructions of groups. Another useful but more complex construction is that of amalgamation. Suppose that $\Gamma_1 = \langle S_1 \mid R_1 \rangle$ and $\Gamma_2 = \langle S_2 \mid R_2 \rangle$ are groups that both have a subgroup isomorphic to Γ'. We want to take the free product of Γ_1 and Γ_2 while identifying the copies of Γ'. More precisely, suppose that $\phi_i : \Gamma' \to \Gamma_i$ are monomorphisms for $i = 1, 2$ and that $S_1 \cap S_2 = \varnothing$. Let $R := \{ \phi_1(\gamma)\phi_2^{-1}(\gamma) ; \ \gamma \in \Gamma' \}$. The relations in R allow us to identify the copies $\phi_i(\Gamma')$ in forming the new group $\langle S_1 \cup S_2 \mid R_1 \cup R_2 \cup R \rangle$, which is called the *amalgamation of* Γ_1 *and* Γ_2 *over* Γ' and denoted $\Gamma_1 *_{\Gamma'} \Gamma_2$. The name and notation do not reflect the role of the maps ϕ_i, even though they are crucial. For example, $\mathbb{Z} *_{2\mathbb{Z}} \mathbb{Z} = \langle a, b \mid a^2 b^{-2} \rangle$ has a Cayley graph that, if its edges are not labeled,* looks just like the usual square lattice; see Figure 7.2. However, $\mathbb{Z} *_{3\mathbb{Z}} \mathbb{Z} = \langle a, b \mid a^3 b^{-3} \rangle$ is quite different: it is nonamenable by Exercise 6.42, as it has the quotient $\langle a, b \mid a^3 b^{-3}, a^3, b^3 \rangle = \mathbb{Z}_3 * \mathbb{Z}_3$. Of course, both $2\mathbb{Z}$ and $3\mathbb{Z}$ are isomorphic to $\mathbb{Z}$; our notation evokes inclusion as the appropriate maps ϕ_i.

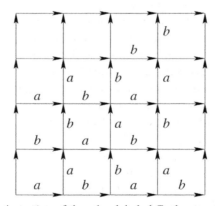

Figure 7.2. A portion of the edge-labeled Cayley graph of $\mathbb{Z} *_{2\mathbb{Z}} \mathbb{Z}$.

In this chapter, we will use $\Phi_{\mathsf{E}}(G)$ *to mean always the expansion constant* $\Phi_{\mathsf{E}}(G, 1, 1)$ *and* $\Phi_{\mathsf{V}}(G)$ *to mean always* $\Phi_{\mathsf{V}}(G, 1)$. This also makes the latter directly comparable to the anchored expansion constant $\Phi_{\mathsf{V}}^*(G)$. Since $|\partial_{\mathsf{E}} K| \geq |\partial_{\mathsf{V}} K|$, we have $\Phi_{\mathsf{E}}(G) \geq \Phi_{\mathsf{V}}(G)$. In the other direction, for any graph of degree bounded by d, we have $\Phi_{\mathsf{V}}(G) \geq \Phi_{\mathsf{E}}(G)/d$. For a tree T, it is clear that $\Phi_{\mathsf{E}}(T) = \Phi_{\mathsf{V}}(T)$. See Section 6.5 for the calculation of $\Phi_{\mathsf{E}}(G)$ when G arises from certain hyperbolic tessellations; for example, $\Phi_{\mathsf{E}}(G) = \sqrt{5}$ for the graph in Figure 2.4. Many of the edge graphs of hyperbolic tessellations are not Cayley graphs but are

---

* When the edges are oriented and labeled by the generators, we get the *Cayley diagram*.

still transitive graphs. The graph in Figure 2.4 is a Cayley graph; see Chaboud and Kenyon (1996) for an analysis of which regular tessellations are Cayley graphs. Another Cayley graph is shown in Figure 6.1.

Besides the hyperbolic graphs just mentioned, some other transitive graphs that are not Cayley graphs can be constructed as follows.

Example 7.1. (Grandparent Graph) Let ξ be a fixed end of a regular tree T of degree at least 3. Ends in graphs will be defined more generally in Section 7.3, but for trees, the definition is simpler. Namely, an **end** is an equivalence class of rays (that is, infinite, simple paths), where rays may start from any vertex and two rays are equivalent if they share all but finitely many vertices. Thus, given an end ξ, for every vertex x in T, there is a unique ray $x_\xi := \langle x_0, x_1, x_2, \ldots \rangle$ in the class ξ that starts from $x = x_0$; for each pair x, y, the rays x_ξ and y_ξ differ by only finitely many vertices. Call x_2 in this ray the ξ-**grandparent** of x. Let G be the graph obtained from T by adding, for every x, an edge $[x, x_2]$ between x and its ξ-grandparent. This is shown in Figure 7.3. Then G is a transitive graph that is not the

Figure 7.3. The grandparent graph, with the new edges shown in gray.

Cayley graph of any group. A proof that this is not a Cayley graph is given in Section 8.2. To see that G is transitive, let x and y be two of its vertices. Let $\langle x_n ; n \in \mathbb{Z} \rangle$ be a bi-infinite, simple path in T that extends x_ξ to negative integer indices. We have redrawn the grandfather graph using this line in Figure 7.4. There is some k such that $x_k \in y_\xi$. We claim that there is

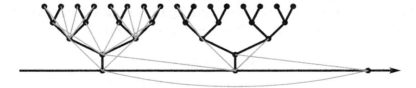

Figure 7.4. The grandparent graph redrawn.

an automorphism of G that takes x to x_k; this implies transitivity since there is then also an automorphism that takes x_k to y. Indeed, the graph $G \setminus \{x_n ; n \in \mathbb{Z}\}$ consists of components that are isomorphic, one for each $n \in \mathbb{Z}$. Seeing this, we also see that it is easy to shift along the path $\langle x_n ; n \in \mathbb{Z} \rangle$ by any integer amount, in particular, by k, via an automorphism of G. This proves the claim. These examples were described by Trofimov (1985).

Example 7.2. (Diestel–Leader Graph) Let $T(1)$ be a 3-regular tree with edges oriented toward some distinguished end, and let $T(2)$ be a 4-regular tree with edges oriented toward some distinguished end. (An edge $\langle x, y \rangle$ is oriented toward an end in a tree if y is on the ray starting from x that belongs to that end.) Let V be the Cartesian product of the vertices of $T(1)$ and the vertices of $T(2)$. Join $(x_1, x_2), (y_1, y_2) \in$ V by an edge if $\langle x_1, y_1 \rangle$ is an edge of $T(1)$ and $\langle x_2, y_2 \rangle$ is an edge of $T(2)$, and precisely one of these edges goes against the orientation. The resulting graph G has infinitely many components, any two components being isomorphic. Each component is a transitive graph, but not a Cayley graph for the same (still-to-be-explained) reason as in Example 7.1. To see that each component of G is transitive, note that if ϕ_i is an automorphism of $T(i)$ that preserves its distinguished end, then $\phi_1 \times \phi_2 : (x_1, x_2) \mapsto (\phi_1(x_1), \phi_2(x_2))$ is an automorphism of G. We saw in Example 7.1 that such automorphisms ϕ_i act transitively. Therefore, G is transitive, whence so is each of its components (and all components are isomorphic). One way to describe this graph is as a family graph: Suppose that there is an infinity of individuals, each of which has two parents and three children. The children are shared by the parents, as is the case in the real world. If an edge is drawn between each individual and his parent, then one obtains this graph for certain parenthood relations (with one component if each individual is related to every other individual). This graph is not a tree: if, say, John and Jane are both parents of Alice, Betty, and Carl, then one cycle in the family graph is from John to Alice to Jane to Betty to John. This example of a transitive graph was first discovered by Diestel and Leader (2001) for the purpose of providing a potential example of a transitive graph that is not roughly isometric to any Cayley graph. The question whether there was any such transitive graph was asked by Woess (1991). Finally, Eskin, Fisher, and Whyte (2012) showed that indeed the Diestel–Leader graph is not roughly isometric to any Cayley graph.

7.2 Tolerance and Ergodicity

This section is devoted to some quite general and elementary properties of the product measures $\mathbf{P}_p$. Notation we will use throughout is that 2^A denotes the collection of subsets of A. It also denotes $\{0, 1\}^A$, the set of functions from A to $\{0, 1\}$. These two spaces are identified by identifying a subset with its indicator function.

The first general property that we treat is that $\mathbf{P}_p$ is *insertion tolerant*, which means the following. Given a configuration $\omega \in 2^\mathsf{E}$ and an edge $e \in \mathsf{E}$, write

$$\Pi_e \omega := \omega \cup \{e\},$$

where we regard ω as the subset of open edges. The notation is chosen to evoke the phrase "put e in ω." Extend this notation to events A by

$$\Pi_e A := \{\Pi_e \omega \, ; \; \omega \in A\}.$$

Also, for any set F of edges, write

$$\Pi_F \omega := \omega \cup F$$

and extend to events in the same way. A bond percolation process **P** on G is **insertion tolerant** if $\mathbf{P}(\Pi_e A) > 0$ for every $e \in \mathsf{E}$ and every measurable $A \subset 2^{\mathsf{E}}$ satisfying $\mathbf{P}(A) > 0$. Here, we use the Borel σ-field generated by the product topology on 2^{E}. For Bernoulli(p) percolation, we have the stronger inequality

$$\mathbf{P}_p(\Pi_e A) \geq p\,\mathbf{P}_p(A). \tag{7.2}$$

▷ **Exercise 7.1.**
Show that if A is measurable, then so is $\Pi_e A$ and prove (7.2).

Likewise, a bond percolation process **P** on G is **deletion tolerant** if $\mathbf{P}(\Pi_{\neg e} A) > 0$ whenever $e \in \mathsf{E}$ and $\mathbf{P}(A) > 0$, where $\Pi_{\neg e}\omega := \omega \setminus \{e\}$. We extend this notation to sets F by $\Pi_{\neg F}\omega := \omega \setminus F$. By symmetry, Bernoulli percolation is also deletion tolerant, with

$$\mathbf{P}_p(\Pi_{\neg e} A) \geq (1-p)\,\mathbf{P}_p(A).$$

Similar definitions hold for site percolation processes.*

▷ **Exercise 7.2.**
Let G be a connected graph and $x, y \in \mathsf{V}(G)$. Consider Bernoulli(p) percolation on G. Prove that if $\theta_y(p) > 0$, then also $\theta_x(p) > 0$.

We will use insertion and deletion tolerance often. Another crucial pair of properties is invariance and ergodicity. Suppose that Γ is a group of automorphisms of a graph G. A measure **P** on 2^{E}, on 2^{V}, or on $2^{\mathsf{E} \cup \mathsf{V}}$ is called a Γ-**invariant** percolation if $\mathbf{P}(\gamma A) = \mathbf{P}(A)$ for all $\gamma \in \Gamma$ and all events A. In the case of a measure on $2^{\mathsf{E} \cup \mathsf{V}}$, we assume that **P** is concentrated on subgraphs of G, that is, whenever an edge lies in a configuration, so do both of its endpoints. Let $\mathcal{I}_\Gamma$ denote the σ-field of events that are invariant under all elements of Γ. The measure **P** is called Γ-**ergodic** if, for each $A \in \mathcal{I}_\Gamma$, we have either $\mathbf{P}(A) = 0$ or $\mathbf{P}(\neg A) = 0$. Bernoulli percolation on any infinite Cayley graph is both invariant and ergodic with respect to translations. The invariance is obvious, and ergodicity is proved in the following proposition.

Proposition 7.3. (Ergodicity of Bernoulli Percolation) *If Γ acts on a connected, locally finite graph G in such a way that each vertex has an infinite orbit, then $\mathbf{P}_p$ is Γ-ergodic.*

Note that if some vertex in G has an infinite Γ-orbit, then every vertex has an infinite Γ-orbit. Recall that we identify a subset $\omega \subseteq \mathsf{E}$ with its indicator function, so that $\omega(e)$ takes the value 0 or 1 depending on whether e lies in the subset. Also, recall that a **cylinder** event B is one for which there is a finite set $F \subset \mathsf{E}$ with the property that for every pair $\omega_1, \omega_2 \in 2^{\mathsf{E}}$ that agree on F, we have either both ω_1, ω_2 are in B or neither are in B. In this case, we say that B **depends only on** F.

---

* In the statistical physics literature, a measure that is both insertion and deletion tolerant is said to have *finite energy*, whereas one that is assumed only insertion tolerant has *positive finite energy*. Of course, these terms have nothing to do with the energy of flows in electrical networks. Rather, they arise by analogy to other processes where a different form of energy plays a role.

Proof. Our notation will be for bond percolation. The proof for site percolation is identical. Let $A \in \mathcal{I}_\Gamma$. The idea is to show that A is almost independent of $\gamma A = A$ for "large" γ. To make this precise, we approximate A by a cylinder event. Thus, let $\epsilon > 0$. Because A is measurable, there is a cylinder event B that depends only on some finite set F such that $\mathbf{P}_p(A \bigtriangleup B) < \epsilon$. For all $\gamma \in \Gamma$, we have $\mathbf{P}_p(\gamma A \bigtriangleup \gamma B) = \mathbf{P}_p[\gamma(A \bigtriangleup B)] < \epsilon$. By the assumption that vertices have infinite orbits, there is some γ such that F and γF are disjoint (because their graph distance can be made arbitrarily large). Since γB depends only on γF, it follows that B and γB are independent. Now for any events C_1, C_2, D, we have

$$\left| \mathbf{P}_p(C_1 \cap D) - \mathbf{P}_p(C_2 \cap D) \right| \le \mathbf{P}_p\left[(C_1 \cap D) \bigtriangleup (C_2 \cap D) \right] \le \mathbf{P}_p(C_1 \bigtriangleup C_2) .$$

Therefore,

$$
\begin{aligned}
\left| \mathbf{P}_p(A) - \mathbf{P}_p(A)^2 \right| &= \left| \mathbf{P}_p(A \cap \gamma A) - \mathbf{P}_p(A)^2 \right| \\
&\le \left| \mathbf{P}_p(A \cap \gamma A) - \mathbf{P}_p(B \cap \gamma A) \right| + \left| \mathbf{P}_p(B \cap \gamma A) - \mathbf{P}_p(B \cap \gamma B) \right| \\
&\quad + \left| \mathbf{P}_p(B \cap \gamma B) - \mathbf{P}_p(B)^2 \right| + \left| \mathbf{P}_p(B)^2 - \mathbf{P}_p(A)^2 \right| \\
&\le \mathbf{P}_p(A \bigtriangleup B) + \mathbf{P}_p(\gamma A \bigtriangleup \gamma B) + \left| \mathbf{P}_p(B)\, \mathbf{P}_p(\gamma B) - \mathbf{P}_p(B)^2 \right| \\
&\quad + \left| \mathbf{P}_p(B) - \mathbf{P}_p(A) \right| \left(\mathbf{P}_p(B) + \mathbf{P}_p(A) \right) \\
&< \epsilon + \epsilon + 0 + 2\epsilon .
\end{aligned}
$$

It follows that $\mathbf{P}_p(A) \in \{0, 1\}$, as desired. ◄

Another way we could have proved Proposition 7.3 is to use Kolmogorov's zero-one law in combination with the fact that every invariant event is a tail event up to a set of probability 0. Recall what the tail events are: For a set of edges $K \subseteq \mathsf{E}$, let $\mathscr{F}(K)$ denote the σ-field of events depending only on K. Define the **tail σ-field** to be the intersection of $\mathscr{F}(\mathsf{E} \setminus K)$ over all finite K.

Lemma 7.4. (Invariant Events Are Almost Tail Events) *Let Γ act on a connected, locally finite graph G in such a way that each vertex has an infinite orbit. Let $\mathbf{P}$ be a Γ-invariant percolation on G. Then, for every Γ-invariant event A, there is a tail event B such that $\mathbf{P}(A \bigtriangleup B) = 0$.*

Proof. Since A is measurable, there exist increasing finite sets $K_n \subset \mathsf{E}$ whose union is E and cylinder events B_n that depend only on K_n with $\sum_{n \ge 1} \mathbf{P}(A \bigtriangleup B_n) < \infty$. Choose $\gamma_n \in \Gamma$ so that $\gamma_n K_n \cap K_n = \varnothing$. Since $\mathbf{P}$ and A are Γ-invariant, we have $\mathbf{P}(A \bigtriangleup \gamma_n B_n) = \mathbf{P}(A \bigtriangleup B_n)$, whence $\sum_{n \ge 1} \mathbf{P}(A \bigtriangleup \gamma_n B_n) < \infty$. The Borel–Cantelli lemma implies that $\mathbf{P}\left(\limsup_n (A \bigtriangleup \gamma_n B_n) \right) = 0$, whence $\mathbf{P}(A \bigtriangleup \limsup_n \gamma_n B_n) = 0$. The event $\limsup_n \gamma_n B_n$ is the tail event we seek. ◄

7.3 The Number of Infinite Clusters

The number of infinite clusters in Bernoulli percolation could be any nonnegative integer, or infinity, with positive probability (see Exercise 7.22), but on transitive graphs, it is quite restricted, as shown by the following theorem of Newman and Schulman (1981). The proof is a beautiful combination of insertion tolerance and ergodicity.

Theorem 7.5. *If G is a transitive, connected graph, then the number of infinite clusters is constant* $\mathbf{P}_p$*-a.s. and equal to either 0, 1, or* ∞*. In fact, it suffices that G is connected and that vertices have infinite orbits under the automorphism group of G.*

Proof. Let N_∞ denote the number of infinite clusters and Γ denote the automorphism group of the infinite graph G. The action of any element of Γ on a configuration does not change N_∞. That is, N_∞ is measurable with respect to the σ-field $\mathcal{I}_\Gamma$ of events that are invariant under all elements of Γ. By ergodicity (Proposition 7.3), this means that N_∞ is constant a.s.

Now suppose that $N_\infty \in [2, \infty)$ a.s. Then there must exist two vertices x and y that belong to distinct infinite clusters with positive probability. For bond percolation, let $e_1, e_2, \ldots, e_n$ be a path of edges from x to y. If A denotes the event that x and y belong to distinct infinite clusters and B denotes the event $\Pi_{\{e_1, e_2, \cdots, e_n\}} A$, then by insertion tolerance, we have $\mathbf{P}_p(A) > 0$ and $\mathbf{P}_p(B) > 0$. Yet N_∞ takes a strictly smaller value on B than on A, which contradicts the constancy of N_∞. (The proof for site percolation is parallel.) ◀

The result mentioned in the introduction that there is at most one infinite cluster for percolation on $\mathbb{Z}^d$, due to Aizenman, Kesten, and Newman (1987), was extended and simplified until the following result appeared, due to Burton and Keane (1989) and Gandolfi, Keane, and Newman (1992).

Theorem 7.6. (Amenability Yields at Most One Infinite Cluster) *If G is a connected, transitive, amenable graph, then* $\mathbf{P}_p$*-a.s. there is at most one infinite cluster, no matter the value of p.*

The proof of Theorem 7.6 makes a more complicated use of insertion and deletion tolerance, but it is still quite striking. A major open conjecture is that the converse of Theorem 7.6 holds; see Conjecture 7.31.

To prove Theorem 7.6, we will use the result of the following exercise.

▷ **Exercise 7.3.**
Let T be a tree that has no vertices of degree 1. Let B be the set of vertices of degree at least 3. Let K be a finite, nonempty subset of vertices of T. Show that $|\partial_V K| \geq |K \cap B| + 2$.

We also use the following notation and concept. Let $E_R(x)$ denote the set of edges that have at least one endpoint at distance at most $R - 1$ from x. This is the edge interior of the ball of radius R about x. We denote the component of x in a percolation configuration by $K(x)$. Call a vertex x a *furcation* of a configuration ω if closing all edges incident to x would split $K(x)$ into at least three infinite clusters.* Similarly, call x a *site-furcation* of a (site) configuration ω if closing all vertices adjacent to x would split $K(x)$ into at least three infinite clusters.

---

* In Burton and Keane (1989), these vertices were called *encounter points*. When $K(x)$ is split into exactly three infinite clusters, x is called a *trifurcation* by Grimmett (1999).

Proof of Theorem 7.6. We first do the proof for bond percolation, then state the modifications needed for the case of site percolation. Let $0 < p < 1$ and fix a vertex $o \in V$. Let $\Lambda = \Lambda(\omega)$ denote the set of furcations of ω. We claim that if there is more than one infinite cluster with positive probability, then there are sufficiently many furcations as to create tree-like structures that force G to be nonamenable. To be precise, we claim first that

$$c := \mathbf{P}_p[o \in \Lambda] > 0$$

and, second, that for each finite set $K \subset V$,

$$|\partial_V K| \geq c|K|. \tag{7.3}$$

These two claims together imply that G is nonamenable.

Now our assumption that there is more than one infinite cluster with positive probability implies that there are, in fact, infinitely many infinite clusters a.s. by Theorem 7.5. We turn these into furcations as follows. Choose some $R > 0$ such that the ball of radius R about o intersects at least three infinite clusters with positive probability. When this event occurs and then $E_R(o)$ is closed, this event still occurs. Thus, by deletion tolerance, we may choose x, y, z at the same distance, R, from o such that $\mathbf{P}_p(A) > 0$, where A is the event that x, y, z belong to distinct infinite clusters and $E_R(o)$ is closed. Join x to y by a geodesic $\mathcal{P}_1$ in the graph induced by G on the ball $B_R(o)$. Then join z to $\mathcal{P}_1$ by a geodesic $\mathcal{P}_2$ in $B_R(o)$. Let the vertex where $\mathcal{P}_2$ meets $\mathcal{P}_1$ be u. By insertion tolerance, $\mathbf{P}_p(A') > 0$, where $A' := \Pi_{E(\mathcal{P}_1) \cup E(\mathcal{P}_2)} A$. Furthermore, u is a furcation on the event A'. Hence $c = \mathbf{P}_p[o \in \Lambda] = \mathbf{P}_p[u \in \Lambda] \geq \mathbf{P}_p(A') > 0$.

It follows that for every finite $K \subset V$,

$$\mathbf{E}_p\big[|K \cap \Lambda|\big] = \sum_{x \in K} \mathbf{P}_p[x \in \Lambda] = c|K|. \tag{7.4}$$

We next claim that

$$|\partial_V K| \geq |K \cap \Lambda|. \tag{7.5}$$

Taking the expectation of (7.5) and using (7.4) shows (7.3).

To see why (7.5) is true, let T be a spanning tree of an infinite component η of ω. Remove all vertices of degree 1 from T and iterate until there are no more vertices of degree 1; call the resulting tree $T(\eta)$. Note that every furcation in η has degree ≥ 3 in $T(\eta)$. Apply Exercise 7.3 to conclude that

$$\big|(\partial_V K) \cap T(\eta)\big| \geq \big|\partial_{V(T(\eta))}(K \cap T(\eta))\big| \geq |K \cap \Lambda \cap T(\eta)| = |K \cap \Lambda \cap \eta|.$$

If we sum this over all components η of ω, we arrive at (7.5).

Now we describe the modifications for site percolation. Let Λ_s denote the set of site-furcations of ω. We choose $R, x, y,$ and z as before. Note, however, that now we close the vertices of $B_R(o)$ (except for $x, y,$ and z), rather than the edges, to get the event A. Let $x', y',$ and z' be vertices at distance $R - 1$ from o that are adjacent to $x, y,$ and z, respectively. Let $\mathcal{P}_1$ be a geodesic from x' to y' in $B_{R-1}(o)$ (which, if $x' = y'$, is just that vertex x') and $\mathcal{P}_2$ be a geodesic from z' to $\mathcal{P}_1$. If z' is on $\mathcal{P}_1$, then let $u := z'$. Otherwise, let u be the vertex of $\mathcal{P}_2$

adjacent to $\mathcal{P}_1$; this is unique, since $\mathcal{P}_2$ is a shortest path to $\mathcal{P}_1$. Then u is a site-furcation on the event $A' := \Pi_{V(\mathcal{P}_1) \cup V(\mathcal{P}_2)} A$, where $\mathbf{P}_p(A') > 0$ by insertion tolerance. This gives us that $\mathbf{P}_p[o \in \Lambda_s] \geq \mathbf{P}_p(A') > 0$.

We claim that for every finite $K \subset \mathsf{V}$, we have $|\partial_V(K \cup \partial_V K)| \geq |K \cap \Lambda_s|/(d+1)$, where d is the degree of G, which is sufficient to complete the proof as before. To prove this, we again use the trees $T(\eta)$. It is not necessarily the case any longer that every site-furcation in η has degree at least 3 in $T(\eta)$. However, if u is a site-furcation of η, then either u or one of its neighbors has degree at least 3 in $T(\eta)$. This proves the desired inequality. ◄

There is an extension of Theorem 7.6 to invariant, insertion-tolerant percolation that was proved in the original papers: we used deletion tolerance in one place of our proof. Because this theorem is so important, we give a second proof that yields this more general theorem. Our proof uses a technique from BLPS (1999b), which will itself be important in Chapter 8. The method uses trees as well as the important notion of ends of graphs.

Let $G = (\mathsf{V}, \mathsf{E})$ be any graph. The ***number of ends*** of G is defined to be the supremum of the number of infinite components of $G \setminus K$ over all finite subsets K of G. In particular, G has no ends iff all components of G are finite and only one end iff the complement of each finite set in G has exactly one infinite component. This definition suffices for our purposes, but nevertheless, what is an end? To define an end, we make two preliminary definitions. First, an infinite set of vertices $V_0 \subset \mathsf{V}$ is ***end convergent*** if, for every finite $K \subset \mathsf{V}$, there is a component of $G \setminus K$ that contains all but finitely many vertices of V_0. Second, two end-convergent sets V_0, V_1 are ***equivalent*** if $V_0 \cup V_1$ is end convergent. Now an ***end*** of G is an equivalence class of end-convergent sets. For example, $\mathbb{Z}$ has two ends, which we could call $+\infty$ and $-\infty$. For a tree, this definition agrees with the one in Example 7.1, and the ends are in natural one-to-one correspondence with the boundary with respect to any fixed root.

▷ **Exercise 7.4.**
Show that for any finitely generated group, the number of ends is the same for all of its Cayley graphs. Thus, we may speak of the number of ends of a group, not merely of a Cayley graph. In fact, show that if two graphs are roughly isometric, then they have the same number of ends.

▷ **Exercise 7.5.**
Show that if G and G' are any two infinite, connected graphs, then the Cartesian product graph $G \,\square\, G'$ (defined on p. 1) has only one end.

▷ **Exercise 7.6.**
Show that if Γ and Γ' are any two finitely generated groups with $|\Gamma| \geq 2$ and $|\Gamma'| \geq 3$, then $\Gamma * \Gamma'$ has infinitely many ends.

Lemma 7.7. (Forests in Percolation) *Let $\mathbf{P}$ be a Γ-invariant percolation process on a graph G. If with positive probability there is a component of ω with at least three ends, then (on a larger probability space) there is a random forest $\mathfrak{F} \subset \omega$ such that the distribution of* *the pair* $(\mathfrak{F}, \omega)$ *is Γ-invariant and with positive probability, there is a component of $\mathfrak{F}$ that has at least three ends.*

Proof. Assign independent, uniform $[0, 1]$ random variables to the edges (independently of ω). Define the ***free minimal spanning forest*** $\mathfrak{F}$ of ω by taking an edge $e \in \omega$ to be present in $\mathfrak{F}$ iff there is no cycle in ω containing e in which e is assigned the maximum value. There is no cycle in $\mathfrak{F}$, since the edge with the largest label in any cycle could not belong to $\mathfrak{F}$. In addition, all components (trees) of $\mathfrak{F}$ that lie in infinite components of ω are a.s. infinite, since if K were a finite component, then consider the element of $\partial_\mathsf{E} K$ with smallest label. Provided all labels are distinct, which occurs a.s., this element would be in $\mathfrak{F}$ by definition of $\mathfrak{F}$, a contradiction. Thus, each vertex in an infinite component of ω belongs to some infinite tree of $\mathfrak{F}$. (Minimal spanning forests are studied in Chapter 11.)

Suppose that $K(x)$ has at least three ends with positive probability. Choose any finite tree T containing x so that with positive probability, $T \subset K(x)$ and $K(x) \setminus \mathsf{V}(T)$ has at least three infinite components. Then with positive probability, all of the following four events occur: (1) $T \subset K(x)$; (2) $K(x) \setminus \mathsf{V}(T)$ has at least three infinite components; (3) all edges in T are assigned values less than $1/2$; and (4) all edges incident to $\mathsf{V}(T)$ but not in T are assigned values greater than $1/2$. On this event, $\mathfrak{F}$ contains T, and T is part of a tree in $\mathfrak{F}$ with at least three ends. ◄

We next show that a version of Theorem 7.5 holds for insertion-tolerant, invariant percolation.

Theorem 7.8. *If $\mathbf{P}$ is an insertion-tolerant, invariant percolation on a transitive, connected graph, then the number of infinite clusters is $\mathbf{P}$-a.s. equal to 0, 1, or ∞.*

Proof. The proof of Theorem 7.5 would apply if we knew that the number of infinite clusters were constant a.s. However, this does not follow from our hypotheses. Yet if we condition on the number of infinite clusters being a fixed number k, and if we are able to show that the conditioned percolation is still insertion tolerant and invariant, then the old proof will work just fine. Luckily, such conditioning does indeed preserve these properties.

To see this, we show that actually, conditioning $\mathbf{P}$ on any invariant event A of positive probability preserves insertion tolerance and invariance. The fact that $\mathbf{P}(\,\cdot\,|\,A)$ is invariant is immediate from the invariance of $\mathbf{P}$ and A. To show insertion tolerance, let B be any event with $\mathbf{P}(B \mid A) > 0$ and e be any edge. By Lemma 7.4, we may assume that A is a tail event, so that $\Pi_e A \cup \Pi_{\neg e} A = A$. Thus,

$$\mathbf{P}(\Pi_e B \mid A) = \frac{\mathbf{P}(\Pi_e B \cap A)}{\mathbf{P}(A)} = \frac{\mathbf{P}(\Pi_e(B \cap A))}{\mathbf{P}(A)} > 0$$

since $\mathbf{P}(B \cap A) > 0$. In other words, $\mathbf{P}(\,\cdot\,|\,A)$ is insertion tolerant. ◄

Here is the promised extension of Theorem 7.6:

Theorem 7.9. (Amenability Yields at Most One Infinite Cluster) *If Γ acts transitively on a connected, amenable graph G and $\mathbf{P}$ is a Γ-invariant, insertion-tolerant percolation process on G, then $\mathbf{P}$-a.s. there is at most one infinite cluster.*

Proof. We prove the case of bond percolation, as the case of site percolation is virtually the same. By Theorem 7.8, it suffices to show that if there are infinitely many infinite clusters with positive probability, then G is not amenable. Let x, y, and z be three vertices such that the event A that they belong to distinct infinite clusters has positive probability. Choose a finite set of edges F that connect x, y, and z. By insertion tolerance, $\mathbf{P}(\Pi_F A) > 0$. On the event $\Pi_F A$, there is an infinite cluster with at least three ends. Thus, the hypothesis of Lemma 7.7 holds. Let $\mathfrak{F}$ be as in Lemma 7.7. Since $\mathfrak{F}$ has a component with at least three ends with positive probability, there is a furcation of $\mathfrak{F}$ with positive probability. This shows (7.4) for some $c > 0$, where Λ is the set of furcations of $\mathfrak{F}$. Exercise 7.3 then gives (7.5), which completes the proof. ◀

7.4 Inequalities for p_c

In Section 6.9, we saw that p_c can be bounded above by using an anchored expansion constant, and we also saw how p_c behaves under covering maps. How do p_c^{bond} and p_c^{site} compare to each other? When is there a real phase transition in the sense that $0 < p_c < 1$? These are the questions we address in this section, but, as we will see, there is still a basic hole in our knowledge.

Our first two results show that p_c^{bond} and p_c^{site} are comparable in a uniform sense depending on the maximum degree.

Proposition 7.10. (Hammersley, 1961a) *For any graph G, any $p \in (0, 1)$, and any vertex $o \in G$, we have $\theta_o^{\text{site}}(p) \leq \theta_o^{\text{bond}}(p)$. Therefore $p_c^{\text{site}}(G) \geq p_c^{\text{bond}}(G)$.*

Proof. As we have done to prove other percolation inequalities, we will construct a coupling of the two percolation measures. That is, given $\xi \in 2^{\mathsf{V}}$, we will construct $\omega \in 2^{\mathsf{E}}$ in such a way that, first, if ξ has distribution $\mathbf{P}_p^{\text{site}}$ on G, then ω has distribution $\mathbf{P}_p^{\text{bond}}$ on G; and, second, if $K_\xi(o)$ is infinite, then so is $K_\omega(o)$, where the subscript indicates which configuration determines the cluster of o. This implies the desired inequalities.

Choose any ordering $\langle x_1, x_2, \ldots \rangle$ of V with $x_1 = o$. Let $\langle Y_e \rangle_{e \in \mathsf{E}}$ be Bernoulli(p) random variables.

We will look at a finite or infinite subsequence of vertices $\langle x_{n_j} \rangle$ via a recursive procedure. If $\xi(o) = 0$, then stop. Otherwise, let $V_1 := \{o\}$, $W_1 := \varnothing$, and set $n_1 := 1$. Suppose that V_k and W_k have been selected. Let n_{k+1} be the smallest index of a vertex in $\mathsf{V} \setminus (V_k \cup W_k)$ that neighbors some vertex in V_k, if any. If there is such a vertex, then let x'_{k+1} be the vertex in V_k that neighbors $x_{n_{k+1}}$ and that has smallest index, and set $\omega([x'_{k+1}, x_{n_{k+1}}]) := \xi(x_{n_{k+1}})$. If $\xi(x_{n_{k+1}}) = 1$, then put $V_{k+1} := V_k \cup \{x_{n_{k+1}}\}$ and $W_{k+1} := W_k$, whereas if $\xi(x_{n_{k+1}}) = 0$, then put $V_{k+1} := V_k$ and $W_{k+1} := W_k \cup \{x_{n_{k+1}}\}$. When n_{k+1} is not defined, stop; $K_\xi(o)$ is finite, and we set $\omega(e) := Y_e$ for the remaining edges $e \in \mathsf{E}$ for which we have not yet specified $\omega(e)$.

If this procedure never ends, then both $K_\xi(o)$ and $K_\omega(o)$ are infinite; assigning $\omega(e) := Y_e$ for any remaining edges $e \in \mathsf{E}$ gives a fair sample of Bernoulli(p) bond percolation on G when $\xi \sim \mathbf{P}_p^{\text{site}}$. This gives the desired coupling. ◀

In the preceding proof, we constructed a certain coupling of two percolation measures. Another kind of coupling that is important is the following. Given two percolation measures $\mathbf{P}$ and $\mathbf{P}'$ on G, we say that $\mathbf{P}$ *stochastically dominates* $\mathbf{P}'$, written $\mathbf{P} \succcurlyeq \mathbf{P}'$, if there are random variables ω and ω' with laws $\mathbf{P}$ and $\mathbf{P}'$, respectively, such that $\omega \geq \omega'$ a.s., which we could also write as $\omega \supseteq \omega'$ a.s.* We use this idea to prove inequalities in the other direction to those in Proposition 7.10. For an intuitive understanding of the next inequality, think about p close to 1.

Proposition 7.11. (Grimmett and Stacey, 1998) *For any graph G of maximal degree d, any $p \in (0, 1)$, and any vertex $o \in G$ of degree d_o, we have*

$$\theta_o^{\mathrm{site}}\left(1 - (1 - p)^{d-1}\right) \geq \left[1 - (1 - p)^{d_o}\right]\theta_o^{\mathrm{bond}}(p).$$

Therefore

$$p_{\mathrm{c}}^{\mathrm{site}}(G) \leq 1 - \left(1 - p_{\mathrm{c}}^{\mathrm{bond}}(G)\right)^{d-1}.$$

Proof. We will again construct a coupling of the two percolation measures. This time, given $\omega \in 2^{\mathsf{E}}$, we will construct $\xi \in 2^{\mathsf{V}}$ in such a way that, first, if ω has distribution $\mathbf{P}_p^{\mathrm{bond}}$ on G, then ξ has a distribution that is stochastically dominated by $\mathbf{P}_q^{\mathrm{site}}$ on G conditioned on $\xi(o) = 1$, where $q := 1 - (1 - p)^{d-1}$; and, second, if $K_\omega(o)$ is infinite, then so is $K_\xi(o)$.

Choose any ordering $\langle x_1, x_2, \ldots \rangle$ of V with $x_1 = o$. Let $\langle Y_x \rangle_{x \in \mathsf{V}}$ be Bernoulli(q) random variables.

We will look at a finite or infinite subsequence of vertices $\langle x_{n_j} \rangle$ via a recursive procedure. If $\omega(e) = 0$ for all edges e incident to o, then stop. Otherwise, let $V_1 := \{o\}$, $W_1 := \varnothing$, $\xi(o) := 1$, and $n_1 := 1$. Note that the probability that some edge incident to o is open is $1 - (1 - p)^{d_o}$.

Suppose that V_k and W_k have been selected. Let n_{k+1} be the smallest index of a vertex in $\mathsf{V} \setminus (V_k \cup W_k)$ that neighbors some vertex in V_k, if any. Define $\xi(x_{n_{k+1}})$ to be the indicator that there is some vertex x *not* in $V_k \cup W_k$ for which $\omega([x_{n_{k+1}}, x]) = 1$. Note that the conditional probability that $\xi(x_{n_{k+1}}) = 1$ is $1 - (1 - p)^r \leq q$, where r is the degree of $x_{n_{k+1}}$ in $G \setminus (V_k \cup W_k)$. If $\xi(x_{n_{k+1}}) = 1$, then put $V_{k+1} := V_k \cup \{x_{n_{k+1}}\}$ and $W_{k+1} := W_k$; otherwise put $V_{k+1} := V_k$ and $W_{k+1} := W_k \cup \{x_{n_{k+1}}\}$. When n_{k+1} is not defined, stop; $K_\omega(o)$ is finite and we set $\xi(x) := Y_x$ for the remaining vertices $x \in \mathsf{V} \setminus V_k$.

If this procedure never ends, then $K_\xi(o)$ (though perhaps not $K_\omega(o)$) is infinite; assigning $\xi(x) := Y_x$ for any remaining vertices $x \in \mathsf{V}$ gives a law of ξ that is dominated by Bernoulli(q) site percolation on G conditioned to have $\xi(o) = 1$. This gives the desired inequality. ◄

We now discuss whether $0 < p_{\mathrm{c}} < 1$. It turns out that a lower bound for p_{c} is rather simple to obtain and implies that $p_{\mathrm{c}} > 0$ for all graphs of bounded degree. (It is easy to see that there are graphs of unbounded degree for which $p_{\mathrm{c}} = 0$.) The idea for our lower bound was used already in Chapter 5, where we calculated p_{c} for general trees. Namely, we use a simple, first-moment argument as in Proposition 5.8. We can apply a similar argument on a general graph by using the associated tree of self-avoiding walks. Recall that in Example 3.6, we described a tree formed from the self-avoiding walks in $\mathbb{Z}^2$. If G is any graph, not necessarily transitive, we may again form the tree T^{SAW} of self-avoiding walks of G, where all walks begin at some fixed base point, o. The lower growth rate of this tree, $\mu(G) := \underline{\mathrm{gr}}\, T^{\mathrm{SAW}}$, is called the **connective constant** of G. This does not depend on choice of o, although that will not matter for us. If G is transitive, then T^{SAW} is 0-subperiodic, whence $\mu(G) = \mathrm{br}\, T^{\mathrm{SAW}}$ by Theorem 3.8.

---

* See Exercise 4.46(a) and Section 10.2 for more on this concept.

Proposition 7.12. *For any connected, infinite graph G, we have $p_c(G) \geq 1/\mu(G)$. In particular, if G has bounded degree, then $p_c(G) > 0$.*

Proof. In view of Proposition 7.10, it suffices to prove the inequality for bond percolation.

Let $K(o)$ be the component of o in Bernoulli(p) percolation. Write $K_n(o)$ for the self-avoiding walks of length n within $K(o)$. Thus, $K_n(o) \subseteq T_n^{SAW}$. If $K(o)$ is infinite, then $K_n(o) \neq \varnothing$ for each n. It follows that

$$\theta_o^{bond}(p) \leq \mathbf{P}_p[K_n(o) \neq \varnothing] \leq \mathbf{E}_p\big[|K_n(o)|\big] = |T_n^{SAW}|p^n \,.$$

Taking nth roots and letting $n \to \infty$, we obtain $1 \leq \mu(G)p$ whenever $\theta_o^{bond}(p) > 0$. In particular, this holds for $p > p_c^{bond}$.   ◄

Upper bounds for p_c are more difficult. In fact, we have seen examples of graphs (trees) that have exponential growth, bounded degree, and $p_c = 1$. The transitive case is thought to be better behaved, according to the following conjecture of Benjamini and Schramm (1996b):

Conjecture 7.13. (Quadratic Growth Yields a Phase Transition) *If G is a transitive graph with at least quadratic growth (that is, the ball of radius n grows at least as fast as some quadratic function of n), then $p_c(G) < 1$.*

This conjecture was finally proved by Duminil-Copin, Goswami, Raoufi, Severo, and Yadin (2020), who showed the stronger result that $p_c(G) < 1$ for every G of bounded degree that satisfies

$$\sup_{\substack{x \in V(G), \\ n \in \mathbb{N}}} p_n(x, x) \cdot n^{d/2} < \infty \tag{7.6}$$

for some $d > 4$, where $p_n(x, x)$ is the probability that simple random walk starting from x is back at x after exactly n steps. See Corollary 6.32 for an expansion condition that guarantees (7.6). The fact that this expansion condition holds on transitive graphs of at least quintic growth rate follows from Lemma 10.46 and Proposition 8.14. The proof of Duminil-Copin, Goswami, Raoufi, Severo, and Yadin (2020) relies on a novel connection to the wired canonical Gaussian field (Exercise 10.31) and percolation in a random environment.

We will prove "most" cases of this conjecture. Given a group, certainly the value of p_c depends on which generators are used. Nevertheless, whether $p_c < 1$ does not depend on the generating set chosen. To prove this, consider two generating sets, S_1 and S_2, of a group Γ. Let G_1 and G_2 be the corresponding Cayley graphs. We will transfer Bernoulli percolation on G_1 to a dependent percolation on G_2 by making an edge of G_2 open iff a certain corresponding path in G_1 is open. We then want to compare this dependent percolation on G_2 to Bernoulli percolation on G_2. To do this, we use a weak form of a comparison result of Liggett, Schonmann, and Stacey (1997). That, in turn, will rely on the following general coupling principle:

▷ **Exercise 7.7.**
Suppose that $\mathbf{P}_i$ ($i = 1, 2, 3$) are three percolation measures on 2^A such that $\mathbf{P}_1 \succcurlyeq \mathbf{P}_2$ and $\mathbf{P}_2 \succcurlyeq \mathbf{P}_3$. Show that there exist random variables ω_i on a common probability space such that $\omega_i \sim \mathbf{P}_i$ for all i and $\omega_1 \geq \omega_2 \geq \omega_3$ a.s.

In the following comparison principle, Bernoulli percolation on a set A is transferred to a dependent percolation on a set B, where the dependency is given by a set $D \subseteq A \times B$. The resulting dependent percolation on B then dominates a Bernoulli percolation on B whose parameter depends on the amount of dependencies introduced.

Proposition 7.14. *Write $\mathbf{P}_p^A$ for the Bernoulli(p) product measure on a set 2^A. Let A and B be two sets and $D \subseteq A \times B$. Write $D_a := (\{a\} \times B) \cap D$ and $D^b := (A \times \{b\}) \cap D$. Suppose that $m := \sup_{a \in A} |D_a| < \infty$ and that $n := \sup_{b \in B} |D^b| < \infty$. Given $0 < p < 1$, let*

$$q := \left(1 - (1 - p)^{1/m}\right)^n .$$

Given $\omega \in 2^A$, define $\omega' \in 2^B$ by

$$\omega'(b) := \begin{cases} \min\{\omega(a) ;\ (a, b) \in D^b\} & \text{if } D^b \neq \varnothing \\ 1 & \text{if } D^b = \varnothing. \end{cases}$$

Let $\mathbf{P}$ be the law of ω' when ω has the law of $\mathbf{P}_p^A$. Then $\mathbf{P}$ stochastically dominates $\mathbf{P}_q^B$.

Proof. Let η have law $\mathbf{P}_{q^{1/n}}^{A \times B}$. Define

$$\zeta(a) := \begin{cases} \max \eta \restriction D_a & \text{if } D_a \neq \varnothing \\ 0 & \text{if } D_a = \varnothing \end{cases} \qquad \text{and} \qquad \zeta'(b) := \begin{cases} \min \eta \restriction D^b & \text{if } D^b \neq \varnothing \\ 1 & \text{if } D^b = \varnothing. \end{cases}$$

Then the collection ζ' has a law μ' that dominates $\mathbf{P}_q^B$, since $\zeta'(b)$ are mutually independent for $b \in B$ and

$$\mathbf{P}_{q^{1/n}}^{A \times B}[\zeta'(b) = 1] = (q^{1/n})^{|D^b|} \geq q .$$

Similarly, $\mathbf{P}_p^A$ dominates the law μ of ζ, since $\zeta(a)$ are mutually independent for $a \in A$ and

$$\mathbf{P}_{q^{1/n}}^{A \times B}[\zeta(a) = 0] = (1 - q^{1/n})^{|D_a|} = (1 - p)^{|D_a|/m} \geq 1 - p .$$

Therefore, if $\omega \sim \mathbf{P}_p^A$, then we may couple ω and ζ so that $\omega \geq \zeta$. Since $\zeta(a) \geq \eta(a, b)$, we have that for each fixed b with $D^b \neq \varnothing$, under our coupling (as extended by Exercise 7.7),

$$\omega'(b) = \min\{\omega(a) ;\ (a, b) \in D^b\} \geq \min\{\zeta(a) ;\ (a, b) \in D^b\} \geq \min \eta \restriction D^b = \zeta'(b) ,$$

whereas $\omega'(b) = 1 = \zeta'(b)$ if $D^b = \varnothing$. It follows that $\mathbf{P} \succcurlyeq \mu' \succcurlyeq \mathbf{P}_q^B$, as desired. ◀

Theorem 7.15. *Let S_1 and S_2 be two finite generating sets for a countable group Γ, yielding corresponding Cayley graphs G_1 and G_2. Then $p_c(G_1) < 1$ iff $p_c(G_2) < 1$.*

Proof. Left and right Cayley graphs with respect to a given set of generators are isomorphic via $x \mapsto x^{-1}$, so we consider only right Cayley graphs. We also give the proof only for bond percolation, as site percolation is treated analogously.

Express each element $s \in S_2$ in terms of a minimal-length word $\varphi(s)$ using letters from S_1. If $s^{-1} \in S_2$, then we take $\varphi(s^{-1}) = \varphi(s)^{-1}$. Let ω_1 be Bernoulli(p) percolation on G_1, and define ω_2 on G_2 by letting $[x, xs] \in \omega_2$ (for $s \in S_2$) iff the path from x to xs in G_1 determined by $\varphi(s)$ lies in ω_1. Then we may apply Proposition 7.14 with A the set of edges of G_1, B the set of edges of G_2, and D the set of pairs (e, e') such that e is used in a path between the endpoints of e' determined by some $\varphi(s)$ for an appropriate $s \in S_2$. Clearly $n = \max\{|\varphi(s)| ;\ s \in S_2\}$ and m is finite, whence we may conclude that ω_2 stochastically dominates Bernoulli(q) percolation on G_2. Since o lies in an infinite cluster with respect to ω_1 if it lies in an infinite cluster with respect to ω_2, it follows that if $q > p_c(G_2)$, then $p \geq p_c(G_1)$, showing that $p_c(G_2) < 1$ implies $p_c(G_1) < 1$. ◀

It is easy to see that the preceding proof extends beyond Cayley graphs to cover the case of two graphs that are roughly isometric and have bounded degrees.

We now show that the Euclidean lattices $\mathbb{Z}^d$ ($d \geq 2$) have a true phase transition for Bernoulli percolation in that $p_c < 1$. The argument again relies on a first-moment calculation, but the new ingredient is planar duality, which introduces "contours."

Theorem 7.16. *For all $d \geq 2$, we have $p_c(\mathbb{Z}^d) < 1$.*

Proof. We use the standard generators of $\mathbb{Z}^d$. In view of Proposition 7.11, it suffices to consider bond percolation. Also, since $\mathbb{Z}^d$ contains a copy of $\mathbb{Z}^2$, it suffices to prove the result for $d = 2$. Consider the plane dual lattice $(\mathbb{Z}^2)^\dagger$. (See Section 6.5 for the definition.) To each configuration ω in 2^{E}, we associate the dual configuration $\omega^\times$ in $2^{\mathsf{E}^\dagger}$ by $\omega^\times(e^\dagger) := 1 - \omega(e)$. Those edges of $(\mathbb{Z}^2)^\dagger$ that lie in $\omega^\times$ we call "open." If $K(o)$ is finite, then $\partial_{\mathsf{E}} K(o)^\dagger$ contains a simple cycle of open edges in $(\mathbb{Z}^2)^\dagger$ that surrounds o. Now each edge in $(\mathbb{Z}^2)^\dagger$ is open with probability $1 - p$. Furthermore, the number of simple cycles in $(\mathbb{Z}^2)^\dagger$ of length n surrounding o is at most $n3^n$, since each one must intersect the x-axis somewhere in $(0, n)$. Let $M(n)$ be the total number of open, simple cycles in $(\mathbb{Z}^2)^\dagger$ of length n surrounding o. Then

$$1 - \theta(p) = \mathbf{P}_p\big[|K(o)| < \infty\big] = \mathbf{P}_p\bigg[\sum_n M(n) \geq 1\bigg]$$

$$\leq \mathbf{E}_p\bigg[\sum_n M(n)\bigg] = \sum_n \mathbf{E}_p[M(n)] \leq \sum_n (n3^n)(1 - p)^n .$$

By choosing p sufficiently close to 1, we may make this last sum less than 1, and that makes $\theta(p) > 0$. ◀

The preceding proof introduced the dual percolation $\omega^\times$, which is another Bernoulli percolation. When $p = 1/2$, the dual percolation has the same law as the original percolation. Some thought may suggest from this that $p_c^{\text{bond}}(\mathbb{Z}^2) = 1/2$. Although we will not prove the full theorem that $p_c^{\text{bond}}(\mathbb{Z}^2) = 1/2$ and $\theta(p_c, \mathbb{Z}^2) = 0$, we will give a beautiful proof due to Zhang, taken from Grimmett (1999), of Harris's theorem that $\theta(1/2, \mathbb{Z}^2) = 0$. This implies that $p_c^{\text{bond}}(\mathbb{Z}^2) \geq 1/2$; a short proof of the converse inequality is given by Duminil-Copin and Tassion (2016b). The proof we are about to give of Harris's theorem uses the self-duality of Bernoulli(1/2) percolation and the uniqueness of the infinite cluster, if there were one.

Theorem 7.17. *For bond percolation on $\mathbb{Z}^2$, we have $\theta(1/2) = 0$.*

Proof. Assume for a contradiction that $\theta(1/2) > 0$. By Theorem 7.6, there is a unique infinite cluster a.s. Let B be a square box with sides parallel to the axes in $\mathbb{Z}^2$. Let A be the event that B intersects the infinite cluster. Then $\mathbf{P}_{1/2}(A)$ is arbitrarily close to 1 provided we take B large enough. Let A_i ($1 \leq i \leq 4$) be the event that there is an infinite path in ω whose only intersection with B is on the ith side of B. These events are increasing and have equal probability. Since $A^c = \bigcap_{i=1}^4 A_i^c$, it follows from Exercise 5.18 that $\mathbf{P}_{1/2}(A_i)$ are also arbitrarily close to 1 when B is large. As in the proof of Theorem 7.16, to each configuration ω in 2^{E}, we associate the dual configuration $\omega^\times$ in $2^{\mathsf{E}^\dagger}$ by $\omega^\times(e^\dagger) := 1 - \omega(e)$. Let B' be the smallest square box in the dual lattice $(\mathbb{Z}^2)^\dagger$ that contains B in its interior. Then similar statements apply to the sides of B' with respect to $\omega^\times$. In particular, the probability is close to 1 when B is large that there are infinite paths in ω from the left and right sides of B and simultaneously infinite paths in $\omega^\times$ from the top and bottom sides of B'. See Figure 7.5. However, on the event that these four things occur simultaneously, there cannot be a unique infinite cluster in both $\mathbb{Z}^2$ and its dual, which is the contradiction we sought.　◄

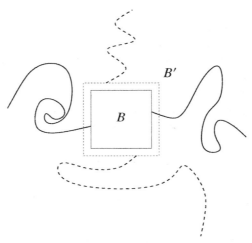

Figure 7.5. Infinite open paths from two sides of B and infinite closed paths from two sides of B'.

In Corollary 7.19 and Theorem 7.20, we establish that $p_c < 1$ for the "majority" of groups. The following language will be useful. We say that a group **almost** has a property if it has a subgroup of finite index that has the property in question. We will use the following fact:

Theorem 7.18. *If Γ is a finitely generated group of at most polynomial growth, then either Γ is almost (isomorphic to) $\mathbb{Z}$ or Γ contains a subgroup isomorphic to $\mathbb{Z}^2$.*

See Section 7.9 for the proof, which relies on a deep theorem of Gromov (1981a). Combining this with Theorems 7.16 and 7.15, we may immediately deduce what we want for groups of polynomial growth:

Corollary 7.19. (Polynomial Growth Yields a Phase Transition) *If G is a Cayley graph of a group Γ with at most polynomial growth, then either Γ is almost (isomorphic to) $\mathbb{Z}$ or $p_c(G) < 1$.*

A proof that avoids Gromov's theorem and that holds for roughly transitive graphs of polynomial growth was given by Candellero and Teixeira (2018).

Now we continue with groups of exponential growth, which are quite a bit easier to handle given what we know about trees. This result is due to Lyons (1995).

Theorem 7.20. (Exponential Growth Yields a Phase Transition) *If G is a Cayley graph of a group with exponential growth, then $p_c(G) < 1$.*

Proof. Let b be the exponential growth rate of balls in G. Recall the construction in Section 3.4 (p. 90) of the lexicographically-minimal spanning tree, $T^{\text{lexmin}}(G)$. This subperiodic tree is isomorphic to a subgraph of G, whence

$$p_c(G) \leq p_c(T^{\text{lexmin}}) = 1/\text{br}(T^{\text{lexmin}}) = 1/\text{gr}(T^{\text{lexmin}}) = 1/b < 1 \qquad (7.7)$$

by Theorems 5.15 and 3.8. ◀

▷ **Exercise 7.8.**
Extend Theorem 7.20 to quasi-transitive graphs with exponential growth.

The reason that Corollary 7.19 and Theorem 7.20 do not cover all groups is that there exist groups that have superpolynomial growth but subexponential growth. These are called groups of "intermediate growth." The examples constructed by Grigorchuk (1983) also have $p_c < 1$ (Muchnik and Pak, 2001). Until Conjecture 7.13 was proved by Duminil-Copin, Goswami, Raoufi, Severo, and Yadin (2020), the only cases among Cayley graphs of Conjecture 7.13 that had remained unresolved were those of intermediate growth.

7.5 Merging Infinite Clusters and Invasion Percolation

As p is increased, Bernoulli(p) percolation adds more edges, whence if there is an infinite cluster a.s. at p_0, then there is also an infinite cluster a.s. at each $p > p_0$. This intuition was made precise by using the standard coupling defined in Section 5.2. How does the number of infinite clusters vary as p is increased through the range where there is at least one infinite cluster? There are two competing features of increasing p: the fact that *finite* clusters can join up means that new infinite clusters could form, while the fact that infinite clusters can join up means that the number of infinite clusters could decrease.

For general graphs, either one of these two features could be dominant. For example, if the Cayley graph of $\mathbb{Z}^2$ is joined by an edge to a tree T with branching number in $\left(1, 1/p_c(\mathbb{Z}^2)\right)$, then for $p_c(\mathbb{Z}^2) < p < 1/\text{br}\,T$, there is a unique infinite cluster, while for $1/\text{br}\,T < p < 1$, there are infinitely many infinite clusters (see Exercise 7.37). On the other hand, we will see in Theorem 7.37 that for graphs that are the product of $\mathbb{Z}$ with a regular tree of large degree, the number of infinite clusters is first 0 in an interval of p, then ∞, and then 1. In Theorem 8.24, we will see the same for the self-dual, hyperbolic pentagonal-tessellation graph shown in Figure 2.4. By combining examples such as these, one can obtain more complicated behavior of the number of infinite clusters.

However, on quasi-transitive graphs, it turns out that once there is a unique infinite cluster, then that remains the case for all larger values of p. To prove this, we use the standard coupling and prove something stronger, due to Schonmann (1999b) and Häggström, Peres, and Schonmann (1999). It shows that the dominant feature of percolation clusters is that the infinite ones grow faster than finite ones can create new infinite clusters. Let $\Theta(G)$ be the set of $p \in [0, 1]$ for which there is an infinite cluster $\mathbf{P}_p$-a.s. on G (so that $p_c(G) = \inf \Theta(G)$). It is conjectured that $\Theta(G) = \left(p_c(G), 1\right]$ when G is quasi-transitive and $p_c(G) < 1$: see Conjecture 8.15.

Theorem 7.21. (Merging Infinite Clusters) *Let G be a quasi-transitive graph. If $p_1 \in \Theta(G)$ is such that there exists a unique infinite cluster $\mathbf{P}_{p_1}$-a.s., then for all $p_2 > p_1$, there is a unique infinite cluster $\mathbf{P}_{p_2}$-a.s. Furthermore, in the standard coupling of Bernoulli percolation processes, a.s. simultaneously for all $p_1, p_2 \in \Theta(G)$ with $p_2 > p_1$, every infinite p_2-cluster contains an infinite p_1-cluster.*

If we define

$$p_u(G) := \inf\{p; \text{ there is a.s. a unique infinite cluster in Bernoulli}(p) \text{ percolation}\}, \quad (7.8)$$

then it follows from Theorem 7.21 that when G is a quasi-transitive graph,

$$p_u(G) = \sup\{p; \text{ there is a.s. not a unique infinite cluster in Bernoulli}(p) \text{ percolation}\}. \quad (7.9)$$

Of course, $p_c(G) \le p_u(G) \le 1$ and two new questions immediately arise: When is $p_u < 1$? When is $p_c < p_u$? We address these questions in Sections 7.6 and 7.7.

Our first order of business, however, is to prove Theorem 7.21. For this, we introduce invasion percolation, which will also be important in our study of minimal spanning forests in Chapter 11. There are two types of invasion percolation, bond and site. We describe first invasion bond percolation. Fix distinct labels $U(e) \in [0, 1]$ for $e \in \mathsf{E}$, usually a sample of independent, uniform random variables. Fix a vertex x. Imagine $U(e)$ as the height of e and the height of x as 0. If we pour water gradually on x, then when the water reaches the height of the lowest edge incident to x, water will flow along that edge. As we keep pouring, the water will continue to "invade" the lowest edge that it touches. This is invasion percolation. More precisely, let $I_1(x)$ be the lowest edge incident to x and define $I_n(x)$ recursively thereafter by letting $I_{n+1}(x) := I_n(x) \cup \{e\}$, where e is the lowest among the edges that are not in $I_n(x)$ but are adjacent to (some edge of) $I_n(x)$. Finally, define the ***invasion basin*** of x to be $I(x) := \bigcup_n I_n(x)$. This is the set of all edges invaded from x. Invasion site percolation is similar, but the vertices are labeled rather than the edges and the invasion basin is a set of vertices rather than a set of edges. For invasion site percolation, we start with $I_1(x) := \{x\}$.

Given the labels $U(e)$, we have the usual subgraphs $\omega_p := \{e; U(e) < p\}$ that form the standard coupling of Bernoulli percolation. The first connection of invasion percolation to Bernoulli percolation is that if x belongs to an infinite cluster η of ω_p, then $I(x) \subseteq \eta$ (this is clear). Similarly, if there is some edge e adjacent to $I(x)$ but not in $I(x)$ that satisfies $U(e) < p$, then $|I(x) \cap \eta| = \infty$ for some infinite cluster η of ω_p. A much deeper connection is the following result of Häggström, Peres, and Schonmann (1999):

Theorem 7.22. (Invasion of Infinite Clusters) *Let G be an infinite, quasi-transitive graph. Then in the standard coupling of Bernoulli percolation processes, a.s. for all $p > p_c(G)$ and all $x \in \mathsf{V}$, there is some infinite p-cluster that intersects $I(x)$.*

The proof of Theorem 7.22 is rather tricky. The main steps will be to show that $I(x)$ contains arbitrarily large balls, that it comes infinitely often within distance 1 of infinite p-clusters, and finally that it actually invades some infinite p-cluster. We present the proof for invasion bond percolation, as the site case is similar.

The first step does not require quasi-transitivity. Recall that $E_R(x)$ denotes the set of edges that have at least one endpoint at distance at most $R - 1$ from x. From now on, invasion percolation uses the edge labels from the standard coupling of Bernoulli percolation.

Lemma 7.23. (Invasion of Large Balls) *Let G be any infinite graph with bounded degrees, $x \in V$, and $R \in \mathbb{N}$. Then a.s. the invasion basin of x contains $E_R(y)$ for some $y \in V$.*

Proof. The idea is that as the invasion from x proceeds, it will encounter balls of radius R for the first time infinitely often. Each time, the ball might have all its labels fairly small, in which case that ball would eventually be contained in the invasion basin. Since the encountered balls can be chosen far apart, these events are independent, and therefore one of them will eventually happen.

To make this precise, fix an enumeration of V. Let d be the maximum degree in G,

$$S_r := \{ y \,;\; \mathrm{dist}_G(x, y) = r \} ,$$

and

$$\tau_n := \inf \left\{ k \,;\; \mathrm{dist}_G\big(I_k(x), S_{2nR}\big) = R \right\} .$$

Since $I(x)$ is infinite, $\tau_n < \infty$ for all $n \in \mathbb{N}$. Let Y_n be the first vertex in the enumeration of V such that $Y_n \in S_{2nR}$ and $\mathrm{dist}_G\big(I_{\tau_n}(x), Y_n\big) = R$. Let A be the event of probability 1 that there is no infinite p-cluster for any $p < p_c(G)$ and consider the events

$$A_n := \big[\forall e \in E_R(Y_n) \;\; U(e) < p_c(G) \big] .$$

On the event $A \cap A_n$, we have $E_R(Y_n) \subset I(x)$, so that it suffices to show that a.s. some A_n occurs. The sets $E_R(Y_n)$ are disjoint for different n. Also, the invasion process up to time τ_n gives no information about the labels of $E_R(Y_n)$. Since $|E_R(y)| \le d^R$ for all y, it follows that $\mathbf{P}(A_n \mid Y_n, A_1, A_2, \ldots, A_{n-1}) = p_c(G)^{|E_R(Y_n)|} \ge p_c(G)^{d^R}$, so that $\mathbf{P}(A_n \mid A_1, A_2, \ldots, A_{n-1}) \ge p_c(G)^{d^R}$. Since $p_c(G) > 0$ by Proposition 7.12, it follows that a.s. some A_n occurs. ◄

We next need an extension of insertion and deletion tolerance that holds for the edge labels. It allows us to change the labels on a finite set of edges while maintaining the positivity of the probability of events.

Lemma 7.24. (Change Tolerance of Labels) *Let X be a denumerable set and A be an event in $[0, 1]^X$. Let $\mathcal{L}_X$ be the product of Lebesgue measures on $[0, 1]^X$. Given a finite $Y \subset X$ and a continuously differentiable, injective map $\varphi \colon [0, 1]^Y \to [0, 1]^Y$ with a.e. nonzero Jacobian determinant, let*

$$A' := \left\{ \omega \in [0, 1]^X \,;\; \exists \eta \in A \;\; \omega{\restriction}(X \setminus Y) = \eta{\restriction}(X \setminus Y) \text{ and } \omega{\restriction}Y = \varphi(\eta{\restriction}Y) \right\} .$$

Then $\mathcal{L}_X(A') > 0$ if $\mathcal{L}_X(A) > 0$.

Proof. Regard $[0, 1]^X$ as $[0, 1]^Y \times [0, 1]^{X \setminus Y}$. Apply Fubini's theorem to this product to see that it suffices to prove the case where $X = Y$. But this case follows from the usual change-of-variable formula

$$\mathcal{L}_X(A') = \int_A |J| \, d\mathcal{L}_X ,$$

where J is the Jacobian determinant of φ. ◄

We now come to the second step, proving that $I(x)$ comes infinitely often within distance 1 of infinite p-clusters. Let $\xi_p(x)$ be the number of edges $[y, z]$ with $y \in I(x)$ and z in some infinite p-cluster. (It is irrelevant to the definition whether $z \in I(x)$, but if $z \in I(x)$ and in some infinite p-cluster, then $\xi_p(x) = \infty$.)

Lemma 7.25. *If G is a quasi-transitive graph, $p \in \Theta(G)$, and $x \in \mathsf{V}$, then $\xi_p(x) = \infty$ a.s.*

Proof. Let $p \in \Theta(G)$ and $\epsilon > 0$. Because G is quasi-transitive, there is some R so that

$$\forall y \in \mathsf{V} \quad \mathbf{P}\big[\text{some infinite } p\text{-cluster comes within distance } R \text{ of } y\big] \geq 1 - \epsilon. \qquad (7.10)$$

This is the only way that quasi-transitivity will enter the proof. Indeed, the rest of the section will not use quasi-transitivity at all but only that $\xi_p(x) = \infty$ a.s. However, we need to reformulate (7.10). Given a set of edges F, write $V(F)$ for the set of its endpoints. By (7.10), if $F \subset \mathsf{E}$ and F contains some set $E_R(y)$, then with probability at least $1 - \epsilon$, the set F is adjacent to some infinite p-cluster. Let $A(F)$ be the event that there is an infinite path in ω_p that starts at distance 1 from $V(F)$ and that does not use any vertex in $V(F)$. Our reformulation of (7.10) is that if $F \subset \mathsf{E}$ is *finite* and contains some set $E_R(y)$, then $\mathbf{P}[A(F)] \geq 1 - \epsilon$.

By Lemma 7.23, there is a.s. a smallest finite k for which $I_k(x)$ contains some $E_R(y)$. Call this smallest time τ. Since the invasion process up to time τ gives no information about the labels of edges that are not adjacent to $I_\tau(x)$, we have $\mathbf{P}\big[A(I_\tau(x)) \mid I_\tau(x)\big] \geq 1 - \epsilon$, whence $\mathbf{P}\big[A(I_\tau(x))\big] \geq 1 - \epsilon$. Since $\xi_p(x) \geq 1$ on the event $A(I_\tau(x))$ and ϵ was arbitrary, we obtain $\mathbf{P}[\xi_p(x) \geq 1] = 1$.

Now suppose for a contradiction that $\mathbf{P}[\xi_p(x) = n] > 0$ for some finite n. Then there is a set F_1 of n edges such that $\mathbf{P}(A_1) > 0$ for the event A_1 that F_1 is precisely the set of edges joining $I(x)$ to an infinite p-cluster. Among the edges adjacent to F_1, there is a nonempty set F_2 of edges such that $\mathbf{P}(A_1 \cap A_2) > 0$ for the event A_2 that F_2 is precisely the set of edges adjacent to F_1 that belong to an infinite p-cluster. Now changing the labels $U(e)$ for $e \in F_2$ to be $p + (1-p)U(e)$ changes $A_1 \cap A_2$ to an event A_3 where $\xi_p(x) = 0$, since $I(x)$ on $A_1 \cap A_2$ is the same as $I(x)$ on the corresponding configuration in A_3. By Lemma 7.24, $\mathbf{P}(A_3) > 0$, so $\mathbf{P}[\xi_p(x) = 0] > 0$. This contradicts the preceding conclusion. Therefore, $\xi_p(x) = \infty$ a.s. ◄

Proof of Theorem 7.22. First fix p_1 and p_2 with $p_2 > p_1 > p_c(G)$ and fix $x \in \mathsf{V}$. Given a labeling of the edges, color an edge blue if it lies in an infinite p_1-cluster. Color an edge red if it is adjacent to some blue edge but is not itself blue. Observe that for red edges e, the labels $U(e)$ are independent and uniform on $[p_1, 1]$.

Now consider invasion from x given this coloring information. We claim that a.s., some edge of $I(x)$ is adjacent to some colored edge e with label $U(e) < p_2$. To see this, note that by Lemma 7.25, $\xi_{p_1}(x) = \infty$ a.s., so that infinitely many edges of $I(x)$ are adjacent to colored edges. When invasion from x first becomes adjacent to a colored edge, e, the distribution of $U(e)$, conditional on the colors of *all* edges and on the invasion process so far, is still concentrated on $[0, p_1)$ if e is blue or is uniform on $[p_1, 1]$ if e is red. Hence $U(e) < p_2$ with conditional probability at least $(p_2 - p_1)/(1 - p_1) > 0$. Since the same holds *every* time a colored edge is encountered, there must be one colored edge with label less than p_2 a.s. that is adjacent to some edge of $I(x)$. This proves the claim.

Now the claim we have just established implies that $I(x)$ a.s. intersects some infinite p_2-cluster by the observation preceding the statement of Theorem 7.22.

Apply this result to a sequence of values of p_2 approaching $p_c(G)$ to get the theorem. ◀

Before we can deduce Theorem 7.21, we need one more lemma.

Lemma 7.26. *If G is a quasi-transitive graph, $p_1 < p_2$ are fixed with $p_1, p_2 \in \Theta(G)$, then a.s. every infinite p_2-cluster contains an infinite p_1-cluster.*

Proof. Consider a labeling U of the edges. Because of Lemma 7.25, we know that for each vertex x in an infinite p_2-cluster C, there are a.s. infinitely many edges $[y, z]$ with $y \in I(x) \subseteq C$ and z in some infinite p_1-cluster. If C does not contain an infinite p_1-cluster, then it must be that C is adjacent via infinitely many edges e to infinite p_1-clusters a.s. Those edges e must then have labels $U(e) \geq p_2$.

To show that this is impossible, condition on the set F of all edges e having label $U(e) < p_1$. All other edges have i.i.d. distribution uniform on $[p_1, 1]$ conditional on F. Given any vertex x that does not belong to an infinite cluster in F (that is, an infinite p_1-cluster), grow the p_2-cluster of x one edge at a time: Choose any deterministic ordering $\langle e_1, e_2, \ldots \rangle$ of E. Define $E_0 := F_0 := \varnothing$. We will look at a finite or infinite subsequence of edges $\langle e_{n_j} \rangle$ via a recursive procedure and discover the p_2-cluster C of x at the end. Suppose that the edges $E_k := \langle e_{n_1}, \ldots, e_{n_k} \rangle$ have been looked at and that F_k has been defined. Let V_k be the union of $\{x\}$ and the endpoints of the edges of E_k whose labels are less than p_2. Let n_{k+1} be the smallest index of an edge in $E \setminus E_k$ that has an endpoint in V_k, if any. If there are none, then stop; C is finite with vertex set V_k. Otherwise, define

$$F_{k+1} := \begin{cases} F_k \cup \{n_{k+1}\} & \text{if } e_{n_{k+1}} \text{ has exactly one endpoint in an infinite } F\text{-cluster} \\ F_k & \text{otherwise.} \end{cases}$$

Define $F_\infty := \bigcup F_k$. Let $\langle Y_i ; i \geq 1 \rangle$ be the random variables $U(e_j)$ for $j \in F_\infty$ ordered in the same order as F_∞. The random variables Y_i are i.i.d. uniform on $[p_1, 1]$ (consider the distribution of Y_i given all Y_m with $m < i$). If C is infinite and does not contain an infinite F-cluster, then F_∞ is infinite, as we said at the start of the proof. If F_∞ is infinite, then $Y_i < p_2$ a.s. for some $i \geq 1$. But this implies that C does contain an infinite F-cluster. So this proves the lemma.   ◀

Proof of Theorem 7.21. We give the proof for bond percolation, as site percolation is treated in an identical manner. We consider first only pairs $p_c(G) < p_1 < p_2$ and use Theorem 7.22. Let A be the event of probability 1 that all edge labels are distinct and that for all $p > p_c(G)$ and all $x \in V$, there is some infinite p-cluster that intersects $I(x)$. On A, for each infinite cluster η_2 of ω_{p_2} and each $x \in \eta_2$, there is some infinite cluster η_1 of ω_{p_1} that intersects $I(x)$. Since $I(x) \subseteq \eta_2$, it follows that η_1 intersects η_2, whence $\eta_1 \subseteq \eta_2$ on A, as desired.

To finish the proof, we need to consider pairs $p_c(G) = p_1 < p_2$ in case $p_c(G) \in \Theta(G)$. But for each p_2, this is the statement of Lemma 7.26, whence it holds simultaneously for any countable collection of p_2. So let $\langle p_n ; n \geq 2 \rangle$ be converging to p_1 from above. We have that a.s. every pair (p_1, p_n) satisfies the conclusion, and by what we have shown in the preceding paragraph, also simultaneously every pair (p_n, p) with $n \geq 2$ and $p_n < p$ satisfies the conclusion. However, whenever a p-cluster contains an infinite p_n-cluster and that p_n-cluster in turn contains an infinite p_1-cluster, then the p-cluster also contains an infinite p_1-cluster, so we are done.   ◀

7.6 **Upper Bounds for** p_u

If G is a regular tree, then it is easy to see that $p_u(G) = 1$. (In fact, this is true for any tree; see Exercise 7.37.) What about the Cayley graph of a free group with respect to a non-free generating set? A special case of the following exercise (combined with Exercise 7.4) shows that such a Cayley graph still has $p_u(G) = 1$.

▷ **Exercise 7.9.**
    (a) Show that if ω is an a.s. nonempty, $\mathrm{Aut}(G)$-invariant percolation on a quasi-transitive graph G, then a.s. every end of G contains an end-convergent subset of ω.
    (b) Show that if G is a quasi-transitive graph with more than one end, then $p_u(G) = 1$.

In fact, one can show that the property $p_u(G) < 1$ does not depend on the generating set for any group (Lyons and Schramm, 1999), but this would be subsumed by the following conjecture, suggested by a question of Benjamini and Schramm (1996b):

Conjecture 7.27. *If G is a quasi-transitive graph with one end, then $p_u(G) < 1$.*

This conjecture has been verified in many cases, most notably, when G is the Cayley graph of a finitely *presented* group. The present section is devoted to proving this case. (See Section 7.9 for a discussion of other known cases.) A key ingredient is the following combinatorial fact.

Lemma 7.28. *Let $k \geq 1$. Suppose that G is a graph that has a set K of cycles of length at most k such that every cycle belongs to the linear span (in $\ell^2_-(\mathsf{E})$) of K. Fix $y, z \in \mathsf{V}$. Let Π be a (possibly infinite) cutset of edges that separates y and z and that is minimal with respect to inclusion. Then, for each nontrivial partition $\Pi = \Pi_1 \cup \Pi_{-1}$, there exist vertices $x_i \in \mathsf{V}(\Pi_i)$ $(i = 1, -1)$ such that the distance between x_1 and x_{-1} is at most $k/2$.*

Note that when a presentation $\langle S \mid R \rangle$ has relators (elements of R) with length at most k, then the associated Cayley graph satisfies the hypothesis of Lemma 7.28. Timár (2007) showed that the hypothesis is invariant under rough isometries (up to changing the value of k). In particular, every Cayley graph of a finitely presentable group satisfies this hypothesis for some k.

Proof. It suffices to show that there is some cycle of K that intersects both Π_i. By the assumption of minimality, for $i = 1, -1$, there is a path $\mathcal{P}_i$ from y to z that does not intersect Π_{-i}. Considering $\mathcal{P}_i$ as elements of $\ell^2_-(\mathsf{E})$, we may write the cycle $\mathcal{P}_1 - \mathcal{P}_{-1}$ as a linear combination of cycles $\sum_{C \in K'} \alpha_C C$ for some finite set $K' \subset K$ with $\alpha_C \in \mathbb{R}$. Let K'_1 denote the cycles of K' that intersect Π_1 and $K'_{-1} := K' \setminus K'_1$. We have

$$\theta := \mathcal{P}_1 - \sum_{C \in K'_1} \alpha_C C = \mathcal{P}_{-1} + \sum_{C \in K'_{-1}} \alpha_C C.$$

The right-hand side is composed of a path and cycles that do not intersect Π_1, whence $\theta(e) = 0$ for all $e \in \Pi_1$. Since θ is a unit flow from y to z, it follows that $\theta(e) \neq 0$ for some $e \in \Pi_{-1}$. Since $\mathcal{P}_1$ does not include an edge from Π_{-1}, we deduce that some cycle in K'_1 does. This provides precisely the sort of cycle we sought. ◀

Write $\widehat{\mathsf{E}} := \{(e, f) \in \mathsf{E} \times \mathsf{E} \; ; \; e, f \text{ are adjacent}\}$ to define the ***line graph****** of G as $\widehat{G} := (\mathsf{E}, \widehat{\mathsf{E}})$. Write $\mathsf{E}^k := \{(x, y) \in \mathsf{V} \times \mathsf{V} \; ; \; 1 \le \text{dist}_G(x, y) \le k\}$ to define the k-***fuzz*** of G as $G^k := (\mathsf{V}, \mathsf{E}^k)$. Finally, write $\widehat{G}^k := (\widehat{G})^k$. If every cycle in G can be written as a (finite) linear combination of cycles with length $< 2k$, then according to Lemma 7.28,

$$\text{every minimal cutset in } G \text{ is the vertex set of a connected subgraph of } \widehat{G}^k. \qquad (7.11)$$

Here, a cutset means one that separates two vertices of G.

Theorem 7.29. (Babson and Benjamini, 1999) *If G is the Cayley graph of a nonamenable, finitely presented group with one end, then $p_u(G) < 1$. In fact, the same holds for any nonamenable, quasi-transitive graph G with one end such that there is a set of cycles with bounded length whose linear span contains all cycles.*

An extension to the amenable case (where $p_u = p_c$) is given in Exercise 7.42. In particular, $p_c < 1$ for finitely presented, amenable Cayley graphs. However, since it is unknown whether there are any finitely presented groups of intermediate growth, it is not clear that this extension adds to what we know already (that $p_c < 1$ for groups other than possibly those of intermediate growth).

Proof. Let $2k$ be strictly larger than the lengths of cycles of some generating set. We do the case of bond percolation and prove that

$$p_u^{\text{bond}}(G) \le \max\{p_c^{\text{bond}}(G), 1 - p_c^{\text{site}}(\widehat{G}^k)\}.$$

The case of site percolation is left as Exercise 7.38. Choose $p < 1$ so that $p > p_c^{\text{bond}}(G)$ and $p > 1 - p_c^{\text{site}}(\widehat{G}^k)$. This is possible by Theorem 7.20, (6.7), and Proposition 7.12. Because G has only one end, any two open infinite clusters must be separated by a closed infinite minimal cutset (indeed, one within the edge boundary of one of the open infinite clusters). Combining this with (7.11), we obtain

$$\mathbf{P}_{p,G}^{\text{bond}}[\exists \text{ at least two open infinite clusters}] \le \mathbf{P}_{p,G}^{\text{bond}}[\exists \text{ a closed infinite minimal cutset in } G]$$

$$\le \mathbf{P}_{p,\widehat{G}^k}^{\text{site}}[\exists \text{ a closed infinite cluster in } \widehat{G}^k]$$

$$= \mathbf{P}_{1-p,\widehat{G}^k}^{\text{site}}[\exists \text{ an open infinite cluster in } \widehat{G}^k] = 0.$$

On the other hand, $\mathbf{P}_{p,G}^{\text{bond}}[\exists \text{ at least 1 open infinite cluster}] = 1$ since $p > p_c^{\text{bond}}(G)$. This proves the result. ◄

Remark 7.30. The proof shows that for every graph G with one end such that there is a set of cycles with bounded length whose linear span contains all cycles, if $p_c(G) < 1$, then $p_u(G) < 1$. For example, this applies to every plane graph with a bounded number of sides on its faces.

---

* Since site percolation on $\widehat{G}$ is equivalent to bond percolation on G, one might wonder why we study bond percolation separately. There are three principal reasons for this: One is that other sorts of bond percolation processes have no natural site analogue. Another is that planar duality is quite different for bonds than for sites. Third, various quantities for graphs would have differing values for line graphs; for example, in Theorem 6.46, $\Phi_{\mathsf{E}}^*(G) \ge \Phi_{\mathsf{V}}^*(\widehat{G})$, so using site percolation on $\widehat{G}$ might give a worse bound for $p_c^{\text{bond}}(G)$.

7.7 **Lower Bounds for** p_u

According to Theorem 7.6, when G is amenable and transitive, there can never be infinitely many infinite clusters, whence $p_c(G) = p_u(G)$. Behavior that is truly different from the amenable case arises when there *are* infinitely many infinite clusters. This has been conjectured always to be the case on nonamenable, transitive graphs for some interval of p:

Conjecture 7.31. (Benjamini and Schramm, 1996b) *If G is a quasi-transitive, nonamenable graph, then $p_c(G) < p_u(G)$.*

It is not easy even to give examples of quasi-transitive graphs where $0 < p_c < p_u < 1$, and thus where we have three phases in Bernoulli percolation. But we will, and we will even prove that every nonamenable group has *some* Cayley graph where $p_c^{\text{bond}} < p_u^{\text{bond}}$. Our first tool will be the following lower bound for p_u. Recall that a **simple cycle** is a cycle that does not use any vertex or edge more than once. Let $a_n(G)$ be the number of simple cycles of length n in G that contain o and

$$\gamma(G) := \limsup_{n\to\infty} a_n(G)^{1/n}, \tag{7.12}$$

where a cycle is counted only as a set without regard to ordering or orientation. (We do not know whether $\lim_{n\to\infty} a_{2n}(G)^{1/2n}$ exists for every transitive, or even every Cayley, graph.)

Theorem 7.32. *If G is a transitive graph, then*

$$p_u(G) \geq 1/\gamma(G). \tag{7.13}$$

Proof. We give the proof for site percolation, the proof for bond percolation being similar. Let ω_p for $0 \leq p \leq 1$ be the standard coupling associated to independent, uniform $[0, 1]$ random variables $U(x)$ indexed by V. Consider $p > p_u$. To show that $p_u(G) \geq \liminf_{n\to\infty} a_n(G)^{-1/n}$, we will show that $\sum_n a_n(G)p^n = \infty$.

First, since ω_p contains a.s. a unique infinite cluster, that infinite cluster K has only one end: otherwise, removing a finite number of sites would create more than one infinite cluster and deletion tolerance would give more than one infinite cluster with positive probability.

Second, with positive probability, there are two (edge- and vertex-) disjoint, infinite rays in K. To see this, suppose not. Then by Menger's theorem (Exercise 3.16), for every vertex $x \in K$, a.s. there would be infinitely many vertices x_n, each of whose removal would leave x in a finite open component. Now the law of $U(y)$ given ω_p is uniform on $[0, p]$ for $y \in \omega_p$. Since K is independent of all $U(y)$ given ω_p, it follows that the law of $U(y)$ given ω_p and that $y \in K$ is still uniform on $[0, p]$. The same holds for the law of $U(x_n)$, since the x_n are determined simply by x and K. Choose any $p' \in (p_c, p)$. Then given ω_p, given any such vertex x as earlier, and given any such vertices x_n, we have that $U(x_n) > p'$ a.s. for some n. This means that the cluster of x in $\omega_{p'}$ is finite a.s. The fact that this holds for all $x \in K$ contradicts $p' > p_c$.

Therefore, with positive probability, there are two infinite rays in ω_p starting at o that are disjoint except at o. Since K has only one end, the two rays may be connected by paths in ω_p that stay outside arbitrarily large balls. In particular, there are an infinite number of simple cycles in ω_p through o, whence the expected number of simple cycles through o in ω_p must be infinite. That is, $\sum_n a_n(G)p^n = \infty$.   ◄

It seems difficult to calculate $\gamma(G)$, although Hammersley (1961b) showed that $\gamma(\mathbb{Z}^d) = \mu(\mathbb{Z}^d)$ for all $d \geq 2$, where μ is the connective constant defined in Section 7.4; in particular, $\gamma(\mathbb{Z}^2) \approx 2.64$ (Madras and Slade, 1993). However, although it is crude, quite a useful estimate of $\gamma(G)$ arises from the spectral radius of G, which is much easier to calculate. For any graph G, recall that the ***spectral radius*** of (simple random walk on) G is defined to be

$$\rho(G) := \limsup_{n \to \infty} \mathbf{P}_o[X_n = o]^{1/n},$$

where $(\langle X_n \rangle, \mathbf{P}_o)$ denotes simple random walk on G starting at o.

▷ **Exercise 7.10.**
Show that for any graph G, the spectral radius satisfies $\rho(G) = \lim_{n \to \infty} \mathbf{P}_o[X_{2n} = o]^{1/2n}$ and $\mathbf{P}_o[X_n = o] \leq \rho(G)^n$ for all n.

Lemma 7.33. *For any graph G of maximum degree d, we have $\gamma(G) \leq \rho(G)d$.*

Proof. For any simple cycle $o = x_0, x_1, \ldots, x_{n-1}, x_n = o$, one way that simple random walk could return to o at time n is by following this cycle. That event has probability at least $1/d^n$. Therefore $\mathbf{P}_o[X_n = o] \geq a_n(G)/d^n$, which gives the result. ◄

Since every simple cycle is a nonbacktracking cycle, we also have the better inequality

$$\gamma(G) \leq \operatorname{cogr}(G). \tag{7.14}$$

Recall that Theorem 6.10 gives a formula for $\operatorname{cogr}(G)$ in terms of $\rho(G)$ when G is a transitive graph. It would be useful to have a still better estimate of $\gamma(G)$, even for $G = \mathbb{T}_{b+1} \square \mathbb{Z}$.

Proposition 7.34. (Spectral Radius of Products) *Let G and G' be transitive graphs of degrees d_G and $d_{G'}$. We have*

$$\rho(G \square G') = \frac{d_G}{d_G + d_{G'}} \rho(G) + \frac{d_{G'}}{d_G + d_{G'}} \rho(G').$$

Proof. We first sketch the calculation, then show how to make it rigorous. We will use superscripts on $\mathbf{P}_o$ to denote on which graph the walk is taking place. The degree in $G \square G'$ is $d_G + d_{G'}$, whence each step of simple random walk has chance $d_G/(d_G + d_{G'})$ to be along an edge coming from G. Take o in $G \square G'$ to be the product of the basepoints in the two graphs. Then a walk of n steps from o in $G \square G'$ that takes k steps along edges from G and the rest from G' is back at o iff the k steps in G lead back to o in G and likewise for the $n-k$ steps in G'. Thus,

$$\mathbf{P}_o^{G \square G'}[X_n = o] = \sum_{k=0}^{n} \binom{n}{k} \left(\frac{d_G}{d_G + d_{G'}} \right)^k \left(\frac{d_{G'}}{d_G + d_{G'}} \right)^{n-k} \mathbf{P}_o^G[X_k = o] \, \mathbf{P}_o^{G'}[X_{n-k} = o]$$

$$\approx \sum_{k=0}^{n} \binom{n}{k} \left(\frac{d_G}{d_G + d_{G'}} \right)^k \left(\frac{d_{G'}}{d_G + d_{G'}} \right)^{n-k} \rho(G)^k \rho(G')^{n-k}$$

$$= \left(\frac{d_G \rho(G) + d_{G'} \rho(G')}{d_G + d_{G'}} \right)^n.$$

Taking nth roots gives the result.

To make this argument rigorous, we will replace the vague approximation by inequalities in both directions.

For an upper bound, we may use the inequality of Exercise 7.10.

For a lower bound, we note that according to Exercise 7.10, it suffices to consider only even n, which we now do. Then we may sum over even k only (for a lower bound) and use $\mathbf{P}_o^G[X_k = o] \geq C(\rho(G) - \epsilon)^k$ and $\mathbf{P}_o^{G'}[X_{n-k} = o] \geq C(\rho(G') - \epsilon)^{n-k}$, where ϵ is chosen arbitrarily small and then C is chosen sufficiently small. Note that for any $0 < p < 1$, we have

$$\lim_{n \to \infty} \sum_{k=0}^{n} \binom{2n}{2k} p^{2k}(1 - p)^{2n-2k} = 1/2 .$$

Putting together these ingredients with $p := d_G(\rho(G) - \epsilon)/\big(d_G(\rho(G) - \epsilon) + d_{G'}(\rho(G') - \epsilon)\big)$ completes the proof. ◄

Since $\rho(\mathbb{Z}) = 1$, it follows that $\rho(\mathbb{Z}^d) = 1$ as well, a fact that is also easy to verify directly. Another particular value of ρ that is useful for us is that of regular trees:

Proposition 7.35. (Spectral Radius of Trees) *For all $b \geq 1$, we have*

$$\rho(\mathbb{T}_{b+1}) = \frac{2\sqrt{b}}{b + 1} .$$

In fact, we proved this in Exercise 6.9, as well as in (6.25); see also Exercises 7.45 and 7.46. We give a more direct proof here.

Proof. If X_n is simple random walk on $\mathbb{T}_{b+1}$, then $|X_n|$ is a biased random walk on $\mathbb{N}$, with probability $p := b/(b + 1)$ to increase by 1 and $q := 1/(b + 1)$ to decrease by 1 when $X_n \neq 0$. Let $\tau_0^+ := \inf\{n \geq 1 ; |X_n| = 0\}$. Now the number of paths of length $2n$ from 0 to 0 in $\mathbb{N}$ that do not visit 0 in between is equal to the number of paths of length $2n - 2$ from 1 to 1 in $\mathbb{Z}$ minus the number of the latter that visit 0. The former number is clearly $\binom{2n-2}{n-1}$. The latter number is $\binom{2n-2}{n}$, since reflection after the first visit to 0 of a path from 1 to 1 that visits 0 yields a bijection to the set of paths from 1 to -1 of length $2n - 2$. Hence, the difference is

$$\binom{2n-2}{n-1} - \binom{2n-2}{n} = \frac{1}{n}\binom{2n-2}{n-1} .$$

(This is the sequence of **Catalan numbers**, which appear in many combinatorial problems.) Each path of length $2n$ from 0 to 0 in $\mathbb{N}$ that does not visit 0 has probability $p^{n-1}q^n$, whence we have

$$\mathbf{P}[\tau_0^+ = 2n] = \frac{1}{n}\binom{2n-2}{n-1}p^{n-1}q^n = \frac{-1}{2}(-4)^n\binom{1/2}{n}p^{n-1}q^n .$$

We have written this probability in a second way because to find the asymptotics as $n \to \infty$, we will use a generating function, and this will turn out to fit just right. That is, provided $|z|$ is sufficiently small,

$$g(z) := \mathbf{E}[z^{\tau_0^+}] = \sum_{n \geq 1} \frac{-1}{2}(-4)^n\binom{1/2}{n}p^{n-1}q^n z^{2n} = (1 - \sqrt{1 - 4pqz^2})/(2p) . \qquad (7.15)$$

Since the square root function cannot be analytically continued in a neighborhood of $z = 1/(2\sqrt{pq}\,)$, it follows that the radius of convergence of the power series for g is $1/(2\sqrt{pq}\,)$. We now use a generating function to find the spectral radius as well. Since

$$G(z) := \sum_{n \geq 0} \mathbf{P}\big[|X_n| = 0\big] z^n = \frac{1}{1 - g(z)} \,,$$

it follows that the radius of convergence of G is also $1/(2\sqrt{pq})$, which shows that the spectral radius is $2\sqrt{pq}$, as desired. ◀

Combining this with Proposition 7.34, we get:

Corollary 7.36. *For all $b \geq 1$, we have*

$$\rho(\mathbb{T}_{b+1} \,\square\, \mathbb{Z}) = \frac{2\sqrt{b} + 2}{b + 3} \,.$$

◀

We now have enough tools at our disposal to give our first example of a graph with two phase transitions. This was also historically the first example ever given, when Grimmett and Newman (1990) gave a version of the following result.

Theorem 7.37. *For all $b \geq 6$, we have $0 < p_c(\mathbb{T}_{b+1} \,\square\, \mathbb{Z}) < p_u(\mathbb{T}_{b+1} \,\square\, \mathbb{Z}) < 1$.*

Proof. The first inequality follows from Proposition 7.12. Since $\mathbb{T}_{b+1}$ and $\mathbb{Z}$ are both finitely presented, so is their product, whence the last inequality is a consequence of Theorem 7.29. To prove the middle inequality, note that by Theorem 7.32 and (7.14),

$$p_c(\mathbb{T}_{b+1} \,\square\, \mathbb{Z}) \leq p_c(\mathbb{T}_{b+1}) = \frac{1}{b} < \frac{1}{\mathrm{cogr}(\mathbb{T}_{b+1} \,\square\, \mathbb{Z})} \leq p_u(\mathbb{T}_{b+1} \,\square\, \mathbb{Z})$$

if $\mathrm{cogr}(\mathbb{T}_{b+1} \,\square\, \mathbb{Z}) < b$. Combining Theorem 6.10 and Corollary 7.36, we obtain that $\mathrm{cogr}(\mathbb{T}_{b+1} \,\square\, \mathbb{Z}) < b$ iff $\rho(\mathbb{T}_{b+1} \,\square\, \mathbb{Z})(b + 3) = 2\sqrt{b} + 2 < b + 1 + 2/b$. It is easy to check that this holds for $b \geq 6$. ◀

Using Exercise 7.47, the proof of Theorem 7.37 extends immediately to all $b \geq 4$. On the other hand, using the weaker bound Lemma 7.33 in place of (7.14), the proof would work only for $b \geq 8$.

To establish some more cases of Conjecture 7.31, we use the following consequence of (6.16). (Recall that in this chapter, $\Phi_E(G)$ means $\Phi_E(G, \mathbf{1}, \mathbf{1})$, whereas in (6.16), it meant $\Phi_E(G, c, \pi)$.)

Theorem 7.38. *If G is a regular graph of degree d, then*

$$\rho(G)^2 + \left(\frac{\Phi_E(G)}{d}\right)^2 \leq 1 \tag{7.16}$$

and

$$\rho(G) + \frac{\Phi_E(G)}{d} \geq 1 \,. \tag{7.17}$$

◀

The following corollary is due to Schonmann (2001) (parts (i) and (ii)) and Pak and Smirnova-Nagnibeda (2000) (part (iii)).

Corollary 7.39. *Let G be a transitive graph of degree d.*
  (i) *If $\Phi_E(G)/d \geq 1/\sqrt{2}$, then $p_c^{bond}(G) < p_u^{bond}(G)$.*
  (ii) *If $\Phi_V(G)/d \geq 1/\sqrt{2}$, then $p_c^{site}(G) < p_u^{site}(G)$.*
  (iii) *If $\rho(G) \leq 1/2$, then $p_c^{bond}(G) < p_u^{bond}(G)$.*

Proof. We will use the following implications:

$$\rho(G)d \leq \Phi_E(G) \quad \Longrightarrow \quad p_c^{bond}(G) < p_u^{bond}(G) \tag{7.18}$$

$$\rho(G)d \leq \Phi_V(G) \quad \Longrightarrow \quad p_c^{site}(G) < p_u^{site}(G). \tag{7.19}$$

To prove (7.18), use Theorem 6.46, Lemma 7.33, and Theorem 7.32 to see that when $\rho(G)d \leq \Phi_E(G)$, we have

$$p_c^{bond}(G) \leq \frac{1}{1 + \Phi_E(G)} < \frac{1}{\Phi_E(G)} \leq \frac{1}{\rho(G)d} \leq \frac{1}{\gamma(G)} \leq p_u^{bond}(G).$$

Similarly, when $\rho(G)d \leq \Phi_V(G)$, we have

$$p_c^{site}(G) \leq \frac{1}{1 + \Phi_V(G)} < \frac{1}{\Phi_V(G)} \leq \frac{1}{\rho(G)d} \leq \frac{1}{\gamma(G)} \leq p_u^{site}(G).$$

Part (i) is now an immediate consequence of (7.16) and (7.18). Part (ii) follows similarly from (7.16) and (7.19), with the observation that $\Phi_E(G) \geq \Phi_V(G)$, so that if $\Phi_V(G)/d \geq 1/\sqrt{2}$, then also $\Phi_E(G)/d \geq 1/\sqrt{2}$, so that $\rho(G)d \leq d/\sqrt{2} \leq \Phi_V(G)$. Part (iii) follows from (7.17) and (7.18). ◀

To get a sense of the strength of the hypotheses of Corollary 7.39, note that by Exercise 6.1, we have $\Phi(\mathbb{T}_{b+1})/d_{\mathbb{T}_{b+1}} = (b-1)/(b+1)$. Also, for a given degree, the regular tree maximizes $\Phi(\cdot)$ over all graphs by Exercise 6.43.

It turns out that any nonamenable group has a generating set that gives a Cayley graph satisfying the hypotheses of Corollary 7.39(iii). To show this, consider the following construction. Let G be a graph and $k \geq 1$. Define a new multigraph $G^{[k]}$ to have vertex set $V(G)$ and to have one edge joining $x, y \in V(G)$ for every path in G of length *exactly* k that joins x and y. Thus $G^{[1]} = G$. Furthermore, if G is the Cayley graph of Γ corresponding to a generating set S that is closed under inverses, then for any odd k, we have that $G^{[k]}$ is the Cayley graph of the same group Γ but presented as $\langle S^{[k]} \mid R^{[k]} \rangle$, where

$$S^{[k]} := \{x_{w_1, w_2, \ldots, w_k} \; ; \; w_i \in S\}$$

and $R^{[k]}$ is the set of products $x_{w_{1,1}, w_{1,2}, \ldots, w_{1,k}} x_{w_{2,1}, w_{2,2}, \ldots, w_{2,k}} \cdots x_{w_{n,1}, w_{n,2}, \ldots, w_{n,k}}$ of elements from $S^{[k]}$ such that $w_{1,1} w_{1,2} \cdots w_{1,k} w_{2,1} w_{2,2} \cdots w_{2,k} \cdots w_{n,1} w_{n,2} \cdots w_{n,k}$ is the identity in Γ. (To see that $G^{[k]}$ is connected, note that for every $s \in S$, we have $S^{[k]} \ni x_{s, s^{-1}, s, s^{-1}, \ldots, s} = s$. If G is not bipartite, then $G^{[k]}$ is connected also for even k.) Of course, this does have the disadvantage that $S^{[k]}$ is not a subset of Γ. That disadvantage can be removed by a result of Thom (2015).

Corollary 7.40. (Pak and Smirnova-Nagnibeda, 2000) *If G is any nonamenable, transitive graph, then* $p_c^{\text{bond}}(G^{[k]}) < p_u^{\text{bond}}(G^{[k]})$ *for all odd* $k \geq -\log 2/\log \rho(G)$. *In particular, every finitely generated, nonamenable group has a generating set whose Cayley graph satisfies* $p_c^{\text{bond}} < p_u^{\text{bond}}$.

Proof. Let k be as given. Then $\rho(G)^k \leq 1/2$. Now simple random walk on $G^{[k]}$ is the same as simple random walk on G sampled every k steps. Therefore $\rho(G^{[k]}) = \rho(G)^k$ by Exercise 7.10. This means that $\rho(G^{[k]}) \leq 1/2$, whence the result follows from Corollary 7.39. ◀

Corollary 7.40 would confirm Conjecture 7.31 if the following conjecture were established:

Conjecture 7.41. (Benjamini and Schramm, 1996b) *If G and G′ are roughly isometric, quasi-transitive graphs and* $p_c(G) < p_u(G)$, *then* $p_c(G') < p_u(G')$.

More detailed information about percolation requires additional tools, which we develop in the next chapter.

7.8 Bootstrap Percolation on Regular Trees

The last section in this chapter is devoted to a different model, bootstrap percolation, which, like ordinary percolation, was introduced by physicists. Our limited study of it will be restricted to regular trees. We will combine some ideas from Sections 5.7 and 5.8. ***Bootstrap percolation*** on an infinite graph G begins with a random initial configuration where each vertex is occupied with probability p, independently of each other – in other words, with Bernoulli(p) site percolation. It then evolves at positive integer times according to a *deterministic* spreading rule with a fixed parameter k: if a vacant site has at least k occupied neighbors at a certain time step, then it becomes occupied in the next step. Perhaps all sites are eventually occupied; the probability of this happening depends on p and k. If we fix k, then we define the critical probability $p(G, k)$ as the infimum of the initial probabilities p that make $\mathbf{P}_p[\text{complete occupation}] > 0$. If $k = 1$, then clearly everything becomes occupied when G is connected and $p > 0$, so $p(G, 1) = 0$, whereas if k is the maximum degree of G, then clearly not everything becomes occupied when $p < 1$, and so $p(G, k) = 1$.

▷ **Exercise 7.11.**
Show that $p(\mathbb{Z}^2, 2) = 0$ for the usual square lattice graph.

Bootstrap percolation is well studied on $\mathbb{Z}^d$ and on finite boxes where some surprising "metastability" phenomena occur (see, for example, Aizenman and Lebowitz (1988), Schonmann (1992), Holroyd (2003), Adler and Lev (2003), Gravner and Holroyd (2009), Balogh, Bollobás, and Morris (2009), Balogh, Bollobás, Duminil-Copin, and Morris (2012)). Balogh, Peres, and Pete (2006) investigated bootstrap percolation on regular and general infinite trees and on graphs with anchored expansion (such as nonamenable Cayley graphs).

For example, Balogh, Peres, and Pete (2006) showed that if Γ is a finitely generated group that contains a free subgroup on two elements, then Γ has a Cayley graph G such that $0 < p(G, k) < 1$ for some k. In contrast, Schonmann (1992) proved that $p(\mathbb{Z}^d, k) = 0$ if $k \leq d$ and $= 1$ if $k > d$. These results raise the following open question:

Question 7.42. (Balogh, Peres, and Pete, 2006) Is a group amenable if and only if for any finite generating set, the resulting Cayley graph G has $p(G, k) \in \{0, 1\}$ for every k?

It turns out that finding the critical probability $p(\mathbb{T}_{d+1}, k)$ on a $(d + 1)$-regular tree is equivalent to the problem of finding certain regular subtrees in a Bernoulli percolation process, so the results of Section 5.7 are directly applicable to this case.

To see this, we introduce the following notion:

Definition 7.43. A finite or infinite, connected subset $F \subseteq V$ of vertices is called a k-***fort*** if, for each $x \in F$, there are at most k edges that join x to (some vertex in) $V \setminus F$.

What's the relationship of forts to the occupation question? If there is an initially *vacant* $(k - 1)$-fort, then it will remain vacant and so there will never be complete occupation. On the other hand, the unoccupied sites in the final configuration are a $(k - 1)$-fort and thus form an initial vacant $(k - 1)$-fort. Therefore, the failure of complete occupation by the k-neighbor rule for a given initial configuration ω is equivalent to the existence of a nontrivial vacant $(k - 1)$-fort in ω.

We can now analyze bootstrap percolation on the $(d + 1)$-regular tree, $\mathbb{T}_{d+1}$, where $d \geq 2$ is a fixed integer. Let $1 \leq k \leq d$. Consider Bernoulli(p) *bond* percolation on $\mathbb{T}_{d+1}$; does the cluster of a fixed vertex contain a $(k + 1)$-regular subtree with positive probability? Define $\pi(d, k)$ to be the infimum of probabilities p for which the answer is yes. Harris's inequality (Section 5.8) implies that this critical probability is the same as the one for having a k-ary subtree at the root in a Galton–Watson tree with offspring distribution Bin(d, p), which we analyzed in Section 5.7. Ergodicity also shows that the probability is either 0 or 1 that somewhere there is a $(k + 1)$-regular subtree in Bernoulli(p) percolation on $\mathbb{T}_{d+1}$; this probability is monotonic in p and changes at $\pi(d, k)$. The relationship of the critical probabilities $\pi(d, k)$ to those for bootstrap percolation are as follows. Asymptotics for large d and k lead to the very cute asymptotic $p(\mathbb{T}_{d+1}, k) \sim k/d$:

Proposition 7.44. (Balogh, Peres, and Pete, 2006) *Let $1 \leq k \leq d$, and consider k-neighbor bootstrap percolation on the $(d + 1)$-regular tree, $\mathbb{T}_{d+1}$. We have the following equality of critical probabilities:*

$$p(\mathbb{T}_{d+1}, k) = 1 - \pi(d, d + 1 - k). \tag{7.20}$$

In particular, for any constant $\gamma \in [0, 1]$ and sequence of integers k_d with $\lim_{d \to \infty} k_d/d = \gamma$, we have

$$\lim_{d \to \infty} p(\mathbb{T}_{d+1}, k_d) = \gamma. \tag{7.21}$$

Proof. The tree $\mathbb{T}_{d+1}$ has no finite $(k - 1)$-forts, and it is easy to see that any infinite $(k - 1)$-fort of $\mathbb{T}_{d+1}$ contains a complete $(d + 2 - k)$-regular subtree. Hence, unsuccessful complete occupation for the k-neighbor rule is equivalent to the existence of a $(d + 2 - k)$-regular, vacant subtree in the initial configuration. Furthermore, the set of initial configurations that lead to complete occupation on $\mathbb{T}_{d+1}$ is invariant under the automorphism group of $\mathbb{T}_{d+1}$, hence has probability 0 or 1: see Proposition 7.3. So incomplete occupation has probability 1 if and only if a fixed origin is contained in a $(d + 2 - k)$-regular, vacant subtree with positive probability. Since the vacant vertices in $\mathbb{T}_{d+1}$ form a Bernoulli($1 - p$) percolation process, we find that (7.20) holds.

Note that if $\lim_{d\to\infty} k_d/d = \gamma$, then $\lim_{d\to\infty}(d + 1 - k_d)/d = 1 - \gamma$, so equation (5.29) of Proposition 5.31 implies (7.21).   ◀

Proposition 5.30 showed that for $k \geq 2$ we already have a k-ary subtree at the critical probability $\pi(d, k)$. For bootstrap percolation, this means that the probability of complete occupation is still 0 at $p = p(\mathbb{T}_{d+1}, k)$ if $k < d$.

7.9 Notes

Results similar to (7.1) are part of a general conjecture due to Schramm, first mentioned in print by Benjamini, Nachmias, and Peres (2011):

Conjecture 7.45. *Suppose that G_n are infinite, transitive graphs that converge to G in the sense that for every $r > 0$, the balls of radius r in G_n are isomorphic to those in G for all large n. If $\sup_n p_c(G_n) < 1$, then $\lim_{n\to\infty} p_c(G_n) = p_c(G)$.*

Other known cases include nonamenable G_n with girth tending to infinity (Benjamini, Nachmias, and Peres, 2011); G is a Cayley graph of an abelian group (Martineau and Tassion, 2017); and Exercise 8.43. That last result was greatly extended by Hutchcroft (2020b) to G_n with a uniform exponential lower bound on their volume growth.

In the nonamenable case, two Cayley graphs are roughly isometric iff there is a bijective rough isometry between them, which is the same as a bi-Lipschitz map: see Whyte (1999). However, in the amenable case, this is not so, and there are lamplighter groups that are roughly isometric but not bi-Lipschitz equivalent: see Dymarz (2010). These lamplighter groups are of the form $\sum_{x\in\mathbb{Z}} \mathbb{Z}_m \rtimes \mathbb{Z}$; other lamplighter groups of interest in probability replace the base space $\mathbb{Z}$ by $\mathbb{Z}^d$ (see Section 14.2).

The grandparent graph G of Example 7.1 is built on a tree T with a fixed end, ξ. The automorphism group of G is the same as the subgroup of automorphisms of T that fix ξ, called the ***affine group*** of T. The reasons for this name are given on p. 1252 of Cartwright, Kaimanovich, and Woess (1994). The grandparent graph has another interesting property: it is the only transitive graph with the same ball of radius 1, as proved by Frisch and Tamuz (2016).

The proof in this chapter of Proposition 5.27 is due to Pemantle and Peres (1996), Lemma 4.2(i). A modification of this proof is used by Peres, Pete, and Scolnicov (2006).

Special cases of Proposition 7.10 were proved by Fisher (1961).

Proposition 7.12 was first observed for bond percolation on $\mathbb{Z}^2$ by Broadbent and Hammersley (1957) and Hammersley (1959).

Proposition 7.14 comes from Lyons and Schramm (1999), Remark 6.2.

The contour argument used to prove Theorem 7.16 is often referred to as a "Peierls argument," after Peierls (1936), who used such an argument to prove that the two-dimensional Ising model has a phase transition.

Inequality (7.7), that $p_c(G) \leq 1/b$, also follows from the following result of Aizenman and Barsky (1987):

Theorem 7.46. (Expected Cluster Size) *If G is any transitive graph and $p < p_c(G)$, then we have $\mathbf{E}_p\big[|K(o)|\big] < \infty$.*

Aizenman and Barsky (1987) worked only on $\mathbb{Z}^d$, but their proof works in greater generality. A very short proof, inspired by theirs, is given by Duminil-Copin and Tassion (2016a).

To prove Theorem 7.18, we will use the following result:

Proposition 7.47. *A subgroup of finite index in a finitely generated group is itself finitely generated.*

Proof. Let Γ be generated by the set S. Given a subgroup $\Gamma' < \Gamma$, choose a set $A \subset \Gamma$ such that A intersects each coset of Γ' exactly once. We may assume that $o \in A$, where o is the identity of Γ. Thus, for each $\gamma \in \Gamma$, there is a unique $k(\gamma) \in A$ and a unique $h(\gamma) \in \Gamma'$ such that $\gamma = k(\gamma)h(\gamma)$. Now if $\gamma = ah(\gamma)$ and $\gamma' = a'h(\gamma')$, then $\gamma\gamma' = \gamma a'h(\gamma') = a''h(\gamma a')h(\gamma')$ for some $a'' \in A$, whence $h(\gamma\gamma') = h(\gamma k(\gamma'))h(\gamma')$. It follows by induction that

$$h(\gamma_1 \cdots \gamma_m) = h(\gamma_1 k(\gamma_2 \cdots \gamma_m))h(\gamma_2 k(\gamma_3 \cdots \gamma_m)) \cdots h(\gamma_{m-1}k(\gamma_m))h(\gamma_m).$$

Therefore, given any choice of $s_i \in S \cup S^{-1}$, the element $h(s_1 \cdots s_m)$ is a product of elements from $S' := \{h(sa) ; s \in S \cup S^{-1}, a \in A\}$. Since $o \in A$, $h(\gamma) = \gamma$ for all $\gamma \in \Gamma'$, whence if $s_1 \cdots s_m \in \Gamma'$, then $h(s_1 \cdots s_m) = s_1 \cdots s_m$. This means that Γ' is generated by S'. ◄

Proof of Theorem 7.18. The principal fact we need is the difficult theorem of Gromov (1981a), who showed that Γ is almost a nilpotent group, as conjectured by Carlitz, Wilansky, Milnor, Struble, Felsinger, Simoes, Power, Shafer, and Maas (1968). (The converse is true as well; see Wolf (1968). See Kleiner (2010) or Ozawa (2018) for other proofs of Gromov's theorem.) Now every finitely generated, nilpotent group is almost torsion-free,[*] while the upper central series of a torsion-free, finitely generated, nilpotent group has all factors isomorphic to free abelian groups (Kargapolov and Merzljakov (1979), Theorem 17.2.2 and its proof). Thus, Γ has a subgroup Γ' of finite index that is torsion-free and either Γ' equals its center, C, or there is a subgroup C' of Γ' such that C'/C is the center of Γ'/C and C'/C is free abelian of rank at least 1. If the rank of C is at least 2, then C already contains $\mathbb{Z}^2$, so suppose that $C \approx \mathbb{Z}$. If $\Gamma' = C$, then Γ is almost $\mathbb{Z}$. If not, then let C'' be a subgroup of C' such that C''/C is isomorphic to $\mathbb{Z}$. We claim that C'' is isomorphic to $\mathbb{Z}^2$, and thus that C'' provides the subgroup we seek. Choose $\gamma \in C''$ such that γC generates C''/C. Let D be the group generated by γ. Then clearly $D \approx \mathbb{Z}$. Since $C'' = DC$ and C lies in the center of C'', it follows that $C'' \approx D \times C \approx \mathbb{Z}^2$. ◄

Transitive graphs of polynomial growth are very close to Cayley graphs of the same growth rate. This was proved by Trofimov (1984). A cleaner way to say that they are "close" to Cayley graphs was noted by Godsil, Imrich, Seifter, Watkins, and Woess (1989) as a consequence of results by Trofimov (1984) and Sabidussi (1964): every transitive graph G of polynomial growth has the property that for some $n \geq 1$, the graph nG is a Cayley graph, where nG has vertex set $\mathsf{V}(G) \times \{1, \ldots, n\}$ and edges connecting pairs (x, j) to (y, k) when $[x, y] \in \mathsf{E}(G)$. In fact, Sabidussi (1964) shows that if Γ is any subgroup of $\mathrm{Aut}(G)$ that acts transitively, n is the cardinality of the Γ-stabilizer of a vertex, o, and $S := \{\gamma \in \Gamma ; \gamma o \sim o\}$, then nG is the Cayley graph of Γ with respect to S. The fact that n and S are necessarily finite follows from Trofimov (1984).

Alexander (1995b) proved Theorem 7.21 in the case of $\mathbb{Z}^d$. The extension of his result to the first part of Theorem 7.21 and a weakening of the second part, when p_1 and p_2 are fixed in advance, was shown by Häggström and Peres (1999) for Cayley graphs and other unimodular, transitive graphs, then by Schonmann (1999b) in general. This answered affirmatively a question of Benjamini and Schramm (1996b). Finally Häggström, Peres, and Schonmann (1999) proved Theorem 7.21 in general. They also showed that Theorem 7.21 holds for **semitransitive** graphs, which are those graphs G for which there exists a finite set F of vertices with the property that for every $x \in \mathsf{V}(G)$, there is some injective homomorphism of G that maps some $y \in F$ to x. For example, $\mathbb{Z}^d \square \mathbb{Z}^+$ is semitransitive, as is every superperiodic tree. Since semitransitive graphs satisfy (7.10), our proof gives this result.

Invasion percolation was introduced by Wilkinson and Willemsen (1983). Chayes, Chayes, and Newman (1985) proved Theorem 7.22 in the case of $\mathbb{Z}^d$. A detailed study of invasion percolation

---

[*] A group is torsion-free if all its elements have infinite order, other than the identity element.

on regular trees is given by Angel, Goodman, den Hollander, and Slade (2008), with its scaling limit studied by Angel, Goodman, and Merle (2013).

Lemma 7.28 is due to Babson and Benjamini (1999), who used topological definitions and proofs. Our proof is due to Timár (2007). If instead of hypothesizing that K spans the cycles with real coefficients, one assumes that K spans the cycles with $\mathbb{F}_2$ coefficients, where $\mathbb{F}_2$ is the field of two elements, then the same statement is true with a very similar proof.

That $p_u < 1$ for planar, quasi-transitive graphs with one end follows from Theorem 7.29 and Exercise 7.42. An alternative proof is given in Section 8.5. Other techniques have established Conjecture 7.27 for the Cartesian product of two infinite graphs (Häggström, Peres, and Schonmann, 1999) and for Cayley graphs of Kazhdan groups, that is, groups with Kazhdan's property (T) (Lyons and Schramm, 1999).

The definition of Kazhdan's property is the following. Let Γ be a countable group and S be a finite subset of Γ. A *unitary representation* of Γ is an action π of Γ by unitary maps on a Hilbert space. An *invariant vector* of π is a vector v such that $\pi(\gamma)v = v$ for all $\gamma \in \Gamma$. Let $\mathscr{U}_\Gamma(\mathscr{H})$ denote the set of unitary representations of Γ on $\mathscr{H}$ that have no invariant vectors except 0. Set

$$\kappa(\Gamma, S) := \max\{\epsilon \,;\, \forall \mathscr{H}\ \forall \pi \in \mathscr{U}_\Gamma(\mathscr{H})\ \forall v \in \mathscr{H}\ \exists s \in S\ \|\pi(s)v - v\| \geq \epsilon\|v\|\}.$$

Then Γ is called *Kazhdan* (or has Kazhdan's property (T)) if $\kappa(\Gamma, S) > 0$ for all finite S (or, equivalently, for a single S that generates Γ). The only amenable Kazhdan groups are the finite ones. Examples of Kazhdan groups include $SL(n, \mathbb{Z})$ for $n \geq 3$. See de la Harpe and Valette (1989) for background; in particular, every Kazhdan group is finitely generated (p. 11 there) but not necessarily finitely presentable (as shown by examples of Gromov; see p. 43 there). Every infinite Kazhdan group has only one end. See Żuk (1996) for examples of Kazhdan groups arising as fundamental groups of finite simplicial complexes. There is a beautiful probabilistic characterization of Kazhdan groups. Let $\mathbf{P}_*$ be the probability measure on subsets of Γ that is the empty set half the time and all of Γ half the time. Recall that Γ acts by translation on the probability measures on 2^Γ.

Theorem 7.48. (Glasner and Weiss, 1997) *A countable infinite group Γ is Kazhdan iff $\mathbf{P}_*$ is not in the weak* closure of the Γ-invariant, ergodic probability measures on 2^Γ.*

The inequality $p_u(G) \geq 1/(\rho(G)d)$ for quasi-transitive graphs of maximum degree d, which follows from Theorem 7.32 combined with Lemma 7.33, was proved earlier by Benjamini and Schramm (1996b).

Theorem 7.32 is due to Schramm; his proof was published by Lyons (2000). A strengthening of Theorem 7.32 and of a result in the proof of Theorem 4 of Benjamini and Schramm (1996b) is the following:

Theorem 7.49. *If G is a transitive graph and $p < 1/\gamma(G)$, then $\lim_{x \to \infty} \mathbf{P}_p[o \leftrightarrow x] = 0$.*

The reason that this implies Theorem 7.32 is that when there is a unique infinite p-cluster, we have $\mathbf{P}_p[o \leftrightarrow x] \geq \mathbf{P}_p[|K(o)| = \infty, |K(x)| = \infty] \geq \theta(p)^2$ by Harris's inequality (Section 5.8).

Proof. Consider first site percolation. Let $p < p_+ < 1/\gamma(G)$. We use the standard coupling of Bernoulli percolation. A vertex z is called an (x, y)-*cutpoint* if x and y belong to the same p_+-open cluster but would not if z were closed. Let $u_k(r)$ denote the probability that for some x outside the ball $B(o, r)$, we have that o and x are connected via a p_+-open path with at most k (o, x)-cutpoints (for the p_+-open vertices). When there are no (o, x)-cutpoints, there are two disjoint paths joining o to x by Menger's theorem (Exercise 3.16), in other words, there is a simple cycle containing o and x. Thus, $u_0(r) \leq \sum_{n > 2r} a_n(G)p_+^n \to 0$ as $r \to \infty$. For $k \geq 0$ and $r \leq R$, consider $x \notin B(o, R)$. If there is a p_+-open path from o to x that has at most $k + 1$ (o, x)-cutpoints, then either this path contains no cutpoints in $B(o, r)$ or it has at most k cutpoints outside $B(o, r)$. In the latter case, it intersects the sphere of radius r in one of at most d^r points, where d is the degree of G. Therefore, we obtain the bound

$$u_{k+1}(R) \leq u_0(r) + d^r u_k(R - r). \tag{7.22}$$

The inequality (7.22) implies inductively that (for any fixed k) $u_k(R)$ tends to zero as $R \to \infty$. (Given ϵ, first choose r such that $u_0(r) < \epsilon$ and then choose R' so that $d^r u_k(R - r) < \epsilon$ for all $R \geq R'$.)

Next, let $\tau_p(R)$ denote the maximum over $x \notin B(o, R)$ of the probability that o and x are in the same p-cluster. Then, for every k, we have

$$\tau_p(R) \leq (p/p_+)^{k+1} + u_k(R) \tag{7.23}$$

by considering whether there is a p_+-open path from o to x with at most k cutpoints; in the latter case, there are $k + 1$ cutpoints that lie on every p_+-open path from o to x, and they must also be p-open. Finally (7.23) implies that $\tau_p(R)$ tends to zero as $R \to \infty$. (Given ϵ, first choose k such that $(p/p_+)^{k+1} < \epsilon$, then choose R' so that $u_k(R) < \epsilon$ for all $R \geq R'$.)

The proof for bond percolation is similar but needs just the following modification: replace $(p/p_+)^{k+1}$ by $\left(1 - (1 - p/p_+)^d\right)^{(k+1)/2}$ in (7.23), because if o and x are in the same p-cluster and there is a no p_+-open path from o to x with at most k cutpoints, then at least one edge incident to each cutpoint must also be p-open. Furthermore, the cutpoints must appear in a fixed order in every p_+-open path from o to x, and two cutpoints can share at most one edge.   ◀

According to Schonmann (1999a), techniques of that paper combined with those of Stacey (1996) can be used to show the same inequality as that of Theorem 7.37 for all $b \geq 2$, that is, $0 < p_c(\mathbb{T}_{b+1} \square \mathbb{Z}) < p_u(\mathbb{T}_{b+1} \square \mathbb{Z}) < 1$, but the proof "is quite technical." Yamamoto (2017) has shown that $p_c(\mathbb{T}_{b+1} \square \mathbb{Z}) \leq 1/(b + 1)$ for all $b \geq 2$, which implies $p_c(\mathbb{T}_{b+1} \square \mathbb{Z}) < p_u(\mathbb{T}_{b+1} \square \mathbb{Z})$ for $b \geq 3$ by the method of proof of Theorem 7.37. Hutchcroft (2020c) has shown that $p_c(\mathbb{T}_{b+1} \square \mathbb{Z}^k) < p_u(\mathbb{T}_{b+1} \square \mathbb{Z}^k)$ for all $b \geq 2$ and $k \geq 0$ by a method that does not rely on estimates of these critical probabilities.

The only groups for which it is known that *all* their Cayley graphs satisfy $p_c < p_u$ are the nonamenable, Gromov hyperbolic groups (Hutchcroft, 2019b) and the groups of "cost" larger than 1. This concept is defined in Section 10.2, but we give a few classes of examples here. This includes, first, free groups of rank at least 2 and fundamental groups of compact surfaces of genus larger than 1. Second, let Γ_1 and Γ_2 be two groups of finite cost with Γ_1 having cost larger than 1. Then every amalgamation of Γ_1 and Γ_2 over an amenable group has cost larger than 1. Third, every so-called HNN extension of Γ_1 over an amenable group has cost larger than 1. Also, every HNN extension of an infinite group over a finite group has cost larger than 1. For proofs that these groups have cost larger than 1, see Gaboriau (1998, 2000, 2002). The proof that $p_c(G) < p_u(G)$ when G is a Cayley graph of a group Γ with cost larger than 1 follows fairly easily from Theorem 8.21 in the next chapter, as noted by Lyons (2000); see Lyons (2013b) for the details. If Γ is a group with a Cayley graph G and the free and wired uniform spanning forest measures on G differ (equivalently, there exist nonconstant harmonic Dirichlet functions), then Γ has cost larger than 1. These are the same as the groups with strictly positive first ℓ^2-Betti number; see Sections 10.2 and 10.8. Possibly these are exactly the groups of cost larger than 1; see Question 10.12. Although it has long been known (Bekka and Valette, 1997) that Kazhdan groups have first ℓ^2-Betti number equal to 0, it was not proved until Hutchcroft and Pete (2020) that they have cost 1; their argument used invariant percolation. Likewise, it is not known that all Cayley graphs of Kazhdan groups have $p_c < p_u$. If the free and wired uniform spanning forests differ on a unimodular, quasi-transitive graph G, then again $p_c(G) < p_u(G)$: see Corollary 4.5 of Gaboriau (2005).

In a different direction towards solving Conjecture 7.31, Hutchcroft (2020c) proved that $p_c(G) < p_u(G)$ for every graph G such that $\mathrm{Aut}(G)$ has a nonunimodular subgroup acting quasi-transitively. (The definition of "unimodular" is in the next chapter.) In particular, this implies the conclusion of Theorem 7.37 for every $b \geq 2$, as mentioned two paragraphs prior.

The following results concerning the uniqueness phase of Bernoulli percolation are also known.

Theorem 7.50. *Let G be a quasi-transitive graph.*
  (i) **(Schonmann, 1999b)** *We have*

$$p_{\mathrm{u}}(G) = \inf\left\{p\, ;\; \sup_{R}\inf_{x} \mathbf{P}_p\big[B(o, R) \leftrightarrow B(x, R)\big] = 1\right\}. \tag{7.24}$$

  (ii) **(Lyons and Schramm (1999), Tang (2019))** *If* $\inf_x \mathbf{P}_p[o \leftrightarrow x] > 0$, *then there is a unique infinite cluster* $\mathbf{P}_p$*-a.s. Therefore,*

$$p_{\mathrm{u}}(G) = \inf\left\{p\, ;\; \inf_{x} \mathbf{P}_p[o \leftrightarrow x] > 0\right\}. \tag{7.25}$$

It is not known which Cayley graphs have the property that there is a unique infinite cluster $\mathbf{P}_{p_{\mathrm{u}}}$-a.s. It is known that there is a unique infinite cluster at p_{u} when G is planar, nonamenable, and transitive (Theorem 8.24). On the other hand, Schonmann (1999a) proved that this does *not* happen on $\mathbb{T}_{b+1} \,\square\, \mathbb{Z}$ with $b \geq 2$, which Peres (2000) extended to all nonamenable Cartesian products of infinite, transitive graphs. Likewise, there are infinitely many infinite clusters at p_{u} when G is a Cayley graph of a Kazhdan group (due to Peres; see Lyons and Schramm (1999)). For further groups with this property, see Gaboriau and Tucker-Drob (2016).

▷ **Exercise 7.12.**
Show that for every transitive graph G, $p_{\mathrm{u}}^{\mathrm{site}}(G) \geq p_{\mathrm{u}}^{\mathrm{bond}}(G)$.

▷ **Exercise 7.13.**
Show that if G and G' are roughly isometric, quasi-transitive graphs, then $p_{\mathrm{u}}(G) < 1$ iff $p_{\mathrm{u}}(G') < 1$.

▷ **Exercise 7.14.**
Let G be a unimodular, transitive graph. Suppose that there are constants $C < \infty$ and $\zeta < 1$ such that for all vertices x in G, there is a probability measure μ on paths from o to x that satisfies

$$(\mu \times \mu)\big[|\xi \cap \xi'| = n\big] \leq C\zeta^n$$

for all n. Show that $p_{\mathrm{u}}(G) \leq \zeta$.

An analogue of nonamenable graphs in the world of finite graphs is a sequence of finite graphs whose expansion constants are bounded away from 0; such a sequence is usually called an expander sequence. Properties of percolation on an expander sequence that are analogues of the existence or uniqueness of infinite percolation clusters for infinite graphs are studied by Alon, Benjamini, and Stacey (2004) and Angel and Benjamini (2007). The analogue of the existence of an infinite cluster is the existence of a "giant" component, that is, a cluster that contains a positive proportion of all vertices; if there is such a giant component, then it is unique. It is somewhat of a mystery, then, what uniqueness of the infinite cluster could correspond to. The answer may be this: the analogue of the uniqueness of infinite clusters is that for a random pair of neighboring vertices, conditioned that they belong to the giant component, the distributions of their distance within the giant component form a tight family. This is similar to whether the limit of uniform spanning trees in an exhaustion of an infinite graph (that is, the free spanning forest) has only one component, a topic studied in Chapter 10. Much remains to be understood for percolation in this finite-analogue world.

7.10 Collected In-Text Exercises

7.1. Show that if A is measurable, then so is $\Pi_e A$ and prove (7.2).

7.2. Let G be a connected graph and $x, y \in V(G)$. Consider Bernoulli(p) percolation on G. Prove that if $\theta_y(p) > 0$, then also $\theta_x(p) > 0$.

7.3. Let T be a tree that has no vertices of degree 1. Let B be the set of vertices of degree at least 3. Let K be a finite, nonempty subset of vertices of T. Show that $|\partial_V K| \geq |K \cap B| + 2$.

7.4. Show that for any finitely generated group, the number of ends is the same for all of its Cayley graphs. Thus, we may speak of the number of ends of a group, not merely of a Cayley graph. In fact, show that if two graphs are roughly isometric, then they have the same number of ends.

7.5. Show that if G and G' are any two infinite, connected graphs, then the Cartesian product graph $G \mathbin{\square} G'$ (defined on p. 1) has only one end.

7.6. Show that if Γ and Γ' are any two finitely generated groups with $|\Gamma| \geq 2$ and $|\Gamma'| \geq 3$, then $\Gamma * \Gamma'$ has infinitely many ends.

7.7. Suppose that $\mathbf{P}_i$ ($i = 1, 2, 3$) are three percolation measures on 2^A such that $\mathbf{P}_1 \succcurlyeq \mathbf{P}_2$ and $\mathbf{P}_2 \succcurlyeq \mathbf{P}_3$. Show that there exist random variables ω_i on a common probability space such that $\omega_i \sim \mathbf{P}_i$ for all i and $\omega_1 \geq \omega_2 \geq \omega_3$ a.s.

7.8. Extend Theorem 7.20 to quasi-transitive graphs with exponential growth.

7.9. (a) Show that if ω is an a.s. nonempty, Aut(G)-invariant percolation on a quasi-transitive graph G, then a.s. every end of G contains an end-convergent subset of ω.
  (b) Show that if G is a quasi-transitive graph with more than one end, then $p_u(G) = 1$.

7.10. Show that for any graph G, the spectral radius satisfies $\rho(G) = \lim_{n \to \infty} \mathbf{P}_o[X_{2n} = o]^{1/2n}$ and $\mathbf{P}_o[X_n = o] \leq \rho(G)^n$ for all n.

7.11. Show that $p(\mathbb{Z}^2, 2) = 0$ for the usual square lattice graph.

7.12. Show that for every transitive graph G, $p_u^{\text{site}}(G) \geq p_u^{\text{bond}}(G)$.

7.13. Show that if G and G' are roughly isometric, quasi-transitive graphs, then $p_u(G) < 1$ iff $p_u(G') < 1$.

7.14. Let G be a unimodular, transitive graph. Suppose that there are constants $C < \infty$ and $\zeta < 1$ such that for all vertices x in G, there is a probability measure μ on paths from o to x that satisfies

$$(\mu \times \mu)\big[|\xi \cap \xi'| = n\big] \leq C \zeta^n$$

for all n. Show that $p_u(G) \leq \zeta$.

7.11 Additional Exercises

7.15. Show that if G' is a transitive representation of a quasi-transitive graph G, then G is amenable iff G' is amenable.

7.16. Show that a tree is quasi-transitive iff it is a directed cover of a finite, directed multigraph (as on p. 83 of Section 3.3).

7.17. Let x and y be two vertices of any graph and define $\tau_p(x, y) := \mathbf{P}_p[x \leftrightarrow y]$. Show that τ is continuous from the left as a function of p.

7.18. Find the Cayley graph of the lamplighter group $\mathbb{Z}^\circ$ with respect to the four generators $(\mathbf{0}, 1)$, $(\mathbf{0}, -1)$, $(\mathbf{1}_{\{0\}}, 1)$, and $(\mathbf{1}_{\{0\}}, -1)$.

7.19. Show that if b_n is the size of a ball of radius n in the graph $\mathbb{T}_{b+1} \,\square\, \mathbb{Z}^d$, then $\lim_{n \to \infty} b_n^{1/n} = b$ for all $b \geq 1$ and $d \geq 0$.

7.20. Use Harris's inequality to do Exercise 7.2.

7.21. Let T be a finite tree. Call a vertex x a *k-**branch point*** of T if $T \setminus x$ has at least three components that each have at least k vertices. Let B_k be the set of k-branch points of T. Show that if $|B_k| \geq 1$, then $|\mathsf{V}(T) \setminus B_k| \geq k(|B_k| + 2)$.

7.22. Give an example of a connected graph such that for some p, the number of infinite clusters of Bernoulli(p) percolation is finite and at least 2 with positive probability.

7.23. Suppose that ω is an invariant percolation on a transitive graph G. Let Λ be the set of furcations of ω. Show that $\Phi_\mathsf{V}(G, \mathbf{1}, \mathbf{1}) \geq \mathbf{P}[o \in \Lambda]$.

7.24. Suppose that $\mathbf{P}$ is an invariant percolation on an amenable, transitive graph. Show that all infinite clusters have at most two ends a.s.

7.25. Let $\mathfrak{F}$ be an invariant random forest on a transitive graph G. Show that

$$\mathbf{E}[\deg_{\mathfrak{F}} o] \leq \Phi_\mathsf{V}(G, \mathbf{1}, \mathbf{1}) + 2 .$$

7.26. Show that if G is a quasi-transitive graph with at least three ends, then G has infinitely many ends.

7.27. Extend Theorem 7.9 to quasi-transitive, amenable graphs.

7.28. Refine Proposition 7.10 to prove that $\theta_o^{\text{site}}(p) \leq p\theta_o^{\text{bond}}(p)$.

7.29. Extend Proposition 7.10 to show that there is a coupling of $\xi \sim \mathbf{P}_p^{\text{site}}$ and $\omega \sim \mathbf{P}_p^{\text{bond}}$ such that every ξ-cluster is contained in some ω-cluster.

7.30. Strengthen Proposition 7.12 to show that for any connected graph G, we have the lower bound $p_\text{c}(G) \geq 1/\text{br}\, T^{\text{SAW}}$.

7.31. Suppose that G is a graph such that for some constant $a < \infty$ and all $n \geq 1$, we have

$$\left|\left\{\Pi \subset \mathsf{E} \,;\; \Pi \text{ is a minimal cutset separating } o \text{ from } \infty, \; |\Pi| = n\right\}\right| \leq ae^{an} , \qquad (7.26)$$

where o is a fixed vertex in G. Show that $p_\text{c}^{\text{bond}}(G) < 1$.

7.32. Prove that $p_\text{c}^{\text{bond}}(\mathbb{Z}^2) \leq 1 - 1/\mu(\mathbb{Z}^2)$.

7.33. Let G be an infinite graph and $x \in \mathsf{V}(G)$. Write $K(x)$ for the cluster of x in percolation; as usual, $\theta_x(p) := \mathbf{P}_p[|K(x)| = \infty]$.

  (a) Show that $\theta_x(p)$ is continuous from the right at all $p \in [0, 1)$.

  (b) For $0 < p \le 1$, show that $\theta_x(p) - \theta_x(p^-) = \mathbf{P}_p[|K(x)| = \infty, \ p_c(K(x)) = 1]$.

  (c) Show that if G is quasi-transitive, then $\theta_x(p)$ is continuous from the left at all $p \in (0, 1]$ with an exception at $p = p_c$ iff $\theta(p_c) > 0$.

7.34. Let $\langle U(e) \rangle$ be distinct labels on an infinite graph, yielding invasion basins $I(x)$. Show that if $\inf\{U(e) ; \ e \in \partial_E I(x)\} < p$, then $|I(x) \setminus \eta| < \infty$ for some infinite cluster η of ω_p.

7.35. Give an example of a graph with a vertex x whose invasion basin $I(x)$ satisfies

$$\mathbf{P}[\exists p > p_c \ \ I(x) \cap \omega_p \text{ has no infinite component}] > 0.$$

7.36. Let G be an infinite, quasi-transitive graph with $p_u < 1$. Show that in the standard coupling of Bernoulli percolation, a.s. for all $p > p_u$, every infinite cluster in ω_p has one end.

7.37. Show that if T is any tree and $p < 1$, then Bernoulli(p) percolation on T has either no infinite clusters a.s. or infinitely many infinite clusters a.s.

7.38. Prove Theorem 7.29 for site percolation.

7.39. Let $G = (\mathsf{V}, \mathsf{E})$ be a graph with a fixed base point o. A finite subset of vertices that includes o and that induces a connected subgraph is called a ***site animal***. Let a_n be the number of n-vertex site animals. Show that if all vertices have degrees at most d, then $\limsup_{n \to \infty} a_n^{1/n} < ed$, where e is the base of natural logarithms.

7.40. Let $G = (\mathsf{V}, \mathsf{E})$ be a graph with a fixed base point o. A finite set of edges that includes some edge incident to o and that form a connected subgraph is called a ***bond animal***. Let b_n be the number of n-edge bond animals. Show that if all vertices have degrees at most d, then $\limsup_{n \to \infty} b_n^{1/n} < e(d - 1)$, where e is the base of natural logarithms.

7.41. A sequence of vertices $\langle x_k \rangle_{k \in \mathbb{Z}}$ is called a ***bi-infinite geodesic*** if $\mathrm{dist}_G(x_j, x_k) = |j - k|$ for all $j, k \in \mathbb{Z}$. Show that an infinite, transitive graph contains a bi-infinite geodesic passing through o.

7.42. Show that if G is a transitive graph with one end such that there is a set of cycles with bounded length whose linear span contains all cycles, then $p_c(G) < 1$. For example, this holds when G is the Cayley graph of a finitely presented group.

7.43. Let G be any connected graph. Let $a_n(G, x)$ be the number of simple cycles of length n in G from x to x and

$$\gamma(G) := \sup_x \limsup_{n \to \infty} a_n(G, x)^{1/n}.$$

Show that if there is a unique infinite cluster $\mathbf{P}_p$-a.s., then $p \ge 1/\gamma(G)$. Moreover, show that if $p < 1/\gamma(G)$, then $\lim_{x \to \infty} \mathbf{P}_p[o \leftrightarrow x] = 0$.

7.44. Let G and G' be transitive networks with conductance sums at vertices π and π', respectively. (These are constants by transitivity.) Define the Cartesian product network as in Exercise 6.25. Show that

$$\rho(G \,\square\, G') = \frac{\pi}{\pi + \pi'} \rho(G) + \frac{\pi'}{\pi + \pi'} \rho(G').$$

7.45. Combine Exercises 5.34 and 5.35(a) to give another proof of (7.15).

7.46. Give another proof of (7.15) by conditioning on the second step to get a recursion for $g(z)$.

7.47. Prove that $p_c^{\text{bond}}(\mathbb{T}_{b+1} \,\square\, \mathbb{N}) \leq (b + 1 - \sqrt{b^2 + 2b - 3}\,)/2$.

7.48. Prove that $p_c^{\text{bond}}(\mathbb{T}_{b+1} \,\square\, \mathbb{Z})$ is at least the smaller positive solution of $(b + 2)^2 p^2 - (b + 1)p^4 = 1$.

7.49. Prove that for every transitive graph G, if b is sufficiently large, then we have the inequality $p_c(\mathbb{T}_{b+1} \,\square\, G) < p_u(\mathbb{T}_{b+1} \,\square\, G)$.

7.50. Show that in Corollary 7.39, the hypotheses of (i)–(iii) may be replaced by, respectively, $\Phi_E(G)/(d - 1) \geq 1/\sqrt{2}$; $\Phi_V(G)/(d - 1) \geq 1/\sqrt{2}$; and $\rho(G) \leq 1/2 + \sqrt{d - 1}/d$.

7.51. Show that Corollary 7.39 holds for quasi-transitive graphs when d is replaced by the maximum degree in G.

7.52. Consider bootstrap percolation with the k-neighbor rule on a $2k$-regular graph, G. Show that if $\Phi_E^*(G) > 0$, then $p(G, k) > 0$. *Hint:* Suppose not. For small p, given the initial configuration ω_p, find an arbitrarily large finite set K such that K becomes occupied even if the outside of K were to be made vacant. Count the increase of the boundary throughout the evolution of the process. Use Exercise 7.39.

7.53. Show that for bootstrap percolation on the d-regular tree $\mathbb{T}_{d+1}$, we have

$$p(\mathbb{T}_{d+1}, d) = 1 - \frac{1}{d} \qquad \text{and} \qquad p(\mathbb{T}_{d+1}, 2) = 1 - \frac{(d - 1)^{2d-3}}{d^{d-1}(d - 2)^{d-2}} \sim \frac{1}{2d^2}. \qquad (7.27)$$

7.54. For bootstrap percolation, define another critical probability, $b(G, k)$, as the infimum of initial probabilities for which, following the k-neighbor rule on G, there will be an infinite, connected component of occupied vertices in the final configuration with positive probability. Clearly, we have $b(G, k) \leq p(G, k)$. Show that for any integers d and k with $2 \leq k \leq d$, if T is an infinite tree with maximum degree $d + 1$, then $p(T, k) \geq b(\mathbb{T}_{d+1}, k) > 0$.

7.55. Consider bootstrap percolation on the family tree T_ξ of a Galton–Watson process with offspring distribution ξ.
   (a) Show that $p(T_\xi, k)$ is a constant almost surely, given nonextinction.
   (b) Prove that $p(T_\xi, k) \geq p(T_\eta, k)$ if η stochastically dominates ξ.

7.56. Consider the Galton–Watson tree T_ξ with offspring distribution $\mathbf{P}(\xi = 2) = \mathbf{P}(\xi = 4) = 1/2$. Then $\text{br}(T_\xi) = \mathbf{E}\xi = 3$ a.s., there are no finite 1-forts in T_ξ, and $0 < p(T_\xi, 2) < 1$ is an almost sure constant by Exercise 7.55. Prove that $p(T_\xi, 2) < p(\mathbb{T}_{3+1}, 2) = 1/9$.

8 | The Mass-Transport Technique and Percolation

Interchanging the order of summation in a double sum is one of the most useful techniques in a mathematician's toolkit. It can always be tried when the sums are finite but requires assumptions when the sums are infinite. This chapter develops a technique that is quite similar to interchange of summation but applies to certain infinite settings where pure interchange of summation would not make sense. This technique, based on the so-called mass-transport principle, is so powerful that its applications often seem magical. The technique has become indispensable in the context of percolation on nonamenable graphs, whose study is continued in this chapter. We will use it again in Chapters 10 and 11 as well. Recall that for us, **percolation** means simply a probability measure on subgraphs of a given graph. There are two main types, **bond percolation**, wherein each subgraph has all the vertices, and **site percolation**, where each subgraph is the graph induced by some of the vertices. In this chapter, all graphs are assumed to be locally finite without explicit mention.

Our main topic is invariant percolation on Cayley graphs. The importance of this topic is, as we will see, that it has many applications to Bernoulli percolation and to random spanning forests. This is not only because these percolations are themselves invariant; it is also because we will construct auxiliary invariant percolations from Bernoulli percolation that will inform us about Bernoulli percolation. In this way, we will show that there is no infinite cluster a.s. at p_c on nonamenable Cayley graphs and show that there is a unique infinite cluster a.s. at p_u on nonamenable, planar Cayley graphs. Likewise, we will also understand what the infinite clusters look like in Bernoulli percolation when there are infinitely many of them.

8.1 The Mass-Transport Principle for Cayley Graphs

Early forms of the mass-transport technique were used by Liggett (1987), Adams (1990), and van den Berg and Meester (1991). It was introduced in the study of percolation by Häggström (1997) and developed further in BLPS (1999b). This method is useful far beyond Bernoulli percolation. The principle on which it depends is best stated in a form that does not mention percolation or even probability at all. However, to motivate it, we first consider probabilistic processes that are invariant with respect to a *countable* group Γ, meaning, in its greatest generality, a probability measure $\mathbf{P}$ on a space Ω on which Γ acts in such a way as to preserve $\mathbf{P}$. For example, we could take $\mathbf{P} := \mathbf{P}_p^{\text{bond}}$ and $\Omega := 2^{\mathsf{E}}$, where E is the edge set of a Cayley graph of Γ. Let $F(x, y; \omega) \in [0, \infty]$ be a function of $x, y \in \Gamma$ and $\omega \in \Omega$. Suppose that F is invariant under the diagonal action of Γ; that is, $F(\gamma x, \gamma y; \gamma \omega) = F(x, y, \omega)$ for all $\gamma \in \Gamma$. We think of giving each element $x \in \Gamma$ some initial mass, possibly depending on ω, then redistributing it so that x sends y the mass $F(x, y; \omega)$. With this terminology, one hopes for "conservation" of mass, at least in expectation. Of course, the total amount of mass is usually infinite. Nevertheless, it turns out that there is a sense in which mass is conserved: the expected mass at an element before transport equals the expected mass at an element afterward. Since F enters this equation only in expectation, it is convenient to set $f(x, y) := \mathbf{E}F(x, y; \omega)$. Then f is also diagonally invariant, that is, $f(\gamma x, \gamma y) = f(x, y)$ for all $\gamma, x, y \in \Gamma$, because $\mathbf{P}$ is invariant. This function f satisfies the following principle:

The Mass-Transport Principle for Countable Groups. *Let Γ be a countable group and o be its identity element. If $f : \Gamma \times \Gamma \to [0, \infty]$ is diagonally invariant, then*

$$\sum_{x \in \Gamma} f(o, x) = \sum_{x \in \Gamma} f(x, o).$$

Proof. Just note that $f(o, x) = f(x^{-1}o, x^{-1}x) = f(x^{-1}, o)$ and that summation of $f(x^{-1}, o)$ over all x^{-1} is the same as $\sum_{x \in \Gamma} f(x, o)$, since inversion is a permutation of Γ.   ◄

Before we use the mass-transport principle in a significant way, we examine a few simple questions to illustrate where it is needed and how it is different from simpler principles. Let G be a Cayley graph of the infinite group Γ, and let $\mathbf{P}$ be an invariant percolation, that is, an invariant measure on 2^V, on 2^E, or even on $2^{V \cup E}$. Let ω be a configuration with distribution $\mathbf{P}$.

Example 8.1. Could it be that ω is a single vertex a.s.? In other words, is there an invariant way to pick a vertex at random?

No: if there were, the assumptions would imply that the probability p of picking x is the same for all x, whence an infinite sum of p would equal 1, an impossibility.

Example 8.2. Could it be that ω is a finite, nonempty vertex set a.s.? In other words, is there an invariant way to pick a finite set of vertices at random?

No: if there were, then we could pick one of the vertices of the finite set at random (uniformly), thereby obtaining an invariant probability measure on singletons.

Recall that *cluster* means connected component of the percolation subgraph.

Example 8.3. The number of *finite* clusters is $\mathbf{P}$-a.s. 0 or ∞. For if not, then we could condition on the number of finite clusters being finite and positive, then take the set of their vertices and arrive at an invariant probability measure on 2^V that is concentrated on finite sets.

Recall that a vertex x is a *furcation* of a configuration ω if removing x would split the cluster containing x into at least three infinite clusters.

Example 8.4. The number of furcations is $\mathbf{P}$-a.s. 0 or ∞. For if not, the set of furcations conditioned to be finite and nonempty has an invariant distribution on 2^V.

Example 8.5. $\mathbf{P}$-a.s. *each* cluster has 0 or ∞ furcations.

This does not follow from elementary considerations as the previous examples do but requires the mass-transport principle. (See Exercise 8.14 for a proof that elementary considerations do not suffice.) To prove the assertion about furcations, for a vertex x

and a configuration ω, let $N(x, \omega)$ be the number of furcations in the cluster $K(x) = K(x, \omega)$. Define $F(x, y; \omega)$ to be 0 if $N(x, \omega) \in \{0, \infty\}$ and otherwise to be $1/N(x, \omega)$ if y is one of the furcations of $K(x)$. Then F is diagonally invariant, whence the mass-transport principle applies to $f(x, y) := \mathbf{E} F(x, y; \omega)$. Since $\sum_y F(x, y; \omega) \le 1$, we have

$$\sum_x f(o, x) \le 1. \tag{8.1}$$

If any cluster has a finite, positive number of furcations, then each of them receives infinite mass. More precisely, if o is one of a finite number of furcations of $K(o)$, then $\sum_x F(x, o; \omega) = \infty$. Therefore, if with positive probability some cluster has a finite, positive number of furcations, then with positive probability o is one of a finite number of furcations of $K(o)$, and therefore $\mathbf{E}\left[\sum_x F(x, o; \omega) \right] = \infty$. That is, $\sum_x f(x, o) = \infty$, which contradicts the mass-transport principle and (8.1).

This is a typical application of the mass-transport principle in that it is qualitative, not quantitative.

A generalization of Examples 8.2 and 8.5 is the following:

Example 8.6. If there are infinite clusters with positive probability, then there is no invariant way to pick a finite, nonempty subset from one or more infinite clusters, whether deterministically or with additional randomness. More precisely, there is no invariant measure on pairs (ω, ϕ) such that ω has infinite clusters with positive probability and with the properties that $\phi \colon \mathsf{V} \to 2^{\mathsf{V}}$ is a function such that $\phi(x) \subseteq K(x)$ is finite for all x, is nonempty for at least one x with $|K(x)| = \infty$, and such that whenever x and y lie in the same ω-cluster, $\phi(x) = \phi(y)$.

To illustrate a deeper use of the mass-transport principle, we now give another proof that in the standard coupling of Bernoulli bond percolation on a Cayley graph, G, for all $p_c(G) < p_1 < p_2$, a.s. every infinite p_2-cluster contains some infinite p_1-cluster. (This was Lemma 7.26. This lemma implies that the two natural definitions of p_u are equivalent, (7.8) and (7.9); the full Theorem 7.21 that we used to deduce (7.9) is not needed.)

Let $\omega_1 \subseteq \omega_2$ be the configurations in the standard coupling of the two Bernoulli(p_i) percolations. Note that (ω_1, ω_2) is invariant. Write $K_2(x)$ for the cluster of x in ω_2.

Let η denote the union of all infinite clusters of ω_1. Define $F\big(x, y; (\omega_1, \omega_2)\big)$ to be 1 if x and y belong to the same ω_2-cluster and y is the unique vertex in η that is closest in ω_2 to x; otherwise, define $F\big(x, y; (\omega_1, \omega_2)\big) := 0$. Note that if there is not a unique such y, then x does not send any mass anywhere. Now F is diagonally invariant and $\sum_y F\big(x, y; (\omega_1, \omega_2)\big) \le 1$.

Suppose that with positive probability there is an infinite cluster of ω_2 that is disjoint from η. Let $A(z, y, e_1, e_2, \ldots, e_n)$ be the event that $K_2(z)$ is infinite and disjoint from η, that $y \in \eta$, and that $e_1, e_2, \ldots, e_n$ form a path of edges from z to y that lies outside $K_2(z) \cup \eta$. Whenever there is an infinite cluster of ω_2 that is disjoint from η, there must exist two vertices z and y and some edges $e_1, e_2, \ldots, e_n$ for which $A(z, y, e_1, e_2, \ldots, e_n)$ holds. Hence, there exists $z, y, e_1, \ldots, e_n$ such that $\mathbf{P}\big(A(z, y, e_1, \ldots, e_n)\big) > 0$. Let $h \colon [0, 1] \to [p_1, p_2]$ be affine and surjective. If B denotes the event obtained by replacing each label $U(e_k)$ ($k = 1, \ldots, n$) by $h(U(e_k))$ on each configuration in $A(z, y, e_1, \ldots, e_n)$, then $\mathbf{P}(B) > 0$ by Lemma 7.24. On the event B, we have $F\big(x, y; (\omega_1, \omega_2)\big) = 1$ for all $x \in K_2(z)$, whence $\sum_x \mathbf{E} F\big(x, y; (\omega_1, \omega_2)\big) = \infty$. This contradicts the mass-transport principle.

8.2 Beyond Cayley Graphs: Unimodularity

It would be nice if the mass-transport principle, in the form given in Section 8.1, held for all transitive graphs. But it does not. For example, if G is the grandparent graph of Example 7.1 with T having degree 3, then consider the function $f(x, y)$ that is the indicator of y being the ξ-grandparent of x. To analyze this, we need the result of the following exercise:

▷ **Exercise 8.1.**
Show that every automorphism of the grandparent graph G fixes the end ξ and therefore that f is diagonally invariant under $\mathrm{Aut}(G)$.

Although f is diagonally invariant, we have that $\sum_{x \in \mathsf{V}(G)} f(o, x) = 1$, yet the other way, $\sum_{x \in \mathsf{V}(G)} f(x, o) = 4$. In particular, G is not a Cayley graph.

▷ **Exercise 8.2.**
Show that the Diestel–Leader graph of Example 7.2 is not a Cayley graph.

Nevertheless, there are many transitive graphs for which the mass-transport principle does hold, the so-called unimodular graphs, and there is a generalization of the mass-transport principle that holds for all graphs. The case of unimodular (quasi-transitive) graphs is the most important case and the one that we will focus on in the rest of this chapter. Even if we were interested only in Cayley graphs, the use of planar duality for planar Cayley graphs would force us to consider quasi-transitive graphs, as we will see in Section 8.5.

We will use the following notation throughout the rest of this chapter. Let G be a connected, locally finite graph and Γ be a group of automorphisms of G. Denote the **stabilizer** of $x \in G$ by $S(x) := \{\gamma \in \Gamma ; \; \gamma x = x\}$. Since all points in $S(x)y := \{\gamma y ; \; \gamma \in S(x)\}$ are at the same distance from x and G is connected and locally finite, the set $S(x)y$ is finite for all x and y.

Theorem 8.7. (Mass-Transport Principle) *If Γ is a group of automorphisms of a connected, locally finite graph $G = (\mathsf{V}, \mathsf{E})$, $f : \mathsf{V} \times \mathsf{V} \to [0, \infty]$ is invariant under the diagonal action of Γ, and $u, w \in \mathsf{V}$, then*

$$\sum_{z \in \Gamma w} f(u, z) = \sum_{y \in \Gamma u} f(y, w) \frac{|S(y)w|}{|S(w)y|} \,. \tag{8.2}$$

This formula is too complicated to remember, but note the form it takes when Γ acts transitively:

Corollary 8.8. *If Γ is a transitive group of automorphisms of a connected, locally finite graph $G = (\mathsf{V}, \mathsf{E})$, $f : \mathsf{V} \times \mathsf{V} \to [0, \infty]$ is invariant under the diagonal action of Γ, and $o \in \mathsf{V}$, then*

$$\sum_{x \in \mathsf{V}} f(o, x) = \sum_{x \in \mathsf{V}} f(x, o) \frac{|S(x)o|}{|S(o)x|} \,. \tag{8.3}$$

Note how this works to restore "conservation of mass" for the grandparent graph of Example 7.1: Suppose again that the graph is based on a tree of degree 3. If o is the ξ-grandparent of x, then $|S(x)o| = 1$ and $|S(o)x| = 4$, so that the left-hand side of (8.3) is the sum of one term equal to 1, whereas the right-hand side is the sum of four terms, each equal to 1/4.

We say* that Γ is **unimodular** if $|S(x)y| = |S(y)x|$ for all pairs (x, y) such that $y \in \Gamma x$. We also say that a graph is **unimodular** when its full automorphism group is. If Γ is unimodular and also transitive, then (8.3) simplifies to

$$\sum_{x \in \mathsf{V}} f(o, x) = \sum_{x \in \mathsf{V}} f(x, o). \tag{8.4}$$

Thus, all the applications of the mass-transport principle in Section 8.1, which were qualitative, apply (with the same proofs) to transitive unimodular graphs. In fact, they apply as well to all quasi-transitive unimodular graphs, since for such graphs, $|S(x)y|/|S(y)x|$ is bounded over all pairs (x, y) by the upcoming Theorem 8.10, and we may sum (8.2) over all pairs (u, w) chosen from a complete set of orbit representatives.

The proof of Theorem 8.7 is not particularly pleasant, but its applications will more than make up for our trouble. To prove Theorem 8.7, let

$$\Gamma_{x,y} := \{\gamma \in \Gamma \,;\; \gamma x = y\}\,.$$

Note that for any $\gamma \in \Gamma_{x_1,x_2}$, we have

$$\Gamma_{x_1,x_2} = \gamma S(x_1) = S(x_2)\gamma\,.$$

Therefore, for all x_1, x_2, y_1 and any $\gamma \in \Gamma_{x_1,x_2}$,

$$|\Gamma_{x_1,x_2} y_1| = |\gamma S(x_1) y_1| = |S(x_1) y_1|\,, \tag{8.5}$$

and, writing $y_2 := \gamma y_1$, we have

$$|\Gamma_{x_1,x_2} y_1| = |S(x_2)\gamma y_1| = |S(x_2) y_2|\,. \tag{8.6}$$

Proof of Theorem 8.7. Let $z \in \Gamma w$, so that $w = \gamma z$ for some $\gamma \in \Gamma_{z,w}$. If $y := \gamma u$, then $f(u, z) = f(\gamma u, \gamma z) = f(y, w)$. That is, $f(u, z) = f(y, w)$ whenever $y \in \Gamma_{z,w} u$. Therefore,

$$\sum_{z \in \Gamma w} f(u, z) = \sum_{z \in \Gamma w} \frac{1}{|\Gamma_{z,w} u|} \sum_{y \in \Gamma_{z,w} u} f(y, w) = \sum_{y \in \Gamma u} f(y, w) \sum_{z \in \Gamma_{y,u} w} \frac{1}{|\Gamma_{z,w} u|}\,,$$

where our interchange of the order of summation in the last step used the calculation

$$\begin{aligned}
\{(z, y)\,;\; z \in \Gamma w,\ y \in \Gamma_{z,w} u\} &= \{(z, y)\,;\; \exists \gamma \in \Gamma \ \ y = \gamma u,\ w = \gamma z\} \\
&= \{(z, y)\,;\; \exists \gamma \in \Gamma \ \ u = \gamma^{-1} y,\ z = \gamma^{-1} w\} \\
&= \{(z, y)\,;\; y \in \Gamma u,\ z \in \Gamma_{y,u} w\}\,.
\end{aligned}$$

In particular, for such (z, y), there is some $\gamma \in \Gamma_{z,w}$ such that $\gamma u = y$. Therefore, using (8.6) and then (8.5), we may rewrite this as

$$\sum_{y \in \Gamma u} f(y, w) \frac{|\Gamma_{y,u} w|}{|S(w) y|} = \sum_{y \in \Gamma u} f(y, w) \frac{|S(y) w|}{|S(w) y|}\,. \qquad \blacktriangleleft$$

▷ **Exercise 8.3.**
Let Γ be a transitive group of automorphisms of a graph that satisfies (8.4) for all Γ-invariant f. Show that Γ is unimodular.

---

  * Although we use the term "unimodular" for a group when our definition is in terms of its action on a graph, we will see that it really depends only on the group, once we are given a natural topology on the group that comes from its action on the graph. See the paragraph after (8.10).

▷ **Exercise 8.4.**

Show that if Γ is a transitive, unimodular group of automorphisms and Γ' is a larger group of automorphisms of the same graph, then Γ' is also transitive and unimodular.

By Exercise 8.3 and the mass-transport principle for countable groups, every Cayley graph is unimodular. How do we find other unimodular graphs? Call a group Γ of automorphisms *discrete* if all stabilizers are finite. For example, when a group acts on its Cayley graph, the stabilizers are singletons. We now show that all discrete groups are unimodular. Recall that $[\Gamma : \Gamma']$ means the index of a subgroup Γ' in a group Γ, that is, the number of cosets of Γ' in Γ.

▷ **Exercise 8.5.**

Show that for all x and y, we have $|S(x)y| = [S(x) : S(x) \cap S(y)]$.

Proposition 8.9. *If Γ is a discrete group of automorphisms, then Γ is unimodular.*

Proof. Suppose that $y = \gamma x$. Then $S(x)$ and $S(y)$ are conjugate subgroups since $\gamma_1 \mapsto \gamma\gamma_1\gamma^{-1}$ is a bijection of $S(x)$ to $S(y)$. Thus, $|S(x)| = |S(y)|$, whence

$$\frac{|S(x)y|}{|S(y)x|} = \frac{[S(x) : S(x) \cap S(y)]}{[S(y) : S(x) \cap S(y)]} = \frac{|S(x)|/|S(x) \cap S(y)|}{|S(y)|/|S(x) \cap S(y)|} = \frac{|S(x)|}{|S(y)|} = 1. \qquad \blacktriangleleft$$

Sometimes we can make additional use of the mass-transport principle by employing the following fact.

Theorem 8.10. *If Γ is a group of automorphisms of any connected, locally finite graph G, then there are nonzero numbers μ_x ($x \in V$) that are unique up to a constant multiple such that for all x and y,*

$$\frac{\mu_x}{\mu_y} = \frac{|S(x)y|}{|S(y)x|} . \tag{8.7}$$

Proof. Recall* that subgroup index is multiplicative: if Γ_3 is a subgroup of Γ_2, which in turn is a subgroup of Γ_1, then

$$[\Gamma_1 : \Gamma_2][\Gamma_2 : \Gamma_3] = [\Gamma_1 : \Gamma_3] . \tag{8.8}$$

Applying (8.8) to $\Gamma_3 := S(x) \cap S(y) \cap S(z)$, $\Gamma_2 := S(x) \cap S(y)$, and Γ_1 equal to either $S(x)$ or $S(y)$, we get

$$\frac{[S(x) : S(x) \cap S(y)]}{[S(y) : S(x) \cap S(y)]} = \frac{[S(x) : S(x) \cap S(y) \cap S(z)]}{[S(y) : S(x) \cap S(y) \cap S(z)]} .$$

There are three forms of this equation arising from the three cyclic permutations of the ordered triple (x, y, z). Combining them with Exercise 8.5, we obtain the "cocycle" identity

$$\frac{|S(x)y|\,|S(y)z|}{|S(y)x|\,|S(z)y|} = \frac{|S(x)z|}{|S(z)x|} \tag{8.9}$$

_____

* Here is a proof: if $A \subset \Gamma_1$ is a set of coset representatives of Γ_1/Γ_2, then the map $(a, \Gamma_2/\Gamma_3) \mapsto a\Gamma_2/\Gamma_3$ is a bijection of $A \times \Gamma_2/\Gamma_3 \to \Gamma_1/\Gamma_3$.

for all x, y, z.

Thus, if we fix $o \in V$, choose $\mu_o \in \mathbb{R} \setminus \{0\}$, and define $\mu_x := \mu_o |S(x)o|/|S(o)x|$, then for all x, y, we get

$$\frac{\mu_x}{\mu_y} = \frac{\mu_o |S(x)o|/|S(o)x|}{\mu_o |S(y)o|/|S(o)y|} = \frac{|S(x)y|}{|S(y)x|}$$

by (8.9). This shows the claimed existence of the numbers μ_x. On the other hand, if numbers μ'_x ($x \in V$) also satisfy (8.7), then

$$\frac{\mu_x}{\mu'_x} = \frac{\mu_o |S(x)o|/|S(o)x|}{\mu'_o |S(x)o|/|S(o)x|} = \frac{\mu_o}{\mu'_o}$$

for all x, so $\mu'_x = C\mu_x$ for all x, where $C := \mu'_o/\mu_o$.   ◀

Note that Γ is unimodular iff $\mu_y = \mu_x$ whenever $y \in \Gamma x$.

The following exercises makes it easier to check the definition of unimodularity.

▷ **Exercise 8.6.**
Show that if Γ acts transitively, then Γ is unimodular iff $|S(x)y| = |S(y)x|$ for all edges $[x, y]$.

▷ **Exercise 8.7.**
Show that if for all edges $[x, y]$, there is some $\gamma \in \Gamma$ such that $\gamma x = y$ and $\gamma y = x$, then Γ is unimodular.

Using the numbers μ_x, we may write (8.2) as

$$\sum_{z \in \Gamma w} f(u, z)\mu_w = \sum_{y \in \Gamma u} f(y, w)\mu_y \,. \tag{8.10}$$

We now state the relevance of Haar measure for the interested reader. Consider the closure $\bar{\Gamma}$ of Γ if Γ is not already closed, where the topology is defined in Exercise 8.20. This is a locally compact topological group. As such, it has both a left-invariant Borel measure and a right-invariant Borel measure, each one being finite on compact sets, called *left Haar* and *right Haar* measures. On countable groups, counting measure serves as Haar measure. See Exercise 8.22 for a simple construction of Haar measures on automorphism groups. If there is a left Haar measure that is also right invariant, then the group is classically called *unimodular*. This agrees with our definition in the sense that Γ is unimodular in our sense iff $\bar{\Gamma}$ is unimodular in the classical sense. It is not hard to prove this by imitating the proof of Proposition 8.9 to show that μ_x is the left-invariant Haar measure of $S(x)$ in $\bar{\Gamma}$. Since discrete groups are countable, this also makes clear why discrete groups are unimodular. One can also use Haar measure to give a simple proof of Theorem 8.7; see Section 8.9. In the amenable, quasi-transitive case, Exercise 8.31 gives another interpretation of the weights μ_x.

Given that there is such a simple form (8.4) of the mass-transport principle for transitive, unimodular graphs, we might hope for almost as simple a form that applies to quasi-transitive, unimodular graphs. Our hopes will be met by (8.12), which is explicitly written out in our next corollary:

Corollary 8.11. *Let* Γ *be a quasi-transitive group of automorphisms of a connected, locally finite graph,* G. *Choose a complete set* $\{o_1, \ldots, o_L\}$ *of representatives in* V *of the orbits of* Γ. *Let* μ_i *be the weight of* o_i *as given by Theorem 8.10. If* Γ *is unimodular, then whenever* $f \colon \mathsf{V} \times \mathsf{V} \to [0, \infty]$ *is invariant under the diagonal action of* Γ, *we have*

$$\sum_{i=1}^{L} \mu_i^{-1} \sum_{z \in \mathsf{V}} f(o_i, z) = \sum_{j=1}^{L} \mu_j^{-1} \sum_{y \in \mathsf{V}} f(y, o_j). \tag{8.11}$$

Proof. Since Γ is unimodular, $\mu_y = \mu_i$ for $y \in \Gamma o_i$. Thus, for each i and j, (8.10) gives

$$\sum_{z \in \Gamma o_j} f(o_i, z)\mu_j = \sum_{y \in \Gamma o_i} f(y, o_j)\mu_i,$$

in other words,

$$\mu_i^{-1} \sum_{z \in \Gamma o_j} f(o_i, z) = \mu_j^{-1} \sum_{y \in \Gamma o_i} f(y, o_j).$$

Adding these equations over all i and j gives the desired result. ◀

Because of this result, in the quasi-transitive, unimodular case, we will always assume that the weights are chosen so that $\sum_i \mu_i^{-1} = 1$. It then makes sense to think of o_i being picked randomly with probability μ_i^{-1}. If we denote such a random root by $\hat{o}$, then (8.11) assumes a very simple form:

$$\mathbf{E}\left[\sum_x f(\hat{o}, x)\right] = \mathbf{E}\left[\sum_x f(x, \hat{o})\right]. \tag{8.12}$$

This will be the usual way we apply the mass-transport principle on quasi-transitive, unimodular graphs. We will call such a random root $\hat{o}$ **normalized**. For example, the *inverse* weights of the two types of vertices in the graph of Figure 8.1 are 1/5 and 4/5.

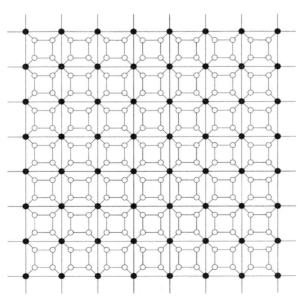

Figure 8.1. A quasi-transitive graph, 1/5 of whose vertices are black, and 4/5 are white.

To make it even more useful, a converse can be added to Corollary 8.11:

Proposition 8.12. *Let Γ be a quasi-transitive group of automorphisms of a connected, locally finite graph, G. Choose a complete set $\{o_1, \ldots, o_L\}$ of representatives in V of the orbits of Γ. Let μ_i be the weight of o_i as given by Theorem 8.10. If there exist numbers $v_x \geq 0$ for $x \in \mathsf{V}(G)$ such that $0 < \sum_x v_x < \infty$ and*

$$\sum_{x \in \mathsf{V}} v_x \sum_{z \in \mathsf{V}} f(x, z) = \sum_{x \in \mathsf{V}} v_x \sum_{y \in \mathsf{V}} f(y, x) \tag{8.13}$$

whenever $f: \mathsf{V} \times \mathsf{V} \to [0, \infty]$ is invariant under the diagonal action of Γ, then Γ is unimodular and

$$\mu_i^{-1} = \sum_{x \in \Gamma o_i} v_x \tag{8.14}$$

for $1 \leq i \leq L$.

Note that (8.11) is the special case of (8.13) where $v_x = \mu_i^{-1}$ if $x = o_i$, and $v_x = 0$ otherwise.

Proof. Assume that v_x are as stated. Define

$$a_i := \sum_{x \in \Gamma o_i} v_x .$$

We first show that $a_i > 0$. Let each vertex x send mass 1 to each vertex in Γo_i that is nearest to x. Since the left-hand side of (8.13) is positive, so is the right-hand side. Since only vertices in Γo_i receive mass, it follows that $a_i > 0$, as desired.

To see that Γ is unimodular, consider any j, k and any $u \in \Gamma o_j$ and $v \in \Gamma o_k$. Let $f(x, y) := \mathbf{1}_{\Gamma_{u,x} v}(y)$. It is straightforward to check that f is diagonally invariant under Γ. Note that

$$|S(x)z| \, \mathbf{1}_{\Gamma x}(y) = |\Gamma_{x,y} z|$$

for all $x, y, z \in \mathsf{V}(G)$ and that

$$z \in \Gamma_{u,x} v \iff x \in \Gamma_{v,z} u . \tag{8.15}$$

Therefore, we have

$$\begin{aligned}
|S(u)v| a_j &= \sum_y v_y |\Gamma_{u,y} v| = \sum_y v_y \sum_x \mathbf{1}_{\Gamma_{u,y} v}(x) \\
&= \sum_y v_y \sum_x f(y, x) = \sum_z v_z \sum_x f(x, z) \qquad \text{[by (8.13)]} \\
&= \sum_z v_z \sum_x \mathbf{1}_{\Gamma_{u,x} v}(z) = \sum_z v_z \sum_x \mathbf{1}_{\Gamma_{v,z} u}(x) \qquad \text{[by (8.15)]} \\
&= \sum_z v_z |\Gamma_{v,z} u| = |S(v)u| a_k .
\end{aligned}$$

In other words,

$$|S(u)v| \, a_j = |S(v)u| \, a_k . \tag{8.16}$$

If we take $j = k$, then we see that Γ is unimodular. In general, comparison of (8.16) with (8.7) shows (8.14). ◀

Remark 8.13. The proof shows that unimodularity and (8.14) follow from verifying (8.13) only for those f taking values 0 and 1 and for which $f(x, y) = 0$ whenever $\mathrm{dist}_G(x, y) > M$, where M is a constant that depends on f.

▷ **Exercise 8.8.**
Extend Exercise 8.4 to the quasi-transitive case: show that if Γ is a quasi-transitive, unimodular group of automorphisms and Γ' is a larger group of automorphisms of the same graph, then Γ' is also quasi-transitive and unimodular.

It is nice to know that when one is in the amenable setting, unimodularity is automatic, as shown by Soardi and Woess (1990):

Proposition 8.14. (Amenability Implies Unimodularity) *Any transitive group Γ of automorphisms of an amenable graph G is unimodular.*

Proof. The idea is the same as that which prevents Ponzi schemes from working: mass must be conserved approximately on Følner sets, and hence exactly in the whole graph in the sense of the mass-transport principle.

Let $\langle F_n \rangle$ be a sequence of finite sets of vertices in G such that $|\overline{F_n}|/|F_n| \to 1$ as $n \to \infty$, where $\overline{F_n} := F_n \cup \partial_V F_n$ is the union of F_n with its exterior vertex boundary.

Fix neighboring vertices $x \sim y$. We count the number of pairs (z, w) such that $z \in F_n$ and $w \in \Gamma_{x,z} y$ (or equivalently, $z \in \Gamma_{y,w} x$) in two ways: by summing over z first or over w first. In view of (8.5), this gives

$$|F_n| \, |S(x)y| = \sum_{z \in F_n} \sum_{w \in \Gamma_{x,z} y} 1 \le \sum_{w \in \overline{F_n}} \sum_{z \in \Gamma_{y,w} x} 1 = |\overline{F_n}| \, |S(y)x| \,.$$

Dividing both sides by F_n and taking a limit, we get $|S(x)y| \le |S(y)x|$. By symmetry and Exercise 8.6, we are done. ◀

Likewise, quasi-transitive amenable graphs are unimodular: see Exercise 8.30.

8.3 Infinite Clusters in Invariant Percolation

What happens in Bernoulli percolation at p_c itself, that is, is there an infinite cluster a.s.? Extending the classical conjecture for Euclidean lattices, Benjamini and Schramm (1996b) made the following conjecture:

Conjecture 8.15. (No Infinite Clusters at Criticality) *If G is any quasi-transitive graph with $p_c(G) < 1$, then $\theta(p_c(G)) = 0$.*

In the next section, we will establish this conjecture under the additional hypotheses that G is nonamenable and unimodular. This will utilize in a crucial way properties of infinite clusters in invariant percolation processes beyond Bernoulli percolation. This is the topic of the present section. What is the advantage conferred by nonamenability that allows a resolution of this conjecture in that case but not in the amenable case? In essence, it is that *all* finite sets have a comparatively large boundary; since clusters are chosen at random in percolation, this is an advantage compared to knowing, in the amenable case, that *certain* large sets have comparatively small boundary. This may appear to be a tautology at the moment, but our next theorem should begin to clarify its meaning.

We begin with a simple but powerful result on a threshold for having infinite clusters. Write $\deg_K(x)$ for the degree of x as a vertex in a subgraph K. **As in the preceding chapter, we will use $\Phi_\mathsf{E}(G)$ to mean the expansion constant $\Phi_\mathsf{E}(G, 1, 1)$.**

Theorem 8.16. (Thresholds for Finite Clusters) *Let G be a transitive, unimodular graph of degree d_G and $\mathbf{P}$ be an automorphism-invariant probability measure on 2^E. If $\mathbf{P}$-a.s. all clusters are finite, then*

$$\mathbf{E}[\deg_\omega o] \leq d_G - \Phi_\mathsf{E}(G).$$

This was first proved for regular trees by Häggström (1997) and then extended by BLPS (1999b). Häggström (1997) showed that the threshold is sharp for regular trees. Of course, it is useless when G is amenable. For Bernoulli percolation, the bound it gives on p_c is worse than Theorem 6.46. But it is extremely useful for a wide variety of other percolation processes, including those that we will use to study Bernoulli percolation.

We could have stated the theorem more generally with a unimodular group $\Gamma \subseteq \mathrm{Aut}(G)$ acting transitively on G and $\mathbf{P}$ being assumed invariant under Γ. In fact, all such results that we present have similar generalizations, but for simplicity of language, we will refer just to "invariant percolation" on a transitive or quasi-transitive graph.

To prove Theorem 8.16, define, for finite subgraphs $K = \big(\mathsf{V}(K), \mathsf{E}(K)\big) \subset G$,

$$\alpha_K := \frac{1}{|\mathsf{V}(K)|} \sum_{x \in \mathsf{V}(K)} \deg_K(x),$$

the average (internal) degree of K. Then set*

$$\alpha(G) := \sup \{\alpha_K \,;\; K \subset G \text{ is finite}\}.$$

If G is a regular graph of degree d_G, then

$$\alpha(G) + \Phi_\mathsf{E}(G) = d_G. \tag{8.17}$$

This is because we may restrict attention to induced subgraphs, and for induced subgraphs,

$$\sum_{x \in \mathsf{V}(K)} \deg_K(x) + |\partial_{\mathsf{E}(G)} \mathsf{V}(K)| = \sum_{x \in \mathsf{V}(K)} \deg_G(x) = d_G |\mathsf{V}(K)|.$$

Dividing by $|\mathsf{V}(K)|$ gives (8.17).

In view of (8.17), we may write the conclusion of Theorem 8.16 as

$$\mathbf{E}[\deg_\omega o] \leq \alpha(G). \tag{8.18}$$

This inequality is actually rather intuitive: it says that random finite clusters have average degree no more than the supremum average degree of arbitrary finite subgraphs. Of course, the first sense of "average" is "expectation," whereas the second is "arithmetic mean." This is reminiscent of the ergodic theorem, which says that a spatial average (expectation) is the limit of time averages (arithmetic means).

---

* We remark that $\beta(G) = 2/\alpha(G)$, with $\beta(G)$ defined as in Section 6.5, except that β was defined with a liminf, rather than an infimum.

Proof of Theorem 8.16. We use the mass-transport principle (8.4). Start with mass $\deg_\omega x$ at each vertex x and redistribute it equally among the vertices in its cluster $K(x)$ (including x itself). After transport, the mass at x is $\alpha_{K(x)}$. This is an invariant transport, so that if $f(x, y)$ denotes the expected mass taken from x and transported to y, then we have

$$\mathbf{E}[\deg_\omega o] = \sum_x f(o, x) = \sum_x f(x, o) = \mathbf{E}[\alpha_{K(o)}].$$

By definition, $\alpha_{K(o)} \le \alpha(G)$, whence (8.18) follows.   ◀

One version of Theorem 8.16 for quasi-transitive graphs is as follows.

Corollary 8.17. *Let G be a quasi-transitive, nonamenable, unimodular graph. There is some $\epsilon > 0$ such that if $\mathbf{P}$ is an automorphism-invariant probability measure on 2^E with all clusters finite $\mathbf{P}$-a.s., then for some $x \in V$,*

$$\mathbf{E}[\deg_\omega x] \le \deg_G x - \epsilon.$$

Proof. Let G' be a transitive representation of G. Then $\mathbf{P}$ induces an invariant percolation $\mathbf{P}'$ on G' by letting an edge $[x, y]$ in G' be open iff x and y are joined by a path of open edges in G of length at most $2r + 1$ (where r is as in the definition of transitive representation). If $\mathbf{P}$ gives a high expected degree to all x, then so does $\mathbf{P}'$, whence the conclusion follows easily from Theorem 8.16.   ◀

Some variations on Theorem 8.16 are contained in the following exercises as well as in others at the end of the chapter.

▷ **Exercise 8.9.**
Show that for any invariant percolation $\mathbf{P}$ on subgraphs of a transitive, unimodular graph that has only finite clusters a.s., $\mathbf{E}[\deg_\omega o \mid o \in \omega] \le \alpha(G)$. More generally, let G be a quasi-transitive, unimodular graph with a normalized random root $\hat{o}$. Show that if $\mathbf{P}$ is an invariant percolation on subgraphs of G such that all clusters are finite a.s., then $\mathbf{E}[\deg_\omega \hat{o} \mid \hat{o} \in \omega] \le \alpha(G)$.

▷ **Exercise 8.10.**
Let $\mathbf{P}$ be an invariant percolation on subgraphs of a transitive, unimodular graph such that all clusters are finite trees a.s. Show that $\mathbf{E}[\deg_\omega o \mid o \in \omega] < 2$. More generally, let G be a quasi-transitive, unimodular graph with a normalized random root $\hat{o}$. Show that if $\mathbf{P}$ is an invariant percolation on subgraphs of G such that all clusters are finite trees a.s., then $\mathbf{E}[\deg_\omega \hat{o} \mid \hat{o} \in \omega] < 2$.

We now take a look at forest percolations with infinite trees. Trees have proved their worth to us before in studying Cayley graphs, and they will continue to do so.

Proposition 8.18. *Let G be a quasi-transitive, unimodular graph with a normalized random root $\hat{o}$. Let $\mathfrak{F}$ be the configuration of an invariant percolation on G, independent of $\hat{o}$, such that $\mathfrak{F}$ is a forest, all of whose trees are infinite a.s. The following dichotomy holds:*
   (i) *if each tree in $\mathfrak{F}$ has one or two ends a.s., then $\mathbf{E}\big[\deg_{\mathfrak{F}} \hat{o} \mid \hat{o} \in \mathfrak{F}\big] = 2$;*
   (ii) *if some tree in $\mathfrak{F}$ has ≥ 3 ends with positive probability, then $\mathbf{E}\big[\deg_{\mathfrak{F}} \hat{o} \mid \hat{o} \in \mathfrak{F}\big] > 2$.*

Proof. We use a mass transport that is subtler than our earlier ones. To set it up, let $\xi(x, y; \mathfrak{F})$ be the indicator that there is a ray in $\mathfrak{F}$ starting at x whose first vertex after x is y. Let

$$F(x, y; \mathfrak{F}) := \begin{cases} 2\xi(x, y; \mathfrak{F}) & \text{if } \xi(y, x; \mathfrak{F}) = 0 \\ \xi(x, y; \mathfrak{F}) & \text{otherwise,} \end{cases}$$

and $f(x, y) := \mathbf{E}[F(x, y; \mathfrak{F})]$. Now $\sum_x (F(o, x; \mathfrak{F}) + F(x, o; \mathfrak{F})) = 2(\deg_{\mathfrak{F}} o)\mathbf{1}_{[o \in \mathfrak{F}]}$, so that $\sum_x (f(o, x) + f(x, o)) = 2 \mathbf{E}[\deg_{\mathfrak{F}} o \,;\, o \in \mathfrak{F}]$. By (8.12), we obtain that

$$\mathbf{E}[\deg_{\mathfrak{F}} \hat{o} \,;\, \hat{o} \in \mathfrak{F}] = \frac{1}{2}\,\mathbf{E}\Big[\sum_x (f(\hat{o}, x) + f(x, \hat{o}))\Big] = \mathbf{E}\Big[\sum_x f(\hat{o}, x)\Big]. \tag{8.19}$$

Now in case (i), $\sum_x F(o, x; \mathfrak{F}) = 2 \cdot \mathbf{1}_{[o \in \mathfrak{F}]}$, whence by (8.19), we have $\mathbf{E}[\deg_{\mathfrak{F}} \hat{o} \,;\, \hat{o} \in \mathfrak{F}] = 2\,\mathbf{P}[\hat{o} \in \mathfrak{F}]$. This gives the desired result.

On the other hand, in case (ii), $\sum_x F(o, x; \mathfrak{F}) \geq 2$ for all $o \in \mathfrak{F}$ and $\sum_x F(o, x; \mathfrak{F}) \geq 3$ for all furcations $o \in \mathfrak{F}$; note that $\hat{o}$ is a furcation with positive probability, since $\mathfrak{F}$ has trees with at least three ends. Therefore, (8.19) yields $\mathbf{E}[\deg_{\mathfrak{F}} \hat{o} \,;\, \hat{o} \in \mathfrak{F}] > 2\,\mathbf{P}[\hat{o} \in \mathfrak{F}]$.   ◄

As a consequence of this, the number of ends is related to the critical value for Bernoulli percolation on trees that arise in invariant percolation:

Theorem 8.19. *Let G be a quasi-transitive, unimodular graph and $\hat{o}$ a normalized random root. Let $\mathfrak{F}$ be the configuration of an invariant percolation on G, independent of $\hat{o}$, such that $\mathfrak{F}$ is a forest a.s. Then the following are equivalent:*
  (i) *some component of $\mathfrak{F}$ has at least three ends with positive probability;*
  (ii) *some component of $\mathfrak{F}$ has $p_c < 1$ with positive probability;*
 (iii) $\mathbf{E}[\deg_{\mathfrak{F}} \hat{o} \mid |K(\hat{o})| = \infty] > 2$.

Of course, by Theorem 5.15, (ii) is equivalent to saying that some component of $\mathfrak{F}$ has branching number > 1 with positive probability.

Proof. Let $\mathfrak{F}'$ be the mixed (site and bond) percolation obtained from $\mathfrak{F}$ by retaining only those vertices and edges that belong to an infinite cluster. Then we may rewrite (iii) as $\mathbf{E}[\deg_{\mathfrak{F}'} \hat{o} \mid \hat{o} \in \mathfrak{F}'] > 2$.

The implication (i) implies (iii) is immediate from Proposition 8.18.

Now assume (iii). Let p be sufficiently close to 1 that independent Bernoulli(p) bond percolation on $\mathfrak{F}'$ yields a configuration $\mathfrak{F}''$ with $\mathbf{E}[\deg_{\mathfrak{F}''} \hat{o} \mid \hat{o} \in \mathfrak{F}''] > 2$. (Note that $\mathsf{V}(\mathfrak{F}'') = \mathsf{V}(\mathfrak{F}')$, so that $\hat{o} \in \mathfrak{F}''$ iff $\hat{o} \in \mathfrak{F}'$.) According to Exercise 8.10, we have that $\mathfrak{F}''$ contains infinite clusters with positive probability, whence (ii) follows.

Finally, (ii) implies (i) trivially.   ◄

Corollary 8.20. (BLPS, 1999b) *Let G be a quasi-transitive, unimodular graph. Let $\mathfrak{F}$ be the configuration of an invariant percolation on G such that $\mathfrak{F}$ is a forest a.s. Then almost surely every component that has at least three ends has $p_c < 1$.*

Proof. If not, condition on having some tree with at least three ends and $p_c = 1$. Then the collection of all such components gives an invariant percolation that contradicts Theorem 8.19.   ◄

8.4 Critical Percolation on Nonamenable Transitive Unimodular Graphs

Here we establish Conjecture 8.15 for nonamenable, quasi-transitive, unimodular graphs. This was shown by BLPS (1999b), with a more direct proof given in Benjamini, Lyons, Peres, and Schramm (1999a). The proof here is similar but partly new.

Theorem 8.21. (No Infinite Clusters at Criticality) *If G is a nonamenable, quasi-transitive, unimodular graph, then $\theta(p_c(G)) = 0$.*

We remark that this also implies that $p_c(G) < 1$; see also Exercise 8.41.

Proof. The proof for site percolation is similar to that for bond, so we treat only bond percolation.

In light of Theorem 7.5, we must rule out the possibilities that the number of infinite clusters at criticality is 1 or ∞. We do these separately and use the standard coupling of percolation arising from i.i.d. uniform $[0, 1]$-valued random variables $U(e)$ on the edges. As usual, let ω_p consist of the edges e with $U(e) < p$.

First suppose that there is a unique infinite cluster in ω_{p_c}. Let ω' be that infinite cluster. The idea is that ω' is fragile in the sense that $p_c(\omega') = 1$; we extend this to break up all of G into finite clusters with high marginal, contradicting Corollary 8.17. To do this, first pick independently for each $x \in V(G)$ uniformly at random one of its nearest vertices $W(x) \in \omega'$, where distance is in G. For $\epsilon > 0$, let ξ_ϵ consist of those edges $[x, y] \in E$ such that $W(x)$ and $W(y)$ belong to the same cluster of $\omega_{p_c - \epsilon}$. Since all clusters of $\omega_{p_c - \epsilon}$ are finite a.s., the mass-transport principle gives that all clusters of ξ_ϵ are also finite a.s. (for each x, transport mass 1 from x to $W(x)$). But $\xi_\epsilon \subseteq \xi_{\epsilon'}$ for $\epsilon > \epsilon'$ and $\bigcup_\epsilon \xi_\epsilon = E$ a.s., whence $\lim_{\epsilon \downarrow 0} \mathbf{P}[[x, y] \in \xi_\epsilon] = 1$. Thus, Corollary 8.17 implies that G is amenable.

Now suppose that there are infinitely many infinite clusters in ω_{p_c} a.s. By insertion tolerance, as in the proof of Theorem 7.9, some infinite cluster has at least three ends a.s. By Lemma 7.7, there is a random forest $\mathfrak{F} \subseteq \omega_{p_c}$ such that the distribution of the pair $(\mathfrak{F}, \omega_{p_c})$ is invariant and such that with positive probability, there is a component of $\mathfrak{F}$ that has at least three ends. Since $p_c(\omega_{p_c(G)}) = 1$ a.s. by Exercise 5.7 and $\mathfrak{F} \subseteq \omega_{p_c}$, we have $p_c(\mathfrak{F}) = 1$ a.s. This contradicts Theorem 8.19. ◄

The method of proof of the first part of Theorem 8.21 shows the following extension:

Proposition 8.22. (BLPS, 1999b) *If $\mathbf{P}$ is an invariant percolation on a nonamenable, quasi-transitive, unimodular graph that has a unique infinite cluster ω' a.s., then $p_c(\omega') < 1$ $\mathbf{P}$-a.s.* ◄

Sometimes the following generalization of Theorem 8.21 is useful. Given a family of probability measures μ_p $(0 \le p \le 1)$ on, say, 2^E, we say that a ***smooth monotone coupling*** of the family, if it exists, is a random field $Z : E \to [0, 1]$ such that for each p, the law of $\{e ; Z(e) < p\}$ is μ_p. (Here, a random field is just a random variable whose values are functions from E to $[0, 1]$. Equivalently, it is a collection of random variables $Z(e)$ for $e \in E$.) Also, let us call a probability measure $\mathbf{P}$ ***weakly insertion tolerant*** if there is a function

$f : E \times 2^E \to 2^E$ such that
  (i) for all e and all ω, we have $\omega \cup \{e\} \subseteq f(e, \omega)$;
 (ii) for all e and all ω, the difference $f(e, \omega) \setminus [\omega \cup \{e\}]$ is finite; and
(iii) for all e and each event A of positive probability, the image of A under $f(e, \cdot)$ is an event of positive probability.

Of course, an insertion-tolerant probability measure satisfies this definition with $f(e, \omega) := \omega \cup \{e\}$. **Weak deletion tolerance** has a similar definition.

Theorem 8.23. *Let G be a nonamenable, quasi-transitive, unimodular graph. Let $p \mapsto \mu_p$ be a family of ergodic, weakly insertion-tolerant probability measures on 2^E. Suppose that the family has a smooth monotone coupling by a random field Z whose law is invariant under automorphisms. Let $\mathscr{A}$ be the event that all clusters are finite. If*

$$p_c := \sup \{ p \,;\; \mu_p(\mathscr{A}) = 1 \} > 0 \,,$$

then $\mu_{p_c}(\mathscr{A}) = 1$.

▷ **Exercise 8.11.**
Prove Theorem 8.23.

8.5 Bernoulli Percolation on Planar, Quasi-transitive Graphs

For planar nonamenable Cayley graphs, we can answer all the most basic questions about percolation. The case of infinitely many ends is easy by what we have already established, since $p_c < 1$ by Theorem 7.20 and (6.7), $p_u = 1$ by Exercise 7.9, and $\theta(p_c) = 0$ by Theorem 8.21. The case of two ends is trivial, since $p_c = 1$. This section, adapted from Benjamini and Schramm (2001a), is devoted to the case of one end.

▷ **Exercise 8.12.**
Show that a plane, (properly embedded, locally finite) quasi-transitive graph with one end has no face with an infinite number of sides.

Planarity is used to exploit properties of percolation on the plane dual of the original graph. However, this dual is not necessarily a Cayley graph; indeed, it is not necessarily even transitive. For this reason, we will need to go beyond Cayley graphs. In the setting of Exercise 8.12, the dual of a Cayley graph is always locally finite, quasi-transitive, and unimodular, as we will see. Also, recall from Exercise 6.32 that if G is nonamenable, then so is its plane dual, $G^\dagger$. Thus, the natural setting preserved under duality is that of nonamenable plane quasi-transitive graphs. The main theorem of this section is the following.

Theorem 8.24. (Double Phase Transition) *Let G be a nonamenable, planar, quasi-transitive graph with one end. Then $0 < p_c(G) < p_u(G) < 1$ and Bernoulli(p_u) percolation on G has a unique infinite cluster a.s.*

We need the following fundamental fact whose proof is given in the appendix to this chapter, Section 8.8.

Theorem 8.25. (Planar Unimodularity) *If G is a planar, quasi-transitive graph with one end, then G is unimodular. Furthermore, there is some plane embedding of G such that $G^\dagger$ is quasi-transitive.*

We will assume for the rest of this section that

$$\left. \begin{array}{l} \textit{G is a nonamenable, plane, quasi-transitive graph with one} \\ \textit{end, G is embedded in such a way that } G^\dagger \textit{ is quasi-transitive,} \\ \textit{and } \omega \textit{ is the configuration of an invariant percolation on } G. \end{array} \right\} \qquad (8.20)$$

When more restrictions are needed, we will be explicit about them. We define the dual configuration $\omega^\times$ on the plane dual graph $G^\dagger$ by

$$\omega^\times(e^\dagger) := 1 - \omega(e). \qquad (8.21)$$

For a set A of edges, we let $A^\dagger$ denote the set of edges $e^\dagger$ for $e \in A$. Write N_∞ for the number of infinite clusters of ω and $\mathsf{N}_\infty^\times$ for the number of infinite clusters of $\omega^\times$. Duality will severely constrain the possible values of the pair $(\mathsf{N}_\infty, \mathsf{N}_\infty^\times)$. We begin by showing they cannot both be 0.

Lemma 8.26. *If (8.20) holds, then $\mathsf{N}_\infty + \mathsf{N}_\infty^\times \geq 1$ a.s.*

Proof. Suppose that $\mathsf{N}_\infty + \mathsf{N}_\infty^\times = 0$ with positive probability. Then by conditioning on that event, we may assume that it holds surely. We are going to create a new invariant percolation from ω that has very high expected degree, whence has infinite clusters, but yet cannot have infinite clusters, leading to our desired contradiction.

Now our assumptions tell us that for each (open) cluster K of ω, there is a unique infinite component of $G \backslash K$, since K is finite and G has only one end. This means that there is a unique open component of $(\partial_E K)^\dagger$ that "surrounds" K; this component of $(\partial_E K)^\dagger$ is contained in some cluster, K', of $\omega^\times$. Since our assumption implies that K' is also finite, this same procedure yields a cluster K'' of ω that surrounds K' and hence surrounds K. This allows us to classify clusters by a "rank" as follows.

Let $\mathscr{K}_0$ denote the collection of all clusters of ω. Define recursively the sequence of collections $\mathscr{K}_{j+1} := \{K''; \ K \in \mathscr{K}_j\}$ for $j \geq 0$. Since no cluster surrounds infinitely many other clusters, it follows that $\limsup_j \mathscr{K}_j = \varnothing$. Thus, we may define

$$r(K) := \max\{j; \ K \in \mathscr{K}_j\}$$

for all clusters $K \in \mathscr{K}_0$. Given $N \geq 0$, let ω_N be the set of those edges whose endpoints belong to (possibly different) clusters K with $r(K) \leq N$, in other words, the "interiors" of the clusters in $\mathscr{K}_{N+1}$. Then $\omega_N \subseteq \omega_{N+1}$ for all N and $\bigcup_N \omega_N = \mathsf{E}$. Also, ω_N is an invariant percolation for each N. Since $\deg_{\omega_N} o \to \deg_G o$, it follows from Corollary 8.17 that for sufficiently large N, with positive probability ω_N has infinite clusters. Yet the interiors of the clusters of $\mathscr{K}_{N+1}$ are finite and disjoint. This is a contradiction. ◄

Our next result is like the Newman–Schulman Theorem 7.5, with planarity substituting for insertion tolerance.

Lemma 8.27. (BLPS, 1999b) *If (8.20) holds, then* $N_\infty \in \{0, 1, \infty\}$ *a.s.*

Proof. If not, we may condition on $2 \le N_\infty < \infty$. In this case, we may pick uniformly at random two distinct infinite clusters of ω, call them K_1 and K_2. Let τ consist of those edges $[x, y]$ such that $x \in K_1$ and y belongs to the component of $G \setminus K_1$ that contains K_2. Then $\tau^\dagger$ is a bi-infinite path in $G^\dagger$ and is an invariant percolation on $G^\dagger$. But the fact that $p_c(\tau^\dagger) = 1$ contradicts Proposition 8.22. ◄

This lemma has the following consequence:

Corollary 8.28. *If (8.20) holds, then* $N_\infty + N_\infty^\times \in \{0, 1, \infty\}$ *a.s.*

Proof. The idea is that the infinite clusters of ω together with the infinite clusters of $\omega^\times$ form infinite clusters of an invariant percolation on another plane graph. Then we may apply Lemma 8.27.

To create this new graph from G and $G^\dagger$, draw G and $G^\dagger$ in the plane in such a way that every edge e intersects $e^\dagger$ in one point, v_e, and there are no other intersections of G and $G^\dagger$. For $e \in G$, write $\hat{e}$ for the pair of edges that result from the subdivision of e by v_e, and likewise for $\hat{e^\dagger}$. This defines a new graph $\hat{G}$, whose vertices are $V(G) \cup V(G^\dagger) \cup \{v_e \; ; \; e \in E(G)\}$ and whose edges are $\bigcup_{e \in E(G)} (\hat{e} \cup \hat{e^\dagger})$. The proof of Theorem 8.25 shows that $\hat{G}$ is quasi-transitive.

Consider the percolation on $\hat{G}$ given by

$$\omega' := \bigcup_{e \in \omega} \hat{e} \cup \bigcup_{e^\dagger \in \omega^\times} \hat{e^\dagger} .$$

This percolation is invariant under the quasi-transitive action of the unimodular group $\mathrm{Aut}(G)$ on $\hat{G}$. (The proof of Theorem 8.25 shows how $\mathrm{Aut}(G)$ acts on $\hat{G}$.) The number of infinite clusters of ω' is $N_\infty + N_\infty^\times$. Applying Lemma 8.27 to ω', we obtain our desired conclusion. ◄

Putting these results together, with just a bit more duality, leads to the following conclusion:

Theorem 8.29. *If (8.20) holds, then* $(N_\infty, N_\infty^\times) \in \{(1, 0), (0, 1), (1, \infty), (\infty, 1), (\infty, \infty)\}$ *a.s.*

Proof. Lemma 8.27 gives $N_\infty, N_\infty^\times \in \{0, 1, \infty\}$. Lemma 8.26 rules out $(N_\infty, N_\infty^\times) = (0, 0)$. Corollary 8.28 rules out $(N_\infty, N_\infty^\times) = (1, 1)$. Since every two infinite clusters of ω must be separated by at least one infinite cluster of $\omega^\times$ (namely, the one containing the path τ in the proof of Lemma 8.27), the case $(N_\infty, N_\infty^\times) = (\infty, 0)$ is impossible. Dual reasoning shows that $(N_\infty, N_\infty^\times) = (0, \infty)$ cannot happen. This leaves the five cases mentioned. ◄

We now come to the place where more assumptions on the percolation are needed. In particular, insertion and deletion tolerance become crucial. To treat site percolation as well as bond percolation, we will use the bond percolation ω^ξ associated to a site percolation ξ as follows:

$$\omega^\xi := \{[x, y] \; ; \; x, y \in \xi\} . \tag{8.22}$$

The ω^N used in the proof of Lemma 8.26 are examples of such associated bond percolations. Note that even when ξ is Bernoulli percolation, ω^ξ is neither insertion tolerant nor deletion tolerant. However, ω^ξ is still weakly insertion tolerant. More generally, it is clear that whenever ξ is insertion [resp., deletion] tolerant, then ω^ξ is weakly insertion [resp., deletion] tolerant. It is also clear that whenever ξ is ergodic, ω^ξ is ergodic: if $\mathscr{A}$ is an invariant set, then $\mathbf{P}(\{\omega^\xi \; ; \; \omega^\xi \in \mathscr{A}\}) = \mathbf{P}(\{\omega \; ; \; \omega^\xi \in \mathscr{A}\}) \in \{0, 1\}$.

Theorem 8.30. *Assume (8.20) and that ω is ergodic, weakly insertion tolerant, and weakly deletion tolerant. Then a.s.*

$$(\mathsf{N}_\infty, \mathsf{N}_\infty^\times) \in \{(1,0), (0,1), (\infty, \infty)\}.$$

Proof. By Theorem 8.29, it is enough to rule out the cases $(1, \infty)$ and $(\infty, 1)$. Let K be a finite, connected subgraph of G. If K intersects distinct infinite clusters of ω, then planarity yields that $\omega^\times \setminus \{e^\dagger \; ; \; e \in \mathsf{E}(K)\}$ must have at least two infinite clusters. Define $A(K)$ to be the event that K intersects distinct infinite clusters of ω. If $\mathsf{N}_\infty = \infty$ with positive probability, then there is some finite subgraph K such that $A(K)$ has positive probability. Fix some such K and write its edge set as $\{e_1, \ldots, e_n\}$. Let $f(e, \omega)$ be a function witnessing the weak insertion tolerance. Define $f(e, A) := \{f(e, \omega) \; ; \; \omega \in A\}$. Let $B(K) := f\big(e_1, f(e_2, \cdots f(e_n, A(K)) \cdots)\big)$. Then $B(K)$ has positive probability and we have $\mathsf{N}_\infty^\times > 1$ on $B(K)$ by our earlier observation. But ergodicity implies that $(\mathsf{N}_\infty, \mathsf{N}_\infty^\times)$ is an a.s. constant. Hence it is (∞, ∞). A dual argument rules out $(1, \infty)$; note that weak insertion tolerance of $\omega^\dagger$ arises from weak deletion tolerance of ω. ◄

Now we are almost done. Theorem 8.30 allows us to deduce precisely the value of N_∞ from the value of $\mathsf{N}_\infty^\times$ and thus, for bond percolation, critical values on G from those on $G^\dagger$. Site percolation will require a little extra arguing.

Theorem 8.31. *If (8.20) holds, then $p_c^{\mathrm{bond}}(G^\dagger) + p_u^{\mathrm{bond}}(G) = 1$ and $\mathsf{N}_\infty = 1$ $\mathbf{P}_{p_u}$-a.s.*

Proof. Let ω_p be Bernoulli(p) bond percolation on G. Then $\omega_p^\times$ is Bernoulli($1-p$) bond percolation on $G^\dagger$. It follows from Theorem 8.30 that

$$p > p_u^{\mathrm{bond}}(G) \quad \Longrightarrow \quad \mathsf{N}_\infty = 1 \quad \Longrightarrow \quad \mathsf{N}_\infty^\times = 0 \quad \Longrightarrow \quad 1 - p \le p_c^{\mathrm{bond}}(G^\dagger)$$

and

$$p < p_u^{\mathrm{bond}}(G) \quad \Longrightarrow \quad \mathsf{N}_\infty \ne 1 \quad \Longrightarrow \quad \mathsf{N}_\infty^\times \ne 0 \quad \Longrightarrow \quad 1 - p \ge p_c^{\mathrm{bond}}(G^\dagger).$$

This gives us $p_c^{\mathrm{bond}}(G^\dagger) + p_u^{\mathrm{bond}}(G) = 1$. Furthermore, according to Theorem 8.21, we have $\mathsf{N}_\infty^\times = 0$ $\mathbf{P}_{p_c^{\mathrm{bond}}(G^\dagger)}^{\mathrm{bond},G^\dagger}$-a.s., whence Theorem 8.30 tells us that $\mathsf{N}_\infty = 1$ $\mathbf{P}_{p_u^{\mathrm{bond}}(G)}^{\mathrm{bond},G}$-a.s.

For site percolation, we must prove that $\mathsf{N}_\infty = 1$ $\mathbf{P}_{p_u}^{\mathrm{site}}$-a.s. Let ξ_p be the standard coupling of site percolation and ω^{ξ_p} the corresponding bond percolation processes. As earlier, we may conclude that for $p > p_u^{\mathrm{site}}(G)$, we have that $(\omega^{\xi_p})^\times$ has no infinite clusters a.s., whereas for $p < p_u^{\mathrm{site}}(G)$, we have that $(\omega^{\xi_p})^\times$ has infinite clusters a.s. Because $p \mapsto (\omega^{\xi_{1-p}})^\times$ satisfies all the hypotheses of Theorem 8.23, it follows that for $p = p_u^{\mathrm{site}}(G)$, we have that $(\omega^{\xi_p})^\times$ has no infinite clusters a.s. Therefore ω^{ξ_p} has a unique infinite cluster a.s., whence so does ξ_p. ◄

Proof of Theorem 8.24. We already know that $p_c > 0$ from Proposition 7.12. Applying this to $G^\dagger$ and using Theorem 8.31, we obtain that $p_u^{\mathrm{bond}} < 1$. Fix $q \in [p_u^{\mathrm{bond}}, 1)$. In light of Proposition 7.14, there is some $p < 1$ such that for $\xi \sim \mathbf{P}_p^{\mathrm{site}}$, we have ω^ξ stochastically dominates $\mathbf{P}_q^{\mathrm{bond}}$, whence for such p, $(\omega^\xi)^\times$ has no infinite clusters a.s. and thus ω^ξ has a unique infinite cluster a.s. This proves that $p_u^{\mathrm{site}} < 1$ too. (The result that $p_u < 1$ also follows from Theorem 7.29.) Comparing Theorems 8.21 and 8.31, we see that it is impossible that $p_c = p_u$, whence $p_c < p_u$. ◄

8.6 Properties of Infinite Clusters

In the proof of Theorem 7.6, we saw that if Bernoulli(p) percolation produces infinitely many infinite clusters a.s., then a.s. at least one of them has a furcation, whence at least three ends. This was a consequence of insertion and deletion tolerance. Likewise, deletion tolerance implies that if there is a unique infinite cluster (and $p < 1$), then that cluster has a unique end a.s.

As the reader might suspect from Example 8.5, one can derive stronger results from the mass-transport principle. Indeed, the following properties are true quite generally:

Theorem 8.32. *Let* **P** *be an invariant, weakly insertion-tolerant percolation process on a nonamenable, quasi-transitive, unimodular graph G. If there are infinitely many infinite clusters a.s., then a.s. every infinite cluster has continuum many ends, no isolated end, and is transient for simple random walk.*

Here, we are using a topology on the set of ends of a graph, defined as follows. Let G be a graph with fixed base point o. Let B_n be the ball of radius n about o. Let Ends be the set of ends of G. Define a metric on Ends by putting

$$d(\xi, \zeta) := \inf\Big\{1/n \, ; \; n = 1 \text{ or } \forall X \in \xi \; \forall Z \in \zeta \; \exists \text{ a component } C \text{ of } G \setminus B_n$$

$$|X \setminus C| + |Z \setminus C| < \infty\Big\}.$$

It is easy to verify that this is a metric; in fact, it is an ultrametric, that is, for any $\xi_1, \xi_2, \xi_3 \in$ Ends, we have $d(\xi_1, \xi_3) \leq \max\{d(\xi_1, \xi_2), d(\xi_2, \xi_3)\}$. Since G is locally finite, it is easy to check that Ends is compact in this metric. Finally, it is easy to see that the topology on Ends does not depend on choice of base point. A **vertex-neighborhood** of $\xi \in$ Ends is a set of vertices that, for some n, contains the component of $G \setminus B_n$ that has an infinite intersection with every set in ξ.

The set of isolated points in any compact metric space is countable; the nonisolated points form a perfect subset (the Cantor–Bendixson theorem), whence, if there are nonisolated points, they have the cardinality of the continuum (see, for example, Kuratowski (1966)).

In the case of site percolation ξ, by transience of ξ we mean transience of ω^ξ, the induced subgraph defined in (8.22).

Now that we have explained the terms in Theorem 8.32, we devote the remainder of this section to proving it. We begin with the following proposition.

Proposition 8.33. *Let* **P** *be an invariant percolation process on a nonamenable, quasi-transitive, unimodular graph G. Almost surely each infinite cluster that has at least three ends has no isolated ends.*

Proof. The idea is to send unit mass from each vertex in an isolated end to a nearest point not in that end, so that some vertex receives infinite mass, which contradicts the mass-transport principle. However, this is too imprecise to make sense. Thus, for each $n = 1, 2, \ldots$, we create a new invariant percolation A_n from ω as follows: A_n is the union of all vertex sets A such that there is some cluster K of ω with the properties that $K \supset A$, the diameter of A is at most n in the metric of K, and $K \setminus A$ has at least three infinite components. Note that if ξ is an isolated end of a percolation cluster K, then for each n, some vertex-neighborhood of ξ in K is disjoint from A_n. Also observe that if K is a cluster with at least three ends, then K intersects A_n for some n.

Fix some $n \geq 1$. Consider the mass transport that sends one unit of mass from each vertex x in a percolation cluster that intersects A_n and distributes it equally among the vertices in A_n that are closest to x in the metric of $K(x)$. In other words, let $C(x)$ be the set of vertices in $K(x) \cap A_n$ that are closest to x in the metric of ω, and set $F(x, y; \omega) := |C(x)|^{-1}$ if $y \in C(x)$ and otherwise $F(x, y; \omega) := 0$. Then $F(x, y; \omega)$ is invariant under the diagonal action. If ξ is an isolated end of an infinite cluster K that intersects A_n, then there is a finite set of vertices B that gets all the mass from all the vertices in some vertex-neighborhood of ξ. But the mass-transport principle tells us that the expected mass transported to a vertex is at most 1. Hence, a.s. clusters that intersect A_n do not have isolated ends. Since this holds for all n, we gather that a.s. infinite clusters with isolated ends do not intersect $\bigcup_n A_n$, whence they have at most two ends. ◄

Corollary 8.34. *Let* **P** *be an invariant, weakly insertion-tolerant bond percolation process on a nonamenable, quasi-transitive, unimodular graph* G. *If there are infinitely many infinite clusters a.s., then a.s. every infinite cluster has continuum many ends and no isolated end.*

Proof. It suffices to prove that there are no isolated ends of clusters. To prove this in turn, observe that if some cluster has an isolated end with positive probability, then because of weak insertion tolerance, with positive probability, some cluster will have at least three ends with one of them being isolated. Hence Corollary 8.34 follows from Proposition 8.33. ◄

We proved a weak form of the following principle in Lemma 7.7.

Lemma 8.35. (BLPS, 1999b) *Let* **P** *be an invariant bond percolation process on a nonamenable, quasi-transitive, unimodular graph* G. *If a.s. there is a component of* ω *with at least three ends, then (on a larger probability space) there is a random forest* $\mathfrak{F} \subset \omega$ *such that the distribution of the pair* $(\mathfrak{F}, \omega)$ *is invariant and such that a.s. for each component* K *of* ω *with at least three ends, there is a component of* $K \cap \mathfrak{F}$ *that has infinitely many ends.*

Proof. We begin as in the proof of Lemma 7.7. Assign independent, uniform $[0, 1]$ random variables to the edges (independently of ω). Define the free minimal spanning forest $\mathfrak{F}$ of ω by taking an edge $e \in \omega$ to be present in $\mathfrak{F}$ iff there is no cycle in ω containing e in which e is assigned the maximum value. As before, we have (a.s.) that $\mathfrak{F}$ is a forest with each tree having the same vertex-cardinality as the cluster of ω in which it lies.

Suppose that $K(x)$ has at least three ends with positive probability. Choose any finite tree T containing x with edge set $\mathsf{E}(T)$ so that with positive probability, $T \subset K(x)$ and $K(x) \setminus \mathsf{E}(T)$ has at least three infinite components. Then with positive probability, $T \subset K(x)$, $K(x) \setminus \mathsf{E}(T)$ has at least three infinite components, all edges in T are assigned values less than $1/2$, and all edges in $\partial_{\mathsf{E}} T$ are assigned values greater than $1/2$. On this event, $\mathfrak{F}$ contains T and T is part of a spanning tree in $\mathfrak{F}$ with at least three ends.

To convert this event of positive probability to an event of probability 1, let $r_x = r(x, \omega)$ be the least cardinality r of a tree T in G such that $K(x) \setminus T$ has at least three infinite components, if such an r exists. If not, set $r(x, \omega) := \infty$. Note that $r(x, \omega) < \infty$ iff $K(x)$ has at least three ends. By Example 8.6, if r_x is finite, then there are a.s. infinitely many such trees T in

$K(x)$. Therefore, given that $K(x)$ has at least three ends, there are a.s. infinitely many trees $T \subset K(x)$ of size r_x at pairwise distance at least 2 from each other such that $K(x) \setminus T$ has at least three infinite components. For such T, the events that all edges in T are assigned values less than 1/2 and all edges in $\partial_E T$ are assigned values greater than 1/2 are independent and have probability bounded below, whence an infinite number of these events occur a.s. Therefore, there is a component of $K(x) \cap \mathfrak{F}$ that has at least three ends a.s., whence, by Proposition 8.33, has infinitely many ends a.s. ◄

Proposition 8.36. *Let* **P** *be an invariant, weakly insertion-tolerant percolation process on a nonamenable, quasi-transitive, unimodular graph G. If there are infinitely many infinite clusters a.s., then a.s. each infinite cluster is transient.*

Proof. It suffices to consider bond percolation. By Corollary 8.34, every infinite cluster of ω has infinitely many ends. Consequently, there is an invariant random forest $\mathfrak{F} \subset \omega$ such that a.s. each infinite cluster K of ω contains a tree of $\mathfrak{F}$ with infinitely many ends by Lemma 8.35. By Corollary 8.20, we know that any such tree has $p_c < 1$. Since it has branching number > 1, it follows that such a tree is transient by Theorem 3.5. The Rayleigh monotonicity law then implies that K is transient. ◄

Theorem 8.32 now follows from Corollary 8.34 and Proposition 8.36.

8.7 Invariant Percolation on Amenable Graphs

The mass-transport principle is most useful in the nonamenable setting. In this section, we give two results that show that nonamenability is in fact necessary for some of our previous work.

We will use Haar measure for the proofs since it would be artificial and somewhat cumbersome to avoid it. In particular, given vertices x and y, the set of automorphisms that take x to y is compact in $\mathrm{Aut}(G)$, whence it has finite Haar measure. This means that we can choose one of its elements at random (via normalized Haar measure). See Exercise 8.22 for a simple construction of left Haar measure on automorphism groups.

▷ **Exercise 8.13.**
Let G be a transitive, unimodular graph and $o \in \mathsf{V}$. For each $x \in \mathsf{V}$, choose a Haar-random $\gamma_x \in \mathrm{Aut}(G)$ that takes o to x. Show that for every finite set $L \subset \mathsf{V}$, we have

$$\mathbf{E}\big|\{x \in \mathsf{V} \,;\; o \in \gamma_x L\}\big| = |L| \,.$$

Our first result is for site percolation; for the bond version, see Exercise 8.58.

Theorem 8.37. (Amenability and Finite Percolation) *Let G be a quasi-transitive, unimodular graph. Then G is amenable iff for all $\alpha < 1$, there is an invariant site percolation ω on G with no infinite clusters and such that $\mathbf{P}[x \in \omega] > \alpha$ for all $x \in \mathsf{V}(G)$.*

Proof. We assume that G is transitive and leave the quasi-transitive case to Exercise 8.59. One direction follows from the site version of Theorem 8.16 on thresholds, which is Exercise 8.38.

Now we prove the converse. Suppose that G is amenable. If G were $\mathbb{Z}^d$, we could randomly center a tiling of G by large cubes and remove the boundaries of the cubes. Since we do not have such a convenient tiling in general, we instead remove the boundaries of randomly placed copies of a large set with small boundary. In fact, to make sure that the remaining clusters are finite, we will remove such boundaries for larger and larger sets that are placed more and more rarely.

Fix $o \in V(G)$. For a finite set $F \subset V$, consider the following percolation. For each $x \in V$, choose a random $\gamma_x \in \text{Aut}(G)$ that takes o to x, and let ζ be a Bernoulli$(1/|F|)$ site percolation on G. Choose all γ_x and ζ independently. Remove the vertices

$$\bigcup_{x \in \zeta} \partial_V(\gamma_x F);$$

that is, consider the percolation subgraph

$$\omega_F := V \setminus \bigcup_{x \in \zeta} \partial_V(\gamma_x F).$$

Then the distribution of ω_F is an invariant percolation on G.

We claim that

$$\mathbf{P}[o \notin \omega_F] \leq |\partial_V F|/|F| \tag{8.23}$$

and

$$\mathbf{P}\big[|K(o, \omega_F)| < \infty\big] \geq 1 - 1/e, \tag{8.24}$$

where $K(o, \omega_F)$ denotes the component of o in ω_F and e is the base of natural logarithms. Both of these will be proved by using the following calculation: For any $L \subset V$, we have

$$\mathbf{E}\big|\{x \in \zeta ;\ o \in \gamma_x L\}\big| = |L|/|F|. \tag{8.25}$$

Indeed,

$$\mathbf{E}\big|\{x \in \zeta ;\ o \in \gamma_x L\}\big| = \sum_{x \in V} \mathbf{P}[x \in \zeta,\ o \in \gamma_x L] = \sum_{x \in V} \mathbf{P}[x \in \zeta]\,\mathbf{P}[o \in \gamma_x L]$$

$$= \sum_{x \in V} \mathbf{P}[o \in \gamma_x L]/|F| = \mathbf{E}\big|\{x \in V ;\ o \in \gamma_x L\}\big|/|F| = |L|/|F|$$

by Exercise 8.13. (Recall that G is unimodular by Proposition 8.14.)

To prove (8.23), note that the probability that $o \notin \omega_F$ is at most the expected number of $x \in \zeta$ such that $o \in \partial_V(\gamma_x F)$. But this expectation is exactly the right-hand side of (8.23) by (8.25) applied to $L := \partial_V F$. To prove (8.24), use the independence to calculate that

$$\mathbf{P}\big[|K(o, \omega_F)| < \infty\big] \geq \mathbf{P}[\exists x\ x \in \zeta,\ o \in \gamma_x F] = 1 - \mathbf{P}\big[\forall x\ \neg(x \in \zeta,\ o \in \gamma_x F)\big]$$

$$= 1 - \prod_{x \in V}(1 - \mathbf{P}[x \in \zeta,\ o \in \gamma_x F])$$

$$\geq 1 - \exp\Big\{-\sum_{x \in V} \mathbf{P}[x \in \zeta,\ o \in \gamma_x F]\Big\}$$

$$= 1 - \exp\Big\{-\mathbf{E}\big|\{x \in \zeta ;\ o \in \gamma_x F\}\big|\Big\} = 1 - 1/e$$

by (8.25) again, applied to $L := F$.

Now, since G is amenable, there is a sequence $\langle F_n \rangle$ of finite sets of vertices such that $\sum_n |\partial_V F_n|/|F_n| < 1 - \alpha$. For each n, let $\omega_n = \omega_{F_n}$ be the random subgraph from the percolation just described based on the set F_n. Choose ω_n to be independent and consider the percolation with configuration $\omega := \bigcap \omega_n$. By (8.23), we have $\mathbf{P}[o \in \omega] > \alpha$ and by (8.24), we have $\mathbf{P}[|K(o, \omega)| < \infty] = 1$. ◄

We now use Theorem 8.37 to establish a converse to Theorem 6.48 on anchored expansion constants in percolation. This result is due to Häggström, Schonmann, and Steif (2000), Theorem 2.4(ii), with a rather indirect proof. Our proof is due to Oded Schramm (personal communication).

Corollary 8.38. *Let G be a quasi-transitive, amenable graph and ω be an invariant bond percolation on G. Then a.s. every component of ω has anchored expansion constant equal to 0.*

Proof. Again, we do the transitive case and leave the quasi-transitive case to the reader. For simplicity, we use the anchored expansion constant defined using the inner vertex boundary, where we define the ***inner vertex boundary*** of a set K as

$$\partial_V^{\text{in}} K := \{x \in K \,;\, \exists y \notin K \;\; y \sim x\}.$$

Choose a sequence $\alpha_n < 1$ with $\sum_n (1 - \alpha_n) < \infty$. Choose a sequence of independent invariant site percolations ω_n, also independent of ω, with no infinite component and such that $\mathbf{P}[x \in \omega_n] > \alpha_n$ for all $x \in \mathsf{V}(G)$. This exists by Theorem 8.37. Let $K(x)$ denote the component of x in ω and $K_n(x)$ the component of x in ω_n. Since $\partial_{\mathsf{V}(K(o))}^{\text{int}}(K(o) \cap K_n(o)) \subseteq \partial_V^{\text{in}} K_n(o) \setminus \partial_V^{\text{in}} K(o)$, it suffices to prove that a.s.,

$$\lim_{n \to \infty} \frac{|\partial_V^{\text{in}} K_n(o) \setminus \partial_V^{\text{in}} K(o)|}{|K(o) \cap K_n(o)|} = 0.$$

Now

$$\mathbf{E}\left[\frac{|\partial_V^{\text{in}} K_n(o) \setminus \partial_V^{\text{in}} K(o)|}{|K(o) \cap K_n(o)|} \right] = \mathbf{P}\left[o \in \partial_V^{\text{in}} K_n(o) \setminus \partial_V^{\text{in}} K(o)\right] \leq \mathbf{P}\left[o \in \partial_V^{\text{in}} K_n(o)\right]$$

$$\leq \deg_G(o)\,\mathbf{P}[o \notin \omega_n] < \deg_G(o)(1 - \alpha_n)$$

by the mass-transport principle. (For the equality, every point $x \in \partial_V^{\text{in}} K_n(x) \setminus \partial_V^{\text{in}} K(x)$ sends mass 1 split equally among the vertices of its component $K_n(x) \cap K(x)$. For the second inequality, x sends mass 1 to y when $x \sim y$, $x \in \omega_n$, and $y \notin \omega_n$.) Therefore

$$\sum_n \mathbf{E}\left[\frac{|\partial_V^{\text{in}} K_n(o) \setminus \partial_V^{\text{in}} K(o)|}{|K(o) \cap K_n(o)|} \right] < \infty,$$

whence

$$\sum_n \frac{|\partial_V^{\text{in}} K_n(o) \setminus \partial_V^{\text{in}} K(o)|}{|K(o) \cap K_n(o)|} < \infty \text{ a.s.,}$$

which gives the result. ◄

8.8 Appendix: Unimodularity of Planar, Quasi-transitive Graphs

We prove here Theorem 8.25. Our approach is based on some ideas we heard from O. Schramm. The transitive case is easier to prove, as this proof simplifies to show that the graph is 3-connected.

We need several lemmas. A graph $G = (V, E)$ is called k-**connected** if $|V| \geq k + 1$ and whenever at most $k - 1$ vertices are removed from G (together with their incident edges), the resulting graph is connected. For the next lemmas, let $Q(x)$ denote the set of vertices that lie in finite components of $G \setminus \{x\}$, and let $Q(x, y)$ denote the set of vertices, *other than* $Q(x) \cup Q(y)$, that lie in finite components of $G \setminus \{x, y\}$. Note that $x \notin Q(x)$ and $x \notin Q(x, y)$.

Lemma 8.39. *If G is a quasi-transitive graph, then $\sup_{x \in V} |Q(x)| < \infty$.*

Proof. Since x has finite degree, $G \setminus \{x\}$ has a finite number of components, whence $Q(x)$ is finite. If x and y are in the same orbit, then $|Q(x)| = |Q(y)|$. Since there are only finitely many orbits, the result follows.   ◄

Lemma 8.40. *If G is a quasi-transitive graph with one end, then $\sup_{x, y \in V} |Q(x, y)| < \infty$.*

Proof. Suppose not. Since there are only a finite number of orbits, there must be some x and y_n such that $d(x, y_n) \to \infty$ and $Q(x, y_n) \neq \varnothing$ for all n. Because of Lemma 8.39, for all large n, we have $x \notin Q(y_n)$, whence for all large n, there are neighbors a_n, b_n of x that cannot be joined by a path in $G \setminus \{x, y_n\}$, but such that a_n lies in an infinite component of $G \setminus \{x, y_n\}$ and b_n can be joined to y_n using only vertices in $Q(x, y_n) \cup \{y_n\}$. There is some pair of neighbors a, b of x for which $a = a_n$ and $b = b_n$ for infinitely many n. But then a and b lie in distinct infinite components of $G \setminus \{x\}$, contradicting the assumption that G has one end.   ◄

We partially order the collection of sets $Q(x, y)$ by inclusion.

Lemma 8.41. *Let G be any graph. If $Q(x, y)$ and $Q(z, w)$ are maximal and nonempty, then either $\{x, y\} = \{z, w\}$ or $Q(x, y) \cap Q(z, w) = \varnothing$. Also,*

$$\{x, y, z, w\} \cap (Q(x, y) \cup Q(z, w)) = \varnothing. \tag{8.26}$$

Proof. Suppose that $z \in Q(x, y)$. Then we cannot have $w \in Q(x, y)$, since that would imply $Q(z, w) \subsetneq Q(x, y)$. Now there is a path from z to w using only vertices in $Q(z, w) \cup \{z, w\}$. Since $z \in Q(x, y)$ and $w \notin Q(x, y)$, this path must include either x or y – say, y. Therefore $y \in Q(z, w) \cup \{w\}$. We cannot have $y = w$, since that would imply $Q(z, w) \subsetneq Q(x, y)$.

Thus $y \in Q(z, w)$. Consider any infinite, simple path starting from any vertex in the union $Q(x, y) \cup Q(z, w)$. We claim it must visit x or w. For if not, then it must visit z or y. If the last vertex among $\{z, y\}$ that it visits is z, then it must visit x, since $z \in Q(x, y)$, a contradiction. If the last vertex among $\{z, y\}$ that it visits is y, then it must visit w, since $y \in Q(z, w)$, a contradiction. This proves our claim, whence $Q(x, w) \supsetneq Q(x, y) \cup Q(z, w)$. But this contradicts maximality of $Q(x, y)$.

Thus, we have proved that $z \notin Q(x, y)$. By symmetry, we deduce (8.26).

Now suppose that $Q(x, y) \cap Q(z, w) \neq \varnothing$. Choose some $a \in Q(x, y) \cap Q(z, w)$. Then every infinite, simple path from a must pass through $\{x, y\}$ and through $\{z, w\}$. Consider an infinite, simple path from a. Without loss of generality, suppose that the first point among

$\{x, y, z, w\}$ that it visits is z. Then $z \in Q(x, y) \cup \{x, y\}$. By the preceding, z is equal to x or y — say, x. Now there is also an infinite, simple path from a such that the first point among $\{x, y, w\}$ is not x. Say it is w. Then $w \in Q(x, y) \cup \{y\}$. By the preceding, $w = y$, and this proves the lemma. ◄

Every embedding ϕ of a planar graph into the plane induces a cyclic ordering $(\phi(x))$ of the edges incident to any vertex x by looking at the clockwise ordering of these edges after embedding. Two cyclic orderings are considered the same if they differ only by a cyclic permutation. Two cyclic orderings are inverses if they can be written in opposite order from each other.

The following extends a theorem of Whitney (1932):

Lemma 8.42. (Imrich, 1975) *If G is a planar, 3-connected graph and ϕ and ψ are two embeddings of G in the plane, then either for all x, we have $(\phi(x)) = (\psi(x))$ or for all x, we have that $(\phi(x))$ and $(\psi(x))$ are inverses.*

Proof. Imrich (1975) gives a proof that is valid for graphs that are not necessarily properly embedded nor locally finite. To simplify the proof, we will assume that the graph is not only properly embedded and locally finite but also quasi-transitive and has only one end. This is the only case we will use.

Given any vertices $x \neq y$, Menger's theorem (Exercise 3.16) shows that we can find three paths joining x and y that are disjoint except at x and y. Comparison of the first edges e_1, e_2, e_3 and the last edges f_1, f_2, f_3 of these paths shows that the cyclic ordering of $\phi(e_i)$ is opposite to that of $\phi(f_i)$, and likewise for ψ. Therefore, it suffices to prove that for each x separately, we have either $(\phi(x)) = (\psi(x))$ or $(\phi(x))$ and $(\psi(x))$ are inverses. For this, it suffices to show that if e_1 and e_2 are two edges incident to x and $\phi(e_1)$ is adjacent to $\phi(e_2)$ in the cyclic ordering $(\phi(x))$, then $\psi(e_1)$ is adjacent to $\psi(e_2)$ in the cyclic ordering $(\psi(x))$.

So let $\phi([x, y])$ and $\phi([x, z])$ be adjacent in $(\phi(x))$. Assume that $\psi([x, y])$ and $\psi([x, z])$ are not adjacent in $(\psi(x))$. Let C be the cycle such that $\phi(C)$ is the border of the (unique) face having sides that include both $\phi([x, y])$ and $\phi([x, z])$. (This exists by Exercise 8.12.)

Now by our assumption, there are two neighbors v and w of x such that the cyclic ordering $(\psi(x))$ induces the cyclic order $\psi(y), \psi(v), \psi(z), \psi(w)$ on these latter four points. The Jordan curve theorem tells us that the ψ-image of every path from v to w must intersect $\psi(C)$, in other words, that every path from v to w must intersect C.

However, if we return to the picture provided by ϕ, then, since there are three paths joining v to w that are disjoint except at their endpoints, at least two of these paths do not contain x, and hence at least one, $\mathcal{P}'$, is mapped by ϕ to a curve that does not intersect $\phi(C)$. But this means $\mathcal{P}'$ does not intersect C, a contradiction. ◄

Lemma 8.43. *If G is a planar, 3-connected graph, then $\mathrm{Aut}(G)$ is discrete.*

Proof. Let x be any vertex. Note that the degree of x is at least 3. By Exercise 8.19, it suffices to show that only the identity fixes x and all its neighbors. Let ϕ be an embedding of G in the plane, and let $\gamma \in \mathrm{Aut}(G)$ fix x and all its neighbors. Then $\phi \circ \gamma$ is an embedding of G in the plane that induces the same cyclic ordering of the edges of x as does ϕ. By Lemma 8.42, it follows that $((\phi \circ \gamma)(y)) = (\phi(y))$ for all y. By induction on the distance of y to x, it is easy to deduce that $\gamma(y) = y$ for all y, which proves the claim. ◄

Because of this and the next corollary, every automorphism of a planar 3-connected graph can be characterized as either orientation preserving or orientation reversing.

Corollary 8.44. *If G is a planar, 3-connected graph, $\gamma \in \mathrm{Aut}(G)$, and ϕ is an embedding of G in the plane, then either for all x, we have $(\phi(x)) = (\phi(\gamma x))$ or for all x, we have that $(\phi(x))$ and $(\phi(\gamma x))$ are inverses.*

Proof. Define another embedding ψ of G as follows: Let $\psi(x) := \phi(\gamma x)$ for $x \in \mathsf{V}(G)$ and $\psi(e) := \phi(\gamma e)$ for $e \in \mathsf{E}(G)$. The conclusion follows from Lemma 8.42. ◀

The next corollary implies that every $\gamma \in \mathrm{Aut}(G)$ induces an element of $\mathrm{Aut}(G^\dagger)$.

Corollary 8.45. *If G is a plane 3-connected graph and $\gamma \in \mathrm{Aut}(G)$, then γ maps every facial cycle to a facial cycle.*

Proof. Let $x_0, x_1, \ldots, x_n = x_0$ be the vertices in counterclockwise order of a facial cycle. Then, for each $i \in [0, n-1]$, the edge $[x_i, x_{i+1}]$ is the edge following $[x_{i-1}, x_i]$ in the cyclic order (x_i). Consider the cycle C formed by $\gamma x_0, \gamma x_1, \ldots, \gamma x_n$. If γ is orientation preserving, then these vertices traverse C in counterclockwise order, so for each $i \in [0, n-1]$, the edge $[\gamma x_i, \gamma x_{i+1}]$ is the edge following $[\gamma x_{i-1}, \gamma x_i]$ in the cyclic order (γx_i). This means that C is facial. The argument is similar when γ is orientation reversing. ◀

Proof of Theorem 8.25. Fix an embedding of G. If G is 3-connected, then let $G' := G$. Otherwise, let G' be the graph formed from G by removing all vertices in the union $W := \bigcup_{x, y \in \mathsf{V}}(Q(x) \cup Q(x, y))$ and by adding new edges $[x, y]$ between each pair of vertices $x, y \notin W$ for which $Q(x, y)$ is maximal and nonempty. Because of Lemmas 8.39 and 8.40, the graph G' is not empty and has one end. Since each new edge may be placed along the trace of a path of edges in G, Lemma 8.41 guarantees that G' is planar. Furthermore, by construction, G' is 3-connected (and quasi-transitive with only one end). Lemma 8.43 shows that $\mathrm{Aut}(G')$ is discrete, whence every subgroup of $\mathrm{Aut}(G')$ is unimodular by Proposition 8.9. Now the restriction to $\mathsf{V}(G')$ of an automorphism γ of G induces an automorphism of G'. Let Γ be the subgroup of $\mathrm{Aut}(G')$ given by such restrictions. Since Γ acts quasi-transitively, it is unimodular, whence $|S(x)y| = |S(y)x|$ for $x \in \mathsf{V}(G')$ and $y \in \Gamma x$, where S denotes the stabilizer in Γ. But in addition, for any pair $x, y \in \mathsf{V}(G')$, the set $S(x)y$ is the same for the stabilizer in Γ as for the stabilizer in $\mathrm{Aut}(G)$, and $\Gamma x = \mathrm{Aut}(G)x$. Therefore, $\mathrm{Aut}(G)$ is unimodular by Exercise 8.28.

We claim next that $(G')^\dagger$ is quasi-transitive. Fix one vertex x_i from each orbit of $\mathrm{Aut}(G)$. Given any face, let x be one of its vertices. Then, for some i, there is a $\gamma \in \mathrm{Aut}(G)$ that maps x to x_i, whence, by Corollary 8.45, maps the face to a face containing x on its boundary. That is, there is an induced $\gamma' \in \mathrm{Aut}((G')^\dagger)$ that maps the face to one of a finite set of faces. This proves our claim.

Finally, if G is not 3-connected, then temporarily add a vertex to each face of G' and connect such a vertex to each of the vertices on the boundary of that face. Call the new temporary graph G''. This graph G'' produces a triangulation of the region spanned by all the faces. There is a triangulation of either the Euclidean plane or the hyperbolic plane using geodesic line segments that is isomorphic to G''; one way to get this is to use circle packing: see Beardon and Stephenson (1990), He and Schramm (1995), or Babai (1997). Now use this new embedding of G' (and forget G''). We can extend each automorphism of G' to an isometry of the (Euclidean or hyperbolic) plane. By Selberg's lemma, there is a torsion-free finite-index normal subgroup Γ of $\text{Aut}(G')$ (see, for example, Corollary 7.6.4 of Ratcliffe (2006)). Since Γ has finite index in $\text{Aut}(G')$, it also acts quasi-transitively on G'. The quotient of the plane by Γ is a compact orientable surface inheriting a finite graph H from G'. Now replace each edge in H by a corresponding $Q(x, y)$ from G and add $Q(x)$ to each vertex in H that is an image of $x \in V(G)$. We may do this in a way that results in a graph L embedded in the surface. Finally, we lift L to the plane by taking the universal cover of the surface; this is an embedding of G. Furthermore, a subgroup of isometries acts quasi-transitively on this embedding, whence the dual of G is quasi-transitive. ◀

Remark 8.46. If G is transitive and has one end, then the construction of G' in the proof of Theorem 8.25 must yield all of G (because if x is omitted, then so is the entire orbit of x). Thus, G is 3-connected. Additional results on connectivity were proved by Mader (1970) and Watkins (1970).

The last part of the proof of Theorem 8.25 is not self-contained, which means that our proofs of several results from Section 8.5 are also not self-contained. In order that Section 8.5 be self-contained, one can avoid that last part by using an induced percolation on the graph G' that occurs in the proof of Theorem 8.25 and by appealing more to Theorem 8.23.

8.9 Notes

The proof of uniqueness monotonicity given in Section 8.1 is essentially that of Häggström and Peres (1999).

The approach we take to unimodularity for automorphism groups of graphs uses the equivalence first observed by Schlichting (1979) and Trofimov (1985). The graph structure is not important; it suffices that Γ act on a set in such a way that $S(x)y$ is finite for all x, y in the set. The approach used in BLPS (1999b) was the classical notion of unimodularity. There, the proof (due to Woess) of Theorem 8.7 was the following:

Second proof of Theorem 8.7. We may assume that Γ is closed. Let μ denote a left Haar measure on Γ. We prove (8.2) in the form (8.10) with $\mu_x = \mu(S(x))$. We have

$$
\sum_{z \in \Gamma w} f(u, z)\mu(S(w)) = \sum_{z \in \Gamma w} f(u, z)\mu(\Gamma_{w,z}) = \int_\Gamma f(u, \gamma w)\, d\mu(\gamma)
$$

$$
= \int_\Gamma f(\gamma^{-1}u, w)\, d\mu(\gamma) = \sum_{y \in \Gamma u} f(y, w)\mu(\{\gamma \in \Gamma \,;\; \gamma^{-1}u = y\})
$$

$$
= \sum_{y \in \Gamma u} f(y, w)\mu(\Gamma_{y,u}) = \sum_{y \in \Gamma u} f(y, w)\mu(S(y)). \qquad ◀
$$

The idea behind the preceding proof was to average over random elements from Γ that take a given vertex u to another given vertex y. The proof in the text of Theorem 8.7 can be seen as analogous to the one for Cayley graphs in Section 8.1; instead of using the bijection $x \mapsto x^{-1}$, we use the fact that for any u and w, the map $S(u)z \mapsto S(w)\gamma^{-1}u$ from $\{S(u)z \,;\; z \in \Gamma w\} \to \{S(w)y \,;\; y \in \Gamma u\}$ is independent of the element $\gamma \in \Gamma_{w,z}$ that takes w to z, is well defined, and is a bijection.

Theorem 8.19 is from BLPS (1999b) and Aldous and Lyons (2007). Using a normalized random root $\hat{o}$ for a quasi-transitive unimodular graph G allows us to treat the randomly rooted graph $(G, \hat{o})$

virtually the same as a transitive unimodular graph. In fact, this idea holds much more generally: The graph G can be random as well, not only its root. A probability measure on rooted graphs is called **unimodular** if it satisfies the mass-transport principle in the sense of (8.12), appropriately generalized. Such probability measures have almost all the properties for percolation that we have seen here, as well as those that we will study in Chapters 10 and 11; see Aldous and Lyons (2007). These measures arise naturally, not only as clusters of the root in invariant percolation on a fixed graph or as modified Galton–Watson trees, but also as limits of uniformly rooted finite graphs. The idea that the mass-transport principle should be defined in this generality is due to Benjamini and Schramm (2001b). Another approach to the same idea is taken in ergodic theory, where the focus is more on the actual measure space, now given a measure-preserving graphed equivalence relation; see Example 9.9 of Aldous and Lyons (2007) for the definitions and how these ideas relate to each other.

A version of the mass-transport principle holds on continuous spaces as well, such as Euclidean space or hyperbolic space. Namely, let G be a unimodular locally compact topological group and H be a compact subgroup. For example, G could be the isometry group of a Euclidean or hyperbolic space and H the stabilizer of a point. (To see that such isometry groups are unimodular, note that every group generated by involutions is unimodular, since the value of the modular function at an involution has square equal to 1 and thus must itself be 1. That the isometries of Euclidean space form such a group is well known and is proved, for example, as Theorem 4.1.2 of Ratcliffe (2006). The corresponding result for hyperbolic space is also proved in Chapter 4 of Ratcliffe (2006).) If μ denotes Haar measure on G, then let ν be the push-forward of μ under the quotient map $G \to G/H$. Up to a constant factor, ν is the unique G-invariant Borel measure on the topological homogeneous space G/H that is finite on compact sets (Nachbin (1965), Theorem 1 and Corollary 4 of Chapter III, Section 4, or Royden (1988), Theorem 14.25; uniqueness does not depend on unimodularity). The following was proved by Benjamini and Schramm (2001a).

Theorem 8.47. *With the notation of the preceding paragraph, let θ be a Borel measure on $G/H \times G/H$ that is invariant under the diagonal action of G. Then $\theta(A \times G/H) = \theta(G/H \times A)$ for all Borel $A \subseteq G/H$. If $\theta(A \times G/H) < \infty$ for some open, nonempty A, then there is a constant c such that $\theta(A \times G/H) = c\nu(A)$ for all Borel $A \subseteq G/H$.*

Proof. Since G is unimodular, $\mu(B) = \mu(B^{-1})$ for all Borel $B \subseteq G$ and $\mu\{g \in G ;\ x \in gA\} = \nu(A)$ is the same for all $x \in G/H$ given any fixed Borel $A \subseteq G/H$. For $A, C \subseteq G/H$, write $\theta_A(C) := \theta(C \times A)$. Fubini's theorem yields that

$$\int \theta\big((gA) \times A\big)\, d\mu(g) = \iint \mathbf{1}_{[x \in gA]}\, d\theta_A(x)\, d\mu(g) = \theta(G/H \times A)\nu(A) .$$

Since $\theta\big((gA) \times A\big) = \theta\big(A \times (g^{-1}A)\big)$ by diagonal invariance, a similar computation gives

$$\theta(G/H \times A)\nu(A) = \theta(A \times G/H)\nu(A) .$$

When $0 < \nu(A) < \infty$, this shows that $\theta(G/H \times A) = \theta(A \times G/H)$. In general, this last equation holds also for $\nu(A) = 0$ by writing such A as a decreasing intersection of sets of positive, finite measure and for $\nu(A) = \infty$ by writing such A as an increasing union of sets of positive, finite measure. Finally, $A \mapsto \theta(A \times G/H)$ is a G-invariant measure, whence if it is finite on some open, nonempty set, it equals ν up to a constant factor. ◀

For a simple illustration of Theorem 8.47, consider an invariant (discrete) point process, Ξ, on G/H, such as a Poisson process. Since $A \mapsto \mathbf{E}|\Xi \cap A|$ is an invariant measure on G/H, it is equal to a constant times ν; that constant is called the **density** of Ξ. Suppose that we associate, in a measurable, equivariant way, to each discrete set W in G/H a measurable partition $f(W)$ of G/H such that the relation $w \in A$ ($w \in W$, $A \in f(W)$) is a bijection of W to $f(W)$; for example, this could be the associated Voronoi partition. Then the expected v-measure of the part associated to o given that $o \in \Xi$ is equal to the reciprocal of the density of Ξ. More precisely, let $N(A)$ be the v-measure of the parts of G/H associated to the points of $\Xi \cap A$, and let λ be the density of Ξ. Then

$$\mathbf{E}[N(A)]/\mathbf{E}|\Xi \cap A| = 1/\lambda, \tag{8.27}$$

whence if A is such that $\mathbf{P}[|\Xi \cap A| \geq 2] = o(v(A))$ as $v(A) \to 0$, then $\mathbf{E}[N(A) \mid \Xi \cap A \neq \varnothing] = 1/\lambda + o(1)$. To see this, let $P(x)$ be the part of G/H associated to x (which is empty if $x \notin \Xi$), and let $\theta(A \times B) := \mathbf{E} \sum_{x \in \Xi \cap A} v(P(x) \cap B)$. Then $\theta(A \times G/H) = \mathbf{E}[N(A)] = \theta(G/H \times A) = v(A)$, from which the identity (8.27) follows.

Exercise 8.10, Proposition 8.18, and Theorem 8.19 are from BLPS (1999b) in the transitive case.

Using Theorem 7.46 and the main result of Timár (2006c), Hutchcroft (2016a) extended Theorem 8.21 to all quasi-transitive graphs of exponential growth. Some other cases of Conjecture 8.15 for groups of intermediate growth were established by Hermon and Hutchcroft (2019a). Theorem 8.21 was used by Schramm (published by Kozma (2011)) to show that if $\langle X_n ; n \geq 0 \rangle$ is simple random walk on a unimodular, nonamenable, transitive graph G and ω is critical Bernoulli percolation on G, then the probability that X_0 is connected to X_n in ω decays exponentially fast in n. It is open whether more generally, $\mathbf{P}[x \leftrightarrow y]$ decays exponentially fast in the distance between x and y for critical Bernoulli percolation on nonamenable, transitive graphs. Hutchcroft (2020a) has made some progress on this.

Theorem 8.24 was proved in the transitive case by Benjamini and Schramm (2001a). The quasi-transitive case is new but not essentially different. This result was proved for certain planar Cayley graphs earlier by Lalley (1998).

Theorem 8.25 was known in the transitive case (Benjamini and Schramm, 2001a), but the quasi-transitive case is due to R. Lyons and is published here for the first time.

The history of Theorem 8.32 is as follows. Benjamini and Schramm (1996b) conjectured that for Bernoulli percolation on any quasi-transitive graph, if there are infinitely many infinite clusters, then a.s. every infinite cluster has continuum many ends. This was proved by Häggström and Peres (1999) for transitive, unimodular graphs and then by Häggström, Peres, and Schonmann (1999) in general. This last paper also shows that in the standard coupling of Bernoulli percolation, a.s. simultaneously all infinite clusters in the nonuniqueness regime have continuum many ends. Proposition 8.33, which implies that each infinite cluster in that more general setting has one, two, or infinitely many ends, is from BLPS (1999b). Our proof of Theorem 8.32 is from Lyons and Schramm (1999), which also proved the statement about transience. It is also true that for any invariant, insertion-tolerant percolation process on a nonamenable, quasi-transitive, unimodular graph with a unique infinite cluster a.s., that cluster is transient, but this is more difficult; see Benjamini, Lyons, and Schramm (1999) for a proof. (In the case of Bernoulli percolation, if Conjecture 7.31 were proven, then this would also follow from Theorem 8.32 and the Rayleigh monotonicity law.) For amenable, transient, quasi-transitive graphs, Benjamini, Lyons, and Schramm (1999) conjectured that infinite clusters are transient a.s. for Bernoulli percolation, following the theorem of Grimmett, Kesten, and Zhang (1993), who established it in $\mathbb{Z}^d$, $d \geq 3$. (We gave a proof of part of this result in Section 5.5. Stronger conjectures were stated in Section 6.9.)

The conclusion of Lemma 8.35, getting an invariant forest whose trees have at least three ends, is useful in geometric group theory: It is analogous to finding a free subgroup (of rank at least 2) in a nonamenable group. However, it is better than the latter in that the latter does not always exist. Such a theorem was proved by Gaboriau and Lyons (2009). An exposition of its use is given by Houdayer (2012).

The following theorem was proved by Benjamini, Lyons, and Schramm (1999):

Theorem 8.48. *Let G be a nonamenable, transitive, unimodular graph. Let ω be an invariant percolation on G that has infinite clusters a.s. Then in each of the following cases (on a larger probability space) there is a percolation $\omega' \subset \omega$ such that $\omega' \neq \emptyset$, $\Phi_E(\omega') > 0$ a.s., and the distribution of the pair (ω', ω) is invariant:*

(i) *ω is Bernoulli percolation;*

(ii) *ω has a unique infinite cluster a.s.;*

(iii) *ω has a cluster with at least three ends a.s.;*

(iv) *$\mathbf{E}[\deg_\omega o \mid o \in \omega] > \alpha(G)$ and ω is ergodic.*

A slight generalization of Theorem 4.1 and Lemma 4.3 of Benjamini and Schramm (2001a) is the following:

Theorem 8.49. *Let $\mathbf{P}$ be an invariant percolation process on a quasi-transitive graph G embedded in hyperbolic space $\mathbb{H}^d$ in such a way that the embedding is a rough isometry. Assume that one of the four conditions in Theorem 8.48 holds. If there are infinite clusters $\mathbf{P}$-a.s., then $\mathbf{P}$-a.s. the set of points z in the ideal boundary $\partial \mathbb{H}^d$ for which there is an open path with limit z is dense in $\partial \mathbb{H}^d$.*

Theorem 8.37 was proved earlier by Ornstein and Weiss (1987) in a more specialized form. Theorem 5.1 of BLPS (1999b) is somewhat more general than our form of it here, Theorem 8.37. The reader may have noticed that each time we prove a property of infinite clusters of insertion-tolerant, invariant percolation, when there is more than one infinite cluster all of the clusters have that property; it was never the case that some do and some don't. Of course, for some properties, this is not true, like the property that a cluster contains o. Clearly the interesting properties for this purpose are those that are automorphism-invariant: a class of subgraphs of G is called ***automorphism-invariant*** if the class is preserved by applying any element of $\text{Aut}(G)$ to each of its member subgraphs. Then we say that a probability measure $\mathbf{P}$ on subgraphs of G has ***indistinguishable infinite clusters*** if, for *every* automorphism-invariant property $\mathcal{A}$ of subgraphs, $\mathbf{P}$-a.s. either all components satisfy $\mathcal{A}$ or they all do not. The main result of Lyons and Schramm (1999) is that every insertion-tolerant, invariant bond percolation on a quasi-transitive, unimodular graph has indistinguishable infinite clusters. The same proof works for site percolation and for weakly insertion-tolerant percolation. In fact, a somewhat stronger version of indistinguishability is proved. In the context of the ergodic theory of equivalence relations mentioned earlier, it turns out that indistinguishability is equivalent to ergodicity of the equivalence relation given by the infinite clusters; see Gaboriau and Lyons (2009).

Indistinguishability has a number of consequences, as shown by Lyons and Schramm (1999), besides such obvious ones as equality of asymptotic growth rates of clusters. For example, it can be used to give an extremely short proof of Lemma 7.26 in the unimodular case. It is used to prove Theorem 7.50(ii) that long-range order is equivalent to uniqueness for Bernoulli percolation. This is then used to prove that $p_u < 1$ for Kazhdan groups and lamplighter groups. Other uses of indistinguishability appear in Benjamini, Lyons, and Schramm (1999). An ingenious use of the mass-transport principle, together with indistinguishability, was made by Timár (2006b), who proved that two infinite clusters in Bernoulli percolation on a quasi-transitive, unimodular graph can come within distance 1 of each other only finitely many times a.s. Häggström, Peres, and Schonmann (1999) asked whether this holds for all quasi-transitive graphs; it is still open in the nonunimodular case.

Indistinguishability does not hold for deletion-tolerant, invariant percolation processes, nor for insertion-tolerant, invariant percolation processes on nonunimodular graphs. However, Häggström, Peres, and Schonmann (1999) show that Bernoulli percolation on all quasi-transitive graphs satisfies a version of indistinguishability restricted to so-called robust properties. Also, Tang (2019) has shown that so-called heavy clusters are indistinguishable in Bernoulli percolation on nonunimodular, quasi-transitive graphs.

For additional results and questions concerning percolation on nonunimodular graphs, see Question 3.17 of Lyons and Schramm (1999), Peres, Pete, and Scolnicov (2006), Timár (2006c), Hutchcroft (2020c), and Tang (2019).

8.10 Collected In-Text Exercises

8.1. Let G be the grandparent graph of Example 7.1 with T having degree 3 and $f(x, y)$ be the indicator that y is the ξ-grandparent of x. Show that every automorphism of G fixes the end ξ and therefore that f is diagonally invariant under $\text{Aut}(G)$.

8.2. Show that the Diestel–Leader graph of Example 7.2 is not a Cayley graph.

8.3. Let Γ be a transitive group of automorphisms of a graph that satisfies (8.4) for all Γ-invariant f. Show that Γ is unimodular.

8.4. Show that if Γ is a transitive, unimodular group of automorphisms and Γ' is a larger group of automorphisms of the same graph, then Γ' is also transitive and unimodular.

8.5. Show that for all x and y, we have $|S(x)y| = [S(x) : S(x) \cap S(y)]$.

8.6. Show that if Γ acts transitively, then Γ is unimodular iff $|S(x)y| = |S(y)x|$ for all edges $[x, y]$.

8.7. Show that if for all edges $[x, y]$, there is some $\gamma \in \Gamma$ such that $\gamma x = y$ and $\gamma y = x$, then Γ is unimodular.

8.8. Extend Exercise 8.4 to the quasi-transitive case: show that if Γ is a quasi-transitive, unimodular group of automorphisms and Γ' is a larger group of automorphisms of the same graph, then Γ' is also quasi-transitive and unimodular.

8.9. Show that for any invariant percolation $\mathbf{P}$ on subgraphs of a transitive, unimodular graph that has only finite clusters a.s., $\mathbf{E}[\deg_\omega o \mid o \in \omega] \le \alpha(G)$. More generally, let G be a quasi-transitive, unimodular graph with a normalized random root $\hat{o}$. Show that if $\mathbf{P}$ is an invariant percolation on subgraphs of G such that all clusters are finite a.s., then $\mathbf{E}[\deg_\omega \hat{o} \mid \hat{o} \in \omega] \le \alpha(G)$.

8.10. Let $\mathbf{P}$ be an invariant percolation on subgraphs of a transitive, unimodular graph such that all clusters are finite trees a.s. Show that $\mathbf{E}[\deg_\omega o \mid o \in \omega] < 2$. More generally, let G be a quasi-transitive, unimodular graph with a normalized random root $\hat{o}$. Show that if $\mathbf{P}$ is an invariant percolation on subgraphs of G such that all clusters are finite trees a.s., then $\mathbf{E}[\deg_\omega \hat{o} \mid \hat{o} \in \omega] < 2$.

8.11. Prove Theorem 8.23.

8.12. Show that a plane, (properly embedded, locally finite) quasi-transitive graph with one end has no face with an infinite number of sides.

8.13. Let G be a transitive, unimodular graph and $o \in \mathsf{V}$. For each $x \in \mathsf{V}$, choose a Haar-random $\gamma_x \in \text{Aut}(G)$ that takes o to x. Show that for every finite set $L \subset \mathsf{V}$, we have

$$\mathbf{E}\big|\{x \in \mathsf{V} \,;\, o \in \gamma_x L\}\big| = |L| \,.$$

8.11 Additional Exercises

8.14. Prove that the result of Example 8.5 cannot be proved by invariance alone. In other words, give an example of an invariant percolation on a transitive graph such that the number of furcations in some cluster is a.s. finite and positive.

8.15. Show that an invariant nonempty percolation on a regular tree that is connected a.s. is the entire tree a.s.

8.16. Show that there is no automorphism-invariant probability measure on the set of ends of a regular tree with degree at least 3.

8.17. Let Ξ be a random closed subset of ends of a regular tree of degree at least 3. Show that if the law of Ξ is automorphism invariant, then a.s. either Ξ is empty or its complement is empty.

8.18. Show that the universal cover of any finite, undirected graph is a unimodular, quasi-transitive tree. Give an example of a quasi-transitive tree that is not unimodular.

8.19. Show that if Γ is a group of automorphisms of a connected graph and the intersection of stabilizers $S(x_1) \cap S(x_2) \cap \cdots \cap S(x_n)$ is finite for some $x_1, x_2, \ldots, x_n$, then Γ is discrete.

8.20. Give $\mathrm{Aut}(G)$ the weak topology generated by its action on G, in other words, a base of open sets at $\gamma \in \mathrm{Aut}(G)$ consists of the sets $\{\gamma' \in \mathrm{Aut}(G) \,;\, \gamma' {\restriction} K = \gamma {\restriction} K\}$ for finite $K \subset \mathsf{V}$. Show that this topology is metrizable, that a subgroup is discrete in this topology iff it is discrete in the sense of Section 8.2, and that if Γ is a closed countable subgroup of $\mathrm{Aut}(G)$, then Γ is discrete.

8.21. Let G be a transitive graph with weights μ_x as in Theorem 8.10.
   (a) Show that the function $x \mapsto \mu_x$ is harmonic on this graph.
   (b) Show that if $f : \mathsf{V} \times \mathsf{V} \to [0, \infty]$ is invariant, then

$$\sum_{x \in \mathsf{V}} \sqrt{\mu_x} f(o, x) = \sum_{x \in \mathsf{V}} \sqrt{\mu_x} f(x, o).$$

   (c) Assign conductances $\sqrt{\mu_{e^-} \cdot \mu_{e^+}}$ to the edges $e \in \mathsf{E}$. Show that the function $x \mapsto \log \mu_x$ is harmonic on this network.

8.22. If Γ is a topological group and μ is a Borel measure on Γ, then we write $L_\gamma \mu$ for the measure $F \mapsto \mu(\gamma^{-1} F)$ and $R_\gamma \mu$ for the measure $F \mapsto \mu(F \gamma)$. We call a Borel measure μ that is finite on compact sets a **left Haar measure** if $L_\gamma \mu = \mu$ for all $\gamma \in \Gamma$ and a **right Haar measure** if $R_\gamma \mu = \mu$ for all $\gamma \in \Gamma$. Here we show how to construct Haar measures on automorphism groups of graphs.
   (a) Let X be a compact metric space. Given $\epsilon > 0$ and $A \subseteq X$, write $B_\epsilon(A)$ for the union of the closed balls of radius ϵ with centers in A. Suppose that A_i $(i = 1, 2)$ are each subsets of X of minimal cardinality with $B_\epsilon(A_i) = X$. Show that there is a bijection $f : A_1 \to A_2$ such that $\mathrm{dist}(x, f(x)) \le 2\epsilon$ for all $x \in A_1$. *Hint:* Use Hall's theorem, Exercise 3.17.
   (b) Given a compact metrized group Γ and $n \ge 1$, choose $A_n \subseteq \Gamma$ of minimal cardinality so that $B_{1/n}(A_n) = \Gamma$. Define $\mu_n := \sum_{x \in A_n} \delta(x)/|A_n|$, where $\delta(x)$ is the unit point mass at x. Show that there is a weak*-limit point μ of $\langle \mu_n \rangle$ and that every such limit point is both a left and right Haar probability measure.
   (c) Show that left and right Haar probability measures are unique. Show that left and right Haar probability measures are equal on every compact group. *Hint:* Use Fubini's theorem.
   (d) Let G be a graph and Γ be a closed subgroup of $\mathrm{Aut}(G)$. Fix $o \in \mathsf{V}(G)$. Choose a maximal set $H \subseteq \Gamma$ so that $\gamma_1 o \ne \gamma_2 o$ for $\gamma_1 \ne \gamma_2 \in H$. Write μ for the Haar probability measure on the stabilizer $S(o)$ in Γ. Show that $\nu := \sum_{\gamma \in H} L_\gamma \mu$ is a left Haar measure on Γ.

8.23. A *perfect matching* of a graph G is a subset M of its edges such that each vertex of G belongs to exactly one edge in M. A graph is called *bipartite* if its vertex set can be partitioned into two parts, A and B, such that all edges have one endpoint in A and one in B. A bipartite graph is called (a, b)-*biregular* if the degrees in one part are a while the degrees in the other part are b. Show that if a biregular, bipartite, quasi-transitive, unimodular graph (for example, a biregular tree) has an invariant percolation that is a perfect matching a.s., then the graph is regular.

8.24. Let G be a quasi-transitive, unimodular graph and $\omega \subseteq \mathfrak{F} \subseteq G$ be random with the law of $(\omega, \mathfrak{F})$ invariant under $\operatorname{Aut}(G)$. Show that if a.s. $\mathfrak{F}$ is a forest all of whose trees are infinite and $\omega \cap T$ is connected for each tree T of $\mathfrak{F}$, then a.s. for each tree T of $\mathfrak{F}$, we have either $\omega \cap T = \varnothing$ or $\omega \cap T = T$.

8.25. Let G be a quasi-transitive, unimodular graph and $(\Xi, \mathfrak{F})$ be random with $\operatorname{Aut}(G)$-invariant law, where $\mathfrak{F}$ is a forest in G all of whose trees have at least three ends and Ξ is a closed set of ends of (trees in) $\mathfrak{F}$. Show that a.s. for each tree T of $\mathfrak{F}$, we have either $\Xi \cap \partial T = \varnothing$ or $\Xi \cap \partial T = \partial T$.

8.26. Give another proof of Proposition 8.14 by using Exercise 8.3.

8.27. Show that for every d, we have $\inf \Phi_{\mathsf{V}}(G) > 0$, where the infimum is taken over transitive, nonunimodular G with degree at most d.

8.28. Show that if Γ is not unimodular and μ_x are the weights of Theorem 8.10, then $\sup_x \mu_x = \infty$ and $\inf_x \mu_x = 0$. In fact, the supremum and infimum may each be taken over any single orbit.

8.29. Let G' be a transitive representation of a quasi-transitive graph G.
  (a) Show that $\operatorname{Aut}(G')$ is unimodular iff $\operatorname{Aut}(G)$ is unimodular.
  (b) Let Γ act quasi-transitively on G. Show that Γ is unimodular iff $\operatorname{Aut}(G')$ is unimodular.

8.30. Show that Proposition 8.14 is also valid for quasi-transitive automorphism groups.

8.31. Let G be an amenable graph and $\Gamma \subseteq \operatorname{Aut}(G)$ be a quasi-transitive subgroup. Choose a complete set $\{o_1, \ldots, o_L\}$ of representatives in V of the orbits of Γ. Choose the weights μ_{o_i} of Theorem 8.10 so that $\sum_i \mu_{o_i}^{-1} = 1$. Show that if K_n is any sequence of finite subsets of vertices such that $|\partial_{\mathsf{V}} K_n|/|K_n| \to 0$, then for all i,

$$\lim_{n \to \infty} \frac{|\Gamma o_i \cap K_n|}{|K_n|} = \mu_{o_i}^{-1} .$$

See Figure 8.1 for an example.

8.32. Show that if Γ is a compact group of automorphisms of a graph, then Γ is unimodular.

8.33. Call two rooted graphs *(rooted) isomorphic* if there is a bijection of their vertex sets preserving adjacency and mapping one root to the other. Our notation for a rooted graph will be (G, o), where $o \in \mathsf{V}(G)$ designates the root. For a rooted graph (G, o), let $[G, o]$ denote the set of rooted graphs that are isomorphic to (G, o). Suppose $\langle X_n \rangle$ is random walk on a quasi-transitive network G whose orbit representatives are $o_1, \ldots, o_L$. Consider the quotient Markov chain $\langle [G, X_n] ; n \geq 0 \rangle$ on the finite state space $\{[G, o_i] ; 1 \leq i \leq L\}$.
  (a) Show that if G is unimodular and X_0 has the distribution of a normalized root biased by π, that is, $\mathbf{P}[X_0 = o_i] \propto \mu_i^{-1} \pi(o_i)$, then the quotient Markov chain $\langle [G, X_n] \rangle$ is stationary and reversible. For example, if G is the graph in Figure 8.1, then the quotient chain on $\{[G, o_1], [G, o_2]\}$ has a loop at $[G, o_1]$, an edge between the two vertices, and two loops at $[G, o_2]$, where all edges have conductance 1. Thus, the stationary distribution is $\langle 2/5, 3/5 \rangle$. The present claim leads to this stationary distribution as follows: $\hat{o} = o_1$ with probability $1/5$ and this gets biased by the degree, 8, whereas $\hat{o} = o_2$ with probability $4/5$, which gets biased by the degree, 3, giving finally relative weights $8/5$ and $3 \cdot 4/5 = 12/5$, which are indeed in the ratio $2 : 3$.

(b) Give an example of a nonunimodular, quasi-transitive network where the quotient chain has a reversible measure.

(c) Give an example of a nonunimodular, quasi-transitive network where the quotient chain does not have a reversible measure.

8.34. Let G be a transitive graph and $o \in \mathsf{V}$. Given a set $S \subseteq \mathsf{V}$ of "seeds," the *Voronoi cell* $V(s)$ of $s \in S$ is the set of vertices x that are closer to s than to any other element of S. Ties are broken at random, with each seed given equal chance. (The exact random mechanism will not matter, as long as it is $\mathrm{Aut}(G)$-invariant.) When $x \in V(s)$ for $s \in S$, we write $V(x) := V(s)$. Given $p \in (0, 1)$, let S be given by Bernoulli(p) site percolation or, more generally, any invariant site percolation with site marginal p.

(a) Show that if G is unimodular, then $\mathbf{E}\big[|V(o)| \mid o \in S\big] = 1/p$.

(b) Show that if G is unimodular, then $\mathbf{E}\big[|V(o)|\big] \geq 1/p$, with equality iff $|V(o)|$ is constant a.s.

(c) Show that if G is unimodular, then the distribution of $|V(o)|$ equals the size-biased conditional distribution of $|V(o)|$ given that $o \in S$, that is, for every bounded $f \colon \mathbb{N} \to \mathbb{R}$, we have

$$\mathbf{E}\big[f(|V(o)|)\big] = \frac{\mathbf{E}\big[f(|V(o)|)\,|V(o)| \mid o \in S\big]}{\mathbf{E}\big[|V(o)| \mid o \in S\big]}.$$

(d) Show that in the case of Bernoulli seeds and where ties are broken independently, even if G is not unimodular, then $\mathbf{E}\big[|V(o)| \mid o \in S\big] = 1/p$ and $\mathbf{E}\big[|V(o)|\big] > 1/p$.

(e) Give an example where $\mathbf{E}\big[|V(o)| \mid o \in S\big] \neq 1/p$ and where the distribution of $|V(o)|$ does not equal the size-biased conditional distribution of $|V(o)|$ given that $o \in S$.

8.35. Sharpen Theorem 8.16 to conclude that $\mathbf{E}[\deg_\omega o] < d_G - \Phi_\mathsf{E}(G)$ by showing that there is some invariant $\mathbf{P}'$ with all clusters finite and with $\mathbf{E}[\deg_\omega o] < \mathbf{E}'[\deg_{\omega'} o]$.

8.36. A subset K of the vertices of a graph is called *dominating* if every vertex is in K or is adjacent to some vertex of K. Suppose that an invariant site percolation on a transitive, unimodular graph of degree d is a dominating set a.s. Show that o belongs to the percolation with probability at least $1/(d+1)$.

8.37. Show that for any invariant percolation $\mathbf{P}$ on a transitive, unimodular graph that has finite clusters with positive probability, $\mathbf{E}\big[\deg_\omega o \mid |K(o)| < \infty\big] \leq \alpha(G)$.

8.38. Let $\mathbf{P}$ be an invariant site percolation on a transitive, unimodular graph of degree d such that all clusters are finite a.s. Show that $\mathbf{P}[o \in \omega] < d/(d + \Phi_\mathsf{V}(G))$.

8.39. Let G be a quasi-transitive, unimodular graph and $\hat{o}$ a normalized random root. Let $\mathfrak{F}$ be the configuration of an invariant random spanning forest on G such that a.s. each tree has one end. ("Spanning" means that the forest includes all vertices of G. We will see important examples of such spanning forests in Chapters 10 and 11.) For a vertex x, denote by $\xi(x) = \langle \xi_n(x) ; n \geq 0 \rangle$ the unique infinite, simple path starting at x. If $y \in \xi(x)$, call x a *descendant* of y. Let $D(x)$ be the (finite) set of all descendants of x.

(a) Show that $\mathbf{E}\big[|D(\hat{o})|\big] = \infty$.

(b) Show that $\mathbf{E}\big[|\{y ; \hat{o} = \xi_n(y)\}|\big] = 1$ for each $n \geq 0$.

(c) Show that $\mathbf{E}\big[\sum_{n \geq 0} 1/|D(\xi_n(\hat{o}))|\big] = 1$.

(d) Show that $\mathbf{E}\big[|D(\xi_n(\hat{o})) \setminus D(\xi_{n-1}(\hat{o}))|\big] = \infty$ for each $n \geq 1$.

(e) Show that $\mathbf{E}\big[|D(\hat{o})|\,(\deg_{\mathfrak{F}} \hat{o} - 2)\big] = \infty$.

8.40. Let T be a regular tree of degree at least 3. Show that for every $p \in (0, 1)$, there is an $\mathrm{Aut}(T)$-invariant probability measure on $\{0, 1\}^{\mathsf{V}(T)}$ such that for all $[x, y] \in \mathsf{E}(T)$, the probability that x is assigned 0 equals $1/2$, the probability that x and y are assigned the same label is at least p, and the probability that two vertices have the same label tends to $1/2$ as their mutual distance tends to infinity.

8.41. Extract from the proof of Theorem 8.21 just enough to prove that $p_c < 1$ for nonamenable, quasi-transitive, unimodular graphs.

8.42. Show that if an invariant percolation on a nonamenable, quasi-transitive, unimodular graph has a finite number of infinite clusters a.s., then a.s. each of those infinite clusters has $p_c < 1$.

8.43. Let G be a nonamenable, unimodular, transitive graph. Write P_n for the graph that is a path of n vertices and C_n for the cycle of n vertices.
    (a) Show that $p_c(G \,\square\, P_n) \to p_c(G \,\square\, \mathbb{Z})$ as $n \to \infty$.
    (b) Show that $p_c(G \,\square\, C_n) \to p_c(G \,\square\, \mathbb{Z})$ as $n \to \infty$.

8.44. Let G be the edge graph of the $(2, 3, 7)$-triangle tessellation of Figure 6.1, that is, the plane dual of the Cayley graph there. By Theorem 8.25 and Remark 8.46, it is quasi-transitive and unimodular. It has three vertex orbits. Find the weight of each vertex that is given by Theorem 8.10.

8.45. Let G be a plane, transitive graph with one end. By Theorem 8.25 and Remark 8.46, its dual $G^\dagger$ is quasi-transitive and unimodular. Find the distribution of a normalized random root of $G^\dagger$.

8.46. Let G be a plane, transitive graph with one end. Every edge $e \in \mathsf{E}(G)$ intersects $e^\dagger \in \mathsf{E}(G^\dagger)$ in one point, v_e. (These are the only intersections of G and $G^\dagger$.) For $e \in \mathsf{E}(G)$, write $\hat{e}$ for the pair of edges that result from the subdivision of e by v_e, and likewise for $e^\dagger$. This defines a new quasi-transitive graph $\hat{G}$, whose vertices are $\mathsf{V}(G) \cup \mathsf{V}(G^\dagger) \cup \{v_e \; ; \; e \in \mathsf{E}(G)\}$ and whose edges are $\bigcup_{e \in \mathsf{E}(G)} (\hat{e} \cup \hat{e}^\dagger)$. Show that $\hat{G}$ is unimodular and find the distribution of a normalized random root of $\hat{G}$.

8.47. Let G be the usual Cayley graph of the (p, q, r)-triangle group and $G^\dagger$ be its dual, where $1/p + 1/q + 1/r \leq 1$. This Cayley graph for $(p, q, r) = (2, 3, 7)$ was shown in Figure 6.1. In general, the group is generated by reflections in the infinitely extended geodesic sides of a Euclidean or hyperbolic triangle whose interior angles measure π/p, π/q, and π/r. The edge graph of the tessellation by such triangles is $G^\dagger$. Let $\mathfrak{F}$ be an invariant random spanning forest of $G^\dagger$ such that all of its trees are infinite and have at most two ends a.s. ("Spanning" means that the forest includes all vertices of $G^\dagger$.) Let the edges of $G^\dagger$ opposite to the angles of measure π/p, π/q, π/r have probabilities $\alpha_p, \alpha_q, \alpha_r$ of belonging to $\mathfrak{F}$, respectively. Show that $\alpha_p + \alpha_q + \alpha_r = 1/p + 1/q + 1/r$. Show also that if $\mathfrak{F}^\times$ is defined on G as in (8.21), then $\mathbf{E}[\deg_o \mathfrak{F}^\times] = 3 - 1/p - 1/q - 1/r$.

8.48. Let G be a plane, transitive graph with one end. Show that

$$\frac{1}{\mu(G^\dagger)} + \frac{1}{\gamma(G)} \leq 1 \,,$$

where μ is the connective constant and γ is defined as in (7.12). Deduce that if $G^\dagger$ is regular of degree $d^\dagger$, then

$$\gamma(G) \geq \frac{d^\dagger - 1}{d^\dagger - 2} \,.$$

8.49. Let **P** be an invariant bond percolation on a quasi-transitive, unimodular graph such that all clusters are infinite a.s. Show that $\mathbf{E}[\deg_\omega \hat{o}] \geq 2$ when $\hat{o}$ is a normalized random root.

8.50. Let ω be the configuration of an invariant percolation on a transitive, unimodular graph G. Show that if
 (i) some component of ω has at least three ends with positive probability,
then
 (ii) a.s. every component of ω with at least three ends has $p_c < 1$ and
 (iii) $\mathbf{E}\big[\deg_\omega o \mid |K(o)| = \infty\big] > 2$.

8.51. Give an example of an invariant random forest on a transitive graph where each component has three ends, but the expected degree of each vertex is smaller than 2.

8.52. Give an example of an invariant random forest on a transitive graph where each tree has one end and the expected degree of each vertex is greater than 2.

8.53. Let ω be an invariant percolation on a quasi-transitive, unimodular graph. Suppose that all components of ω have at least three ends a.s. Show that the probability space cannot be enlarged so as to pick exactly one end from some of the components of ω.

8.54. Let **P** be an insertion-tolerant, invariant bond percolation on a quasi-transitive, unimodular graph that has infinitely many infinite clusters a.s. Show that a.s. for each infinite cluster, there are infinitely many edges with one endpoint in that cluster and one endpoint in a different infinite cluster.

8.55. Let G be a quasi-transitive, unimodular graph with $p_c(G) < p_u(G)$. Show that in the standard coupling $p \mapsto \omega_p$ of Bernoulli percolation, a.s. for each pair (p_1, p_2) with $p_c < p_1 < p_2 < p_u$, we have that every infinite cluster of ω_{p_2} contains infinitely many infinite clusters of ω_{p_1}.

8.56. Let T be a 3-regular tree and fix an end ξ of T. Define $W(e)$ to be independent, symmetric $\{-1, 1\}$-valued random variables for $e \in E(T)$. Let $Y(x) := \max_{e \sim x} W(e)$. If x_ξ denotes the first edge on the ray from x that belongs to ξ, then let $Z(x) := \max_{e \sim x, \, e \neq x_\xi} W(e)$. Clearly $Y(x) \geq Z(x)$ for all x and the distribution of Y is Aut(T)-invariant.
  (a) Show that the distribution of Z is i.i.d.
  (b) Show that the distribution of (Y, Z) is not Aut(T)-invariant.
  (c) Show that there is a Z' equal to Z in distribution such that $Y \geq Z'$ and the distribution of (Y, Z') is Aut(T)-invariant.

8.57. Let ω be an invariant bond percolation on a transitive graph, G, such that all clusters are finite a.s. Fix $o \in V$. Let $K(o)$ be the cluster of o in ω. Let Z be a uniformly random vertex of $K(o)$. Recall the notion of rooted isomorphism from Exercise 8.33.
  (a) Suppose that G is unimodular. Show that the law of the rooted isomorphism class of $\big(K(o), Z\big)$ is the same as the law of the rooted isomorphism class of $\big(K(o), o\big)$.
  (b) Give an example on a nonunimodular graph where the result of (a) fails.

8.58. Let G be a quasi-transitive, unimodular graph. Show that G is amenable iff for all $\alpha < 1$, there is an invariant bond percolation ω on G with no infinite clusters and such that $\mathbf{E}[\deg_\omega x] > \alpha \deg_G x$ for all $x \in V(G)$.

8.59. Theorem 8.37 was proved only in the transitive case. Prove it in the general quasi-transitive case.

9 | Infinite Electrical Networks and Dirichlet Functions

In Chapters 2 and 3, we looked at current flows from a vertex to infinity to analyze transience and recurrence. We never looked at current flows from one vertex to another in an infinite network, except briefly on recurrent networks in Exercises 2.71 and 2.72. We had no need for that; it also turns out to be more complicated. But those complications are also quite interesting, as we will see in this chapter. Moreover, we will ultimately use our work here to answer a question of recurrence again. Namely, given a transient network, consider the subnetwork formed by the edges that are crossed at least once by a random walk on the original network. Is this random subnetwork transient or recurrent? The answer is in the last section of this chapter.

Even more importantly, our work here will be the foundation in the next chapter for extending the study of uniform spanning trees on finite networks to so-called uniform spanning forests on infinite networks.

9.1 Free and Wired Electrical Currents

We begin by taking a look at currents from one vertex to another on infinite networks. It turns out that there are two natural ways of defining such currents that correspond to two ways of taking limits over finite networks. In some sense, these two ways may differ due to the possibility of current "passing via infinity." Our approach in this section will be to give definitions of both these currents using Hilbert space; then we will show how they correspond to limits of currents over finite graphs.

Let G be a connected network whose conductances c satisfy the usual condition that $\sum_{e^- = x} c(e) < \infty$ for each vertex $x \in G$. Note that this condition guarantees that the stars $\sum_{e^- = x} c(e) \chi^e$ of G have finite energy. We assume this condition is satisfied for all networks in this chapter. We also assume that networks are connected. Let ★ denote the closure of the linear span of the stars and ◇ the closure of the linear span of the cycles of a graph $G = (V, E)$, both of these closures taking place in the Hilbert space of antisymmetric edge functions of finite energy,

$$\ell_-^2(E, r) := \left\{ \theta : E \to \mathbb{R} ; \ \forall e \quad \theta(-e) = -\theta(e) \text{ and } \sum_{e \in E} \theta(e)^2 r(e) < \infty \right\}.$$

Recall that by (2.15), we have $\sum_{e^- = x} |\theta(e)| < \infty$ for all $x \in V$ and all $\theta \in \ell_-^2(E, r)$.

The exhaustions we considered in Chapter 2 by finite, induced networks led to currents in ★. Indeed, for a finite, induced subnetwork H of G, recall that H^W is formed by identifying the complement of H to a single vertex and that we identify $\mathsf{E}(H^{\mathrm{W}})$ with a subset of $\mathsf{E}(G)$. The star space of H^{W} lies in the star space of G, since for all $x \in \mathsf{V}(H)$, the star of x in H^{W} coincides with the star of x in G, while the star of the new vertex of H^{W} equals the negative of the sum of the other stars of H^{W}. Therefore, the unit current flow i_a in G from a vertex a to infinity (defined in Proposition 2.12) lies in the star space of G.

Since each star is orthogonal to each cycle, it is still true (as for finite networks) that $\bigstar \perp \Diamond$. However, it is *no longer necessarily the case* that $\ell_-^2(\mathsf{E}, r) = \bigstar \oplus \Diamond$. Thus, we are led to define two possibly different currents:

$$i_{\mathrm{F}}^e := P_{\Diamond}^\perp \chi^e, \tag{9.1}$$

the ***unit free current*** between the endpoints of e (also called the ***limit current***), and

$$i_{\mathrm{W}}^e := P_{\bigstar} \chi^e, \tag{9.2}$$

the ***unit wired current*** between the endpoints of e (also called the ***minimal current***).

▷ **Exercise 9.1.**
Calculate i_{F}^e and i_{W}^e in a regular tree.

The names for these currents are explained by the following two propositions. Recall that for a subnetwork $G_n \subset G$, we identify $\mathsf{E}(G_n)$ as a subset of $\mathsf{E}(G)$ and also identify $\mathsf{E}(G_n^{\mathrm{W}})$ as a subset of $\mathsf{E}(G)$.

Proposition 9.1. (Free Currents as Limit Currents) *Let G be an infinite network exhausted by finite subnetworks $\langle G_n \rangle$. Let e be an edge in G_1 and i_n be the unit current flow in G_n from e^- to e^+. Then $\|i_n - i_{\mathrm{F}}^e\|_r \to 0$ as $n \to \infty$ and $\mathscr{E}(i_{\mathrm{F}}^e) = i_{\mathrm{F}}^e(e)r(e)$.*

Proof. Decompose $\ell_-^2(\mathsf{E}_n, r) = \bigstar_n \oplus \Diamond_n$ on G_n into the spaces spanned by the stars and cycles in G_n and recall that $i_n = \chi^e - P_{\Diamond_n}\chi^e$. We may regard the spaces $\Diamond_n$ as lying in $\ell_-^2(\mathsf{E}, r)$, where they form an increasing sequence. (Each cycle in G_n lies in G_{n+1}, but the same is not true of the stars.) The closure of $\bigcup_n \Diamond_n$ is $\Diamond$. Note that the orthogonal projection on $\Diamond_n$ of any element of $\ell_-^2(\mathsf{E}_n, r)$ is the same as its projection on $\Diamond_n$ in $\ell_-^2(\mathsf{E}, r)$ since for $\theta \in \ell_-^2(\mathsf{E}_n, r)$, if $\theta \perp \Diamond_n$ in $\ell_-^2(\mathsf{E}_n, r)$, then also $\theta \perp \Diamond_n$ in $\ell_-^2(\mathsf{E}, r)$. Thus, the fact that $\|i_n - i_{\mathrm{F}}^e\|_r \to 0$ follows from the standard result given in the upcoming Exercise 9.2. Also, we have that

$$i_{\mathrm{F}}^e(e)r(e) = (i_{\mathrm{F}}^e, \chi^e)_r = (i_{\mathrm{F}}^e, P_{\Diamond}^\perp \chi^e)_r = \mathscr{E}(i_{\mathrm{F}}^e). \qquad \blacktriangleleft$$

▷ **Exercise 9.2.**
Let H_n be increasing closed subspaces of a Hilbert space H and P_n be the orthogonal projection on H_n. Let P be the orthogonal projection on the closure of $\bigcup H_n$. Show that for all $u \in H$, we have $\|P_n u - Pu\| \to 0$ as $n \to \infty$.

Proposition 9.2. (Wired Currents as Minimal Currents) *Let G be an infinite network exhausted by finite, induced subnetworks $\langle G_n \rangle$. Form G_n^{W} by identifying the complement of G_n to a single vertex. Let e be an edge in G_1 and i_n be the unit current flow in G_n^{W} from e^- to e^+. Then $\|i_n - i_{\mathrm{W}}^e\|_r \to 0$ as $n \to \infty$ and $\mathscr{E}(i_{\mathrm{W}}^e) = i_{\mathrm{W}}^e(e)r(e)$, which is the minimum energy among all unit flows from e^- to e^+.*

▷ **Exercise 9.3.**
Prove Proposition 9.2.

Since $\bigstar \subseteq \Diamond^\perp$, we have $\mathcal{E}(i_W^e) \leq \mathcal{E}(i_F^e)$ with equality iff $i_W^e = i_F^e$. Therefore,

$$i_W^e(e) \leq i_F^e(e) \text{ with equality iff } i_W^e = i_F^e. \tag{9.3}$$

Proposition 9.3. *For any network, $i_W^e = i_F^e$ for every edge e iff $\ell_-^2(\mathsf{E}, r) = \bigstar \oplus \Diamond$.*

Proof. Note that $\ell_-^2(\mathsf{E}, r) = \bigstar \oplus \Diamond$ is equivalent to $P_\bigstar = P_\Diamond^\perp$. Since $\{\chi^e ; e \in \mathsf{E}_{1/2}\}$ spans $\ell_-^2(\mathsf{E}, r)$, this is also equivalent to $P_\bigstar \chi^e = P_\Diamond^\perp \chi^e$ for all edges e, as desired. ◄

Given two vertices a and z, define the **free** and **wired unit currents** from a to z as $i_F^{a,z} := \sum_{k=1}^n i_F^{e_k}$ and $i_W^{a,z} := \sum_{k=1}^n i_W^{e_k}$, where $e_1, e_2, \ldots, e_n$ is an oriented path from a to z.

▷ **Exercise 9.4.**
Show that the choice of path in the definition of the free and wired currents from a to z does not influence their values.

We call $\mathcal{R}^F(a \leftrightarrow z) := \mathcal{E}(i_F^{a,z})$ and $\mathcal{R}^W(a \leftrightarrow z) := \mathcal{E}(i_W^{a,z})$ the **free** and **wired effective resistance**, respectively, between a and z. Note that these are equal to the limits of $\mathcal{E}(i_n)$, where i_n is the unit current flow from a to z on G_n or G_n^W, respectively, since for any sequence of vectors u_n converging in norm to a vector u, we have $\|u_n\| \to \|u\|$; indeed, $\big| \|u_n\| - \|u\| \big| \leq \|u_n - u\|$ by the triangle inequality. Of course, the reciprocals of the effective resistances are called the free and wired effective conductances.

9.2 Planar Duality

In this section, we recall from Section 6.5 the basic notions of duality for planar graphs and show how the dual graphs give related electrical networks.

A **planar** graph is one that can be drawn in the plane in such a way that edges do not cross; an actual such embedding is called a **plane** graph. If G is a plane graph such that each bounded set in the plane contains only finitely many vertices of G, then G is said to be **properly embedded** in the plane. We will always assume without further mention that plane graphs are properly embedded. A **face** of a plane graph is a connected component of the complement of the graph in the plane. If G is a plane (multi)graph, then the **plane dual** $G^\dagger$ of G is the (multi)graph formed as follows: The vertices of $G^\dagger$ are the faces formed by G. Two faces of G are joined by an edge in $G^\dagger$ precisely when they share an edge in G. Thus, $\mathsf{E}(G)$ and $\mathsf{E}(G^\dagger)$ are in a natural one-to-one correspondence. Furthermore, if one draws each vertex of $G^\dagger$ in the interior of the corresponding face of G and each edge of $G^\dagger$ so that it crosses only the corresponding edge of G, then the dual of $G^\dagger$ is G.

We choose orientations of the edges so that for $e \in \mathsf{E}$, the corresponding edge $e^\dagger$ of the dual crosses e from right to left as viewed from the direction of e. Thus, the orientation of the pair $(e, e^\dagger)$ is the same as the usual counterclockwise orientation of the plane.

If conductances $c(e)$ are assigned to the edges e of G, then we define the conductance $c(e^\dagger)$ of $e^\dagger$ to be the *resistance* $r(e)$. In this case, we will assume without mention that G is such that $\sum_{(e^\dagger)^-=u} c(e^\dagger) < \infty$ for each vertex $u \in G^\dagger$, so that the stars of $G^\dagger$ have finite energy.

The bijection $e \mapsto e^\dagger$ between edge sets provides a natural isometric isomorphism of Hilbert spaces $^\dagger\colon \ell^2_-(E(G), r) \to \ell^2_-(E(G^\dagger), r)$ via $\theta^\dagger(e^\dagger) := r(e)\theta(e)$. That is, $\theta \mapsto \theta^\dagger$ is a surjective linear map such that for all $\theta, \psi \in \ell^2_-(E(G), r)$, we have $\theta^\dagger, \psi^\dagger \in \ell^2_-(E(G^\dagger), r)$ and

$$(\theta, \psi)_r = \sum_{e \in E(G)} \theta(e)\psi(e)r(e) = \sum_{e^\dagger \in E(G^\dagger)} \theta^\dagger(e^\dagger)\psi^\dagger(e^\dagger)r(e^\dagger) = (\theta^\dagger, \psi^\dagger)_r .$$

Note that $(\theta^\dagger)^\dagger = -\theta$.

It is clear that if θ is a star in $\ell^2_-(E(G), r)$, then $\theta^\dagger$ is a cycle in $\ell^2_-(E(G^\dagger), r)$. Moreover, it is easy to see that † induces an isomorphism from the star space on G to the cycle space on $G^\dagger$ and from the cycle space on G to the star space on $G^\dagger$. That's cute. What is the implication of this for currents? For an edge $e \in G$, consider the orthogonal decomposition

$$\chi^e = i^e_W + \theta ,$$

where $i^e_W \in \bigstar(G)$ and $\theta \in \bigstar(G)^\perp$. Applying the map †, we obtain

$$r(e)\chi^{e^\dagger} = (\chi^e)^\dagger = (i^e_W)^\dagger + \theta^\dagger ,$$

whence

$$\chi^{e^\dagger} = c(e)(i^e_W)^\dagger + c(e)\theta^\dagger ,$$

where the first term on the right is a vector in $\Diamond(G^\dagger)$ and the second is in $\Diamond(G^\dagger)^\perp$. It follows from this and the definition (9.1) that

$$i^{e^\dagger}_F = c(e)\theta^\dagger = \chi^{e^\dagger} - c(e)(i^e_W)^\dagger .$$

Likewise, one can check that

$$i^{e^\dagger}_W = \chi^{e^\dagger} - c(e)(i^e_F)^\dagger .$$

In particular, we obtain

$$i^{e^\dagger}_F(e^\dagger) = \chi^{e^\dagger}(e^\dagger) - c(e)(i^e_W)^\dagger(e^\dagger) = 1 - c(e)r(e)i^e_W(e) = 1 - i^e_W(e) \tag{9.4}$$

and

$$i^{e^\dagger}_W(e^\dagger) = 1 - i^e_F(e) .$$

This has the following curious consequence:

Proposition 9.4. *Let G be a plane network and $[a, z]$ be an edge of G. Let the dual edge be $[b, y]$. Suppose that the graph G' obtained by deleting the edge $[a, z]$ from G is connected and that the graph $(G^\dagger)'$ obtained by deleting the edge $[b, y]$ is connected. Then the free effective resistance between a and z in G' equals the wired effective conductance between b and y in $(G^\dagger)'$.*

▷ **Exercise 9.5.**
Prove Proposition 9.4.

9.3 Harmonic Dirichlet Functions

As we saw from the definitions, wired and free currents are equal iff $\ell^2_-(E, r) = \bigstar \oplus \lozenge$, or said another way, iff the orthogonal complement of $\bigstar \oplus \lozenge$ in $\ell^2_-(E, r)$ is $\mathbf{0}$. What is this orthogonal complement? We will identify it and use this identification to develop criteria for its vanishing. In preparation for applications in the next sections, we then analyze the limiting behavior of certain functions along a random walk path.

Recall that the ***gradient*** of a function f on V is the antisymmetric function

$$\nabla f := c \, df$$

on E. Ohm's law in this notation is $\nabla v = i$. Define the space of ***Dirichlet*** functions

$$\mathbf{D} := \left\{ f \,;\, \nabla f \in \ell^2_-(E, r) \right\}.$$

Given a vertex $o \in V$, we use the inner product on $\mathbf{D}$

$$\langle f, g \rangle := f(o)g(o) + (\nabla f, \nabla g)_r = f(o)g(o) + (df, dg)_c.$$

This makes $\mathbf{D}$ a Hilbert space, whose norm we denote by $\|\cdot\|_\mathbf{D}$. The choice of o does not matter in the sense that changing it leads to an equivalent norm: for any x, take a path of edges $\langle e_j \,;\, 1 \le j \le n \rangle$ leading from x to o and note that by the Cauchy–Schwarz inequality,

$$f(x)^2 = \left[f(o) + \sum_j df(e_j) \right]^2 \le \left[1 + \sum_j r(e_j) \right] \left[f(o)^2 + \sum_j c(e_j) df(e_j)^2 \right]$$

$$\le \left[1 + \sum_j r(e_j) \right] \langle f, f \rangle,$$

whence

$$f(x)^2 + (df, df)_c \le \left[2 + \sum_j r(e_j) \right] \langle f, f \rangle.$$

The quantity $\mathscr{D}(f) := \|\nabla f\|^2_r = \|df\|^2_c$ is called the ***Dirichlet energy*** of f.* Of course, the constant functions, which we identify as $\mathbb{R}$, lie in $\mathbf{D}$. Since it is the gradient of a function that matters most here, we usually work with $\mathbf{D}/\mathbb{R}$ using the inner product

$$\langle f + \mathbb{R}, g + \mathbb{R} \rangle := (df, dg)_c.$$

Then $\mathbf{D}/\mathbb{R}$ is a Hilbert space.

If $\phi: \mathbb{R} \to \mathbb{R}$ is a contraction (that is, $|\phi(x) - \phi(y)| \le |x - y|$ for $x, y \in \mathbb{R}$), then $f \mapsto \phi \circ f$ maps $\mathbf{D}$ to $\mathbf{D}$ by decreasing the energy: $\mathscr{D}(\phi \circ f) \le \mathscr{D}(f)$. Useful examples include $\phi(s) := |s|$ and the truncation maps

$$\phi_N(s) := \begin{cases} s & \text{if } |s| \le N \\ sN/|s| & \text{if } |s| > N. \end{cases}$$

---

* The classical Dirichlet energy of a smooth function f in a domain Ω is $\int_\Omega |\nabla f|^2 \, d\lambda$, where λ is Lebesgue measure.

Thus, for $f \in \mathbf{D}$, there is a sequence $\langle \phi_N \circ f \rangle$ of bounded functions in $\mathbf{D}$ that converge to f in norm by Lebesgue's dominated convergence theorem. If $f, g \in \mathbf{D}$ are both bounded functions, then

$$\|d(fg)\|_c \le \|f\|_\infty \|dg\|_c + \|df\|_c \|g\|_\infty \tag{9.5}$$

since

$$\begin{aligned}
|(fg)(x) - (fg)(y)| &= \big| f(x)[g(x) - g(y)] + [f(x) - f(y)]g(y) \big| \\
&\le \|f\|_\infty |g(x) - g(y)| + \|g\|_\infty |f(x) - f(y)| \\
&= |\widetilde{g}(x) - \widetilde{g}(y)| + |\widetilde{f}(x) - \widetilde{f}(y)|,
\end{aligned}$$

where $\widetilde{g} := \|f\|_\infty g$ and $\widetilde{f} := \|g\|_\infty f$, whence

$$\|d(fg)\|_c \le \big\| |d\widetilde{g}| + |d\widetilde{f}| \big\|_c \le \|d\widetilde{g}\|_c + \|d\widetilde{f}\|_c = \|f\|_\infty \|dg\|_c + \|df\|_c \|g\|_\infty.$$

Recall that $\bigstar$ denotes the closed span of the stars in $\ell^2_-(\mathsf{E}, r)$ and $\Diamond$ the closed span of the cycles in $\ell^2_-(\mathsf{E}, r)$. The gradient map $\nabla: \mathbf{D}/\mathbb{R} \to \Diamond^\perp$ is an isometric isomorphism (since G is connected). Just as we reasoned in Section 2.4, an element $\theta \in (\bigstar \oplus \Diamond)^\perp$ is the gradient of a harmonic function $f \in \mathbf{D}$. Thus, if $\mathbf{HD}$ denotes the set of $f \in \mathbf{D}$ that are harmonic, we have the orthogonal decomposition

$$\ell^2_-(\mathsf{E}, r) = \bigstar \oplus \Diamond \oplus \nabla \mathbf{HD}. \tag{9.6}$$

Since $\ell^2_-(\mathsf{E}, r) = \bigstar \oplus \Diamond$ iff $\nabla \mathbf{HD} = \mathbf{0}$ iff $\mathbf{HD} = \mathbb{R}$, we may add the condition that there are no nonconstant harmonic Dirichlet functions to those in Proposition 9.3:

Theorem 9.5. (Doyle, 1988) *Let G be a denumerable network. We have $\mathbf{HD} = \mathbb{R}$ iff $i^e_{\mathrm{W}} = i^e_{\mathrm{F}}$ for each $e \in \mathsf{E}$.*

Doyle's theorem is often stated another way as a criterion for uniqueness of currents. We call the elements i of $\Diamond^\perp$ **currents** (with sources where $d^* i > 0$ and sinks where $d^* i < 0$). We say that **currents are unique** if whenever i, i' are currents with $d^* i = d^* i'$, we have $i = i'$. Observe that by subtraction and by (2.8), this is the same as saying that $\Diamond^\perp \cap \bigstar^\perp = \mathbf{0}$, in other words, $\mathbf{HD} = \mathbb{R}$. For this reason, the class of networks with unique currents is often denoted O_{HD}.

Let $\mathbf{D}_0$ be the closure in $\mathbf{D}$ of the set of $f \in \mathbf{D}$ with finite support.

▷ **Exercise 9.6.**

(a) Show that $\nabla \mathbf{D}_0 = \bigstar$.

(b) Show that $\mathbf{D}/\mathbb{R} = \widetilde{\mathbf{D}_0}/\mathbb{R} \oplus \mathbf{HD}/\mathbb{R}$, where $\widetilde{\mathbf{D}_0} := \mathbf{D}_0 + \mathbb{R}$.

(c) Show that currents are unique iff $\mathbf{D}/\mathbb{R} = \widetilde{\mathbf{D}_0}/\mathbb{R}$.

(d) Show that $\|\mathbf{1} - \mathbf{D}_0\|^2_{\mathbf{D}} = \mathscr{C}(o \leftrightarrow \infty)/[1 + \mathscr{C}(o \leftrightarrow \infty)]$, where o is the vertex used to define the inner product on $\mathbf{D}$.

(e) Show that G is recurrent iff $\mathbf{1} \in \mathbf{D}_0$.

(f) **(Royden Decomposition)** Show that if G is transient, then every $f \in \mathbf{D}$ has a unique decomposition $f = g + h$ with $g \in \mathbf{D}_0$ and $h \in \mathbf{HD}$. Note that this is not an orthogonal decomposition.

(g) With the assumption and notation of part (f), show that $g(x) = (\nabla f, i_x)_r$ and that $g(x)^2 \le \mathscr{D}(f) \mathscr{G}(x, x)/\pi(x)$, where i_x is the unit current flow from x to infinity (from Proposition 2.12) and $\mathscr{G}(\cdot, \cdot)$ is the Green function.

(h) Show that if G is transient, then $\nabla: \mathbf{D}_0 \to \bigstar$ is invertible with bounded inverse.

This exercise allows us to show that recurrent networks have unique currents. (This also follows from Exercise 2.72.) Since current cannot go to infinity at all in recurrent networks, this fits with our intuition that nonunique currents require the ability of current to pass via infinity. An extension is in Exercise 9.23.

Corollary 9.6. (Recurrence Yields Unique Currents) *A network G is recurrent iff* $\mathbf{D} = \mathbf{D}_0$, *in which case currents are unique.*

Proof. If $\mathbf{D} = \mathbf{D}_0$, then $\mathbf{1} \in \mathbf{D}_0$, whence G is recurrent by Exercise 9.6(e). Conversely, suppose that G is recurrent. By Exercise 9.6(e), we have $\mathbf{1} \in \mathbf{D}_0$. Thus, there exist $g_n \to \mathbf{1}$ with g_n having finite support and $\mathbf{0} \leq g_n \leq \mathbf{1}$. (We use the contraction $\phi(s) := s\mathbf{1}_{[0,1]}(s) + \mathbf{1}_{(1,\infty)}(s)$ if necessary to get the values of g_n to be in $[0, 1]$.) Let $f \in \mathbf{D}$ be bounded. Then $fg_n \in \mathbf{D}$ by (9.5) and $fg_n \to f$ in $\mathbf{D}$ by the dominated convergence theorem. Hence $f \in \mathbf{D}_0$, in other words, $\mathbf{D}_0$ contains all bounded Dirichlet functions. Since these functions are dense in $\mathbf{D}$, we get $\mathbf{D}_0 = \mathbf{D}$. Finally, when this happens, currents are unique by Exercise 9.6(c). ◀

Some transient networks also have unique currents. To exhibit some of these, we will use the following criterion, which generalizes a result of Thomassen (1989). It is analogous to the Nash-Williams criterion. It shows that high connectedness of cutsets, rather than their small size, can force currents to be unique. This is reasonable given that the difference between free and wired currents is a matter of whether wiring cutsets matters in the limit. Let $\mathscr{R}(x \leftrightarrow y; A)$ denote the effective resistance between vertices x and y in a finite network A. We will use this when A is a subnetwork of G; in this case, the effective resistance is computed purely in terms of the network A. Let

$$\mathrm{RD}(A) := \sup\{\mathscr{R}(x \leftrightarrow y; A); \ x, y \in \mathsf{V}(A)\}$$

be the "effective-resistance diameter" of A. Note that in the case of unit conductances on the edges, $\mathrm{RD}(A)$ is at most the graph diameter of A. We say that a subnetwork W *separates* x *from* ∞ if every simple infinite path starting at x intersects W in some vertex.

Theorem 9.7. (Unique Currents from Internal Connectivity) *If $\langle W_n \rangle$ is a sequence of pairwise edge-disjoint, finite subnetworks of a locally finite network G such that each (equivalently, some) vertex is separated from ∞ by all but finitely many W_n and such that*

$$\sum_n \frac{1}{\mathrm{RD}(W_n)} = \infty, \tag{9.7}$$

then G has unique currents.

Proof. Let f be any nonconstant harmonic function. We will show that $f \notin \mathbf{D}$ by bounding the left-hand side of (9.7) in terms of f. Take an edge e_0 whose endpoints have different values for f, that is, $df(e_0) \neq 0$. Let n_0 be such that W_n separates e_0 from infinity for $n \geq n_0$. Let H_n be the set of vertices that W_n separates from infinity, including the vertices of W_n. Because G is locally finite, H_n is finite. Let x_n and y_n be points of H_n where f takes its maximum and minimum, respectively, on H_n. By the maximum principle, we may assume that $x_n, y_n \in W_n$. Thus, for $n \geq n_0$, we have $f(x_n) - f(y_n) \geq |df(e_0)|$. Normalize f to take the value 1 at x_n

and 0 at y_n, that is, define F_n on V to be the function $F_n := (f - f(y_n))/(f(x_n) - f(y_n))$. Then $|dF_n| \leq |df|/|df(e_0)|$. By Dirichlet's principle (Exercise 2.13), we have

$$1/\mathrm{RD}(W_n) \leq \mathscr{C}(x_n \leftrightarrow y_n \ ; \ W_n) \leq \sum_{e \in W_n} c(e)\, dF_n(e)^2 \leq \sum_{e \in W_n} c(e)\, df(e)^2/df(e_0)^2 \, .$$

Since edges of the networks W_n are disjoint, it follows that

$$\sum_{n \geq n_0} 1/\mathrm{RD}(W_n) \leq \sum_{n \geq n_0} \sum_{e \in W_n} c(e)\, df(e)^2/df(e_0)^2 \leq \langle f, f \rangle / df(e_0)^2 \, .$$

Therefore our hypothesis implies that f is not Dirichlet. Thus, $\mathbf{HD} = \mathbb{R}$, so the conclusion follows from Theorem 9.5. ◄

▷ **Exercise 9.7.**
One can define the product of two networks in various ways. For example, given two networks $G_i = (V_i, E_i)$ with conductances c_i ($i = 1, 2$), define the ***Cartesian product*** $G = (V, E)$ with conductances c by $V := V_1 \times V_2$,

$$E := \left\{ ((x_1, x_2), (y_1, y_2)) ; \ (x_1 = y_1, \ (x_2, y_2) \in E_2) \text{ or } ((x_1, y_1) \in E_1, \ x_2 = y_2) \right\} ,$$

and

$$c((x_1, x_2), (y_1, y_2)) := \begin{cases} c(x_2, y_2) & \text{if } x_1 = y_1 \\ c(x_1, y_1) & \text{if } x_2 = y_2. \end{cases}$$

We write $G = G_1 \square G_2$. Show that if G_i are infinite, locally finite graphs with unit conductances, then G has unique currents.

It follows that the usual nearest-neighbor graph on $\mathbb{Z}^d$ has unique currents for all $d \geq 1$.

If one network is "similar" to another, must they both have unique currents or both not? One such case that is easy to decide is a graph with two "similar" assignments on conductances, c and c':

Proposition 9.8. *Let G be a graph with two assignments of conductances, c and c'. If $c \asymp c'$, meaning that the ratio c/c' is bounded and bounded away from 0, then (G, c) has unique currents iff (G, c') does.*

Proof. From Exercise 9.6(c), currents are unique iff the functions with finite support span a dense subspace of $\mathbf{D}/\mathbb{R}$. Since the two norms on $\mathbf{D}/\mathbb{R}$ for the different conductances c and c' are equivalent, density is the same for both. ◄

This has a very useful extension due to Soardi (1993), analogous to a combination of Proposition 2.18 and Theorem 2.17.

Theorem 9.9. (Rough Isometry Preserves Unique Currents) *Let G and G' be two infinite, roughly isometric networks with conductances c and c'. If c, c', c^{-1}, c'^{-1} are all bounded and the degrees in G and G' are all bounded, then G has unique currents iff G' does.*

Proof. Since $c \asymp 1$ and $c' \asymp 1$, we may assume that actually $c = 1$ and $c' = 1$ by Proposition 9.8.

Let $\phi: \mathsf{V} \to \mathsf{V}'$ be a rough isometry. Suppose first that ϕ is a bijection. We claim that the map $\Phi: f + \mathbb{R} \mapsto f \circ \phi^{-1} + \mathbb{R}$ from $\mathbf{D}/\mathbb{R}$ to $\mathbf{D}'/\mathbb{R}$ is an isomorphism of Banach spaces, where $\mathbf{D}'$ is the space of Dirichlet functions on G'. Since the minimum distance between distinct vertices is 1, the fact that ϕ is a bijective rough isometry implies that ϕ is actually bi-Lipschitz: for some constant γ_1, we have

$$\gamma_1^{-1} d(x, y) \le d'(\phi(x), \phi(y)) \le \gamma_1 d(x, y)$$

for $x, y \in \mathsf{V}$. For $e' = (\phi(x), \phi(y)) \in \mathsf{E}'$, let $\mathcal{P}(e')$ be a path of $d(x, y) \le \gamma_1$ edges in E that joins x to y. Given $e \in \mathsf{E}$ and $e' \in \mathsf{E}'$ with $e \in \mathcal{P}(e')$, we know that the endpoints of e' are the ϕ-images of vertices that are within distance γ_1 of the endpoints of e. Since the degrees of G are bounded, the number of possibilities for such pairs of endpoints of e' is no more than some constant, γ_2. Therefore no edge in E appears in more than γ_2 paths of the form $\mathcal{P}(e')$ for $e' \in \mathsf{E}'$. Thus, for $f \in \mathbf{D}$, we have by the Cauchy–Schwarz inequality

$$\|f \circ \phi^{-1} + \mathbb{R}\|^2 = \frac{1}{2} \sum_{e' \in \mathsf{E}'} \nabla(f \circ \phi^{-1})(e')^2 = \frac{1}{2} \sum_{e' \in \mathsf{E}'} \left(\sum_{e \in \mathcal{P}(e')} \nabla f(e) \right)^2$$

$$\le \frac{\gamma_1}{2} \sum_{e' \in \mathsf{E}'} \sum_{e \in \mathcal{P}(e')} \nabla f(e)^2 \le \frac{\gamma_1 \gamma_2}{2} \sum_{e \in \mathsf{E}} \nabla f(e)^2 = \gamma_1 \gamma_2 \|f + \mathbb{R}\|^2 .$$

This shows that Φ is a bounded map; symmetry gives the boundedness of Φ^{-1}, establishing our claim.

Now clearly Φ is a bijection between the subspaces of functions with finite support. Hence Φ also gives an isomorphism between $\mathbf{D}_0/\mathbb{R}$ and $\mathbf{D}_0'/\mathbb{R}$. Therefore, the result follows from Exercise 9.6(c).

Now consider the case that ϕ is not a bijection. We will "fluff up" the graphs G and G' to extend ϕ to a bijection so as to use the result we have just established. Because the image of V comes within some fixed distance β of every vertex in G' and because G' has bounded degrees, V' can be partitioned into subsets $N'(\phi(x))$ ($x \in \mathsf{V}$) of bounded cardinality in such a way that every vertex in $N'(\phi(x))$ lies within distance β of $\phi(x)$ and so that $\phi(x) \in N'(\phi(x))$. Also, because ϕ does not shrink distances too much and the degrees in G are bounded, the cardinalities of the preimages $\phi^{-1}(x')$ ($x' \in \mathsf{V}'$) are bounded. For each $x' \in \phi(\mathsf{V})$, let $\psi(x')$ denote some vertex in $\phi^{-1}(x')$. Create a new graph G_* by joining each vertex $\psi(x')$ ($x' \in \phi(\mathsf{V})$) to new vertices $v_1(x'), \ldots, v_{|N'(x')|-1}(x')$ by new edges. Also, create a new graph G_*' by joining each vertex $\phi(x)$ ($x \in \mathsf{V}$) to new vertices $w_1(x), \ldots, w_{|\phi^{-1}(\phi(x))|-1}(x)$ by new edges. Then G_* and G_*' have bounded degrees. Define $\phi_*: G_* \to G_*'$ as follows: For $x' \in \phi(\mathsf{V})$, let $\phi_*(\psi(x')) := x'$, and let ϕ_* be a bijection from $v_1(x'), \ldots, v_{|N'(x')|-1}(x')$ to $N'(x') \setminus \{x'\}$. For $x \in \mathsf{V}$, let ϕ_* be a bijection from $\phi^{-1}(\phi(x)) \setminus \{\psi(\phi(x))\}$ to $w_1(x), \ldots, w_{|\phi^{-1}(\phi(x))|-1}(x)$. Then ϕ_* is a bijective rough isometry. By the first part of the proof, G_* has unique currents iff G_*' does. Since every harmonic function on G_* has the same value on $v_i(x')$ as on $\psi(x')$ ($x' \in \mathsf{V}'$), it follows that G_* has unique currents iff G does. The same holds for G_*' and G', which proves the theorem.   ◀

Here's a simple application. Every finitely generated, abelian group is isomorphic to $\mathbb{Z}^d \times \Gamma$ for some d and some finite abelian group Γ. Therefore, each of its Cayley graphs is roughly isometric to the usual graph on $\mathbb{Z}^d$, whence has unique currents, according to our observation after Exercise 9.7. We record this conclusion:

Corollary 9.10. *Every Cayley graph of a finitely generated, abelian group has unique currents. More generally, every bounded-degree graph roughly isometric to a Euclidean space has unique currents.*

We'll see in Exercise 10.11 that amenable, transitive graphs also have unique currents.

In the next section, we examine whether currents are unique in graphs roughly isometric to hyperbolic spaces. To do that, we will use the fact that $\langle f(X_n) \rangle$ converges a.s. for all $f \in \mathbf{D}$ on planar, transient networks. This is true not only on planar, transient networks but on all transient networks, as we show next. We begin with the following exercise, which illustrates a basic technique in the theory of harmonic functions.

▷ **Exercise 9.8.**
Let G be transient and let $f \in \mathbf{D}_0$. Show that there is a unique $g \in \mathbf{D}_0$ having minimal energy such that $g \geq |f|$. Show that this g is **superharmonic**, meaning that for all vertices x,

$$g(x) \geq \frac{1}{\pi(x)} \sum_{y \sim x} c(x, y) g(y).$$

Theorem 9.11. (Dirichlet Functions along Random Walks) *If G is a transient network, $\langle X_n \rangle$ the corresponding random walk, and $f \in \mathbf{D}$, then $\langle f(X_n) \rangle$ has a finite limit a.s. and in L^2. Furthermore, if $f = f_{\mathbf{D}_0} + f_{\mathbf{HD}}$ is the Royden decomposition of f, then $\lim f(X_n) = \lim f_{\mathbf{HD}}(X_n)$ a.s.*

This is due to Ancona, Lyons, and Peres (1999).

Proof. A vague idea of the proof is that fluctuations of $f(X_n)$ use energy of f; since f has finite energy, the fluctuations must tend to 0. To make this rigorous, we use the Royden decomposition of f to reduce the problem to one involving martingales and supermartingales.

The following notation will be handy:

$$\mathscr{E}_f(x) := \sum_y p(x, y) [f(y) - f(x)]^2 = \mathbf{E}_x \left[|f(X_1) - f(X_0)|^2 \right].$$

Thus, we have

$$\mathscr{D}(f) = \frac{1}{2} \sum_{x \in \mathsf{V}} \pi(x) \mathscr{E}_f(x).$$

Since the vertex o used in defining the norm on $\mathbf{D}$ is arbitrary, we may take $o = X_0$. (We assume that X_0 is nonrandom, without loss of generality.) We first observe that for any $f \in \mathbf{D}$, it is easy to bound the sum of squared increments along the random walk: For any Markov chain, we have

$$\mathscr{G}(y, o) = \sum_{n \geq 0} \mathbf{P}_y[\tau_o = n] \, \mathbf{E}_y \left[\sum_{k \geq 0} \mathbf{1}_{\{X_{n+k} = o\}} \, \middle| \, \tau_o = n \right] = \mathbf{P}_y[\tau_o < \infty] \, \mathscr{G}(o, o) \leq \mathscr{G}(o, o).$$

In our reversible case, Exercise 2.1(e) tells us that $\pi(o)\mathscr{G}(o, y) = \pi(y)\mathscr{G}(y, o) \leq \pi(y)\mathscr{G}(o, o)$, whence

$$\sum_{k=1}^{\infty} \mathbf{E}\big[|f(X_k) - f(X_{k-1})|^2\big] = \sum_{y \in \mathsf{V}} \sum_{k=1}^{\infty} \mathbf{E}\big[|f(X_k) - f(X_{k-1})|^2 \mid X_{k-1} = y\big]\mathbf{P}[X_{k-1} = y]$$

$$= \sum_{y \in \mathsf{V}} \mathscr{E}_f(y)\mathscr{G}(o, y) \leq \frac{\mathscr{G}(o, o)}{\pi(o)} \sum_{y \in \mathsf{V}} \pi(y)\mathscr{E}_f(y)$$

$$= 2\frac{\mathscr{G}(o, o)}{\pi(o)}\mathscr{D}(f). \tag{9.8}$$

Also, we have

$$\mathbf{E}\big[f(X_k)^2 - f(X_{k-1})^2\big] = \mathbf{E}\big[|f(X_k) - f(X_{k-1})|^2\big] + 2\,\mathbf{E}\big[(f(X_k) - f(X_{k-1}))f(X_{k-1})\big]$$

$$\leq \mathbf{E}\big[|f(X_k) - f(X_{k-1})|^2\big]$$

in case f is harmonic (since then $\langle f(X_n)\rangle$ is a martingale) or f is superharmonic and nonnegative (since then $\langle f(X_n)\rangle$ is a supermartingale). In either of these two cases, we obtain by summing these inequalities for $k = 1, \ldots, n$ that

$$\mathbf{E}\big[f(X_n)^2 - f(X_0)^2\big] \leq \sum_{k=1}^{n} \mathbf{E}\big[|f(X_k) - f(X_{k-1})|^2\big] \leq 2\frac{\mathscr{G}(o, o)}{\pi(o)}\mathscr{D}(f)$$

by (9.8). That is,

$$\mathbf{E}\big[f(X_n)^2\big] \leq 2\frac{\mathscr{G}(o, o)}{\pi(o)}\mathscr{D}(f) + f(o)^2 \tag{9.9}$$

in case f is harmonic or f is superharmonic and nonnegative.

It follows from (9.9) applied to $f_{\mathbf{HD}}$ that $\langle f_{\mathbf{HD}}(X_n)\rangle$ is a martingale bounded in L^2, whence by Doob's theorem it converges a.s. and in L^2. It remains to show that $f_{\mathbf{D}_0}$ converges to 0 a.s. and in L^2. Given $\epsilon > 0$, write $f_{\mathbf{D}_0} = f_1 + f_2$, where f_1 is finitely supported and $\mathscr{D}(f_2) < \pi(o)\epsilon/(3\,\mathscr{G}(o, o))$. Exercise 9.8 applied to $f_2 \in \mathbf{D}_0$ yields a superharmonic function $g \in \mathbf{D}_0$ that satisfies $g \geq |f_2|$ and $\mathscr{D}(g) \leq \mathscr{D}(|f_2|) \leq \mathscr{D}(f_2)$. Also, $g(o)^2 \leq \mathscr{G}(o, o)\mathscr{D}(g)/\pi(o)$ by Exercise 9.6(g) applied to $g \in \mathbf{D}_0$ (that is, $g = g + \mathbf{0}$ is its Royden decomposition). Combining this with (9.9) applied to g, we get that $\mathbf{E}\big[g(X_n)^2\big] \leq \epsilon$ for all n. Since $\langle g(X_n)\rangle$ is a nonnegative supermartingale, it converges a.s. and in L^2 to a limit whose second moment is at most ϵ. Since $|f_2(X_n)| \leq g(X_n)$, it follows that both $\mathbf{E}\big[\limsup_{n\to\infty} f_2(X_n)^2\big]$ and $\mathbf{E}\big[f_2(X_n)^2\big]$ are at most ϵ. Now transience implies that $f_1(X_n) \to 0$ a.s. and (by the bounded convergence theorem) $\mathbf{E}\big[f_1(X_n)^2\big] \to 0$ as $n \to \infty$. Therefore, it follows that $\mathbf{E}\big[\limsup_{n\to\infty} f_{\mathbf{D}_0}(X_n)^2\big] \leq \epsilon$ and $\limsup_{n\to\infty} \mathbf{E}\big[f_{\mathbf{D}_0}(X_n)^2\big] \leq \epsilon$. Since ϵ was arbitrary, $\langle f_{\mathbf{D}_0}(X_n)\rangle$ must tend to 0 a.s. and in L^2. This completes the proof. ◄

9.4 Planar Graphs and Hyperbolic Graphs

There is a surprising phase transition between dimensions 2 and 3 in hyperbolic space $\mathbb{H}^d$: in graphs that are roughly isometric to $\mathbb{H}^d$, currents are not unique when $d = 2$, but they are unique when $d \geq 3$. We first treat the case of $\mathbb{H}^2$. It turns out that this nonuniqueness of currents is vastly more general: it holds for *virtually all* transient, planar graphs!

Theorem 9.12. (Transient Planar Networks Have Nonunique Currents) *Suppose that G is a transient, planar network. Let $\pi(x)$ denote the sum of the conductances of the edges incident to x. If $\pi(\cdot)$ is bounded, then currents are not unique.*

This theorem is due to Benjamini and Schramm (1996a, 1996c). For simplicity, we will assume from now on that G *is a proper, simple, plane, transient network all of whose faces have a finite number of sides*. Since we can always add edges of conductance 0, the last assumption here does not lose any generality.

To prove Theorem 9.12, we show that in some sense, random walk on G is like Brownian motion in the hyperbolic disc. Benjamini and Schramm showed this in two geometric senses: one used circle packing (1996a) and the other used square tiling (1996c). We will show this in a combinatorial sense that is essentially the same as the approach with square tiling.

Our first goal is to establish a (polar) coordinate system on V. Fix a vertex $o \in$ V; this, of course, will be our origin. We will use voltages and currents to assign radii and angles to the other vertices. Let i_o be the unit current flow on G from o to ∞, and let v be the voltage function that is 0 at o and 1 at ∞, in other words, $v(x)$ is the probability that a random walk started at x will never visit o. Note that the voltage function corresponding to i_o is not v but rather $\mathscr{E}(i_o)(1 - v)$.

Recall our conventions for plane dual graphs from Section 9.2. Define $i_o^\times(e^\dagger) := i_o(e)$. Now any face of $G^\dagger$ contains a vertex of G in its interior. If $i_o^\times$ is summed counterclockwise around a face of $G^\dagger$ surrounding x, then we obtain $d^* i_o(x)$, which is 0 unless $x = o$, in which case it is 1. Since any cycle in $G^\dagger$ can be written as a sum of cycles surrounding faces, it follows that the sum of $i_o^\times$ along any cycle is an integer. Therefore, we may define $\alpha : G^\dagger \to \mathbb{R}/\mathbb{Z}$ by picking any vertex $o^\dagger \in G^\dagger$ and, for $x^\dagger \in$ V$(G^\dagger)$, setting $\alpha(x^\dagger)$ to be the sum (mod 1) of $i_o^\times$ along any path in $G^\dagger$ from $o^\dagger$ to $x^\dagger$.

We now have the essence of the polar coordinates on V, with v giving the radial distance and α giving the angle; however, α is defined on V†, not on V. In fact, we prefer to assign to each $x \in$ V an arc $J(x)$ of angles to get all angles of $\mathbb{R}/\mathbb{Z}$. To do this, let $\mathsf{Out}(x) := \{e \,; \; e^- = x, \; i_o(e) > 0\}$ and $\mathsf{In}(x) := \{e \,; \; e^+ = x, \; i_o(e) > 0\}$. For example, $\mathsf{In}(o)$ is empty.

Lemma 9.13. *For every $x \in$ V, the sets $\mathsf{Out}(x)$ and $\mathsf{In}(x)$ do not interleave, that is, their union can be ordered counterclockwise so that no edge of $\mathsf{In}(x)$ precedes any edge of $\mathsf{Out}(x)$.*

Proof. Consider any x and any two edges $\langle y, x \rangle, \langle z, x \rangle \in \mathsf{In}(x)$. We have $v(y) < v(x)$ and $v(z) < v(x)$ by definition of $\mathsf{In}(x)$. By Corollary 3.3, there are paths from o to y and o to z using only vertices with $v < v(x)$. Extend these paths to x by adjoining the edges $\langle y, x \rangle$ and $\langle z, x \rangle$, respectively. These two paths from o to x bound one or more regions in the plane. By the maximum principle, any vertices inside these regions also have $v < v(x)$. In particular, there can be none that are endpoints of edges in $\mathsf{Out}(x)$. This implies what we want. ◄

If $J(e)$ denotes the closed counterclockwise arc on $\mathbb{R}/\mathbb{Z}$ from $\alpha((e^\dagger)^-)$ to $\alpha((e^\dagger)^+)$, then we let

$$J(x) := \bigcup_{e \in \mathsf{Out}(x)} J(e).$$

By Lemma 9.13, we also have $J(x) = \bigcup_{e \in \mathsf{In}(x)} J(e)$ for $x \neq o$.

This assignment $x \mapsto J(x)$ of arcs to vertices gives a nice collection of "rays": Consider some ϑ not in the range of α. If $\vartheta \in J(e)$ for some $e \in \mathsf{In}(x)$, then there is exactly one $e \in \mathsf{Out}(x)$ with $\vartheta \in J(e)$ by Exercise 3.1. Thus, there is a unique infinite path $\mathcal{P}_\vartheta$ in G starting at o and containing only vertices x with $\vartheta \in J(x)$. Such a path corresponds to a radial line in the hyperbolic disc. We claim that for every edge e with $i_o(e) > 0$, the Lebesgue measure of $\{\vartheta \, ; \, e \in \mathcal{P}_\vartheta\}$ is the length of $J(e)$, that is, $i_o(e)$. Indeed, consider the flow ϕ defined to be the expectation of the path $\mathcal{P}_\vartheta$ when ϑ is chosen uniformly at random in $\mathbb{R}/\mathbb{Z}$, just as we created flows from random paths in Section 2.5. Then ϕ is a unit flow from o and $0 \leq \phi(e) \leq i_o(e)$ when $i_o(e) > 0$. Therefore, $\mathscr{E}(\phi) \leq \mathscr{E}(i_o)$, yet i_0 has the minimum energy among all unit flows from o (Proposition 2.12). Hence $\phi = i_o$, which implies our claim.

We illustrate these coordinates for the edge graph G of the (2,3,7)-triangle tessellation of the hyperbolic plane, that is, the dual of the Cayley graph shown in Figure 6.1. This embedding of G is shown in Figure 9.1; in the left-hand part, each edge of G corresponds to an annular region, while each vertex of G corresponds to a collection of adjacent arcs. This is a little confusing, so to make it look like a standard graph, we can place a vertex in the middle of the arc collection to which it corresponds; this gives the right-hand part of Figure 9.1.

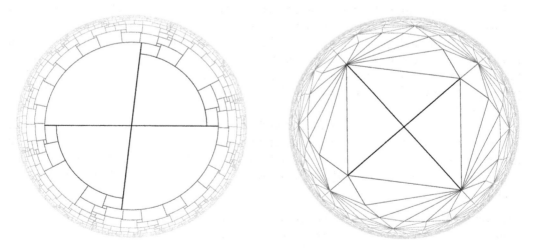

Figure 9.1. The polar embeddings of the (2,3,7)-triangle tessellation of the hyperbolic plane.

We can also illustrate the coordinates with the square lattice. Although the square lattice is recurrent, if we make the conductance of each edge $\langle x, y \rangle$ equal to the maximum of the distance of x and y to the origin, then it becomes transient. The polar embeddings of this network are shown in Figure 9.2.

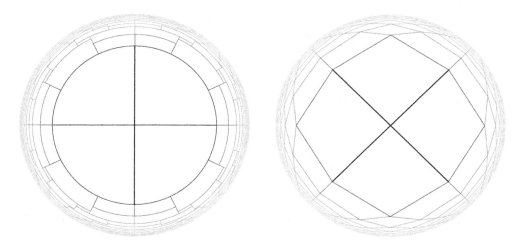

Figure 9.2. The polar embeddings of the distance-weighted square lattice.

The "rays" $\mathcal{P}_\vartheta$ that we have defined earlier go from the origin to "radius" 1:

Lemma 9.14. *For Lebesgue-almost every $\vartheta \in \mathbb{R}/\mathbb{Z}$, we have $\sup\{v(x)\,;\ x \in \mathcal{P}_\vartheta\} = 1$.*

Proof. Let $h(\vartheta) := \sup\{v(x)\,;\ x \in \mathcal{P}_\vartheta\}$. Then $h \le 1$ everywhere. Also,

$$\int_{\mathbb{R}/\mathbb{Z}} h(\vartheta)\,d\vartheta = \int_{\mathbb{R}/\mathbb{Z}} \sum_{e \in \mathcal{P}_\vartheta} |dv(e)|\,d\vartheta = \sum_{i_o(e)>0} |dv(e)|\,i_o(e) = 1$$

since $|dv(e)| = i_o(e)r(e)/\mathscr{E}(i_o)$ (recall that v is not exactly the voltage function corresponding to i_o). Therefore, $h = 1$ a.e. ◄

It follows that we can go the other way and assign a vertex to a point in the unit disc: for $0 \le \rho < 1$ and for a.e. ϑ, Lemma 9.14 allows us to define $x(\rho, \vartheta)$ as the vertex $x \in \mathcal{P}_\vartheta$ where $v(x) \le \rho$ and $v(x)$ is maximum. We now come to the key calculation made possible by our coordinate system.

Lemma 9.15. *For every $f \in \mathbf{D}$,*

$$\overline{f}(\vartheta) := \lim_{\rho \uparrow 1} f(x(\rho, \vartheta))$$

exists for Lebesgue-almost every ϑ and satisfies

$$\|\overline{f}\|_{L^1} \le \sqrt{1 + \mathscr{E}(i_o)}\,\|f\|_{\mathbf{D}}\,.$$

Proof. The Cauchy–Schwarz inequality yields the bound

$$\int_{\mathbb{R}/\mathbb{Z}} \sum_{e \in \mathcal{P}_\vartheta} |df(e)|\,d\vartheta = \sum_{i_o(e)>0} |df(e)|\,i_o(e) \le \|df\|_c\|i_o\|_r < \infty\,,$$

whence the integrand is finite a.e. This proves that

$$\lim_{\rho \uparrow 1} f(x(\rho, \vartheta)) = f(o) - \lim_{\rho \uparrow 1} \sum_{\substack{e \in \mathcal{P}_\vartheta, \\ v(e^+) \le \rho}} df(e)$$

exists a.e. and has L^1 norm at most

$$|f(o)| + \|df\|_c \|i_o\|_r \le \sqrt{1 + \mathscr{E}(i_o)} \, \|f\|_{\mathbf{D}}$$

by the Cauchy–Schwarz inequality.   ◀

▷ **Exercise 9.9.**
With $f \in \mathbf{D}$ and $\bar{f}$ defined as in Lemma 9.15, show that $\left\| f(x(\rho, \cdot)) - \bar{f} \right\|_{L^1} \to 0$ as $\rho \uparrow 1$.

If f has finite support, then of course $\bar{f} \equiv 0$. Since the map $f \mapsto \bar{f}$ from $\mathbf{D}$ to L^1 is continuous in light of Lemma 9.15, it follows that $\bar{f} = 0$ a.e. for $f \in \mathbf{D}_0$. Therefore, writing $f = f_{\mathbf{D}_0} + f_{\mathbf{HD}}$ for the Royden decomposition of f (see Exercise 9.6(f)), we have $\bar{f} = \bar{f}_{\mathbf{HD}}$ a.e. To show that $\mathbf{HD} \neq \mathbb{R}$, then, it suffices to show that there is a Dirichlet function f such that $\bar{f}$ is not an a.e. constant. An evident candidate for such an f is the angle function, α. This doesn't quite work because α is defined on $\mathsf{V}^\dagger$ rather than on V and because, moreover, α takes values in $\mathbb{R}/\mathbb{Z}$. Thus, we make the following modifications. Let F_x be any face of G with x as one of its vertices. For $\vartheta \in \mathbb{R}/\mathbb{Z}$, let $|\vartheta|$ denote the distance of (any representative of) ϑ to the integers. Set $\psi(x) := |\alpha(F_x)|$.

Lemma 9.16. *If $\pi(\cdot)$ is bounded, then $\psi \in \mathbf{D}$.*

Proof. Let $\pi \le M$. Given adjacent vertices x, y in G, there is a path of edges $e_1^\dagger, \ldots, e_j^\dagger$ in $G^\dagger$ from F_x to F_y with each e_k incident to either x or y (see Figure 9.3). Therefore,

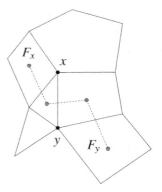

Figure 9.3. A dual path.

$$d\psi(x, y)^2 \le |\alpha(F_x) - \alpha(F_y)|^2 \le \left(\sum_{k=1}^j |i_o(e_k)| \right)^2$$

$$\le \sum_{k=1}^j i_o(e_k)^2 r(e_k) \sum_{k=1}^j c(e_k)$$

$$\le 2M \sum_{e^- \in \{x, y\}} i_o(e)^2 r(e)$$

by the Cauchy–Schwarz inequality. Rewrite this as

$$d\psi(e)^2 \le M \sum_{e' \sim e^\pm} i_o(e')^2 r(e')$$

for every edge e, where $e' \sim e^{\pm}$ denotes that e' is incident to at least one of the endpoints of e, including the possibility that $e' = e$. (Both orientations of e' are included, which is why we lost a factor of 2.) It follows that

$$\mathscr{D}(\psi) = \frac{1}{2} \sum_{e \in E} d\psi(e)^2 c(e) \leq \frac{M}{2} \sum_{e \in E} \sum_{e' \sim e^{\pm}} i_o(e')^2 r(e') c(e)$$

$$= \frac{M}{2} \sum_{e' \in E} i_o(e')^2 r(e') \sum_{e \sim e'^{\pm}} c(e) \leq 4M^2 \mathscr{E}(i_o) < \infty . \quad \blacktriangleleft$$

Since $\alpha(F_x) \in J(x)$ and $J(x)$ has length tending to zero as the distance from x to o tends to infinity (by the first inequality of (2.15)), we have that $\lim_{\rho \uparrow 1} \psi(x(\rho, \vartheta)) = |\vartheta|$ for every ϑ for which $\sup\{v(x) ; x \in \mathcal{P}_\vartheta\} = 1$, in other words, for a.e. ϑ. Thus, ψ is the sought-for Dirichlet function with $\bar{\psi}$ not an a.e. constant. This proves Theorem 9.12.

We can use our polar coordinates to prove another wonderful result. By Theorem 9.11, $\psi(X_n) = |\alpha(F_{X_n})|$ converges a.s. when $\pi(\cdot)$ is bounded. Since the length of $J(X_n) \to 0$ a.s. and $J(X_n) \cap J(X_{n+1}) \neq \varnothing$, it follows that $\alpha(F_{X_n})$ also converges a.s. Benjamini and Schramm (1996c) showed that its limiting distribution is Lebesgue measure. Of course, the choice of faces F_x has no effect, since the lengths of the intervals $J(X_n)$ tend to 0 a.s. This angle convergence is what we meant by saying that random walk on G is similar to Brownian motion in the disc.

Theorem 9.17. (Benjamini and Schramm, 1996c) *If G is a transient, simple, plane network with $\pi(\cdot)$ bounded, then $J(X_n)$ tends to a point on the circle $\mathbb{R}/\mathbb{Z}$ a.s. The distribution of the limiting point is Lebesgue measure when $X_0 = o$.*

Proof. That $J(X_n)$ tends to a point a.s. follows from the a.s. convergence of $\alpha(F_{X_n})$. To say that the limiting distribution is Lebesgue measure is to say that for any arc A,

$$\mathbf{P}_o[\lim \alpha(F_{X_n}) \in A] = \int \mathbf{1}_A(\vartheta) \, d\vartheta . \tag{9.10}$$

This is implied by the statement

$$\mathbf{E}_o[\lim f(X_n)] = \int \bar{f}(\vartheta) \, d\vartheta \tag{9.11}$$

for every Lipschitz function $\bar{f}$ on $\mathbb{R}/\mathbb{Z}$, where f is defined by

$$f(x) := \bar{f}(\alpha(F_x)) . \tag{9.12}$$

The reason is that if (9.11) holds for all Lipschitz f, then it also holds for $\mathbf{1}_A$, since we can sandwich $\mathbf{1}_A$ arbitrarily closely by Lipschitz functions. This sandwiching gives

$$\int \mathbf{1}_{A^\circ}(\vartheta) \, d\vartheta \leq \mathbf{P}_o\left[\lim \mathbf{1}_{A^\circ}(\alpha(F_{X_n}))\right] \leq \mathbf{P}_o\left[\lim \mathbf{1}_{\bar{A}}(\alpha(F_{X_n}))\right] \leq \int \mathbf{1}_{\bar{A}}(\vartheta) \, d\vartheta ,$$

where $A°$ is the interior of A and $\bar{A}$ is the closure of A. In particular, the chance that $\lim \alpha(F_{X_n})$ is an endpoint of A is 0, so that (9.10) holds.

To show (9.11), note that the same arguments as we used to prove Lemma 9.16 and Theorem 9.12 yield that all f as in (9.12) are Dirichlet functions and

$$\lim_{\rho \uparrow 1} f(x(\rho, \vartheta)) = \bar{f}(\vartheta)$$

for all ϑ for which $\sup\{v(x); \ x \in \mathcal{P}_\vartheta\} = 1$. Now for any $f \in \mathbf{D}$, Theorem 9.11 gives that

$$\mathbf{E}_o\big[\lim f(X_n)\big] = \mathbf{E}_o\big[\lim f_{\mathbf{HD}}(X_n)\big] = \lim \mathbf{E}_o\big[f_{\mathbf{HD}}(X_n)\big] = f_{\mathbf{HD}}(o),$$

where we use the convergence in L^1 in the middle step. On the other hand, Exercise 9.9 gives

$$\int \bar{f}(\vartheta)\, d\vartheta = \lim_{\rho \uparrow 1} \int f(x(\rho, \vartheta))\, d\vartheta = \lim_{\rho \uparrow 1} \int \left[f(o) - \sum_{\substack{e \in \mathcal{P}_\vartheta, \\ v(e^+) \le \rho}} df(e) \right] d\vartheta$$

$$= f(o) - (\nabla f, i_o)_r = f(o) - f_{\mathbf{D}_0}(o) = f_{\mathbf{HD}}(o)$$

by Exercise 9.6(g). Comparing these two results gives (9.11).   ◀

We may now completely analyze which graphs roughly isometric to a hyperbolic space have unique currents.

Theorem 9.18. *If G is a bounded-degree graph that is roughly isometric to $\mathbb{H}^d$ for some $d \ge 2$, then currents are unique on G iff $d \ge 3$.*

Proof. That graphs roughly isometric to $\mathbb{H}^d$ are transient was established in Theorem 2.19; an example appeared in Figure 2.4. Thus, by Theorems 9.12 and 9.9, they do not have unique currents when $d = 2$.

Suppose now that $d \ge 3$. By Theorem 9.9, it suffices to prove unique currents for a particular graph. Choose an origin $o \in \mathbb{H}^d$. For each $n \ge 1$, choose a maximal 1-separated set A_n in the sphere S_n of radius n centered at o. Let $G := (\mathsf{V}, \mathsf{E})$ with $\mathsf{V} := \bigcup_n A_n$ and E the set of pairs of vertices with mutual distance at most 3. Clearly G is roughly isometric to $\mathbb{H}^d$. Consider the subgraphs W_n of G induced on A_n. We claim that they satisfy the hypotheses of Theorem 9.7.

Elementary hyperbolic geometry shows that S_n is isometric to a Euclidean sphere of radius $r_n := \alpha_n e^{\beta_n n}$ for some numbers α_n and β_n (depending on d) that have positive, finite limits as $n \to \infty$. Let $\mathrm{RD}(W_n)$ be the effective resistance diameter of W_n. The random path method shows, just as in the proof of Proposition 2.15, that $\mathrm{RD}(W_n)$ is comparable to $\log r_n$ if $d = 3$ and to a constant if $d \ge 4$. Therefore, we have $\mathrm{RD}(W_n) \le Cn$ for some constant C. Theorem 9.7 completes the proof.   ◀

9.5 Random-Walk Traces

Consider the network random walk on a locally finite, transient network (G, c) when it starts from some fixed vertex o. The *trace* of the random walk is the random set of edges traversed at least once by the random walk. How big can the trace be? We show that it cannot be very large in that the trace forms a.s. a recurrent graph (for simple random walk). This result is due to Benjamini, Gurel-Gurevich, and Lyons (2007), from which this section is taken.

Our proof will demonstrate the following stronger results. Let $N(x, y)$ denote the number of traversals of the edge $[x, y]$.

Theorem 9.19. (Recurrence of Traces) *The network* $(G, \mathbf{E}[N])$ *is recurrent. The networks* (G, N) *and* $(G, \mathbf{1}_{[N>0]})$ *are a.s. recurrent.*

By Exercise 2.97, if $(G, \mathbf{E}[N])$ is recurrent, then so a.s. is (G, N). Furthermore, Rayleigh's monotonicity principle implies that when (G, N) is recurrent, so is $(G, \mathbf{1}_{[N>0]})$.

Thus, it remains to prove that $(G, \mathbf{E}[N])$ is recurrent.

Recall from Proposition 2.12 that the effective resistance from o to infinity in the network (G, c) equals

$$R := \mathscr{G}(o, o)/\pi(o). \tag{9.13}$$

Let the voltage function be $v(\cdot)$ throughout this section, where $v(o) = 1$ and $v(\cdot)$ is 0 at ∞. Then $v(x)$ is the probability of ever visiting o for a random walk starting at x.

Note that

$$\mathbf{E}[N(x, y)] = \mathscr{G}(o, x)p(x, y) + \mathscr{G}(o, y)p(y, x) = \big(\mathscr{G}(o, x)/\pi(x) + \mathscr{G}(o, y)/\pi(y)\big)\, c(x, y)$$

and, by Exercise 2.1 and Proposition 2.12,

$$\pi(o)\,\mathscr{G}(o, x) = \pi(x)\,\mathscr{G}(x, o) = \pi(x)v(x)\,\mathscr{G}(o, o).$$

Thus, we have (from the definition (9.13))

$$\mathbf{E}[N(x, y)] = Rc(x, y)\big[v(x) + v(y)\big] \tag{9.14}$$

$$\leq 2R \max\{v(x), v(y)\}c(x, y). \tag{9.15}$$

In a finite network (H, c), we write $\mathscr{C}(A \leftrightarrow z; H, c)$ for the effective conductance between a subset A of its vertices and a vertex z. Clearly, the inclusions $A \subset B \subset \mathsf{V}$ imply the inequality $\mathscr{C}(A \leftrightarrow z; H, c) \leq \mathscr{C}(B \leftrightarrow z; H, c)$. The effective conductance to infinity from an infinite set A of vertices is defined to be the supremum of the effective conductance from B to infinity over all finite subsets $B \subset A$.

Lemma 9.20. *Let* (H, c) *be a finite network and* $a, z \in \mathsf{V}(H)$. *Let* v_H *be the voltage function that is 1 at* a *and 0 at* z. *For* $0 < t < 1$, *let* A_t *be the set of vertices* x *with* $v_H(x) \geq t$. *Then* $\mathscr{C}(A_t \leftrightarrow z; H, c) \leq \mathscr{C}(a \leftrightarrow z; H, c)/t$. *More generally, for every* $A \subset \mathsf{V}(H) \setminus \{z\}$, *we have*

$$\mathscr{C}(A \leftrightarrow z; H, c) \leq \frac{\mathscr{C}(a \leftrightarrow z; H, c)}{\min\left(v_H \restriction A\right)}.$$

Proof. It is easy to prove the result when the voltages at all the inner boundary vertices of A_t are equal to t. To compare to this case, we subdivide edges as follows. If any edge (x, y) is such that $v_H(x) > t$ and $v_H(y) < t$, then subdividing the edge (x, y) with a vertex z by giving resistances

$$r(x, z) := \frac{v_H(x) - t}{v_H(x) - v_H(y)} r(x, y)$$

and

$$r(z, y) := \frac{t - v_H(y)}{v_H(x) - v_H(y)} r(x, y)$$

will result in a network such that $v_H(z) = t$ while no other voltages change (this is the series law). Doing this for all such edges gives a possibly new graph H' and a new set of vertices $A'_t \supseteq A_t$ whose inner vertex boundary is a set W'_t on which the voltage is identically equal to t. We have $\mathscr{C}(A_t \leftrightarrow z; H, c) = \mathscr{C}(A_t \leftrightarrow z; H', c) \leq \mathscr{C}(A'_t \leftrightarrow z; H', c)$. Now $\mathscr{C}(A'_t \leftrightarrow z; H', c) = \mathscr{C}(a \leftrightarrow z; H, c)/t$, since the current flowing in (H', c) from a to z induces a current from A'_t to z with strength $\mathscr{C}(a \leftrightarrow z; H, c)$ and voltage difference t. Therefore, $\mathscr{C}(A_t \leftrightarrow z; H, c) \leq \mathscr{C}(a \leftrightarrow z; H, c)/t$, as desired.

For a general A, let $t := \min v_H \restriction A$. Then $\mathscr{C}(A \leftrightarrow z; H, c) \leq \mathscr{C}(A_t \leftrightarrow z; H, c)$ since $A \subset A_t$. Combined with the previous inequality, this yields the final conclusion. ◄

For $t \in (0, 1)$, let $V_t := \{x \in V ; v(x) < t\}$. Let W_t be the outer vertex boundary of V_t, that is, the set of vertices outside V_t that have a neighbor in V_t. Write G_t for the subgraph of G induced by $V_t \cup W_t$.

We will refer to the conductances c as the ***original*** ones and the conductances $\mathbf{E}[N]$ as the ***new*** ones for convenience.

Lemma 9.21. *For $t \in (0, 1)$, the effective conductance from W_t to ∞ in the network $(G_t, \mathbf{E}[N])$ is at most 2.*

Proof. Again, we want to compare to the case where $v \restriction W_t \equiv t$, so we subdivide. If any edge (x, y) is such that $v(x) > t$ and $v(y) < t$, then subdividing the edge (x, y) with a vertex z as in the proof of Lemma 9.20 and consequently adding z to W_t has the effect of replacing the edge (x, y) by an edge (z, y) with conductance $c(z, y) = c(x, y)[v(x) - v(y)]/[t - v(y)] > c(x, y)$ in the original network and, by (9.14), with larger conductance in the corresponding new network:

$$\begin{aligned} \mathbf{E}[N(z, y)] &= Rc(z, y)[t + v(y)] = Rc(z, y)[t - v(y) + 2v(y)] \\ &= Rc(x, y)[v(x) - v(y)] + 2Rc(z, y)v(y) \\ &> Rc(x, y)[v(x) - v(y)] + 2Rc(x, y)v(y) = \mathbf{E}[N(x, y)]. \end{aligned}$$

Since raising edge conductances clearly raises effective conductance, it suffices to prove the lemma in the case that $v(x) = t$ for all $x \in W_t$. Thus, we assume this case for the remainder of the proof.

Suppose that $\langle (H_n, c \restriction \mathsf{E}(H_n)); \; n \geq 1 \rangle$ is an increasing exhaustion of (G, c) by finite, induced networks that include o. Identify the boundary (in G) of H_n to a single vertex, z_n. Let v_n be the corresponding voltage functions with $v_n(o) = 1$ and $v_n(z_n) = 0$. Then we have the limits $\mathscr{C}(o \leftrightarrow z_n; H_n, c) \downarrow 1/R$ and $v_n(x) \uparrow v(x)$ as $n \to \infty$ for all $x \in \mathsf{V}(G)$. Let A be a finite subset of W_t. By Lemma 9.20, as soon as $A \subset \mathsf{V}(H_n)$, we have that the effective conductance from A to z_n in H_n is at most $\mathscr{C}(o \leftrightarrow z_n; H_n, c)/\min\{v_n(x);\, x \in A\}$. Therefore, by Rayleigh's monotonicity principle, $\mathscr{C}(A \leftrightarrow \infty; G_t, c) \leq \mathscr{C}(A \leftrightarrow \infty; G, c) = \lim_{n\to\infty} \mathscr{C}(A \leftrightarrow z_n; H_n, c) \leq 1/(Rt)$. Since this holds for all such A, we have

$$\mathscr{C}(W_t \leftrightarrow \infty; G_t, c) \leq 1/(Rt). \tag{9.16}$$

By (9.15), the new conductances on G_t are obtained by multiplying the original conductances by factors that are at most $2Rt$. Combining this with (9.16), we obtain that the new effective conductance from W_t to infinity in G_t is at most 2. ◀

When the complement of V_t is finite for all t, which is the case for "most" networks, this completes the proof by the following lemma (and by the fact that $\bigcap_{t>0} \mathsf{V}_t = \varnothing$):

Lemma 9.22. *If H is a transient, locally finite network, then for all $m > 0$, there exists a finite subset $K \subset \mathsf{V}(H)$ such that for all finite $K' \supseteq K$, the effective conductance from K' to infinity is more than m.*

▷ **Exercise 9.10.**
Prove this lemma.

Even when the complement of V_t is not finite for all t, this is enough to show that the network (G, N) is a.s. recurrent. For if X_n denotes the position of the random walk on (G, c) at time n, then $v(X_n) \to 0$ a.s.: it converges a.s., since it is a nonnegative supermartingale, and its expectation tends to the probability that the random walk visits o infinitely often, that is, to 0. Thus, the path is a.s. contained in V_t after some time, no matter the value of $t > 0$. Let B_n be the ball of radius n about o. By Lemma 9.22, if (G, N) is transient with probability $p > 0$, then $\mathscr{C}(B_n \leftrightarrow \infty; G, N)$ tends in probability, as $n \to \infty$, to a random variable that is infinite with probability p. In particular, this effective conductance is at least $6/p$ with probability at least $p/2$ for all large n. Fix n with this property. Let $t > 0$ be such that $\mathsf{V}_t \cap B_n = \varnothing$. Write D for the (random finite) set of endpoints of edges $e \notin G_t$ with $N(e) > 0$. Then $\mathscr{C}(W_t \leftrightarrow \infty; G_t, N) = \mathscr{C}(W_t \cup D \leftrightarrow \infty; G, N) \geq \mathscr{C}(B_n \leftrightarrow \infty; G, N)$. However, in combination with Exercise 2.75, this implies that $\mathscr{C}(W_t \leftrightarrow \infty; G_t, \mathbf{E}[N]) \geq \mathbf{E}[\mathscr{C}(W_t \leftrightarrow \infty; G_t, N)] \geq (p/2)(6/p) = 3$, which contradicts Lemma 9.21.

We now complete the proof that $(G, \mathbf{E}[N])$ is recurrent in general. This depends on the following extension of Lemma 9.22. This lemma will approximate a current flow by a flow that is 0 on many edges that we'd prefer to ignore.

Lemma 9.23. *Let (H, c) be a transient, locally finite network, $o \in \mathsf{V}(H)$, and $B \subset \mathsf{V}(H)$ be such that $o \in B$, the induced subgraph $H \restriction B$ is connected, and the network random walk on H starting at o visits B only finitely many times a.s. Let i be the unit current flow on H from o to ∞ and $\epsilon > 0$. Then there is a unit flow θ on H from o to ∞ such that $\theta(e) \neq 0$ for only finitely many edges e incident to B and*

$$\sum_{e \in \mathsf{E}(H) \setminus (B \times B)} \theta(e)^2 r(e) \leq \epsilon \mathscr{E}(i) + \sum_{e \in \mathsf{E}(H) \setminus (B \times B)} i(e)^2 r(e).$$

Proof. The idea is to take θ to be the unit current flow corresponding to the random walk conditioned not to visit a very large subset of B that is very far from o. Let $H_1 := H{\restriction}B$. Let $B' \subset B$ be finite with $B' \ni o$ and

$$\sum_{e \in \mathsf{E}(H_1) \cap (B' \times B')} i(e)^2 r(e) \geq (1 - \epsilon/3) \sum_{e \in \mathsf{E}(H_1)} i(e)^2 r(e). \tag{9.17}$$

Given $D \subset B$, let $h_D(x)$ be the probability that when the random walk on H starts from x, it never visits D. Since the random walk visits B only finitely often a.s., there is some (possibly empty) set $D \subset B \setminus B'$ such that $B \setminus D$ is finite and $h_D(x) \geq 1 - \delta$ for all $x \in B'$, where we will choose $\delta > 0$ later. Enlarge D if necessary to $h_D^{-1}(0)$ so that $h_D(x) > 0$ for all $x \notin D$. Then h_D is positive and harmonic off D, so by the solution to Exercise 2.32, the random walk conditioned never to visit D corresponds to the conductances $c'(e) := c(e) h_D(e^-) h_D(e^+)$ on the graph $K := H{\restriction}(\mathsf{V}(H) \setminus D)$. Let θ be the unit current flow on (K, c') from o to ∞. We claim that θ satisfies the desired conclusions. It is clear that $\theta(e) \neq 0$ for only finitely many edges e incident to B, so it remains to verify the inequality.

Write $\pi_H(o) := \sum_{e \sim o} c(e)$ and $\pi_K(o) := \sum_{e \sim o} c'(e)$. Since $c' \leq c$ everywhere, we may choose D so that $\pi_K(o)/\pi_H(o) \in (1 - \delta, 1]$. We may couple random walks on (K, c') and on (H, c) as follows: if the random walk on H never visits D, then the random walks are the same; otherwise, the random walks are independent. Thus, the random walks starting at o are identical with probability at least $h_D(o) \geq 1 - \delta$. Therefore, the escape probability from o in K differs from the escape probability in H by at most δ. It follows from (2.4) that $\mathscr{E}'(\theta)/\mathscr{E}(i) \in (1 - \epsilon/3, 1 + \epsilon/3)$ for an appropriate choice of δ, where $\mathscr{E}$ denotes energy for the conductances c and $\mathscr{E}'$ denotes energy for the conductances c'.

Now $c(e) \geq c'(e) \geq (1 - \delta)^2 c(e)$ for $e \in \mathsf{E}(K) \cap (B' \times B')$. If v denotes the voltage function on H that is 1 at o and 0 at ∞, and v' denotes the similar voltage function on K, then $|v'(x) - v(x)| \leq \delta$ for all $x \in B'$ (by the same coupling as we used in the previous paragraph, except starting from x this time). Since the voltage function for the unit current flow θ is $\mathscr{E}'(\theta)v'$, we have $\theta(e)^2 r'(e) = \mathscr{E}'(\theta)^2 [v'(e^-) - v'(e^+)]^2 c'(e)$ for every edge e. A similar relation holds for i on H. Therefore, we may choose δ so that

$$\sum_{e \in \mathsf{E}(K) \cap (B' \times B')} \theta(e)^2 r'(e) \quad \geq \quad \sum_{e \in \mathsf{E}(K) \cap (B' \times B')} i(e)^2 r(e) - \epsilon \mathscr{E}(i)/3.$$

We may now put together all our inequalities to achieve the conclusion:

$$\sum_{e \in \mathsf{E}(H) \setminus (B \times B)} \theta(e)^2 r(e) \quad \leq \quad \sum_{e \in \mathsf{E}(H) \setminus (B' \times B')} \theta(e)^2 r(e)$$

$$\leq \quad \sum_{e \in \mathsf{E}(H) \setminus (B' \times B')} \theta(e)^2 r'(e)$$

$$= \quad \mathscr{E}'(\theta) \; - \sum_{e \in \mathsf{E}(K) \cap (B' \times B')} \theta(e)^2 r'(e)$$

$$\leq \quad (1 + \epsilon/3)\mathscr{E}(i) \; - \sum_{e \in \mathsf{E}(K) \cap (B' \times B')} i(e)^2 r(e) + \epsilon \mathscr{E}(i)/3$$

$$\leq (1 + \epsilon/3)\mathscr{E}(i) - (1 - \epsilon/3) \sum_{e \in E(H_1)} i(e)^2 r(e) + \epsilon\mathscr{E}(i)/3$$

[by (9.17)]

$$\leq \epsilon\mathscr{E}(i) + \sum_{e \in E(H) \setminus (B \times B)} i(e)^2 r(e). \qquad \blacktriangleleft$$

Proof of Theorem 9.19. The function $x \mapsto v(x)$ has finite Dirichlet energy for the original network, hence for the new (since conductances are multiplied by a bounded factor). Assume (for a contradiction) that the new random walk is transient. Then by Theorem 9.11, $\langle v(X_n) \rangle$ converges a.s., where v is the original voltage function and $\langle X_n \rangle$ is the new random walk. Consider $t > 0$. By Exercise 2.98, the complement of V_t induces a recurrent network for the original conductances; since the new conductances are at most $2R$ times the original ones by (9.15), this recurrence holds also for the new conductances. By Exercise 2.81, it follows that $\langle v(X_n) \rangle$ a.s. cannot have a limit $> t$. Thus, it converges to 0 a.s.

We can therefore apply Lemma 9.23 with H being $(G, \mathbf{E}[N])$ and B being the complement of V_t. Here, we choose t so that the unit current flow i on H from o to ∞ satisfies $\sum_{e \in E(G_t)} i(e)^2 / \mathbf{E}[N] < \epsilon$. We obtain in this way a unit flow θ_t from o to ∞ that is nonzero on only finitely many edges incident to B, whence θ_t restricts to a unit flow on G_t from (a finite subset of) W_t to ∞. The energy of this restriction is at most $\epsilon\mathscr{R}(o \leftrightarrow \infty; H) + \epsilon$ by Lemma 9.23. For sufficiently small ϵ, this contradicts Lemma 9.21. ◀

9.6 Notes

Propositions 9.1 and 9.2 go back in some form to Flanders (1971) and Zemanian (1976).

Equation (9.6) is an elementary form of the Hodge decomposition of L^2 1-cochains. To see this, add an oriented 2-cell with boundary θ for each cycle θ in some spanning set of cycles.

Proposition 9.4 is classical in the case of a finite, plane network and its dual. The finite case can also be proved using Kirchhoff's laws and Ohm's law, combined with the max-flow min-cut theorem: the cycle law for i implies the node law for $i^\dagger$, while the node law for i implies the cycle law for $i^\dagger$.

Theorem 9.7 is due to the authors and is published here for the first time. Thomassen (1989) proved the weaker result for (unweighted) graphs where RD(A) is replaced by the diameter of A.

The proof we have given of Theorem 9.9 was communicated to us by O. Schramm.

In light of Exercise 9.7 and Theorem 9.9, Cayley graphs of groups that are direct products of infinite groups have unique currents. An extension is the following: If G is a Cayley graph of a group that has an infinite, normal, infinite-index subgroup that is finitely generated, then G has unique currents: see Theorem 6.8 of Gaboriau (2002). Also, Cayley graphs of infinite Kazhdan groups have unique currents: see Bekka and Valette (1997).

Ancona, Lyons, and Peres (1999) also prove a crossing inequality related to Theorem 9.11.

Our proof of Theorem 9.12 was influenced by the proof of a related result by Kenyon (1998). The tiling associated by Benjamini and Schramm (1996c) to a transient, plane network is the following. We use the notation at the beginning of Section 9.4. Let $R := \mathscr{R}(o \leftrightarrow \infty)$. If $i_o(e) > 0$, then let $S(e) := J(e) \times [Rv(e^-), Rv(e^+)]$ in the cylinder $\mathbb{R}/\mathbb{Z} \times [0, R]$. Each such $S(e)$ is a square and the set of all such squares tiles $\mathbb{R}/\mathbb{Z} \times [0, R]$. For the (2,3,7)-triangle tessellation of the hyperbolic place, the result is shown in Figure 9.4. This works on finite networks too, of course. For example, a square tiling of a cylinder arising from a 21×21 grid in the plane is shown in Figure 9.4. When current flows from one vertex to another on the same face (such as the outer face), then one can unroll the

cylinder to a rectangle, as in Figure 9.5. The polar embeddings of large pieces of the square lattice
have an interesting structure, as shown in Figure 9.6. See Section II.2 of Bollobás (1998) for more
on square tilings, following the original work of Brooks, Smith, Stone, and Tutte (1940). There are
also connections to Riemann's mapping theorem; see Cannon, Floyd, and Parry (1994). In the case
of a planar triangulation, more information about the set of harmonic Dirichlet functions is given by
Hutchcroft (2019a).

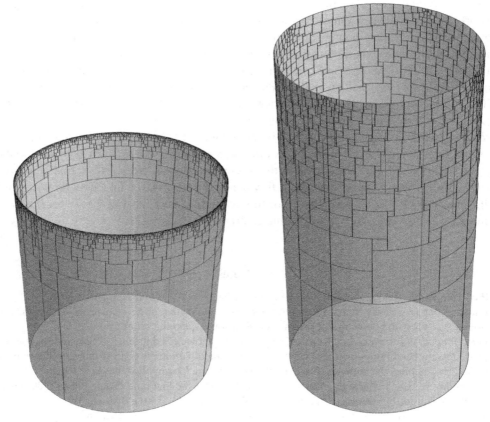

color
plate **Figure 9.4.** Tilings of cylinders by squares corresponding to the (2,3,7)-
triangle tessellation of the hyperbolic plane on the left and the 21×21 grid
with current from its center vertex to its boundary vertices on the right.

▷ **Exercise 9.11.**
We have seen that in square tilings of cylinders corresponding to planar graphs, the squares correspond
to edges and horizontal segments correspond to vertices. What correspond to the faces of the planar
graphs?

▷ **Exercise 9.12.**
Given distinct $a_i > 0$ with $\sum_{i=1}^{k} a_i = 1$, let $N(a_1, \ldots, a_k)$ be the number of tilings of a unit square by
rectangles whose areas are $a_1, \ldots, a_k$. Show that $N(a_1, \ldots, a_k)$ depends on k but not otherwise on a_i.

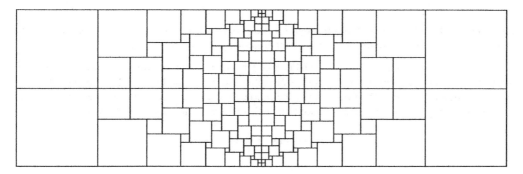

Figure 9.5. A tiling of a rectangle by squares corresponding to the 10×10 grid with current from one corner vertex to its opposite corner vertex.

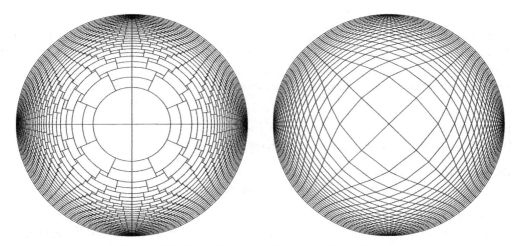

Figure 9.6. The polar embeddings of a 41×41 square grid graph from its center vertex to its boundary vertices.

Carmesin and Georgakopoulos (2020) show that every nonamenable, planar graph has nonunique currents. The proof relies on a result of Carmesin (2012), which says that a network (G, c) has nonunique currents iff there are disjoint $W_1, W_2 \subset V(G)$ such that $G \restriction W_i$ are both transient and there is a function $f: V(G) \to \mathbb{R}$ with finite Dirichlet energy and with $f \restriction W_i$ each being constant, different for $i = 1$ than for $i = 2$.

Theorem 9.18 also follows immediately from a theorem of Holopainen and Soardi (1997), which says that the property $\mathbf{HD} = \mathbb{R}$ is preserved under rough isometries between graphs and manifolds, together with a theorem of Dodziuk (1979), which says that $\mathbb{H}^d$ satisfies $\mathbf{HD} = \mathbb{R}$ iff $d \geq 3$.

We say that a Markov chain has the ***Liouville property*** if all bounded harmonic functions for that chain are constant. In contrast to Proposition 9.8, the Liouville property of a network (G, c) is not stable under change of conductances by a bounded factor; see Exercise 14.19.

Determining which random walks on groups have the Liouville property is very important, and is a central theme of Chapter 14. Here we discuss only the abelian case. The following is due to Blackwell

(1955), later generalized by Choquet and Deny (1960). We say a function f is μ-*harmonic* if, for all x, we have $f(x) = \sum_{\mu(g)>0} \mu(g)f(xg)$.

Theorem 9.24. *If Γ is an abelian group and μ is a probability measure on Γ with countable support that generates Γ, then there are no nonconstant bounded μ-harmonic functions.*

Proof. (Due to Dynkin and Maljutov (1961).) Let f be a harmonic function. For any element g of the support of μ, the function $w_g(x) := f(x) - f(xg)$ is also harmonic because Γ is abelian. If f is not constant, then for some g, the function w_g is not identically 0, whence it takes, say, a positive value. Let $M := \sup w_g$. If $M = \infty$, then f is not bounded. Otherwise, for any x, we have

$$w_g(x) = \sum_{\mu(h)>0} \mu(h)w_g(xh) \leq M\big(1 - \mu(g)\big) + \mu(g)w_g(xg),$$

which is to say that

$$M - w_g(xg) \leq \big(M - w_g(x)\big)/\mu(g).$$

Iterating this inequality gives for all $n \geq 1$,

$$M - w_g(xg^n) \leq \big(M - w_g(x)\big)/\mu(g)^n.$$

Choose x so that $M - w_g(x) < M\mu(g)^n/2$. Then $w_g(xg^k) > M/2$ for $k = 0, \ldots, n$, whence $\sum_{k=0}^{n} w_g(xg^k) = f(x) - f(xg^{n+1}) > M(n+1)/2$. Since n is arbitrary, f is not bounded. ◀

By Exercise 9.44, if a graph has no nonconstant bounded harmonic functions, then it also has no nonconstant Dirichlet harmonic functions. Thus, Theorem 9.24 strengthens Corollary 9.10 for abelian groups.

▷ **Exercise 9.13.**
Give another proof of Theorem 9.24 using the Kreĭn–Milman theorem.

▷ **Exercise 9.14.**
Give another proof of Theorem 9.24 using the Hewitt–Savage theorem.

▷ **Exercise 9.15.**
Complete the following alternative proof of Theorem 9.24. Let f be a bounded μ-harmonic function. Define

$$u_n(x) := \sum_{y_1, \ldots, y_n} \left(f\left(x + \sum_{i=1}^{n} y_i\right) - f\left(x + \sum_{i=2}^{n} y_i\right) \right)^2 \prod_{i=1}^{n} \mu(y_i).$$

(a) Show that $u_n(x) \leq u_{n+1}(x)$.
(b) Show that

$$u_n(x) = \sum_{y_1, \ldots, y_n} f\left(x + \sum_{i=1}^{n} y_i\right)^2 \prod_{i=1}^{n} \mu(y_i) - \sum_{y_2, \ldots, y_n} f\left(x + \sum_{i=2}^{n} y_i\right)^2 \prod_{i=2}^{n} \mu(y_i).$$

(c) Show that $\sum_n u_n(x) < \infty$.
(d) Show that $u_1(x) = 0$.

▷ **Exercise 9.16.**
Show that if G is a Cayley graph of a finitely generated, abelian group Γ and μ is a symmetric probability measure on Γ with finite support that generates Γ, then there are no nonconstant μ-harmonic functions h whose growth is sublinear in distance, that is, such that $h(x)/\text{dist}_G(o, x) \to 0$ as $\text{dist}_G(o, x) \to \infty$.

Benjamini, Gurel-Gurevich, and Lyons (2007) suggest that a Brownian analogue of Theorem 9.19 may be true, that is, given Brownian motion on a transient Riemannian manifold, the 1-neighborhood of its trace is recurrent for Brownian motion. For background on recurrence in the Riemannian context, see, for example, Section 2.9. It would be interesting to prove similar theorems for other processes. For example, consider the trace of a branching random walk on a graph G. Then Benjamini, Gurel-Gurevich, and Lyons (2007) conjecture that almost surely the trace is recurrent for branching random walk with the same branching law. Partial results have been proved by Benjamini and Müller (2012).

9.7 Collected In-Text Exercises

9.1. Calculate i_F^e and i_W^e in a regular tree.

9.2. Let H_n be increasing closed subspaces of a Hilbert space H and P_n be the orthogonal projection on H_n. Let P be the orthogonal projection on the closure of $\bigcup H_n$. Show that for all $u \in H$, we have $\|P_n u - Pu\| \to 0$ as $n \to \infty$.

9.3. Prove Proposition 9.2.

9.4. Show that the choice of path in the definition of the free and wired currents from a to z does not influence their values.

9.5. Prove Proposition 9.4.

9.6. **(a)** Show that $\nabla \mathbf{D}_0 = \bigstar$.
 (b) Show that $\mathbf{D}/\mathbb{R} = \widetilde{\mathbf{D}}_0/\mathbb{R} \oplus \mathbf{HD}/\mathbb{R}$, where $\widetilde{\mathbf{D}}_0 := \mathbf{D}_0 + \mathbb{R}$.
 (c) Show that currents are unique iff $\mathbf{D}/\mathbb{R} = \widetilde{\mathbf{D}}_0/\mathbb{R}$.
 (d) Show that $\|\mathbf{1} - \mathbf{D}_0\|_{\mathbf{D}}^2 = \mathscr{C}(o \leftrightarrow \infty)/[1 + \mathscr{C}(o \leftrightarrow \infty)]$, where o is the vertex used to define the inner product on $\mathbf{D}$.
 (e) Show that G is recurrent iff $\mathbf{1} \in \mathbf{D}_0$.
 (f) **(Royden Decomposition)** Show that if G is transient, then every $f \in \mathbf{D}$ has a unique decomposition $f = g + h$ with $g \in \mathbf{D}_0$ and $h \in \mathbf{HD}$. Note that this is not an orthogonal decomposition.
 (g) With the assumption and notation of part (f), show that $g(x) = (\nabla f, i_x)_r$ and that $g(x)^2 \le \mathscr{D}(f)\mathscr{G}(x, x)/\pi(x)$, where i_x is the unit current flow from x to infinity (from Proposition 2.12) and $\mathscr{G}(\cdot, \cdot)$ is the Green function.
 (h) Show that if G is transient, then $\nabla: \mathbf{D}_0 \to \bigstar$ is invertible with bounded inverse.

9.7. One can define the product of two networks in various ways. For example, given two networks $G_i = (\mathsf{V}_i, \mathsf{E}_i)$ with conductances c_i $(i = 1, 2)$, define the **Cartesian product** $G = (\mathsf{V}, \mathsf{E})$ with conductances c by $\mathsf{V} := \mathsf{V}_1 \times \mathsf{V}_2$,

$$\mathsf{E} := \left\{\big((x_1, x_2), (y_1, y_2)\big);\ \big(x_1 = y_1,\ (x_2, y_2) \in \mathsf{E}_2\big) \text{ or } \big((x_1, y_1) \in \mathsf{E}_1,\ x_2 = y_2\big)\right\},$$

and

$$c\big((x_1, x_2), (y_1, y_2)\big) := \begin{cases} c(x_2, y_2) & \text{if } x_1 = y_1 \\ c(x_1, y_1) & \text{if } x_2 = y_2. \end{cases}$$

We write $G = G_1 \,\square\, G_2$. Show that if G_i are infinite, locally finite graphs with unit conductances, then G has unique currents.

9.8. Let G be transient and let $f \in \mathbf{D}_0$. Show that there is a unique $g \in \mathbf{D}_0$ having minimal energy such that $g \ge |f|$. Show that this g is **superharmonic**, meaning that for all vertices x,

$$g(x) \ge \frac{1}{\pi(x)} \sum_{y \sim x} c(x, y)g(y).$$

9.9. With $f \in \mathbf{D}$ and $\bar{f}$ defined as in Lemma 9.15, show that $\|f(x(\rho, \cdot)) - \bar{f}\|_{L^1} \to 0$ as $\rho \uparrow 1$.

9.10. Prove Lemma 9.22.

9.11. We have seen that in square tilings of cylinders corresponding to planar graphs, the squares correspond to edges and horizontal segments correspond to vertices. What correspond to the faces of the planar graphs?

9.12. Given distinct $a_i > 0$ with $\sum_{i=1}^k a_i = 1$, let $N(a_1, \ldots, a_k)$ be the number of tilings of a unit square by rectangles whose areas are $a_1, \ldots, a_k$. Show that $N(a_1, \ldots, a_k)$ depends on k but not otherwise on a_i.

9.13. Give another proof of Theorem 9.24 using the Kreĭn–Milman theorem.

9.14. Give another proof of Theorem 9.24 using the Hewitt–Savage theorem.

9.15. Complete the following alternative proof of Theorem 9.24. Let f be a bounded μ-harmonic function. Define

$$u_n(x) := \sum_{y_1, \ldots, y_n} \left(f\left(x + \sum_{i=1}^n y_i \right) - f\left(x + \sum_{i=2}^n y_i \right) \right)^2 \prod_{i=1}^n \mu(y_i).$$

(a) Show that $u_n(x) \le u_{n+1}(x)$.
(b) Show that

$$u_n(x) = \sum_{y_1, \ldots, y_n} f\left(x + \sum_{i=1}^n y_i \right)^2 \prod_{i=1}^n \mu(y_i) - \sum_{y_2, \ldots, y_n} f\left(x + \sum_{i=2}^n y_i \right)^2 \prod_{i=2}^n \mu(y_i).$$

(c) Show that $\sum_n u_n(x) < \infty$.
(d) Show that $u_1(x) = 0$.

9.16. Show that if G is a Cayley graph of a finitely generated, abelian group Γ and μ is a symmetric probability measure on Γ with finite support that generates Γ, then there are no nonconstant μ-harmonic functions h whose growth is sublinear in distance, that is, such that $h(x)/\mathrm{dist}_G(o, x) \to 0$ as $\mathrm{dist}_G(o, x) \to \infty$.

9.8 Additional Exercises

9.17. Let G be an infinite network exhausted by finite, induced subnetworks $\langle G_n \rangle$. Form G_n^{W} by identifying the complement of G_n to a single vertex.
   (a) Given $\theta \in \ell_-^2(\mathsf{E}, r)$, define $f := d^*\theta$. Let i_n be the current on G_n^{W} such that $d^* i_n \restriction \mathsf{V}(G_n) = f \restriction \mathsf{V}(G_n)$. Show that $i_n \to P_\star \theta$ in $\ell_-^2(\mathsf{E}, r)$.
   (b) Let $f: \mathsf{V} \to \mathbb{R}$ and i_n be the current on G_n^{W} such that $d^* i_n \restriction \mathsf{V}(G_n) = f \restriction \mathsf{V}(G_n)$. Show that $\sup \mathscr{E}(i_n) < \infty$ iff there is some $\theta \in \ell_-^2(\mathsf{E}, r)$ such that $f = d^*\theta$.

9.18. Let G be a network and, if G is recurrent, $z \in \mathsf{V}$. Let $\mathscr{H}$ be the Hilbert space of functions f on V with $\sum_{x,y} \pi(x)\mathscr{G}(x, y)f(x)f(y) < \infty$. Give $\mathscr{H}$ the inner product $\langle f, g \rangle := \sum_{x,y} \pi(x)\mathscr{G}(x, y)f(x)g(y)$, where $\mathscr{G}(\cdot, \cdot)$ is the Green function for random walk, absorbed at z if G is recurrent. Define the **divergence** operator by $\mathrm{div}\,\theta := \pi^{-1}d^*\theta$. Show that $\mathrm{div}: \bigstar \to \mathscr{H}$ is an isometric isomorphism.

9.19. Let G be a transient network. Define the space $\mathscr{H}$ as in Exercise 9.18 and $\mathscr{G}$, I, and P as in Exercise 2.27. Show that $I - P$ is a bounded operator from $\mathbf{D}_0$ to $\mathscr{H}$ with bounded inverse $\mathscr{G}$.

9.20. Let G be a transient network. Show that if $u \in \mathbf{D}_0$ is superharmonic, then $u \geq 0$.

9.21. Let G be a transient network and $f \in \mathbf{HD}$. Show that there exist nonnegative $u_1, u_2 \in \mathbf{HD}$ such that $f = u_1 - u_2$.

9.22. Let $u \in \mathbf{D}$. Show that u is superharmonic iff $\mathscr{D}(u) \leq \mathscr{D}(u + f)$ for all nonnegative $f \in \mathbf{D}_0$.

9.23. Show that if G is recurrent, then the only superharmonic Dirichlet functions are the constants. *Hint:* If u is superharmonic, then use $f := u - Pu$ in Exercise 9.17, where P is the transition operator defined in Exercise 2.27. Use Exercise 2.60 to show that $f = \mathbf{0}$.

9.24. Suppose that there is a finite set K of vertices such that $G \backslash K$ has at least two transient components, where this notation indicates that K and all edges incident to K are deleted from G. Show that G has a nonconstant harmonic Dirichlet function.

9.25. Find the free and wired effective resistances between arbitrary pairs of vertices in regular trees.

9.26. Let r and r' be two assignments of resistances to a graph with $r \leq r'$ everywhere. Let a and z be two vertices in the graph. Prove the two inequalities $\mathscr{R}^{\mathrm{F}}(a \leftrightarrow z; r) \leq \mathscr{R}^{\mathrm{F}}(a \leftrightarrow z; r')$ and $\mathscr{R}^{\mathrm{W}}(a \leftrightarrow z; r) \leq \mathscr{R}^{\mathrm{W}}(a \leftrightarrow z; r')$.

9.27. Let a and z be distinct vertices in a network. Show that the wired effective resistance between a and z equals

$$\min \left\{ \mathscr{E}(\theta) \, ; \, \theta \text{ is a unit flow from } a \text{ to } z \right\},$$

whereas the free effective resistance between a and z equals

$$\min \left\{ \mathscr{E}(\theta) \, ; \, \theta - \sum_{j=1}^{k} \chi^{e_j} \in \Diamond \right\}$$

for any oriented path $e_1, \ldots, e_k$ from a to z.

9.28. Let G be a network with an induced exhaustion $\langle G_n \rangle$. Suppose that $a, z \in \mathsf{V}(G_n)$ for all n. Show that the effective resistance between a and z in G_n is monotone decreasing with limit the free effective resistance in G, whereas the effective resistance between a and z in G_n^{W} is monotone increasing with limit the wired effective resistance in G.

9.29. Let G be an infinite network and $x, y \in \mathsf{V}(G)$. Show that

$$\mathscr{R}^{\mathrm{W}}(x \leftrightarrow y) \leq \mathscr{R}(x \leftrightarrow \infty) + \mathscr{R}(y \leftrightarrow \infty).$$

9.30. Let G be a finite, plane network. In the notation of Proposition 9.4, show that the maximum flow from a to z in G' when the conductances are regarded as capacities is equal to the distance between b and y in $(G^{\dagger})'$ when the resistances are regarded as edge lengths.

9.31. Suppose that G is a plane graph with bounded degree and bounded number of sides of its faces. Show that G is transient iff $G^{\dagger}$ is transient. On the other hand, give an example of a recurrent, plane graph of bounded degree and with one end whose plane dual is transient.

9.32. Let G be transient, $a \in \mathsf{V}$, and $f(x) := \mathscr{G}(x, a)/\pi(a)$. Show that $f \in \mathbf{D}_0$ and $\nabla f = i_a$, the unit current flow from a to ∞.

9.33. Let $\mathbf{BD}$ denote the space of bounded Dirichlet functions with the norm $\|f\| := \|f\|_\infty + \|df\|_c$. Show that $\mathbf{BD}$ is a commutative Banach algebra (with respect to the pointwise product) and that $\mathbf{BD} \cap \mathbf{D}_0$ is a closed ideal.

9.34. Let G be a network.

(a) Show that G is recurrent iff every (or some) star lies in the closed span of the other stars.

(b) Show that if G is transient, then the current flow from any vertex a to infinity corresponding to unit voltage at a and zero voltage at infinity is the orthogonal projection of the star at a on the orthocomplement of the other stars. (Here, the current flow is $i_a / \mathscr{E}(i_a)$.)

9.35. Show that no transient tree has unique currents; use Exercise 2.49 instead of Theorem 9.12.

9.36. Let G be a recurrent network. Show that if $\theta \in \ell^2_-(\mathsf{E}, r)$ satisfies $\sum_x |d^*\theta(x)| < \infty$, then $\sum_x d^*\theta(x) = 0$.

9.37. Show that $\|d(Pf)\|_c \leq \|df\|_c$ for all $f \in \mathbf{D}$, where P is the transition operator defined in Exercise 2.27.

9.38. Let (G, c) be an infinite network. Show that $\Phi_{\mathsf{E}}(G, c, \pi) > 0$ iff $\ell^2(\mathsf{V}, \pi) = \mathbf{D}_0$.

9.39. Let G be a transient network and i_x be the unit current flow from x to infinity. Given $\theta \in \ell^2_-(\mathsf{E}, r)$, define $F(x) := (\theta, i_x)_r$. Show that $F \in \mathbf{D}_0$ and $\nabla F = P_\star \theta$.

9.40. Let G be a transient network and i_x be the unit current flow from x to infinity. Show that $i_x - i_y = i_{\mathsf{W}}^{x,y}$ for all $x \neq y \in \mathsf{V}$.

9.41. Let $a \neq z \in \mathsf{V}$.

(a) Show that if $F \in \mathbf{D}$, then $(\nabla F, i_{\mathsf{F}}^{a,z})_r = F(a) - F(z)$.

(b) Show that if $F \in \mathbf{D}_0$, then $(\nabla F, i_{\mathsf{W}}^{a,z})_r = F(a) - F(z)$.

9.42. Let a and z be distinct vertices in a network. Show that the free effective conductance between a and z equals

$$\min\{\mathscr{D}(F);\ F \in \mathbf{D},\ F(a) = 1,\ F(z) = 0\},$$

whereas the wired effective conductance between a and z equals

$$\min\{\mathscr{D}(F);\ F \in \mathbf{D}_0,\ F(a) = 1,\ F(z) = 0\}.$$

9.43. Let a and z be distinct vertices in a network. Show that the free effective conductance between a and z equals

$$\min\left\{\sum_{e \in \mathsf{E}_{1/2}} c(e)\ell(e)^2\right\},$$

where ℓ is an assignment of nonnegative lengths so that the associated shortest-path distance from a to z is 1, whereas the wired effective conductance between a and z equals

$$\inf\left\{\sum_{e \in \mathsf{E}_{1/2}} c(e)\ell(e)^2\right\},$$

where ℓ is an assignment of nonnegative lengths so that the distance from a to z is 1, where in measuring the distance, we also allow "extended paths" from a to z that pass via ∞, that is, the union of an infinite path from a and an infinite path from z.

9.44. Show that the bounded harmonic Dirichlet functions are dense in $\mathbf{HD}$.

9.45. Let G be an infinite graph and H be a finite graph. Consider the Cartesian product graph $G \,\square\, H$.

(a) Show that every $f \in \mathbf{HD}(G \,\square\, H)$ has the property that it does not depend on the second coordinate, that is, $f(x, y) = f(x, z)$ for all $x \in \mathsf{V}(G)$ and all $y, z \in \mathsf{V}(H)$.

(b) Show that if e is an edge in $G \,\square\, H$ that connects (x, y) and (x, z), then $Y_{\mathsf{F}}(e, e) = Y_{\mathsf{W}}(e, e)$.

9.46. Let $W \subseteq V$ be such that for all $x \in V$, random walk started at x eventually visits $V \setminus W$, that is, $\mathbf{P}_x[\tau_{V \setminus W} < \infty] = 1$. Show that if $f \in \mathbf{D}$ is supported on W and is harmonic at all vertices in W, then $f \equiv 0$.

9.47. Prove the following variant of Theorem 9.7. If G is a network such that for every pair $\mathcal{P}_1, \mathcal{P}_2$ of disjoint simple paths, there exist vertices $x_n \in \mathcal{P}_1$ and $y_n \in \mathcal{P}_2$ and edge-disjoint subgraphs H_n of G containing both x_n and y_n with the property that $\sum_n \mathscr{C}(x_n \leftrightarrow y_n; H_n) = \infty$, then G has unique currents.

9.48. Give an example of a graph with unique currents and with nonconstant bounded harmonic functions.

9.49. Extend Theorem 9.9 to show that under the same assumptions, $\mathbf{HD}$ and $\mathbf{HD}'$ have the same dimensions.

9.50. Let G and G' be networks with bounded conductances, resistances, and degrees. Suppose that there is a rough embedding from G to G' such that each vertex of G' is within some constant distance of the image of $V(G)$.
    (a) Find an example where G has unique currents but G' does not.
    (b) Find an example where G' has unique currents but G does not.

9.51. Show that the result of Exercise 9.8 also holds when G is recurrent.

9.52. Let G be a transient network. Suppose that $f \in \mathbf{D}_0$, h is harmonic, and $|h| \leq |f|$. Prove that $h = \mathbf{0}$.

9.53. Give a transient, planar network with unique currents.

9.54. Complete an alternative proof of Theorem 9.17 as follows. Show that by subdividing (adding vertices to) edges as necessary, we may assume that for each k, there is a set of vertices Π_k where $v = 1 - 1/k$ and such that the random walk visits Π_k a.s. Show that the harmonic measure on Π_k "converges" to Lebesgue measure on the circle.

9.55. Give an example of two recurrent graphs (V, E_i) $(i = 1, 2)$ on the same vertex set V whose union $(V, E_1 \cup E_2)$ is transient. On the other hand, show that on any transient network, the union of finitely many traces is a.s. recurrent, even if the random walks that produce the traces are dependent.

10 | Uniform Spanning Forests

In Chapter 4, we looked at the remarkable model of uniform spanning trees in finite (or recurrent) graphs. In this chapter, we extend this model to infinite, transient graphs.

We will discover several fascinating things. Besides the intimate connections between spanning trees, random walks, and electric networks exposed in Chapter 4 and deepened here, we will also find an intimate connection to harmonic Dirichlet functions. This leads to an unexpected constancy of expected degree when different generators are used to form a Cayley graph of a given group. We will see some amazing phase transitions of uniform spanning forests in Euclidean space as the dimension increases, and then again in hyperbolic space. Many interesting questions remain open, some of which are collected in the last section.

It turns out that there are two important ways to extend the ideas of Chapter 4 to connected, infinite graphs. In each case, sometimes we end up with spanning trees, but other times, with spanning forests, all of whose trees are infinite. Recall that a *forest* is a graph all of whose connected components are trees. Some of the results in Sections 10.5 and 10.6 are the most difficult to prove in the entire book, but we hope their interest will reward the reader's efforts. Most unattributed results in this chapter are from Benjamini, Lyons, Peres, and Schramm (2001), hereinafter referred to as BLPS (2001).

10.1 Limits over Exhaustions

How can one define a "uniform" spanning tree on an infinite graph? One natural way to try is to take the uniform spanning tree on each of a sequence of finite subgraphs that grow larger and larger so as to exhaust the whole infinite graph, hoping all the while that the measures on spanning trees have some sort of limit. We saw in Section 4.2 that this works for recurrent graphs. Luckily for us, this works also on transient graphs. In fact, there are two ways to make it work. Here are the details.

Let G be an infinite, connected, locally finite network; in fact, ***throughout this chapter, assume our graphs are locally finite***. Let $G_n = (V_n, E_n)$ be finite, connected subgraphs that ***exhaust*** G, that is, $G_n \subseteq G_{n+1}$ and $G = \bigcup G_n$. Let μ_n^F be the (weighted) uniform spanning tree probability measure on G_n that is described in Chapter 4 (the superscript F stands for "free" and will be explained in the paragraph after next). Given a finite set B of edges, we have $B \subseteq E_n$ for large enough n and, for such n, we claim that $\mu_n^F[B \subseteq T]$ is decreasing in n, where T denotes the random spanning tree. To see this, let $i(e; H)$ be the current that flows along e when a unit current is imposed in the network H from the tail to the head of e. Write

$B = \{e_1, \ldots, e_m\}$. By Kirchhoff's effective-resistance formula, we have that

$$\mu_n^F[B \subseteq T] = \prod_{k=1}^m \mu_n^F[e_k \in T \mid \forall j < k \quad e_j \in T] = \prod_{k=1}^m i(e_k; G_n/\{e_j \,;\, j < k\})$$

$$\geq \prod_{k=1}^m i(e_k; G_{n+1}/\{e_j \,;\, j < k\}) \tag{10.1}$$

$$= \mu_{n+1}^F[B \subseteq T],$$

where the inequality is from Rayleigh's monotonicity principle. In particular, $\mu^F[B \subseteq \mathfrak{F}] := \lim_{n \to \infty} \mu_n^F[B \subseteq T]$ exists. (Here, we denote a random forest by $\mathfrak{F}$ and a random tree by T to avoid prejudice about whether the forest is a tree. We say "forest" since if B contains a cycle, then by our definition, $\mu^F[B \subseteq \mathfrak{F}] = 0$.) It follows that we may define μ^F on all *elementary cylinder* sets (that is, sets of the form $\{\mathfrak{F} \,;\, B_1 \subseteq \mathfrak{F}, B_2 \cap \mathfrak{F} = \varnothing\}$ for finite, disjoint sets $B_1, B_2 \subset \mathsf{E}$) by the inclusion-exclusion formula:

$$\mu^F[B_1 \subseteq \mathfrak{F}, B_2 \cap \mathfrak{F} = \varnothing] := \sum_{S \subset B_2} \mu^F[B_1 \cup S \subseteq \mathfrak{F}](-1)^{|S|}$$

$$= \sum_{S \subset B_2} \lim_{n \to \infty} \mu_n^F[B_1 \cup S \subseteq \mathfrak{F}](-1)^{|S|}$$

$$= \lim_{n \to \infty} \mu_n^F[B_1 \subseteq T, B_2 \cap T = \varnothing].$$

This lets us define μ^F on *cylinder* sets, that is, finite (disjoint) unions of elementary cylinder sets. Again, $\mu^F(\mathscr{A}) = \lim_{n \to \infty} \mu_n^F(\mathscr{A})$ for cylinder sets $\mathscr{A}$, so these probabilities are consistent and so, by Kolmogorov's theorem, uniquely determine a probability measure μ^F on subgraphs of G. We call μ^F the *free (uniform) spanning forest* measure on G, denoted FSF or FUSF, since clearly it is carried by the set of spanning forests of G. We say that μ_n^F *converges weakly* to FSF. The term "free" will be explained in a moment.

How can it be that the limit FSF is not concentrated on the set of spanning trees of G? This happens if, for some $x, y \in \mathsf{V}$, the μ_n^F-distributions of the distance in T between x and y do not form a tight family. For example, if G is the lattice graph $\mathbb{Z}^d$, Pemantle (1991) proved the remarkable theorem that FSF is concentrated on spanning trees iff $d \leq 4$ (see Theorem 10.30).

Now there is another possibility for taking similar limits. In disregarding the complement of G_n, we are (temporarily) disregarding the possibility that a spanning tree or forest of G may connect the boundary vertices of G_n in ways that would affect the possible connections within G_n itself. An alternative approach takes the opposite view and forces *all* connections outside of G_n: As in the proof of Theorem 2.11 and in Chapter 9, let G_n^W be the graph obtained from G_n by identifying all the vertices of $G \setminus G_n$ to a single vertex, z_n. (The superscript W stands for "wired," since we think of G_n^W as having its boundary "wired" together.) Assume now that each subgraph G_n is an induced subgraph. Let μ_n^W be the random spanning tree measure on G_n^W. For any finite $B \subset \mathsf{E}$ and any n with $B \subset \mathsf{E}(G_n)$, we have $\mu_n^W[B \subseteq T] \leq \mu_{n+1}^W[B \subseteq T]$: this is proved just like the inequality (10.1), with the key difference that

$$\prod_{k=1}^m i(e_k; G_n^W/\{e_j \,;\, j < k\}) \leq \prod_{k=1}^m i(e_k; G_{n+1}^W/\{e_j \,;\, j < k\})$$

since G_n^W may be obtained from G_{n+1}^W by contracting still more edges. Thus, we may again define the limiting probability measure μ^W, called the ***wired (uniform) spanning forest*** and denoted WSF or WUSF. In the case where G is itself a tree, the free spanning forest is trivially concentrated on just $\{G\}$, whereas the wired spanning forest is usually more interesting (see Exercise 10.5). In statistical mechanics, measures on infinite configurations are also defined using limiting procedures similar to those employed here to define FSF and WSF. One needs to specify boundary conditions, just as we did. The terms "free" and "wired" have analogous uses there.

As we will see, the wired spanning forest is much better understood than the free one. For example, there is a direct construction of it that avoids exhaustions: Let G be a transient network. Define $\mathfrak{F}_0 = \varnothing$. Inductively, for each $n = 1, 2, \ldots$, pick a vertex x_n and run a network random walk starting at x_n. Stop the walk when it hits $\mathfrak{F}_{n-1}$, if it does, but otherwise let it run indefinitely. Let $\mathcal{P}_n$ denote this walk. Since G is transient, with probability 1, $\mathcal{P}_n$ visits no vertex infinitely often, so its loop erasure $\mathsf{LE}(\mathcal{P}_n)$ is well defined. Set $\mathfrak{F}_n := \mathfrak{F}_{n-1} \cup \mathsf{LE}(\mathcal{P}_n)$ and $\mathfrak{F} := \bigcup_n \mathfrak{F}_n$. Assume that the choices of the vertices x_n are made in such a way that $\{x_1, x_2, \ldots\} = \mathsf{V}$. The same reasoning as in Section 4.1 shows that the resulting distribution on forests is independent of the order in which we choose starting vertices. This also follows from the result we are about to prove. We will refer to this method of generating a random spanning forest as ***Wilson's method rooted at infinity*** (although it was introduced by BLPS (2001)).

Proposition 10.1. *The wired spanning forest on any transient network G is the same as the random spanning forest generated by Wilson's method rooted at infinity.*

Proof. For any path $\langle x_k \rangle$ that goes to ∞, $\mathsf{LE}(\langle x_k ;\ k \le K \rangle)$ converges to $\mathsf{LE}(\langle x_k ;\ k \ge 0 \rangle)$ as $K \to \infty$. That is, if $\mathsf{LE}(\langle x_k ;\ k \le K \rangle) = \langle u_i^K ;\ i \le m_K \rangle$ and $\mathsf{LE}(\langle x_k ;\ k \ge 0 \rangle) = \langle u_i ;\ i \ge 0 \rangle$, then for each i and all large K, we have $u_i^K = u_i$; this follows from the definition of loop erasure. Since G is transient, it follows that if $\langle X_k \rangle$ is a random walk starting from any fixed vertex, then $\mathsf{LE}(\langle X_k ;\ k \le K \rangle) \to \mathsf{LE}(\langle X_k ;\ k \ge 0 \rangle)$ as $K \to \infty$ a.s.

Let G_n be an exhaustion of G by induced subgraphs and G_n^W the graph formed by contracting the vertices outside G_n to a vertex z_n. Let $T(n)$ be a random spanning tree on G_n^W and $\mathfrak{F}$ the limit of $T(n)$ in law. Given $e_1, \ldots, e_M \in \mathsf{E}$, let $\langle X_k(u_i) \rangle$ be independent random walks starting from the endpoints $u_1, \ldots, u_L$ of $e_1, \ldots, e_M$. Run Wilson's method rooted at z_n from the vertices $u_1, \ldots, u_L$ in that order; actually, we do not use z_n but simply stop the random walks once they leave G_n. In this way, we can couple all the random walks and spanning trees by using the same (infinite) random walk paths from each u_j regardless of n. Let τ_j^n be the time that $\langle X_k(u_j) \rangle$ reaches the portion of the spanning tree created by the preceding random walks $\langle X_k(u_l) \rangle$ $(l < j)$. Then

$$\mathbf{P}[e_i \in T(n) \text{ for } 1 \le i \le M] = \mathbf{P}\left[e_i \in \bigcup_{j=1}^L \mathsf{LE}(\langle X_k(u_j) ;\ k \le \tau_j^n \rangle) \text{ for } 1 \le i \le M\right].$$

Let τ_j be the stopping times corresponding to Wilson's method rooted at infinity. We use the same random walks as we did in G_n. By induction on j, we see that $\tau_j^n \to \tau_j$ as $n \to \infty$, so that

$$\mathbf{P}[e_i \in \mathfrak{F} \text{ for } 1 \le i \le M] = \mathbf{P}\left[e_i \in \bigcup_{j=1}^{L} \text{LE}(\langle X_k(u_j) ; \ k \le \tau_j \rangle) \text{ for } 1 \le i \le M\right] ;$$

in other words, $\mathfrak{F}$ has the same law as the random spanning forest generated by Wilson's method rooted at infinity. ◄

We will see that in many important cases, such as $\mathbb{Z}^d$, the free and wired spanning forests agree. In fact, such agreement will be crucial to understanding the free spanning forest.

▷ **Exercise 10.1.**

The choice of exhaustion $\langle G_n \rangle$ does not change the resulting measure WSF by the proof of Proposition 10.1. Show that the choice also does not change the resulting measure FSF.

An ***automorphism*** of a network is an automorphism of the underlying graph that preserves edge weights.

▷ **Exercise 10.2.**

Show that FSF and WSF are invariant under any automorphisms that the network may have.

▷ **Exercise 10.3.**

Show that if G is an infinite, recurrent network, then the wired spanning forest on G is the same as the free spanning forest, that is, the random spanning tree of Section 4.2.

▷ **Exercise 10.4.**

Let G be a network such that there is a finite subset of edges whose removal from G leaves at least two transient components. Show that the free and wired spanning forests are different on G.

▷ **Exercise 10.5.**

Let G be a tree with unit conductances. Show that FSF = WSF iff G is recurrent.

Proposition 10.2. *Let G be a locally finite network. For both* FSF *and* WSF, *all trees are a.s. infinite.*

Proof. A finite tree, if it occurs, must occur with positive probability at some specific location, meaning that certain specific edges are present and certain other specific edges surrounding the edges of the tree are absent. But every such event has probability 0 for the approximations μ_n^{F} and μ_n^{W}, provided n is large enough, and there are only countably many such events. ◄

▷ **Exercise 10.6.**
Let G be an edge-amenable, infinite graph as witnessed by the vertex sets $\langle V_n \rangle$ (see Section 4.3). Let G_n be the subgraph induced by V_n.

(a) Let F be any spanning forest all of whose components (trees) are infinite. Show that if k_n denotes the number of trees of $F \cap G_n$, then $k_n = o(|V_n|)$.

(b) Let $\mathfrak{F}$ be a random spanning forest all of whose components (trees) are infinite. Show that the average degree, in two senses, of vertices is 2:

$$\lim_{n \to \infty} |V_n|^{-1} \sum_{x \in V_n} \deg_{\mathfrak{F}}(x) = 2 \quad \text{a.s.}$$

and

$$\lim_{n \to \infty} |V_n|^{-1} \sum_{x \in V_n} \mathbf{E}[\deg_{\mathfrak{F}}(x)] = 2.$$

In particular, if G is a transitive graph such as $\mathbb{Z}^d$, then every vertex has expected degree 2 in both the free spanning forest and the wired spanning forest.

▷ **Exercise 10.7.**
Let (T, c) be a network on a tree and $e \in T$. Show that $\mathsf{WSF}[e \in \mathfrak{F}] < 1$ iff both components of $T \backslash e$ are transient.

10.2 Coupling, Harmonic Dirichlet Functions, and Expected Degree

Often FSF = WSF; investigating when this happens will leads us to some quite interesting phenomena. In all cases, though, there is a simple inequality between these two probability measures, namely,

$$\forall e \in \mathsf{E} \quad \mathsf{FSF}[e \in \mathfrak{F}] \geq \mathsf{WSF}[e \in \mathfrak{F}], \tag{10.2}$$

since given an induced exhaustion G_n, by Rayleigh's monotonicity principle, this is true for μ_n^{F} and μ_n^{W} as soon as n is large enough that $e \in \mathsf{E}(G_n)$. Alternatively, we can write

$$\mathsf{FSF}[e \in \mathfrak{F}] = i_{\mathrm{F}}^e(e) \quad \text{and} \quad \mathsf{WSF}[e \in \mathfrak{F}] = i_{\mathrm{W}}^e(e) \tag{10.3}$$

by Propositions 9.1 and 9.2 combined with Kirchhoff's effective-resistance formula. In (9.3), we saw the inequality (10.2) for these currents.

More generally, we claim that for every increasing cylinder set $\mathscr{A}$, we have

$$\mathsf{FSF}(\mathscr{A}) \geq \mathsf{WSF}(\mathscr{A}). \tag{10.4}$$

Recall that $\mathscr{A}$ is called *increasing* if $\omega \cup \{e\} \in \mathscr{A}$ whenever $\omega \in \mathscr{A}$. We therefore say that FSF *stochastically dominates* WSF, and we write FSF $\succcurlyeq$ WSF. To show (10.4), it suffices to show that for each increasing cylinder event $\mathscr{A}$, we have $\mu_n^{\mathrm{F}}(\mathscr{A}) \geq \mu_n^{\mathrm{W}}(\mathscr{A})$ as soon as n is large enough that $\mathscr{A}$ depends only on the edge set E_n. Note that G_n is a subgraph of G_n^{W}. Thus, what we want is a consequence of the following more general result:

Lemma 10.3. *Let G be a connected subgraph of a finite, connected graph H. Let μ_G and μ_H be the corresponding uniform spanning tree measures. Then $\mu_G(\mathscr{A}) \geq \mu_H(\mathscr{A})$ for every increasing event $\mathscr{A}$ in the edge set $\mathsf{E}(G)$.*

Proof. By induction, it suffices to prove this when H has only one more edge, e, than G. Now

$$\mu_H(\mathscr{A}) = \mu_H[e \in T]\mu_H[\mathscr{A} \mid e \in T] + \mu_H[e \notin T]\mu_H[\mathscr{A} \mid e \notin T].$$

If $\mu_H[e \notin T] = 0$, then $\mu_H(\mathscr{A}) = \mu_G(\mathscr{A})$, whereas if $\mu_H[e \notin T] > 0$, then

$$\mu_H[\mathscr{A} \mid e \in T] \leq \mu_H[\mathscr{A} \mid e \notin T] = \mu_G(\mathscr{A})$$

by Theorem 4.6. This gives the result.   ◄

▷ **Exercise 10.8.**
Let G be a graph obtained by identifying some vertices of a finite, connected graph H, keeping all edges of H, though some may become loops. Let μ_G and μ_H be the corresponding uniform spanning tree measures. Show that $\mu_G(\mathscr{A}) \leq \mu_H(\mathscr{A})$ for every increasing event $\mathscr{A}$ depending on the edges of G.

The stochastic inequality FSF $\succcurlyeq$ WSF implies that the two measures, FSF and WSF, can be **monotonically coupled**. What this means is that there is a probability measure on the set

$$\left\{ (\mathfrak{F}_1, \mathfrak{F}_2) ; \ \mathfrak{F}_i \text{ is a spanning forest of } G \text{ and } \mathfrak{F}_1 \subseteq \mathfrak{F}_2 \right\}$$

that projects in the first coordinate to WSF and in the second to FSF. It is easy to see that the existence of a monotonic coupling implies the stochastic domination inequality. The converse is surprising. This equivalence between existence of a monotonic coupling and stochastic domination is quite a general result. In fact, we've encountered the notion of stochastic domination before in Exercise 4.46 and Section 7.4. All these notions can be unified by considering probability measures on partially ordered sets. In this generality, we present the two equivalent definitions of stochastic domination on finite sets in the following theorem. Extension to infinite sets is often straightforward.

Theorem 10.4. (Strassen, 1965) *Let $(X, \preccurlyeq)$ be a partially ordered finite set with two probability measures, μ_1 and μ_2. Call a subset $A \subseteq X$ **increasing** if whenever $x \in A$ and $x \preccurlyeq y$, also $y \in A$. The following are equivalent:*
  (i) *There is a probability measure ν on $\{(x, y) \in X \times X ; \ x \preccurlyeq y\}$ whose coordinate projections are μ_i.*
  (ii) *We have $\mu_1(A) \leq \mu_2(A)$ for each increasing subset $A \subseteq X$.*

In case these properties hold, we write $\mu_1 \preccurlyeq \mu_2$ and we say that μ_1 is **stochastically dominated** by μ_2. We are interested here in the case where X consists of the subsets of edges of a finite graph, ordered by $\subseteq$.

Proof. It turns out that this is a special case of the fecund max-flow min-cut theorem. The coupling ν of (i) is a way to distribute the μ_1-mass of each point $x \in X$ among the points $y \geq x$ in such a way as to obtain the distribution of μ_2. This is just a more graphic way of expressing the requirements of (i) that

$$\forall x \quad \sum_{y \geq x} \nu(x, y) = \mu_1(x)$$

$$\forall y \quad \sum_{x \leq y} \nu(x, y) = \mu_2(y) .$$

If we make a directed graph whose vertices are $(X \times \{1\}) \cup (X \times \{2\}) \cup \{\Delta_1, \Delta_2\}$ with edges from Δ_1 to each vertex of $X \times \{1\}$, from each vertex of $X \times \{2\}$ to Δ_2, and from $(x, 1)$ to $(y, 2)$ whenever $x \leq y$, then we can think of ν as a flow from Δ_1 to Δ_2 by letting $\nu(x, y)$ be the amount of flow on the edge from $(x, 1)$ to $(y, 2)$. To put this into the framework of the max-flow min-cut theorem, let the capacity of the edge joining Δ_i with (x, i) be $\mu_i(x)$ for $i = 1, 2$, while we set the capacity of all other edges to be 2.

It is evident that the condition (i) is that the maximum flow from Δ_1 to Δ_2 is 1. We claim that the condition (ii) is that the minimum cutset sum is 1, from which the theorem follows. To see this, note that any cutset of minimum sum does not use any edges of the form $\langle(x, 1), (y, 2)\rangle$, since these all have capacity 2. Thus, given a minimum cutset sum, let $B \times \{1\}$ be the set of vertices in $X \times \{1\}$ that are *not* separated from Δ_1 by the cutset. By minimality, we have that the set of vertices in $X \times \{2\}$ that *are* separated from Δ_2 is $A \times \{2\}$, where

$$A := \{y \in X ; \exists x \in B \ x \leq y\} .$$

Thus, the cutset sum is $\mu_2(A) + 1 - \mu_1(B)$. Note that A is an increasing set that contains B. Thus, we may increase B to A if $B \neq A$ while not increasing the cutset sum. That is, minimality again shows that, without loss of generality, $B = A$, whence the cutset sum is $\mu_2(A) + 1 - \mu_1(A)$. Conversely, every increasing A yields a cutset sum equal to $\mu_2(A) + 1 - \mu_1(A)$. Thus, the minimum cutset sum equals 1 iff (ii) holds, as claimed. ◄

Corollary 10.5. *On every infinite network G, we have FSF $\geq$ WSF, and there is a monotone coupling $(\mathfrak{F}_1, \mathfrak{F}_2) \in 2^{E(G)} \times 2^{E(G)}$ with $\mathfrak{F}_1 \sim$ FSF, $\mathfrak{F}_2 \sim$ WSF, and $\mathfrak{F}_1 \supseteq \mathfrak{F}_2$ a.s.*

Proof. We have already proved that FSF $\geq$ WSF, whence by Theorem 10.4, we may monotonically couple the measures induced by FSF and WSF on any finite subgraph of G. By taking an exhaustion of G and a weak limit point of these couplings, we obtain a monotone coupling of FSF and WSF on all of G. ◄

The coupling shows that $\mathsf{FSF}(\mathscr{A}) \geq \mathsf{WSF}(\mathscr{A})$ not only for all increasing cylinder events $\mathscr{A}$ but for all increasing events $\mathscr{A}$.

Question 10.6. The proof of Corollary 10.5 implicitly involves making a choice of a coupling for each finite graph in an exhaustion of G. Since there is no "canonical" choice, we do not necessarily get an invariant coupling in the limit, even though FSF and WSF are automorphism invariant by Exercise 10.2. Is there a "natural" monotone coupling of FSF and WSF? In particular, is there a monotone coupling that is invariant under all graph automorphisms? As we will see soon, FSF = WSF on amenable Cayley graphs, so in that case, there is nothing to do. Bowen (2004) has proved there is an invariant monotone coupling for all so-called residually amenable groups. This was extended by Lyons and Thom (2016) to all so-called sofic groups. On the other hand, there do exist invariant percolation processes that have monotone couplings but no invariant monotone coupling, as shown by Mester (2013).

▷ **Exercise 10.9.**
Show that the number of trees in the free spanning forest on a network is stochastically dominated by the number in the wired spanning forest on the network. Show that if the number of trees in the free spanning forest is a.s. finite, then, in distribution, it equals the number in the wired spanning forest iff FSF = WSF.

So when are the free and wired spanning forests the same? Here is one test.

Proposition 10.7. *If* $E[\deg_{\mathfrak{F}}(x)]$ *is the same under* FSF *and* WSF *for every* $x \in V$, *then* FSF = WSF.

Proof. In the monotone coupling described earlier, the set of edges adjacent to a vertex x in the WSF is a subset of those adjacent to x in the FSF. The hypothesis implies that for each x, these two sets coincide a.s.   ◄

Note that this proof works for any pair of measures where one stochastically dominates the other.

Remark 10.8. It follows that if FSF and WSF agree on single-edge probabilities, that is, if equality holds in (10.2) for all $e \in E$, then FSF = WSF. This is due to Häggström (1995).

We may now deduce that FSF = WSF for many graphs, including Cayley graphs of abelian groups such as $\mathbb{Z}^d$. Call a graph or network ***transitive*** if, for every pair of vertices x and y, there is an automorphism of the graph or network that takes x to y.

The following is essentially due to Häggström (1995).

Corollary 10.9. *On any transitive, amenable network,* FSF = WSF, *and for both measures, the expected degree of every vertex is 2.*

Proof. By transitivity and Exercise 10.6, $E[\deg_{\mathfrak{F}}(x)] = 2$ for both FSF and WSF. Apply Proposition 10.7.   ◄

▷ **Exercise 10.10.**
Give an amenable graph on which FSF ≠ WSF.

The amenability assumption gave not only equality of the two forests but restricted their expected degree to 2. What are the expected degrees otherwise?

Proposition 10.10. *If* G *is a transitive network, the* WSF-*expected degree of every vertex is* 2.

Proof. If G is recurrent, then it is amenable by Theorem 6.7, and the result follows from Corollary 10.9. So assume that G is transient. Think of the wired spanning forest as oriented toward infinity from Wilson's method rooted at infinity; that is, orient each edge of the forest in the direction it is last traversed by the associated random walk. We claim that the law of the orientation does not depend on the choices in Wilson's method rooted at infinity. Indeed, since this obviously holds for finite graphs when orienting the tree toward a fixed root, it follows by taking an exhaustion of G and using the proof of Proposition 10.1. Alternatively, one can modify the proof of Theorem 4.1 to prove it directly.

Now the out-degree of every vertex in this orientation is 1. Fix a vertex o. We need to show that the expected in-degree of o is 1. For this, it suffices to prove that for every neighbor x of o, the probability that the edge $\langle x, o \rangle$ belongs to the oriented forest is the same as the probability of the edge $\langle o, x \rangle$.

Now the probability of the edge $\langle o, x \rangle$ is $\left(\pi(o) \mathbf{P}_o[\tau_o^+ = \infty] \right)^{-1} c(o, x) \mathbf{P}_x[\tau_o = \infty]$ by Exercise 4.26. Since $\pi(x) = \pi(o)$ and $\mathscr{G}(x, x) = \mathscr{G}(o, o)$ by transitivity, it follows from Exercise 2.26 that

$$\mathbf{P}_x[\tau_o < \infty] = \mathbf{P}_o[\tau_x < \infty]. \tag{10.5}$$

Combined with the previous sentence, this gives the result.   ◄

Now what about the expected degree of a vertex in the FSF? This turns out to be even more interesting than in the WSF. One can show (Lyons, 2009) from (10.3) and Definition 2.9 of Gaboriau (2005) that the expected degree in an infinite, transitive graph G equals $2 + 2\beta_1(G)$, where $\beta_1(G)$ is the so-called *first ℓ^2-Betti number* of G. This is a nonnegative real number whose definition was made originally by Atiyah (1976) in a continuous context, was extended to a discrete context by Dodziuk (1977), was considerably extended by Cheeger and Gromov (1986), and was made in this context by Gaboriau (2005). Consider the special case where G is a Cayley graph of an infinite group, Γ. A marvelous fact is that $\beta_1(G)$ is the same for all Cayley graphs of Γ, so we normally write $\beta_1(\Gamma)$ instead. Thus,

$$\mathbf{E}_{\mathsf{FSF}}[\deg_{\widetilde{\mathfrak{F}}} o] = 2 + 2\beta_1(\Gamma). \tag{10.6}$$

For our purposes, we may take (10.6) as a definition of $\beta_1(\Gamma)$, but we also mention that it equals the von Neumann dimension of $\nabla\mathbf{HD}(G)$ with respect to Γ. See Section 10.8 for definitions and a derivation of (10.6), including the fact that this does not depend on the generators. Here are some known values of first ℓ^2-Betti numbers (see, for example, Gaboriau (2002), Cheeger and Gromov (1986), and Lück (2009)):

- $\beta_1(\Gamma) = 0$ if Γ is finite or amenable
- $\beta_1(\Gamma_1 * \Gamma_2) = \beta_1(\Gamma_1) + \beta_1(\Gamma_2) + 1 - \frac{1}{|\Gamma_1|} - \frac{1}{|\Gamma_2|}$ if Γ_i are not trivial
- $\beta_1(\Gamma_1 *_{\Gamma_3} \Gamma_2) = \beta_1(\Gamma_1) + \beta_1(\Gamma_2)$ if Γ_3 is amenable and infinite
- $\beta_1(\Gamma_2) = [\Gamma_1 : \Gamma_2] \beta_1(\Gamma_1)$ if Γ_2 has finite index in Γ_1
- $\beta_1(\Gamma) = 2g - 2$ if Γ is the fundamental group of an orientable surface of genus g
- $\beta_1(\Gamma) = s - 2$ if Γ is torsion free and can be presented with $s \geq 2$ generators and one nontrivial relation

If we take $\beta_1(\Gamma) = 0$ as a definition for finite Γ and (10.6) as a definition for infinite Γ, then the fact that $\beta_1(\Gamma) = 0$ for amenable Γ follows from Exercise 10.6, while the formula $\beta_1(\Gamma_1 * \Gamma_2) = \beta_1(\Gamma_1) + \beta_1(\Gamma_2) + 1 - \frac{1}{|\Gamma_1|} - \frac{1}{|\Gamma_2|}$ follows from taking a generating set for $\Gamma_1 * \Gamma_2$

by combining generating sets for Γ_1 and Γ_2: the restriction of the FSF of $\Gamma_1 * \Gamma_2$ to each copy of the Cayley graph of, say, Γ_1 in the Cayley graph of $\Gamma_1 * \Gamma_2$ has the same law as the FSF of just Γ_1, which is the uniform spanning tree if Γ_1 is finite.

Question 10.11. If G is a Cayley graph of a finitely presented group Γ, is the FSF-expected degree of o a rational number? Atiyah (1976) has asked whether all the ℓ^2-Betti numbers (not only the first one discussed here) are rational if Γ is finitely presented. Various generalizations of this question have negative answers; see, for example, Grabowski (2014).

An important open question is whether $\beta_1(\Gamma)$ is equal to one less than the **cost** of Γ, which is one-half of the infimum of the expected degree in any random connected spanning graph of Γ whose law is Γ-invariant (see Gaboriau (2000)). Gaboriau (2002) proved that $\beta_1(\Gamma)$ is always at most one less than the cost of Γ. To establish equality, it would therefore suffice to show that for every $\epsilon > 0$, there is some probability space with two 2^{E}-valued random variables, $\mathfrak{F}$ and $\mathfrak{X}$, such that the law of $\mathfrak{F}$ is FSF, the union $\mathfrak{F} \cup \mathfrak{X}$ is connected and has a Γ-invariant law, and the expected degree of a vertex in $\mathfrak{X}$ is less than ϵ. In Chapter 11, we will see that an analogue of this does hold for the free *minimal* spanning forest.

Question 10.12. If G is a Cayley graph, $\mathfrak{F}$ is the FSF on G, and $\epsilon > 0$, is there an invariant connected percolation ω containing $\mathfrak{F}$ such that for every edge e, the probability that $e \in \omega \setminus \mathfrak{F}$ is less than ϵ? This question was asked by Damien Gaboriau (personal communication, 2001). It is known that whenever G is nonamenable, the union of WSF and independent Bernoulli(ϵ) percolation is *not* connected a.s. for small ϵ (see BLPS (2001)). Therefore, when G is nonamenable and WSF = FSF, we cannot connect the FSF by adding a small percolation. In Proposition 10.14, we describe when WSF = FSF.

One use of the invariance of $\beta_1(\Gamma)$ with respect to generating sets is to prove uniform exponential growth of certain Cayley graphs, a question we discussed briefly in Section 3.4. Recall from that discussion that if a Cayley graph has exponential growth, then so does every Cayley graph of the same group, yet for some groups, the infimum of the growth rates is 1, that is, those groups do not have uniform exponential growth. Here, we show that if FSF $\neq$ WSF for a given Cayley graph of a group Γ, then Γ does have uniform exponential growth. Note that FSF $\neq$ WSF iff $\beta_1(\Gamma) > 0$ by (10.6), Proposition 10.10, and Proposition 10.7. The argument is an elaboration of Exercise 10.6 and was observed by Lyons, Pichot, and Vassout (2008). It shows even more than uniform exponential growth, namely, a uniform expansion property.

Theorem 10.13. (Uniform Expansion) *Let G be a Cayley graph of a finitely generated infinite group Γ with respect to a finite generating set S. For every finite $K \subset \Gamma$, we have*

$$\frac{|\partial_\vee K|}{|K|} > 2\beta_1(\Gamma).$$

In particular, this implies that finitely generated groups Γ with $\beta_1(\Gamma) > 0$ have uniform exponential growth: if B_n denotes the ball of radius n centered at o, then

$$|B_{n+1}|/|B_n| > 1 + 2\beta_1(\Gamma),$$

so

$$|B_n| > [1 + 2\beta_1(\Gamma)]^n .$$

Proof. Let $\mathfrak{F} \sim \mathsf{FSF}$. Let $\mathfrak{F}'$ be the part of $\mathfrak{F}$ that touches K, that is, the edges that are incident to some vertex of K, together with their endpoints. Let $L := \mathsf{V}(\mathfrak{F}') \setminus K$. Since $\mathfrak{F}'$ is a forest,

$$\sum_{x \in K} \deg_{\mathfrak{F}} x \le \sum_{x \in K \cup L} \deg_{\mathfrak{F}'} x - |L| = 2|\mathsf{E}(\mathfrak{F}')| - |L| < 2|\mathsf{V}(\mathfrak{F}')| - |L|$$

$$= 2|K| + |L| \le 2|K| + |\partial K| .$$

Take the expectation, use the formula (10.6), and divide by $|K|$ to get the result. ◀

We turn now to an electrical criterion for the equality of FSF and WSF. As we saw in Chapter 9, there are two natural ways of defining currents between vertices of a graph, corresponding to the two ways of defining spanning forests.

We may add to Proposition 9.3 and Theorem 9.5 as follows:

Proposition 10.14. *For any network, the following are equivalent:*
  (i) $\mathsf{FSF} = \mathsf{WSF}$;
  (ii) $i_W^e = i_F^e$ *for every edge* e;
  (iii) $\ell_-^2(\mathsf{E}) = \bigstar \oplus \Diamond$;
  (iv) $\mathbf{HD} = \mathbb{R}$.

Proof. Use Remark 10.8, (10.3), and (9.3) to deduce that (i) and (ii) are equivalent. The other equivalences were proved already. ◀

Combining this with Theorem 9.7, we get another proof that $\mathsf{FSF} = \mathsf{WSF}$ on $\mathbb{Z}^d$.

▷ **Exercise 10.11.**
Show that every transitive, amenable network has unique currents.

Let $Y_F(e, e') := i_F^e(e')$ and $Y_W(e, e') := i_W^e(e')$ be the free and wired transfer-current matrices.

Theorem 10.15. (Free and Wired Transfer-Current Theorems) *Given any network G and any distinct edges $e_1, \ldots, e_k \in G$, we have*

$$\mathsf{FSF}[e_1, \ldots, e_k \in \mathfrak{F}] = \det[Y_F(e_i, e_j)]_{1 \le i, j \le k}$$

and

$$\mathsf{WSF}[e_1, \ldots, e_k \in \mathfrak{F}] = \det[Y_W(e_i, e_j)]_{1 \le i, j \le k} .$$

Proof. This is immediate from the transfer-current theorem of Section 4.2 and Propositions 9.1 and 9.2. ◀

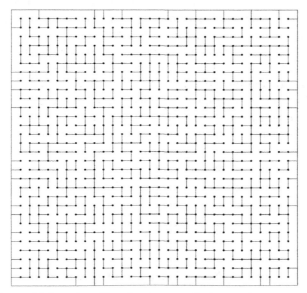

Figure 10.1. A uniformly chosen wired spanning tree on a subgraph of $\mathbb{Z}^2$, drawn by Wilson (see Propp and Wilson (1998)).

10.3 Planar Networks and Euclidean Lattices

Consider now plane graphs. Examination of Figure 10.1 reveals two spanning trees: one in white, the other in black on the plane dual graph. (See Section 9.2 for the definition of dual.) Note that in the dual, the outer boundary of the grid is identified to a single vertex. In general, suppose that G is a proper, simple, plane network whose plane dual $G^\dagger$ is locally finite. Consider an exhaustion of G by subnetworks G_n such that $G \backslash V(G_n)$ has only one (infinite) component. Let $G_n^\dagger$ be the plane dual of G_n. Note that $G_n^\dagger$ can be regarded as a finite subnetwork of $G^\dagger$ but with its outer boundary vertices identified to a single vertex. Spanning trees T of G_n are in one-to-one correspondence with spanning trees $T^\times$ of $G_n^\dagger$ in the same way as in Figure 10.1:

$$e \in T \iff e^\dagger \notin T^\times. \tag{10.7}$$

Furthermore, this correspondence preserves the relative weights of Section 4.1: we have

$$\Xi(T) = \prod_{e \in T} c(e) = \prod_{e^\dagger \notin T^\times} r(e^\dagger) = \frac{\prod_{e^\dagger \in T^\times} c(e^\dagger)}{\prod_{e^\dagger \in G_n^\dagger} c(e^\dagger)} = \frac{\Xi(T^\times)}{\prod_{e^\dagger \in G_n^\dagger} c(e^\dagger)}.$$

Taking $n \to \infty$, we find that the FSF of G is "dual" to the WSF of $G^\dagger$; that is, the relation (10.7) transforms one to the other. As a consequence, (10.3) and the definition (10.7) explain (9.4). Let's look at this a little more closely for the graph $\mathbb{Z}^2$. By recurrence and Exercise 10.3, the free and wired spanning forests are the same as the uniform spanning tree. In particular, $\mathbf{P}[e \in T] = \mathbf{P}[e^\dagger \in T^\times]$. Since these add to 1, they are equal to 1/2, as we derived in another fashion in (4.14).

Let's see what Theorem 10.15 says more explicitly for unit conductances on the hypercubic lattice $\mathbb{Z}^d$. The case $d = 2$ was treated already in Section 4.3 and Exercise 4.54. The transient case $d \geq 3$ is actually easier, but the approach is the same as in Section 4.3. To find the transfer currents $Y(e_1, e_2)$, we will first find voltages, then use $i = dv$. When i is a unit flow from x to y, we have $d^* i = \mathbf{1}_{\{x\}} - \mathbf{1}_{\{y\}}$. Hence the voltages satisfy $\Delta v := d^* dv = \mathbf{1}_{\{x\}} - \mathbf{1}_{\{y\}}$. We are interested in solving this equation when $x := e_1^-$ and $y := e_1^+$ and then computing $v(e_2^-) - v(e_2^+)$. Unlike in the recurrent case of Section 4.3, however, we can solve $\Delta v = \mathbf{1}_{\{x\}}$, in other words, the voltage v for a current from x to infinity. Again, we begin with a formal (that is, heuristic) derivation of the solution, and then prove that the resulting formula is actually correct.

Let $\mathbb{T}^d := (\mathbb{R}/\mathbb{Z})^d$ be the d-dimensional torus. For $(x_1, \ldots, x_d) \in \mathbb{Z}^d$ and $(\alpha_1, \ldots, \alpha_d) \in \mathbb{T}^d$, write $(x_1, \ldots, x_d) \cdot (\alpha_1, \ldots, \alpha_d) := x_1\alpha_1 + \cdots + x_d\alpha_d \in \mathbb{R}/\mathbb{Z}$. For a function f on $\mathbb{Z}^d$, define the function $\widehat{f}$ on $\mathbb{T}^d$ by

$$\widehat{f}(\alpha) := \sum_{x \in \mathbb{Z}^d} f(x) e^{-2\pi i x \cdot \alpha} \,.$$

For example, $\widehat{\mathbf{1}_{\{x\}}}(\alpha) = e^{-2\pi i x \cdot \alpha}$. Now a formal calculation shows that

$$\widehat{\Delta f}(\alpha) = \widehat{f}(\alpha)\varphi(\alpha) \,,$$

where

$$\varphi((\alpha_1, \ldots, \alpha_d)) := 2d - \sum_{k=1}^{d}(e^{2\pi i \alpha_k} + e^{-2\pi i \alpha_k}) = 2d - 2\sum_{k=1}^{d} \cos 2\pi\alpha_k \,. \tag{10.8}$$

Hence, to solve $\Delta f = g$, we may try to solve $\widehat{\Delta f} = \widehat{g}$ by using $\widehat{f} := \widehat{g}/\varphi$ and then finding f from $\widehat{f}$. In fact, a formal calculation shows that we may recover f from $\widehat{f}$ by the formula

$$f(x) = \int_{\mathbb{T}^d} \widehat{f}(\alpha) e^{2\pi i x \cdot \alpha} \, d\alpha \,,$$

where the integration is with respect to Lebesgue measure. What makes the transient case easier is that $1/\varphi \in L^1(\mathbb{T}^d)$, because

$$\varphi(\alpha) = (2\pi)^2 |\alpha|^2 + O(|\alpha|^4) \tag{10.9}$$

as $|\alpha| \to 0$ and $r \mapsto 1/r^2$ is integrable in $\mathbb{R}^d$ for $d \geq 3$; here, $|\alpha|$ is the minimal distance in $\mathbb{R}^d$ between the coset α and $\mathbb{Z}^d$.

$\triangleright$ **Exercise 10.12.**
(The Riemann–Lebesgue Lemma)   Show that if $F \in L^1(\mathbb{T}^d)$ and

$$f(x) = \int_{\mathbb{T}^d} F(\alpha) e^{2\pi i x \cdot \alpha} \, d\alpha \,,$$

then $\lim_{|x| \to \infty} f(x) = 0$. *Hint:* This is obvious if F is a ***trigonometric polynomial***, that is, a finite linear combination of functions $\alpha \mapsto e^{2\pi i x \cdot \alpha}$. The Stone–Weierstrass theorem implies that such functions are dense in $L^1(\mathbb{T}^d)$.

Proposition 10.16. (Voltages on $\mathbb{Z}^d$) *Let $d \geq 3$. Define φ as in (10.8). The voltage at x when a unit current flows from $\mathbf{0}$ to infinity in $\mathbb{Z}^d$ and the voltage is 0 at infinity is*

$$v(x) = \int_{\mathbb{T}^d} \frac{e^{2\pi i x \cdot \alpha}}{\varphi(\alpha)} \, d\alpha \, . \tag{10.10}$$

Proof. Define $v'(x)$ to be the integral on the right-hand side of (10.10). By the analogue of Exercise 4.9 for $d \geq 3$, we have

$$\Delta v'(x) = \int_{\mathbb{T}^d} \frac{e^{2\pi i x \cdot \alpha}}{\varphi(\alpha)} \varphi(\alpha) \, d\alpha = \int_{\mathbb{T}^d} e^{2\pi i x \cdot \alpha} \, d\alpha = \mathbf{1}_{\{0\}}(x) \, .$$

That is, $\Delta v' = \mathbf{1}_{\{0\}}$. Since v satisfies the same equation, we have $\Delta(v' - v) = 0$. In other words, $v' - v$ is harmonic at every point in $\mathbb{Z}^d$. Furthermore, v' is bounded in absolute value by the L^1 norm of $1/\varphi$, whence $v' - v$ is bounded. Since the only bounded harmonic functions on $\mathbb{Z}^d$ are the constants (by, say, Theorem 9.24), this means that $v' - v$ is constant. Now the Riemann–Lebesgue lemma* (Exercise 10.12) implies that v' tends to 0 at infinity. Since v also tends to 0 at infinity (by either Exercise 2.91 and that $v' - v$ is constant, or else by Exercise 2.92 and symmetry), this constant is 0. Therefore, $v' = v$, as desired. ◄

Now one can compute

$$Y(e_1, e_2) = v(e_2^- - e_1^-) - v(e_2^+ - e_1^-) - v(e_2^- - e_1^+) + v(e_2^+ - e_1^+)$$

using (10.10).

10.4 Tail Triviality

How much does the configuration of a uniform spanning forest in one region influence the configuration in a faraway region? Are they asymptotically independent? It turns out that they are indeed, in several senses.

For a set of edges $K \subseteq \mathsf{E}$, let $\mathscr{F}(K)$ denote the σ-field of events depending only on K. Define the ***tail σ-field*** to be the intersection of $\mathscr{F}(\mathsf{E} \setminus K)$ over all finite K. Triviality of the tail σ-field means that every tail-measurable event has probability 0 or 1. For example, Kolmogorov's zero-one law says that the tail σ-field is trivial for independent percolation. More generally, tail triviality is equivalent to a strong form of asymptotic independence:

Proposition 10.17. (Tail Triviality and Asymptotic Independence) *For any probability measure $\mathbf{P}$, tail triviality is equivalent to*

$$\forall \mathscr{A}_1 \in \mathscr{F}(\mathsf{E}) \ \forall \epsilon > 0 \ \exists K \text{ finite } \forall \mathscr{A}_2 \in \mathscr{F}(\mathsf{E} \setminus K) \quad \left| \mathbf{P}(\mathscr{A}_1 \cap \mathscr{A}_2) - \mathbf{P}(\mathscr{A}_1) \mathbf{P}(\mathscr{A}_2) \right| < \epsilon \, . \tag{10.11}$$

---

* One could avoid the Riemann–Lebesgue lemma by solving the more complicated equation $\Delta v = \mathbf{1}_{\{x\}} - \mathbf{1}_{\{y\}}$ instead, as in Section 4.3.

Proof. Suppose first that (10.11) holds. Let $\mathscr{A}$ be a tail event. Then we may take $\mathscr{A}_1 := \mathscr{A}_2 :=$ $\mathscr{A}$ in (10.11), no matter what ϵ is. This shows that $\mathscr{A}$ is independent of itself, so is trivial.

For the converse, let $\mathbf{P}$ be tail trivial. Let $\langle K_n \rangle$ be an exhaustion of E. Then $\mathcal{T} :=$ $\bigcap_n \mathscr{F}(\mathsf{E} \setminus K_n)$ is the tail σ-field. The reverse-martingale convergence theorem shows that for every $\mathscr{A}_1 \in \mathscr{F}(\mathsf{E})$, we have

$$Z_n := \mathbf{P}\big[\mathscr{A}_1 \mid \mathscr{F}(\mathsf{E} \setminus K_n)\big] \to \mathbf{P}(\mathscr{A}_1 \mid \mathcal{T})$$

not only a.s. but in $L^1(\mathbf{P})$. Since the tail is trivial, the limit equals $\mathbf{P}(\mathscr{A}_1)$ a.s. Thus, given $\epsilon > 0$, there is some n such that $\mathbf{E}\big[|Z_n - \mathbf{P}(\mathscr{A}_1)|\big] < \epsilon$. Hence, for all $\mathscr{A}_2 \in \mathscr{F}(\mathsf{E} \setminus K_n)$, we have

$$\big|\mathbf{P}(\mathscr{A}_1 \cap \mathscr{A}_2) - \mathbf{P}(\mathscr{A}_1)\,\mathbf{P}(\mathscr{A}_2)\big| = \Big|\mathbf{E}\big[(Z_n - \mathbf{P}(\mathscr{A}_1))\mathbf{1}_{\mathscr{A}_2}\big]\Big| \le \mathbf{E}\big[|(Z_n - \mathbf{P}(\mathscr{A}_1))\mathbf{1}_{\mathscr{A}_2}|\big] < \epsilon. \quad \blacktriangleleft$$

This equivalence shows why the following theorem is interesting.

Theorem 10.18. *The* WSF *and* FSF *have trivial tail on every network.*

Proof. This is a consequence of the fact that the boundary conditions defining WSF and FSF are each extremal. To see this, let G be an infinite network exhausted by finite, connected, induced subnetworks $\langle G_n \rangle$. Write $G_n = (\mathsf{V}_n, \mathsf{E}_n)$. Recall from Section 10.1 that μ_n^{F} denotes the uniform spanning tree measure on G_n, and μ_n^{W} denotes the uniform spanning tree measure on the "wired" graph G_n^{W}. Let ν_n be any "partially wired" measure, that is, the uniform spanning tree measure on a graph G_n^* obtained from a finite network G_n' satisfying $G_n \subset G_n' \subset G$ by contracting some of the edges in G_n' that are not in G_n. Lemma 10.3 and Exercise 10.8 give that if $\mathscr{B} \in \mathscr{F}(\mathsf{E}_n)$ is increasing, then

$$\mu_n^{\mathrm{W}}(\mathscr{B}) \le \nu_n(\mathscr{B}) \le \mu_n^{\mathrm{F}}(\mathscr{B}). \tag{10.12}$$

This is what we meant by "extremality" earlier.

To see its consequences, let $M > n$, and let $\mathscr{A} \in \mathscr{F}(\mathsf{E}_M \setminus \mathsf{E}_n)$ be a cylinder event such that $\mu_M^{\mathrm{W}}(\mathscr{A}) > 0$. If we condition separately on each possible configuration of edges of $G_M \setminus G_n$ that is in $\mathscr{A}$ and use (10.12) on each of these configurations, then we get that for each increasing $\mathscr{B} \in \mathscr{F}(\mathsf{E}_n)$, we have

$$\mu_n^{\mathrm{W}}(\mathscr{B}) \le \mu_M^{\mathrm{W}}(\mathscr{B} \mid \mathscr{A}). \tag{10.13}$$

Fixing $\mathscr{A}$ and letting $M \to \infty$ in (10.13) gives

$$\mu_n^{\mathrm{W}}(\mathscr{B}) \le \mathsf{WSF}(\mathscr{B} \mid \mathscr{A}). \tag{10.14}$$

This applies to all cylinder events $\mathscr{A} \in \mathscr{F}(\mathsf{E} \setminus \mathsf{E}_n)$ with $\mathsf{WSF}(\mathscr{A}) > 0$, and therefore the assumption that $\mathscr{A}$ is a cylinder event can be dropped. In particular, (10.14) holds for all tail events $\mathscr{A}$ of positive probability. Taking $n \to \infty$ there gives

$$\mathsf{WSF}(\mathscr{B}) \le \mathsf{WSF}(\mathscr{B} \mid \mathscr{A}), \tag{10.15}$$

where $\mathscr{B}$ is any increasing cylinder event and $\mathscr{A}$ is any tail event. Thus, (10.15) also applies to the complement $\mathscr{A}^c$. Since $\mathsf{WSF}(\mathscr{B}) = \mathsf{WSF}(\mathscr{A})\,\mathsf{WSF}(\mathscr{B} \mid \mathscr{A}) + \mathsf{WSF}(\mathscr{A}^c)\,\mathsf{WSF}(\mathscr{B} \mid \mathscr{A}^c)$, it follows that $\mathsf{WSF}(\mathscr{B}) = \mathsf{WSF}(\mathscr{B} \mid \mathscr{A})$. Therefore, every tail event $\mathscr{A}$ is independent of every increasing cylinder event. By inclusion-exclusion, such $\mathscr{A}$ is also independent of every elementary cylinder event, whence of every cylinder event, whence of every event. That is, $\mathscr{A}$ is trivial. The argument for the FSF is similar. $\quad \blacktriangleleft$

Tail triviality has additional consequences when the underlying graph is a Cayley graph. More generally, consider graphs that have automorphisms that can move points arbitrarily far from where they start. If Γ is a group of automorphisms of a network or graph G, we say that Γ *acts* on G. Any action extends in the obvious way to an action on the collection of subnetworks or subgraphs of G. We now want to take limits at infinity of functions on Γ. What does this mean? A "large" element of Γ is one that moves points far. To be precise, for an action of Γ on G and a real-valued function f on Γ, we write $\lim_{\gamma \to \infty} f(\gamma) = a$ to mean that for every $\epsilon > 0$ and every $x \in V$, there is some N such that for all γ satisfying $\mathrm{dist}(x, \gamma x) > N$, we have $|f(\gamma) - a| < \epsilon$. Similarly, $\limsup_{\gamma \to \infty} f(\gamma) \le a$ means that for every $\epsilon > 0$ and every $x \in V$, there is some N such that whenever $\mathrm{dist}(x, \gamma x) > N$, we have $f(\gamma) < a + \epsilon$. We say that the action is *mixing* for a Γ-invariant probability measure $\mathbf{P}$ on subnetworks or subgraphs of G if, for any pair of events $\mathscr{A}$ and $\mathscr{B}$,

$$\lim_{\gamma \to \infty} \mathbf{P}(\mathscr{A}, \gamma \mathscr{B}) = \mathbf{P}(\mathscr{A}) \mathbf{P}(\mathscr{B}). \tag{10.16}$$

We call an action *ergodic* for $\mathbf{P}$ if the only $(\Gamma, \mathbf{P})$-invariant events are trivial. Here, an event $\mathscr{A}$ is called $(\Gamma, \mathbf{P})$-*invariant* if $\mathbf{P}(\mathscr{A} \triangle \gamma \mathscr{A}) = 0$ for all $\gamma \in \Gamma$. As usual, as long as there is some infinite orbit, mixing implies ergodicity: if $\mathscr{A}$ is an invariant event, then just take $\mathscr{B} = \mathscr{A}$ in (10.16). In addition, tail triviality implies mixing: Let $\mathscr{A}$ and $\mathscr{B}$ be two events. The equation (10.16) follows immediately from (10.11) in case $\mathscr{B}$ is a cylinder. For general $\mathscr{B}$, let $\epsilon > 0$, and let $\mathscr{D}$ be a cylinder such that $\mathbf{P}(\mathscr{B} \triangle \mathscr{D}) < \epsilon$. Since $\mathbf{P}$ is invariant under Γ, we have

$$|\mathbf{P}(\mathscr{A}, \gamma \mathscr{B}) - \mathbf{P}(\mathscr{A}) \mathbf{P}(\mathscr{B})| \le |\mathbf{P}(\mathscr{A}, \gamma \mathscr{D}) - \mathbf{P}(\mathscr{A}) \mathbf{P}(\mathscr{D})| + \mathbf{P}[\mathscr{A}, \gamma(\mathscr{B} \triangle \mathscr{D})] + \mathbf{P}(\mathscr{A}) \mathbf{P}(\mathscr{B} \triangle \mathscr{D})$$
$$< |\mathbf{P}(\mathscr{A}, \gamma \mathscr{D}) - \mathbf{P}(\mathscr{A}) \mathbf{P}(\mathscr{D})| + 2\epsilon.$$

Taking γ to infinity, we get that $\limsup_{\gamma \to \infty} |\mathbf{P}(\mathscr{A}, \gamma \mathscr{B}) - \mathbf{P}(\mathscr{A}) \mathbf{P}(\mathscr{B})| \le 2\epsilon$. Since ϵ is arbitrary, the action is mixing.

An interesting consequence of ergodicity is the following: distinct ergodic, invariant probability measures under any group action are always mutually singular. To see this, suppose that μ_1 and μ_2 are both invariant and ergodic probability measures for an action of a group Γ on a network G. According to the Lebesgue decomposition theorem, there is a *unique* pair of measures ν_1, ν_2 such that $\mu_1 = \nu_1 + \nu_2$ with $\nu_1 \ll \mu_2$ and $\nu_2 \perp \mu_2$. Applying any element of Γ, we see that ν_1 and ν_2 are both Γ-invariant. Choose an event $\mathscr{A}$ with $\nu_1(\mathscr{A}^c) = 0$ and $\nu_2(\mathscr{A}) = 0$. Then $\mathscr{A}$ is (Γ, ν_1)-invariant and (Γ, ν_2)-invariant, whence (Γ, μ_1)-invariant, whence μ_1-trivial. If $\mu_1(\mathscr{A}) = 0$, then $\mu_1 = \nu_2 \perp \mu_2$. On the other hand, if $\mu_1(\mathscr{A}^c) = 0$, then $\mu_1 = \nu_1 \ll \mu_2$. Let f be the Radon–Nikodým derivative of μ_1 with respect to μ_2. Then f is measurable with respect to the σ-field of (Γ, μ_2)-invariant events, which is trivial, so f is constant μ_2-a.e. That is, $\mu_1 = \mu_2$.

Thus, by Theorem 10.18 and Exercise 10.2, we obtain the following consequences:

Corollary 10.19. *Let Γ be a group acting on a network G so that every vertex has an infinite orbit. Then the action is mixing and ergodic for* FSF *and for* WSF. *If* WSF *and* FSF *on G are distinct, then they are singular measures on the space 2^{E}.*

We do not know if the preceding singularity assertion holds without the hypothesis that every vertex has an infinite orbit under $\mathrm{Aut}(G)$; see Question 10.58.

10.5 The Number of Trees

We've not yet said much about how the uniform spanning forests look. For example, when is the free spanning forest or the wired spanning forest of a network a.s. a single tree, as in the case of recurrent networks? There are few cases where we know the answer for the free spanning forest; almost all of them come from knowing the answer for the wired spanning forest and then using some relationship between the two forests, such as their being equal or being dual. So we begin with the wired spanning forest. In that case, we have the following answer for the wired spanning forest due to Pemantle (1991).

Proposition 10.20. *Let G be any network. The wired spanning forest is a single tree a.s. iff from every (or some) vertex, random walk and independent loop-erased random walk intersect infinitely often a.s. Moreover, the chance that two given vertices x and y belong to the same tree equals the probability that random walk from x intersects independent loop-erased random walk from y.*

This is obvious from Proposition 10.1 (which wasn't available to Pemantle at the time) but is otherwise quite striking. How, then, do we decide whether a random walk and a loop-erased random walk intersect a.s.? Pemantle (1991) used results of Lawler (1986, 1988) to answer this for simple random walk in $\mathbb{Z}^d$. However, Lyons, Peres, and Schramm (2003) later showed that for any transient Markov chain, two independent paths started at any pair of states intersect infinitely often (i.o.) with probability 1 iff the loop erasure of one intersects the other i.o. with probability 1. In fact, more is true:

Theorem 10.21. *Let $\langle X_m \rangle$ and $\langle Y_n \rangle$ be independent, transient Markov chains on the same state space V that have the same transition probabilities but possibly different initial states. Then, given the event that $\big|\{X_m\} \cap \{Y_n\}\big| = \infty$, almost surely $\big|\mathsf{LE}\langle X_m \rangle \cap \{Y_n\}\big| = \infty$.*

This makes it considerably easier to decide whether the wired spanning forest is a single tree. Thus:

Theorem 10.22. *Let G be any network. The wired spanning forest is a single tree a.s. iff two independent random walks started at any different states intersect with probability 1.* ◄

And how do we decide whether two random walks intersect a.s.? We will give a useful test in the transitive case. In fact, this will work for Markov chains that may not be reversible, so we should say what we mean by "transitive" Markov chain:

Definition 10.23. Let $p(\cdot, \cdot)$ be a transition kernel on a state space V. Suppose that there is a group Γ of permutations of V that acts transitively (that is, with a single orbit) and satisfies $p(\gamma x, \gamma y) = p(x, y)$ for all $\gamma \in \Gamma$ and $x, y \in \mathsf{V}$. Then we call the Markov chain with transition kernel $p(\cdot, \cdot)$ *transitive*.

We will use $\mathbf{P}_{x_0, y_0}$ and $\mathbf{E}_{x_0, y_0}$ to denote probability and expectation when independent Markov chains $\langle X_m \rangle$ and $\langle Y_n \rangle$ start at x_0 and y_0, respectively.

In the transitive case, we may use the Green function for our test:

Theorem 10.24. (Intersections of Transitive Markov Chains) *Let $p(\cdot, \cdot)$ be the transition kernel of a transient, transitive Markov chain on the countable state space V. Let X and Y be* *two independent copies of the Markov chain with initial states x_0 and y_0. Let o be a fixed element of* V. *If*

$$\sum_{z \in V} \mathscr{G}(o, z)^2 = \infty, \tag{10.17}$$

then $\mathbf{P}_{x_0, y_0}\big[|\{X_m\} \cap \{Y_n\}| = \infty\big] = \mathbf{P}_{x_0, y_0}\big[|\mathsf{LE}\langle X_m \rangle \cap \{Y_n\}| = \infty\big] = 1$, *whereas if (10.17) fails, then* $\mathbf{P}_{x_0, y_0}\big[|\{X_m\} \cap \{Y_n\}| < \infty\big] = 1$.

If we specialize still further, we obtain:

Corollary 10.25. (Intersections of Random Walks in Transitive Graphs) *Let G be an infinite, locally finite, vertex-transitive graph. Denote by V_n the number of vertices in G at distance at most n from a fixed vertex o.*

  (i) *If* $\sup_n V_n/n^4 = \infty$, *then two independent sample paths of simple random walk in G have only finitely many intersections a.s.*

  (ii) *Conversely, if* $\sup_n V_n/n^4 < \infty$, *then two independent sample paths of simple random walk in G intersect infinitely often a.s.*

These results tell us when the WSF has a single tree. How many trees are there otherwise? Usually there are infinitely many a.s., but there can be only finitely many:

▷ **Exercise 10.13.**
Join two copies of the usual nearest-neighbor graph of $\mathbb{Z}^3$ by an edge at their origins. How many trees does the free uniform spanning forest have? How many does the wired uniform spanning forest have?

To give the general answer, we use the following quantity: let $\alpha(w_1, \dots, w_K)$ be the probability that independent random walks started at $w_1, \dots, w_K$ have no pairwise intersections.

Theorem 10.26. (Constant Number of Trees in the WSF**)** *Let G be a connected network. The number of trees of the* WSF *is a.s.*

$$\sup\{K ; \exists w_1, \dots, w_K \quad \alpha(w_1, \dots, w_K) > 0\}. \tag{10.18}$$

Moreover, if the probability is 0 that two independent random walks from every (or some) vertex x intersect infinitely often, then the number of trees of the WSF *is a.s. infinite.*

Corollary 10.27. (WSF **in Nonamenable Networks)** *If (G, c) is a network such that $\Phi_E(G, c, \pi) > 0$ and $\sup_{x \in V} \pi(x) < \infty$, then the number of trees of the* WSF *is a.s. infinite.*

Theorem 10.26 tells us that in particular, on any network, the number of trees of the WSF is equal a.s. to a constant. The case of the free spanning forest (when it differs from the wired) is quite a puzzle. In particular, we do not know whether the number of components is deterministic or random:

Question 10.28. Let G be an infinite network. Is the number of trees of the FSF a.s. constant?

If Aut(G) has an infinite orbit, then the number is deterministic, since the number is invariant under automorphisms and the invariant σ-field is trivial by Corollary 10.19. Motivated by Theorem 7.5, we can ask whether, in this case, the number of trees is either 1 or ∞:

Question 10.29. (O. Häggström) Let G be a transitive network. By ergodicity, the number of trees of the FSF is a.s. constant. Is it 1 or ∞ a.s.?

See Theorem 10.53 for one case that is understood completely. For the unimodular case, Question 10.29 was answered positively and independently by Hutchcroft and Nachmias (2017) and Timár (2018).

Returning to the WSF, we deduce the following wonderful result of Pemantle (1991), which is truly stunning without an understanding of the approach using random walks:

Theorem 10.30. (Phase Transition in $\mathbb{Z}^d$) *The uniform spanning forest on $\mathbb{Z}^d$ has one tree a.s. for $d \leq 4$ and infinitely many trees a.s. for $d \geq 5$.* ◄

(Recall that by Corollary 10.9, FSF = WSF on $\mathbb{Z}^d$.)

We now prove all the preceding claims, though sometimes only in special cases. The special cases always include simple random walk on $\mathbb{Z}^d$. In particular, we prove Theorem 10.24 but not Theorem 10.21.

Nevertheless, we begin with a heuristic argument for Theorem 10.21. On the event that $X_m = Y_n$, the continuation paths $X' := \langle X_j \rangle_{j \geq m}$ and $Y' := \langle Y_k \rangle_{k \geq n}$ have the same distribution, whence the chance is at least $1/2$ that Y' intersects $L := \mathsf{LE}\langle X_0, \ldots, X_m \rangle$ at an earlier point than X' ever does, where "earlier" means in the clock of L. On this event, the earliest intersection point of Y' and L will remain in $\mathsf{LE}\langle X_j \rangle_{j \geq 0} \bigcap \{Y_k\}_{k \geq 0}$. The difficulty in making this heuristic precise lies in selecting a pair (m, n) such that $X_m = Y_n$, given that such pairs exist. The natural rules for selecting such a pair (for example, lexicographic ordering) affect the law of at least one of the continuation paths, and invalidate the argument above; R. Pemantle (private communication, 1996) showed that this holds for *all* selection rules. Our solution to this difficulty is based on applying a second-moment argument to a count of intersections. In the cases we will prove here (Theorem 10.24), we will show that there is a second-moment bound for intersections of X and Y. The general case, Theorem 10.21, also has a similar second-moment bound, as shown by Lyons, Peres, and Schramm (2003), but this is a little too long to prove here. We will then transfer the second-moment bound for intersections of X and Y to one for intersections of $\mathsf{LE}\langle X \rangle$ and Y.

Ultimately, the second-moment argument relies on the following widely used inequality. It allows one to deduce that a random variable has a reasonable chance to be large from knowing that its first moment is large compared to its second moment. We saw this inequality in Section 5.5.

The Paley–Zygmund Inequality (1932). *Let Z be a random variable with $\mathbf{E}[Z] \geq 0$ and $\mathbf{P}[Z = 0] \neq 1$. Let $0 < \epsilon < 1$. Then*

$$\mathbf{P}\big[Z \geq \epsilon\, \mathbf{E}[Z]\big] > (1 - \epsilon)^2 \frac{\mathbf{E}[Z]^2}{\mathbf{E}[Z^2]}.$$

Proof. Let $\mathscr{A}$ be the event that $Z \geq \epsilon\, \mathbf{E}[Z]$. The Cauchy–Schwarz inequality gives

$$\mathbf{E}[Z^2]\, \mathbf{P}(\mathscr{A}) = \mathbf{E}[Z^2]\, \mathbf{E}[\mathbf{1}_{\mathscr{A}}^2] \geq \mathbf{E}[Z\mathbf{1}_{\mathscr{A}}]^2 = \big(\mathbf{E}[Z] - \mathbf{E}[Z\mathbf{1}_{\mathscr{A}^c}]\big)^2 > \big(\mathbf{E}[Z] - \epsilon\, \mathbf{E}[Z]\big)^2. \quad ◄$$

We begin with a second-moment bound on intersections of two Markov chains that start at the same state. The calculation leading to (10.20) below follows Le Gall and Rosen (1991), Lemma 3.1. Denote $\mathscr{G}_N(o, x) := \sum_{m=0}^{N} \mathbf{P}_o[X_m = x]$. Let

$$I_N := \sum_{m=0}^{N} \sum_{n=0}^{N} \mathbf{1}_{[X_m = Y_n]} \tag{10.19}$$

be the number of intersections of X and Y by time N. We'll be interested in whether $\mathbf{E}[I_N] \to \infty$, and the Paley–Zygmund inequality will be used to show that $I_N \to \infty$ with reasonable probability when $\mathbf{E}[I_N] \to \infty$. For that, we'll need that the first moment of I_N is large compared to its second moment, which is the purpose of the following lemma.

Lemma 10.31. *Let $p(\cdot, \cdot)$ be the transition kernel of a transitive Markov chain on a countable state space V. Start both Markov chains X, Y at $o \in \mathsf{V}$. Then*

$$\frac{\mathbf{E}[I_N]^2}{\mathbf{E}[I_N^2]} \geq \frac{1}{4} . \tag{10.20}$$

Proof. By transitivity,

$$\sum_{w \in \mathsf{V}} \mathscr{G}_N(z, w)^2 = \sum_{w \in \mathsf{V}} \mathscr{G}_N(o, w)^2 \tag{10.21}$$

for all $z \in \mathsf{V}$. We have

$$I_N = \sum_{z \in \mathsf{V}} \sum_{m,n=0}^{N} \mathbf{1}_{[X_m = z = Y_n]} .$$

Thus,

$$\begin{aligned} \mathbf{E}[I_N] &= \sum_{z \in \mathsf{V}} \sum_{m=0}^{N} \sum_{n=0}^{N} \mathbf{P}[X_m = z = Y_n] \\ &= \sum_{z \in \mathsf{V}} \sum_{m=0}^{N} \mathbf{P}[X_m = z] \cdot \sum_{n=0}^{N} \mathbf{P}[Y_n = z] \\ &= \sum_{z \in \mathsf{V}} \mathscr{G}_N(o, z)^2 . \end{aligned} \tag{10.22}$$

To estimate the second moment of I_N, observe that

$$\begin{aligned} \sum_{m,j=0}^{N} \mathbf{P}[X_m = z, X_j = w] &= \sum_{m=0}^{N} \sum_{j=m}^{N} \mathbf{P}[X_m = z] \, \mathbf{P}[X_j = w \mid X_m = z] \\ &\quad + \sum_{j=0}^{N} \sum_{m=j+1}^{N} \mathbf{P}[X_j = w] \, \mathbf{P}[X_m = z \mid X_j = w] \\ &\leq \mathscr{G}_N(o, z) \, \mathscr{G}_N(z, w) + \mathscr{G}_N(o, w) \, \mathscr{G}_N(w, z) . \end{aligned}$$

Therefore

$$
\begin{aligned}
\mathbf{E}[I_N^2] &= \sum_{z,w\in\mathsf{V}} \sum_{m,n=0}^{N} \sum_{j,k=0}^{N} \mathbf{P}[X_m = z = Y_n, X_j = w = Y_k] \\
&= \sum_{z,w\in\mathsf{V}} \sum_{m,j=0}^{N} \mathbf{P}[X_m = z, X_j = w] \cdot \sum_{n,k=0}^{N} \mathbf{P}[Y_n = z, Y_k = w] \\
&\le \sum_{z,w\in\mathsf{V}} \big[\mathscr{G}_N(o,z)\,\mathscr{G}_N(z,w) + \mathscr{G}_N(o,w)\,\mathscr{G}_N(w,z)\big]^2 \\
&\le \sum_{z,w\in\mathsf{V}} 2\big[\mathscr{G}_N(o,z)^2\,\mathscr{G}_N(z,w)^2 + \mathscr{G}_N(o,w)^2\,\mathscr{G}_N(w,z)^2\big] \\
&= 4 \sum_{z,w\in\mathsf{V}} \mathscr{G}_N(o,z)^2\,\mathscr{G}_N(z,w)^2 \,.
\end{aligned}
$$

Summing first over w and using (10.21), then (10.22), we deduce that

$$
\mathbf{E}[I_N^2] \le 4\Big(\sum_{z\in\mathsf{V}} \mathscr{G}_N(o,z)^2\Big)^2 = 4\,\mathbf{E}[I_N]^2\,. \qquad \blacktriangleleft
$$

We now extend this to any pair of starting points when (10.17) holds.

Corollary 10.32. *Let $p(\cdot,\cdot)$ be the transition kernel of a transitive Markov chain on a countable state space V. Start the Markov chains X, Y at o, o'. If (10.17) holds, then*

$$
\liminf_{N\to\infty} \frac{\mathbf{E}_{o,o'}[I_N]^2}{\mathbf{E}_{o,o'}[I_N^2]} \ge \frac{1}{4}\,. \tag{10.23}
$$

Proof. Let $a_N := \mathbf{E}_{o,o}[I_N]$. As in the proof of Lemma 10.31, we have for any pair o, o',

$$
\begin{aligned}
\mathbf{E}_{o,o'}[I_N^2] &= \sum_{z,w\in\mathsf{V}} \sum_{m,j=0}^{N} \mathbf{P}_o[X_m = z, X_j = w] \cdot \sum_{n,k=0}^{N} \mathbf{P}_{o'}[Y_n = z, Y_k = w] \\
&\le \sum_{z,w\in\mathsf{V}} \big[\mathscr{G}_N(o,z)\,\mathscr{G}_N(z,w) + \mathscr{G}_N(o,w)\,\mathscr{G}_N(w,z)\big] \\
&\qquad\qquad \cdot \big[\mathscr{G}_N(o',z)\,\mathscr{G}_N(z,w) + \mathscr{G}_N(o',w)\,\mathscr{G}_N(w,z)\big] \\
&\le \sum_{z,w\in\mathsf{V}} \big[\mathscr{G}_N(o,z)^2\,\mathscr{G}_N(z,w)^2 + \mathscr{G}_N(o,w)^2\,\mathscr{G}_N(w,z)^2 \\
&\qquad\qquad + \mathscr{G}_N(o',z)^2\,\mathscr{G}_N(z,w)^2 + \mathscr{G}_N(o',w)^2\,\mathscr{G}_N(w,z)^2\big] \\
&= 4a_N^2 \tag{10.24}
\end{aligned}
$$

by the elementary inequality

$$
(a+b)(c+d) \le a^2 + b^2 + c^2 + d^2\,.
$$

We also have

$$\mathbf{E}_{o,o'}[I_N] = \sum_{m,n \leq N} \sum_{z \in V} p_m(o, z) p_n(o', z) = \sum_{z \in V} \mathcal{G}_N(o, z) \mathcal{G}_N(o', z) \leq a_N \qquad (10.25)$$

by the Cauchy–Schwarz inequality, transitivity, and (10.22).

By (10.17), we have $a_N \to \infty$. Let $u_N(x, y) := \mathbf{E}_{x,y}[I_N]/a_N$. Thus, $u_N(x, x) = 1$ for all x and $u_N(x, y) \leq 1$ for all x, y. Fix r such that $p_r(o, o') > 0$. Now

$$
\begin{aligned}
a_{r+N} = a_{r+N} u_{r+N}(o, o) &= \sum_{m,n \leq r+N} \mathbf{P}_{o,o}[X_m = Y_n] \\
&= \sum_{m,n \leq N} \mathbf{P}_{o,o}[X_m = Y_{r+n}] \\
&\quad + \sum_{m \leq r+N} \sum_{n < r} \mathbf{P}_{o,o}[X_m = Y_n] \qquad (10.26) \\
&\quad + \sum_{1 \leq m \leq r} \sum_{n \leq N} \mathbf{P}_{o,o}[X_{N+m} = Y_{r+n}].
\end{aligned}
$$

The Markov property and (10.25) show that the first of the three latter sums is equal to

$$\mathbf{E}_{o,o}[a_N u_N(o, Y_r)] \leq a_N[p_r(o, o') u_N(o, o') + 1 - p_r(o, o')].$$

The second sum in (10.26) is

$$\sum_{z \in V} \mathcal{G}_{r+N}(o, z) \mathcal{G}_{r-1}(o, z) \leq \sqrt{a_{r+N} a_{r-1}}$$

by the Cauchy–Schwarz inequality. Similarly, by conditioning on (X_{N+1}, Y_r) and using transitivity and the Cauchy–Schwarz inequality, we see that the third sum in (10.26) is

$$\mathbf{E}_{o,o}\left[\sum_{z \in V} \mathcal{G}_{r-1}(X_{N+1}, z) \mathcal{G}_N(Y_r, z)\right] \leq \sqrt{a_{r-1} a_N}.$$

Substituting these bounds in (10.26) yields

$$
\begin{aligned}
1 &\leq (a_N/a_{r+N})[p_r(o, o') u_N(o, o') + 1 - p_r(o, o')] + \sqrt{a_{r-1}/a_{r+N}} + \sqrt{a_{r-1} a_N}/a_{r+N} \\
&\leq p_r(o, o') u_N(o, o') + 1 - p_r(o, o') + o(1)
\end{aligned}
$$

as $N \to \infty$. This implies that $\liminf_{N \to \infty} u_N(o, o') \geq 1$. Combining this with (10.24) gives the result. ◄

We now give the crucial step that converts intersections of random walks to intersections when one of the paths is loop erased.

Lemma 10.33. *Let $p(\cdot, \cdot)$ be a transition kernel on a countable state space V that gives a transient, transitive Markov chain. Fix $k \geq 0$ and a path $\langle x_j \rangle_{j=-k}^0$ in V. Let $\langle X_m \rangle_{m \geq 0}$ and $\langle Y_n \rangle_{n \geq 0}$ be independent Markov chains on V with initial states x_0 and y_0. Set $X_j := x_j$ for $-k \leq j \leq -1$. If (10.17) holds, then the probability that $|\mathsf{LE}\langle X_m \rangle_{m \geq -k} \cap \{Y_n\}| = \infty$ is at least $1/16$.*

Proof. Denote

$$\langle L_j^m \rangle_{j=0}^{J(m)} := \mathsf{LE}\langle X_{-k}, X_{-k+1}, \ldots, X_m \rangle \,.$$

When it happens that $X_m = Y_n$, we want to see which of the continuations of X and Y intersects $\langle L_j^m \rangle$ earlier. On the event $[X_m = Y_n]$, define

$$j(m, n) := \min \{ j \geq 0 \,;\, L_j^m \in \{X_m, X_{m+1}, X_{m+2}, \ldots\} \} \tag{10.27}$$

$$i(m, n) := \min \{ i \geq 0 \,;\, L_i^m \in \{Y_n, Y_{n+1}, Y_{n+2}, \ldots\} \} \,. \tag{10.28}$$

Note that the sets on the right-hand sides of (10.27) and (10.28) both contain $J(m)$ if $X_m = Y_n$. Define $j(m, n) := i(m, n) := 0$ on the event $[X_m \neq Y_n]$. When the continuation of Y has an earlier intersection than the continuation of X does, then that intersection will be an intersection of the loop erasure of X with Y. Thus, let $\chi(m, n) := \mathbf{1}_{[i(m,n) \leq j(m,n)]}$. Given $[X_m = Y_n = z]$, the continuations $\langle X_m, X_{m+1}, X_{m+2}, \ldots \rangle$ and $\langle Y_n, Y_{n+1}, Y_{n+2}, \ldots \rangle$ are exchangeable with each other, so for every $z \in \mathsf{V}$,

$$\mathbf{E}[\chi(m, n) \mid X_m = Y_n = z] = \mathbf{P}[i(m, n) \leq j(m, n) \mid X_m = Y_n = z] \geq \frac{1}{2} \,. \tag{10.29}$$

As we said, if $X_m = Y_n$ and $i(m, n) \leq j(m, n)$, then $L_{i(m,n)}^m$ is in $\mathsf{LE}\langle X_r \rangle_{r=-k}^\infty \cap \{Y_\ell\}_{\ell=0}^\infty$.

Consider the random variables

$$\Upsilon_N := \sum_{m=0}^N \sum_{n=0}^N \mathbf{1}_{[X_m = Y_n]} \chi(m, n)$$

for $N \geq 1$. Obviously $\Upsilon_N \leq I_N$ everywhere, so

$$\frac{1}{\mathbf{E}[\Upsilon_N^2]} \geq \frac{1}{\mathbf{E}[I_N^2]} \,. \tag{10.30}$$

On the other hand, by conditioning on X_m and Y_n and by applying (10.29), we see that

$$\mathbf{E}[\Upsilon_N] = \sum_{m=0}^N \sum_{n=0}^N \mathbf{E}\Big[\mathbf{1}_{[X_m = Y_n]} \mathbf{E}[\chi(m, n) \mid X_m, Y_n] \Big] \geq \frac{1}{2} \mathbf{E}[I_N] \,. \tag{10.31}$$

By the Paley–Zygmund inequality, we have, for every $\epsilon > 0$,

$$\mathbf{P}[\Upsilon_N \geq \epsilon \mathbf{E}[\Upsilon_N]] \geq (1 - \epsilon)^2 \frac{\mathbf{E}[\Upsilon_N]^2}{\mathbf{E}[\Upsilon_N^2]} \,.$$

By (10.30), (10.31), and Corollary 10.32, we conclude that for every $\epsilon > 0$, we have for large enough N that

$$\mathbf{P}[\Upsilon_N \geq \epsilon \mathbf{E}[\Upsilon_N]] \geq (1 - \epsilon)^2 \frac{\mathbf{E}[I_N]^2}{4 \mathbf{E}[I_N^2]} \geq \frac{(1 - \epsilon)^2}{16} - \epsilon \,.$$

Since $\mathbf{E}[\Upsilon_N] \to \infty$ by (10.31) and (10.17), it follows that $\Upsilon_N \to \infty$ with probability at least $1/16$. On the event that $\Upsilon_N \to \infty$, we have $|\mathsf{LE}\langle X_m \rangle_{m \geq -k} \cap \{Y_n\}| = \infty$ by the observation following (10.29) and by transience. This finishes the proof. ◀

Proof of Theorem 10.24. Suppose first that (10.17) holds. Denote by Λ the event of interest, $|\mathsf{LE}\langle X_m \rangle \cap \{Y_n\}| = \infty$. We will show that $\mathbf{P}(\Lambda) = 1$. By virtue of Lévy's zero-one law, we have that $\lim_{n \to \infty} \mathbf{P}_{x_0, y_0}(\Lambda \mid X_1, \ldots, X_n, Y_1, \ldots, Y_n) = \mathbf{1}_\Lambda$ a.s. On the other hand,

$$\mathbf{P}_{x_0, y_0}(\Lambda \mid X_1, \ldots, X_n, Y_1, \ldots, Y_n) = \mathbf{P}_{x_0, Y_n}(\Lambda \mid X_1, \ldots, X_n) \geq 1/16$$

by the Markov property and by Lemma 10.33. Thus, $\mathbf{1}_\Lambda \geq 1/16$ a.s., which means that $\mathbf{P}_{x_0, y_0}(\Lambda) = 1$, as desired.

Now suppose that (10.17) fails. Then the expected number of pairs (m, n) such that $X_m = Y_n$ is (by the monotone convergence theorem) $\lim_N \mathbf{E}[I_N] = \sum_z \mathscr{G}(o, z)^2 < \infty$. ◀

▷ **Exercise 10.14.**
(Parseval's Identity) Suppose that $F \in L^1(\mathbb{T}^d)$ and that $f(x) := \int_{\mathbb{T}^d} F(\alpha) e^{2\pi i x \cdot \alpha} \, d\alpha$ for $x \in \mathbb{Z}^d$. Show that $\int_{\mathbb{T}^d} |F(\alpha)|^2 \, d\alpha = \sum_{\mathbb{Z}^d} |f(x)|^2$. *Hint:* Prove that the functions $G_x(\alpha) := e^{-2\pi i x \cdot \alpha}$ for $x \in \mathbb{Z}^d$ form an orthonormal basis of $L^2(\mathbb{T}^d)$.

Proof of Corollary 10.25. For $\mathbb{Z}^d$, the result is easy: We need treat only the transient case. By Proposition 2.1, (10.17) is equivalent to the voltage function v of Proposition 10.16 not being square summable. By Exercise 10.14, this in turn is equivalent to $1/\varphi \notin L^2(\mathbb{T}^d)$, where φ is defined in (10.8). By (10.9), this holds iff $d \leq 4$.

The general statement requires certain facts that are beyond the scope of this book but otherwise is not hard. For independent simple random walks, reversibility and regularity of G imply that

$$\sum_z \mathscr{G}(o, z)^2 = \lim_{n \to \infty} \mathbf{E}_{o,o}[I_N] = \sum_{m=0}^\infty \sum_{n=0}^\infty \mathbf{P}_{o,o}[X_m = Y_n]$$

$$= \sum_{m=0}^\infty \sum_{n=0}^\infty \mathbf{P}_o[X_{m+n} = o] = \sum_{n=0}^\infty (n+1) \mathbf{P}_o[X_n = o]. \qquad (10.32)$$

In case (i), the assumption that $\sup_n V_n/n^4 = \infty$ implies that $V_n \geq cn^5$ for some $c > 0$ and all n: see Kleiner (2010). By Corollary 6.32, this yields $\mathbf{P}_x[X_n = x] \leq Cn^{-5/2}$. Thus the sum in (10.32) converges.

In case (ii), combining the results (14.5), (14.12), and (14.19) in Woess (2000), we infer that the assumption $V_n = O(n^4)$ implies that $P_x[X_{2n} = x] \geq cn^{-2}$ for some $c > 0$ and all $n \geq 1$. Thus the series (10.32) diverges.

Hence, both assertions follow from Theorem 10.24. ◀

Proof of Theorem 10.26. Of course, the recurrent case is a consequence of our work in Chapter 4, so we restrict ourselves to the transient case.

Let $\langle X_n(u); \ n \geq 0, u \in \mathsf{V} \rangle$ be a collection of independent random walks, one starting at each $u \in \mathsf{V}$. Denote the event that $X_n(w_i) \neq X_m(w_k)$ for all $i \neq k$ and $n, m \geq j$ by $\mathscr{B}_j(w_1, \ldots, w_K)$. Thus, $\alpha(w_1, \ldots, w_K) = \mathbf{P}[\mathscr{B}_0(w_1, \ldots, w_K)]$ and $\mathscr{B}(w_1, \ldots, w_K) :=$

$\bigcup_j \mathscr{B}_j(w_1, \ldots, w_K)$ is the event that there are only finitely many pairwise intersections among the random walks starting at $w_1, \ldots, w_K$. For every $j \geq 0$, we have

$$\liminf_{n \to \infty} \alpha\big(X_n(w_1), \ldots, X_n(w_K)\big) \geq \lim_{n \to \infty} \mathbf{P}\big[\mathscr{B}_j(w_1, \ldots, w_K) \,\big|\, \langle X_m(w_i)\,;\; m \leq n,\, i \leq K\rangle\big]$$

$$= \mathbf{1}_{\mathscr{B}_j(w_1, \ldots, w_K)} \text{ a.s.}$$

by Lévy's zero-one law. It follows that

$$\liminf_{n \to \infty} \alpha\big(X_n(w_1), \ldots, X_n(w_K)\big) \geq \mathbf{1}_{\mathscr{B}(w_1, \ldots, w_K)} \text{ a.s.} \tag{10.33}$$

First suppose that $\alpha(w_1, \ldots, w_K) > 0$ for some $w_1, \ldots, w_K$. Then by (10.33), for every $\epsilon > 0$, there is an $n \in \mathbb{N}$ such that $\alpha\big(X_n(w_1), \ldots, X_n(w_K)\big) > 1 - \epsilon$ with positive probability. In particular, there are $w_1', \ldots, w_K'$ such that $\alpha(w_1', \ldots, w_K') > 1 - \epsilon$. Using Wilson's method rooted at infinity starting with the vertices $w_1', \ldots, w_K'$, this implies that with probability greater than $1 - \epsilon$, the number of trees for WSF is at least K. As $\epsilon > 0$ was arbitrary, this implies that the number of trees is WSF-a.s. at least (10.18).

For the converse, suppose that $\alpha(w_1, \ldots, w_K) = 0$ for all $w_1, \ldots, w_K$. By (10.33), we find that there exist a.s. $i \neq j$ with infinitely many intersections between $\langle X_n(w_i)\rangle$ and $\langle X_n(w_j)\rangle$, whence also between $\text{LE}\langle X_n(w_i)\rangle$ and $\langle X_n(w_j)\rangle$ by Theorem 10.21 in general and by Theorem 10.24 in the transitive case. By Wilson's method rooted at infinity, the probability that all $w_1, \ldots, w_K$ belong to different trees is 0. Since this holds for all $w_1, \ldots, w_K$, it follows that the number of trees is WSF-a.s. at most (10.18).

Moreover, if the probability is zero that two independent random walks X^1, X^2 intersect i.o. starting at some $w \in \mathsf{V}$, then $\lim_{n \to \infty} \alpha(X_n^1, X_n^2) = 1$ a.s. by (10.33). Therefore

$$\lim_{n \to \infty} \alpha\big(X_n^1, \ldots, X_n^k\big) = 1$$

a.s. for any independent random walks $X^1, \ldots, X^k$. This implies that the number of components of WSF is a.s. infinite. ◄

Proof of Corollary 10.27. By Theorem 6.7, the spectral radius ρ of random walk in G is strictly less than 1. Let d be an upper bound for $\pi(x)$. Fix $o \in \mathsf{V}$ and consider independent random walks X and Y starting at o. Then by reversibility and (6.13), $\mathbf{P}[X_m = Y_n] \leq d\,\mathbf{P}[X_{m+n} = o] \leq d\rho^{m+n}$. Summing on m and n gives that the expected number of pairs of times where the two random walks intersect is finite, whence the number of intersections is finite a.s. Thus, the result follows from the last part of Theorem 10.26. ◄

▷ **Exercise 10.15.**
Let (T, c) be a network on a tree. Show that $\text{WSF}[\mathfrak{F} \neq T] = 1$ iff there is some edge $e \in T$ such that both components of $T \setminus e$ are transient iff $\text{WSF}[\mathfrak{F} \neq T] > 0$.

10.6 The Size of the Trees

In the preceding section, we discovered how many trees there are in the wired spanning forest. Now we ask how big these trees are. Of course, there is more than one way to define "big." The most obvious probabilistic notion of "big" is "transient." In this sense, all the trees are small:

Theorem 10.34. (Morris, 2003) *Let (G, c) be a network. For WSF-a.e. $\mathfrak{F}$, all trees T in $\mathfrak{F}$ have the property that the wired spanning forest of (T, c) equals T. In particular, if $c(\cdot)$ is bounded, then (T, c) is recurrent.*

To prove Morris's theorem, we will use the following observation:

Lemma 10.35. *Let (G, c) be a transient network and $\langle e_n \rangle$ an enumeration of its edges. Write G_n for the network obtained by contracting the edges $e_1, e_2, \ldots, e_n$. For any vertex x, we have $\lim_{n \to \infty} \mathscr{R}(x \leftrightarrow \infty; G_n) = 0$.*

Proof. Let θ be a unit flow in G from x to ∞ of finite energy. The restriction of θ to $\langle e_k ; \; k > n \rangle$ is a unit flow in G_n from x to infinity with energy tending to 0 as $n \to \infty$. Since the energy of any unit flow is an upper bound on the effective resistance by Thomson's principle, the result follows. ◀

We will also use the result of the following exercise:

▷ **Exercise 10.16.**
Let (T, c) be a transient network on a tree with bounded conductances $c(\cdot)$. Show that there is some $e \in T$ such that both components of $T \backslash e$ are transient (with respect to c).

Proof of Theorem 10.34. Consider any edge e of G. Let $\mathscr{A}_e$ be the event that both endpoints of e lie in transient components of $\mathfrak{F} \backslash e$ (with respect to the conductances $c(\cdot)$). We'll prove that $\mathsf{WSF}[e \in \mathfrak{F}, \mathscr{A}_e] = 0$. The first part of the theorem then follows from Exercise 10.7, while the last part follows from Exercise 10.16.

Enumerate the edges of $G \backslash e$ as $\langle e_n \rangle$. Let $\mathscr{F}_n$ be the σ-field generated by the events $[e_k \in \mathfrak{F}]$ for $k < n$, where $n \leq \infty$. Recall that conditioning on an edge being in the spanning forest is equivalent to contracting the edge from the original network, whereas conditioning on the edge being absent is equivalent to deleting the edge from the original network. Thus, for $n < \infty$, let G_n be the random network obtained from G by contracting e_k when $e_k \in \mathfrak{F}$ and deleting e_k otherwise, where we do this for all $k < n$. Let i_n be the wired unit current flow in G_n from the tail to the head of e. Then $\mathsf{WSF}[e \in \mathfrak{F} \mid \mathscr{F}_n] = i_n(e)$ by (10.3). This is $c(e)$ times the wired effective resistance between e^- and e^+ in G_n. By Lemma 10.35 and Exercise 9.29, this tends to 0 on the event $\mathscr{A}_e$. By Lévy's martingale convergence theorem, we obtain $\mathsf{WSF}[e \in \mathfrak{F} \mid \mathscr{F}_\infty] \mathbf{1}_{\mathscr{A}_e} = 0$ a.s. Taking the expectation of this equation gives $\mathsf{WSF}[e \in \mathfrak{F}, \mathscr{A}_e] = 0$, as desired. ◀

We now look at another notion of size of trees. Call an infinite path in a tree that starts at any vertex and does not backtrack a **ray**. Call two rays **equivalent** if they have infinitely many vertices in common. An equivalence class of rays is called an **end**. How many ends do the trees of a uniform spanning forest have?

Let's begin by thinking about the case of the wired uniform spanning forest on a regular tree of degree $d + 1$. Choose a vertex, o. Begin Wilson's method rooted at infinity from o. We obtain a ray ξ from o to start our forest. Now o has d neighbors not in ξ, $x_1, \ldots, x_d$. By beginning random walks at each of them in turn, we see that the events $\mathscr{A}_i := [x_i$ connected to $o]$ are independent. Furthermore, the resistance from x_i to ∞ in the descendant subtree of x_i (we think of o as the parent of x_i) is $1/d + 1/d^2 + 1/d^3 + \cdots = 1/(d-1)$, whence the probability that random walk started at x_i ever hits o is $1/d$. This is the probability of $\mathscr{A}_i$. On the event $\mathscr{A}_i$, we add only the edge $[o, x_i]$ to the forest, and then we repeat the analysis from x_i. Thus, the tree containing o includes, apart from the ray ξ, a critical Galton–Watson tree with binomial offspring distribution $\mathrm{Bin}(d, 1/d)$. In addition, each vertex on ξ has another random subtree attached to it; its first generation has distribution $\mathrm{Bin}(d - 1, 1/d)$, but subsequent generations yield Galton–Watson trees with distribution $\mathrm{Bin}(d, 1/d)$. In particular, a.s. every tree added to ξ is finite. This means that the tree containing o has only one end, the equivalence class of ξ. This analysis is easily extended to form a complete description of the entire wired spanning forest. In fact, if we work on the d-ary tree instead of the $(d + 1)$-regular tree, the description is slightly simpler: each tree in the forest is a size-biased Galton–Watson tree grown from its lowest vertex, generated from the offspring distribution $\mathrm{Bin}(d, 1/d)$ (see Section 12.1). The resulting description is due to Häggström (1998), whose work predated Wilson's algorithm.

Now consider the WSF on general graphs. If we begin Wilson's algorithm at a vertex o in a graph, it immediately generates one end of the tree containing o. For this tree to have more than one end, however, we need a succession of "coincidences" to occur, building up other ends by gradually adding on finite pieces. This is possible (see Exercises 10.17, 10.48, or 10.49), but it suggests that "usually," the wired spanning forest has trees with only one end each. Indeed, we will show that this is very often the case.

▷ **Exercise 10.17.**
Give a transient graph such that the wired uniform spanning forest has a single tree with more than one end.

First, we consider the planar recurrent case, which has an amazingly simple analysis:

Theorem 10.36. *Suppose that G is a proper, simple, plane, recurrent network and $G^\dagger$ its plane dual. Assume that $G^\dagger$ is locally finite and recurrent. Then the uniform spanning tree on G has only one end a.s.*

Proof. Because both networks are recurrent, their free and wired spanning forests coincide and are a single tree a.s. We observed in Section 10.3 that the uniform spanning tree T of G is "dual" to that of $G^\dagger$. If T had at least two ends, the bi-infinite path in T joining two of them would separate $T^\times$ into at least two trees. Since we know that $T^\times$ is a single tree, this is impossible. ◄

In particular, we have verified the claim in Section 4.3 that the maze in $\mathbb{Z}^2$ is connected and has exactly one way to get to infinity from any square. If we want to get a somewhat more global picture of the uniform spanning tree in $\mathbb{Z}^2$ incorporating the fact that it has only one end, then we can look at the uniform spanning tree in a wired piece of the square grid in a similar fashion to Figure 4.5. That is, we can plot the distance to the outer boundary vertex: see Figure 10.2. Near the middle, this should be similar to what we would see in the infinite grid, where "distance" is replaced by "horodistance" to the unique end. Horodistance is defined only up to an additive constant; if we fix the horodistance to be 0, say, at the origin, then all other horodistances $h(x)$ are determined by $h(x) := d_T(x, x_n) - d_T(\mathbf{0}, x_n)$ for all large n, where d_T indicates distance in the uniform spanning tree T and $\langle x_n \rangle$ is a ray that represents the unique end of T.

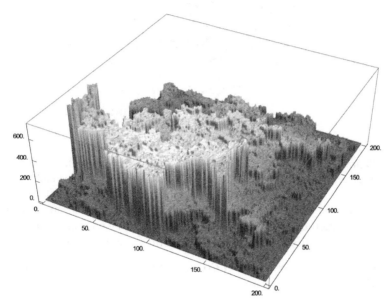

color plate **Figure 10.2.** The distances to the outer boundary in a uniform spanning tree of a wired 200×200 square grid.

Similar reasoning shows the following:

Proposition 10.37. *Suppose that G is a proper, simple, plane network and $G^\dagger$ its plane dual. Assume that $G^\dagger$ is locally finite. If each tree of the WSF of G has only one end a.s., then the FSF of $G^\dagger$ has only one tree a.s. If, in addition, the WSF of G has infinitely many trees a.s., then the tree of the FSF of $G^\dagger$ has infinitely many ends a.s.* ◀

This is illustrated for the Cayley graph of Figure 6.1 and its dual in Figure 10.3.

Now we go beyond the planar cases. By Proposition 10.10 and Theorem 8.19, it follows that if G is a transient, transitive, unimodular network, almost surely each tree of the wired spanning forest has one or two ends. It it one or two? This is less easy to decide. The answer is that each tree of the WSF has only one end a.s. in every quasi-transitive, transient network as well as in a host of other natural networks. The first result of this kind was due to Pemantle (1991), who proved it for $\mathbb{Z}^d$ with $d = 3, 4$ and also showed that there are at most two ends per tree for $d \geq 5$. BLPS (2001) extended and completed this for all unimodular, transitive networks, showing that each tree has only one end. This was then extended to

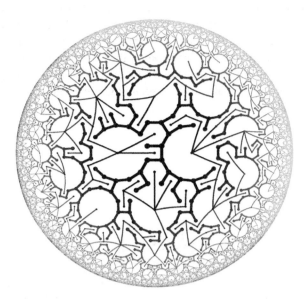

Figure 10.3. The FSF of a Cayley graph in the hyperbolic disc, which
is a tree, and its thinner dual WSF, each of whose trees has one end.

all quasi-transitive, transient networks and more by Lyons, Morris, and Schramm (2008),
who found a simpler method of proof that worked in greater generality. This is the proof we
present here. We prove it first for the case of $\mathbb{Z}^d$ before giving more general results.

Lemma 10.38. *Let $\mathscr{A}$ be an event of positive probability in a probability space and $\mathscr{F}$ be a
σ-field. Then $\mathbf{P}(\mathscr{A} \mid \mathscr{F}) > 0$ a.s. given $\mathscr{A}$.*

Proof. Let $\mathscr{B}$ be the event where $\mathbf{P}(\mathscr{A} \mid \mathscr{F}) = 0$. We want to show that $\mathbf{P}(\mathscr{B} \cap \mathscr{A}) = 0$. Since
$\mathscr{B} \in \mathscr{F}$, we have by definition of conditional probability that

$$\mathbf{P}(\mathscr{B} \cap \mathscr{A}) = \mathbf{E}\big[\mathbf{1}_{\mathscr{B}}\,\mathbf{P}(\mathscr{A} \mid \mathscr{F})\big] = 0\,. \qquad \blacktriangleleft$$

Lemma 10.39. *Let $B(n)$ be a box in $\mathbb{Z}^d$ for some $d \geq 3$ whose sides are parallel to the axes
and of length n. Then*

$$\lim_{N \to \infty} \inf\big\{\mathscr{C}\big(K \leftrightarrow \infty; \mathbb{Z}^d \backslash B(n)\big)\,;\; K \subset \mathbb{Z}^d \backslash B(n),\; |K| = N,\; n \geq 0\big\} = \infty\,.$$

Proof. Since the effective conductance is monotone increasing in K, it suffices to bound it
from below when all points in K are at pairwise distance at least N from each other and lie
on one side of a hyperplane that includes a face of $B(n)$. Since the effective conductance is
also monotone increasing in the rest of the network (Rayleigh's monotonicity principle), it
suffices to bound from below the conductance from K to ∞ in $G := \mathbb{N} \,\square\, \mathbb{Z}^{d-1}$. That G is
transient follows from the proof of Pólya's theorem, or can be deduced from Pólya's theorem
by reflection. Thus, there exists a unit flow θ of finite energy from the origin to ∞ in G. Write
θ_x for the image of θ under the translation of the origin to $x \in G$. Then $\theta_K := \sum_{x \in K} \theta_x / |K|$

is a unit flow between K and ∞, so it suffices to bound its energy from above. We claim that the inner product (θ_x, θ_y) is small when x and y are far from each other. Indeed, given $\epsilon > 0$, choose a finite set $F \subset E(G)$ such that $\sum_{e \notin F} \theta(e)^2 < \epsilon^2$. Write $\theta_x = \theta_x^{(1)} + \theta_x^{(2)}$, where $\theta_x^{(1)} := \theta_x \restriction (F + x)$ and $\theta_x^{(2)} := \theta_x \restriction (F + x)^c$. Then if the distance between x and y is so large that $(F + x) \cap (F + y) = \varnothing$, we have

$$(\theta_x, \theta_y) = (\theta_x^{(1)}, \theta_y^{(2)}) + (\theta_x^{(2)}, \theta_y^{(1)}) + (\theta_x^{(2)}, \theta_y^{(2)}) \leq 2\epsilon \mathcal{E}(\theta)^{1/2} + \epsilon^2$$

by the Cauchy–Schwarz inequality. Therefore, we have

$$\mathcal{E}(\theta_K) = \sum_{x \in K} \mathcal{E}(\theta_x / |K|) + \sum_{x \neq y \in K} (\theta_x, \theta_y)/|K|^2 = \mathcal{E}(\theta)/|K| + o(1) = o(1),$$

as desired. ◄

We need one more simple lemma before proving that the uniform spanning forest in $\mathbb{Z}^d$ has one end in each tree a.s. for $d \geq 2$.

Lemma 10.40. *Let G be a network and $F \subset E(G)$. For $e \in F$, let us abbreviate $r_e := \mathscr{R}^{\mathrm{W}}(e^- \leftrightarrow e^+; G \backslash F)$. Then*

$$\mathsf{WSF}[F \cap \mathfrak{F} = \varnothing] \geq \prod_{e \in F} \frac{1}{1 + c(e)r_e}.$$

Proof. It suffices to prove the analogous statement for finite networks, so we write T for the random spanning tree rather than $\mathfrak{F}$. For $e \in F$, we have

$$\mathbf{P}[e \in T] = c(e)\mathscr{R}(e^- \leftrightarrow e^+; G) = \frac{c(e)}{c(e) + \mathscr{C}(e^- \leftrightarrow e^+; G \backslash e)} \leq \frac{c(e)}{c(e) + \mathscr{C}(e^- \leftrightarrow e^+; G \backslash F)}$$

by Kirchhoff's effective-resistance formula and Rayleigh's monotonicity principle. Thus,

$$\mathbf{P}[e \notin T] \geq \frac{1}{1 + c(e)\mathscr{R}(e^- \leftrightarrow e^+; G \backslash F)}.$$

If we order F and use this bound one at a time, conditioning each time that the prior edges of F are not in T and deleting them from G, we get the desired bound. ◄

Recall that the ***inner vertex boundary*** of a set K is

$$\partial_{\mathsf{V}}^{\mathrm{in}} K := \{x \in K ; \exists y \notin K \ y \sim x\}.$$

Theorem 10.41. (One End for Trees in $\mathbb{Z}^d$) *In $\mathbb{Z}^d$ with $d \geq 2$, every tree in the uniform spanning forest has only one end a.s.*

Proof. The case $d = 2$ is part of Theorem 10.36, so assume that $d \geq 3$ and, thus, that the graph is transient by Pólya's theorem. Let $\mathscr{A}_e$ be the event that $\mathfrak{F} \backslash e$ has a finite component.

The way we will show that every tree in $\mathfrak{F}$ has one end a.s. is to show that every edge $e \in \mathsf{E}(G)$ has the property that $\mathsf{WSF}[\mathscr{A}_e \mid e \in \mathfrak{F}] = 1$.

Fix e. Let $\langle G_n \rangle$ be an exhaustion by boxes that contain e. Let $\mathscr{F}_n$ be the σ-field generated by the events $[f \in \mathfrak{F}]$ for $f \in \mathsf{E}(G_n) \setminus \{e\}$. Let G'_n be $\mathbb{Z}^d$ after we delete the edges of $G_n \setminus (\mathfrak{F} \cup \{e\})$ and contract each edge of $G_n \cap (\mathfrak{F} \setminus \{e\})$. Note that $e \in G'_n$. As in the proof of Theorem 10.34,

$$\mathsf{WSF}[e \in \mathfrak{F} \mid \mathscr{F}_n] = \mathscr{R}^{\mathrm{W}}(e^- \leftrightarrow e^+; G'_n) = \frac{1}{1 + \mathscr{C}^{\mathrm{W}}(e^- \leftrightarrow e^+; G'_n \setminus e)} \,.$$

Since $\mathsf{WSF}[e \in \mathfrak{F} \mid \mathscr{F}_n]$ has a nonzero limit a.s. given $e \in \mathfrak{F}$ by the martingale convergence theorem and Lemma 10.38, it follows that $\mathscr{C}^{\mathrm{W}}(e^- \leftrightarrow e^+; G'_n \setminus e)$ is bounded a.s. given $e \in \mathfrak{F}$. Let T be the component of $\mathfrak{F}$ that contains e when $e \in \mathfrak{F}$, and let $T := \varnothing$ otherwise. By combining the previous bound on the wired effective conductance with Lemma 10.39 and Exercise 9.29, it follows that there is some finite random N with the following property: for all n, conditional on the event that $e \in \mathfrak{F}$, a.s. at least one of the two components of $T \setminus e$ has at most N vertices on the inner vertex boundary of the box G_n. Suppose that the component T' of e^- is such a component, so that e^- has degree at most $2dN$ in $G'_n \setminus e$. By Lemma 10.40, it follows that on the event $e \in \mathfrak{F}$, the conditional probability that $T' = T' \cap G_n$ given $\mathscr{F}_n$ is at least ϵ^{2dN} for some $\epsilon > 0$ and all large n. In particular, $\mathsf{WSF}[\mathscr{A}_e \mid \mathscr{F}_n] \geq \epsilon^{2dN}$ on the event that $e \in \mathfrak{F}$. Since $\mathscr{A}_e$ lies in the σ-field generated by $\bigcup \mathscr{F}_n$, the limit of these conditional probabilities as $n \to \infty$ is a.s. the indicator of $\mathscr{A}_e$, whence $\mathbf{1}_{\mathscr{A}_e} \geq \epsilon^{2dN} \mathbf{1}_{[e \in \mathfrak{F}]}$ a.s., whence $\mathbf{1}_{\mathscr{A}_e} \geq \mathbf{1}_{[e \in \mathfrak{F}]}$ a.s., as desired. ◀

We now prove the same for networks with a "reasonable isoperimetric profile," as defined in Section 6.8. Recall that $|F|_c := \sum_{e \in F} c(e)$ for $F \subseteq \mathsf{E}$ and $|K|_\pi := \sum_{x \in K} \pi(x)$ for $K \subseteq \mathsf{V}$. Write

$$\psi(G, t) := \inf \{ |\partial_{\mathsf{E}} K|_c \; ; \; t \leq |K|_\pi < \infty \} \,.$$

We need one more lemma, which states that if G has a good isoperimetric profile (in terms of ψ), then G can be exhausted by finite subgraphs whose complements still have good isoperimetric profiles.

Lemma 10.42. *Let G be an infinite, connected network such that*

$$\psi(G, 0^+) := \lim_{t \downarrow 0} \psi(G, t) > 0 \text{ and } \lim_{t \to \infty} \psi(G, t) = \infty \,. \tag{10.34}$$

Then G has an exhaustion $\langle G_n \rangle$ by finite, connected, induced subgraphs such that

$$\left| \partial_{\mathsf{E}} U \setminus \partial_{\mathsf{E}} \mathsf{V}(G_n) \right|_c \geq \left| \partial_{\mathsf{E}} U \right|_c / 2 \tag{10.35}$$

for all n and all finite $U \subset \mathsf{V}(G) \setminus \mathsf{V}(G_n)$ and

$$\psi\big(G \setminus \mathsf{V}(G_n), t\big) \geq \psi(G, t)/2 \tag{10.36}$$

for all n and $t > 0$.

Proof. Given $K \subset \mathsf{V}(G)$ such that its induced subgraph $G \restriction K$ is connected, let $W(K)$ be a set L that minimizes $|\partial_{\mathsf{E}} L|_c$ over all finite sets L that contain $K \cup \partial_{\mathsf{V}} K$; such a set $W(K)$ exists by our assumptions (10.34) on G. Furthermore, its induced subgraph $G \upharpoonright W(K)$ is connected. Let $G' := G \backslash W(K)$ and write ∂'_E for the edge-boundary operator in G'. If U is a finite subset of vertices in G', then $\left| \partial'_\mathsf{E} U \right|_c \geq \left| \partial_\mathsf{E} U \right|_c / 2$, which is the same as $\left| \partial_\mathsf{E} U \backslash \partial_\mathsf{E} W(K) \right|_c \geq \left| \partial_\mathsf{E} U \right|_c / 2$, since if not, $W(K) \cup U$ would have a smaller edge boundary but larger size than $W(K)$, contradicting the definition of $W(K)$. Thus, $\psi(G', t) \geq \psi(G, t)/2$ for all $t > 0$. It follows that we may define an exhaustion having the desired properties by the recursion $K_1 := W(\{o\})$ and $K_{n+1} := W(K_n)$, where o is a fixed vertex of G, and $G_n := G \upharpoonright K_n$. ◀

The following theorem shows that isoperimetric growth at a rate faster than square root guarantees one end per tree in the WSF.

Theorem 10.43. (Isoperimetric Condition for One-Ended Trees) *Suppose that G is an infinite network with $\pi_0 := \inf_{x \in \mathsf{V}} \pi(x) > 0$. If $\psi(t) := \psi(G, t)$ has the property that $\psi(t) \geq f(t)$ for some increasing function f on $[\pi_0, \infty)$ that satisfies $0 \leq f(t) \leq t$ and $f(2t) \leq \alpha f(t)$ for some α and*

$$\int_{\pi_0}^\infty \frac{dt}{f(t)^2} < \infty ,$$

then WSF-a.s. every tree has only one end.

Just as we saw that Theorem 6.41 was an easy consequence of Theorem 6.42, so Theorem 10.43 is easy to deduce from the following version that does not assume any regularity on $\psi(G, t)$:

Theorem 10.44. *Suppose that G is an infinite network. Let $\psi(t) := \psi(G, t)$. Define $s_1 := \psi(0^+)/2$ and $s_{k+1} := s_k + \psi(s_k)/2$ inductively for $k \geq 1$. If*

$$\psi(0^+) > 0 \text{ and } \sum_{k \geq 1} \frac{1}{\psi(s_k)} < \infty , \tag{10.37}$$

then WSF-a.s. every tree has only one end.

Proof. Let $\mathscr{A}_e$ be the event that $\mathfrak{F} \backslash e$ has a finite component. Just as we did in the proof of Theorem 10.41, we will show that every edge $e \in \mathsf{E}(G)$ satisfies $\mathsf{WSF}[\mathscr{A}_e \mid e \in \mathfrak{F}] = 1$. Unfortunately, the proof will be rather lengthier, but it is the same in spirit.

Our assumptions (10.37) imply those (10.34) of Lemma 10.42. Let $\langle G_n \rangle$ be an exhaustion by finite, connected, induced subgraphs that contain e and satisfy (10.35) and (10.36). Let G'_n be G after deleting $\mathsf{E}(G_n) \backslash (\mathfrak{F} \cup \{e\})$, contracting $\mathsf{E}(G_n) \cap (\mathfrak{F} \backslash \{e\})$, and removing any resulting loops. Note that $e \in G'_n$. Let $H_n := G'_n \backslash e$ and write π_n for the corresponding vertex weights in H_n. As in the proof of Theorem 10.41,

$$\mathsf{WSF}[e \in \mathfrak{F} \mid G'_n] = c(e) \mathscr{R}^\mathsf{W}(e^- \leftrightarrow e^+; G'_n) = \frac{c(e)}{c(e) + \mathscr{C}^\mathsf{W}(e^- \leftrightarrow e^+; H_n)} .$$

Since $\mathsf{WSF}[e \in \mathfrak{F} \mid G'_n]$ has a nonzero limit a.s. given $e \in \mathfrak{F}$ by the martingale convergence theorem and Lemma 10.38, it follows that

$$\sup_n \mathscr{C}^\mathsf{W}(e^- \leftrightarrow e^+; H_n) < \infty \text{ a.s. given } e \in \mathfrak{F} . \tag{10.38}$$

If either of the endpoints of e is isolated in H_n, then $\mathscr{A}_e$ occurs. If not, then H_n is obtained from $L_n := G \backslash V(G_n)$ by adding new vertices and edges. Write $\pi'_n(x)$ for the sum of edge weights incident to x in L_n. We claim that for some fixed $R < \infty$, all $n \geq 1$, and all $x, y \in V(L_n)$, we have

$$\mathscr{R}^{\mathrm{W}}(x \leftrightarrow y; L_n) \leq R. \tag{10.39}$$

To see this, given $x \in V(L_n)$, define $r_0 := \pi'_n(x)$ and $r_{k+1} := r_k + \psi(L_n, x, r_k)/2$ inductively. Here, as in (6.57), we use the definition

$$\psi(L_n, x, t) := \inf\{|\partial_{\mathsf{E}} K|_c \; ; \; x \in K \subset V(L_n), \; L_n{\upharpoonright}K \text{ is connected}, t \leq |K|_\pi < \infty\}.$$

We claim that $r_{2k} \geq s_k$ for all k. We prove this by induction on k. Obviously $r_0 \geq s_0$. Assume that $r_{2k} \geq s_k$ for some $k \geq 0$. By (10.36), we have

$$r_{2k+2} \geq r_{2k+1} + \psi(G, r_{2k+1})/4 \geq r_{2k} + \psi(G, r_{2k})/4 + \psi(G, r_{2k})/4 \geq s_k + \psi(G, s_k)/2 = s_{k+1},$$

which completes the proof. Since $r_{2k+1} \geq s_k$ as well, the bound of Theorem 6.42 (which applies, since $\psi(0^+) > 0$ implies $|V(G)|_\pi = \infty$) shows that

$$\mathscr{R}(x \leftrightarrow \infty; L_n) \leq \sum_k \frac{2}{\psi(L_n, x, r_k)} \leq \sum_k \frac{4}{\psi(G, r_k)} \leq \sum_k \frac{8}{\psi(G, s_k)}.$$

In combination with Exercise 9.29, this proves (10.39) with $R = \sum_k 16/\psi(G, s_k)$. In addition, the same proof shows that if it happens that $\pi'_n(x) \geq s_m$ for some m, then we get $\mathscr{R}(x \leftrightarrow \infty; L_n) \leq \sum_{k \geq m} 8/\psi(G, s_k)$. This tends to 0 as $m \to \infty$.

Let x_n and y_n be the endpoints of e in H_n. Consider now

$$\mathscr{R}^{\mathrm{W}}(x_n \leftrightarrow \infty; H_n) \leq \mathscr{R}^{\mathrm{W}}(x_n \leftrightarrow \infty; J_n),$$

where J_n is the network formed by adding to L_n the vertex x_n together with the edges joining it to its neighbors in H_n. By (10.35), $\psi(J_n, x_n, t) = \pi_n(x_n)$ for $t = \pi_n(x_n)$. We claim that

$$\psi(J_n, x_n, t) \geq \psi(L_n, t/3) \tag{10.40}$$

for all $t \geq (3/2)\pi_n(x_n)$. Let ∂'_{E} denote the edge-boundary operator in L_n. Let π''_n denote the vertex weights in J_n. Consider a finite, connected set K of vertices in J_n that strictly contains x_n and with $|K|_{\pi''_n} \geq (3/2)\pi_n(x_n)$. Let $K' := K \setminus \{x_n\}$. Then $|K'|_{\pi'_n} \geq |K|_{\pi''_n}/3$, while $|\partial'_{\mathsf{E}} K'|_c \leq |\partial_{\mathsf{E}} K|_c$. This proves the claim. An argument similar to the preceding shows, therefore, that $\mathscr{R}(x_n \leftrightarrow \infty; J_n)$ is small if $\pi_n(x_n)$ is large because the first two terms in the series of Theorem 6.42 are $2/\pi_n(x_n)$ and $2/\psi(J_n, x_n, (3/2)\pi_n(x_n))$, and the latter is at most $2/\psi(L_n, \pi_n(x_n)/2)$ by (10.40). The same holds for y_n. If both resistances were small, then $\mathscr{R}^{\mathrm{W}}(x_n \leftrightarrow y_n; H_n)$ would also be small by Exercise 9.29, which would contradict (10.38). Therefore, $\{\min\{\pi_n(x_n), \pi_n(y_n)\} \; ; \; n \geq 1\}$ is bounded a.s. For simplicity, let's say that $\{\pi_n(x_n) \; ; \; n \geq 1\}$ is bounded a.s.

Form G''_n from G'_n by contracting the edges incident to y_n (other than e) that lie in $\mathfrak{F}$ and by deleting the others. If none lie in $\mathfrak{F}$, then $\mathscr{A}_e$ occurs, so assume at least one does belong to

$\mathfrak{F}$. If F denotes the set of edges of H_n incident to x_n, then by Lemma 10.40, (10.39), and Rayleigh's monotonicity principle, we have

$$\mathsf{WSF}[F \cap \mathfrak{F} = \varnothing \mid G_n''] \geq \exp\left\{-\sum_{f \in F} c(f)(r(e) + R)\right\} = \exp\left\{-\pi_n(x_n)(r(e) + R)\right\}$$

if neither endpoint of e is isolated in H_n. This is bounded away from 0. In particular, $\mathsf{WSF}[\mathscr{A}_e \mid G_n'']$ is bounded away from 0 a.s. on the event that $e \in \mathfrak{F}$. Since the limit of these conditional probabilities as $n \to \infty$ is a.s. the indicator of $\mathscr{A}_e$, the proof is completed as for Theorem 10.41. ◄

Which (unweighted) graphs satisfy the hypothesis of Theorem 10.43? Of course, all nonamenable graphs do – with a linear function f. Combining this with Corollary 10.27, we get a good description of the WSF on bounded-degree nonamenable graphs:

Theorem 10.45. *If G is a connected, nonamenable graph of bounded degree, then the* WSF *a.s. has infinitely many trees, each with one end.*

Although it is not obvious, all quasi-transitive, transient graphs also satisfy the hypothesis of Theorem 10.43. To show this, we begin with the following lemma due to Coulhon and Saloff-Coste (1993), Saloff-Coste (1995), and Lyons, Morris, and Schramm (2008). It extends Theorem 6.29. (Recall from Proposition 8.14 that nonunimodular, transitive graphs are nonamenable and so satisfy a better inequality than the following one.)

Lemma 10.46. *Let G be a unimodular, transitive graph. Let $R(m)$ be the smallest radius of a ball in G that contains at least m vertices. Then, for all finite $K \subset V$, we have*

$$\frac{|\partial_V^{\mathrm{in}} K|}{|K|} \geq \frac{1}{2R(2|K|)}.$$

Proof. Fix a finite set K and let $R := R(2|K|)$. Let $B'(x, r)$ be the punctured ball of radius r about x, that is, the ball but missing x itself, and let $b := |B'(x, R)|$. For $x, y, z \in V(G)$, define $f_k(x, y, z)$ as the proportion of geodesic (that is, shortest) paths from x to z whose kth vertex is y. Let $S(x, r)$ be the sphere of radius r about x. Write $q_r := |S(x, r)|$. Let $F_{r,k}(x, y) := \sum_{z \in S(x,r)} f_k(x, y, z)$. Clearly, $\sum_y F_{r,k}(x, y) = q_r$ for every $x \in V(G)$ and $r \geq 1$. Since $F_{r,k}$ is invariant under the diagonal action of the automorphism group of G, the mass-transport principle gives $\sum_x F_{r,k}(x, y) = q_r$ for every $y \in V(G)$ and $r \geq 1$. Now we consider the sum

$$Z_r := \sum_{x \in K} \sum_{z \in S(x,r) \setminus K} \sum_{y \in \partial_V^{\mathrm{in}} K} \sum_{k=0}^{r-1} f_k(x, y, z).$$

If we fix $x \in K$ and $z \in S(x, r) \setminus K$, then the inner double sum is at least 1, since if we fix any geodesic path from x to z, it must pass through $\partial_V^{\mathrm{in}} K$. It follows that

$$Z_r \geq \sum_{x \in K} |S(x, r) \setminus K|,$$

whence

$$Z := \sum_{r=1}^{R} Z_r \geq \sum_{x \in K} |B'(x, R) \setminus K| \geq \sum_{x \in K} |B'(x, R)|/2 = |K|b/2 .$$

On the other hand, if we do the summation in another order, we find

$$Z_r = \sum_{y \in \partial_V^{in} K} \sum_{k=0}^{r-1} \sum_{x \in K} \sum_{z \in S(x,r) \setminus K} f_k(x, y, z)$$

$$\leq \sum_{y \in \partial_V^{in} K} \sum_{k=0}^{r-1} \sum_{x \in V(G)} \sum_{z \in S(x,r)} f_k(x, y, z)$$

$$= \sum_{y \in \partial_V^{in} K} \sum_{k=0}^{r-1} \sum_{x \in V(G)} F_{r,k}(x, y)$$

$$= \sum_{y \in \partial_V^{in} K} \sum_{k=0}^{r-1} q_r = |\partial_V^{in} K| r q_r .$$

Therefore,

$$Z \leq \sum_{r=1}^{R} |\partial_V^{in} K| r q_r \leq |\partial_V^{in} K| R b .$$

Comparing these upper and lower bounds for Z, we get the desired result. ◀

An immediate consequence of Lemma 10.46 (and Proposition 8.14) is the following bound:

Corollary 10.47. *If G is a transitive graph with balls of radius n having at least cn^3 vertices for some constant c, then*

$$\psi(G, t) \geq c' t^{2/3}$$

for some constant c' and all $t \geq 1$. ◀

As in Theorem 6.40, we may deduce transience:

Corollary 10.48. *If G is a transitive graph with balls of radius n having at least cn^3 vertices for some constant c, then simple random walk is transient on G.* ◀

Since all quasi-transitive, transient graphs have at least cubic volume growth by a theorem of Gromov (1981a) and Trofimov (1985) (see also the discussion of polynomial growth in Section 7.9), we obtain one-endedness in the WSF:

Theorem 10.49. *If G is a transient, transitive network, then WSF-a.s. every tree has only one end.* ◀

▷ **Exercise 10.18.**
Show that Theorem 10.49 holds in the quasi-transitive case as well.

Question 10.50. If G and G' are two Cayley graphs of the same group and FSF_G is connected a.s., then is $\mathsf{FSF}_{G'}$ also connected a.s.?

In contrast to Theorem 10.49, we have:

Proposition 10.51. *If G is a unimodular, transitive network with $\mathsf{WSF} \neq \mathsf{FSF}$, then FSF-a.s., there is a tree with infinitely many ends – in fact, with branching number > 1.*

Proof. We have that $\mathbf{E}_{\mathsf{WSF}}[\deg_{\mathfrak{F}} x] = 2$ for all x, whence $\mathbf{E}_{\mathsf{FSF}}[\deg_{\mathfrak{F}} x] > 2$ for all x. Apply Theorem 8.19 and ergodicity (Corollary 10.19). ◀

Question 10.52. Let G be a transitive network with $\mathsf{WSF} \neq \mathsf{FSF}$. Must *all* components of the FSF have infinitely many ends a.s.?

In view of Proposition 10.51, this would follow in the unimodular case from a proof of the following conjecture from BLPS (2001):

Conjecture. The components of the FSF on a unimodular, transitive graph are indistinguishable in the sense that for every automorphism-invariant property $\mathcal{A}$ of subgraphs, either a.s. all components satisfy $\mathcal{A}$ or a.s. they all do not. The same holds for the WSF.

This conjecture was finally proved by Hutchcroft and Nachmias (2017). Independently, it was proved by Timár (2018) for the FSF with the additional assumption that $\mathsf{FSF} \neq \mathsf{WSF}$. It fails in the nonunimodular setting, however; see Exercise 10.51.

To sum up our results for graphs roughly isometric to hyperbolic space, we need the fact that such graphs are nonamenable; see Proposition F.6.12 of Benedetti and Petronio (1992).

Taking stock, we arrive at the following surprising results:

Theorem 10.53. *Let G be a graph of bounded degree that is roughly isometric to $\mathbb{H}^d$.*

(i) *If G is a plane graph such that both G and its plane dual are roughly isometric to $\mathbb{H}^2$, then the WSF of G has infinitely many trees a.s., each having one end a.s., whereas the FSF of G has one tree a.s. with infinitely many ends a.s.*

(ii) *If $d \geq 3$, then the $\mathsf{WSF} = \mathsf{FSF}$ of G has infinitely many trees a.s., each having one end a.s.*

Proof. By Theorem 10.45 and nonamenability of G, the descriptions of the WSF follow for all $d \geq 2$. In case (i), the description of the FSF follows from Proposition 10.37. In case (ii), it follows from Theorem 9.18 and Proposition 10.14. In the planar case, it suffices merely to have both G and its dual locally finite and nonamenable. ◀

An example of a self-dual plane Cayley graph roughly isometric to $\mathbb{H}^2$ was shown in Figure 2.4.

Here is a summary of the phase transitions. It is quite surprising that the free spanning forest undergoes three phase transitions as the dimension increases. (We think of hyperbolic space as having infinitely many Euclidean dimensions, since volume grows exponentially in hyperbolic space but only polynomially in Euclidean space.)

	$\mathbb{Z}^d$		$\mathbb{H}^d$	
d	2–4	≥ 5	2	≥ 3
FSF: trees	1	∞	1	∞
ends	1	1	∞	1
WSF: trees	1	∞	∞	∞
ends	1	1	1	1

To go beyond hyperbolic graphs, let G be a proper, transient, plane graph with bounded degree and a bounded number of sides to its faces. Recall that Theorem 9.12 and Proposition 10.14 imply that WSF $\neq$ FSF. By Theorem 10.26 and Theorem 9.17, WSF has infinitely many trees. If the wired spanning forest has only one end in each tree, then the end containing any given vertex has a limiting direction that is distributed according to Lebesgue measure in the parametrization of Section 9.4. Indeed this hypothesis holds:

Theorem 10.54. *Let G be a proper, transient, plane graph with bounded degree and a bounded number of sides to its faces. Each tree in the wired spanning forest has only one end a.s., and the free spanning forest is a single tree a.s.*

The recurrent case follows by combining Theorem 10.36 with the fact that $G^\dagger$ is roughly isometric to G, and thus that $G^\dagger$ is also recurrent by Theorem 2.17. The transient case was proved (in somewhat greater generality) by Hutchcroft and Nachmias (2019), thus answering a question of BLPS (2001).

Finally, we mention the following beautiful extension of Pemantle's Theorem 10.30; see Benjamini, Kesten, Peres, and Schramm (2004).

Theorem 10.55. *Identify each tree in the uniform spanning forest on $\mathbb{Z}^d$ to a single point. In the new graph metric, the diameter of the resulting (locally infinite) graph is a.s. $\lfloor (d-1)/4 \rfloor$.*

10.7 Loop-Erased Random Walk and Harmonic Measure from Infinity

Infinite loop-erased random walk is defined in any transient network by chronologically erasing cycles from the random walk path. This was the first stage in constructing the WSF via Wilson's method rooted at infinity. Is there a sensible way to define infinite loop-erased random walk in a recurrent network? Suppose we try this: run random walk until it first reaches distance n from its starting point, erase cycles, and take a weak limit as $n \to \infty$.

▷ **Exercise 10.19.**
Show that on a general recurrent network, such a weak limit need not exist.

Despite the discouraging results of that exercise, this approach does work in many cases. In $\mathbb{Z}^2$, weak convergence of these measures was established by Lawler (1988) using Harnack inequalities (see Lawler (1991), Proposition 7.4.2). Lawler's approach yields explicit estimates of the rate of convergence but is difficult to extend to other networks. Using spanning trees, however, we can often prove that the limit exists, as shown in the following exercise.

▷ **Exercise 10.20.**
Let $\langle G_n \rangle$ be an induced exhaustion of a recurrent network G. Consider the network random walk $\langle X_k \; ; \; k \geq 0 \rangle$ started from $o \in G$. Denote by τ_n the first exit time of G_n, and let L_n be the loop erasure of the path $\langle X_k \; ; \; 0 \leq k \leq \tau_n \rangle$. Show that if the uniform spanning tree T_G in G has one end a.s., then the random paths L_n converge weakly to the law of the unique ray from o in T_G. In particular, show that this applies if G is a proper, plane network with a locally finite, recurrent dual.

Now we go the other way: Instead of walking to infinity, we walk from infinity. More precisely, we do not try to find a path coming from infinity but only the hitting distribution on a finite set of that imagined path. Let A be a finite set of vertices in a recurrent network G. Denote by τ_A the hitting time of A, and by h_u^A the harmonic measure from u on A:

$$\forall B \subseteq A \quad h_u^A(B) := \mathbf{P}_u\big[X_{\tau_A} \in B\big].$$

If the measures h_u^A converge as $\text{dist}(u, A) \to \infty$, then it is natural to refer to the limit as **harmonic measure from infinity on** A. This convergence fails in some recurrent networks (for example, in $\mathbb{Z}$), but it does hold in $\mathbb{Z}^2$; see Lawler (1991), Theorem 2.1.3. On transient networks, wired harmonic measure from infinity always exists: see Exercise 2.50. Here, we show that on recurrent networks with one end in their uniform spanning tree, harmonic measure from infinity also exists:

Theorem 10.56. *Let G be a recurrent network and A be a finite set of vertices. Suppose that the uniform spanning tree T_G in G has one end a.s. Then the harmonic measures h_u^A converge as $\text{dist}(u, A) \to \infty$.*

Proof. Essentially, we would like to run Wilson's algorithm with all of A as root. To do this, let (A, F) be a tree with $F \cap \mathsf{E}(G) = \varnothing$, and form the graph $G' := \big(\mathsf{V}(G), \mathsf{E}(G) \cup F\big)$. Assign unit conductances to the edges of F. Note that the event that T has one end belongs to the tail σ-field of Section 10.4. Now $T_{G'}$ conditioned on $T_{G'} \cap F = \varnothing$ has the same distribution as T_G. Therefore, tail triviality (Theorem 10.18) tell us that a.s. $T_{G'}$ has one end. By this fact and similar reasoning, because $T_{G'}$ conditioned on the event $\big[T_{G'} \cap F = F\big]$ has the same distribution as $T_{G'/F}$, also $T_{G'/F}$ has one end a.s.

The path from u to A in $T_{G'/F}$ is constructed by running a random walk from u until it hits A and then loop erasing. Thus, when $\text{dist}(u, A) \to \infty$, the measures h_u^A must tend to the conditional distribution, given $\big[T_{G'} \cap F = F\big]$, of the point in A that is closest (in $T_{G'}$) to the unique end of $T_{G'}$. ◄

10.8 Appendix: Von Neumann Dimension and ℓ^2-Betti Numbers

Our goal is to show that the expected degree of the FSF in a Cayley graph depends only on the group and not on the set of generators. To accomplish this, we will explain what (10.6) means and why it is true. In particular, we show that the first ℓ^2-Betti number of a Cayley graph depends only on the group.

Let G be the Cayley graph of an infinite group Γ with respect to a finite generating set S closed under inverses. Write e_s for the edge joining the identity o to s. By (10.3), we can express the expected degree in the FSF as

$$\mathbf{E}_{\mathsf{FSF}}[\deg_{\widetilde{\mathfrak{F}}} o] = \sum_{s \in S} \mathsf{FSF}[e_s \in \mathfrak{F}] = \sum_{s \in S} \left(P_\diamond^\perp \chi^{e_s}, \chi^{e_s} \right). \qquad (10.41)$$

The idea now is to identify $\diamond^\perp$ with a Γ-invariant subspace of $\ell^2(\Gamma \times S) \cong \ell^2(\Gamma)^{|S|}$ and to define von Neumann dimension for Γ-invariant subspaces of $\ell^2(\Gamma)^n$ in general. Note that Γ acts on $\ell^2(\Gamma)^n$ by $(\gamma f)(\gamma_1, \ldots, \gamma_n) := f(\gamma^{-1}\gamma_1, \ldots, \gamma^{-1}\gamma_n)$. To show that $\mathbf{E}_{\mathsf{FSF}}[\deg_{\widetilde{\mathfrak{F}}} o]$ does not depend on S, we define von Neumann dimension still more generally and then prove its invariance under embeddings.

The identification of a Γ-invariant, closed subspace $H \subseteq \ell^2(\mathsf{E})$, such as $\diamond^\perp$, with a Γ-invariant subspace of $\ell^2(\Gamma \times S)$ goes as follows. Let the standard basis elements of $\ell^2(\Gamma \times S)$ be $\{f_{\gamma,s} \,;\, \gamma \in \Gamma, s \in S\}$. Identify $\ell^2(\mathsf{E})$ with the range in $\ell^2(\Gamma \times S)$ of the map defined by sending $\chi^{\langle \gamma, \gamma s \rangle}$ to the vector $(f_{\gamma,s} - f_{\gamma s, s^{-1}})/\sqrt{2}$. This is well defined, since it respects the identity $\chi^{\langle \gamma, \gamma s \rangle} = -\chi^{\langle \gamma s, \gamma \rangle}$. These image vectors form an orthonormal basis of the range, so the map is an isometry. Then H becomes identified with a Γ-invariant, closed subspace H_S. Write Q for the orthogonal projection of $\ell^2(\Gamma \times S)$ onto H_S. Since $(f_{\gamma,s} + f_{\gamma s, s^{-1}}) \perp \ell^2(\mathsf{E})_S$, and hence $(f_{\gamma,s} + f_{\gamma s, s^{-1}}) \perp H_S$, we have

$$Q f_{o,s} = -Q f_{s,s^{-1}}.$$

Therefore, the sum we encountered in (10.41) can be written in this more general context as

$$\sum_{s \in S} \left(P_H \chi^{e_s}, \chi^{e_s} \right) = \sum_{s \in S} \| Q(f_{o,s} - f_{s,s^{-1}})/\sqrt{2} \|^2 = \sum_{s \in S} \| \sqrt{2} Q f_{o,s} \|^2 = 2 \sum_{s \in S} \left(Q f_{o,s}, f_{o,s} \right).$$
$$(10.42)$$

To motivate the definition of von Neumann dimension, recall first that the ordinary dimension is the minimal number of vectors needed to span (in an appropriate sense) a vector or Banach space. When we have a group acting that preserves a vector space, then the von Neumann dimension has the same intuitive definition as this, but in spanning, we are allowed to act via the group. Thus, the von Neumann dimension of $\ell^2(\Gamma)$ (with respect to Γ) is 1 and the von Neumann dimension of $\ell^2(\Gamma)^2$ is 2. However, this does not help us to see clearly what the von Neumann dimension of a subspace of $\ell^2(\Gamma)$ is. So let us reformulate our intuitive definition as the minimal number of vectors needed in the ordinary sense, but per group element. When Γ is finite, this makes it quite easy to rigorously define the von Neumann dimension of an invariant subspace H of $\ell^2(\Gamma)$, namely, $\dim_\Gamma H := (\dim H)/|\Gamma|$. In particular, note that von Neumann dimension is not always an integer. To extend this idea to infinite groups, make the key observation that dim $H = \operatorname{tr} P_H$, the trace of the orthogonal projection to H. When H is invariant, we will see in a moment that all the diagonal elements of the matrix of P_H are the same (in the standard orthonormal basis of $\ell^2(\Gamma)$), whence $\dim_\Gamma H$ is simply the common value on the diagonal. This is how we define it in general: when H is a closed, Γ-invariant subspace of $\ell^2(\Gamma)$, we put $\dim_\Gamma H := (P_H \mathbf{1}_{\{o\}}, \mathbf{1}_{\{o\}})$, which is the same as $(P_H \mathbf{1}_{\{\gamma\}}, \mathbf{1}_{\{\gamma\}})$ for all $\gamma \in \Gamma$.

To see that all the diagonal elements of the matrix of P_H are the same, let H be a closed, Γ-invariant subspace of $\ell^2(\Gamma)$. Then $H^\perp$ is also closed and Γ-invariant, since Γ acts on $\ell^2(\Gamma)$ by isometries. Thus, for all $f \in \ell^2(\Gamma)$ and all $\gamma \in \Gamma$, the equation $f = P_H(f) + P_H^\perp(f)$ implies that $\gamma f = \gamma P_H(f) + \gamma P_H^\perp(f)$ is the orthogonal decomposition of γf with respect to $H \oplus H^\perp$, that is,

$$P_H(\gamma f) = \gamma P_H(f). \tag{10.43}$$

In particular, $(P_H \mathbf{1}_{\{\gamma\}}, \mathbf{1}_{\{\gamma\}}) = (P_H \gamma \mathbf{1}_{\{o\}}, \gamma \mathbf{1}_{\{o\}}) = (P_H \mathbf{1}_{\{o\}}, \mathbf{1}_{\{o\}})$, as desired.

For example, if $\Gamma = \mathbb{Z}$, then $\ell^2(\mathbb{Z})$ is isometrically isomorphic to $L^2([0, 1])$ (with Lebesgue measure) by the Fourier map $L^2([0, 1]) \ni f \mapsto \hat{f} \in \ell^2(\mathbb{Z})$ (the inverse of the map used in Section 10.3, as in Exercise 10.39), and it can be shown that $\mathbb{Z}$-invariant subspaces of $\ell^2(\mathbb{Z})$ correspond under this isomorphism to spaces of the form $L^2(A)$ for A a measurable subset of $[0, 1]$. Then $\dim_\mathbb{Z} \widehat{L^2(A)} = |A|$ (the Lebesgue measure of A), since $P_{L^2(A)} f = f \mathbf{1}_A$, so that $\dim_\mathbb{Z} \widehat{L^2(A)} = \int_0^1 (\mathbf{1}\mathbf{1}_A)\mathbf{1} = |A|$.

We need to extend this notion to closed, invariant subspaces H of $\ell^2(\Gamma)^n$. Note that $\ell^2(\Gamma)^n \cong \ell^2(\Gamma \times \{1, \ldots, n\})$. The matrix (with respect to the standard orthonormal basis) of any linear operator on this space can be written in block form, where an $n \times n$-block corresponds to fixing $\gamma, \gamma' \in \Gamma$ and taking all rows corresponding to (γ, i) and all columns corresponding to (γ', j), where $i, j \in \{1, \ldots, n\}$. The diagonal blocks are those with $\gamma = \gamma'$. When H is invariant, the diagonal blocks of the matrix of P_H are all the same, whence we define the **von Neumann dimension** of H with respect to Γ to be the common trace, $\dim_\Gamma H := \sum_{i=1}^n (P_H \mathbf{1}_{\{(o, i)\}}, \mathbf{1}_{\{(o, i)\}})$. Thus, the sum in (10.42) equals $2 \dim_\Gamma H_S$.

Next we need to show that von Neumann dimension is an intrinsic number, not dependent on the particular representation as a subspace of any $\ell^2(\Gamma)^n$. To what, then, is this dimension intrinsic?* Recall that a **Fréchet space** is a locally convex topological vector space whose topology is induced by a complete, translation-invariant metric. Define a Fréchet space M to be a **Hilbertable Γ-space** if Γ acts on M by continuous linear transformations and there exists a Γ-equivariant continuous linear injection $\alpha: M \to \ell^2(\Gamma)^n$ for some finite n whose image, $\operatorname{img} \alpha$, is closed. Here, to say that α is Γ-*equivariant* means that α intertwines the (left) Γ-actions, that is, $\alpha(\gamma f) = \gamma \alpha(f)$ for all $f \in M$ and all $\gamma \in \Gamma$. In this case, we define $\dim_\Gamma M := \dim_\Gamma \operatorname{img} \alpha$. Clearly, we must show that this does not depend on the choice of α.

This leads us to revisit the earlier use of the trace. Let $N(\Gamma)$ be the **von Neumann algebra** of Γ, that is, the set of Γ-equivariant bounded linear operators on $\ell^2(\Gamma)$. This includes orthogonal projections on closed, invariant subspaces, as we saw in (10.43).

For $\phi \in N(\Gamma)$, define $\operatorname{tr}_\Gamma(\phi) := (\phi(\mathbf{1}_{\{o\}}), \mathbf{1}_{\{o\}})$. We'll need the following little device: given $f \in \ell^2(\Gamma)$, write $\widetilde{f}$ for the element of $\ell^2(\Gamma)$ defined by $\gamma \mapsto \overline{f(\gamma^{-1})}$. (We're taking complex

---

* One could avoid this question by using the first part of the proof of Theorem 9.9, but one would still need most of the mathematics that follows.

Hilbert spaces to be general, but real ones suffice for our application.) Note that for all $f, g \in \ell^2(\Gamma)$, we have $(f, g) = (\widetilde{g}, \widetilde{f})$.

As usual, write ϕ^* for the adjoint of ϕ. We have that

$$\phi^*(\mathbf{1}_{\{o\}}) = \widetilde{\phi(\mathbf{1}_{\{o\}})}, \tag{10.44}$$

since for all $\gamma \in \Gamma$,

$$\begin{aligned}
\left(\phi^*(\mathbf{1}_{\{o\}}), \mathbf{1}_{\{\gamma\}}\right) &= \left(\mathbf{1}_{\{o\}}, \phi(\mathbf{1}_{\{\gamma\}})\right) = \left(\mathbf{1}_{\{o\}}, \gamma\phi(\mathbf{1}_{\{o\}})\right) = \left(\gamma^{-1}\mathbf{1}_{\{o\}}, \phi(\mathbf{1}_{\{o\}})\right) \\
&= \left(\mathbf{1}_{\{\gamma^{-1}\}}, \phi(\mathbf{1}_{\{o\}})\right) = \left(\widetilde{\mathbf{1}_{\{\gamma\}}}, \phi(\mathbf{1}_{\{o\}})\right) = \left(\widetilde{\phi(\mathbf{1}_{\{o\}})}, \mathbf{1}_{\{\gamma\}}\right).
\end{aligned}$$

This leads to the key property of a trace:

Proposition 10.57. *For all* $\phi, \psi \in N(\Gamma)$*, we have* $\mathrm{tr}_\Gamma(\phi\psi) = \mathrm{tr}_\Gamma(\psi\phi)$.

Proof. By definition and (10.44), we have

$$\begin{aligned}
\mathrm{tr}_\Gamma(\phi\psi) &= \left(\phi(\psi(\mathbf{1}_{\{o\}})), \mathbf{1}_{\{o\}}\right) = \left(\psi(\mathbf{1}_{\{o\}}), \phi^*(\mathbf{1}_{\{o\}})\right) = \left(\psi(\mathbf{1}_{\{o\}}), \widetilde{\phi(\mathbf{1}_{\{o\}})}\right) \\
&= \left(\phi(\mathbf{1}_{\{o\}}), \widetilde{\psi(\mathbf{1}_{\{o\}})}\right) = \left(\phi(\mathbf{1}_{\{o\}}), \psi^*(\mathbf{1}_{\{o\}})\right) \\
&= \left(\psi(\phi(\mathbf{1}_{\{o\}})), \mathbf{1}_{\{o\}}\right) = \mathrm{tr}_\Gamma(\psi\phi).
\end{aligned}$$
◀

We now extend the trace to the algebra $M_n(N(\Gamma))$ of Γ-equivariant bounded linear operators on $\ell^2(\Gamma)^n$. For $\phi \in M_n(N(\Gamma))$ and $(f_1, \ldots, f_n) \in \ell^2(\Gamma)^n$, write

$$\phi(f_1, \ldots, f_n) = \left(\sum_{j=1}^n \phi_{1,j} f_j, \ldots, \sum_{j=1}^n \phi_{n,j} f_j\right) \in \ell^2(\Gamma)^n.$$

That is, think of ϕ as an $n \times n$-matrix $[\phi_{i,j}]$, where $\phi_{i,j} \in N(\Gamma)$. Define $\mathrm{tr}_\Gamma(\phi) := \sum_{i=1}^n \mathrm{tr}_\Gamma(\phi_{i,i}) = \mathrm{tr}_\Gamma\left(\sum_{i=1}^n \phi_{i,i}\right)$. From Proposition 10.57, a short calculation reveals that

$$\mathrm{tr}_\Gamma(\phi\psi) = \mathrm{tr}_\Gamma(\psi\phi) \tag{10.45}$$

for all $\phi, \psi \in M_n(N(\Gamma))$.

Given a Hilbertable Γ-space M, choose, as the definition allows us, a Γ-equivariant continuous linear injection $\alpha \colon M \to \ell^2(\Gamma)^n$ with closed image. Note that $\mathrm{img}\,\alpha$ is Γ-invariant. Define $\dim_\Gamma M := \mathrm{tr}_\Gamma(P_{\mathrm{img}\,\alpha}) = \dim_\Gamma \mathrm{img}\,\alpha$; we must verify that this does not depend on the choice of α. For example, consider the map $\alpha \oplus \mathbf{0} \colon M \to \ell^2(\Gamma)^{n+k}$ for fixed $k \geq 1$ induced by the inclusion $f \mapsto (f, \mathbf{0})$ of $\ell^2(\Gamma)^n$ into $\ell^2(\Gamma)^{n+k}$. Here, it is easy to see from the definition that $\mathrm{tr}_\Gamma(P_{\mathrm{img}\,\alpha}) = \mathrm{tr}_\Gamma(P_{\mathrm{img}(\alpha \oplus \mathbf{0})})$. Thus, if we have another Γ-equivariant continuous linear injection $\beta \colon M \to \ell^2(\Gamma)^m$ with closed image, then to show that $\mathrm{tr}_\Gamma(P_{\mathrm{img}\,\beta}) = \mathrm{tr}_\Gamma(P_{\mathrm{img}\,\alpha})$, we may assume that $n = m$. Now define $T := \beta\alpha^{-1} \colon \mathrm{img}\,\alpha \to \mathrm{img}\,\beta$. This is a Γ-equivariant continuous linear bijection by the open mapping theorem, whence its polar decomposition gives a Γ-equivariant unitary operator $U := T(T^*T)^{-1/2}$ that maps $\mathrm{img}\,\alpha \to \mathrm{img}\,\beta$ (see Theorem 12.35 of Rudin (1991), for example). Extend U to $\phi := U \oplus \mathbf{0} \in M_n(N(\Gamma))$, where $\phi \restriction \text{img}\, \alpha = U$ and $\phi \restriction (\text{img}\, \alpha)^\perp = \mathbf{0}$. It is easy to check that $\phi^* = U^* \oplus \mathbf{0}$ in that $\phi^* \restriction \text{img}\, \beta = U^*$ and $\phi^* \restriction (\text{img}\, \beta)^\perp = \mathbf{0}$. From this, it is easy to see that $P_{\text{img}\, \alpha} = \phi^* \phi$ and $P_{\text{img}\, \beta} = \phi \phi^*$, whence (10.45) gives the desired identity.

We have now shown how to define von Neumann dimension for general Hilbertable Γ-spaces. Clearly $\dim_\Gamma M = \dim_\Gamma M'$ if M and M' are *isomorphic* as Hilbertable Γ-spaces, that is, if there is a Γ-equivariant continuous linear bijection from M to M'. The following properties of von Neumann dimension are not hard to verify, but we will not need them: $\dim_\Gamma M \geq 0$ with equality iff $M = \mathbf{0}$ (use the fact that an orthogonal projection P is equal to P^*P); $\dim_\Gamma(M \oplus M') = \dim_\Gamma M + \dim_\Gamma M'$; and if Γ' is a subgroup of Γ of index $m < \infty$, then $\dim_{\Gamma'} M = m \dim_\Gamma M$.

The final step is to define $\diamondsuit^\perp$ as a Hilbertable Γ-space in a way that does not depend on the group generators, S. For a function $F: \Gamma \to \mathbb{R}$ and $\gamma \in \Gamma$, write $\rho(\gamma)F$ for the function $\gamma' \mapsto F(\gamma'\gamma)$ for $\gamma' \in \Gamma$. Let

$$D(\Gamma) := \left\{ F: \Gamma \to \mathbb{R} \, ; \, \forall \gamma \ (\rho(\gamma)F - F) \in \ell^2(\Gamma) \right\} .$$

Then define $Z(\Gamma)$ to be the vector space of *cocycles*, that is,

$$Z(\Gamma) := \left\{ z: \Gamma \to \ell^2(\Gamma) \, ; \, \exists F \in D(\Gamma) \ \forall \gamma \ z(\gamma) = \rho(\gamma)F - F \right\} .$$

Here, we may always choose F so that $F(o) = 0$ if we wish, in which case we may recover F from z by $F(\gamma) = -z(\gamma^{-1})(\gamma)$. To define a metric on $Z(\Gamma)$, choose an ordering $\langle \gamma_i \, ; \, i \geq 1 \rangle$ of Γ, and let the distance between z and z' be the infimum of $\epsilon > 0$ such that $\|z(\gamma_i) - z'(\gamma_i)\| < \epsilon$ for all $i < 1/\epsilon$. It is not hard to check that $Z(\Gamma)$ is a Fréchet space. The (left) action of Γ on $Z(\Gamma)$ is $(\gamma z)(\gamma'): \gamma'' \mapsto z(\gamma')(\gamma^{-1}\gamma'')$. It is easy to see that $z \in Z(\Gamma)$ satisfies

$$z(\gamma\gamma') = z(\gamma) + \rho(\gamma)z(\gamma')$$

for all $\gamma, \gamma' \in \Gamma$. (In fact, $Z(\Gamma)$ equals the space of all maps $z: \Gamma \to \ell^2(\Gamma)$ that satisfy this *cocycle identity*.) Thus, every cocycle is determined by its values on a generating set. In other words, the map $\alpha: Z(\Gamma) \to \ell^2(\Gamma)^S$ given by $z \mapsto z \restriction S$ is an injection. It is obviously Γ-equivariant and continuous. It is not hard to check by using the cocycle identity that $\text{img}\, \alpha$ is closed, whence $Z(\Gamma)$ is a Hilbertable Γ-space.

It remains to show that $Z(\Gamma)$ is isomorphic as a Hilbertable Γ-space to $\diamondsuit^\perp$, where the latter is a Hilbertable Γ-space by virtue of being a Γ-invariant Hilbert subspace of $\ell^2(E)$. Note that $\mathbf{D} = D(\Gamma)$. Also, recall that $\diamondsuit^\perp = \nabla \mathbf{D}$. Thus, define $T: \diamondsuit^\perp \to Z(\Gamma)$ by $T(\theta)(\gamma) := \rho(\gamma)F - F$ when $\theta = dF$. (Recall that all conductances are 1 here.) This is clearly an isomorphism.

From our earlier calculation (10.41)–(10.42), it follows that $\mathbf{E}_{\text{FSF}}[\deg_{\mathfrak{F}} o] = 2 \dim_\Gamma Z(\Gamma)$ does not depend on S. This finishes our main task.

We now explain the connection to Betti numbers. Ordinary Betti numbers are the dimensions of the cohomology groups with real coefficients. In certain cases of infinite CW-complexes acted on by a group, Γ, one defines ℓ^2-Betti numbers as the von Neumann dimensions of reduced ℓ^2-cohomology Hilbert Γ-spaces. We explain part of this and refer to Eckmann (2000) for more; there are also higher-dimensional analogues of spanning forests that corresponding to the higher ℓ^2-Betti numbers, as shown by Lyons (2009). Define the space of **coboundaries**

$$B(\Gamma) := \left\{z : \Gamma \to \ell^2(\Gamma) \, ; \; \exists F \in \ell^2(\Gamma) \; \forall \gamma \; z(\gamma) = \rho(\gamma)F - F\right\}.$$

The only difference between $Z(\Gamma)$ and $B(\Gamma)$ is that in the latter, F is required to belong to $\ell^2(\Gamma)$, not just to $D(\Gamma)$. Since Γ is infinite, this implies that for $z \in B(\Gamma)$, there is a *unique* $F \in \ell^2(\Gamma)$ with $z(\gamma) = \rho(\gamma)F - F$. Make $B(\Gamma)$ into a Hilbert space by defining the inner product $(z, z') := (F, F')$, where $F, F' \in \ell^2(\Gamma)$ correspond to $z, z' \in B(\Gamma)$ as in the definition. Thus, $B(\Gamma)$ is isomorphic to $\ell^2(\Gamma)$. On the other hand, the closure of the inclusion of $B(\Gamma)$ in $Z(\Gamma)$ is isomorphic to $\bigstar$ as a Hilbertable Γ-space, whence also $B(\Gamma)$ is isomorphic to $\bigstar$; this relies on the polar decomposition and is proved as Lemma 2.5.3 of Eckmann (2000). Thus, $1 = \dim_{\Gamma} \ell^2(\Gamma) = \dim_{\Gamma} B(\Gamma) = \dim_{\Gamma} \bigstar$. (This gives another way to prove Proposition 10.10 for Cayley graphs.) Since $\Diamond^{\perp} = \nabla\mathbf{HD} \oplus \bigstar$, it follows that

$$\mathbf{E}_{\mathsf{FSF}}[\deg_{\widetilde{\mathfrak{F}}} o] = 2 \dim_{\Gamma} \nabla\mathbf{HD} + 2 = 2 \dim_{\Gamma} Z(\Gamma)/\overline{B(\Gamma)} + 2 = 2\beta_1(\Gamma) + 2 \, ,$$

where $Z(\Gamma)/\overline{B(\Gamma)}$ is the **first reduced ℓ^2-cohomology** of Γ and $\beta_1(\Gamma) := \dim_{\Gamma} Z(\Gamma)/\overline{B(\Gamma)}$ is the **first ℓ^2-Betti number** of Γ. The reason that $Z(\Gamma) \cong \Diamond^{\perp}$ has the name of cocycles is that if one adds a 2-cell to the Cayley graph for every relation in the group, then $\Diamond^{\perp}$ is the kernel of the natural coboundary map defined on $\ell^2_-(\mathsf{E})$. Similarly, $\bigstar$ is the closure of the image of the coboundary map defined on $\ell^2(\mathsf{V})$, which accounts for the name "coboundaries." (In the case of finite groups, $\nabla\mathbf{HD} = \mathbf{0}$, which is why $\beta_1(\Gamma) = 0$ when Γ is finite.)

▷ **Exercise 10.21.**
Let G be the Cayley graph of Γ with respect to S. Assign positive conductances to the edges in such a way that they depend only on the generators associated to the edges. Show that still $\mathbf{E}_{\mathsf{FSF}}[\deg_{\widetilde{\mathfrak{F}}} o] = 2 + 2\beta_1(\Gamma)$.

An extension of the invariance of expected degree of the FSF can be proved by using the major result of Gaboriau (2002). Namely, let G be a *random* connected graph whose vertex set is Γ and whose law is Γ-invariant. Then still $\mathbf{E}\big[\mathbf{E}_{\mathsf{FSF}(G)}[\deg_{\widetilde{\mathfrak{F}}} o \mid G]\big] = 2 + 2\beta_1(\Gamma)$. One immediate consequence of this is the resolution of a conjecture made by Benjamini, Lyons, and Schramm (1999), namely, that if G is a Cayley graph of Γ and ω is an invariant percolation with a unique infinite cluster, then G has no nonconstant harmonic Dirichlet functions iff ω has no nonconstant harmonic Dirichlet functions a.s. See Gaboriau (2005) for this and for an extension to unimodular, quasi-transitive graphs G.

10.9 Notes

The free spanning forest was first suggested by Lyons, but its existence was proved by Pemantle (1991). Pemantle also implicitly proved the existence of the wired spanning forest and showed that the free and wired uniform spanning forests are the same on Euclidean lattices $\mathbb{Z}^d$. The first explicit construction of the wired spanning forest is due to Häggström (1995).

We have no need for the general theory of weak convergence in defining or analyzing the free or wired spanning forests. However, since $\{0, 1\}^{\mathsf{E}}$ is compact, it is easy to verify that our notion of weak convergence coincides with the usual notion for Borel probability measures on topological spaces.

The version we present of Strassen's theorem, Theorem 10.4, is only a special case of what he proved.

The main results in this chapter that come from BLPS (2001) are Theorems 10.15, 10.18, 10.26, 10.36, 10.49, 10.45, 10.53, and 10.56 and Propositions 10.1, 10.10, 10.14, 10.37, and 10.51. Theorems 10.49, 10.45, and 10.53 were proved there in more restricted versions. The extended versions proved here are from Lyons, Morris, and Schramm (2008).

Theorems 10.21 and 10.24, Corollary 10.25, and Lemma 10.31 are from Lyons, Peres, and Schramm (2003). We have modified the more general results of Lyons, Peres, and Schramm (2003) and simplified their proofs for the special cases presented here in Theorem 10.24, Corollary 10.32, and Lemma 10.33. For estimates of the transition kernel for group-invariant random walks, such as those used in the proof of Corollary 10.25, see also Hebisch and Saloff-Coste (1993). The last part of Theorem 10.34 was conjectured by BLPS (2001).

Theorem 10.36 was first stated without proof for $\mathbb{Z}^2$ by Pemantle (1991). It is shown in BLPS (2001) that the same is true on any recurrent transitive network.

The appendix is based on Lyons (2009) and Eckmann (2000). Although it was known (Bekka and Valette, 1997) that Kazhdan groups have first ℓ^2-Betti number equal to 0, it was not proved until Hutchcroft and Pete (2020) that they have cost 1; their argument used invariant percolation.

Uniform spanning forests help in studying sandpiles on infinite graphs; see Járai and Redig (2008) and Járai and Werning (2014).

In Section 10.6, we gave several families of examples of graphs for which the WSF has one end in each tree. This has been extended to various random (rooted) transient graphs that satisfy the property of unimodularity mentioned in Section 8.9; when the degree is bounded, this was done by Aldous and Lyons (2007), whereas the case of bounded expected degree was done by Hutchcroft (2016b) and then the completely general case was done by Hutchcroft (2018). One consequence is that each component in the wired spanning forest on an infinite supercritical Galton–Watson tree has one end a.s., a question that was asked by BLPS (2001). A key new tool introduced by Hutchcroft (2018) generates the WSF by extending the Aldous–Broder algorithm to transient graphs via Sznitman's random interlacement process (for which see Černý and Teixeira (2012) and Drewitz, Ráth, and Sapozhnikov (2014)). Hutchcroft (2018) also showed that there are roughly isometric graphs G and G' of bounded degree such that the wired spanning forest in G has only one end in each tree a.s., while a.s., some tree of the wired spanning forest in G' has uncountably many ends; this answered a question of Lyons, Morris, and Schramm (2008). Additional information on the size of the trees in the WSF is given by Hutchcroft (2020d).

There are many intriguing open questions related to spanning forests. Besides the ones we have already given, here are a few more.

Question 10.58. (BLPS, 2001) Let G be an infinite network such that WSF $\neq$ FSF on G. Does it follow that WSF and FSF are mutually singular measures?

This question has a positive answer for trees (there is exactly one component FSF-a.s. on a tree, while the number of components is a constant WSF-a.s. by Theorem 10.26) and for networks G where $\mathrm{Aut}(G)$ has an infinite orbit (Corollary 10.19).

Question 10.59. (Lyons, Morris, and Schramm, 2008) Is the probability that each tree has only one end equal to either 0 or 1 for both WSF and FSF? This is true on trees by Theorem 11.1 of BLPS (2001).

Conjecture 10.60. (BLPS, 2001) *Let T_o be the component of the identity o in the* WSF *on a Cayley graph, and let $\xi = \langle x_n ; n \geq 0 \rangle$ be the unique ray from o in T_o. The sequence of "bushes" $\langle b_n \rangle$ observed along ξ converges in distribution. (Formally, b_n is the connected component of x_n in $T \setminus \{x_{n-1}, x_{n+1}\}$, multiplied on the left by x_n^{-1}.)*

Conjecture 10.60 is now almost completely established. A positive answer for Cayley graphs of at least quintic volume growth follows from Proposition 3.2 of Lawler (1983), which shows that $x_n^{-1}\xi = \langle x_n^{-1} x_{n+k} ; k \geq -n \rangle$ converges in distribution (to a bi-infinite loop-erased random walk). Combined with Wilson's algorithm rooted at infinity, this ensures the existence of the limiting distribution of $x_n^{-1}T_o$ rooted at x_n. This was extended to $\mathbb{Z}^4$ by Lawler, Sun, and Wu (2019) and to $\mathbb{Z}^3$ and $\mathbb{Z}^2$ by Lawler (2020).

10.10 Collected In-Text Exercises

10.1. The choice of exhaustion $\langle G_n \rangle$ does not change the resulting measure WSF by the proof of Proposition 10.1. Show that the choice also does not change the resulting measure FSF.

10.2. Show that FSF and WSF are invariant under any automorphisms that the network may have.

10.3. Show that if G is an infinite, recurrent network, then the wired spanning forest on G is the same as the free spanning forest, that is, the random spanning tree of Section 4.2.

10.4. Let G be a network such that there is a finite subset of edges whose removal from G leaves at least two transient components. Show that the free and wired spanning forests are different on G.

10.5. Let G be a tree with unit conductances. Show that FSF = WSF iff G is recurrent.

10.6. Let G be an edge-amenable, infinite graph as witnessed by the vertex sets $\langle V_n \rangle$ (see Section 4.3). Let G_n be the subgraph induced by V_n.
    (a) Let F be any spanning forest all of whose components (trees) are infinite. Show that if k_n denotes the number of trees of $F \cap G_n$, then $k_n = o(|V_n|)$.
    (b) Let $\mathfrak{F}$ be a random spanning forest all of whose components (trees) are infinite. Show that the average degree, in two senses, of vertices is 2:

$$\lim_{n \to \infty} |V_n|^{-1} \sum_{x \in V_n} \deg_{\mathfrak{F}}(x) = 2 \quad \text{a.s.}$$

and

$$\lim_{n \to \infty} |V_n|^{-1} \sum_{x \in V_n} \mathbf{E}\big[\deg_{\mathfrak{F}}(x)\big] = 2 .$$

In particular, if G is a transitive graph such as $\mathbb{Z}^d$, then every vertex has expected degree 2 in both the free spanning forest and the wired spanning forest.

10.7. Let (T, c) be a network on a tree and $e \in T$. Show that $\mathsf{WSF}[e \in \mathfrak{F}] < 1$ iff both components of $T \setminus e$ are transient.

10.8. Let G be a graph obtained by identifying some vertices of a finite, connected graph H, keeping all edges of H, though some may become loops. Let μ_G and μ_H be the corresponding uniform spanning tree measures. Show that $\mu_G(\mathscr{A}) \leq \mu_H(\mathscr{A})$ for every increasing event $\mathscr{A}$ depending on the edges of G.

10.9. Show that the number of trees in the free spanning forest on a network is stochastically dominated by the number in the wired spanning forest on the network. Show that if the number of trees in the free spanning forest is a.s. finite, then, in distribution, it equals the number in the wired spanning forest iff FSF = WSF.

10.10. Give an amenable graph on which FSF $\neq$ WSF.

10.11. Show that every transitive, amenable network has unique currents.

10.12. (**The Riemann–Lebesgue Lemma**) Show that if $F \in L^1(\mathbb{T}^d)$ and

$$f(x) = \int_{\mathbb{T}^d} F(\alpha) e^{2\pi i x \cdot \alpha} \, d\alpha \,,$$

then $\lim_{|x| \to \infty} f(x) = 0$. *Hint:* This is obvious if F is a ***trigonometric polynomial***, that is, a finite linear combination of functions $\alpha \mapsto e^{2\pi i x \cdot \alpha}$. The Stone–Weierstrass theorem implies that such functions are dense in $L^1(\mathbb{T}^d)$.

10.13. Join two copies of the usual nearest-neighbor graph of $\mathbb{Z}^3$ by an edge at their origins. How many trees does the free uniform spanning forest have? How many does the wired uniform spanning forest have?

10.14. (**Parseval's Identity**) Suppose that $F \in L^1(\mathbb{T}^d)$ and that $f(x) := \int_{\mathbb{T}^d} F(\alpha) e^{2\pi i x \cdot \alpha} \, d\alpha$ for $x \in \mathbb{Z}^d$. Show that $\int_{\mathbb{T}^d} |F(\alpha)|^2 \, d\alpha = \sum_{\mathbb{Z}^d} |f(x)|^2$. *Hint:* Prove that the functions $G_x(\alpha) := e^{-2\pi i x \cdot \alpha}$ for $x \in \mathbb{Z}^d$ form an orthonormal basis of $L^2(\mathbb{T}^d)$.

10.15. Let (T, c) be a network on a tree. Show that $\mathsf{WSF}[\mathfrak{F} \neq T] = 1$ iff there is some edge $e \in T$ such that both components of $T \setminus e$ are transient iff $\mathsf{WSF}[\mathfrak{F} \neq T] > 0$.

10.16. Let (T, c) be a transient network on a tree with bounded conductances $c(\cdot)$. Show that there is some $e \in T$ such that both components of $T \backslash e$ are transient (with respect to c).

10.17. Give a transient graph such that the wired uniform spanning forest has a single tree with more than one end.

10.18. Show that Theorem 10.49 holds in the quasi-transitive case as well.

10.19. Consider a random walk until it first reaches distance n from its starting point and erase cycles. Show that on a general recurrent network, a weak limit of these random paths need not exist.

10.20. Let $\langle G_n \rangle$ be an induced exhaustion of a recurrent network G. Consider the network random walk $\langle X_k \,;\, k \geq 0 \rangle$ started from $o \in G$. Denote by τ_n the first exit time of G_n, and let L_n be the loop erasure of the path $\langle X_k \,;\, 0 \leq k \leq \tau_n \rangle$. Show that if the uniform spanning tree T_G in G has one end a.s., then the random paths L_n converge weakly to the law of the unique ray from o in T_G. In particular, show that this applies if G is a proper, plane network with a locally finite, recurrent dual.

10.21. Let G be the Cayley graph of Γ with respect to S. Assign positive conductances to the edges in such a way that they depend only on the generators associated to the edges. Show that still $\mathbf{E}_{\mathsf{FSF}}[\deg_{\mathfrak{F}} o] = 2 + 2\beta_1(\Gamma)$.

10.11 Additional Exercises

10.22. Let $\langle G_n \rangle$ be an exhaustion of G by finite, induced subgraphs. Write F_n for the set of spanning forests of G_n such that each component tree includes exactly one vertex of the inner vertex boundary of G_n. Show that WSF is the limit as $n \to \infty$ of the uniform measure on F_n.

10.23. Show that the FSF of the usual Cayley graph of $\mathbb{Z}_2 * \mathbb{Z}_3$ (shown in Figure 3.3) is a tree whose branching number equals 1.35^+.

10.24. Let G be exhausted by finite, induced subgraphs G_n and μ_n^{F} be the uniform spanning tree measure on G_n. Let also μ_n^{W} be the uniform spanning tree measure on G_n^{W}. Show that for each n, we have $\mu_n^{\mathrm{F}} \geqslant \mu_{n+1}^{\mathrm{F}} \upharpoonright 2^{\mathsf{E}(G_n)}$ and $\mu_n^{\mathrm{W}} \leqslant \mu_{n+1}^{\mathrm{W}} \upharpoonright 2^{\mathsf{E}(G_n^{\mathrm{W}})}$.

10.25. Consider the free or wired uniform spanning forest measure on an infinite, transient network G. Let X and Y be increasing random variables on $\{0, 1\}^{\mathsf{E}(G)}$ with finite second moments that depend on disjoint sets of edges. Show that $\mathbf{E}[XY] \leq \mathbf{E}[X]\,\mathbf{E}[Y]$.

10.26. Let $\langle X_n \rangle$ be the network random walk on a transitive network. Show that if the speed is 0, that is, $\lim_{n\to\infty} \mathrm{dist}(X_0, X_n)/n = 0$ a.s., then all harmonic Dirichlet functions on the network are constant and WSF = FSF.

10.27. Let G be a planar Cayley graph. Show that simple random walk on G is transient iff G is nonamenable iff G has exponential growth.

10.28. Let G be the edge graph of a degree-3 hyperbolic tessellation all of whose faces have the same number of sides. Show that the probability that the degree of a vertex in the wired uniform spanning forest on G is 1, 2, or 3 is, respectively, $1/4$, $1/2$, and $1/4$.

10.29. Let (G, c) be a denumerable network with an exhaustion by finite, induced subnetworks G_n. Fix $o \in \mathsf{V}(G_1)$. Let Z_n be the canonical Gaussian field on G_n from Exercise 2.137 (with $W = \{o\}$ and $u(o) = 0$ there). Show that the weak limit of Z_n exists; it is called the *free canonical Gaussian field*. Show that if Z_n^{W} denotes the canonical Gaussian field on the wired network G_n^{W}, then the weak limit of Z_n^{W} also exists, called the *wired canonical Gaussian field* (pinned at o). Let $X(e)$ be independent, normal random variables with variance $r(e)$ for $e \in \mathsf{E}_{1/2}$. Show that if $i_{\mathrm{F}}^{x,o}$ denotes the free unit current flow from x to o, then $x \mapsto \sum_{e \in \mathsf{E}_{1/2}} i_{\mathrm{F}}^{x,o}(e)X(e)$ has the law of the free canonical Gaussian field, where the sum converges in L^2 and a.s. Similarly, show that if $i_{\mathrm{W}}^{x,o}$ denotes the wired unit current flow from x to o, then $x \mapsto \sum_{e \in \mathsf{E}_{1/2}} i_{\mathrm{W}}^{x,o}(e)X(e)$ has the law of the wired canonical Gaussian field. Finally, show that the wired and free canonical Gaussian fields coincide in distribution iff $\mathbf{HD} = \mathbb{R}$.

10.30. Consider the canonical Gaussian field Z on $\mathbb{Z}^d$ with $Z(\mathbf{0}) = 0$, as defined in Exercise 10.29. Show that $\{Z(x)\,;\ x \in \mathbb{Z}^d\}$ is a tight collection of random variables iff $d \geq 3$.

10.31. Let (G, c) be a transient network with an exhaustion by finite, induced subnetworks G_n. Show that if Z_n^{W} denotes the canonical Gaussian field on the wired network G_n^{W} that is pinned to be 0 at z_n, the vertex resulting from identifying the complement of G_n, then the weak limit of Z_n^{W} exists, called the *wired canonical Gaussian field*, which we can think of as "pinned at infinity." Show that if i_x denotes the unit current flow from x to infinity and $X(e)$ are independent, normal random variables with variance $r(e)$ for $e \in \mathsf{E}_{1/2}$, then $x \mapsto \sum_{e \in \mathsf{E}_{1/2}} i_x(e)X(e)$ has the law of the wired canonical Gaussian field, where the sum converges in L^2 and a.s. Show that the wired canonical Gaussian field has a law that is invariant under all automorphisms of (G, c) and that has a trivial tail σ-field.

10.32. Let (G, c) be a transient network with an exhaustion by finite, induced subnetworks G_n. Let $X(e)$ be independent, normal random variables with variance $r(e)$ for $e \in \mathsf{E}_{1/2}$; put $X(-e) := -X(e)$ for $e \notin \mathsf{E}_{1/2}$. Given random walks starting at each $x \in \mathsf{V}(G)$, define $S_n(x)$ to be the sum of $X(e)$ over the edges e traversed by the random walk starting at x until it exits G_n for the first time. Show that $x \mapsto \lim_{n\to\infty} \mathbf{E}[S_n(x) \mid X]$ exists and has the law of the wired canonical Gaussian field.

10.33. Let G be an infinite graph and H be a finite graph. Consider the Cartesian product graph $G \square H$. Fix $x \in V(G)$ and $y \in V(H)$. Show that the free spanning forests $\mathfrak{F}_1^{\mathrm{F}}$ in $G \square H$ and $\mathfrak{F}_2^{\mathrm{F}}$ in G and the wired spanning forests $\mathfrak{F}_1^{\mathrm{W}}$ in $G \square H$ and $\mathfrak{F}_2^{\mathrm{W}}$ in G satisfy

$$\left(\mathbf{E}\big[\deg_{\mathfrak{F}_1^{\mathrm{F}}}(x,y)\big] - \mathbf{E}\big[\deg_{\mathfrak{F}_1^{\mathrm{W}}}(x,y)\big]\right)|V(H)| = \mathbf{E}\big[\deg_{\mathfrak{F}_2^{\mathrm{F}}} x\big] - \mathbf{E}\big[\deg_{\mathfrak{F}_2^{\mathrm{W}}} x\big].$$

10.34. Let G be a plane network all of whose faces have a finite number of sides. Show that G has unique currents iff its dual $G^\dagger$ has unique currents.

10.35. Let G be the edge graph of a degree-3 hyperbolic tessellation all of whose faces have d sides. Show that the probability that the degree of a vertex in the free uniform spanning forest on G is 1, 2, or 3 is, respectively, $9/d^2$, $6/d - 18/d^2$, and $(1 - 3/d)^2$.

10.36. Let G be a plane, regular graph of degree d with regular dual of degree $d^\dagger$. Show that the FSF-expected degree of each vertex in G is $d(1 - 2/d^\dagger)$.

10.37. Let G be the usual Cayley graph of the (p, q, r)-triangle group, where $1/p + 1/q + 1/r \le 1$, shown in Figure 6.1 for $(2, 3, 7)$ and defined in Exercise 8.47. It has three generators, which are reflections in the sides of a fundamental triangle. Show that the expected degree of the FSF of G is $3 - 1/p - 1/q - 1/r$.

10.38. Complete the following outline of an alternative proof of Proposition 10.16. Let $\psi(\alpha) := 1 - \varphi(\alpha)/(2d)$, where φ is as defined in (10.8). For all $n \in \mathbb{N}$ and $x \in \mathbb{Z}^d$, we have $p_n(\mathbf{0}, x) = \int_{\mathbb{T}^d} \psi(\alpha)^n e^{2\pi i x \cdot \alpha} \, d\alpha$. Therefore, $\mathscr{G}(\mathbf{0}, x)/(2d)$ equals the right-hand side of (10.10). Now apply Proposition 2.1.

10.39. For a function $f \in L^1(\mathbb{T}^d)$ and $x \in \mathbb{Z}^d$, define $\widehat{f}(x) := \int_{\mathbb{T}^d} f(\alpha) e^{-2\pi i x \cdot \alpha} \, d\alpha$. Let Y be the transfer-current matrix for the hypercubic lattice $\mathbb{Z}^d$. Write $u_k := \mathbf{1}_{\{k\}}$ for the vector with a 1 in the kth place and 0s elsewhere. Let $e_x^k := [x, x + u_k]$. Show that $Y(e_{\mathbf{0}}^1, e_x^k) = \widehat{f_k}(x)$, where

$$f_k(\alpha_1, \ldots, \alpha_d) := \frac{(1 - e^{2\pi i \alpha_1})(1 - e^{-2\pi i \alpha_k})}{4 \sum_{j=1}^d \sin^2 \pi \alpha_j} \qquad (1 \le k \le d).$$

10.40. Let $d \ge 3$, and let e be any edge of $\mathbb{Z}^d$. For $n \in \mathbb{Z}$, let X_n be the indicator that $e + (n, n, \ldots, n)$ lies in the uniform spanning forest of $\mathbb{Z}^d$. Show that X_n are i.i.d. The same holds for the other 2^{d-1} collections of translates given by changing the signs of some of the last $d - 1$ coordinates.

10.41. Show that if FSF = WSF, then for all cylinder events $\mathscr{A}$ and $\epsilon > 0$, there is a finite set of edges K such that

$$\sup\{|\mathrm{WSF}[\mathscr{A} \mid \mathscr{B}] - \mathrm{WSF}[\mathscr{A}]| \; ; \; \mathscr{B} \in \mathscr{F}(\mathsf{E} \setminus K)\} < \epsilon.$$

10.42. For $k \ge 1$, we say that an action of Γ on $(\Omega, \mathbf{P})$ is ***mixing of order*** k (or k-***mixing***) if, for any events $A_1, \ldots, A_{k+1}$, we have

$$\lim_{\substack{\gamma_1, \ldots, \gamma_k \to \infty, \\ \forall i \ne j \; \gamma_i \gamma_j^{-1} \to \infty}} \mathbf{P}(A_1, \gamma_1 A_2, \gamma_2 A_3, \ldots, \gamma_k A_{k+1}) = \mathbf{P}(A_1) \mathbf{P}(A_2) \cdots \mathbf{P}(A_{k+1}).$$

Let $\Gamma \subseteq \mathrm{Aut}(G)$ have an infinite orbit and $\mathbf{P}$ be a Γ-invariant probability measure on $(\Omega, 2^{\mathsf{E}(G)})$ with a trivial tail. Show that $\mathbf{P}$ is mixing of all orders.

10.43. Define I_N as in (10.19). Show that $\mathbf{E}[I_N^k] \le (k!)^2 (\mathbf{E} I_N)^k$ for every $k \ge 1$.

10.44. Let $\alpha(w_1, \ldots, w_K)$ be the probability that independent random walks started at $w_1, \ldots, w_K$ have no pairwise intersections. Let $\mathscr{B}(w_1, \ldots, w_K)$ be the event that the number of pairwise intersections among the random walks $\langle X_n(w_i) \rangle$ is finite. Show that

$$\lim_{n \to \infty} \alpha\big(X_n(w_1), \ldots, X_n(w_K)\big) = \mathbf{1}_{\mathscr{B}(w_1, \ldots, w_K)} \quad \text{a.s.}$$

10.45. Suppose that the number of components in the WSF is a.s. $k < \infty$ and that $w_1, \ldots, w_k$ are vertices such that with positive probability, independent random walks started at $w_1, \ldots, w_k$ have no pairwise intersections. Let $\{X(w_i);\ 1 \le i \le k\}$ be independent random walks indexed by their initial states. Consider the random functions

$$h_i(w) := \mathbf{P}\big[Y(w) \text{ intersects } X(w_i) \text{ i.o.} \mid X(w_1), \ldots, X(w_k)\big],$$

where the random walk $Y(w)$ starts at w and is independent of all $X(w_i)$. Show that a.s. on the event that $X(w_1), \ldots, X(w_k)$ have pairwise disjoint paths, the functions $\{h_1, \ldots, h_k\}$ form a basis for $\mathbf{BH}(G)$, the vector space $\mathbf{BH}(G)$ of bounded harmonic functions on G. Deduce that if the number of components of the WSF is finite a.s., then it a.s. equals the dimension of $\mathbf{BH}(G)$.

10.46. It follows from Corollary 10.25 that if two transitive graphs are roughly isometric, then the a.s. number of trees in the wired spanning forest of one graph is the same as in the other. Show that this is false without the assumption of transitivity.

10.47. Find an example of a graph G of bounded degree such that
   (a) FSF_G has two components a.s.;
   (b) FSF_G has two components a.s. and $\mathsf{FSF}_G \ne \mathsf{WSF}_G$.

10.48. Consider a d-ary tree (T, c) with conductances corresponding to the random walk RW_λ. Show that a.s. all the trees in the wired spanning forest on (T, c) have branching number λ for $1 \le \lambda \le d$.

10.49. Let (T, c) be an arbitrary tree with arbitrary conductances and root o. For a vertex $x \in T$, let $\alpha(x)$ denote the effective conductance of T^x (from x to infinity). Consider the independent percolation on T that keeps the edge $e(x)$ preceding x with probability $c(e(x))/\big[c(e(x)) + \alpha(x)\big]$.
   (a) Show that the WSF on T has a tree with more than one end with positive probability iff this percolation on T has an infinite cluster a.s.
   (b) In the case of WSF on a spherically symmetric tree with unit conductances, show that the conditions in (a) are equivalent to

$$\sum_n \frac{1}{|T_n|^2 \rho_n} \prod_{k=1}^{n-1} \Big(1 + \frac{1}{|T_k| \rho_k}\Big) < \infty,$$

where $\rho_n := \sum_{k>n} 1/|T_k|$. In particular, show that this holds if $|T_n| \approx n^a$ for some $a > 1$.
   (c) Show that for a spherically symmetric tree with unit conductances, a.s. either every tree in the WSF has a single end or a.s. every tree has infinitely many ends.

10.50. Show that if G is a transitive graph such that the balls of radius r have cardinality asymptotic to αr^d for some positive finite α and d, then the lower bound of Lemma 10.46 is optimal up to a constant factor. In other words, show that in this case,

$$\liminf_{|K| \to \infty} R(2|K|) \frac{|\partial_V^{\text{in}} K|}{|K|} < \infty.$$

10.51. Let G be the grandparent graph of Example 7.1 defined using the end ξ of a regular tree. Consider the FSF on $G \square \mathbb{Z}$. Show that the components are distinguishable by an automorphism-invariant property.

10.52. Let G be the Cayley graph of the free product $\mathbb{Z}^d * \mathbb{Z}_2$, where $\mathbb{Z}_2$ is the group with two elements, with the obvious generating set. Show that the FSF is connected iff $d \le 4$ and FSF $\ne$ WSF for all $d > 0$.

11 | Minimal Spanning Forests

In Chapters 4 and 10, we looked at spanning trees chosen uniformly at random and their analogues for infinite graphs. We saw that they are intimately connected to random walks. Another measure on spanning trees and forests has been studied a great deal, especially in optimization theory. This measure, minimal spanning trees and forests, turns out to be connected to percolation rather than to random walks. In fact, one of the measures on forests is closely tied to critical bond percolation and invasion percolation, whereas the other is related to percolation at the uniqueness threshold. However, many fundamental questions remain open for both types of minimal spanning forest measures.

On the whole, minimal spanning forests share many similarities with uniform spanning forests. In some cases, we know a result for one measure whose analogue is only conjectured for the other. Occasionally, there are striking differences between the two settings.

The standard coupling of Bernoulli bond percolation will be ubiquitous in this chapter, so much that we need some special notation for it. Namely, given labels $U: E(G) \to \mathbb{R}$, we'll write $G[p]$ for the subgraph formed by the edges $\{e ; \ U(e) < p\}$. (In previous chapters, we denoted $G[p]$ by ω_p, but in this chapter, both G and p often assume more complicated expressions.) Our treatment is drawn from Lyons, Peres, and Schramm (2006), as are most of the results.

11.1 Minimal Spanning Trees

Let $G = (V, E)$ be a finite, connected (multi)graph. Since loops cannot belong to trees, we will ignore any loops that G may have. Given an injective function, $U: E \to \mathbb{R}$, we'll refer to $U(e)$ as the *label* of e. The labeling U induces a total ordering on E, where $e < e'$ if $U(e) < U(e')$. We'll say that e is *lower* than e' and that e' is *higher* than e when $e < e'$.

Define T_U to be the subgraph whose vertex set is V and whose edge set consists of all edges $e \in E$ whose endpoints cannot be joined by a path containing only edges lower than e. We claim that T_U is a spanning tree. First, the largest edge in any cycle of G is not in T_U, whence T_U is a forest. Second, if $\varnothing \neq A \subsetneq V$, then the lowest edge of G connecting A with $V \setminus A$ must belong to T_U, which shows that T_U is connected. Thus, it is a spanning tree.

▷ **Exercise 11.1.**
Show that among all spanning trees of a finite graph, T_U is the unique one that has minimum edge-label sum, $\sum_{e \in T} U(e)$.

When $\langle U(e)\,;\ e \in \mathsf{E}\rangle$ are independent, uniform $[0,1]$ random variables, U is a.s. injective and the law of the corresponding spanning tree T_U is called simply the ***minimal spanning tree (measure)***. It is a probability measure on 2^{E}. Note that this independent labeling U induces a uniform random ordering on E.

There is an easy monotonicity principle for the minimal spanning tree measure, which is analogous to a similar principle, Lemma 10.3 and Exercise 10.8, for uniform spanning trees:

Proposition 11.1. *Let G and H be connected, finite graphs. Denote by T_G and T_H the corresponding minimal spanning trees. If G is a subgraph of H, then T_G stochastically dominates $T_H \cap \mathsf{E}(G)$. On the other hand, if G is obtained by identifying some vertices in H, then T_G is stochastically dominated by $T_H \cap \mathsf{E}(G)$.*

Proof. We'll prove the first part, as the second part is virtually identical. Let $U(e)$ be i.i.d. uniform $[0,1]$ random variables for $e \in \mathsf{E}(H)$. We use these labels to construct both T_G and T_H. In this coupling, if $[x, y] \in \mathsf{E}(G)$ is contained in T_H, in other words, there is no path in $\mathsf{E}(H)$ joining x and y that uses only lower edges, then a fortiori there is no such path in $\mathsf{E}(G)$, so that $[x, y]$ is also contained in T_G. That is, $T_H \cap \mathsf{E}(G) \subseteq T_G$, which proves the result. ◄

The reader might expect next to see negative correlations for the minimal spanning tree, as we saw for uniform spanning trees. Surprisingly, however, the presence of two edges can be positively correlated! To see this, we will use the following formula for computing probabilities of spanning trees. Let MST denote the minimal spanning tree measure on a finite, connected graph.

Proposition 11.2. *Let G be a finite, connected graph. Given a set F of edges, let $N(F)$ be the number of edges of G that do not become loops when each edge in F is contracted. Note that $N(\varnothing)$ is the number of edges of G that are not loops. Let $N'(e_1, \ldots, e_k) := \prod_{j=0}^{k-1} N(\{e_1, \ldots, e_j\})$ (which does not depend on e_k). Let $T = \{e_1, \ldots, e_n\}$ be a spanning tree of G. Then*

$$\mathsf{MST}(T) = \sum_{\sigma} N'(e_{\sigma(1)}, \ldots, e_{\sigma(n)})^{-1},$$

where the sum is over all permutations σ of $\{1, 2, \ldots, n\}$.

Proof. To make the dependence on G explicit, we write $N(F) = N(G; F)$. Note that $N(G/F; \varnothing) = N(G; F)$, where G/F is the graph G with each edge in F contracted. Given the edge labels U, one way to find the minimal spanning tree T_U is to choose the lowest edge e_1 that is not a loop, put this in T_U, then choose the lowest edge e_2 that is not a loop in G/e_1 and put this in T_U, and so forth. This is known as Prim's algorithm. Our proof involves simply keeping track of the probabilities as we follow that algorithm.

Given an edge e that is not a loop, the chance that e is the lowest edge in the minimal spanning tree of G equals $N(G; \varnothing)^{-1}$. Furthermore, given that this is the case, the ordering on the nonloops of the edge set of G/e is uniform. Thus, if f is not a loop in G/e, then the chance that f is the next lowest edge in the minimal spanning tree of G given that e is the lowest edge in the minimal spanning tree of G equals $N(G/e; \varnothing)^{-1} = N(G; \{e\})^{-1}$. Hence we may easily condition, contract, and repeat.

Thus, the probability that the minimal spanning tree is T *and* that $e_{\sigma(1)} < \cdots < e_{\sigma(n)}$ is equal to

$$\prod_{j=0}^{n-1} N\big(G/\{e_{\sigma(1)}, e_{\sigma(2)}, \ldots, e_{\sigma(j)}\}; \varnothing\big)^{-1} = N'\big(e_{\sigma(1)}, \ldots, e_{\sigma(n)}\big)^{-1}.$$

Summing this over all possible induced orderings of T gives $\mathsf{MST}(T)$.   ◀

An example of a graph where MST has positive correlations is provided by the following exercise.

▷ **Exercise 11.2.**
Construct G as follows. Begin with the complete graph, K_4. Let e and f be two of its edges that do not share endpoints. Replace e by three edges in parallel, e_1, e_2, and e_3, that have the same endpoints as e. Likewise, replace f by three parallel edges f_i. Show that in G,

$$\mathsf{MST}[e_1, f_1 \in T] > \mathsf{MST}[e_1 \in T]\,\mathsf{MST}[f_1 \in T].$$

The following difference from the uniform spanning tree must also be kept in mind:

▷ **Exercise 11.3.**
Show that given an edge, e, the minimal spanning tree measure on G conditioned on the event not to contain e need not be the same as the minimal spanning tree measure on $G \backslash e$, the graph G with e deleted.

In Figure 11.1, we show the distances to the lower left vertex in a minimal spanning tree on a 100×100 square grid as well as the path in the tree that joins the opposite corners. However, unlike the case of uniform spanning trees (Figure 4.5), it does not seem simple to sample from the minimal spanning tree distribution conditional on a given distance profile. Also, the distances seem to be generally lower than for the uniform spanning tree: see the value for "D_f" reported by Wieland and Wilson (2003), Table III.

▷ **Exercise 11.4.**
Find an example of a finite graph G with vertex $o \in G$ such that there are two spanning trees T and T' of G having the properties that for all x, the distance from x to o in T is the same as in T', yet T and T' are not equally likely under the minimal spanning tree measure.

This is all the theory of minimal spanning trees that we'll need! We can move directly to infinite graphs.

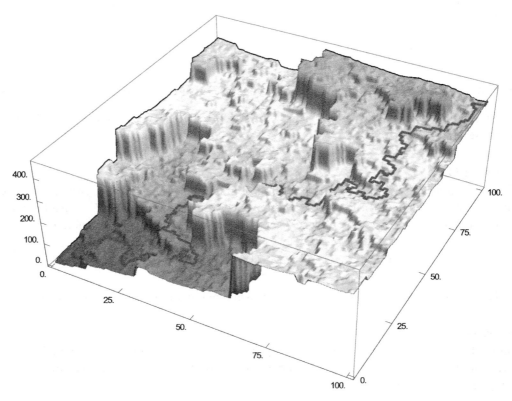

color
plate **Figure 11.1.**   The distances to a vertex in a minimal spanning tree
in a 100×100 grid and the path between the opposite corners.

11.2 Deterministic Results

Just as there were for the uniform spanning trees, there are free and wired extensions to infinite graphs of the minimal spanning trees. Unlike the uniform case, however, the minimal case can be done without weak limits and, indeed, without probability whatsoever. We give these deterministic definitions and associated results in this section.

Let $G = (\mathsf{V}, \mathsf{E})$ be an infinite, connected, locally finite graph and $U \colon \mathsf{E} \to \mathbb{R}$ be an injective labeling of the edges. Let $\mathfrak{F}_{\mathrm{f}} = \mathfrak{F}_{\mathrm{f}}(U) = \mathfrak{F}_{\mathrm{f}}(U, G)$ be the set of edges $e \in \mathsf{E}$ such that in every path in G connecting the endpoints of e, there is at least one edge e' with $U(e') \geq U(e)$. When $\langle U(e) \,; \ e \in \mathsf{E} \rangle$ are independent, uniform random variables in $[0, 1]$, the law of $\mathfrak{F}_{\mathrm{f}}$ (or sometimes, $\mathfrak{F}_{\mathrm{f}}$ itself) is called the ***free minimal spanning forest*** on G and is denoted by FMSF or FMSF(G).

An ***extended path*** joining two vertices $x, y \in \mathsf{V}$ is either a simple path in G joining them or the union of a simple infinite path starting at x and a disjoint simple infinite path starting at y. (The latter possibility may be considered as a simple path connecting x and y through ∞.) Let $\mathfrak{F}_{\mathrm{w}} = \mathfrak{F}_{\mathrm{w}}(U) = \mathfrak{F}_{\mathrm{w}}(U, G)$ be the set of edges $e \in \mathsf{E}$ such that in every extended path joining the endpoints of e, there is at least one edge e' with $U(e') \geq U(e)$. Again, when U is chosen according to the product measure on $[0, 1]^E$, we call $\mathfrak{F}_w$ the *wired minimal spanning forest* on G. The law of $\mathfrak{F}_w$ is denoted WMSF or WMSF(G).

▷ **Exercise 11.5.**
Show that $\mathfrak{F}_w(U)$ consists of those edges e for which there is a finite set of vertices $W \subset V$ such that e is the least edge joining W to $V \setminus W$.

Clearly, $\mathfrak{F}_w(U) \subset \mathfrak{F}_f(U)$. Note that $\mathfrak{F}_w(U)$ and $\mathfrak{F}_f(U)$ are indeed forests, since in every cycle of G, the edge e with $U(e)$ maximal is not present in either $\mathfrak{F}_f(U)$ or in $\mathfrak{F}_w(U)$. In addition, all the connected components in $\mathfrak{F}_f(U)$ and in $\mathfrak{F}_w(U)$ are infinite. Indeed, the lowest edge joining any finite vertex set to its complement belongs to both forests.

One of the nice properties that minimal spanning forests have is that there are these direct definitions on infinite graphs. Although we won't need this property, one can also describe them as weak limits, parallel to the definitions for uniform spanning forests:

▷ **Exercise 11.6.**
Consider an increasing sequence of finite, nonempty, connected (not necessarily induced) subgraphs $G_n \subset G$ ($n \in \mathbb{N}$) such that $\bigcup_n G_n = G$. For $n \in \mathbb{N}$, let G_n^W be the graph obtained from G by identifying the vertices outside of G_n to a single vertex, then removing all resulting loops based at that vertex. Let $T_n(U)$ and $T_n^W(U)$ denote the minimal spanning trees on G_n and G_n^W, respectively, that are induced by the labels U. Show that $\mathfrak{F}_f(U) = \lim_{n \to \infty} T_n(U)$ and that $\mathfrak{F}_w(U) = \lim_{n \to \infty} T_n^W(U)$ in the sense that for every $e \in \mathfrak{F}_f(U)$, we have $e \in T_n(U)$ for every sufficiently large n, for every $e \notin \mathfrak{F}_f(U)$ we have $e \notin T_n(U)$ for every sufficiently large n, and similarly for $\mathfrak{F}_w(U)$. Deduce that $T_n(U) \Rightarrow$ FMSF and $T_n^W(U) \Rightarrow$ WMSF.

It will be useful to make more explicit the comparisons that determine which edges belong to the two spanning forests. Define

$$Z_f(e) = Z_f^U(e) := \inf_{\mathcal{P}} \max\{U(e') ;\ e' \in \mathcal{P}\},$$

where the infimum is over simple paths $\mathcal{P}$ in $G \setminus e$ that connect the endpoints of e; if there are none, the infimum is defined to be ∞. Thus, $\mathfrak{F}_f(U) = \{e ;\ U(e) \le Z_f(e)\}$. In the random case, since $Z_f(e)$ is independent of $U(e)$ and $U(e)$ is a continuous random variable, we can also write $\mathfrak{F}_f(U) = \{e ;\ U(e) < Z_f(e)\}$ a.s. Similarly, define

$$Z_w(e) = Z_w^U(e) := \inf_{\mathcal{P}} \sup\{U(e') ;\ e' \in \mathcal{P}\},$$

where the infimum is over extended paths $\mathcal{P}$ in $G \setminus e$ that join the endpoints of e. Again, if there are no such extended paths, then the infimum is defined to be ∞. Thus,

$$\{e ;\ U(e) < Z_w(e)\} \subseteq \mathfrak{F}_w(U) \subseteq \{e ;\ U(e) \le Z_w(e)\}$$

and, in the random case, $\mathfrak{F}_w(U) = \{e ;\ U(e) < Z_w(e)\}$ a.s. The infimum in the definition of $Z_w(e)$ is attained whenever there is some extended path in $G \setminus e$ that connects the endpoints of e, as we'll see in the course of proving the next lemma.

It turns out that there are also dual definitions for Z_f and Z_w. To state these, we use the following terminology: if $W \subseteq V$, then the set of edges $\partial_E W$ joining W to $V \setminus W$ is a *cut*.

Lemma 11.3. (Dual Criteria) *For any injection* $U: \mathsf{E} \to \mathbb{R}$ *on any locally finite graph* G, *we have*

$$Z_{\mathrm{f}}(e) = \sup_{\Pi} \inf\{U(e');\ e' \in \Pi \setminus \{e\}\}, \tag{11.1}$$

where the supremum is over all cuts Π *that contain* e; *also, this supremum is attained. Similarly,*

$$Z_{\mathrm{w}}(e) = \sup_{\Pi} \inf\{U(e');\ e' \in \Pi \setminus \{e\}\}, \tag{11.2}$$

where now the supremum is over all cuts Π *containing* e *such that* $\Pi = \partial_{\mathsf{E}} W$ *for some finite* $W \subset \mathsf{V}$.

Proof. We first verify (11.1). If $\mathcal{P}$ is a simple path in $G \backslash e$ that connects the endpoints of e and Π is a cut that contains e, then $\mathcal{P} \cap (\Pi \setminus \{e\}) \neq \varnothing$, so

$$\max\{U(e');\ e' \in \mathcal{P}\} \geq \inf\{U(e');\ e' \in \Pi \setminus \{e\}\}.$$

This proves one inequality ($\geq$) in (11.1). To prove the reverse inequality, fix one endpoint x of e, and let W be the vertex set of the component of x in $(G \backslash e)[Z_{\mathrm{f}}(e)]$. Then $\Pi := \partial_{\mathsf{E}} W$ is a cut that contains e by definition of Z_{f}. Using this Π in the right-hand side of (11.1) yields the inequality $\leq$ in (11.1) and shows that the supremum there is achieved.

The inequality $\geq$ in (11.2) is proved in the same way as in (11.1). For the other direction, we dualize the preceding proof. Let Z denote the right-hand side in (11.2); we may assume that $Z < \infty$. Let W be the vertex set of the connected component of one of the endpoints of e in the set of edges $e' \neq e$ such that $U(e') \leq Z$. We clearly have $U(e') > Z$ for each $e' \in \partial_{\mathsf{E}} W \setminus \{e\}$. Thus, by the definition of Z, the other endpoint of e is in W if W is finite, in which case there is a path in $G \backslash e$ connecting the endpoints of e that uses only edges with labels at most Z. The same argument applies with the roles of the endpoints of e switched. If the sets W corresponding to both endpoints of e are infinite, then there is an extended path $\mathcal{P}$ connecting the endpoints of e in $G \backslash e$ with $\sup\{U(e');\ e' \in \mathcal{P}\} \leq Z$. This completes the proof of (11.2) and also shows that the infimum in the definition of $Z_{\mathrm{w}}(e)$ is attained (when $Z < \infty$). ◄

These lead to interesting and useful representations of the two minimal spanning forests, the wired one as an increasing union of "small" subtrees and the free one as a decreasing intersection of "large" supergraphs. (If there are edges with $U(e) = Z_{\mathrm{w}}(e)$, then the following representation misses them; but they will a.s. not occur when U is random.)

Proposition 11.4. *Let* $U: \mathsf{E} \to \mathbb{R}$ *be an injective labeling of the edges of a locally finite graph* $G = (\mathsf{V}, \mathsf{E})$. *For each* $p < 1$, *the edges* e *with* $U(e) < p Z_{\mathrm{w}}(e)$ *form subtrees of* $\mathfrak{F}_{\mathrm{w}}(U)$ *with at most one end each, whereas the edges* e *with* $1 - U(e) > p[1 - Z_{\mathrm{f}}(e)]$ *form a supergraph of* $\mathfrak{F}_{\mathrm{f}}(U)$ *that is connected.*

Proof. Fix $p < 1$. Write η_p for the graph formed by the edges e with $U(e) < p Z_{\mathrm{w}}(e)$. For the assertion on $\mathfrak{F}_{\mathrm{w}}(U)$, it suffices to show that η_p does not contain any simple, bi-infinite path. Let $\mathcal{P}$ be a simple, bi-infinite path and $\alpha := \sup_{e \in \mathcal{P}} U(e)$. If $\mathcal{P} \subset \eta_p$, then we would have

$$\forall e \in \mathcal{P} \quad U(e) < p Z_{\mathrm{w}}(e) \leq p \sup_{\mathcal{P}} U = p \alpha.$$

This would imply that $\alpha \leq p \alpha$, whence $\alpha = 0$, which is clearly impossible. So $\mathcal{P}$ is not a subset of η_p, as desired.

Write ξ_p for the graph formed by the edges e with $1 - U(e) > p[1 - Z_f(e)]$. Consider any nonempty cut Π in G, and let $\alpha := \inf_{e \in \Pi} U(e)$. Then $1 - \alpha = \sup_{\Pi}(1 - U)$, so we may choose $e \in \Pi$ to satisfy $1 - U(e) > p(1 - \alpha)$. By (11.1), $Z_f(e) \geq \inf_{\Pi \setminus \{e\}} U \geq \alpha$, whence $e \in \xi_p$. Since ξ_p intersects every nonempty cut, it is connected.   ◄

The ***invasion tree*** $T(x) = T_U(x)$ of a vertex x is defined as the increasing union of the trees t_n, where $t_0 := \{x\}$ and t_{n+1} is t_n together with the least edge joining t_n to a vertex not in t_n. (If G is finite, we stop when t_n contains V.) Invasion trees play a role in the wired minimal spanning forest similar to the role played by Wilson's method rooted at infinity in the wired uniform spanning forest:

▷ **Exercise 11.7.**
Let $U: \mathsf{E} \to \mathbb{R}$ be an injective labeling of the edges of a locally finite graph $G = (\mathsf{V}, \mathsf{E})$. Show that the union $\bigcup_{x \in \mathsf{V}} T_U(x)$ of all the invasion trees is equal to $\mathfrak{F}_{\mathrm{w}}(U)$.

Recall from Section 7.5 that the ***invasion basin*** $I(x)$ of a vertex x is defined as the union of the subgraphs G_n, where $G_0 := \{x\}$ and G_{n+1} is G_n together with the lowest edge not in G_n but incident to some vertex in G_n. Note that $I(x)$ has the same vertices as $T(x)$ but may have additional edges.

While the invasion basins (and trees) starting from different vertices can certainly differ, they can differ in only finitely many edges when they belong to the same component of $\mathfrak{F}_{\mathrm{w}}$.

Proposition 11.5. *Let $U: \mathsf{E} \to \mathbb{R}$ be an injective labeling of the edges of a locally finite graph $G = (\mathsf{V}, \mathsf{E})$. If x and y are vertices in the same component of $\mathfrak{F}_{\mathrm{w}}(U)$, then the symmetric differences $I(x) \bigtriangleup I(y)$ and $T_U(x) \bigtriangleup T_U(y)$ are finite.*

Proof. We prove only that $|I(x) \bigtriangleup I(y)| < \infty$, since the proof for $T_U(x) \bigtriangleup T_U(y)$ is essentially the same. It suffices to prove this when $e := [x, y] \in \mathfrak{F}_{\mathrm{w}}(U)$. Consider the connected components $C(x)$ and $C(y)$ of x and y in $G[U(e)]$. Not both $C(x)$ and $C(y)$ can be infinite, since $e \in \mathfrak{F}_{\mathrm{w}}(U)$. If both are finite, then invasion from each x and y will fill $C(x) \cup C(y) \cup \{e\}$ before invading elsewhere, and therefore $I(x) = I(y)$ in this case. Finally, if, say, $C(x)$ is finite and $C(y)$ is infinite, then $I(x) = C(x) \cup \{e\} \cup I(y)$.   ◄

Similarly, changing the edge labels on a finite set can change the free and wired spanning forests by only finitely many edges. This will be a very useful property for us.

Lemma 11.6. *Let G be any infinite, locally finite graph with distinct fixed labels $U(e)$ on its edges. Let $\mathfrak{F}$ be the corresponding free or wired minimal spanning forest. If the label $U(e)$ is changed at a single edge e, then the forest changes at most at e and at one other edge (an edge f with $U(f) = Z_f(e)$ or $Z_{\mathrm{w}}(e)$, respectively). More generally, if $\mathfrak{F}'$ is the corresponding forest when labels only in $K \subset \mathsf{E}$ are changed, then $|(\mathfrak{F} \bigtriangleup \mathfrak{F}') \setminus K| \leq |K|$.*

Proof. Consider first the free minimal spanning forest. Suppose the two values of $U(e)$ are u_1 and u_2 with $u_1 < u_2$. Let $\mathfrak{F}_1$ and $\mathfrak{F}_2$ be the corresponding free minimal spanning forests. Then $\mathfrak{F}_1 \setminus \mathfrak{F}_2 \subseteq \{e\}$. Suppose that $f \in \mathfrak{F}_2 \setminus \mathfrak{F}_1$. Then there must be a path $\mathcal{P} \subset G \setminus e$ joining the endpoints of e and including f such that $U(f) = \max_{\mathcal{P}} U > u_1$. Suppose that there were a path $\mathcal{P}' \subset G \setminus e$ joining the endpoints of e such that $\max_{\mathcal{P}'} U < U(f)$. Then $\mathcal{P} \cup \mathcal{P}'$ would contain a cycle including f but not e on which f has the maximum label. This contradicts $f \in \mathfrak{F}_2$. Therefore, $Z_f(e) = U(f)$. Since the labels are distinct, there is at most one such f.

For the wired minimal spanning forest, the proof is the same, only with "extended path" replacing "path" and "$Z_w(e)$" replacing "$Z_f(e)$."

The second conclusion in the lemma follows by induction from the first. ◄

11.3 Basic Probabilistic Results

Here are some of the easier analogues of several results on uniform spanning forests:

▷ **Exercise 11.8.**

Let G be a connected, locally finite graph. Prove the following.

(a) If G is edge-amenable, then the average degree of vertices in both the free and wired minimal spanning forests on G is a.s. 2.

(b) The free and wired minimal spanning forests on G are the same if they have a.s. the same finite number of trees or if the expected degree of every vertex is the same for both measures.

(c) The free and wired minimal spanning forests on G are the same on any transitive, amenable graph and have expected degree 2.

(d) If $\mathfrak{F}_w$ is connected a.s. or if each component of $\mathfrak{F}_f$ has a.s. one end, then WMSF(G) = FMSF(G).

(e) The measures WMSF(G) and FMSF(G) are invariant under all automorphisms of G, as is the law of $(\mathfrak{F}_w, \mathfrak{F}_f)$.

(f) If G is unimodular and transitive with WMSF(G) ≠ FMSF(G), then a.s. $\mathfrak{F}_f$ has a component with uncountably many ends and, in fact, with $p_c < 1$.

When are the free and wired minimal spanning forests the same? Say that a graph G has ***almost everywhere uniqueness*** (of the infinite cluster) if, for almost every $p \in (0, 1)$ in the sense of Lebesgue measure, there is a.s. at most one infinite cluster for Bernoulli(p) percolation on G. This is the analogue for minimal spanning forests of uniqueness of currents for uniform spanning forests (Proposition 10.14):

Proposition 11.7. *On any connected graph G, we have* FMSF = WMSF *iff G has almost everywhere uniqueness.*

Proof. Let $A(e)$ be the event that the two endpoints of e are in distinct infinite components of $(G \backslash e)[U(e)]$. When U is injective, $A(e)$ is the same as the event that $e \in \mathfrak{F}_f \setminus \mathfrak{F}_w$. Since $\mathfrak{F}_w \subset \mathfrak{F}_f$ and E is countable, FMSF = WMSF is equivalent to the statement that $\mathbf{P}[A(e)] = 0$ for all edges e. Write $f(e, p)$ for the probability that the endpoints of e belong to distinct infinite clusters with positive probability in Bernoulli(p) percolation on $G \backslash e$. Then we have $\mathbf{P}[A(e) \mid U(e)] = f(e, U(e))$ a.s., whence $\mathbf{P}[A(e)] = \int_0^1 f(e, p) \, dp$. Let $B(e) \subseteq (0, 1)$ be the set of p for which $f(e, p) > 0$. Then the preceding identity yields that $B(e)$ has measure zero iff $\mathbf{P}[A(e)] = 0$. Insertion and deletion tolerance show that $\bigcup_e B(e)$ is the set of p for which Bernoulli(p) percolation gives more than one infinite cluster with positive probability, whence $\bigcup_e B(e)$ has Lebesgue measure zero iff almost everywhere uniqueness holds. It follows that almost everywhere uniqueness holds iff $\mathbf{P}[A(e)] = 0$ for all e. ◄

Corollary 11.8. *On any graph G, if almost everywhere uniqueness fails, then* WMSF *is not a.s. a tree.*

Proof. By Exercise 11.8(b), if WMSF is a tree a.s., then WMSF = FMSF. ◄

It is not easy to give a graph on which the FMSF is not a tree a.s., especially if the graph is transitive. This will be done in Section 11.6.

Another corollary of Proposition 11.7 is the following result of Häggström (1998).

Corollary 11.9. *If G is a tree, then the free and wired minimal spanning forests are the same iff* $p_c(G) = 1$.

Proof. By Exercise 7.37, only at $p = 1$ can $G[p]$ have a unique infinite cluster a.s. Thus, almost everywhere uniqueness is equivalent to $p_c(G) = 1$. ◄

What if we specialize to quasi-transitive graphs? In that setting, Proposition 11.7 and Theorem 7.21 show that FMSF = WMSF iff $p_c = p_u$; according to Conjecture 7.31, this holds iff G is amenable. Of course, for a quasi-transitive, amenable G and *every* $p \in [0, 1]$, there is a.s. at most one infinite cluster in $G[p]$; see Theorem 7.6. This is slightly stronger than $p_c = p_u$, and gives another proof that for quasi-transitive, amenable graphs, FMSF = WMSF (compare Exercise 11.8(c)).

Recall that tail triviality is a strong form of asymptotic independence (Section 10.4). It holds for minimal spanning forests, just as it does for uniform spanning forests (Theorem 10.18):

Theorem 11.10. *Both measures* WMSF *and* FMSF *have a trivial tail σ-field on every graph.*

Proof. This is really a nonprobabilistic result in the following sense. Let $\mathscr{F}(K)$ be the σ-field generated by $U(e)$ for $e \in K$. We will show that the tail σ-field is contained in the tail σ-field of the labels of the edges, $\bigcap_{K \text{ finite}} \mathscr{F}(E \setminus K)$. This implies the desired result by Kolmogorov's zero-one law.

Let $\phi: [0, 1]^E \to 2^E$ be the map that assigns the (free or wired) minimal spanning forest to a configuration of labels. (Actually, ϕ is defined only on the configurations of distinct labels.) Let A be a tail event of 2^E. We claim that $\phi^{-1}(A)$ lies in the tail σ-field $\bigcap_{K \text{ finite}} \mathscr{F}(E \setminus K)$. Indeed, for any finite set K of edges and any two labelings U_1 and U_2 that differ only on K, we know by Lemma 11.6 that $\phi(U_1)$ and $\phi(U_2)$ differ at most on $2|K|$ edges, whence both $\phi(U_i)$ are in A or neither are, in other words, both U_i are in $\phi^{-1}(A)$ or neither are. In other words, $\phi^{-1}(A) \in \mathscr{F}(E \setminus K)$. ◄

11.4 Tree Sizes

Here we prove analogues of results from Section 10.6. There, we gave very general sufficient conditions for each tree in the wired uniform spanning forest to have one end a.s. We do not know such a general theorem for the wired minimal spanning forest. Even in the transitive case, we do not know how to prove this without assuming unimodularity (see Section 8.2 for its definition) and $\theta(p_c) = 0$; recall that $\theta(p)$ is the probability that a vertex belongs to an infinite cluster in Bernoulli(p) percolation and that when G is nonamenable and unimodular, it is known that $\theta(p_c) = 0$ (Theorem 8.21). On the other hand, we will be able to answer the analogue of Question 10.52 in the unimodular case and to prove analogues of Theorem 10.34 for both the wired and free minimal spanning forests and in great generality.

We begin by showing that the trees in the WMSF have at most two ends each.

Theorem 11.11. (WMSF Expected Degree 2) *Let G be a unimodular, transitive graph. Then the WMSF-expected degree of each vertex is 2 and each tree has at most two ends a.s.*

Proof. Fix $p < 1$. Write η_p for the graph formed by the edges e with $U(e) < pZ_w(e)$. By Proposition 11.4, the trees in η_p have at most one end each, whence by Exercise 8.10 and Proposition 8.18, the expected degree in η_p is at most 2. Since $\mathfrak{F}_w = \bigcup_p \eta_p$ a.s., the same follows for $\mathfrak{F}_w$. This in turn implies that the trees in $\mathfrak{F}_w$ have at most two ends each a.s. Finally, Proposition 8.18 implies that the expected degree in WMSF is at least 2.   ◄

With an additional hypothesis on critical Bernoulli percolation, we can assert that the trees have only one end.

Theorem 11.12. (One End) *Let G be a unimodular, transitive graph. If $\theta(p_c, G) = 0$, then a.s. each component of the WMSF has one end.*

Proof. Suppose that $\theta(p_c) = 0$. Fix a vertex x. Let $e_1, e_2, \ldots$ be the edges in the invasion tree of x, in the order they are added. Then $\sup_{n \geq k} U(e_n) > p_c$ for every k. By Theorem 7.22, $\limsup U(e_n) = p_c$. Therefore, there are infinitely many k such that $U(e_k) = \sup_{n \geq k} U(e_n)$. For each such k, we have that $U(e_n) < U(e_k)$ for all $n > k$, whence e_n is on the other side of e_k from x. Thus, the edge e_k separates x from ∞ in the invasion tree of x. It follows that the invasion tree of x a.s. has one end. It also follows that the limsup of the labels U along that end is equal to p_c.

Thus, all invasion trees have one end a.s. Since each pair of invasion trees is either disjoint or shares all but finitely many vertices by Proposition 11.5, there is a well-defined special end for each component of $\mathfrak{F}_w$, namely, the end of any invasion tree contained in that component (by Exercise 11.7).

Suppose that with positive probability, the event A occurs that some component has two ends. Let the *trunk* of a component with two ends be the unique bi-infinite path that it contains. Enumerate the vertices of the trunk as x_n ($n \in \mathbb{Z}$), with $\langle x_n, x_{n+1} \rangle$ being the edges of the trunk, oriented toward the special end. Since $\theta(p_c) = 0$, we have

$$\epsilon := \left[\sup_{n \in \mathbb{Z}} U([x_n, x_{n+1}]) - p_c \right] / 2 > 0$$

a.s. on A. By the first paragraph, $\limsup_{n \to \infty} U([x_n, x_{n+1}]) = p_c$ a.s. Thus, a.s. on A, there is a largest $m \in \mathbb{Z}$ such that $U([x_m, x_{m+1}]) > p_c + \epsilon$. We can then transport mass 1 from each vertex in such a component to the vertex x_m. The vertex x_m would then receive infinite mass, contradicting the mass-transport principle. Therefore, all components have only one end a.s.

◄

Question 11.13. Let G be a transitive graph whose automorphism group is not unimodular. Does every tree of the WMSF on G have one end a.s.? Are there reasonably general conditions to guarantee one-ended trees in the WMSF without any homogeneity of the graph?

Theorem 11.12 gives one relation between the WMSF and critical Bernoulli percolation. Another is an immediate consequence of Theorem 7.22 and Exercise 11.7:

Proposition 11.14. *Provided G is quasi-transitive, a.s. for every $p > p_c(G)$ and for every component T of WMSF(G), there is some infinite cluster K of $G[p]$ such that $T \cap K$ has an infinite connected component.* ◄

This is not true for general graphs:

▷ **Exercise 11.9.**
Give an infinite, connected graph G such that for some $p > p_c(G)$, with positive probability, there is some component of the WMSF that intersects no infinite cluster of $G[p]$.

Now we answer the analogue of Question 10.52 in the unimodular case. This result is due to Timár (2006a).

Theorem 11.15. (Infinitely Many Ends) *If G is a quasi-transitive, unimodular graph and WMSF $\neq$ FMSF, then a.s. every tree in $\mathfrak{F}_f$ has infinitely many ends and contains infinitely many trees of $\mathfrak{F}_w$.*

Proof. Suppose that WMSF $\neq$ FMSF. Then G is nonamenable by Exercise 11.8. Since $\mathfrak{F}_w \subseteq \mathfrak{F}_f$, each tree of $\mathfrak{F}_f$ consists of trees of $\mathfrak{F}_w$ together with edges joining them. By Example 8.6, the number of edges in a tree of $\mathfrak{F}_f$ that do not belong to $\mathfrak{F}_w$ is either 0 or ∞. By Theorems 11.12 (as extended to quasi-transitive, unimodular graphs in Exercise 11.23) and 8.21, each tree of $\mathfrak{F}_w$ has one end, so it remains to show that no tree of $\mathfrak{F}_f$ is a tree of $\mathfrak{F}_w$. Call a tree of $\mathfrak{F}_f$ that is a tree of $\mathfrak{F}_w$ *lonely*. All other trees of $\mathfrak{F}_f$ have infinitely many ends. Suppose for a contradiction that there is a lonely tree with positive probability.

By the discussion in Section 11.3, we have $p_c < p_u$, so we may choose $p \in (p_c, p_u)$. In view of Proposition 11.14, we may then choose a finite, simple path $\mathcal{P}$ with vertices ordered as $\langle x_1, x_2, \dots, x_n \rangle$ such that $\mathbf{P}(A) > 0$ for the event A that x_1 and x_n belong to distinct infinite clusters of $G[p]$, that x_1 belongs to a lonely tree, T, and that T does not intersect $\mathcal{P}$ except at x_1. Let $F \subset \mathsf{E}$ be the set of edges not in $\mathcal{P}$ that have an endpoint in $\{x_2, \dots, x_{n-1}\}$. Since T has one end, so does $T \cap G[p]$.

Define A' to be the event that results from A by changing the labels $U(e)$ for $e \in F$ to $p + (1 - p)U(e)$. This increases all the labels in F and also makes them larger than p. Let $\mathfrak{F}'_f$ be the new free minimal spanning forest and T' be the component of $\mathfrak{F}'_f$ that contains x_1. By Lemma 11.6, $\mathfrak{F}'_f \bigtriangleup \mathfrak{F}_f$ is finite. Furthermore, $T' \supseteq T$. Therefore, if T' is not lonely on A', then it contains infinitely many ends and an isolated end (the end of T). Since $\mathbf{P}(A') > 0$ by Lemma 7.24, this is impossible by Proposition 8.33. Hence, T' is lonely on A'.

Define A'' to be the event that results from A' by changing the labels $U(e)$ for $e \in \mathcal{P}$ to $pU(e)$. This decreases all the labels in $\mathcal{P}$ and also makes them smaller than p. Let $\mathfrak{F}''_f$ be the new free minimal spanning forest and T'' be the component of $\mathfrak{F}''_f$ that contains x_1. By Lemma 11.6, $\mathfrak{F}''_f \triangle \mathfrak{F}_f$ is finite. On A'', the path $\mathcal{P}$ belongs to a single infinite cluster of $G[p]$. Furthermore, $\mathcal{P} \subset \mathfrak{F}''_f$ on A'', since all cycles containing an edge of $\mathcal{P}$ must contain an edge with label larger than p. In addition, $T' \cap G[p] \subseteq T''$ for the same reason: Let $e \in T' \cap G[p]$, and let C be a cycle containing e. If $C \cap \mathcal{P} = \varnothing$, then the labels on C are the same on A'' as on A', in which case e is not the highest edge on C, since $e \in T'$; whereas if $C \cap \mathcal{P} \neq \varnothing$, then C contains an edge labeled more than p, in which case again e is not the highest edge on C, since $e \in G[p]$. Thus, $e \in T''$. By symmetry, the intersection of $G[p]$ with the component of $\mathfrak{F}'_f$ that contains x_n has all its edges in $\mathfrak{F}''_f$, which means that T'' has at least two and hence (by our first paragraph) infinitely many ends and an isolated end (the end of T'). Since $\mathbf{P}(A'') > 0$ by Lemma 7.24, this is again impossible by Proposition 8.33. This contradiction proves the theorem.   ◀

We believe that unimodularity is not needed for this property:

Conjecture 11.16. *If G is a quasi-transitive graph and* WMSF $\neq$ FMSF, *then* FMSF-*a.s. every tree has infinitely many ends.*

Theorem 11.12 and Proposition 11.14 give two relations between the WMSF and critical Bernoulli percolation. The next result, though not much related to the title of this section, gives a relation between the FMSF and Bernoulli(p_u) percolation. Recall that

$$p_u(G) := \inf\{p \,; \text{ there is a.s. a unique infinite cluster in Bernoulli}(p) \text{ percolation}\}$$

for a general graph, G.

Proposition 11.17. *Under the standard coupling, a.s. each component of* FMSF(G) *intersects at most one infinite cluster of $G[p_u]$. Thus, the number of trees in* FMSF(G) *is a.s. at least the number of infinite clusters in $G[p_u]$. If G is quasi-transitive with $p_u(G) > p_c(G)$, then a.s. each component of* FMSF(G) *intersects exactly one infinite cluster of $G[p_u]$.*

Proof. Let $\langle p_j \rangle$ be a sequence satisfying $\lim_{j\to\infty} p_j = p_u$ and that is contained in the set of $p \in [p_u, 1]$ such that there is a.s. a unique infinite cluster in $G[p]$. Let $\mathcal{P}$ be a finite, simple path in G, and let $\mathscr{A}$ be the event that $\mathcal{P} \subset \mathfrak{F}_f$ and the endpoints of $\mathcal{P}$ are in distinct infinite p_u-clusters. Since a.s. for every $j = 1, 2, \ldots$ there is a unique infinite cluster in $G[p_j]$, a.s. on $\mathscr{A}$ there is a path joining the endpoints of $\mathcal{P}$ in $G[p_j]$. Because $\mathcal{P} \subset \mathfrak{F}_f$ on $\mathscr{A}$, we have $\max_\mathcal{P} U \leq p_j$ a.s. on $\mathscr{A}$. Thus, $\max_\mathcal{P} U \leq p_u$ a.s. on $\mathscr{A}$. On the other hand, $\max_\mathcal{P} U \geq p_u$ a.s. on $\mathscr{A}$, since on $\mathscr{A}$, the endpoints of $\mathcal{P}$ are in distinct p_u components. This implies $\mathbf{P}[\mathscr{A}] \leq \mathbf{P}[\max_\mathcal{P} U = p_u] = 0$, and the first statement follows.

The second sentence follows from the fact that every vertex belongs to some component of $\mathfrak{F}_f$. Finally, the third sentence follows from Theorem 7.22 and the fact that invasion trees are contained in the wired minimal spanning forest, which, in turn, is contained in the free minimal spanning forest.   ◀

There is a related conjecture of Benjamini and Schramm (personal communication, 1998):

Conjecture 11.18. *Let G be a quasi-transitive, nonamenable graph with $p_u(G) < 1$. Then* FMSF *is a single tree a.s. iff there is a unique infinite cluster in $G[p_u]$ a.s.*

We can strengthen this conjecture to say that the number of trees in the FMSF equals the number of infinite clusters at p_u. An even stronger conjecture would be that in the natural coupling of Bernoulli percolation and the FMSF, each infinite cluster at p_u intersects exactly one component of the FMSF and each component of the FMSF intersects exactly one infinite cluster at p_u.

Question 11.19. Must the number of trees in the FMSF and the WMSF in a quasi-transitive graph be either 1 or ∞ a.s.? This question for $\mathbb{Z}^d$ is due to Newman (1997).

When FMSF $\neq$ WMSF and the quasi-transitive graph is unimodular, a positive answer was given by Timár (2018).

We now prove an analogue of Theorem 10.34, which showed that the trees in the wired uniform spanning forest of a graph are all a.s. recurrent. What is the analogue of recurrence? It is $p_c = 1$: see Exercise 11.14 for one reason to consider this as a proper analogue. We show $p_c = 1$, in fact, for something even larger than the trees of the wired minimal spanning forest. Namely, define the ***invasion basin of infinity***, $I(\infty) = I^U(\infty)$, as the set of edges $[x, y]$ such that there do not exist disjoint, infinite, simple paths from x and y consisting only of edges e satisfying $U(e) < U([x, y])$. Thus, we have

$$I(\infty) \supset \bigcup_{x \in \mathsf{V}} I(x) \supset \bigcup_{x \in \mathsf{V}} T_U(x) = \mathfrak{F}_w(U).$$

In addition, $I(\infty)$ does not contain any edge that joins different trees in $\mathfrak{F}_w$. For an edge e, define

$$Z_\infty^U(e) := Z_\infty(e) := \inf_{\mathcal{P}} \sup\{U(f) \,;\, f \in \mathcal{P} \setminus \{e\}\},$$

where the infimum is over bi-infinite, simple paths that contain e; if there is no such path $\mathcal{P}$, define $Z_\infty(e) := 1$. Similarly to the expression for $\mathfrak{F}_w$ in terms of Z_w, we have

$$\{e \,;\, U(e) < Z_\infty(e)\} \subseteq I(\infty) \subseteq \{e \,;\, U(e) \leq Z_\infty(e)\}$$

and a.s. $I(\infty) = \{e \,;\, U(e) < Z_\infty(e)\}$.

Theorem 11.20. *Let $G = (\mathsf{V}, \mathsf{E})$ be a graph of bounded degree. Then $p_c\big(I(\infty)\big) = 1$ a.s. Therefore $p_c(\mathfrak{F}_w) = 1$ a.s.*

To prove this, we begin with the following lemma that will provide a coupling between percolation and invasion that is different from the usual one we work with.

Lemma 11.21. *Let $G = (\mathsf{V}, \mathsf{E})$ be a locally finite, infinite graph and $\langle U(e) \,;\, e \in \mathsf{E} \rangle$ be i.i.d. uniform $[0, 1]$ random variables. Let $A \subset \mathsf{E}$ be finite. Conditioned on $A \subset I(\infty)$, the random variables*

$$\frac{U(e)}{Z_\infty(e)} \qquad (e \in A)$$

are i.i.d. uniform $[0, 1]$.

The heart of the lemma is that a uniform random variable conditioned on a set of possible values is uniform on that set. The form of this that we use is in the following exercise:

▷ **Exercise 11.10.**
Consider two independent random vectors $U = \langle U_i ; 1 \leq i \leq k \rangle$ and $Z = \langle Z_i ; 1 \leq i \leq k \rangle$, where U is distributed uniformly in $[0, 1]^k$ and Z has an arbitrary distribution in $(0, 1]^k$. Show that, given $U_i < Z_i$ for all $1 \leq i \leq k$, the conditional law of the vector $\langle U_i/Z_i ; 1 \leq i \leq k \rangle$ is uniform in $[0, 1]^k$.

Proof of Lemma 11.21. Given Exercise 11.10, the essence of the proof is the plausible fact that $Z_\infty \restriction I(\infty)$ gives no information about $U \restriction I(\infty)$ other than $U \leq Z_\infty$ on $I(\infty)$. This is reasonable because no edge in $I(\infty)$ can be the highest edge in any bi-infinite, simple path, whence its label cannot determine the value of Z_∞ anywhere. Let $A \subset \mathsf{E}$ be finite. Define $\widetilde{U}(e) := 0$ for $e \in A$ and $\widetilde{U}(e) := U(e)$ for $e \notin A$, and let $Z_A^U := Z_\infty^U \restriction A$ denote the restriction of Z_∞^U to A. Certainly $Z_A^{\widetilde{U}}$ is independent of $U \restriction A$.

We claim that on the event $\big[A \subset I^U(\infty)\big]$, we have $Z_A^U = Z_A^{\widetilde{U}}$. Indeed, consider any bi-infinite, simple path $\mathcal{P}$. If $e \in I^U(\infty) \cap \mathcal{P}$, then $U(e) < \sup\{U(e') ; e \neq e' \in \mathcal{P}\}$. Hence, for every such $\mathcal{P}$,

$$\sup_{\mathcal{P}} U = \sup_{\mathcal{P} \setminus A} U = \sup_{\mathcal{P} \setminus A} \widetilde{U} = \sup_{\mathcal{P}} \widetilde{U}$$

on the event $\big[A \subset I^U(\infty)\big]$. This proves the claim.

One consequence is that the symmetric difference of the two events $\big[A \subset I^U(\infty)\big]$ and $\big[U \restriction A < Z_A^{\widetilde{U}}\big]$ has probability 0. Indeed, the second event is contained in the first event since $Z_A^{\widetilde{U}}(e) \leq Z_A^U(e)$ for all $e \in A$. For the converse, $A \subset I^U(\infty)$ implies that $U \restriction A < Z_A^U$ a.s. and $Z_A^U = Z_A^{\widetilde{U}}$, as we saw in the preceding paragraph. Together, these give the converse.

Thus, the distribution of $\langle U(e)/Z_A^U(e) ; e \in A \rangle$ conditional on $\big[A \subset I^U(\infty)\big]$ is a.s. the same as the distribution of $\langle U(e)/Z_A^{\widetilde{U}}(e) ; e \in A \rangle$ conditional on $\big[U \restriction A < Z_A^{\widetilde{U}}\big]$. By Exercise 11.10, this distribution is uniform on $[0, 1]^A$, as desired. ◀

We also need the following fact.

Lemma 11.22. *If a graph H of bounded degree does not contain a simple, bi-infinite path, then $p_c^{\mathrm{bond}}(H) = 1$.*

Proof. Suppose that H does not contain a simple, bi-infinite path. Let x be a vertex of H. Since every pair of infinite paths from x must share another vertex besides x, it follows by Menger's theorem (Exercise 3.16) that there is a vertex $y \neq x$ that belongs to every infinite path from x. That is, removal of y from H leaves x in a finite component. We can then repeat the argument with infinite paths starting at y and eventually find infinitely many vertices z such that x is in a finite component of $H \setminus \{z\}$. Since H has bounded degree, it follows that $p_c^{\mathrm{bond}}(H) = 1$. (Even without bounded degree, we get that $p_c^{\mathrm{site}}(H) = 1$.) ◀

Proof of Theorem 11.20. A random subset ω of E is Bernoulli(p) percolation on $I(\infty)$ iff $\omega \subseteq I(\infty)$ a.s. and for all finite $A \subset \mathsf{E}$, the probability that $A \subseteq \omega$ given that $A \subset I(\infty)$ is $p^{|A|}$. Let η_p be the set of edges e satisfying $U(e) < p Z_\infty(e)$. Lemma 11.21 implies that η_p has the law of Bernoulli(p) percolation on $I(\infty)$. Thus, by Lemma 11.22, it suffices to show that η_p does not contain any simple, bi-infinite path. In fact, this was already shown for a larger set of edges in Proposition 11.4. ◀

Thinking about planar duality leads one to suspect that a dual statement holds for the FMSF. (Planar duality is explored in the next section.) And indeed, a dual argument shows that the FMSF is almost connected in the following sense.

Theorem 11.23. *Let G be a locally finite, connected graph and $\epsilon \in (0, 1)$. Let $\mathfrak{F}_f$ have the law* FMSF *and ω be an independent copy of $G[\epsilon]$. Then $\mathfrak{F}_f \cup \omega$ is connected a.s.*

For this, we use a lemma dual to Lemma 11.21; it will provide a coupling of $\mathfrak{F}_f$ and $\mathfrak{F}_f \cup \omega$.

Lemma 11.24. *Let $G = (\mathsf{V}, \mathsf{E})$ be a locally finite, infinite graph and $\langle U(e); \ e \in \mathsf{E}\rangle$ be i.i.d. uniform $[0, 1]$ random variables. Let $A \subset \mathsf{E}$ be a finite set such that $\mathbf{P}[A \cap \mathfrak{F}_f = \varnothing] > 0$. Let $B \subset \mathsf{E}$ be a finite set disjoint from A. Conditioned on $A \cap \mathfrak{F}_f = \varnothing$ and $B \subset \mathfrak{F}_f$, the random variables*

$$\frac{1 - U(e)}{1 - Z_f(e)} \qquad (e \in A)$$

are i.i.d. uniform $[0, 1]$.

Proof. Let $\widetilde{U}(e) := 1$ for $e \in A$ and $\widetilde{U}(e) := U(e)$ for $e \notin A$, and let Z_A^U denote the restriction of Z_f^U to A. Consider any cut Π. If $e \in \Pi \setminus \mathfrak{F}_f(U)$, then

$$U(e) > Z_f^U(e) \ge \inf\{U(e'); \ e' \in \Pi \setminus \{e\}\}$$

by (11.1). Hence, if $A \cap \mathfrak{F}_f = \varnothing$, then for every cut Π,

$$\inf_{\Pi} U = \inf_{\Pi \setminus A} U = \inf_{\Pi \setminus A} \widetilde{U} = \inf_{\Pi} \widetilde{U},$$

and therefore (still assuming that $A \cap \mathfrak{F}_f = \varnothing$ and using (11.1) again) $Z_A^U = Z_A^{\widetilde{U}}$. Hence $A \cap \mathfrak{F}_f(U) = \varnothing$ implies $U > Z_A^{\widetilde{U}}$ on A. Moreover, $A \cap \mathfrak{F}_f(U) = \varnothing$ is actually equivalent to $U > Z_A^{\widetilde{U}}$ on A, because $Z_A^{\widetilde{U}} \ge Z_A^U$. In addition, $\mathfrak{F}_f(U) = \mathfrak{F}_f(\widetilde{U})$ on the event $A \cap \mathfrak{F}_f(U) = \varnothing$. Thus, conditioned on $A \cap \mathfrak{F}_f(U) = \varnothing$ and $B \subset \mathfrak{F}_f(U)$, which is the same as conditioning on $U{\restriction}A > Z_A^{\widetilde{U}}$ and $B \subset \mathfrak{F}_f(\widetilde{U})$, the random variables $\langle(1 - U(e))/(1 - Z_f(e)); \ e \in A\rangle = \langle(1 - U(e))/(1 - Z_A^{\widetilde{U}}(e)); \ e \in A\rangle$ are i.i.d. uniform in $[0, 1]$ by Exercise 11.10. ◀

Proof of Theorem 11.23. According to Lemma 11.24, $\mathfrak{F}_f \cup \omega$ has the same law as $\xi := \{e; \ 1 - U(e) \ge (1 - \epsilon)[1 - Z_f(e)]\}$. We saw in Proposition 11.4 that ξ is connected. ◀

Recall that an analogous question about the FUSF, Question 10.12, is open. One difficulty in trying to provide an analogous solution is to imagine what corresponds to Bernoulli(ϵ) percolation in the FUSF world.

11.5 Planar Graphs

When we add planar duality to our tools, it will be easy to deduce all the major properties of both the free and wired minimal spanning forests on planar, quasi-transitive graphs.

Recall the definition (10.7)

$$e \in \omega \iff e^\dagger \notin \omega^\times.$$

Proposition 11.25. *Let G and $G^\dagger$ be proper, locally finite, dual plane graphs. For any injection $U\colon \mathsf{E} \to \mathbb{R}$, let $U^\dagger(e^\dagger) := 1 - U(e)$. We have*

$$\big(\mathfrak{F}_{\mathrm{f}}(U, G)\big)^\times = \big\{ e^\dagger \;;\; U^\dagger(e^\dagger) < Z_{\mathrm{w}}^{U^\dagger}(e^\dagger) \big\},$$

whence $\big(\mathfrak{F}_{\mathrm{f}}(U, G)\big)^\times = \mathfrak{F}_{\mathrm{w}}(U^\dagger, G^\dagger)$ if $U^\dagger(e^\dagger) \neq Z_{\mathrm{w}}^{U^\dagger}(e^\dagger)$ for all $e^\dagger \in \mathsf{E}^\dagger$.

Proof. The Jordan curve theorem implies that a set $\mathcal{P} \subset \mathsf{E} \setminus \{e\}$ is a simple path between the endpoints of e iff the set $\Pi := \{f^\dagger \;;\; f \in \mathcal{P}\} \cup \{e^\dagger\}$ is a cut of a finite set. Thus $Z_{\mathrm{f}}^{U}(e) = 1 - Z_{\mathrm{w}}^{U^\dagger}(e^\dagger)$ by (11.2). This means that $e^\dagger \in \big(\mathfrak{F}_{\mathrm{f}}(U, G)\big)^\times$ iff $e \notin \mathfrak{F}_{\mathrm{f}}(U, G)$ iff $U(e) > Z_{\mathrm{f}}^{U}(e)$ iff $U^\dagger(e^\dagger) < Z_{\mathrm{w}}^{U^\dagger}(e^\dagger)$. ◀

The following corollary is proved in the same way that Proposition 10.37 is proved.

Corollary 11.26. *Let G be a proper, plane graph with $G^\dagger$ locally finite. If each tree of the WMSF of G has only one end a.s., then the FMSF of $G^\dagger$ has only one tree a.s. If, in addition, the WMSF of G has infinitely many trees a.s., then the tree of the FMSF of $G^\dagger$ has infinitely many ends a.s.* ◀

This allows us decide what happens on $\mathbb{Z}^2$, a result of Alexander and Molchanov (1994).

Corollary 11.27. *The minimal spanning forest of $\mathbb{Z}^2$ is a.s. a tree with one end.*

Proof. The hypothesis $\theta(p_{\mathrm{c}}) = 0$ of Theorem 11.12 applies by Harris (1960) and Kesten (1980). Therefore, each tree in the WMSF has one end. By Corollary 11.26, this means that the FMSF has one tree. On the other hand, the wired and free measures are the same by Exercise 11.8.

However, we can get by in our proof with less than Kesten's theorem, namely, with only Harris's Theorem 7.17. To see this, consider the labels $U^\dagger(e^\dagger) := 1 - U(e)$ on $(\mathbb{Z}^2)^\dagger$. Let U be an injective $[0, 1]$-labeling where all the $(1/2)$-clusters in $\mathbb{Z}^2$ and $(\mathbb{Z}^2)^\dagger$ are finite and $U(e) \neq Z_{\mathrm{w}}^{U}(e)$ for all $e \in \mathsf{E}$, which happens a.s. for the standard labeling. Suppose that e is an edge where $U(e) \leq 1/2$. We claim that the endpoints of e belong to the same tree in $\mathfrak{F}_{\mathrm{w}}(U, \mathbb{Z}^2)$. Indeed, the invasion basin of e^- contains e by our assumption, whence the invasion tree of e^- contains e^+.

Therefore, if $\mathfrak{F}_{\mathrm{w}}(U, \mathbb{Z}^2)$ contains more than one tree, all edges joining two of its components have labels larger than $1/2$. If F is the edge boundary of one of the components, then $F^\dagger$ contains an infinite path with labels all less than $1/2$, which contradicts our assumption. This proves that $\mathfrak{F}_{\mathrm{w}}(U, \mathbb{Z}^2)$ is one tree. So is $\mathfrak{F}_{\mathrm{f}}\big(U^\dagger, (\mathbb{Z}^2)^\dagger\big)$, whence $\mathfrak{F}_{\mathrm{w}}(U, \mathbb{Z}^2) = \big(\mathfrak{F}_{\mathrm{f}}(U^\dagger, G^\dagger)\big)^\times$ has just one end. ◀

Similar reasoning can be applied in the nonamenable case:

Proposition 11.28. *Let G be a connected, nonamenable, quasi-transitive, planar graph with one end. Then the* FMSF *on G is a.s. a tree.*

The nonamenability assumption can be replaced by the assumption that the planar dual of G satisfies $\theta(p_c) = 0$. The latter assumption is known to hold in many amenable cases (see Kesten (1982)).

Proof. Let G be such a graph. By Theorem 8.25, $\text{Aut}(G)$ is unimodular, and we may embed G so that $\text{Aut}(G^\dagger)$ is also unimodular. By Exercise 6.32, the graph $G^\dagger$ is also nonamenable. Thus, we may apply Theorem 8.21 to $G^\dagger$ to see that $\theta(p_c, G^\dagger) = 0$. Theorem 11.12 and Corollary 11.26 now yield the desired conclusion.   ◀

It is amusing to see how we can use minimal spanning forests now to give another proof of the bond percolation part of Theorem 8.24.

Corollary 11.29. *If G is a connected, nonamenable, quasi-transitive, planar graph with one end, then for bond percolation, $p_c(G) < p_u(G)$. In addition, there is a unique infinite cluster in Bernoulli($p_u(G)$) bond percolation.*

Proof. Again, by Theorem 8.25, $\text{Aut}(G)$ is unimodular and we may embed G so that $\text{Aut}(G^\dagger)$ is also unimodular. By Theorem 7.21 and Proposition 11.7, for the first part, it suffices to show that WMSF $\neq$ FMSF on G. Now if the forests were the same, then they would also be the same on $G^\dagger$, so that each would be one tree with one end, as in the proof of Proposition 11.28. But this is impossible by Proposition 8.22.

Furthermore, by Proposition 11.28, the FMSF is a tree on G, whence by the second sentence of Proposition 11.17, there is a unique infinite cluster in Bernoulli(p_u) percolation on G.   ◀

11.6 Nontreeable Groups

We don't know any good way to tell when the free minimal spanning forest is a.s. a single tree, even for $\mathbb{Z}^d$ when $d \geq 3$. So far in this chapter, we have not presented a single example of a Cayley graph where it is not a tree. One way to present such an example is to use Proposition 11.17 and give a Cayley graph where there are infinite many infinite clusters in Bernoulli(p_u) percolation. In Section 7.9, some examples of such Cayley graphs were mentioned. As it happens, all of those examples have another surprising property: they don't admit *any* invariant random spanning tree! That, of course, also implies that the FMSF is a.s. not a tree. This is the method we use here: we give examples of Cayley graphs that have no invariant random spanning tree. The following theorem is due to Pemantle and Peres (2000).

Theorem 11.30. (Nontreeable Products) *Let Γ and Δ be infinite countable groups and G be a Cayley graph of $\Gamma \times \Delta$. If there is a random invariant spanning tree of G, then G is amenable.*

Proof. Let H be a Cayley graph of Γ. We'll use the spanning tree measure assumed to exist to create an invariant percolation on H with finite clusters and with expected degree arbitrarily close to the degree of H; amenability of H (and hence of Γ) then follows from Theorem 8.16. By symmetry, Δ is also amenable, whence so is G by Exercise 6.25.

To create this percolation on H, we first define an equivalence relation on Γ. Consider a fixed spanning tree T of G. Write o for the identities of Γ and Δ. Write Γ_n for the points in G that lie within distance n of $\Gamma \times \{o\}$. Let H_r denote the points in H that are within H-distance r of o and let $\Delta_r := H_r \times \Delta$. For $n \geq 1$, let $\delta_n \in \Delta$ be any element such that the distance in G from (o, o) to (o, δ_n) is at least $2n + 1$. Given n and r, let $C(T, n, r) \in \Gamma \times \Gamma$ consist of the pairs (γ, γ') such that either $\gamma = \gamma'$ or all of the following four properties hold:

  (i) the path in T from (γ, o) to (γ', o) lies in Γ_n;
  (ii) the path in T from (γ, δ_n) to (γ', δ_n) lies in $(o, \delta_n)\Gamma_n$;
  (iii) the path in T from (γ, o) to (γ, δ_n) lies in $(\gamma, o)\Delta_r$; and
  (iv) the path in T from (γ', o) to (γ', δ_n) lies in $(\gamma', o)\Delta_r$.

It is easy to see that $C(T, n, r)$ is an equivalence relation for each n and r.

Now let T be random with a $\Gamma \times \Delta$-invariant law. Since $C((\gamma, o)T, n, r) = \gamma C(T, n, r)$ for every $\gamma \in \Gamma$, the law of $C(T, n, r)$ is Γ-invariant. The probabilities of the events in (i) and (ii), which are the same, tend to 1 as $n \to \infty$. Given n, the probabilities of the events in (iii) and (iv) tend to 1 as $r \to \infty$. Thus, given any pair $\gamma \neq \gamma'$, we may choose n and r large enough that $\mathbf{P}\big[(\gamma, \gamma') \in C(T, n, r)\big]$ is as close as desired to 1. On the other hand, when $(\gamma, \gamma') \in C(T, n, r)$, we may concatenate the paths in T from (γ, o) to (γ', o) to (γ', δ_n) to (γ, δ_n) to (γ, o). Since T contains no simple cycles, this means that $(\gamma, o)\Delta_r \cap (\gamma', o)\Delta_r \neq \varnothing$, whence γ and γ' lie within distance $2r$ of each other. Thus, the equivalence classes of $C(T, n, r)$ are finite. To make a percolation out of them, just take $C(T, n, r) \cap \mathsf{E}(H)$.   ◀

We do not know much about the number of infinite clusters in Bernoulli(p_{u}) percolation on Cayley graphs. For example, the following is open:

Question 11.31. Let G be a Cayley graph with $p_{\mathrm{u}}(G) < 1$. Both of the following properties imply that the FMSF on G is a.s. not a tree:

  (i) Bernoulli(p_{u}) percolation has more than one infinite cluster a.s.;
  (ii) there is no random invariant spanning tree of G.
Is there any implication between (i) and (ii)?

11.7 Notes

Proposition 11.5 was first proved by Chayes, Chayes, and Newman (1985) for $\mathbb{Z}^2$, then by Alexander (1995a) for all $\mathbb{Z}^d$, and finally by Lyons, Peres, and Schramm (2006) for general graphs.

Lemma 11.6 is a strengthening due to Lyons, Peres, and Schramm (2006) of Theorem 5.1(i) of Alexander (1995a).

All other results in this chapter that are not explicitly attributed are due to Lyons, Peres, and Schramm (2006), although Proposition 11.4 was only implicit there.

A curious comparison of FUSF and FMSF was found by Lyons, Peres, and Schramm (2006), extending work of Lyons (2000) and Gaboriau (2005): Let $\overline{\deg}(\mu)$ denote the expected degree of a vertex under an automorphism-invariant percolation μ on a transitive graph (so that it is the same for all vertices).

Proposition 11.32. *Let $G = (\mathsf{V}, \mathsf{E})$ be a transitive, unimodular, connected, infinite graph of degree d. Then*

$$\overline{\deg}(\mathsf{FUSF}) \leq \overline{\deg}(\mathsf{FMSF}) \leq 2 + d \int_{p_c}^{p_u} \theta(p)^2 \, dp \, .$$

It is not known whether the unimodularity assumption is needed.

Some examples of particular behaviors of minimal spanning forests are given by Lyons, Peres, and Schramm (2006); they are somewhat hard to prove but worth recounting here. Namely, there are examples of the following: a planar graph whose free and wired minimal spanning forests are equal and have two components; a planar graph such that the number of trees in the wired minimal spanning forest is not an a.s. constant; a planar graph such that the number of trees in the free minimal spanning forest is not an a.s. constant; and a graph for which $\mathsf{WUSF} \neq \mathsf{FUSF}$ and $\mathsf{WMSF} = \mathsf{FMSF}$ (unlike the situation in Proposition 11.32).

Theorem 11.30 is essentially taken from Pemantle and Peres (2000), which contains generalizations. A precursor is in Adams (1988). Gaboriau (2000) (Corollary VI.22) shows that nonamenable groups of cost 1 are not treeable, that is, they have no invariant random spanning tree; this implies Theorem 11.30. Cayley graphs of Kazhdan groups also are not treeable; a version of this result appears in Adams and Spatzier (1990). More generally, a nonamenable group Γ with first ℓ^2-Betti number $\beta_1(\Gamma) = 0$ is nontreeable by Gaboriau (2002), Proposition 6.10. Gaboriau (2000) (Proposition VI.18) shows that if a group contains a nontreeable subgroup, then the group itself is not treeable. Note that Gaboriau calls a group "arborable" if *all* its free actions are treeable and "anti-arborable" if none of them are. It is unknown if there are any other kinds of groups. An example given by Gaboriau (2000) (Remark VI.11) of a nontreeable group is $(A \oplus B) *_B (B \oplus C)$, where A, B, C are all isomorphic to $\mathbb{Z}$; this group is nonamenable and has cost 1, yet is an amalgamated product of treeable groups over $\mathbb{Z}$.

▷ **Exercise 11.11.**
Call a graph G *almost treeable* if there exists a sequence of $\mathrm{Aut}(G)$-invariant spanning forests $\mathfrak{F}_n$ on G with the property that for all $x, y \in \mathsf{V}(G)$, we have $\lim_{n\to\infty} \mathbf{P}[x \leftrightarrow y \text{ in } \mathfrak{F}_n] = 1$. Use Theorem 7.48 to show that Cayley graphs of Kazhdan groups are not almost treeable.

A striking use of the minimal spanning tree on a Poisson process in $\mathbb{R}^2$ (minimizing Euclidean distance) is made by Krikun (2007).

We end with a few more open questions and conjectures, mostly from Lyons, Peres, and Schramm (2006).

Conjecture 11.33. *The components of the* FMSF *on a unimodular, transitive graph are indistinguishable in the sense that for every automorphism-invariant property $\mathcal{A}$ of subgraphs, either a.s. all components satisfy $\mathcal{A}$ or a.s. all do not. The same holds for the* WMSF.

This was proved for the FMSF by Timár (2018) in the special case that $\mathsf{FMSF} \neq \mathsf{WMSF}$.

This surely does not extend to nonunimodular, transitive graphs, though we do not have an example.

Conjecture 11.34. *Let T_o be the component of the identity o in the* WMSF *on a Cayley graph not roughly isometric to $\mathbb{Z}$, and let $\xi = \langle x_n \, ; \, n \geq 0 \rangle$ be a ray from o in T_o. (This ray is conjectured to be unique, since $\theta(p_c) = 0$ by Conjecture 8.15.) The sequence of "bushes" $\langle b_n \rangle$ observed along ξ converges in distribution. (Formally, b_n is the connected component of x_n in $T \setminus \{x_{n-1}, x_{n+1}\}$, multiplied on the left by x_n^{-1}.)*

Question 11.35. For which d is the minimal spanning forest of $\mathbb{Z}^d$ a.s. a tree? This question is due to Newman and Stein (1996), who conjecture that the answer is $d < 8$ or $d \leq 8$. Jackson and Read (2010a, 2010b) suggest instead that the answer is $d < 6$ or $d \leq 6$. This is related to the number of ground states of the Edwards–Anderson model: see Newman and Stein (2006).

Question 11.36. One may consider the minimal spanning tree on $\epsilon \mathbb{Z}^2 \subset \mathbb{R}^2$ and let $\epsilon \to 0$. It would be interesting to show that the limit exists in various senses. Aizenman, Burchard, Newman, and Wilson (1999) have shown that a subsequential limit exists. According to simulations of Wilson (2004a), the scaling limit in a simply connected domain with free or wired boundary conditions does not have the conformal invariance property one might expect. This contrasts with the situation of the uniform spanning forest, where the limit exists and is conformally invariant, as was proved by Lawler, Schramm, and Werner (2004a). For the triangular lattice, Garban, Pete, and Schramm (2018) establish existence of the limit.

Question 11.37. If G is a graph that is roughly isometric to a tree, then is the free minimal spanning forest on G a.s. a tree?

11.8 Collected In-Text Exercises

11.1. Show that among all spanning trees of a finite graph, T_U is the unique one that has minimum edge-label sum, $\sum_{e \in T} U(e)$.

11.2. Construct G as follows. Begin with the complete graph, K_4. Let e and f be two of its edges that do not share endpoints. Replace e by three edges in parallel, e_1, e_2, and e_3, that have the same endpoints as e. Likewise, replace f by three parallel edges f_i. Show that in G,

$$\mathsf{MST}[e_1, f_1 \in T] > \mathsf{MST}[e_1 \in T]\,\mathsf{MST}[f_1 \in T]\,.$$

11.3. Show that given an edge, e, the minimal spanning tree measure on G conditioned on the event not to contain e need not be the same as the minimal spanning tree measure on $G \backslash e$, the graph G with e deleted.

11.4. Find an example of a finite graph G with vertex $o \in G$ such that there are two spanning trees T and T' of G having the properties that for all x, the distance from x to o in T is the same as in T', yet T and T' are not equally likely under the minimal spanning tree measure.

11.5. Show that $\mathfrak{F}_{\mathrm{w}}(U)$ consists of those edges e for which there is a finite set of vertices $W \subset \mathsf{V}$ such that e is the least edge joining W to $\mathsf{V} \setminus W$.

11.6. Consider an increasing sequence of finite, nonempty, connected (not necessarily induced) subgraphs $G_n \subset G$ ($n \in \mathbb{N}$) such that $\bigcup_n G_n = G$. For $n \in \mathbb{N}$, let G_n^{W} be the graph obtained from G by identifying the vertices outside of G_n to a single vertex, then removing all resulting loops based at that vertex. Let $T_n(U)$ and $T_n^{\mathrm{W}}(U)$ denote the minimal spanning trees on G_n and G_n^{W}, respectively, that are induced by the labels U. Show that $\mathfrak{F}_{\mathrm{f}}(U) = \lim_{n \to \infty} T_n(U)$ and that $\mathfrak{F}_{\mathrm{w}}(U) = \lim_{n \to \infty} T_n^{\mathrm{W}}(U)$ in the sense that for every $e \in \mathfrak{F}_{\mathrm{f}}(U)$, we have $e \in T_n(U)$ for every sufficiently large n, for every $e \notin \mathfrak{F}_{\mathrm{f}}(U)$ we have $e \notin T_n(U)$ for every sufficiently large n, and similarly for $\mathfrak{F}_{\mathrm{w}}(U)$. Deduce that $T_n(U) \Rightarrow \mathsf{FMSF}$ and $T_n^{\mathrm{W}}(U) \Rightarrow \mathsf{WMSF}$.

11.7. Let $U: \mathsf{E} \to \mathbb{R}$ be an injective labeling of the edges of a locally finite graph $G = (\mathsf{V}, \mathsf{E})$. Show that the union $\bigcup_{x \in \mathsf{V}} T_U(x)$ of all the invasion trees is equal to $\mathfrak{F}_{\mathrm{w}}(U)$.

11.8. Let G be a connected, locally finite graph. Prove the following.

    (a) If G is edge-amenable, then the average degree of vertices in both the free and wired minimal spanning forests on G is a.s. 2.

    (b) The free and wired minimal spanning forests on G are the same if they have a.s. the same finite number of trees or if the expected degree of every vertex is the same for both measures.

(c) The free and wired minimal spanning forests on G are the same on any transitive, amenable graph and have expected degree 2.

(d) If $\mathfrak{F}_w$ is connected a.s. or if each component of $\mathfrak{F}_f$ has a.s. one end, then $\mathsf{WMSF}(G) = \mathsf{FMSF}(G)$.

(e) The measures $\mathsf{WMSF}(G)$ and $\mathsf{FMSF}(G)$ are invariant under all automorphisms of G, as is the law of $(\mathfrak{F}_w, \mathfrak{F}_f)$.

(f) If G is unimodular and transitive with $\mathsf{WMSF}(G) \neq \mathsf{FMSF}(G)$, then a.s. $\mathfrak{F}_f$ has a component with uncountably many ends and, in fact, with $p_c < 1$.

11.9. Give an infinite, connected graph G such that for some $p > p_c(G)$, with positive probability, there is some component of the WMSF that intersects no infinite cluster of $G[p]$.

11.10. Consider two independent random vectors $U = \langle U_i \; ; \; 1 \leq i \leq k \rangle$ and $Z = \langle Z_i \; ; \; 1 \leq i \leq k \rangle$, where U is distributed uniformly in $[0, 1]^k$ and Z has an arbitrary distribution in $(0, 1]^k$. Show that, given $U_i < Z_i$ for all $1 \leq i \leq k$, the conditional law of the vector $\langle U_i/Z_i \; ; \; 1 \leq i \leq k \rangle$ is uniform in $[0, 1]^k$.

11.11. Call a graph G *almost treeable* if there exists a sequence of $\mathrm{Aut}(G)$-invariant spanning forests $\mathfrak{F}_n$ on G with the property that for all $x, y \in \mathsf{V}(G)$, we have $\lim_{n \to \infty} \mathbf{P}[x \leftrightarrow y \text{ in } \mathfrak{F}_n] = 1$. Use Theorem 7.48 to show that Cayley graphs of Kazhdan groups are not almost treeable.

11.9 Additional Exercises

11.12. Let G be the graph in Figure 11.2. There are 11 spanning trees of G. Show that under the minimal spanning tree measure, they are not all equally likely and calculate their probabilities. Show, however, that there are conductances such that the corresponding weighted uniform spanning tree measure equals the minimal spanning tree measure.

11.13. Let G be a complete graph on four vertices (that is, all pairs of vertices are joined by an edge). Calculate the minimal spanning tree measure and show that there are no conductances that give the minimal spanning tree measure as a weighted uniform spanning tree.

Figure 11.2. See Exercise 11.12.

11.14. Let G be an infinite, locally finite graph. Show that if $p_c(G) = 1$, then for all x, we have $I(x) = G$ a.s., whereas if $p_c(G) < 1$, then for all x, we have $I(x) \neq G$ a.s. Here, $I(x)$ denotes the invasion basin of x.

11.15. Show that the FMSF of the usual Cayley graph of $\mathbb{Z}_2 * \mathbb{Z}_3$ (as in Figure 3.3) is a tree a.s. whose branching number equals 1.35^+.

11.16. Let G be a transitive graph, possibly with loops. Let $\rho = \rho(G)$ be the spectral radius (6.14). Show that the expected degree of FMSF_G is at least $\max\{1/\rho - 1 + \log\rho, (1 - \log 2)(1/\rho - 1)\}$. Show that if G has no loops, then the expected degree of FMSF_G is at least $1/\rho - \log(1 + 1/\rho)$. *Hint:* Consider the edges incident to o. For each, bound above the probability that it is the largest in some cycle.

11.17. As discussed in Section 10.2, for each finitely generated group, the degree of the FUSF does not depend on the Cayley graph chosen.

(a) Show that there is a finitely generated group Γ such that the expected degree of a vertex in the FMSF of a Cayley graph of Γ does depend on which Cayley graph is used.

(b) Let Γ be a finitely generated, nonamenable group. Show that for every $c > 0$, there is a Cayley graph G of Γ such that the expected degree of FMSF_G is at least c.

11.18. Let T be a 3-regular tree. Calculate the chance that a given vertex is a leaf in the wired minimal spanning forest on T.

11.19. Show that for the ladder graph of Exercise 4.2, the minimal spanning forest is a tree a.s. and calculate the chance that the bottom rung of the ladder is in the minimal spanning tree.

11.20. Let $f(p)$ be the probability that two given neighbors in $\mathbf{Z}^d$ are in different components in Bernoulli(p) percolation. Show that

$$\int_0^1 \frac{f(p)}{1-p}\,dp = \frac{1}{d}\,.$$

11.21. Let G be a connected graph. Let $\alpha(x_1, \ldots, x_K)$ be the probability that the invasion basins $I(x_1), \ldots, I(x_K)$ are pairwise vertex-disjoint. Show that the WMSF-essential supremum of the number of trees is

$$\sup\{K\,;\,\exists x_1, \ldots, x_K \in \mathsf{V} \quad \alpha(x_1, \ldots, x_K) > 0\}\,.$$

11.22. Show that if we sum the number of ends over all trees in the free minimal spanning forest of a graph, then we get an a.s. constant, and likewise for the wired minimal spanning forest.

11.23. Show that if G is a unimodular, quasi-transitive graph and $\theta(p_c, G) = 0$, then a.s. each component of the WMSF has one end.

11.24. Show that if G is a quasi-transitive, nonamenable, unimodular graph, then there are infinitely many trees in the WMSF a.s.

11.25. Theorem 11.20 was stated for bounded-degree graphs. Prove that if $G = (\mathsf{V}, \mathsf{E})$ is an infinite graph, then the WMSF $\mathfrak{F}_w$ satisfies $p_c(\mathfrak{F}_w) = 1$ a.s., and moreover, $p_c(\bigcup_{v \in \mathsf{V}} I(v)) = 1$ a.s.

11.26. Show that if G is a unimodular, transitive, locally finite, connected graph, then $p_c(G) < p_u(G)$ iff $p_c(\mathfrak{F}_f) < 1$ a.s.

11.27. Let G be a plane, regular graph of degree d with regular dual of degree $d^\dagger$. Show that the FMSF-expected degree of each vertex in G is $d(1 - 2/d^\dagger)$.

11.28. Let G be the usual Cayley graph of the (p, q, r)-triangle group, where $1/p + 1/q + 1/r \leq 1$, shown in Figure 6.1 for $(2, 3, 7)$ and defined in Exercise 8.47. It has three generators, which are reflections in the sides of a fundamental triangle. Show that the expected degree of the FMSF of G is $3 - 1/p - 1/q - 1/r$.

11.29. Consider the Cayley graph corresponding to the presentation $\langle a, b, c, d \mid a^2, b^2, c^2, abd^{-1} \rangle$. Show that the expected degree of a vertex in the FMSF is 3.

11.30. Give a Cayley graph with $p_u = 1$ on which the FMSF a.s. is not a single tree.

12 | Limit Theorems for Galton–Watson Processes

How quickly does a supercritical Galton–Watson branching process grow? In Section 5.1, we gave an answer via the Kesten–Stigum theorem, but its proof was postponed to the present chapter. How quickly does the survival probability to generation n decay as $n \to \infty$ in critical and subcritical processes? These are the major questions we answer in this chapter. A special biasing of the branching process, called size-biasing, transforms these questions to much easier ones. Size-biased branching processes turn out to be an example of branching processes with immigration, so we also study those a bit.

This chapter is adapted from Lyons, Pemantle, and Peres (1995a).

12.1 Size-Biased Trees and Immigration

Recall from Section 5.1

The Kesten–Stigum Theorem (1966). *Let L be the offspring random variable of a Galton–Watson process with mean $m \in (1, \infty)$ and martingale limit W. The following are equivalent:*

(i) $\mathbf{P}[W = 0] = q$;

(ii) $\mathbf{E}[W] = 1$;

(iii) $\mathbf{E}[L \log^+ L] < \infty$.

Although condition (iii) appears technical, there is a conceptual proof of the theorem that uses only the crudest estimates. The dichotomy of Corollary 5.7 as expanded in the Kesten–Stigum theorem turns out to arise from the following elementary dichotomy:

Lemma 12.1. *Let $X, X_1, X_2, \ldots$ be nonnegative, i.i.d. random variables. Then a.s.*

$$\limsup_{n \to \infty} \frac{1}{n} X_n = \begin{cases} 0 & \text{if } \mathbf{E}[X] < \infty \\ \infty & \text{if } \mathbf{E}[X] = \infty. \end{cases}$$

▷ **Exercise 12.1.**

Prove this by using the Borel–Cantelli lemma.

We will use Lemma 12.1 mainly via the following consequence:

▷ **Exercise 12.2.**

Given X, X_n as in Lemma 12.1, show that if $\mathbf{E}[X] < \infty$, then $\sum_n e^{X_n} c^n < \infty$ a.s. for all $c \in (0, 1)$, whereas if $\mathbf{E}[X] = \infty$, then $\sum_n e^{X_n} c^n = \infty$ a.s. for all $c \in (0, 1)$.

This dichotomy will be applied to an auxiliary random variable. Let $\widehat{L}$ be a random variable whose distribution is that of *size-biased L*; that is,

$$\mathbf{P}[\widehat{L} = k] = \frac{kp_k}{m},$$

where, as usual, $p_k := \mathbf{P}[L = k]$. Note that

$$\mathbf{E}[\log \widehat{L}] = \frac{1}{m}\,\mathbf{E}[L \log^+ L].$$

Lemma 12.1 will be applied to $\log \widehat{L}$.

▷ **Exercise 12.3.**
Let X be a nonnegative random variable with $0 < \mathbf{E}[X] < \infty$. We say that $\widehat{X}$ has the *size-biased distribution* of X if $\mathbf{P}[\widehat{X} \in A] = \mathbf{E}[X\mathbf{1}_A(X)]/\mathbf{E}[X]$ for intervals $A \subseteq [0, \infty)$. Show that this is equivalent to $\mathbf{E}[f(\widehat{X})] = \mathbf{E}[Xf(X)]/\mathbf{E}[X]$ for all Borel $f \colon [0, \infty) \to [0, \infty)$.

▷ **Exercise 12.4.**
Suppose that $X_n \geq 0$ satisfy $0 < \mathbf{E}[X_n] < \infty$ and $\mathbf{P}[X_n > 0]/\mathbf{E}[X_n] \to 0$. Show that the size-biased random variables $\widehat{X}_n$ tend to infinity in probability.

We will now define certain "size-biased" random trees, called *size-biased Galton–Watson* trees. Note that this process, as well as the usual Galton–Watson process, will be a way of putting a measure on the space of trees, which we think of as rooted and labeled, as in Section 5.1. The law of the size-biased random tree will be denoted $\widehat{\mathbf{GW}}$, whereas the law of an ordinary Galton–Watson tree is denoted $\mathbf{GW}$. We will show that the Kesten–Stigum dichotomy is equivalent to the following: these two measures on the space of trees are either mutually absolutely continuous or mutually singular.

How can we size bias in a probabilistic manner? Suppose that we have an urn of balls such that when we reach into the urn and choose a ball uniformly at random, the probability of picking a ball numbered k (with "size" k) is q_k. If, for each k, we replace each ball numbered k with k balls numbered k, then the new probability of picking a ball numbered k is the size-biased probability. Thus, a probabilistic way of biasing $\mathbf{GW}$ according to Z_n is as follows: Imagine an urn containing trees up to generation n, the number copies of each tree being proportional to its $\mathbf{GW}$ probability. If we count trees according to the sizes of their nth generations, however, then we find that they are size-biased. Thus, reach in the urn and pick a vertex uniformly at random from among all vertices in the nth generation of some tree. We can think of the result as a tree with a path from the root to the nth generation (ending in the chosen vertex). We will couple these measures for all n, giving a random infinite tree with an infinite path. The resulting joint distribution will be called $\widehat{\mathbf{GW}}_*$. This motivates the following definitions.

For a tree t with Z_n vertices at level n, write $W_n(t) := Z_n/m^n$. For any rooted tree t and any $n \geq 0$, denote by $[t]_n$ the set of rooted trees whose first n levels agree with those of t. (In particular, if the height of t is less than n, then $[t]_n = \{t\}$.) If v is a vertex at the nth level of

t, then let $[t; v]_n$ denote the set of **trees with distinguished paths** such that the tree is in $[t]_n$ and the path starts from the root, does not backtrack, and goes through v.

To construct $\widehat{\mathbf{GW}}$, we will construct a measure $\widehat{\mathbf{GW}}_*$ on the set of infinite trees with infinite distinguished paths; this measure will satisfy

$$\widehat{\mathbf{GW}}_*[t; v]_n = \frac{1}{m^n} \mathbf{GW}[t]_n \tag{12.1}$$

for all n and all $[t; v]_n$ as earlier. By using the branching property and the fact that the expected number of children of v is m, it is easy to verify consistency of these finite-height distributions. Kolmogorov's existence theorem thus provides such a measure $\widehat{\mathbf{GW}}_*$. However, this verification may be skipped, as we will give a more useful, direct construction of a measure with these marginals in a moment.

Note that *if* a measure $\widehat{\mathbf{GW}}_*$ satisfying (12.1) exists, then its projection to the space of trees, which is denoted simply by $\widehat{\mathbf{GW}}$, automatically satisfies

$$\widehat{\mathbf{GW}}[t]_n = W_n(t) \mathbf{GW}[t]_n \tag{12.2}$$

for all n and all trees t. It is for this reason that we call $\widehat{\mathbf{GW}}$ "size-biased."

How do we define $\widehat{\mathbf{GW}}_*$? Assuming still that (12.1) holds, note that the recursive structure of Galton–Watson trees yields a recursion for $\widehat{\mathbf{GW}}_*$. Assume that t is a tree of height at least $n + 1$ and that the root of t has k children with descendant trees $t^{(1)}, t^{(2)}, \ldots, t^{(k)}$. Any vertex v in level $n + 1$ of t is in one of these – say, $t^{(i)}$. Now

$$\mathbf{GW}[t]_{n+1} = p_k \prod_{j=1}^{k} \mathbf{GW}[t^{(j)}]_n = kp_k \cdot \frac{1}{k} \cdot \mathbf{GW}[t^{(i)}]_n \cdot \prod_{j \neq i} \mathbf{GW}[t^{(j)}]_n .$$

Thus any measure $\widehat{\mathbf{GW}}_*$ that satisfies (12.1) must satisfy the recursion

$$\widehat{\mathbf{GW}}_*[t; v]_{n+1} = \frac{kp_k}{m} \cdot \frac{1}{k} \cdot \widehat{\mathbf{GW}}_*[t^{(i)}; v]_n \cdot \prod_{j \neq i} \mathbf{GW}[t^{(j)}]_n . \tag{12.3}$$

Conversely, if a probability measure $\widehat{\mathbf{GW}}_*$ on the set of trees with distinguished paths satisfies this recursion, then induction shows that the measure satisfies (12.1); this observation leads to the following direct construction of $\widehat{\mathbf{GW}}_*$.

Recall that $\widehat{L}$ is a random variable whose distribution is that of size-biased L, in other words, $\mathbf{P}[\widehat{L} = k] = kp_k/m$. To construct a size-biased Galton–Watson tree $\widehat{T}$, start with an initial particle v_0. Give it a random number $\widehat{L}_1$ of children, where $\widehat{L}_1$ has the law of $\widehat{L}$. Pick one of these children at random, v_1. Give the *other* children independently ordinary Galton–Watson descendant trees and give v_1 an independent size-biased number $\widehat{L}_2$ of children. Again, pick one of the children of v_1 at random, call it v_2, and give the others ordinary Galton–Watson descendant trees. Continue in this way indefinitely. (See Figure 12.1.) Note that since $\widehat{L} \geq 1$, size-biased Galton–Watson trees are always infinite (there is no extinction).

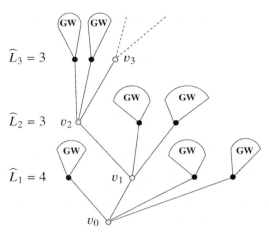

Figure 12.1. Schematic of size-biased Galton–Watson trees.

Now we can finally **define** the measure $\widehat{\mathbf{GW}}_*$ as the joint distribution of the random tree $\widehat{T}$ and the random path $\langle v_0, v_1, v_2, \ldots \rangle$. This measure clearly satisfies the recursion (12.3), and hence also (12.1).

Note that, given the first n levels of the tree $\widehat{T}$, the measure $\widehat{\mathbf{GW}}_*$ makes the vertex v_n in the random path $\langle v_0, v_1, \ldots \rangle$ uniformly distributed on the nth level of $\widehat{T}$; this is not obvious from the explicit construction of this random path, but it is immediate from the formula (12.1) in which the right-hand side does not depend on v.

▷ **Exercise 12.5.**
Define $\widehat{\mathbf{GW}}_*$ formally on a space analogous to the space $\mathscr{T}$ of Exercise 5.2 and define $\widehat{\mathbf{GW}}$ formally on $\mathscr{T}$.

The vertices off the distinguished path $\langle v_0, v_1, \ldots \rangle$ of the size-biased tree form a **branching process with immigration**. In general, such a process is defined by two distributions, an offspring distribution and an immigration distribution. The process starts with no particles, say, and at every generation $n \geq 1$, there is an immigration of Y_n particles, where Y_n are i.i.d. with the given immigration law. Meanwhile, each particle has, independently, an ordinary Galton–Watson descendant tree with the given offspring distribution.

Thus, the $\widehat{\mathbf{GW}}$-law of $Z_n - 1$ is the same as that of the generation sizes of an immigration process with $Y_n = \widehat{L}_n - 1$. Since $\mathbf{E}\big[\log^+(\widehat{L} - 1)\big] = m^{-1}\,\mathbf{E}\big[L \log^+(L - 1)\big]$, the condition (iii) of the Kesten–Stigum theorem is equivalent to $\mathbf{E}\big[\log^+(\widehat{L} - 1)\big] < \infty$. The finiteness of this last expectation determines the behavior of the variables $\langle \log^+ Y_n \rangle$ via Lemma 12.1.

We will need to take conditional expectations of random variables that might not have finite means, so let's recall how this works. Note that if $\nu \ll \mu$ and μ is σ-finite, then ν has a Radon–Nikodým derivative with respect to μ without any further assumption on ν (see, for example, Problem 32.8 of Billingsley (1995)). Suppose that $X \geq 0$ is a random variable and $\mathscr{F}$ is a σ-field. We allow X to take the value $+\infty$. We define $\mathbf{E}[X \mid \mathscr{F}]$ to be the Radon–Nikodým derivative of $(X\,\mathbf{P}){\restriction}\mathscr{F}$ with respect to $\mathbf{P}{\restriction}\mathscr{F}$. Uniqueness of the derivative shows that

$$\mathbf{E}[X \mid \mathscr{F}] = \mathbf{E}[X\mathbf{1}_{[X<\infty]} \mid \mathscr{F}] + \infty \cdot \mathbf{P}[X = \infty \mid \mathscr{F}].$$

If $\mathbf{E}[X \mid \mathscr{F}] < \infty$ a.s., then $X < \infty$ a.s. The conditional monotone convergence theorem says that if $0 \le X_n \uparrow X$, then $\lim_{n \to \infty} \mathbf{E}[X_n \mid \mathscr{F}] = \mathbf{E}[X \mid \mathscr{F}]$ a.s. To see this, let Y be the limit of the increasing sequence $\langle \mathbf{E}[X_n \mid \mathscr{F}] \rangle$ and use the ordinary monotone convergence theorem to deduce that for every $F \in \mathscr{F}$, we have $\mathbf{E}[Y\mathbf{1}_F] = \mathbf{E}[X\mathbf{1}_F]$, as desired. One can then deduce the conditional version of Fatou's lemma: if $X_n \ge 0$, then $\mathbf{E}[\liminf_{n \to \infty} X_n \mid \mathscr{F}] \le \liminf_{n \to \infty} \mathbf{E}[X_n \mid \mathscr{F}]$ a.s.

12.2 Supercritical Processes: Proof of the Kesten–Stigum Theorem

We mentioned that the Kesten–Stigum dichotomy arises from a dichotomy concerning mutual singularity of measures. The latter dichotomy involves the following lemma, which is more or less standard.

Lemma 12.2. *Let μ be a finite measure and ν be a probability measure on a σ-field $\mathscr{F}$. Suppose that $\mathscr{F}_n$ are increasing sub-σ-fields whose union generates $\mathscr{F}$ and that $(\mu \!\restriction\! \mathscr{F}_n)$ is absolutely continuous with respect to $(\nu \!\restriction\! \mathscr{F}_n)$ with Radon–Nikodým derivative X_n. Set $X := \limsup_{n \to \infty} X_n$. Then*

$$\mu \ll \nu \iff X < \infty \quad \mu\text{-a.e.} \iff \int X \, d\nu = \int d\mu$$

and

$$\mu \perp \nu \iff X = \infty \quad \mu\text{-a.e.} \iff \int X \, d\nu = 0.$$

Proof. Since $\langle (X_n, \mathscr{F}_n) \rangle$ is a nonnegative martingale with respect to ν (see Exercise 12.17), it converges to X ν-a.s. and $X < \infty$ ν-a.s. We claim that μ has the following decomposition into a ν-absolutely continuous part and a ν-singular part:

$$\mu = X\nu + \mathbf{1}_{[X=\infty]}\mu. \tag{12.4}$$

Given this, the lemma follows: For if $\mu \ll \nu$, then $X < \infty$ μ-a.e.; if $X < \infty$ μ-a.e., then by (12.4), $\int X \, d\nu = \int d\mu$; and if $\int X \, d\nu = \int d\mu$, then by (12.4), $X < \infty$ μ-a.e. and $\mu \ll \nu$. On the other hand, if $\mu \perp \nu$, then by (12.4), $\mu = \mathbf{1}_{[X=\infty]}\mu$, whence $X = \infty$ μ-a.e.; if $X = \infty$ μ-a.e., then by (12.4), $\int X \, d\nu = 0$; and if $\int X \, d\nu = 0$, then by (12.4), $X = \infty$ μ-a.e., whence $\mu \perp \nu$.

To establish (12.4), suppose first that $\mu \ll \nu$ with Radon–Nikodým derivative $\widetilde{X}$. Then X_n is (a version of) the conditional expectation of $\widetilde{X}$ given $\mathscr{F}_n$ (with respect to ν), whence $X_n \to \widetilde{X}$ ν-a.s. by the martingale convergence theorem. In particular, $X = \widetilde{X}$ ν-a.s., so the decomposition is simply the definition of Radon–Nikodým derivative.

To treat the general case, we use a common trick: Define the probability measure $\rho := (\mu + \nu)/C$, where $C := \int d(\mu + \nu)$. Then $\mu, \nu \ll \rho$, so that we may apply what we have just shown to the variables $U_n := d(\mu \!\restriction\! \mathscr{F}_n)/d(\rho \!\restriction\! \mathscr{F}_n)$ and $V_n := d(\nu \!\restriction\! \mathscr{F}_n)/d(\rho \!\restriction\! \mathscr{F}_n)$. Let $U := \limsup U_n$ and $V := \limsup V_n$. Since $U_n + V_n = C$ ρ-a.s., we have $\rho[U = V = 0] = 0$ and thus ρ-a.s.

$$U/V = \lim U_n / \lim V_n = \lim(U_n/V_n) = \lim X_n = X.$$

Therefore, using three times what we established in the preceding paragraph, we obtain

$$\mu = U\rho = XV\rho + \mathbf{1}_{[V=0]}U\rho = X\nu + \mathbf{1}_{[X=\infty]}\mu. \qquad \blacktriangleleft$$

The Kesten–Stigum theorem will be an immediate consequence of the following theorem on the growth rate of immigration processes.

Theorem 12.3. (Seneta, 1970) *Let Z_n be the generation sizes of a Galton–Watson process with immigration Y_n. Let $m := \mathbf{E}[L] \in (1, \infty)$ be the mean of the offspring law, and let Y have the same law as Y_n. If $\mathbf{E}[\log^+ Y] < \infty$, then $\lim Z_n/m^n$ exists and is finite a.s., whereas if $\mathbf{E}[\log^+ Y] = \infty$, then $\limsup Z_n/c^n = \infty$ a.s. for every constant $c > 0$.*

Proof. Assume first that $\mathbf{E}[\log^+ Y] = \infty$. By Lemma 12.1, $\limsup Y_n/c^n = \infty$ a.s. Since $Z_n \geq Y_n$, the result follows.

Now assume that $\mathbf{E}[\log^+ Y] < \infty$. Let $\mathscr{G}$ be the σ-field generated by $\{Y_k \, ; \; k \geq 1\}$. Let $Z_{n,k}$ be the number of descendants at level n of the Y_k particles that immigrated in generation k. Thus, the total number of vertices at level n is $\sum_{k=1}^{n} Z_{n,k}$. With our notation, we have

$$\mathbf{E}[Z_n/m^n \mid \mathscr{G}] = \mathbf{E}\left[\frac{1}{m^n} \sum_{k=1}^{n} Z_{n,k} \,\middle|\, \mathscr{G} \right] = \sum_{k=1}^{n} \frac{1}{m^k} \, \mathbf{E}\left[\frac{Z_{n,k}}{m^{n-k}} \,\middle|\, \mathscr{G} \right].$$

Note that conditioning on $\mathscr{G}$ is just fixing values for all Y_k; since Y_k are independent of all other random variables, conditioning on $\mathscr{G}$ amounts to using Fubini's theorem. Now for $k \leq n$, the random variable $Z_{n,k}/m^{n-k}$ is the $(n-k)$th element of the ordinary Galton–Watson martingale sequence starting, however, with Y_k particles. Therefore, its conditional expectation is just Y_k, and so

$$\mathbf{E}[Z_n/m^n \mid \mathscr{G}] = \sum_{k=1}^{n} \frac{Y_k}{m^k}.$$

Our assumption gives, by Exercise 12.2, that this series converges a.s. This implies by the conditional Fatou lemma that $\mathbf{E}[\liminf Z_n/m^n \mid \mathscr{G}] < \infty$ a.s., whence $\liminf Z_n/m^n < \infty$ a.s. Finally, since $\langle Z_n/m^n \rangle$ is a submartingale when conditioned on $\mathscr{G}$ with bounded expectation (given $\mathscr{G}$), it converges a.s. ◄

Proof of the Kesten–Stigum theorem. (Lyons, Pemantle, and Peres, 1995a) Let $\mathscr{F}_n$ be the σ-field generated by the first n levels of trees. Then (12.2) is the same as

$$\frac{d(\widehat{\mathbf{GW}} \!\restriction\! \mathscr{F}_n)}{d(\mathbf{GW} \!\restriction\! \mathscr{F}_n)}(t) = W_n(t). \tag{12.5}$$

To define W for every infinite tree t, set

$$W(t) := \limsup_{n \to \infty} W_n(t).$$

From (12.5) and Lemma 12.2 follows the key dichotomy:

$$\int W \, d\mathbf{GW} = 1 \iff \widehat{\mathbf{GW}} \ll \mathbf{GW} \iff W < \infty \quad \widehat{\mathbf{GW}}\text{-a.s.}, \tag{12.6}$$

whereas

$$W = 0 \quad \mathbf{GW}\text{-a.s.} \iff \mathbf{GW} \perp \widehat{\mathbf{GW}} \iff W = \infty \quad \widehat{\mathbf{GW}}\text{-a.s.} \tag{12.7}$$

This is key because it allows us to change the problem from one about the $\mathbf{GW}$-behavior of W to one about the $\widehat{\mathbf{GW}}$-behavior of W. Indeed, since the $\widehat{\mathbf{GW}}$-behavior of W is described by Theorem 12.3, the theorem is immediate: if $\mathbf{E}[L \log^+ L] < \infty$, that is, $\mathbf{E}[\log \widehat{L}] < \infty$, then $W < \infty$ $\widehat{\mathbf{GW}}$-a.s. by Theorem 12.3, whence $\int W \, d\mathbf{GW} = 1$ by (12.6); whereas if $\mathbf{E}[L \log^+ L] = \infty$, then $W = \infty$ $\widehat{\mathbf{GW}}$-a.s. by Theorem 12.3, whence $W = 0$ $\mathbf{GW}$-a.s. by (12.7). ◄

12.3 Subcritical Processes

When a Galton–Watson process is subcritical or critical, the questions we asked in Section 5.1 about rate of growth are inappropriate. Other questions come to mind, however, such as, How quickly does the process die out? One way to make this question precise is to ask for the decay rate of $\mathbf{P}[Z_n > 0]$. An easy estimate in the subcritical case is

$$\mathbf{P}[Z_n > 0] \leq \mathbf{E}[Z_n] = m^n. \tag{12.8}$$

We determine in this section when m^n is the right decay rate (up to some factor). In the next section, we treat the critical case.

Theorem 12.4. (Heathcote, Seneta, and Vere-Jones, 1967) *For any Galton–Watson process with $0 < m < \infty$, the sequence $\langle \mathbf{P}[Z_n > 0]/m^n \rangle$ is decreasing. If $m < 1$, then the following are equivalent:*
  (i) $\lim_{n \to \infty} \mathbf{P}[Z_n > 0]/m^n > 0$;
  (ii) $\sup \mathbf{E}[Z_n \mid Z_n > 0] < \infty$;
  (iii) $\mathbf{E}[L \log^+ L] < \infty$.

The fact that (i) holds when $\mathbf{E}[L^2] < \infty$ was proved by Kolmogorov (1938).

To prove Theorem 12.4, we use an approach analogous to that in the preceding section: we combine a general lemma with a result on immigration.

Lemma 12.5. *Let $\langle v_n \rangle$ be a sequence of probability measures on the positive integers with finite means a_n. Let $\widehat{v}_n$ be size biased, that is, $\widehat{v}_n(k) = k v_n(k)/a_n$. If $\{\widehat{v}_n\}$ is tight, then $\sup a_n < \infty$, whereas if $\widehat{v}_n \to \infty$ in distribution, then $a_n \to \infty$.*

▷ **Exercise 12.6.**
Prove Lemma 12.5.

Theorem 12.6. (Heathcote, 1966) *Let Z_n be the generation sizes of a Galton–Watson process with immigration Y_n. Let Y have the same law as Y_n. Suppose that the mean m of the offspring random variable L is less than 1. If $\mathbf{E}[\log^+ Y] < \infty$, then Z_n converges in distribution to a proper* random variable, whereas if $\mathbf{E}[\log^+ Y] = \infty$, then Z_n converges in probability to infinity.*

Proof. Let $\mathscr{G}$ be the σ-field generated by $\{Y_k ; \ k \geq 1\}$. For any n, let $Z_{n,k}$ be the number of descendants at level n of the Y_k vertices that immigrated in generation k. Thus, the total number of vertices at level n is $Z_n = \sum_{k=1}^{n} Z_{n,k}$. Since the distribution of $Z_{n,k}$ depends only on $n - k$, this total Z_n has the same distribution as $\sum_{k=1}^{n} Z_{2k-1,k}$. This latter sum increases in n to some limit Z'_∞. By Kolmogorov's zero-one law, Z'_∞ is a.s. finite or a.s. infinite. Hence, we need only to show that $Z'_\infty < \infty$ a.s. iff $\mathbf{E}[\log^+ Y] < \infty$.

Assume that $\mathbf{E}[\log^+ Y] < \infty$. Now $\mathbf{E}[Z'_\infty \mid \mathscr{G}] = \sum_{k=1}^{\infty} Y_k m^{k-1}$. By Exercise 12.2, this sum converges a.s. Therefore, Z'_∞ is finite a.s.

---

* That is, finite a.s.

Now assume that $Z'_\infty < \infty$ a.s. Writing $Z_{2k-1,k} = \sum_{i=1}^{Y_k} \zeta_k(i)$, where $\zeta_k(i)$ are the sizes of generation $k-1$ of i.i.d. Galton–Watson branching processes with one initial particle, we have $Z'_\infty = \sum_{k=1}^{\infty} \sum_{i=1}^{Y_k} \zeta_k(i)$ written as a random sum of independent, integer-valued random variables; the latter are still independent conditioned on $\mathscr{G}$. Almost surely only a finite number of the $\zeta_k(i)$ are at least one, whence by the conditional Borel–Cantelli lemma with respect to $\mathscr{G}$, we get $\sum_{k=1}^{\infty} Y_k \mathbf{GW}[Z_{k-1} \geq 1] < \infty$ a.s. Since $\mathbf{GW}[Z_{k-1} \geq 1] \geq \mathbf{P}[L > 0]^{k-1}$, it follows from Exercise 12.2 that $\mathbf{E}[\log^+ Y] < \infty$.   ◀

Proof of Theorem 12.4. (Lyons, Pemantle, and Peres, 1995a) Let μ_n be the law of Z_n conditioned on $Z_n > 0$. For any tree t with $Z_n(t) > 0$, let $\xi_n(t)$ be the lowest-labeled child of the root that has at least one descendant in generation n. (Recall we are using the labeling of Section 5.1.) If $Z_n(t) > 0$, let $H_n(t)$ be the number of descendants of $\xi_n(t)$ in generation n; otherwise, let $H_n(t) := 0$. It is easy to see that

$$\mathbf{P}[H_n = k \mid Z_n > 0] = \mathbf{P}[H_n = k \mid Z_n > 0, \xi_n = v] = \mathbf{P}[Z_{n-1} = k \mid Z_{n-1} > 0]$$

for all children v of the root. Since $H_n \leq Z_n$, this shows that $\langle \mu_n \rangle$ increases stochastically as n increases. Now

$$\mathbf{P}[Z_n > 0] = \frac{\mathbf{E}[Z_n]}{\mathbf{E}[Z_n \mid Z_n > 0]} = \frac{m^n}{\int x \, d\mu_n(x)} .$$

Therefore, $\langle \mathbf{P}[Z_n > 0]/m^n \rangle$ is decreasing and (i) $\Leftrightarrow$ (ii).

Now $\widehat{\mu}_n$ is not only the size-biased version of μ_n but also the law of the size-biased random variable $\widehat{Z}_n$. Thus, from Section 12.1, we know that $\langle \widehat{\mu}_n \rangle$ describes the generation sizes plus 1 of a process with immigration $\widehat{L} - 1$. Suppose that $m < 1$. If (ii) holds, that is, the means of μ_n are bounded, then by Lemma 12.5, the laws $\widehat{\mu}_n$ do not tend to infinity. Applying Theorem 12.6 to the associated immigration process, we see that (iii) holds. Conversely, if (iii) holds, then by Theorem 12.6, $\langle \widehat{\mu}_n \rangle$ converges, whence is tight. In light of Lemma 12.5, (ii) follows.   ◀

12.4 Critical Processes

In the critical case, the easy estimate (12.8) is useless. What then is the rate of decay?

Theorem 12.7. (Kesten, Ney, and Spitzer, 1966) *Suppose that* $m = 1$, *and let* $\sigma^2 := \mathrm{Var}(L) = \mathbf{E}[L^2] - 1$. *Then we have*

(i) **Kolmogorov's estimate:** $\lim_{n\to\infty} n \, \mathbf{P}[Z_n > 0] = 2/\sigma^2$;

(ii) **Yaglom's limit law:** *If* $\sigma < \infty$, *then the conditional distribution of* Z_n/n *given* $Z_n > 0$ *converges as* $n \to \infty$ *to an exponential law with mean* $\sigma^2/2$.

Under the assumption that $\mathbf{E}[L^3] < \infty$, parts (i) and (ii) of this theorem are due to Kolmogorov (1938) and Yaglom (1947), respectively. The case where $\sigma = \infty$ in (ii) appears to be open. We give a proof that combines ideas of Lyons, Pemantle, and Peres (1995a) and Geiger (1999). The exponential limit law in part (ii) will arise from the following characterization of exponential random variables due to Pakes and Khattree (1992):

▷ **Exercise 12.7.**
Let A be a nonnegative random variable with a positive, finite mean, and let $\widehat{A}$ have the corresponding size-biased distribution. Denote by U a uniform random variable in $[0, 1]$ that is independent of $\widehat{A}$. Prove that $U \cdot \widehat{A}$ and A have the same distribution iff A has an exponential distribution.

The size-biased offspring random variable $\widehat{L}$ will arise in the following way:

▷ **Exercise 12.8.**
Let L be a random variable taking nonnegative integer values with $0 < \mathbf{E}[L] < \infty$, and let $\widehat{L}$ be its size-biased version. Suppose that for each n, there are events $H_1^{(n)}, \ldots, H_L^{(n)}$ that, given L, are independent with probability $p_n > 0$ each and that $p_n \to 0$ as $n \to \infty$. Let $Y_n := \sum_{i=1}^{L} \mathbf{1}_{H_i^{(n)}}$ be the number of events $H_i^{(n)}$ that occur. Show that the following hold:
   (a) $\lim_{n \to \infty} \mathbf{P}[Y_n = 1 \mid Y_n > 0] = 1$;
   (b) $\lim_{n \to \infty} \mathbf{P}[L \le k \mid Y_n > 0] = \mathbf{P}[\widehat{L} \le k]$;
   (c) $\lim_{n \to \infty} \mathbf{P}[H_i^{(n)} \mid Y_n > 0, L = k] = 1/k$ for $1 \le i \le k$.

We will also use the results of the following exercises.

▷ **Exercise 12.9.**
Suppose that A, A_n are nonnegative random variables with positive, finite means such that $A_n \to A$ in law and $\widehat{A}_n \to B$ in law. Show that if B is a proper random variable, then B has the law of $\widehat{A}$.

▷ **Exercise 12.10.**
Suppose that $0 \le A_i \le B_i$ are random variables, that $A_i \to 0$ in probability, and that B_i are identically distributed with finite mean. Show that $\sum_{i=1}^{n} A_i/n \to 0$ in probability. Show that if, in addition, $C_{i,j}$ are random variables with $\mathbf{E}\big[|C_{i,j}| \;\big|\; A_i\big] \le 1$ and A_i takes integer values, then $\sum_{i=1}^{n} \sum_{j=1}^{A_i} C_{i,j}/n \to 0$ in probability.

▷ **Exercise 12.11.**
Let A be a random variable independent of the random variables B and C. Suppose that the function $x \mapsto \mathbf{P}[C \le x]/\mathbf{P}[B \le x]$ is increasing for $x > \operatorname{ess\,inf} B$, that $\mathbf{P}[A \ge B] > 0$, and that $\mathbf{P}[A \ge C] > 0$. Show that the law of A given that $A \ge B$ is stochastically dominated by the law of A given that $A \ge C$. Show that the hypothesis on B and C is satisfied when they are geometric random variables with B having a larger parameter than C.

Proof of Theorem 12.7. It will be convenient to refer to the lexicographic ordering of labeled vertices as increasing from left to right. Let Y_n be the number of individuals of the first generation that have a descendant in generation n. Since $\mathbf{P}[Z_n > 0] \to 0$, it follows from Exercise 12.8 that $\mathbf{P}[Y_n = 1 \mid Z_n > 0] \to 1$, that the conditional distribution of Z_1 given $Z_n > 0$ tends to the distribution of the size-biased random variable $\widehat{L}$, and that the conditional distribution of the left-most individual of the first generation that has a descendant in generation n tends to a uniform pick among the individuals of the first generation. Since each child of the initial progenitor independently has a descendant in generation n with probability $\mathbf{P}[Z_{n-1} > 0]$, we have that the law of Z_1 given that $Z_n > 0$ is the law of Z_1 given that $Z_1 \geq D_n$, where D_n is a geometric random variable independent of Z_1 with mean $1/\mathbf{P}[Z_{n-1} > 0]$. Since $\mathbf{P}[Z_{n-1} > 0] > \mathbf{P}[Z_n > 0]$, Exercise 12.11 implies that the conditional distribution of Z_1 given $Z_n > 0$ stochastically increases with n. Since $\mathbf{P}[Z_1 \geq k \mid Z_n > 0]$ increases with n, the tail formula for expectation and the monotone convergence theorem yield $\mathbf{E}[Z_1 \mid Z_n > 0] \to \mathbf{E}[\widehat{L}] = \sigma^2 + 1$.

Let u_n^n be the left-most individual in generation n when $Z_n > 0$. Let its ancestors back to the initial progenitor be $u_{n-1}^n, \ldots, u_0^n$, where u_i^n is in generation i. Let X_i' denote the number of descendants of u_i^n in generation n that are not descendants of u_{i+1}^n. Let X_i be the number of children of u_i^n that are to the right of u_{i+1}^n. Then $Z_n = 1 + \sum_{i=0}^{n-1} X_i'$ and $\mathbf{E}[X_i' \mid Z_n > 0] = \mathbf{E}[X_i \mid Z_n > 0]$, since each of these X_i individuals generates an independent *critical* Galton–Watson descendant tree (with offspring law the same as that of the original process). Therefore,

$$\frac{1}{n\,\mathbf{P}[Z_n > 0]} = \frac{\mathbf{E}[Z_n]}{n\,\mathbf{P}[Z_n > 0]} = \frac{1}{n}\,\mathbf{E}[Z_n \mid Z_n > 0] = \frac{1}{n} + \frac{1}{n}\sum_{i=0}^{n-1}\mathbf{E}[X_i \mid Z_n > 0]. \quad (12.9)$$

In the first paragraph of the proof, we have seen that the conditional distribution of X_i given that $Z_n > 0$ tends to that of $\lfloor U \cdot \widehat{L} \rfloor$, where U denotes a uniform $[0, 1]$-random variable that is independent of $\widehat{L}$. Thus, $\lim_{n\to\infty} \mathbf{E}[X_i \mid Z_n > 0] = \mathbf{E}[\lfloor U \cdot \widehat{L} \rfloor] = \mathbf{E}[\widehat{L} - 1]/2 = \sigma^2/2$, which gives Kolmogorov's estimate. Actually, we need to justify this passage to the limit for the expectations. When $\sigma < \infty$, it follows from the fact that the conditional distributions of Z_1 given $Z_n > 0$ are uniformly integrable (a consequence of the first paragraph) and that the conditional distribution of X_i given $Z_n > 0$ is dominated by the conditional distribution of Z_1 given that $Z_{n-i+1} > 0$, whence the conditional distributions of X_i given $Z_n > 0$ are uniformly integrable as well. When $\sigma = \infty$, the passage to the limit follows from Fatou's lemma.

Now suppose that $\sigma < \infty$. We are going to compare the conditional distribution of Z_n/n given $Z_n > 0$ with the law of R_n/n, where R_n is the number of individuals in generation n to the right of v_n in the size-biased tree with distinguished path $v_0, v_1, \ldots$. Recall that X_i denotes the number of children of u_i^n to the right of u_{i+1}^n. Since we are interested in its distribution as $n \to \infty$, we will be explicit and write $X_i^{(n)} := X_i$. For $1 \leq j \leq X_i^{(n)}$, let $S_{i,j}^{(n)}$ be the number of descendants in generation n of the jth child of u_i^n to the right of u_{i+1}^n. On the other hand, in the size-biased tree, let X_i'' be the number of children of v_i to the right of v_{i+1}, and let $V_{i,j}^{(n)}$ be the number of descendants in generation n of the jth child of v_i to the right of v_{i+1}. For each n and i, $S_{i,j}^{(n)}$ and $V_{i,j}^{(n)}$ are identically distributed with mean 1 (since they pertain to $n - i - 1$ generations of independent critical Galton–Watson trees). In virtue of the first paragraph, we may couple them so that $S_{i,j}^{(n)} = V_{i,j}^{(n)}$ for $j \leq X_i^{(n)}$ and $0 \leq X_i'' - X_i^{(n)} \leq \widehat{L}_{i+1}$. Then $X_i^{(n)} - X_i'' \to 0$ in measure as $n \to \infty$. Since

$$Z_n = 1 + \sum_{i=0}^{n-1}\sum_{j=1}^{X_i^{(n)}} S_{i,j}^{(n)} \qquad \text{and} \qquad R_n = 1 + \sum_{i=0}^{n-1}\sum_{j=1}^{X_i''} V_{i,j}^{(n)},$$

it follows from Exercise 12.10 that in this coupling, $Z_n/n - R_n/n \to 0$ in measure as $n \to \infty$.

Now we prove that the limit of R_n/n exists in law and identify it. The $\widehat{\mathbf{GW}}$ laws of Z_n/n have uniformly bounded means by Exercise 5.27 and so are tight. This implies the tightness of $\{\mu_n\}$, where μ_n is the $\mathbf{GW}$-conditional distribution of Z_n/n given that $Z_n > 0$, and of the $\widehat{\mathbf{GW}}_*$ laws of R_n/n. Therefore, we can find $n_k \to \infty$ so that μ_{n_k} and the $\widehat{\mathbf{GW}}_*$ laws of R_{n_k}/n_k converge to the law of a (proper) random variable A and the $\widehat{\mathbf{GW}}$ laws of Z_{n_k}/n_k converge to the law of a (proper) random variable B. Note that the $\widehat{\mathbf{GW}}$ law of Z_n/n can also be gotten by size biasing μ_n. By virtue of Exercise 12.9, therefore, the variables $\widehat{A}$ and B are identically distributed. Since R_n is a uniform pick from $\{0, 1, \ldots, Z_n - 1\}$, we also have that A has the same law as $U \cdot B$, in other words, as $U \cdot \widehat{A}$. By Exercise 12.7, it follows that A is an exponential random variable whose mean, by (12.9) and part (i), is $\sigma^2/2$. In particular, the limit of μ_{n_k} is independent of the sequence $\langle n_k \rangle$, and hence we actually have convergence in law of the whole sequence μ_n to A, as desired. ◀

12.5 Notes

Ideas related to size-biased Galton–Watson trees occur in Hawkes (1981), Joffe and Waugh (1982), Waymire and Williams (1996), and Chauvin, Rouault, and Wakolbinger (1991). There have been very many uses since then of these ideas. The sequence of generation sizes of size-biased Galton–Watson trees is known as a Q-process in the case $m \le 1$; see Athreya and Ney (1972), pp. 56–60.

The proof of Theorem 12.3 is that of Asmussen and Hering (1983), pp. 50–51. The proof of Theorem 12.6 is a slight improvement on Asmussen and Hering (1983), pp. 52–53.

The proof that the law of Z_1 given $Z_n > 0$ stochastically increases in n that appears at the beginning of the proof of Theorem 12.7 is due to Matthias Birkner (personal communication, 2000). A rate of convergence in Yaglom's limit law is given by Peköz and Röllin (2011), using Stein's method and ideas from the proof of Theorem 12.7 given in Lyons, Pemantle, and Peres (1995a).

12.6 Collected In-Text Exercises

12.1. Prove Lemma 12.1 by using the Borel–Cantelli lemma.

12.2. Given X, X_n as in Lemma 12.1, show that if $\mathbf{E}[X] < \infty$, then $\sum_n e^{X_n} c^n < \infty$ a.s. for all $c \in (0, 1)$, whereas if $\mathbf{E}[X] = \infty$, then $\sum_n e^{X_n} c^n = \infty$ a.s. for all $c \in (0, 1)$.

12.3. Let X be a nonnegative random variable with $0 < \mathbf{E}[X] < \infty$. We say that $\widehat{X}$ has the *size-biased distribution* of X if $\mathbf{P}[\widehat{X} \in A] = \mathbf{E}[X \mathbf{1}_A(X)]/\mathbf{E}[X]$ for intervals $A \subseteq [0, \infty)$. Show that this is equivalent to $\mathbf{E}[f(\widehat{X})] = \mathbf{E}[X f(X)]/\mathbf{E}[X]$ for all Borel $f: [0, \infty) \to [0, \infty)$.

12.4. Suppose that $X_n \ge 0$ satisfy $0 < \mathbf{E}[X_n] < \infty$ and $\mathbf{P}[X_n > 0]/\mathbf{E}[X_n] \to 0$. Show that the size-biased random variables $\widehat{X}_n$ tend to infinity in probability.

12.5. Define $\widehat{\mathbf{GW}}_*$ formally on a space analogous to the space $\mathscr{T}$ of Exercise 5.2 and define $\widehat{\mathbf{GW}}$ formally on $\mathscr{T}$.

12.6. Prove Lemma 12.5.

12.7. Let A be a nonnegative random variable with a positive, finite mean, and let $\widehat{A}$ have the corresponding size-biased distribution. Denote by U a uniform random variable in $[0, 1]$ that is independent of $\widehat{A}$. Prove that $U \cdot \widehat{A}$ and A have the same distribution iff A has an exponential distribution.

12.8. Let L be a random variable taking nonnegative integer values with $0 < \mathbf{E}[L] < \infty$, and let $\widehat{L}$ be its size-biased version. Suppose that for each n, there are events $H_1^{(n)}, \ldots, H_L^{(n)}$ that, given L, are independent with probability $p_n > 0$ each and that $p_n \to 0$ as $n \to \infty$. Let $Y_n := \sum_{i=1}^{L} \mathbf{1}_{H_i^{(n)}}$ be the number of events $H_i^{(n)}$ that occur. Show that the following hold:

    (a) $\lim_{n \to \infty} \mathbf{P}[Y_n = 1 \mid Y_n > 0] = 1$;

    (b) $\lim_{n \to \infty} \mathbf{P}[L \le k \mid Y_n > 0] = \mathbf{P}[\widehat{L} \le k]$;

    (c) $\lim_{n \to \infty} \mathbf{P}[H_i^{(n)} \mid Y_n > 0, L = k] = 1/k$ for $1 \le i \le k$.

12.9. Suppose that A, A_n are nonnegative random variables with positive, finite means such that $A_n \to A$ in law and $\widehat{A}_n \to B$ in law. Show that if B is a proper random variable, then B has the law of $\widehat{A}$.

12.10. Suppose that $0 \le A_i \le B_i$ are random variables, that $A_i \to 0$ in probability, and that B_i are identically distributed with finite mean. Show that $\sum_{i=1}^{n} A_i/n \to 0$ in probability. Show that if, in addition, $C_{i,j}$ are random variables with $\mathbf{E}\big[|C_{i,j}| \mid A_i\big] \le 1$ and A_i takes integer values, then $\sum_{i=1}^{n} \sum_{j=1}^{A_i} C_{i,j}/n \to 0$ in probability.

12.11. Let A be a random variable independent of the random variables B and C. Suppose that the function $x \mapsto \mathbf{P}[C \le x]/\mathbf{P}[B \le x]$ is increasing for $x > \operatorname{ess\,inf} B$, that $\mathbf{P}[A \ge B] > 0$, and that $\mathbf{P}[A \ge C] > 0$. Show that the law of A given that $A \ge B$ is stochastically dominated by the law of A given that $A \ge C$. Show that the hypothesis on B and C is satisfied when they are geometric random variables with B having a larger parameter than C.

12.7 Additional Exercises

12.12. Let $X_1, \ldots, X_n$ be independent, nonnegative random variables with finite, positive mean. Define $X := \sum_{i=1}^{n} X_i$ and $Y := \prod_{i=1}^{n} X_i$. Let I be a random variable independent of $X_1, \ldots, X_n$ with $\mathbf{P}[I = i] = \mathbf{E}[X_i]/\mathbf{E}[X]$.

    (a) Show that $\widehat{X} \overset{\mathscr{D}}{=} \sum_{i=1}^{n} \big(X_i \mathbf{1}_{[I \ne i]} + \widehat{X}_i \mathbf{1}_{[I=i]}\big)$.

    (b) Show that $\widehat{Y} \overset{\mathscr{D}}{=} \prod_{i=1}^{n} \widehat{X}_i$.

    (c) Show that if X_i are identically distributed, then $\widehat{X} \overset{\mathscr{D}}{=} \widehat{X}_1 + \sum_{i=2}^{n} X_i$.

12.13. Show that if $X \sim \operatorname{Bin}(n, p)$ and $p > 0$, then $\widehat{X} \sim 1 + \operatorname{Bin}(n-1, p)$, whereas if $X \sim \operatorname{Pois}(\lambda)$ and $\lambda > 0$, then $\widehat{X} \sim 1 + \operatorname{Pois}(\lambda)$.

12.14. Let X be a mixed binomial random variable, that is, there are independent events $A_1, \ldots, A_n$ such that $X = \sum_{i=1}^{n} \mathbf{1}_{A_i}$. Suppose that $\mathbf{P}[X > 0] > 0$. Let I be a random variable independent of $A_1, \ldots, A_n$ such that $\mathbf{P}[I = i] = \mathbf{P}[A_i]/\sum_{j=1}^{n} \mathbf{P}[A_j]$ for all i. Show that $\widehat{X} \sim 1 + \sum_{i=1}^{n} \mathbf{1}_{A_i} \mathbf{1}_{[I \ne i]}$.

12.15. Let $X, X_1, \ldots, X_k$ be i.i.d. nonnegative random variables for some $k \ge 0$. Suppose that $0 < \mathbf{E}[X] < \infty$, and let $\widehat{X}$ be an independent random variable with the size-biased distribution of X. Show that

$$\mathbf{E}\left[\frac{k+1}{\widehat{X} + X_1 + \cdots + X_k}\right] = \frac{1}{\mathbf{E}[X]}\,.$$

12.16. Show that if $X \ge 0$ and $0 < \mathbf{E}[X] < \infty$, then X is stochastically dominated by $\widehat{X}$. Deduce the arithmetic mean–quadratic mean inequality, that $\mathbf{E}[X]^2 \le \mathbf{E}[X^2]$, and determine when equality occurs. Deduce the Cauchy–Schwarz inequality from this.

12.17. In the notation of Lemma 12.2, show that $\langle (X_n, \mathscr{F}_n) \rangle$ is a martingale with respect to ν. Deduce that if μ is a probability measure and $\int X_n^2 \, d\nu < \infty$, then $\langle (X_n, \mathscr{F}_n) \rangle$ is a submartingale with respect to μ.

12.18. The simplest proof of the Kesten–Stigum theorem along traditional lines is due to Tanny (1988). Complete the following outline of this proof. A branching process in varying environments (BPVE) is one in which the offspring distribution depends on the generation. Namely, if $\langle L_i^{(n)}; \; n, i \geq 1 \rangle$ are independent random variables with values in $\mathbb{N}$ such that for each n, the variables $L_i^{(n)}$ are identically distributed, then set $Z_0 := 1$ and, inductively, $Z_{n+1} := \sum_{i=1}^{Z_n} L_i^{(n+1)}$. Let $m_n := \mathbf{E}[L_i^{(n)}]$ and $M_n := \prod_{k=1}^{n} m_k$. Show that $M_n = \mathbf{E}[Z_n]$ and that Z_n/M_n is a martingale. Its limit is denoted W.

Given a Galton–Watson branching process $\langle Z_n \rangle$ and a number $A > 0$, define a BPVE $\langle Z_n(A) \rangle$ by letting the offspring random variables $L_i^{(n)}(A)$ in generation n have the distribution of $L\mathbf{1}_{[L < Am^n]}$. Write $W(A)$ for the martingale limit of this BPVE. Use the fact that $W < \infty$ a.s. to show that for any $\epsilon > 0$, one can choose A sufficiently large that $\mathbf{P}[\forall n \; Z_n = Z_n(A)] > 1 - \epsilon$. Show that when $Z_n = Z_n(A)$ for all n, we have

$$W = W(A) \prod_{n \geq 1} (1 - \mathbf{E}[L; \; L \geq Am^n]/m) \,.$$

Show that this product is 0 iff $\mathbf{E}[L \log^+ L] = \infty$. Conclude that if $\mathbf{E}[L \log^+ L] = \infty$, then $W = 0$ a.s.

For the converse, define a BPVE $\langle Z_n(B) \rangle$ by letting the offspring random variables $L_i^{(n)}(B)$ in generation n have the distribution of $L\mathbf{1}_{[L < Bm^{3n/4}]}$. Choose B large enough that $M_n(B) := \mathbf{E}[Z_n(B)] > 0$ for all n. Show that $Z_n(B)/M_n(B)$ is bounded in L^2, whence its limit $W(B)$ has expectation 1. From $Z_n \geq Z_n(B)$, conclude that $\mathbf{E}[W] \geq \lim M_n(B)/m^n$. Show that by appropriate choice of B, if $\mathbf{E}[L \log^+ L] < \infty$, then this last limit can be made arbitrarily close to 1.

12.19. Let **GW** be a subcritical or critical Galton–Watson measure. Show that the limit in distribution as $n \to \infty$ of **GW** conditioned on $Z_n > 0$ is $\widehat{\mathbf{GW}}$.

12.20. Let **GW** be a critical Galton–Watson measure whose offspring distribution has variance $\sigma^2 < \infty$. Let X be an exponential random variable with mean $\sigma^2/2$. Show that the $\widehat{\mathbf{GW}}$-laws of Z_n/n tend to the law of $\widehat{X}$ as $n \to \infty$.

12.21. Let $G_n := \sup\{|u|; \; T_n \subseteq T^u\}$ be the generation of the most recent common ancestor of all individuals in generation n. Show that for a critical Galton–Watson branching process whose offspring random variable satisfies $\mathrm{Var}(L) < \infty$, the conditional distribution of G_n/n given $Z_n > 0$ tends to the uniform distribution on $[0, 1]$.

A traditional proof of Theorem 12.7 observes that $\mathbf{P}[Z_n > 0] = 1 - f^{(n)}(0)$ and analyzes the rate at which the iterates of f tend to 1. The following exercises outline such a proof.

12.22. Show that if $m = 1$, then $\lim_{s \uparrow 1} f''(s) = \sigma^2$.

12.23. Suppose that $m = 1$, $p_1 \neq 1$, and $\sigma < \infty$.
  (a) Define $\delta(s) := [1 - f(s)]^{-1} - [1 - s]^{-1}$. Show that $\lim_{s \uparrow 1} \delta(s) = \sigma^2/2$.
  (b) Let $s_n \in [0, 1)$ be such that $n(1 - s_n) \to \alpha \in [0, \infty]$. Show that

$$\lim_{n \to \infty} n[1 - f^{(n)}(s_n)] = \frac{1}{\sigma^2/2 + \alpha^{-1}} \,.$$

(Recall that $f^{(n)}$ denotes the nth iterate of f, not its nth derivative.)

12.24. Use Exercise 12.23 and Laplace transforms to prove Theorem 12.7 in the case that $\sigma < \infty$.

13 | Escape Rate of Random Walks and Embeddings

If a random walk on a network is transient, how quickly does the walk increase its distance from its starting point? In particular, the limit of the distance divided by the time (when it exists) is called *speed* of the random walk. If this speed is 0, that is, if the rate of escape is sublinear, then one may ask for the right order of magnitude of the escape rate. In many cases, the growth of the graph restricts the rate of escape; a key theorem in this direction is the Varopoulos–Carne bound, Theorem 13.4. We'll also see some relations to problems of embedding finite metric spaces, particularly graphs, in Euclidean space. In Chapters 14 and 17, we study random walk on groups (respectively, on Galton–Watson trees) and, in particular, their speed.

13.1 Basic Examples

If $\langle S_n \rangle$ is a sum of i.i.d. real-valued random variables, then $\lim_n S_n/n$ could be called its speed (on the real line) when it exists. Of course, the strong law of large numbers (SLLN) says that this limit does exist a.s. and equals the mean increment of S_n – when this mean is well defined. Actually, the independence of the increments is not needed if we have some other control of the increments. We present two such general results. Recall that X and Y are called ***uncorrelated*** if they have finite variance and $\mathbf{E}\big[(X - \mathbf{E}[X])(Y - \mathbf{E}[Y])\big] = 0$.

Theorem 13.1. (SLLN for Uncorrelated Random Variables) *Let $\langle X_n \rangle$ be a sequence of uncorrelated random variables with $\sup_n \mathrm{Var}(X_n) < \infty$. Then*

$$\frac{1}{n} \sum_{k=1}^{n} (X_k - \mathbf{E}[X_k]) \to 0$$

a.s. as $n \to \infty$.

Proof. We may clearly assume that $\mathbf{E}[X_n] = 0$ and $\mathbf{E}[X_n^2] \le 1$ for all n. Write $S_n := \sum_{k=1}^{n} X_k$.

We begin with the simple observation that if $\langle Y_n \rangle$ is a sequence of random variables such that

$$\sum_n \mathbf{E}[Y_n^2] < \infty,$$

then $\mathbf{E}\big[\sum_n Y_n^2\big] < \infty$, whence $\sum_n Y_n^2 < \infty$ a.s. and $Y_n \to 0$ a.s.

Using this, it is easy to verify the SLLN for $n \to \infty$ along the sequence of squares. Indeed,

$$\mathbf{E}\big[(S_n/n)^2\big] = \frac{1}{n^2}\,\mathbf{E}\big[S_n^2\big] = \frac{1}{n^2}\sum_{k=1}^{n}\mathbf{E}\big[X_k^2\big] \le \frac{1}{n}\,.$$

This is not summable, but if we set $Y_n := S_{n^2}/n^2$, we get $\mathbf{E}\big[Y_n^2\big] \le 1/n^2$, which is summable. Therefore, the preceding observation implies $Y_n \to 0$ a.s., that is, $S_{n^2}/n^2 \to 0$ a.s.

To deal with the limit over all the integers, consider $m^2 \le n < (m+1)^2$. Then

$$\mathbf{E}\left[\left|\frac{S_n}{m^2} - \frac{S_{m^2}}{m^2}\right|^2\right] = \frac{1}{m^4}\,\mathbf{E}\left[\left|\sum_{k=m^2+1}^{n} X_k\right|^2\right] = \frac{1}{m^4}\,\mathbf{E}\left[\sum_{k=m^2+1}^{n} X_k^2\right]$$

$$= \frac{1}{m^4}\sum_{k=m^2+1}^{n}\mathbf{E}\big[X_k^2\big] \le \frac{2}{m^3}\,,$$

since the sum has at most $2m$ terms, each of size at most 1. Thus, write

$$Z_n := \frac{S_n}{m(n)^2} - \frac{S_{m(n)^2}}{m(n)^2}\,,$$

where $m(n) := \lfloor\sqrt{n}\rfloor$. Then, since each $m = m(n)$ is associated to at most $2m+1$ different values of n, we get

$$\sum_{n=1}^{\infty}\mathbf{E}\big[Z_n^2\big] \le \sum_{n=1}^{\infty}\frac{2}{m(n)^3} \le \sum_{m}(2m+1)\frac{2}{m^3} < \infty\,,$$

so by the initial observation, $Z_n \to 0$ a.s. This implies $S_n/m(n)^2 \to 0$ a.s., which in turn implies $S_n/n \to 0$ a.s., as desired. ◀

As an example, note that martingale increments (that is, the differences between successive terms of a martingale) are uncorrelated when they are square integrable.

More refined information for sums of i.i.d. real-valued random variables is given, of course, by the central limit theorem or by Chernoff–Cramér's theorem on large deviations. For the case of simple random walk on $\mathbb{Z}$, the latter implies that given $0 < s < 1$, the chance that the location at time n is at least sn is at most $e^{-nI(s)}$, where

$$I(s) := \frac{(1+s)\log(1+s) + (1-s)\log(1-s)}{2} \tag{13.1}$$

(see Billingsley (1995), p. 151, or Dembo and Zeitouni (1998), Exercise 2.2.23(b)). Note that for small $|s|$,

$$I(s) = \frac{s^2}{2} + O(s^4)\,. \tag{13.2}$$

In other situations, one does not have i.i.d. random variables. An extension (though not as sharp) of Chernoff–Cramér's theorem is a large deviation inequality due to Hoeffding (1963) and rediscovered by Azuma (1967). As an upper bound, the Hoeffding–Azuma inequality is just as sharp to the first two orders in the exponent as the Chernoff–Cramér theorem is for simple random walk on $\mathbb{Z}$.

Theorem 13.2. (Hoeffding–Azuma Inequality) *Let $\langle X_1, \ldots, X_n \rangle$ be bounded, nonconstant random variables such that*

$$\mathbf{E}[X_{i_1} \cdots X_{i_k}] = 0 \quad \text{whenever } 1 \le i_1 < \cdots < i_k \le n$$

(for instance, independent variables with zero mean, or martingale differences). Then, for all $L > 0$,

$$\mathbf{P}\left[\sum_{i=1}^{n} X_i \ge L\right] \le \exp\left(-\frac{2L^2}{\sum_{i=1}^{n}(\text{ess sup } X_i - \text{ess inf } X_i)^2}\right).$$

Proof. We will use the hypothesis in the following form: for any sequences of constants $\langle a_i \rangle$ and $\langle b_i \rangle$, we have

$$\mathbf{E}\left[\prod_{i=1}^{n}(a_i + b_i X_i)\right] = \prod_{i=1}^{n} a_i . \tag{13.3}$$

Write $m_i := \text{ess inf } X_i \le 0$ and $M_i := \text{ess sup } X_i \ge 0$.

The convexity of the function $x \mapsto e^{tx}$ implies that a.s.,

$$e^{tX_i} \le \frac{M_i - X_i}{M_i - m_i} e^{m_i} + \frac{X_i - m_i}{M_i - m_i} e^{M_i} .$$

Multiplying over all i, taking the expectation, and using (13.3), we obtain

$$\mathbf{E}\left[\exp\left(t \sum_{i=1}^{n} X_i\right)\right] \le \prod_{i=1}^{n}\left(\frac{M_i}{M_i - m_i} e^{tm_i} - \frac{m_i}{M_i - m_i} e^{tM_i}\right). \tag{13.4}$$

We claim that the ith factor is at most $\exp\left(t^2(M_i - m_i)^2/8\right)$. Assuming this, we obtain

$$\mathbf{E}\left[\exp\left(t \sum_{i=1}^{n} X_i\right)\right] \le \exp\left(\frac{t^2}{8} \sum_{i=1}^{n}(M_i - m_i)^2\right).$$

By Markov's inequality, it follows that for all $t > 0$,

$$\mathbf{P}\left[\sum_{i=1}^{n} X_i \ge L\right] = \mathbf{P}\left[\exp\left(t \sum_{i=1}^{n} X_i\right) \ge e^{Lt}\right] \le e^{-Lt} \exp\left(\frac{t^2}{8} \sum_{i=1}^{n}(M_i - m_i)^2\right).$$

By making the (optimal) choice $t := 4L\left(\sum_{i=1}^{n}(M_i - m_i)^2\right)^{-1}$, we obtain the required result.

It remains to prove our claimed bound on the ith factor of (13.4). Define Y_i to be the random variable that takes the value m_i with probability $M_i/(M_i - m_i)$ and the value M_i otherwise. Then the ith factor equals $\mathbf{E}[e^{tY_i}]$. Calculation shows that if $g_i(t) := \log \mathbf{E}[e^{tY_i}]$, then $g_i(0) = 0$, $g_i'(0) = \mathbf{E}[Y_i] = 0$, and $g_i''(t) = \text{Var}(\widehat{Y_{i,t}}) \le \mathbf{E}[(\widehat{Y_{i,t}} - (M_i + m_i)/2)^2] = (M_i - m_i)^2/4$, where $\widehat{Y_{i,t}}$ has the distribution of Y_i biased by e^{tY_i}. Therefore, $|g_i'(t)| \le |t|(M_i - m_i)^2/4$ and $g_i(t) \le t^2(M_i - m_i)^2/8$, as desired. ◄

For example, consider simple random walk $\langle X_n \rangle$ on an infinite tree T starting at its root, o. When the walk is at x, it has a "push" away from the root equal to

$$f(x) := \begin{cases} (\deg x - 2)/\deg x & \text{if } x \neq o \\ 1 & \text{if } x = o. \end{cases}$$

What we mean by "push" is that $\langle |X_n| - |X_{n-1}| - f(X_{n-1}) \rangle$ is a martingale-difference sequence. Thus, by either of the preceding theorems, this sequence obeys the SLLN,

$$\lim_{n \to \infty} \frac{1}{n} \left(|X_n| - \sum_{k=1}^{n-1} f(X_k) \right) = 0 \quad \text{a.s.}$$

Now the density of times at which the walk visits the root is 0, since the tree is infinite and has a stationary measure proportional to the degree (see also Exercise 2.46), whence we may write the preceding equation as

$$\lim_{n \to \infty} \left(\frac{|X_n|}{n} - \frac{1}{n} \sum_{k=1}^{n-1} (1 - 2/\deg X_k) \right) = 0 \quad \text{a.s.} \tag{13.5}$$

For example, if T is regular of degree d, then the random walk has a speed of $1 - 2/d$ a.s. For a more interesting example, suppose that T is the universal cover of a finite, connected graph G with at least one cycle (so that T is infinite). Denote the covering map by $\varphi : T \to G$. Then $\langle \varphi(X_n) \rangle$ is simple random walk on G, whence the density of times at which $\varphi(X_k) = y \in \mathsf{V}(G)$ equals $\deg_G y / (2|\mathsf{E}(G)|)$ a.s. Of course, $\deg_G \varphi(x) = \deg_T x$. Thus, (13.5) tells us that the speed on T is a.s.

$$1 - \sum_{y \in \mathsf{V}(G)} \frac{\deg y}{2|\mathsf{E}(G)|} (2/\deg y) = 1 - |\mathsf{V}(G)|/|\mathsf{E}(G)| = 1 - 2/\bar{d}(G), \tag{13.6}$$

where $\bar{d}(G) := 2|\mathsf{E}(G)|/|\mathsf{V}(G)|$ is the average degree in G. Furthermore, for any $\epsilon > 0$, the probability that the speed after n steps differs from $1 - 2/\bar{d}(G)$ by more than ϵ decays exponentially in n; this can be shown by combining the Hoeffding–Azuma inequality with the exponential convergence of occupation times (divided by n) on G to the stationary measure and the geometric distribution of the number of visits to the root of T. Note that the speed is positive iff $|\mathsf{E}(G)| > |\mathsf{V}(G)|$, in other words, iff G contains at least two distinct simple cycles, which is also equivalent to $\mathrm{br}\, T > 1$.

How does this speed compare to $1 - 2/(\mathrm{br}\, T + 1)$, which is the speed when T is regular? To answer this, we need an estimate of $\mathrm{br}\, T$ so that we can compare $\bar{d}(G)$ to $\mathrm{br}\, T + 1$. Let H be the graph obtained from G by iteratively removing all vertices (if any) of degree 1, and let T' be the universal cover of H. Then clearly $\mathrm{br}\, T' = \mathrm{br}\, T$, while $\bar{d}(H) \geq \bar{d}(G)$. By Theorem 3.8, we know that $\mathrm{br}\, T' = \mathrm{gr}\, T'$. Now $\mathrm{gr}\, T'$ equals the growth rate of the number $N(L)$ of nonbacktracking paths in H of length L (from any starting point) as $L \to \infty$:

$$\mathrm{gr}\, T' = \lim_{L \to \infty} N(L)^{1/L}.$$

To estimate this limit, write B for the matrix indexed by the oriented edges of H such that $B((x, y), (y, z)) = 1$ when $(x, y), (y, z) \in \mathsf{E}(H)$ and $x \neq z$, and all other entries of B are 0. Consider a stationary Markov chain on the oriented edges $\mathsf{E}(H)$ with stationary probability measure σ and transition probabilities $p(e, f)$ such that $p(e, f) > 0$ only if $B(e, f) > 0$. Such a chain gives a probability measure on the paths of length L whose entropy is at most that of the uniform measure, in other words, at most $\log N(L)$ (see (6.40)). On the other hand, this path entropy equals $-\sum_e \sigma(e) \log \sigma(e) - L \sum_{e,f} \sigma(e) p(e, f) \log p(e, f)$ (see Exercise 6.75). Thus, $\log \operatorname{gr} T'$ is at least the Markov-chain entropy:

$$\log \operatorname{gr} T' \geq -\sum_{e,f} \sigma(e) p(e, f) \log p(e, f) \,. \tag{13.7}$$

Here, given any Markov chain with transition probabilities $p_{i,j}$ and stationary probabilities π_i, its **entropy** is defined to be

$$-\sum_{i,j} \pi_i p_{i,j} \log p_{i,j} \,.$$

Now choose $p((x, y), (y, z)) = 1/(\deg y - 1)$ when $B((x, y), (y, z)) > 0$, in other words, simple nonbacktracking random walk. It is easy to verify that $\sigma(x, y) = 1/D(H)$ is a stationary probability measure. To calculate the entropy of this Markov chain, suppose $e^+ = y$. Then

$$-\sum_f p(e, f) \log p(e, f) = \log(\deg y - 1) \,.$$

Since there are $\deg y$ such edges e, we get that the entropy equals

$$D(H)^{-1} \sum_{y \in \mathsf{V}(H)} (\deg y) \log(\deg y - 1) = \bar{d}(H)^{-1} \frac{1}{|\mathsf{V}(H)|} \sum_{y \in \mathsf{V}(H)} (\deg y) \log(\deg y - 1) \,.$$

Because the function $t \mapsto t \log (t - 1)$ is convex for $t \geq 2$, it follows that the entropy is at least $\bar{d}(H)^{-1} \bar{d}(H) \log(\bar{d}(H) - 1) = \log(\bar{d}(H) - 1)$. Therefore,

$$\operatorname{br} T = \operatorname{gr} T' \geq \bar{d}(H) - 1 \geq \bar{d}(G) - 1 \,.$$

This result is due to David Wilson (personal communication, 1993) but first appeared in print in an article by Alon, Hoory, and Linial (2002).

Substitute this bound in (13.6) to obtain that the speed on T is at most $1 - 2/(\operatorname{br} T + 1)$. For example, if the branching number is an integer, then this shows that the regular tree of that branching number has the greatest speed among all covering trees of the same branching number.

Return now to the qualitative results: For covering trees of finite graphs, we have seen that simple random walk has positive speed iff the tree has growth rate or branching number larger than 1. For more general trees, are any of these implications still valid? We consider these questions now.

If every vertex has at least two children, then by (13.5), the **liminf speed** of the random walk, that is, $\liminf_{n \to \infty} |X_n|/n$, is positive a.s. A more general sufficient condition is given in Exercise 13.25. However, it does not suffice that $\operatorname{br} T > 1$ for the speed of simple random walk on T to be positive:

▷ **Exercise 13.1.**

Show that simple random walk has speed 0 on the tree T formed from a binary tree by joining a unary tree to every vertex, as in Figure 13.1.

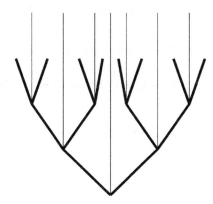

Figure 13.1. A binary tree with unary trees attached.

In the other direction, $\mathrm{br}\,T > 1$ is necessary for positive speed. The following bound was proved by Peres (1999), Theorem 5.4; see Section 13.9 for a better result due to Virág (2000b). Note that $s \mapsto I(s)/s$ is monotonic increasing on $(0, 1)$ (where I is defined in (13.1)) since its derivative is $-(2s^2)^{-1}\log(1 - s^2)$.

Proposition 13.3. *If simple random walk on T escapes at a linear rate, then $\mathrm{br}\,T > 1$. More precisely, if $\langle X_n \rangle$ is simple random walk on T and*

$$\liminf_{n\to\infty} \frac{|X_n|}{n} \geq s > 0$$

with positive probability, then $\mathrm{br}\,T \geq e^{I(s)/s}$.

Proof. We may assume that T has no leaves, since leaves only slow the random walk and do not change the branching number. (The slowing effect of leaves can be proved rigorously by coupling a random walk $\langle X_n \rangle$ on T with a random walk $\langle X_n' \rangle$ on T', where T' is the result of iteratively removing the leaves from T. The walks can be coupled so that $\liminf |X_n'|/n \geq \liminf |X_n|/n$ by letting $\langle X_n' \rangle$ take only the moves of $\langle X_n \rangle$ that do not enter $T \setminus T'$.) Given $0 < s' < s$, there is some L such that

$$q := \mathbf{P}\big[\forall n \geq L \ |X_n| > s'n\big] > 0. \tag{13.8}$$

Define a general percolation on T by keeping all edges $e(x)$ with $|x| \leq s'L$ as well as those edges $e(x)$ such that $X_n = x$ for some $n < |x|/s'$. According to (13.8), the component of the root in this percolation is infinite with probability at least q. On the other hand, if $|x| > s'L$, then $\mathbf{P}[o \leftrightarrow x]$ is bounded above by the probability that simple random walk $\langle S_k \rangle$ on $\mathbb{Z}$ moves distance at least $|x|$ in fewer than $|x|/s'$ steps:

$$\mathbf{P}[o \leftrightarrow x] \leq \mathbf{P}\Big[\max_{n < |x|/s'} |S_n| \geq |x|\Big]. \tag{13.9}$$

(This is proved rigorously by coupling the random walk on T to a random walk $\langle Y_n \rangle$ on $\mathbb{N}$ by letting $\langle Y_n \rangle$ take only the moves of $\langle X_n \rangle$ that lie on the shortest path between o and x.) Now by the reflection principle,

$$\mathbf{P}\left[\max_{n \leq N} |S_n| \geq j\right] \leq 2\,\mathbf{P}\left[\max_{n \leq N} S_n \geq j\right] \leq 4\,\mathbf{P}[S_N \geq j]\,.$$

We have seen the basic Chernoff–Cramér bound that $\mathbf{P}[S_N \geq j] \leq e^{-NI(j/N)}$. Therefore, with $j := |x|$ and N the largest integer less than $|x|/s'$, (13.9) gives $\mathbf{P}[o \leftrightarrow x] \leq 4e^{-|x|I(s')/s'}$. Here, we used the monotonicity of $I(t)/t$. In light of Proposition 5.8, this means that for any cutset Π with all edges at level $> s'L$, we have

$$q \leq \sum_{e(x) \in \Pi} 4e^{-|x|I(s')/s'}\,.$$

Therefore, $\operatorname{br} T \geq e^{I(s')/s'}$. Since this holds for all $s' < s$, the result follows.   ◀

We return to random walks on trees in Chapter 17, where we walk on Galton–Watson trees and, among other things, calculate the speed exactly, as we did here for covering trees.

13.2 The Varopoulos–Carne Bound

Consider a network random walk. Recall from Section 6.2 that the transition operator

$$(Pf)(x) := \sum_y p(x, y)f(y)$$

is a bounded self-adjoint operator on $\ell^2(\mathsf{V}, \pi)$. We saw in (6.13) that

$$\forall n \quad p_n(x, y) \leq \sqrt{\pi(y)/\pi(x)}\,\|P\|_\pi^n\,. \tag{13.10}$$

Of course, if the distance between x and y is larger than n, then $p_n(x, y) = 0$. How can (13.10) be modified to show how $p_n(x, y)$ depends on $d(x, y)$?

The following fundamental inequality does that. It is a generalization and improvement by Carne (1985) of a result of Varopoulos (1985b), which we have improved a bit more by adding the factor $\|P\|_\pi^n$.

Theorem 13.4. (Varopoulos–Carne Bound)   *For any reversible random walk, we have*

$$p_n(x, y) \leq 2\sqrt{\pi(y)/\pi(x)}\,\|P\|_\pi^n\,e^{-d(x,y)^2/(2n)}\,.$$

Compare this bound to the following bound for simple random walk on $\mathbb{Z}$:

$$\sum_{|k| \geq d} q_n(k) \leq 2e^{-d^2/(2n)}\,, \tag{13.11}$$

where $q_n(k)$ denotes the probability that simple random walk on $\mathbb{Z}$ starting at 0 is at k after the nth step. The inequality (13.11) is an immediate consequence of the Hoeffding–Azuma inequality specialized to simple random walk on $\mathbb{Z}$. As it happens, we will use (13.11) to establish Theorem 13.4.

To prove Theorem 13.4, we need some standard facts from analysis.

Suppose S is a bounded self-adjoint operator on a Hilbert space $\mathscr{H}$ and Q is a polynomial with real coefficients. If $\mathscr{H}$ is finite-dimensional, then by diagonalizing S, we see that all eigenvalues of $Q(S)$ can be written as $Q(\lambda)$, where λ is an eigenvalue of S. It follows that

$$\|Q(S)\| \le \max_{s \in [-\|S\|, \|S\|]} |Q(s)| . \tag{13.12}$$

If $\mathscr{H}$ is infinite-dimensional, this inequality still holds. An elementary reduction to the finite-dimensional case is outlined in Exercise 13.30. Alternatively, one can invoke the infinite-dimensional spectral theorem; see Section 13.9.

Next, we need the ***Chebyshev polynomials***, T_k for $k \in \mathbb{Z}$. These are the unique polynomials of degree $|k|$ such that

$$T_k(\cos \theta) = \cos k\theta \qquad (\theta \in \mathbb{R}) . \tag{13.13}$$

The existence of such polynomials follows by induction from the identity

$$\cos(k + 1)\theta + \cos(k - 1)\theta = 2 \cos \theta \cos k\theta .$$

Alternatively, expand $\cos k\theta + i \sin k\theta = (\cos \theta + i \sin \theta)^k$, take the real part, and replace every occurrence of $\sin^2 \theta$ by $1 - \cos^2 \theta$.

Since every $s \in [-1, 1]$ is of the form $\cos \theta$,

$$|T_k(s)| \le 1 \text{ whenever } -1 \le s \le 1 . \tag{13.14}$$

We are now ready to prove the central formula that allows one to deduce information about P from information about simple random walk on $\mathbb{Z}$. The following formula is a modification of that proved by Carne (1985).

Lemma 13.5. *Let T_k be the Chebyshev polynomials. For any reversible random walk, we have*

$$P^n = \|P\|_\pi^n \sum_{k \in \mathbb{Z}} q_n(k) T_k\big(P/\|P\|_\pi\big) . \tag{13.15}$$

Furthermore, $\big\|T_k\big(P/\|P\|_\pi\big)\big\|_\pi \le 1$ for all k.

Proof. For $z = \cos \theta$ and $w = e^{i\theta}$, the binomial expansion gives

$$z^n = \big[(w + w^{-1})/2\big]^n = \sum_{k \in \mathbb{Z}} q_n(k) w^k = \sum_{k \in \mathbb{Z}} q_n(k)(w^k + w^{-k})/2 = \sum_{k \in \mathbb{Z}} q_n(k) T_k(z) .$$

Since this identity holds for all $z \in [-1, 1]$, the two polynomials in z must coincide. Thus we may apply the identity to operators, substituting $P/\|P\|_\pi$ for z to get (13.15). The final estimate derives from (13.12) (with $S := P/\|P\|_\pi$) and (13.14). ◄

Proof of Theorem 13.4. Fix $x, y \in V$ and write $d := d(x, y)$. Consider the unit vectors $f_x := \mathbf{1}_{\{x\}}/\sqrt{\pi(x)}$ and $f_y := \mathbf{1}_{\{y\}}/\sqrt{\pi(y)}$. We have

$$p_n(x, y) = \sqrt{\pi(y)/\pi(x)}\,(f_x, P^n f_y)_\pi \,. \tag{13.16}$$

When we substitute (13.15) for P^n here, we may exploit the fact that

$$\left(f_x, T_k(P/\|P\|_\pi)f_y\right)_\pi = 0 \quad \text{for } |k| < d\,,$$

since T_k has degree $|k|$ and $p_i(x, y) = 0$ for $i < d$. Furthermore, we may use the bound

$$\left|\left(f_x, T_k(P/\|P\|_\pi)f_y\right)_\pi\right| \leq \|T_k(P/\|P\|_\pi)\|_\pi \|f_x\|_\pi \|f_y\|_\pi \leq 1\,.$$

We obtain

$$p_n(x, y) \leq \sqrt{\pi(y)/\pi(x)}\,\|P\|_\pi^n \sum_{|k| \geq d} q_n(k)\,.$$

Now use (13.11) to complete the proof.   ◄

13.3 An Application to Mixing Time

The basic convergence theorem for finite, irreducible, aperiodic Markov chains ensures that the distribution at time t of such a chain approaches the stationary distribution as $t \to \infty$; the time t required for these distributions to be close (that is, within some prescribed distance ϵ) is known as the *mixing time* (which depends on ϵ). One intuitive interpretation of the mixing time is that it is the time needed for the chain to "forget" its initial state, that is, the time t required for X_t to become approximately independent of X_0, no matter what the distribution of X_0. See the formal definition later.

Mixing time is one of the most important parameters describing a finite Markov chain. It is studied in computer science, where it is often the main component in randomized algorithms for sampling and counting combinatorial structures, as well as in statistical physics; we refer the reader to the books by Aldous and Fill (2002) and Levin and Peres (2017) for more information.

The Varopoulos–Carne theorem has a striking consequence for the mixing time of a random walk on a graph: it implies a lower bound that is close to the square of the diameter. To make this precise, we define the notions of total variation distance and mixing time. For any two probability measures μ and ν on the same measurable space $(\Omega, \mathscr{F})$, their **total variation distance** is

$$\|\mu - \nu\|_{\mathrm{TV}} := \sup_{A \in \mathscr{F}} |\mu(A) - \nu(A)|\,.$$

(This definition is convenient in probability theory; in analysis, the variation of the signed measure $\mu - \nu$ is defined to be twice the preceding value.) For finite Ω, we take $\mathscr{F}$ to be the collection of all subsets.

▷ **Exercise 13.2.**

Let Ω be finite or countable, and let μ and v be probability measures on Ω.

(a) Show that $\|\mu - v\|_{TV} = \frac{1}{2}\sum_{z \in \Omega}|\mu(z) - v(z)|$ and $1 - \|\mu - v\|_{TV} = \sum_{z \in \Omega}(\mu(z) \wedge v(z))$, where $x \wedge y := \min\{x, y\}$.

(b) Deduce that
$$\|\mu - v\|_{TV} = \min_{X \sim \mu,\, Y \sim v} \mathbf{P}[X \neq Y].$$

(c) Prove that $\|\mu P - v P\|_{TV} \leq \|\mu - v\|_{TV}$ for every transition matrix P on Ω.

Let V be a finite state space, and let $\langle X_t \rangle$ be an irreducible Markov chain on V with transition probabilities $p(x, y)$ and stationary **probability** measure π. For $t \geq 0$, the distance to stationarity at time t is
$$\delta(t) := \max_{x \in V} \|p_t(x, \cdot) - \pi\|_{TV}.$$

Sometimes it is more convenient to work with

$$\bar{\delta}(t) := \max_{x, y} \|p_t(x, \cdot) - p_t(y, \cdot)\|_{TV}.$$

Both $\delta(t)$ and $\bar{\delta}(t)$ are decreasing in t by Exercise 13.2(c). The following exercise relates these two notions of distance to stationarity to each other and to the maximum relative distance considered in Section 6.4.

▷ **Exercise 13.3.**

(a) Show that the two notions of distance to stationarity, $\delta(\cdot)$ and $\bar{\delta}(\cdot)$, satisfy

$$\delta(t) \leq \bar{\delta}(t) \leq 2\delta(t) \text{ for } t \geq 0.$$

(b) Show that $\delta(t) = \max_{x \in V} \frac{1}{2}\sum_{y \in V}|p_t(x, y) - \pi(y)| \leq \max_{x, y \in V}\left|\frac{p_t(x, y) - \pi(y)}{2\pi(y)}\right|.$

For $0 < \epsilon < 1$, define the **mixing time**

$$t_{\text{mix}}(\epsilon) := \min\{t \geq 0 ;\ \delta(t) \leq \epsilon\}.$$

Since $\bar{\delta}(t) \leq 2\epsilon$ at time $t = t_{\text{mix}}(\epsilon)$, we see that if ϵ is small, then at time $t_{\text{mix}}(\epsilon)$, the chain has indeed almost "forgotten" its initial state.

Let $\pi_{\min} := \min_x \pi(x)$. Recall the **absolute spectral gap** $g_* := 1 - \max_{j > 1}|\lambda_j|$, where $1 = \lambda_1 > \lambda_2 \geq \cdots \geq \lambda_n$ are the eigenvalues of the transition matrix P. (Recall that for lazy chains, $g_* = 1 - \lambda_2$.) Theorem 6.13 and Exercise 13.3 imply that $\delta(t) \leq e^{-g_* t}/(2\pi_{\min})$, and this yields a useful upper bound for mixing time.

Corollary 13.6. *In every connected network,* $t_{\text{mix}}(\epsilon) \leq 1 + \lceil\log(2\epsilon\pi_{\min})\rceil/g_*$ *for* $0 < \epsilon < 1$.

◀

A bound in the other direction is given in the following exercise.

▷ **Exercise 13.4.**

For a reversible chain, define $\lambda_* := \max_{j>1} |\lambda_j|$. Show that $\lambda_*^t \leq 2\delta(t)$ for all $t \geq 1$, and deduce that $t_{\mathrm{mix}}(\epsilon) \geq [g_*^{-1} - 1] \log 1/(2\epsilon)$.

As usual, we define a graph G on V where $[x, y]$ is an edge iff $p(x, y) > 0$. Let $d(\cdot, \cdot)$ denote the corresponding graph distance, and let D be the ***diameter*** of G, that is, the largest distance between pairs of vertices of G. Fix $a, z \in \mathsf{V}$ such that $d(a, z) = D$. Then, for $t < D/2$, the distributions $p_t(a, \cdot)$ and $p_t(z, \cdot)$ have disjoint supports, so

$$\bar{\delta}(t) \geq \|p_t(a, \cdot) - p_t(z, \cdot)\|_{\mathrm{TV}} = 1 .$$

By Exercise 13.3, $\delta(t) \geq \frac{1}{2}$. This gives a crude lower bound of the mixing time for $\epsilon < \frac{1}{2}$:

$$t_{\mathrm{mix}}(\epsilon) \geq \frac{D}{2} .$$

Using the Varopoulos–Carne bound, we can derive a sharper inequality under some extra conditions. A ***lazy*** simple random walk on a graph is obtained from simple random walk by averaging the transition matrix with the identity matrix. In many finite, transitive graphs, for example, the discrete torus, the mixing time of lazy simple random walk can be bounded above by a constant multiple of the diameter squared times the degree. It is an open problem whether such a bound holds for all transitive graphs.* Next, we prove a related general lower bound.

Proposition 13.7. *Consider random walk on a network G with n vertices and diameter D. Let π denote the stationary distribution. Then, for*

$$t < \frac{D^2}{12 \log n + 4 |\log \pi_{\min}|} ,$$

we have $\bar{\delta}(t) > 1 - 4/\sqrt{n}$. Consequently, given any $\epsilon < \frac{1}{2}$, if $n \geq 16/(1 - 2\epsilon)^2$, then

$$t_{\mathrm{mix}}(\epsilon) \geq \frac{D^2}{12 \log n + 4 |\log \pi_{\min}|} .$$

In particular, for simple random walk (or lazy simple random walk) on an n-vertex simple graph,

$$t_{\mathrm{mix}}(\epsilon) \geq \frac{D^2}{20 \log n} .$$

Proof. Choose $a, z \in \mathsf{V} = \mathsf{V}(G)$ such that $d(a, z) = D$. Consider the set of vertices $A := \{y \in \mathsf{V} ; d(a, y) \leq d(z, y)\}$. Then it is easy to see that

$$y \in A^c \implies d(a, y) \geq \frac{D}{2}$$

---

* This conjecture was posed by the second author in lectures; it had already been proved for certain graphs of moderate growth by Diaconis and Saloff-Coste (1994).

and

$$y \in A \implies d(z, y) \geq \frac{D}{2}.$$

Applying the Varopoulos–Carne bound, we obtain that for every t,

$$p_t(a, A^c) = \sum_{y \in A^c} p_t(a, y) \leq \sum_{y \in A^c} 2\sqrt{\frac{\pi(y)}{\pi(a)}} \exp\left\{-\frac{d^2(a, y)}{2t}\right\} \leq \frac{2n}{\sqrt{\pi_{\min}}} \exp\left\{-\frac{D^2}{8t}\right\},$$

where in the last step we used that $|A^c| \leq n$. Now let $t < \dfrac{D^2}{12 \log n + 4 |\log \pi_{\min}|}$. We get

$$p_t(a, A^c) < \frac{2n}{\sqrt{\pi_{\min}}} \exp\left\{\frac{-3 \log n - |\log \pi_{\min}|}{2}\right\} = \frac{2}{\sqrt{n}}.$$

Similarly, for this value of t, we have $p_t(z, A) < 2/\sqrt{n}$, so

$$p_t(z, A^c) > 1 - \frac{2}{\sqrt{n}}.$$

Now

$$\bar{\delta}(t) \geq p_t(z, A^c) - p_t(a, A^c) > 1 - \frac{2}{\sqrt{n}} - \frac{2}{\sqrt{n}} = 1 - \frac{4}{\sqrt{n}},$$

and the first two assertions of the theorem follow. For the last assertion, note that $\pi_{\min} \geq n^{-2}$ in every simple graph. ◄

The bound in Proposition 13.7 is sharp (up to a constant factor) for lazy simple random walk on expander graphs, as shown in Exercise 13.33.

Combining Proposition 13.7 (for $\epsilon := 1/4$) with Corollary 13.6 yields an upper bound for the absolute spectral gap g_* (see Corollary 13.24 for an alternative bound).

Corollary 13.8. *For every connected network with $n \geq 64$ vertices and diameter D,*

$$g_* \leq \frac{|\log(\pi_{\min}/2)| \cdot (24 \log n + 8 |\log \pi_{\min}|)}{D^2}.$$

In particular, for simple random walk on a simple graph,

$$g_* \leq \frac{160 \log^2 n}{D^2}.$$
◄

The dependence on $\pi_{\min}$ in Proposition 13.7 is not just an artifact of the proof. For example, given $n > 1$, consider the Markov chain on the states $\{0, 1, \ldots, n-1\}$ where $p(x, x) = 1/2$ for all x and $p(x, x+1) = 2p(x, x-1) = 1/3$ if $0 < x < n-1$, while $p(0, 1) = p(n-1, n-2) = 1/2$; otherwise, $p(x, y) = 0$. Then $D = n - 1$, and the mixing time satisfies $t_{\mathrm{mix}} \asymp n$. Note that for this sequence of chains, $\pi_{\min}$ decays exponentially in n.

Nonetheless, there is an alternative lower bound for mixing time (the upcoming Proposition 13.11) that does not depend on $\pi_{\min}$, where the squared diameter is replaced by the *average squared distance*

$$\widehat{D}^2 := \sum_{x, y} \pi(x)\pi(y)d^2(x, y).$$

▷ **Exercise 13.5.**
Prove that in every transitive network, we have $D^2 \le 4\widehat{D}^2$.

For transitive networks, D^2 and $\widehat{D}^2$ are comparable by Exercise 13.5. For nontransitive networks, the ratio $D/\widehat{D}$ may be arbitrarily large, as in the case of the chain on $\{0, 1, \ldots, n-1\}$ described earlier, where $\widehat{D} \asymp 1$ and $D = n - 1$.

To prove our next lower bound on mixing time, we require the following proposition. It implies, in particular, that a stationary, reversible chain on n states can escape from its starting point at linear rate for at most $O(\log n)$ steps. Contrast this with biased random walk on the n-cycle, a nonreversible chain where linear rate of escape is maintained for order n steps.

Proposition 13.9. *For an n-vertex network with stationary distribution π, we have*

$$\mathbf{E}_\pi\big[d(X_0, X_t)^2\big] \le 3t \log n \tag{13.17}$$

if $n > e^4$, where we use the notation $\mathbf{E}_\pi := \sum_{x \in V} \pi(x)\,\mathbf{E}_x$.

Proof. Fix t. For $\beta > 0$, let $A_\beta := \{(x, y);\ d^2(x, y) \ge \beta t \log n\}$. Using the Varopoulos–Carne bound, we have for every $\beta > 0$ that

$$\mathbf{P}_\pi\big[d^2(X_0, X_t) \ge \beta t \log n\big] = \sum_{(x,y) \in A_\beta} \pi(x) p_t(x, y) \le \sum_{(x,y) \in A_\beta} 2\sqrt{\pi(x)\pi(y)}\,n^{-\beta/2}$$

$$\le \sum_{x,y} \big(\pi(x) + \pi(y)\big) n^{-\beta/2} = 2n^{1-\beta/2}\,.$$

Thus we have

$$\mathbf{E}_\pi\left[\frac{d^2(X_0, X_t)}{t \log n}\right] = \int_0^\infty \mathbf{P}_\pi\big[d^2(X_0, X_t) \ge \beta t \log n\big]\, d\beta \le \int_0^\infty \big(2n^{1-\beta/2} \wedge 1\big)\, d\beta$$

$$\le 2 + 2n \int_2^\infty n^{-\beta/2}\, d\beta = 2 + \frac{4}{\log n} < 3$$

if $n > e^4$.   ◀

We will also need the following key property of mixing in reversible Markov chains, due to Aldous and Fill (2002).

Lemma 13.10. *For every connected network and any two vertices x and y, we have for all t*

$$\frac{p_{2t}(x, y)}{\pi(y)} \ge \big(1 - \overline{\delta}(t)\big)^2\,.$$

Proof. Sum over all possible positions at time t, then use reversibility and Cauchy–Schwarz:

$$\frac{p_{2t}(x, y)}{\pi(y)} = \sum_z p_t(x, z)\frac{p_t(z, y)}{\pi(y)} = \left(\sum_z p_t(x, z)\frac{p_t(y, z)}{\pi(z)}\right) \cdot \sum_z \pi(z)$$

$$\ge \left(\sum_z \sqrt{p_t(x, z)p_t(y, z)}\right)^2 \ge \left(\sum_z [p_t(x, z) \wedge p_t(y, z)]\right)^2$$

$$= \left(1 - \|p_t(x, \cdot) - p_t(y, \cdot)\|_{\mathrm{TV}}\right)^2 \ge \big(1 - \overline{\delta}(t)\big)^2,$$

where the penultimate step used Exercise 13.2(a).   ◀

We deduce the following consequence for mixing times.

Proposition 13.11. *Let $n > e^4$. For every n-vertex network, we have*

$$t_{\text{mix}}(\epsilon) \geq \frac{(1-2\epsilon)^2}{6} \frac{\widehat{D}^2}{\log n} \tag{13.18}$$

for all $\epsilon < 1/2$.

At an intuitive level, it is easy to derive (13.18) from (13.17), since for $t = t_{\text{mix}}(\epsilon)$, the random variables X_0 and X_t should be roughly independent, and therefore $\mathbf{E}_\pi[d^2(X_0, X_t)]$ should be roughly $\widehat{D}^2$. The rigorous proof uses Lemma 13.10 as well.

Proof. Let $t := t_{\text{mix}}(\epsilon)$. On the one hand, (13.17) tells us that

$$\mathbf{E}_\pi d^2(X_0, X_{2t}) \leq 6t \log n \tag{13.19}$$

for $n > e^4$. On the other hand, $\bar{\delta}(t) \leq 2\epsilon < 1$ by Exercise 13.3, so Lemma 13.10 implies that

$$\mathbf{E}_\pi d^2(X_0, X_{2t}) = \sum_{x,y} \pi(x)\pi(y)\frac{p_{2t}(x, y)}{\pi(y)} d^2(x, y)$$

$$\geq \sum_{x,y} \pi(x)\pi(y)\big(1 - \bar{\delta}(t)\big)^2 d^2(x, y) \geq (1-2\epsilon)^2 \widehat{D}^2. \tag{13.20}$$

Combining (13.19) and (13.20) gives (13.18). ◄

13.4 Markov Type of Metric Spaces

The next three sections of this chapter are about sublinear rates of escape. They show how the escape rate can be constrained by the combinatorial structure of a graph. Also, the escape rate can force large distortion of Euclidean embeddings of the graph. The third of these sections shows that Cayley graphs have escape rates at least as large as the rate of simple random walk on $\mathbb{Z}$, in other words, they are at least diffusive.

In this section, we consider Markov chains in metric spaces and see how quickly they increase their squared distance in expectation. That is, given a metric space (X, d) and a finite-state, reversible stationary Markov chain $\langle Z_t \, ; \, t \in \mathbb{N} \rangle$ whose state space is a subset of X, how quickly can $\mathbf{E}[d(Z_t, Z_0)^2]$ increase in t? Since the chain is stationary, the proper normalization for the distance is $\mathbf{E}[d(Z_1, Z_0)^2]$. It turns out that this notion, invented by Ball (1992), is connected to many interesting questions in functional analysis. It is more convenient to relax the notion slightly from Markov chains to functions of Markov chains. Thus, with Ball, we make the following definition, where the "2" in the name refers to the exponent.

Definition 13.12. Given a metric space (X, d), we say that X has ***Markov type*** 2 if there exists a constant $M < \infty$ such that for every positive integer n, every stationary, reversible Markov chain $\langle Z_t \rangle_{t=0}^{\infty}$ on $\{1, \dots, n\}$, every mapping $f \colon \{1, \dots, n\} \to X$, and every time $t \in \mathbb{N}$,

$$\mathbf{E}\left[d(f(Z_t), f(Z_0))^2\right] \le Mt\, \mathbf{E}\left[d(f(Z_1), f(Z_0))^2\right]. \tag{13.21}$$

We will prove that the real line has Markov type 2 (see Exercise 13.6 for a space that does not have Markov type 2). Since adding the *squared* coordinates gives squared distance in higher dimensions, even in Hilbert space, it follows that Hilbert space also has Markov type 2. This result is due to Ball (1992).

Theorem 13.13. $\mathbb{R}$ *has Markov type* 2 *with constant* $M = 1$ *in (13.21).*

Proof. As in Section 6.2, let P be the transition operator of the Markov chain on $\{1, \dots, n\}$ with stationary probability measure π. We saw in that section that P is a self-adjoint operator in $\ell^2(\{1, \dots, n\}, \pi)$. This implies that $\ell^2(\{1, \dots, n\}, \pi)$ has an orthogonal basis of eigenfunctions of P with real eigenvalues. We also saw (Exercise 6.6) that $\|P\|_\pi \le 1$, whence all eigenvalues lie in $[-1, 1]$.

The first step is to reexpress the left-hand side of (13.21) in terms of the operator P. Note that P^t is also reversible with respect to π. By virtue of (6.17), we have

$$\mathbf{E}d(f(Z_t), f(Z_0))^2 = \sum_{i,j} \pi_i p_{ij}^{(t)} [f(i) - f(j)]^2 = 2((I - P^t)f, f)_\pi. \tag{13.22}$$

In particular,

$$\mathbf{E}d(f(Z_1), f(Z_0))^2 = 2((I - P)f, f)_\pi.$$

Thus, we want to prove that

$$((I - P^t)f, f)_\pi \le t((I - P)f, f)_\pi$$

for all functions f. Note that if f is an eigenfunction with eigenvalue λ, this reduces to the inequality $(1 - \lambda^t) \le t(1 - \lambda)$. Since $|\lambda| \le 1$, this in turn reduces to

$$1 + \lambda + \dots + \lambda^{t-1} \le t,$$

which is obviously true.

The claim follows for functions f that are not eigenfunctions by taking $f = \sum_{j=1}^{n} a_j f_j$, where $\{f_j\}$ is an orthonormal basis of eigenfunctions:

$$((I - P^t)f, f)_\pi = \sum_{j=1}^{n} a_j^2 ((I - P^t)f_j, f_j)_\pi \le \sum_{j=1}^{n} a_j^2 t((I - P)f_j, f_j)_\pi = t((I - P)f, f)_\pi. \blacktriangleleft$$

▷ **Exercise 13.6.**

A collection of metric spaces has ***uniform Markov type*** 2 if there exists a constant $M < \infty$ such that each space in the collection has Markov type 2 with constant M. Prove that the set of k-dimensional hypercube graphs $\{0, 1\}^k$ does not have uniform Markov type 2. From this, deduce that ℓ^1 does not have Markov type 2.

Our next goal is to extend Theorem 13.13 to trees, by showing that tree metrics have uniform Markov type 2. The proof will involve a martingale decomposition of a function of a stationary, reversible Markov chain; this decomposition has multiple uses. A tree with prescribed edge lengths defines a metric on the vertices: the distance between two vertices v and w is the sum of the edge lengths in the unique path from v to w.

Theorem 13.14. *Trees (with arbitrary edge lengths) have uniform Markov type* 2.

The key to Theorem 13.14 is the following lemma, a maximal-inequality version of Theorem 13.13.

Lemma 13.15. *Consider a stationary, reversible Markov chain* $\langle Z_t \rangle_{t=0}^\infty$ *on* $\{1, \ldots, n\}$ *and a function* $f : \{1, \ldots, n\} \to \mathbb{R}$. *Then, for every time* $t > 0$,

$$\mathbf{E}\left[\max_{0 \le s \le t} \left(f(Z_s) - f(Z_0) \right)^2 \right] \le 7t\, \mathbf{E}\left[\left(f(Z_1) - f(Z_0) \right)^2 \right].$$

Proof. Define $\alpha : \{1, \ldots, n\} \to \mathbb{R}$ by $\alpha(i) := \mathbf{E}[f(Z_1) \mid Z_0 = i] - f(i) = \sum_{j=1}^n p_{ij} f(j) - f(i)$. Notice that

$$M_s := f(Z_s) - f(Z_0) - \sum_{u=0}^{s-1} \alpha(Z_u)$$

is a martingale (with respect to the natural filtration of $Z_0, \ldots, Z_t$). Write $\widetilde{Z}_s := Z_{t-s}$ for $0 \le s \le t$. Because the distribution of $(Z_0, \ldots, Z_t)$ is the same as that of $(\widetilde{Z}_0, \ldots, \widetilde{Z}_t)$,

$$\widetilde{M}_s := f(\widetilde{Z}_s) - f(\widetilde{Z}_0) - \sum_{u=0}^{s-1} \alpha(\widetilde{Z}_u)$$

is a martingale (with respect to the natural filtration of $\widetilde{Z}_0, \ldots, \widetilde{Z}_t$). Note that

$$\widetilde{M}_{t-s} - \widetilde{M}_t = f(\widetilde{Z}_{t-s}) - f(\widetilde{Z}_t) + \sum_{u=t-s}^{t-1} \alpha(\widetilde{Z}_u) = f(Z_s) - f(Z_0) + \sum_{j=1}^s \alpha(Z_j).$$

Averaging this with the definition of M_s and rearranging yields

$$f(Z_s) - f(Z_0) = \frac{1}{2}\left(M_s + \widetilde{M}_{t-s} - \widetilde{M}_t + \alpha(Z_0) - \alpha(Z_s) \right).$$

Therefore,

$$2 \max_{0 \le s \le t} |f(Z_s) - f(Z_0)| \le \max_{0 \le s \le t} |M_s| + \max_{0 \le s \le t} |\widetilde{M}_s| + |\widetilde{M}_t| + |\alpha(Z_0)| + \max_{1 \le s \le t} |\alpha(Z_s)|.$$

It follows that

$$2\left\| \max_{0 \le s \le t} |f(Z_s) - f(Z_0)| \right\| \le \left\| \max_{0 \le s \le t} |M_s| \right\| + \left\| \max_{0 \le s \le t} |\widetilde{M}_s| \right\| + \left\| \widetilde{M}_t \right\| + \left\| \alpha(Z_0) \right\| + \left\| \max_{1 \le s \le t} |\alpha(Z_s)| \right\|,$$

where all norms are in $L^2(\mathbf{P})$.

We will use Doob's L^2-maximal inequality for martingales $\langle N_s \rangle$ (see, for example, Theorem 4.4.4 in Durrett (2019)),

$$\mathbf{E}\left[\max_{0 \le s \le t} N_s^2 \right] \le 4\mathbf{E}[N_t^2] = 4\left(\mathbf{E}[N_0^2] + \sum_{s=1}^t \mathbf{E}\left[(N_s - N_{s-1})^2 \right] \right).$$

In our case, $M_0 = 0$ and for $s > 0$,

$$M_s - M_{s-1} = f(Z_s) - f(Z_{s-1}) - \alpha(Z_{s-1}) = f(Z_s) - f(Z_{s-1}) - \mathbf{E}\big[f(Z_s) - f(Z_{s-1}) \mid Z_{s-1}\big]$$

is orthogonal to $\mathbf{E}\big[f(Z_s) - f(Z_{s-1}) \mid Z_{s-1}\big]$ in $L^2(\mathbf{P})$, whence

$$\mathbf{E}\big[(M_s - M_{s-1})^2\big] + \mathbf{E}\big[\alpha(Z_{s-1})^2\big] = \mathbf{E}\big[|f(Z_s) - f(Z_{s-1})|^2\big] =: V\,,$$

where V does not depend on $s > 0$ by stationarity. Write V_1 and V_2 for the expectations on the left-hand side. A similar equation holds for $\widetilde{M}$. This gives us

$$\Big\|\max_{0 \le s \le t} |M_s|\Big\| + \Big\|\max_{0 \le s \le t} |\widetilde{M}_s|\Big\| + \big\|\widetilde{M}_t\big\| \le 2\|M_t\| + 3\|\widetilde{M}_t\| = 5\|M_t\| = 5\sqrt{tV_1}\,.$$

In addition,

$$\big\|\alpha(Z_0)\big\| + \Big\|\max_{1 \le s \le t} |\alpha(Z_s)|\Big\| = \sqrt{V_2} + \mathbf{E}\Big[\max_{1 \le s \le t} \alpha(Z_s)^2\Big]^{1/2}$$

$$\le \sqrt{V_2} + \mathbf{E}\Big[\sum_{s=1}^{t} \alpha(Z_s)^2\Big]^{1/2} = \big(\sqrt{t} + 1\big)\sqrt{V_2}\,.$$

Summing up, dividing by 2, squaring, and using the Cauchy–Schwarz inequality, we obtain

$$\mathbf{E}\Big[\max_{0 \le s \le t} |f(Z_s) - f(Z_0)|^2\Big] \le \big(5\sqrt{tV_1} + (\sqrt{t} + 1)\sqrt{V_2}\big)^2/4 \le (25t + t + 2\sqrt{t} + 1)(V_1 + V_2)/4$$

$$= (26t + 2\sqrt{t} + 1)V/4 \le 7tV$$

for $t \ge 2$, as desired. ◀

Proof of Theorem 13.14. Let T be a weighted tree, $\langle Z_j \rangle$ be a stationary, reversible Markov chain on $\{1, \ldots, n\}$, and $F: \{1, \ldots, n\} \to T$. Let $X_j := F(Z_j)$. Choose an arbitrary root, o, and set $\psi(u) := d(o, u)$ for all vertices u. For $x, y \in T$, write $\overline{x, y}$ for the shortest path in T joining x and y. Let w be the closest vertex on $\overline{X_0, X_t}$ to o. Concatenating all paths $\overline{X_i, X_{i+1}}$ gives a path between X_0 and X_t. Let j be such that w is on $\overline{X_j, X_{j+1}}$; this exists because $\overline{X_0, X_t} \subseteq \bigcup_{i=0}^{t-1} \overline{X_i, X_{i+1}}$. Drawing a picture reveals that

$$d(X_0, X_t) \le |\psi(X_0) - \psi(X_j)| + |\psi(X_{j+1}) - \psi(X_t)| + d(X_j, X_{j+1})\,,$$

whence

$$d(X_0, X_t) \le 2\max_{0 \le j \le t} |\psi(X_0) - \psi(X_j)| + \max_{0 \le j < t} d(X_j, X_{j+1})\,.$$

Using norms in $L^2(\mathbf{P})$ and Lemma 13.15 with $f := \psi \circ F$, we obtain

$$\|d(X_0, X_t)\| \le 2\Big\|\max_{0 \le j \le t} |\psi(X_0) - \psi(X_j)|\Big\| + \Big\|\max_{0 \le j < t} d(X_j, X_{j+1})\Big\|$$

$$\le 2\sqrt{7t}\|\psi(X_0) - \psi(X_1)\| + \mathbf{E}\Big[\sum_{j=0}^{t-1} d(X_j, X_{j+1})^2\Big]^{1/2}$$

$$\le 2\sqrt{7t}\|d(X_0, X_1)\| + \sqrt{t}\|d(X_0, X_1)\| = (2\sqrt{7} + 1)\sqrt{t}\|d(X_0, X_1)\|$$

by the triangle inequality $|\psi(X_1) - \psi(X_0)| \le d(X_0, X_1)$. This gives the result. ◀

13.5 Embeddings of Finite Metric Spaces

If we map one metric space into another, distances can change in various ways. For example, a homothety merely multiplies all distances by the same constant. Thus, a homothety does not change the "shape" of the domain space. We can measure changes in shape, or distortion, by how much some distances change compared to the change in other distances. This motivates the following definition.

Definition 13.16. Given metric spaces (X, d_X) and (Y, d_Y), a mapping $f : X \to Y$ has *distortion at most* C if there exists a number $r > 0$ such that for all $x_1, x_2 \in X$,

$$r\, d_X(x_1, x_2) \le d_Y\big(f(x_1), f(x_2)\big) \le Cr\, d_X(x_1, x_2). \tag{13.23}$$

The infimum of such numbers C is called the *distortion of* f.

We will consider only the case where Y is Hilbert space and X is finite. In this case, the infimum of the distortions of all embeddings is a minimum and is called the *Hilbert distortion* of X, also called the *Euclidean distortion*. Usually X will be a finite graph with the shortest-path metric.

For example, consider X to be a hypercube graph. The distance in the graph is an ℓ^1-metric, whereas an embedded image in Hilbert space gets an ℓ^2-metric. These metrics are generally incomparable: as norms, they do not induce the same topology (in infinite dimensions). A finitistic and quantitative version of this inequivalence is that there must be a fair amount of distortion. The obvious embedding of the hypercube $\{0, 1\}^k$ has distortion $\sqrt{k}$. Enflo (1969) proved that $\sqrt{k}$ is indeed the minimum distortion. We first prove a weaker version of this result because our method, via random walks and Markov type, applies in much greater generality. It is inspired by Linial, Magen, and Naor (2002).

Proposition 13.17. (Distortion of Hypercubes) *There exists $c > 0$ such that for all k, the Hilbert distortion of the hypercube $\{0, 1\}^k$ is at least $c\sqrt{k}$.*

Proof. In the solution to Exercise 13.6, we showed that if $\langle Z_t \rangle$ is a simple random walk in the hypercube, then

$$\mathbf{E}\big[d(Z_0, Z_t)^2\big] \ge \frac{t^2}{4} \qquad \forall t \le k/4.$$

Take $t := \lfloor k/4 \rfloor$. Now let $f : \{0, 1\}^k \to \ell^2(\mathbb{N})$ be a map. Assume that (13.23) holds; we may take $r = 1$ there. We saw that Hilbert space has Markov type 2 with constant $M = 1$. Therefore, if we take Z_0 to be uniform on the vertices of the hypercube, we obtain

$$\mathbf{E}d^2\big(f(Z_0), f(Z_t)\big) \le t\,\mathbf{E}d^2\big(f(Z_0), f(Z_1)\big) \le C^2 t\,\mathbf{E}d^2(Z_0, Z_1) = C^2 t,$$

where C is the distortion of f. We conclude

$$C^2 t \ge \mathbf{E}d^2\big(f(Z_0), f(Z_t)\big) \ge \mathbf{E}d^2(Z_0, Z_t) \ge t^2/4,$$

whence $C \ge \sqrt{t}/2$, which implies the result. ◀

We give another beautiful approach to prove lower bounds on distortion that is flexible in a different way and also yields Enflo's sharp result. This approach uses a spectral gap, similar to that defined in Section 6.4. The method appears first in Gromov (1983), Section 9.1, where the author thanks V. Milman for these ideas. It was refined by Jolissaint and Valette (2014). The essential difference between Section 6.4 and the approach we are about to explain is that the Rayleigh quotient, Exercise 6.7, was used in Section 6.4 to bound the spectral gap, whereas here we *use* the spectral gap to bound the distortion. This will require a simple observation that extends the Rayleigh quotient from real-valued to Hilbert-space-valued functions, as well as another observation that relates norms of functions to norms of differences.

For a finite, connected network (G, c) with corresponding transition operator P, write $\lambda_2(P)$ for the second-largest eigenvalue of P. Thus, $1 - \lambda_2(P)$ is what we called the spectral gap in Section 6.4. As we did there, we write $\pi(x)$ for the stationary **probability** measure on $V(G)$. However, we do not want to assume that the conductances are normalized. If we write $\gamma := \sum_{e \in E_{1/2}} c(e)$, then $\pi(x) = \sum_{e^- = x} c(e)/(2\gamma)$.

Lemma 13.18. *Let (G, c) be a finite, connected network with transition operator P and stationary probability measure π. Then, for every map $f : V \to \mathscr{H}$, where $\mathscr{H}$ is a Hilbert space, we have*

$$\left(1 - \lambda_2(P)\right) \sum_{x, y \in V} \pi(x)\pi(y)\left|f(x) - f(y)\right|^2 \leq \frac{1}{\sum_{e \in E_{1/2}} c(e)} \sum_{e \in E_{1/2}} c(e)\left\|f(e^+) - f(e^-)\right\|^2.$$

Proof. Recall (6.32) for real-valued functions $f : V \to \mathbb{R}$:

$$1 - \lambda_2(P) = \min_{f \perp \mathbf{1}} \frac{\left((I - P)f, f\right)_\pi}{(f, f)_\pi}.$$

The numerator, as we saw earlier, can also be written as $\sum_{e \in E_{1/2}} c(e)|f(e^+) - f(e^-)|^2/(2\gamma)$ (whether $f \perp \mathbf{1}$ or not). The condition that $f \perp \mathbf{1}$ is inconvenient here, but we may easily remove it by using $f - \bar{f}$ in place of f, where $\bar{f} := \sum_x \pi(x)f(x)$. Now $2\|f - \bar{f}\|_\pi^2 = \sum_{x, y \in V} \pi(x)\pi(y)|f(x) - f(y)|^2$, whence we can write

$$\left(1 - \lambda_2(P)\right) \sum_{x, y \in V} \pi(x)\pi(y)|f(x) - f(y)|^2 \leq \sum_{e \in E_{1/2}} c(e)|f(e^+) - f(e^-)|^2/\gamma$$

for all $f : V \to \mathbb{R}$. By considering the coordinates with respect to an orthonormal basis of $\mathscr{H}$, we see that the same holds for all $f : V \to \mathscr{H}$, provided we replace absolute values by norms. ◄

Recall that $\widehat{D}^2$ is the π-weighted average squared distance between pairs of vertices in G. The preceding inequality says that if the π-weighted average squared distance between *images* of vertices is at least $\widehat{D}^2$ (for example, if (13.23) holds with $r = 1$), then the c-weighted average squared distance between images of *neighboring* vertices is at least $\widehat{D}^2(1 - \lambda_2(P))$. In particular, a lower bound for the distortion is $\widehat{D}\sqrt{1 - \lambda_2(P)}$:

Corollary 13.19. (Spectral Bound on Distortion) *Let (G, c) be a finite, connected network with transition operator P and stationary probability measure π. Then the Hilbert distortion of (G, c) is at least*

$$\left[\left(1 - \lambda_2(P) \right) \sum_{x,y \in V} \pi(x) \pi(y) \, d(x, y)^2 \right]^{1/2}. \qquad \blacktriangleleft$$

For example, suppose that G is a graph with n vertices and all degrees at most $k \geq 3$. The ball of radius r about any vertex x has at most $\left(k(k-1)^r - 2 \right)/(k-2)$ vertices, which is $< n/2$ when $r \leq \log(n/2)/\log(k-1)$. Therefore $\widehat{D} > \log(n/2)/(2 \log(k-1))$. This gives the following result of Linial, London, and Rabinovich (1995):

Corollary 13.20. (Distortion of Expanders) *Let $k \geq 3$ be an integer and $\lambda < 1$. There is a constant $C_{k,\lambda}$ such that for each graph G on n vertices with $\lambda_2(P) \leq \lambda$ and maximal degree k, where P stands for the simple random walk transition operator, the Hilbert distortion of G is at least $C_{k,\lambda} \log n$.* $\blacktriangleleft$

Recall that such graphs do exist for arbitrarily large n: see Theorems 6.17 and 6.15.

Corollary 13.19 could instead be phrased in terms of the expansion constants: using Theorem 6.15, we have that the distortion of a network with expansion constant Φ_* is at least $\widehat{D}\Phi_*/\sqrt{2}$, where $\widehat{D}$ is as previously.

In the special case of transitive networks, such as unweighted Cayley graphs, we can refine the preceding bounds to be sometimes sharp. Denote by $\operatorname{diam}(G)$ the diameter of G. For a finite Cayley graph G, there is a permutation σ of the vertex set (not necessarily an automorphism of G) such that $d(x, \sigma(x)) = \operatorname{diam}(G)$ for all $x \in V$: if $d(x, x\gamma)$ equals the diameter for some x and some group element γ, then it equals the diameter for all x and the same γ, so we may choose $\sigma(x) := x\gamma$. There is also such a permutation for every transitive graph G: see Exercise 13.42.

Theorem 13.21. (Spectral Bound on Distortion – Transitive Case) *Let (G, c) be a finite, connected, transitive network with transition operator P. Then the Hilbert distortion of G is at least* $\operatorname{diam}(G)\sqrt{(1 - \lambda_2(P))/2}$.

Proof. Transitivity of the network implies that $\pi(x) = 1/|V|$ for all vertices x. In the proof of Lemma 13.18, we saw that for real-valued $f \perp \mathbf{1}$, we have

$$\left(1 - \lambda_2(P) \right) \frac{1}{|V|} \sum_{x \in V} |f(x)|^2 = \left(1 - \lambda_2(P) \right) \|f\|_\pi^2 \leq \sum_{e \in E_{1/2}} c(e) \, |f(e^+) - f(e^-)|^2/(2\gamma).$$

Given a permutation σ of V, consider the unitary operator U on $\ell^2(V)$ defined by $U(f) := f \circ \sigma$. Let I be the identity operator. The triangle inequality tells us that $\|I - U\| \leq 2$, whence

$$\sum_{x \in V} \left| f(x) - f(\sigma(x)) \right|^2 \leq 4 \sum_{x \in V} |f(x)|^2.$$

Therefore, we have

$$\frac{1 - \lambda_2(P)}{|V|} \sum_{x \in V} \left| f(x) - f(\sigma(x)) \right|^2 \leq \frac{2}{\gamma} \sum_{e \in E_{1/2}} c(e) \, |f(e^+) - f(e^-)|^2.$$

At this point, it no longer matters whether $f \perp \mathbf{1}$. Furthermore, as before, this inequality extends to $\mathscr{H}$-valued f if we replace absolute values by norms:

$$\frac{1 - \lambda_2(P)}{|V|} \sum_{x \in V} \left|f(x) - f(\sigma(x))\right|^2 \le \frac{2}{\gamma} \sum_{e \in E_{1/2}} c(e) \|f(e^+) - f(e^-)\|^2 . \tag{13.24}$$

Now take σ so that $d(x, \sigma(x)) = \operatorname{diam}(G)$ for all $x \in V$. Assume (13.23) holds with $r = 1$. Then the left-hand side of (13.24) is at least $(1 - \lambda_2(P)) \operatorname{diam}(G)^2$, while the right-hand side is at most $2C^2$. This gives the result. ◀

Some interesting special cases are detailed in the exercises, as is an extension to variable edge lengths. We do one example here. The identity embedding of $\{0, 1\}^k \to \mathbb{R}^k$ shows that the Hilbert distortion of the k-dimensional hypercube G is at most $\sqrt{k}$. To apply our spectral bound, we must find $\lambda_2(P)$. The hypercube is a Cartesian product graph, so let's consider the general problem of the eigenvalues of Cartesian product Markov chains. In other words, let P_i be transition matrices for Markov chains on finite state spaces V_i for $i = 1, \ldots, k$. Also, let α_i be probabilities summing to 1. Then consider the Markov chain on the product state space $V_1 \times \cdots \times V_k$ with each transition occurring in one coordinate only; the ith coordinate changes with probability α_i, and when it does, it uses the transition matrix P_i. Write P for the transition operator of this Markov chain. It is easy to see that if f_i is an eigenvector of P_i with eigenvalue λ_i, then $(f_1, \ldots, f_k)$ is an eigenvector of P with eigenvalue $\sum_i \alpha_i \lambda_i$. It follows that the eigenvalues of P include the α-averages of eigenvalues of P_i, with geometric multiplicities accordingly. If each P_i is reversible, then this gives all eigenvalues of P. For the k-dimensional hypercube, simple random walk is a product Markov chain of this form, where P_i is simple random walk on an edge and all $\alpha_i = 1/k$. Since the eigenvalues of P_i are ± 1, we obtain that $1 - \lambda_2(P) = 2/k$. Since $\operatorname{diam}(G) = k$, Theorem 13.21 gives Enflo's result that the distortion of the k-dimensional hypercube is exactly $\sqrt{k}$.

We now consider metric spaces more general than graphs. What if we know nothing about the finite metric space X other than its cardinality: how little can we distort X by embedding in Euclidean space, and how small can we take the dimension of the Euclidean space to be? Of course, if the space has n points, then its image in Hilbert space spans a subspace of Hilbert space with dimension at most n, so we may always embed it in $\mathbb{R}^n$ just as well as in Hilbert space. It turns out that the dimension of the codomain can always be made much smaller than n without increasing the distortion much: the dimension need be only $O(\log n)$. It turns out that also the distortion need never be more than $O(\log n)$. Both of these theorems are proved via probability: a certain clever random embedding will give the distortion result, while once the original space of n points is embedded in $\mathbb{R}^n$, the reduction in dimension can be effectuated by a random linear map, shown to have the desired property with positive probability.

The dimension-reduction proposition is due to Johnson and Lindenstrauss (1984). It is now widely used in theoretical computer science. Embeddings of networks, regarded as metric spaces given by shortest-path distance using the edge labels as lengths, allow one to use Euclidean geometry and analysis to solve (at least, approximately) discrete optimization problems. Although the proofs of many theorems, including Proposition 13.22, are probabilistic, producing a desired object only with positive probability, that still allows one to construct probabilistic algorithms that work with probability as close to 1 as desired. One such optimization problem is to find a "sparsest cut," that is, to determine a subset of a network whose edge boundary is as small as possible relative to the product of the size of the set and the size of its complement; see Naor (2010) for a discussion of its relationship to embeddings.

Proposition 13.22. (Dimension Reduction) *For any $0 < \epsilon < 1/2$ and any finite number of distinct points $v_1, \ldots, v_n \in \mathbb{R}^l$ with the Euclidean metric, there exists a linear map $A \colon \mathbb{R}^l \to \mathbb{R}^k$, where $k := \lceil 24 \log n / \epsilon^2 \rceil$, that has distortion at most $(1 + \epsilon)/(1 - \epsilon)$ when restricted to the n-point space $\{v_1, \ldots, v_n\}$.*

Proof. Let $A := \frac{1}{\sqrt{k}} (X_{i,j})_{1 \le i \le k, 1 \le j \le l}$ be a $k \times l$ matrix where the entries $X_{i,j}$ are independent, standard normal random variables. We prove that with probability at least $1/n$, this map has distortion at most $(1 + \epsilon)/(1 - \epsilon)$.

Consider a pair $v_p \ne v_q$. Since A is a linear map, the distortion of A on the pair v_p, v_q is measured by $\|Au\|$, where $u := (v_p - v_q)/\|v_p - v_q\|$ has norm 1. Denote the coordinates of u by $u = (u_1, \ldots, u_l)$. Clearly,

$$Au = \frac{1}{\sqrt{k}} \left(\sum_{j=1}^l u_j X_{1,j}, \cdots, \sum_{j=1}^l u_j X_{k,j} \right),$$

so

$$\|Au\|^2 = \frac{1}{k} \sum_{i=1}^k \left(\sum_{j=1}^l u_j X_{i,j} \right)^2.$$

Note that for any i the sum $\sum_{j=1}^l u_j X_{i,j}$ is distributed as a standard normal random variable. So $\|Au\|^2$ is distributed as $\frac{1}{k} \sum_{i=1}^k Y_i^2$, where $Y_1, \ldots, Y_k$ are independent, standard normal random variables. We wish to show that Au is quite concentrated around its mean. To achieve that, we compute the moment generating function of Y^2, where $Y \sim N(0, 1)$. For any $\lambda \in (0, 1/2)$, we have

$$\mathbf{E} e^{\lambda Y^2} = \frac{1}{\sqrt{2\pi}} \int_{-\infty}^{\infty} e^{\lambda y^2} e^{-y^2/2} dy = \frac{1}{\sqrt{1 - 2\lambda}},$$

and using Taylor expansion, we get

$$\varphi(\lambda) := \left| \log \mathbf{E} e^{\lambda(Y^2 - 1)} \right| = \left| -\frac{1}{2} \log(1 - 2\lambda) - \lambda \right|$$

$$= \sum_{k=2}^{\infty} \frac{2^{k-1} \lambda^k}{k} \le 2\lambda^2 [1 + 2\lambda + (2\lambda)^2 + \cdots] = \frac{2\lambda^2}{1 - 2\lambda}.$$

Now,

$$\mathbf{P}[\|Au\|^2 > 1 + \epsilon] = \mathbf{P}\left[e^{\lambda \sum_{j=1}^k (Y_j^2 - 1)} > e^{\lambda \epsilon k} \right] \le e^{-\lambda \epsilon k} e^{k \varphi(\lambda)} \le \exp\left(-\lambda \epsilon k + \frac{2\lambda^2 k}{1 - 2\lambda} \right).$$

Choose $\lambda := \epsilon/4$; this gives that the exponent

$$-\lambda \epsilon k + \frac{2\lambda^2 k}{1 - 2\lambda} = -\frac{\epsilon^2 k}{4} \frac{1/2 - \epsilon/2}{1 - \epsilon/2} < -\frac{\epsilon^2 k}{12}$$

since $\epsilon < 1/2$. With our definition of $k := \lceil 24 \log n/\epsilon^2 \rceil$, this yields

$$\mathbf{P}\big[\|Au\|^2 > 1 + \epsilon\big] \le \exp(-\epsilon^2 k/12) \le n^{-2}.$$

One can prove similarly that

$$\mathbf{P}\big[\|Au\|^2 < 1 - \epsilon\big] \le n^{-2}.$$

Since we have $\binom{n}{2}$ pairs of vectors v_p, v_q, it follows that with probability at least $1/n$, for all $p \ne q$,

$$(1 - \epsilon)\|v_p - v_q\| \le \|Av_p - Av_q\| \le (1 + \epsilon)\|v_p - v_q\|.$$

In this case, the distortion of A is no more than $(1 + \epsilon)/(1 - \epsilon)$.   ◀

We now prove a theorem of Bourgain (1985) that any metric space on n points can be embedded in Hilbert space with distortion $O(\log n)$; the proof gives an embedding into Euclidean space of dimension $O(\log n)^2$. Corollary 13.20 showed that there are graphs on n vertices whose distortion is at least a constant times $\log n$.

Theorem 13.23. (Log Upper Bound on Distortion) *For all $n \ge 2$, every n-point metric space (X, d) can be embedded in Hilbert space with distortion at most $52 \log n$.*

Proof. We may obviously assume that $n \ge 4$. Let $\alpha \ge 60$ and put $L := \lceil \alpha \log n \rceil$. For each integer $k \le n$ that is a positive power of 2, randomly pick L sets $A \subseteq X$ independently, by including in A independently each $x \in X$ with probability $1/k$; in other words, each such set A is a Bernoulli$(1/k)$ site percolation on X. Write $m := \lfloor \log_2 n \rfloor$. Then altogether, we obtain Lm random sets $A_1, \ldots, A_{Lm}$; they are independent but not identically distributed. Define the mapping $f : X \to \mathbb{R}^{Lm}$ by

$$f(x) := \big(d(x, A_1), d(x, A_2), \ldots, d(x, A_{Lm})\big). \tag{13.25}$$

Here, we interpret $d(x, \varnothing) := 0$. We will show that with probability at least $1 - n^{2-\alpha/20} \log n > 0$, the distortion of f is at most $52 \log n$.

For $x, y \in X$ and $A_i \subset X$, the triangle inequality gives $|d(x, A_i) - d(y, A_i)| \le d(x, y)$, so

$$\|f(x) - f(y)\|_2^2 = \sum_{i=1}^{Lm} |d(x, A_i) - d(y, A_i)|^2 \le Lm\, d(x, y)^2.$$

For the lower bound, let $B(x, r) := \{z \in X \, ; \, d(x, z) \le r\}$ denote the closed ball of radius r centered at x and $B^\circ(x, r) := \{z \in X \, ; \, d(x, z) < r\}$ denote the open ball. Fix two points $x \ne y \in X$. Set $r_0 := 0$. Consider an integer $t \ge 0$. If there is some $r \le d(x, y)/4$ such that both $|B(x, r)| \ge 2^t$ and $|B(y, r)| \ge 2^t$, then define r_t to be the least such r. This gives us a sequence of radii $r_0, r_1, \ldots, r_{\hat{t}}$, where $\hat{t} \in [0, m]$ is the largest index for which such an r exists. Define one more radius, $r_{\hat{t}+1} := d(x, y)/4$. Note that $B(x, r_i)$ and $B(y, r_j)$ are always disjoint for $i, j \le \hat{t} + 1$.

Let $1 \le t \le \hat{t} + 1$. By definition, either $|B^\circ(x, r_t)| < 2^t$ or $|B^\circ(y, r_t)| < 2^t$. Let us assume the former, without loss of generality. Now for any set $A \subseteq X$, we have that

$$A \cap B^\circ(x, r_t) = \varnothing \iff d(x, A) \ge r_t$$

and

$$A \cap B(y, r_{t-1}) \ne \varnothing \implies d(y, A) \le r_{t-1}.$$

Therefore, if both conditions hold, then $|d(x, A) - d(y, A)| \ge r_t - r_{t-1}$. Now, we also have $|B(y, r_{t-1})| \ge 2^{t-1}$. Suppose A is a Bernoulli(2^{-t}) percolation on X. Then

$$\mathbf{P}\big[A \text{ misses } B^\circ(x, r_t)\big] = (1 - 2^{-t})^{|B^\circ(x, r_t)|} > (1 - 2^{-t})^{2^t} \ge \frac{1}{4}$$

and

$$\mathbf{P}\big[A \text{ hits } B(y, r_{t-1})\big] = 1 - (1 - 2^{-t})^{|B(y, r_{t-1})|} \ge 1 - (1 - 2^{-t})^{2^{t-1}} \ge 1 - e^{-1/2} \ge \frac{1}{3}.$$

Since these two balls are disjoint, these two events are independent, whence such an A has probability $> 1/12$ to simultaneously miss $B^\circ(x, r_t)$ and intersect $B(y, r_{t-1})$. Since we choose L such sets independently at random, the probability that fewer than $L/81$ of them have that last property is less than

$$e^{-L I_{1/12}(1/81)} < e^{-L/20} \le n^{-\alpha/20}$$

by (6.61). So with probability at least $1 - n^{2-\alpha/20} \log n$, simultaneously for every pair $x \ne y \in X$ and every $t \in [1, \hat{t} + 1]$, we have at least $L/81$ sets that satisfy the condition. In this case,

$$\|f(x) - f(y)\|_2^2 \ge \sum_{t=1}^{\hat{t}+1} \frac{L}{81} (r_t - r_{t-1})^2.$$

Since $\sum_{t=1}^{\hat{t}+1} (r_t - r_{t-1}) = r_{\hat{t}+1} = d(x, y)/4$ and $2 \cdot 2^{\hat{t}} \le n$, we then have

$$\|f(x) - f(y)\|_2^2 \ge \frac{L}{81} \left(\frac{d(x, y)}{4(\hat{t}+1)} \right)^2 (\hat{t} + 1) = \frac{L \, d(x, y)^2}{36^2(\hat{t}+1)} \ge \frac{L \, d(x, y)^2}{36^2 m}.$$

In this case, the distortion of f is at most $36m < 52 \log n$, which proves our claim. ◄

If we combine this upper bound on distortion with the lower bound for networks that we proved earlier (Corollary 13.19) in terms of average distance and spectral gap, we obtain the following upper bound on the gap:

Corollary 13.24. *Let (G, c) be an n-vertex, connected network with transition operator P and stationary probability measure π. Let $\widehat{D}^2$ be the π-weighted average squared distance between pairs of vertices in G. Then*

$$1 - \lambda_2(P) \le \frac{2704 \log^2 n}{\widehat{D}^2}. \qquad ◄$$

The advantage of this bound over Corollary 13.8 is that it does not involve $\pi_{\min}$; however, for simple random walk on graphs, the bound in the earlier corollary is better.

13.6 A Diffusive Lower Bound for Cayley Graphs

As we know from a trivial calculation of variance, simple random walk on $\mathbb{Z}$ has variance n at time n. Given that $\mathbb{Z}$ is the smallest infinite Cayley graph, it is reasonable to conjecture that every infinite Cayley graph has the property that the distance of simple random walk at time n has second moment at least of order n. Indeed this is true; proving it is the goal of this section. In fact, the same holds for finite Cayley graphs as long as the time n is not too large, as suggested by simple random walk on a cycle.

Let Γ be a (finite or countable) group, with a finite generating set S of size d. We assume that $S = S^{-1}$ and denote by G the right Cayley graph determined by S on the vertex set Γ. Let dist denote graph distance in G, and let P be the transition matrix for simple random walk $\langle X_n \rangle_{n \geq 0}$ on G. In this section, we prove the following lower bounds on the rate of escape.

Theorem 13.25. *If $|\Gamma| < \infty$, then $\mathbf{E}\big[\mathrm{dist}(X_0, X_n)^2\big] \geq \dfrac{n}{2d}$ for $n \leq \dfrac{1}{1 - \lambda}$, where $\lambda = \lambda_2(P)$ is the second-largest eigenvalue of P.*

Theorem 13.26. *If $|\Gamma| = \infty$ and Γ is amenable, then $\mathbf{E}\big[\mathrm{dist}(X_0, X_n)^2\big] \geq \dfrac{n}{d}$ for all $n \geq 1$.*

Note that the time bound $(1 - \lambda)^{-1}$ in Theorem 13.25 is of order k^2 for a k-cycle, as we would expect.

Why are nonamenable groups not included in Theorem 13.26? One answer is that we know something even stronger for them: If Γ is nonamenable, then $\rho(G) < 1$ by Theorem 6.7. Proposition 6.9 then implies that for some constant $c_G > 0$ and all n, we have $\mathbf{E}\big[\mathrm{dist}(X_0, X_n)\big] \geq c_G n$, so that $\mathbf{E}\big[\mathrm{dist}(X_0, X_n)^2\big] \geq c_G^2 n^2$. Unfortunately, however, this does not establish a universal lower bound of the form in Theorem 13.26 for such groups, as c_G might be arbitrary small. Indeed, such a universal bound is not known, which is curious.

To prove these theorems, we use the following lemma. For each $n \geq 1$, define the ***Dirichlet form*** $\mathscr{D}_n \colon \ell^2(\Gamma) \to \mathbb{R}$ by $\mathscr{D}_n(f) := \langle (I - P^n)f, f \rangle$. (The form $\mathscr{D}_1$ differs from the Dirichlet energy $\mathscr{D}$ in Section 9.3 by a factor of d: see (6.17).)

Lemma 13.27. *Let G be a d-regular Cayley graph of a (finite or countable) group Γ and suppose that $f \in \ell^2(\Gamma)$ is not constant. Then simple random walk on G satisfies, for every n,*

$$\mathbf{E}\big[\mathrm{dist}(X_0, X_n)^2\big] \geq \frac{1}{d} \frac{\mathscr{D}_n(f)}{\mathscr{D}_1(f)} \, .$$

Before proving the lemma, we show it implies Theorem 13.25.

Proof of Theorem 13.25. Let $f \in \ell^2(\Gamma)$ be an eigenfunction such that $Pf = \lambda f$ with $\|f\|_2 = 1$. Then $\mathscr{D}_n(f) = 1 - \lambda^n$, whence the lemma gives that for $n \leq 1/(1 - \lambda)$,

$$d \cdot \mathbf{E}\big[\mathrm{dist}(X_0, X_n)^2\big] \geq \frac{1 - \lambda^n}{1 - \lambda} = \sum_{j=0}^{n-1} \lambda^j \geq \sum_{j=0}^{n-1} \Big(1 - \frac{1}{n}\Big)^j = n\Big[1 - \Big(1 - \frac{1}{n}\Big)^n\Big] \geq n\Big(1 - \frac{1}{e}\Big). \ \blacktriangleleft$$

Now we prove the lemma itself.

Proof of Lemma 13.27. Given a nonconstant $f \in \ell^2(\Gamma)$, we use an associated Hilbert embedding $F : \Gamma \to \ell^2(\Gamma)$, defined by $F(x)(\gamma) := f(\gamma x)$. For every possible starting point $X_0 := x_0$, use (6.17) to compute (as in (13.22))

$$\mathbf{E}\|F(X_0) - F(X_n)\|^2 = \mathbf{E}\sum_{\gamma \in \Gamma} \big[f(\gamma X_0) - f(\gamma X_n)\big]^2 = \sum_{x \in \Gamma}\sum_{y \in \Gamma} \big[f(x) - f(y)\big]^2 p_n(x, y)$$

$$= 2\langle(I - P^n)f, f\rangle = 2\,\mathscr{D}_n(f). \tag{13.26}$$

Observe that (13.26) with $n = 1$ implies that $\frac{1}{d}\|F(x_0) - F(y)\|^2 \leq 2\,\mathscr{D}_1(f)$ for any $x_0, y \in \Gamma$ that are neighbors in G. Thus, F is Lipschitz with $\mathrm{Lip}(F) \leq \sqrt{2d\,\mathscr{D}_1(f)}$. Therefore,

$$2\,\mathscr{D}_n(f) = \mathbf{E}\|F(X_n) - F(X_0)\|^2 \leq \mathrm{Lip}(F)^2 \cdot \mathbf{E}\big[\mathrm{dist}(X_0, X_n)^2\big] \leq 2d\,\mathscr{D}_1(f)\,\mathbf{E}\big[\mathrm{dist}(X_0, X_n)^2\big].$$

◀

The proof of Theorem 13.25 does not extend as is to the infinite case because we may not have the relevant eigenfunctions. However, use of the infinite-dimensional spectral theorem enables a proof to go through that is virtually identical. The notes, Section 13.9, review the spectral theorem and use it to give a short proof of Theorem 13.26.

To give a direct proof without using the spectral theorem, we use the following exercise and lemma.

▷ **Exercise 13.7.**
Let P be the transition matrix for simple random walk on an infinite graph (V, E). Prove that $\langle P^i\psi, \psi\rangle \to 0$ as $i \to \infty$ for every $\psi \in \ell^2(\mathsf{V})$.

Lemma 13.28. *Given $f \in \ell^2(\Gamma)$ and $n \geq 1$,*

$$\mathbf{E}\big[\mathrm{dist}(X_0, X_n)^2\big] \geq \frac{n}{d} - \frac{n^2}{2d}\frac{\|(I - P)f\|^2}{\mathscr{D}_1(f)}.$$

Proof. We use Lemma 13.27. We need to lower bound $\mathscr{D}_n(f)$ and show it grows almost linearly. For this, we bound second differences. First,

$$\Delta_j := \mathscr{D}_{j+1}(f) - \mathscr{D}_j(f) = \langle P^j f - P^{j+1}f, f\rangle = \langle(I - P)P^j f, f\rangle = \langle P^j f, (I - P)f\rangle.$$

Second, if $\delta := \|(I - P)f\|$, then by Cauchy–Schwarz,

$$|\Delta_j - \Delta_{j-1}| = |\langle P^{j-1}(I - P)f, (I - P)f\rangle| \leq \|P^{j-1}(I - P)f\| \cdot \|(I - P)f\| \leq \delta^2$$

since P is a contraction (Exercise 6.6). Since $\Delta_0 = \mathscr{D}_1(f)$, it follows that $\Delta_j \geq \Delta_0 - j\delta^2$, whence

$$\mathscr{D}_n(f) = \sum_{j=0}^{n-1} \Delta_j \geq n\Delta_0 - \frac{n(n-1)}{2}\delta^2 \geq n\,\mathscr{D}_1(f) - \frac{n^2\delta^2}{2}.$$

Thus,

$$\frac{\mathscr{D}_n(f)}{\mathscr{D}_1(f)} \geq n - \frac{n^2\|(I - P)f\|^2}{2\,\mathscr{D}_1(f)},$$

and the lemma follows from Lemma 13.27. ◀

Proof of Theorem 13.26. We first prove the theorem assuming that G is transient and that the Green function $\mathscr{G}(x_0, x) = \sum_{j=0}^{\infty} p_j(x_0, x)$ satisfies $\mathscr{G}(x_0, \cdot) \in \ell^2(\Gamma)$, and later remove those assumptions.

Let $\langle A_k \rangle$ be a sequence of Følner sets, that is, $\delta_k := |\partial_V A_k|/|A_k| \to 0$ as $k \to \infty$. Write $\psi_k := \mathbf{1}_{A_k}$ and $f_k := \sum_{j=0}^{\infty} P^j \psi_k$. The assumptions of the preceding paragraph imply that $f_k \in \ell^2(G)$. Note that $(I - P)f_k = \psi_k$ and $f_k(x) = \mathbf{E}_x\left[\sum_{j=0}^{\infty} \mathbf{1}_{[X_j \in A_k]}\right]$. If $\text{dist}(x, A_k^c) \geq r$, then $f_k(x) \geq r$, whence

$$\mathscr{D}_1(f_k) = \langle (I - P)f_k, f_k \rangle = \langle \psi_k, f_k \rangle = \sum_{x \in A_k} f_k(x) \geq r\left|\{x \in A_k \, ; \, \text{dist}(x, A_k^c) \geq r\}\right|$$

$$\geq r\left[|A_k| - d^r |\partial_V A_k|\right] = r|A_k|(1 - d^r \delta_k)$$

for every $r > 0$. Letting $k \to \infty$ gives $\liminf_{k \to \infty} \mathscr{D}_1(f_k)/|A_k| \geq r$, so $\mathscr{D}_1(f_k)/|A_k| \to \infty$. By Lemma 13.28,

$$\mathbf{E}\left[\text{dist}(X_0, X_n)^2\right] \geq \frac{n}{d} - \frac{n^2}{2d} \frac{|A_k|}{\mathscr{D}_1(f_k)} \, .$$

Letting $k \to \infty$ proves the theorem under the assumptions of the preceding paragraph.

Next, we prove the theorem without those assumptions. Since Γ is amenable, for every $\theta \in (0, 1/2)$ there exists $\psi \in \ell^2(\Gamma)$ such that $\|\psi\| = 1$ and $\|P\psi - \psi\| \leq \theta$ (see Exercise 6.7 or take $\psi = \mathbf{1}_A/\sqrt{|A|}$ for a suitable Følner set, A). By Exercise 13.7, $\langle P^i \psi, \psi \rangle \to 0$ as $i \to \infty$, so the upcoming Lemma 13.29 yields a function $\varphi \in \ell^2(\Gamma)$ such that $\|(I - P)\varphi\|^2 \leq 32\,\theta \mathscr{D}_1(\varphi)$. Thus, by Lemma 13.28, we conclude that

$$\mathbf{E}\left[\text{dist}(X_0, X_n)^2\right] \geq \frac{n}{d} - \frac{32\theta n^2}{2d} \, .$$

Letting $\theta \to 0$ completes the proof.   ◀

Lemma 13.29. *Suppose that P is a self-adjoint contraction on $\ell^2(\mathsf{V})$ and $\psi \in \ell^2(\mathsf{V})$ has $\|\psi\| = 1$ and $\langle P^i \psi, \psi \rangle \to 0$ as $i \to \infty$. If $\|P\psi - \psi\| \leq \theta \in (0, 1/2)$, then there exists $k \geq 1$ such that $\varphi_k = \sum_{i=0}^{k-1} P^i \psi$ satisfies*

$$\frac{\|(I - P)\varphi_k\|^2}{\langle \varphi_k, (I - P)\varphi_k \rangle} \leq 32\theta. \tag{13.27}$$

Proof. Since $(I - P)\varphi_k = (I - P^k)\psi$ and $I - P^k$ is self-adjoint,

$$\langle \varphi_k, (I - P)\varphi_k \rangle = \langle \varphi_k, (I - P^k)\psi \rangle = \left\langle (I - P^k) \sum_{i=0}^{k-1} P^i \psi, \psi \right\rangle = \langle 2\varphi_k - \varphi_{2k}, \psi \rangle \, .$$

Combining this with $\|(I - P)\varphi_k\| = \|(I - P^k)\psi\| \leq 2$ yields

$$\frac{\|(I - P)\varphi_k\|^2}{\langle \varphi_k, (I - P)\varphi_k \rangle} \leq \frac{4}{\langle 2\varphi_k - \varphi_{2k}, \psi \rangle} \, . \tag{13.28}$$

Thus, our proof will be concluded once we demonstrate the existence of $k \in \mathbb{N}$ such that

$$\langle 2\varphi_k - \varphi_{2k}, \psi \rangle \geq \frac{1}{8\theta} . \tag{13.29}$$

To find such a k, note first that since P is a contraction, $\|P^j\psi - P^{j-1}\psi\| \leq \theta$ for every $j \geq 1$, so by the triangle inequality, $\|P^j\psi - \psi\| \leq j\theta$ for every $j \geq 1$. Cauchy–Schwarz then gives $\langle \psi, (I - P^j)\psi \rangle \leq j\theta$, in other words, $\langle \psi, P^j\psi \rangle \geq 1 - j\theta$. Thus for every $j \geq 1$, we have $\langle \varphi_{2^j}, \psi \rangle \geq 2^j(1 - 2^j\theta)$.

Fix $\ell \in \mathbb{N}$ so that $2^\ell\theta \leq 1/2 \leq 2^{\ell+1}\theta$, yielding

$$\langle \varphi_{2^\ell}, \psi \rangle \geq \frac{1}{8\theta} . \tag{13.30}$$

Let $a_m := \langle \varphi_{2^m}, \psi \rangle$, and write, for $N > \ell$,

$$a_\ell - \frac{a_N}{2^{N-\ell}} = \sum_{m=\ell}^{N-1} \frac{2a_m - a_{m+1}}{2^{m-\ell+1}} . \tag{13.31}$$

The hypothesis that $\langle P^i\psi, \psi \rangle \to 0$ as $i \to \infty$ implies that $\lim_{N\to\infty} a_N/2^N = 0$. Using (13.30) and taking $N \to \infty$ in (13.31) yields

$$\frac{1}{8\theta} \leq a_\ell = \sum_{m=\ell}^{\infty} \frac{2a_m - a_{m+1}}{2^{m-\ell+1}} .$$

Since $\sum_{m=\ell}^{\infty} 1/2^{m-\ell+1} = 1$, there must exist some $m \geq \ell$ with $2a_m - a_{m+1} \geq \frac{1}{8\theta}$. This establishes the property (13.29) for $k = 2^m$. ◀

13.7 Branching Number of a Graph

In Proposition 13.3, we gave an upper bound for the liminf speed of simple random walk on a tree in terms of the branching number of the tree. Here we do the same for general graphs. For a connected, locally finite graph G, choose some vertex $o \in G$ and define the **branching number** $\mathrm{br}\, G$ of G as the supremum of those $\lambda \geq 1$ such that there is a positive flow from o to infinity when the edges have capacities $\lambda^{-|e|}$, where distance is measured from o. By the max-flow min-cut theorem, this is the same as

$$\mathrm{br}\, G = \inf\left\{\lambda \geq 1 ;\ \inf\left\{\sum_{e\in\Pi} \lambda^{-|e|} ;\ \Pi \text{ separates } o \text{ from } \infty\right\} = 0\right\} .$$

Let $\widetilde{M}_n$ be the number of edges that lead from a vertex at distance $n - 1$ from o to a vertex at distance n. Then consideration of spherical cutsets shows that

$$\mathrm{br}\, G \leq \liminf_{n\to\infty} \widetilde{M}_n^{1/n} .$$

We have seen already that when G is a tree, strict inequality may hold here; we can even have that the left-hand side is 1, while the right-hand side is not.

▷ **Exercise 13.8.**

Show that br G does not depend on the choice of vertex o.

▷ **Exercise 13.9.**

Show that if G' is a subgraph of G, then br $G' \leq$ br G.

If G is nonamenable, then br $G > 1$ by this exercise combined with the result of Benjamini and Schramm (1997) mentioned at the end of Section 6.1.

Our bound for trees, Proposition 13.3, can be extended to general graphs as follows, but it is not as good as the one obtained for trees. See Section 13.9 for a better result.

Proposition 13.30. *If G is a connected, locally finite graph with bounded degree on which simple random walk has positive speed, then* br $G > 1$. *More precisely, if*

$$\liminf_{n \to \infty} \frac{|X_n|}{n} \geq s$$

with positive probability, then br $G \geq e^{s/2}$.

Proof. For s with the property stated, consider $0 < s' < s'' < s$. There is some L such that

$$q := \mathbf{P}\big[\forall n \geq L \quad |X_n| > s''n\big] > 0. \tag{13.32}$$

Orient all edges e of G so that e^+ is farther from o than is e^-. As in the proof of Proposition 13.3, define a general percolation on G by keeping all edges e with $|e^+| \leq s''L$ as well as those edges e such that $X_n = e^+$ for some $n < |e^+|/s''$. According to (13.32), the component of the root in this percolation is infinite with probability at least q. On the other hand, if $|x| > s''L$, then by the Varopoulos–Carne bound,

$$\mathbf{P}[o \leftrightarrow x] \leq \mathbf{P}\big[\exists n < |x|/s'' \quad X_n = x\big] \leq \sum_{n=1}^{|x|/s''} p_n(o,x) \leq \sum_{n=1}^{|x|/s''} 2\sqrt{\deg(x)/\deg(o)}\, e^{-|x|^2/(2n)}$$

$$\leq 2\big(|x|/s''\big)\sqrt{\deg(x)/\deg(o)}\, e^{-s''|x|/2} < Ce^{-s'|x|/2}$$

for some constant C. In light of Proposition 5.8, this means that for any cutset Π, we have

$$q < \sum_{e \in \Pi} Ce^{-s'|e^+|/2} \,.$$

Therefore, br $G \geq e^{s'/2}$. Since this holds for all $s' < s$, the result follows.   ◀

We saw in Exercise 13.1 that the converse of Proposition 13.30 fails for trees. It fails even for Cayley graphs, which is rather surprising. Recall from Section 3.4 that we call a subtree T of G rooted at o **geodesic** if for every vertex $x \in T$, the distance from x to o is the same in T as in G. Note that for such trees, br $T \leq$ br G by Exercise 13.9. Let $\mathbb{Z}^\circ$ be the lamplighter group on $\mathbb{Z}$ described in Section 3.4. We showed there that $\mathbb{Z}^\circ$ has exponential growth rate $(1 + \sqrt{5})/2$; in fact, the sets T_n there are parts of the nth levels of a geodesic subtree of $\mathbb{Z}^\circ$, whence br $\mathbb{Z}^\circ = (1 + \sqrt{5})/2$. On the other hand, simple random walk has speed 0 on $\mathbb{Z}^\circ$, as mentioned in Section 3.4 (see Corollary 14.21 for a proof).

Of course, for trees T, we earlier established that $\mathrm{br}\,T$ is the critical value of λ separating transience from recurrence for RW_λ on T. Things are not so nice in general (Exercise 13.55), but we do have some useful relationships. Recall that $\lambda_c(G)$ denotes the critical value of λ separating transience from recurrence for RW_λ on G.

▷ **Exercise 13.10.**
Show that $\lambda_c(G) \le \mathrm{br}\,G$.

It follows from Section 3.4 that when G is a Cayley graph of a finitely generated group of growth rate b, there is a geodesic spanning tree T of G with $\mathrm{br}\,T = b = \lambda_c(G)$. Hence, we also have $\mathrm{br}\,G = b = \lambda_c(G)$. Most of this also holds for many planar graphs, as we show now. The proof will show how planar graphs are like trees. (For other resemblances of planar graphs to trees, see Theorems 9.12 and 10.53 and Proposition 11.28.)

Theorem 13.31. *Let G be an infinite, connected, plane graph of bounded degree such that only finitely many vertices lie in any bounded region of the plane. Suppose that G has a geodesic spanning tree T with no leaves. Then $\lambda_c(G) = \mathrm{br}\,G = \mathrm{br}\,T$.*

Proof. We first prove that $\mathrm{br}\,T = \mathrm{br}\,G$. Since $\mathrm{br}\,T \le \mathrm{br}\,G$, it suffices to show that for $\lambda > \mathrm{br}\,T$, we have $\lambda \ge \mathrm{br}\,G$. Given a cutset Π of T, we will define a cutset Π^* of G whose corresponding cutset sum is not much larger than that of Π. We may assume that o is at the origin of the plane and that all vertices in T_n are on the circle of radius n in the plane. Now every vertex $x \in T$ has a descendant subtree $T^x \subseteq T$. For $n \ge |x| > 0$, this subtree cuts off an arc of the circle of radius n; in the clockwise order of $T^x \cap T_n$, there is a least element $\underline{x}_n$ and a greatest element $\bar{x}_n$. Each edge in Π has two endpoints; collect the ones farther from o in a set W. Define Π^* to be the collection of edges incident to the set of vertices

$$W^* := \left\{ \underline{x}_n, \bar{x}_n \; ; \; x \in W, \, n \ge |x| \right\}.$$

We claim that, since T has no leaves, Π^* is a cutset of G. For if $o = y_1, y_2, \dots$ is a path in G with an infinite number of distinct vertices, let y_k be the first vertex belonging to T^x for some $x \in W$. Planarity implies that $y_k \in W^*$, whence the path intersects Π^*, as desired.

Now let c be the maximum degree of vertices in G. We have

$$\sum_{e \in \Pi^*} \lambda^{-|e|} \le c \sum_{x \in W^*} \lambda^{-|x|+1} \le c \sum_{x \in W} \sum_{n \ge |x|} 2\lambda^{-n+1}$$

$$= \frac{2c\lambda}{\lambda - 1} \sum_{x \in W} \lambda^{-|x|+1} = \frac{2c\lambda}{\lambda - 1} \sum_{e \in \Pi} \lambda^{-|e|}.$$

This makes it evident that $\mathrm{br}\,G \le \mathrm{br}\,T$. It also implies that $\lambda_c(G) \le \mathrm{br}\,G$, which we knew as well from Exercise 13.10.

To finish, it remains to show that $\lambda_c(G) \ge \mathrm{br}\,G$. Let $\lambda < \mathrm{br}\,G$. Since $\lambda < \mathrm{br}\,T$, it follows from Theorem 3.5 that RW_λ is transient on T. Therefore, RW_λ is also transient on G. ◀

13.8 Tree-Indexed Random Walks

Label the vertices of a tree T by a collection of i.i.d. real random variables $\langle X_v \rangle_{v \in V(T)}$. Define the **tree-indexed random walk** $\langle S_v \rangle_{v \in V(T)}$ by

$$S_v := \sum_{w \leq v} X_w \,,$$

where $w \leq v$ means that v is a descendant of w. Throughout this section, we will assume that each variable X_v satisfies, for some $\lambda > 0$,

$$X_v \text{ is not a.s. constant, } \mathbf{E}[X_v] = 0, \text{ and } \mathbf{E}[e^{\lambda X_v}] < \infty \,. \tag{13.33}$$

We want to determine the speed of tree-indexed random walks, or at least recognize when the speed is positive. Here are three natural notions of speed for such walks:

$$\textbf{Cloud Speed} \quad l_{\text{cloud}} := \limsup_n \max_{|v|=n} S_v/n \,;$$

$$\textbf{Burst Speed} \quad l_{\text{burst}} := \sup_{\xi \in \partial T} \limsup_{v \in \xi} S_v/|v| \,;$$

$$\textbf{Sustainable Speed} \quad l_{\text{sust}} := \sup_{\xi \in \partial T} \liminf_{v \in \xi} S_v/|v| \,.$$

Each of these speeds is a.s. constant by Kolmogorov's zero-one law.

In general, $l_{\text{cloud}} \geq l_{\text{burst}} \geq l_{\text{sust}}$. Exercises 13.11 and 13.15 show that the inequalities may be strict.

▷ **Exercise 13.11.**
Show that if the 1–3 tree in Example 1.2 is labeled by ± 1-valued random variables $\langle X_v \rangle$ of mean zero, then $l_{\text{cloud}} > 0$, but $l_{\text{burst}} = l_{\text{sust}} = 0$.

Denote by $\langle \widetilde{S}_n \rangle_{n \geq 0}$ the ordinary random walk indexed by the nonnegative integers with i.i.d. increments distributed like X_v. Since

$$\mathbf{P}\big[\widetilde{S}_{n+m} \geq (n+m)a\big] \geq \mathbf{P}\big[\widetilde{S}_n \geq na\big] \mathbf{P}\big[\widetilde{S}_m \geq ma\big] \,,$$

Fekete's lemma (Exercise 3.9) implies that for all $a \in \mathbb{R}$, the limit

$$I(a) := \lim_{n \to \infty} -\frac{1}{n} \log \mathbf{P}\big[\widetilde{S}_n \geq na\big]$$

exists. This limiting function of a is known as the **rate function** for the random walk $\langle \widetilde{S}_n \rangle_{n \geq 0}$. By Exercise 3.9(a), we have that $\mathbf{P}\big[\widetilde{S}_n \geq na\big] \leq e^{-nI(a)}$ for all n and a. Write $s_\infty := \mathrm{ess\,sup}\, X_o \in [0, \infty]$. Then $I(a) < \infty$ for $a < s_\infty$ and $I(a) = \infty$ for $a > s_\infty$. We also have $I(s_\infty) = -\log \mathbf{P}[X_o = s_\infty]$. By Chernoff–Cramér's theorem on large deviations (see Durrett (2019), (2.7.2) and Lemma 2.7.2, and Dembo and Zeitouni (1998), (2.2.6)), (13.33) implies that $I(a) > 0$ for all $a > 0$, and $I(\cdot)$ is convex on $\mathbb{R}$ where it is finite. Since $I(0) = 0$ by the central limit theorem, $I(\cdot)$ is strictly increasing on $[0, s_\infty]$. For $\alpha > 0$, consider

$$s^* := s^*(\alpha) := \sup\{s \,;\, I(s) \leq \alpha\} \,.$$

If $I(s^*) \neq \alpha$, then $s^* = s_\infty$.

▷ **Exercise 13.12.**
Verify that if the law of X_v is N(0, 1), then $I(a) = a^2/2$, whereas if X_v are ± 1 valued with mean zero, then $I(a) = \frac{(1+a)}{2} \log(1+a) + \frac{(1-a)}{2} \log(1-a)$. (That this rate function is an upper bound for simple random walk is noted in (13.1); for biased walks, a similar inequality is (6.62).)

If the underlying tree T is a family tree of a Galton–Watson process, then the speed is determined by the mean of the offspring distribution and the rate function.

Theorem 13.32. (Biggins, 1977) *Let T be a Galton–Watson tree with mean $m > 1$. Suppose that the vertices of T are labeled by random variables X_v that satisfy (13.33). Given the event that T is infinite, a.s. all three speeds coincide and equal $s^* = s^*(\log m)$, where*

$$I(s^*) = \log m$$

unless $\mathbf{P}[X_o = s_\infty] > m^{-1}$, *in which case* $s^* = s_\infty$.

Proof. We first show that $l_{\text{cloud}} \leq s^*$. By the definition of s^*, for any $\epsilon > 0$ there is $\delta > 0$ such that $I(s^* + \epsilon) > \log m + \delta$. Therefore,

$$\mathbf{P}\big[\widetilde{S}_n \geq n(s^* + \epsilon)\big] \leq e^{-n(\log m + \delta)} = m^{-n} e^{-n\delta}.$$

Consequently,

$$\mathbf{P}\big[S_v \geq n(s^* + \epsilon) \text{ for some } v \in T_n \mid \text{nonextinction}\big] \leq \frac{m^n}{1 - q} m^{-n} e^{-n\delta},$$

where q is the probability of extinction. The proof is concluded by invoking the Borel–Cantelli lemma.

To prove the reverse inequality, $l_{\text{sust}} \geq s^*$, let $a < s^*$ be given. Using the strict monotonicity of the rate function and the definition of s^*, choose ϵ so that $I(a) + 2\epsilon < \log m$. For each $k \geq 1$ and $M \in [1, \infty]$, we define a new embedded branching process as follows: start from the root of T, and take the set of offspring $T(v, k, M)$ of a vertex v to consist of all its descendants w in T that satisfy

- $|w| = |v| + k$ in T,
- $S_w \geq S_v + ka$, and
- $S_u > S_v - M$ for all u on the path from v to w.

(If $M = \infty$, then the last requirement holds automatically.) It is not hard to calculate that $\mathbf{E}|T(v, k, \infty)| = m^k \mathbf{P}\big[\widetilde{S}_k \geq ka\big]$, so the definition of I yields that for sufficiently large k,

$$\mathbf{E}|T(v, k, \infty)| \geq m^k e^{-k[I(a)+\epsilon]} > e^{k\epsilon} > 2.$$

By choosing M large, we can ensure that the embedded process has mean offspring

$$\mathbf{E}|T(v, k, M)| \geq \frac{1}{2} \mathbf{E}|T(v, k, \infty)| > 1.$$

Thus for large k and M, this embedded process is supercritical. Therefore $l_{\text{sust}} > a$ with positive probability. Since

$$\{T \,;\, T \text{ finite or } l_{\text{sust}} \leq a \text{ on } T\}$$

is an inherited property, Proposition 5.6 implies that $\mathbf{P}[l_{\text{sust}} > a \mid \text{survival}] = 1$. Hence, given survival, we have that a.s.,

$$s^* \geq l_{\text{cloud}} \geq l_{\text{burst}} \geq l_{\text{sust}} \geq s^* . \qquad \blacktriangleleft$$

Theorem 13.32 was extended by Lyons and Pemantle (1992), who determined the sustainable speed for a general tree-indexed walk. For the other speed notions, there is no analogous exact formula, but there is a characterization for positivity of speed; see the notes.

Theorem 13.33. *Suppose that T is an infinite, locally finite tree, labeled with random variables $\langle X_v \rangle$ satisfying (13.33). Let $I(\cdot)$ be the corresponding rate function. Then $l_{\text{sust}} = s^*(\log \text{br} \, T)$, so*

$$I(l_{\text{sust}}) = \log \text{br}(T)$$

unless $\mathbf{P}[X_o = s_\infty] > (\text{br} \, T)^{-1}$, in which case $l_{\text{sust}} = s_\infty$.

The proof will use Exercise 3.25, repeated here:

▷ **Exercise 13.13.**
Given a tree T and $k \geq 1$, form the tree $T^{[k]}$ by taking the vertices x of T for which $|x|$ is a multiple of k and joining x and y by an edge in $T^{[k]}$ when their distance is k in T. Show that $\text{br} \, T^{[k]} = (\text{br} \, T)^k$.

Proof of Theorem 13.33. Suppose that $I(a) > \log \text{br} \, T$. Choose a sequence of cutsets Π_k such that the cutset sums $c_k := \sum_{e(v) \in \Pi_k} e^{-I(a)|v|} \to 0$ as $k \to \infty$. Since $\mathbf{P}[\widetilde{S}_n \geq na] \leq e^{-nI(a)}$ for all n, we have that $\mathbf{P}[\exists v \in \Pi_k \ S_v \geq |v|a] \leq c_k$, whence $l_{\text{sust}} \leq a$ a.s. Hence, $l_{\text{sust}} \leq s^*(\log \text{br} \, T)$.

For the other inequality, fix a so that $I(a) < \log \text{br} \, T$. Choose an integer k that satisfies $\mathbf{P}[\widetilde{S}_k \geq ka] > (\text{br} \, T)^k$ and then select $M < \infty$ such that

$$\mathbf{P}[\widetilde{S}_k \geq ka \text{ and } \widetilde{S}_j > -M \text{ for all } j \in (0, k)] > (\text{br} \, T)^k .$$

Define a general percolation on the compressed tree $T^{[k]}$ in which, for v the parent of w, the edge $[v, w]$ is retained if $S_w - S_v \geq ka$ and $S_u > S_v - M$ for all u on the path in T from v to w. This general percolation process is not independent; however, for each fixed k, it is quasi-independent. By Theorem 5.19 and Exercise 13.13, this percolation survives with positive probability, whence $l_{\text{sust}} \geq a$ a.s by the Kolmogorov zero-one law. It follows that $l_{\text{sust}} \geq s^*(\log \text{br} \, T)$ a.s. $\qquad \blacktriangleleft$

13.9 Notes

The spectral theorem on Hilbert spaces can be used in this chapter to greatly shorten some proofs. We first summarize what that theorem says for bounded self-adjoint operators on a (real or complex) Hilbert space, $\mathscr{H}$. For more details, see, for example, Rudin (1991), Chapter 12. Let $\mathscr{B}$ be the Borel σ-field in $\mathbb{R}$. A ***resolution of the identity*** $E(\cdot)$ is a map from $\mathscr{B}$ to the set of orthogonal projections on $\mathscr{H}$ that satisfies properties similar to a probability measure, namely, $E(\varnothing) = 0$; $E(\mathbb{R}) = I$; for all $B_1, B_2 \in \mathscr{B}$, we have $E(B_1 \cap B_2) = E(B_1)E(B_2)$ and, if $B_1 \cap B_2 = \varnothing$, then $E(B_1 \cup B_2) = E(B_1) + E(B_2)$; and for all $f, g \in \mathscr{H}$, the map $B \mapsto \langle E(B)f, g \rangle$ is a finite (real or complex) measure on $\mathscr{B}$. Note that $B \mapsto \langle E(B)f, f \rangle = \|E(B)f\|^2$ is a positive measure of norm $\|f\|^2$.

Now suppose that $T: \mathscr{H} \to \mathscr{H}$ is a bounded self-adjoint operator. The ***spectrum*** of T is the set of $\lambda \in \mathbb{R}$ such that $T - \lambda I$ does not have an inverse on $\mathscr{H}$; the spectrum is compact and contained in $[-\|T\|, \|T\|]$. Isolated points λ of the spectrum are necessarily eigenvalues, that is, values where $T - \lambda I$ has a nonzero kernel. The ***spectral theorem*** says that there is a unique resolution of the identity, $E_T(\cdot)$, such that $T = \int \lambda \, dE_T(\lambda)$ in the sense that for all $f, g \in \mathscr{H}$, we have $\langle Tf, g \rangle = \int \lambda \, d\langle E_T(\lambda)f, g \rangle$. Furthermore, E_T is supported on the spectrum of T. For a bounded, Borel-measurable function $h: \mathbb{R} \to \mathbb{R}$, one can define (via what is called the ***functional calculus***) a bounded self-adjoint operator $h(T)$ by $h(T) := \int h(\lambda) \, dE_T(\lambda)$, with integration meant in the same sense as earlier. The operator $h(T)$ commutes with T. As an instance of this, we have $\mathbf{1}_B(T) = E_T(B)$.

If $\mathscr{H}$ is finite-dimensional and T has spectrum σ, then $E_T(B) = \sum_{\lambda \in \sigma \cap B} P_\lambda$, where P_λ is the orthogonal projection onto the λ-eigenspace. Writing $T = \sum_{\lambda \in \sigma} \lambda P_\lambda$ amounts to diagonalizing T. Here we have $h(T) = \sum_{\lambda \in \sigma} h(\lambda)P_\lambda$ for any function h; because only finitely many values of h are used, we may take h to be a polynomial.

We now use the spectral theorem to prove (13.12) when $\mathscr{H}$ is infinite-dimensional. Let E be the resolution of the identity for S:

$$S = \int_{-\|S\|}^{\|S\|} s \, dE(s) \,.$$

If Q is a polynomial with real coefficients, then

$$Q(S) = \int_{-\|S\|}^{\|S\|} Q(s) \, dE(s) \,.$$

Applying this to any $f \in \mathscr{H}$, we obtain

$$\left| \langle Q(S)f, f \rangle \right| = \left| \int_{-\|S\|}^{\|S\|} Q(s) \, d\langle E(s)f, f \rangle \right| \leq \max_{-\|S\| \leq s \leq \|S\|} |Q(s)| \cdot \int d\langle E(s)f, f \rangle$$

$$= \max_{-\|S\| \leq s \leq \|S\|} |Q(s)| \cdot \|f\|^2 \,.$$

Since $Q(S)$ is self-adjoint, this proves (13.12).

We next give a second proof of Theorem 13.26 via the spectral theorem.

Second proof of Theorem 13.26. Since Γ is amenable, there are finite vertex sets K_n that satisfy $\|P\mathbf{1}_{K_n} - \mathbf{1}_{K_n}\|/\|\mathbf{1}_{K_n}\| \to 0$, whence 1 lies in the spectrum of P, which is self-adjoint. However, 1 is not an eigenvalue since Γ is infinite. Therefore, $E(1 - \epsilon, 1) \neq 0$ for every $\epsilon > 0$, where $E(\cdot)$ is the resolution of the identity for P. Consider $\epsilon > 0$. Choose $f \in \ell^2(\Gamma) \setminus \{\mathbf{0}\}$ in the image of $E(1 - \epsilon, 1)$. Write $\mu(\cdot) := \langle E(\cdot)f, f \rangle$ for the spectral measure of f. Then

$$\mathscr{D}_n(f) = \int_{1-\epsilon}^1 (1 - \lambda^n) \, d\mu(\lambda) = \int_{1-\epsilon}^1 (1 - \lambda) \sum_{j=0}^{n-1} \lambda^j \, d\mu(\lambda)$$

$$\geq \frac{1 - (1-\epsilon)^n}{\epsilon} \int_{1-\epsilon}^1 (1 - \lambda) \, d\mu(\lambda) = \frac{1 - (1-\epsilon)^n}{\epsilon} \mathscr{D}_1(f) \neq 0 \,.$$

Letting $\epsilon \to 0$, we obtain $\inf_{f \neq 0} \mathscr{D}_n(f)/\mathscr{D}_1(f) \geq n$. Use of Lemma 13.27 now completes the proof. ◀

Another version of Hoeffding's inequality for martingales says this:

Theorem 13.34. *Let $\langle (Y_1, \mathscr{F}_1), \ldots, (Y_n, \mathscr{F}_n) \rangle$ be a martingale and*

$$c_i := \inf\{\|Y_i - Z\|_\infty \,;\, Z \in \mathscr{F}_{i-1}\},$$

where $\mathscr{F}_0 := \{\varnothing, \Omega\}$ is the trivial σ-field. Then, for all $L > 0$,

$$\mathbf{P}[Y_n - \mathbf{E}[Y_n] \geq L] \leq \exp\left(-\frac{L^2}{2\sum_{i=1}^n c_i^2}\right).$$

A very useful special case is the following inequality of McDiarmid (1989):

Theorem 13.35. (Bounded-Differences Inequality) *Let $\langle Z_1, \ldots, Z_n \rangle$ be independent random variables and $f(z_1, \ldots, z_n)$ be a real-valued function such that*

$$|f(\mathbf{z}) - f(\mathbf{z}')| \leq a_i$$

when the vectors $\mathbf{z}$ and $\mathbf{z}'$ differ only in the ith coordinate. Write $Y := f(Z_1, \ldots, Z_n)$. Then, for all $L > 0$,

$$\mathbf{P}[Y - \mathbf{E}[Y] \geq L] \leq \exp\left(-\frac{2L^2}{\sum_{i=1}^n a_i^2}\right).$$

Proof. Put $Y_i := \mathbf{E}[f(Z_1, \ldots, Z_n) \mid Z_1, \ldots, Z_i]$ for $0 \leq i \leq n$, so that $Y_n = Y$. We may apply Theorem 13.34 to this martingale with $c_i \leq a_i/2$. To see this bound, write $Y_i = f_i(Z_1, \ldots, Z_i)$. Define $g_i^{\pm}(z_1, \ldots, z_{i-1}) := \sup f_i(z_1, \ldots, z_{i-1}, z_i) \pm \inf f_i(z_1, \ldots, z_{i-1}, z_i)$. Since $0 \leq g_i^- \leq a_i$, it follows that $\|Y_i - g_i^+(Z_1, \ldots, Z_{i-1})/2\|_\infty \leq a_i/2$. ◀

See McDiarmid (1989) for a proof of Theorem 13.34, variations, and applications. An extension to the case where the hypothesis holds only on an event of probability less than 1 is given by Combes (2015).

The first statement of (13.6) was in Lyons, Pemantle, and Peres (1996b). An extension to the much harder case of (biased) random walks on directed covers was achieved by Takacs (1997), with a further extension to certain random trees by Takacs (1998). More general weights on the edges were treated by Nagnibeda and Woess (2002) and Gilch and Müller (2011). On standard Cayley graphs of free products of copies of $\mathbb{Z}_2$, such results go back to Sawyer and Steger (1987).

A probabilistic proof of the Varopoulos–Carne bound, Theorem 13.4, was given by Peyre (2008). The application of this bound to mixing time was prompted by discussions with Itai Benjamini, James Lee, and Manor Mendel.

Spielman and Teng (2007) used circle packing to give a (short) proof that on a planar graph G with n vertices and maximum degree d, the spectral gap of simple random walk satisfies $1 - \lambda_2 \leq 8d/n$. Consider now lazy simple random walk on such planar graphs. Write D for the diameter of G. Jian Ding and Asaf Nachmias (personal communication) noted that in conjunction with Proposition 13.7, the above bound implies that $t_{\text{mix}}(\epsilon) \geq c_d D^2/\log D$, where c_d is a constant. This can be seen by Exercise 13.4 and by considering whether $D \leq \sqrt{n}$. In view of Corollary 13.6, again considering whether $D \leq \sqrt{n}$, it follows that $1 - \lambda_2 \leq c_d'(\log D)^2/D^2$, as first proved by Louder and Souto (2012). In that paper and in Lee and Qin (2012), examples were given of planar graphs for which this lower bound on the spectral gap is sharp.

Naor, Peres, Schramm, and Sheffield (2006) proved Theorem 13.14. They also showed that L^p spaces for $p > 2$ have Markov type 2, as conjectured by Ball, as do word-hyperbolic groups and simply connected Riemannian manifolds of pinched negative curvature. A key technique is a decomposition of a function of a stationary, reversible Markov chain into a difference of a forward and a backward martingale. This decomposition idea was discovered independently by Kesten (1986) and T. Lyons and Zheng (1988) and employed later by Barlow and Perkins (1989) and T. Lyons and Zhang (1994). The same martingale decomposition was also used by Ding, Lee, and Peres (2013), who proved that planar graphs and doubling metric spaces have Markov type 2. In fact, the result of Kesten (1986) that simple random walk on supercritical percolation clusters of $\mathbb{Z}^d$ is at most diffusive in the metric of $\mathbb{Z}^d$ can be proved by using the fact that $\mathbb{R}^d$ has Markov type 2; see Peres, Stauffer, and Steif (2015) for a more general result proved this way.

Corollary 13.20 was proved by Linial, London, and Rabinovich (1995) not for embeddings in Hilbert space but in L^1 spaces. That is harder, since ℓ^2 embeds isometrically in L^1: Let Z_n ($n \geq 0$) be i.i.d. standard normal random variables. Given $u = \langle u_n \; ; \; n \geq 0 \rangle \in \ell^2(\mathbb{N})$, define the random variable $Z(u) := \sqrt{\pi/2} \sum_{n \geq 0} u_n Z_n$. Then $Z(u)$ is a centered normal random variable with standard deviation equal to $\sqrt{\pi/2} \, \|u\|_2$. Therefore, $\mathbf{E}\big[|Z(u)|\big] = \|u\|_2$, so $u \mapsto Z(u)$ is an isometry.

Theorem 13.21 and Exercise 13.46 are straightforward extensions of a special case of the main result of Jolissaint and Valette (2014).

Our proof of Proposition 13.22 is a small variant of the original proof; it was known to many people shortly after the original paper appeared. A version of it was published by Indyk and Motwani (1999). For more variants and history, see Matoušek (2008). A lower bound on k of the same order as in Proposition 13.22 is known for almost the entire range of n compared to l and ϵ, even for nonlinear maps: see Larsen and Nelson (2016, 2017).

An easier implementation of dimension reduction, Proposition 13.22, uses random variables $X_{i,j}$ that independently take the values ± 1 with probability $1/2$ each. That this is possible was first noted by Achlioptas and McSherry (2001, 2007). Actually, it is possible to use random variables that have the distribution of a random variable X for which there exists a constant $C > 0$ such that $\mathbf{E}e^{\lambda X} \leq e^{C\lambda^2}$ (for $X = \pm 1$ with probability $1/2$ each, we have $\mathbf{E}e^{\lambda X} = \cosh(\lambda) \leq e^{\lambda^2/2}$). This is proved by the following argument due to Assaf Naor (personal communication, 2004):

Let X_i be i.i.d. with the same distribution as X, and set $Y := \sum_{i=1}^{k} u_i X_i$ with $\sum_{j=1}^{k} u_j^2 = 1$. Take Z to be a standard normal random variable independent of $\{X_i\}$. Recall that for all real α, we have $\mathbf{E}e^{\alpha Z} = e^{\alpha^2/2}$. Since Y and Z are independent, using Fubini's theorem, we get that for any $\lambda < C/4$,

$$\mathbf{E}e^{\lambda Y^2} = \mathbf{E}e^{(\sqrt{2\lambda}Y)^2/2} = \mathbf{E}e^{\sqrt{2\lambda}YZ} = \mathbf{E}e^{\sum_{i=1}^{k} \sqrt{2\lambda}u_i X_i Z} = \mathbf{E}\mathbf{E}\big[e^{\sum_{i=1}^{k} \sqrt{2\lambda}u_i X_i Z} \mid Z\big]$$

$$\leq \mathbf{E}e^{C\sum_{i=1}^{k} 2\lambda u_i^2 Z^2} = \mathbf{E}e^{2C\lambda Z^2} = \frac{1}{\sqrt{1 - 4C\lambda}} \, .$$

Then the rest of the argument is the same as Proposition 13.22.

Our exposition of the proof of Theorem 13.23 follows Linial, London, and Rabinovich (1995), but the idea is the same as the original proof of Bourgain. Embeddings of the sort (13.25) are known as **_Fréchet embeddings_**. For more on the use of embeddings in computer science, see Chapter 15 of Matoušek (2002).

The following theorem was proved by Benjamini, Lyons, and Schramm (1999):

Theorem 13.36. (Speed on Percolation Clusters) *Let G be a nonamenable, transitive, unimodular graph. Let ω be an invariant percolation on G. Then simple random walk on some infinite cluster of ω has positive speed with positive probability in each of the following cases:*

 (i) *ω is Bernoulli percolation that has infinite components a.s.;*

 (ii) *ω has a unique infinite cluster a.s.;*

 (iii) *ω has a cluster with at least three ends with positive probability;*

 (iv) *$\mathbf{E}[\deg_\omega o \mid o \in \omega] > \alpha(G)$.*

Equivariant maps $f : \Gamma \to \mathscr{H}$, where Γ is a (finitely generated) group and $\mathscr{H}$ is a Hilbert Γ-space, that is, a Hilbert space on which Γ acts by isometries, are also related to speed of random walks. Fix a

Cayley graph G of Γ (it doesn't matter which one). Using the metric of G, we see that all such maps f are Lipschitz with Lipschitz constant $\max_{x \sim o} \|f(x) - f(o)\|$. Let $\alpha^{\#}(\Gamma)$ be the supremum of α for which there exists some equivariant f and $c > 0$ with $\|f(x) - f(y)\| \geq c \, d_G(x, y)^{\alpha}$. We call $\alpha^{\#}(\Gamma)$ the **equivariant Hilbert compression exponent** of Γ. Let $\beta^*(\Gamma)$ be the supremum of β for which simple random walk $\langle X_n \rangle$ on some Cayley graph G of Γ satisfies $\mathbf{E}[d_G(X_0, X_n)] \geq cn^{\beta}$ for some $c > 0$ and all $n \geq 0$. Then $\alpha^{\#}(\Gamma)\beta^*(\Gamma) \leq 1/2$; see Naor and Peres (2008).

Theorem 13.26 was discovered by Anna Erschler (private communication), who noted that it follows from a result of Mok (1995) and Korevaar and Schoen (1997) that ensures the existence of a nonconstant equivariant harmonic map from Γ to a Hilbert Γ-space $\mathcal{H}$, provided Γ is not Kazhdan. Kleiner (2010) presents a short proof of the existence of such a map. The proof we present of Theorems 13.25 and 13.26 follows Lee and Peres (2013), where these results were extended to random walks on transitive graphs. That paper also establishes the existence of an automorphism-equivariant harmonic map to $\mathcal{H}$ in this setting, provided the automorphism group of the graph is not Kazhdan. Lee, Peres, and Smart (2016) deduced that for every infinite, vertex-transitive graph G, there is a constant $C_G > 0$ such that, if $\langle X_t \rangle$ is the simple random walk on G, then for every $\epsilon > 0$ and $t \geq 1/\epsilon^2$,

$$\mathbf{P}\left[\mathrm{dist}(X_t, X_0) \leq \epsilon \sqrt{t}\,\right] \leq C_G \, \epsilon \,.$$

Although simple random walk on nonamenable Cayley graphs escapes at positive (linear) speed, the inequality $\mathbf{E}[\mathrm{dist}(X_0, X_t)^2] \geq t/d$ of Theorem 13.26 without a group-dependent constant is not known for such graphs (except for Cayley graphs of non-Kazhdan groups, which are precisely the infinite groups that admit a nonconstant equivariant harmonic map to a Hilbert space, $\mathcal{H}$).

Proposition 13.30 is due to Peres (1997), unpublished. The inequalities of Propositions 13.3 and 13.30 were sharpened in remarkable work of Virág (2000b, 2002). In the first paper, he proved that if RW_λ is considered instead of simple random walk, then we have the bound

$$\frac{\mathrm{br}\, G - \lambda}{\mathrm{br}\, G + \lambda} \geq s \tag{13.34}$$

for all graphs G, where s is the essential supremum of the liminf speed. In other words, the liminf speed is a.s. at most the left-hand side of (13.34). In fact, he defined the branching number for networks and proved that

$$\frac{\mathrm{br}\, G - 1}{\mathrm{br}\, G + 1} \geq s \tag{13.35}$$

holds for all networks. Here, for a network (G, c), its **branching number** is

$$\mathrm{br}(G, c) = \inf\left\{\lambda > 0 \, ; \; \inf\left\{\sum_{e \in \Pi} c(e)\lambda^{-|e|} \, ; \; \Pi \text{ separates } o \text{ from } \infty\right\} = 0\right\}.$$

Thus, if G_λ is the network on the graph G corresponding to RW_λ, then $\mathrm{br}\, G_\lambda = (\mathrm{br}\, G)/\lambda$, so that (13.35) includes (13.34). In this same first paper, Virág defined a more refined notion of branching number, the **essential branching number**

$$\mathrm{ess}\,\mathrm{br}\, G = \inf\left\{\lambda > 0 \, ; \; \inf\left\{\sum_{e \in \partial_E W} c(e)\lambda^{-|e|} \, ; \; \text{the random walk from } o \text{ a.s. leaves } W\right\} = 0\right\}, \tag{13.36}$$

where W is a set of vertices, possibly infinite. When W is restricted to be finite, this is simply another form of the definition of branching number for a network. Virág (2000b) showed that

$$\frac{\mathrm{ess}\,\mathrm{br}\, G - 1}{\mathrm{ess}\,\mathrm{br}\, G + 1} \vee 0 \geq s \,. \tag{13.37}$$

In his second paper, Virág proved an inequality analogous to (13.35) that relates limsup speed to a notion of upper growth for general networks.

Theorem 13.31 is due to Lyons (1996).

The first reference we know to tree-indexed random walk is Dubins and Freedman (1967). They considered the case where T is a binary tree and $X_v = \pm 1$ with probabilities p and $1 - p$.

Biggins (1977) gave a result more general than our Theorem 13.32, as he allowed dependencies among siblings, as in Section 5.9.

Cloud speed and burst speed of tree-indexed walks were studied in Benjamini and Peres (1994b), while sustainable speed was studied earlier by Lyons and Pemantle (1992).

To describe when a tree-indexed walk has positive burst speed, we will need one more definition, due (in greater generality) to Tricot (1982). The **packing dimension** of ∂T is defined by

$$\dim_P(\partial T) := \inf\{\sup_i \log \overline{\mathrm{gr}}\, T^{(i)}\},$$

where the infimum extends over all countable collections $\{T^{(i)}\}$ of subtrees of T such that $\partial T = \bigcup_i \partial T^{(i)}$. In particular, $\dim_P(\partial T) \le \log \overline{\mathrm{gr}}\, T$. For the 1–3 tree of Example 1.2, we have $\dim_P(\partial T) = 0$ and $\log \overline{\mathrm{gr}}\, T = \log 2$.

Theorem 13.37. (Benjamini and Peres, 1994b) *Suppose that the vertices of an infinite, locally finite tree T are labeled by random variables X_v that satisfy (13.33). Then*
  (i) $l_{\mathrm{cloud}} > 0 \Leftrightarrow \log \overline{\mathrm{gr}}(T) > 0$, *provided that T has no leaves, and*
  (ii) $l_{\mathrm{burst}} > 0 \Leftrightarrow \dim_P(\partial T) > 0$.

A quantitative version of part (i) of this theorem can be given if the labeling variables X_v are Gaussian.

▷ **Exercise 13.14.**
Suppose that T is an infinite tree without leaves and that its vertices are labeled by i.i.d. variables $X_v \sim N(0, 1)$. Denote $d^* := \log \overline{\mathrm{gr}}(T)$. Prove that

$$\sqrt{d^*/2} \le l_{\mathrm{cloud}} \le \sqrt{2d^*}$$

and both bounds can be achieved.

Part (ii) of the preceding theorem, combined with Theorem 13.33, can be used to show that on some trees, burst speed is strictly greater than sustainable speed:

▷ **Exercise 13.15.**
Let $\langle n_k \rangle_{k \ge 1}$ be an increasing sequence of positive integers. Construct a tree T as follows: The first n_1 levels of T are as in the 1–3 tree. To each vertex v in level n_1 of T, attach a copy of the first $n_2 - n_1$ levels of the 1–3 tree, with v as its root. Continue iteratively by attaching a copy of the first $n_{k+1} - n_k$ levels of the 1–3 tree to each vertex at level n_k of T. Show that for every choice of $\langle n_k \rangle$ (and for every distribution of the variables X_v that satisfies (13.33)), we have $l_{\mathrm{burst}} > 0$, but if the n_k increase sufficiently quickly, then $l_{\mathrm{sust}} = 0$.

For further probabilistic applications of packing dimension, see Khoshnevisan, Peres, and Xiao (2000).

Let T be a tree without leaves. Instead of a T-indexed random walk, one can consider a T-indexed Markov chain determined by a transition kernel $p(\cdot, \cdot)$ on a countable state space V. Along any infinite ray of T, this reduces to the trajectory of a standard Markov chain on V with transition kernel $p(\cdot, \cdot)$. Let ρ be the spectral radius of the latter chain, defined as in (6.14). Benjamini and Peres (1994a) proved that there exists a ray in T with a bounded trajectory iff $\rho \cdot \mathrm{br}\, T > 1$. (This is closely related to Example 5.20.) Call a T-indexed Markov chain **recurrent** if infinitely many vertices of T are mapped to the image of the root a.s. Given a tree T without leaves and a transition matrix $p(\cdot, \cdot)$, Benjamini and Peres (1994a) also showed that $\overline{\mathrm{gr}}\, T > 1$ iff there exists a nonamenable graph where the resulting T-indexed chain is recurrent.

13.10 Collected In-Text Exercises

13.1. Show that simple random walk has speed 0 on the tree T formed from a binary tree by joining a unary tree to every vertex, as in Figure 13.1.

13.2. Let Ω be finite or countable, and let μ and ν be probability measures on Ω.

    (a) Show that $\|\mu - \nu\|_{\mathrm{TV}} = \frac{1}{2}\sum_{z \in \Omega}|\mu(z) - \nu(z)|$ and $1 - \|\mu - \nu\|_{\mathrm{TV}} = \sum_{z \in \Omega}(\mu(z) \wedge \nu(z))$, where $x \wedge y := \min\{x, y\}$.

    (b) Deduce that
$$\|\mu - \nu\|_{\mathrm{TV}} = \min_{X \sim \mu,\, Y \sim \nu} \mathbf{P}[X \neq Y].$$

    (c) Prove that $\|\mu P - \nu P\|_{\mathrm{TV}} \leq \|\mu - \nu\|_{\mathrm{TV}}$ for every transition matrix P on Ω.

13.3. **(a)** Show that the two notions of distance to stationarity, $\delta(\cdot)$ and $\bar{\delta}(\cdot)$, satisfy
$$\delta(t) \leq \bar{\delta}(t) \leq 2\delta(t) \text{ for } t \geq 0.$$

    (b) Show that $\delta(t) = \max_{x \in V} \frac{1}{2}\sum_{y \in V}|p_t(x, y) - \pi(y)| \leq \max_{x,y \in V}\left|\dfrac{p_t(x, y) - \pi(y)}{2\,\pi(y)}\right|.$

13.4. For a reversible chain, define $\lambda_* := \max_{j > 1}|\lambda_j|$. Show that $\lambda_*^t \leq 2\delta(t)$ for all $t \geq 1$, and deduce that $t_{\mathrm{mix}}(\epsilon) \geq [g_*^{-1} - 1]\log 1/(2\epsilon)$.

13.5. Prove that in every transitive network, we have $D^2 \leq 4\widehat{D}^2$.

13.6. A collection of metric spaces has **uniform Markov type** 2 if there exists a constant $M < \infty$ such that each space in the collection has Markov type 2 with constant M. Prove that the set of k-dimensional hypercube graphs $\{0, 1\}^k$ does not have uniform Markov type 2. From this, deduce that ℓ^1 does not have Markov type 2.

13.7. Let P be the transition matrix for simple random walk on an infinite graph (V, E). Prove that $\langle P^i\psi, \psi \rangle \to 0$ as $i \to \infty$ for every $\psi \in \ell^2(V)$.

13.8. Show that $\mathrm{br}\, G$ does not depend on the choice of vertex o.

13.9. Show that if G' is a subgraph of G, then $\mathrm{br}\, G' \leq \mathrm{br}\, G$.

13.10. Show that $\lambda_c(G) \leq \mathrm{br}\, G$.

13.11. Show that if the 1–3 tree in Example 1.2 is labeled by ± 1-valued random variables $\langle X_v \rangle$ of mean zero, then $l_{\mathrm{cloud}} > 0$, but $l_{\mathrm{burst}} = l_{\mathrm{sust}} = 0$.

13.12. Verify that if the law of X_v is $N(0, 1)$, then $I(a) = a^2/2$, whereas if X_v are ± 1 valued with mean zero, then $I(a) = \frac{(1+a)}{2}\log(1 + a) + \frac{(1-a)}{2}\log(1 - a)$. (That this rate function is an upper bound for simple random walk is noted in (13.1); for biased walks, a similar inequality is (6.62).)

13.13. Given a tree T and $k \geq 1$, form the tree $T^{[k]}$ by taking the vertices x of T for which $|x|$ is a multiple of k and joining x and y by an edge in $T^{[k]}$ when their distance is k in T. Show that $\mathrm{br}\, T^{[k]} = (\mathrm{br}\, T)^k$.

13.14. Suppose that T is an infinite tree without leaves and that its vertices are labeled by i.i.d. variables $X_v \sim N(0, 1)$. Denote $d^* := \log \overline{\mathrm{gr}}(T)$. Prove that
$$\sqrt{d^*/2} \leq l_{\mathrm{cloud}} \leq \sqrt{2d^*}$$

and both bounds can be achieved.

13.15. Let $\langle n_k \rangle_{k \geq 1}$ be an increasing sequence of positive integers. Construct a tree T as follows: The first n_1 levels of T are as in the 1–3 tree. To each vertex v in level n_1 of T, attach a copy of the first $n_2 - n_1$ levels of the 1–3 tree, with v as its root. Continue iteratively by attaching a copy of the first $n_{k+1} - n_k$ levels of the 1–3 tree to each vertex at level n_k of T. Show that for every choice of $\langle n_k \rangle$ (and for every distribution of the variables X_v that satisfies (13.33)), we have $I_{\text{burst}} > 0$, but if the n_k increase sufficiently quickly, then $I_{\text{sust}} = 0$.

13.11 Additional Exercises

13.16. Does expected distance from the starting point always increase monotonically for a random walk? If the whole graph is a single edge, then the answer is no; perhaps we want to make an assumption on the walk, such as that it is *lazy*, that is, for every vertex x we have that $p(x, x) \geq 1/2$. Still, if we take a binary tree of large height and then identify the leaves to a single vertex which we join to the root by a new edge, then for (lazy) simple random walk starting at the root, the expected distance from the root will increase for a while but will eventually be small, approximately $\sum_{n \geq 1} n/2^n = 2$. This suggests that we might want to take expectation over a stationary starting distribution. However, this is still not enough, though it does suffice for the distance not to go down by a factor of more than 2.

(a) Show that the expected distance of simple random walk on the usual Cayley graph of $\mathbb{Z}^d$ is monotonic.

(b) Consider a cycle on N points to which we add new edges of conductance 2^{-N} between every pair of points whose distance in the cycle is at least $\sqrt{N}$. If c_N denotes the sum of conductances at a vertex in this network, then add a loop at each vertex of conductance c_N. For this new (transitive) network, where N is large and fixed, show that the expected distance of this lazy random walk is not monotone in the number of steps.

(c) Given a finite, connected network, let $X = \langle X_n \rangle$ be the associated network random walk with X_0 having the stationary probability distribution. Write $d(x, y)$ for the graph distance between two vertices x and y in the network. Define $D(n) := \mathbf{E}[d(X_0, X_n)]$. Show that for $j, k \geq 0$, we have $D(j + k) = \mathbf{E}[d(X_j, X_k') \mid X_0 = X_0']$ when X' is an independent copy of X.

(d) In the setting of part (c), show that if $0 \leq k \leq \min\{m, n\}$, then $D(n) \leq D(m) + D(m + n - 2k)$. In particular, if $n = 2k$, then $D(n) \leq 2D(m)$ and if $n = 2k + 1$, then $D(n) \leq D(m) + D(m + 1)$ and $D(n) \leq 2D(m) + 1$.

(e) In the setting of part (c), show that if X is lazy, then there is a stationary random walk Y on another network so that for all $n \geq 0$, we have $X_n = Y_{2n}$.

(f) Deduce that in the setting of part (c), if X is lazy, then $D(n) \leq 2D(m)$ whenever $m \geq n/2$.

13.17. Show that every submartingale $\langle Y_n \rangle$ with bounded increments satisfies $\liminf_{n \to \infty} Y_n/n \geq 0$.

13.18. Extend Theorem 13.1 beyond the uncorrelated case as follows.

(a) Show that if $a_n \geq 0$ satisfy $\sum_n a_n/n < \infty$, then there exists an increasing sequence of integers n_k such that $\sum_k a_{n_k} < \infty$ and $\lim_{k \to \infty} n_{k+1}/n_k = 1$.

(b) Show that if X_n are random variables with $|X_n| \leq 1$ a.s. and

$$\sum_n \frac{1}{n} \mathbf{E}\left[\left|\frac{1}{n}\sum_{k=1}^{n} X_k\right|^2\right] < \infty,$$

then $\sum_{k=1}^{n} X_k/n \to 0$ a.s.

13.19. Extend Theorem 13.1 beyond the uncorrelated case as follows.

(a) Show that if $a_n \geq 0$ satisfy $\sum_n a_n/n < \infty$, then there exists an increasing sequence of integers n_k such that $\sum_k a_{n_k}^2 < \infty$ and $\sum_k (n_{k+1}/n_k - 1)^2 < \infty$.

(b) Show that if X_n are random variables with $\mathbf{E}[X_n^2] \leq 1$ and

$$\sum_n \frac{1}{n} \mathbf{E}\left[\left|\frac{1}{n} \sum_{k=1}^{n} X_k\right|^2\right]^{1/2} < \infty,$$

then $\sum_{k=1}^{n} X_k/n \to 0$ a.s.

13.20. Extend Theorem 13.2 to show that under the same hypotheses,

$$\mathbf{P}(\mathscr{A}) \leq e^{-L^2/\left(2\sum_{i=1}^{n} \|X_i\|_\infty^2\right)}$$

whenever $\mathscr{A}$ is an event such that $\mathbf{E}\left[\sum_{i=1}^{n} X_i \mid \mathscr{A}\right] \geq L$.

13.21. Let $\langle X_1, \ldots, X_n \rangle$ be bounded random variables such that their partial sums $S_k = \sum_{i=1}^{k} X_i$ form a supermartingale.

(a) Show that the Hoeffding–Azuma inequality holds, that is, for all $L > 0$,

$$\mathbf{P}[S_n \geq L] \leq \exp\left(-\frac{2L^2}{\sum_{i=1}^{n}(\text{ess sup } X_i - \text{ess inf } X_i)^2}\right).$$

(b) Show that for all $L > 0$,

$$\mathbf{P}\left[\max_{1 \leq k \leq n} S_k \geq L\right] \leq \exp\left(-\frac{2L^2}{\sum_{i=1}^{n}(\text{ess sup } X_i - \text{ess inf } X_i)^2}\right).$$

13.22. Let G be a finite, directed graph such that at every vertex x, the number of edges going out from x equals the number of edges coming into x, which we denote by $d(x)$. Let $b(G)$ denote the maximum growth rate of the directed covers of G (the maximum is taken over possible starting vertices of paths). Show that $b(G) \geq \sum_{x \in V} d(x)/|V|$.

13.23. Suppose that G is a finite, directed, simple graph such that there is a (directed) path from each vertex to each other vertex. Show that there is a stationary Markov chain on the vertices of G such that the transition probability from x to y is positive only if $(x, y) \in \mathsf{E}(G)$ and such that the entropy of the Markov chain equals the log of the growth rate of the directed cover of G.

13.24. One version of continuous-time random walk on a network can be described in either of the following two equivalent ways: Let $X := \langle X_n ; n \in \mathbb{N} \rangle$ be the usual discrete-time network random walk. Let S_n be exponential random variables that are independent given X, with the parameter of S_n being $\pi(X_n)$. For $t \in [0, \infty)$, write $n(t) := \inf\{k ; \sum_{j=0}^{k} S_j \geq t\}$. Now define $Y_t := X_{n(t)}$. Alternatively, there are independent Poisson processes associated to all edges, where the process associated to e has rate $c(e)$. When the process Y is at x, it moves across the edge incident to x at the next arrival time for the Poisson processes associated to the edges incident to x.

Consider the 3-regular tree and a perfect matching F of the tree. Let $c(e) = 1$ for every edge $e \notin F$ and $c(e) = \alpha$ for all $e \in F$. Calculate $\lim_{t \to \infty} |Y_t|/t$.

13.25. Let T be a tree and N be an integer such that for every vertex u, there is some vertex $x \geq u$ with at least two children and with $|x| < |u| + N$. Show that if $\langle X_n \rangle$ is simple random walk on T, then

$$\liminf_{n \to \infty} \frac{|X_n|}{n} \geq \frac{1}{3N} \quad \text{a.s.}$$

13.26. One might expect that the speed of RW_λ would be monotonic decreasing in λ. But this is not the case, even in simple examples: Let T be a binary tree to every vertex of which is joined a unary tree as in Figure 13.1. Show that for $1 \leq \lambda \leq 2$, the speed is

$$\frac{(2 - \lambda)(\lambda - 1)}{\lambda^2 + 3\lambda - 3},$$

which is maximized at $\lambda = (5 + \sqrt{7})/6 = 1.27^+$.

13.27. Find a finite, connected graph whose universal cover has the property that the speed of RW_λ is not monotonic.

13.28. Consider the directed graph shown in Figure 13.2. Let T be one of its directed covers (shown in Figure 3.2). Why is T called the **Fibonacci tree**? Show that the speed of RW_λ on T is

$$\frac{(\sqrt{\lambda + 1} + 2)(\sqrt{\lambda + 1} - \lambda)}{\sqrt{\lambda + 1}\,(2 + \lambda + \sqrt{\lambda + 1}\,)}$$

for $0 \leq \lambda < (\sqrt{5} + 1)/2$.

Figure 13.2. A directed graph whose cover is the Fibonacci tree.

13.29. Identify the binary tree with the set of all finite sequences of 0s and 1s. Let the conductance of an edge be $x + 1$ when its vertex farthest from the root ends in x, where $x \in \{0, 1\}$. Calculate the speed of the corresponding random walk.

13.30. Prove (13.12) by reducing to the spectral theorem from linear algebra on finite-dimensional spaces as follows. Let $\mathscr{H}_n$ be finite-dimensional subspaces increasing to the entire Hilbert space. Let A_n be the orthogonal projection onto $\mathscr{H}_n$. Then A_n converges to the identity operator in the strong operator topology, that is, $A_n f$ converges to f in norm for every f. Since A_n has norm 1, it follows that $A_n S A_n$ converges to S in the strong operator topology. Furthermore, $A_n S A_n$ is really a self-adjoint operator of norm at most $\|S\|$ on $\mathscr{H}_n$. Since $Q(A_n S A_n)$ converges to $Q(S)$ in the strong operator topology and thus $\|Q(A_n S A_n)\| \to \|Q(S)\|$, the result (13.12) can be applied to $A_n S A_n$ and then deduced for S.

13.31. Suppose that (G, c) is a network with $\pi(\cdot)$ bounded above. Suppose that for all large r, the balls $B(o, r)$ have cardinality at most $A r^d$ for some finite constants A, d. Write $|x| := d(o, x)$. Show that the network random walk $\langle X_n \rangle$ obeys

$$\limsup_{n \to \infty} \frac{|X_n|}{\sqrt{n \log n}} \leq \sqrt{d + 2} \quad \text{a.s.}$$

13.32. Suppose that (G, c) is a network with $\pi(\cdot)$ bounded above. Suppose that for all large r, the balls $B(o, r)$ have cardinality at most $A r^d e^{B r^\alpha}$ for some finite constants A, B, d, α, with $0 < \alpha \leq 1$. Write $|x| := d(o, x)$. Show that the network random walk $\langle X_n \rangle$ obeys

$$\limsup_{n \to \infty} \frac{|X_n|}{n^{1/(2-\alpha)}} \leq (2B)^{1/(2-\alpha)} \quad \text{a.s.}$$

13.33. Let $\langle G_n \rangle$ be a (k, c)-expander family as defined before Theorem 6.17, that is, G_n is a regular graph of degree k with n vertices and expansion at least c. We use the notation $x_n \asymp y_n$ to mean that the ratio y_n/x_n is bounded above and below by positive constants. Show that $\text{diam}(G_n) \asymp \log n$ and the mixing time of the lazy simple random walk on G_n satisfies $t_{\text{mix}}(G_n) \asymp \log n$, where the implied constants depend only on k and c.

13.34. Show that for any finite network and any vertex a, there is a random time τ such that the network random walk $\langle X_t \rangle$ satisfies $\mathbf{P}_a[X_\tau = \bullet] = \pi(\bullet)$, the stationary distribution, and $\mathbf{E}_a[\tau] \le 4\, t_{\mathrm{mix}}(1/4)$.

13.35. Show that a metric space (X, d) has Markov type 2 iff there exists a constant $M < \infty$ such that for every reversible Markov chain $\langle Z_t \rangle_{t=0}^{\infty}$ with a stationary probability measure on a countable state space W, every mapping $f : W \to X$, and every time $t \in \mathbb{N}$,

$$\mathbf{E}\big[d\big(f(Z_t), f(Z_0)\big)^2\big] \le M t\, \mathbf{E}\big[d\big(f(Z_1), f(Z_0)\big)^2\big].$$

13.36. Combine the martingale decomposition in the proof of Lemma 13.15 with the Hoeffding–Azuma inequality to obtain a large deviation bound for functions of stationary, reversible Markov chains taking values in $\mathbb{R}^d$ with bounded increments.

13.37. Show that the lamplighter group $\mathbb{Z}^\odot$ defined in Section 3.4 has Markov type 2.

13.38. Consider a finite lamplighter group $\mathbb{Z}_n^3{}^\odot$ where the lamps are attached to vertices of the three-dimensional discrete torus, $\mathbb{Z}_n^3$. Show that $\mathbb{Z}_n^3{}^\odot$ has Hilbert distortion at least $cn^{3/2}$ for some absolute constant $c > 0$.

13.39. For simple random walk in $\mathbb{Z}^2$, show that the probability not to return to the starting point for k steps is of order $(\log k)^{-1}$. Deduce that the expected range of simple random walk on $\mathbb{Z}^2$ after t steps is of order $t/(\log t)$. *Hint:* Consider the effective conductance from $\mathbf{0}$ to the circles of radii k or $k^{1/4}$.

13.40. Suppose G is an amenable, transitive graph where the graph metric has Markov type 2 with constant M. Prove that simple random walk on G satisfies $\mathbf{E}[d(X_0, X_t)^2] \le M^2 t$ for all $t \in \mathbb{N}$, and show that the amenability assumption is needed. Deduce that for $d \ge 2$, the Cayley graph of $\mathbb{Z}^d{}^\odot$, the lamplighter group over the lattice $\mathbb{Z}^d$, does not have Markov type 2.

13.41. Show that there is a constant $c > 0$ such that for all graphs G with all degrees at least 3 and girth at least g, the Hilbert distortion of G is at least $c\sqrt{g}$. Here, **girth** is the length of the smallest simple cycle.

13.42. Let G be a finite, connected, transitive graph of diameter D. Show that there is a permutation σ of V such that for all $x \in \mathsf{V}$, we have $d(x, \sigma(x)) = D$.

13.43. Let G be a finite, connected graph with maximum degree $k \ge 3$. Show that there exists a permutation σ of V such that for all $x \in \mathsf{V}$, we have $d(x, \sigma(x)) > \log(|\mathsf{V}|/2)/(2\log(k-1))$. *Hint:* Choose σ at random, not quite uniformly.

13.44. Let (G, c) be a finite, connected, simple network. Define the **network Laplacian** Δ_G to be the $\mathsf{V} \times \mathsf{V}$ matrix whose (x, y) entry is $-c(x, y)$ if $x \ne y$ and is $\pi(x)$ if $x = y$. Another way to define Δ_G is that for a function f on V, we have $\Delta_G(f) = d^* c\, df$. In particular, $\big(\Delta_G(f), f\big) = \|df\|_c^2 \ge 0$ and Δ_G is self-adjoint on $\ell^2(\mathsf{V})$. Thus, the eigenvalues of Δ_G are nonnegative. Since each row sum of Δ_G is 0, the smallest eigenvalue is 0; denote the next smallest by $\lambda_1(G, c)$. Write $\bar{c} := 2\sum_{e \in \mathsf{E}_{1/2}} c(e)/|\mathsf{V}|$. Show that for every permutation σ of V and every map $f : \mathsf{V} \to \mathscr{H}$, where $\mathscr{H}$ is a Hilbert space, we have

$$\frac{\lambda_1(G, c)}{2\bar{c}} \cdot \frac{1}{|\mathsf{V}|} \sum_{x \in \mathsf{V}} \big|f(x) - f(\sigma(x))\big|^2 \le \frac{1}{\sum_{e \in \mathsf{E}_{1/2}} c(e)} \sum_{e \in \mathsf{E}_{1/2}} c(e) \,\|f(e^+) - f(e^-)\|^2. \tag{13.38}$$

13.45. Let (G, c) be a finite, connected, simple network. Define $\lambda_1(G, c)$ and $\bar{c}$ as in Exercise 13.44. Show that for every map $f : \mathsf{V} \to \mathscr{H}$, where $\mathscr{H}$ is a Hilbert space, we have

$$\frac{\lambda_1(G, c)}{2\bar{c}} \cdot \frac{1}{|\mathsf{V}|^2} \sum_{x, y \in \mathsf{V}} \big|f(x) - f(y)\big|^2 \le \frac{1}{\sum_{e \in \mathsf{E}_{1/2}} c(e)} \sum_{e \in \mathsf{E}_{1/2}} c(e) \,\|f(e^+) - f(e^-)\|^2.$$

Show that if $\widehat{D}^2$ is the average squared distance between pairs of vertices in G, then the distortion is at least $\widehat{D}\sqrt{\lambda_1(G, c)/(2\bar{c})}$.

13.46. Let (G, c) be a finite, connected, simple network. Define $\lambda_1(G, c)$ and $\bar{c}$ as in Exercise 13.44. Consider a permutation σ of V; define the **displacement** of σ to be

$$\widehat{\mathsf{d}}(\sigma) := \sqrt{\frac{1}{|\mathsf{V}|} \sum_{x \in \mathsf{V}} d(x, \sigma(x))^2}.$$

Then define the **maximal displacement** of G to be

$$\widehat{\mathsf{d}}(G) := \max_{\sigma} \widehat{\mathsf{d}}(\sigma).$$

Show that the Hilbert distortion of G is at least $\widehat{\mathsf{d}}(G)\sqrt{\lambda_1(G, c)/(2\bar{c})}$.

13.47. Let (G, c) be a finite, connected, simple network. Define $\lambda_1(G, c)$ as in Exercise 13.44. Give each edge e the length $\ell(e)$ and use the induced shortest-path metric d_ℓ on V. Define the associated displacement of a permutation σ by

$$\widehat{\mathsf{d}}(\sigma, \ell) := \sqrt{\frac{1}{|\mathsf{V}|} \sum_{x \in \mathsf{V}} d_\ell(x, \sigma(x))^2}$$

and the maximal displacement $\widehat{\mathsf{d}}(G, \ell) := \max_{\sigma} \widehat{\mathsf{d}}(\sigma, \ell)$ accordingly. Show that the Hilbert distortion of (V, d_ℓ) is at least $\widehat{\mathsf{d}}(G, \ell)\sqrt{\lambda_1(G, c)/(2\widetilde{c})}$, where $\widetilde{c} := 2 \sum_{e \in \mathsf{E}_{1/2}} c(e)\ell(e)^2/|\mathsf{V}|$.

13.48. Let G be the k-dimensional hypercube graph, but give all edges in direction i the length ℓ_i ($1 \le i \le k$). Show that the Hilbert distortion of G (in the sense of Exercise 13.47) has the value $(\ell_1 + \cdots + \ell_k)/\sqrt{\ell_1^2 + \cdots + \ell_k^2}$.

13.49. Show that the Hilbert distortion of the cycle on $2n$ vertices equals $n \sin(\pi/(2n))$ for every integer $n \ge 1$.

13.50. Suppose that the Hilbert distortion of G equals $C_G := \widehat{\mathsf{d}}(G)\sqrt{\lambda_1(G)/(2\bar{c}(G))}$, with notation as in Exercise 13.46. Show that the Hilbert distortion of the n-fold Cartesian product $G^{\square n}$ of G with itself equals $C_G \sqrt{n}$.

13.51. Let G be the Cartesian product of a 3-cycle and an edge. Show that the Hilbert distortion of the n-fold Cartesian product $G^{\square n}$ of G with itself equals $\sqrt{n}$.

13.52. Suppose that G_i for $i = 1, 2$ are transitive networks on two Cayley graphs of the same diameter δ that have Hilbert distortion $C_i := \delta\sqrt{(1 - \lambda_2(P_i))/2}$. Show that the Hilbert distortion of $G_1 \square G_2$ equals $2\sqrt{C_1 C_2/(C_1 + C_2)}$.

13.53. Let G be the Cartesian product of a 6-cycle and the three-dimensional hypercube. Calculate the Hilbert distortion of the n-fold Cartesian product $G^{\square n}$ of G with itself.

13.54. Show that the hypothesis of bounded degree in Proposition 13.30 is necessary.

13.55. Give an example of a graph G such that $\lambda_c(G) < \mathrm{br}(G)$.

14 | Random Walks on Groups and Poisson Boundaries

In this chapter we examine group-invariant random walks and, more generally, transitive Markov chains. What can we say about their asymptotic behavior? It turns out that a key particular asymptotic, the speed of the random walk, determines whether there is any nontrivial asymptotic behavior. Asymptotic behavior is also closely related to the existence of nonconstant bounded harmonic functions and to entropy.

Consider, first, simple random walk on a Cayley graph. For instance, in $\mathbb{Z}^d$, simple random walk has speed zero; also, every bounded harmonic function is constant, as shown in Section 9.6. On the other hand, on free groups other than $\mathbb{Z}$, simple random walk has positive speed; also, it is easy to construct nonconstant bounded harmonic functions: the value at x can be the probability that the random walk is in some given part of the Cayley graph for all sufficiently large times. Although there are no nonconstant Dirichlet harmonic functions on the product of $\mathbb{Z}$ with a free group, there *are* nonconstant bounded harmonic functions there. Very interesting examples intermediate between abelian and free groups and that mix features of both are provided by the lamplighter groups $\mathbb{Z}^{d\odot}$. These groups have the same definition as $\mathbb{Z}^\odot$ from Section 3.4, but the base group is $\mathbb{Z}^d$ for $d \geq 1$. All of them are amenable groups of exponential growth. But, surprisingly, the speed of simple random walk on $\mathbb{Z}^{d\odot}$ is positive iff $d \geq 3$; see Corollary 14.21.

Returning to simple random walks on general Cayley graphs, let $H(X_n)$ be the entropy of the distribution of X_n. The sequence $\langle H(X_n)\,;\ n \geq 0 \rangle$ is subadditive (see Section 14.1), whence the limit

$$\boldsymbol{h} := \lim_{n\to\infty} \frac{H(X_n)}{n}$$

exists; it is the **Avez entropy** of the random walk. We will prove the following fundamental theorem, which is a consequence of work of Avez, Derriennic, Kaimanovich and Vershik, and Varopoulos.

Theorem 14.1. (Speed, Entropy, Tail, and Liouville) *For a simple random walk $\langle X_n \rangle$ on a Cayley graph, the following four properties are equivalent:*

(i) *The random walk has positive speed, that is, there exists some $\boldsymbol{l} > 0$ such that*

$$\frac{d(X_0, X_n)}{n} \to \boldsymbol{l} \quad \text{almost surely.}$$

(ii) *The Avez entropy $\boldsymbol{h} > 0$.*
(iii) *The random walk has a nontrivial tail σ-field.*
(iv) *The Cayley graph admits nonconstant bounded harmonic functions.*

Recall that a σ-field $\mathscr{F}$ is **P**-trivial iff every $A \in \mathscr{F}$ satisfies $\mathbf{P}(A) \in \{0, 1\}$. Since we may interpret "asymptotic behavior" as the tail σ-field, this theorem does indeed show that the speed determines whether there is any nontrivial asymptotic behavior. Here, "nontrivial" means behavior that has probability strictly between 0 and 1; it doesn't mean that the σ-field is uninteresting. For example, the law of the iterated logarithm is highly nontrivial and describes asymptotic behavior of random walks on $\mathbb{Z}$, but it describes almost sure behavior.

The preceding theorem is valid in a more general setting; see Theorem 14.20. In particular, we will show that the equivalence of (ii), (iii), and (iv) does not require symmetry of $p(\cdot, \cdot)$. In contrast, biased nearest-neighbor random walk on $\mathbb{Z}$ has positive speed but zero Avez entropy.

Although the above theorem provides powerful dichotomies, it does not describe the tail σ-field when it is nontrivial, nor does it describe the set of bounded harmonic functions. In fact, one can easily describe the tail σ-field in terms of the set of bounded harmonic functions and vice versa: see Proposition 14.12 and Theorem 14.18. Furstenberg gave a framework for concrete descriptions of the bounded harmonic functions in terms of what he called Poisson boundaries, and also gave a criterion for identifying them. It turns out that an extension of Theorem 14.1 due to Kaimanovich provides a more powerful tool for the purpose of identification; see Section 14.4.

14.1 Tail, Entropy, and Speed for Transitive Markov Chains

A Markov chain is determined by a countable state space V, a transition matrix $P = \left(p(x, y)\right)_{x, y \in \mathsf{V}}$, and an initial distribution. Endow the sequence space $\mathsf{V}^{\mathbb{N}}$ with the usual Borel σ-field $\mathscr{B}$ (the minimal σ-field that contains all the elementary cylinders $\langle y_0, \dots, y_k \rangle \times \mathsf{V}^{\{k+1, k+2, \dots\}}$ for all k and $y_0, \dots, y_k \in \mathsf{V}$). The left-shift S acts on $\mathsf{V}^{\mathbb{N}}$ via $S(x_0, x_1, \dots) := (x_1, x_2, \dots)$. For each $x \in \mathsf{V}$, let $\mathbf{P}_x$ denote the law of the Markov chain $X_0, X_1, \dots$ with transition matrix P and $X_0 = x$. For any probability measure θ on V, write $\mathbf{P}_\theta := \sum_{x \in \mathsf{V}} \theta(x) \mathbf{P}_x$ for the law of the chain with initial distribution θ. The **tail σ-field** on $\mathsf{V}^{\mathbb{N}}$ is defined by $\mathcal{T} := \bigcap_{n=0}^{\infty} S^{-n} \mathscr{B}$. A function $f \colon \mathsf{V}^{\mathbb{N}} \to \mathbb{R}$ is called a **tail function** if it is $\mathcal{T}$-measurable.

It is sometimes convenient to think of tail events and functions via the corresponding equivalence relation, although we will not use this interpretation explicitly.

▷ **Exercise 14.1.**
For $\mathbf{y}, \mathbf{z} \in \mathsf{V}^{\mathbb{N}}$, write $\mathbf{y} \overset{\mathcal{T}}{\sim} \mathbf{z}$ iff $\exists n \; \forall m \geq n \; y_m = z_m$. Let $A \subseteq \mathsf{V}^{\mathbb{N}}$ be a Borel set. Show that $A \in \mathcal{T}$ iff A is a union of tail equivalence classes (that is, $\mathbf{y} \in A$, $\mathbf{y} \overset{\mathcal{T}}{\sim} \mathbf{z} \Rightarrow \mathbf{z} \in A$). Similarly, show that a Borel function $f \colon \mathsf{V}^{\mathbb{N}} \to \mathbb{R}$ is a tail function iff $\mathbf{y} \overset{\mathcal{T}}{\sim} \mathbf{z}$ implies $f(\mathbf{y}) = f(\mathbf{z})$.

Example 14.2. Consider the 3-regular infinite tree $\mathbb{T}_3$ with a distinguished vertex, o. Removing o yields three connected components, each a binary tree, denoted by T_1, T_2, T_3. Then the three events $E_i := \{\mathbf{y} \in \mathsf{V}^{\mathbb{N}} \, ; \; \exists N \; \forall n \geq N \; y_n \in T_i\}$ are in $\mathcal{T}$, and for simple random walk on $\mathbb{T}_3$ started at any vertex x, we have $0 < \mathbf{P}_x(E_i) < 1$.

▷ **Exercise 14.2.**
Let (T, c) be a transient network on a tree with bounded conductances. Show that the network walk has a nontrivial tail.

Let Γ be a countable group, and let μ be a probability measure on Γ. A random walk on Γ with step distribution μ, also called a μ-*walk*, is a Markov chain $\langle X_n \rangle$ with transition probabilities $p(x, y) := \mu(x^{-1}y)$. We often assume that the support of μ generates Γ as a semigroup; this ensures that the μ-walk is irreducible on Γ. If the walk is started at the identity o, then X_n can be represented as a random product $g_1 g_2 \cdots g_n$, where the increments g_i are i.i.d. with law μ. A special case is simple random walk on the right Cayley graph of a finitely generated group. When $X \sim \mu$ and $Y \sim \nu$ are independent, the law of XY is usually denoted by $\mu * \nu$, which is called the **convolution** of μ and ν. For this reason, we refer to μ-walks as **convolution random walks**.

Example 14.3. For a convolution random walk on an abelian group started at x, every tail event is invariant under finite permutations of the increments. Since the exchangeable σ-field of the i.i.d. increments is trivial by the Hewitt–Savage zero-one law, it follows that the tail σ-field of a random walk on an abelian group is $\mathbf{P}_x$-trivial. Note, however, that the tail σ-field need not be $\mathbf{P}_\theta$-trivial for other initial distributions θ: for simple random walk on $\mathbb{Z}$, the tail event $A := \bigcup_{k \geq 1} \bigcap_{n \geq k} \{\mathbf{y} ; y_n - n \in 2\mathbb{Z}\}$ has $0 < \mathbf{P}_\theta(A) < 1$ if $\theta(0)\,\theta(1) > 0$.

Definition 14.4. Let P be a transition matrix on V, and let γ be a permutation of V. We say that γ is an **automorphism** of (V, P), and write $\gamma \in \text{Aut}(V, P)$, if $p(\gamma x, \gamma y) = p(x, y)$ for all x, y in V. A Markov chain with transition matrix P is called **transitive** (or **spatially homogeneous**) if $\text{Aut}(V, P)$ acts transitively on V, that is, for all z, w in V there exists $\gamma \in \text{Aut}(V, P)$ such that $\gamma(z) = w$. Simple random walks on transitive graphs are, of course, transitive chains; so are convolution random walks on groups.

We will show that for transitive chains, tail triviality may be characterized via entropy. Indeed, entropy will be a key tool throughout this chapter. Recall that for a discrete random variable X, the **entropy** of X is defined by $H(X) = -\sum_x \mathbf{P}[X = x] \log \mathbf{P}[X = x]$. More generally, given a σ-field $\mathscr{F}$, the **conditional entropy** of X is

$$H(X \mid \mathscr{F}) := -\mathbf{E}\left[\sum_x \mathbf{P}[X = x \mid \mathscr{F}] \cdot \log \mathbf{P}[X = x \mid \mathscr{F}] \right]. \tag{14.1}$$

▷ **Exercise 14.3.**
Show that if a discrete random variable X satisfies $H(X \mid \mathscr{F}) = H(X)$ a.s., then X is independent of $\mathscr{F}$.

▷ **Exercise 14.4.**
Prove that we always have $H(X, Y \mid \mathscr{F}) \geq H(X \mid \mathscr{F})$.

The conditional entropy is monotone decreasing in $\mathscr{F}$, which extends (6.42):

Lemma 14.5. (Monotonicity of Conditional Entropy) *Let X be a discrete random variable defined on a probability space $(\Omega, \mathscr{F}, \mathbf{P})$.*
 (i) *If $\mathscr{G} \subset \mathscr{F}$ is a smaller σ-field, then $H(X \mid \mathscr{G}) \geq H(X \mid \mathscr{F})$. Equality holds if and only if $\mathbf{P}[X = \cdot \mid \mathscr{F}] = \mathbf{P}[X = \cdot \mid \mathscr{G}]$ a.s.*
 (ii) *Suppose that $\langle \mathscr{G}_n \rangle$ is a decreasing sequence of sub-σ-fields of $\mathscr{F}$, with $\bigcap_n \mathscr{G}_n = \mathscr{G}$. Then $H(X \mid \mathscr{G}_n) \uparrow H(X \mid \mathscr{G})$.*

Proof. (i) The tower property of conditional expectations (conditioning on $\mathscr{G}$ in the right-hand side below) yields that

$$H(X \mid \mathscr{G}) = -\mathbf{E}\Big[\sum_x \mathbf{P}[X = x \mid \mathscr{F}] \cdot \log \mathbf{P}[X = x \mid \mathscr{G}]\Big].$$

Therefore

$$H(X \mid \mathscr{F}) - H(X \mid \mathscr{G}) = \mathbf{E}\Big[\sum_x \mathbf{P}[X = x \mid \mathscr{F}] \cdot \log \frac{\mathbf{P}[X = x \mid \mathscr{G}]}{\mathbf{P}[X = x \mid \mathscr{F}]}\Big]. \tag{14.2}$$

Observe that $\sum_x \mathbf{P}[X = x \mid \mathscr{F}] = \mathbf{E}[\mathbf{1} \mid \mathscr{F}] = \mathbf{1}$ a.s. Invoking the inequality $\log r \le r - 1$ implies that the right-hand side of (14.2) is nonpositive. For the right-hand side to vanish, the ratio $\mathbf{P}[X = x \mid \mathscr{G}]/\mathbf{P}[X = x \mid \mathscr{F}]$ must equal 1 a.s. for all x with $\mathbf{P}[X = x \mid \mathscr{F}] > 0$, since the inequality $\log r \le r - 1$ is strict for positive $r \ne 1$. Equivalently, we can rewrite (14.2) in the form

$$H(X \mid \mathscr{G}) - H(X \mid \mathscr{F}) = \mathbf{E}\Big[D_{\mathrm{KL}}\big(\mathbf{P}[X = \cdot \mid \mathscr{F}] \,\big\|\, \mathbf{P}[X = \cdot \mid \mathscr{G}]\big)\Big],$$

and deduce the assertion of (i) from nonnegativity of the relative entropy, (6.39).

(ii) The Lévy zero-one law implies that $\lim_n \mathbf{P}[X = x \mid \mathscr{G}_n] = \mathbf{P}[X = x \mid \mathscr{G}]$ a.s. for every x, so by Fatou's lemma, $H(X \mid \mathscr{G}) \le \liminf_n H(X \mid \mathscr{G}_n) = \lim_n H(X \mid \mathscr{G}_n)$. The reverse inequality follows from (i). ◀

Consider a transitive chain $\langle X_n \rangle$ on V started at $X_0 = o$. Transitivity gives us the identity $H(X_{m+n} \mid X_m) = H(X_n)$, so $H(X_{m+n}) \le H(X_m) + H(X_n)$ by (6.41) and (6.43). In particular, if $H(X_1) < \infty$, then $H(X_n) < \infty$ for all n. Also, the special case $H(X_{n+1} \mid X_1) = H(X_n)$ shows that the sequence $\langle H(X_n) \rangle$ is (weakly) increasing. By Fekete's lemma (Exercise 3.9), the *Avez (asymptotic) entropy*

$$\boldsymbol{h} := \lim_{n \to \infty} \frac{H(X_n)}{n} \tag{14.3}$$

exists.

Avez (1976) proved a lower bound for entropy in terms of the spectral radius, ρ.

Proposition 14.6. *For any symmetric Markov chain $(p(x, y) = p(y, x)$ for all $x, y)$ and any initial state o, we have $\liminf_{n \to \infty} H(X_n)/n \ge 2\log(1/\rho)$. In particular, transitive symmetric chains with $\rho < 1$ have $\boldsymbol{h} > 0$.*

Proof. By Jensen's inequality and symmetry,

$$-H(X_n) = \sum_x p_n(o, x) \log p_n(o, x) \le \log \sum_x p_n^2(o, x) = \log \sum_x p_n(o, x) p_n(x, o)$$
$$= \log p_{2n}(o, o).$$

Dividing by $-n$ and taking $\liminf$ yield the claim. ◀

The next theorem already indicates the usefulness of the Avez entropy.

Theorem 14.7. (Entropy and Tail Triviality) *For a transitive Markov chain with initial state $X_0 = o$ and $H(X_1) < \infty$, the tail field $\mathcal{T}$ is $\mathbf{P}_o$-trivial if and only if $\boldsymbol{h} = 0$.*

In particular, this theorem readily implies that for any transitive graph with subexponential growth, simple random walk has a trivial tail field. Since abelian groups have polynomial growth, we obtain another proof of Example 14.3. Note that transitivity is crucial here, since simple random walk on a transient tree has a nontrivial tail by Exercise 14.2, and there exist transient trees of polynomial growth.

Proof of Theorem 14.7. For $k < n$, decompose $H(X_k, X_n)$ in two ways:

$$H(X_k \mid X_n) + H(X_n) = H(X_k, X_n) = H(X_k) + H(X_n \mid X_k) = H(X_k) + H(X_{n-k}). \quad (14.4)$$

By the Markov property, we have that

$$H(X_k \mid \langle X_j \rangle_{j \geq n}) = H(X_k \mid X_n).$$

Thus (14.4) implies

$$H(X_k \mid \langle X_j \rangle_{j \geq n}) = H(X_k) + H(X_{n-k}) - H(X_n). \quad (14.5)$$

The left-hand side is increasing in n and converges to $H(X_k \mid \mathcal{T})$ by Lemma 14.5. Considering $k = 1$, we see that $H(X_n) - H(X_{n-1})$ converges as $n \to \infty$; since the averages $H(X_n)/n$ tend to $\boldsymbol{h}$, we infer that

$$H(X_n) - H(X_{n-1}) \to \boldsymbol{h}. \quad (14.6)$$

By taking $n \to \infty$ in (14.5), we conclude that

$$H(X_k \mid \mathcal{T}) = H(X_k) - k\boldsymbol{h}. \quad (14.7)$$

Thus $\boldsymbol{h} = 0$ if $\mathcal{T}$ is $\mathbf{P}_o$-trivial.

To see the converse, note that by the Markov property,

$$H(X_1, \ldots, X_{k-1} \mid X_k, \mathcal{T}) = H(X_1, \ldots, X_{k-1} \mid X_k);$$

adding this to (14.7) yields

$$H(X_1, \ldots, X_k \mid \mathcal{T}) = H(X_1, \ldots, X_k) - k\boldsymbol{h}. \quad (14.8)$$

Therefore, if $\boldsymbol{h} = 0$, then the sequence $X_1, \ldots, X_k$ is independent of $\mathcal{T}$ for every k by Exercise 14.3, whence $\mathcal{T}$ must be $\mathbf{P}_o$-trivial. ◀

To study the speed of random walk, a suitable metric is needed.

Definition 14.8. Suppose that P is a transition matrix on V and $G = (\mathsf{V}, \mathsf{E})$ is a graph, with corresponding graph metric d. If for all z, w in V there exists $\gamma \in \mathrm{Aut}(\mathsf{V}, P) \cap \mathrm{Aut}(G)$ such that $\gamma(z) = w$, then we call the chain with transition matrix P a ***transitive Markov chain with an invariant graph metric***.

The simplest example is simple random walk on a transitive graph. Another example is a convolution walk on a countable group Γ, where the metric d arises from a right Cayley graph G on Γ. In this case, transitions need not be restricted to edges of G, yet the action of the group on itself by left multiplication defines automorphisms of (V, P) and of G.

Next, consider a transitive chain on V with an invariant graph metric d determined by the transitive graph $G = (\mathsf{V}, \mathsf{E})$. Fix a basepoint $o \in \mathsf{V}$. By transitivity, the balls in G satisfy $|B(o, r + s)| \le |B(o, r)| \cdot |B(o, s)|$. Therefore, Fekete's lemma (Exercise 3.9) implies that the ***volume growth exponent*** $\boldsymbol{v} := \lim_{r \to \infty} \frac{1}{r} \log |B(o, r)|$ exists. Similarly, if $\langle X_n \rangle$ is a transitive Markov chain on V started at o, then transitivity implies that $\mathbf{E}|X_{n+m}| \le \mathbf{E}|X_n| + \mathbf{E}|X_m|$, where we write $|x| := d(o, x)$ for $x \in \mathsf{V}$. Thus, by Fekete's lemma, if $\mathbf{E}|X_1| < \infty$, then the asymptotic ***speed*** $\boldsymbol{l} := \lim_n \mathbf{E}|X_n|/n$ exists. The next lemma relates these quantities to the entropy $\boldsymbol{h}$ and the spectral radius ρ of the chain.

Theorem 14.9. (Entropy and Speed) *Let $\langle X_n \rangle$ be a transitive Markov chain on V with an invariant graph metric d determined by a transitive graph $G = (\mathsf{V}, \mathsf{E})$. Fix an initial state o.*
  (i) *If $\mathbf{E}|X_1| < \infty$, then the entropy, speed, and volume growth are related by $\boldsymbol{h} \le \boldsymbol{l}\boldsymbol{v}$.*
  (ii) *Suppose that $\langle X_n \rangle$ is a network random walk for some transitive network on G. (In particular, $p(x, y) = p(y, x)$ for all x and y by transitivity, and $\boldsymbol{l} < 1$.) Then we have $\boldsymbol{h} \ge \boldsymbol{l}^2/2 + \log(1/\rho)$.*

Part (i) is intuitive, since X_n is likely to be in $B(o, n(\boldsymbol{l}+\epsilon))$, so $H(X_n)$ should be bounded by the logarithm of the volume of this ball. This intuition can be rigorized using Theorem 14.10; see Exercise 14.26. Here we give a direct, elementary (though less intuitive) proof.

Proof. Abbreviate $p_n(x) := p_n(o, x)$.

(i) We may assume that $\boldsymbol{v} < \infty$, since the inequality is vacuous otherwise. Define the spheres $S_k := \{x \in \mathsf{V} ; |x| = k\}$. Given $\epsilon > 0$, consider the probability measure ν_ϵ on V defined by $\nu_\epsilon(x) := e^{-z_\epsilon - k\epsilon} |S_k|^{-1}$ for $x \in S_k$ (where $e^{-z_\epsilon} = 1 - e^{-\epsilon}$ is a normalizing constant). By the nonnegativity of Kullback–Leibler divergence, (6.39),

$$0 \le D_{\mathrm{KL}}(p_n \| \nu_\epsilon) = \sum_{k=0}^{\infty} \sum_{x \in S_k} p_n(x) \log\big(p_n(x) e^{z_\epsilon + k\epsilon} |S_k|\big).$$

By the definition of $\boldsymbol{v}$, there exists $C_\epsilon < \infty$ such that $\log |S_k| \le k(\boldsymbol{v} + \epsilon) + C_\epsilon$ for all $k \ge 0$. Therefore,

$$H(X_n) \le \sum_{k=0}^{\infty} \mathbf{P}_o[X_n \in S_k](z_\epsilon + k\epsilon + \log |S_k|) \le \sum_{k=0}^{\infty} \mathbf{P}_o[X_n \in S_k](z_\epsilon + C_\epsilon + k(\boldsymbol{v} + 2\epsilon))$$
$$= \mathbf{E}\big[|X_n|\big](\boldsymbol{v} + 2\epsilon) + z_\epsilon + C_\epsilon .$$

Dividing by n and taking limits yields $\boldsymbol{h} \le \boldsymbol{l}(\boldsymbol{v} + 2\epsilon)$. This concludes the proof.

(ii) Since π is constant, the Varopoulos–Carne bound, Theorem 13.4, states that for every $x \in V$,

$$p_n(x) \leq 2\rho^n \exp\left(-\frac{|x|^2}{2n}\right), \quad \text{or equivalently,} \quad \log \frac{1}{p_n(x)} \geq \frac{|x|^2}{2n} - n \log \rho - \log 2.$$

Therefore,

$$H(X_n) \geq \sum_{x \in V} p_n(x) \frac{|x|^2}{2n} - n \log \rho - \log 2 = \frac{\mathbf{E}\big[|X_n|^2\big]}{2n} - n \log \rho - \log 2.$$

Divide by n and use the arithmetic mean–quadratic mean inequality to get

$$\frac{H(X_n)}{n} \geq \frac{\big[\mathbf{E}|X_n|\big]^2}{2n^2} - \log \rho - \frac{\log 2}{n}.$$

Passing to the limit yields the claimed inequality.   ◀

▷ **Exercise 14.5.**

   (a) Show that simple random walk on a $(b+1)$-regular tree $\mathbb{T}_{b+1}$ satisfies $\boldsymbol{h} = \boldsymbol{lv}$.

   (b) Find a Cayley graph where simple random walk satisfies $\boldsymbol{h} < \boldsymbol{lv}$. *Hint:* Consider a Cartesian product of trees.

The following theorem gives pathwise interpretations of the asymptotic speed $\boldsymbol{l}$ and the Avez entropy $\boldsymbol{h}$, which were defined via expectations.

Theorem 14.10. (Pathwise Limits) *Let $\langle X_n \rangle$ be a transitive Markov chain starting at o.*

   (i) *If the chain is endowed with an invariant graph metric d and $\mathbf{E}\big[d(o, X_1)\big] < \infty$, then almost surely*

$$\lim_n \frac{1}{n} d(o, X_n) = \boldsymbol{l}.$$

   (ii) *If $H(X_1) < \infty$, then almost surely*

$$\lim_n \frac{1}{n} \log p_n(o, X_n) = -\boldsymbol{h}.$$

Proof. (i) Since $d(o, X_{n+m}) \leq d(o, X_n) + d(X_n, X_{n+m})$, almost sure existence of the limit $\lim_n \frac{1}{n} d(o, X_n)$ follows from the subadditive ergodic theorem (Theorem 14.44). The expectation of this limit must be $\boldsymbol{l}$ by the same theorem, so it remains only to check that the limit is a.s. constant. If $\langle X_n \rangle$ is a random walk on a group, then (taking o to be the identity) $X_n = \prod_{i=1}^{n} Z_i$ with Z_i i.i.d., whence the assertion follows from the Kolmogorov zero-one law. For a general transitive graph, invoke the upcoming Lemma 14.11.

   (ii) Since $p_{n+m}(o, X_{n+m}) \geq p_n(o, X_n) \cdot p_m(X_n, X_{n+m})$, almost sure existence of the limit $\Lambda(X_0, X_1, \ldots) := \lim_n \frac{1}{n} \log p_n(o, X_n)$ follows from the subadditive ergodic theorem. To see that Λ is almost surely constant, observe that $p_n(X_0, X_n) \geq p(X_0, X_1)p_{n-1}(X_1, X_n)$; taking logarithms, we see that $\Lambda(X_0, X_1, \ldots) \geq \Lambda(X_1, X_2 \ldots)$, so all the hypotheses of Lemma 14.11 are satisfied.   ◀

The next lemma was invoked in the preceding proof to establish that certain parameters for transitive chains are a.s. constant.

Lemma 14.11. *Let $\langle X_n \rangle_{n \geq 0}$ be a Markov chain on V and let $\Lambda \colon V^{\mathbb{N}} \to \mathbb{R}$ be a Borel function. If the law of $\Lambda(X_0, X_1, \ldots)$ does not depend on the initial state X_0 and $\Lambda(X_1, X_2, \ldots) \leq \Lambda(X_0, X_1, \ldots)$ a.s., then $\Lambda(X_0, X_1, \ldots)$ is a.s. constant.*

Proof. We will show that for all $t \in \mathbb{R}$, we have $\mathbf{P}[\Lambda(X_0, X_1, \ldots) \leq t] \in \{0, 1\}$. Indeed, the two sides of $\Lambda(X_1, X_2, \ldots) \leq \Lambda(X_0, X_1, \ldots)$, have the same law, so they have the same probability to be at most t. As this holds for all $t \in \mathbb{R}$, it follows that $\Lambda(X_1, X_2, \ldots) = \Lambda(X_0, X_1, \ldots)$ a.s. Then, by using this shift-invariance of Λ, the independence of Λ from X_0, and the Markov property, then again the shift-invariance of Λ, we get that

$$\mathbf{P}[\Lambda(X_0, X_1, \ldots) \leq t] = \mathbf{P}[\Lambda(X_k, X_{k+1}, \ldots) \leq t \mid \mathscr{F}_k] = \mathbf{P}[\Lambda(X_0, X_1, \ldots) \leq t \mid \mathscr{F}_k],$$

which tends to 0 or 1 as $k \to \infty$ by the Lévy zero-one law.   ◀

A result similar to Theorem 14.10 about the behavior of Markov chains, which we will use later, is the following.

▷ **Exercise 14.6.**
Given a transitive Markov chain $\langle X_j \rangle$ on a state space V, denote the number of distinct states among its first n by $R_n := |\{X_0, X_1, \ldots, X_{n-1}\}|$. Fix $o \in V$, and show that $\lim_{n \to \infty} \mathbf{E}_o[R_n]/n = \mathbf{P}_o[\forall j \geq 1 \ X_j \neq o]$ as $n \to \infty$. Deduce that R_n/n converges $\mathbf{P}_o$-a.s. to the same limit.

14.2 Harmonic Functions and the Liouville Property

Recall that a function $u \colon V \to \mathbb{R}$ is ***harmonic*** for a transition matrix $P = p(\cdot, \cdot)$ on V if $Pu = u$, where $(Pu)(x) := \sum_{y \in V} p(x, y)u(y)$. We say that a Markov chain has the ***Liouville property*** if all bounded harmonic functions for that chain are constant. The main goal of this section is to characterize which transitive Markov chains have this property. (Some of our theorems will not require transitivity, so we will state explicitly whenever transitivity is assumed.)

To this end, we define another important σ-field on the sequence space $V^{\mathbb{N}}$, the ***invariant σ-field*** $\mathcal{I} := \{A \in \mathscr{B}; \ S^{-1}(A) = A\}$. (This is also sometimes called the ***stationary σ-field***.) As we will see, $\mathcal{I}$ is closely related to harmonic functions on V; in particular, $x \mapsto \mathbf{P}_x(A)$ is harmonic for every $A \in \mathcal{I}$.

One might be inclined to interpret "asymptotic behavior" as $\mathcal{I}$ rather than $\mathcal{T}$. How are these two σ-fields related? Clearly $\mathcal{I} \subseteq \mathcal{T}$. For simple random walk on $\mathbb{Z}$, the event $A_* := \{\mathbf{y}; \ \forall n \ y_n \equiv n \bmod 2\}$ is a tail event but is not invariant. On the other hand, this event A_* has probability 0 or 1 for every $\mathbf{P}_x$. This is not an accident: for transitive chains, the $\mathbf{P}_x$-completions of $\mathcal{I}$ and $\mathcal{T}$ coincide for every x; see Section 14.6. Here, we will prove this also holds for ***lazy*** chains, that is, ones for which $p(x, x) \geq 1/2$ for all $x \in V$.

We generally assume that our Markov chains are irreducible but will include that assumption when used.

Our first proposition exhibits the promised connection between the invariant σ-field and the space $\mathbf{BH}(\mathsf{V}, P)$ of bounded harmonic functions on V. A function $f: \mathsf{V}^{\mathbb{N}} \to \mathbb{R}$ is called an **invariant function** if it is $\mathcal{I}$-measurable (equivalently, if $f = f \circ S$). Say that two functions $f, g: \mathsf{V}^{\mathbb{N}} \to \mathbb{R}$ are $\mathbf{P}_x$**-equivalent** if $\mathbf{P}_x[f = g] = 1$. Recall that $L^{\infty}(\mathsf{V}^{\mathbb{N}}, \mathcal{I}, \mathbf{P}_o)$ denotes the space of $\mathbf{P}_o$-equivalence classes of bounded, invariant functions on $\mathsf{V}^{\mathbb{N}}$. We are going to exhibit an isometric isomorphism between the Banach spaces $L^{\infty}(\mathsf{V}^{\mathbb{N}}, \mathcal{I}, \mathbf{P}_o)$ and $\mathbf{BH}(\mathsf{V}, P)$ (with the sup norm). We invite the reader to warm up by doing the following exercise.

▷ **Exercise 14.7.**
Suppose that u is a bounded harmonic function on an irreducible Markov chain (V, P) and c is a constant such that $\lim_n u(X_n) = c$ almost surely-$\mathbf{P}_o$. Show that $u(x) = c$ for all $x \in \mathsf{V}$.

Proposition 14.12. (Equivalence of $\mathcal{I}$ and BH) *For every bounded, invariant function* $f: \mathsf{V}^{\mathbb{N}} \to \mathbb{R}$, *the function* $U_f: \mathsf{V} \to \mathbb{R}$ *defined by*

$$U_f(x) := \mathbf{E}_x\big[f(X_0, X_1, \ldots)\big] \tag{14.9}$$

is harmonic. Moreover, if the chain is irreducible, the mapping $f \mapsto U_f$ is a linear isometry from $L^{\infty}(\mathsf{V}^{\mathbb{N}}, \mathcal{I}, \mathbf{P}_o)$ onto $\mathbf{BH}(\mathsf{V}, P)$. In this case, its inverse is given by $u \mapsto F_u$, where

$$F_u(x_0, x_1, \ldots) := \limsup_{n \to \infty} u(x_n). \tag{14.10}$$

Proof. Let $f: \mathsf{V}^{\mathbb{N}} \to \mathbb{R}$ be a bounded, invariant function. For notational clarity, let both $\langle X_j \rangle$ and $\langle Y_j \rangle$ denote realizations of the Markov chain. Then $f(X_0, X_1, \ldots) = f(X_1, X_2, \ldots)$ by invariance of f, so for all $x \in \mathsf{V}$,

$$U_f(x) = \mathbf{E}_x f(X_1, X_2, \ldots) = \sum_{y \in \mathsf{V}} p(x, y)\, \mathbf{E}_y f(Y_0, Y_1, \ldots) = (PU_f)(x),$$

showing that U_f is harmonic. To check that the map $f \mapsto U_f$ is well defined on the space $L^{\infty}(\mathsf{V}^{\mathbb{N}}, \mathcal{I}, \mathbf{P}_o)$ of equivalence classes, consider two $\mathbf{P}_o$-equivalent bounded, invariant functions f and g on $\mathsf{V}^{\mathbb{N}}$. Then $\mathbf{1}_{[f \neq g]}$ is invariant. Therefore, $x \mapsto \mathbf{P}_x[f \neq g]$ is a nonnegative harmonic function vanishing at o, whence it must vanish on all of V by irreducibility and the maximum principle. Thus $U_f = U_g$ identically on V, as desired. Linearity of the map $f \mapsto U_f$ is clear. We will verify that its inverse is given by $u \mapsto F_u$ defined in (14.10).

First, let $u: \mathsf{V} \to \mathbb{R}$ be bounded and harmonic. Then $F_u: \mathsf{V}^{\mathbb{N}} \to \mathbb{R}$ is clearly an invariant function. Writing $g := F_u$, we have, for every $x \in \mathsf{V}$,

$$U_g(x) = \mathbf{E}_x \limsup_{n \to \infty} u(X_n) = \mathbf{E}_x \lim_{n \to \infty} u(X_n) = u(x)$$

by the martingale convergence theorem and the bounded convergence theorem.
Conversely, consider an invariant, bounded function $f: \mathsf{V}^{\mathbb{N}} \to \mathbb{R}$. Since f is invariant,

$$\mathbf{E}_o\big[f(X_0, X_1, \ldots) \mid X_0, \ldots, X_n\big] = \mathbf{E}_o\big[f(X_n, X_{n+1}, \ldots) \mid X_0, \ldots, X_n\big].$$

By the Markov property, this equals $U_f(X_n)$. Lévy's zero-one law then gives

$$\lim_{n \to \infty} U_f(X_n) = f(X_0, X_1, \ldots) \qquad \mathbf{P}_o\text{-a.s.},$$

in other words, $F_{U_f} = f$ a.s.-$\mathbf{P}_o$.
Finally, both $f \mapsto U_f$ and $u \mapsto F_u$ are clearly contractions. Being inverses of each other, they must be isometries. ◀

Since the only functions measurable with respect to a trivial σ-field are the constants, we obtain the following corollary.

Corollary 14.13. *Consider an irreducible Markov chain. Then the invariant σ-field $\mathcal{I}$ on $\mathsf{V}^{\mathbb{N}}$ is $\mathbf{P}_o$-trivial (that is, $\mathbf{P}_o(A) \in \{0, 1\}$ for all $A \in \mathcal{I}$) if and only if all bounded harmonic functions on V are constant.* ◀

Example 14.14. Consider again the 3-regular infinite tree $\mathbb{T}_3$, with a distinguished vertex o. Removing o gives three connected components denoted by T_1, T_2, T_3. The three events $E_i := \{\mathbf{y} \in \mathsf{V}^{\mathbb{N}} ; \ \exists N \ \forall n \geq N \ y_n \in T_i\}$ are in $\mathcal{I}$, and for simple random walk on $\mathbb{T}_3$, the three functions $u_i(x) := \mathbf{P}_x(E_i)$ are nonconstant bounded harmonic functions of $x \in \mathbb{T}_3$.

The next example shows that for general Markov chains, the completions of $\mathcal{I}$ and $\mathcal{T}$ need not coincide.

Example 14.15. Consider the network walk $\langle X_j \rangle$ on $\mathbb{N}$ obtained by assigning conductance $2^{\binom{k}{2}}$ to the edge between $k - 1$ and k for every $k \geq 1$. If τ_k denotes the hitting time of k, then $\mathbf{P}_0[\tau_{k+1} \neq \tau_k + 1] = (1 + 2^k)^{-1}$. By Borel–Cantelli, the sequence $j - X_j$ is eventually constant $\mathbf{P}_0$-a.s. Therefore shifts of $\langle X_j \rangle$ all look the same eventually, so $\mathcal{I}$ is $\mathbf{P}_0$-trivial. However, for every $m \geq 0$, the tail event $A_m := [\limsup_{j \to \infty}(j - X_j) = 2m]$ satisfies $0 < \mathbf{P}_0(A_m) < 1$.

To show that for lazy Markov chains, the invariant σ-field coincides with the tail σ-field up to sets of $\mathbf{P}_x$-measure 0, we need the following lemma. Recall that $\|\mu - \nu\|_{\mathrm{TV}} = \frac{1}{2}\|\mu - \nu\|_1 = \sup_A |\mu(A) - \nu(A)|$.

Lemma 14.16. *Let $\widetilde{P}$ be the transition matrix for a lazy Markov chain on V. Then, for every $x \in \mathsf{V}$,*

$$\lim_{n \to \infty} \|\widetilde{p}^n(x, \cdot) - \widetilde{p}^{n+1}(x, \cdot)\|_{\mathrm{TV}} = 0. \tag{14.11}$$

Proof. Laziness implies that $\widetilde{P} = (I + P)/2$ for some transition matrix P. Since

$$2(I + P)^n - (I + P)^{n+1} = \sum_{k=0}^{n+1}\left[2\binom{n}{k} - \binom{n+1}{k}\right]P^k = \sum_{k=0}^{n+1}\left[\binom{n}{k} - \binom{n}{k-1}\right]P^k$$

(with the convention $\binom{n}{n+1} = \binom{n}{-1} = 0$), we have

$$\|\widetilde{p}^n(x, \cdot) - \widetilde{p}^{n+1}(x, \cdot)\|_1 \leq 2^{-n-1}\sum_{k=0}^{n+1}\left|\binom{n}{k} - \binom{n}{k-1}\right|$$

$$= 2^{-n}\sum_{k=0}^{\lfloor n/2 \rfloor}\left|\binom{n}{k} - \binom{n}{k-1}\right| = 2^{-n}\binom{n}{\lfloor n/2 \rfloor}.$$

(To see the penultimate identity, consider separately n odd and n even.) ◀

We saw that invariant functions can be described in terms of harmonic functions. It turns out that tail functions can be described using the following auxiliary notion, which we will then use to show the equivalence of $\mathcal{I}$ and $\mathcal{T}$ for lazy chains. A function $\Lambda\colon \mathsf{V} \times \mathbb{N} \to \mathbb{R}$ is called **space-time harmonic** if

$$\forall x \in \mathsf{V} \;\; \forall n \in \mathbb{N} \quad \Lambda(x, n) = \sum_{y \in \mathsf{V}} p(x, y)\Lambda(y, n + 1)\,,$$

in other words, Λ is a harmonic function for the space-time chain $\langle (X_n, n)\,;\; n \geq 0 \rangle$. To see a relationship of these functions to the tail field, consider a bounded tail function $f\colon \mathsf{V}^{\mathbb{N}} \to \mathbb{R}$. If the Markov chain is either lazy, irreducible, or transitive, then for any state $x \in \mathsf{V}$ and any $n \in \mathbb{N}$, we may choose $z \in \mathsf{V}$ such that $p_n(z, x) > 0$ and define

$$\Lambda_f(x, n) := \mathbf{E}_z\big[f(X_0, X_1, \ldots) \mid X_n = x\big]\,; \qquad (14.12)$$

this definition does not depend on the choice of z, since f is a tail function. By considering all possibilities for X_{n+1}, we see that Λ_f is space-time harmonic.

Next we show that for lazy chains, bounded space-time harmonic functions do not depend on time, and so coincide with harmonic functions on V.

Proposition 14.17. *Consider a lazy chain with transition matrix $\widetilde{P}$ on V, and let $\Lambda\colon \mathsf{V} \times \mathbb{N} \to \mathbb{R}$ be a bounded space-time harmonic function.*
  (i) *For every $x \in \mathsf{V}$ and $\ell \geq 0$, we have $\Lambda(x, \ell) = \Lambda(x, \ell + 1)$.*
  (ii) *There exists a bounded harmonic function $u\colon \mathsf{V} \to \mathbb{R}$ such that for all $x \in \mathsf{V}$ and $n \in \mathbb{N}$,*

$$u(x) = \Lambda(x, n)\,. \qquad (14.13)$$

Proof. (i) Since Λ is space-time harmonic, for every $n \in \mathbb{N}$

$$\Lambda(x, \ell) = \mathbf{E}_x \Lambda(X_{n+1}, \ell + n + 1) \quad \text{and} \quad \Lambda(x, \ell + 1) = \mathbf{E}_x \Lambda(X_n, \ell + n + 1)\,.$$

Therefore,

$$|\Lambda(x, \ell) - \Lambda(x, \ell + 1)| \leq 2 \sup |\Lambda| \cdot \|\widetilde{p}^{\,n}(x, \cdot) - \widetilde{p}^{\,n+1}(x, \cdot)\|_{\mathrm{TV}} \to 0 \text{ as } n \to \infty\,.$$

(ii) For every $x \in \mathsf{V}$ and $n \in \mathbb{N}$, define $u(x) := \Lambda(x, n)$. Part (i) implies this does not depend on the choice of n. Harmonicity of u follows. ◀

We can now relate the tail and invariant fields, as promised.

Theorem 14.18. *Consider a lazy chain on V, and let $o \in \mathsf{V}$. Then, for every $x \in \mathsf{V}$, the $\widetilde{\mathbf{P}}_x$-completions of the corresponding tail σ-field $\widetilde{\mathcal{T}}$ and the corresponding invariant σ-field $\widetilde{\mathcal{I}}$ coincide.*

Proof. Since $\widetilde{\mathcal{I}} \subseteq \widetilde{\mathcal{T}}$, it suffices to show that for any bounded tail function $f\colon \mathsf{V}^{\mathbb{N}} \to \mathbb{R}$, there is an invariant function that coincides with it $\widetilde{\mathbf{P}}_o$-a.s. Define $\Lambda := \Lambda_f$ as in (14.12), and let u be the corresponding harmonic function in (14.13). In particular,

$$p_n(o, y) > 0 \quad \Longrightarrow \quad u(y) = \Lambda_f(y, n) := \mathbf{E}_o\big[f(X_0, X_1, \ldots) \mid X_n = y\big]\,. \qquad (14.14)$$

The invariant function F_u corresponding to u was described in (14.10); it will be the sought-for invariant function. Indeed,

$$F_u(X_0, X_1, \ldots) = \limsup_n u(X_n) = \limsup_n \mathbf{E}_o\big[f(X_0, X_1, \ldots) \mid X_n\big]$$

$$= \limsup_n \mathbf{E}_o\big[f(X_0, X_1, \ldots) \mid X_0, \ldots, X_n\big]\,,$$

by the Markov property of $\langle X_i \rangle$. The Lévy zero-one law thus yields $\widetilde{\mathbf{P}}_o[F_u = f] = 1$, as desired. ◀

Of course, we are not interested only in chains that are lazy. Luckily, the speed and entropy of a transitive chain are easily related to the speed and entropy of the corresponding lazy chain.

Lemma 14.19. *Let P be the transition matrix for a transitive Markov chain $\langle X_n \rangle$ on V, and let $\langle \widetilde{X}_n \rangle$ be the corresponding lazy chain, with transition matrix $\widetilde{P} = (P + I)/2$. Then the Avez entropy $\widetilde{h}$ of the lazy chain satisfies $\widetilde{h} = h/2$. If, moreover, V is equipped with an invariant graph metric d, then the speed $\widetilde{l}$ of the lazy chain satisfies $\widetilde{l} = l/2$.*

Proof. Let $B_0 = 0$ and suppose that $\langle B_{n+1} - B_n \rangle_{n \geq 0}$ are independent Bernoulli$(1/2)$ random variables, so we can write $\widetilde{X}_n = X_{B_n}$ for all $n \geq 0$. Since $H(X_k) = k h(1 + o(1))$ as $k \to \infty$, it follows that $H(X_{B_n} \mid B_n) = \sum_{k=0}^{n} \mathbf{P}[B_n = k] H(X_k) = n h(1/2 + o(1))$ as $n \to \infty$. By (6.42), we have $0 \leq H(X_{B_n}) - H(X_{B_n} \mid B_n) \leq H(B_n) < \log(n + 1)$; dividing by n yields $\widetilde{h} = h/2$. The statement concerning speed is even easier. ◄

Putting together the results of the preceding two sections, we arrive at the following striking equivalence.

Theorem 14.20. (Speed, Entropy, and Harmonic Functions for Transitive Chains) *For an irreducible, transitive Markov chain with transition matrix P on a countable state space V and $H(X_1) < \infty$, the following four properties are equivalent:*

(i) *The invariant σ-field is $\mathbf{P}_o$-trivial for some (hence for every) $o \in \mathsf{V}$.*

(ii) *The Liouville property holds (all bounded harmonic functions are constant).*

(iii) *The Avez entropy $h = 0$.*

(iv) *The chain has a $\mathbf{P}_o$-trivial tail for some (hence for every) $o \in \mathsf{V}$.*

Moreover, suppose that the transitive chain (V, P) is endowed with an invariant graph metric. If the asymptotic speed $l = \lim_n \frac{1}{n} \mathbf{E}[d(X_0, X_n)]$ vanishes and $v < \infty$, then (i)–(iv) hold; conversely, for network walks, (i)–(iv) imply that $l = 0$.

Proof. That (i) $\Leftrightarrow$ (ii) follows from Proposition 14.12. Next, assume (ii). Then the lazy version of the chain also has the Liouville property, since $Pu = u$ iff $\widetilde{P}u = u$. Therefore $\widetilde{P}$ has a trivial invariant σ-field and, by Theorem 14.18, a trivial tail. By Theorem 14.7, the Avez entropy of the lazy chain is $\widetilde{h} = 0$, so Lemma 14.19 yields that $h = 0$. Thus (ii) $\Rightarrow$ (iii). The converse (iii) $\Rightarrow$ (ii) is proved similarly. The equivalence (iii) $\Leftrightarrow$ (iv) is Theorem 14.7.

Finally, if the chain has an invariant graph metric with $v < \infty$ and $l = 0$, then $h = 0$ by Theorem 14.9(i); for transitive network walks, part (ii) of the same theorem yields the implication $h = 0 \Rightarrow l = 0$. ◄

To illustrate the preceding theorem, we consider lamplighter graphs, a generalization of $\mathbb{Z}^{\circ}$ from Section 3.4. Given any graph G, the direct sum $\sum_{x \in \mathsf{V}(G)} \mathbb{Z}_2$ is the collection of maps $\psi \colon \mathsf{V}(G) \to \mathbb{Z}_2$ such that $\psi^{-1}(\{1\})$ is a finite set. Now define the **lamplighter graph** G° as follows. First, let $\mathsf{V}(G^{\circ}) := \left(\sum_{x \in \mathsf{V}(G)} \mathbb{Z}_2 \right) \times \mathsf{V}(G)$. Second, declare that two configurations, (ψ, x) and (φ, y) in $\mathsf{V}(G^{\circ})$, form an edge in $\mathsf{E}(G^{\circ})$ if either $x = y$ and ψ, φ differ only at x, or if $\psi = \varphi$ and $[x, y] \in \mathsf{E}(G)$. We regard a configuration (ψ, x) as a collections of lamps, one at each vertex of G, and a lamplighter at x. We call the lamp at y "on" when $\psi(y) = 1$. Thus, crossing an edge on G° means either changing the state of the lamp at the location of the lamplighter, or moving the lamplighter to a neighboring location. Observe that if G is a transitive graph, then so is $G^{\odot}$. The next corollary is due to Kaimanovich and Vershik (1983).

Corollary 14.21. *Let G be an infinite, transitive graph. Then $G^{\odot}$ has the Liouville property if and only if simple random walk on G is recurrent. In particular, $\mathbb{Z}^{d\,\odot}$ has the Liouville property iff $d \leq 2$.*

Proof. Denote the simple random walk on $G^{\odot}$ by $\langle(\Psi_n, X_n)\rangle$. Fix $o \in \mathsf{V}(G)$. If G is transient, then define $u(\psi, x) := \mathbf{P}_{(\psi,x)}[\lim_n \Psi_n(o) = 1]$. Clearly,

$$u(\mathbf{0}, o) \geq \mathbf{P}_o[X_1 = o \text{ and } \forall n \geq 2 \ X_n \neq o] > 0 .$$

(This event says that first the lamp at o is turned on, then the lamplighter never returns to o.) On the other hand,

$$\lim_k \mathbf{E}_o\big[u(\mathbf{0}, X_k)\big] \leq \lim_k \mathbf{P}_o[\exists n \geq k \ X_n = o] = 0 .$$

Thus u is a nonconstant bounded harmonic function on $G^{\odot}$.

Conversely, if G is recurrent, then $R_n := |\{X_0, X_1, \ldots, X_{n-1}\}|$ satisfies $\mathbf{E}_o[R_n] = o(n)$ by Exercise 14.6. For any finite, connected graph G_0, there is a cyclic tour in G_0 of length at most $2\mathsf{V}(G_0) - 2$ that visits all vertices of G_0 (to see this, consider depth-first search on a spanning tree of G_0). Applying this to the induced subgraph of G on the vertex set $\{X_0, X_1, \ldots, X_{n-1}\}$, it follows that the distance in $G^{\odot}$ satisfies $d\big((\Psi_0, X_0), (\Psi_{n-1}, X_{n-1})\big) \leq 4|R_n|$, so $\mathbf{E}\big[d\big((\Psi_0, X_0), (\Psi_{n-1}, X_{n-1})\big)\big] = o(n)$. ◄

The proof of the last implication in Theorem 14.20 (Liouville $\Rightarrow l = 0$) relied on the Varopoulos–Carne bound, so it was restricted to network walks. Karlsson and Ledrappier (2007) proved the same implication for symmetric random walks on groups with steps of finite mean distance, as we now show. Let Γ be a finitely generated group with identity o, endowed with a Cayley-graph metric $d(\cdot, \cdot)$. Thus $|x| = d(o, x)$ is the minimal length of a word in the generators or their inverses that represents x.

Proposition 14.22. (Symmetry and Liouville Implies Speed 0) *Let μ be a probability measure on Γ that satisfies $\int_\Gamma |x| \, d\mu(x) < \infty$. If (Γ, μ) is Liouville, then there is a group homomorphism $\varphi\colon \Gamma \to \mathbb{R}$ such that $\int_\Gamma \varphi(x) \, d\mu(x) = l$, the speed of the μ-walk $\langle X_n\rangle$. If, in addition, μ is symmetric (that is, $\mu(x) = \mu(x^{-1})$ for all $x \in \Gamma$), then $l = 0$.*

Proof. Let $X_0 := o$, the identity in Γ. Using the Cantor diagonal method, we can find a sequence $n_j \to \infty$ such that the limit

$$\varphi(\gamma) := \lim_j \frac{1}{n_j} \sum_{k=0}^{n_j - 1} \mathbf{E}\big[|\gamma X_k| - |X_k|\big] \tag{14.15}$$

exists for every $\gamma \in \Gamma$. Suppose Z_1 is independent of $\langle X_k\rangle$ and has law μ. Then $Z_1 X_k$ has the same law as X_{k+1} for each k, so

$$\mathbf{E}\big[\varphi(\gamma Z_1)\big] = \lim_j \frac{1}{n_j} \sum_{k=0}^{n_j - 1} \mathbf{E}\big[|\gamma X_{k+1}| - |X_k|\big]$$

$$= \varphi(\gamma) + \lim_j \frac{1}{n_j} \mathbf{E}\big[|\gamma X_{n_j}| - |\gamma|\big] = \varphi(\gamma) + l . \tag{14.16}$$

For $\gamma, x \in \Gamma$, define $u_\gamma(x) := \varphi(\gamma x) - \varphi(x)$. Since $\big| |\gamma x X_k| - |x X_k| \big| \le |\gamma|$, it follows that u_γ is bounded. Moreover,

$$\mathbf{E}[u_\gamma(xZ_1)] = \mathbf{E}[\varphi(\gamma x Z_1)] - \mathbf{E}[\varphi(xZ_1)] = \varphi(\gamma x) - \varphi(x) = u_\gamma(x),$$

in other words, u_γ is μ-harmonic; by the Liouville hypothesis, u_γ is constant. Clearly $u_\gamma(o) = \varphi(\gamma)$, so $u_\gamma(x) = \varphi(\gamma)$ for all $x \in \Gamma$. Thus φ is indeed a homomorphism; it satisfies $\int_\Gamma \varphi(x)\,d\mu = l$ by (14.16) with $\gamma := o$. Finally, if μ is symmetric, then $\int_\Gamma \varphi(x^{-1})\,d\mu = l$, and adding this to the previous identity shows that $l = 0$. ◀

14.3 Harmonic Functions and the Poisson Boundary

Given a continuous function u on the closed unit disc $\overline{\mathbb{D}} = \{z \in \mathbb{C};\ |z| \le 1\}$ that is harmonic on the open unit disc $\mathbb{D}$, the classical Poisson formula (see, for example, Rudin (1987), Theorem 11.8) represents $u(z)$ for $z \in \mathbb{D}$ in terms of boundary values on the circle $\partial\mathbb{D}$:

$$u(z) = \frac{1}{2\pi} \int_0^{2\pi} \mathcal{P}(z, e^{i\theta})\, u(e^{i\theta})\, d\theta, \tag{14.17}$$

where $\mathcal{P}(z, w) := \frac{1 - |z|^2}{|w - z|^2}$ is the **Poisson kernel**. Moreover, *every* bounded harmonic function u on $\mathbb{D}$ has a representation of the form (14.17). In this case the boundary values $u(e^{i\theta})$ are not given; rather, they have to be defined, for example, as radial limits,

$$u(e^{i\theta}) := \lim_{r\uparrow 1} u(re^{i\theta}), \tag{14.18}$$

that exist for a.e. θ. See Rudin (1987), Theorems 11.23 and 11.30, for such results. The "harmonic" measures $d\nu_z(e^{i\theta}) := \mathcal{P}(z, e^{i\theta})\frac{d\theta}{2\pi}$ that appear in the Poisson formula have a probabilistic meaning: ν_z is the hitting measure on the unit circle for Brownian motion started at z; see, for example, Bass (1995) or Mörters and Peres (2010). Moreover, (14.17) defines an isometric isomorphism $u|_{\partial\mathbb{D}} \mapsto u|_{\mathbb{D}}$ from $L^\infty(\partial\mathbb{D}, \mathscr{B}_{\partial\mathbb{D}}, d\theta)$ to the Banach space of bounded harmonic functions $\mathbf{BH}(\mathbb{D})$; the inverse mapping is given by (14.18).

The utility of this Poisson representation led Furstenberg (1963) to define the notion of "Poisson boundary" for random walks on groups. We proceed to discuss this concept in the more general setting of Markov chains. As a running example, consider simple random walk on the four-regular tree $\mathbb{T}_4$, the Cayley graph of the free group on two letters $\mathbb{F}_2 = \langle a, b \mid \rangle$. In this case, as we will see, there is a natural analogue of (14.17), with the circle $\partial\mathbb{D}$ replaced by the space $\partial\mathbb{T}_4$ of ends of $\mathbb{T}_4$ and the harmonic measure ν_z replaced by the hitting measure of random walk.

For a general Markov chain, a coarse analogue of (14.17) is provided by Proposition 14.12: the measures $\langle \mathbf{P}_x ;\ x \in \mathsf{V} \rangle$ on $\mathsf{V}^\mathbb{N}$ satisfy

$$u(x) = \mathbf{E}_x[F_u(X_0, X_1, \dots)] = \int_{\mathsf{V}^\mathbb{N}} F_u\, d\mathbf{P}_x \tag{14.19}$$

for all u in the space $\mathbf{BH}(\mathsf{V}, P)$ of bounded harmonic functions. Here, $F_u(\mathbf{x}) = \limsup u(x_n)$ from (14.10). Moreover, the correspondence $F_u \mapsto u$ given by (14.19) (or (14.9)) is a linear isometry from $L^\infty(\mathsf{V}^{\mathbb{N}}, \mathcal{I}, \mathbf{P}_o)$ to $\mathbf{BH}(\mathsf{V}, P)$ (the choice of basepoint o is immaterial).

However, the measurable space $(\mathsf{V}^{\mathbb{N}}, \mathcal{I})$ is too large to serve as a proper boundary, since $\mathcal{I}$ does not separate points in $\mathsf{V}^{\mathbb{N}}$. As an example, for random walks on $\mathbb{T}_4$, we might like to use as a boundary the quotient of $\mathsf{V}^{\mathbb{N}}$ obtained by identifying any two paths that converge to the same end.

Definition 14.23. Let P be the transition matrix for an irreducible Markov chain on V and recall that $\mathsf{V}^{\mathbb{N}}$ is endowed with the Borel σ-field, $\mathscr{B}_{\mathsf{V}^{\mathbb{N}}}$. A (measure theoretic) **boundary** $(\Theta, \mathscr{F}_\Theta, \mathbf{b}_\Theta)$ of the Markov chain (V, P) is a measurable space $(\Theta, \mathscr{F}_\Theta)$ with a measurable **boundary map** $\mathbf{b} = \mathbf{b}_\Theta \colon (\mathsf{V}^{\mathbb{N}}, \mathscr{B}_{\mathsf{V}^{\mathbb{N}}}) \to (\Theta, \mathscr{F}_\Theta)$ satisfying the following two properties:

(i) $\mathscr{F}_\Theta$ is countably generated and separates points: there is a collection of sets $\{A_n\}_{n \geq 1}$ in $\mathscr{F}_\Theta$ such that $\mathscr{F}_\Theta = \sigma\{A_n\}_{n \geq 1}$, and for every pair $\theta \neq \xi$ of points in Θ, there exists n such that $\mathbf{1}_{A_n}(\theta) \neq \mathbf{1}_{A_n}(\xi)$.

(ii) $\mathbf{b}_\Theta$ is shift-invariant: $\mathbf{b}_\Theta(x_0, x_1, \ldots) = \mathbf{b}_\Theta(x_1, x_2, \ldots)$ for all $\mathbf{x} = (x_0, x_1, \ldots) \in \mathsf{V}^{\mathbb{N}}$.

Every boundary $(\Theta, \mathscr{F}_\Theta, \mathbf{b}_\Theta)$ of (V, P) is naturally endowed with the family of **harmonic measures** $\nu_x := \mathbf{P}_x \circ \mathbf{b}_\Theta^{-1}$ for $x \in \mathsf{V}$.

We call the elements of the family $\langle \nu_x \,;\; x \in \mathsf{V} \rangle$ "harmonic measures" because they vary harmonically in the sense that for every $A \in \mathscr{F}_\Theta$, the function $x \mapsto \nu_x(A)$ is harmonic on (V, P). By irreducibility of P, these measures are all mutually absolutely continuous. Thus, if one wishes, one can fix a basepoint o and define an analogue of the Poisson kernel to be the collection of Radon–Nikodým derivatives $(x, \theta) \mapsto \frac{d\nu_x}{d\nu_o}(\theta)$.

For any boundary, the harmonic measures yield a solution of the Dirichlet problem, which is half of what we want from a Poisson boundary:

Proposition 14.24. *Given a boundary $(\Theta, \mathscr{F}_\Theta, \mathbf{b})$ of (V, P) and $\varphi \in L^\infty(\Theta, \mathscr{F}_\Theta, \nu_o)$, the function*

$$u(x) := \int_\Theta \varphi(\theta) \, d\nu_x(\theta) \tag{14.20}$$

is harmonic and bounded on V and satisfies $\lim_n u(X_n) = \varphi(\mathbf{b}(\mathbf{X}))$ a.s.-$\mathbf{P}_o$ as $n \to \infty$ for all $o \in \mathsf{V}$, where $\mathbf{X} := (X_0, X_1, \ldots)$.

Proof. Define $f = \varphi \circ \mathbf{b}$ and fix $o \in \mathsf{V}$. Then f is an invariant function on $\mathsf{V}^{\mathbb{N}}$ since $\mathbf{b}$ is shift-invariant, so by Proposition 14.12,

$$U_f(x) = \mathbf{E}_x f(X_0, X_1, \ldots) = \int_{\mathsf{V}^{\mathbb{N}}} (\varphi \circ \mathbf{b}) \, d\mathbf{P}_x = \int_\Theta \varphi \, d\nu_x = u(x)$$

satisfies $\lim_n U_f(X_n) = f(\mathbf{X}) = \varphi(\mathbf{b}(\mathbf{X}))$ a.s.-$\mathbf{P}_o$. ◀

We will look at the other half of being a Poisson boundary later, that is, whether *every* harmonic function can be represented on the boundary.

We say that a boundary is **trivial** if one of the measures ν_x is concentrated at a point; in such a case, every ν_x is concentrated at the same point. Condition (a) in the definition of boundary imparts meaning to individual points in Θ, as illustrated in the following proof. Without condition (a), the introduction of boundaries is pointless – one could just consider σ-fields of invariant sets in $\mathsf{V}^{\mathbb{N}}$ instead.

Proposition 14.25. (Boundary Triviality) *The irreducible chain* (V, P) *is Liouville if and only if all boundaries of* (V, P) *are trivial.*

Proof. First suppose that (V, P) is Liouville and $(\Theta, \mathscr{F}_\Theta, \mathbf{b})$ is a boundary. By Proposition 14.12, $\nu_o(A) = \mathbf{P}_o \mathbf{b}^{-1}(A) \in \{0, 1\}$ for every $A \in \mathscr{F}_\Theta$. In particular, this applies to the sets A_n in part (a) of the definition. Replacing A_n by its complement when necessary, we may assume that $\nu_o(A_n) = 1$ for all $n \geq 1$. Then $\nu_o\left(\bigcap_n A_n\right) = 1$. On the other hand, this intersection must be a singleton, since if it contained two distinct points, the point separation assumption in (a) would yield a contradiction.

Conversely, if (V, P) is not Liouville, then by Proposition 14.12, there is an invariant set $A \subset \mathsf{V}^{\mathbb{N}}$ such that $0 < \mathbf{P}_o(A) < 1$. Then $\left(\{0, 1\}, 2^{\{0,1\}}, \mathbf{1}_A\right)$ is a nontrivial boundary of (V, P) since $\nu_o(1) = \mathbf{P}_o(A)$. ◀

Alternatively, to prove the preceding converse more directly, suppose that all boundaries of (V, P) are trivial and $u \in \mathbf{BH}(\mathsf{V}, P)$. Construct a boundary $(\mathbb{R}, \mathscr{B}_{\mathbb{R}}, \mathbf{b})$ by letting $\mathbf{b}(\mathbf{x}) := \limsup_n u(x_n)$. Since $\nu_o = \mathbf{P}_o \mathbf{b}^{-1}$ is concentrated at a point, $\lim_n u(X_n)$ is $\mathbf{P}_o$-a.s. constant, so u must be constant by Exercise 14.7.

An appealing way to ensure condition (i) in Definition 14.23, used by Furstenberg (1971b), is to consider compactifications.

Definition 14.26. We say that $(\Theta, \mathscr{F}_\Theta, \mathbf{b})$ is a ***compactification boundary*** of the Markov chain (V, P) if there is a metric on $\mathsf{V} \cup \Theta$ such that

(i) $\mathsf{V} \cup \Theta$ is a compact space in which V is discrete (that is, each point in V is an open set) and dense, with $\Theta = \partial \mathsf{V}$ in this metric;

(ii) $\mathscr{F}_\Theta = \mathscr{B}_\Theta$ is the Borel σ-field;

(iii) $\forall x \in \mathsf{V}$ the chain $\langle X_n \rangle$ converges to an element of Θ a.s.-$\mathbf{P}_x$; and

(iv) $\mathbf{b}(\mathbf{x}) = \lim_n x_n \in \Theta$ for all $\mathbf{x} \in \mathsf{V}^{\mathbb{N}}_{\text{conv}}$, the set of sequences in $\mathsf{V}^{\mathbb{N}}$ that converge to an element of Θ. On $\mathsf{V}^{\mathbb{N}} \setminus \mathsf{V}^{\mathbb{N}}_{\text{conv}}$, we just require that $\mathbf{b}$ is a shift-invariant and measurable map to Θ (for instance, $\mathbf{b}(\mathbf{x}) = \eta$ for all $x \notin \mathsf{V}^{\mathbb{N}}_{\text{conv}}$, where $\eta \in \Theta$ is fixed).

Observe that conditions (i) and (ii) in Definition 14.23 follow from (i)–(iv) in Definition 14.26, so a compactification boundary is indeed a boundary as defined there.

Example 14.27. Suppose that T is a locally finite tree with prescribed edge conductances such that the corresponding network walk (T, P) is transient. Recall that the space of ends of T (described in Example 7.1) can be identified with the set ∂T of rays emanating from a fixed root o; the latter set was equipped in Section 1.8 with a natural metric $d(\xi, \theta) = e^{-|\xi \cap \theta|}$ (considering each ray as a set of edges) that makes it a compact metric space. The same metric extends to $T \cup \partial T$ if we identify each vertex $x \in T$ with the shortest path $[o, x]$ from o to x, that is, for $x, y \in \mathsf{V}$ and $\theta \in \partial T$, we let $d(x, \theta) := e^{-|[o,x] \cap \theta|}$ and $d(x, y) := e^{-|[o,x] \cap [o,y]|}$. Then (i), (ii), and (iii) in Definition 14.26 hold, so if we define $\mathbf{b}$ as in (iv) (in particular, $\mathbf{b}(\mathbf{x}) = \lim_n x_n$ for $\mathbf{x} \in \mathsf{V}^{\mathbb{N}}_{\text{conv}}$), then $(\partial T, \mathscr{B}_{\partial T}, \mathbf{b})$ is a compactification boundary of (T, P). (If T is not locally finite, then ∂T is a metric boundary that is not compact.)

Returning for a moment to the classical setting of the unit disc $\mathbb{D}$, we note that the group of the Möbius transformations γ preserving the unit disc acts naturally on the space $\mathbf{BH}(\mathbb{D})$ (via $u \mapsto u \circ \gamma$) and also on the boundary $\partial \mathbb{D}$. They transform harmonic measures via

$v_{\gamma z} = v_z \circ \gamma^{-1}$ for all $z \in \mathbb{D}$. Similarly, in the setting of discrete Markov chains, it is natural to require that any symmetries of the chain also be reflected in the boundary. This leads to the next definition.

Definition 14.28. Let $(\Theta, \mathscr{F}_\Theta, \mathbf{b})$ be a boundary of (V, P). Suppose that Γ is a group of automorphisms of (V, P) that also acts on Θ. Note that Γ acts diagonally on $V^{\mathbb{N}}$ via $\gamma(x_0, x_1, \ldots) := (\gamma x_0, \gamma x_1, \ldots)$. Suppose that $\mathbf{b}$ is Γ-*equivariant*, that is, $\gamma \circ \mathbf{b} = \mathbf{b} \circ \gamma$ for all $\gamma \in \Gamma$.

(i) We say that $(\Theta, \mathscr{F}_\Theta, \mathbf{b})$ is a Γ-*boundary* of the Markov chain (V, P) if Γ acts $\mathscr{F}_\Theta$-measurably on Θ.

(ii) If $(\Theta, \mathscr{F}_\Theta, \mathbf{b})$ is a compactification boundary and Γ acts continuously on Θ, then we call $(\Theta, \mathscr{F}_\Theta, \mathbf{b})$ a *compactification Γ-boundary* of (V, P).

In particular, for a μ-walk on a group Γ where $p(x, y) = \mu(x^{-1}y)$ for all $x, y \in \Gamma$, every $\gamma \in \Gamma$ acts as an automorphism of (Γ, P) via $x \mapsto \gamma x$. In this case, we will focus on describing Γ-boundaries of (Γ, P).

Recall that boundaries can be used to represent harmonic functions: we used (14.20) to define u, but we could have regarded the same formula as a Poisson representation for u on Θ. However, for some boundaries, not all bounded harmonic functions have such a representation. For instance, consider a one-point boundary for a non-Liouville chain. Boundaries that support a Poisson representation for *all* bounded harmonic functions (that is, *Poisson boundaries*) are the subject of the next theorem/definition.

Theorem 14.29. *Let $(\Theta, \mathscr{F}_\Theta, \mathbf{b})$ be a boundary of the irreducible chain (V, P) with harmonic measures $\langle v_x \rangle_{x \in V}$. Fix $o \in V$. Then the following are equivalent:*

(i) **(Poisson Representation)** *For every bounded harmonic $u \colon V \to \mathbb{R}$, there is a bounded measurable function $\widetilde{u} \colon \Theta \to \mathbb{R}$ such that $u(x) = \int_\Theta \widetilde{u}(\theta)\, dv_x(\theta)$ for all $x \in V$.*

(ii) **(Harmonic Limits Determined)** *For every bounded harmonic $u \colon V \to \mathbb{R}$, there is a bounded measurable $\widetilde{u} \colon \Theta \to \mathbb{R}$ such that $\lim_n u(X_n) = \widetilde{u}(\mathbf{b}(\mathbf{X}))$ $\mathbf{P}_o$-a.s., where $\mathbf{X} = (X_0, X_1, \ldots)$.*

(iii) **(Invariant σ-field Saturation)** $\mathbf{b}^{-1}(\mathscr{F}_\Theta) = \mathcal{I} \bmod \mathbf{P}_o$, *that is, for every $A \in \mathcal{I}$, there is an $E \in \mathscr{F}_\Theta$ such that $\mathbf{P}_o(A \triangle \mathbf{b}^{-1}(E)) = 0$.*

(iv) **(Maximality)** *For every other boundary $(\Theta', \mathscr{F}_{\Theta'}, \mathbf{b}')$ of (V, P), there is a measurable map $\pi \colon \Theta \to \Theta'$ that satisfies $\mathbf{b}' = \pi \circ \mathbf{b}$ almost surely-$\mathbf{P}_o$.*

*A boundary $(\Theta, \mathscr{F}_\Theta, \mathbf{b})$ that satisfies these conditions is called a **Poisson boundary** of (V, P).*

We will need a simple equivalence from measure theory.

▷ **Exercise 14.8.**

(Saturation) Let $(M_1, \mathscr{F}_1, Q_1)$ and $(M_2, \mathscr{F}_2, Q_2)$ be two probability spaces, and suppose that $\Phi \colon (M_1, \mathscr{F}_1) \to (M_2, \mathscr{F}_2)$ is measurable with $Q_2 = Q_1 \circ \Phi^{-1}$. Show that the following are equivalent:

(i) The Q_1-completions of $\mathscr{F}_1$ and $\Phi^{-1}\mathscr{F}_2$ coincide.

(ii) The mapping $f_2 \mapsto f_2 \circ \Phi$ from $L^\infty(M_2, \mathscr{F}_2, Q_2)$ to $L^\infty(M_1, \mathscr{F}_1, Q_1)$ is surjective.

Proof of Theorem 14.29. The implication (i) $\Rightarrow$ (ii) is part of Proposition 14.24, and the equivalence (ii) $\Leftrightarrow$ (iii) follows from Proposition 14.12.

(iii) $\Rightarrow$ (i): Given $u \in \mathbf{BH}(\mathsf{V}, P)$, Proposition 14.12 yields a bounded, invariant function $f: \mathsf{V}^{\mathbb{N}} \to \mathbb{R}$ such that $u(x) = \mathbf{E}_x f(X_0, X_1, \ldots)$ for all $x \in \mathsf{V}$. By (iii) and Exercise 14.8, f coincides mod $\mathbf{P}_o$ with $\widetilde{u} \circ \mathbf{b}$ for some (bounded) $\widetilde{u}: \Theta \to \mathbb{R}$ that is $\mathscr{F}_\Theta$-measurable. Thus, for all $x \in \mathsf{V}$,

$$u(x) = \int_{\mathsf{V}^{\mathbb{N}}} \widetilde{u} \circ \mathbf{b} \, d\mathbf{P}_x = \int_\Theta \widetilde{u} \, d\nu_x \,.$$

(iv) $\Rightarrow$ (iii): For $A \in \mathcal{I}$, consider the corresponding 2-point boundary $(\{0, 1\}, 2^{\{0,1\}}, \mathbf{1}_A)$. By (iv), there is a measurable $\pi: \Theta \to \{0, 1\}$ such that $\mathbf{1}_A = \pi \circ \mathbf{b}$ a.s.-$\mathbf{P}_o$. This yields (iii) with $E := \pi^{-1}(1)$.

(iii) $\Rightarrow$ (iv): Suppose $\{A'_n; \ n \geq 1\}$ separates points in Θ' and generates $\mathscr{F}_{\Theta'}$. By (iii), there exist sets $E_n \in \mathscr{F}_\Theta$ such that $\mathbf{P}_o\big[\mathbf{1}_{A'_n} \circ \mathbf{b}' = \mathbf{1}_{E_n} \circ \mathbf{b}\big] = 1$ for all n. Let

$$\Omega_1 := \bigcap_n \Big\{ \mathbf{x} \in \mathsf{V}^{\mathbb{N}} ; \ \mathbf{1}_{A'_n}(\mathbf{b}'(\mathbf{x})) = \mathbf{1}_{E_n}(\mathbf{b}(\mathbf{x})) \Big\} \,.$$

Since $\mathbf{P}_o(\Omega_1) = 1$, it suffices to define π on $\mathbf{b}(\Omega_1)$: Here, we extend ν_o to the σ-field generated by $\mathscr{F}_\Theta$ and those $E \subset \Theta$ for which $\mathbf{b}^{-1}(E)$ lies in a subset of $\mathscr{B}_{\mathsf{V}^{\mathbb{N}}}$ that has $\mathbf{P}_o$-measure 0. With this extension, $\nu_o(\mathbf{b}(\Omega_1)) = 1$. If $\theta = \mathbf{b}(\mathbf{x})$ and $\mathbf{x} \in \Omega_1$, define $\pi(\theta) := \mathbf{b}'(\mathbf{x})$. To see this does not depend on the choice of $\mathbf{x}$, suppose that $\mathbf{x}, \mathbf{y} \in \Omega_1$ satisfy $\mathbf{b}(\mathbf{x}) = \mathbf{b}(\mathbf{y})$, yet $\mathbf{b}'(\mathbf{x}) \neq \mathbf{b}'(\mathbf{y})$. Then $\mathbf{1}_{A'_n}(\mathbf{b}'(\mathbf{x})) \neq \mathbf{1}_{A'_n}(\mathbf{b}'(\mathbf{y}))$ for some n by the separation assumption, which implies that $\mathbf{1}_{E_n}(\mathbf{b}(\mathbf{x})) \neq \mathbf{1}_{E_n}(\mathbf{b}(\mathbf{y}))$, a contradiction. Clearly, $\mathbf{b}' = \pi \circ \mathbf{b}$ almost surely-$\mathbf{P}_o$. ◄

The utility of the maximality condition is shown in the next exercise.

▷ **Exercise 14.9.**

(a) Prove that the map π in Theorem 14.29(iv) is unique up to a set of ν_o-measure 0.

(b) Prove that the Poisson boundary of a Markov chain (V, P) is unique up to a pointwise isomorphism π of measure spaces, defined ν_o-a.e., that respects boundary maps. In other words, show that if $(\Theta, \mathscr{F}_\Theta, \mathbf{b})$ and $(\Theta', \mathscr{F}_{\Theta'}, \mathbf{b}')$ are both Poisson boundaries of (V, P), then the map π in Theorem 14.29(iv) is one-to-one on a set $\Theta_1 \in \mathscr{F}_\Theta$ of full ν_o-measure, and π^{-1} is measurable.

(c) Show that if both of these Poisson boundaries in (b) are Γ-boundaries for some countable Γ, then π is Γ-equivariant (that is, $\pi \circ \gamma(\theta) = \gamma \circ \pi(\theta)$ for ν_o-a.e. $\theta \in \Theta$ and all $\gamma \in \Gamma$).

Next, we apply Theorem 14.29 to transient trees.

Example 14.30. (Poisson Boundary of a Tree) Let T be a locally finite tree with given edge conductances such that the corresponding network walk (T, P) is transient. In Example 14.27 we described a compactification boundary $(\partial T, \mathscr{B}_{\partial T}, \mathbf{b})$ for (T, P). Now we prove that it is a Poisson boundary by using criterion (ii) in Theorem 14.29. The network walk $\langle X_n \rangle$, started at o, almost surely converges to a random ray $\mathbf{Z} = \langle Z_0, Z_1, \ldots \rangle \in \partial T$. Observe that $\mathbf{Z}$ is obtained by iteratively erasing backtracks from $\langle X_n \rangle$; in particular, the former is a subsequence of the latter. Let u be a bounded harmonic function on T. For $\mathbf{z} = \langle z_j \rangle \in \partial T$, denote $\widetilde{u}(\mathbf{z}) := \lim \sup u(z_j)$. Then $\lim_n u(X_n)$ a.s. exists and equals $\lim_j u(Z_j) = \widetilde{u}(\mathbf{Z})$. This verifies criterion (ii), as promised. If Γ is a group of automorphisms of (T, P), then for each $\gamma \in \Gamma$ and every ray $(x_0, x_1, \ldots)$ in ∂T, the path $(\gamma x_0, \gamma x_1, \ldots)$ eventually coincides (up to a time-shift) with a unique ray $(y_0, y_1, \ldots)$ in ∂T. It is easy to check that defining $\gamma(x_0, x_1, \ldots) := (y_0, y_1, \ldots)$ makes $(\partial T, \mathscr{B}_{\partial T}, \mathbf{b})$ a Γ-boundary.

We turn now to proving the existence of Poisson boundaries in general. Let us write $L^1(\mathsf{V}^{\mathbb{N}}, \mathscr{F}, \mathbf{P}, \{0, 1\})$ for the collection of equivalence classes of $\{0, 1\}$-valued functions in $L^1(\mathsf{V}^{\mathbb{N}}, \mathscr{F}, \mathbf{P})$. We will use the following measure-theory exercise.

▷ **Exercise 14.10.**
Suppose that $(\Omega, \mathscr{F}, \mathbf{P})$ is a probability space. Show that if $\{A_n ; n \geq 1\} \subset \mathscr{F}$ is such that $\{\mathbf{1}_{A_n}\}$ is dense in $L^1(\Omega, \mathscr{F}, \mathbf{P}, \{0, 1\})$, then the $\mathbf{P}$-completions of $\mathscr{F}$ and of $\sigma(\{A_n ; n \geq 1\})$ coincide.

Theorem 14.31. (Poisson Boundary Existence) *Every irreducible chain (V, P) has a Poisson boundary. If Γ is a countable group of automorphisms of (V, P), then the Poisson boundary can be taken to be a Γ-boundary.*

Proof. Since the Borel σ-field $\mathscr{B}_{\mathsf{V}^{\mathbb{N}}}$ in the sequence space $\mathsf{V}^{\mathbb{N}}$ is countably generated (by the cylinder sets), the metric space $L^1(\mathsf{V}^{\mathbb{N}}, \mathscr{B}_{\mathsf{V}^{\mathbb{N}}}, \mathbf{P}_o, \{0, 1\})$ is separable, whence so is the subspace $L^1(\mathsf{V}^{\mathbb{N}}, \mathcal{I}, \mathbf{P}_o, \{0, 1\})$ of invariant indicators. Let $\langle I_n \rangle_{n \in \mathbb{N}}$ be a countable dense sequence in this subspace. Define $\mathbf{b} \colon \mathsf{V}^{\mathbb{N}} \to \{0, 1\}^{\mathbb{N}}$ by $\mathbf{b}(\mathbf{x}) := \langle I_n(\mathbf{x}) \rangle_{n \in \mathbb{N}}$ and let $\Theta := \overline{\mathbf{b}(\mathsf{V}^{\mathbb{N}})}$ be the closure, in the product topology, of the image of $\mathbf{b}$. Recall that the product topology on $\{0, 1\}^{\mathbb{N}}$ can be metrized by

$$\rho(\xi, \eta) = \sum_k 2^{-k} |\xi_k - \eta_k| .$$

The Borel σ-field $\mathscr{B}_{\Theta}$ in Θ is generated by the elementary cylinder sets $\{\theta \in \Theta ; \theta_n = 1\}$; clearly these separate points in Θ. Since I_n are invariant functions, $\mathbf{b} \circ S = \mathbf{b}$. Thus $(\Theta, \mathscr{B}_{\Theta}, \mathbf{b})$ is a boundary of (V, P). To see that it is a Poisson boundary, we will verify condition (iii) in Theorem 14.29. Let $\pi_n \colon \Theta \to \{0, 1\}$ be the coordinate projections. By construction, $(\pi_n \circ \mathbf{b})^{-1}$ is dense in $L^1(\mathsf{V}^{\mathbb{N}}, \mathcal{I}, \mathbf{P}_o, \{0, 1\})$, whence saturation follows from Exercise 14.10.

Next, suppose that Γ is a countable group of automorphisms of (V, P). Then Γ also acts on $\mathsf{V}^{\mathbb{N}}$ by the diagonal action $\gamma(x_0, x_1, \ldots) := (\gamma x_0, \gamma x_1, \ldots)$. Enlarging the collection $\langle I_n \rangle_{n \in \mathbb{N}}$, we may assume that for every $k \in \mathbb{N}$ and $\gamma \in \Gamma$, the composition $I_k \circ \gamma$ is also in $\langle I_n \rangle_{n \in \mathbb{N}}$. Therefore, if $\mathbf{x}, \mathbf{y} \in \mathsf{V}^{\mathbb{N}}$ satisfy $I_k(\mathbf{x}) = I_k(\mathbf{y})$ for all k, then also $I_k(\gamma \mathbf{x}) = I_k(\gamma \mathbf{y})$ for all k. Thus, we may define the action of $\gamma \in \Gamma$ on $\mathbf{b}(\mathsf{V}^{\mathbb{N}})$ by $\gamma(\mathbf{b}(\mathbf{x})) := \mathbf{b}(\gamma \mathbf{x})$. This defines a uniformly continuous self-map of $\mathbf{b}(\mathsf{V}^{\mathbb{N}})$ with respect to the metric ρ, so it extends uniquely to a continuous self-map of $\Theta = \overline{\mathbf{b}(\mathsf{V}^{\mathbb{N}})}$. The resulting action of Γ on Θ is equivariant by construction on $\mathbf{b}(\mathsf{V}^{\mathbb{N}})$ and so by continuity on all of Θ. ◀

If Γ is an uncountable subgroup of $\mathrm{Aut}(\mathsf{V}, P)$, then we can find a countable dense subgroup $\Gamma_* \subset \Gamma$ (in the weak topology described in Exercise 8.20) and verify that the Γ_*-boundary of (V, P) constructed above can also serve as a Γ-boundary; see Exercise 14.33.

Remark 14.32. (The Poisson Boundary as a Compactification Boundary) Suppose that (V, P) is not Liouville. Let $u_n(x) := \mathbf{E}_x[I_n]$ be the harmonic function that corresponds to the invariant indicator I_n in the preceding proof. We may extend the metric ρ to $V \cup \Theta$ by setting $\rho(v, \xi) := \sum_n 2^{-n}|u_n(v) - \xi_n|$ for $v \in V$ and $\xi \in \Theta$, and $\rho(v, w) := \inf_{\xi \in \Theta}(\rho(v, \xi) + \rho(w, \xi))$ for distinct $v, w \in V$. Thus, $v \in V$ is close to $\xi \in \Theta$ iff $u_n(v)$ is close to ξ_n for all n. Since there is some nonconstant u_n, the maximum principle guarantees that $\rho(v, \Theta) > 0$ for all $v \in V$. In particular, $\rho(v, w) \geq \rho(v, \Theta) + \rho(w, \Theta) > 0$ for $v \neq w$. Since $\lim_k u_n(x_k) = I_n(\mathbf{x})$ for $\mathbf{P}_o$-a.e. $\mathbf{x} \in V^{\mathbb{N}}$ and every n, it follows that the set of limit points $\Theta' := \partial V \subset \Theta$, in the metric ρ, has full ν_o-measure in Θ. Define $\mathbf{b}'$ as in Definition 14.26, and observe that it coincides with $\mathbf{b}$ a.e.-ν_o. The other conditions of Definition 14.26 are satisfied, so $(\Theta', \mathcal{B}_{\Theta'}, \mathbf{b}')$ is a compactification boundary of (V, P).

Let us unravel this development and see how this meets our original goal of describing the asymptotic behavior of Markov chains. Since $\mathcal{I}$ describes the asymptotic behavior, so does $\langle I_n ; \ n \geq 1 \rangle$, and therefore so does Θ'. That is, the points of Θ' tell us all the possible asymptotic behaviors of the Markov chain. Furthermore, if $v_n \to \xi \in \Theta'$, then as far as the Markov chain can distinguish, ξ tells us everything about the way that v_n tends to infinity.

In Example 14.30, we showed that the Poisson boundary is the natural geometric boundary. That is, asymptotic behavior of the random walk is determined entirely by erasing all backtracks.

In general, one wants not merely to know that the Poisson boundary exists but to identify it. Such an identification would allow us to determine all possible asymptotic behavior. Normally one can easily observe certain kinds of asymptotic behavior; one then wants to establish that such kinds are the only kinds. Techniques for doing so are the subject of the next section.

14.4 Identifying the Poisson Boundary

Here, we develop a powerful method, due to Kaimanovich (1985, 1994, 2000), to identify when a boundary $(\Theta, \mathscr{F}_\Theta, \mathbf{b})$ of a transitive Markov chain is, in fact, a Poisson boundary.

We are going to study the Markov chain conditioned on a subfield of the invariant σ-field, $\mathcal{I}$. We begin this study with some elementary aspects of such conditioning. Given a Markov chain $\langle X_n \rangle$ and a σ-field $\mathcal{J}$ in $V^{\mathbb{N}}$, write

$$p_n^{\mathcal{J}}(x, y) := \mathbf{P}_x[X_n = y \mid \mathcal{J}].$$

For instance, if $m > n$ and $\mathcal{J} = \sigma(X_m)$, then

$$p_n^{\mathcal{J}}(x, y) = \frac{p_n(x, y)p_{m-n}(y, X_m)}{p_m(x, X_m)}$$

by the Markov property.

▷ **Exercise 14.11.**
Suppose that the Markov chain (V, P) is simple random walk on a $(d + 1)$-regular tree, and $\mathcal{J}$ is the σ-field determined by the boundary mapping $\mathbf{b}$ in Example 14.27. For any two vertices x, y and every $\mathbf{z} \in V_{\text{conv}}^{\mathbb{N}}$, there is a unique first meeting point of the rays from x and y that belong to the same end as $\mathbf{b}(\mathbf{z})$. Show that $p_n^{\mathcal{J}}(x, y)(\mathbf{z}) = p_n(x, y) \cdot d^{k_x - k_y}$ for $\mathbf{P}_x$-a.e. $\mathbf{z} \in V^{\mathbb{N}}$, where $k_x = k_x(x, y, \mathbf{z})$ is the distance from x to this meeting point.

Lemma 14.33. *Given an irreducible (but possibly periodic) chain* (V, P), *a vertex* $x \in V$ *and a* σ-*field* $\mathcal{J} \subseteq \mathcal{I}$, *denote by* Q_x *the restriction of* $\mathbf{P}_x$ *to* $\mathcal{J}$. *Then, for all* $x, y \in V$, *the measures* Q_x *and* Q_y *are mutually absolutely continuous and the Radon–Nikodým derivative satisfies*

$$p_n^{\mathcal{J}}(x, y) = p_n(x, y) \cdot \frac{dQ_y}{dQ_x} \quad \mathbf{P}_x\text{-}a.e. \tag{14.21}$$

Proof. We first show absolute continuity. By irreducibility, there is some k for which $p_k(x, y) > 0$. Suppose $A \in \mathcal{J}$ and $Q_y(A) > 0$. Then

$$Q_x(A) = Q_x(S^{-k}A) \geq p_k(x, y)Q_y(A) > 0,$$

as desired.

Now for $A \in \mathcal{J}$, the Markov property yields

$$\int_A \mathbf{1}_{[X_n=y]} \, d\mathbf{P}_x = p_n(x, y)Q_y(A) = \int_A p_n(x, y) \cdot \frac{dQ_y}{dQ_x} \, d\mathbf{P}_x \,.$$

The integrand on the right-hand side is clearly $\mathcal{J}$-measurable, so it is a version of the conditional expectation $\mathbf{E}_x[\mathbf{1}_{\{X_n=y\}} \mid \mathcal{J}]$. ◀

A similar statement holds when we consider the chain at two different times:

▷ **Exercise 14.12.**
In the setting of Lemma 14.33, show that for all $x, y, z \in V$, we have $\mathbf{P}_x$-a.s.

$$\mathbf{P}_x[X_n = y, X_{n+m} = z \mid \mathcal{J}] = p_n(x, y)p_m(y, z)\frac{dQ_z}{dQ_x} = p_n^{\mathcal{J}}(x, y)p_m^{\mathcal{J}}(y, z) \,. \tag{14.22}$$

Suppose that Γ is a group acting on $V^{\mathbb{N}}$. We say that a σ-field $\mathcal{J}$ in $V^{\mathbb{N}}$ is Γ-*closed* (also called "Γ-invariant") if, for all $A \in \mathcal{J}$ and $\gamma \in \Gamma$, we have $\gamma(A) \in \mathcal{J}$.

The main properties we need of conditioning a Markov chain on $\mathcal{J}$ are given in the following proposition. These properties are analogues of those we saw already for the unconditioned chain.

Proposition 14.34. *Let* Γ *be a transitive group of automorphisms of* (V, P), *acting on* $V^{\mathbb{N}}$ *via the diagonal action. If* $\langle X_n \rangle$ *is a Markov chain determined by* (V, P) *with initial state* $X_0 = o$ *and* $H(X_1) < \infty$ *and* $\mathcal{J} \subset \mathcal{I}$ *is a* Γ-*closed* σ-*field, then for every* $y \in V$ *and* $m \geq 1$, *we have*

$$-\mathbf{E}_y \sum_{z \in V} p_m^{\mathcal{J}}(y, z) \log p_m^{\mathcal{J}}(y, z) = H(X_m \mid \mathcal{J}) \,. \tag{14.23}$$

Therefore

$$H(X_{m+n} \mid \mathcal{J}) \leq H(X_n, X_{m+n} \mid \mathcal{J}) = H(X_n \mid \mathcal{J}) + H(X_m \mid \mathcal{J}) \,, \tag{14.24}$$

and the limit

$$h^{\mathcal{J}} := \lim_n \frac{H(X_n \mid \mathcal{J})}{n} \tag{14.25}$$

exists. Moreover,

$$\lim_n \frac{1}{n} \log p_n^{\mathcal{J}}(o, x_n) = -h^{\mathcal{J}} \quad \text{for } \mathbf{P}_o\text{-a.e. } \mathbf{x} \in \mathsf{V}^{\mathbb{N}} . \tag{14.26}$$

Proof. Choose $\gamma \in \Gamma$ that maps y to o. For every $z \in \mathsf{V}$, we have $p_m^{\mathcal{J}}(y, z) = p_m^{\gamma \mathcal{J}}(o, \gamma z) \circ \gamma = p_m^{\mathcal{J}}(o, \gamma z) \circ \gamma$, since $\mathcal{J}$ is Γ-closed, so

$$-p_m^{\mathcal{J}}(y, z) \log p_m^{\mathcal{J}}(y, z) = -\left[p_m^{\mathcal{J}}(o, \gamma z) \log p_m^{\mathcal{J}}(o, \gamma z) \right] \circ \gamma .$$

Summing over $z \in \mathsf{V}$, then applying $\mathbf{E}_y$ to both sides, and using the identity $\mathbf{E}_y[f \circ \gamma] = \mathbf{E}_o[f]$ for measurable functions $f \colon \mathsf{V}^{\mathbb{N}} \to \mathbb{R}_+$, we obtain (14.23). By (14.22),

$$H(X_n, X_{m+n} \mid \mathcal{J}) = -\mathbf{E}_o \Bigg[\sum_{y,z \in \mathsf{V}} p_n^{\mathcal{J}}(o, y) p_m^{\mathcal{J}}(y, z) \log(p_n^{\mathcal{J}}(o, y) p_m^{\mathcal{J}}(y, z)) \Bigg] . \tag{14.27}$$

Rewrite the logarithm of a product as a sum of logarithms in (14.27) and use the identity $\sum_z p_m^{\mathcal{J}}(y, z) = 1$ and Lemma 14.33 to deduce that

$$H(X_n, X_{m+n} \mid \mathcal{J}) = -\mathbf{E}_o \Bigg[\sum_{y \in \mathsf{V}} p_n^{\mathcal{J}}(o, y) \log p_n^{\mathcal{J}}(o, y) \Bigg]$$
$$- \sum_y p_n(o, y) \sum_z \mathbf{E}_o \Bigg[\frac{dQ_y}{dQ_o} p_m^{\mathcal{J}}(y, z) \log p_m^{\mathcal{J}}(y, z) \Bigg] .$$

By definition, $\mathbf{E}_o \Big[f \cdot \frac{dQ_y}{dQ_o} \Big] = \mathbf{E}_y[f]$ for any $\mathcal{J}$-measurable function f. Thus,

$$H(X_n, X_{m+n} \mid \mathcal{J}) - H(X_n \mid \mathcal{J}) = - \sum_y p_n(o, y) \mathbf{E}_y \Bigg[\sum_z p_m^{\mathcal{J}}(y, z) \log p_m^{\mathcal{J}}(y, z) \Bigg] ,$$

which equals $H(X_m \mid \mathcal{J})$ by (14.23). Together with Exercise 14.4, this proves (14.24), and (14.25) follows using Fekete's lemma. The proof of (14.26) is similar to the proof of Theorem 14.10(ii): the existence of the limit follows from the subadditive ergodic theorem, and this limit is a.e. constant by Lemma 14.11. The subadditivity required for (14.26) follows from the inequality $p_{n+m}^{\mathcal{J}}(x, z) \geq p_n^{\mathcal{J}}(x, y) \cdot p_m^{\mathcal{J}}(y, z)$, a consequence of Exercise 14.12. ◀

The next theorem is the main goal of this section. Recall that $\mathcal{T}$ is the tail σ-field in $\mathsf{V}^{\mathbb{N}}$.

Theorem 14.35. (Kaimanovich's Conditional Entropy Criterion) *Let Γ be a transitive group of automorphisms of (V, P), acting on $\mathsf{V}^{\mathbb{N}}$ via the diagonal action. Suppose that $\langle X_n \rangle$ is a Markov chain determined by (V, P) with initial state $X_0 = o$ and $H(X_1) < \infty$ and that $\mathcal{J} \subset \mathcal{I}$ is a Γ-closed σ-field. Then $\mathcal{J} = \mathcal{T}$ mod $\mathbf{P}_o$ if and only if $h^{\mathcal{J}} = 0$.*

Since $\mathcal{I} \subset \mathcal{T}$, the equality $\mathcal{J} = \mathcal{T}$ mod $\mathbf{P}_o$ immediately implies that $\mathcal{J} = \mathcal{I}$ mod $\mathbf{P}_o$. The converse follows from the identity $\mathcal{I} = \mathcal{T}$ mod $\mathbf{P}_o$, proved in Theorem 14.18 (for lazy chains) and in Theorem 14.47 (in general).

Proof. For $k < n$, by (14.24),

$$H(X_k \mid X_n, \mathcal{J}) + H(X_n \mid \mathcal{J}) = H(X_k, X_n \mid \mathcal{J}) = H(X_k \mid \mathcal{J}) + H(X_{n-k} \mid \mathcal{J}). \qquad (14.28)$$

Since $\mathcal{J} \subset \sigma\{X_j\}_{j \geq n}$, the Markov property implies that

$$H(X_k \mid X_n, \mathcal{J}) = H(X_k \mid X_n) = H(X_k \mid \{X_j\}_{j \geq n}).$$

This allows us to rewrite (14.28) as

$$H(X_k \mid \{X_j\}_{j \geq n}) = H(X_k \mid \mathcal{J}) + H(X_{n-k} \mid \mathcal{J}) - H(X_n \mid \mathcal{J}). \qquad (14.29)$$

The left-hand side is increasing in n and converges to $H(X_k \mid \mathcal{T})$ by Lemma 14.5. Considering $k = 1$, we see that $H(X_n \mid \mathcal{J}) - H(X_{n-1} \mid \mathcal{J})$ converges as $n \to \infty$. Since the averages $n^{-1}H(X_n \mid \mathcal{J})$ tend to $\boldsymbol{h}^{\mathcal{J}}$, we infer that

$$H(X_n \mid \mathcal{J}) - H(X_{n-1} \mid \mathcal{J}) \to \boldsymbol{h}^{\mathcal{J}}. \qquad (14.30)$$

By taking $n \to \infty$ in (14.29), we conclude that

$$H(X_k \mid \mathcal{T}) = H(X_k \mid \mathcal{J}) - k\boldsymbol{h}^{\mathcal{J}}. \qquad (14.31)$$

Thus $\boldsymbol{h}^{\mathcal{J}} = 0$ if $\mathcal{J} = \mathcal{T} \bmod \mathbf{P}_o$.

To see the converse, note that by the Markov property,

$$H(X_1, \ldots, X_{k-1} \mid X_k, \mathcal{T}) = H(X_1, \ldots, X_{k-1} \mid X_k, \mathcal{J});$$

adding this to (14.31) yields

$$H(X_1, \ldots, X_k \mid \mathcal{T}) = H(X_1, \ldots, X_k \mid \mathcal{J}) - k\boldsymbol{h}^{\mathcal{J}}. \qquad (14.32)$$

Suppose that $\boldsymbol{h}^{\mathcal{J}} = 0$. Then (14.32) and Lemma 14.5(i) imply that

$$\mathbf{P}_o\big[(X_1, \ldots X_k) = \cdot \mid \mathcal{J}\big] = \mathbf{P}_o\big[(X_1, \ldots X_k) = \cdot \mid \mathcal{T}\big]$$

for all k, whence

$$\mathbf{E}_o[f \mid \mathcal{J}] = \mathbf{E}_o[f \mid \mathcal{T}] \qquad (14.33)$$

for every Borel measurable function $f \colon \mathsf{V}^{\mathbb{N}} \to \mathbb{R}$ that depends on finitely many coordinates. Such functions are dense in $L^1(\mathsf{V}^{\mathbb{N}}, \mathscr{B}_{\mathsf{V}^{\mathbb{N}}}, \mathbf{P}_o)$, so (14.33) holds for all f in this L^1 space. Considering indicator functions of events in $\mathcal{T}$ shows that $\mathcal{J} = \mathcal{T} \bmod \mathbf{P}_o$. ◀

Although Theorem 14.35 provides a necessary and sufficient condition to identify $\mathcal{T}$, it is not so clear how one might actually use it. For example, how does one understand the Markov chain conditioned on $\mathcal{J}$? The following corollary is aimed at making this easier. In its statement, we use random finite subsets of V; here, the collection $\mathcal{W}$ of all finite subsets of V is itself countable, whence $\mathcal{W}$ comes with the σ-field consisting of all subsets of $\mathcal{W}$.

Corollary 14.36. (Kaimanovich's Enumeration Criterion) *Let Γ be a transitive group of automorphisms of (V, P), and let $\mathcal{J} \subset \mathcal{I}$ be a Γ-closed σ-field in the sequence space $\mathsf{V}^{\mathbb{N}}$. Denote by $\langle X_n \rangle$ a Markov chain determined by (V, P) with initial state $X_0 = o$ and $H(X_1) < \infty$. If for every n there is a finite $\mathcal{J}$-measurable random set $W_n \subset \mathsf{V}$ such that*

$$\limsup_{n\to\infty} \mathbf{P}_o[X_n \in W_n] > 0 \tag{14.34}$$

and $|W_n|^{1/n} \to 1$ a.s.-$\mathbf{P}_o$ as $n \to \infty$, then $\mathcal{J} = \mathcal{T}$ mod $\mathbf{P}_o$. More generally, it suffices that for every $\epsilon > 0$, there is a sequence $\langle W_n \rangle_{n\geq 1} = \langle W_n^{\epsilon} \rangle_{n\geq 1}$ of $\mathcal{J}$-measurable random sets that satisfies (14.34) and

$$\limsup_{n} \frac{1}{n} \log |W_n| \leq \epsilon \quad \mathbf{P}_o\text{-a.s.} \tag{14.35}$$

Proof. We prove the more general criterion. It is enough to show that $h^{\mathcal{J}} = 0$. Suppose that $h^{\mathcal{J}} > 0$ and define

$$S_n := \left\{ x \in \mathsf{V} \, ; \, p_n^{\mathcal{J}}(o, x) \leq \exp(-nh^{\mathcal{J}}/2) \right\}.$$

By (14.26), $\mathbf{P}_o[X_n \notin S_n] \to 0$ as $n \to \infty$. Take $\epsilon := h^{\mathcal{J}}/4$ and find sets W_n that satisfy (14.34) and (14.35). Then

$$\mathbf{P}_o[X_n \in W_n \cap S_n \mid \mathcal{J}] \leq |W_n| \cdot \exp(-nh^{\mathcal{J}}/2) \to 0 \quad \mathbf{P}_o\text{-a.s. as } n \to \infty.$$

Taking expectations and using bounded convergence, we get that $\mathbf{P}_o[X_n \in W_n \cap S_n] \to 0$ as $n \to \infty$, whence also $\mathbf{P}_o[X_n \in W_n] \to 0$, contradicting (14.34). ◄

We may also express the preceding theorem and corollary in terms of boundaries.

Corollary 14.37. *Let Γ be a transitive group of automorphisms of (V, P). Suppose that $\mathbf{X} = \langle X_n \rangle$ is a Markov chain determined by (V, P) with initial state $X_0 = o$ and $H(X_1) < \infty$. Given a Γ-boundary $(\Theta, \mathscr{F}_\Theta, \mathbf{b})$ of (V, P), it is a Poisson boundary iff $H(X_n \mid \mathbf{b}^{-1}\mathscr{F}_\Theta) = o(n)$. To verify the latter condition, it suffices to find a sequence of $\mathscr{F}_\Theta$-measurable mappings W_n, from Θ to the collection of finite subsets of V, such that $\limsup_n \mathbf{P}_o[X_n \in W_n(\mathbf{b}(\mathbf{X}))] > 0$ and $|W_n(\mathbf{b}(\mathbf{X}))|^{1/n} \to 1$ a.s.-$\mathbf{P}_o$.* ◄

To illustrate the power of Corollary 14.36, we apply it to several examples. The simplest is random walks with drift on lamplighter groups. Looking again at the proof of Corollary 14.21 that simple random walk on $\mathbb{Z}^{d\circledcirc}$ is non-Liouville for $d \geq 3$, we see that, in fact, an entire sub-σ-field of $\mathcal{I}$ is apparent, namely, the final configuration of the lamps. Indeed, for any random walk $\langle (\Psi_n, X_n) \rangle$ on a lamplighter graph $G^\circledcirc$ where $\langle X_n \rangle$ is transient, there is a limiting lamp configuration $\Psi_\infty := \lim_{n\to\infty} \Psi_n$ a.s. Clearly $\sigma(\Psi_\infty) \subseteq \mathcal{I}$, and we may ask whether these are equal mod $\mathbf{P}_o$. We will show that this is indeed the case in at least certain situations.

Consider a μ-walk on $\mathbb{Z}^{d\circledcirc}$ with μ supported on $\{(\overline{\mathbf{0}}, x)\}_{x \in \mathbb{Z}^d} \cup \{(\delta_{\mathbf{0}}, \mathbf{0})\}$, where $\overline{\mathbf{0}}$ is the function that is 0 at every site in $\mathbb{Z}^d$. Such a support entails that in every step of the walk, the only lamp that can change is at the current location of the lamplighter.

Corollary 14.38. (Lamplighter Boundaries: Biased Walks) *Fix $d \geq 1$, and let μ be a probability measure on $\mathbb{Z}^{d\odot}$ that is supported on $\{(\overline{\mathbf{0}}, x)\}_{x \in \mathbb{Z}^d} \cup \{(\delta_0, \mathbf{0})\}$ with support generating $\mathbb{Z}^{d\odot}$ as a semigroup. Denote by $\langle(\Psi_n, X_n)\rangle$ the corresponding convolution walk. Suppose that $\mathbf{E}_\mu[|X_1|] = \int |x| \, d\mu(\psi, x) < \infty$ and $\mathbf{E}_\mu[X_1] = \int x \, d\mu(\psi, x) \neq \mathbf{0}$. Then the tail σ-field of $\langle(\Psi_n, X_n) ; n \geq 0\rangle$ is generated by $\Psi_\infty : \mathbb{Z}^d \to \{0, 1\}$, where $\Psi_\infty(x) := \lim_n \Psi_n(x)$.*

Proof. The hypothesis implies that for some $i \leq d$, the standard basis vector $\mathbf{e}_i$ satisfies $\mathbf{E}_\mu[\langle X_1, \mathbf{e}_i \rangle] \neq 0$. Fix such an i, and without loss of generality, assume this expectation is positive. For each n, the events A_n and B_n are clearly independent, where

$$A_n := \left[\forall k < n \; \langle X_k - X_n, \mathbf{e}_i \rangle < 0\right] \qquad \text{and} \qquad B_n := \left[\forall m > n \; \langle X_m - X_n, \mathbf{e}_i \rangle > 0\right].$$

By transience of the convolution random walk $\langle \langle X_n, \mathbf{e}_i \rangle ; n \geq 0\rangle$ on $\mathbb{Z}$, there exists some $\delta > 0$ such that $\mathbf{P}(A_n) \geq \delta$ and $\mathbf{P}(B_n) \geq \delta$ for all $n \geq 1$. Let $D_n := \left[|X_n| \leq n^2\right]$, so that $\lim_{n\to\infty} \mathbf{P}(D_n) = 1$. Consider the half-space $\mathbb{H}_i := \{v \in \mathbb{Z}^d ; \langle v, \mathbf{e}_i \rangle \leq 0\}$. On the event $A_n \cap B_n \cap D_n$, the random walk satisfies $(\Psi_n, X_n) \in W_n$, where

$$W_n = W_n(\Psi_\infty) := \{(\psi, x) ; |x| \leq n^2, \; \psi(z) = \Psi_\infty(z)\mathbf{1}_{\mathbb{H}_i}(z - x)\}.$$

Since $|W_n|^{1/n} \to 1$ and $\liminf_n \mathbf{P}(A_n \cap B_n \cap D_n) \geq \delta^2$, Corollary 14.36 implies that $\sigma(\Psi_\infty)$ coincides with the tail $\mathcal{T}$ mod $\mathbf{P}_o$.   ◄

The next result we present was conjectured by Vershik and Kaimanovich (1979, 1983); it was proved by Erschler (2011, 2010) for $d \geq 5$ and by Lyons and Peres (2021) for $d = 3, 4$.

Theorem 14.39. (Lamplighter Boundaries: Unbiased Walks) *Fix $d \geq 3$. Let $\langle(\Psi_n, X_n)\rangle$ denote simple random walk on the Cayley graph of $\mathbb{Z}^{d\odot}$, endowed with the standard $2d + 1$ generators $\{(\overline{\mathbf{0}}, \pm\mathbf{e}_i)\}_{i=1}^d \cup \{(\delta_0, \mathbf{0})\}$. Then the tail σ-field of $\langle(\Psi_n, X_n) ; n \geq 0\rangle$ is generated by $\Psi_\infty : \mathbb{Z}^d \to \{0, 1\}$, where $\Psi_\infty(x) := \lim_n \Psi_n(x)$.*

In the next two proofs, C_d will denote a constant that depends only on d and could vary from line to line.

Lemma 14.40. *Let $\langle X_n \rangle$ be simple random walk on $\mathbb{Z}^d$, except for a little laziness: for each n, we have $\mathbf{P}[X_n = X_{n+1}] = \alpha \in (0, 1)$. Then, for all $n, k, r \in \mathbb{N}$ with $k \leq n$,*

$$\mathbf{E}\left[|\{m > n ; |X_k - X_m| \leq r\}|\right] \leq C_d r^d (n + 1 - k)^{-(d-2)/2}.$$

Proof. Such a random walk in $\mathbb{Z}^d$ satisfies $p_t(x, y) = \mathbf{P}_x[X_t = y] \leq C_d t^{-d/2}$ for all $x, y \in \mathbb{Z}^d$ and $t \geq 1$ (see Exercise 2.100 and Exercise 6.40, or Corollary 6.32). Thus, for $k \leq n < m$, we have

$$\mathbf{P}\left[|X_k - X_m| \leq r\right] = \mathbf{P}_0\left[|X_{m-k}| \leq r\right] \leq C_d r^d (m - k)^{-d/2}.$$

Summing this bound over all $m > n$ gives the result.   ◄

We will also need the elementary inequality

$$k \leq n/3 \quad \Longrightarrow \quad \sum_{j=0}^k \binom{n}{j} \leq 2n^k. \tag{14.36}$$

(Proof: $\binom{n}{j+1} \geq 2\binom{n}{j}$ for $j < n/3$.)

Proof of Theorem 14.39. For $x \in \mathbb{Z}^d$, denote by $B(x, r) := \{y \in \mathbb{Z}^d \, ; \, |x - y| \leq r\}$ the ball of radius r in the graph metric, centered at x. Set $r_n := \lfloor \log n \rfloor$ and $s_n := \lceil n/r_n \rceil$. Define $W_n(\psi_\infty)$ to be the set of all (ψ_n, x) such that there are S and U satisfying

(i) $S = \langle x_0, x_1, \ldots, x_{s_n} \rangle \in (\mathbb{Z}^d)^{s_n + 1}$ with $x_{i+1} \in B(x_i, r_n)$ for $0 \leq i < s_n$, $x_0 = \mathbf{0}$, and $x_{s_n} = x$,

(ii) $U \subseteq B(S, r_n) := \bigcup_{i=0}^{s_n} B(x_i, r_n)$ with $|U| \leq n^{3/4}$,

and

(iii) $\psi_n(y) = \begin{cases} \psi_\infty(y) & \text{for } y \in B(S, r_n) \setminus U \\ 0 & \text{for } y \notin B(S, r_n). \end{cases}$

The number of sequences S satisfying (i) is at most $(C_d r_n^d)^{s_n}$, which is subexponential in n. For each such S, the number of sets U satisfying (ii) is at most $\left(C_d s_n r_n^d\right)^{n^{3/4}}$ by (14.36), and this bound is also subexponential in n. Given S and U, the number of choices of ψ_n satisfying (iii) is at most $2^{|U|} \leq 2^{n^{3/4}}$. Multiplying these bounds, we obtain that $|W_n(\psi_\infty)|$ is subexponential.

We will prove that $\lim_{n \to \infty} \mathbf{P}\big[(\Psi_n, X_n) \in W_n(\Psi_\infty)\big] = 1$.

Write $S := \langle X_0, X_{r_n}, X_{2r_n}, \ldots, X_{(s_n - 1)r_n}, X_n \rangle$. Obviously S satisfies (i).

Let $U := B(S, r_n) \cap \{X_m \, ; \, m > n\}$. Write $A_n := \big[|U| \leq n^{3/4}\big]$. It follows from Lemma 14.40 that $\mathbf{E}\big[|U|\big] \leq C_d r_n^d n^{1/2} = o(n^{3/4})$, whence $\mathbf{P}(A_n) \to 1$ as $n \to \infty$ by Markov's inequality.

Now $\Psi_n(x) = \Psi_\infty(x)$ for every $x \in B(S, r_n) \setminus U$, and $\Psi_n(x) = 0$ for all $x \notin B(S, r_n)$. That is, $\widehat{X}_n \in W_n(\Psi_\infty)$ on the event A_n. ◀

In Example 14.30, we described the Poisson boundary for nearest-neighbor walks on a tree. Our next application of Corollary 14.36 is to general convolution random walks on a free group. The most natural questions for such walks are transience and convergence to an end. For simplicity, we will continue to identify ends of a tree with rays from o. Recall from Section 7.3 that a set S of vertices of a tree T *converges* to a ray $\xi = \langle \xi_0, \xi_1, \ldots \rangle \in \partial T$, where $\xi_0 = o$, if for all n, all but finitely many vertices in S are separated from o by ξ_n. We will then write $\lim S = \xi$. Let $\mathbb{T}_4$ be the 4-regular tree, regarded as the usual Cayley graph of the free group $\mathbb{Z} * \mathbb{Z}$ on two letters. Then $\mathbb{Z} * \mathbb{Z}$ acts on the boundary $\partial \mathbb{T}_4$ by $(\gamma, \xi) \mapsto \lim\{\gamma \xi_n \, ; \, n \geq 0\}$. Let μ be a measure on $\mathbb{Z} * \mathbb{Z}$ with support that generates $\mathbb{Z} * \mathbb{Z}$ as a group (not necessarily as a semigroup). A startling argument of Furstenberg (1973), as described in Exercise 14.36, shows – with no other assumptions – that the μ-walk is transient and, with a little further work, converges a.s. to an end. (Transience also follows from Theorem 14.48.) Building on these results, Kaimanovich (2000) proved the following.

Proposition 14.41. (Long-Range Walks on Free Groups) *Let $\mathbb{T}_4$ be the 4-regular tree, regarded as the usual Cayley graph of the free group $\mathbb{Z} * \mathbb{Z}$ on two letters. Denote the identity by o and graph distance of $x \in \mathbb{T}_4$ from the identity by $|x|$. Let μ be a measure on $\mathbb{Z} * \mathbb{Z}$ with support that generates that group as a semigroup. Suppose that μ has finite entropy $H(\mu)$ and finite logarithmic moment $\int \log(1 + |x|) \, d\mu$. Define $\mathbf{b}(S) := \lim S$ when S is end-convergent and to be some fixed end ξ' otherwise. Then $(\partial \mathbb{T}_4, \mathscr{B}_{\partial \mathbb{T}_4}, \mathbf{b})$ is a Poisson boundary for the μ-walk $\langle X_n \rangle$ on $\mathbb{T}_4$, and also a compactification $\mathbb{Z} * \mathbb{Z}$-boundary.*

Proof. As stated earlier, the μ-walk $\langle X_n \rangle$ converges $\mathbf{P}_o$-a.s. to a random ray $\mathbf{Z} = \langle Z_j \rangle_{j \geq 0}$ with $Z_0 = o$. Observe that every orbit of the action of $\mathbb{Z} * \mathbb{Z}$ on $\partial \mathbb{T}_4$ is dense. Since the support of μ generates $\mathbf{Z} * \mathbf{Z}$, it follows that the law of $\mathbf{Z}$ has full support in $\partial \mathbb{T}_4$. This implies that $(\partial \mathbb{T}_4, \mathscr{B}_{\partial \mathbb{T}_4}, \mathbf{b})$ is a compactification $\mathbf{Z} * \mathbf{Z}$-boundary for $\langle X_n \rangle$. It remains to identify it as a Poisson boundary, or equivalently, that the σ-field generated by $\mathbf{Z}$ coincides with the tail of $\langle X_n \rangle$ mod $\mathbf{P}_o$. By the preceding, there is some $\delta > 0$ so that each of the four neighbors y of o in $\mathbb{T}_4$ satisfies $\mathbf{P}[Z_1 = y] \le 1 - \delta$. Because the increments of the μ-walk are i.i.d., it follows that given $X_n \ne o$, the first step on the ray from X_n toward the end $\mathbf{Z}$ will coincide with the first step on the geodesic from X_n toward o with probability at most $1 - \delta$. Therefore, $\mathbf{P}[X_n \in \mathbf{Z}] \ge \delta$ for all n. Write $X_n = Y_1 \cdot Y_2 \cdots Y_n$, where $\langle Y_j \rangle$ are i.i.d. random elements of $\mathbf{Z} * \mathbf{Z}$ with distribution μ. Given $\epsilon > 0$, for large n we have

$$\mathbf{P}\big[|X_n| \ge e^{n\epsilon}\big] \le n\,\mathbf{P}\big[|Y_1| \ge e^{n\epsilon}/n\big] \le n\,\mathbf{P}\big[\log|Y_1| \ge n\epsilon/2\big] \to 0$$

as $n \to \infty$. Thus, the sets $W_n = W_n^\epsilon(\mathbf{Z}) := \langle Z_j \,;\, 0 \le j \le e^{n\epsilon} \rangle$ satisfy the hypotheses (14.34) and (14.35) of Corollary 14.36, so $\sigma(\mathbf{Z})$ generates $\mathcal{T}$ mod $\mathbf{P}_o$. ◀

We conclude this section with the following enhanced version of Corollary 14.36 from Lyons and Peres (2021). For applications, see Exercises 14.13 and 14.38.

Proposition 14.42. *Let Γ be a transitive group of automorphisms of (V, P), and let $\mathcal{J} \subset \mathcal{I}$ be a Γ-closed σ-field in the sequence space $\mathsf{V}^{\mathbb{N}}$. Denote by $\langle X_n \rangle$ a Markov chain determined by (V, P) with initial state $X_0 = o$ and $H(X_1) < \infty$. If, for every $\epsilon > 0$, there is a sequence of finite $\mathcal{J}$-measurable random sets $\langle \mathbf{W}_n^\epsilon \rangle_{n \ge 1}$ in V such that*

$$\limsup_{n \to \infty} \mathbf{P}_o[\exists m \ge n \;\; X_m \in \mathbf{W}_n^\epsilon] > 0 \tag{14.37}$$

and

$$\limsup_{n \to \infty} \frac{1}{n} \log |\mathbf{W}_n^\epsilon| \le \epsilon \quad \mathbf{P}_o\text{-a.s.,}$$

then $\mathcal{J} = \mathcal{T}$ mod $\mathbf{P}_o$.

Proof. It suffices to show that $\boldsymbol{h}^{\mathcal{J}} = 0$. Suppose that $\boldsymbol{h}^{\mathcal{J}} > 0$ and define

$$S_m := \big\{ x \in \mathsf{V} \,;\, p_m^{\mathcal{J}}(o, x) \le \exp(-m\boldsymbol{h}^{\mathcal{J}}/2) \big\}.$$

For $\epsilon > 0$,

$$\mathbf{P}_o[X_m \in \mathbf{W}_n^\epsilon \cap S_m \mid \mathcal{J}] \le |\mathbf{W}_n^\epsilon| \cdot \exp(-m\boldsymbol{h}^{\mathcal{J}}/2).$$

Summing over $m \ge n$, we deduce that for $0 < \epsilon < \boldsymbol{h}^{\mathcal{J}}/2$,

$$\mathbf{P}_o[\exists m \ge n \;\; X_m \in \mathbf{W}_n^\epsilon \cap S_m \mid \mathcal{J}] \le |\mathbf{W}_n^\epsilon| \cdot C \exp(-n\boldsymbol{h}^{\mathcal{J}}/2) \to 0 \tag{14.38}$$

almost surely as $n \to \infty$, where $C = C(\boldsymbol{h}^{\mathcal{J}})$ is a constant. Therefore,

$$\mathbf{P}_o[\exists m \ge n \;\; X_m \in \mathbf{W}_n^\epsilon \cap S_m] \to 0 \quad \text{as } n \to \infty. \tag{14.39}$$

By (14.26), $\mathbf{P}_o[\exists m \ge n \;\; X_m \notin S_m] \to 0$ as $n \to \infty$, since the intersection of these events has probability 0. In conjunction with (14.39), this implies that

$$\mathbf{P}_o[\exists m \ge n \;\; X_m \in \mathbf{W}_n^\epsilon] \to 0 \quad \text{as } n \to \infty,$$

contradicting the hypothesis (14.37). ◀

▷ **Exercise 14.13.**

Let $\mathbb{T}_3$ be a 3-regular tree, and suppose that ξ is an end of $\mathbb{T}_3$. Consider the Markov chain $\langle (\Psi_n, X_n) \rangle$ on the lamplighter graph $\mathbb{T}_3^{\circ}$ where $p\big((\psi, x), (\phi, y)\big)$ is $1/3$ if $x = y$ and ψ, ϕ differ only at x or if $\psi = \phi$ and y is the parent of x (the unique neighbor of x in the direction of ξ); $p\big((\psi, x), (\phi, y)\big) = 1/6$ if $\psi = \phi$ and x is the parent of y; and otherwise $p\big((\psi, x), (\phi, y)\big) = 0$. The projected Markov chain $\langle X_n \rangle$ is invariant under the transitive group of automorphisms of $\mathbb{T}_3$ that preserve ξ. Fix $o \in \mathbb{T}_3$.

  (a) Show that the projected chain $\langle X_n \rangle$ tends to ξ.

  (b) Show that the Avez entropy is 0 for the projected chain $\langle X_n \rangle$.

  (c) For the Markov chain $\langle (\Psi_n, X_n) \rangle$ started at $(\overline{\mathbf{0}}, o)$, show that the tail is generated by the final configuration of the lamps, Ψ_∞.

14.5 Appendix: Ergodic Theorems

In this section, we give proofs of Birkhoff's pointwise ergodic theorem and Kingman's subadditive ergodic theorem. Kingman's theorem was used in this chapter to show that random walks on groups have an asymptotic speed and to prove a Shannon theorem for entropy. Birkhoff's theorem is used in Chapter 17 to analyze speed and harmonic measure for random walks on Galton–Watson trees. The proof we give of Birkhoff's theorem is one that easily extends to prove Kingman's.

Suppose that $(\Omega, \mathscr{F}, \mathbf{P})$ is a probability space. Recall that $T \colon (\Omega, \mathscr{F}) \to (\Omega, \mathscr{F})$ means that T is measurable, that is, T^{-1} maps $\mathscr{F}$ to itself, and that $T \colon (\Omega, \mathscr{F}, \mathbf{P}) \to (\Omega, \mathscr{F}, \mathbf{P})$ means that T is also measure preserving, that is, $\mathbf{P} \circ T^{-1} = \mathbf{P}$. In this case, we call $(\Omega, \mathscr{F}, \mathbf{P}, T)$ a ***probability measure-preserving system***. The system is ***ergodic*** if every $\mathscr{F}$-measurable $f \colon \Omega \to \mathbb{R}$ that satisfies $f = f \circ T$ is constant almost everywhere. If $\mathscr{I}$ is the T-invariant σ-field $\{A \in \mathscr{F} \,;\, T^{-1}(A) = A\}$, then ergodicity is equivalent to $\mathscr{I}$ being $\mathbf{P}$-trivial.

▷ **Exercise 14.14.**

Often a system is defined to be ergodic if the only f that satisfy $f = f \circ T$ a.s. are constant a.s. Show that this is equivalent to the preceding definition. Show also that if $f \geq f \circ T$ a.s., then $f = f \circ T$ a.s.

Let $\langle X_n \rangle_{n \geq 0}$ be a real-valued stochastic process defined on $(\Omega, \mathscr{F}, \mathbf{P})$. If T is measure preserving and $X_n \circ T = X_{n+1}$ for all n, in other words, $X_n = X_0 \circ T^n$, then the process $\langle X_n \rangle_{n \geq 0}$ is ***stationary***, that is, for every n, the distribution of $\langle X_0, X_1, \ldots, X_n \rangle$ is the same as that of $\langle X_1, X_2, \ldots, X_{n+1} \rangle$. If we are not interested $(\Omega, \mathscr{F}, \mathbf{P})$ beyond the random variables X_n, then we can use instead the measure-preserving system $\big(\mathbb{R}^{\mathbb{N}}, \mathscr{B}_{\mathbb{R}^{\mathbb{N}}}, S, \mu\big)$, where μ is the law of $\langle X_n \,;\, n \geq 0 \rangle$ and S is the left shift. When this latter system is ergodic, then we also say that $\langle X_n \,;\, n \geq 0 \rangle$ is ergodic. Note that ergodicity here is equivalent to the invariant σ-field of $\langle X_n \rangle$ being trivial.

▷ **Exercise 14.15.**

(a) Prove that if the random variables X_n are i.i.d., then $\langle X_n \rangle$ is ergodic.

(b) Let $(\Omega, \mathscr{F}, \mathbf{P}, T)$ be a probability measure-preserving system. Let $(\Lambda, \mathscr{G})$ be a measurable space, and suppose there are measurable maps $\Psi: \Omega \to \Lambda$ and $\widetilde{T}: \Lambda \to \Lambda$ such that $\Psi \circ T = \widetilde{T} \circ \Psi$. Prove that $(\Lambda, \mathscr{G}, \mathbf{P} \circ \Psi^{-1}, \widetilde{T})$ is a probability measure-preserving system. Moreover, prove that if $(\Omega, \mathscr{F}, \mathbf{P}, T)$ is ergodic, then so is $(\Lambda, \mathscr{G}, \mathbf{P} \circ \Psi^{-1}, \widetilde{T})$.

(c) Use (b) to deduce the following statement: if $\langle X_n \rangle$ is ergodic, $g: \mathbb{R}^{\mathbb{N}} \to \mathbb{R}$ is measurable with respect to the product σ-field, and $Y_n = g(X_n, X_{n+1}, \ldots)$, then the process $\langle Y_n \rangle_{n \geq 0}$ is also ergodic.

For $0 \leq k < m$, define

$$S[k, m] := \sum_{j=k}^{m-1} X_j \, .$$

Theorem 14.43. (Birkhoff's Ergodic Theorem) *Suppose $\langle X_n \rangle_{n \geq 0}$ is stationary and satisfies $\mathbf{E}[|X_0|] < \infty$. Then almost surely and in L^1,*

$$\lim_{n \to \infty} \frac{1}{n} S[0, n] = \mathbf{E}[X_0 \mid \mathscr{I}] \, .$$

Proof. We will assume ergodicity and leave the general case to Exercise 14.16. We may assume that $X_n \geq 0$. The general case then follows by writing $X_n = X_n^+ - X_n^-$. Let

$$\bar{A} := \limsup_{n \to \infty} \frac{1}{n} S[0, n] \, .$$

Since $\bar{A} = \bar{A} \circ T$, ergodicity of T implies that $\bar{A} \in [0, \infty]$ is constant a.s. It remains to show that

$$\bar{A} = \liminf_{n \to \infty} \frac{1}{n} S[0, n] \text{ a.s.} \tag{14.40}$$

The basic approach is to look at partial averages $\ell^{-1} S[k, k + \ell)$ that are close to $\bar{A}$; these should occur regularly enough to imply (14.40).

Given $\epsilon > 0$, let $\alpha := (\bar{A} \wedge 1/\epsilon) - \epsilon$, where $x \wedge y := \min\{x, y\}$. For each k, define

$$L(k) := \min\left\{\ell \geq 1 \, ; \; \frac{S[k, k + \ell)}{\ell} \geq \alpha\right\} < \infty \, .$$

Note that by stationarity, the distribution of $L(k)$ does not depend on k.

First we consider the case where $L(k)$ is uniformly bounded by a constant, M. The key idea is to break the sum $\sum_{j=0}^{n-1} X_j$ into sums over blocks $[k_i, k_{i+1})$ where the average in each block is at least α, plus a remainder block of length less than M. Let $k_1 := 0$, and for $i \geq 1$, define

$$k_{i+1} := k_i + L(k_i) \, .$$

Discarding the remainder block, we obtain for $n > M$ that

$$\sum_{j=0}^{n-1} X_j \geq \sum_{i=1}^{m} S[k_i, k_i + L(k_i)) \geq \sum_{i=1}^{m} L(k_i)\alpha \geq (n - M)\alpha,$$

where $m := \max\{i \; ; \; k_i + L(k_i) < n\}$. The last inequality follows from our assumption that $L(k) \leq M$. Dividing by n and taking liminf on both sides, we deduce that

$$\liminf_{n \to \infty} \frac{1}{n} \sum_{j=0}^{n-1} X_j \geq \liminf_{n \to \infty} \frac{n - M}{n}\alpha = \alpha = (\bar{A} \wedge 1/\epsilon) - \epsilon.$$

Letting $\epsilon \downarrow 0$, it follows that $\liminf_{n \to \infty} S[0, n)/n \geq \bar{A}$, whence $\lim_{n \to \infty} S[0, n)/n = \bar{A}$.

Now suppose that $\langle L(k) \rangle$ is unbounded. Pick $M \geq 1$ so large that $\mathbf{P}[L(k) > M] < \epsilon$ for all k (the probability does not depend on k). For each k, define

$$X_k^* := \begin{cases} X_k & \text{if } L(k) \leq M \\ \alpha & \text{if } L(k) > M. \end{cases}$$

Note that if $L(k) > M$, then $X_k = S[k, k+1) < \alpha$, whence $X_k^* \geq X_k$. Also, let

$$S^*[k, m) := \sum_{j=k}^{m-1} X_j^* \quad \text{and} \quad L^*(k) := \min\left\{\ell \geq 1 \; ; \; \frac{S^*[k, k+\ell)}{\ell} \geq \alpha\right\}.$$

Then $L^*(k) \leq M$, since $X_j^* \geq X_j$ for all j. Splitting the sum as in the previous case, we have

$$\sum_{j=0}^{n-1} X_j^* \geq (n - M)\alpha. \tag{14.41}$$

Hence $n\,\mathbf{E}[X_0^*] \geq (n - M)\alpha$. Note that $\mathbf{E}[X_0^*] \leq \mathbf{E}[X_0] + \alpha\epsilon$ by definition of X^*. It follows that

$$\mathbf{E}[X_0] \geq \frac{n - M}{n}\alpha - \alpha\epsilon.$$

Letting $n \to \infty$ gives $\mathbf{E}[X_0] \geq \alpha - \alpha\epsilon$; since $\alpha = (\bar{A} \wedge 1/\epsilon) - \epsilon$, by taking $\epsilon \downarrow 0$ we deduce that a.s.

$$\mathbf{E}[X_0] \geq \bar{A}. \tag{14.42}$$

Next, define $Z_n := X_n^* - X_n \geq 0$ for all n. Since $\langle Z_n \rangle$ is ergodic, we can apply the conclusions already obtained for $\langle X_n \rangle$ to $\langle Z_n \rangle$. Thus, by (14.42), we find that a.s.

$$\alpha\epsilon \geq \mathbf{E}[Z_0] \geq \limsup_{n \to \infty} \frac{1}{n} \sum_{j=0}^{n-1} Z_j.$$

Taking liminf in (14.41), we obtain that a.s.

$$\liminf_{n\to\infty} \frac{1}{n}\sum_{j=0}^{n-1} X_j + \limsup_{n\to\infty} \frac{1}{n}\sum_{j=0}^{n-1} Z_j \geq \liminf_{n\to\infty} \frac{1}{n}\left(\sum_{j=0}^{n-1} X_j + \sum_{j=0}^{n-1} Z_j\right) \geq \alpha\,.$$

Combining the two preceding displays of inequalities, we get

$$\liminf_{n\to\infty} \frac{S[0,n)}{n} + \alpha\epsilon \geq \alpha \text{ a.s.;}$$

letting $\epsilon \downarrow 0$ gives $\liminf_{n\to\infty} S[0,n)/n \geq \bar{A}$ a.s.

This completes the proof of a.s. convergence, except for identifying $\bar{A}$ as $\mathbf{E}[X_0]$. If $\langle X_n \rangle$ is bounded, then $\bar{A} = \mathbf{E}[X_0]$ by the bounded convergence theorem. In general, for each $C > 0$, the bounded case we just established shows that a.s.

$$\bar{A} \geq \lim_{n\to\infty} \frac{1}{n}\sum_{j=0}^{n-1} (X_i \wedge C) = \mathbf{E}[X_0 \wedge C]\,.$$

Letting $C \to \infty$ and using the monotone convergence theorem yields $\bar{A} \geq \mathbf{E}[X_0]$. On the other hand, Fatou's lemma and (14.40) show the reverse inequality.

Finally, the convergence in L^1 of $S[0,n)/n$ to $\bar{A}$ follows from Theorem 4.6.3 in Durrett (2019). ◀

▷ **Exercise 14.16.**
Prove the general version of the Birkhoff ergodic theorem that does not require ergodicity.

The method we used to prove Birkhoff's ergodic theorem can also be applied to prove the subadditive ergodic theorem. This theorem has many applications, for example, in first-passage percolation and random walk on groups. The version we present is due to Kingman (1968). (In fact, what we state is the superadditive version.) This includes Birkhoff's theorem: take $Y(m,n) := S[m,n)$.

Theorem 14.44. (Kingman's Superadditive Ergodic Theorem) *Let $(\Omega, \mathscr{F}, \mathbf{P}, T)$ be a probability measure-preserving system. Let $\langle Y(m,n)\,;\ 0 \leq m \leq n \rangle$ be a stochastic process such that for all $m \leq n$,*
   (i) $Y(0,n) \geq Y(0,m) + Y(m,n)$,
   (ii) $Y(m,n) \circ T = Y(m+1,n+1)$, and
   (iii) $\mathbf{E}\big[|Y(m,n)|\big] < \infty$.
Then almost surely the following limit exists:

$$\beta := \lim_{n\to\infty} \frac{Y(0,n)}{n} \in (-\infty, \infty]\,.$$

Moreover, $\beta = \lim_{n\to\infty} \mathbf{E}\big[Y(0,n) \mid \mathscr{I}\big]/n$ a.s. and if $\mathbf{E}[\beta] < \infty$, then $\mathbf{E}\Big[\Big|\frac{Y(0,n)}{n} - \beta\Big|\Big] \to 0$.

Proof. We will assume ergodicity and leave the general case to Exercise 14.17. By the superadditivity assumption (i), $Y(0,n) \geq Y(0,n-1) + Y(n-1,n)$, which inductively yields that $Y(0, n) \geq \sum_{k=1}^{n} Y(k - 1, k)$. Applying this to $n - m$ in place of n and using (ii), we deduce that

$$\widetilde{Y}(m, n) := Y(m, n) - \sum_{k=m+1}^{n} Y(k - 1, k) \geq 0.$$

Since the process $S(m, n) := \sum_{k=m+1}^{n} Y(k - 1, k)$ is additive, the process $\widetilde{Y}(m, n)$ satisfies (i), (ii), and (iii). Birkhoff's ergodic theorem may be applied to $\langle Y(k - 1, k) \rangle$. This reduction allows us to assume that $Y(m, n) \geq 0$, since otherwise we may replace Y by $\widetilde{Y}$.

Let $\beta := \limsup_{n \to \infty} Y(0, n)/n$. Since $\beta \geq \beta \circ T$, ergodicity ensures that β is constant almost surely. Given $\epsilon > 0$, let $\alpha := (\beta \wedge 1/\epsilon) - \epsilon$ and define

$$L(k) := \min\{\ell \geq 1 ; \; Y(k, k + \ell) \geq \ell\alpha\}.$$

Choose M so large that $\mathbf{P}[L(k) > M] < \epsilon$ for all k. Define

$$L^*(k) := \begin{cases} L(k) & \text{if } L(k) \leq M \\ 1 & \text{if } L(k) > M. \end{cases}$$

Let $k_1 := L(0)$ and $k_{i+1} := k_i + L^*(k_i)$. Analogously to the proof of Birkhoff's ergodic theorem, we have

$$Y(0, n) \geq Y(0, k_1) + Y(k_1, k_2) + \cdots + Y(k_{m-1}, k_m) \geq \left[(n - M) - \sum_{k=0}^{n} \mathbf{1}_{[L(k)>M]}\right]\alpha,$$

where $m := \max\{i ; \; k_i + L^*(k_i) \leq n\}$. Since $\langle \mathbf{1}_{[L(k)>M]} \rangle_k$ is ergodic, by Birkhoff's ergodic theorem, we have that a.s.

$$\liminf_{n \to \infty} \frac{Y(0, n)}{n} \geq \alpha(1 - \mathbf{P}[L(0) > M]) \geq \alpha(1 - \epsilon).$$

Letting $\epsilon \downarrow 0$ and recalling the definition of α, we conclude that $\liminf_{n \to \infty} Y(0, n)/n \geq \beta$ a.s., whence $\lim_{n \to \infty} Y(0, n)/n = \beta$ a.s.

We next prove that $\beta = \lim_{n \to \infty} \mathbf{E}[Y(0, n)]/n$. First, note that $\lim_{n \to \infty} \mathbf{E}[Y(0, n)]/n$ exists by superadditivity (see Exercise 3.9). By Fatou's lemma,

$$\beta \leq \liminf_{n \to \infty} \frac{\mathbf{E}[Y(0, n)]}{n} = \lim_{n \to \infty} \frac{\mathbf{E}[Y(0, n)]}{n}.$$

If $\beta < \infty$, then we still need to prove the other direction. For every integer $n > 0$, by Birkhoff's ergodic theorem and superadditivity, almost surely

$$\frac{1}{n} \mathbf{E}[Y(0, n)] = \lim_{k \to \infty} \frac{1}{kn} \sum_{j=1}^{k} Y((j - 1)n, jn) \leq \lim_{k} \frac{1}{kn} Y(0, kn) = \beta.$$

The convergence in L^1 of $Y(0, n)/n$ to β now follows from Theorem 4.6.3 in Durrett (2019). ◀

▷ **Exercise 14.17.**
Prove the more general version of Kingman's subadditive ergodic theorem that does not require ergodicity.

Just as Birkhoff's ergodic theorem applies to stationary sequences of random variables, Kingman's subadditive ergodic theorem applies to stationary triangular arrays of random variables. That is, let $\blacktriangle := \{(m, n) \, ; \, 0 \leq m \leq n\}$ and suppose $\langle Y(m, n) \, ; \, (m, n) \in \blacktriangle \rangle$ is a triangular array of real-valued random variables whose law on the space $\mathbb{R}^{\blacktriangle}$ is unchanged under the shift $T \colon \langle y(m, n) \rangle \mapsto \langle y(m + 1, n + 1) \rangle$. Then (i) and (iii) of Theorem 14.44 imply the conclusions of that theorem.

14.6 Appendix: The Zero-Two Law for Transitive Markov Chains

In this section, we show that for transitive Markov chains, the invariant σ-field coincides with the tail σ-field up to sets of $\mathbf{P}_o$-measure 0. The first step is the following theorem. Recall the notation $\|\mu - \nu\|_{\mathrm{TV}} := \sup_{A \in \mathscr{F}} |\mu(A) - \nu(A)|$.

Theorem 14.45. *Let $p_n(\cdot, \cdot)$ be the n-step transition probabilities for a transitive Markov chain on a countable set V. Fix a basepoint o. Then, for any integer $k > 0$,*

$$\lim_{n \to \infty} \|p_{n+k}(o, \cdot) - p_n(o, \cdot)\|_{\mathrm{TV}} \in \{0, 1\} \, . \tag{14.43}$$

Because of transitivity, the norm in (14.43) does not depend on the choice of o.

The distance $\|p_{n+k}(o, \cdot) - p_n(o, \cdot)\|_{\mathrm{TV}}$ is decreasing in n by Exercise 13.2(c). Therefore, if the limit in (14.43) is 1, then

$$\forall n \quad \|p_{n+k}(o, \cdot) - p_n(o, \cdot)\|_{\mathrm{TV}} = 1 \, .$$

Theorem 14.45 as stated does not hold without the assumption of transitivity (as demonstrated in the following exercise), but there is an extension valid for general chains; see Theorem 14.53 in the notes.

▷ **Exercise 14.18.**
Let G denote $\mathbb{Z}^3$ with a loop added at o. Show that simple random walk on G satisfies $0 < \lim_{n \to \infty} \|p_{n+1}(o, \cdot) - p_n(o, \cdot)\|_{\mathrm{TV}} < 1$.

As a warm-up to the general proof, it will be useful to consider first a special case.

Proof of Theorem 14.45 for random walks on groups. We first consider the case where V is a countable group Γ and $p(x, y) = \mu(x^{-1}y)$ for some probability measure μ on Γ. Take the basepoint o to be the identity and abbreviate $p_\ell(\cdot) := p_\ell(o, \cdot)$.

Suppose there exists an integer $\ell > 0$ such that $\|p_{\ell+k} - p_\ell\|_{\mathrm{TV}} < 1$. This implies that there exists $\epsilon > 0$ and $\gamma_* \in \Gamma$ such that $p_{\ell+k}(\gamma_*) \geq \epsilon$ and $p_\ell(\gamma_*) \geq \epsilon$. Let $\mu^{\otimes \ell}$ be the ℓ-fold product probability measure on the Cartesian product Γ^ℓ. In this notation, for every $\gamma \in \Gamma$, we have

$$\mu^{\otimes \ell}\big(\{(\gamma_1, \ldots, \gamma_\ell) \, ; \, \gamma_1 \cdots \gamma_\ell = \gamma\}\big) = p_\ell(\gamma) \, . \tag{14.44}$$

Let v_ℓ denote $\mu^{\otimes\ell}$ conditioned on $\{(\gamma_1, \ldots, \gamma_\ell); \gamma_1 \cdots \gamma_\ell = \gamma_*\}$. Then $\mu^{\otimes\ell} \geq \epsilon v_\ell$, since $p_\ell(\gamma_*) \geq \epsilon$. Thus, we can define a probability measure θ_ℓ on Γ^ℓ by

$$\mu^{\otimes\ell} = \epsilon v_\ell + (1 - \epsilon)\theta_\ell . \tag{14.45}$$

We can similarly write $\mu^{\otimes(\ell+k)} = \epsilon v_{\ell+k} + (1 - \epsilon)\theta_{\ell+k}$, where $v_{\ell+k}$ is $\mu^{\otimes(\ell+k)}$ conditioned on $\{(\gamma_1, \ldots, \gamma_{\ell+k}); \gamma_1 \cdots \gamma_{\ell+k} = \gamma_*\}$ and $\theta_{\ell+k}$ is a probability measure on $\Gamma^{\ell+k}$.

Next, we will construct a sequence of i.i.d. random variables $W_1, W_2, \ldots$ taking values in the union $\Gamma^\ell \cup \Gamma^{\ell+k}$ with distribution

$$\frac{1}{2}\left(\mu^{\otimes\ell} + \mu^{\otimes(\ell+k)}\right). \tag{14.46}$$

(To pick a variable W from the distribution (14.46), toss a fair coin and pick W from $\mu^{\otimes\ell}$ if it falls heads and from $\mu^{\otimes(\ell+k)}$ otherwise.) We will couple this sequence with another sequence $W_1', W_2', \ldots$ that has the same law.

We start by describing the joint law of one pair (W, W'). Toss a coin with probability ϵ of heads. If it falls heads, sample the pair (W, W') from $\frac{1}{2}(v_\ell \otimes v_{\ell+k} + v_{\ell+k} \otimes v_\ell)$; if tails, sample W from $\frac{1}{2}(\theta_\ell + \theta_{\ell+k})$ and set $W' := W$. As a formula, the joint distribution of (W, W') is

$$\frac{\epsilon}{2}(v_\ell \otimes v_{\ell+k}) + \frac{\epsilon}{2}(v_{\ell+k} \otimes v_\ell) + \frac{1-\epsilon}{2}\text{diag}_*(\theta_\ell) + \frac{1-\epsilon}{2}\text{diag}_*(\theta_{\ell+k}), \tag{14.47}$$

where $\text{diag}_*(\theta)$ is the distribution of the random variable (U, U) when U is distributed according to θ. Clearly, the marginal distributions of both W and W' in (14.47) are the law specified in (14.46).

We define (W_i, W_i') inductively. First, pick the pair (W_1, W_1') from the joint distribution (14.47). For the inductive step, suppose that (W_i, W_i') have been defined for all $i < m$. If

$$\forall j < m \qquad \sum_{i=1}^{j} \text{length}(W_i') \neq k + \sum_{i=1}^{j} \text{length}(W_i),$$

then let (W_m, W_m') be an independent pick from (14.47); otherwise, let W_m be an independent pick from μ^ℓ and set $W_m' := W_m$.

Observe that $t \mapsto \sum_{i=1}^{t}\left[\text{length}(W_i') - \text{length}(W_i)\right]$ is a symmetric, nondegenerate random walk on $\mathbb{Z}$ with increments $\{-k, 0, k\}$, started at 0 and absorbed at k; this walk is recurrent, so the first time τ that it reaches k is a.s. finite. The concatenation $W_1 W_2 \ldots$ is an infinite word in which the first n letters have the distribution $\mu^{\otimes n}$ for all n. Thus, the *product* of the first n letters, for any n, gives an element $X_n \in \Gamma$ distributed according to p_n. The same holds for the product $X_n' \in \Gamma$ of the first n letters of the concatenation $W_1' W_2' \ldots$.

This coupling ensures that for each i, either $W_i = W_i'$, or W_i and W_i' are two words with lengths differing by k but both representing the same group element γ_*. Therefore, on the event $\sum_{i=1}^{\tau} \text{length}(W_i) \leq n$, the first n letters of the concatenation $W_1 W_2 \ldots$ and the first $n + k$ letters of the concatenation $W_1' W_2' \ldots$ yield products X_n and X_{n+k}' that are equal in Γ because of the coupling. Thus, by the coupling interpretation of total variation, Exercise 13.2(b), we have

$$\|p_{n+k} - p_n\|_{\text{TV}} \leq \mathbf{P}[X_{n+k}' \neq X_n] \leq \mathbf{P}\left[\sum_{i=1}^{\tau} \text{length}(W_i) > n\right] \to 0$$

as $n \to \infty$, since $\tau < \infty$ a.s. ◀

Next we prove Theorem 14.45 in general. The proof is similar to the group case but involves paths rather than words.

Proof of Theorem 14.45 for general transitive chains. Suppose that

$$\|p_{\ell+k}(o, \cdot) - p_\ell(o, \cdot)\|_{\mathrm{TV}} < 1.$$

By transitivity, there exists $\epsilon > 0$ such that for every $x \in \mathsf{V}$, there is a vertex $y = y(x) \in \mathsf{V}$ that satisfies $p_{\ell+k}(x, y) \geq \epsilon$ and $p_\ell(x, y) \geq \epsilon$. Denote by $p_{[1,\ell]}^{x_0}(\cdot)$ the distribution on paths of length ℓ in V given by

$$p_{[1,\ell]}^{x_0}(x_1, \ldots, x_\ell) = \prod_{i=1}^{\ell} p(x_{i-1}, x_i).$$

For every $x, z \in \mathsf{V}$, we have

$$p_{[1,\ell]}^x\big(\{(x_1, \ldots, x_\ell);\ x_\ell = z\}\big) = p_\ell(x, z). \tag{14.48}$$

For $x, x_1, \ldots, x_\ell \in \mathsf{V}$, define $v_\ell^x(x_1, \ldots, x_\ell) := p_{[1,\ell]}^x(x_1, \ldots, x_\ell)/p_\ell(x_\ell)$ if $x_\ell = y(x)$, and $v_\ell^x(x_1, \ldots, x_\ell) := 0$ otherwise. Then, for every $x \in \mathsf{V}$, we have $p_{[1,\ell]}^x(\cdot) \geq \epsilon v_\ell^x$, so we can define a probability measure θ_ℓ^x on V^ℓ by

$$p_{[1,\ell]}^x(\cdot) = \epsilon v_\ell^x + (1 - \epsilon)\theta_\ell^x. \tag{14.49}$$

We can similarly write $p_{[1,\ell+k]}^x(\cdot) = \epsilon v_{\ell+k}^x + (1 - \epsilon)\theta_{\ell+k}^x$.

Given $x \in \mathsf{V}$, let (W, W') be a pair of random paths in V with joint law

$$\Upsilon^x := \frac{\epsilon}{2}(v_\ell^x \otimes v_{\ell+k}^x) + \frac{\epsilon}{2}(v_{\ell+k}^x \otimes v_\ell^x) + \frac{1-\epsilon}{2}\mathrm{diag}_*(\theta_\ell^x) + \frac{1-\epsilon}{2}\mathrm{diag}_*(\theta_{\ell+k}^x). \tag{14.50}$$

Clearly, the marginal distributions of both W and W' in (14.50) are

$$\frac{1}{2}\big(p_{[1,\ell]}^x(\cdot) + p_{[1,\ell+k]}^x(\cdot)\big). \tag{14.51}$$

Observe that under Υ^x, the paths W and W' have the same final vertex a.s.

Next, we define inductively a sequence of pairs (W_i, W_i') of paths in V. First, pick the pair (W_1, W_1') from the joint distribution Υ^o in (14.50). For the inductive step, suppose that (W_i, W_i') has been defined for all $i < m$ and that z_m is the final vertex of W_{m-1}. If

$$\forall j < m \quad \sum_{i=1}^{j} \mathrm{length}(W_i') \neq k + \sum_{i=1}^{j} \mathrm{length}(W_i),$$

then let (W_m, W_m') be an independent pick from Υ^{z_m}; otherwise, let W_m be an independent pick from $p_{[1,\ell]}^{z_m}(\cdot)$ and set $W_m' := W_m$. As before, the first time τ that the random walk $t \mapsto \sum_{i=1}^{t}\big[\mathrm{length}(W_i') - \mathrm{length}(W_i)\big]$ reaches k is a.s. finite. The concatenation $W_1 W_2 \ldots$ is an infinite path where for every n, the nth vertex X_n is distributed according to $p_n(o, \cdot)$. The same holds for the nth vertex X_n' on the infinite path $W_1' W_2' \ldots$. Therefore,

$$\|p_{n+k}(o, \cdot) - p_n(o, \cdot)\|_{\mathrm{TV}} \leq \mathbf{P}[X_{n+k}' \neq X_n] \leq \mathbf{P}\Big[\sum_{i=1}^{\tau}\mathrm{length}(W_i) > n\Big] \to 0$$

as $n \to \infty$, since $\tau < \infty$ a.s. ◄

Recall from Section 14.2 that a function $\Lambda\colon \mathsf{V} \times \mathbb{N} \to \mathbb{R}$ is called **space-time harmonic** if it is a harmonic function for the space-time chain $\langle (X_n, n)\,;\ n \geq 0 \rangle$.

The following consequence of the zero-two law shows that bounded space-time harmonic functions coincide with harmonic functions on the Markov chain trajectory. The meaning of this latter phrase is spelled out in (14.52). Note that it need not be the case that $u(x) = \Lambda(x, n)$ for all x and n: to see this, consider simple random walk on $\mathbb{Z}$ and define $\Lambda(x, n)$ to be the indicator that x and n have the same parity.

Proposition 14.46. *Consider a transitive chain with transition kernel $p(\cdot, \cdot)$ on V, and let $\Lambda\colon \mathsf{V} \times \mathbb{N} \to \mathbb{R}$ be a bounded space-time harmonic function.*

  (i) *Suppose that $x \in \mathsf{V}$, $\ell \geq 0$, and $k \geq 1$. If $p_\ell(o, x) > 0$ and $p_{\ell+k}(o, x) > 0$, then*
$$\Lambda(x, \ell) = \Lambda(x, \ell + k).$$

  (ii) *There exists a bounded harmonic function $u\colon \mathsf{V} \to \mathbb{R}$ such that for all $x \in \mathsf{V}$ and $n \in \mathbb{N}$,*

$$p_n(o, x) > 0 \quad \Longrightarrow \quad u(x) = \Lambda(x, n). \tag{14.52}$$

Proof. (i) The hypothesis implies that $\|p_\ell(o, \cdot) - p_{\ell+k}(o, \cdot)\|_{\mathrm{TV}} < 1$, so by the zero-two law and transitivity, $\|p_n(x, \cdot) - p_{n+k}(x, \cdot)\|_{\mathrm{TV}} \to 0$ as $n \to \infty$. Since Λ is space-time harmonic, for every $n \in \mathbb{N}$,

$$\Lambda(x, \ell) = \mathbf{E}_x \Lambda(X_{n+k}, \ell + n + k) \quad \text{and} \quad \Lambda(x, \ell + k) = \mathbf{E}_x \Lambda(X_n, \ell + n + k).$$

Therefore,

$$|\Lambda(x, \ell) - \Lambda(x, \ell + k)| \leq 2 \sup |\Lambda| \cdot \|p_n(x, \cdot) - p_{n+k}(x, \cdot)\|_{\mathrm{TV}} \to 0 \quad \text{as} \quad n \to \infty.$$

(ii) For every $x \in \mathsf{V}$ and $n \in \mathbb{N}$ with $p_n(o, x) > 0$, define $u(x) := \Lambda(x, n)$. Part (i) implies this does not depend on the choice of n. Harmonicity of u follows: Given x and n as earlier, if $p(x, y) > 0$, then $p_{n+1}(o, y) \geq p_n(o, x)p(x, y) > 0$, so $u(y) = \Lambda(y, n + 1)$. Thus

$$u(x) = \Lambda(x, n) = \sum_{y \in \mathsf{V}} p(x, y)\Lambda(y, n + 1) = \sum_{y \in \mathsf{V}} p(x, y)u(y). \qquad \blacktriangleleft$$

We can now deduce the main goal of this section. Given the preceding proposition, the proof is identical to the proof of Theorem 14.18, so we do not repeat it. Although (14.52) is weaker than what was established for lazy chains, namely, (14.13), it was all that was used: see (14.14).

Theorem 14.47. *Consider a transitive chain on V, and let $o \in \mathsf{V}$. Then the $\mathbf{P}_o$-completions of the tail σ-field $\mathcal{T}$ and the invariant σ-field $\mathcal{I}$ coincide.* $\qquad \blacktriangleleft$

14.7 Notes

The theory developed in this chapter originates with the pioneering works of Furstenberg (1963, 1971a, 1971b). Given a discrete group Γ and a measure μ on it, recall that (Γ, μ) is Liouville if all bounded μ-harmonic functions on Γ are constant. Furstenberg (1973) conjectured that Γ is amenable iff there is a symmetric μ whose support generates Γ such that (Γ, μ) is Liouville. This was established by Rosenblatt (1981) and Kaimanovich and Vershik (1983). These authors also proved the following.

Theorem 14.48. *Let Γ be a nonamenable discrete group. Then every distribution μ on Γ is non-Liouville.*

For symmetric μ of finite entropy, this follows from Proposition 14.6 and Theorem 14.20. If we just assume finite entropy, then we can replace Proposition 14.6 by Corollary 6.32(i) to deduce that $h > 0$ (using the observation that P is Liouville iff $(P + I)/2$ is Liouville). For other μ, this can be deduced from Exercise 14.22; see Kaimanovich and Vershik (1983), Theorem 4.2. The following elegant, general argument was given by Björklund (2014).

Proof of Theorem 14.48. Since Γ is nonamenable, there exists a compact Hausdorff space K on which Γ acts by homeomorphisms without an invariant measure (this follows from one of the standard characterizations of amenability; see, for example, Paterson (1988)). By Exercise 14.36(a), there is a probability measure ν on K that satisfies $\mu * \nu = \nu$, that is,

$$\int_\Gamma \int_K \varphi(\gamma x) \, d\nu(x) \, d\mu(\gamma) = \int_K \varphi \, d\nu \tag{14.53}$$

for all $\varphi \in C(K)$. Since ν is not invariant, there exists $\psi \in C(K)$ such that $u(g) := \int_K \psi(gx) \, d\nu(x)$ is not constant on Γ. Clearly u is bounded. Applying (14.53) with $\varphi = \psi \circ g$ yields

$$\int_\Gamma u(g\gamma) \, d\mu(\gamma) = \int_\Gamma \int_K \psi(g\gamma x) \, d\nu(x) \, d\mu(\gamma) = \int_K \psi(gx) \, d\nu(x) = u(g) \,;$$

in other words, u is μ-harmonic, so (Γ, μ) is not Liouville. ◄

Another open question had been whether there exists an amenable group Γ with a symmetric non-Liouville measure whose support generates Γ as a group. To answer this, Vershik and Kaimanovich (1979, 1983) utilized the lamplighter groups $\mathbb{Z}^{d^\odot} = \mathbb{Z}_2 \wr \mathbb{Z}^d$, which are key examples in this chapter; they denoted $\mathbb{Z}^{d^\odot}$ by G_d.

Every convolution walk on every nilpotent group is Liouville, as shown by Dynkin and Maljutov (1961). By a celebrated theorem of Gromov (1981a), it follows that the same holds for all groups of polynomial growth, as noted by Kaimanovich and Vershik (1983) (compare Exercise 14.30). As described in Section 7.9, transitive graphs of polynomial growth are very close to Cayley graphs of the same growth rate. This was proved by Trofimov (1984). A cleaner way to say that they are "close" to Cayley graphs was noted by Godsil, Imrich, Seifter, Watkins, and Woess (1989) as a consequence of results by Trofimov (1984) and Sabidussi (1964): every transitive graph G of polynomial growth has the property that for some $n \geq 1$, the graph nG is a Cayley graph of a transitive subgroup of $\mathrm{Aut}(G)$, where nG has vertex set $\mathsf{V}(G) \times \{1, \ldots, n\}$ and edges connecting pairs (x, j) to (y, k) when $[x, y] \in \mathsf{E}(G)$. Thus, we have the following theorem:

Theorem 14.49. *Every transitive Markov chain with invariant graph metric on a graph of polynomial growth is Liouville.*

Proof. Let n be such that nG is a Cayley graph of a transitive subgroup, Γ, of the automorphism group preserving the Markov chain and the graph, G. In fact, Sabidussi (1964) shows that for $\gamma \in \Gamma$,

$1 \le j \le n$, and $x \in V(G)$, there is some $k \in [1, n]$ such that $\gamma(x, j) = (\gamma x, k)$. In other words, the left translations of Γ on nG project to the action of Γ on G as a subgroup of $\mathrm{Aut}(G)$. If $p(\cdot, \cdot)$ are the transition probabilities on $V(G)$, then let $q((x, j), (y, k)) := p(x, y)/n$ to define a transitive Markov chain on nG. It follows from the above relationship of the actions of Γ that this chain is invariant under Γ, whence it is a convolution walk. Since Γ has polynomial growth, this chain on nG is Liouville. Exercise 14.28 then gives that the original chain is also Liouville. ◀

When $\mathbf{E}_o[\log(1 + d(o, X_1))] < \infty$, Theorem 14.49 is also a consequence of Exercise 14.23 and Theorem 14.7. Frisch, Hartman, Tamuz, and Vahidi Ferdowsi (2019) provide a converse to Theorem 14.49, namely, that every finitely generated group that is not of polynomial growth admits a symmetric probability measure with finite entropy that is not Liouville.

Kaimanovich and Vershik (1983) asked whether the Liouville property for simple random walk on a Cayley graph of a finitely generated group is stable under change of generators. More generally, it is natural to ask the following: suppose a transitive network (G, c) with the Liouville property is roughly isometric to another transitive network (G, c'); must the second network be Liouville as well?

If the transitivity assumption is omitted, then the answer is negative; indeed, the Liouville property of a network (G, c) is not even stable under change of conductances by a bounded factor (compare Proposition 9.8). The first example of this is due to T. Lyons (1987); the next exercise indicates a simpler example from Benjamini (1991). The latter paper also describes graphs of polynomial growth where the Liouville property is unstable under bounded conductance perturbations.

▷ **Exercise 14.19.**
Let T be an infinite binary tree, where every vertex except the root has an address consisting of a finite sequence of left and right turns. Let $\langle v_j \rangle$ be an enumeration of all vertices of T with $|v_j| \ge 100$ that have more left turns than right turns in their address. Construct a graph G from T and $\mathbb{Z}^4$ by gluing v_j to the node $(j, 0, 0, 0)$ of $\mathbb{Z}^4$ for every $j \ge 1$. Show that simple random walk on G is Liouville, but if all the tree edges in G leading right are assigned conductance 2 (while all other edges have conductance 1), then the resulting network (G, c) is not Liouville.

As noted in (14.3), for a transitive Markov chain of finite entropy, a subadditivity argument implies that $\lim_n H(X_n)/n$ exists. However, when $H(X_n)$ grows sublinearly, it can fluctuate wildly; indeed, Brieussel (2013) gives examples of random walks on groups where

$$\liminf \frac{\log H(X_n)}{\log n} = 1/2 \quad \text{and} \quad \limsup \frac{\log H(X_n)}{\log n} = 1.$$

It can also grow more regularly with $\lim \dfrac{\log H(X_n)}{\log n}$ existing and equal to any prescribed value in $[1/2, 1)$: see Amir and Virág (2017). Similar results hold for the speed: Brieussel (2013) gives examples of random walks on groups where

$$\liminf \frac{\log \mathbf{E}|X_n|}{\log n} = 1/2 \quad \text{and} \quad \limsup \frac{\log \mathbf{E}|X_n|}{\log n} = 1,$$

while between the two papers Amir and Virág (2017) and Brieussel and Zheng (2021), we obtain examples where $\lim \dfrac{\log \mathbf{E}|X_n|}{\log n}$ exists and equals any prescribed value in $[1/2, 1)$.

Another interesting result similar to Theorem 14.10 is the following: Recall the notation $\tau_x := \inf\{n \ge 0; X_n = x\}$. Denote by $\zeta_{\mathrm{Gr}}(x, y) := -\log \mathbf{P}_x[\tau_y < \infty]$, the negative log of the probability that the chain, started at x, ever visits $y \in V$. This is not a metric since it need not be symmetric; however, it is nonnegative and satisfies the triangle inequality. These properties are actually enough for the proof of Theorem 14.9(i) to hold when $\mathbf{E}[\zeta_{\mathrm{Gr}}(o, X_1)] < \infty$ and the chain is transitive, as is easily checked. The following result is due to Benjamini and Peres (1994b) and Blachère, Haïssinsky, and Mathieu (2008):

Theorem 14.50. *Let $\langle X_n \rangle$ be a transitive Markov chain with an invariant graph metric on a graph of finite degree. Suppose that $X_0 = o$ and $H(X_1) < \infty$. Then $\mathbf{P}_o$-almost surely and in L^1,*

$$\lim_n -\frac{1}{n} \log \zeta_{\mathrm{Gr}}(o, X_n) = \boldsymbol{h} \,.$$

Proof. Because of the triangle inequality, the subadditive ergodic theorem ensures the existence of the desired limit, both a.s. and in L^1. Denote the limit by α. Since $\mathbf{P}_x[\tau_y < \infty] \cdot \mathscr{G}(y, y) = \mathscr{G}(x, y)$ and $\mathscr{G}(y, y)$ does not depend on y, the claim is equivalent to saying that $\lim_n -n^{-1} \mathbf{E}_o \log \mathscr{G}(o, X_n) = \boldsymbol{h}$. Now

$$-\mathbf{E}_o \log \mathscr{G}(o, X_n) = -\sum_x p_n(x) \log \mathscr{G}(o, x) \leq -\sum_x p_n(x) \log p_n(x) = H(X_n) \,,$$

whence $\alpha \leq \boldsymbol{h}$.

To show that $\boldsymbol{h} \leq \alpha$, we may assume that $\boldsymbol{h} > 0$. In particular, by Theorem 14.49, we may assume that the graph has superpolynomial growth, that is, for every $d \in (0, \infty)$, there is some $c > 0$ such that the graph balls of radius r have at least cr^d points. In addition, just as in Lemma 14.19, we may assume that the chain is lazy. By Lemma 10.46 and the upcoming Exercise 14.20, it follows that for some c_1, all $x \in \mathsf{V}$, and all $k \geq 1$, we have $p_k(o, x) \leq c_1 k^{-d/2}$. Consider the sets $A(r) := \{x \in \mathsf{V} \,;\, -\log \mathscr{G}(o, x) \leq r\} = \{x \in \mathsf{V} \,;\, \mathscr{G}(o, x) \geq e^{-r}\}$. Define s by $s^{2-d} = e^{-r}$. Choose $c_2 > 0$ such that $\sum_{k > c_2 s^2} c_1 k^{-d/2} \leq s^{2-d}/2 = e^{-r}/2$; note that c_2 does not depend on r. Then, for all $x \in A(r)$, we have $\sum_{k \leq c_2 s^2} p_k(o, x) \geq e^{-r}/2$, whence

$$c_2 s^2 + 1 \geq \sum_{x \in \mathsf{V}} \sum_{k \leq c_2 s^2} p_k(o, x) \geq \sum_{x \in A(r)} \sum_{k \leq c_2 s^2} p_k(o, x) \geq |A(r)| e^{-r}/2 = |A(r)| s^{2-d}/2 \,.$$

In other words, $|A(r)| \leq 2c_2 s^d + 2s^{d-2} \leq c_3 e^{rd/(d-2)}$ for some c_3 that does not depend on r. Since $A(r)$ is the "ball" of "radius" $r + \log \mathscr{G}(o, o)$ in the "metric" ζ_{Gr}, this means that in the "metric" ζ_{Gr}, we have $\boldsymbol{v} \leq d/(d-2)$. As this holds for all d, we deduce that $\boldsymbol{v} \leq 1$. Notice that in this "metric," $\mathbf{E}_o[\zeta_{\mathrm{Gr}}(o, X_1)] = -\sum_x p(o, x) \log \mathbf{P}_o[\tau_x < \infty] \leq -\sum_x p(o, x) \log p(o, x) = H(X_1) < \infty$. Therefore Theorem 14.9(i) yields $\boldsymbol{h} \leq \alpha$, as desired. ◀

▷ **Exercise 14.20.**
Consider a transitive, lazy Markov chain on (V, P) with an invariant graph metric on a graph G of finite degree. Suppose that for some constants $c, d > 0$ and all $r \geq 1$, the balls of radius r have at least cr^d points in them. Show that for some c_1, all $x \in \mathsf{V}$, and all $k \geq 1$, we have $p_k(o, x) \leq c_1 k^{-d/2}$.

The proof method of Lemma 14.11 is from Benjamini, Lyons, and Schramm (1999). Proposition 14.12, a cornerstone of the theory, is due to Blackwell (1955).

For symmetric transitive chains, Ledrappier (1992) proved the inequality $\boldsymbol{h} \geq 4(1 - \rho)$ relating the entropy and the spectral radius. For ρ near 1, this improves on Avez's bound $\boldsymbol{h} \geq -2 \log \rho$ in Proposition 14.6. An estimate that improves on both of these bounds was established in Gouëzel, Mathéus, and Maucourant (2015). Avez (1972, 1974) proved that for finitely supported random walks on groups, $\boldsymbol{h} = 0$ implies the Liouville property. Theorem 14.7 for random walks on groups is due to Vershik and Kaimanovich (1979, 1983) and Derriennic (1980). The equivalence of (i)–(iv) in Theorem 14.20 was also proved in those papers. These results were later generalized to transitive chains by Kaimanovich and Woess (2002). Theorem 14.9(i), due to Guivarc'h (1980), is sometimes referred to as the "fundamental inequality"; our exposition follows Karlsson and Ledrappier (2007). That paper also showed that the implication $\boldsymbol{h} = 0 \Rightarrow \boldsymbol{l} = 0$ holds for all symmetric μ-walks with jumps in L^1: see

Proposition 14.22. Part (ii) of Theorem 14.9 is a streamlined version of an inequality from Varopoulos (1985b). These two inequalities yield the statements on speed in Theorem 14.20. In free products and certain related groups, speed and entropy can be calculated explicitly; see, for example, Mairesse and Mathéus (2007a, 2007b) and Gilch (2007, 2011). Theorem 14.10(i) is due to Guivarc'h (1980), p. 174, and Woess (2000), Theorem 8.14, and Theorem 14.10(ii) is due to Vershik and Kaimanovich (1979, 1983), Derriennic (1980), and Kaimanovich and Woess (2002).

Another way to prove part of Theorem 14.20 is from the following inequality of Benjamini, Duminil-Copin, Kozma, and Yadin (2015b), which makes some results of Erschler and Karlsson (2010) more explicit:

Theorem 14.51. *Let μ be a symmetric measure of finite entropy on a discrete group Γ, and let $\langle X_n \rangle$ be the corresponding μ-walk. For every harmonic function f on Γ, we have*

$$\left(\mathbf{E}|f(X_1) - f(o)|\right)^2 \le 4\,\mathbf{E}\big[|f(X_n) - f(o)|^2\big]\big(H(X_n) - H(X_{n-1})\big).$$

For the promised application, note that if the harmonic function f is not constant and the support of μ generates Γ, then the left-hand side is not 0, whence if f is also bounded, then $H(X_n) - H(X_{n-1})$ is bounded away from 0, whence $h \neq 0$. This proves that (iii) implies (ii) in Theorem 14.20.

A related inequality is due to Erschler and Karlsson (2010), as sharpened by Benjamini, Duminil-Copin, Kozma, and Yadin (2015a) (who also provide an extension to a wider setting): let $\mu_n = p_n(o, \cdot)$ denote the distribution of X_n.

Theorem 14.52. *Let $\langle X_n \rangle$ be simple random walk on the Cayley graph of a group Γ determined by the symmetric finite generating set S. Then, for $n \ge 1$,*

$$\sum_{s \in S} \sum_{x \in \Gamma} \frac{(\mu_n(sx) - \mu_n(x))^2}{\mu_n(sx) + \mu_n(x)} \le 2|S|\big(H(X_n) - H(X_{n-1})\big).$$

A related inequality giving a lower bound on $H(\mu * \nu) - H(\nu)$ appears at the end of Section 3 of Ozawa (2018).

Proposition 14.22 is due to Karlsson and Ledrappier (2007). The short proof we give is due to Björklund (2014), who was inspired by Erschler and Karlsson (2010).

The equivalence of parts (i), (ii), and (iv) of Theorem 14.29 in the context of random walks on groups is due to Furstenberg (1963, 1971b). The extension to more general chains and the equivalence with (iii) are due to Kaimanovich and Vershik (1983). The construction of the Poisson boundary in Theorem 14.31 is a streamlined version of the construction in Furstenberg (1971b). The passage from $(V^{\mathbb{N}}, \mathcal{I}, \mathbf{P}_o)$ to $(\Theta, \mathscr{B}_\Theta, \nu_o)$ is part of a theory of Rokhlin (1949) concerning standard probability spaces. Moreover, it is automatic that $\mathbf{b}(\Omega_1)$ is ν_o-measurable: $(\Theta, \mathscr{B}_\Theta, \nu_o)$ is a Lebesgue–Rokhlin space, so we can assume it is a standard Borel space. The set $\mathbf{b}(\Omega_1)$ is then analytic, so is universally measurable, that is, measurable with respect to every Borel measure.

Furstenberg (1971b) and Kaimanovich and Vershik (1983) gave entropy criteria for identifying the Poisson boundary. Two notable papers by Ledrappier (1984, 1985) used this criterion to determine the Poisson boundary for discrete matrix groups. Ballmann and Ledrappier (1994) developed further the entropy method in the context of rank-one manifolds. Kaimanovich (1985, 1994, 2000) refined the entropy method more generally and proved the key criteria Theorem 14.35 and Corollary 14.36. (More precisely, those papers were written for random walks on groups; the extension to transitive chains was made later in Kaimanovich and Woess (2002).) One of Kaimanovich's important observations was that the sets W_n in Corollary 14.36 can often be defined geometrically, using "rays" or "strips."

Kaimanovich's criteria led to much progress in identifying Poisson boundaries, such as the works by Kaimanovich and Masur (1996, 1998), Karlsson (2003), Malyutin (2003), Karlsson and Woess (2007),

Sava (2010b), Brofferio and Schapira (2011), Gautero and Mathéus (2012), Malyutin, Nagnibeda, and Serbin (2017), Nevo and Sageev (2013), Maher and Tiozzo (2018), and Malyutin and Svetlov (2014).

Vershik and Kaimanovich (1979, 1983) asked for a description of the Poisson boundary for finitely supported μ on the lamplighter groups $\mathbb{Z}^{d\circleddash}$ when it is nontrivial, in other words, when the projected walk on $\mathbb{Z}^d$ is transient. Moreover, they suggested a natural candidate, namely, $(\mathbb{Z}_2)^{\mathbb{Z}^d}$ with the probability measure given by the final configuration of lamps, Ψ_∞.

Kaimanovich (2001) made some progress on this lamplighter question by proving that for μ whose projection on the base group, $\mathbb{Z}^d$, has nonzero mean, the final lamps do indeed give the Poisson boundary; in particular, this includes Corollary 14.38.

In 2008, Erschler (2011, 2010) achieved a breakthrough, proving that the conjecture of Vershik and Kaimanovich (1979) holds when $d \geq 5$ and allowing infinitely supported μ with a finite third moment. This was extended by Lyons and Peres (2021) to all $d \geq 3$, allowing μ with finite second moment.

The result of Exercise 14.13 is due to Sava (2010a), where an extension to walks with bounded jumps was conjectured; this conjecture was proved by Lyons and Peres (2021) using the enhanced criterion, Proposition 14.42.

Let G be a one-ended, planar graph of bounded degree. The Poisson boundary for simple random walk on G was conjectured by Benjamini and Schramm (1996c) to be a circle, endowed with the harmonic measure obtained from convergence of the walk in its square tiling representation, as in Section 9.4. This conjecture was proved by Georgakopoulos (2016). An alternative representation of the Poisson boundary, using circle packings, was obtained in Angel, Barlow, Gurel-Gurevich, and Nachmias (2016). A simplified proof for both representations, and an extension to graphs that are roughly isometric to one-ended, planar graphs of bounded degree, are given by Hutchcroft and Peres (2017).

Our proof of Theorem 14.43 follows the lines of Katznelson and Weiss (1982) and Keane (1995), who were in turn inspired by Kamae (1982); see also Shields (1987). The proof of Kingman's theorem that we present follows Steele (1997) and is modeled after the proofs of Birkhoff's theorem mentioned earlier. An alternative proof of Kingman's theorem with slightly weaker hypotheses was found by Liggett (1985). A different concise proof of the Birkhoff theorem is given by Ross and Peköz (2007); that proof does not generalize as easily to the subadditive case.

Zero-two laws for positive operators were first proved by Orstein and Sucheston (1970) and Foguel (1971, 1976). The version in Theorem 14.45 is due to Derriennic (1976). The reason for the name "zero-two law" is that in these papers, the L_1 distance between measures, $\|\mu - \nu\|_1 = \sum_{x\in V} |\mu(x) - \nu(x)|$, which is twice the total variation distance $\|\mu - \nu\|_{\mathrm{TV}}$, was used (see Exercise 13.2).

As noted after the zero-two law, Theorem 14.45, that theorem requires the transitivity assumption. Nevertheless, Derriennic (1976) did prove the following extension.

Theorem 14.53. (Derriennic's General Zero-Two Law) *Let $p_n(\cdot, \cdot)$ be the n-step transition probabilities for a Markov chain on a countable set V. Then, for any integer $k > 0$,*

$$\sup_{x\in V} \lim_{n\to\infty} \|p_{n+k}(x, \cdot) - p_n(x, \cdot)\|_{\mathrm{TV}} \in \{0, 1\}. \qquad (14.54)$$

It can be proved similarly to our proof of Theorem 14.45.

Kaimanovich and Vershik (1983) pointed out that the zero-two law implies the coincidence of the tail and invariant σ-fields (Theorems 14.18 and 14.47).

The theory of this chapter can be extended to random walks in random environments: see Kaimanovich (1990), Benjamini and Curien (2012), and Benjamini, Duminil-Copin, Kozma, and Yadin (2015a).

There is also a theory of positive harmonic functions, involving the so-called Martin boundary. For an exposition, see Dynkin (1969) or Sawyer (1997). The latter explains the relationship of the Martin boundary to the Poisson boundary. Kaimanovich (1996) surveys the topic of this chapter and discusses symmetric spaces and Martin boundaries as well.

14.8 Collected In-Text Exercises

14.1. For $\mathbf{y}, \mathbf{z} \in \mathsf{V}^{\mathbb{N}}$, write $\mathbf{y} \overset{\mathcal{T}}{\sim} \mathbf{z}$ iff $\exists n \; \forall m \geq n \; y_m = z_m$. Let $A \subseteq \mathsf{V}^{\mathbb{N}}$ be a Borel set. Show that $A \in \mathcal{T}$ iff A is a union of tail equivalence classes (that is, $\mathbf{y} \in A, \; \mathbf{y} \overset{\mathcal{T}}{\sim} \mathbf{z} \Rightarrow \mathbf{z} \in A$). Similarly, show that a Borel function $f \colon \mathsf{V}^{\mathbb{N}} \to \mathbb{R}$ is a tail function iff $\mathbf{y} \overset{\mathcal{T}}{\sim} \mathbf{z}$ implies $f(\mathbf{y}) = f(\mathbf{z})$.

14.2. Let (T, c) be a transient network on a tree with bounded conductances. Show that the network walk has a nontrivial tail.

14.3. Show that if a discrete random variable X satisfies $H(X \mid \mathscr{F}) = H(X)$ a.s., then X is independent of $\mathscr{F}$.

14.4. Prove that we always have $H(X, Y \mid \mathscr{F}) \geq H(X \mid \mathscr{F})$.

14.5. **(a)** Show that simple random walk on a $(b+1)$-regular tree $\mathbb{T}_{b+1}$ satisfies $\boldsymbol{h} = \boldsymbol{l}\boldsymbol{v}$.
(b) Find a Cayley graph where simple random walk satisfies $\boldsymbol{h} < \boldsymbol{l}\boldsymbol{v}$. *Hint:* Consider a Cartesian product of trees.

14.6. Given a transitive Markov chain $\langle X_j \rangle$ on a state space V, denote the number of distinct states among its first n by $R_n := |\{X_0, X_1, \ldots, X_{n-1}\}|$. Fix $o \in \mathsf{V}$, and show that $\lim_{n \to \infty} \mathbf{E}_o[R_n]/n = \mathbf{P}_o[\forall j \geq 1 \; X_j \neq o]$ as $n \to \infty$. Deduce that R_n/n converges $\mathbf{P}_o$-a.s. to the same limit.

14.7. Suppose that u is a bounded harmonic function on an irreducible Markov chain (V, P) and c is a constant such that $\lim_n u(X_n) = c$ almost surely-$\mathbf{P}_o$. Show that $u(x) = c$ for all $x \in \mathsf{V}$.

14.8. **(Saturation)** Let $(M_1, \mathscr{F}_1, Q_1)$ and $(M_2, \mathscr{F}_2, Q_2)$ be two probability spaces, and suppose that $\Phi \colon (M_1, \mathscr{F}_1) \to (M_2, \mathscr{F}_2)$ is measurable with $Q_2 = Q_1 \circ \Phi^{-1}$. Show that the following are equivalent:
(i) The Q_1-completions of $\mathscr{F}_1$ and $\Phi^{-1}\mathscr{F}_2$ coincide.
(ii) The mapping $f_2 \mapsto f_2 \circ \Phi$ from $L^{\infty}(M_2, \mathscr{F}_2, Q_2)$ to $L^{\infty}(M_1, \mathscr{F}_1, Q_1)$ is surjective.

14.9. **(a)** Prove that the map π in Theorem 14.29(iv) is unique up to a set of ν_o-measure 0.
(b) Prove that the Poisson boundary of a Markov chain (V, P) is unique up to a pointwise isomorphism π of measure spaces, defined ν_o-a.e., that respects boundary maps. In other words, show that if $(\Theta, \mathscr{F}_{\Theta}, \mathbf{b})$ and $(\Theta', \mathscr{F}_{\Theta'}, \mathbf{b}')$ are both Poisson boundaries of (V, P), then the map π in Theorem 14.29(iv) is one-to-one on a set $\Theta_1 \in \mathscr{F}_{\Theta}$ of full ν_o-measure, and π^{-1} is measurable.
(c) Show that if both of these Poisson boundaries in (b) are Γ-boundaries for some countable Γ, then π is Γ-equivariant (that is, $\pi \circ \gamma(\theta) = \gamma \circ \pi(\theta)$ for ν_o-a.e. $\theta \in \Theta$ and all $\gamma \in \Gamma$).

14.10. Suppose that $(\Omega, \mathscr{F}, \mathbf{P})$ is a probability space. Show that if $\{A_n \; ; \; n \geq 1\} \subset \mathscr{F}$ is such that $\{\mathbf{1}_{A_n}\}$ is dense in $L^1(\Omega, \mathscr{F}, \mathbf{P}, \{0, 1\})$, then the $\mathbf{P}$-completions of $\mathscr{F}$ and of $\sigma(\{A_n \; ; \; n \geq 1\})$ coincide.

14.11. Suppose that the Markov chain (V, P) is simple random walk on a $(d+1)$-regular tree, and $\mathcal{J}$ is the σ-field determined by the boundary mapping $\mathbf{b}$ in Example 14.27. For any two vertices x, y and every $\mathbf{z} \in \mathsf{V}_{\mathrm{conv}}^{\mathbb{N}}$, there is a unique first meeting point of the rays from x and y that belong to the same end as $\mathbf{b}(\mathbf{z})$. Show that $p_n^{\mathcal{J}}(x, y)(\mathbf{z}) = p_n(x, y) \cdot d^{k_x - k_y}$ for $\mathbf{P}_x$-a.e. $\mathbf{z} \in \mathsf{V}^{\mathbb{N}}$, where $k_x = k_x(x, y, \mathbf{z})$ is the distance from x to this meeting point.

14.12. In the setting of Lemma 14.33, show that for all $x, y, z \in \mathsf{V}$, we have $\mathbf{P}_x$-a.s.

$$\mathbf{P}_x[X_n = y, \; X_{n+m} = z \mid \mathcal{J}] = p_n(x, y) p_m(y, z) \frac{dQ_z}{dQ_x} = p_n^{\mathcal{J}}(x, y) p_m^{\mathcal{J}}(y, z). \tag{14.22}$$

14.13. Let $\mathbb{T}_3$ be a 3-regular tree, and suppose that ξ is an end of $\mathbb{T}_3$. Consider the Markov chain $\langle (\Psi_n, X_n) \rangle$ on the lamplighter graph $\mathbb{T}_3^{\circ}$ where $p((\psi, x), (\phi, y))$ is $1/3$ if $x = y$ and ψ, ϕ differ only at x or if $\psi = \phi$ and y is the parent of x (the unique neighbor of x in the direction of ξ); $p((\psi, x), (\phi, y)) = 1/6$

if $\psi = \phi$ and x is the parent of y; and otherwise $p\big((\psi, x), (\phi, y)\big) = 0$. The projected Markov chain $\langle X_n \rangle$ is invariant under the transitive group of automorphisms of $\mathbb{T}_3$ that preserve ξ. Fix $o \in \mathbb{T}_3$.

  (a) Show that the projected chain $\langle X_n \rangle$ tends to ξ.

  (b) Show that the Avez entropy is 0 for the projected chain $\langle X_n \rangle$.

  (c) For the Markov chain $\langle (\Psi_n, X_n) \rangle$ started at $(\overline{\mathbf{0}}, o)$, show that the tail is generated by the final configuration of the lamps, Ψ_∞.

14.14. Often a system is defined to be ergodic if the only f that satisfy $f = f \circ T$ a.s. are constant a.s. Show that this is equivalent to the definition in the appendix. Show also that if $f \geq f \circ T$ a.s., then $f = f \circ T$ a.s.

14.15. **(a)** Prove that if the random variables X_n are i.i.d., then $\langle X_n \rangle$ is ergodic.

  (b) Let $(\Omega, \mathscr{F}, \mathbf{P}, T)$ be a probability measure-preserving system. Let $(\Lambda, \mathscr{G})$ be a measurable space, and suppose there are measurable maps $\Psi: \Omega \to \Lambda$ and $\widetilde{T}: \Lambda \to \Lambda$ such that $\Psi \circ T = \widetilde{T} \circ \Psi$. Prove that $(\Lambda, \mathscr{G}, \mathbf{P} \circ \Psi^{-1}, \widetilde{T})$ is a probability measure-preserving system. Moreover, prove that if $(\Omega, \mathscr{F}, \mathbf{P}, T)$ is ergodic, then so is $(\Lambda, \mathscr{G}, \mathbf{P} \circ \Psi^{-1}, \widetilde{T})$.

  (c) Use (b) to deduce the following statement: if $\langle X_n \rangle$ is ergodic, $g: \mathbb{R}^{\mathbb{N}} \to \mathbb{R}$ is measurable with respect to the product σ-field, and $Y_n = g(X_n, X_{n+1}, \ldots)$, then the process $\langle Y_n \rangle_{n \geq 0}$ is also ergodic.

14.16. Prove the general version of the Birkhoff ergodic theorem that does not require ergodicity.

14.17. Prove the more general version of Kingman's subadditive ergodic theorem that does not require ergodicity.

14.18. Let G denote $\mathbb{Z}^3$ with a loop added at o. Show that simple random walk on G satisfies $0 < \lim_{n \to \infty} \| p_{n+1}(o, \cdot) - p_n(o, \cdot) \|_{\mathrm{TV}} < 1$.

14.19. Let T be an infinite binary tree, where every vertex except the root has an address consisting of a finite sequence of left and right turns. Let $\langle v_j \rangle$ be an enumeration of all vertices of T with $|v_j| \geq 100$ that have more left turns than right turns in their address. Construct a graph G from T and $\mathbb{Z}^4$ by gluing v_j to the node $(j, 0, 0, 0)$ of $\mathbb{Z}^4$ for every $j \geq 1$. Show that simple random walk on G is Liouville, but if all the tree edges in G leading right are assigned conductance 2 (while all other edges have conductance 1), then the resulting network (G, c) is not Liouville.

14.20. Consider a transitive, lazy Markov chain on (V, P) with an invariant graph metric on a graph G of finite degree. Suppose that for some constants $c, d > 0$ and all $r \geq 1$, the balls of radius r have at least $c r^d$ points in them. Show that for some c_1, all $x \in \mathsf{V}$, and all $k \geq 1$, we have $p_k(o, x) \leq c_1 k^{-d/2}$.

14.9 Additional Exercises

14.21. Given a transitive Markov chain, prove that $\langle H(X_n) \rangle$ is a concave sequence, that is, $H(X_n) \geq \big(H(X_{n-1}) + H(X_{n+1})\big)/2$ for all $n > 0$.

14.22. Consider an irreducible Markov chain, $\langle X_n \rangle$. Show that its tail σ-field is trivial iff $\forall y$

$$p(o, y) > 0 \quad \Longrightarrow \quad \lim_{n \to \infty} \frac{p_{n-1}(y, X_n)}{p_n(o, X_n)} = 1 \quad \mathbf{P}_o\text{-a.s.}$$

14.23. Consider a transitive Markov chain $\langle X_n \rangle$ with an invariant metric on a graph G of at most polynomial growth. Suppose that $\mathbf{E}_o\big[\log\big(1 + d(o, X_1)\big)\big] < \infty$. Show that $H(X_1) < \infty$ and that $\boldsymbol{h} = 0$.

14.24. Let K be a finite subset of a denumerable group, Γ. Let $\langle X_n \rangle$ be a convolution random walk on Γ. Define $R_n(K) := |\{X_j \gamma; \ \gamma \in K, \ 0 \le j < n\}|$. Define the *capacity* of K to be $C(K) := \sum_{\gamma \in K} \mathbf{P}_{\gamma^{-1}}[\forall n \ge 1 \ X_n \notin K^{-1}]$. Extend Exercise 14.6 to show that $\mathbf{E}_o[R_n(K)/n] \to C(K)$ as $n \to \infty$ and $R_n(K)/n \to C(K)$ a.s.

14.25. For a transitive Markov chain $\langle X_n \rangle$ on a countable set V with $H(X_1) < \infty$, show that the following are equivalent:
 (i) The Avez entropy vanishes: $\boldsymbol{h} = 0$.
 (ii) There exist finite sets $W_n \subset V$ with $\log |W_n| = o(n)$ and $\mathbf{P}_o[X_n \in W_n] \to 1$.
 (iii) There exist finite sets $W_n \subset V$ with $\log |W_n| = o(n)$ and $\limsup_n \mathbf{P}_o[X_n \in W_n] > 0$.
 (iv) There exist finite sets $W_n \subset V$ with $\log |W_n| = o(n)$ and $\limsup_n \mathbf{P}_o[\exists m \ge n \ X_m \in W_n] > 0$.

14.26. Fix $o \in V$. Given a transitive Markov chain $\langle X_j \rangle$ on V endowed with an invariant graph metric d, denote $\boldsymbol{l}_* := \liminf d(o, X_n)/n$ and $\boldsymbol{l}^* := \limsup d(o, X_n)/n$. Recall from Theorem 14.10 that if $\mathbf{E}_o d(o, X_1) < \infty$, then $\boldsymbol{l}_* = \boldsymbol{l}^*$ a.s.
 (a) Show that $\boldsymbol{l}_*$ and $\boldsymbol{l}^*$ are $\mathbf{P}_o$-a.s. constant.
 (b) Show that if $\mathbf{E}_o d(o, X_1) = \infty$, then $\boldsymbol{l}^* = \infty$.
 (c) Find a transitive chain where $\mathbf{E}_o d(o, X_1) = \infty$ yet $\boldsymbol{l}_* < \infty$.
 (d) Suppose that $H(X_1) < \infty$. Show that $\boldsymbol{h} \le \boldsymbol{v} \boldsymbol{l}_*$ always holds, strengthening Theorem 14.9(i). (*Hint:* Use Theorem 14.10.) In particular, if $\boldsymbol{l}_* < \infty$ and $\boldsymbol{v} = 0$, then $\boldsymbol{h} = 0$.

14.27. Let Γ be a group with a finite symmetric set of generators, S. For any probability measure μ on S, denote by $\boldsymbol{h}(\mu)$ and $\boldsymbol{l}(\mu)$ the Avez entropy and the speed of the μ-walk on Γ.
 (a) Show that there exists a symmetric μ of maximal speed and a symmetric μ of maximal Avez entropy.
 (b) Give an example of a group Γ and a set of generators S where no measure μ on S simultaneously maximizes $\boldsymbol{h}(\cdot)$ and $\boldsymbol{l}(\cdot)$.

14.28. Let P be a transition matrix on V.
 (a) Suppose that for every x, y in V there is a coupling of a chain $\langle X_n \rangle$ started at x and a chain $\langle Y_n \rangle$ started at y, both with transition matrix P, such that $\mathbf{P}[\exists m \ \forall n \ge m \ X_n = Y_n] = 1$. Show that the tail σ-field $\mathcal{T}$ is $\mathbf{P}_o$-trivial for every $o \in V$.
 (b) Suppose that for every x, y in V there is a coupling of a chain $\langle X_n \rangle$ started at x and a chain $\langle Y_n \rangle$ started at y, both with transition matrix P, such that $\mathbf{P}[\exists m, k \ \forall n \ge m \ X_n = Y_{n+k}] = 1$ (this is known as a *shift coupling*). Show that the invariant σ-field $\mathcal{I}$ is $\mathbf{P}_o$-trivial for every $o \in V$.

14.29. For $\mathbf{y}, \mathbf{z} \in V^{\mathbb{N}}$, write $\mathbf{y} \overset{\mathcal{I}}{\sim} \mathbf{z}$ if $\exists k, n \ \forall m \ge n \ y_m = z_{m+k}$. Let $A \subseteq V^{\mathbb{N}}$ be a Borel set. Show that $A \in \mathcal{I}$ iff A is a union of invariant equivalence classes (that is, $\mathbf{y} \in A$, $\mathbf{y} \overset{\mathcal{I}}{\sim} \mathbf{z} \Rightarrow \mathbf{z} \in A$). Similarly, show that a Borel function $f : V^{\mathbb{N}} \to \mathbb{R}$ is an invariant function iff $\mathbf{y} \overset{\mathcal{I}}{\sim} \mathbf{z}$ implies $f(\mathbf{y}) = f(\mathbf{z})$.

14.30. Let (V_1, P_1) and (V_2, P_2) be two Markov chains with respective tail σ-fields $\mathcal{T}_1$ and $\mathcal{T}_2$. Consider the Cartesian product chain $(V_1 \times V_2, (P_1 + P_2)/2)$.
 (a) Show that if (V_2, P_2) is Liouville, then every bounded harmonic function $u(x, y)$ for the product chain is constant in the second coordinate. (Compare with Exercise 9.45.)
 (b) Show that the tail σ-field of the product chain is equal to $\mathcal{T}_1 \times \mathcal{T}_2$ mod 0. (Compare with Exercise 9.7.)

14.31. Show that for every graph G, the lamplighter graph $G^{\odot}$ has the Liouville property iff G is recurrent. Thus, the transitivity assumption in Corollary 14.21 can be dropped.

14.32. Show that the multiplication operation $(u, v) \mapsto u \circledast v$ on $\mathbf{BH}(V, P)$ defined by $(u \circledast v)(x) := \lim_n \mathbf{E}_x[u(X_n)v(X_n)]$ makes $\mathbf{BH}(V, P)$ into a commutative C^*-algebra whose Gelfand spectrum Ξ allows a Poisson representation in the sense that there is a family $\{v_x ; \ x \in V\}$ of probability measures on Ξ such that the Gelfand isomorphism $\varphi : \mathbf{BH}(V, P) \to C(\Xi)$ satisfies $u(x) = \int_{\Xi} \varphi(u) \, dv_x$ for all $u \in \mathbf{BH}(V, P)$ and $x \in V$.

14.33. (Uncountable Automorphism Groups and the Poisson Boundary) Let Γ be an uncountable subgroup of $\mathrm{Aut}(V, P)$, endowed with the weak topology described in Exercise 8.20. Observe that Γ acts on $V^{\mathbb{N}}$ via the diagonal action.

(a) Show that there exists a countable dense subgroup $\Gamma_* \subset \Gamma$.

(b) Prove that for every $\gamma \in \Gamma$ and $f \in L^1(V^{\mathbb{N}}, \mathscr{B}_{V^{\mathbb{N}}}, \mathbf{P}_o)$, there is a sequence $\langle \gamma_j \rangle$ in Γ_* such that $f(\gamma_j \mathbf{x}) \to f(\gamma \mathbf{x})$ for $\mathbf{P}_o$-a.e. $\mathbf{x} \in V^{\mathbb{N}}$.

(c) Let $(\Theta, \mathscr{B}_{\Theta}, \mathbf{b})$ be the Γ_*-boundary of (V, P) constructed in Theorem 14.31. Using (b), define an action of Γ on Θ and verify that $(\Theta, \mathscr{B}_{\Theta}, \mathbf{b})$ is also a Γ-boundary of (V, P).

14.34. Show that the assumption that $\mathcal{J}$ is Γ-closed is crucial in Theorem 14.35.

14.35. Prove the following strong converse to Corollary 14.36: Let Γ be a transitive group of automorphisms of (V, P), and let $\mathcal{J} \subset \mathcal{I}$ be a Γ-closed σ-field. Fix $o \in V$. If $\mathcal{J} = \mathcal{I}$ mod $\mathbf{P}_o$, then there exists a sequence of finite, $\mathcal{J}$-measurable sets $\langle W_n \rangle$ in V such that $\mathbf{P}_o[X_n \in W_n] \to 1$ and $|W_n|^{1/n} \to 1$ a.s. as $n \to \infty$.

14.36. Here we develop a martingale method due to Furstenberg and use it to show that very general random walks on free groups converge to an end. Let Γ be any countable group and K be a compact Hausdorff space on which Γ acts via continuous maps. Let μ be a probability measure on Γ whose support generates Γ as a group. Let $\langle X_n \rangle$ be the μ-walk on Γ. If ν is a probability measure on K, we define $\mu * \nu$ to be the law of $X_1 Y$, where $Y \sim \nu$.

(a) Show that there is a Borel probability measure ν on K with $\mu * \nu = \nu$. We say that such a ν is μ-**stationary**. *Hint:* Let ν be a weak limit point of $(1/n) \sum_{k=1}^{n} \mu^{*k} \nu_0$ for some ν_0.

(b) Show that if ν is μ-stationary and $A \subseteq K$ is Borel, then $\langle (X_n \nu)(A) \rangle = \langle \nu(X_n^{-1} A) \rangle$ is a martingale and a.s. converges.

(c) Show that if there is no Borel probability measure on K that is invariant under Γ, then $\langle X_n \rangle$ is transient. *Hint:* Take A so that $\nu(\gamma^{-1} A) \neq \nu(A)$ for some $\gamma \in \Gamma$.

(d) Let $\mathbb{T}_4$ be the 4-regular tree, regarded as the usual Cayley graph of the free group $\mathbb{Z} * \mathbb{Z}$ on two letters. Recall from Section 14.4 that $\mathbb{Z} * \mathbb{Z}$ acts on the boundary $\partial \mathbb{T}_4$ by $(\gamma, \xi) \mapsto \lim\{\gamma \xi_n ; n \geq 1\}$, where $\xi = (\xi_1, \xi_2, \ldots)$. Show that there is no Borel probability measure on $\partial \mathbb{T}_4$ that is invariant under the action of $\mathbb{Z} * \mathbb{Z}$. Deduce that if the support of μ generates $\mathbb{Z} * \mathbb{Z}$, then the μ-walk is transient.

(e) Prove that if μ generates $\mathbb{Z} * \mathbb{Z}$ as in (d), then every μ-stationary measure ν on $\partial \mathbb{T}_4$ is nonatomic.

(f) Show that if ν is a nonatomic measure on $\partial \mathbb{T}_4$ and the sequence $\langle y_k \rangle$ in $\mathbb{T}_4$ tends $\xi \in \partial \mathbb{T}_4$ as $k \to \infty$, then $y_k \nu$ converges weakly to the Dirac measure δ_{ξ}.

(g) Prove that if μ generates $\mathbb{Z} * \mathbb{Z}$ as in (d), then the μ-walk on $\mathbb{Z} * \mathbb{Z}$ a.s. converges to a point in $\partial \mathbb{T}_4$.

(h) Prove that if μ generates $\mathbb{Z} * \mathbb{Z}$ as in (d) and ν is μ-stationary, then ν is the law of $\lim_{n \to \infty} X_n$ in $\partial \mathbb{T}_4$.

14.37. Let $d \geq 3$. For simple random walk $\langle X_n \rangle$ in $\mathbb{Z}^d$ and $r > 1$, consider the events

$$A_r(m) := \left[\forall k < m \ \|X_k\|_{\infty} < r \quad \text{and} \quad \forall j > m \ \|X_j\|_{\infty} > r \right],$$

and write $A_r := \bigcup_{m \geq 1} A_r(m)$. Show that $\liminf_{n \to \infty} \mathbf{P}_o\left[\bigcup_{r=n}^{n^2} A_r \right] > 0$.

14.38. Give a short alternative proof of Theorem 14.39 by using Proposition 14.42 and Exercise 14.37.

14.39. Let $\langle X_n \rangle$ be a transitive Markov chain starting at o with $H(X_1) < \infty$. Define $\zeta_n(x) := -\log \mathbf{P}_o[\tau_x \leq n]$. Show that $\mathbf{P}_o$-a.s. and in L^1,

$$\lim_{n \to \infty} n^{-1} \zeta_n(X_n) = \boldsymbol{h} .$$

Hint: $-\log p_n(x) \geq \zeta_n(x) \geq -\log\big(\nu_n(x)(n + 1)\big)$, where ν_n is the probability measure $\nu_n(x) := (n + 1)^{-1} \sum_{k=0}^{n} p_k(x)$.

15 | Hausdorff Dimension

How do we capture the intrinsic dimension of a geometric object (such as a curve, a surface, or a body)? What if the intrinsic dimension seems not to be an integer? There are various notions of dimension that can answer these questions; for many purposes, Hausdorff dimension is the most important one. In Chapter 1, we saw that the Hausdorff dimension of the boundary of a tree is merely the logarithm of the branching number of the tree. We also saw how we can use trees to represent closed sets in Euclidean space and thereby capture their Hausdorff dimension. In this chapter, we develop these ideas in greater depth and give several applications. For example, we will use our work on Galton–Watson networks from Section 5.9 to analyze random fractals, such as the zero-set of one-dimensional Brownian motion.

15.1 Basics

Suppose that we have an open bounded subset of Euclidean space. One way we can infer its dimension is via scaling: when we scale the subset by a homothety with factor r, its volume changes by a factor r^d, where d is the dimension. A disadvantage of this approach is that it requires moving outside the object itself: it is not measured intrinsically.

Another approach is to cover the object by small sets, leaving the object itself unchanged. If E is a bounded set in Euclidean space, let $N(E, \epsilon)$ be the number of closed balls of diameter at most ϵ required to cover E. Then when E is d-dimensional, $N(E, \epsilon) \approx C/\epsilon^d$. To get at d, then, look at $\log N(E, \epsilon)/\log(1/\epsilon)$ and take some kind of limit as $\epsilon \to 0$. This gives the **upper** and **lower Minkowski dimensions**:

$$\dim^{M} E := \limsup_{\epsilon \to 0} \frac{\log N(E, \epsilon)}{\log(1/\epsilon)}, \tag{15.1}$$

$$\dim_{M} E := \liminf_{\epsilon \to 0} \frac{\log N(E, \epsilon)}{\log(1/\epsilon)}. \tag{15.2}$$

The lower Minkowski dimension is often called simply the Minkowski dimension. Note that we could use cubes instead of balls, or even b-adic cubes

$$[a_1 b^{-n}, (a_1 + 1)b^{-n}] \times [a_2 b^{-n}, (a_2 + 1)b^{-n}] \times \cdots \times [a_d b^{-n}, (a_d + 1)b^{-n}] \quad (a_i \in \mathbb{Z}, n \in \mathbb{N}),$$

and not change these dimensions. In the future, we will call n the **order** of a b-adic cube if the sides of the cube have length b^{-n}.

Since balls are defined in any metric space, these definitions (15.1) and (15.2) make sense in any metric space E. Thus, they are intrinsic. It is clear that the Minkowski dimension of a singleton is 0.

Example 15.1. Let E be the standard middle-thirds Cantor set,

$$E := \left\{ \sum_{n \geq 1} x_n 3^{-n} \, ; \; x_n \in \{0, 2\} \right\}.$$

To calculate the Minkowski dimensions of E, it is convenient to use triadic intervals, of course. We have that $N(E, 3^{-n}) = 2^n$, so for $\epsilon = 3^{-n}$, we have

$$\frac{\log N(E, \epsilon)}{\log (1/\epsilon)} = \frac{\log 2}{\log 3}.$$

Thus, the upper and lower Minkowski dimensions are both $\log 2/\log 3$.

One problem with Minkowski dimension is that an unbounded set will have infinite Minkowski dimension. But things are even worse than that:

Example 15.2. Let $E := \{1/n \, ; \; n \geq 1\}$. Given $\epsilon \in (0, 1)$, let k be such that $1/k^2 \approx \epsilon$. Then it takes about $1/\sqrt{\epsilon}$ intervals of length ϵ to cover $E \cap [0, 1/k]$ and about $1/\sqrt{\epsilon}$ more to cover the k points in $E \cap [1/k, 1]$. It can be shown that, indeed, $N(E, \epsilon) \approx 2/\sqrt{\epsilon}$, so that $\dim_M E = 1/2$.

This example shows that a bounded countable union of sets of Minkowski dimension 0 may have positive Minkowski dimension, even when the union consists entirely of isolated points. For this reason, Minkowski dimensions are not entirely suitable for measuring the dimension of a set.

So we consider yet another approach. In what dimension should we measure the *size* of E? Note that a surface has infinite one-dimensional measure but zero three-dimensional measure. How do we measure size in an arbitrary dimension? Define ***Hausdorff α-dimensional (outer) measure*** by

$$\mathcal{H}_\alpha(E) := \lim_{\epsilon \to 0} \inf \left\{ \sum_{i=1}^{\infty} (\operatorname{diam} E_i)^\alpha \, ; \; E \subseteq \bigcup_{i=1}^{\infty} E_i, \, \forall i \;\; \operatorname{diam} E_i < \epsilon \right\};$$

here, the sets E_i are unrestricted except for their diameter. The limit in this definition exists because the infimum is taken over smaller classes of sets as ϵ decreases. Note that, up to a bounded factor, the restriction of $\mathcal{H}_d$ to Borel sets in $\mathbb{R}^d$ is d-dimensional Lebesgue measure, $\mathcal{L}_d$.

▷ **Exercise 15.1.**
Show that for any $E \subseteq \mathbb{R}^d$, there exists a real number α_0 such that $\alpha < \alpha_0 \Rightarrow \mathcal{H}_\alpha(E) = +\infty$ and $\alpha > \alpha_0 \Rightarrow \mathcal{H}_\alpha(E) = 0$.

The number α_0 of the preceding exercise is called the ***Hausdorff dimension*** of E, denoted $\dim_H E$ or simply $\dim E$. We can also write

$$\dim E = \inf \left\{ \alpha \, ; \; \inf \left\{ \sum_{i=1}^{\infty} (\operatorname{diam} E_i)^\alpha \, ; \; E \subseteq \bigcup_i E_i \right\} = 0 \right\}$$

since the only way the sum of $(\text{diam } E_i)^\alpha$ can be small is when all $\text{diam } E_i$ are small. Again, these definitions make sense in any metric space (in some circumstances, you may need to recall that the infimum of the empty set is $+\infty$). In $\mathbb{R}^d$, we could restrict ourselves to covers by open sets, spheres or b-adic cubes; this would change $\mathcal{H}_\alpha$ by at most a bounded factor, and so would leave the dimension unchanged. Also, $\dim E \le \dim_M E$ for any E: if $E \subseteq \bigcup_{i=1}^{N(E,\epsilon)} E_i$ with $\text{diam } E_i \le \epsilon$, then

$$\sum_{i=1}^{N(E,\epsilon)} (\text{diam } E_i)^\alpha \le \sum_{i=1}^{N(E,\epsilon)} \epsilon^\alpha = N(E,\epsilon)\epsilon^\alpha \,.$$

Thus, if $\alpha > \dim_M E$, we get $\mathcal{H}_\alpha(E) = 0$. Indeed, Minkowski dimension is useful mostly as an upper bound on Hausdorff dimension.

Example 15.3. Let E be the standard middle-thirds Cantor set again. For an upper bound on its Hausdorff dimension, we use the Minkowski dimension, $\log 2/\log 3$. For a lower bound, let μ be the Cantor–Lebesgue measure, that is, the law of $\sum_{n\ge 1} X_n 3^{-n}$ when X_n are i.i.d. with $\mathbf{P}[X_n = 0] = \mathbf{P}[X_n = 2] = 1/2$. Let E_i be triadic intervals whose union covers E. If $\mu(E_i) \ne 0$ and E_i has diameter 3^{-n}, we have

$$(\text{diam } E_i)^{\log 2/\log 3} = 2^{-n} = \mu(E_i) \,.$$

Therefore,

$$\sum_i (\text{diam } E_i)^{\log 2/\log 3} \ge \sum_i \mu(E_i) \ge \mu(E) = 1 \,.$$

(Except for wasteful covers, these inequalities are equalities.) This shows that $\dim E \ge \log 2/\log 3$. Therefore, the Hausdorff dimension is in fact equal to the Minkowski dimension in this example.

Example 15.4. If $E := \{1/n \,;\, n \ge 1\}$, then $\dim E = 0$: for any $\alpha, \epsilon > 0$, we may cover E by the sets $E_i := [1/i, 1/i + (\epsilon/2^i)^{1/\alpha}]$, showing that $\mathcal{H}_\alpha(E) < \epsilon$.

Example 15.5. Let E be the Cantor middle-thirds set. The set $E \times E$ is called the planar Cantor set. We have $\dim(E \times E) = \dim^M(E \times E) = \log 4/\log 3$. To prove this, note that it requires 4^n triadic squares of order n to cover $E \times E$, whence $\dim^M(E \times E) = \log 4/\log 3$. On the other hand, if μ is the Cantor–Lebesgue measure, then $\mu \times \mu$ is supported by $E \times E$. If E_i are triadic squares covering $E \times E$, then, as in Example 15.3,

$$\sum_i (\text{diam } E_i)^{\log 4/\log 3} \ge \sum_i (\mu \times \mu)(E_i) \ge (\mu \times \mu)(E \times E) = 1 \,.$$

Hence $\dim(E \times E) \ge \log 4/\log 3$.

▷ **Exercise 15.2.**
The *Sierpinski carpet* is the set

$$E := \left\{ \left(\sum_n x_n 3^{-n}, \sum_n y_n 3^{-n} \right) \,;\, (x_n, y_n) \in \{0, 1, 2\}^2 \setminus \{(1, 1)\} \right\}.$$

That is, the unit square is divided into its nine triadic subsquares of order 1 and the interior of the middle one is removed. This process is repeated on each of the remaining eight squares, and so on. Show that $\dim E = \dim^M E = \log 8/\log 3$.

▷ **Exercise 15.3.**

The *Sierpinski gasket* is the set obtained by partitioning an equilateral triangle into four congruent pieces and removing the interior of the middle one, then repeating this process ad infinitum on the remaining pieces, as in Figures 15.1 and 15.2. Show that its Hausdorff dimension is $\log 3/\log 2$.

Figure 15.1. The first three stages of the construction of the Sierpinski gasket.

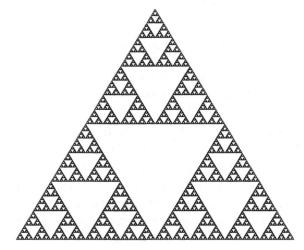

Figure 15.2. The Sierpinski gasket, drawn by O. Schramm.

▷ **Exercise 15.4.**

Show that for any sets E_n, $\dim\left(\bigcup_{n=1}^{\infty} E_n\right) = \sup \dim E_n$.

▷ **Exercise 15.5.**

Show that if $E_1 \supseteq E_2 \supseteq \cdots$, then $\dim \bigcap E_n \leq \lim \dim E_n$.

One can extend the notion of Hausdorff measures to other *gauge functions* $h: \mathbb{R}^+ \to \mathbb{R}^+$ in place of $h(t) := t^{\alpha}$. One requires merely that h be continuous and increasing and that $h(0^+) = 0$. This allows for finer examination of the size of E in cases when $\mathcal{H}_{\dim E}(E) \in \{0, \infty\}$.

15.2 Coding by Trees

There are general ways to associate rooted trees to bounded sets in $\mathbb{R}^d$, and vice versa. We start with the simplest case, closed sets in $[0, 1]$. We described this case in Section 1.10, following Furstenberg (1970). Namely, consider the system of b-adic subintervals of $[0, 1]$. Given a tree T, suppose that we associate to each $x \in T_n$ a b-adic interval $I_x \subseteq [0, 1]$ of order n in such a way that (1) $|I_x \cap I_y| \leq 1$ for $|x| = |y|$, $x \neq y$, and (2) I_x is contained in I_z when z is the parent of x. Then the tree T *codes (in base b)* the closed set $E := \bigcap_{n \geq 0} \bigcup_{x \in T_n} I_x$. Each ray in T corresponds to a point in E, though this correspondence might not be injective. Nevertheless, we can think of the boundary of T (the set of all rays of T) as representing E.

Figure 15.3. The Furstenberg coding of the Cantor middle-thirds set in base 3.

Figure 15.4. The Furstenberg coding of the Cantor middle-thirds set in base 2.

For example, the Cantor middle-thirds set for the base $b = 3$ is associated to the binary tree, shown in Figure 15.3. Note that if, instead of always taking out the middle third in the construction of the set, we took out the last third, we could still code it by the binary tree but with different 3-adic intervals associated to the vertices of the tree. However, if we code in base 2, we get a different tree, as in Figures 15.4 and 15.5.

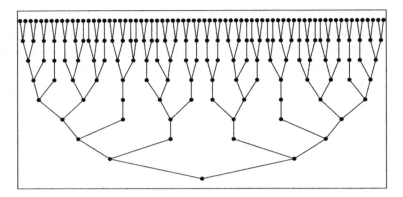

Figure 15.5. The tree of the Furstenberg coding of the Cantor middle-thirds set in base 2. The tree has branching number $2^{\log 2/\log 3} = 1.55^-$.

If T codes E in base b, then we claim that

$$\operatorname{br} T = b^{\dim E}, \tag{15.3}$$

$$\underline{\operatorname{gr}} T = b^{\dim_M E}, \tag{15.4}$$

$$\overline{\operatorname{gr}} T = b^{\dim^M E}. \tag{15.5}$$

For note that a cover of E by b-adic intervals is essentially the same as a cutset of T. If the interval I_x is associated to the vertex $x \in T$, then diam $I_x = b^{-|x|}$. Thus,

$$\dim E = \inf\left\{\alpha\,;\; \inf_\Pi \sum_{e \in \Pi} b^{-\alpha|e|} = 0\right\}.$$

Comparing this with the formula (3.4) for the branching number gives (15.3).

▷ **Exercise 15.6.**
Deduce (15.4) and (15.5) in the same way.

   Therefore, a tree that codes a set determines the dimension of the set, and the placement of particular digits in the coding doesn't influence the dimension. Recall from Section 1.10 that there is a maximal tree $T_{[b]}(E)$ that codes a given closed set, E. Its boundary differs from the boundary of any tree T that codes E by at most countably many rays, whence the two boundaries have the same Hausdorff dimension and the two trees have the same branching number. Since the branching number of a tree is a kind of average number of children per vertex, and since vertices of a tree that codes E correspond to b-adic intervals that intersect E, we may say that if a given b-adic interval intersects E, then on "average," the number of b-adic subintervals intersecting E of order one greater than the given interval is $b^{\dim E}$.

Example 15.6. We may easily rederive the Hausdorff and Minkowski dimensions of the Cantor middle-thirds set, E. Since E is coded by a binary tree, it has branching number and growth 2. From (15.3) and (15.4), it follows that the Hausdorff and Minkowski dimensions of E are both $\log 2/\log 3$.

In the context of ergodic theory, important examples are given by closed subsets $E \subseteq [0, 1]$ that are invariant under the map $s \mapsto bs$ (mod 1), that is, $bE \subseteq E$ (mod 1). This condition of invariance is equivalent to the condition that $T_{[b]}(E)$ be 0-subperiodic; indeed, this was Furstenberg's original motivation for studying such trees.

▷ **Exercise 15.7.**
Reinterpret Theorems 3.5 and 5.15 as theorems about random walks on the collection of b-adic intervals intersecting E and about random b-adic possible coverings of E.

There is a more general way than b-adic coding to relate trees to certain types of sets. Let T be a tree with root o; suppose that to each vertex $x \in V(T)$ there is associated a compact nonempty set $I_x \subseteq \mathbb{R}^d$. Denote the interior of a set I by int I and the parent of x by $\overleftarrow{x}$. Suppose that the following properties hold:

$$I_x = \overline{\text{int } I_x}, \tag{15.6}$$

$$x \neq o \quad \Longrightarrow \quad I_x \subseteq I_{\overleftarrow{x}}, \tag{15.7}$$

$$\overleftarrow{y} = \overleftarrow{x} \text{ and } y \neq x \quad \Longrightarrow \quad \text{int } I_y \cap \text{int } I_x = \varnothing, \tag{15.8}$$

$$\forall \xi \in \partial T \quad \lim_{x \in \xi} \text{diam } I_x = 0, \tag{15.9}$$

$$C_1 := \inf_{x \neq o} \frac{\text{diam } I_x}{\text{diam } I_{\overleftarrow{x}}} > 0, \tag{15.10}$$

$$C_2 := \inf_x \frac{\mathcal{L}_d(\text{int } I_x)}{(\text{diam } I_x)^d} > 0. \tag{15.11}$$

For example, if the sets I_x are b-adic cubes of order $|x|$, this is just a multidimensional version of the previous coding. For the Sierpinski gasket (Exercise 15.3), no b-adic coding is natural, but equilateral triangles give natural sets I_x. Similarly, for the von Koch snowflake (Figure 15.6), natural sets to use are again equilateral triangles (see Figure 15.7).

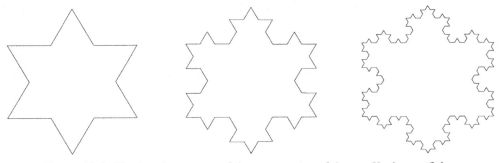

Figure 15.6. The first three stages of the construction of the von Koch snowflake.

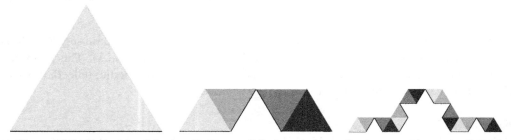

Figure 15.7. Each side of the von Koch snowflake is coded by a 4-ary tree, each vertex of which is associated to an equilateral triangle. The children of a vertex have triangles on the base of the parent triangle. The sets I_x for one side are shown for the root and the next two generations.

▷ **Exercise 15.8.**
Prove that under the conditions (15.6)–(15.11),

$$\lim_{n\to\infty} \max_{|x|=n} \operatorname{diam} I_x = 0 \,.$$

We may associate the following set to T and $I_\cdot$:

$$I_T := \bigcup_{\xi\in\partial T} \bigcap_{x\in\xi} I_x \,.$$

If T is locally finite (as it must be when C_1 and C_2 are positive), then we also have

$$I_T = \bigcap_{n\geq 1} \bigcup_{|x|=n} I_x \,. \tag{15.12}$$

For example, $I_{T_{[b]}(E)} = E$.

▷ **Exercise 15.9.**
Prove (15.12) (if T is locally finite).

The sets $\langle I_x \,;\; x \in T \rangle$ are actually the only ones we need consider in determining $\dim I_T$ or even, up to a bounded factor, $\mathcal{H}_\alpha(I_T)$:

Theorem 15.7. (Transference of Hausdorff Dimension and Measure) *If (15.6)–(15.11) hold, then*

$$\dim I_T = \inf\left\{\alpha \,;\; \inf_\Pi \sum_{e(x)\in\Pi} (\operatorname{diam} I_x)^\alpha = 0\right\} \,. \tag{15.13}$$

In fact, for $\alpha > 0$, we have

$$\mathcal{H}_\alpha(I_T) \leq \liminf_{d(0,\Pi)\to\infty} \sum_{e(x)\in\Pi} (\operatorname{diam} I_x)^\alpha \leq C_3\,\mathcal{H}_\alpha(I_T)\,, \tag{15.14}$$

where

$$C_3 := \frac{4^d}{C_2 C_1^d} \,.$$

To prove Theorem 15.7, we need a little geometric fact about Euclidean space:

Lemma 15.8. *Let E and O_i $(1 \le i \le n)$ be subsets of $\mathbb{R}^d$ such that O_i are open and disjoint, $\overline{O_i} \cap E \ne \varnothing$, $\operatorname{diam} O_i \le \operatorname{diam} E$, and $\mathcal{L}_d(O_i) \ge C(\operatorname{diam} E)^d$. Then $n \le 4^d/C$.*

Proof. Fix $x_0 \in E$ and let B be the closed ball centered at x_0 with radius $2 \cdot \operatorname{diam} E$. Then $\overline{O_i} \subseteq B$, whence

$$(4 \operatorname{diam} E)^d \ge \mathcal{L}_d(B) \ge \sum_{i=1}^n \mathcal{L}_d(O_i) \ge nC(\operatorname{diam} E)^d \,. \qquad \blacktriangleleft$$

▷ **Exercise 15.10.**
Prove the left-hand inequality of (15.14).

Proof of Theorem 15.7. For the right-hand inequality of (15.14), consider a cover by sets of positive diameter

$$I_T \subseteq \bigcup_{i=1}^\infty E_i \,.$$

Without loss of generality, we may assume that $\operatorname{diam} E_i < \operatorname{diam} I_o$ for each i. Thus, the set

$$\Pi_i := \left\{ e(x) \,;\ \operatorname{diam} I_x \le \operatorname{diam} E_i < \operatorname{diam} I_{\overleftarrow{x}} \right\}$$

is not empty. In fact, Π_i is a cutset, so

$$I_T \cap E_i \subseteq I_T \subseteq \bigcup_{e(x) \in \Pi_i} I_x \,.$$

Also, for $e(x) \in \Pi_i$, we have

$$\mathcal{L}_d(\operatorname{int} I_x) \ge C_2(\operatorname{diam} I_x)^d \ge C_2 C_1^d(\operatorname{diam} I_{\overleftarrow{x}})^d > C_2 C_1^d(\operatorname{diam} E_i)^d \,,$$

whence by Lemma 15.8,

$$\Pi_i' := \left\{ e(x) \in \Pi_i \,;\ I_x \cap E_i \ne \varnothing \right\}$$

has at most $C_3 = 4^d/(C_2 C_1^d)$ elements. Now

$$I_T \subseteq \bigcup_{i=1}^\infty \left(E_i \cap \bigcup_{e(x) \in \Pi_i'} I_x \right) \subseteq \bigcup_i \bigcup_{e(x) \in \Pi_i'} I_x \,,$$

and the cutset $\Pi := \bigcup_{i=1}^\infty \Pi_i'$ satisfies

$$\sum_{e(x) \in \Pi} (\operatorname{diam} I_x)^\alpha \le \sum_{i,\, e(x) \in \Pi_i'} (\operatorname{diam} I_x)^\alpha \le \sum_{i,\, e(x) \in \Pi_i'} (\operatorname{diam} E_i)^\alpha \le C_3 \sum_i (\operatorname{diam} E_i)^\alpha \,. \qquad \blacktriangleleft$$

Example 15.9. For two-dimensional coding by b-adic squares, we have $\operatorname{diam} I_x = \sqrt{2}\, b^{-|x|}$. In this case, (15.13) says that

$$\dim I_T = \frac{\dim \partial T}{\log b} \,,$$

which is actually the same as (15.3), the one-dimensional coding formula. The random fractal percolation set $Q_b(p)$ studied in Theorem 5.33 is coded by the cluster of the root under Bernoulli(p) percolation on a b^2-ary tree, provided that cluster is infinite (so that $Q_b(p) \ne \varnothing$). Therefore, Theorem 15.7 and Corollary 5.10 give that $\dim Q_b(p) = \log_b(pb^2) = 2 + \log_b p$ a.s. given that $Q_b(p) \ne \varnothing$.

15.3 Galton–Watson Fractals

Example 15.9 showed how to compute the Hausdorff dimension of fractal percolation in the plane, a simple process corresponding to Bernoulli percolation. In this section, we combine the transference Theorem 15.7 with Falconer's Theorem 5.35 on Galton–Watson networks to determine the dimension of more general random fractals. Thus, suppose that the sets I_x are randomly assigned to a Galton–Watson tree in such a way that the regularity conditions

$$I_x = \overline{\operatorname{int} I_x}\,,$$

$$x \neq o \quad \Longrightarrow \quad I_x \subseteq I_{\overleftarrow{x}}\,,$$

$$\overleftarrow{y} = \overleftarrow{x} \text{ and } y \neq x \quad \Longrightarrow \quad \operatorname{int} I_y \cap \operatorname{int} I_x = \varnothing\,,$$

$$\inf_x \mathcal{L}_d(\operatorname{int} I_x)/(\operatorname{diam} I_x)^d > 0$$

are satisfied a.s. and the normalized diameters $c(e(x)) := \operatorname{diam} I_x/\operatorname{diam} I_0$ are the capacities of a Galton–Watson network with generating random variable $\mathscr{L} := (L, A_1, \ldots, A_L)$, where $0 < A_i < 1$ a.s. Because Galton–Watson trees can have leaves, we denote by T' the subtree of T that consists of vertices with infinite lines of descent. (You might want to glance at the examples following the proof of the theorem at this point.) The following result is due to Falconer (1986) and Mauldin and Williams (1986).

Theorem 15.10. (Dimension of Galton–Watson Fractals) *Almost surely on nonextinction,*

$$\dim I_{T'} = \min\left\{\alpha\,;\ \mathbf{E}\left[\sum_{i=1}^{L} A_i^{\alpha}\right] \leq 1\right\}.$$

Proof. Since $I_o \supseteq \bigcup_{|x|=1} I_x$ and $\langle \operatorname{int} I_x\,;\ |x| = 1\rangle$ are disjoint, $\mathcal{L}_d(I_o) \geq \sum \mathcal{L}_d(\operatorname{int} I_x)$, whence

$$(\operatorname{diam} I_o)^d \geq \mathcal{L}_d(I_o) \geq C_2 \sum (\operatorname{diam} I_x)^d = C_2(\operatorname{diam} I_o)^d \sum_{|x|=1} A_x^d\,,$$

and so $\mathbf{E}\left[\sum_{i=1}^{L} A_i^d\right] \leq 1/C_2$. By Lebesgue's dominated convergence theorem, the function $\alpha \mapsto \mathbf{E}\left[\sum_1^{L} A_i^{\alpha}\right]$ is continuous on $[d, \infty]$ and has limit 0 at ∞, so there is some α such that $\mathbf{E}\left[\sum_1^{L} A_i^{\alpha}\right] \leq 1$. Since $\alpha \mapsto \mathbf{E}\left[\sum_1^{L} A_i^{\alpha}\right]$ is continuous from the right on $[0, \infty)$ by the monotone convergence theorem, it follows that the minimum written in the theorem statement exists.

To prove that this minimum is the Hausdorff dimension of $I_{T'}$, assume first that for some $\epsilon > 0$, we have $\forall i\ \epsilon \leq A_i \leq 1 - \epsilon$ a.s. Then the last regularity conditions, (15.9) and (15.10), are satisfied a.s.: $\forall \xi \in \partial T'\ \lim_{x \in \xi} \operatorname{diam} I_x = 0$, and

$$\inf_{x \neq o} \frac{\operatorname{diam} I_x}{\operatorname{diam} I_{\overleftarrow{x}}} = \inf_{x \neq o} A_x \geq \epsilon.$$

Hence Theorem 15.7 assures us that

$$\dim I_{T'} = \inf \left\{ \alpha \; ; \; \inf_{\Pi} \sum_{e(x) \in \Pi} (\operatorname{diam} I_x)^\alpha = 0 \right\}$$

$$= \inf \left\{ \alpha \; ; \; \inf_{\Pi} \sum_{e \in \Pi} c(e)^\alpha = 0 \right\}.$$

Now the capacities $\langle c(e)^\alpha \rangle$ come from a Galton–Watson network based on the random vector $\mathscr{L}^{(\alpha)} := (L, A_1^\alpha, \ldots, A_L^\alpha)$, whence Falconer's Theorem 5.35 says that unless $\sum_1^L A_i^\alpha = 1$ a.s.,

$$\inf_{\Pi} \sum_{e \in \Pi} c(e)^\alpha = 0 \text{ a.s. if } \mathbf{E}\left[\sum_1^L A_i^\alpha \right] \leq 1,$$

whereas

$$\inf_{\Pi} \sum_{e \in \Pi} c(e)^\alpha > 0 \text{ a.s. if } \mathbf{E}\left[\sum_1^L A_i^\alpha \right] > 1.$$

This gives the result, since there is at most one exceptional value of α where $\sum_1^L A_i^\alpha = 1$ a.s.
For the general case, note that

$$\mathbf{E}\left[\sum_1^L A_i^\alpha \right] \leq 1 \quad \Longrightarrow \quad \inf_{\Pi} \sum c(e)^\alpha = 0 \text{ a.s.} \quad \Longrightarrow \quad \mathcal{H}_\alpha(I_{T'}) = 0 \text{ a.s.},$$

except possibly for one exceptional value of α that we may ignore. For the other direction, consider the Galton–Watson subnetwork $T_{(\epsilon)}$ consisting of those branches whose ratios satisfy $\epsilon \leq A_i \leq 1 - \epsilon$. Thus $I_{T'_{(\epsilon)}} \subseteq I_{T'}$. From the preceding, we have

$$\mathbf{E}\left[\sum_1^L A_i^\alpha \mathbf{1}_{\epsilon \leq A_i \leq 1-\epsilon} \right] > 1 \quad \Longrightarrow \quad \mathcal{H}_\alpha(I_{T'_{(\epsilon)}}) > 0 \text{ a.s. on nonextinction of } T_{(\epsilon)}$$

$$\Longrightarrow \quad \mathcal{H}_\alpha(I_{T'}) > 0 \text{ a.s. on nonextinction of } T_{(\epsilon)}.$$

Since $\mathbf{P}[T'_{(\epsilon)} \neq \varnothing] \to \mathbf{P}[T' \neq \varnothing]$ as $\epsilon \to 0$ by Exercise 5.22 and

$$\mathbf{E}\left[\sum_1^L A_i^\alpha \right] = \lim_{\epsilon \to 0} \mathbf{E}\left[\sum_1^L A_i^\alpha \mathbf{1}_{\epsilon \leq A_i \leq 1-\epsilon} \right]$$

by the monotone convergence theorem, it follows that $\mathbf{E}\left[\sum A_i^\alpha \right] > 1 \Rightarrow \mathcal{H}_\alpha(I_{T'}) > 0$ a.s. on $T' \neq \varnothing$. ◀

Remark 15.11. By Falconer's theorem, $\mathcal{H}_\alpha(I_{T'}) = 0$ a.s. unless $\sum_1^L A_i^\alpha = 1$ a.s., where $\alpha := \dim I_{T'}$, in which case $\mathcal{H}_\alpha(I_{T'}) = \left(\operatorname{diam} I_o \right)^\alpha$ a.s.

Example 15.12. Divide $[0, 1]$ into three equal parts and keep each independently with probability p. Repeat this with the remaining intervals ad infinitum. The Galton–Watson network comes from the random variable $\mathscr{L} = (L, A_1, \ldots, A_L)$, where L is a Bin$(3, p)$-random variable, so that $\mathbf{E}[s^L] = (1 - p + ps)^3$, and $A_i \equiv 1/3$. Thus extinction is a.s. iff $p \leq 1/3$, whereas for $p > 1/3$, the probability of extinction q satisfies $(1 - p + pq)^3 = q$ and a.s. on nonextinction,

$$\dim I_{T'} = \min \left\{ \alpha \; ; \; \mathbf{E}\left[\sum_1^L A_i^\alpha\right] \leq 1 \right\}$$

$$= \min\{\alpha \; ; \; (1/3)^\alpha 3p \leq 1\}$$

$$= 1 + \log p / \log 3 \, .$$

Example 15.13. Remove from $[0, 1]$ a central portion leaving two intervals of random length $A_1 = A_2 \in (0, 1/2)$. Repeat on the remaining intervals ad infinitum. Here $L \equiv 2$. There is no extinction and a.s. $\dim I_T$ is the root of

$$1 = \mathbf{E}[A_1^\alpha + A_2^\alpha] = 2\,\mathbf{E}[A_1^\alpha] \, .$$

For example, if A_1 is uniform on $(0, 1/2)$, then

$$\mathbf{E}[A_1^\alpha] = 2 \int_0^{1/2} t^\alpha \, dt = \frac{1}{2^\alpha(\alpha + 1)} \, ,$$

whence $\dim I_T \approx 0.457$.

Example 15.14. Suppose M and N are random integers with $M \geq 2$ and $0 \leq N \leq M^2$. Divide the unit square into M^2 equal squares and keep N of them (in some manner). Repeat on the remaining squares ad infinitum. The probability of extinction of the associated Galton–Watson network is a root of $\mathbf{E}[s^N] = s$ and a.s. on nonextinction,

$$\dim I_{T'} = \min\{\alpha \; ; \; \mathbf{E}[N/M^\alpha] \leq 1\} \, .$$

By selecting the squares appropriately, one can have infinite connectivity of $I_{T'}$.

Example 15.15. Here, we consider the zero set E of Brownian motion B_t on $\mathbb{R}$. That is, $E := \{t \; ; \; B_t = 0\}$. To find $\dim E$, we may restrict to an interval $[0, t_0]$ where $B_{t_0} = 0$, such as the largest $t \leq 2$ with $B_t = 0$. Of course, t_0 is random. By scale invariance of B_t and of Hausdorff dimension and by the strong Markov property of B_t, the time t_0 makes no difference. So let's condition on $B_1 = 0$. This is known as ***Brownian bridge***. Calculation of finite-dimensional marginals shows B_t given $B_1 = 0$ to have the same law as $B_t^0 := B_t - tB_1$. One way to construct E is as follows. From $[0, 1]$, remove the interval (τ_1, τ_2), where $\tau_1 := \max\{t \leq \frac{1}{2} \; ; \; B_t^0 = 0\}$ and $\tau_2 := \min\{t \geq \frac{1}{2} \; ; \; B_t^0 = 0\}$. Independence and scaling shows that E is obtained as the iteration of this process. That is, $E = I_T$, where T is the binary tree and $A_1 := \tau_1$ and $A_2 := 1 - \tau_2$ are the lengths of the surviving intervals. One can show that the joint density of A_1 and A_2 is

$$\mathbf{P}[A_1 \in da_1, A_2 \in da_2] = \frac{1}{2\pi} \frac{1}{\sqrt{a_1 a_2 (1 - a_1 - a_2)^3}} \, da_1 \, da_2 \qquad (a_1, a_2 \in [0, 1/2]) \, .$$

Straightforward calculus then shows that $\mathbf{E}[A_1^{1/2} + A_2^{1/2}] = 1$, so $\dim E = 1/2$ a.s. This result is due to Taylor (1955), but this method is due to Graf, Mauldin, and Williams (1988).

More examples of the use of Theorem 15.10 can be found in Mauldin and Williams (1986).

15.4 Hölder Exponent

There is an intrinsic metric on the boundary of a tree; when we use it to compute the Hausdorff dimension of the boundary, we find a simple relationship to the Hausdorff dimension of a set coded by T with b-adic intervals. This ability to work purely on the tree for much of the calculation has advantages and also suggests interesting extensions, which we begin to study here.

For any tree T with root o, consider $\xi, \eta \in \partial T$. Define $\xi \wedge \eta$ to be the vertex common to both ξ and η that is farthest from o if $\xi \neq \eta$ and $\xi \wedge \eta := \xi$ if $\xi = \eta$. Write $|\xi| = \infty$. Now use these notations to define a metric on the boundary ∂T by $d(\xi, \eta) := e^{-|\xi \wedge \eta|}$ ($\xi \neq \eta$). With this metric, we can consider $\dim E$ for $E \subseteq \partial T$. To compute Hausdorff dimensions of subsets of ∂T, it suffices to consider covers by sets of the form

$$\beta_x := \{\xi \in \partial T \,;\; x \in \xi\}, \tag{15.15}$$

since any subset of ∂T is contained in such a special set of the same diameter. (Namely, if $F \subseteq \partial T$ and $x := \bigwedge_{\xi \in F} \xi$, in the obvious extension of the $\wedge$ notation, then $F \subseteq \beta_x$ and $\operatorname{diam} F = \operatorname{diam} \beta_x$.)

▷ **Exercise 15.11.**
What is $\operatorname{diam} \beta_x$?

In the particular case $E = \partial T$, we have

$$\dim \partial T = \log \operatorname{br} T \tag{15.16}$$

as in Section 1.8. More generally, when $E \subseteq \partial T$ is closed and $T(E)$ denotes the set of vertices that belong to some ray in E, then

$$\dim E = \log \operatorname{br} T(E).$$

By the transference Theorem 15.7, we obtain the following:

Corollary 15.16. *If the sets I_x of Section 15.2 are b-adic cubes of order $|x|$ in $\mathbb{R}^d$, then*

$$\dim I_T = \dim \partial T / \log b. \qquad \blacktriangleleft$$

When this corollary is applied to a tree T that b-adically codes a set E, we obtain (15.3) again by (15.16).

We now consider extensions of the ideas of Hausdorff dimension. Let T be an infinite, locally finite rooted tree and μ be a Borel probability measure on ∂T. A subset $E \subseteq \partial T$ is a *carrier* of μ if $\mu(E) = 1$. Even if the support of μ is ∂T, there may still be "much smaller" carriers of μ. That is, there may be sets $E \subseteq \partial T$ with $\dim E < \dim \partial T$ and $\mu(E) = 1$. This suggests defining the *dimension* of μ as

$$\dim \mu := \min\{\dim E \,;\; \mu(E) = 1\}. \tag{15.17}$$

▷ **Exercise 15.12.**
Show that this minimum exists.

▷ **Exercise 15.13.**

Let T be an infinite, locally finite tree. Show that there is a one-to-one correspondence between unit flows θ from o to ∂T and Borel probability measures μ on ∂T satisfying

$$\theta(x) = \mu(\beta_x) \,.$$

Here and in the sequel, we write $\theta(x)$ for $\theta(\bar{x}, x)$.

The dimension of a measure is most often computed not by using the definition but by calculating pointwise information and then using the following theorem of Billingsley (1965). For a ray $\xi = \langle \xi_1, \xi_2, \ldots \rangle \in \partial T$, define the **Hölder exponent** of μ at ξ to be

$$\text{Hö}(\mu)(\xi) := \liminf_{n \to \infty} \frac{1}{n} \log \frac{1}{\mu(\beta_{\xi_n})} \,.$$

For example, if T is the m-ary tree, $\theta(x) := m^{-|x|}$, and μ is the probability measure corresponding to the unit flow θ, then the Hölder exponent of μ is everywhere $\log m$.

Theorem 15.17. (Dimension and Hölder Exponent) *For every Borel probability measure μ on the boundary of a tree,*

$$\dim \mu = \mu\text{-ess sup Hö}(\mu) \,.$$

Proof. Let $d := \mu\text{-ess sup Hö}(\mu)$. We first exhibit a carrier of μ whose dimension is at most d. Then we show that every carrier of μ has dimension at least d. Let θ be the unit flow that corresponds to μ as in Exercise 15.13.

Given $k \in \mathbb{N}$ and $\alpha > d$, let

$$E(k, \alpha) := \left\{ \xi \in \partial T \,;\; \exists n \geq k \quad (1/n) \log(1/\theta(\xi_n)) \leq \alpha \right\}$$
$$= \left\{ \xi \in \partial T \,;\; \exists n \geq k \quad e^{-n\alpha} \leq \theta(\xi_n) \right\} \,.$$

Then clearly $\bigcap_\alpha \bigcap_k E(k, \alpha)$ is a carrier of μ. To show that it has Hausdorff dimension at most d, it suffices, by Exercise 15.5, to show that $\dim \bigcap_k E(k, \alpha) \leq \alpha$. Clearly $E(k, \alpha)$ is open, so is a countable union of disjoint sets of the form (15.15), call them β_{x_i}, with all $|x_i| \geq k$ and

$$\left(\text{diam } \beta_{x_i} \right)^\alpha \leq e^{-|x_i|\alpha} \leq \mu(\beta_{x_i}) \,.$$

Since

$$\sum_i \left(\text{diam } \beta_{x_i} \right)^\alpha \leq \sum_i \mu(\beta_{x_i}) \leq 1$$

and the sets β_{x_i} also cover $\bigcap_k E(k, \alpha)$, it follows that $\mathcal{H}_\alpha\left(\bigcap_k E(k, \alpha)\right) \leq 1$. This implies our desired inequality, $\dim \bigcap_k E(k, \alpha) \leq \alpha$.

For the other direction, suppose that F is a carrier of μ. For $k \in \mathbb{N}$ and $\alpha < d$, let

$$F(k, \alpha) := \left\{ \xi \in F \,;\; \forall n \geq k \quad (1/n) \log(1/\theta(\xi_n)) \geq \alpha \right\} \,.$$

Then $\mu\big(F(k, \alpha)\big) > 0$ for sufficiently large k. Fix such a k. To show that $\dim F \geq d$, it suffices to show that $\dim F(k, \alpha) \geq \alpha$. Indeed, reasoning similar to that in the preceding paragraph shows that $\mathcal{H}_\alpha\big(F(k, \alpha)\big) \geq \mu\big(F(k, \alpha)\big)$, which completes the proof. ◀

Example 15.18. For an example of the calculation of Hölder exponent, let T be an infinite rooted tree without leaves (except possibly the root). Consider the harmonic measure of **simple nonbacktracking random walk** on T, which is the random walk that starts at the root and chooses randomly (uniformly) among the children of the present vertex as the next vertex. The corresponding harmonic measure on ∂T is called **visibility measure**, denoted VIS_T, and corresponds to the **equally-splitting flow**. Suppose now that T is a Galton–Watson tree starting, as usual, with one particle and having L children, where $L > 0$ a.s. Identifying measure and flow, we have that VIS_T is a flow on the random tree T. Let $\mathbf{GW}$ denote the distribution of Galton–Watson trees. The probability measure that corresponds to choosing a Galton–Watson tree T at random followed by a ray of ∂T chosen according to VIS_T will be denoted $\mathsf{VIS} \times \mathbf{GW}$. Formally, this is

$$(\mathsf{VIS} \times \mathbf{GW})(F) := \iint \mathbf{1}_F(\xi, T)\, d\mathsf{VIS}_T(\xi)\, d\mathbf{GW}(T)\,.$$

Since

$$\frac{1}{n} \log \frac{1}{\mathsf{VIS}_T(\xi_n)} = \frac{1}{n} \sum_{k=0}^{n-1} \log \frac{\mathsf{VIS}_T(\xi_k)}{\mathsf{VIS}_T(\xi_{k+1})}$$

and the random variables $\mathsf{VIS}_T(\xi_{k-1})/\mathsf{VIS}_T(\xi_k)$ are $\mathsf{VIS} \times \mathbf{GW}$-i.i.d. with the same distribution as L, the strong law of large numbers gives

$$\mathsf{H\ddot{o}}(\mathsf{VIS}_T)(\xi) = \mathbf{E}[\log L] \quad \text{for } \mathsf{VIS} \times \mathbf{GW}\text{-a.e. } (\xi, T)\,.$$

Thus $\dim \mathsf{VIS}_T = \mathbf{E}[\log L]$ for $\mathbf{GW}$-a.e. tree T. Jensen's inequality (or the arithmetic mean–geometric mean inequality) shows that this dimension is less than $\log m$, except in the deterministic case $L = m$ a.s. Recall that the Hausdorff dimension of the full boundary ∂T is $\log m$ a.s. by Corollary 5.10.

15.5 Derived Trees

This section is based on some ideas of Furstenberg (1970). In the context of Euclidean sets coded by trees, it is important to view trees as (rooted and) labeled, as in Section 5.1. However, since our interest centers more on the trees themselves, we will not label the trees; that is, we consider rooted trees equal if they are rooted isomorphic.

Most of the time, for the sake of compactness, we need to restrict our trees to have uniformly bounded degree. Thus, fix an integer r and let $\mathbf{T}$ be the r-ary tree. We will consider only subtrees of $\mathbf{T}$ rooted at the root of $\mathbf{T}$ and that have no leaves (other than possibly the root). Recall that for a tree T and vertex $v \in T$, we denote by T^v the subtree of T formed from the descendants of v. We view T^v as a rooted subtree of $\mathbf{T}$ with v identified with the root of T^v. Given a unit flow θ on T and a vertex $v \in T$, the conditional flow through v is defined to be the unit flow

$$\theta^v(x) := \theta(x)/\theta(v) \quad \text{for } x \geq v$$

on T^v when $\theta(v) \neq 0$. The class of subtrees of $\mathbf{T}$ is given the natural topology as in Exercise 5.2. This class is compact. For a subtree T, let $\mathscr{D}(T)$ denote the closure of the set of its descendant trees, $\{T^v \,;\ v \in T\}$. We call these trees the **derived trees** of T. For example, $\mathscr{D}(\mathbf{T}) = \{\mathbf{T}\}$.

▷ **Exercise 15.14.**

Let T code the set $\{0\} \cup \{1/n \,;\, n \geq 1\}$ in base r. Show that $\mathbf{T} \in \mathscr{D}(T)$.

▷ **Exercise 15.15.**

Show that if $T^* \in \mathscr{D}(T)$, then $\mathscr{D}(T^*) \subseteq \mathscr{D}(T)$.

Define

$$T_n^v := \{x \in T^v \,;\, |x| \leq |v| + n\},$$

$$M_n(v) := |\partial_{\mathrm{L}} T_n^v| = \big|\{x \in T^v; |x| = |v| + n\}\big|,$$

$$\dim \sup \partial T := \lim_{n \to \infty} \max_{v \in T} \frac{1}{n} \log M_n(v), \tag{15.18}$$

$$\dim \inf \partial T := \lim_{n \to \infty} \min_{v \in T} \frac{1}{n} \log M_n(v). \tag{15.19}$$

▷ **Exercise 15.16.**

Show that these limits exist.

Clearly,

$$\dim \partial T \leq \dim_{\mathrm{M}} \partial T = \liminf_{n \to \infty} \frac{1}{n} \log M_n(o) \leq \dim \sup \partial T. \tag{15.20}$$

Also, if $\alpha < \dim \inf \partial T$, then there is some n such that

$$\min_{v \in T} \frac{1}{n} \log M_n(v) > \alpha,$$

in other words, $M_n(v) \geq \lceil e^{\alpha n} \rceil$ for all $v \in T$. Thus, in the notation of Exercise 3.25, $T^{[n]}$ contains a $\lceil e^{\alpha n} \rceil$-ary subtree. Therefore $\dim \partial T^{[n]} \geq \log \lceil e^{\alpha n} \rceil \geq \alpha n$, whence $\dim \partial T \geq \alpha$. Thus, we obtain

$$\dim \partial T \geq \dim \inf \partial T. \tag{15.21}$$

Since for any $T^* \in \mathscr{D}(T)$, $v \in T^*$, and $n \geq 0$, there is some $w \in T$ such that $T_n^w = (T^*)_n^v$, we have $\dim \inf \partial T \leq \dim \inf \partial T^*$ and $\dim \sup \partial T^* \leq \dim \sup \partial T$. Combining these inequalities with (15.20) and (15.21) as applied to T^*, we arrive at the following:

Proposition 15.19. *For any $T^* \in \mathscr{D}(T)$, we have*

$$\dim \inf \partial T \leq \dim \partial T^* \leq \dim \sup \partial T. \qquad \blacktriangleleft$$

▷ **Exercise 15.17.**

Show that if T is subperiodic, then $\dim \partial T = \dim \sup \partial T$, whereas if T is superperiodic, then $\dim \partial T = \dim \inf \partial T$.

Proposition 15.19 is sharp in a strong sense, as shown by the following theorem adapted from Furstenberg (1970) (compare Theorem 5.1 of Furstenberg (2008)):

Theorem 15.20. *If T is a tree of uniformly bounded degree, then there exist $T^* \in \mathscr{D}(T)$ and a unit flow θ^* on T^* such that*

$$\dim \partial T^* = \dim \sup \partial T, \tag{15.22}$$

$$\forall x \in T^* \setminus \{o\} \quad \frac{1}{|x|} \log \frac{1}{\theta^*(x)} \geq \dim \sup \partial T, \tag{15.23}$$

and for θ^*-a.e. $\xi \in \partial T^*$,

$$\mathrm{H\ddot{o}}(\theta^*)(\xi) = \dim \sup \partial T. \tag{15.24}$$

Similarly, there exist $T^{**} \in \mathscr{D}(T)$ and a unit flow θ^{**} on T^{**} such that

$$\dim \partial T^{**} = \dim \inf \partial T \tag{15.25}$$

and

$$\forall x \in T^{**} \setminus \{o\} \quad \frac{1}{|x|} \log \frac{1}{\theta^{**}(x)} \leq \dim \inf \partial T. \tag{15.26}$$

Proof. We concentrate first on (15.23): the idea is to find T^* as the support of θ^* and to find θ^* as the limit of flows on finite trees. We claim that for any positive integer L and any $\alpha < \dim \sup \partial T$, there is some vertex $v \in T$ and some unit flow θ on T_L^v (from v to $\partial_L T_L^v$) such that

$$\forall x \in T_L^v \setminus \{v\} \quad \frac{1}{|x| - |v|} \log \frac{1}{\theta(x)} \geq \alpha. \tag{15.27}$$

Suppose for a contradiction that this is not the case for some L and α. By choice of α, there are v and n so that $M_n(v)^{1/n} > e^\alpha$. Since $r^{L/n} \to 1$ as $n \to \infty$, we may choose v and n so that n is a multiple of L and

$$M_n(v)^{1/n} > e^\alpha r^{L/n}. \tag{15.28}$$

(Recall that r is the degree of the root of **T**.) Let θ be the flow on T_n^v such that

$$\forall x \in \partial_L T_n^v \setminus \{v\} \quad \theta(x) = 1/M_n(v).$$

Then θ restricts to a flow on T_L^v, whence, by our assumption that (15.27) fails,

$$\exists x_1 \in T_L^v \quad \theta(x_1) > e^{-(|x_1| - |v|)\alpha}.$$

By similar reasoning, define a finite sequence $\langle x_k \rangle$ inductively as follows. Provided $|x_k| - |v| \leq n - L$, choose $x_{k+1} \in T_L^{x_k}$ so that $\theta^{x_k}(x_{k+1}) > e^{-(|x_{k+1}| - |x_k|)\alpha}$. Let the last index thereby achieved on $\langle x_k \rangle$ be K. Then, using $x_0 := v$, we have

$$\theta(x_K) = \prod_{k=0}^{K-1} \theta^{x_k}(x_{k+1}) > e^{-(|x_K| - |v|)\alpha} \geq e^{-n\alpha} > \frac{r^L}{M_n(v)},$$

where in the last step we have used (15.28). On the other hand, since $|v| + n - |x_K| < L$, we have

$$\theta(x_K) = \frac{\left| \partial_L T^{x_K}_{|v|+n-|x_K|} \right|}{M_n(v)} < \frac{r^L}{M_n(v)} \,.$$

As these two inequalities contradict each other, our claim is established.

For $j \geq 1$, there is thus some unit flow θ_j on some $T_j^{v_j}$ such that

$$\forall x \in T_j^{v_j} \setminus \{v_j\} \qquad \frac{1}{|x| - |v_j|} \log \frac{1}{\theta_j(x)} \geq \left(1 - \frac{1}{j} \right) \dim \sup \partial T \,. \tag{15.29}$$

Identifying $T_j^{v_j}$ with a rooted subtree of $\mathbf{T}$ identifies θ_j with a unit flow on $\mathbf{T}$. By taking a subsequence if necessary, we may assume that these flows θ_j have an edgewise limit θ^*. Those edges where $\theta_j^* > 0$ for infinitely many j form a tree $T^* \in \mathscr{D}(T)$ such that θ^* is a unit flow on T^*. Because of (15.29), we obtain (15.23). By definition of Hölder exponent, this means that $\mathrm{H\ddot{o}}(\theta^*) \geq \dim \sup \partial T$. On the other hand, Theorem 15.17 gives $\mathrm{H\ddot{o}}(\theta^*) \leq \dim \sup \partial T^* \leq \dim \sup \partial T$, whence (15.22) and (15.24) follow.

The proof of (15.26) is parallel to that of (15.23), and we omit it. To deduce (15.25), let

$$M_n^{**} := \left| \{ x \in T^{**} ; \ |x| = n \} \right|$$

and $\alpha := \dim \inf \partial T$. By (15.26), we have

$$1 = \theta^{**}(0) = \sum_{|v|=n} \theta^{**}(v) \geq \sum_{|v|=n} e^{-\alpha n} = M_n^{**} e^{-\alpha n} \,,$$

so that

$$\frac{1}{n} \log M_n^{**} \leq \alpha \,.$$

Therefore, $\dim \partial T^{**} \leq \alpha$. The other inequality follows from Proposition 15.19. (We deduce that also $\dim^{\mathrm{M}} \partial T = \alpha$.) ◄

The first part of Theorem 15.20 allows us to give another proof of Furstenberg's important Theorem 3.8: If T is subperiodic, then every $T^* \in \mathscr{D}(T)$ is isomorphic to a subtree of a descendant subtree T^v. In particular,

$$\dim \partial T^* \leq \dim \partial T \,.$$

On the other hand, if $T^* \in \mathscr{D}(T)$ satisfies (15.22), then

$$\dim \partial T \geq \dim \partial T^* = \dim \sup \partial T \geq \dim^{\mathrm{M}} \partial T \geq \dim \partial T \,.$$

Therefore, $\dim \partial T = \dim^{\mathrm{M}} \partial T$, as desired.

15.6 Notes

There is much more one can say about Hausdorff dimension. Some good books to consult are by Edgar (1990), Falconer (1990), Mattila (1995), Pesin (1997), and Barreira (2008).

As noted by Virág (2000b), the essential branching number $\operatorname{ess\,br} T$ of a tree T, defined in (13.36), is related to the Hausdorff dimension of harmonic measure μ for simple random walk on the tree by the inequality

$$\operatorname{ess\,br} T \le e^{\dim \mu} \,.$$

Strict inequality can hold.

See Furstenberg (2008) for a development of the ideas of Section 15.5 in the context of Euclidean sets.

The proof we give of Theorem 15.20 is due to F. Ledrappier and Y. Peres, 1990, previously unpublished.

15.7 Collected In-Text Exercises

15.1. Show that for any $E \subseteq \mathbb{R}^d$, there exists a real number α_0 such that $\alpha < \alpha_0 \Rightarrow \mathcal{H}_\alpha(E) = +\infty$ and $\alpha > \alpha_0 \Rightarrow \mathcal{H}_\alpha(E) = 0$.

15.2. The *Sierpinski carpet* is the set

$$E := \left\{ \left(\sum_n x_n 3^{-n}, \sum_n y_n 3^{-n} \right); \ (x_n, y_n) \in \{0, 1, 2\}^2 \setminus \{(1, 1)\} \right\}.$$

That is, the unit square is divided into its nine triadic subsquares of order 1 and the interior of the middle one is removed. This process is repeated on each of the remaining eight squares, and so on. Show that $\dim E = \dim^{\mathrm{M}} E = \log 8 / \log 3$.

15.3. The *Sierpinski gasket* is the set obtained by partitioning an equilateral triangle into four congruent pieces and removing the interior of the middle one, then repeating this process ad infinitum on the remaining pieces, as in Figures 15.1 and 15.2. Show that its Hausdorff dimension is $\log 3 / \log 2$.

15.4. Show that for any sets E_n, $\dim \left(\bigcup_{n=1}^\infty E_n \right) = \sup \dim E_n$.

15.5. Show that if $E_1 \supseteq E_2 \supseteq \cdots$, then $\dim \bigcap E_n \le \lim \dim E_n$.

15.6. Deduce (15.4) and (15.5) in the same way as we deduced (15.3).

15.7. Reinterpret Theorems 3.5 and 5.15 as theorems about random walks on the collection of b-adic intervals intersecting E and about random b-adic possible coverings of E.

15.8. Prove that under the conditions (15.6)–(15.11),

$$\lim_{n \to \infty} \max_{|x|=n} \operatorname{diam} I_x = 0 \,.$$

15.9. Prove (15.12) (if T is locally finite).

15.10. Prove the left-hand inequality of (15.14).

15.11. What is $\operatorname{diam} \beta_x$ of (15.15)?

15.12. Show that the minimum of (15.17) exists.

15.13. Let T be an infinite, locally finite tree. Show that there is a one-to-one correspondence between unit flows θ from o to ∂T and Borel probability measures μ on ∂T satisfying

$$\theta(x) = \mu(\beta_x).$$

15.14. Let T code the set $\{0\} \cup \{1/n ; \ n \geq 1\}$ in base r. Show that $\mathbf{T} \in \mathscr{D}(T)$.

15.15. Show that if $T^* \in \mathscr{D}(T)$, then $\mathscr{D}(T^*) \subseteq \mathscr{D}(T)$.

15.16. Show that the limits in (15.18) and (15.19) exist.

15.17. Show that if T is subperiodic, then $\dim \partial T = \dim \sup \partial T$, whereas if T is superperiodic, then $\dim \partial T = \dim \inf \partial T$.

15.8 Additional Exercises

15.18. Let $\Omega \subseteq \mathbb{R}^2$ be open and nonempty and $f: \Omega \to \mathbb{R}$ be Lipschitz. Show that the Hausdorff dimension of the graph of f is 2.

15.19. Let $E := \{\sum_{n \geq 1} x_n 2^{-n} ; \ x_n \in \{0, 1\}, \ \forall n \ x_n x_{n+1} = 0\}$. What is $\dim E$?

15.20. Suppose that $E \subseteq [0, 1]$ is closed and $bE \subseteq E \pmod 1$. Show that $\dim_{\mathrm{H}} E = \dim^{\mathrm{M}} E$ and $\mathcal{H}_{\dim_{\mathrm{H}} E}(E) > 0$.

15.21. Is there a subperiodic tree of exponential growth whose boundary has infinite Hausdorff measure in its dimension?

15.22. Consider fractal percolation in the unit square with parameters b and $p > b^{-2}$. Recall from Example 15.9 that the fractal percolation set $Q_b(p)$ has Hausdorff dimension $\log_b(pb^2)$ a.s. on the event that it is nonempty. Let Γ denote the union of all connected components of $Q_b(p)$ that have at least two points.

(a) Show that $\dim \Gamma$ is a.s. constant on the event that $Q_b(p)$ is nonempty.

(b) Show that $\dim \Gamma < \log_b(pb^{-2})$ a.s. *Hint:* For the case $b = 3$, consider removing every square at level k such that at level $k + 1$ only the central subsquare is retained.

15.23. Let $m > 1$. Show that for $\gamma < 1$,

$$\max\left\{\dim I_{T'} ; \ \|L\|_\infty < \infty, \mathbf{E}[L] = m, \mathbf{E}\left[\sum_1^L A_i\right] = \gamma\right\} = \left(1 - \frac{\log \gamma}{\log m}\right)^{-1} < 1,$$

where the maximum is over all Galton–Watson fractals $I_{T'}$ based on random vectors $(L, A_1, \ldots, A_L)$ as in Section 15.3, whereas for $m > \gamma > 1$,

$$\min\left\{\dim I_{T'} ; \ \|L\|_\infty < \infty, \mathbf{E}[L] = m, \mathbf{E}\left[\sum_1^L A_i\right] = \gamma\right\} = \left(1 - \frac{\log \gamma}{\log m}\right)^{-1} > 1,$$

with equality in each case iff $\forall i \ A_i = \gamma/m$ a.s. What if $\gamma = 1$?

15.24. To create (the top side of) a random von Koch curve E, begin with the unit interval. Replace the middle portion of random length $\in (0, 1/3)$ by the other two sides of an equilateral triangle, as in Figure 15.8. Repeat this process proportionally and independently on each of the remaining four pieces, and so on. Of course, here, the associated Galton–Watson network has no extinction. Show that if the length is uniform on $(0, 1/3)$, then $\dim E \approx 1.144$ a.s.

Figure 15.8. Random side replacement.

15.25. Create a random Cantor set E in $[0, 1]$ by removing a middle interval between two points chosen independently uniformly on $[0, 1]$. Repeat indefinitely proportionally and independently on the remaining intervals. Show that $\dim E = (\sqrt{17} - 3)/2$ a.s.

15.26. Let T be a Galton–Watson tree with offspring random variable L having mean $m > 1$. Show that the Hausdorff measure of ∂T in dimension $\log m$ is 0 a.s., except when L is constant.

15.27. Fix a positive integer d and an integer $b \geq 2$. Consider d-dimensional b-adic codings of closed sets $E \subseteq [0, 1]^d$. For a subtree T of the b^d-ary tree that contains the root, write Q_T for the closed subset of $[0, 1]^d$ coded by the rays of T. For $0 \leq p \leq 1$, write $Q_{d,b}(p)$ for the corresponding random closed set Q_T when T is the component of the root for Bernoulli(p) percolation on the b^d-ary tree. Consider any closed subset $\Lambda \subseteq [0, 1]^d$ and any $\beta > 0$.

    (a) Show that if $\dim_H \Lambda < \beta$, then $\Lambda \cap Q_{d,b}(b^{-\beta}) = \varnothing$ a.s.

    (b) Show that if $\dim_H \Lambda > \beta$, then $\Lambda \cap Q_{d,b}(b^{-\beta}) \neq \varnothing$ with positive probability.

    (c) Show that if $\dim_H \Lambda > \beta$, then $\dim_H(\Lambda \cap Q_{d,b}(b^{-\beta})) \leq \dim_H \Lambda - \beta$ a.s. and that the essential supremum of the left-hand side is equal to the right-hand side.

15.28. Let $1 \leq k < d$ be integers and $\gamma > 0$ be real. Let L be a linear transformation from $\mathbb{R}^d$ onto $\mathbb{R}^k$, and let $Q = Q_{d,b}(b^{-\gamma})$ be as in Exercise 15.27.

    (a) Show that if $d - \gamma > k$, then the image $L(Q)$ has positive k-dimensional Lebesgue measure a.s. given that $Q \neq \varnothing$.

    (b) Show that if $0 < d - \gamma \leq k$, then $\dim_H L(Q) = d - \gamma$ a.s. given that $Q \neq \varnothing$.

15.29. **(a)** Suppose that μ is a probability measure on $\mathbb{R}^d$ such that the measure of every b-adic cube of order n is at most $Cb^{-n\alpha}$ for some constants C and α. Show that there is a constant C' such that the μ-measure of every ball of radius r is at most $C'r^\alpha$.

    (b) The *Frostman exponent* of a probability measure μ on $\mathbb{R}^d$ is

$$\mathrm{Frost}(\mu) := \sup\left\{\alpha \in \mathbb{R}; \sup_{x \in \mathbb{R}^d, r > 0} \mu(B_r(x))/r^\alpha < \infty\right\},$$

where $B_r(x)$ is the ball of radius r centered at x. Show that the Hausdorff dimension of every compact set is the supremum of the Frostman exponents of the probability measures it supports.

15.30. Given a probability vector $\langle p_i; 1 \leq i \leq k \rangle$, the capacities of the Galton–Watson network generated on a k-ary tree by the nonrandom vector $(k, p_1, p_2, \ldots, p_k)$ as in Section 5.9 define a unit flow θ where the flow along every edge equals the capacity of that edge. The flow θ corresponds to a probability measure μ on the boundary. Show that

$$\dim \mu = \sum_i p_i \log \frac{1}{p_i}.$$

15.31. Let $p(\cdot, \cdot)$ be the transition probabilities of a finite-state, irreducible Markov chain. Let $\pi(\cdot)$ be the stationary probability distribution. The directed graph G associated to this chain has for vertices the states and for edges all (x, y) for which $p(x, y) > 0$. Let T be the directed cover of G (see Section 3.3). Define the unit flow

$$\theta(\langle x_1, \ldots, x_{n+1} \rangle) := \pi(x_1) \prod_{i=1}^n p(x_i, x_{i+1}).$$

Let μ be the corresponding probability measure on ∂T. Show that

$$\dim \mu = \sum_x \pi(x) \sum_y p(x, y) \log \frac{1}{p(x, y)}.$$

15.32. Let $p_k > 0$ ($1 \leq k \leq r$) satisfy $\sum_{k=1}^r p_k = 1$. Let **GW** be the Galton–Watson measure on trees with offspring distribution $\langle p_k \rangle$. Show that for **GW**-a.e. T, $\mathscr{D}(T)$ is equal to the set of *all* subtrees of the r-ary tree **T**.

15.33. Let T be the tree of Exercise 3.33 with $N = 0$ and $\alpha = 1/3$. Show that $\dim\inf \partial T = \frac{1}{3}\log 3 + \frac{2}{3}\log \frac{3}{2}$ and $\dim\sup \partial T = \log 2$.

15.34. Let $p_k \geq 0$ ($k \geq 1$) satisfy $\sum_{k=1}^{\infty} p_k = 1$. Let **GW** be the Galton–Watson measure on trees with offspring distribution $\langle p_k \rangle$. Show that for **GW**-a.e. T, $\dim\sup \partial T = \log\sup\{k\,;\ p_k > 0\}$ and $\dim\inf \partial T = \log\min\{k\,;\ p_k > 0\}$. Here, we use the same definitions of $\dim\sup$ and $\dim\inf$ as in (15.18) and (15.19), even though the degrees may not be bounded.

15.35. Is there a tree T of uniformly bounded degree such that $\operatorname{br} T > 1$ and such that every derived tree of T is amenable?

15.36. Let T code the Cantor middle-thirds set in base 2. Show that $\dim\sup \partial T = (\log 2)^2/\log 3$.

16 | Capacity and Stochastic Processes

One refinement of Hausdorff dimension is, of course, Hausdorff measure. Another is capacity. The latter turns out to be more widely related to probability than the former. The notion of capacity will lead to important reformulations of our theorems concerning random walk and percolation on trees. As one consequence, we will be able to deduce a classical relationship of Hausdorff dimension and capacity in Euclidean space. Since capacity also is intimately related to Brownian motion, we will be able to solve problems about Brownian motion by using our work on percolation on trees: capacity translates one domain to the other.

16.1 Definitions

We call a function $\Psi \geq 0$ on a rooted tree T such that $x \to y \Rightarrow \Psi(x) \leq \Psi(y)$ a *gauge*. For a gauge Ψ, extend Ψ to ∂T by $\Psi(\xi) := \lim_{x \in \xi} \Psi(x)$ and define the *kernel* $K := \partial T \times \partial T \to [0, \infty]$ by

$$K(\xi, \eta) := \Psi(\xi \wedge \eta),$$

where, as in Section 15.4, $\xi \wedge \eta$ is the vertex common to both ξ and η that is farthest from o if $\xi \neq \eta$ and $\xi \wedge \eta := \xi$ if $\xi = \eta$.

A common gauge is $\Psi(x) := \lambda^{|x|}$ or $\Psi(x) := \lambda^{|x|}/(\lambda - 1)$ for $\lambda > 1$. Using $r(e(x)) := \lambda^{|x|-1}(\lambda - 1)$ or $r(e(x)) := \lambda^{|x|-1}$, we see that these gauges are special cases of the following assignment of a gauge to conductances (and resistances):

$$\Psi(x) := \Psi(o) + \sum_{o < u \leq x} r\big(e(u)\big). \tag{16.1}$$

For another example, $r \equiv 1$ corresponds to $\Psi(x) = \Psi(o) + |x|$, which gives the kernel $K(\xi, \eta) = \Psi(o) - \log d(\xi, \eta)$ in terms of the distance on ∂T defined in Section 15.4. Furthermore, given a gauge Ψ, we can implicitly define conductances by (16.1).

Write $\mathsf{Prob}(\partial T)$ for the set of Borel probability measures on ∂T. Fix a kernel, K. Given $\mu \in \mathsf{Prob}(\partial T)$, its *potential* is the function

$$V_\mu(\xi) := \int_{\partial T} K(\xi, \eta) \, d\mu(\eta)$$

and its *energy* is the number

$$\mathscr{E}(\mu) := \int_{\partial T \times \partial T} K \, d(\mu \times \mu) = \int_{\partial T} V_\mu(\xi) \, d\mu(\xi).$$

The *capacity* of $E \subseteq \partial T$ measures inversely how small the energy can be of a probability measure on E, that is,

$$\operatorname{cap} E := \left[\inf \{ \mathscr{E}(\mu) \, ; \ \mu \in \operatorname{Prob}(\partial T), \ \mu(\partial T \setminus E) = 0 \} \right]^{-1}.$$

When $E = \varnothing$, then we define $\operatorname{cap} E := 0$. ***These definitions are made similarly for any topological space X in place of ∂T and any Borel function $K \colon X \times X \to [0, \infty]$.***

The concepts of potential, energy, and capacity come from physics. For example, if μ is a distribution of electric charge in space and $K(x, y)$ is one over the distance between x and y (we ignore physical constants and units), then V_μ is the electric potential generated by μ (its negative gradient is the electric field) and $\mathscr{E}(\mu)$ is the electrostatic potential energy inherent in this configuration. If a unit charge is placed on a perfect conductor, then it will distribute itself to minimize its energy; the reciprocal of that energy is the capacitance of the conductor.

Recall from Exercise 15.13 that Borel probabilities μ on ∂T are in 1-1 correspondence with unit flows θ on T via

$$\theta(x) = \mu(\{ \xi \, ; \ x \in \xi \}).$$

Here, as in Chapter 15, we will abbreviate $\theta(e(x))$ as $\theta(x)$. For convenience, set $\theta(o) := 1$. We may express V_μ and $\mathscr{E}(\mu)$ in terms of θ as follows. Set

$$\Phi(x) := \begin{cases} \Psi(x) - \Psi(\overleftarrow{x}) & \text{if } x \neq o \\ \Psi(o) & \text{if } x = o. \end{cases}$$

(Recall that $\overleftarrow{x}$ is the parent of x.) Thus, $\Phi(x)$ is the resistance of $e(x)$ for $x \neq o$ when (16.1) holds.

Proposition 16.1. (Lyons, 1990) *With the preceding notation, we have*

$$\forall \xi \in \partial T \qquad V_\mu(\xi) = \sum_{x \in \xi} \Phi(x) \theta(x)$$

and

$$\mathscr{E}(\mu) = \sum_{x \in T} \Phi(x) \theta(x)^2.$$

Proof. Since $\Psi(x) = \sum_{u \leq x} \Phi(u)$ and $\Psi(\xi) = \sum_{x \in \xi} \Phi(x)$, we have

$$V_\mu(\xi) = \int_{\partial T} K(\xi, \eta) \, d\mu(\eta) = \int_{\partial T} \Psi(\xi \wedge \eta) \, d\mu(\eta) = \int_{\partial T} \sum_{x \leq \xi \wedge \eta} \Phi(x) \, d\mu(\eta)$$

$$= \int_{\partial T} \sum_{x \in \xi} \Phi(x) \mathbf{1}_{[x \in \eta]} \, d\mu(\eta) = \sum_{x \in \xi} \Phi(x) \int_{\partial T} \mathbf{1}_{[x \in \eta]} \, d\mu(\eta) = \sum_{x \in \xi} \Phi(x) \theta(x).$$

Now integrate to get

$$\mathscr{E}(\mu) = \int_{\partial T} V_\mu \, d\mu = \int_{\partial T} \sum_{x \in \xi} \Phi(x) \theta(x) \, d\mu(\xi) = \sum_{x \in T} \Phi(x) \theta(x) \mu(\{ \xi \, ; \ x \in \xi \}). \qquad \blacktriangleleft$$

When conductances determine Ψ through (16.1) and the equation $\Psi(o) := 0$, we have $\Phi(x) = r(e(x))$ and thus $\mathscr{E}(\mu) = \mathscr{E}(\theta)$, as we defined energy of flows in Section 2.4. In particular, $\mathrm{cap}\,\partial T = \mathscr{C}(o \leftrightarrow \infty)$; when this is positive, $\mathscr{E}(\mu)$ is minimum when $\mathscr{E}(\theta)$ is minimum and, hence, $\mathscr{E}(\cdot)$ has a unique minimum corresponding to unit current flow. If i denotes unit current flow and μ^i the corresponding measure on ∂T, then μ^i gives the distribution of current "outflow" on ∂T. Probabilistically, this is **harmonic measure** for the random walk, that is, the "hitting" distribution on ∂T: by Proposition 2.12, given an edge $e(x)$, we have

$$\mu^i(\{\xi\,;\ x \in \xi\}) = i(e(x))$$
$$= \mathbf{E}[\text{number of transitions from } \tilde{x} \text{ to } x$$
$$- \text{ number of transitions from } x \text{ to } \tilde{x}\,]$$
$$= \mathbf{P}[\text{random walk enters and eventually stays in } T^x]\,.$$

To summarize:

Theorem 16.2. (Random Walk and Capacity) *Given conductances on a tree T, the associated random walk is transient iff the capacity of ∂T is positive in the associated gauge. The unit flow corresponds to the harmonic measure on ∂T, which minimizes the energy.*

In this context, Theorem 3.5 becomes

$$\mathrm{br}\,T = \inf\left\{\lambda > 1\,;\ \mathrm{cap}\,\partial T = 0 \text{ in the gauge } \Psi(u) = \lambda^{|u|}\right\}.$$

Although $\mathrm{br}\,T$ has a similar definition via Hausdorff measures, it is capacity, not Hausdorff measure, in the critical gauge that determines the type of critical random walk, $\mathrm{RW}_{\mathrm{br}\,T}$.

Write $v(\xi) := \lim_{x \in \xi} v(x)$. Then from Proposition 16.1, we have

$$V_{\mu^i}(\xi) = \sum_{o < u \in \xi} i(e(u))r(e(u)) = \sum_{o < u \in \xi} \left[v(\tilde{u}) - v(u)\right] = v(o) - v(\xi)$$
$$= \mathscr{E}(i) - v(\xi) = \mathscr{E}(\mu^i) - v(\xi) = (\mathrm{cap}\,\partial T)^{-1} - v(\xi)\,.$$

In particular, $\forall \xi\ \ V_{\mu^i}(\xi) \le (\mathrm{cap}\,\partial T)^{-1}$. Since $\int V_{\mu^i}\,d\mu^i = (\mathrm{cap}\,\partial T)^{-1}$, it follows that $V_{\mu^i}(\xi) = (\mathrm{cap}\,\partial T)^{-1}$ for μ^i-a.e. ξ. One can show more, that $V_{\mu^i}(\xi) = (\mathrm{cap}\,\partial T)^{-1}$, in other words, $v(\xi) = 0$, except for a set of ξ of capacity 0 (Lyons, 1990). This further justifies thinking of the electrical network T as "grounded at ∞."

16.2 Percolation on Trees

Let p_x be survival probabilities for independent percolation, as in Section 5.3. Consider the gauge

$$\Psi(x) := \mathbf{P}[o \leftrightarrow x]^{-1} = \prod_{o < u \leq x} p_u^{-1} \, .$$

Note that

$$\Psi(o) = 1 \, .$$

If we define resistances by

$$r(e(u)) := \Psi(u) - \Psi(\overleftarrow{u}) \, ,$$

then $\Psi(x)$ is 1 plus the resistance between o and x. Thus, (5.12) holds. Because $\Psi(o) = 1$ but $i(o)$ does not enter in the sum defining $\mathscr{E}(i)$, we have

$$\mathscr{E}(\mu^i) = 1 + \mathscr{E}(i) = 1 + \mathscr{C}(o \leftrightarrow \infty)^{-1} \, ,$$

whence

$$\operatorname{cap} \partial T = \mathscr{E}(\mu^i)^{-1} = \frac{\mathscr{C}(o \leftrightarrow \infty)}{1 + \mathscr{C}(o \leftrightarrow \infty)} \, .$$

We may therefore rewrite Theorem 5.24 as

$$\operatorname{cap} \partial T \leq \mathbf{P}[o \leftrightarrow \partial T] \leq 2 \operatorname{cap} \partial T \, . \tag{16.2}$$

These inequalities are easily generalized to subsets of ∂T (Lyons, 1992):

Theorem 16.3. (Tree Percolation and Capacity) *For Borel $E \subseteq \partial T$, we have*

$$\operatorname{cap} E \leq \mathbf{P}[o \leftrightarrow E] \leq 2 \operatorname{cap} E.$$

Proof. If E is closed, let $T_{(E)} := \{u ; \ \exists \xi \in E \ \ u \in \xi\}$. Then $T_{(E)}$ is a subtree of T with $\partial T_{(E)} = E$. Hence the result follows from (16.2). The general case follows from the theory of Choquet capacities, which we omit (see Lyons (1992)). ◀

▷ **Exercise 16.1.**
Show that if $p_x \equiv p \in (0, 1)$ and T is spherically symmetric, then

$$\operatorname{cap} \partial T = \left(1 + (1 - p) \sum_{n=1}^{\infty} \frac{1}{p^n |T_n|} \right)^{-1} \, .$$

Thus, $\mathbf{P}[o \leftrightarrow \partial T] > 0$ iff $\sum_{n=1}^{\infty} \frac{1}{p^n |T_n|} < \infty$.

If we specialize to trees coding closed sets, as in Sections 1.10 and 15.2, and take $p_u \equiv p$, the result that $\mathbf{P}[o \leftrightarrow \partial T] > 0$ iff $\operatorname{cap} \partial T > 0$ is due to Fan (1989, 1990).

16.3 Euclidean Space

Like Hausdorff dimension in Euclidean space, capacity in Euclidean space is also related to trees. We treat first the case of $\mathbb{R}^1$.

Consider a closed set $E \subseteq [0, 1]$ and a kernel

$$K(x, y) := f(|x - y|),$$

where $f \geq 0$ and f is decreasing. Here, f is called the *gauge* function. We denote the capacity of E with respect to this kernel by $\mathrm{cap}_f(E)$.

▷ **Exercise 16.2.**
Suppose that f and g are two gauge functions such that $f/c_1 - c_2 \leq g \leq c_1 f + c_2$ for some constants c_1 and c_2. Show that for all E, $\mathrm{cap}_f(E) > 0$ iff $\mathrm{cap}_g(E) > 0$.

The gauge functions $f(z) = z^{-\alpha}$ $(\alpha > 0)$ or $f(z) = \log^+ 1/z$ are so frequently used that the corresponding capacities have their own notation, cap_α and cap_0, respectively. For any set E, there is a critical α_0 such that $\mathrm{cap}_\alpha(E) > 0$ for $\alpha < \alpha_0$ and $\mathrm{cap}_\alpha(E) = 0$ for $\alpha > \alpha_0$. In fact, $\alpha_0 = \dim E$. This result is due to Frostman (1935); we will deduce it from our work on trees. We will also show the following result.

Theorem 16.4. *Let $E \subseteq [0, 1]$ be closed. If T codes E in base b, then critical homesick random walk on T is transient iff $\mathrm{cap}_{\dim E}(E) > 0$. In particular, this does not depend on b.*

This result is a slight generalization of one due to Benjamini and Peres (1992), who stated this for simple random walk and $\mathrm{cap}_0 E$. That case bears a curious relation to a result of Kakutani (1944): $\mathrm{cap}_0 E > 0$ iff Brownian motion in $\mathbb{R}^2$ hits E a.s.; see Lemma 16.10.

To prove Theorem 16.4, we will use the following important proposition that relates energy on trees to energy in Euclidean space. It is due to Benjamini and Peres (1992), as extended by Pemantle and Peres (1995b).

Proposition 16.5. (Energy and Capacity Transference in $\mathbb{R}$) *Let $E \subseteq [0, 1]$ be closed and T code E in base b. Let $r_n > 0$. Give T the resistances $r(e(u)) := r_{|u|}$ for $u \in \mathsf{V}(T)$ and suppose that $f \geq 0$ is decreasing and satisfies*

$$\forall n \quad f(b^{-n}) = \sum_{1 \leq k \leq n} r_k \quad \text{and} \quad f(0) = \sum_{n=1}^{\infty} r_n.$$

If $v \in \mathsf{Prob}(E)$ and $\mu \in \mathsf{Prob}(\partial T)$ satisfy $\theta(u) = v(I_u)$ for $u \in \mathsf{V}(T)$, where θ is the flow corresponding to μ and I_u is the b-ary interval corresponding to u, then

$$\frac{1}{2}\mathscr{E}_c(\theta) \leq \mathscr{E}_f(v) \leq 3b\,\mathscr{E}_c(\theta).$$

Hence, (T, c) is transient iff $\mathrm{cap}_f(E) > 0$.

Proof of Theorem 16.4. Critical random walk uses the resistances $r(e(u)) = \lambda_c^{|u|}$, where $\lambda_c = \mathrm{br}\,T = b^{\dim E}$. Hence, set $r_n := b^{n\,\dim E}$ and choose any f as in Proposition 16.5. According to the proposition, RW_{λ_c} is transient iff $\mathrm{cap}_f E > 0$. Since there are constants c_1 and c_2 such that $f(z)/c_1 - c_2 \leq z^{-\dim E} \leq c_1 f(z) + c_2$ when $\dim E > 0$ and $f(z)/c_1 - c_2 \leq \log^+ 1/z \leq c_1 f(z) + c_2$ when $\dim E = 0$, we have $\mathrm{cap}_f E > 0$ iff $\mathrm{cap}_{\dim E}(E) > 0$ by Exercise 16.2. ◀

Corollary 16.6. (Frostman, 1935) *For $E \subseteq [0, 1]$ closed, $\dim E = \inf\{\alpha \,; \, \mathrm{cap}_\alpha(E) = 0\}$.*

▷ **Exercise 16.3.**
Prove Corollary 16.6.

Although we have just shown that the base used for coding E does not affect transience of critical random walk, the following question is open:

Question 16.7. Let E be the Cantor middle-thirds set, μ_b harmonic measure of simple random walk on $T_{[b]}(E)$, and ν_b the corresponding measure on E. Thus ν_3 is Cantor–Lebesgue measure. Is $\nu_2 \perp \nu_3$?

Proof of Proposition 16.5. If $t \le b^{-N}$, then $f(t) \ge f(b^{-N})$, whence

$$f(t) \ge \sum_{n\ge 1} r_n \mathbf{1}_{\{b^{-n} \ge t\}},$$

so

$$\mathscr{E}_f(\nu) = \iint f(|x-y|)\, d\nu(x)\, d\nu(y) \ge \iint \sum_{n\ge 1} r_n \mathbf{1}_{\{|x-y|\le b^{-n}\}}\, d\nu(x)\, d\nu(y)$$

$$= \sum_{n\ge 1} r_n (\nu\times\nu)\{|x-y|\le b^{-n}\}.$$

Now

$$\{|x-y|\le b^{-n}\} \supseteq \bigcup_{k=0}^{b^n-1} I_k^n \times I_k^n, \tag{16.3}$$

where $I_k^n := \left[\frac{k}{b^n}, \frac{k+1}{b^n}\right]$, and each point in the right-hand side lies in at most two of the terms in the union, whence

$$\mathscr{E}_f(\nu) \ge \frac{1}{2}\sum_{n\ge 1} r_n \sum_k (\nu I_k^n)^2 = \frac{1}{2}\sum_{n\ge 1} r_n \sum_{|u|=n} \theta(u)^2 = \frac{1}{2}\mathscr{E}_c(\theta).$$

On the other hand, $f(t) \le f(b^{-N})$ when $t \ge b^{-N}$, in other words,

$$f(t) \le \sum_{n\ge 1} r_n \mathbf{1}_{\{b^{-n+1}>t\}}.$$

Therefore

$$\mathscr{E}_f(\nu) = \iint f(|x-y|)\, d\nu(x)\, d\nu(y) \le \iint \sum_{n\ge 1} r_n \mathbf{1}_{\{|x-y|<b^{-n+1}\}}\, d\nu(x)\, d\nu(y)$$

$$= \sum_{n\ge 1} r_n (\nu\times\nu)\{|x-y|< b^{-n+1}\}.$$

Now

$$\{|x - y| \le b^{-n}\} \subseteq \bigcup_{k=0}^{b^n-1} \left(I_k^n \times (I_k^n \cup I_{k-1}^n \cup I_{k+1}^n) \right). \quad (16.4)$$

(See Figure 16.1.) This gives the bound

$$(\nu \times \nu)\{|x - y| \le b^{-n}\} \le \sum_{k=0}^{b^n-1} (\nu I_k^n)(\nu I_k^n + \nu I_{k-1}^n + \nu I_{k+1}^n).$$

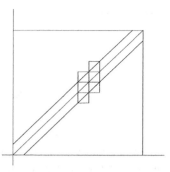

The cross terms are estimated by the following inequality:

$$AB \le \frac{A^2 + B^2}{2}.$$

Figure 16.1. Covering by b-adic squares.

Thus

$$(\nu \times \nu)\{|x - y| \le b^{-n}\} \le \sum_{k=0}^{b^n-1} (\nu I_k^n)^2 + \sum_{k=0}^{b^n-1} (\nu I_k^n)^2 + \sum_{k=0}^{b^n-1} \frac{(\nu I_{k-1}^n)^2 + (\nu I_{k+1}^n)^2}{2}$$

$$\le 3 \sum_{|u|=n} (\nu I_u)^2 = 3 \sum_{|u|=n} \theta(u)^2,$$

whence by the arithmetic mean–quadratic mean inequality (or the Cauchy–Schwarz inequality),

$$\mathscr{E}_f(\nu) \le 3 \sum_{n\ge 1} r_n \sum_{|u|=n-1} \theta(u)^2 = 3 \sum_{n\ge 1} r_n \sum_{|u|=n-1} \left(\sum_{u\to x} \theta(x) \right)^2$$

$$\le 3b \sum_{n\ge 1} r_n \sum_{|u|=n-1} \sum_{u\to x} \theta(x)^2 = 3b \, \mathscr{E}_c(\theta). \qquad \blacktriangleleft$$

The situation in higher-dimensional Euclidean space is very similar. Denote the Euclidean distance between x and y by $|x - y|$. Consider a closed set $E \subseteq [0, 1]^d$ and a kernel $K(x, y) := f(|x - y|)$, where $f \ge 0$ and f is decreasing. Again, f is called the **gauge** function, and we denote the capacity of E with respect to f by $\mathrm{cap}_f(E)$. To code E by a tree T, we now use b-adic cubes in $[0, 1]^d$. Again, the transference principle is due to Pemantle and Peres (1995b).

Proposition 16.8. (Energy and Capacity Transference in $\mathbb{R}^d$) *Let $E \subseteq [0, 1]^d$ be closed and T code E in base b. Let $r_n > 0$ satisfy $\sum r_n = \infty$. Give T the resistances $r(e(u)) := r_{|u|}$ and suppose that f is decreasing and satisfies*

$$\forall n \quad f(b^{-n}) = \sum_{1 \le k \le n} r_k \quad and \quad f(0) = \sum_{n=1}^{\infty} r_n.$$

If $\nu \in \mathrm{Prob}(E)$ and $\mu \in \mathrm{Prob}(\partial T)$ satisfy $\theta(u) = \nu(I_u)$, where θ is the flow corresponding to μ and I_u is the b-adic cube corresponding to u, then

$$2^{-d} b^{-d\ell} \mathscr{E}_c(\theta) \le \mathscr{E}_f(\nu) \le 3^d b^d \mathscr{E}_c(\theta),$$

where $\ell := \lceil (1/2) \log_b d \rceil$.

▷ **Exercise 16.4.**
Prove Proposition 16.8.

16.4 Fractal Percolation and Brownian Intersections

In this section, we will consider a special case, known as fractal percolation, of the Galton–Watson fractals defined in Example 15.14. We will see how the use of capacity can help us understand intersection properties of Brownian motion traces in $\mathbb{R}^d$ by transferring the question to these much simpler fractal percolation sets. Moreover, the fractal percolation sets will help analyze capacity. In Section 5.7, we studied connectivity properties of fractal percolation sets in the plane. Here we consider them in greater generality, both in dimension and in survival probability at each generation.

Given integers $d, b \geq 2$ and numbers $p_n \in [0, 1]$ ($n \geq 1$), consider the natural tiling of the unit cube $[0, 1]^d$ by b^d closed cubes of side $1/b$. Let $\mathcal{K}_1$ be a random subcollection of these cubes, where each cube has probability p_1 of belonging to $\mathcal{K}_1$, and these events are mutually independent. (Thus, the cardinality $|\mathcal{K}_1|$ of $\mathcal{K}_1$ is a binomial random variable.) In general, if $\mathcal{K}_n$ is a collection of cubes of side b^{-n}, tile each cube $Q \in \mathcal{K}_n$ by b^d closed subcubes of side b^{-n-1} (with disjoint interiors) and include each of these subcubes in $\mathcal{K}_{n+1}$ with probability p_{n+1} (independently). Finally, define

$$A_n := A_{n,d,b}(p_n, \mathcal{K}_{n-1}) := \bigcup \mathcal{K}_n \quad \text{and} \quad Q_d(\langle p_n \rangle) := Q_{d,b}(\langle p_n \rangle) := \bigcap_{n=1}^{\infty} A_n \,.$$

In the construction of $Q_d(\langle p_n \rangle)$, the cardinalities $|\mathcal{K}_n|$ of $\mathcal{K}_n$ form a branching process in a varying environment, where the offspring distribution at level n is $\mathrm{Bin}(b^d, p_n)$. When $p_n \equiv p$, the cardinalities form a Galton–Watson branching process. Alternatively, the successive subdivisions into b-ary subcubes define a natural mapping from a b^d-ary tree to the unit cube; the construction of $Q_d(\langle p_n \rangle)$ corresponds to performing independent percolation with parameter p_n at level n on this tree and considering the set of infinite open paths emanating from the root. The process $\langle A_n ; \; n \geq 1 \rangle$ is called *fractal percolation*, whereas $Q_d(\langle p_n \rangle)$ is the *limit set*. When $p_n \equiv p$, we write $Q_d(p)$ for $Q_d(\langle p_n \rangle)$.

▷ **Exercise 16.5.**
  (a) Show that if $p_n \leq b^{-d}$ for all n, then $Q_{d,b}(\langle p_n \rangle) = \varnothing$ a.s.
  (b) Characterize the sequences $\langle p_n \rangle$ for which $Q_d(\langle p_n \rangle)$ has positive volume with positive probability.

The main result in this section, from Peres (1996), is that the Brownian trace is "intersection equivalent" to the limit set of fractal percolation for an appropriate choice of parameters $\langle p_n \rangle$. We will obtain as easy corollaries facts about Brownian motion, for example, that two traces in 3-space intersect but not three traces. Here, the *trace* of Brownian motion $\langle B_t \rangle$ is the set of its values for t belonging to some given interval. This is a random set, by which we mean the following.

Definition 16.9. A *random set* A is a measurable function $\omega \mapsto A(\omega)$ on a measurable space Ω whose values are sets; measurability means that for every closed Λ, the outcomes ω where $A(\omega) \cap \Lambda \neq \varnothing$ form an event (that is, a measurable subset of Ω). Two random (Borel) sets A and B in $\mathbb{R}^d$ are *intersection equivalent* in a set U if there exist constants C_1, C_2 such that for every closed* set $\Lambda \subseteq U$, we have

$$C_1 \, \mathbf{P}[A \cap \Lambda \neq \varnothing] \leq \mathbf{P}[B \cap \Lambda \neq \varnothing] \leq C_2 \, \mathbf{P}[A \cap \Lambda \neq \varnothing], \qquad (16.5)$$

which we write as

$$\mathbf{P}[A \cap \Lambda \neq \varnothing] \asymp \mathbf{P}[B \cap \Lambda \neq \varnothing].$$

We need a few basic facts about Brownian motion. For more information on Brownian motion, see the book by Mörters and Peres (2010). Further references can also be found in the notes at the end of this chapter. If g_d is the d-dimensional *radial potential* function,

$$g_d(r) := \begin{cases} \log^+(r^{-1}) & \text{if } d = 2 \\ r^{2-d} & \text{if } d \geq 3, \end{cases}$$

then the kernel $G_d(x, y) = c_d g_d(|x - y|)$, where $c_d > 0$ is a constant given in Exercise 16.6, is called the **Green kernel** for Brownian motion. This is the continuous analogue of the Green function for Markov chains: As Exercise 16.6 shows, the expected one-dimensional Lebesgue measure of the time that Brownian motion spends in a set turns out to be absolutely continuous with respect to d-dimensional Lebesgue measure, and its density at y when started at x is the Green kernel $G_d(x, y)$. Actually, in two dimensions, (neighborhood) recurrence of Brownian motion means that we need to kill it at some finite time. If we kill it at a random time with an exponential distribution, then it remains a Markov process. It is not hard to compare properties of the exponentially killed process to one that is killed at a fixed time: see Exercise 16.13.

▷ **Exercise 16.6.**
Let $p_t(x, y) := (2\pi t)^{-d/2} \exp\left(-|x - y|^2/(2t)\right)$ be the Brownian transition density function, and for $d \geq 3$, define $G_d(x, y) := \int_0^\infty p_t(x, y) \, dt$ for $x, y \in \mathbb{R}^d$.

(a) Show that $G_d(x, y) = c_d g_d(|x - y|)$ with a constant $0 < c_d < \infty$.

(b) Define $F_A(x) := \int_A G_d(x, z) \, dz$ for Borel sets $A \subseteq \mathbb{R}^d$. Show that $F_A(x)$ is the expected time the Brownian motion started at x spends in A.

(c) Show that $x \mapsto G_d(\mathbf{0}, x)$ is harmonic on $\mathbb{R}^d \setminus \{\mathbf{0}\}$, in other words, has zero Laplacian there. Equivalently, if $B \subset \mathbb{R}^d \setminus \{\mathbf{0}\}$ is a ball, then the average value (with respect to Lebesgue measure) of $G_d(\mathbf{0}, x)$ over $x \in B$ is equal to the value at the center of the ball.

(d) Consider Brownian motion $\langle B_t \rangle$ in $\mathbb{R}^2$, killed at a random time with an exponential distribution with parameter 1. In other words, let τ be an Exponential(1) random time, independent of the Brownian motion. The expected occupation measure ν_x for B_t started at x and killed at time τ is defined by $\nu_x(A) := \mathbf{E} \int_0^\tau \mathbf{1}_A(B_t) \, dt$ for Borel sets A in $\mathbb{R}^2$. Show that $\nu_x(A) = \int_A G_2^*(x, y) \, dy$, where $G_2^*(x, y) := \int_0^\infty p_t(x, y) e^{-t} \, dt$.

(e) Show that in two dimensions, $G_2^*(x, y) \sim -(1/\pi) \log |x - y|$ for $|x - y| \downarrow 0$.

---

* Intersection equivalence of A and B implies that (16.5) holds for all Borel sets Λ by the Choquet capacitability theorem (see Carleson (1967), p. 3, or Dellacherie and Meyer (1978), III.28).

The following is a classical result relating the hitting probability of a set Λ by Brownian motion to the capacity of Λ in the Green kernel. It is the key to our understanding in this section of Brownian motion. A proof using the so-called Martin kernel and the second-moment method was found by Benjamini, Pemantle, and Peres (1995); see Section 16.6. We will write $\mathrm{cap}(\,\cdot\,;f) := \mathrm{cap}_f(\,\cdot\,)$ to make it easier to see the gauge function f, which will be crucial in what follows and which will change many times.

Lemma 16.10. *If $\langle B_t \rangle$ is Brownian motion (killed at an exponential time for $d = 2$), and the initial distribution v of B_0 has a bounded density on $\mathbb{R}^d$, then*

$$\mathbf{P}_v[\exists t \geq 0 \ \ B_t \in \Lambda] \asymp \mathrm{cap}(\Lambda; g_d) \tag{16.6}$$

for all Borel sets $\Lambda \subset [0,1]^d$. (The implicit constants depend on v.)

A picture of Brownian motion in the plane is convincing that there are double points a.s., that is, points that are visited at least twice by the Brownian motion. What about in higher dimensions? What about points of greater multiplicity? Which sets contain double points with positive probability? Because of the independence of increments in Brownian motion, it is intuitive that the questions have the same answers as asking whether two (or more) independent Brownian motions intersect. We will later prove (partly) that these are indeed equivalent. So let B and B' be two independent Brownian motions, and write $[B]$ for the trace $\{B_t \; ; \; t \geq 0\}$.

For which Λ is $\mathbf{P}\big[\Lambda \cap [B] \cap [B'] \neq \varnothing\big] > 0$?

Evans (1987) and Tongring (1988) gave a sufficient condition:

If $\mathrm{cap}(\Lambda; g_d^2) > 0$, then $\mathbf{P}\big[\Lambda \cap [B] \cap [B'] \neq \varnothing\big] > 0$.

Later Fitzsimmons and Salisbury (1989) showed that $\mathrm{cap}(\Lambda; g_d^2) > 0$ is also necessary. Moreover, they showed that in dimension $d = 2$, if $B^{(i)}$ are independent Brownian motions, then

$$\mathrm{cap}(\Lambda; g_2^k) > 0 \ \text{ if and only if } \ \mathbf{P}\big[\Lambda \cap [B^{(1)}] \cap \cdots \cap [B^{(k)}] \neq \varnothing\big] > 0.$$

Chris Bishop (personal communication, 1994) then made the following conjecture in every dimension:

$$\mathbf{P}\big[\mathrm{cap}(\Lambda \cap [B]; f) > 0\big] > 0 \ \text{ if and only if } \ \mathrm{cap}(\Lambda; fg_d) > 0 \tag{16.7}$$

whenever f is a nonnegative, decreasing function. Iteration of (16.7) implies the preceding results on multiple intersections. We will establish all of this.

To do so, we will relate gauge functions used for kernels on Euclidean space to fractal percolation in the following way. If $f \geq 0$ is a decreasing function and $p_n \in [0,1]$ satisfy

$$p_1 \cdots p_n = \min\{f(b^{-n})^{-1}, 1\}, \tag{16.8}$$

then the limit set of fractal percolation with retention probability p_n at level n will be denoted by $Q_d[f]$.

Note that the numbers $\langle p_n \rangle$ satisfying (16.8) for $f = g_d$ are

$$\begin{cases} p_n = b^{2-d} & \text{if } d \geq 3, \\ p_1 = 1/\log b, \ p_n = (n-1)/n \text{ for } n \geq 2 & \text{if } d = 2 \text{ and } b \geq 3, \\ p_1 = 1, \ p_2 = 1/(2\log 2), \ p_n = (n-1)/n \text{ for } n \geq 3 & \text{if } d = 2 \text{ and } b = 2. \end{cases}$$

Consider the canonical map Υ from the boundary ∂T of the b^d-ary tree T onto $[0,1]^d$. Let ω be the percolation on T that has probability p_n of retaining an edge e when e joins vertices at level $n-1$ to vertices at level n. Write $\partial_o \omega$ for the set of open rays from the root of T. Then the image of $\partial_o \omega$ under Υ has the same law as $Q_d(\langle p_n \rangle)$. Moreover, for every closed $\Lambda \subseteq [0,1]^d$, the probability that $Q_d(\langle p_n \rangle) \cap \Lambda \neq \varnothing$ equals the probability that $\Upsilon^{-1}(\Lambda) \cap \partial_o \omega \neq \varnothing$. This fact, along with the percolation result Theorem 16.3 and the transfer result Proposition 16.5, are the ingredients behind the following theorem.

Theorem 16.11. (Peres, 1996) *Let $f \geq 0$ be a decreasing function. Then, for all closed sets $\Lambda \subseteq [0,1]^d$,*

$$\mathrm{cap}(\Lambda; f) \asymp \mathbf{P}\big[\Lambda \cap Q_d[f] \neq \varnothing\big]. \tag{16.9}$$

For $f = g_d$ in particular, $Q_d[g_d]$ is intersection equivalent in $[0,1]^d$ to the Brownian trace: If B is Brownian motion with initial distribution v that has a bounded density, stopped at an exponential time when $d = 2$, and $[B] := \{B_t \ ; \ t \geq 0\}$, then for all closed $\Lambda \subseteq [0,1]^d$,

$$\mathbf{P}\big[\Lambda \cap Q_d[g_d] \neq \varnothing\big] \asymp \mathbf{P}\big[\Lambda \cap [B] \neq \varnothing\big]. \tag{16.10}$$

Proof. Let $\langle p_n \rangle$ be determined by (16.8). Let $\widetilde{\mathbf{P}}$ be the independent percolation on T with retention probability p_n for each edge joining level $n-1$ to level n. Define the gauge Ψ on T by $\Psi(x) := \max\{f(b^{-|x|}), 1\}$. By Theorem 16.3,

$$\mathbf{P}\big[Q_d[f] \cap \Lambda \neq \varnothing\big] = \widetilde{\mathbf{P}}\big[o \leftrightarrow \Upsilon^{-1}(\Lambda)\big] \asymp \mathrm{cap}(\Upsilon^{-1}(\Lambda); \Psi), \tag{16.11}$$

where the constants in $\asymp$ on the right-hand side are 1 and 2. On the other hand, Proposition 16.5 says that

$$\mathrm{cap}(\Upsilon^{-1}(\Lambda); \Psi) \asymp \mathrm{cap}(\Lambda; f). \tag{16.12}$$

Combining (16.11) and (16.12) yields (16.9). From (16.9) and (16.6), we get (16.10). ◀

To apply the preceding theorem to intersections of several random sets, we use the following two lemmas.

Lemma 16.12. *Suppose that $A_1, \ldots, A_k, F_1, \ldots, F_k$ are independent random closed* sets, with A_i intersection equivalent to F_i for $1 \leq i \leq k$. Then $A_1 \cap A_2 \cap \cdots \cap A_k$ is intersection equivalent to $F_1 \cap F_2 \cap \cdots \cap F_k$.*

Proof. By induction, reduce to the case $k = 2$. It clearly suffices to show that $A_1 \cap A_2$ is intersection equivalent to $F_1 \cap A_2$, and this is done by conditioning on A_2:

$$\begin{aligned} \mathbf{P}[A_1 \cap A_2 \cap \Lambda \neq \varnothing] &= \mathbf{E}\big[\mathbf{P}[A_1 \cap A_2 \cap \Lambda \neq \varnothing \mid A_2]\big] \\ &\asymp \mathbf{E}\big[\mathbf{P}[F_1 \cap A_2 \cap \Lambda \neq \varnothing \mid A_2]\big] = \mathbf{P}[F_1 \cap A_2 \cap \Lambda \neq \varnothing]. \end{aligned}$$ ◀

Lemma 16.13. *For any $0 < p, q < 1$, if $Q_d(p)$ and $Q_d'(q)$ are independent, then their intersection $Q_d(p) \cap Q_d'(q)$ has the same distribution as $Q_d(pq)$.*

Proof. This is immediate from the construction of $Q_d(p)$. ◀

---

* In fact, A_i and F_i may be taken to be Borel by the Choquet capacitability theorem, as before.

Now we can reap the corollaries. In what follows, Brownian paths will be started either from arbitrary fixed points or from initial distributions that have bounded densities in the unit cube $[0, 1]^d$. The proof of the first corollary was first completed by Dvoretzky, Erdős, Kakutani, and Taylor (1957), following earlier work of Dvoretzky, Erdős, and Kakutani (1950). A proof using the renormalization group method was given by Aizenman (1985). (In fact, the earlier authors phrased their results in terms of multiple points, but their proofs show these results about intersections as well.) Compare the following to Corollary 10.25 and note the difference in four dimensions. The reason for the difference is analogous to the difference between recurrence of two-dimensional random walk and neighborhood recurrence of two-dimensional Brownian motion: although two four-dimensional Brownian motions do get arbitrarily close to each other (see Jain and Taylor (1973), Remark 4.2), they do not meet, whereas each time two random walks get close, then they will meet with positive probability.

Corollary 16.14. (Dvoretzky, Erdős, Kakutani, Taylor)

  (i) *For all $d \geq 4$, two independent Brownian traces in $\mathbb{R}^d$ are disjoint a.s., except, of course, at their starting point if they are identical.*

 (ii) *In $\mathbb{R}^3$, two independent Brownian traces intersect a.s., but three traces a.s. have no points of mutual intersection (except, possibly, their starting point).*

(iii) *In $\mathbb{R}^2$, any finite number of independent Brownian traces have nonempty mutual intersection almost surely.*

Proof. (i) It suffices to prove the case $d = 4$, since for $d \geq 5$, the first four coordinates of Brownian motion in $\mathbb{R}^d$ give Brownian motion in $\mathbb{R}^4$. Fix $\epsilon > 0$. The distribution of B_ϵ has bounded density. By Theorem 16.11 and Lemma 16.12, the intersection of two independent copies of $\{B_t ; t \geq \epsilon\}$ is intersection equivalent in $[0, 1]^d$ to the intersection of two independent copies of $Q_{4,b}(b^{-2})$; by Lemma 16.13, this latter intersection is intersection equivalent to $Q_{4,b}(b^{-4})$. But $Q_{4,b}(b^{-4})$ is a.s. empty because critical branching processes die out. Thus two independent copies of $\{B_t ; t \geq \epsilon\}$ are a.s. disjoint in $[0, 1]^d$. This argument actually worked for fractal percolation in any cube of any size (not just the unit cube), whence two independent copies of $\{B_t ; t \geq \epsilon\}$ are a.s. disjoint. As ϵ is arbitrary, the claim follows.

(ii) Since $\{B_t ; t \geq \epsilon\}$ is intersection equivalent to $Q_{3,b}(b^{-1})$ in the unit cube, the intersection of three independent Brownian traces (from any time $\epsilon > 0$ on) is intersection equivalent in the cube to the intersection of three independent copies of $Q_3(b^{-1})$, which has the same distribution as $Q_3(b^{-3})$. Again, a critical branching process is obtained, and hence the triple intersection is a.s. empty.

On the other hand, the intersection of two independent copies of $\{B_t ; t \geq \epsilon\}$ is intersection equivalent to $Q_{3,b}(b^{-2})$ in the unit cube for any positive ϵ. Since $Q_3(b^{-2})$ is defined by a supercritical branching process, we have $p(\epsilon, \infty) > 0$, where

$$p(u, v) := \mathbf{P}\big[\{B_t ; u < t < v\} \cap \{B'_s ; u < s < v\} \neq \varnothing\big].$$

Here, B' is an independent copy of B. Suppose now first that the two Brownian motions are started from the origin; we will show that $p(1, \infty) = 1$. Note that $p(\epsilon, \infty) \uparrow p(0, \infty)$ as $\epsilon \downarrow 0$, hence $p(0, \infty) > 0$. Furthermore, $p(0, v) \to p(0, \infty)$ as $v \to \infty$. On the other hand, by Brownian scaling,* $p(0, v)$ is independent of $v > 0$ and $p(u, \infty)$ is independent of $u > 0$. Therefore, $p(0, v) = p(u, \infty) = p(0, \infty) > 0$ for all $u, v > 0$, and

$$\lim_{v \downarrow 0} p(0, v) = \mathbf{P}\Big[0 = \inf\big\{v > 0 \,;\, \{B_t \,;\, 0 < t < v\} \cap \{B'_s \,;\, 0 < s < v\} \neq \varnothing\big\}\Big] > 0 \,.$$

This last positive probability, by Blumenthal's zero-one law (see Durrett (2019), Section 7.2, or Mörters and Peres (2010), Section 2.1), has to be 1, thus $p(0, 1) = p(1, \infty) = p(0, \infty) = 1$.

When the Brownian motions are started at other points, or from any initial distributions, then at unit time, their joint distribution is absolutely continuous with respect to the joint distribution of Brownian motions at unit time, started from the origin. For these latter Brownian motions, we know already that $p(1, \infty) = 1$, so we have this for the original motions as well. ◄

▷ **Exercise 16.7.**
Prove part (iii) of the preceding Corollary.

We now show how Theorem 16.11 leads to a proof of (16.7).

Corollary 16.15. *Let f and h be nonnegative decreasing functions. If a random closed set A in $[0, 1]^d$ satisfies*

$$\mathbf{P}[A \cap \Lambda \neq \varnothing] \asymp \operatorname{cap}(\Lambda; h) \tag{16.13}$$

for all closed $\Lambda \subseteq [0, 1]^d$, then

$$\mathbf{E}\big[\operatorname{cap}(A \cap \Lambda; f)\big] \asymp \operatorname{cap}(\Lambda; fh) \tag{16.14}$$

for all closed $\Lambda \subseteq [0, 1]^d$. In particular, for each such Λ,

$$\mathbf{P}\big[\operatorname{cap}(A \cap \Lambda; f) > 0\big] > 0 \quad \text{if and only if } \operatorname{cap}(\Lambda; fh) > 0 \,,$$

and (16.7) follows by putting $A := [B]$ and $h := g_d$.

Proof. Enlarge the probability space on which A is defined to include two independent fractal percolations, $Q_d[f]$ and $Q_d[h]$. Because

$$\mathbf{P}\big[A \cap \Lambda \cap Q_d[f] \neq \varnothing\big] = \mathbf{E}\big[\mathbf{P}[A \cap \Lambda \cap Q_d[f] \neq \varnothing \mid A]\big] \,,$$

and by Theorem 16.11,

$$\mathbf{P}\big[A \cap \Lambda \cap Q_d[f] \neq \varnothing | A\big] \asymp \operatorname{cap}(A \cap \Lambda; f) \,,$$

we have

$$\mathbf{E}\big[\operatorname{cap}(A \cap \Lambda; f)\big] \asymp \mathbf{P}\big[A \cap \Lambda \cap Q_d[f] \neq \varnothing\big] \,. \tag{16.15}$$

Conditioning on $Q_d[f]$ and then using (16.13) with $\Lambda \cap Q_d[f]$ in place of Λ gives

$$\mathbf{P}\big[A \cap \Lambda \cap Q_d[f] \neq \varnothing\big] \asymp \mathbf{E}\big[\operatorname{cap}(\Lambda \cap Q_d[f]; h)\big] \,. \tag{16.16}$$

Conditioning on $Q_d[f]$ and then applying Theorem 16.11 yields

$$\mathbf{E}\big[\operatorname{cap}(\Lambda \cap Q_d[f]; h)\big] \asymp \mathbf{P}\big[\Lambda \cap Q_d[f] \cap Q_d[h] \neq \varnothing\big] \,. \tag{16.17}$$

Note that $Q_d[f] \cap Q_d[h]$ has the same distribution as $Q_d[fh]$ if $f, h \geq 1$; otherwise, this is true of the functions $f \vee 1$ and $h \vee 1$. In either case, Theorem 16.11 implies that

$$\mathbf{P}\big[\Lambda \cap Q_d[f] \cap Q_d[h] \neq \varnothing\big] \asymp \operatorname{cap}(\Lambda; fh) \,. \tag{16.18}$$

Combining (16.15), (16.16), (16.17), and (16.18) proves (16.14). ◄

---

* That is, for $a > 0$, the distribution of $t \mapsto a B_{t/a^2}$ is the same as that of $t \mapsto B_t$.

Remark 16.16. We have assumed in the statement of Corollary 16.15 that the function $A \mapsto \operatorname{cap}(A \cap \Lambda; f)$ is measurable. When f is continuous, this follows from the Fekete–Szegő theorem to come later in Theorem 16.18(ii). In the notation there, $\operatorname{cap}(A \cap \Lambda; f) = \lim_{n \to \infty} D_n(A \cap \Lambda)^{-1}$, and it is not hard to see that $A \mapsto D_n(A \cap \Lambda)$ is continuous. Without any real loss of generality, one can assume that f is continuous, since we are interested in capacity only up to a bounded factor. However, one can show measurability in general.

We finish this section by showing how the intersection of several independent copies of Brownian motion is related to multiple points of a single Brownian motion.

Corollary 16.17. *Consider Brownian motion in dimension $d \geq 2$. Then almost surely*

    (i) *if $d \geq 4$, no double points exist, in other words, Brownian motion is injective;*

  (ii) *if $d = 3$, double points exist, but triple points fail to exist;*

 (iii) *if $d = 2$, points of multiplicity at least k exist for every finite k.*

Proof. We prove (i) and the first part of (ii). The rest is plausible from Corollary 16.14, and a full proof can be found in Mörters and Peres (2010), Theorem 9.22.

Let $\langle B_t ; t \geq 0 \rangle$ be a Brownian motion. To show part (i), it suffices to show that for any rational $\alpha > 0$, almost surely, there exist no times (t_1, t_2) with $0 \leq t_1 < \alpha < t_2$ and $B_{t_1} = B_{t_2}$. Fix such an α. The Brownian motions $\langle B_t^{(1)} ; t > 0 \rangle$ and $\langle B_t^{(2)} ; 0 < t \leq \alpha \rangle$ given by

$$B_t^{(1)} := B_{\alpha+t} - B_\alpha \quad \text{and} \quad B_t^{(2)} := B_{\alpha-t} - B_\alpha$$

are independent and hence, by Corollary 16.14, a.s. they do not intersect. This proves the statement.

To show existence of double points in $d \leq 3$, we consider the independent Brownian motions $\langle B_t^{(1)} ; 0 < t \leq 1 \rangle$ and $\langle B_t^{(2)} ; 0 < t \leq 1 \rangle$ given by

$$B_t^{(1)} = B_{1+t} - B_1 \quad \text{and} \quad B_t^{(2)} = B_{1-t} - B_1$$

to see that the two traces intersect with positive probability; in fact, this probability is one by Blumenthal's zero-one law. ◄

A remarkable extension of Corollary 16.17(c) was proved by Le Gall (1987) concerning infinite multiplicity. Namely, for each totally disconnected compact set $K \subset \mathbb{R}$, there is a.s. some $x \in \mathbb{R}^2$ such that the time set $A(x) := \{t \geq 0 ; B_t = x\}$ has the same order type as K, that is, $A(x)$ is homeomorphic to K via a monotonic, increasing function.

16.5 Generalized Diameters and Average Meeting Height on Trees

There are amusing alternative definitions of capacity that we present in some generality and then specialize to trees. Suppose that X is a compact Hausdorff* space and $K : X \times X \to [0, \infty]$ is continuous and symmetric. Let $\mathrm{Prob}(X)$ denote the set of Borel probability measures on X. For $\mu \in \mathrm{Prob}(X)$, set

$$V_\mu(x) := \int_X K(x, y) \, d\mu(y), \qquad \mathscr{E}(\mu) := \int_{X \times X} K \, d(\mu \times \mu),$$

$$\mathscr{E}(X) := \inf\{\mathscr{E}(\mu) \, ; \; \mu \in \mathrm{Prob}(X)\}, \qquad \mathrm{cap}(X) := \mathscr{E}(X)^{-1}$$

as before, and define the **Chebyshev constants**

$$M_n(X) := \max_{x_1, \dots, x_n \in X} \; \min_{x \in X} \frac{1}{n} \sum_{k=1}^n K(x, x_k)$$

and the **generalized diameters**

$$D_n(X) := \min_{x_1, \dots, x_n \in X} \frac{1}{\binom{n}{2}} \sum_{1 \le j < k \le n} K(x_j, x_k).$$

To see where generalized diameters get their name, consider the case $n = 2$ and $K(x, y) = 1/d(x, y)$ for a metric space. When generalized diameters were first named, however, the kernel was $\log(1/d(x, y))$. See p. 37 of Carleson (1967) for the classical case.

Theorem 16.18. *Let X be compact and Hausdorff and K be continuous and symmetric.*
 (i) *We have*

$$\mathscr{E}(X) = \inf\{\|V_\mu\|_{L^\infty(\mu)} \, ; \; \mu \in \mathrm{Prob}(X)\}.$$

If $\mathscr{E}(X) < \infty$, then for some $\mu \in \mathrm{Prob}(X)$, we have $V_\mu = \mathscr{E}(X)$ μ-a.e.
 (ii) **(Fekete–Szegő Theorem)** *We have*

$$D_n(X) \uparrow \mathscr{E}(X) = \lim_{n \to \infty} \inf\{M_n(\hat{X}) \, ; \; \hat{X} \subseteq X \text{ is compact}\}.$$

If $\mathscr{E}(X) < \infty$, then for some $\mu \in \mathrm{Prob}(X)$, there is a compact set $\hat{X} \subseteq X$ such that $V_\mu {\restriction} \hat{X} \le \mathscr{E}(X)$ and $\mu(\hat{X}) = 1$; for any such μ and $\hat{X}$, we have $M_n(\hat{X}) \le \mathscr{E}(X)$ for all n and $\mathscr{E}(X) = \lim_{n \to \infty} M_n(\hat{X})$.

See the notes at the end of this chapter for a proof.

We are interested in the case that X is the boundary of a tree T and the kernel $K(\xi, \eta) = \Psi(\xi \wedge \eta)$ corresponds to conductances on T as in Section 16.1. In that case, there is a measure μ, namely, harmonic measure for the network random walk on T, such that $V_\mu \le \mathscr{E}(\partial T)$ everywhere (see the end of Section 16.1). Hence, by the Fekete–Szegő theorem,

$$\mathscr{E}(\partial T) = \lim D_n(\partial T) = \lim M_n(\partial T),$$

with $D_n(\partial T) \le \mathscr{E}(\partial T)$ and $M_n(\partial T) \le \mathscr{E}(\partial T)$. We will transfer this from the boundary to the vertices of T to obtain the following theorem.

---

* We will have need only of metric spaces, but the proofs are no simpler than for Hausdorff spaces.

Theorem 16.19. (Benjamini and Peres, 1992) *If $\forall \xi \in \partial T$ $\sum_{o \neq x \in \xi} r(e(x)) = \infty$ and T is locally finite, then the following are equivalent:*

(i) $\mathcal{R}(o \leftrightarrow \infty) < \infty$;

(ii) $\exists A < \infty$ $\forall n \geq 1$ $\exists$ *distinct* $x_1, \ldots, x_n \in T$ *such that*

$$\binom{n}{2}^{-1} \sum_{1 \leq j < k \leq n} \mathcal{R}(o \leftrightarrow x_j \wedge x_k) \leq A ;$$

(iii) $\exists A < \infty$ $\forall n \geq 1$ $\forall x_1, \ldots, x_n \in T$ $\exists u \in T$ $\forall k$ $u \not\leq x_k$ *and*

$$\frac{1}{n} \sum_{k=1}^{n} \mathcal{R}(o \leftrightarrow u \wedge x_k) \leq A .$$

Here, the effective resistance between vertices is the free effective resistance, which is just the sum of the resistances of the edges between the vertices.

In the case of simple random walk, that is, $c \equiv 1$, we have $\mathcal{R}(o \leftrightarrow x \wedge u) = |x \wedge u|$. Thus, the quantities in (ii) and (iii) are average meeting heights.

For example, consider simple random walk on the unary tree. This is recurrent, whence the three conditions in Theorem 16.19 are false. Indeed, it is not hard to see that given n distinct vertices, the average in (ii) is at least $(n + 1)/2$, whereas even for $n = 1$, there is no bound on the quantity in (iii).

For a more interesting example, consider simple random walk on the binary tree, T. Now the three conditions are true. To illustrate (ii), suppose that $n = 2^\ell$ is a power of 2. Choose $x_1, \ldots, x_n$ to be the ℓth level of T. Then the quantity in (ii) is approximately equal to 1. To see that (iii) holds with $A = 1$, suppose that $H := \langle x_1, \ldots, x_n \rangle$ is given. Choose a child u_1 of the root of T such that T^{u_1} contains at most half of H. Then choose a child u_2 of u_1 such that T^{u_2} contains at most half of $H \cap T^{u_1}$. Continue in this way until we reach a vertex u_j with $H \cap T^{u_j} = \varnothing$. Let $u := u_j$. Since $u \wedge x = o$ for $x \in H \setminus T^{u_1}$, these vertices x contribute 0 to the left-hand side of (iii). Similarly, the vertices in $H \setminus T^{u_2}$ contribute at most $1/4$ to the left-hand side of (iii), and so on.

Remark 16.20. The proof will show that the smallest possible A in (ii), as well as in (iii), is $\mathcal{R}(o \leftrightarrow \infty)$.

Proof of Theorem 16.19. Assume (i). Use $K(\xi, \eta) := \mathcal{R}(o \leftrightarrow \xi \wedge \eta)$. Then $D_n(\partial T) \leq \mathcal{E}(\partial T)$, so $\exists \xi_1, \ldots, \xi_n \in \partial T$ such that

$$\binom{n}{2}^{-1} \sum_{j < k} \mathcal{R}(o \leftrightarrow \xi_j \wedge \xi_k) \leq \mathcal{E}(\partial T) .$$

From the hypothesis that $\forall \xi \in \partial T$ $\mathcal{R}(o \leftrightarrow \xi) = \infty$, the ξ_k are distinct. Hence there exist distinct $x_k \in \xi_k$ such that $x_j \wedge x_k = \xi_j \wedge \xi_k$, namely, pick $x_k \in \xi_k$ such that $|x_k| > \max_{i \neq j} |\xi_i \wedge \xi_j|$. This gives (ii) with $A = \mathcal{E}(\partial T)$.

For the remainder of the proof of the theorem, we will need new trees T^* and T^{**}, created by adding a ray to each leaf or vertex of T, respectively, with all new edges having conductance 1. Since $\partial T^* \setminus \partial T$ and $\partial T^{**} \setminus \partial T$ are countable and $K(\xi, \xi) \equiv \infty$, we have $\mathscr{E}(\partial T^*) = \mathscr{E}(\partial T^{**}) = \mathscr{E}(\partial T)$.

Assume (i) again. Given $x_1, \ldots, x_n \in T$, let $\xi_k \in \partial T^*$ be such that $x_k \in \xi_k$. Since, as we noted before the statement of the theorem, $M_n(\partial T^*) \leq \mathscr{E}(\partial T^*) = \mathscr{E}(\partial T)$, there exists $\eta \in \partial T^*$ such that

$$\frac{1}{n} \sum_{k=1}^{n} \mathscr{R}(o \leftrightarrow \eta \wedge \xi_k) \leq \mathscr{E}(\partial T).$$

Then $\forall k \quad \eta \neq \xi_k$, so $\exists u \in \eta \cap T$ such that $\forall k \quad u \wedge x_k \leq \eta \wedge \xi_k$ and $u \nleq x_k$. This gives (iii).

Now assume (iii). Given $\xi_1, \ldots, \xi_n \in \partial T^*$, choose $x_k \in \xi_k \cap T^*$ so that in case $\xi_k \in \partial T^* \setminus \partial T$, then $x_k \in T^* \setminus T$, whereas if not, then $|x_k|$ is so large that $\mathscr{R}(o \leftrightarrow x_k) > nA$. Let u be as asserted in (iii). Then $\forall k \quad u \nleq x_k$. Choose $\eta \in \partial T^*$ so that $u \in \eta$. Then $\forall k \quad \eta \wedge \xi_k = u \wedge x_k$, so

$$\frac{1}{n} \sum_{k=1}^{n} \mathscr{R}(o \leftrightarrow \eta \wedge \xi_k) \leq A.$$

Therefore $\mathscr{E}(\partial T^*) \leq A$, so $\mathscr{R}(o \leftrightarrow \infty; T) < \infty$.

Finally, assume (ii). Given n, let $x_1, \ldots, x_n$ be as in (ii). Choose $\xi_k \in \partial T^{**} \setminus \partial T$ so that $x_k \in \xi_k$. Then $\forall j < k \quad \xi_j \wedge \xi_k = x_j \wedge x_k$, whence

$$\binom{n}{2}^{-1} \sum_{j<k} \mathscr{R}(o \leftrightarrow \xi_j \wedge \xi_k) \leq A.$$

Therefore $\mathscr{E}(\partial T^{**}) \leq A$, so $\mathscr{R}(o \leftrightarrow \infty; T) < \infty$. ◀

16.6 Notes

For general background on Brownian motion, see Mörters and Peres (2010). Other nice references on potential theory are Bass (1995), Port and Stone (1978), and Sznitman (1998).

As noted by Benjamini, Pemantle, and Peres (1995), capacity in the ***Martin kernel***

$$K_d(x, y) := G_d(x, y)/G_d(\mathbf{0}, y)$$

is better suited for studying Brownian hitting probabilities than the Green kernel $G_d(x, y)$. (In two dimensions, we use $G_2^*(x, y)$, but we will not indicate this by our notation. Also, K_d is not symmetric, but we will use it only in a symmetric fashion, so it could be replaced by $(K_d(x, y) + K_d(y, x))/2$.) The reason is that whereas the Green kernel, and hence the corresponding capacity, are translation invariant, the hitting probability of a set Λ by standard d-dimensional Brownian motion is not translation invariant but is invariant under scaling for $d \geq 3$. This scale invariance is shared by the Martin kernel $K_d(x, y)$. In particular, using the second-moment method, Benjamini, Pemantle, and Peres (1995) proved the following result, where the constants ($1/2$ and 1) are the best possible. The case $d = 2$ is spelled out in Mörters and Peres (2010).

Theorem 16.21. *Let B_t be Brownian motion started at $\mathbf{0}$ in $\mathbb{R}^d$ for $d \geq 3$, or in $\mathbb{R}^2$ but killed at an exponential time, and let K_d be the corresponding Martin kernel. Then, for any closed set Λ in $\mathbb{R}^d$,*

$$\frac{1}{2}\mathrm{cap}(\Lambda; K_d) \leq \mathbf{P}[\exists t \geq 0 \ \ B_t \in \Lambda] \leq \mathrm{cap}(\Lambda; K_d). \tag{16.19}$$

If the Brownian motion is started according to the measure ν, then (16.19) remains true with the Martin kernel $K_d(x, y) := G_d(x, y)/G_d(\nu, y)$, where

$$G_d(\nu, y) := \int G_d(x, y)\, d\nu(x). \tag{16.20}$$

See Exercise 16.10 for a discrete version. From Theorem 16.21, Lemma 16.10 is immediate:

Proof of Lemma 16.10. By Theorem 16.21, it is enough to show that the ratio of the Martin kernel to the Green kernel is bounded above and below. By definition of the Martin kernel, it suffices to check that the Greenian potential $G_d(\nu, y)$ defined in (16.20) is bounded. This is clearly the case when ν has a bounded density.   ◄

▷ **Exercise 16.8.**
Let $[B]$ be the trace of Brownian motion started at $\mathbf{0}$ in $\mathbb{R}^d$ for $d \geq 3$, or in $\mathbb{R}^2$ but killed at an exponential time. Let $\Lambda \subset \mathbb{R}^d \setminus \{\mathbf{0}\}$ be closed. Show that $\mathbf{P}[|[B] \cap \Lambda| \in \{0, 2^{\aleph_0}\}] = 1$.

The proof of Bishop's conjecture (16.7) is published here for the first time (Corollary 16.15). Now we prove Theorem 16.18.

Proof of Theorem 16.18. (i) Since $\mathscr{E}(\mu) = \|V_\mu\|_{L^1(\mu)} \leq \|V_\mu\|_{L^\infty(\mu)}$, we have $\mathscr{E}(X) \leq \inf_\mu \|V_\mu\|_{L^\infty(\mu)}$. On the other hand, suppose that $\mathscr{E}(X) < \infty$ (since otherwise certainly $\mathscr{E}(X) \geq \inf_\mu \|V_\mu\|_{L^\infty(\mu)}$). The space $\mathrm{Prob}(X)$ is Hausdorff and compact under the weak* topology. The definition of Hausdorff is equivalent to the diagonal $\Delta := \{(\mu, \mu) \in \mathrm{Prob}(X) \times \mathrm{Prob}(X)\}$ being closed in $\mathrm{Prob}(X) \times \mathrm{Prob}(X)$. Since $\forall t \in \mathbb{R} \ \ K \wedge t \in C(X \times X)$, we have

$$\Delta \cap \left\{(\mu, \nu) \in \mathrm{Prob}(X) \times \mathrm{Prob}(X)\,; \ \int K \wedge t\, d(\mu \times \nu) \leq \mathscr{E}(X) + \epsilon\right\}$$

is compact and nonempty for each $\epsilon > 0$. Hence there is a measure μ with (μ, μ) in all these sets for $t < \infty$ and $\epsilon > 0$. This measure μ necessarily satisfies $\mathscr{E}(\mu) = \mathscr{E}(X)$.

We claim that $V_\mu \geq \mathscr{E}(X)$ μ-a.e.; this gives $V_\mu = \mathscr{E}(X)$ μ-a.e., since $\mathscr{E}(X) = \mathscr{E}(\mu) = \int V_\mu\, d\mu$. Suppose there were a set $F \subsetneq X$ and $\delta > 0$ such that $\mu F > 0$ and $V_\mu \restriction F \leq \mathscr{E}(X) - \delta$. Then move a little of μ's mass to F: Let $\nu := \mu(\bullet \mid F)$ be the normalized restriction of μ to F and $\eta > 0$. We have $\mathscr{E}(\nu) < \infty$ and

$$\mathscr{E}\big((1-\eta)\mu + \eta\nu\big) = (1-\eta)^2\mathscr{E}(\mu) + \eta^2\mathscr{E}(\nu) + 2(1-\eta)\eta\int V_\mu\, d\nu$$

$$\leq (1-\eta)^2\mathscr{E}(X) + \eta^2\mathscr{E}(\nu) + 2(1-\eta)\eta\big(\mathscr{E}(X) - \delta\big)$$

$$\leq \mathscr{E}(X) - 2\eta\delta + O(\eta^2).$$

Hence $\mathscr{E}\big((1-\eta)\mu + \eta\nu\big) < \mathscr{E}(X)$ for sufficiently small η, a contradiction.

(ii) Regard $D_{n+1}(X)$ as an average of averages over all n-subsets of $\{x_1, \ldots, x_{n+1}\}$; each n-subset average is $\geq D_n(X)$, whence so is $D_{n+1}(X)$.

Claim: $D_{n+1}(X) \le M_n(X)$. Let $D_{n+1}(X) = \binom{n+1}{2}^{-1} \sum_{1 \le j < k \le n+1} K(x_j, x_k)$. For each j, write this as $f_j(x_1, \ldots, x_{j-1}, x_{j+1}, \ldots, x_{n+1}) + \binom{n+1}{2}^{-1} \sum_{k \ne j} K(x_j, x_k)$. Only the second term depends on x_j, hence x_j is such that

$$\sum_{k \ne j} K(x_j, x_k) = \min_{x \in X} \sum_{k \ne j} K(x, x_k) \le n M_n(X).$$

Therefore

$$D_{n+1}(X) = \frac{1}{n(n+1)} \sum_{j=1}^{n+1} \sum_{k \ne j} K(x_j, x_k) \le \frac{1}{n+1} \sum_{j=1}^{n+1} M_n(X) = M_n(X).$$

Thus, for any compact $\hat{X} \subseteq X$, we have $D_{n+1}(X) \le D_{n+1}(\hat{X}) \le M_n(\hat{X})$, whence $D_{n+1}(X) \le \inf_{\hat{X}} M_n(\hat{X})$.

Claim: $\inf_{\hat{X}} M_n(\hat{X}) \le \mathscr{E}(X)$ and if $V_\mu \le \mathscr{E}(X)$ on $\hat{X}$ with $\mu\hat{X} = 1$, then $M_n(\hat{X}) \le \mathscr{E}(X)$. We may assume $\mathscr{E}(X) < \infty$. From part (i), $\exists \mu \in \mathrm{Prob}(X)$ $V_\mu = \mathscr{E}(X)$ μ-a.e. Since V_μ is lower semicontinuous by Fatou's lemma, the set $\hat{X}_0$ where $V_\mu \le \mathscr{E}(X)$ is compact and $\mu\hat{X}_0 = 1$. For any $\hat{X} \subseteq \hat{X}_0$ with $\mu\hat{X} = 1$ and any $x_k \in \hat{X}$, we have

$$\min_{x \in \hat{X}} \frac{1}{n} \sum_{1}^{n} K(x, x_k) \le \int_{\hat{X}} \frac{1}{n} \sum_{1}^{n} K(x, x_k) \, d\mu(x) = \int_{X} \frac{1}{n} \sum_{1}^{n} K(x, x_k) \, d\mu(x)$$

$$= \frac{1}{n} \sum_{1}^{n} V_\mu(x_k) \le \mathscr{E}(X),$$

in other words, $M_n(\hat{X}) \le \mathscr{E}(X)$.

Claim: $\mathscr{E}(X) \le \lim D_n(X)$. Let $D_n(X) = \binom{n}{2}^{-1} \sum_{1 \le j < k \le n} K(x_j, x_k)$ and $\mu_n := \sum_{j=1}^{n} \frac{1}{n} \delta(x_j)$. Let μ be a weak* limit point of $\{\mu_n\}$. Weak* convergence of μ_{n_j} to μ implies weak* convergence of the squares $\mu_{n_j} \times \mu_{n_j}$ to $\mu \times \mu$, since the linear span of $C(X) \otimes C(X)$ is dense in $C(X \times X)$ by the Stone–Weierstrass theorem. We have for $t \in \mathbb{R}$

$$\int K \wedge t \, d(\mu_n \times \mu_n) \le \frac{n-1}{n} D_n(X) + \frac{t}{n},$$

whence

$$\int K \wedge t \, d(\mu \times \mu) \le \lim_{n \to \infty} D_n(X),$$

and thus

$$\mathscr{E}(\mu) \le \lim_{n \to \infty} D_n(X).$$

Putting together the claims, we see that

$$D_{n+1}(X) \le \inf_{\hat{X}} M_n(\hat{X}) \le \mathscr{E}(X) \le \lim_{n \to \infty} D_n(X),$$

from which the theorem follows. ◀

Remark. We have also seen that when $\mathscr{E}(X) < \infty$, a minimizing measure (called equilibrium measure) on X can be obtained as a weak* limit point of the empirical measures of minimizing sets for $D_n(X)$.

16.7 Collected In-Text Exercises

16.1. Show that if $p_x \equiv p \in (0, 1)$ and T is spherically symmetric, then

$$\operatorname{cap} \partial T = \left(1 + (1 - p) \sum_{n=1}^{\infty} \frac{1}{p^n |T_n|}\right)^{-1}.$$

Thus, $\mathbf{P}[o \leftrightarrow \partial T] > 0$ iff $\sum_{n=1}^{\infty} \frac{1}{p^n |T_n|} < \infty$.

16.2. Suppose that f and g are two gauge functions such that $f/c_1 - c_2 \leq g \leq c_1 f + c_2$ for some constants c_1 and c_2. Show that for all E, $\operatorname{cap}_f(E) > 0$ iff $\operatorname{cap}_g(E) > 0$.

16.3. Prove Corollary 16.6.

16.4. Prove Proposition 16.8.

16.5. **(a)** Show that if $p_n \leq b^{-d}$ for all n, then $Q_{d,b}(\langle p_n \rangle) = \varnothing$ a.s.
   (b) Characterize the sequences $\langle p_n \rangle$ for which $Q_d(\langle p_n \rangle)$ has positive volume with positive probability.

16.6. Let $p_t(x, y) := (2\pi t)^{-d/2} \exp\left(-|x - y|^2/(2t)\right)$ be the Brownian transition density function, and for $d \geq 3$, define $G_d(x, y) := \int_0^\infty p_t(x, y)\, dt$ for $x, y \in \mathbb{R}^d$.
   (a) Show that $G_d(x, y) = c_d g_d(|x - y|)$ with a constant $0 < c_d < \infty$.
   (b) Define $F_A(x) := \int_A G_d(x, z)\, dz$ for Borel sets $A \subseteq \mathbb{R}^d$. Show that $F_A(x)$ is the expected time the Brownian motion started at x spends in A.
   (c) Show that $x \mapsto G_d(0, x)$ is harmonic on $\mathbb{R}^d \setminus \{0\}$, in other words, has zero Laplacian there. Equivalently, if $B \subset \mathbb{R}^d \setminus \{0\}$ is a ball, then the average value (with respect to Lebesgue measure) of $G_d(0, x)$ over $x \in B$ is equal to the value at the center of the ball.
   (d) Consider Brownian motion $\langle B_t \rangle$ in $\mathbb{R}^2$, killed at a random time with an exponential distribution with parameter 1. In other words, let τ be an Exponential(1) random time, independent of the Brownian motion. The expected occupation measure v_x for B_t started at x and killed at time τ is defined by $v_x(A) := \mathbf{E}\int_0^\tau \mathbf{1}_A(B_t)\, dt$ for Borel sets A in $\mathbb{R}^2$. Show that $v_x(A) = \int_A G_2^*(x, y)\, dy$, where $G_2^*(x, y) := \int_0^\infty p_t(x, y)e^{-t}\, dt$.
   (e) Show that in two dimensions, $G_2^*(x, y) \sim -(1/\pi) \log |x - y|$ for $|x - y| \downarrow 0$.

16.7. Prove part (iii) of Corollary 16.14.

16.8. Let $[B]$ be the trace of Brownian motion started at $\mathbf{0}$ in $\mathbb{R}^d$ for $d \geq 3$, or in $\mathbb{R}^2$ but killed at an exponential time. Let $\Lambda \subset \mathbb{R}^d \setminus \{0\}$ be closed. Show that $\mathbf{P}\big[|[B] \cap \Lambda| \in \{0, 2^{\aleph_0}\}\big] = 1$.

16.8 Additional Exercises

16.9. Consider a network on a tree T and a subset $E \subseteq \partial T$. Let μ^i be harmonic measure on ∂T. Show that cap $E \geq \text{cap}(\partial T) \cdot \mu^i(E)^2$, whence if $\mu^i(E) > 0$, then $\mathbf{P}[o \leftrightarrow E] > 0$ for the corresponding percolation. Find T and $E \subseteq \partial T$ so that $\mu^i(E) = 0$ yet $\mathbf{P}[o \leftrightarrow E] > 0$ (so cap $E > 0$).

16.10. Use a method similar to the proof of Theorem 5.24 to prove the following theorem. Let $\langle X_n \rangle$ be a transient Markov chain on a countable state space with initial distribution π and transition probabilities $p(x, y)$. Define the Green function $\mathcal{G}(x, y) := \sum_{n=1}^{\infty} p^{(n)}(x, y)$ and the Martin kernel $K(x, y) := \mathcal{G}(x, y)/\sum_z \pi(z)\mathcal{G}(z, y)$. Note that K may not be symmetric. Then with capacity defined using the kernel K, we have for any set S of states,

$$\frac{1}{2}\,\text{cap}(S) \leq \mathbf{P}[\exists n \geq 0\ X_n \in S] \leq \text{cap}(S).$$

16.11. Prove (1.10) that for $\alpha > 0$,

$$\frac{1}{2}\,\text{cap}_\alpha E \leq \frac{1}{1 - b^{-\alpha}}\text{cap}_{\alpha \log b}\partial T \leq 3b\,\text{cap}_\alpha E$$

when T codes the closed set $E \subseteq [0, 1]$ in base b.

16.12. Show that if μ is a probability measure on $\mathbb{R}^d$ satisfying $\mu(B_r(x)) \leq Cr^\alpha$ for some constants C and α and if $\beta < \alpha$, then the energy of μ in the gauge $z^{-\beta}$ is finite. By using this and previous exercises, give another proof of Corollary 16.6.

16.13. Let $A \subset \mathbb{R}^d$ be a closed set contained in the d-dimensional ball of radius a centered at the origin. Write $B(s, t)$ for the trace of d-dimensional Brownian motion during the time interval (s, t). Let τ be an exponential random variable with parameter 1 independent of B. Show that

$$e^{-1}\,\mathbf{P}_0\big[B(0, 1) \cap A \neq \varnothing\big] \leq \mathbf{P}_0\big[B(0, \tau) \cap A \neq \varnothing\big] \leq C\,\mathbf{P}_0\big[B(0, 1) \cap A \neq \varnothing\big]$$

for some constant C (depending on a and d). *Hint:* For the upper bound, first show the following general lemma: Let f_1, f_2 be probability densities on $[0, \infty)$. Suppose that the likelihood ratio $\psi(r) := f_2(r)/f_1(r)$ is increasing and $h: [0, \infty) \to [0, \infty)$ is decreasing on $[a, \infty)$. Then

$$\frac{\int_0^\infty h(r)f_2(r)\,dr}{\int_0^\infty h(r)f_1(r)\,dr} \leq \psi(a) + \frac{\int_a^\infty f_2(r)\,dr}{\int_a^\infty f_1(r)\,dr}\,.$$

Second, use this by conditioning on $|B_{t_j}|$ for $j = 1, 2$ to get an upper bound on

$$\frac{\mathbf{P}_0\big[B(t_2, t_2 + s) \cap A \neq \varnothing\big]}{\mathbf{P}_0\big[B(t_1, t_1 + s) \cap A \neq \varnothing\big]}$$

for $t_1 \leq t_2$ and $s \geq 0$. Third, bound $\mathbf{P}_0\big[B(0, \tau) \cap A \neq \varnothing\big]$ by summing over intersections in $[j/2, (j+1)/2]$ for $j \in \mathbb{N}$.

16.14. Let $\Lambda \subset \mathbb{R}^d$ be compact. Fitzsimmons and Salisbury (1989) showed that when $d = 2$, if $B^{(i)}$ are independent Brownian motions, then $\text{cap}(\Lambda; g_d^k) > 0$ if and only if $\mathbf{P}[\Lambda \cap [B^{(1)}] \cap \cdots \cap [B^{(k)}] \neq \varnothing] > 0$. Does this also hold for $d \geq 3$?

16.15. Let $\Lambda \subset \mathbb{R}^d$ be a k-dimensional cube of side length a, where $1 \leq k \leq d$. Find the capacity $\text{cap}(\Lambda; g_d)$ up to constant factors depending on k and d only.

16.16. Let $[B]$ be the Brownian trace in dimension d defined in Theorem 16.11. Consider any closed subset $\Lambda \subseteq \mathbb{R}^d$.

(a) Show that if $d \geq 3$ and $\dim \Lambda > d - 2$, then $\dim(\Lambda \cap [B]) \leq \dim \Lambda + 2 - d$ a.s., and that the essential supremum of the left-hand side is equal to the right-hand side.

(b) Show that if $d = 2$ and $[B^{(1)}], \dots, [B^{(k)}]$ are independent Brownian traces for some $k \geq 1$, then $\dim(\Lambda \cap [B^{(1)}] \cap \cdots \cap [B^{(k)}]) = \dim \Lambda$ a.s.

(c) Show that if $d \geq 3$ and $[B^{(1)}], \dots, [B^{(k)}]$ are independent Brownian traces for some $k \in [1, \dim \Lambda/(d - 2))$, then $\dim(\Lambda \cap [B^{(1)}] \cap \cdots \cap [B^{(k)}]) \leq \dim \Lambda - k(d - 2)$ a.s., and that the essential supremum of the left-hand side is equal to the right-hand side.

16.17. Consider the following variant of the fractal percolation process, defined on all of $\mathbb{R}^3$. For each $n \in \mathbb{Z}$, tile space by cubes of side length 3^{-n} with sides parallel to the coordinate axes, each one being a translation of the cube with center $\mathbf{0} = (0, 0, 0)$ by some vector in $\mathbb{Z}^3/3^n$. Generate a random collection $\mathcal{K}_n$ of these cubes by always including the cube containing $\mathbf{0}$, and including each of the remaining cubes independently with probability p. As before, define

$$C_n := \bigcup \mathcal{K}_n \quad \text{and} \quad R(p) := \bigcap_{n \in \mathbb{Z}} C_n \, .$$

Show that for $p = 1/3$, the random set $R(p)$ is intersection equivalent in all of $\mathbb{R}^3$ to the trace of Brownian motion started at $\mathbf{0}$. *Hint:* Use Theorem 16.21.

16.18. Let $A, B \subset [0, 1]^d$ be two random closed sets that are intersection equivalent in $[0, 1]^d$, with α and β being the essential suprema of their Hausdorff dimensions. Prove that $\alpha = \beta$.

16.19. Let $[B]$ and $[B']$ be independent Brownian traces in dimension 3, as defined in Theorem 16.11. Show that $\mathbf{P}\big[[B] \cap [B'] \cap \Lambda \neq \varnothing\big] \asymp \mathrm{cap}_2(\Lambda)$ for all closed subsets $\Lambda \subseteq [0, 1]^3$.

16.20. Can one remove the hypothesis from Theorem 16.19 that T be locally finite?

16.21. Give an example of a kernel on a space X such that $\mathscr{E}(X) < \infty$ yet for all $\mu \in \mathrm{Prob}(X)$, there is some $x \in X$ with $V_\mu(x) = \infty$.

16.22. Suppose that X is a compact Hausdorff space and $K \colon X \times X \to [0, \infty]$ is continuous and symmetric. Give $\mathrm{Prob}(X)$ the weak* topology. Show that $\mu \mapsto \mathscr{E}(\mu)$ is lower semicontinuous.

17 | Random Walks on Galton–Watson Trees

We have analyzed random walks on trees in Chapters 3 and 13; we have seen many examples of interesting trees in several chapters, some of them random. Up to now, however, we have not looked at random walks on random trees. That is, choose a tree at random in an interesting way; then fix the tree and run a random walk on it. What happens?

We will be particularly interested in the case that the tree is chosen at random according to Galton–Watson measure and the walk is simple random walk. Is it transient or recurrent? If transient, how fast does it escape to infinity, and what is the Hausdorff dimension of its harmonic measure on the boundary of the tree?

The setting now is a little more complicated than for random walks on graphs. It will turn out to be crucial that our trees are rooted and that we move the root with the random walker. The set of trees on which the walk will take place will be uncountable, so although the starting rooted tree determines a countable set of possible rooted trees of the later trajectory of the random walk, altogether the Markov chain takes place on an uncountable state space. Thus we start by discussing Markov chains on general state spaces. In particular, we need to understand the relationship of concepts from the theory of Markov chains to concepts from ergodic theory.

We reiterate that all trees in this chapter are rooted.

17.1 Markov Chains and Ergodic Theory

We begin with a brief review of ergodic theory. A ***probability measure-preserving system*** $(X, \mathscr{F}, \mu, S)$ is a probability measure space $(X, \mathscr{F}, \mu)$ together with a measurable map S from X to itself such that for all $A \in \mathscr{F}$, $\mu(S^{-1}A) = \mu(A)$. Fix $A \in \mathscr{F}$ with $\mu(A) > 0$. We denote the ***induced measure*** on A by $\mu_A(C) := \mu(C)/\mu(A)$ for $C \subseteq A$. We also write $\mu(C \mid A)$ for $\mu_A(C)$, since it is a conditional measure. A set $A \in \mathscr{F}$ is called S-***invariant*** if $\mu(A \bigtriangleup S^{-1}A) = 0$. The σ-field of invariant sets is called the ***invariant σ-field***. The system is called ***ergodic*** if the invariant σ-field is trivial (that is, consists only of sets of measure 0 and 1). A sufficient condition for ergodicity is that the system be ***mixing***, that is, that for all $A, B \in \mathscr{F}$, we have $\lim_{n \to \infty} \mu(A \cap S^{-n}B) = \mu(A)\mu(B)$. (To see that this is indeed sufficient, just apply it to $A = B$ in the invariant σ-field.) For a function f on X, we write Sf for the function $f \circ S$. The ***ergodic theorem*** states that for $f \in L^1(X, \mu)$, the limit of the averages $\sum_{k=0}^{n-1} S^k f / n$ exists a.s. and equals the conditional expectation of f with respect to the invariant σ-field; see Section 14.5. In particular, if the system is ergodic, then the limit equals the expectation of f a.s.

Now let $\mathscr{X}$ be a measurable state space and $\mathbf{p}$ be a transition kernel on $\mathscr{X}$ (that is, $\mathbf{p}(x, A)$ is measurable in x for each measurable set A and is a probability measure for each fixed x). The corresponding Markov chain satisfies $\mathbf{P}[X_{n+1} \in A \mid X_n] = \mathbf{p}(X_n, A)$. This gives the usual operator P on bounded* measurable functions f, where

$$(Pf)(x) := \int_{\mathscr{X}} f(y)\, \mathbf{p}(x, dy),$$

and its adjoint P^* on probability measures, where

$$\int_{\mathscr{X}} f\, dP^* v := \int_{\mathscr{X}} Pf\, dv.$$

Let μ be a **p-*stationary*** probability measure, that is, $P^*\mu = \mu$, and let $\mathcal{I}$ be the σ-field of ***invariant*** sets, that is, those measurable sets A such that

$$\mathbf{p}(x, A) = \mathbf{1}_A(x) \quad \text{for } \mu\text{-a.e. } x.$$

The Markov chain is called ***ergodic*** if $\mathcal{I}$ is trivial.[†]

We use these terms in more general situations as well. That is, if μ is a positive measure, perhaps infinite, then we say that μ is **p**-stationary if, for all measurable $f \geq 0$, we have $\int_{\mathscr{X}} f\, d\mu = \int_{\mathscr{X}} Pf\, d\mu$. Even if μ is not **p**-stationary, we define the invariant σ-field to consist of those measurable sets A such that $\mu\{x \,;\, \mathbf{p}(x, A) \neq \mathbf{1}_A(x)\} = 0$ and we call the Markov chain ergodic if every invariant A satisfies $\mu(A) = 0$ or $\mu(A^c) = 0$.

We have now used "invariant" and "ergodic" each in two apparently different senses and will explain why they are actually equivalent (Proposition 17.2 and Corollary 17.3).

A bounded measurable function f is called ***harmonic*** if $Pf = f$ μ-a.s. Let $(\mathscr{X}^\infty, \mathbf{p} \times \mu)$ be the space of (one-sided) sequences of states with the measure induced by choosing the initial state according to μ and making transitions via $\mathbf{p}$. That is, if $\langle X_n \,;\, n \geq 0\rangle$ is the Markov chain on $\mathscr{X}$ with $X_0 \sim \mu$, then $\mathbf{p} \times \mu$ is its law.

Lemma 17.1. *Consider a Markov chain $\langle X_n \,;\, n \geq 0\rangle$ with transition kernel $\mathbf{p}$ and $\mathbf{p}$-stationary probability measure μ on a measurable state space $\mathscr{X}$. If f is a bounded measurable function on $\mathscr{X}$, then the following are equivalent:*

(i) *f is harmonic;*

(ii) *f is $\mathcal{I}$-measurable;*

(iii) *$f(X_0) = f(X_1)$ $\mathbf{p} \times \mu$-a.s.*

The idea is that when there is a finite stationary measure, then, as in the case of a Markov chain with a denumerable number of states, there really aren't any nontrivial bounded harmonic functions.

▷ **Exercise 17.1.**
Show that Lemma 17.1 may not be true if μ is an infinite stationary measure, where (iii) then means that $(\mathbf{p} \times \mu)\big(f(X_0) \neq f(X_1)\big) = 0$.

---

* Everything we will say about bounded functions applies equally to nonnegative functions.

[†] These definitions of invariant and ergodic match the preceding ones for ergodic theory. They are slightly different from the ones we used in Chapter 14, but compare Exercise 14.14.

Proof of Lemma 17.1. *(i) $\Rightarrow$ (ii):* Let f be harmonic. Fix $\alpha \in \mathbb{R}$. Since P is a positive operator (that is, it maps nonnegative functions to nonnegative functions),

$$P(f \wedge \alpha) \leq (Pf) \wedge \alpha = f \wedge \alpha.$$

Also

$$\int P(f \wedge \alpha)\,d\mu = \int f \wedge \alpha\,dP^*\mu = \int f \wedge \alpha\,d\mu.$$

Combining these two equations yields that $P(f \wedge \alpha) = f \wedge \alpha$ μ-a.s., in other words, that $f \wedge \alpha$ is harmonic. Now consider a state x for which $f(x) \geq \alpha$ and also where $(f \wedge \alpha)(x) = (P(f \wedge \alpha))(x)$. These two conditions imply that $\mathbf{p}(x, [f \geq \alpha]) = 1$. Since each of these two conditions holds μ-a.s., we get that for μ-a.e. $x \in [f \geq \alpha]$, we have $\mathbf{p}(x, [f \geq \alpha]) = 1$. Likewise, for μ-a.e. $x \in [f \leq \beta]$, we have $\mathbf{p}(x, [f \leq \beta]) = 1$ for every β. Since $[f < \alpha] = \bigcup_{\beta < \alpha}[f \leq \beta]$, it follows that for μ-a.e. $x \in [f < \alpha]$, we have $\mathbf{p}(x, [f < \alpha]) = 1$, whence $[f \geq \alpha] \in \mathcal{I}$. Thus, f is $\mathcal{I}$-measurable.

(ii) $\Rightarrow$ (iii): For any interval I, we have $\mathbf{p}(x, [f \in I]) = \mathbf{1}_{[f \in I]}(x)$ μ-a.s. Thus, if $f(X_0) \in I$, then $f(X_1) \in I$ a.s.

(iii) $\Rightarrow$ (i): This is immediate from the definition of harmonic. ◄

The ***shift*** map on X^∞ is the map $(x_0, x_1, \ldots) \mapsto (x_1, x_2, \ldots)$. It preserves the measure $\mathbf{p} \times \mu$ since μ is stationary. We now show that functions on $\mathscr{X}^\infty$ that are (a.s.) shift invariant depend only on their first coordinate:

Proposition 17.2. *Consider a Markov chain $\langle X_n;\ n \geq 0 \rangle$ with transition kernel $\mathbf{p}$ and $\mathbf{p}$-stationary probability measure μ on a measurable state space $\mathscr{X}$. A bounded measurable function h on $\mathscr{X}^\infty$ is shift invariant iff there exists a bounded $\mathcal{I}$-measurable function f on $\mathscr{X}$ such that $h(X_0, X_1, \ldots) = f(X_0)$ $\mathbf{p} \times \mu$-a.s. Indeed, $f(x)$ can be determined from h by starting the chain at x and defining $f(x) := \mathbf{E}_x\big[h(x, X_1, \ldots)\big]$.*

Proof. If h has the form given in terms of f, then the lemma shows that h is shift invariant. Conversely, given h, define f as indicated. Then we have a.s.

$$
\begin{aligned}
f(X_0) &= \mathbf{E}\big[\mathbf{E}[h(X_0, X_1, \ldots) \mid X_0, X_1] \mid X_0\big] \\
&= \mathbf{E}\big[\mathbf{E}[h(X_1, X_2, \ldots) \mid X_0, X_1] \mid X_0\big] \quad \text{by shift invariance} \\
&= \mathbf{E}[f(X_1) \mid X_0] \quad \text{by the Markov property and the definition of } f \\
&= (Pf)(X_0) \quad \text{by definition of } P.
\end{aligned}
$$

That is, f is harmonic, so is $\mathcal{I}$-measurable. Now similar reasoning, together with the martingale convergence theorem, gives a.s.

$$
\begin{aligned}
h(X_0, X_1, \ldots) &= \lim_{n \to \infty} \mathbf{E}\big[h(X_0, X_1, \ldots) \mid X_0, \ldots, X_n\big] \\
&= \lim_{n \to \infty} \mathbf{E}\big[h(X_n, X_{n+1}, \ldots) \mid X_0, \ldots, X_n\big] = \lim_{n \to \infty} f(X_n) = f(X_0)
\end{aligned}
$$

by the lemma. ◄

As an immediate corollary, we get a criterion for ergodicity:

Corollary 17.3. *Consider a Markov chain with transition kernel $\mathbf{p}$ and $\mathbf{p}$-stationary probability measure μ. We have that $\mathbf{p} \times \mu$ is ergodic for the shift iff $\mathcal{I}$ is trivial.* ◄

17.2 Stationary Measures on Trees

We now begin to apply the general theory of the preceding section to simple random walk on random trees. We are interested in linear rate of escape, so the case of recurrent trees is not germane. In particular, trees with one or two ends will receive minimal attention. Recall that an **end** of a tree is an equivalence class of rays with arbitrary starting vertices, where two rays are equivalent if their symmetric difference is finite.

We haven't said yet what random trees we consider. The answer will be that the tree is chosen according to some probability measure that gives us a Markov chain that is stationary in an appropriate sense. Stationarity will greatly facilitate the analysis, but it won't hold in the most interesting case of Galton–Watson trees: after all, the root has a smaller degree (by 1) in distribution than do other vertices. We'll be able to fix this problem in a surprisingly easy way, but in this section, we discuss the general theory that we'll invoke when we do have stationarity. To achieve a stationary Markov chain, we need to change our point of view from a random walk on a fixed (though random) tree to a random walk on the space of isomorphism classes of trees.

Actually, there is no good Borel space of unrooted trees. Think, for example, how one would define a distance between two trees. Instead, it is necessary to consider rooted trees. Then two rooted trees are close if they agree, up to isomorphism sending one root to the other, in a large ball around their roots. This gives a topology, and the topology generates a Borel σ-field. What one then does is walk on the space of rooted trees, changing the root to the location of the walker and keeping the underlying unrooted tree the same. However, to have a measure on rooted trees that is stationary with respect to this chain, we need to use isomorphism classes of rooted trees.

The formalism is as follows. Let $\mathscr{T}$ be the space of rooted trees in Exercise 5.2. Call two rooted trees *(rooted) isomorphic* if there is a bijection of their vertex sets preserving adjacency and mapping one root to the other. Since the roots will be changing with the walker, we will often write the root explicitly. Our notation for a rooted tree will be (T, x), where $x \in \mathsf{V}(T)$ designates the root. For $(T, o) \in \mathscr{T}$, let $[T, o]$ denote the set of trees that are isomorphic to (T, o). Let $[\mathscr{T}] := \{[T, o] \; ; \; (T, o) \in \mathscr{T}\}$. Normally, we have a measure μ such as **GW** on rooted trees $\mathscr{T}$; such a measure induces a measure $[\mu]$ on isomorphism classes of rooted trees $[\mathscr{T}]$ in the obvious way.

Consider the Markov chain that moves from a rooted tree (T, x) to the rooted tree (T, y) for a uniform random neighbor y of x. For a fixed tree T, this chain is "isomorphic" to simple random walk on T. Write the transition probabilities as

$$p\big((T, x), (T, y)\big) = \begin{cases} 1/\deg_T(x) & \text{if } y \sim x \\ 0 & \text{otherwise.} \end{cases}$$

As we said, to get stationarity, we are really interested in the Markov chain induced by this chain on isomorphism classes of trees. Thus, define

$$p([T, x], [T', y]) := \frac{1}{\deg_T(x)} \big|\{z \in T \; ; \; z \sim x, \; [T', y] = [T, z]\}\big|.$$

This gives the transition kernel

$$\mathsf{p}_{\text{SRW}}([T, x], A) := \sum_{[T', y] \in A} p([T, x], [T', y]).$$

We will call a Borel measure μ on $\mathcal{T}$ stationary if the induced measure $[\mu]$ on $[\mathcal{T}]$ is. Rather than say that μ is $\mathbf{p}_{\mathrm{SRW}}$-stationary, we will say more simply that μ is SRW-stationary.

▷ **Exercise 17.2.**
Let $G = (\mathsf{V}, \mathsf{E})$ be a finite, connected graph. For $x \in \mathsf{V}$, let T_x be the universal cover of G based at x (see p. 83 of Section 3.3). Define

$$\mu([T, o]) := \frac{1}{2|\mathsf{E}|} \sum \{\deg x \, ; \, x \in \mathsf{V}, \, [T_x, x] = [T, o]\} \, .$$

Show that μ is an SRW-stationary, ergodic probability measure on $[\mathcal{T}]$.

Let $\mathrm{SRW} \times \mu$ (instead of $\mathbf{p}_{\mathrm{SRW}} \times \mu$) denote the probability measure on paths in trees given by choosing a tree according to μ and then conditionally independently running simple random walk on the tree starting at its root.

Theorem 17.4. (Speed for Stationary Measures) *If μ is an SRW-stationary probability measure on the space of rooted trees $\mathcal{T}$ such that μ-a.e. tree is infinite, then the speed (rate of escape) of simple random walk $\langle X_n \rangle$ satisfies*

$$\mathbf{E}\left[\lim_{n \to \infty} \frac{|X_n|}{n} \right] = \sum_{k \geq 1} \mu[\deg_T o = k]\left(1 - \frac{2}{k}\right) \geq 0 \, . \tag{17.1}$$

The speed is positive a.s. iff μ-a.e. tree has at least three ends, in which case μ-a.e. tree has uncountably many ends and branching number > 1. In case μ is ergodic, then the sum in (17.1) is $\mathrm{SRW} \times \mu$-a.s. the speed.

Proof. Since the measure on rooted trees is stationary for simple random walk, the sequence of degrees $\langle \deg_T X_k \rangle$ is a stationary sequence. Thus, (17.1) follows from (13.5) and the ergodic theorem.

Now the speed is of course 0 a.s. on all trees with at most two ends. In the opposite case, we make use of the following construction. This is also where we begin the deeper, fruitful interplay of discrete probability (on trees) with ergodic theory (on the space of trees).

For a tree T, write $T^\diamond$ for the ***bi-infinitary part*** of T consisting of the vertices and edges of T that belong to some bi-infinite, simple path. This is the same as what remains of T after iteratively pruning all its leaves. The bi-infinitary part of T is nonempty iff T has at least two ends; we are now assuming that μ-a.e. tree has at least three ends. The parts of T that are not in its bi-infinitary part are finite trees that we call ***shrubs***. Since simple random walk on a shrub is recurrent, simple random walk on T visits $T^\diamond$ infinitely often and, in fact, takes infinitely many steps on $T^\diamond$. If we observe $\langle X_k \rangle$ only when it makes a transition along an edge of $T^\diamond$, then we see simple random walk on $T^\diamond$ (by the strong Markov property). In particular, if we begin simple random walk on $\mathcal{T}$ with the initial stationary probability measure μ, then a.s. there exists k for which $X_k \in T^\diamond$. By stationarity, it follows that $X_0 \in T^\diamond$ with positive probability. Thus, the set of states $\mathcal{A}_\diamond := \{[T, o] \, ; \, o \in T^\diamond\}$ has positive probability. Likewise, the event $\mathcal{A}'_\diamond := \left[[T, X_0], [T, X_1] \in \mathcal{A}_\diamond \right]$ satisfies $(\mathrm{SRW} \times \mu)(\mathcal{A}'_\diamond) > 0$, so the sequence of elements of $\mathscr{A}_\Diamond$ given by successive returns to $\mathscr{A}'_\Diamond$ is shift-stationary by Exercise 2.30. Let $\mu^\Diamond$ be the law of $[T, X_0]$ when $([T, X_0], [T, X_1])$ has the law (SRW $\times \mu$) conditioned on $\mathscr{A}'_\Diamond$. Then $\mu^\Diamond$ is stationary for simple random walk on $T^\Diamond$, since that is what corresponds to a return to $\mathscr{A}'_\Diamond$.

Now $\mu^\Diamond[\deg_T(o) \geq 3] > 0$, since otherwise the walk would be restricted to a copy of $\mathbb{Z}$. Since $\deg_T(o) \geq 2 \; \mu^\Diamond$-a.s., it follows from (17.1) that the $\mu^\Diamond$-expected speed is positive, whence the speed is positive with positive $\mu^\Diamond$-probability. In fact, the speed is positive a.s., since if not, consider the set A of trees where the speed is 0. Then A is invariant under the random walk, so if $\mu(A) > 0$, then $[\mu]$ conditioned to A is also SRW-stationary, and applying the result we just proved to this conditioned measure would give a contradiction.

Now let $\langle Y_k \rangle$ be the random walk induced on $T^\Diamond$ and Z_k be the number of steps that the random walk on T takes between the kth step on $T^\Diamond$ and the $(k + 1)$th step on $T^\Diamond$. Since the random walk returns to $T^\Diamond$ infinitely often a.s., its speed on T is

$$\lim_{k \to \infty} \frac{|Y_k|}{\sum_{j<k} Z_j}.$$

To show that this is positive a.s., it remains to show that

$$\lim_{k \to \infty} \frac{1}{k} \sum_{j<k} Z_j < \infty \quad \text{a.s.,} \tag{17.2}$$

since the $\mu^\Diamond$-speed on $T^\Diamond$ is $\lim |Y_k|/k$, which we have already shown is positive a.s. Now $\langle Z_j \rangle$ is a stationary nonnegative sequence if we condition on $\mathscr{A}_\Diamond$, so the limit in (17.2) equals the mean of Z_0 by the ergodic theorem. But Z_0 is the time it takes for the random walk to make a step along $T^\Diamond$, in other words, the time it takes to return to $\mathscr{A}'_\Diamond$. This has finite expectation by the Kac lemma (Exercise 2.30).

The fact that μ-a.e. tree has infinitely many ends is a consequence of the transience, and the stronger fact that the branching number is > 1 follows from Proposition 13.3. The last sentence is a consequence of the ergodic theorem. ◄

Where do we get SRW-stationary probability measures? In the next section, we will give an explicit stationary measure that is relevant to Galton–Watson trees. For the rest of this section, however, we describe a general way of finding such measures in the context of Cayley graphs.

Sometimes, it is easier to find a measure that is stationary for delayed simple random walk, rather than for simple random walk. Here, we define ***delayed simple random walk*** on a graph with maximum degree at most D, abbreviated DSRW, to have the transition probabilities

$$p(x, y) := \begin{cases} 1/D & \text{if } x \sim y, \\ 1 - \deg x / D & \text{if } x = y, \\ 0 & \text{otherwise.} \end{cases}$$

(The choice of D will be left implicit but will be clear from context.) Thus, any uniform measure on the vertices is an (infinite) stationary measure for delayed simple random walk on a single infinite graph. But how do we find an invariant probability measure on the space of rooted trees? Before we answer that question, we show how to pass from a DSRW-stationary measure to a SRW-stationary measure:

Lemma 17.5. *If μ is a* DSRW-*stationary probability measure on $\mathcal{T}$, then the degree-biased measure μ' is a* SRW-*stationary probability measure on $\mathcal{T}$; more precisely, μ' is defined as the measure $\mu' \ll \mu$ with*

$$\frac{d\mu'}{d\mu}(T, o) := C^{-1} \deg_T(o),$$

where $C := \int \deg_T(o) \, d\mu(T, o)$. If μ is ergodic, then so is μ'.

Proof. We are given that for all events A,

$$(\text{DSRW} \times \mu)[[T, X_0] \in A] = (\text{DSRW} \times \mu)[[T, X_1] \in A].$$

If we write this out, it becomes

$$\int \mathbf{1}_A([T, X_0]) \, d\text{DSRW} \times \mu(T, X_0) = \int \left(\left(1 - \frac{\deg_T X_0}{D}\right) \mathbf{1}_A([T, X_0]) \right.$$
$$\left. + \frac{1}{D} \left| \{x \sim X_0 \,;\, [T, x] \in A\} \right| \right) d\text{DSRW} \times \mu(T, X_0).$$

Canceling what can be obviously canceled, we get

$$\int (\deg_T X_0) \mathbf{1}_A([T, X_0]) \, d\text{DSRW} \times \mu(T, X_0) = \int \left| \{x \sim X_0 \,;\, [T, x] \in A\} \right| d\text{DSRW} \times \mu(T, X_0).$$

This is the same as

$$(\text{SRW} \times \mu')[[T, X_0] \in A] = (\text{SRW} \times \mu')[[T, X_1] \in A].$$

The ergodicity claim is immediate from the definitions. ◀

▷ **Exercise 17.3.**

Show that if μ is an ergodic, DSRW-stationary probability measure on the space of rooted trees $\mathcal{T}$ such that μ-a.e. tree is infinite, then the rate of escape of simple random walk (not delayed) is SRW $\times$ μ-a.s.

$$\lim_{n \to \infty} \frac{|X_n|}{n} = 1 - \frac{2}{\int \deg_T(o) \, d\mu(T, o)}.$$

Now where do DSRW-stationary probability measures on $[\mathcal{T}]$ come from? One place is from invariant probability measures on forests in Cayley graphs. Recall that an edge $[x, y]$ is present in a Cayley graph iff there is a generator or its inverse s such that $xs = y$. Thus, for any γ in the group, Γ, multiplication by γ on the left is an automorphism of G. Given a percolation on a Cayley graph $G = (V, E)$, that is, a Borel probability measure $\mathbf{P}$ on the subsets of E, write $\omega \subseteq E$ for the random subset given by the percolation. The action of multiplication by γ induces a map

$$\gamma \omega := \{[\gamma x, \gamma y] \,;\, [x, y] \in \omega\}.$$

Thus, γ acts on $\mathbf{P}$ by pushing it forward to $\mathbf{P} \circ \gamma^{-1}$; we call $\mathbf{P}$ *translation-invariant* or Γ-*invariant* if $\gamma \mathbf{P} = \mathbf{P}$ for all $\gamma \in V$. If S denotes the generating set for G, then clearly $\mathbf{P}$ is translation-invariant iff $s \mathbf{P} = \mathbf{P}$ for all $s \in S$. Call the percolation a *random forest* if each component is a tree. For example, the uniform spanning forests FUSF and WUSF are translation-invariant random forests by Exercise 10.2, as are the minimal spanning forests, FMSF and WMSF, by Exercise 11.8.

Example 17.6. Here's a trivial example. Let G be the usual Cayley graph of $\mathbb{Z}$. Consider the three possible spanning forests that have only trees with three vertices. If these three are equally likely, then we get a translation-invariant random forest of G. The measure on the component of 0 is easily seen to be DSRW-stationary but not SRW-stationary.

Let μ denote the law of the component of the identity, o, of a translation-invariant random forest $\mathbf{P}$, rooted at o. Then μ is stationary for a random walk $\langle X_n \rangle$ starting at $X_0 = o$ iff $\mathbf{E}[X_1 \mu] = \mu$ (where $X_1^{-1} \omega$ is rooted at o). The following is essentially due to Häggström (1997).

Theorem 17.7. (Invariance and Stationarity) *If μ is the law of the component of the identity in a translation-invariant random forest on a Cayley graph, then μ is stationary for delayed simple random walk. In fact, μ is globally reversible (defined below).*

Proof. Let $\langle X_k \rangle$ be DSRW on the component of the identity. Let T_x denote the component of x, so $\mu(A) = \mathbf{P}\big[[T_o, o] \in A\big]$. **Global reversibility** is the property that for all Borel $A, B \subseteq [\mathscr{T}]$,

$$\mathbf{P}\big[[T_o, o] \in A, \ [T_{X_1}, X_1] \in B\big] = \mathbf{P}\big[[T_o, o] \in B, \ [T_{X_1}, X_1] \in A\big].$$

We will show more generally (that is, beyond looking at just components) that for Borel $A, B \subseteq \{0, 1\}^{\mathsf{E}}$, we have

$$\mathbf{P}[A, \ X_1^{-1} B] = \mathbf{P}[B, \ X_1^{-1} A].$$

We may assume that S is closed under inverses, so that $|S|$ is the degree of the Cayley graph. Now we may write the event on the left above as a disjoint union

$$A \cap X_1^{-1} B = \big(A \cap B \cap [X_1 = o]\big) \cup \bigcup_{s \in S} \big(A \cap s^{-1} B \cap [X_1 = s]\big).$$

The first union is unchanged when we switch A and B, so it suffices to show that the probability of the second union is unchanged under switching. Now

$$\mathbf{P}\Bigg[\bigcup_{s \in S} \big(A \cap s^{-1} B \cap [X_1 = s]\big)\Bigg] = \sum_{s \in S} \mathbf{P}\big[A, \ s^{-1}B, \ [o, s] \in \omega\big]/|S|.$$

By translation invariance of $\mathbf{P}$, this equals

$$\sum_{s \in S} \mathbf{P}\big[s\big(A \cap s^{-1} B \cap [[o, s] \in \omega]\big)\big]/|S| = \sum_{s \in S} \mathbf{P}\big[sA, \ B, \ [o, s] \in s\omega\big]/|S|$$

$$= \sum_{s \in S} \mathbf{P}\big[sA, \ B, \ [s^{-1}, o] \in \omega\big]/|S|$$

$$= \sum_{s \in S} \mathbf{P}\big[B, \ s^{-1}A, [o, s] \in \omega\big]/|S|,$$

since inversion is a permutation of S. But this amounts to switching A and B, as desired. ◀

The following corollary generalizes results of Häggström (1997). (Part of this was proved in Corollary 8.20.)

Corollary 17.8. *If μ is the component law of a translation-invariant random forest on a Cayley graph, then simple random walk on the infinite trees with at least three ends has positive speed a.s. Hence, these trees have $\mathrm{br} > 1$. In particular, there are a.s. no trees with a finite number, at least three, of ends.*

Proof. We have seen that $[\mu]$ is DSRW-stationary on $[\mathscr{T}]$. If $A \subseteq [\mathscr{T}]$ denotes the set of tree classes with at least three ends, then A is invariant under the random walk, so if $\mu(A) > 0$, then $[\mu]$ conditioned on A is also DSRW-stationary. From Lemma 17.5, we also get a SRW-stationary probability measure on A, to which we may apply Theorem 17.4. This proves the claims for the component of o; but this passes to all the components, because if some vertex has positive probability of belonging to a component with a given property, then so does o by translation invariance. ◄

Since a tree with branching number > 1 also has exponential growth, it follows that if G is a Cayley graph of subexponential growth, then every tree in a translation-invariant random forest has at most two ends a.s. Actually, this holds for all amenable Cayley graphs:

Corollary 17.9. *If G is an amenable Cayley graph, then every tree in a translation-invariant random spanning forest all of whose trees are infinite has at most two ends a.s.*

Proof. By Exercise 10.6, the expected degree of every vertex in the forest is two. Therefore, the speed of simple random walk is 0 a.s. by Exercise 17.3, whence the number of ends cannot be at least three with positive probability. ◄

In fact, this result holds still more generally:

Theorem 17.10. *If G is an amenable Cayley graph, then every component in a translation-invariant percolation has at most two ends a.s.*

This is due to Burton and Keane (1989) in the case where $G = \mathbb{Z}^d$. It was proved in Exercise 7.24.

▷ **Exercise 17.4.**
Use Corollary 17.8 to prove that for a translation-invariant random forest on a Cayley graph, there are a.s. no isolated ends in trees with an infinite number of ends. (An end is isolated if there is a ray $\langle x_0, x_1, \ldots \rangle$ in its equivalence class such that no other ray begins $\langle x_0, x_1 \rangle$.)

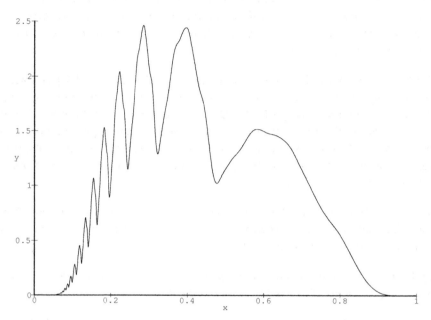

Figure 17.1. The apparent density of the conductance for $p_1 = p_2 = 1/2$.

17.3 Speed on Galton–Watson Trees

When we run random walks on Galton–Watson trees T, the asymptotic properties of the walks reveal information about the structure of T beyond the growth rate and branching number. This is the theme of the rest of the chapter. ***Assume that $p_0 = 0$ and $p_k < 1$ for all k for the rest of this chapter, except when stated otherwise.***

We know by Theorem 3.5 and Corollary 5.10 that simple random walk on a Galton–Watson tree T is almost surely transient. Equivalently, by Theorem 2.3, the effective conductance of T from the root to infinity is a.s. positive when each edge has unit conductance. The effective conductance makes transience quantitative. Since T is random, what can we say about the distribution of its effective conductance? Figure 17.1 shows the apparent density of the effective conductance when an individual has one or two children with equal probability; we will study this interesting graph further in Section 17.10.

Transience means only that the distance of a random walker from the root of T tends to infinity a.s. Is the rate of escape positive? Can we calculate the rate? According to (13.5), it would suffice to know the proportion of time the walk spends at vertices of degree $k + 1$ for each k. As we demonstrate in Theorem 17.13, this proportion turns out to be simply p_k, so that the speed is a.s.

$$l := \sum_{k=1}^{\infty} p_k \frac{k - 1}{k + 1} \,. \tag{17.3}$$

In particular, the speed is positive.

Here's something interesting about this formula: The function $s \mapsto (s - 1)/(s + 1)$ is strictly concave, whence by Jensen's inequality, $l < (m - 1)/(m + 1)$. The latter is the speed on the regular tree of the same growth rate when m is an integer. Thus, the randomness inherent in the tree slows down the simple random walk compared to a regular tree.

Our plan to prove (17.3) is to use Theorem 17.4 rather than (13.5) directly. That is, we will identify a stationary ergodic measure for simple random walk on the space of trees that is sufficiently similar to **GW** that results for it apply also to **GW**. To construct a stationary Markov chain on the space of trees, we will use isomorphism classes of rooted trees, as discussed in the preceding section. The family tree of a Galton–Watson process is rooted at the initial individual. However, Galton–Watson trees are naturally labeled, and we will need to refer to various vertices within them, which would be impossible if we were to use only classes of isomorphic rooted trees. Thus, we will play it both ways: when we need stationarity, we will use isomorphism classes, but otherwise not. To have it both ways, we will use *labeled* trees and introduce a new σ-field, $[\mathscr{F}]$, which contains only the events that are invariant under rooted isomorphisms. Restricting to $[\mathscr{F}]$ will give a stationary measure.

As we noted earlier, the root of a Galton–Watson tree is different from the other vertices, since it has stochastically one fewer neighbor. To remedy this defect, consider *augmented Galton–Watson measure*, **AGW**. This measure is defined just like **GW**, except that the number of children of the root (only) has the law of $Z_1 + 1$; that is, the root has $k + 1$ children with probability p_k, and these children all have independent, standard Galton–Watson descendant trees.

Theorem 17.11. (AGW is SRW-Stationary) *The Markov chain with transition kernel* $\mathbf{p}_{\mathsf{SRW}}$ *and initial distribution* **AGW** *is stationary on the isomorphism-invariant events* $[\mathscr{F}]$.

Proof. The measure **AGW** is stationary, since when the walk takes a step to a neighbor of the root, it goes to a vertex with one neighbor (where it just came from) plus a **GW**-tree; and the neighbor it came from also has attached another independent **GW**-tree. This is the same as **AGW**. ◄

The proof we have just given is not really rigorous, though it is convincing. A rigorous proof is not very enlightening, except for showing how to prove such things with proper notation and technique. Such a proof may be found in Lyons, Pemantle, and Peres (1995b). It also proves more: First, it is clear that the chain is locally reversible, which means that it is reversible on every communicating class. This, however, does not imply (global) reversibility. We will have no need for global reversibility, only for local reversibility (which is, indeed, a consequence of global reversibility). See Theorem 17.7 and Exercise 17.16 for the definitions.

It now follows from Theorem 17.4 that simple random walk has positive speed with positive **AGW**-probability; we want to establish the value of the speed, so we will show ergodicity. In Proposition 7.3, we showed ergodicity for Bernoulli percolation without much difficulty. However, showing ergodicity for measure-preserving systems is, in general, not an easy thing. In the present context, one might hope that, since **AGW** is built on so much independence, it would guarantee that $\mathsf{SRW} \times \mathbf{AGW}$ is ergodic. This is true (Theorem 4.6 of Aldous and Lyons (2007) has a general principle), but it would take us longer to prove this general property than it will to prove ergodicity in this particular case. Besides, we will obtain some interesting facts about $\mathsf{SRW} \times \mathbf{AGW}$ that will be crucial in later sections.

We will find it convenient to work with random walks indexed by $\mathbb{Z}$ rather than by $\mathbb{N}$. We will denote such a bi-infinite path $\ldots, x_{-1}, x_0, x_1, \ldots$ by $\ddot{x}$. Similarly, a path of vertices

$x_0, x_1, \ldots$ in T will be denoted $\vec{x}$ and a path $\ldots, x_{-1}, x_0$ will be denoted $\overleftarrow{x}$. We will regard a ray as either a path $\vec{x}$ or $\overleftarrow{x}$ that starts at the root and doesn't backtrack. The path of simple random walk has the property that a.s., it **converges** to a ray in the sense that there is a unique ray with which it shares infinitely many vertices. If a path $\vec{x}$ converges to a ray ξ in this sense, then we will write $x_{+\infty} = \xi$. Similarly for a limit $x_{-\infty}$ of a path $\overleftarrow{x}$. The space of convergent paths $\vec{x}$ in T will be denoted $\vec{T}$; likewise, $\overleftarrow{T}$ denotes the convergent paths $\overleftarrow{x}$ and $\overleftrightarrow{T}$ denotes the paths $\overleftrightarrow{x}$ for which both $\vec{x}$ and $\overleftarrow{x}$ converge and have distinct limits.

Define the path space (actually, path bundle over the space of trees)

$$\mathsf{PathsInTrees} := \big\{ (\overleftrightarrow{x}, T) \, ; \; \overleftrightarrow{x} \in \overleftrightarrow{T} \big\}.$$

The rooted tree corresponding to $(\overleftrightarrow{x}, T)$ is (T, x_0). Let S be the shift map:

$$(S\overleftrightarrow{x})_n := x_{n+1},$$

$$S(\overleftrightarrow{x}, T) := (S\overleftrightarrow{x}, T).$$

Extend simple random walk to all integer times by letting $\overleftarrow{x}$ be an independent copy of $\vec{x}$. We will reuse the notation SRW $\times$ **AGW** to denote the measure on PathsInTrees associated to this Markov chain, which is stationary when restricted to the isomorphism classes of $(\overleftrightarrow{x}, T)$. We will extend $[\mathscr{F}]$ to denote the events in PathsInTrees that are invariant under isomorphisms.

▷ **Exercise 17.5.**
Show that this chain on PathsInTrees is indeed stationary.

When a random walk traverses an edge for the first and last time simultaneously, we say it **regenerates** since it will now remain in a previously unexplored tree. Thus, we define the set of **regeneration points**

$$\mathsf{Regen} := \big\{ (\overleftrightarrow{x}, T) \in \mathsf{PathsInTrees} \, ; \; \forall n < 0 \; x_n \neq x_0, \; \forall n > 0 \; x_n \neq x_{-1} \big\}.$$

Note that $\mathsf{Regen} \in [\mathscr{F}]$.

Proposition 17.12. (Infinitely Many Regeneration Points) *For* SRW $\times$ **AGW**-*a.e.* $(\overleftrightarrow{x}, T)$, *there are infinitely many $n \geq 0$ for which $S^n(\overleftrightarrow{x}, T) \in \mathsf{Regen}$.*

Proof. Define the set of **fresh points**

$$\mathsf{Fresh} := \big\{ (\overleftrightarrow{x}, T) \in \mathsf{PathsInTrees} \, ; \; \forall n < 0 \; x_n \neq x_0 \big\}.$$

Note that by a.s. transience of simple random walk and the fact that independent simple random walks on a transient tree a.s. converge to distinct ends (Exercise 2.49), there are a.s. infinitely many $n \geq 0$ such that $S^n(\overleftrightarrow{x}, T) \in \mathsf{Fresh}$. Let $\mathscr{F}_n$ be the σ-field generated by $\langle \ldots, x_{-1}, x_0, \ldots, x_n \rangle$. (Since the tree is labeled, $\mathscr{F}_n$ tells us part of the tree but only a small part.) Then by a.s. transience of **GW** trees, $\alpha := (\mathrm{SRW} \times \mathbf{AGW})(\mathsf{Regen} \mid \mathsf{Fresh}) > 0$ and, in fact, $(\mathrm{SRW} \times \mathbf{AGW})(\mathsf{Regen} \mid \mathsf{Fresh}, \mathscr{F}_{-1}) = \alpha$. Fix k_0. Let R be the event that there is at least one $k \geq k_0$ for which $S^k(\overleftrightarrow{x}, T) \in \mathsf{Regen}$. Then, for $n \geq k_0$, the conditional probability $(\text{SRW} \times \mathbf{AGW})(R \mid \mathscr{F}_n)$ is *at least* the conditional probability that the walk comes to a fresh vertex after time n, which is 1, times the conditional probability that at the first such time, the walk regenerates, which is a constant, α. On the other hand, R belongs to the σ-field generated by $\bigcup_n \mathscr{F}_n$, whence the martingale convergence theorem tells us that $(\text{SRW} \times \mathbf{AGW})(R \mid \mathscr{F}_n) \to \mathbf{1}_R$ a.s. Putting together these two facts about the limit, we conclude that the limit must be **1**. That is, R occurs a.s., which completes the proof since k_0 was arbitrary. ◀

Given a rooted tree T and a vertex x in T, recall that the descendant tree of x, the subtree T^x rooted at x, denotes the subgraph of T formed from those edges and vertices that become disconnected from the root of T when x is removed. The sequence of fresh trees $[T \setminus T^{x_{-1}}]$ seen at regeneration points $(\overset{\leftrightarrow}{x}, T)$ is clearly stationary but not i.i.d. However, the part of a tree between regeneration points, together with the path taken through this part of the tree, is independent of the rest of the tree and of the rest of the walk. We call this part a **slab**. To define this notion precisely, we use the return time n_{Regen}, where

$$n_{\text{Regen}}(\overset{\leftrightarrow}{x}, T) := \inf\{n > 0 ; \ S^n(\overset{\leftrightarrow}{x}, T) \in \text{Regen}\}.$$

For $(\overset{\leftrightarrow}{x}, T) \in \text{Regen}$, the associated **slab** (including the path taken through the slab) is

$$\text{Slab}(\overset{\leftrightarrow}{x}, T) := \left[(\langle x_0, x_1, \ldots, x_{n-1}\rangle, T \setminus (T^{x_{-1}} \cup T^{x_n}))\right],$$

where $n := n_{\text{Regen}}(\overset{\leftrightarrow}{x}, T)$ and $[\,\cdot\,]$ again indicates isomorphism class. Let $S_{\text{Regen}} := S^{n_{\text{Regen}}}$ when $(\overset{\leftrightarrow}{x}, T) \in \text{Regen}$. Then the random variables $\text{Slab}(S_{\text{Regen}}^k(\overset{\leftrightarrow}{x}, T))$ are i.i.d. Since the slabs generate the whole tree and the walk through the tree (except for the location of the root), it is easy to see that the system $(\text{PathsInTrees}, \text{SRW} \times \mathbf{AGW}, S)$ is mixing on $[\mathscr{F}]$, hence ergodic.

Theorem 17.13. *The speed (rate of escape) of simple random walk is* $\text{SRW} \times \mathbf{AGW}$*-a.s.*

$$\lim_{n \to \infty} \frac{|x_n|}{n} = \mathbf{E}\left[\frac{Z_1 - 1}{Z_1 + 1}\right]. \tag{17.4}$$

This is immediate now.

▷ **Exercise 17.6.**
Show that the same formula (17.4) holds for the speed of simple random walk on **GW**-a.e. tree.

This accomplishes our main goal for the speed. What about the case when $p_0 > 0$? As usual, let q be the probability of extinction of the Galton–Watson process. Let $\neg\text{Ext}$ be the event of nonextinction of an **AGW** tree and $\mathbf{AGW}_{\neg\text{Ext}}$ be **AGW** conditioned on $\neg\text{Ext}$. Since **AGW** is still stationary and $\neg\text{Ext}$ is an invariant event for the Markov chain, $\mathbf{AGW}_{\neg\text{Ext}}$ is SRW-stationary. The $\mathbf{AGW}_{\neg\text{Ext}}$-distribution of the degree of the root is

$$\mathbf{AGW}[\deg x_0 = k + 1 \mid \neg\text{Ext}] = \frac{\mathbf{AGW}[\neg\text{Ext} \mid \deg x_0 = k + 1]}{\mathbf{AGW}(\neg\text{Ext})} p_k$$
$$= p_k \frac{1 - q^{k+1}}{1 - q^2}. \tag{17.5}$$

▷ **Exercise 17.7.**
Prove this formula.

The proof of Theorem 17.13 on speed is valid when one conditions on nonextinction in the appropriate places. It gives the following formula:

$$\lim_{n \to \infty} \frac{|x_n|}{n} = \mathbf{E}\left[\frac{Z_1 - 1}{Z_1 + 1} \,\Big|\, \neg\mathsf{Ext} \right] = \sum_{k \geq 0} \frac{k - 1}{k + 1} p_k \frac{1 - q^{k+1}}{1 - q^2} \qquad \text{a.s.}$$

For example, the speed when $p_1 = p_2 = 1/2$ is $1/6$, whereas for the offspring distribution of the same unconditional mean $p_0 = p_3 = 1/2$, the speed given nonextinction is only $(7 - 3\sqrt{5})/8 = 0.036^+$: the time spent at leaves is a serious drag.

17.4 Harmonic Measure: The Goal

We've now found the rate of escape of simple random walk on Galton–Watson trees. What about its "direction" of escape? With "direction" interpreted as harmonic measure, this will be studied in the rest of the chapter. That is, since the random walk on a Galton–Watson tree T is transient, it converges to a ray of T a.s. The law of that ray is called **harmonic measure**, denoted HARM(T), which we identify with a unit flow on T from its root to infinity.

Of course, if the offspring distribution is concentrated on a single integer, then the direction is uniform. But we are assuming that this is not the case, that is, the tree is **nondegenerate**. We still assume that $p_0 = 0$ unless otherwise specified, since if we condition on nonextinction, the random walk will always leave any finite descendant subtree, and therefore the harmonic measure lives on the subtree of vertices with infinite lines of descent. (Recall also from Proposition 5.28 that this subtree is still a Galton–Watson tree.) We will see that the random irregularities that recur in a nondegenerate Galton–Watson tree T direct or confine the random walk to an exponentially smaller subtree of T. One aspect of this, and the key tool in its proof, is the dimension of harmonic measure. Now, since br $T = m$ a.s., the boundary ∂T has Hausdorff dimension $\log m$ a.s.

▷ **Exercise 17.8.**
Show that the Hausdorff dimension of harmonic measure is a.s. constant.

Our main goal now is prove that the Hausdorff dimension of harmonic measure is strictly less than that of the full boundary:

Theorem 17.14. (Dimension Drop of Harmonic Measure) *The Hausdorff dimension of harmonic measure on the boundary of a nondegenerate Galton–Watson tree T is a.s. a constant $d < \log m = \dim(\partial T)$, that is, there is a Borel subset of ∂T of full harmonic measure and dimension d.*

This result is established in a sharper form in Theorem 17.27.

With some further work, Theorem 17.14 will yield the following restriction on the range of random walk.

Theorem 17.15. (Confinement of Random Walk) *Fix a nondegenerate offspring distribution with mean m. Let d be as in Theorem 17.14. For any $\epsilon > 0$ and for almost every Galton–Watson tree T, there is a rooted subtree $T(\epsilon)$ of T having growth*

$$\lim_{n\to\infty} |T(\epsilon)_n|^{\frac{1}{n}} = e^d < m$$

such that with probability $1 - \epsilon$, the sample path of simple random walk on T is contained in $T(\epsilon)$. (Here, $|T(\epsilon)_n|$ is the cardinality of the nth level of $T(\epsilon)$.)

See Theorem 17.30 for a restatement and proof.

This theorem gives a partial explanation for the "low" speed of simple random walk on a Galton–Watson tree: the walk is confined to a *much* smaller subtree.

The setting and results of Section 17.3 will be fundamental to our work here. Certain Markov chains on the space of trees (inspired by Furstenberg (1970)) are discussed in Section 17.5 and used in Section 17.6 to compute the dimension of the limit uniform measure, extending a theorem of Hawkes (1981). A general condition for dimension drop is given in Section 17.7 and applied to harmonic measure in Section 17.8, where Theorem 17.14 is proved; its application to Theorem 17.15 is given in Section 17.9. In Section 17.10, we analyze the electrical conductance of a Galton–Watson tree using a functional equation for its distribution. This yields a numerical scheme for approximating the dimension of harmonic measure.

17.5 Flow Rules and Markov Chains on the Space of Trees

These chains are inspired by Furstenberg (1970). In the rest of this chapter, a *flow* on a tree will mean a unit flow from its root to infinity. Given a flow θ on a tree T and a vertex $x \in T$ with $\theta(x) > 0$, we write θ^x for the (conditional) flow on T^x given by

$$\theta^x(y) := \theta(y)/\theta(x) \qquad (y \in T^x).$$

The space of flows on trees can be given a natural topology just as $\mathscr{T}$ is given. We call a Borel function $\Theta: \mathscr{T} \to \{\text{flows on trees}\}$ a *(consistent) flow rule* if $\Theta(T)$ is a flow on T such that

$$x \in T, \ |x| = 1, \ \Theta(T)(x) > 0 \quad \Longrightarrow \quad \Theta(T)^x = \Theta(T^x).$$

A consistent flow rule may also be thought of as a Borel function that assigns to a k-tuple $(T^{(1)}, \ldots, T^{(k)})$ of trees a k-tuple of nonnegative numbers adding to one representing the probabilities of choosing the corresponding trees $T^{(i)}$ in $\bigvee_{i=1}^{k} T^{(i)}$, which is the tree formed by joining the roots of $T^{(i)}$ by single edges to a new vertex, the new vertex being the root of the new tree. It follows from the definition that for all $x \in T$, not only those at distance 1 from the root, $\Theta(T)(x) > 0 \Rightarrow \Theta(T)^x = \Theta(T^x)$. We will usually write Θ_T for $\Theta(T)$.

We will always assume without mention that our flow rules Θ are *equivariant* (as they will be in our particular examples), which means that for any rooted isomorphism $\varphi: T \to \varphi(T)$, we have $\varphi(\Theta(T)) = \Theta(\varphi(T))$. This is important to get events that are isomorphism invariant, and so to apply ergodic theory.

We have already encountered two flow rules: The principal object of interest in the rest of this chapter, harmonic measure, comes from a flow rule, HARM. Another natural example is harmonic measure, HARM$^\lambda$, for homesick random walk RW$_\lambda$; this was studied by Lyons, Pemantle, and Peres (1996a). Visibility measure, encountered in Section 15.4, gives a flow rule, VIS. A final example, UNIF, will be studied in Section 17.6. It is easily verified that all these are flow rules.

Proposition 17.16. *If Θ and Θ' are two flow rules such that for **GW**-a.e. tree T and all vertices $|x| = 1$, $\Theta_T(x) + \Theta'_T(x) > 0$, then* **GW**$(\Theta_T = \Theta'_T) \in \{0, 1\}$.

Proof. By the hypothesis, if $\Theta_T = \Theta'_T$ and $|x| = 1$, then $\Theta_{T^x} = \Theta'_{T^x}$. Thus, the result follows from Proposition 5.6.   ◀

Given a flow rule Θ, there is an associated Markov chain on the space of trees given by the transition probabilities

$$\mathbf{p}_\Theta(T, \{T^x\}) := \Theta_T(x)$$

for $T \in \mathscr{T}$, $x \in T$, and $|x| = 1$. We say that a (possibly infinite) measure μ on the space of trees is Θ-***stationary*** if it is $\mathbf{p}_\Theta$-stationary, or, in other words, for any Borel set $A \subseteq [\mathscr{F}]$ of trees,

$$\mu(A) = \int \mathbf{p}_\Theta(T, A)\, d\mu(T) = \int \sum_{\substack{|x|=1, \\ T^x \in A}} \Theta_T(x)\, d\mu(T)\,.$$

If we denote the vertices along a ray ξ by $\xi_0, \xi_1, \ldots$, then the path of such a Markov chain is a sequence $\langle T^{\xi_n} \rangle_{n=0}^\infty$ for some tree T and some ray $\xi \in \partial T$. Clearly, we may identify the space of such paths with the ray bundle

$$\mathsf{RaysInTrees} := \big\{(\xi, T)\,;\ \xi \in \partial T\big\}\,.$$

For the corresponding path measure on RaysInTrees, write $\Theta \times \mu$ for $\mathbf{p}_\Theta \times \mu$. Likewise, we say Θ-***invariant*** for $\mathbf{p}_\Theta$-*invariant*.

In this setting, Corollary 17.3 says that if μ is a Θ-invariant probability measure, then $\Theta \times \mu$ is ergodic (for the shift map) iff every Θ-invariant (and isomorphism-invariant) set of trees has μ-measure 0 or 1. Moreover, even without ergodicity, Proposition 17.2 says that shift-invariant ([$\mathscr{F}$]-measurable) functions on RaysInTrees correspond to Θ-invariant functions; in particular, they depend (a.s.) only on their second coordinate.

We call two measures ***equivalent*** if they are mutually absolutely continuous.

Proposition 17.17. *Let Θ be a flow rule such that for* **GW**-*a.e. tree T and for all $|x| = 1$, $\Theta_T(x) > 0$. Then the Markov chain with transition kernel $\mathbf{p}_\Theta$ and initial distribution* **GW** *is ergodic, though not necessarily stationary. Hence, if a Θ-stationary measure μ exists that is absolutely continuous with respect to* **GW**, *then μ is equivalent to* **GW** *and the associated stationary Markov chain is ergodic.*

Proof. Let A be a Borel set of trees that is Θ-invariant. It follows from our assumption that for **GW**-a.e. T,

$$T \in A \iff T^x \in A \quad \text{for every } |x| = 1\,.$$

Thus, the first claim follows from Proposition 5.6.

Now let $\mu \ll \mathbf{GW}$ be as in the second claim. If A is a Borel set of trees that is Θ-invariant, then by what we have just shown, $\mu(A) = 0$ or $\mu(A^c) = 0$. That is, μ is ergodic. Finally, if A is any set with $\mu(A) = 0$, then by stationarity, A is Θ-invariant. The preceding paragraph then shows that $\mathbf{GW}(A) \in \{0, 1\}$. Since $\mu(A^c) \neq 0$, it follows that $\mathbf{GW}(A) = 0$. This shows that $\mathbf{GW} \ll \mu$.   ◀

We do not know whether the first hypothesis of Proposition 17.17 is necessary for the final conclusion:

Question 17.18. If a flow rule has a stationary measure equivalent to $\mathbf{GW}$, must the associated Markov chain be ergodic?

Given a Θ-stationary probability measure μ on the space of trees, we define the ***entropy*** of the associated stationary Markov chain as

$$\mathsf{Ent}_\Theta(\mu) := \int \sum_{|x|=1} \mathbf{p}_\Theta(T, T^x) \log \frac{1}{\mathbf{p}_\Theta(T, T^x)} \, d\mu(T)$$

$$= \int \sum_{|x|=1} \Theta_T(x) \log \frac{1}{\Theta_T(x)} \, d\mu(T)$$

$$= \iint \log \frac{1}{\Theta_T(\xi_1)} \, d\Theta_T(\xi) \, d\mu(T)$$

$$= \int \log \frac{1}{\Theta_T(\xi_1)} \, d(\Theta \times \mu)(\xi, T) \,.$$

Write

$$g_\Theta(\xi, T) := \log \frac{1}{\Theta_T(\xi_1)} \,;$$

this function is $[\mathscr{F}]$-measurable by equivariance of Θ. Let S be the shift on RaysInTrees. The ergodic theorem tells us that the Hölder exponent (Section 15.4) of Θ_T is actually a limit a.s.:

$$\mathsf{H\ddot{o}}(\Theta_T)(\xi) = \lim_{n\to\infty} \frac{1}{n} \log \frac{1}{\Theta_T(\xi_n)} = \lim_{n\to\infty} \frac{1}{n} \sum_{k=0}^{n-1} \log \frac{\Theta_T(\xi_k)}{\Theta_T(\xi_{k+1})}$$

$$= \lim_{n\to\infty} \frac{1}{n} \sum_{k=0}^{n-1} \log \frac{1}{\Theta(T)^{\xi_k}(\xi_{k+1})} = \lim_{n\to\infty} \frac{1}{n} \sum_{k=0}^{n-1} S^k g_\Theta(\xi, T)$$

exists $\Theta \times \mu$-a.s., and it satisfies

$$\int \mathsf{H\ddot{o}}(\Theta_T)(\xi) \, d(\Theta \times \mu)(\xi, T) = \mathsf{Ent}_\Theta(\mu) \,.$$

If the Markov chain is ergodic, then

$$\mathsf{H\ddot{o}}(\Theta_T)(\xi) = \mathsf{Ent}_\Theta(\mu) \quad \Theta \times \mu\text{-a.s.} \tag{17.6}$$

This is our principal tool for calculating Hausdorff dimension. Note that even if the Markov chain is not ergodic, the Hölder exponent $\mathsf{H\ddot{o}}(\Theta_T)(\xi)$ is constant Θ_T-a.s. for μ-a.e. T: since $(\xi, T) \mapsto \mathsf{H\ddot{o}}(\Theta_T)(\xi)$ is a shift-invariant $[\mathscr{F}]$-measurable function, it depends only on T (a.s.) (Proposition 17.2).

17.6 The Hölder Exponent of Limit Uniform Measure

For the rest of the chapter, assume that $m < \infty$. By the Seneta–Heyde theorem of Section 5.1, if Z_n and Z'_n are two i.i.d. Galton–Watson processes without extinction, then $\lim_{n\to\infty} Z_n/Z'_n$ exists a.s. Thus, if we fix two vertices x and y at the same level k in a Galton–Watson tree T, then given T up to level k, we have that

$$\lim_{n\to\infty} \frac{|T^x \cap T_n|}{|T^y \cap T_n|}$$

exists a.s., where T_n denotes the vertices of the nth generation of T. This allows us to define (a.s.) a probability measure UNIF_T on the boundary of a Galton–Watson tree by

$$\mathrm{UNIF}_T(x) := \lim_{n\to\infty} \frac{|T^x \cap T_n|}{Z_n} .$$

(We are identifying the measure on the boundary with a unit flow on the tree.) We call this measure ***limit uniform*** since, before the limit is taken, it corresponds to the flow from o to T_n that is uniform on T_n. It is clear that UNIF is a flow rule. Figure 5.1 was drawn by considering the uniform measure on generation 9 and inducing the masses on the preceding generations. Figures 17.2 and 17.3 show this same tree drawn using the uniform measure on generations 14 and 19, respectively.

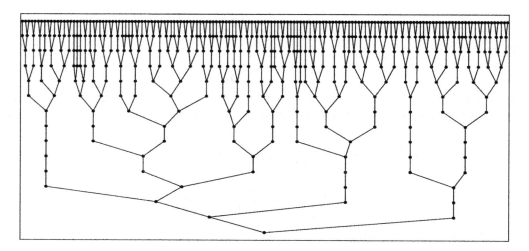

Figure 17.2. Generations 0 to 14 of a typical Galton–Watson tree for $f(s) = (s + s^2)/2$.

We may write limit uniform measure another way: Let c_n be constants with $c_{n+1}/c_n \to m$ such that

$$\widetilde{W}(T) := \lim_{n\to\infty} Z_n/c_n$$

exists and is finite and nonzero a.s.; these constants are provided by the Seneta–Heyde theorem. Then we have

$$\mathrm{UNIF}_T(x) = \frac{\widetilde{W}(T^x)}{m^{|x|}\widetilde{W}(T)} . \tag{17.7}$$

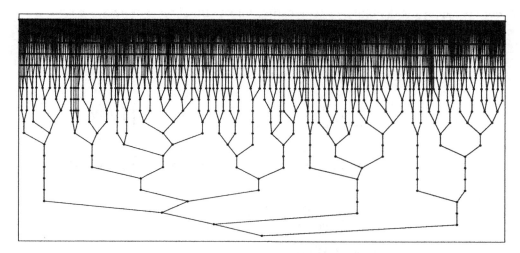

Figure 17.3. Generations 0 to 19 of a typical Galton–Watson tree for $f(s) = (s + s^2)/2$.

Note that

$$\widetilde{W}(T) = \frac{1}{m} \sum_{|x|=1} \widetilde{W}(T^x) \,. \tag{17.8}$$

According to the Kesten–Stigum theorem, when $\mathbf{E}[Z_1 \log Z_1] < \infty$, we may take c_n to be m^n, and so W may be used in place of $\widetilde{W}$ in (17.7) and (17.8). A theorem of Athreya (1971) gives that

$$\int \widetilde{W}(T) \, d\mathbf{GW}(T) < \infty \iff \mathbf{E}[Z_1 \log Z_1] < \infty \,. \tag{17.9}$$

As we mentioned in Section 17.4, $\dim(\partial T) = \log m$ a.s. Now Hawkes (1981) showed that the Hölder exponent of UNIF_T is $\log m$ a.s. provided $\mathbf{E}[Z_1(\log Z_1)^2] < \infty$. One could anticipate Hawkes's result that UNIF_T has full Hausdorff dimension, $\log m$, since limit uniform measure "spreads out" the most possible (at least under some moment condition on Z_1). Furthermore, one might guess that no other measure that comes from a consistent flow rule can have full dimension. We will show that this is indeed true provided that the flow rule has a finite stationary measure for the associated Markov chain that is absolutely continuous with respect to $\mathbf{GW}$. We will then show that such is the case for harmonic measure of simple random walk. Incidentally, this method will allow us to give a simpler proof of Hawkes's theorem, as well as to extend its validity to the case where $\mathbf{E}[Z_1 \log Z_1] < \infty$. On the other hand, Aïdékon (2020) showed that $\dim \mathsf{UNIF}_T = 0$ when $\mathbf{E}[Z_1 \log Z_1] = \infty$.

In this section, we prove and extend the theorem of Hawkes (1981) on the Hölder exponent of limit uniform measure and study further the associated Markov chain. We begin by showing that a (possibly infinite) UNIF-stationary measure on trees is $\widetilde{W} \cdot \mathbf{GW}$; we will use this only when the measure is finite.

Proposition 17.19. *The Markov chain with transition kernel* $\mathbf{p}_{\mathrm{UNIF}}$ *and initial distribution* $\widetilde{W} \cdot \mathbf{GW}$ *is stationary and ergodic.*

Proof. We apply the definition of stationarity: for any Borel set A of trees, we have

$$\int \mathbf{p}_{\text{UNIF}}(T, A) \cdot \widetilde{W}(T) \, d\mathbf{GW}(T) = \int \sum_{\substack{|x|=1, \\ T^x \in A}} \frac{\widetilde{W}(T^x)}{m\widetilde{W}(T)} \cdot \widetilde{W}(T) \, d\mathbf{GW}(T)$$

$$= \sum_{k=1}^{\infty} p_k \frac{1}{m} \int_{T^{(1)}} \cdots \int_{T^{(k)}} \sum_{i=1}^{k} \mathbf{1}_A(T^{(i)}) \widetilde{W}(T^{(i)}) \prod_{j=1}^{k} d\mathbf{GW}(T^{(j)})$$

$$= \sum_{k=1}^{\infty} p_k \frac{1}{m} \sum_{i=1}^{k} \int_{T^{(1)}} \cdots \int_{T^{(k)}} \mathbf{1}_A(T^{(i)}) \widetilde{W}(T^{(i)}) \prod_{j=1}^{k} d\mathbf{GW}(T^{(j)})$$

$$= \sum_{k=1}^{\infty} p_k \frac{1}{m} \sum_{i=1}^{k} \int_A \widetilde{W} \, d\mathbf{GW} = \int_A \widetilde{W} \, d\mathbf{GW}$$

$$= (\widetilde{W} \cdot \mathbf{GW})(A) \,,$$

as desired. Since $\widetilde{W} > 0$ $\mathbf{GW}$-a.s., ergodicity is guaranteed by Proposition 17.17. ◀

▷ **Exercise 17.9.**
This chain is closely connected to the size-biased Galton–Watson trees of Section 12.1. Show that in case $\mathbf{E}[Z_1 \log Z_1] < \infty$, the distribution of (T, ξ) is $\widehat{\mathbf{GW}}_*$ when $T \sim W \cdot \mathbf{GW}$ and ξ is a UNIF_T-path in T.

   To calculate the Hölder exponent of limit uniform measure, we will use the following lemma of ergodic theory:

Lemma 17.20. *If S is a measure-preserving transformation on a probability space and g is finite and measurable, then $\int (g - Sg)^+ = \int (g - Sg)^-$. Therefore, if $g - Sg$ is bounded below by an integrable function, then $g - Sg$ is integrable with integral zero.*

Proof. When g is integrable, this is immediate from the fact that $\int g = \int Sg$. Now if a function $f : \mathbb{R} \to \mathbb{R}$ is an increasing contraction, then $(f(x) - f(y))^+ \le (x - y)^+$ and $(f(x) - f(y))^- \le (x - y)^-$ for $x, y \in R$. Such functions include $F_n(x) := (x \wedge n) \vee (-n)$ for $n \ge 1$. Note that $F_n \circ (Sg) = S(F_n \circ g)$. Therefore $(F_n \circ g - S(F_n \circ g))^+ \le (g - Sg)^+$ and, since $F_n = F_n \circ F_{n+1}$, $(F_n \circ g - S(F_n \circ g))^+ \le (F_{n+1} \circ g - S(F_{n+1} \circ g))^+$. Therefore, the monotone convergence theorem gives us that $\lim_{n \to \infty} \int (F_n \circ g - S(F_n \circ g))^+ = \int (g - Sg)^+$. The same holds for the negative parts. Since $F_n \circ g$ is integrable, the identity holds for g. ◀

Theorem 17.21. (Full Dimension of Limit Uniform Measure) *If $\mathbf{E}[Z_1 \log Z_1] < \infty$, then the Hölder exponent at ξ of limit uniform measure UNIF_T is equal to $\log m$ for UNIF_T-a.e. ray $\xi \in \partial T$ and $\mathbf{GW}$-a.e. tree T. In particular, $\dim \text{UNIF}_T = \log m$ for $\mathbf{GW}$-a.e. T.*

Proof. The hypothesis and Proposition 17.19 ensure that $W \cdot \mathbf{GW}$ is a stationary probability distribution. Let S be the shift on the ray bundle RaysInTrees with the invariant probability measure $\mathsf{UNIF} \times (W \cdot \mathbf{GW})$. Define $g(\xi, T) := \log W(T)$ for a Galton–Watson tree T and $\xi \in \partial T$ (so g does not depend on ξ). Then

$$(g - Sg)(\xi, T) = \log W(T) - \log W(T^{\xi_1}) = \log \frac{m W(T)}{W(T^{\xi_1})} - \log m$$

$$= \log \frac{1}{\mathsf{UNIF}_T(\xi_1)} - \log m .$$

In particular, $g - Sg \geq -\log m$, whence the lemma implies that $g - Sg$ has integral zero.

Now, for $\mathsf{UNIF} \times (W \cdot \mathbf{GW})$-a.e. (ξ, T) (hence for $\mathsf{UNIF} \times \mathbf{GW}$-a.e. (ξ, T)), we have that

$$\mathsf{H\ddot{o}}(\mathsf{UNIF}_T)(\xi) = \mathsf{Ent}_{\mathsf{UNIF}}(W \cdot \mathbf{GW})$$

by ergodicity and (17.6). By definition and the preceding calculation, this in turn is

$$\mathsf{Ent}_{\mathsf{UNIF}}(W \cdot \mathbf{GW}) = \iint \log \frac{1}{\mathsf{UNIF}_T(\xi_1)} \, d\mathsf{UNIF}_T(\xi) \, W(T) \, d\mathbf{GW}(T)$$

$$= \log m + \iint (g - Sg) \, d\mathsf{UNIF}_T(\xi) \, W(T) \, d\mathbf{GW}(T)$$

$$= \log m. \qquad \blacktriangleleft$$

17.7 Dimension Drop for Other Flow Rules

Recall that our goal is to prove that harmonic measure has less than full dimension on $\mathbf{GW}$-a.e. tree. In this section, we give a general condition on flow rules for this dimension drop to hold. Gibbs's inequality will be the tool we use to compare the dimension of measures arising from flow rules to the dimension of the whole boundary: the inequality states that

$$a_i, b_i \in [0, 1], \quad \sum a_i = \sum b_i = 1 \qquad \Longrightarrow \qquad \sum a_i \log \frac{1}{a_i} \leq \sum a_i \log \frac{1}{b_i},$$

with equality iff $a_i \equiv b_i$.

▷ **Exercise 17.10.**
Prove Gibbs's inequality.

Theorem 17.22. *If Θ is a flow rule such that $\Theta_T \neq \mathsf{UNIF}_T$ for $\mathbf{GW}$-a.e. T and there is a Θ-stationary probability measure μ absolutely continuous with respect to $\mathbf{GW}$, then for μ-a.e. T, we have $\mathsf{H\ddot{o}}(\Theta_T) < \log m$ Θ_T-a.s. and $\dim(\Theta_T) < \log m$.*

Proof. Recall that the Hölder exponent of Θ_T is constant Θ_T-a.s. for μ-a.e. T and equal to the Hausdorff dimension of Θ_T. Thus, it suffices to show that the invariant set of trees

$$A := \{T ; \dim \Theta_T = \log m\} = \{T ; \mathsf{H\ddot{o}}(\Theta_T) = \log m \ \Theta_T\text{-a.s.}\}$$

has μ-measure 0. Suppose that $\mu(A) > 0$. Recall that μ_A denotes μ conditioned on A. Now, since $\mu \ll \mathbf{GW}$, the limit uniform measure UNIF_T is defined and satisfies (17.7) for μ_A-a.e. T. Let $g(\xi, T) := \log \widetilde{W}(T)$. As in the proof of Theorem 17.21, $g - Sg \geq -\log m$. Since the entropy is the mean Hölder exponent, we have by Gibbs's inequality and Lemma 17.20,

$$\log m = \mathsf{Ent}_\Theta(\mu_A) = \int \sum_{|x|=1} \Theta_T(x) \log \frac{1}{\Theta_T(x)} \, d\mu_A(T)$$

$$< \int \sum_{|x|=1} \Theta_T(x) \log \frac{1}{\mathsf{UNIF}_T(x)} \, d\mu_A(T)$$

$$= \iint \log \frac{1}{\mathsf{UNIF}_T(\xi_1)} \, d\Theta_T(\xi) \, d\mu_A(T)$$

$$= \log m + \iint (g - Sg) \, d\Theta_T(\xi) \, d\mu_A(T)$$

$$= \log m \,.$$

This contradiction shows that $\mu(A) = 0$, as desired.   ◄

To apply this theorem to harmonic measure, we need to find a stationary measure for the harmonic flow rule with the preceding properties. A general condition for a flow rule to have such a stationary measure is unknown. But it was conjectured by Lyons, Pemantle, and Peres (1995b) that any flow rule other than limit uniform gives boundary measures of dimension less than $\log m$ **GW**-a.s.:

Conjecture 17.23. *If $\Theta \neq \mathsf{UNIF}$ is a flow rule, then $\dim(\Theta_T) < \log m$ for **GW**-a.e. T.*

17.8 Harmonic-Stationary Measure

Consider the set of "last exit points"

$$\mathsf{Exit} := \left\{ (\vec{x}, T) \in \mathsf{PathsInTrees} \,;\ x_{-1} \in x_{-\infty} \,,\ \forall n > 0 \ \ x_n \neq x_{-1} \right\}.$$

This is precisely the event that the path has just exited, for the last time, a horoball centered at $x_{-\infty}$; in other words, x_n is further from $x_{-\infty}$ for $n > 0$ than is x_{-1}. Since regeneration points are exit points, it follows from Proposition 17.12 that the set Exit has positive measure and for a.e. $(\vec{x}, T)$, there is an $n > 0$ such that $S^n(\vec{x}, T) \in \mathsf{Exit}$. (This also follows directly merely from the almost sure transience of simple random walk.) Inducing on this set will yield the measure we need to apply Theorem 17.22.

We recall some more terminology from ergodic theory for this. Let $(X, \mathscr{F}, \mu, S)$ be a probability measure-preserving system. Fix $A \in \mathscr{F}$ with $\mu(A) > 0$. Define the ***return time*** to A by $n_A(x) := \inf\{n \geq 1 \,;\ S^n x \in A\}$ for $x \in A$ and, if $n_A(x) < \infty$, the ***return map*** $S_A(x) := S^{n_A(x)}(x)$. The Poincaré recurrence theorem (Petersen (1983), p. 34) states that $n_A(x) < \infty$ for a.e. $x \in A$. Thus, S_A is defined μ_A-a.e.; $(A, \mathscr{F} \cap A, \mu_A, S_A)$ is a probability measure-preserving system (Petersen (1983), p. 39, or Exercise 2.30), called the ***induced system***. Given two measure-preserving systems, $(X_1, \mathscr{F}_1, \mu_1, S_1)$ and $(X_2, \mathscr{F}_2, \mu_2, S_2)$, the second is called a ***factor*** of the first if there is a measurable map $f: (X_1, \mathscr{F}_1) \to (X_2, \mathscr{F}_2)$ such that $\mu_2 = \mu_1 \circ f^{-1}$ and $f \circ S_1 = S_2 \circ f$ μ_1-a.e.

Theorem 17.24. *There is a unique ergodic,* HARM-*stationary probability measure* μ_{HARM} *equivalent to* **GW**.

Proof. Define $\pi_0(\ddot{x}, T) := x_0$. For $(\ddot{x}, T) \in$ Exit, let $x'_k := \pi_0(S^k_{\text{Exit}}(\ddot{x}, T))$ for $k \geq 0$. The key point is that $\langle x'_k \rangle$ is a sample from the ray generated by the Markov chain associated to $\text{HARM}_{T \setminus T^{x_{-1}}}$. Note that the Markov property of this factor $\langle (x'_k, T) ; \ k \geq 0 \rangle$ of the system induced on Exit is a consequence of the fact that HARM is a consistent flow rule.

Now, since the pushforward of $(\text{SRW} \times \mathbf{AGW})_{\text{Exit}}$ under the map $(\ddot{x}, T) \mapsto T$ is absolutely continuous with respect to **AGW**, we have that the $(\text{SRW} \times \mathbf{AGW})_{\text{Exit}}$-law of $T \setminus T^{x_{-1}}$ is absolutely continuous with respect to **GW**. Since $(\text{SRW} \times \mathbf{AGW})_{\text{Exit}}$ is S_{Exit}-invariant, its pushforward under the map $(\ddot{x}, T) \mapsto T \setminus T^{x_{-1}}$ is HARM-stationary. From Proposition 17.17, it follows that the $(\text{SRW} \times \mathbf{AGW})_{\text{Exit}}$-law of $T \setminus T^{x_{-1}}$ is equivalent to **GW**. (This can also be seen directly: for **AGW**-a.e. T, the SRW_T-probability that $(\ddot{x}, T) \in$ Exit is positive, whence the $(\text{SRW} \times \mathbf{AGW})_{\text{Exit}}$-law of T is equivalent to **AGW**. This gives that the $(\text{SRW} \times \mathbf{AGW})_{\text{Exit}}$-law of $T \setminus T^{x_{-1}}$ is equivalent to **GW**.)

Therefore, the preceding natural factor of the induced measure-preserving system

$$\left(\text{Exit}, (\text{SRW} \times \mathbf{AGW})_{\text{Exit}}, S_{\text{Exit}}\right)$$

obtained by mapping $(\ddot{x}, T) \mapsto \langle t(k) \setminus t(k)^{x'_{k-1}} \rangle_{k \geq 0}$, where $t(k) := (T, x'_k)$ and $x'_{-1} := x_{-1}$, is a HARM-stationary Markov chain on trees with a stationary measure μ_{HARM} equivalent to **GW**.

The fact that $\text{HARM} \times \mu_{\text{HARM}}$ is ergodic follows from our general result on ergodicity, Proposition 17.17. Ergodicity implies that μ_{HARM} is the unique HARM-stationary measure absolutely continuous with respect to **GW**. ◀

Since increases in distance from the root can be considered to come only at exit points, it is natural that the speed is also the probability of being at an exit point:

Proposition 17.25. *The measure of the exit set is the speed:*

$$(\text{SRW} \times \mathbf{AGW})(\text{Exit}) = \mathbf{E}\big[(Z_1 - 1)/(Z_1 + 1)\big].$$

▷ **Exercise 17.11.**
Prove Proposition 17.25.

Given disjoint trees T_1, T_2, define $[\![T_1 \bullet\!-T_2]\!]$ to be the tree rooted at $\text{root}(T_1)$ formed by joining $\text{root}(T_1)$ and $\text{root}(T_2)$ by an edge. Define $T_\Delta := [\![\Delta \bullet\!-T]\!]$, where Δ is a single vertex not in T, to be thought of as representing the past. Let $\gamma(T)$ be the probability that simple random walk started at Δ never returns to Δ:

$$\gamma(T) := \text{SRW}_{T_\Delta}[\forall n > 0 \ \ x_n \neq \Delta].$$

This is also equal to $\text{SRW}_{[\![T \bullet\!-\Delta]\!]}[\forall n > 0 \ \ x_n \neq \Delta]$. Let $\mathscr{C}(T)$ denote the effective conductance of T from its root to infinity when each edge has unit conductance. Clearly,

$$\gamma(T) = \frac{\mathscr{C}(T)}{1 + \mathscr{C}(T)} = \mathscr{C}(T_\Delta).$$

The notation γ is intended to remind us of the word "conductance."

The next proposition is intuitively obvious but crucial.

Proposition 17.26. *For* **GW**-*a.e. T, we have* $\text{HARM}_T \neq \text{UNIF}_T$.

Proof. In view of the zero-one law, Proposition 17.16, we need merely show that we do not have $\text{HARM}_T = \text{UNIF}_T$ a.s. Now, for any tree T and any $x \in T$ with $|x| = 1$, we have

$$\text{HARM}_T(x) = \frac{\gamma(T^x)}{\sum_{|y|=1} \gamma(T^y)},$$

whereas

$$\text{UNIF}_T(x) = \frac{\widetilde{W}(T^x)}{\sum_{|y|=1} \widetilde{W}(T^y)}.$$

Therefore, if $\text{HARM}_T = \text{UNIF}_T$, the vector

$$\left\langle \frac{\gamma(T^x)}{\widetilde{W}(T^x)} \right\rangle_{|x|=1} \tag{17.10}$$

is a multiple of the constant vector $\mathbf{1}$. For Galton–Watson trees, the components of this vector are i.i.d. with the same law as that of $\gamma(T)/\widetilde{W}(T)$. The only way (17.10) can be a (random) multiple of $\mathbf{1}$, then, is for $\gamma(T)/\widetilde{W}(T)$ to be a constant **GW**-a.s. But $\gamma < 1$ and, since Z_1 is not constant, $\widetilde{W}$ is obviously unbounded, so this is impossible. ◀

Taking stock of our preceding results, we obtain our main theorem:

Theorem 17.27. *The dimension of harmonic measure is* **GW**-*a.s. less than* $\log m$. *The Hölder exponent exists a.s. and is constant.*

Proof. The hypotheses of Theorem 17.22 are shown to hold in Theorem 17.24 and Proposition 17.26. The constancy of the Hölder exponent follows from (17.6). ◀

No moment assumptions (other than $m < \infty$) were used in this proof.

Question 17.28. We saw in Section 15.4 that the dimension of visibility measure is a.s. $\mathbf{E}[\log Z_1]$. In the direction of comparison opposite to that of Theorem 17.27, is this a lower bound for $\dim \text{HARM}_T$? This question, due to Ledrappier (personal communication, 1994), was posed in Lyons, Pemantle, and Peres (1995b, 1997). A visual comparison of harmonic measure, uniform measure (these two calculated based on generation 19, not the actual limit), and visibility measure appears in Figure 17.4. A positive answer to this question, showing $\dim \text{HARM}_T > \mathbf{E}[\log Z_1]$ a.s., was given finally by Lin (2019).

Question 17.29. For $0 \leq \lambda < m$, is the dimension of harmonic measure for RW_λ on a Galton–Watson tree T monotonic increasing in the parameter λ? Is it strictly increasing? This was asked by Lyons, Pemantle, and Peres (1997). Rousselin (2018) has a formula that allows numerical computation; the resulting graphs support a positive answer to the question.

▷ **Exercise 17.12.**
Suppose that $p_0 > 0$. Show that given nonextinction, the dimension of harmonic measure is a.s. less than $\log m$.

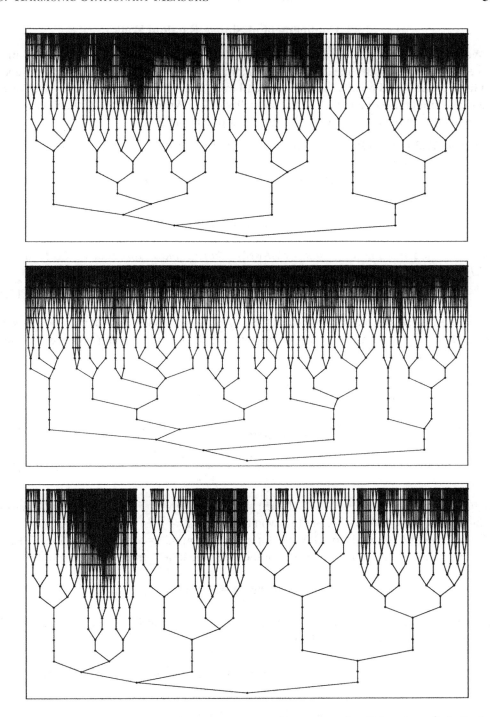

Figure 17.4. Generations 0 to 19 of a typical Galton–Watson tree for $f(s) = (s + s^2)/2$ displayed according to harmonic, uniform, and visibility measure, respectively. This means that if Θ is the flow rule, then the vertex x is centered in an interval of length $\Theta_T(x)$, where the total width of the figure is 1.

17.9 Confinement of Simple Random Walk

We now demonstrate how the drop in dimension of harmonic measure, proved in the preceding section, implies the confinement of simple random walk to a much smaller subtree.

Given a tree T and positive integer n, recall that T_n denotes the set of the vertices of T at distance n from the root and $|T_n|$ is the cardinality of T_n. Write $\mathbf{P}_T$ for the probability measure associated to simple random walk $\langle X_n \rangle$ on T starting at the root.

Theorem 17.30. *For **GW**-almost all trees T and for every $\epsilon \in (0, 1)$, there is a subtree $T(\epsilon) \subseteq T$ such that*

$$\mathbf{P}_T\big[X_n \in T(\epsilon) \text{ for all } n \geq 0\big] \geq 1 - \epsilon \tag{17.11}$$

and

$$\frac{1}{n} \log |T(\epsilon)_n| \to d, \tag{17.12}$$

where $d < \log m$ is the dimension of HARM_T. Furthermore, any subtree $T(\epsilon)$ satisfying (17.11) must have growth

$$\liminf \frac{1}{n} \log |T(\epsilon)_n| \geq d. \tag{17.13}$$

Proof. Let $n_k := 1 + \max\{n; \ |x_n| = k\}$ be the kth exit epoch and $D(x, k)$ be the set of descendants y of x with $|y| \leq |x| + k$. We will use three sample path properties of simple random walk on a fixed tree, T:

$$\textit{Speed:} \quad l := \lim_{n \to \infty} \frac{|X_n|}{n} > 0 \quad \mathbf{P}_T\text{-a.s.} \tag{17.14}$$

$$\textit{Hölder exponent:} \quad \lim_{k \to \infty} \frac{1}{k} \log \frac{1}{\mathsf{HARM}_T(X_{n_k})} = d \quad \mathbf{P}_T\text{-a.s.} \tag{17.15}$$

$$\textit{Neighborhood size:} \quad \forall \delta > 0 \quad \limsup_{n \to \infty} \frac{\log |D(X_n, \delta|X_n|)|}{|X_n|} \leq \delta \log m \quad \mathbf{P}_T\text{-a.s.} \tag{17.16}$$

We have already shown (17.14) and (17.15). In fact, the limit in (17.16) exists and equals the right-hand side for **GW**-a.e. T, but this won't be needed.

To see that (17.16) holds for **GW**-a.e. tree, recall from the proof of Proposition 17.12 that the fresh points are the vertices visited for the first time in a bi-infinite random walk:

$$\mathsf{Fresh} := \big\{(\vec{x}, T) \in \mathsf{PathsInTrees}; \ \forall n < 0 \ \ x_n \neq x_0\big\}.$$

Denote by y_k the kth fresh point visited by simple random walk. Then the statement that (17.16) holds for **GW**-a.e. T can be written as

$$\forall \delta > 0 \quad \limsup_{k \to \infty} |y_k|^{-1} \log |D(y_k, \delta|y_k|)| \leq \delta \log m \quad \mathsf{SRW} \times \mathbf{GW}\text{-a.s.},$$

and since $|y_k|/k$ has a positive a.s. liminf, this is equivalent to

$$\forall \delta^* > 0 \quad \limsup_{k} k^{-1} \log |D(y_k, \delta^* k)| \leq \delta^* \log m \quad \mathsf{SRW} \times \mathbf{GW}\text{-a.s.} \tag{17.17}$$

Now the random variables $|D(y_k, \delta^* k)|$ are identically distributed, though not independent. Indeed, the descendant subtree of y_k has the law of **GW**. Since the expected number of descendants of y_k at generation $|y_k| + j$ is m^j for every j, we have by Markov's inequality that for every $\delta' > 0$,

$$(\text{SRW} \times \text{GW})\big[|D(y_k, \delta^* k)| \geq m^{\delta' k}\big] \leq m^{-\delta' k} \sum_{j=0}^{\delta^* k} m^j \,.$$

If $\delta' > \delta^*$, then the right-hand side decays exponentially in k. The Borel–Cantelli lemma thus yields (17.17), hence (17.16).

Now that we have (17.14)–(17.16), we are ready to prove (17.11)–(17.13). Fix T satisfying (17.14)–(17.16). Then (17.15) alone implies (17.13) (Exercise 17.13).

Applying Egorov's theorem to the two almost sure asymptotics (17.14) and (17.15), we see that for each $\epsilon > 0$, there is a set of paths A_ϵ with $\mathbf{P}_T(A_\epsilon) > 1 - \epsilon$ and such that the convergence in (17.14) and (17.15) is uniform on A_ϵ. Thus, we can choose $\langle \delta_n \rangle$ decreasing to 0 such that on A_ϵ, for all k and all n,

$$\text{HARM}_T(x_{n_k}) > e^{-k(d+\delta_k)} \quad \text{and} \quad \left| \frac{|x_n|}{nl} - 1 \right| < \delta_n \,. \tag{17.18}$$

Now, since δ_n is eventually less than any fixed δ, (17.16) implies that

$$\limsup_{n \to \infty} |x_n|^{-1} \log \big| D(x_n, 3\delta_{|x_n|}|x_n|) \big| = 0 \quad \text{a.s.,}$$

so applying Egorov's theorem once more and replacing A_ϵ by a subset thereof (which we continue to denote A_ϵ), we may assume that there exists a sequence $\langle \eta_n \rangle$ decreasing to 0 such that

$$\big| D(x_n, 3\delta_{|x_n|}|x_n|) \big| \leq e^{|x_n|\eta_n} \quad \text{for all } n \tag{17.19}$$

on A_ϵ.

Define $F_0^{(\epsilon)}$ to consist of all vertices $v \in T$ such that either $\delta_{|v|} \geq 1/3$ or both

$$\text{HARM}_T(v) \geq e^{-|v|(d+\delta_{|v|})} \quad \text{and} \quad \big| D(v, 3\delta_{|v|}|v|) \big| \leq e^{|v|\eta_{|v|}} \,. \tag{17.20}$$

Finally, let

$$F^{(\epsilon)} := \bigcup_{v \in F_0^{(\epsilon)}} D(v, 3\delta_{|v|}|v|)$$

and denote by $T(\epsilon)$ the component of the root in the subforest of T induced by $F^{(\epsilon)}$. Since the number of vertices $v \in T_n$ satisfying $\text{HARM}_T(v) \geq e^{-|v|(d+\delta_{|v|})}$ is at most $e^{n(d+\delta_n)}$, the bounds (17.20) yield for sufficiently large n that

$$|T(\epsilon)_n| \leq \sum_{\substack{v \in F_0^{(\epsilon)}, \\ n-3\delta_{|v|}|v| \leq |v| \leq n}} \big| D(v, 3\delta_{|v|}|v|) \big| \leq \sum_{k=1}^{n} e^{k(d+\delta_k)} e^{k\eta_k} = e^{n(d+\alpha_n)} \,,$$

where $\lim_{n \to \infty} \alpha_n = 0$. Hence

$$\limsup \frac{1}{n} \log |T(\epsilon)_n| \le d \,.$$

In combination with the lower bound (17.13), this gives (17.12).

It remains to establish that the walk stays inside $F^{(\epsilon)}$ forever on the event A_ϵ, since that will imply that the walk is confined to $T(\epsilon)$ on this event. The points visited at exit epochs n_k are in $F_0^{(\epsilon)}$ by the first part of (17.18) and (17.19). Fix a path $\langle x_j \rangle$ in A_ϵ and a time n, and suppose that the last exit epoch before n is n_k, that is, $n_k \le n < n_{k+1}$. Denote by $N := n_{k+1} - 1$ the time preceding the next exit epoch, and observe that $x_N = x_{n_k}$. If $\delta_n \ge 1/3$, then x_n is in $F_0^{(\epsilon)}$ since $\delta_{|x_n|} \ge \delta_n$, so consider the case that $\delta_n < 1/3$. By the second part of (17.18), we have

$$\frac{|x_n|}{nl} < 1 + \delta_n \quad \text{and} \quad \frac{|x_N|}{nl} \ge \frac{|x_N|}{Nl} > 1 - \delta_N \ge 1 - \delta_n \,.$$

Dividing these inequalities, we find that

$$|x_n| < \frac{1 + \delta_n}{1 - \delta_n} |x_N| \le (1 + 3\delta_n)|x_N| \,.$$

It follows that x_n is in $D(x_{n_k}, 3\delta_{|x_{n_k}|} |x_{n_k}|)$. Since $x_{n_k} \in F_0^{(\epsilon)}$, we arrive at our desired conclusion that $x_n \in F^{(\epsilon)}$.   ◄

▷ **Exercise 17.13.**
Prove (17.13).

17.10 Numerical Calculations

Our primary aim in this section is to compute the dimension of harmonic measure, d, numerically. Recall the notation $\gamma(T)$ and $\mathscr{C}(T)$ of Section 17.8. Note that

$$\sum_{|x|=1} \gamma(T^x) = \mathscr{C}(T) = \frac{\gamma(T)}{1 - \gamma(T)} \,,$$

whence for $|x| = 1$,

$$\text{HARM}_T(x) = \frac{\gamma(T^x)}{\sum_{|y|=1} \gamma(T^y)} = \gamma(T^x)(1 - \gamma(T))/\gamma(T) \,. \tag{17.21}$$

Thus, we have

$$
\begin{aligned}
d = \text{Ent}_{\text{HARM}}(\mu_{\text{HARM}}) &= \int \log \frac{1}{\text{HARM}_T(\xi_1)} \, d\text{HARM} \times \mu_{\text{HARM}}(\xi, T) \\
&= \int \log \frac{\gamma(T)}{\gamma(T^{\xi_1})(1 - \gamma(T))} \, d\text{HARM} \times \mu_{\text{HARM}}(\xi, T) \\
&= \int \log \frac{1}{1 - \gamma(T)} \, d\mu_{\text{HARM}}(T) = \int \log(1 + \mathscr{C}(T)) \, d\mu_{\text{HARM}}(T)
\end{aligned}
$$

by stationarity and Lemma 17.20, provided we show that the final integral here is finite; see Exercise 17.14.

To compute such an integral, we use the following expression for the Radon–Nikodým derivative of the HARM-stationary Galton–Watson measure μ_{HARM} with respect to **GW**. Denote by $\mathscr{R}(T)$ the effective resistance of T from its root to infinity.

Proposition 17.31. *The Radon–Nikodým derivative of* μ_{HARM} *with respect to* **GW** *is*

$$\frac{d\mu_{\mathsf{HARM}}}{d\mathbf{GW}}(T) = \frac{1}{l} \int \frac{1}{1 + \mathscr{R}(T) + \mathscr{R}(T')} \, d\mathbf{GW}(T') \,. \tag{17.22}$$

Proof. Since the SRW $\times$ **AGW**-law of $T \setminus T^{x_{-1}}$ is **GW**, we have, for every event A,

$$\mathbf{GW}(A) = (\mathsf{SRW} \times \mathbf{AGW})[T \setminus T^{x_{-1}} \in A]$$

and

$$\mu_{\mathsf{HARM}}(A) = (\mathsf{SRW} \times \mathbf{AGW})[T \setminus T^{x_{-1}} \in A \mid \mathsf{Exit}] \,.$$

Thus, using Proposition 17.25, we have

$$\begin{aligned}
\frac{d\mu_{\mathsf{HARM}}}{d\mathbf{GW}}(t) &= \frac{(\mathsf{SRW} \times \mathbf{AGW})[T \setminus T^{x_{-1}} = t \mid \mathsf{Exit}]}{(\mathsf{SRW} \times \mathbf{AGW})[T \setminus T^{x_{-1}} = t]} \\
&= \frac{1}{l}(\mathsf{SRW} \times \mathbf{AGW})[\mathsf{Exit} \mid T \setminus T^{x_{-1}} = t] \,.
\end{aligned}$$

Of course, the event that $T \setminus T^{x_{-1}} = t$ has probability 0; we should consider events $A_n := [t]_n$, the set of rooted trees whose first n levels agree with those of t, and then take $n \to \infty$. However, we will continue to calculate more informally. Note that on Exit, we have $x_{-1} = (x_{-\infty})_1$. Thus,

$$\frac{d\mu_{\mathsf{HARM}}}{d\mathbf{GW}}(t) = \frac{1}{l}(\mathsf{SRW} \times \mathbf{AGW})\big[\vec{x} \subset t \text{ and } x_{-\infty} \notin \partial t \mid T \setminus T^{x_{-1}} = t\big]. \tag{17.23}$$

Recall that under SRW_T, $\overleftarrow{x}$ and $\vec{x}$ are independent simple random walks starting at $\mathrm{root}(T)$. For a measure μ on trees, let $[\![T_1 \bullet\!\!-\mu]\!]$ denote the law of $[\![T_1 \bullet\!\!-T_2]\!]$ when T_2 has the law of μ; and similarly for other notation. For example, the $(\mathsf{SRW} \times \mathbf{AGW} \mid T \setminus T^{x_{-1}} = t)$-law of $T^{x_{-1}}$ is **GW**, whence the $(\mathsf{SRW} \times \mathbf{AGW} \mid T \setminus T^{x_{-1}} = t)$-law of T is $[\![t \bullet\!\!-\mathbf{GW}]\!]$. Since the conditioning in (17.23) forces $x_{-1} \notin t$, we have

$$\begin{aligned}
\frac{d\mu_{\mathsf{HARM}}}{d\mathbf{GW}}(t) &= \frac{1}{l}\gamma(t) \int \mathsf{HARM}_{[\![T' \bullet\!\!-t]\!]}(\partial T') \, d\mathbf{GW}(T') \\
&= \frac{\gamma(t)}{l} \int \frac{\mathscr{C}(T')}{\gamma(t) + \mathscr{C}(T')} \, d\mathbf{GW}(T') \\
&= \frac{1}{l} \int \frac{1}{\gamma(t)^{-1} + \mathscr{C}(T')^{-1}} \, d\mathbf{GW}(T') \\
&= \frac{1}{l} \int \frac{1}{1 + \mathscr{R}(t) + \mathscr{R}(T')} \, d\mathbf{GW}(T') \,. \quad \blacktriangleleft
\end{aligned}$$

▷ **Exercise 17.14.**
Show that $\int \log(1 + \mathscr{C}(T)) \, d\mu_{\mathsf{HARM}} < \infty$.

Of course, it follows that

$$l = \iint \frac{1}{1 + \mathcal{R}(T) + \mathcal{R}(T')} \, d\mathbf{GW}(T) \, d\mathbf{GW}(T') \, ;$$

the right-hand side may be thought of as the $[\![\mathbf{GW} \bullet \text{-}\mathbf{GW}]\!]$-expected effective conductance from $-\infty$ to $+\infty$, where the boundary of one of the $\mathbf{GW}$ trees is $-\infty$ and the boundary of the other is $+\infty$.

The next computational step is to find the $\mathbf{GW}$-law of $\mathcal{R}(T)$, or, equivalently, of $\gamma(T)$. Since

$$\gamma(T) = \frac{\mathcal{C}(T)}{1 + \mathcal{C}(T)} = \frac{\sum_{|x|=1} \gamma(T^x)}{1 + \sum_{|x|=1} \gamma(T^x)}, \qquad (17.24)$$

we have, for $s \in (0, 1)$,

$$\gamma(T) \le s \iff \sum_{|x|=1} \gamma(T^x) \le \frac{s}{1 - s} \, .$$

Since $\gamma(T^x)$ are i.i.d. under $\mathbf{GW}$ with the same law as γ, the $\mathbf{GW}$-c.d.f. F_γ of γ satisfies

$$F(s) = \begin{cases} \sum_k p_k F^{*k}\left(\dfrac{s}{1-s}\right) & \text{if } s \in (0, 1), \\ 0 & \text{if } s \le 0, \\ 1 & \text{if } s \ge 1. \end{cases} \qquad (17.25)$$

Theorem 17.32. *The functional equation (17.25) has a unique solution, F_γ. Define the operator on c.d.f.'s*

$$\mathcal{K} : F \mapsto \sum_k p_k F^{*k}\left(\frac{s}{1-s}\right) \qquad (s \in (0, 1)) \, .$$

For any initial c.d.f. F with $F(0) = 0$ and $F(1) = 1$, we have weak convergence under iteration to F_γ:

$$\lim_{n \to \infty} \mathcal{K}^n(F) = F_\gamma \, .$$

For a proof, see Lyons, Pemantle, and Peres (1997).

Theorem 17.32 provides a method of calculating F_γ. It is not known that F_γ has a density, but calculations support a conjecture that it does. In the case that the offspring distribution is bounded and always at least 2, Perlin (2001) proved this conjecture and, in fact, that the effective conductance has a bounded density. Some graphs of the apparent $\mathbf{GW}$-density of $\gamma(T)$ for certain progeny distributions appear in Figures 17.5–17.7. They were calculated by iterating a discrete version of the operator $\mathcal{K}$ many times.

These density graphs reflect the stochastic self-similarity of the Galton–Watson trees. Consider, for example, Figure 17.5. Roughly speaking, the peaks represent the number of generations with no branching. For example, note that the full binary tree has conductance 1, whence its γ value is 1/2. Thus, the tree with one child of the root followed by the full binary tree has conductance 1/2 and γ value 1/3. The wide peak at the right of Figure 17.5 is thus

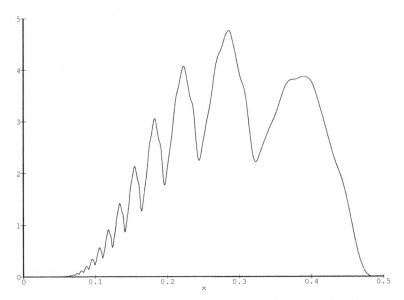

Figure 17.5. The apparent **GW**-density of $\gamma(T)$ for $f(s) = (s + s^2)/2$.

due entirely to those trees that begin with two children of the root. The peak to its left, roughly lying over 0.29, is due, at first approximation, to an unspecified number of generations without branching, while the nth peak to the left of it is due to n generations without branching (and an unspecified continuation). Of course, the next level of approximation deals with further resolution of the peaks; for example, the central peak over 0.29 is actually the sum of two nearby peaks.

Numerical calculations (still for the case $p_1 = p_2 = 1/2$) give the mean of $\gamma(T)$ to be about 0.297, the mean of $\mathscr{C}(T)$ to be about 0.44, and the mean of $\mathscr{R}(T)$ to be about 2.76. This last can be compared with the mean energy of the equally-splitting flow VIS_T, which is exactly 3:

▷ **Exercise 17.15.**
Show that the mean energy of VIS_T is $\left(1 - \mathbf{E}[1/Z_1]\right)^{-1} - 1$.

In terms of F_γ, we have

$$d = -\int \log\left(1 - \gamma(T)\right) d\mu_{\mathsf{HARM}}(T) = -\frac{1}{l} \int_{s=0}^{1} \int_{t=0}^{1} \frac{\log\left(1 - s\right)}{s^{-1} + t^{-1} - 1} \, dF_\gamma(t) \, dF_\gamma(s) .$$

In the case $p_1 = p_2 = 1/2$, it turns out that the dimension of harmonic measure is about 0.38, in other words, about $\log 1.47$, which should be compared with the dimensions of visibility measure, $\log \sqrt{2}$, and of limit uniform measure, $\log 1.5$. We can also calculate that the mean number of children of the vertices visited by a HARM_T path, which is the same as the μ_{HARM}-mean degree of the root, is about 1.58. This is about halfway between the average seen by the entire walk (and by simple forward walk), namely, exactly 1.5, and the average

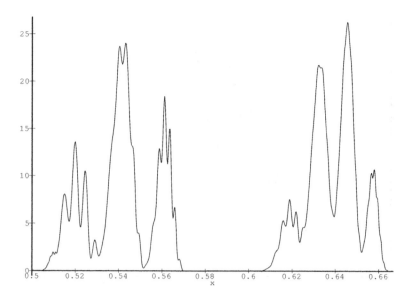

Figure 17.6. The apparent **GW**-density of $\gamma(T)$ for $f(s) = (s^2 + s^3)/2$.

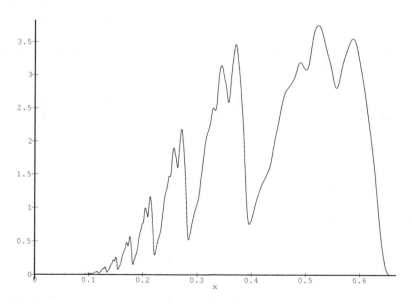

Figure 17.7. The apparent **GW**-density of $\gamma(T)$ for $f(s) = (s + s^2 + s^3)/3$.

seen by a UNIF_T path, 5/3. This last calculation comes from Exercise 17.9 that a UNIF_T-path has the law of $\widehat{\mathbf{GW}}_*$; from Section 12.1, we know that this implies that the number of children of a vertex on a UNIF_T-path has the law of the size-biased variable $\widehat{Z}_1$.

17.11 **Notes**

Our development of the theory of Markov chains on general state spaces is based on Kifer (1986), pp. 19–22; for another approach, see Rosenblatt (1971), especially pp. 96–97. Good references for ergodic theory include Petersen (1983) and Walters (1982).

Theorem 17.7 also holds when "Cayley graph" is replaced by "transitive, unimodular graph" and "translation-invariant" is replaced by "automorphism-invariant." The proof uses the subject of Chapter 8, the mass-transport principle. In fact, Lemma 17.5 and Theorem 17.7 are special cases of unimodular random rooted graphs, as described briefly in Section 8.9; see Aldous and Lyons (2007) for details, where there is also a simpler proof of Theorem 17.4 under the assumption of global reversibility, which is equivalent to unimodularity of an associated probability measure (see Theorems 4.9 and 6.2 there).

In Section 17.3, we gave a short proof that simple random walk on infinite Galton–Watson trees is a.s. transient. This result was first proved by Grimmett and Kesten (1983), Lemma 2. See also Exercises 5.44, 17.15, and 17.34. For another proof that is direct and short, see Collevecchio (2006).

Sections 17.3–17.10 are based on Lyons, Pemantle, and Peres (1995b, 1996a). Unless otherwise attributed, all results here on Galton–Watson trees are from those papers, especially the former.

Since random walk on a random spherically symmetric tree is essentially the same as a special case of random walk in a random environment (RWRE) on the nonnegative integers, we may compare the slowing of speed on Galton–Watson trees to the fact that randomness also slows down random walk for the general RWRE on the integers (Solomon, 1975).

In Proposition 17.17, μ can be an infinite stationary measure. The fact that $\widetilde{W} \cdot \mathbf{GW}$ is UNIF-stationary (Proposition 17.19) was also observed by Hawkes (1981), p. 378. Related ideas occur in Joffe and Waugh (1982). For certain flow rules, Rousselin (2018) has a somewhat explicit formula of a stationary measure that is absolutely continuous with respect to $\mathbf{GW}$.

An analogue to Theorem 17.30 for critical Galton–Watson trees conditioned to survive for many generations is given by Curien and Le Gall (2017). A notable aspect in that case is that the dimension drop is the same for all offspring distributions with finite variance. This universality ultimately arises from a universal scaling limit of the reduced subtrees of critical Galton–Watson trees, as shown by Fleischmann and Siegmund-Schultze (1977). This, in turn, is a consequence of the result of Zubkov (1975) given in Exercise 12.21.

Proposition 17.31 has not appeared in print before, though it was alluded to in Lyons, Pemantle, and Peres (1995b).

Theorems 17.13, 17.14, and 17.15 are applied by Berestycki, Lubetzky, Peres, and Sly (2018) to determine the mixing time of random walk on a random graph started at a typical vertex.

Homesick random walks RW_λ on Galton–Watson trees have been studied as well but are more difficult to study than simple random walks because no explicit stationary measure is known. Nevertheless, Lyons, Pemantle, and Peres (1996a) showed that the speed is a positive constant a.s. when $1 < \lambda < m$. See also Exercise 17.37 for the case $\lambda < 1$. The critical case $\lambda = m$ has been studied by Peres and Zeitouni (2008), who found a stationary measure. Ben Arous, Hu, Olla, and Zeitouni (2013) proved that the derivative (in λ) of the speed at $\lambda = m$ is equal to $-(m^2 - m)/(2\,\mathbf{E}[L^2] - 2m)$ provided that $p_0 = 0$ and $\mathbf{E}[s^L] < \infty$ for some $s > 1$. A stationary measure that is somewhat explicit was found for all λ in the positive-speed regime by Aïdékon (2014).

Other works about random walks on Galton–Watson trees include Kesten (1986), Aldous (1991), Piau (1996), Chen (1997), Piau (1998), Pemantle and Stacey (2001), Dembo, Gantert, Peres, and Zeitouni (2002), Piau (2002), Dembo, Gantert, and Zeitouni (2004), Dai (2005), Collevecchio (2006), Chen and Zhang (2007), Aïdékon (2008), Croydon and Kumagai (2008), Croydon (2008), Aïdékon (2010), Faraud (2011), Faraud, Hu, and Shi (2012), Ben Arous, Friberg, Gantert, and Hammond (2012), and Gantert, Müller, Popov, and Vachkovskaia (2012). However, the following questions from Lyons, Pemantle, and Peres (1996a, 1997) remain open:

Question 17.33. Is the speed of RW_λ on Galton–Watson trees monotonic decreasing in the parameter λ when $p_0 = 0$?

Ben Arous, Fribergh, and Sidoravicius (2014) and Aïdékon (2013) have made some progress on this question. Rousselin (2018) has a formula that allows numerical computation; the resulting graphs support a positive answer to the question.

Question 17.34. Is the speed of RW_λ a real-analytic function of $\lambda \in (0, m)$ for Galton–Watson trees with $p_0 = 0$? Bowditch and Tokushige (2020) finally showed that the answer is yes.

Other uses of some of the ideas in Section 17.5 appear in Furstenberg and Weiss (2003), who show "tree-analogues" of theorems of van der Waerden and Szemerédi on arithmetic progressions.

It was shown in Theorem 17.13 that the speed of simple random walk on a Galton–Watson tree with mean m is strictly smaller than the speed of simple random walk on a deterministic tree where each vertex has m children ($m \in \mathbb{N}$). Since we have also shown that simple random walk is essentially confined to a smaller subtree of growth e^d, it is natural to ask whether its speed is, in fact, smaller than $(e^d - 1)/(e^d + 1)$. This is true and was shown by Virág (2000b).

For a treatment of classical harmonic measure, see Garnett and Marshall (2005).

17.12 Collected In-Text Exercises

17.1. Show that Lemma 17.1 may not be true if μ is an infinite stationary measure, where (iii) then means that $(\mathbf{p} \times \mu)\big(f(X_0) \neq f(X_1)\big) = 0$.

17.2. Let $G = (V, E)$ be a finite, connected graph. For $x \in V$, let T_x be the universal cover of G based at x (see p. 83 of Section 3.3). Define

$$\mu\big([T, o]\big) := \frac{1}{2|E|} \sum \big\{\deg x \,;\; x \in V,\ [T_x, x] = [T, o]\big\}.$$

Show that μ is an SRW-stationary, ergodic probability measure on $[\mathscr{T}]$.

17.3. Show that if μ is an ergodic, DSRW-stationary probability measure on the space of rooted trees $\mathscr{T}$ such that μ-a.e. tree is infinite, then the rate of escape of simple random walk (not delayed) is SRW $\times \mu$-a.s.

$$\lim_{n \to \infty} \frac{|X_n|}{n} = 1 - \frac{2}{\int \deg_T(o)\, d\mu(T, o)}.$$

17.4. Use Corollary 17.8 to prove that for a translation-invariant random forest on a Cayley graph, there are a.s. no isolated ends in trees with an infinite number of ends. (An end is isolated if there is a ray $\langle x_0, x_1, \ldots \rangle$ in its equivalence class such that no other ray begins $\langle x_0, x_1 \rangle$.)

17.5. Show that the Markov chain SRW $\times$ **AGW** on PathsInTrees is stationary.

17.6. Show that the same formula (17.4) holds for the speed of simple random walk on **GW**-a.e. tree.

17.7. Prove (17.5).

17.8. Show that the Hausdorff dimension of harmonic measure is a.s. constant.

17.9. The Markov chain of Proposition 17.19 is closely connected to the size-biased Galton–Watson trees of Section 12.1. Show that in case $\mathbf{E}[Z_1 \log Z_1] < \infty$, the distribution of (T, ξ) is $\widehat{\mathbf{GW}}_*$ when $T \sim W \cdot \mathbf{GW}$ and ξ is a UNIF_T-path in T.

17.10. Prove Gibbs's inequality.

17.11. Prove Proposition 17.25.

17.12. Suppose that $p_0 > 0$. Show that given nonextinction, the dimension of harmonic measure is a.s. less than $\log m$.

17.13. Prove (17.13).

17.14. Show that $\int \log(1 + \mathscr{C}(T)) \, d\mu_{\mathrm{HARM}} < \infty$.

17.15. Show that the mean energy of VIS_T is $\left(1 - \mathbf{E}[1/Z_1]\right)^{-1} - 1$.

17.13 Additional Exercises

In all the exercises about* GW *or* AGW, *assume that* $p_0 = 0$ *unless stated otherwise.

17.16. Since simple random walk on a graph is a reversible Markov chain, the Markov chain on $[\mathscr{T}]$ induced by simple random walk is *locally reversible* in being reversible on each communicating class of states. However, a stationary measure μ is not necessarily *globally reversible*, that is, it is not necessarily the case that for Borel sets $A, B \subseteq [\mathscr{T}]$,

$$\int_A p([T, o], B) \, d\mu(T, o) = \int_B p([T, o], A) \, d\mu(T, o).$$

Give such an example, that is, a stationary measure that is not globally reversible. On the other hand, prove that every globally reversible μ is stationary.

17.17. Show that if μ is a probability measure on rooted trees that is stationary for simple random walk, then $\int_{\mathscr{A}_\diamond} \deg_T(o)/\deg_{T\diamond}(o) \, d\mu(T, o) < \infty$. See the proof of Theorem 17.4 for the notation.

17.18. Deduce from Corollary 17.9 the following: if G is an amenable Cayley graph, then every tree in a translation-invariant random forest has at most two ends a.s.

17.19. For a rooted tree (T, o), define T_Δ to be the tree obtained by adding an edge from o to a new vertex Δ and rooting at Δ. This new vertex Δ is thought of as representing the past. Let $\gamma(T)$ be the probability that simple random walk started at Δ never returns to Δ:

$$\gamma(T) := \mathrm{SRW}_{T_\Delta}[\forall n > 0 \ \ X_n \neq \Delta].$$

Let $\mathscr{C}(T)$ denote the effective conductance of T from its root to infinity when each edge has unit conductance. The series law gives us that

$$\gamma(T) = \frac{\mathscr{C}(T)}{1 + \mathscr{C}(T)} = \mathscr{C}(T_\Delta).$$

The notation γ is intended to remind us of the word "conductance." Compare the remark following Corollary 5.25. For $(\overleftrightarrow{x}, T) \in \mathsf{PathsInTrees}$, write $N(\overleftrightarrow{x}, T) := |\{n \, ; \ x_n = x_0\}|$ for the number of visits to the root of T. For $k \in \mathbb{N}$, let $D_k(\overleftrightarrow{x}, T) := \{j \in \mathbb{Z} \, ; \ \deg x_j = k + 1\}$.

(a) Show that

$$\lim_{n \to \infty} \frac{1}{2n + 1} |D_k(\overleftrightarrow{x}, T) \cap [-n, n]| = p_k \qquad \mathrm{SRW} \times \mathbf{AGW}\text{-a.s.}$$

This means that the proportion of time that simple random walk spends at vertices of degree $k + 1$ is p_k.

(b) For $i \geq 0$, let γ_i be i.i.d. random variables with the distribution of the **GW**-law of $\gamma(T)$. Let

$$\Gamma_k := \mathbf{E}\left[\frac{k + 1}{\gamma_0 + \cdots + \gamma_k}\right].$$

Show that

$$\int N(\overset{\leftrightarrow}{x}, T)\, d\text{SRW} \times \mathbf{AGW}((\overset{\leftrightarrow}{x}, T) \mid \deg x_0 = k + 1) = 2\Gamma_k - 1$$

and that Γ_k are finite and decreasing in k. What is $\lim_{k \to \infty} \Gamma_k$?

(c) The result in (a) says that there is no biasing of visits to a vertex according to its degree, just as Theorem 17.11 says. Yet the result in (b) indicates that there is indeed a biasing. How can these results be compatible?

(d) What is

$$\int N(\overset{\leftrightarrow}{x}, T)\, d\text{SRW} \times \mathbf{AGW}((\overset{\leftrightarrow}{x}, T) \mid \deg x_0 = k + 1, \text{Fresh})?$$

17.20. Show that a labeled tree chosen according to a nondegenerate **GW** measure a.s. has no graph automorphisms other than the identity map. (Automorphisms need not fix the root.)

17.21. Show that the hypothesis of Proposition 17.16 is needed. You may want to use two flow rules that both follow a 2-ray when it exists (see Exercise 17.22) but do different things otherwise.

17.22. Give an example as follows of a flow rule Θ with a Θ-stationary measure that is absolutely continuous with respect to **GW** but whose associated Markov chain is *not* ergodic. Call a ray $\xi \in \partial T$ an *n-ray* if every vertex in the ray has exactly n children, and write $T \in A_n$ if ∂T contains an n-ray. Since the root must have n children if there is an n-ray, the sets A_n are pairwise disjoint. Consider the Galton–Watson process with $p_3 := p_4 := 1/2$. Show that $\mathbf{GW}(A_n) > 0$ for $n = 3, 4$. Define Θ_T to choose equally among all children of the root on $(A_3 \cup A_4)^c$ and to choose equally among all children of the root belonging to an n-ray when $T \in A_n$. Show that $\mathbf{GW}_{A_n}$ is Θ-stationary for both $n = 3, 4$, whence the Θ-stationary measure $(\mathbf{GW}_{A_3} + \mathbf{GW}_{A_4})/2$ gives a nonergodic Markov chain.

17.23. Let T be the Fibonacci tree of Exercise 13.28. Show that the dimension of harmonic measure for RW_λ on T is

$$\frac{1 + \sqrt{\lambda + 1}}{2 + \sqrt{\lambda + 1}} \log\left(1 + \sqrt{\lambda + 1}\right) - \frac{\sqrt{\lambda + 1}}{(2 + \sqrt{\lambda + 1})} \log \sqrt{\lambda + 1}.$$

17.24. Identify the binary tree with the set of all finite sequences of 0s and 1s. Let the conductance of an edge be $x + 1$ when its vertex farthest from the root ends in x, where $x \in \{0, 1\}$. Calculate the dimension of harmonic measure of the corresponding random walk.

*The following sequence of exercises, 17.25–17.33, treats the ideas of Furstenberg (1970) that inspired those of Section 17.5. We adopt the setting and notation of Section 15.5. **In particular, fix an integer r and let $\mathbf{T}$ be the r-ary tree.***

17.25. Let $\mathscr{U}$ be the space of unit flows on $\mathbf{T}$. Give $\mathscr{U}$ a natural compact topology.

17.26. A Markov chain on $\mathscr{U}$ is called *canonical* if it has transition probabilities $p(\theta, \theta^x) = \theta(x)$ for $|x| = 1$. Any Borel probability measure μ on $\mathscr{U}$ can be used as an initial distribution to define a canonical Markov chain on $\mathscr{U}$, denoted $\text{Markov}(\mu)$. Regard $\text{Markov}(\mu)$ as a Borel probability measure on path space $\mathscr{U}^\infty$ (which has the product topology). Show that the set of canonical Markov chains on $\mathscr{U}$ is weak*-compact and convex.

17.27. Let S be the left shift on $\mathscr{U}^\infty$. For a probability measure μ on $\mathscr{U}$, let $\mathsf{Stat}(\mu)$ be the set of weak*-limit points of

$$\frac{1}{N} \sum_{n=0}^{N-1} S^n \mathsf{Markov}(\mu).$$

Show that $\mathsf{Stat}(\mu)$ is nonempty and consists of stationary canonical Markov chains.

17.28. Let f be a continuous function on $\mathscr{U}^\infty$, μ be a probability measure on $\mathscr{U}$, and $\nu \in \mathsf{Stat}(\mu)$. Show that $\int f \, d\nu$ is a limit point of the numbers

$$\frac{1}{N} \sum_{n=0}^{N-1} \int \sum_{|x|=n} \theta(x) \int f(\theta^x, \theta^{x_1}, \theta^{x_2}, \ldots) \, d\theta^x(x_1, x_2, \ldots) \, d\mu(\theta),$$

where we identify $\mathscr{U}$ with the set of Borel probability measures on $\partial \mathbf{T}$.

17.29. For any probability measure μ on $\mathscr{U}$, define its **entropy** as $\mathsf{Ent}(\mu) := \int F(\theta) \, d\mu(\theta)$, where

$$F(\theta) := \sum_{|x|=1} \theta(x) \log \frac{1}{\theta(x)}.$$

Define this also to be the entropy of the associated Markov chain $\mathsf{Markov}(\mu)$. Show that $\mathsf{Ent}(\mu) = \int H \, d\mathsf{Markov}(\mu)$, where

$$H(\theta, \theta^{x_1}, \theta^{x_2}, \ldots) := \log \frac{1}{\theta(x_1)}.$$

Show that if $\nu \in \mathsf{Stat}(\mu)$, then $\int H \, d\nu$ is a limit point of the numbers

$$\frac{1}{N} \int \sum_{|x|=N} \theta(x) \log \frac{1}{\theta(x)} \, d\mu(\theta).$$

17.30. Suppose that the initial distribution is concentrated at a single unit flow θ_0, so that $\mu = \delta_{\theta_0}$. Show that if, for all x with $\theta_0(x) \neq 0$, we have

$$\alpha \leq \frac{1}{|x|} \log \frac{1}{\theta_0(x)} \leq \beta,$$

then for all $\nu \in \mathsf{Stat}(\mu)$, we have $\alpha \leq \mathsf{Ent}(\nu) \leq \beta$.

17.31. Show that if ν is a stationary canonical Markov chain with initial distribution μ, then its entropy is

$$\mathsf{Ent}(\nu) = \lim_{N \to \infty} \int \frac{1}{N} \sum_{|x|=N} \theta(x) \log \frac{1}{\theta(x)} \, d\mu(\theta).$$

17.32. Show that if ν is a stationary canonical Markov chain, then

$$\lim_{n \to \infty} \frac{1}{n} \log \frac{1}{\theta(x_n)}$$

exists for almost every trajectory $\langle \theta^{x_1}, \theta^{x_2}, \ldots \rangle$ and has expectation $\mathsf{Ent}(\nu)$. If the Markov chain is ergodic, then this limit is $\mathsf{Ent}(\nu)$ a.s.

17.33. Show that if T is a tree of uniformly bounded degree, then there is a stationary canonical Markov chain ν such that for almost every trajectory $\langle \theta, \theta^{x_1}, \theta^{x_2}, \ldots \rangle$, the flow θ is carried by a derived tree of T and

$$\lim_{n \to \infty} \frac{1}{n} \log \frac{1}{\theta(x_n)} = \mathrm{Ent}(\nu) = \dim \sup \partial T \, .$$

Show similarly that there is a stationary canonical Markov chain ρ such that for almost every trajectory $\langle \theta, \theta^{x_1}, \theta^{x_2}, \ldots \rangle$, the flow θ is carried by a derived tree of T and

$$\lim_{n \to \infty} \frac{1}{n} \log \frac{1}{\theta(x_n)} = \mathrm{Ent}(\rho) \le \dim \inf \partial T \, .$$

17.34. Let T be a Galton–Watson tree without extinction. Suppose that $\mathbf{E}[L^2] < \infty$. Consider the flow θ on T of strength $W(T)$ given by $\theta(e(x)) := W(T^x)/m^{|x|}$, in other words, the flow corresponding to the measure $W(T)\,\mathrm{UNIF}_T$. Show that $\mathbf{E}[W^2] = 1 + \mathrm{Var}(L)/(m^2 - m)$. Show that if $\lambda < m$, then for the conductances $c(e) := \lambda^{-|e|}$, we have $\mathbf{E}[\mathscr{E}_c(\theta)] = \lambda \mathbf{E}[W^2]/(m - \lambda)$. Show that for these same conductances, the expected effective conductance from the root to infinity is at least $(m - \lambda)/(\lambda \mathbf{E}[W^2])$. Use that $\mathbf{E}[\mathscr{E}_c(\theta)] < \infty$ when $\mathbf{E}[L^2] < \infty$ to give another proof that for every Galton–Watson tree of mean $m > 1$ (without restriction on $\mathbf{E}[L^2]$), RW_λ is transient a.s. given nonextinction for all $\lambda < m$.

17.35. Let μ be an SRW-stationary probability measure on the space of rooted trees $\mathscr{T}$ such that μ-a.e. tree has at least three ends. Show that harmonic measure of simple random walk has positive Hausdorff dimension μ-a.s. This gives another proof that μ-a.e. tree has branching number > 1.

17.36. Consider simple random walk on augmented Galton–Watson trees, T. Define loop-erased simple random walk as the limit x_∞ of a simple random walk path $\vec{x}$. Show that the (expected) probability that the path of a loop-erased simple random walk from the root of T does not intersect the path of an independent (not loop-erased) simple random walk from a uniformly chosen neighbor of the root of T is the speed $\mathbf{E}[(Z_1 - 1)/(Z_1 + 1)]$. Hence, by Proposition 10.20, the chance that the root of T does not belong to the same tree as a uniformly chosen neighbor in the wired uniform spanning forest on T is the speed $\mathbf{E}[(Z_1 - 1)/(Z_1 + 1)]$.

17.37. Suppose that $p_0 > 0$. Let T be a Galton–Watson tree conditioned on nonextinction. Show that the speed of RW_λ is zero if $0 \le \lambda \le f'(q)$.

Comments on Exercises

Chapter 1

1.1. To show that every cover by open sets has a finite subcover, it suffices to consider covers $\{B_x \; ; \; x \in W\}$ by sets of the form in (1.3). Consider the vertices in T that are connected to the root by a path that does not include any $x \in W$. These vertices form a subtree of T that, by definition, contains no ray.

Chapter 2

2.1. **(e)** Use part (b).

(f) A proof of the well-known uniqueness of the stationary distribution is given in Exercise 2.43.

2.8. By symmetry, all vertices at a given distance from o have the same voltage. Therefore, they can be identified without changing any voltages (or currents). This yields the graph $\mathbb{N}$ with multiple edges and loops. We may remove the loops. We see that the parallel edges between $n - 1$ and n are equivalent to a single edge whose conductance is C_n. These new edges are in series.

2.10. The complete bipartite graph $K_{4,4}$.

2.13. Use that the minimum occurs iff F is harmonic at each $x \notin A \cup Z$. Or, let i be the unit current flow from A to Z and use the Cauchy–Schwarz inequality:

$$\sum_{e \in \mathsf{E}_{1/2}} dF(e)^2 c(e) \mathscr{R}(A \leftrightarrow Z) = \sum_{e \in \mathsf{E}_{1/2}} dF(e)^2 c(e) \sum_{e \in \mathsf{E}_{1/2}} i(e)^2 r(e)$$

$$\geq \left(\sum_{e \in \mathsf{E}_{1/2}} i(e) dF(e) \right)^2 = \left(\sum_{x \in \mathsf{V}} d^* i(x) F(x) \right)^2$$

$$= \left(\sum_{x \in A} d^* i(x) F(x) \right)^2 = 1 \,.$$

See Griffeath and Liggett (1982), Theorem 2.1 for essentially the same statement. The minimum is the same even if we allow all F with $F \geq 1$ on A and $F \leq 0$ on Z.

2.14. Since $\{\chi^e \; ; \; e \in \mathsf{E}_{1/2}\}$ form an orthogonal basis of $\ell^2_-(\mathsf{E}, r)$ and the norms of θ_n are bounded, it follows that θ_n tend weakly to θ. Hence the norm of θ is at most $\liminf_n \mathscr{E}(\theta_n)$. Furthermore, as $d^* \theta_n(x)$ is the inner product of θ_n with the star at x, it converges to $d^* \theta$.

2.15. (This is due to T. Lyons (1983).) Let U_1, U_2 be independent, uniform $[0, 1]$ random variables. Take a path in W_f that stays fairly close to the points $(n, U_1 n, U_2 f(n))$ $(n \geq 1)$.

2.21. Let the successive intervals in $\mathbb{N} \setminus S$ have lengths successive powers of 2, whereas the successive intervals of S have lengths successive powers of 5. For more detailed analysis, see Benjamini, Pemantle, and Peres (1996), Proposition 4.1.

2.22. (This is due to Solomon (1975).) Use the Chung–Fuchs theorem when this expectation equals 0. The similar RWRE on $\mathbb{N}$ is a.s. recurrent when $\mathbf{E}[\log A_0 - \log(1 - A_0)] \leq 0$. See Lyons and Pemantle (1992) for an extension to trees.

2.24. (a) Use the fact that $\langle f(X_n) \rangle$ is a submartingale.

2.26. Use Exercise 2.1.

2.28. This is called the ***Riesz decomposition***. The decomposition also exists and is unique if instead of $f \geq 0$, we require $\mathscr{G}|f| < \infty$; but we still have $f \geq 0$.

2.29. The original proof is by Starr (1966). This presentation of it is taken from Norris, Peres, and Zhai (2017).

2.30. (a) The events $[X_k \in A$ and $\forall j > k\ X_j \notin A]$ are disjoint for $k \geq 0$ and all have the same probability. We remark that by applying this result to $\inf\{n \geq 1 ;\ X_{mn} \in A\}$, we conclude that a.s., there is a return to A for some multiple of m for each positive integer m, and thus that there are infinitely many returns a.s.

(b) We may assume that our stationary sequence is bi-infinite. Shifting the sequence to the left preserves the probability measure on sequences. Write $Y := X_{\tau_A^+}$. We want to show that for measurable $B \subseteq A$, we have $\mathbf{P}[Y \in B \mid X_0 \in A] = \mu_A(B)$. Write $\sigma_A^- := \sup\{n \leq -1 ;\ X_n \in A\}$. A proof similar to that in (a) shows that $\mathbf{P}[\sigma_A^- < \infty \mid X_0 \in A] = 1$. Now $\mathbf{P}[Y \in B, X_0 \in A] = \sum_{n \geq 1} \mathbf{P}[X_0 \in A, \tau_A^+ = n, X_n \in B]$. By shifting the nth set here by n to the left, we obtain $\mathbf{P}[Y \in B, X_0 \in A] = \sum_{n \geq 1} \mathbf{P}[\sigma_A^- = -n, X_0 \in B]$, which equals $\mathbf{P}[X_0 \in B]$. This gives (b). The same method shows that shifting the entire sequence $\langle X_n \rangle$ to the left by τ_A^+ preserves the measure given that $X_0 \in A$.

(c) We give two proofs of the Kac lemma, but the second proof assumes that $\langle X_n \rangle$ is ergodic.

First, write $\sigma_A := \sup\{n \leq 0 ;\ X_n \in A\}$. We have $\mathbf{E}[\tau_A^+; X_0 \in A] = \sum_{k \geq 0} \mathbf{P}[X_0 \in A, \tau_A^+ > k]$. If we shift the kth set in this sum to the left by k, then we obtain the set where $\sigma_A = -k$, whence $\mathbf{E}[\tau_A^+; X_0 \in A] = \sum_{k \geq 0} \mathbf{P}[\sigma_A = -k] = 1$.

Second, consider the asymptotic frequency that $X_n \in A$. By the ergodic theorem, this equals $\mathbf{P}[X_0 \in A] = \mu(A)$ a.s. Let τ_A^k be the time between the kth visit to A and the succeeding visit to A. Decomposing the trajectory by visits to A, we see that the asymptotic frequency of visits to A is also the reciprocal of the asymptotic average of τ_A^k. Since the random variables $\langle \tau_A^k \rangle$ are stationary when conditioned on $X_0 \in A$ by the first part of our proof, the ergodic theorem again yields that the asymptotic average of τ_A^k equals $\mathbf{E}[\tau_A^+ \mid X_0 \in A]$ a.s. Equating these asymptotics gives the Kac lemma.

2.31. Express the equations for i_c by Exercise 2.2. Cramer's rule gives that i_c is a rational function of c.

2.32. The corresponding conductances are $c^h(x, y) := c(x, y)h(x)h(y)$.

2.33. (a) Let L be the diameter of G. Consider the random walk every L steps to see whether it has retired. Compare to a geometric random variable.

2.36. Use the craps principle and (2.4).

See Exercise 2.68 for an exact value of the left-hand side.

2.37. *Solution 1.* Let u be the first vertex among $\{x, y\}$ that is visited by a random walk starting from a. Before being absorbed on Z, the walk is as likely to make cycles at u in one direction as in the other by reversibility. This leaves at most one net traversal of the edge between x and y.

Solution 2. Let, say, $v(x) \geq v(y)$. Let $\Pi := \{[u, w] ;\ v(u) \geq v(x),\ v(w) \leq v(x)\}$. Then Π is a cutset separating a from Z, whence $\sum_{[u, w] \in \Pi} i(u, w) = 1$ since $\Pi \setminus \{e ;\ i(e) = 0\}$ is a minimal cutset (see Section 3.1). Since $i(u, w) \geq 0$ for all $[u, w] \in \Pi$, and since $[x, y] \in \Pi$, it follows that $i(x, y) \leq 1$.

2.38. Use Proposition 2.2.

2.39. Consider two networks that differ only by the conductance on (x, y). Couple the two corresponding random walks so that one crosses from x to y at least as many times as the other; one walk may wait for the other to return to x, for example. This extends to nonreversible chains.

2.41. Use the stationary distribution from Exercise 2.1 or Exercise 2.30(c).

2.42. This is due to Aldous and Fill (2002), Proposition 2.3, but is classical for $\tau = \tau_a^+$.

2.43. This is well known. One proof of (a) and (b) is to apply the martingale convergence theorem to $\langle f(X_n) \rangle$, where f is harmonic. For (c), apply (b) to the reversed chain.

2.45. (This result is due to Kemeny and Snell (1960), Theorem 4.4.10, when the state space is finite. Our solution is due to Yiping (Kenneth) Hu.) For an integer $N > 0$, show first that

$$\sum_{n=0}^{N} [p_n(x, x) - \pi(x)] = \sum_{k=0}^{N} \mathbf{P}_\pi[\tau_x = k] \sum_{i=N-k+1}^{N} p_i(x, x) + \mathbf{P}_\pi[\tau_x > N] \sum_{i=0}^{N} p_i(x, x),$$

then take $N \to \infty$, considering separately the cases $\mathbf{E}_\pi[\tau_x] < \infty$ and $= \infty$.

2.46. Assume that $\mathbf{E}_a[\tau_a^+] < \infty$ and use the ideas of Exercise 2.42 to show that there is a stationary probability measure.

2.48. (The first part is due to Kemeny and Snell (1960), Theorem 4.4.10. This proof of constancy is due to Doyle in 1983; see Doyle (2009). The second part is due to Broder and Karlin (1989).) Show that f is harmonic by using Exercise 2.42(b). For the second part, use Exercise 2.45 (noting that $f(x) = \sum_z \pi(z) f(z)$). The common value of f is called the ***Kemeny constant*** of the Markov chain; a formula for it follows from Exercise 2.45.

2.49. Use Fubini's theorem. It suffices that $\sum_{e \in \xi} r(e) = \infty$ for every ray ξ.

2.50. (b) From part (a), we have

$$\mathbf{P}_z[\tau_A < \tau_z^+] \mu(x) = \pi(x) \sum_{\mathcal{P}} \prod_{e \in \mathcal{P}} c(e) \Big/ \prod_{w \in \mathcal{P}} \pi(w) = \pi(x) v(x) / \pi(z),$$

where the sum is over paths $\mathcal{P}$ from z to x that visit z and x just once and do not visit $A \setminus \{x\}$.

2.54. This is known as Parrondo's paradox, as it combines games that are not winning into a winning game. See Parrondo (1996) and Harmer and Abbott (1999). For additional analysis, see Pyke (2003) and Ethier and Lee (2009). For other aspects of turning fair games into unfair ones, see Durrett, Kesten, and Lawler (1991).

2.55. (More such exercises can be found in Doyle and Snell (1984).) **(a)** $29/63$, $29/35$, $17/20$.

2.59. Fix $o \in \mathsf{V}$ and let i_x be the unit current flow from x to o. Use $\mathscr{R}(u \leftrightarrow x) = \|i_u - i_x\|_r^2$, (2.21), or Exercise 2.62(g). This is equivalent to saying that the effective resistance metric (see Exercise 2.67) has ***1-negative type***, that is, $\sum_{x, y \in \mathsf{V}} \mathscr{R}(x \leftrightarrow y) \alpha_x \alpha_y \leq 0$ whenever $\sum_{x \in \mathsf{V}} \alpha_x = 0$: see Schoenberg (1937, 1938). In this latter form, the result is due to Jorgensen and Pearse (2008) (see Theorem 5.1). Moreover, equality holds in this inequality iff all $\alpha_x = 0$, as equality is equivalent to $\sum_x \alpha_x i_x = \mathbf{0}$; one then says that this metric has strict negative type. Metric spaces of negative type include Euclidean spaces; this fact is useful in statistics (Székely and Rizzo (2005a, 2005b, 2005b), Bakirov, Rizzo, and Székely (2006), Székely, Rizzo, and Bakirov (2007), Lyons (2013a)) and in theoretical computer science (Deza and Laurent (1997), Naor (2010)). By Corollary 2.21, we have that the commute-time metric has negative type. This, in fact, holds for Markov chains that are not necessarily reversible: see Doyle and Steiner (2008) and Theorem 15 in Boley, Ranjan, and Zhang (2011).

2.60. Use superposition and $\mathscr{E}(i) = (i, dv) = (d^*i, v) = (f, v)$.

2.62. (a) Multiply $\Delta_G \cdot v_a$ and use properties of voltages, or multiply in the other order and use superposition of currents.

(g) Consider how Δ_G acts on $\ell^2(\mathsf{V}) = \mathbb{R} \oplus \mathbb{R}^\perp$, where we identify $\mathbb{R}$ with the constant vectors.

2.63. (This is noted by Coppersmith, Doyle, Raghavan, and Snir (1993). See also Ponzio (1998).) Use Exercise 2.62.

2.65. (This is due to Foster (1948).) Compute a trace. Such a proof is first due to Flanders (1974). For other short proofs and an extension to nonreversible Markov chains, see Exercise 2.124 and Exercise 4.29.

2.66. With general conductances, we have $r(e)i^e(e) \geq r(e')i^e(e')$.

2.67. There are several solutions. One involves superposition of unit currents, one from u to x and one from x to w. One can also use Exercise 2.68 (which explains the left-hand side minus the right-hand side) or Corollary 2.21.

2.68. There are many interesting proofs. For one, use Exercise 2.62. For another solution, see Tetali (1991), who discovered this formula.

See Exercise 2.36 for an upper bound.

2.69. **(a)** Use superposition, Exercise 2.57. Also, use Exercise 2.68 to see that effective resistances determine voltages. In the notation of Exercise 2.62, we have $\Delta_G f = \Delta_{G'} f$ on W.

(b) Use (a).

(c) Use Exercise 2.67 or part (d).

(d) Let G'' be G' with a loop at each x of conductance $\pi(x)p_W(x,x)$. Then the escape probability from each x to each y is the same in G as in G'', as is the sum of the conductances around x, whence we deduce equality of effective resistances. But the loops do not affect the effective resistances, whence the result. We remark that the network walk on G'' is the walk on G after inducing on W: see Exercises 2.30 and 6.66.

(e) Use (d). Alternatively, show that the voltages are the same on both copies of $\mathsf{V}(G')$ by showing that the current outflows on N via the new edges equal those to z in G. A remarkable use of this transformation is in Caputo, Liggett, and Richthammer (2010). It is also called the ***star-mesh transformation***. It is due to Campbell (1911); the original star-triangle transformation is due to Kennelly (1899).

2.70. Use that $\partial i/\partial r(e)$ is a flow with strength 0.

2.71. **(a)** The probability that the random walk leaves G_n before visiting a or z tends to 0 as $n \to \infty$. We may couple random walks X_k on G_n and Y_k on G as follows. Start at $X_0 := Y_0 := x$. Define Y_k as the usual network random walk on G, stopped when it reaches $\{a, z\}$. Define $\tau := \inf\{k \,;\, Y_k \notin G_n\}$. For $k < \tau$, let $X_k := Y_k$, whereas for $k \geq \tau$, continue the random walk X_k independently on G_n, stopped at $\{a, z\}$. Thus, $v_n(x)$ is the probability that X_k reaches a, whereas $v(x)$ is the probability that Y_k reaches a. For large n, it is likely that $\tau = \infty$, on which event either both random walks reach a or neither does.

(b) In fact, $i_n \to i$ in norm.

(e) Use the coupling in (a) or use Exercise 2.24.

2.73. Let $\langle G_n \rangle$ be an exhaustion by finite, induced subnetworks. Note that the star spaces of G_n^{W} increase to $\bigstar$ and the cycle spaces of G_n increase to $\Diamond$. It follows from Exercises 2.72 and 2.71 that the projections of χ^e onto $\bigstar$ and the orthocomplement of $\Diamond$ agree for each e.

2.74. Use Exercise 2.13 and the fact that the minimum of linear functions is concave.

2.75. Use Exercise 2.13.

2.77. Use Exercises 2.70 and 2.75.

2.78. The expression given by Thomson's principle can also be regarded as an extremal width. This identity is due to Duffin (1962). Given ℓ, find F with $|dF| \leq \ell$ and use Dirichlet's principle. The minimum is the same even if we allow all ℓ with the distance between any point of A and any point of Z to be at least 1.

2.80. This is due to Abrams and Kenyon (2017).

2.81. If G is recurrent, this is obvious. If G is transient and the conclusion does not hold, then let $h(x) := \mathbf{P}_x[\forall n \ X_n \in H]$, so $h(x) > 0$ for all $x \in \mathsf{V}(H)$ and h is harmonic on H. Apply the Doob transform, Exercise 2.32, to get new conductances on H that correspond to the random walk conditioned to stay in H. This new walk is transient, yet the new conductances are less than the old ones, which contradicts Rayleigh's monotonicity principle.

2.82. See the beginning of the proof of Theorem 3.1.

2.83. The tree of Example 1.2 will do. The following analysis is due to Tom Kalvari (personal communication, 2014). Order the cutsets according to their cardinality. For each n, let $A_n := \bigcup_{k=1}^n \Pi_k$. Every infinite path from the root intersects A_n at least n times, hence for each vertex at level $n-1$ in the tree, there exists an edge of A_n above it in the tree. Therefore, $|A_n| \geq 2^{n-1}$, whence $|\Pi_n| \geq 2^{n-1}/n$.

Alternatively, use Exercise 2.84.

There *is* a converse of the Nash-Williams criterion if one allows subdivision of edges; this was proved by Nash-Williams (1959).

2.84. (This is due to Yoram Gat, personal communication, 1997.) Write $A := \{\Pi_n ; n \geq 1\}$ and $S(A) := \sum_n |\Pi_n|^{-1}$. We may clearly assume that each Π_n is minimal. Suppose that there is some edge $e \notin \bigcup_n \Pi_n$. Now there is some Π_k that does not separate the root from e. Let Π'_k be the cutset obtained from Π_k by replacing all the descendants of e in Π_k by e. Then $|\Pi'_k| \leq |\Pi_k|$. Let $A' := \{\Pi_n ; n \neq k\} \cup \{\Pi'_k\}$. Then A' is also a sequence of disjoint cutsets and $S(A') \geq S(A)$.

If we order the edges of T in any fashion that makes $|e|$ increasing and apply the above procedure recursively to all edges not in the current collection of cutsets, we obtain a sequence A_i of collections of disjoint cutsets with $S(A_i)$ increasing in i. Let $A_\infty := \liminf A_i$. Then A_∞ is also a sequence of disjoint cutsets, and $S(A_\infty) \geq S(A)$. Furthermore, each edge of T appears in some element of A_∞.

Call two cutsets **comparable** if one separates the root from the other. Suppose that there are two cutsets of A_∞ that are not comparable, Π_n and Π_m. Then we may create two new cutsets, Π'_n and Π'_m, that are comparable and whose union contains the same edges as $\Pi_n \cup \Pi_m$. Since $|\Pi'_n| + |\Pi'_m| = |\Pi_n| + |\Pi_m|$ and $\min\{|\Pi'_n|, |\Pi'_m|\} \leq \min\{|\Pi_n|, |\Pi_m|\}$, it follows that $|\Pi'_n|^{-1} + |\Pi'_m|^{-1} \geq |\Pi_n|^{-1} + |\Pi_m|^{-1}$. By replacing Π_n and Π_m in A_∞ by Π'_n and Π'_m and repeating this procedure as long as there are at least two cutsets in the collection that are not comparable, we obtain in the limit a collection A'_∞ of disjoint pairwise comparable cutsets containing each edge of T and such that $S(A'_\infty) \geq S(A)$. But the only collection of disjoint pairwise comparable cutsets containing each edge of T is $A'_\infty = \{T_n ; n \geq 1\}$.

2.85. This is due to Itai Benjamini (personal communication, 1996).

(a) Given cutsets, consider the new network where the edges of the cutsets are divided in two by an extra vertex for each edge, each half getting half the resistance.

(b) Reverse time.

(c) Note that A_n are disjoint.

2.86. (This improvement of Lemma 13.5 of BLPS (2001) is due to Yiping (Kenneth) Hu, personal communication, 2016.) Use the Nash-Williams criterion.

2.89. **(a)** Cycles at a are equally likely to be traversed in either direction. Thus, cycles contribute nothing to $\mathbf{E}[S_e - S_{-e}]$.

2.91. By Proposition 2.12, $v(X_n) = \mathbf{P}_{X_n}[\exists k \geq n \; X_k = a]$. Now the intersection of the events $[\exists k \geq n \; X_k = a]$ is the event that a is visited infinitely often, which has probability 0. Since these events are also decreasing, their limiting probability is 0. That is, $\mathbf{E}[v(X_n)] \to 0$. On the other hand, $\langle v(X_n) \rangle$ is a nonnegative supermartingale, whence it converges a.s. and $\mathbf{E}[\lim v(X_n)] = \lim \mathbf{E}[v(X_n)] = 0$.

2.92. The probability that a random walk started at a returns to a after $2d(a, x)$ or more steps is at least

$$\mathbf{P}_a[\tau_x < \infty] \mathbf{P}_x[\tau_a < \infty] = \mathbf{P}_a[\tau_x < \infty] \frac{\mathscr{G}(x, a)}{\mathscr{G}(a, a)} = \mathbf{P}_a[\tau_x < \infty] \frac{\pi(a) \mathscr{G}(a, x)}{\pi(x) \mathscr{G}(a, a)}$$

$$\geq \mathbf{P}_a[\tau_x < \infty]^2 \frac{\pi(a)}{\pi(x) \mathscr{G}(a, a)} .$$

2.93. Use Exercise 2.13.

2.94. Suppose that (G, c) is recurrent. By Exercise 2.93, we may choose an increasing sequence r_n of radii, starting with $r_0 = 0$, and functions $f_n : \mathsf{V}(G) \to \mathbb{R}$ such that $f_n(x) = n$ when $d(o, x) \leq r_n$,

$f_n(x) = n + 1$ when $d(o, x) \geq r_{n+1}$, and $\sum_n \sum_{e \in E} df_n(e)^2 c(e) < \infty$. Now put $f(x) := f_n(x)$ when $r_n \leq d(o, x) < r_{n+1}$.

Conversely, suppose a function f satisfies the conditions. Without loss of generality, we may suppose also that $f(o) = 0$ and $\sum_{e \in E} df(e)^2 c(e) < 1$. Then given $\epsilon > 0$, define $F(x) := 1 - \min\{\epsilon f(x), 1\}$. This new function F has finite support, $F(o) = 1$, and $\sum_{e \in E} dF(e)^2 c(e) < \epsilon^2$.

2.96. Use Exercise 2.74.

2.97. Use Exercise 2.75.

2.98. (This is due to Benjamini, Gurel-Gurevich, and Lyons (2007).) Consider the random walk conditioned to return to o.

2.99. Use Proposition 2.1 and estimate $L_n := \mathscr{G}_{z_n}(0, 0) = 4\mathscr{R}(0 \leftrightarrow z_n) \sim (2/\pi) \log n$. To do this, note that in $\mathbb{Z}^2$, we have $p_{2k}(0, 0) \sim 1/(\pi k)$. Also, the chance of reaching z_n before $n^{2-\epsilon}$ steps is extremely small. Thus, $L_n \geq (1 - \epsilon) \sum_0^{n^{2-\epsilon}} 1/(\pi k)$ for large n. To get an upper bound, define $L_n' := \max_x \mathscr{G}_{z_n}(x, 0)$ and use the fact that the chance for a random walk starting at x not to reach z_n before $n^{2+\epsilon}$ steps is extremely small and that no matter where within distance n the random walk is at that time, the expected number of visits to 0 after that time is at most L_n'. Thus, $L_n \leq L_n'$ and $L_n' \leq \sum_0^{n^{2+\epsilon}} 1/(\pi k) + \epsilon L_n'$ for large n.

2.100. **(a)** Use complex exponentials to evaluate the integral.

(b) According to Glasser and Zucker (1977), Borwein and Zucker (1992), and Zucker (2011), for $d = 3$,

$$\sum_n p_n(0, 0) = \frac{\sqrt{3} - 1}{32\pi^3} \left(\Gamma(1/24) \Gamma(11/24)\right)^2 = 1.516386^+ \,.$$

2.101. This method of proving transience of $\mathbb{Z}^3$ is due to Levin and Peres (2010).

2.106. (This is due to Tom Hutchcroft, personal communication, 2015.) Consider three independent copies of the random walk. The triple is simple random walk on a graph that contains a graph that is roughly isometric to $\mathbb{N}^3$, and so is transient by Pólya's theorem. The bound can be sharpened to $\pi \Gamma(3/4)^{-4} \deg_G(y)^3 = 1.39^+ \deg_G(y)^3$ by using Rayleigh's monotonicity principle.

This argument works as long as G is a network that contains an infinite, simple path with conductances bounded away from 0.

Theorem 6.31 gives $p_n(y, y) \leq 6 \deg_G(y)/\sqrt{n + 1}$, which also gives the result. For the slightly sharper bound $p_n(y, y) \leq 4 \deg_G(y)/\sqrt{n + 1}$, see, for example, Lyons (2005), Remark 3.

It is well known that for every Markov chain that is not positive recurrent, $\lim_{n \to \infty} p_n(x, y) = 0$ for every pair of states x, y. For a proof, see, for example, Billingsley (1995), Theorem 8.7, Thorisson (2000), Theorem 2.3.2, or Grimmett and Stirzaker (2001), Theorem 6.4.17.

2.108. (This is due to Tetali (1991).) Use Proposition 2.20 and Exercise 2.68.

2.109. **(b)** Suppose that $H_A f = 0$. Apply (a) to conclude that for all $z \in A$, we have

$$f(z)\left(\sum_{x \in A} \mu_z^+(x) H(x, z) + \mathbf{E}_z[\tau_A^+]\right) = \mathbf{E}_z[\tau_A^+] \sum_{y \in A} f(y) \,.$$

Deduce that all values of f have the same sign, whence $f = 0$.

If the chain is aperiodic and finite, then the matrix H of hitting times determines the chain: see Theorem 4.4.12 of Kemeny and Snell (1960). For more on the inverse of H, see Neumann and Sze (2011).

2.110. Decompose a path of length k that starts at x and visits y into the part to the last visit of y and the rest (which may be empty). Reverse the first part and append the result of moving the second part to x via a fixed automorphism. The 1-skeleton of the truncated tetrahedron is an example of a transitive graph for which there are two vertices x and y such that no automorphism interchanges x and y.

2.112. The first identity (2.28) follows from a path-reversal argument. For the second, (2.29), use Exercise 2.108. This proof was given by Coppersmith, Tetali, and Winkler (1993), who discovered the result. The first equality of (2.29) does not follow from an easy path-reversal argument: consider, for example, the path $\langle x, y, x, y, z, x \rangle$, where $\tau_{y,z,x} = 6$, but for the reversed path, $\tau_{z,y,x} = 4$.

2.113. (This is due to Coppersmith, Tetali, and Winkler (1993), with a somewhat different approach to (d).) For (a), consider $\mathbf{P}_\pi[\tau_{x,y} > k]$.

2.114. A tricky bijective proof was given by Tanushev and Arratia (1997). A simpler proof proceeds by showing the equality with "$\le k$" in place of "$= k$." Decompose a path of length k into a cycle at x that completes the tour plus a path that does not return to x; reverse the cycle.

2.115. Run the chain from x until it first visits a and then z. This will also be the first visit to z from x, unless $\tau_z < \tau_a$. In the latter case, the path from x to a to z involves an extra commute between z and a beyond time τ_z. Taking expectations yields

$$\mathbf{E}_x[\tau_a] + \mathbf{E}_a[\tau_z] = \mathbf{E}_x[\tau_z] + \mathbf{P}_x[\tau_z < \tau_a]\big(\mathbf{E}_z[\tau_a] + \mathbf{E}_a[\tau_z]\big)$$

This yields the formula. In the reversible case, the cycle identity (2.29) yields

$$\mathbf{E}_x[\tau_a] + \mathbf{E}_a[\tau_z] - \mathbf{E}_x[\tau_z] = \mathbf{E}_a[\tau_x] + \mathbf{E}_z[\tau_a] - \mathbf{E}_z[\tau_x].$$

Adding these two quantities gives a sum of two commute times minus a third. Let γ denote the sum of all edge conductances, summed over all oriented edges. Then, by the commute-time formula of Corollary 2.21, the denominator in (2.30) is $\gamma \mathscr{R}(a \leftrightarrow z)$, and the numerator is $(\gamma/2)[\mathscr{R}(x \leftrightarrow a) + \mathscr{R}(a \leftrightarrow z) - \mathscr{R}(z \leftrightarrow x)]$.

2.117. Use Proposition 2.20. The result extends to trees with loops, where $A_y(z)$ also counts one half the number of loops separated from y.

2.118. This was conjectured by Jim Propp.

Solution 1, due to Y. Peres. The second statement is the same as saying that $\mathbf{E}_x[(1 + s)^{\tau_y}]$ has integer coefficients in s. Writing τ_y as a sum of hitting times over the path from x to y shows that it suffices to consider x and y neighbors. Of course, we may also assume y is a leaf. In fact, we may consider the return time $\mathbf{E}_y[(1 + s)^{\tau_y^+}] = \mathbf{E}_x[(1 + s)^{1+\tau_y}]$. The return time to y is $1 +$ a sum of a random number G of excursion lengths from x in the tree minus y, the random number G having a geometric distribution with parameter $1/d$, where d is the degree of x. Note that, since $\mathbf{P}[G = k] = (1 - 1/d)^k/d$ for $k \ge 0$, we have $\mathbf{E}[(1 + s)^G] = \sum(d - 1)^k s^k$. In the tree where y was deleted, each excursion is to one of the $d - 1$ neighbors of x other than y, each with probability $1/(d - 1)$. Furthermore, the excursion lengths above are return times to x in a smaller tree. Putting all this together gives the result.

Solution 2, due later to Yiping (Kenneth) Hu, personal communication, 2016. Let $G_y(x, z) := \mathbf{E}_x \sum_{k=1}^{\tau_y} \mathbf{1}_{[X_k=z]}$. That $G_y(x, z) \in \mathbb{N}$ can be proved by the same method of solution as that for Exercise 2.117. Then show that $\mathbf{E}_x\big[\binom{\tau_y}{k}\big] = \mathbf{E}_x \sum_{z_1,\ldots,z_k \text{ distinct}} \sum_{1 \le n_1 < \cdots < n_k \le \tau_y} \prod_{i=1}^k \mathbf{1}_{[X_{n_i}=z_i]} = \sum_{z_1,\ldots,z_k \text{ distinct}} \prod_{i=1}^k G_y(z_{i-1}, z_i)$, where $z_0 := x$.

2.119. Use the fact that the expected number of visits to x by time n is at most $\mathbf{P}_o[\tau_x \le n]\mathscr{G}(x, x)$.

2.122. Part (h) is due to Doyle and Steiner (2008), from which this proof is extracted. See also Gaudillière and Landim (2014) for a similar result and extensions to the study of transience and recurrence for nonreversible chains. Part (i) is due to Gaudillière and Landim (2014), Lemma 2.5, and Balázs and Folly (2016), Corollary 3.11.

(c) Note that $\big(L + \widehat{L}\big)/2 = I - \overline{P}$. Alternatively, show directly that

$$(Lf, f)_\pi = \frac{1}{2} \sum_{x,y} \pi(x) p(x, y)[f(x) - f(y)]^2.$$

(d) Use Exercise 2.120.

(g) Write $f = \overline{h} + \psi$ and $g = \overline{h} - \psi$.

(i) Note that the min-max cannot increase if it is required that $f = g = \varphi$.

2.123. Rewrite Exercise 2.122(d) as a matrix equation for matrices indexed by states other than z, namely, $L_z H_z = D_z$, where $L_z(x, y) := L(x, y)$, $H_z(x, y) := \mathbf{P}_x[\tau_y < \tau_z]$, and D_z is the diagonal matrix whose (x, x)-entry is $r(x, z)/\pi(x)$. This gives $D_z^{-1} L_z = H_z^{-1}$. Since the diagonal of L_z is 1, this determines D_z and hence L_z. Using three different choices of z, this allows us to determine L and therefore P. (If there are only two states, the result is trivial.)

2.124. This is due to Tetali (1994b).

2.125. For the original proofs, see Aldous and Fill (2002), Corollary 9.23, and Tetali (1994a), Corollary 2.3.

(a) Use Exercise 2.124 and Exercise 2.122(i).

(b) Use Exercise 2.123.

2.126. Use Propositions 2.1 and 2.2.

2.128. (This formula is due to Coppersmith, Doyle, Raghavan, and Snir (1993).) Use Exercises 2.126 and 2.65.

2.129. (This is due to Thomas Sauerwald, personal communication, 2007.) Let the distance be n, and let Π_k be the edges with one endpoint at distance k from a and the other at distance $k + 1$ from a, where $0 \le k < n$. Write $A_k := \sum_{e \in \Pi_k} c(e)$. Corollary 2.21 and (2.13) give that the commute time is

$$2\mathscr{R}(a \leftrightarrow z) \sum_{e \in \mathsf{E}_{1/2}} c(e) \ge 2 \sum_{k=0}^{n-1} A_k^{-1} \sum_{k=0}^{n-1} A_k \ge 2n^2$$

by the Cauchy–Schwarz inequality.

2.130. (This is due to Aleliunas, Karp, Lipton, Lovász, and Rackoff (1979). An improvement for regular graphs was given by Kahn, Linial, Nisan, and Saks (1989). See Exercise 2.132 for a further improvement.) Take a spanning tree T of G and consider a cycle in G that covers each edge of T twice. For each edge of T, consider the commute time between the endpoints of that edge. Summing these commute times is an upper bound for the expected cover time. This bound can be improved by a factor of what is called the edge toughness of G; though not the definition, the edge toughness equals the maximum of p/q over all (p, q) such that there exist p spanning trees that use each edge no more than q times. This improvement follows from a theorem of Tutte (1961) and Nash-Williams (1961) by replacing each edge with q parallel edges. Another improvement is in Exercise 2.131.

2.132. (These results are due to Coppersmith, Feige, and Shearer (1996).) Use Foster's theorem, Exercise 2.65.

2.133. **(a)** The voltage is constant on the vertices with fixed coordinate sum.

(b) Consider the graph used in (a) where we identify certain vertices. The expected hitting time from k to 0 in this graph is the same as the expected hitting time from a vertex of distance k from $\mathbf{0}$ in the hypercube. When $k = 1$, the expected hitting time can also be computed by using the fact that the expected return time equals 2^n.

2.135. Let $\mathscr{K}$ be the linear span of $Z(w)$ for $w \in W$. Let G/W be the network obtained from G by identifying W to a single vertex. Let Z_W be the associated canonical Gaussian field. Then it suffices to show that $Z_W(x) = P_{\mathscr{K}}^\perp Z(x)$ in light of Proposition 2.24(a). We show this in the form $dZ_W(e) = P_{\mathscr{K}}^\perp dZ(e)$. Now $dZ(e) = P_{\star(\mathscr{H})} X(e)$ and $dZ_W(e) = P_{\mathscr{L}} X(e)$, where $\mathscr{L}$ is the linear span of $\sum_{e^- = x} X(e)/r(e)$ for $x \notin W$. Thus, it suffices to show that $\star(\mathscr{H}) = \mathscr{L} \oplus \mathscr{K}$. The only part of this that is not immediate is the orthogonality; to prove that, write $Z(w) = \sum_{e \in \psi} dZ(e)$ for a path ψ joining o to w.

The result may also be proved by using densities: see Ding, Lee, and Peres (2012), Lemma 2.15.

2.136. The orthogonal projection that gives ∇Z produces a standard normal random vector in ★, which is not concentrated on any subspace of smaller dimension.

2.137. **(a)** This is a deterministic result.

(c) Use Proposition 2.24. This also follows from Exercise 2.62 and the standard result relating the covariance of Gaussians to their density $\left(\text{note that } \|dZ\|_C^2 = (Z, \Delta_G Z)\right)$.

2.138. Add a new vertex and an edge from every vertex to that new vertex. Then apply Exercise 2.137.

Chapter 3

3.1. The random path $\langle Y_n \rangle$ visits x at most once.

3.3. Let $Z := \{e^+ ; e \in \Pi\}$, and let G be the subtree of T induced by those vertices that Π separates from ∞. Then the restriction of θ to G is a flow from o to Z, whence the result is a consequence of Lemma 2.8.

3.4. Consider spherically symmetric trees.

3.9. **(a)** Let $\beta > \inf a_n/n$ and let m be such that $a_m/m < \beta$. Write $a_0 := 0$. For any n, write $n = qm + r$ with $0 \le r \le m - 1$; then we have

$$a_n = a_{qm+r} \le a_m + \cdots + a_m + a_r = qa_m + a_r \,,$$

whence

$$\frac{a_n}{n} \le \frac{qa_m + a_r}{qm + r} = \frac{a_m}{m} \cdot \frac{qm}{qm + r} + \frac{a_r}{n} < \beta \cdot \frac{qm}{qm + r} + \frac{a_r}{n} \,,$$

and so

$$\limsup_{n \to \infty} \frac{a_n}{n} \le \beta \,.$$

(b) Modify the proof so that if an infinite a_r appears, then it is replaced by a_{m+r} instead.

(c) Observe that $\langle \log |T_n| \rangle$ is subadditive.

3.10. Modify the proof of Theorem 3.9. A closer look at the proof shows that Fekete's lemma (Exercise 3.9) was not, in fact, needed.

3.13. Suppose that w_x is broken arbitrarily as the concatenation of two words w_1 and w_2. Let x_i be the product of the generators in w_i ($i = 1, 2$). Then $x = x_1 x_2$. If for some i, we had $w_{x_i} \ne w_i$, then we could substitute the word w_{x_i} for w_i and find another word w whose product was x yet would be either shorter than w_x or come earlier lexicographically than w_x. Either of these circumstances would contradict the definition of w_x as minimal, whence $w_i = w_{x_i}$ for both i.

This fact for $i = 1$ shows that T is a tree; for $i = 2$, it shows that T is subperiodic.

3.14. Replace each vertex x by two vertices x' and x'' together with an edge from x' to x'' of capacity $c(x)$. All edges that led into x now lead into x', whereas all edges that led out of x now lead out of x''. Apply the max-flow min-cut theorem for directed networks (with edge capacities).

3.16. Use Exercise 3.14(a) for part (a). For part (b), let $\langle H_n \rangle$ be an exhaustion of H by finite graphs and b_n be the maximum number of pairwise disjoint paths from a to the complement of H_n. Then $\langle b_n \rangle$ is decreasing and so eventually constant. Now apply part (a).

3.17. Use Menger's theorem. Introduce a source a connected to A and a sink z connected to Z. The same proofs show that there is a matching that covers A if (1) the degree of every vertex in A is at least the degree of every vertex in Z, or if (2) $|A| \le |Z|$ and the rest of the problem statement (b) holds.

3.19. A minimal cutset is the same as a cut, as defined in Exercise 2.61.

3.21. (This is Proposition 13 of Hoffman, Holroyd, and Peres (2006).) Find a flow $\theta \le q$. Consider θ' of Proposition 3.2. Show that $\mathscr{E}(\theta') < \infty$.

3.22. This is due to Y. Peres.

3.23. Let G be the subtree of T induced by those vertices that Π separates from ∞. If G were infinite, then we could find a path $o = x_0, x_1, x_2, \ldots$ of vertices in G such that each x_{i+1} is a child of x_i by choosing x_{i+1} to be a child of x_i with an infinite number of descendants in G. This would produce a path from o to ∞ that did not intersect Π, which contradicts our assumption. Hence G is finite. Let Π' be the set of edges joining a vertex of G to a vertex of $T \setminus G$ that has an infinite number of descendants.

3.24. If T is transient, then the voltage function, normalized to take the value 1 at the root o and vanish at infinity, works as F. Conversely, given F, use the max-flow min-cut theorem to find a nonzero flow θ from the root with $\theta(e) \leq dF(e)c(e)$ for all e. Thus $\sum_{e \in \xi} \theta(e)/c(e) \leq F(o)$ for every path ξ from o. For each path ξ connecting o to x, multiply this inequality by $\theta(e(x))$ and sum over $x \in T_n$. This implies that θ has finite energy.

3.27. This an immediate consequence of the max-flow min-cut theorem. In the form of Exercise 15.29, it is due to Frostman (1935).

3.28. Let $x \in T$ with $|x| > k$. Let u be the ancestor of x which has $|u| = |x| - k$. Then the embedding of T^u into T embeds T^x into T^w for some $w \in T_k$.

3.30. Consider the directed graph on $0, 1, \ldots, k$ with edges $\langle i, j \rangle$ for $i \leq j \leq i + 1$.

3.31. This is part of Theorem 5.1 of Lyons (1990).

3.32. This seems to be new. Let β be an irrational number. Write $\{x\}$ for the fractional part of a real number x. For real x, y and $k \in \mathbb{N}$, set $f_k(x, y) := (\mathbf{1}_{[0,1/2]}(\{x + k\beta\}), \mathbf{1}_{[0,1/2]}(\{y + k\beta\}))$ and let $F_n(x, y)$ be the sequence $\langle f_k(x, y); 0 \leq k < n \rangle$. Let T be the tree of all finite sequences of the form $F_n(x, y)$ for $x, y \in \mathbb{R}$ and $n \in \mathbb{N}$, together with the null sequence, which is the root of T. Join $F_n(x, y)$ to $F_{n+1}(x, y)$ by an edge. Then $|T_n| = 4n^2$. (A sequence $\langle \mathbf{1}_{[0,1/2]}(\{x + k\beta\}); 0 \leq k < n \rangle$ changes as x increases from 0 to 1 exactly when one of the points $\{x + k\beta\}$ passes $1/2$ or 1, so each of these n points contributes to two changes.) Let $\mathcal{L}_2$ be Lebesgue measure on $[0, 1]^2$, and for vertices $w \in T$, set $\theta(w) := \mathcal{L}_2\{(x, y); F_{|w|}(x, y) = w\}$. Now choose β not well approximable by rationals – for example, set $\beta := \sqrt{5}$. Then θ is approximately uniform on T_n: Given $w \in T_n$, we want to show that the set of (x, y) for which $F_n(x, y) = w$ has measure at most c/n^2. This set is a product of two intervals. In fact, all $2n$ intervals determined by $\{k\beta\}$ and $\{k\beta + 1/2\}$ for $0 \leq k < n$ are of length at most $3/n$: see the argument at the bottom of p. 125 of Kuipers and Niederreiter (1974). Therefore θ has finite energy for unit conductances. That is, simple random walk is transient.

3.33. Given $(x_1, \ldots, x_n)$ and $(y_1, \ldots, y_m)$ that correspond to vertices in T, the sequence obtained by concatenating them with N zeros in between also corresponds to a vertex in T. Thus T is N-superperiodic and $\mathrm{br}\, T = \mathrm{gr}\, T$. To rule out $(N-1)$-superperiodicity, use rational approximations to α. If $\alpha > 1/2$ or $\alpha = 1/2$, then $\mathrm{gr}\, T = 2$ by the SLLN or the ballot theorem, respectively. If $\alpha < 1/2$, then Chernoff–Cramér's theorem on large deviations, or Stirling's formula, combined with the idea in the discussion after Theorem 2.1 of Spitzer (1956) imply that $\log \mathrm{gr}\, T$ is the entropy of $(\alpha, 1 - \alpha)$.

3.34. Use Exercise 3.33.

3.35. Product trees were introduced by Lyons (1992) for studying random labeling of trees.

(c) $T(S) \cdot T(\mathbb{N} \setminus S)$ is a binary tree.

3.38. The Cayley graph of $\mathbb{Z}$ with respect to the generators $\{1, n\}$ is a skew cylinder, that is, the strip $\mathbb{Z} \times [0, n]$ with each (x, n) identified with $(x + 1, 0)$. Similarly, the Cayley graphs of $\mathbb{Z}$ with respect to $\{1, n, n^2\}$ approximate the usual Cayley graph of $\mathbb{Z}^3$.

3.39. This is due to Lyons (1995).

Chapter 4

4.1. For the random walk version, this is proved in Sections 6.5 and 7.3 of Lawler (1991). Use the "craps principle" (Pitman (1993), p. 210). First prove equality for the distribution of the first step.

4.3. Use also Rayleigh's monotonicity principle.

4.6. (This is also due to Feder and Mihail (1992).) We follow the proof of Theorem 4.6. We induct on the number of edges of G. Given $\mathscr{A}$ and $\mathscr{B}$ as specified, there is an edge e on which $\mathscr{A}$ depends (and so $\mathscr{B}$ ignores) such that $\mathscr{A}$ is positively correlated with the event $e \in T$ (since $\mathscr{A}$ is negatively correlated with those edges that $\mathscr{A}$ ignores). Thus, $\mathbf{P}[\mathscr{A} \mid e \in T] \geq \mathbf{P}[\mathscr{A} \mid e \notin T]$. Now

$$\mathbf{P}[\mathscr{A} \mid \mathscr{B}] = \mathbf{P}[e \in T \mid \mathscr{B}]\mathbf{P}[\mathscr{A} \mid \mathscr{B}, e \in T] + \mathbf{P}[e \notin T \mid \mathscr{B}]\mathbf{P}[\mathscr{A} \mid \mathscr{B}, e \notin T]. \qquad (18.1)$$

The induction hypothesis implies that (18.1) is at most

$$\mathbf{P}[e \in T \mid \mathscr{B}]\mathbf{P}[\mathscr{A} \mid e \in T] + \mathbf{P}[e \notin T \mid \mathscr{B}]\mathbf{P}[\mathscr{A} \mid e \notin T]. \qquad (18.2)$$

By Theorem 4.6, we have that $\mathbf{P}[e \in T \mid \mathscr{B}] \leq \mathbf{P}[e \in T]$, and we have chosen e so that $\mathbf{P}[\mathscr{A} \mid e \in T] \geq \mathbf{P}[\mathscr{A} \mid e \notin T]$. Therefore, (18.2) is at most

$$\mathbf{P}[e \in T]\mathbf{P}[\mathscr{A} \mid e \in T] + \mathbf{P}[e \notin T]\mathbf{P}[\mathscr{A} \mid e \notin T] = \mathbf{P}[\mathscr{A}].$$

For the second part, note that every increasing random variable X has the form $X = c + \sum_i a_i \mathbf{1}_{\mathscr{A}_i}$ for some constant c, positive a_i, and increasing events $\mathscr{A}_i$.

4.7. (Compare Theorem 3.2 in Thomassen (1990).) The number of components of T when restricted to the subgraph induced by V_n is at most $|\partial_E \mathsf{V}_n|$. A tree with k vertices has $k - 1$ edges. The statement on expectation follows from the bounded convergence theorem.

4.10. You need to use $H(1, 1)$ to find $Y(e, f) = 1/2 - 1/\pi$ for e the edge from $(0, 0)$ to $(1, 0)$ and f the edge from $(0, 0)$ to $(0, 1)$. From this, the values of $Y(e, g)$ for the other edges g incident to the origin follow. Use the transfer-current theorem directly to find $\mathbf{P}[\deg_T(0, 0) = 4]$. Other probabilities can be computed by Exercise 4.41 or by computing $\mathbf{P}[\deg_T(0, 0) \geq 3]$ and using the fact that the expected degree is 2 (Exercise 4.7). See Burton and Pemantle (1993), p. 1346, for some of the details.

It is not needed for the solution to this problem, but here are some values of H. Such a table

4	$80 - \dfrac{736}{3\pi}$	$-49 + \dfrac{160}{\pi}$	$12 - \dfrac{472}{15\pi}$	$-1 + \dfrac{48}{5\pi}$	$\dfrac{704}{105\pi}$
3	$17 - \dfrac{48}{\pi}$	$-8 + \dfrac{92}{3\pi}$	$1 + \dfrac{8}{3\pi}$	$\dfrac{92}{15\pi}$	$-1 + \dfrac{48}{5\pi}$
2	$4 - \dfrac{8}{\pi}$	$-1 + \dfrac{8}{\pi}$	$\dfrac{16}{3\pi}$	$1 + \dfrac{8}{3\pi}$	$12 - \dfrac{472}{15\pi}$
1	1	$\dfrac{4}{\pi}$	$-1 + \dfrac{8}{\pi}$	$-8 + \dfrac{92}{3\pi}$	$-49 + \dfrac{160}{\pi}$
0	0	1	$4 - \dfrac{8}{\pi}$	$17 - \dfrac{48}{\pi}$	$80 - \dfrac{736}{3\pi}$
(x_1, x_2)	0	1	2	3	4

was first constructed by McCrea and Whipple (1940). It is used for studying harmonic measure in that paper, as well as in Spitzer (1976), Section 15. See those references for proofs that

$$\lim_{|x_1|+|x_2|\to\infty}\left[H(x_1,x_2)-\frac{2}{\pi}\log\sqrt{|x_1|^2+|x_2|^2}\right]=\frac{2\gamma+\log 8}{\pi},$$

where γ is Euler's constant. In fact, the convergence is quite rapid. See Kozma and Schreiber (2004) for more precise estimates and higher dimensions. See also Uchiyama (1998) for walks with unbounded jumps and Kazami and Uchiyama (2008) for walks on other Euclidean lattices. Using the preceding table, we can calculate the transfer currents, which we give in two tables. We show $Y(e,f)$ for e the edge from $(0,0)$ to $(1,0)$ and f varying. The first table is for the horizontal edges f, labeled with the left-hand endpoint of f:

4	$-\dfrac{129}{2}+\dfrac{608}{3\pi}$	$\dfrac{95}{2}-\dfrac{746}{5\pi}$	$-\dfrac{37}{2}+\dfrac{872}{15\pi}$	$\dfrac{7}{2}-\dfrac{1154}{105\pi}$	0
3	$-\dfrac{25}{2}+\dfrac{118}{3\pi}$	$\dfrac{17}{2}-\dfrac{80}{3\pi}$	$-\dfrac{5}{2}+\dfrac{118}{15\pi}$	0	$-\dfrac{7}{2}+\dfrac{1154}{105\pi}$
2	$-\dfrac{5}{2}+\dfrac{8}{\pi}$	$\dfrac{3}{2}-\dfrac{14}{3\pi}$	0	$\dfrac{5}{2}-\dfrac{118}{15\pi}$	$\dfrac{37}{2}-\dfrac{872}{15\pi}$
1	$-\dfrac{1}{2}+\dfrac{2}{\pi}$	0	$-\dfrac{3}{2}+\dfrac{14}{3\pi}$	$-\dfrac{17}{2}+\dfrac{80}{3\pi}$	$-\dfrac{95}{2}+\dfrac{746}{5\pi}$
0	$\dfrac{1}{2}$	$\dfrac{1}{2}-\dfrac{2}{\pi}$	$\dfrac{5}{2}-\dfrac{8}{\pi}$	$\dfrac{25}{2}-\dfrac{118}{3\pi}$	$\dfrac{129}{2}-\dfrac{608}{3\pi}$
(x_1,x_2)	0	1	2	3	4

The second table is for the vertical edges f, labeled with the lower endpoint of f:

4	$-138+\dfrac{6503}{15\pi}$	$79-\dfrac{1241}{5\pi}$	$-25+\dfrac{1649}{21\pi}$	$4-\dfrac{1321}{105\pi}$
3	$-26+\dfrac{245}{3\pi}$	$13-\dfrac{613}{15\pi}$	$-3+\dfrac{47}{5\pi}$	$\dfrac{1}{2}-\dfrac{167}{105\pi}$
2	$-5+\dfrac{47}{3\pi}$	$2-\dfrac{19}{3\pi}$	$-\dfrac{1}{2}+\dfrac{23}{15\pi}$	$-3+\dfrac{47}{5\pi}$
1	$-1+\dfrac{3}{\pi}$	$\dfrac{1}{2}-\dfrac{5}{3\pi}$	$2-\dfrac{19}{3\pi}$	$13-\dfrac{613}{15\pi}$
0	$-\dfrac{1}{2}+\dfrac{1}{\pi}$	$-1+\dfrac{3}{\pi}$	$-5+\dfrac{47}{3\pi}$	$-26+\dfrac{245}{3\pi}$
(x_1,x_2)	1	2	3	4

Symmetries of the plane give some other values from these.

4.11. (From Propp and Wilson (1998).) Let T be a tree rooted at some vertex x. Choose any directed path $x = u_0, u_1, \ldots, u_l = x$ from x back to x that visits every vertex. For $1 \le i < l$, let $\mathcal{P}_i$ be the path $u_0, u_1, \ldots, u_i$ followed by the path in T from u_i to x. In the trajectory $\mathcal{P}_{l-1}, \mathcal{P}_{l-2}, \ldots, \mathcal{P}_1$, the last time any vertex $u \neq x$ is visited, it is followed by its parent in T. Therefore, if the chain on spanning trees begins at any spanning tree, following this trajectory (which has positive probability of happening) will lead to T.

4.13. We have
$$\widehat{p}(T, B(T, e)) = \frac{\pi(B(T, e))}{\pi(T)} p(B(T, e), T) = \frac{\Psi(B(T, e))}{\Psi(T)} p(g) = p(e).$$

4.14. A solution using the Aldous–Broder algorithm was noted by Broder (1989).

4.15. This is due to Aldous (1990).

4.16. (This exercise was motivated by Kozdron, Richards, and Stroock (2013).) Cramer's theorem tells us that $\Delta_G[a]^{-1}(x, x) = \det \Delta_G[a, x]/\deg \Delta_G[a]$ in the notation of Exercise 2.62. Here, $\Delta_G[a, x]$ means deleting the rows and columns indexed by both a and x. By Exercise 2.62, this equals $v_a(x, x) = \mathscr{R}(a \leftrightarrow x; G)$. Kirchhoff's effective-resistance formula then gives that if (x, a) is an edge of t, $\mathbf{P}[(x, a) \in T_G] = c(x, a) \det \Delta_G[a, x]/\deg \Delta_G[a]$. Now contract the edge (x, a) and repeat for another edge of t connected to x or a. Multiply all the successive quotients, which telescope to give
$$\frac{\Xi(t)}{\Xi(G)} = \mathbf{P}[T_G = t] = \frac{\Xi(t)}{\det \Delta_G[a]}.$$

4.18. (This is due to Aldous (1990).) Use Exercise 4.15 and torus grids (or square grids).

4.19. There are too many spanning trees.

4.20. This is due to Edmonds (1971); see Corollary 50.7(c) of Schrijver (2003).

4.21. (These results are due to Asadpour, Goemans, Mądry, Oveis Gharan, and Saberi (2010) and depend very little on properties of spanning trees. See Singh and Vishnoi (2014) for an extension and related issues and Gel'fand, Goresky, MacPherson, and Serganova (1987) for a related extension to matroids.) Use Lagrange multipliers for (a). Deduce (b) from (a).

4.23. This is due to Kirchhoff (1847).

4.25. (The first part is from Propp and Wilson (1998).) We sum over the number of transitions out of x that are needed. We may find this by popping cycles at x. The number of such is the number of visits to x starting at x before visiting r. This is $\pi(x)(\mathbf{E}_x[\tau_r] + \mathbf{E}_r[\tau_x])$, as can be seen by considering the frequency of appropriate events in a bi-infinite path of the Markov chain. This also follows from Exercise 2.42 by using the stopping time equal to the first visit to x after the first visit to r. The number of visits to x starting at x before visiting r is $\sum_{n \ge 0} P_r^n(x, x)$, whence the trace formula follows from writing the matrix inverse as an infinite series; this formula is due to Marchal (2000). In the reversible case, we can also use the formulas in Section 2.2 to get
$$\sum_{x \text{ a state}} \sum_{e^- = x} c(e)\mathscr{R}(x \leftrightarrow r),$$
which is the same as what is to be shown.

4.26. Use the craps principle as in the solution to Exercise 4.1.

4.27. This is due to Meir and Moon (1970).

4.29. This is due to Foster (1948).

4.30. Add a (new) edge e from a to z of unit conductance. Then spanning trees of $G \cup \{e\}$ containing e are in 1-1 correspondence preserving $\Xi(\cdot)$ with spanning trees of $G/\{a, z\}$. Thus, the right-hand side of (4.21) is $(1 - \mathbf{P}[e \in T])/\mathbf{P}[e \in T]$ in $G \cup \{e\}$. Now apply Kirchhoff's effective-resistance formula.

4.33. Use Exercise 4.21.

4.34. (This is due to R. Lyons.) For (a), let X be the set of subsets A of vertices of G such that both A and $V \setminus A$ induce connected subgraphs. Identify each $A \in X$ with the subnetwork it induces. On X, put the measure $\mu(A) := \Xi(A)\Xi(A^c)/(2\Xi(G))$. Map a vertex x in G to the function on X given by $A \mapsto \mathbf{1}_A(x)$. Use Exercise 4.30(a) to verify that this is an isometry into $\ell^1(X, \mu)$. Now use this embedding to deduce (b). A metric space that satisfies property (b) is called *hypermetric*; Kelly (1970) noted that this property is a consequence of the existence of an embedding into some L^1 space as indicators, which is always the case when there is some embedding into L^1 (see, for example, Naor (2010)). He also noted that the hypermetric property implies the negative type property of Exercise 2.59.

4.36. (Part (b) is due to Morris (2003).) If there is an edge joining a to z, use Kirchhoff's effective-resistance formula. If there is no edge joining a to z, add one and let its conductance tend to 0.

4.37. Use Wilson's algorithm with a given rung in place to calculate the chance that the next rung (in one direction) is at a given location. Alternatively, use the transfer-current theorem. This problem was originally analyzed by Häggström (1994), who used the theory of subshifts of finite type.

4.39. Use the fact that $i - i' - [i(f) - i'(f)]\chi^f$ is a flow between the endpoints of f.

4.41. For the first part, compare coefficients of monomials in x_i. The second part is Corollary 4.4 in Burton and Pemantle (1993).

4.42. Calculate the chance that all the edges incident to a given vertex are present.

4.45. In the case where all w_e are equal, this is a special case of what is proved by Feder and Mihail (1992). In general, use the same proof as that of Exercise 4.6. An earlier and more straightforward proof of this special case is given by Joag-Dev and Proschan (1983), paragraph 3.1(c). The general case follows from this special case via the implication ULC $\Rightarrow$ CNA$^+$ of Theorem 2.7 in Pemantle (2000).

4.46. **(a)** Let $U \sim \text{Unif}[0, 1]$ and $X_i := \min\{r \,;\, U \le \mu_i((-\infty, r])\}$.

(b) Use the result of Exercise 4.45.

4.48. **(a)** This is Theorem 14.3 of BLPS (2001).

(b) Suppose that $y = x + (1, 0)$. If $\mathbf{E}[L] < \infty$, then we may choose N so that $\sum_{n \ge N} \mathbf{P}[L \ge n] < 1/8$ and M so that the T-distance between $(0, N)$ and $(0, -N)$ is at most M with probability less than $1/4$. Then with positive probability, all the following occur simultaneously: the T-distance between (i, N) and $(i + 1, N)$ is less than $N + |i|$ for all i, the T-distance between $(i, -N)$ and $(i + 1, -N)$ is less than $N + |i|$ for all i, the T-distance between (M, N) and $(M, -N)$ is at most M, and the T-distance between $(-M, N)$ and $(-M, -N)$ is at most M. In this case, there is a path in T that winds around the origin, whence T is not a tree.

The actual rate of decay of the probabilities for the uniform spanning tree is $\mathbf{P}[L \ge n] \asymp 1/n^{3/5}$. This follows from combining Theorem 1.1 of Lawler (2014) with Barlow and Masson (2010); see also Corollary 3.15 of Barlow (2016).

4.49. See Exercise 2.100(a).

4.50. This particular question was asked in the comic strip xkcd, `http://xkcd.com/356/`, "Nerd Sniping." The strip's author first saw it on a Google Labs Aptitude Test.

4.51. Compare this exact result to Proposition 2.15.

4.52. When $x = (\pm n, \pm n)$, this follows from Proposition 4.7 and (4.19). For other $x \ne \mathbf{0}$, use the fact that $x \mapsto \mathcal{R}(\mathbf{0} \leftrightarrow x)$ is harmonic and thus $\mathcal{R}(\mathbf{0} \leftrightarrow x) = \mathbf{E}_x[\mathcal{R}(\mathbf{0} \leftrightarrow Z_\tau)]$, where $\langle Z_k \rangle$ is simple random walk on $\mathbb{Z}^2$ and τ is the first time the random walk visits a diagonal. To estimate the hitting distribution on the diagonals, note that if $Z_k = (X_k, Y_k)$, then $X_{2k} + Y_{2k}$ and $X_{2k} - Y_{2k}$ are independent simple random walks on $\mathbb{Z}$ observed at even times, so it suffices to understand the distribution of one of these walks at the time that the other hits 0 for the first time. One can use the reflection principle to estimate the first hitting time of 0.

Compare to the result in Exercise 2.99.

4.53. This is due to Spitzer (1962); see also Section 15 of Spitzer (1976).

(a) Note that for a starting point y outside the square, μ_y is an average of μ_z for z on the square. For points z and w at odd starting distance, use independent horizontal lazy simple random walks until they have even distance, where ***lazy simple random walk*** has transition probabilities $p(k, k) = 1/2$ and $p(k, k \pm 1) = 1/4$. For another proof of existence of harmonic measure from infinity, see Theorems 10.56 and 10.36.

(b) Apply Exercise 2.111 to a torus graph $(\mathbb{Z}/n\mathbb{Z})^2$, and let $n \to \infty$ in combination with part (a) here.

(c)–(e) Use the calculations in the answer to Exercise 4.10. All matrices that need to be inverted here have the form

$$\begin{bmatrix} 0 & 1 & 4x \\ 1 & 0 & 1 \\ 4x & 1 & 0 \end{bmatrix}.$$

4.55. More generally, consider a network on the vertex set $\mathbb{Z}$ that is invariant under translations, that is, the conductance c_k of an edge $[x, x + k]$ depends on k but not on x. If there are edges $[0, z]$ and $[x, y]$, then the transfer current from the first edge to the second is

$$Y\big([0, z], [x, y]\big) = \int_{\mathbb{T}} \frac{c_{|y-x|}(1 - e^{-2\pi i z\alpha})(e^{2\pi i x\alpha} - e^{2\pi i y\alpha})}{\sum_{k \geq 1} 2c_k(1 - \cos 2\pi k\alpha)}\, d\alpha.$$

Chapter 5

5.1. Various theorems will work. The monotone convergence theorem works even when $m = \infty$.

5.2. Compare Neveu (1986).

5.3. (Compare Neveu (1986).) Let Ω be the probability space on which the random variables $L_i^{(n)}$ are defined. Then **GW** is the law of the random variable $T: \Omega \to \mathscr{T}$ defined by

$$T := \big\{\langle i_1, \ldots, i_n\rangle;\ n \geq 0,\ i_j \in \mathbb{Z}^+\ (1 \leq j \leq n),\ i_{j+1} \leq L_{I(i_1,\ldots,i_j)}^{(j+1)}\ (1 \leq j \leq n-1)\big\},$$

where $I(i_1, \ldots, i_j)$ is the index appropriate to the individual $\langle i_1, \ldots, i_j\rangle$. Actually, any injection I from the set of finite sequences to $\mathbb{Z}^+$ will do, rather than the one implicit in the definition of a Galton–Watson process, that is, (5.1). For example, if $\langle P_n\rangle$ denotes the sequence of prime numbers, then we may use $I(i_1, \ldots, i_j) := \prod_{l=1}^{j} P_{i_l}$.

5.4. We will soon see (Corollary 5.10) that also $\operatorname{br} T = m$ a.s. given nonextinction.

5.5. Show that for each n, the event that the diameter of $K(x)$ is at least n is measurable.

5.6. One way is to let $\langle G_n\rangle$ be an exhaustion of G by finite subgraphs containing x. If

$$\mathbf{P}_p[\exists \text{ infinite-diameter cluster}] > 0,$$

then for some n, we have $\mathbf{P}_p[\exists u \in G_n\ \ u \leftrightarrow \infty] > 0$. This latter event is independent of the event that all the edges of G_n are present. The intersection of these two events is contained in the event that $x \leftrightarrow \infty$.

5.7. Let $p \geq p_c(G)$ and $p' \in [0, 1]$. Given ω_p, the law of $\omega_{pp'}$ is precisely that of percolation on ω_p with survival parameter p'. Therefore, we want to show that if $p' < p_c(G)/p$, that is, if $pp' < p_c(G)$, then $\omega_{pp'}$ a.s. has no infinite components, whereas if $pp' > p_c(G)$, then $\omega_{pp'}$ a.s. has an infinite component. But this follows from the definition of $p_c(G)$.

5.14. For (c), let $h(x) := \theta(e(x))^2 - \sum_{\overleftarrow{y} = x} \theta(e(y))^2 \geq 0$. By (b), we have $\theta(e(x)) = \sum_{z \geq x} h(z)$.

5.15. We have

$$G_d(s) = 1 - (1-s)^{d+1} p^{d+1} - (d+1)(1-s)^d p^d (1 - p + ps).$$

Since $G_d(0) > 0$ and $G_d(1) = 1$, there is a fixed point of G_d in $(0, 1)$ if $G_d(s) < s$ for some $s \in (0, 1)$. Consider

$$g(s) := 1 - (1-s)^{d+1} - (d+1)(1-s)^d s$$

obtained from $G_d(s)$ by taking $p \to 1$. Since $g'(0) = 0$, there is certainly some $s \in (0, 1)$ with $g(s) < s$. Hence the same is true for G_d when p is sufficiently close to 1.

5.19. (Parts (a)–(c) are due to Curien and Le Gall (2011), Lemma 5.5.) For (b), use the optional stopping theorem, considering $\inf\{k \geq 0 ;\ A(\xi_k) < 4\}$. For (c), use Proposition 5.36 to get a contradiction if the result does not hold. More information about the finite paths with labels at least 4 is in Curien and Peres (2011).

5.20. Generate the labels at the same time as you generate the tree. That is, the root is labeled i with probability $1/k$; then there are j children of the root with probability $e^{-c} c^j/j!$ and they are labeled by a subset of $[1, k] \setminus \{i\}$ with probability $\binom{k-1}{j}^{-1}$ each, and so on.

For a more formal proof using induction, let B be the random labeling of T. Let t be a k-vertex rooted tree with a labeling b. Suppose the root of t has j children, with corresponding subtrees of sizes $k_1, \ldots, k_j$. Thus $\sum_{i=1}^{j} k_i = k - 1$. The probability that the root labels match in B and b is $1/k$, and given that, the chance that the set of labels assigned to each of the j subtrees matches in B and b is $j! \dfrac{\prod_{i=1}^{j} k_i!}{(k-1)!}$; the factor of $j!$ is due to the possibility of permuting the children of the root. Thus the probability that (T, B) coincides with (t, b) equals

$$\mathbf{P}[Z_1(T) = j] \frac{1}{k} j! \frac{\prod_{i=1}^{j} k_i!}{(k-1)!} \prod_{i=1}^{j} \frac{e^{-c k_i} c^{k_i - 1}}{k_i!},$$

and this indeed equals $e^{-ck} c^{k-1}/k!$, since $\mathbf{P}[Z_1(T) = j] = e^{-c} c^j/j!$.

5.21. This type of percolation is known as the **Erdős–Rényi random graph**. For (a), use a labeling similar to that used for trees in Section 5.1. For (b) and (c), consider a random total ordering of the vertices of K_n. It induces a relative ordering of the vertices of $C(o)$, which in turn induces a labeling. For (c), calculate the exact value of the left-hand side before the limit is taken.

5.22. Let f_n and f be the corresponding p.g.f.'s. Then $f_n(s) \to f(s)$ for each $s \in [0, 1]$.

5.29. **(a)** By symmetry, we have on the event A that

$$\mathbf{E}\left[\frac{L_i^{(n+1)}}{\sum_{j=1}^{Z_n} L_j^{(n+1)} + \sum_{j=1}^{Z_n'} L_j'^{(n+1)}} \,\middle|\, \mathscr{F}_n \right] = \mathbf{E}\left[\frac{L_i'^{(n+1)}}{\sum_{j=1}^{Z_n} L_j'^{(n+1)} + \sum_{j=1}^{Z_n'} L_j^{(n+1)}} \,\middle|\, \mathscr{F}_n \right]$$

$$= \frac{1}{Z_n + Z_n'}.$$

(b) By part (a), we have

$$\mathbf{E}[Y_{n+1} \mid \mathscr{F}_n] = \frac{Z_n}{Z_n + Z_n'} \mathbf{P}[A \mid \mathscr{F}_n] + Y_n \mathbf{P}[\neg A \mid \mathscr{F}_n]$$

$$= \mathbf{E}\left[\frac{Z_n}{Z_n + Z_n'} \mathbf{1}_A \,\middle|\, \mathscr{F}_n \right] + \mathbf{E}\left[Y_n \mathbf{1}_{\neg A} \,\middle|\, \mathscr{F}_n \right]$$

$$= \mathbf{E}[Y_n \mathbf{1}_A + Y_n \mathbf{1}_{\neg A} \mid \mathscr{F}_n] = Y_n.$$

Since $0 \le Y_n \le 1$, the martingale converges to $Y \in [0, 1]$.

(c) Assume that $1 < m < \infty$. We have

$$\mathbf{E}[Y \mid Z_0, Z_0'] = Y_0 = Z_0/(Z_0 + Z_0'). \tag{18.3}$$

Now $\langle Z_{k+n}' \rangle_{n \ge 0}$ is also a Galton–Watson process, so

$$Y^{(k)} := \lim_n \frac{Z_n}{Z_n + Z_{k+n}'}$$

exists a.s. too with

$$\mathbf{E}[Y^{(k)} \mid Z_0, Z_k'] = Z_0/(Z_0 + Z_k')$$

by (18.3). We have

$$
\begin{aligned}
\mathbf{P}[Y = 1,\ Z_n \not\to 0,\ Z_n' \not\to 0] &= \mathbf{P}\left[\frac{Z_n'}{Z_n} \to 0,\ Z_n \not\to 0,\ Z_n' \not\to 0\right] \\
&= \mathbf{P}\left[\frac{Z_{n+k}'}{Z_n} \to 0,\ Z_n \not\to 0,\ Z_n' \not\to 0\right] \\
&= \mathbf{P}[Y^{(k)} = 1,\ Z_n \not\to 0,\ Z_n' \not\to 0] \\
&\le \mathbf{E}\left[Y^{(k)} \mathbf{1}_{[Z_k' > 0]}\right] = \mathbf{E}\left[\frac{Z_0}{Z_0 + Z_k'} \mathbf{1}_{[Z_k' > 0]}\right] \\
&\to 0 \text{ as } k \to \infty,
\end{aligned}
$$

where the second equality is due to the fact that, by the Seneta–Heyde theorem, $Z_{n+k}'/Z_n' \to m^k$ a.s. given nonextinction. Hence $\mathbf{P}[Y = 1,\ Z_n \not\to 0,\ Z_n' \not\to 0] = 0$. By symmetry, $\mathbf{P}[Y = 0,\ Z_n \not\to 0,\ Z_n' \not\to 0] = 0$ too.

5.31. Take $c_n := Z_n'$ for almost any particular realization of $\langle Z_n' \rangle$ with $Z_n' \not\to 0$. Part (iii) of the Seneta–Heyde theorem, the fact that $c_{n+1}/c_n \to m$, follows from $Z_{n+1}/Z_n \xrightarrow{\mathbf{P}} m$. That is, if $Z_n/c_n \to V$ a.s., $0 < V < \infty$ a.s. on nonextinction, then

$$\frac{Z_{n+1}}{Z_n} \cdot \frac{c_n}{c_{n+1}} \to \frac{V}{V} = 1 \tag{18.4}$$

a.s. on nonextinction. Since $Z_{n+1}/Z_n \xrightarrow{\mathbf{P}} m$, it follows that $c_n/c_{n+1} \to 1/m$.

5.32. This follows either from (18.4) in the solution to Exercise 5.31 or from the Seneta–Heyde theorem.

5.35. There are various ways to prove these; see Pitman (1998) for some of them and the history. For (a), one way to proceed is to replace the tree in stages as follows: replace the initial individuals by their progeny; then replace each of these in turn by their progeny; and so on. Each replacement decreases the total by a copy of $L - 1$. For (b), given L_j with $S_n = -k$, show that among the n cyclic permutations of $\langle L_j ;\ j \le n \rangle$, there are exactly k for which the first time the sum is $-k$ is n (that is, for which the event in (a) occurs). You might want to do this first for $k = 1$.

5.36. The distribution in part (a) is known as the ***Borel distribution***.

5.37. (This is due to David Wilson, personal communication, 2009.) Write $g := g_\infty$ in the notation of Exercise 5.33. Then $g(s) = se^{g(s)-1}$, whence $\log s = \log g(s) + 1 - g(s)$ and $ds/s = (1/g - 1)dg$. Therefore, $\int_0^1 [g(s)/s]ds = \int_0^1 g[1/g - 1]dg = 1/2$. A similar method, but with generating functions of two variables, shows that $\mathbf{P}[|X| \ge k] = 1/(k + 1)$ for every $k \ge 1$ (David Wilson, personal communication, 2010).

5.39. We have br $T(p) = (1 - p)/(p_c(T) - p)$ a.s.

5.41. We have $1 - \theta_n(p) = \left(1 - p\theta_n(p)\right)^n$.

5.42. Define $X(\mu)$ as in (5.8). The bilinear form $(\mu_1, \mu_2) \mapsto \mathbf{E}[X(\mu_1)X(\mu_2)]$ gives the seminorm $\mathscr{E}(\,\boldsymbol{\cdot}\,)^{1/2}$, whence the seminorm satisfies the parallelogram law, which is the desired identity.

5.43. Use Exercise 5.42.

5.44. Use Theorem 5.15. This result is due originally to Grimmett and Kesten (1983), Lemma 2, whose proof was very long. See also Exercises 17.15 and 17.34. For a beautiful proof that is direct and short, see Collevecchio (2006).

5.45. **(a)** Use Exercise 2.75. See Exercise 17.34 for a lower bound on the expected effective conductance.

(b) This is due to Chen (1997).

5.50. Consider oriented paths starting at the origin. For a lower bound on $p_c(d)$, use the first-moment method. For an upper bound, use paths with exponential intersection tails. This result is due to Kesten and published by Cox and Durrett (1983), who gave sharper asymptotics.

5.52. Part (a) is used by Bateman and Katz (2008).

5.53. $\mathscr{E}(o \leftrightarrow \partial T(n))/(1 + \mathscr{E}(o \leftrightarrow \partial T(n))) = 2/(n + 2)$ while $\mathbf{P}_{1/2}[o \leftrightarrow \partial T(n)] \sim 4/n$, so the inequalities with "2" in them are better for large n. To see this asymptotic, let $p_n := \mathbf{P}_{1/2}[o \leftrightarrow \partial T(n)]$. We have $p_{n+1} = p_n - p_n^2/4$. For $\epsilon > 0$ and t_1, t_2, N chosen appropriately, show that $a_n := (4 - \epsilon)/(n - t_1)$ and $b_n := (4 + \epsilon)/(n - t_2)$ satisfy $a_N = b_N = p_N$ and $a_{n+1} < a_n - a_n^2/4$, $b_{n+1} > b_n - b_n^2/4$ for $n > N$. Deduce that $a_n < p_n < b_n$ for $n > N$. Theorem 12.7 determines this kind of asymptotic more generally for critical Galton–Watson processes.

5.54. $\mathscr{E}(o \leftrightarrow \partial T)/(1 + \mathscr{E}(o \leftrightarrow \partial T)) = (2p - 1)/p$ and $\mathbf{P}_p[o \leftrightarrow \partial T] = (2p - 1)/p^2$.

5.55. For related results, see Adams and Lyons (1991).

5.57. See Pemantle and Peres (1996).

5.58. This is due to Lyons (1992).

5.60. Parts (a) and (b) are due to Lyons (1992), and parts (c) and (d) are due to Marchal (1998).

5.62. Use Exercise 5.61 with $n := 1$ and Proposition 5.28(ii).

5.64. These are due to Pakes and Dekking (1991).

5.65. This follows from the facts that $B_{n,d,1}(s) < s$ for $s > 0$ small enough, that the functions $B_{n,d,p}(s)$ converge uniformly to $B_{n,d,1}(s)$ as $p \to 1$, and that $B_{n,d,p}(0) > 0$ for $p > 0$.

5.66. This is due to Balogh, Peres, and Pete (2006).

5.68. A similar statement is as follows. Let $B \subseteq \mathbb{R}^{\mathbb{N}}$ be an increasing set in the sense that if $\langle a_n \rangle \in B$ and $b_n \geq a_n$ for all n, then $\langle b_n \rangle \in B$. Define the associated *target percolation* ω on T to consist of those sites x for which there is some $\langle a_n \rangle \in B$ such that if the path in T from o to x is $x_0 = o, x_1, \ldots, x_k = x$, then $A(x_i) = a_i$ for $0 \leq i \leq k$. For example, if $A(x)$ is uniform on $[0, 1]$ and $B = [1 - p, 1]^{\mathbb{N}}$, then ω is the component of the root for Bernoulli(p) percolation. Claim: if each ray in T has probability 0 of belonging to ω, then the number of rays in ω is a.s. 0 or $2^{\aleph_0}$. This is Lemma 4.1(i) of Pemantle and Peres (1995a), who give more information on target percolation.

5.69. (This is part of Theorem 2.2(ii) of Pemantle and Peres (1995a).) Use Exercise 5.68.

5.70. This and Exercise 5.71 are due to Chayes, Chayes, and Durrett (1988) and Dekking and Meester (1990), whereas our outlines are based on Chayes (1995a).

5.71. **(a)** Let γ be a left-to-right crossing of $[0, 1] \times [0, 2]$ consisting of retained level k squares. If γ intersects A, then of course A had to be retained, and this has probability p^m. Considering the last square in γ that touches $L_A \cup L_{A'}$ (see Figure 5.6) shows that if γ is disjoint from $A \cup A'$, then either γ connects L_A to R_A within $[0, 1] \times [1, 2]$, or γ connects $L_{A'}$ to $R_{A'}$ within $[0, 1] \times [0, 1]$.

5.72. (a) Set $h_a(t) := (t - a)^+$. For one direction, just note that h_a is a nonnegative, increasing convex function. For the other, given h, let g give a line of support of h at the point $(\mathbf{E}Y, \mathbf{E}h(Y))$. Then $\mathbf{E}h(Y) = g(\mathbf{E}Y) = \mathbf{E}g(Y) \le \mathbf{E}g^+(Y) \le \mathbf{E}g^+(X) \le \mathbf{E}h(X)$.

5.73. (a) Condition on X_i and Y_i for $i < n$.

(b) Show that $n \mapsto \mathbf{E}\left[h\left(\sum_{i=1}^n X_i\right)\right]$ is an increasing convex function when h is. See Ross (1996), p. 444 for details.

5.74. (a) Use Exercise 5.73.

(b) Use (a) and Exercise 5.72. See Ross (1996), p. 446.

Chapter 6

6.1. In fact, for every finite, connected subset K of $\mathbb{T}_{b+1}$, we have $|\partial_E K| = (b - 1)|K| + 2$.

6.2. Modify the first part of the proof of Theorem 6.1. Alternatively, increase D.

6.6. Let the transition probabilities be $p(x, y)$. Then the arithmetic mean–quadratic mean inequality gives us

$$
\begin{aligned}
\|Pf\|_\pi^2 &= \sum_{x \in V} \pi(x)(Pf)(x)^2 = \sum_{x \in V} \pi(x)\left[\sum_{y \in V} p(x, y)f(y)\right]^2 \\
&\le \sum_{x \in V} \pi(x) \sum_{y \in V} p(x, y)f(y)^2 = \sum_{y \in V} f(y)^2 \sum_{x \in V} \pi(x)p(x, y) \\
&= \sum_{y \in V} f(y)^2 \pi(y) = \|f\|_\pi^2 \, .
\end{aligned}
$$

6.7. The first equality depends only on the fact that P is self-adjoint. We will omit the subscripts π. Suppose that $|(Pf, f)| \le C(f, f)$ for all f. Then, for all f, g, we have

$$
\begin{aligned}
|(Pf, g)| &= \left|\frac{(P(f + g), f + g) - (P(f - g), f - g)}{4}\right| \\
&\le C[(f + g, f + g) + (f - g, f - g)]/4 = C[(f, f) + (g, g)]/2 \, .
\end{aligned}
$$

Put $g = Pf\|f\|/\|Pf\|$ to get $\|Pf\| \le C\|f\|$. This shows that $\|P\| = \sup\{|(Pf, f)|/(f, f) ; f \in \mathbf{D}_{00} \setminus \{\mathbf{0}\}\}$. This gives the first identity.

6.9. Use the fact that $\Phi_E(\mathbb{T}_{b+1}) = (b - 1)/(b + 1)$ to get an upper bound on $\rho(\mathbb{T}_{b+1})$. To get a lower bound, calculate $\|Pf\|$ for $f := \sum_{n=1}^N b^{-n/2}\mathbf{1}_{S_n}$, where N is arbitrary and S_n is the sphere of radius n.

6.10. In the case of Cayley graphs, this is known as Grigorchuk's criterion for amenability.

6.13. Let x be such that $f_*(x) \ne 0$, where f_* is an eigenfunction with eigenvalue $\pm\lambda_*$. The constants a_i in the proof of Theorem 6.13 equal $f_i(y)/\|f_i\|_\pi^2$.

6.16. $Q(S, V) = \pi(S) = \widehat{Q}(S, V)$, so $Q(S, S^c) = Q(S, V) - Q(S, S) = \widehat{Q}(S, S^c)$.

6.17. The relation between the profiles is clear, since $\widetilde{Q}(x, y) := \pi(x)\widetilde{p}(x, y) = \pi(x)p(x, y)/2$ for $x \ne y$. The identity $\pi(x)p_{2n}(x, x) = \|P^n\mathbf{1}_x\|_\pi$ implies that $p_{2n}(x, x)$ is decreasing. Finally, write $\widetilde{p}_{2n}(x, x) \ge \sum_{k=0}^n \binom{2n}{2k}2^{-2n}p_{2k}(x, x) \ge p_{2n}(x, x)/2$.

6.19. To prove this, do not condition that a vertex belongs to an infinite cluster. Instead, grow the cluster from x and continue to explore even if the cluster turns out finite. Disjoint large balls will be encountered infinitely often and will have a certain probability of containing only a long path. These events are independent and so one will occur a.s. For more details of this kind of proof, see the proof of Lemma 7.23.

6.25. For finite networks and the corresponding definition of expansion constant (see Exercise 6.49), the situation is quite different; see Chung and Tetali (1998).

6.27. (This was the original proof of Theorem 6.2 in BLPS (1999b).) Let θ be an antisymmetric function on E with $|\theta(e)| \le c(e)$ for all edges e and $d^*\theta(x) = \Phi_E(G)D(x)$ for all vertices x. We may assume that θ is acyclic, that is, that there is no cycle of oriented edges on each of which $\theta > 0$. (Otherwise, we modify θ by subtracting appropriate cycles.) Let $K \subset V$ be finite and nonempty. Suppose for a contradiction that $|\partial_E K|_c / |K|_D = \Phi_E(G)$. The proof of Theorem 6.1 shows that for all $e \in \partial_E K$, we have $\theta(e) = c(e)$ if e is oriented to point out of K; in particular, $\theta(e) > 0$. Let $(x_1, x_0) \in \partial_E K$ with $x_1 \in K$ and $x_0 \notin K$, and let γ be an automorphism of G that carries x_0 to x_1. Write x_2 for the image of x_1 and γK for the image of K. Since we also have $|\partial_E \gamma K|_c / |\gamma K|_D = \Phi_E(G)$ and $(x_2, x_1) \in \partial_E \gamma K$, it follows that $\theta(x_2, x_1) > 0$. We may similarly find x_3 such that $\theta(x_3, x_2) > 0$ and so on, until we arrive at some x_k that equals some previous x_j or is outside K. Both lead to a contradiction, the former contradicting the acyclicity of θ and the latter the fact that on all edges leading out of K, we have $\theta > 0$.

6.28. (This is due to R. Lyons.) Use the method of solution of Exercise 6.27.

6.30. This is due to R. Lyons.

6.33. Consider cosets.

6.34. (This was the original proof of Theorem 6.4 in BLPS (1999b).) Let θ be an antisymmetric function on E with $\mathrm{flow}_+(\theta, x) \le 1$ and $d^*\theta(x) = \Phi_V(G)$ for all vertices x. Let $K \subset V$ be finite and nonempty. Suppose for a contradiction that $|\partial_V K|/|K| = \Phi_V(G)$. The proof of Theorem 6.3 shows that for all $x \in \partial_V K$, we have $\mathrm{flow}_+(\theta, x) = 1$ and $\mathrm{flow}_+(-\theta, v) = \Phi_V(G) + 1$; in particular, there is some e leading to x with $\theta(e) \ge 1/d$ and some e leading away from x with $\theta(e) \ge (\Phi_V(G) + 1)/d \ge 1/d$, where d is the degree in G. Since G is transitive, the same is true for all $x \in V$. Therefore, we may find either a cycle or a bi-infinite path with all edges e having the property that $\theta(e) \ge 1/d$. We may then subtract $1/d$ from these edges, yielding another function θ' that satisfies $\mathrm{flow}_+(\theta', v) \le 1$ and $d^*\theta'(x) = \Phi_V(G)$ for all vertices x. But then $\mathrm{flow}_+(\theta', x) < 1$ for some x, a contradiction.

6.35. This is due to R. Lyons.

6.38. This is Exercise $6.17\frac{1}{2}$ in Gromov (1999). Consider the graph G_k formed by adding all edges $[x, y]$ with $\mathrm{dist}_G(x, y) \le k$.

6.39. (This is due to Y. Peres.) Identify the binary tree with the set of all finite sequences of 0s and 1s. Let T be the subtree of the binary tree that contains all rays ξ with the property that for every $k \ge 100$, every path in ξ of length k^2 contains a subpath of 0s of length k. In other words, T contains the vertex corresponding to $(x_1, \ldots, x_n)$ iff for every $k \ge 100$ and $j \ge 1$ with $k^2 + j \le n$, there exists $j' \ge j$ with $k + j' \le k^2 + j$ such that $x_i = 0$ for all $i \in [1 + j', k + j']$.

6.40. **(a)** Use Cauchy–Schwarz.

(b) The spectral theorem guarantees the existence of a square root of P; see Section 13.9 for a review of the spectral theorem. Alternatively, use Exercise 13.16(e).

(c) Adjoin to the usual graph on $\mathbb{N}$ a binary tree at $y_0 := 0$ and a loop at each positive integer. Let $y_n := \lfloor \sqrt{n} \rfloor$. Then $p_n(y_0, y_0) \asymp n^{-3/2}$ and $p_n(y_0, y_n) \asymp 1/n$.

6.45. Only part (b).

6.46. (Prasad Tetali suggested that Theorem 6.1 might be used for this purpose, by analogy with the proof in Alon (1986).) Let θ be an antisymmetric function on E such that $|\theta| \le c$ and $d^*\theta = \Phi_E(G)\pi$. Then, for all $f \in \mathbf{D}_{00}$, we have $(f, f)_\pi = (f^2, \pi) = \Phi_E(G)^{-1}(f^2, d^*\theta) = \Phi_E(G)^{-1}(d(f^2), \theta) \le \Phi_E(G)^{-1} \sum_{e \in E_{1/2}} c(e) |f(e^+)^2 - f(e^-)^2|$.

6.47. This bound for λ_2 is due to Nilli (1991) (a pseudonym for Noga Alon). It is a refinement of the Alon–Boppana bound proved in Alon (1986). See also Section 3 of Murty (2003) for details. A similar proof shows that if G contains s vertices so that the distance between each pair is at least $2k$, then the s largest eigenvalues all satisfy the same bound. A further improvement of Nilli (2004) is that under this same distance condition, the s largest eigenvalues are all at least $2\sqrt{b} \cos(\pi/k)$. This bound is better for $k \ge 7$

and $d \geq 4$. Here, one uses s functions f_j corresponding to s vertices a_j at pairwise distance at least $2k$, where

$$f_j(x) := \begin{cases} \alpha_{i+i_0} & \text{if } \mathrm{dist}(x, a_j) = i \in [0, k - i_0] \\ 0 & \text{otherwise,} \end{cases}$$

$$\alpha_i := b^{-i/2} \sin(\pi i/k),$$

and i_0 is the largest $i < k$ such that $\alpha_i < \alpha_{i+1}$. Each function f_j satisfies $(Af_j, f_j) \geq 2\sqrt{b} \cos(\pi/k) \|f_j\|^2$, as does every linear combination of them. The key algebraic fact that enables this inequality is that $\alpha_{i-1} + b\alpha_{i+1} = 2\sqrt{b} \cos(\pi/k) \alpha_i$.

6.49. This is due to Chung and Tetali (1998).

6.50. Use the Cauchy–Schwarz inequality and Proposition 6.6:

$$\mathbf{P}_x[X_n \in A] = \left(\mathbf{1}_{\{x\}}, P^n \mathbf{1}_A\right)_\pi / \pi(x) \leq \sqrt{\pi(x)} \|P^n \mathbf{1}_A\|_\pi / \pi(x)$$
$$\leq \|P\|_\pi^n \sqrt{|A|_\pi / \pi(x)} = \rho(G)^n \sqrt{|A|_\pi / \pi(x)}.$$

A similar result is Proposition 4.2 of Babai (1991).

6.51. (This confirms a conjecture of Jan Swart (personal communication, 2008), who used it in Swart (2009).) Consider $(P^n \mathbf{1}_A, \mathbf{1}_A)_\pi$.

6.52. Recall from Proposition 2.1 and Exercise 2.1 that $v(x) = \mathscr{G}(o, x)/\pi(x) = \mathscr{G}(x, o)/\pi(o)$. Therefore

$$\sum_{x \in V} \pi(x)v(x)^2 = \sum_{x \in V} \mathscr{G}(o, x)\mathscr{G}(x, o)/\pi(o).$$

Now

$$\mathscr{G}(o, x)\mathscr{G}(x, o) = \sum_{m,n} p_m(o, x)p_n(x, o),$$

whence

$$\sum_{x \in V} \mathscr{G}(o, x)\mathscr{G}(x, o) = \sum_k (k + 1)p_k(o, o).$$

Since $p_k(o, o) \leq \rho^k$, this sum is finite.

6.53. (This is due to Benjamini, Nachmias, and Peres (2011). When A is a singleton, $\{x\}$, it is an easy consequence of (6.13) and the usual formulas for $\mathscr{G}(x, x)$.) Define

$$\tau := \inf\{n \geq 0 \,;\, X_n \in A\} \quad \text{and} \quad \tau^+ := \inf\{n > 0 \,;\, X_n \in A\}.$$

Let

$$f(x) := \mathbf{P}_x[\tau < \infty].$$

This is the voltage function from A to ∞. By Exercise 6.52, we have $f \in \ell^2(V, \pi)$. Observe that $f \equiv 1$ on A. For all $x \in V$,

$$(Pf)(x) = \mathbf{P}_x[\tau^+ < \infty].$$

In particular, $Pf = f$ on $V \setminus A$. Thus $((I - P)f)(x) = \mathbf{P}_x[\tau^+ = \infty]$ for $x \in A$ and $((I - P)f)(x) = 0$ for $x \in V \setminus A$. Therefore

$$(f, (I - P)f)_\pi = \sum_{x \in A} \pi(x)\mathbf{P}_x[\tau^+ = \infty] = \pi(A)\mathbf{P}_{\pi_A}[\tau^+ = \infty].$$

On the other hand, clearly

$$(f, f)_\pi \geq \sum_{x \in A} \pi(x)f(x)^2 = \pi(A).$$

The claim follows by combining the last two displays with Exercise 6.7.

6.54. Use the fact that $\|A\| = \sqrt{\|A^* A\|}$.

6.55. (This appears as Lemma 4.2 in Virág (2000a).) We may apply Exercise 6.50 to bound $\mathbf{P}\big[|X_n| \le 2\alpha n\big]$ for $\alpha < -\log \rho(G)/\log b$.

6.56. (This example is due to Omer Angel, Tom Hutchcroft, Asaf Nachmias, and Gourab Ray, personal communication, 2014.) Let the vertices be $\mathbb{N}$. Join n to $n + 1$ by 2^n edges for $n \ge 0$ and n to $2n$ by one edge for $n \ge 2$.

6.57. Put $z := 1/b$ in (6.23). An extension is given by Guillotin-Plantard (2005).

6.58. We follow the hint. Differentiation k times shows that the kth coefficient is $\sum_{n\ge0} \binom{n}{k} a_n z_0^{n-k}$. This gives that

$$f(R + \epsilon) = \sum_{k\ge0}\sum_{n\ge0} \binom{n}{k} a_n z_0^{n-k} (R + \epsilon - z_0)^k = \sum_{n\ge0} a_n (z_0 + R + \epsilon - z_0)^n$$

by the binomial theorem. For more details and applications, see Theorem IV.6 and elsewhere in Flajolet and Sedgewick (2009).

6.61. (This is due to Lyons and Peres (2015).) For (c), consider concatenating a long cycle that starts with e and ends with $-e$ with a short cycle that does not.

6.62. **(a)** From a nonbacktracking cycle, one can build cycles of various lengths by inserting pure backtracking cycles. The number of pure backtracking cycles of a given length equals the number of cycles of that length on $\mathbb{T}_{b+1}$. In order not to count any result more than once, the steps inserted must not use the last step before the insertion point of the nonbacktracking cycle, which leads to counting cycles on the b-ary tree T, rather than on $\mathbb{T}_{b+1}$, for all insertion points other than the first.

(b) Sum over excursions from the root. The choice of sign in solving the resulting quadratic equation is dictated by the requirement that $H(0) = 1$.

(c) Summing over the crossings of a fixed edge incident to the root gives

$$H_0(z) = \sum_{n\ge0} z^{2n} H(z)^{2n+1} .$$

Use the quadratic equation that H satisfies to simplify the algebra.

6.64. **(a)** $bB^{-1}\theta(x, y) = \sum_{u \sim x} \theta(u, x) - b\theta(y, x)$.

(b) $C\theta(x, y) = \sum_{u \sim x} \theta(u, x) + \sum_{z \sim y} \theta(y, z) - d\theta(y, x)$.

(c) Show that $(C\theta, \psi) = (\psi, C\theta)$. Consideration of the matrix C seems to be new.

(d)–(h) Similar results can be found in Kotani and Sunada (2000) and for nonregular graphs in Angel, Friedman, and Hoory (2015). A complete description of the Jordan form of B in the regular case was obtained by Lubetzky and Peres (2016). For (h) with G infinite, use the fact that $\kappa \in \sigma(B)$ iff $\bar\kappa$ is an eigenvalue of B^* or κ is an approximate eigenvalue of B (that is, for all $\epsilon > 0$, there exists $\theta \in \ell^2(\mathsf{E})$ such that $\|B\theta - \kappa\theta\| < \epsilon\|\theta\|$).

(i) Nonbacktracking random walk is aperiodic, since otherwise B/b would have an eigenvalue that is a nonreal root of unity, contradicting (h). Each real eigenvalue of B is smaller in absolute value than the corresponding eigenvalue of A by (h). By Exercise 6.47, the second-largest eigenvalue of A is larger than $\sqrt{b}$, which equals the absolute value of each nonreal eigenvalue of B other than b. The fact that nonbacktracking random walk mixes faster is due to Alon, Benjamini, Lubetzky, and Sodin (2007).

6.65. Use Exercise 2.30.

6.66. (This fact appears in unpublished notes of David Aldous, 1999, and in Ancona, Lyons, and Peres (1999), where it is stated in a different form as Lemma 3.1.) Let g_A be the spectral gap of the induced chain and g be the original spectral gap. Write π_A for the induced stationary probability measure on A and P_A for the induced transition operator. Choose $\phi: A \to \mathbb{R}$ with $(\phi, \mathbf{1})_{\pi_A} = 0$ and $g_A = \left((I - P_A)\phi, \phi\right)_{\pi_A} / \|\phi\|_{\pi_A}^2$. Let $\psi: V \to \mathbb{R}$ be the harmonic extension of ϕ. Show that $P_A\phi$ is the restriction of $P\psi$ to A, that $\left((I - P_A)\phi, \phi\right)_{\pi_A} = \left((I - P)\psi, \psi\right)_\pi / \pi(A)$, and that $\|\phi\|_{\pi_A}^2 \leq \|\psi - (\psi, \mathbf{1})_\pi\|_\pi^2 / \pi(A)$. We remark that a similar proof shows the same inequality for the gaps of the network Laplacians (Exercise 2.62), which are the gaps for continuous-time random walk. A special case of this (in view of Exercise 2.69) is Proposition 2.1 of Caputo, Liggett, and Richthammer (2010).

6.68. This is from Häggström, Jonasson, and Lyons (2002).

6.69. (This is due to Y. Peres and published in Häggström, Jonasson, and Lyons (2002).) The amenable case is trivial, so assume that G is nonamenable. According to the reasoning of the first paragraph of the proof of Theorem 6.19 and (6.37), we have

$$\frac{|(\widehat{K})'|}{|E((\widehat{K})')|} + \frac{|K|}{|E^*(K)|} \leq \frac{|(\widehat{K})'|}{|E((\widehat{K})')|} + \frac{|\widehat{K}|}{|E^*(\widehat{K})|} \leq 1 + \frac{1}{|E((\widehat{K})')|} . \tag{18.5}$$

Write

$$\kappa_n := \frac{|\partial_E K_n|}{|K_n|} - \Phi'_E(G)$$

and

$$\lambda_n := \frac{|\partial_E L_n|}{|L_n|} - \Phi'_E(G^\dagger) .$$

Also write $d := d_G$, $d^\dagger := d_{G^\dagger}$, $\Phi := \Phi'_E(G)$, and $\Phi^\dagger := \Phi'_E(G^\dagger)$. We may rewrite (18.5) as

$$\frac{2}{d^\dagger - |\partial_E L_n|/|L_n|} + \frac{2}{d + |\partial_E K_n|/|K_n|} \leq 1 + \frac{1}{|E(L_n)|} ,$$

or, again, as

$$\frac{2}{d^\dagger - \Phi^\dagger - \lambda_n} + \frac{2}{d + \Phi + \kappa_n} \leq 1 + \frac{1}{|E(L_n)|} = \frac{2}{d^\dagger - \Phi^\dagger} + \frac{2}{d + \Phi} + \frac{1}{|E(L_n)|} ,$$

whence

$$\frac{2\lambda_n}{(d^\dagger - \Phi^\dagger)(d^\dagger - \Phi^\dagger - \lambda_n)} + \frac{2\kappa_n}{(d + \Phi)(d + \Phi + \kappa_n)} \leq \frac{1}{|E(L_n)|} .$$

Therefore

$$\begin{aligned}
2\lambda_n &\leq \frac{(d^\dagger - \Phi^\dagger)(d^\dagger - \Phi^\dagger - \lambda_n)}{(d + \Phi)(d + \Phi + \kappa_n)}(2\kappa_n) + \frac{(d^\dagger - \Phi^\dagger)(d^\dagger - \Phi^\dagger - \lambda_n)}{|E(L_n)|} \\
&\leq \left(\frac{d^\dagger - \Phi^\dagger}{d + \Phi}\right)^2 2\kappa_n + \frac{(d^\dagger - \Phi^\dagger)^2}{|E(L_n)|} .
\end{aligned}$$

Similarly, we have

$$2\kappa_{n+1} \leq \left(\frac{d - \Phi}{d^\dagger + \Phi^\dagger}\right)^2 2\lambda_n + \frac{(d - \Phi)^2}{|E(K_{n+1})|} .$$

Putting these together, we obtain

$$2\kappa_{n+1} \leq a(2\kappa_n) + b_n ,$$

where

$$a := \left(\frac{(d - \Phi)(d^\dagger - \Phi^\dagger)}{(d + \Phi)(d^\dagger + \Phi^\dagger)}\right)^2$$

and

$$b_n := \left(\frac{(d - \Phi)(d^\dagger - \Phi^\dagger)}{d^\dagger + \Phi^\dagger}\right)^2 \frac{1}{|E(L_n)|} + \frac{(d - \Phi)^2}{|E(K_{n+1})|} .$$

Therefore

$$2\kappa_n \leq 2\kappa_0 a^{n-1} + \sum_{j=0}^{n-2} a^j b_{n-j} .$$

Since $a < 1$ and $b_n \to 0$, we obtain $\kappa_n \to 0$. Hence $\lambda_n \to 0$ too.

6.70. (a) Fix $o \in V(G)$. Let the flow along e be π^{-1} times the area of the triangle defined by the endpoints of $e^\dagger$ and o.

(b) Choose the embedding so that the angles of the faces at a vertex of $G^\dagger$ with degree m are $2\pi/m$. The lower bound this gives is the infimum of $a(F)$ over faces F, where $a(F) := n - 2 - \sum_{i=1}^n 2/d_i$ when F is a face of n sides whose vertices have degrees d_i.

6.73. The function $t \mapsto -t \log t$ is concave. Consider a random permutation of the distribution of X.

6.74. Let $A \subset \{0, \ldots, n-1\}^d$ with $|A| < n^d/2$. Let m be such that $|\mathcal{P}_m(A)|$ is maximal over all projections, and let
$$F = \{a \in \mathcal{P}_m(A);\ |\mathcal{P}_m^{-1}(a)| = n\}\,.$$

Notice that for any $a \notin F$, there is at least one edge in $\partial_{\mathsf{E}} A$, and for different as we get disjoint edges, so that $|\partial_{\mathsf{E}} A| \geq |\mathcal{P}_m(A) \setminus F|$. By Theorem 6.22, we get
$$|A|^{d-1} \leq \prod_{j=1}^{d} |\mathcal{P}_j(A)| \leq |\mathcal{P}_m(A)|^d\,,$$

and so $|\mathcal{P}_m(A)| \geq |A|/|A|^{1/d} \geq 2^{1/d}|A|/n \geq 2^{1/d}|F|$, which together yield
$$|\partial_{\mathsf{E}} A| \geq |\mathcal{P}_m(A) \setminus F| \geq (1 - 2^{-1/d})|\mathcal{P}_m(A)| \geq (1 - 2^{-1/d})|A|^{\frac{d-1}{d}}\,.$$

6.76. Follow the proof of Theorem 6.31, but apply Jensen's inequality instead of Lemma 6.39.

6.77. Considering P^j instead of P reduces to the case $j = 1$. In this case, (6.51) is replaced by
$$\mathbf{E}_S\big[\pi(S') \mid U < \tfrac{1}{2}\big] \geq \pi(S)\Big(1 + \frac{2\eta}{1-\eta}\Phi_S\Big)\,.$$

The rest of the proof is identical; see Morris and Peres (2005).

6.78. For every $v \in [-r, r]$, we have
$$\mathbf{P}_v\Big[\max_{j \leq r^2} |X_j| \leq 2r \ \text{ and } \ |X_{r^2}| \leq r\Big] \geq e^{-c}$$

for some constant c. Repeatedly using the Markov property at times that are multiples of r^2, we deduce that
$$\mathbf{P}_v\Big[\max_{j \leq 2n} |X_j| \leq 2r\Big] \geq e^{-c\lceil n/r^2 \rceil}\,.$$

Adding the requirement that each lamp in $[-2r, 2r]$ is turned off at the last visit to it, and that the lamplighter returns to the origin, we infer that $p_n(o, o) \geq \exp(-cn/r^2 - c_1 r)$. Optimizing this lower bound suggests taking $r = n^{1/3}$. See Pittet and Saloff-Coste (1999).

6.79. (a) Subdivide edges the original network where the voltage would equal $v(z)/2$, identify the vertices where the voltage is $v(z)/2$ to a new vertex b, and apply Lemma 6.43 either to a, b or to b, z. This result is very similar to Benjamini and Kozma (2005).

(b) Note that $\psi \geq 1$.

6.80. These results are due to Benjamini and Schramm (2004). There, they refer to Benjamini and Schramm (2001b) for an example of a tree with balls having cardinality in $[r^d/c, cr^d]$, yet containing arbitrarily large finite subsets with only one boundary vertex. That reference has a minor error; to fix it, Δ on p. 10 there should be assumed to equal the diameter. The hypothesis for the principal result can be weakened to $|B(x, r)| \leq ab^r$ and $\lim_{R \to \infty} |B(x, R)|/|B(x, R - r)| \geq b^r/a$.

6.82. (This is due to McGuinness (1989).) The following simple construction was found by Benjamini and Schramm (1997). Let $B(\ell)$ denote the union of two binary trees of height ℓ, glued at the leaves. Start with an infinite binary tree T, and for each $k \geq 1$, replace every edge e at level k of T by a copy of $B(3^k)$, gluing each of the two roots of $B(3^k)$ to one endpoint of e.

6.83. Use the method of proof of Theorem 3.10. In light of Theorem 8.32, a.s. *every* infinite open cluster is transient. This stronger result is due to Benjamini, Lyons, and Schramm (1999).

6.86. Use Proposition 6.51.

6.87. Conjecture 7.13 says that a proper two-dimensional isoperimetric inequality implies $p_c(G) < 1$. Itai Benjamini (personal communication) has conjectured that the stronger conclusion $|\mathcal{A}_n| \leq e^{Cn}$ for some $C < \infty$ also holds under this assumption. This exercise shows that these conjectures are not true with the weakened assumption of anchored isoperimetry.

6.88. Suppose that the distribution ν of L does not have an exponential tail. Then, for every $c > 0$ and every $\epsilon > 0$, we have $\mathbf{P}\left[\sum_{i=1}^{n} L_i \geq cn\right] \geq \mathbf{P}[L_1 \geq cn] \geq e^{-\epsilon n}$ for infinitely many ns, where $\langle L_i \rangle$ are i.i.d. with law ν. Let G be a binary tree with the root o as the basepoint. Pick a collection of 2^n pairwise disjoint paths from level n to level $2n$. Then

$$\mathbf{P}\left[\text{along at least one of these } 2^n \text{ paths } \sum_{i=1}^{n} L_i \geq cn\right] \geq 1 - (1 - e^{-\epsilon n})^{2^n}$$

$$\geq 1 - \exp(-e^{-\epsilon n} 2^n) \to 1 .$$

With probability very close to 1 (depending on n), there is a path from level n to $2n$ along which $\sum_{i=1}^{n} L_i \geq cn$. Take such a path and extend it to the root, o. Let S be the set of vertices in the extended path from the root o to level $2n$. Then

$$\frac{|\partial_{\mathsf{E}} S|}{\sum_{e \in E(S)} L_e} \leq \frac{2n + 1}{cn} \approx \frac{2}{c} .$$

Since c can be arbitrarily large, $\Phi_{\mathsf{E}}^*(G^\nu) = 0$ a.s.

Chapter 7

7.1. Divide A into two pieces depending on the value of $\omega(e)$.

7.3. Prove this by induction, removing one leaf of $K \cap T$.

7.7. Let μ_{ω_2} and ν_{ω_2} be the conditional distributions of ω_1 and ω_3 given ω_2. Then let $(\omega_1, \omega_2, \omega_3)$ have the law where (ω_1, ω_3) has conditional law $\mu_{\omega_2} \times \nu_{\omega_2}$ given ω_2. See Exercise 18.20 of Billingsley (1995) for a general construction from specified conditional distributions. One can also make a similar construction for finite A and then use Kolmogorov's existence theorem.

7.8. The transitive case is due to Lyons (1995).

7.9. (a) Given $\epsilon > 0$, let R be large enough that the balls of radius R satisfy $\mathbf{P}[B_R(x) \cap \omega \neq \varnothing] > 1 - \epsilon$ for every vertex x. Given any finite $F \subset \mathsf{V}$, every infinite component K of $G \setminus F$ contains a ball of radius R and therefore intersects ω with probability more than $1 - \epsilon$. Since ϵ is arbitrary, the probability that $K \cap \omega \neq \varnothing$ is actually 1. Since this holds for all finite F, the result follows.

7.10. *Solution 1.* This follows from (6.13) and the proof of Proposition 6.6.

Solution 2. A direct proof goes as follows. Let us first suppose that $\mathbf{P}_o[X_k = o]$ is strictly positive for all large k. Note that for any k and n, we have

$$\mathbf{P}_o[X_n = o] \geq \mathbf{P}_o[X_k = o]\,\mathbf{P}_o[X_{n-k} = o] .$$

Thus, by Fekete's lemma (Exercise 3.9), we have $\rho(G) = \lim_{n\to\infty} \mathbf{P}_o[X_n = o]^{1/n}$ as well as $\mathbf{P}_o[X_n = o] \leq \rho(G)^n$ for all n.

On the other hand, if $\mathbf{P}_o[X_k = o] = 0$ for odd k, it is still true that $\mathbf{P}_o[X_k = o]$ is strictly positive for all even k, whence $\lim_{n\to\infty} \mathbf{P}_o[X_{2n} = o]^{1/2n}$ exists and equals $\rho(G)$.

7.11. Suppose that a large square is initially occupied. Then the chance is close to 1 that it will grow to occupy everything.

7.14. This was used by Lyons and Schramm (1999).

7.15. Use Exercise 6.5.

7.17. (This solution was shown to us by Jacob Magnusson, personal communication, 2012.) Use the standard coupling $\omega_p := \{e \,;\, U(e) < p\}$ and show that $\mathbf{1}_{[x\leftrightarrow y]}(\omega_p)$ is continuous from the left.

7.18. (These generators were introduced by Revelle (2001).) It is the same as the graph of Example 7.2 would be if the trees in that example were chosen both to be 3-regular.

7.20. Let $A := \{y \leftrightarrow \infty\}$ and $B := \{x \leftrightarrow y\}$.

7.21. This is due to Angel and Szegedy (2016).

7.24. (This is noted in BLPS (1999b).) Use Lemma 7.7.

7.25. This is due to Lyons, Pichot, and Vassout (2008). Compare the result to Theorem 6.5.

Solution 1. Use Exercise 7.23 and $\mathbf{1}_\Lambda(o) \leq \deg_{\mathfrak{F}} o - 2$.

Solution 2. Here is a direct proof. It is not hard to see that it suffices to establish the case where every tree of $\mathfrak{F}$ is infinite a.s. Let $K \subset \mathsf{V}$ be finite and write $\overline{K} := K \cup \partial_{\mathsf{V}} K$. Let Y be the subgraph of G spanned by those edges of $\mathfrak{F}$ that are incident to some vertex of K. This is a forest with no isolated vertices, whence

$$\sum_{x\in K} \deg_{\mathfrak{F}} x \leq \sum_{x\in \overline{K}} \deg_Y x - |\mathsf{V}(Y) \setminus K| = 2\,|\mathsf{E}(Y)| - |\mathsf{V}(Y) \setminus K|$$
$$< 2\,|\mathsf{V}(Y)| - |\mathsf{V}(Y) \setminus K| = 2\,|K| + |\mathsf{V}(Y) \setminus K| \leq 2\,|K| + |\partial_{\mathsf{V}} K|\,.$$

Take the expectation and divide by $|K|$ to get the result.

7.30. Modify the proof of Theorem 6.47.

7.31. It is conjectured by Babson and Benjamini (1999) that the hypothesis holds for every Cayley graph of at least quadratic growth. Timár (2007) showed that the hypothesis does not depend on the generating set. Prove that $p_c^{\mathrm{bond}}(G) < 1$ as in the proof of Theorem 7.16.

7.32. See Grimmett (1999), pp. 18–19.

7.33. These solutions were shown to us by Jacob Magnusson (personal communication, 2011). Let $C_\infty(p)$ denote the union of the infinite clusters formed by the edges with labels at most p.

(a) We have $C_\infty(p) = \bigcap_{q>p} C_\infty(q)$ since G is locally finite.

(b) (This is due to van den Berg and Keane (1984).) Let $K_p := C_\infty(p) \setminus \bigcup_{q<p} C_\infty(q)$. Consider the event that $K(x)$ is the component of x in K_p.

(c) (This is due to Schonmann (1999b).) Use Theorem 7.21 to show that a.s. $K_p = \varnothing$ in the solution to (b).

7.36. This is Corollary 4.1 of Häggström, Peres, and Schonmann (1999).

7.37. This fact is folklore, published for the first time by Peres and Steif (1998).

Solution 1. Use Proposition 5.27.

Solution 2. If there is an infinite cluster with positive probability, then by Kolmogorov's zero-one law, there is an infinite cluster a.s. Let A be the set of vertices x of T for which $\omega \cap T^x$ contains an infinite cluster a.s. Then A is a subtree of T and clearly cannot have a finite boundary. Furthermore, since A is countable, for a.e. ω, each $x \in A$ has the property that $\omega \cap T^x$ contains an infinite cluster. Also, a.s. for all $x \in A$, $\omega \cap T^x \neq T^x \cap A$. Therefore, a.e. ω has the property

$$\forall x \in A \quad \omega \cap T^x \text{ contains an infinite cluster different from } T^x \cap A. \tag{18.6}$$

For any ω with this property, for all $x \in A$, there is some $y \in T^x \cap A \setminus \omega$. Therefore, such an ω contains infinitely many infinite clusters. Since a.e. ω does have property (18.6), it follows that ω contains infinitely many infinite clusters a.s.

7.39. (This is from Kesten (1982).) Let $a_{n,\ell}$ denote the number of n-vertex site animals K such that $|\partial_V K| = \ell$. Note that for such an animal, $\ell \leq (d-1)|K| = (d-1)n$ provided $n \geq 2$. Let $p := 1/d$ and consider Bernoulli(p) site percolation on G. Writing the fact that 1 is at least the probability that the cluster of o is finite, we obtain

$$1 \geq \sum_{n,\ell} a_{n,\ell} p^n (1-p)^\ell \geq \sum_{n \geq 2} a_n p^n (1-p)^{(d-1)n}.$$

Therefore $\limsup_{n \to \infty} a_n^{1/n} \leq 1/[p(1-p)^{d-1}]$. Putting in the chosen value of p gives the result.

7.40. (This is from Häggström, Jonasson, and Lyons (2002).) Let $b_{n,\ell}$ denote the number of n-edge bond animals (V', E') such that $|\partial_E V'| = \ell$. Note that for such a subgraph,

$$\ell \leq d|V'| - 2n \leq d(n+1) - 2n = (d-2)n + d. \tag{18.7}$$

Let $p := 1/(d-1)$ and consider Bernoulli(p) bond percolation on G. Writing the fact that 1 is at least the probability that the cluster of o is finite and using (18.7), we obtain

$$1 \geq \sum_{n,\ell} b_{n,\ell} p^n (1-p)^\ell \geq \sum_n b_n p^n (1-p)^{(d-2)n+d}.$$

Therefore $\limsup_{n \to \infty} b_n^{1/n} \leq 1/[p(1-p)^{d-2}]$. Putting in the chosen value of p gives the result.

7.41. Since the graph has bounded degree, it contains a bi-infinite geodesic passing through o iff it contains geodesics of length $2k$ for each k with o in the middle. By transitivity, it suffices to find a geodesic of length $2k$ anywhere. But this is trivial. Watkins (1986) proves that if $\mathcal{P}$ is a bi-infinite geodesic in G, then the complement of $\mathcal{P}$ does not contain any finite components.

7.42. (This is from Babson and Benjamini (1999).) We give the solution for bond percolation. Let d be the degree of vertices and $2k$ be an upper bound for the length of cycles in a set spanning all cycles. If $K(o)$ is finite, then $\partial_E K(o)$ is a cutset separating o from ∞. Let $\Pi \subseteq \partial_E K(o)$ be a minimal cutset. All edges in Π are closed, which is an event of probability $(1-p)^{|\Pi|}$. We claim that the number of minimal cutsets separating o from ∞ and with n edges is at most $C n D^n$ for some constants C and D that do not depend on n, which implies that $p_c(G) < 1$ as in the proof of Theorem 7.16 (or use Exercise 7.31). Fix any bi-infinite geodesic $\langle x_i \rangle_{i \in \mathbb{Z}}$ with $x_0 = o$ (see Exercise 7.41). Let Π be any minimal cutset separating o from ∞ with n edges. Since Π separates o from ∞, there are some $j, l \geq 0$ such that $[x_{-j-1}, x_{-j}], [x_l, x_{l+1}] \in \Pi$. Because G has one end, Π also separates o from some vertex in G. Since Π is connected in $\widehat{G}^k$ and $\langle x_i \rangle$ is a geodesic, it follows that $j + l < nk$. By Exercise 7.39, the number of connected subgraphs of $\widehat{G}^k$ that have n vertices and that include the edge $[x_l, x_{l+1}]$ is at most $c D^n$, where c is some constant and D is the degree of $\widehat{G}^k$. Since there are no more than nk choices for l, the bound we want follows with $C := ck$.

Note that transitivity was used only to get a bi-infinite geodesic.

7.43. Follow the method of proof of Theorem 7.32 or of Theorem 7.49.

7.47. Compare to a Galton–Watson process: Fix a root of $\mathbb{T}_{b+1}$. Let (x, n) have progeny $(y, n+k)$ when y is a child of x in $\mathbb{T}_{b+1}$ and $k \geq 0$ is minimal such that all the following edges are open: $[(x, n), (x, n+1)], [(x, n+1), (x, n+2)], \ldots, [(x, n+k-1), (x, n+k)], [(x, n+k), (y, n+k)]$.

7.48. Calculate the expected number of nodes connected to the root by an open path of length 2 from the root in the graph $p_c^{\text{bond}}(\mathbb{T}_{b+1} \,\square\, \mathbb{Z})$ with an edge e adjacent to the root removed (maximized over the choice of e).

7.49. This is due to Benjamini and Schramm (1996b).

7.52. See Theorem 1.4 of Balogh, Peres, and Pete (2006) for details.

7.53. The first formula of (7.27) translates to the statement $\pi(d, 1) = 1/d$, while the second formula follows from Exercise 5.66. The formulae of (7.27) were first given in Chalupa, Reich, and Leath (1979), which was the first paper to introduce bootstrap percolation into the statistical physics literature.

7.54. The first inequality follows immediately from viewing T as a subgraph of $\mathbb{T}_{d+1}$. To prove the positivity of the critical probability $b(\mathbb{T}_{d+1}, k)$, consider the probability that a simple path of length n starting from a fixed vertex x does not intersect any vacant $(k - 1)$-fort of the initial Bernoulli(p) configuration of occupied vertices. Using Exercise 5.65, show that this probability is bounded above by some $O(z(p)^n)$, where $z(p) \to 0$ as $p \to 0$. But there is a fixed exponential bound on the number of simple paths of length n, so we can deduce that for p small enough, every infinite, simple path started at x eventually intersects a vacant $(k - 1)$-fort a.s., hence infinite occupied clusters are impossible.

The main idea of this proof came from Howard (2000). That paper, together with Fontes, Schonmann, and Sidoravicius (2002), used bootstrap percolation to understand the Glauber dynamics of the Ising model at zero temperature.

7.56. Let $R(x, T_\xi)$ be the event [the vertex x of T_ξ is in an infinite vacant 1-fort], and set $r(T_\xi) := \mathbf{P}_p[R(o, T_\xi)]$. This is not an almost sure constant, so let us take expectation over all Galton–Watson trees: $r := \mathbf{E}[r(T_\xi)]$. With a recursion as in Theorem 5.29, one can write the equation $r = \frac{1}{2}(1 - p)(2r - r^2 + 4r^3 - 3r^4)$. So we need to determine the infimum of those p for which there is no solution $r \in (0, 1]$; that infimum will be $p(T_\xi, 2)$. Setting $f(r) = 2 - r + 4r^2 - 3r^3$, an examination of $f'(r)$ gives that $\max\{f(r)\,;\ r \in [0, 1]\} = f((4 + \sqrt{7})/9) = 2.2347\ldots$. So there is no solution $r > 0$ iff $2/(1 - p) > 2.2347\ldots$, which gives $p(T_\xi, 2) = 0.10504\ldots < 1/9$.

Chapter 8

8.3. Fix $u, w \in \mathsf{V}$. Let $f(x, y)$ be the indicator that $y \in \Gamma_{u,x}w$.

8.4. Use Exercise 8.3. Caution: this does not extend to general locally compact groups. For example, given $a, b \in \mathbb{R}$ with $a \neq 0$, define the map $T_{a,b} \colon x \mapsto ax + b$ for $x \in \mathbb{R}$. The collection of all such maps $T_{a,b}$ forms a nonunimodular group under composition acting on $\mathbb{R}$, but it has the unimodular, transitive subgroup $\mathbb{R}$ (where $a = 1$).

8.8. A quantitative strengthening is that the weights μ'_j for Γ' are sums of the weights μ_i for Γ as follows: $\mu'_j = \sum_{o_i \in \Gamma' o'_j} \mu_i$.

8.9. Use the same proof as of Theorem 8.16 but only put mass on vertices in ω.

8.10. (This is from BLPS (1999b).) Note that if K is a finite tree, then $\alpha_K < 2$.

8.12. If there were two faces with an infinite number of sides, then a ball that intersects both of them would contain a finite number of vertices whose removal would leave more than one infinite component. If there were only one face with an infinite number of sides, then let x_n ($n \in \mathbb{Z}$) be the vertices on that face in order. Quasi-transitivity would imply that there is a maximum distance M of any vertex to $A := \{x_n\,;\ n \in \mathbb{Z}\}$. Since the graph has one end, for all large n, there is a path $(y_1, \ldots, y_p)$ from x_{-n} to x_n that avoids the M-neighborhood of x_0. For $1 \le i \le p$, choose $t(i) \in \mathbb{Z}$ so that the distance between $x_{t(i)}$ and y_i is at most M. For some $i < p$, we have $t(i) < 0$ and $t(i + 1) > 0$; choose such an i and denote $r(n) := -t(i)$ and $s(n) := t(i + 1)$. Then the distance between $x_{-r(n)}$ and $x_{s(n)}$ is at most $2M + 1$. If $\sup_n r(n)s(n) = \infty$, then there would be points (either $x_{-r(n)}$ or $x_{s(n)}$) the removal of whose $(2M + 1)$-neighborhood would leave an arbitrarily large finite component (one containing x_0) by planarity. This would contradict the quasi-transitivity. Hence all paths from x_{-n} to x_n intersect some fixed neighborhood of x_0. But this contradicts having just one end.

8.13. By the mass-transport principle, we have

$$\mathbf{E}\big|\{x \in \mathsf{V} ;\ o \in \gamma_x L\}\big| = \sum_{x \in \mathsf{V}} \mathbf{P}[o \in \gamma_x L] = \sum_{x \in \mathsf{V}} \mathbf{P}[x \in \gamma_o L] = \mathbf{E}|\gamma_o L| = |L|.$$

8.14. This is due to BLPS (1999b). See also Häggström (2013).

8.16. There are many ways of proving this. One way is as follows: Suppose one could pick an end at random in an invariant way. Send mass 1 from each vertex to its unique neighbor closer to the end.

8.17. Define the ***convex hull*** of Ξ to be the set of vertices of the tree that lie on a bi-infinite, simple path that converges, in both directions, to an end in Ξ. Apply Exercise 8.15 to the convex hull of Ξ when the hull is nonempty, and apply Exercise 8.16 otherwise.

8.19. For any x, consider the elements of $S(x)$ that fix $x_1, \ldots, x_n$.

8.22. The proof that Haar measure exists on compact groups using parts (a) and (b) is due to Maak (1935).

8.25. Extend the ideas of the solution to Exercise 8.17.

8.26. It is enough to prove (8.4) for bounded f.

8.27. It is unknown whether $\inf \Phi_\mathsf{V}(G) = 0$ without the degree constraint.

8.29. This is adapted from BLPS (1999b).

8.30. This is due to Salvatori (1992).

8.31. (This result is from BLPS (1999b).) By Exercise 8.30, we know that Γ is unimodular, but this will also follow from our proof. We imitate the idea at the beginning of Section 6.1, where it is explained why bounded Ponzi schemes don't work on Euclidean lattices. It suffices to prove the result when

$$\lim_{n \to \infty} \frac{|\Gamma o_i \cap K_n|}{|K_n|} =: \nu_{o_i}$$

exists for each i. Set $\nu_x := 0$ for $x \notin \{o_1, \ldots, o_L\}$. By Remark 8.13, it suffices to show that (8.13) holds when f is an indicator that does not transport mass more than distance M. But it is clear that total mass is preserved within K_n up to the mass that starts or ends within distance M of the boundary of K_n. More precisely, the total over vertices in K_n of mass sent minus mass received is in absolute value at most the number of vertices within distance M of the boundary of K_n. Hence the average mass change per vertex in K_n tends to 0 as $n \to \infty$. Since the mass change is the same for all vertices in a given orbit, this shows that (8.13) holds.

8.32. Use Exercise 8.28 or 8.22.

8.34. Use (c) to prove (b). For (c), transport mass $f(|V(x)|)$ to s if s is the seed of $V(x)$. For (d), calculate the conditional expectation directly as $\sum_x \mathbf{P}[x \in V(o) \mid o \in S]$ and obtain a telescoping sum; use Harris's inequality for the unconditional expectation. It is not clear whether (c) holds in the setting of (d).

8.35. This was noted by Häggström (private communication, 1997). The same sharpening was shown another way in BLPS (1999b) by using Theorem 6.2.

8.38. This is from BLPS (1999b).

8.39. These are due to R. Lyons.

8.40. Consider the clusters of Bernoulli(p) percolation on T. Assign the same label to all the vertices in a given cluster, where that label is chosen independently for different clusters and is equally likely to be 0 or 1. This measure has the stronger property of a trivial tail σ-field (see Section 10.4 for the definition).

8.43. **(a)** If not, then there is some $p > p_c(G \mathbin{\square} \mathbb{Z})$ such that $p < p_c(G \mathbin{\square} P_n)$ for all n. Let $\eta = \langle \eta_k \; ; \; k \geq 1 \rangle$ be an invariant percolation on $G \mathbin{\square} \mathbb{Z}$ with the properties that $\eta_k \subset \eta_{k+1}$ and all components of η_k are isomorphic to $G \mathbin{\square} P_{2k}$ for all k. Bernoulli(p) percolation ω_p on $G \mathbin{\square} \mathbb{Z}$ intersected with η_k has only finite components a.s. The infinite components of ω_p are unions of the finite components of $\omega_p \cap \eta_k$. Use the method of proof of Theorem 8.21 to get a contradiction.

(b) (This establishes a special case of Conjecture 7.45.) Show that $p_c(G \mathbin{\square} \mathbb{Z}) \leq p_c(G \mathbin{\square} C_n) \leq p_c(G \mathbin{\square} P_n)$.

8.45. (This is a special case of Example 9.6 of Aldous and Lyons (2007).) Verify (8.13) for probabilities equal to the reciprocal of the degrees.

8.46. (This is a special case of Example 9.6 of Aldous and Lyons (2007).) Verify (8.13) for the following probability measure: Let o be a vertex of G. Let $Z := 1 + \deg o/2 + \sum_{f \sim o} 1/\deg f$, where f denotes a face of G and $\deg f$ denotes its degree in $G^\dagger$. Choose $\hat{o}$ equal to o with probability $1/Z$, equal to v_e with probability $1/(2Z)$ for each $e \sim o$, and equal to f with probability $1/(Z \deg f)$ for each $f \sim o$.

8.49. More general versions are in BLPS (1999b) and Levitt (1995). Take Bernoulli(ϵ) site percolation and send mass 1 from each vertex to a random ω-neighbor that is closer in ω to an open vertex, if any. Then the expected mass sent is $1 - \epsilon$, so the expected degree in ω is at least $2 - 2\epsilon$. Or take the free minimal spanning forest of ω and use Proposition 8.18.

8.50. (This is from BLPS (1999b).) Combine Lemma 8.35 with Theorem 8.19. See also Corollary 8.20.

8.51. (This is from BLPS (1999b).) Let $\mathbb{Z}_3 := \mathbb{Z}/3\mathbb{Z}$ be the group of order 3, and let T be the 3-regular tree with a distinguished end, ξ. On T, let every vertex be connected to precisely one of its offspring (as measured from ξ), each with probability 1/2. Then every component is a ray. Let ω_1 be the preimage of this configuration under the coordinate projection $T \times \mathbb{Z}_3 \to T$. For every vertex x in T that has distance 5 from the root of the ray containing x, add to ω_1 two edges at random in the $\mathbb{Z}_3$ direction to connect the three preimages of x in $T \times \mathbb{Z}_3$. The resulting configuration is a stationary spanning forest with three ends per tree and expected degree $(1/2)1 + (1/2)2 + 2^{-(5+1)}((2/3)1 + (1/3)2)$.

8.52. (This is from BLPS (1999b).) Use the notation of the solution to Exercise 8.51. For every vertex x that is a root of a ray in T, add to ω_1 two edges at random in the $\mathbb{Z}_3$ direction to connect the three preimages of x, and for every edge e in a ray in T that has distance 5 from the root of the ray, delete two of its preimages at random.

8.53. This result is essentially due to Epstein and Hjorth (2009) in a different setting (and with a different proof). We begin with a deterministic result: if H is a graph, ξ is an end of H, and K is a subset of $\mathsf{V}(H)$ such that no sequence from K converges to ξ, then there is a "canonical" way to associate to ξ and K a finite set $A = A(K, \xi) \subset \mathsf{V}(H)$ that separates K from ξ, where the meaning of "canonical" will become clear when we apply this to a random situation. To prove this, let n be the smallest integer such that there is a set of cardinality n that separates K from ξ. There are only finitely many sets of cardinality n that separate K from ξ and whose maximal distance to K is minimal; let A be their union.

Now suppose for a contradiction that we could pick some ends Ξ as described in the exercise statement. Consider a forest $\mathfrak{F}$ as in Lemma 8.35. By Corollary 8.20, at least some trees in $\mathfrak{F}$ have $p_c < 1$. Thus, we may choose $p \in (0, 1)$ so that Bernoulli(p) percolation on $\mathfrak{F}$ leaves a subgraph η with infinite clusters with positive probability. Clearly a.s. not all infinite clusters of η have subsets that converge to an end in Ξ. To each infinite cluster K of η that belongs to a component of ω with an end $\xi \in \Xi$, but such that no subset of K converges to ξ, we associate the set $A(K, \xi)$ as earlier. Let each vertex in K send mass $1/|A(K, \xi)|$ to each point in $A(K, \xi)$. This is an invariant mass transport that sends out mass at most 1 from each point, but some points receive infinite mass with positive probability. Thus, we arrive at our desired contradiction.

8.54. Compare Proposition 7.1 of Häggström, Peres, and Schonmann (1999).

8.55. (This is Theorem 1.5 of Häggström, Peres, and Schonmann (1999).) Use Exercise 8.54 and Theorem 7.21.

8.56. **(c)** Let $e_1(x)$ be a random uniform edge incident to x, independent for different x. Now let $Z'(x) := \max_{e \sim x, \; e \neq e_1(x)} -W(e)W(e_1(x))$.

If max is replaced by sum, then part (c) is open.

It had been asked by Lyons and Schramm in 1997 whether invariant processes that can be monotonically coupled can also be monotonically coupled in an invariant way. The construction here with sum was proposed as a counterexample by Lalley in 1998, whereas the solution with max was given by Peres in 1998. The answer to the question of Lyons and Schramm was finally provided by Mester (2013); it is no.

8.57. Fix (H, a). For every x such that $(K(x), x)$ is rooted isomorphic to (H, a), let x send total mass 1 split equally among the vertices of $K(x)$.

Chapter 9

9.1. Find $\theta \perp \star$ such that $\mathcal{X}^e = i_W^e + \theta$.

9.2. It suffices to do the case $H = \overline{\bigcup H_n}$. Let $K_1 := H_1$ and define K_n for $n > 1$ by $H_n = H_{n-1} \oplus K_n$. Then

$$H = \bigoplus_{n=1}^{\infty} K_n := \left\{ \sum u_n \,;\, u_n \in K_n, \sum \|u_n\|^2 < \infty \right\}.$$

This makes the result obvious.

9.3. Follow the proof of Proposition 9.1, but use the stars more than the cycles.

9.5. Let $e := \langle a, z \rangle$. The free effective resistance in G' between a and z equals $r(e)i_F^e(e)/\left[1 - i_F^e(e)\right]$ when i_F^e is the free current in G.

9.6. **(a)** If $x \in \mathsf{V}$, then $\nabla \mathbf{1}_{\{x\}}$ is the star at x.

(d) Use Exercise 2.93.

(g) Show that $(\nabla f, i_x)_r = (\nabla g, i_x)_r$, and then show that $g(x) = (\nabla g, i_x)_r$ when g has finite support. (This is an example where (df, i_x) might not equal (f, d^*i_x).)

(h) Use part (g) or the open mapping theorem.

9.7. (This is due to Thomassen (1989).) Fix vertices $a_i \in \mathsf{V}_i$. Use Theorem 9.7 with

$$W_n := \left\{ (x_1, x_2) \,;\, |x_1 - a_1| \vee |x_2 - a_2| = n \right\}.$$

9.8. The set of $g \geq |f|$ in $\mathbf{D}_0$ is closed and convex. Use Exercise 9.6(h).

9.10. Let θ be a unit flow of finite energy from a vertex o to ∞. Since θ has finite energy, there is some finite $K \subset \mathsf{V}(G)$ such that the energy of θ on the edges with some endpoint not in K is less than $1/m$. That is, the effective resistance from K to infinity is less than $1/m$.

9.12. (This is due to Abrams and Kenyon (2017).) In fact, the number of such tilings with prescribed "combinatorics" also depends only on k. Use Exercise 2.80.

9.14. If f is bounded and harmonic, let $\langle X_n \rangle$ be the random walk whose increments are i.i.d. with distribution μ. Then $\lim f(X_n)$ exists and belongs to the exchangeable σ-field.

9.15. This proof is due to Raugi (2004).

9.16. To say that μ is symmetric means that $\mu(g) = \mu(g^{-1})$ for all $g \in \Gamma$. Examine the solution of Exercise 9.15 closely.

9.17. **(a)** $P_{\star_n}\big(\theta \!\restriction\! \mathsf{E}(G_n^W)\big) = i_n$.

(b) Take θ to be a limit point of $\langle i_n \rangle$.

9.18. Use Exercise 2.60.

9.19. $\operatorname{div} \nabla = I - P$; use Exercises 9.6(g–h) and 9.18.

9.20. Use Exercise 9.19.

9.21. Let $f_1 := (-f) \vee \mathbf{0}$ and $f_2 := f \vee \mathbf{0}$. Apply Exercise 9.6(f) to f_1 and f_2 and use Exercise 9.20.

9.23. After following the hint, we get that u is harmonic. That u is constant then follows from Corollary 9.6. One can avoid Corollary 9.6 as follows: For any a, if u is harmonic, then $u \wedge a$ is superharmonic. But we just proved that this means $u \wedge a$ is harmonic. Since this holds for all a, u is constant.

9.24. Let the function values be certain probabilities.

9.25. The wired effective resistance is $2(b^n - 1)/[b^{n-1}(b^2 - 1)]$ if the branching number is b and the distance between the vertices is n.

9.31. Use Theorem 2.17. For the example in the second part, make G recurrent by giving it infinitely many cut-edges.

9.33. The space **BD** is called the ***Dirichlet algebra***. The maximal ideal space of **BD** is called the ***Royden compactification*** of the network.

9.34. (**a**) Use Exercise 2.64.

(**b**) Use Exercise 2.64 and Proposition 2.12. Alternatively, use the fact that the subspace onto which we are projecting is effectively one-dimensional.

9.35. Another proof uses Exercise 9.24 (and Exercise 10.16).

9.36. Use Exercise 9.6(e).

9.37. By Exercise 9.6(e–f), it suffices to prove this for $f \in \mathbf{D}_0$. Therefore, it suffices to prove it for $f \in \mathbf{D}_{00}$. Let $g := (I - P)f$. By (6.17), we have $\|df\|_c^2 - \|dPf\|_c^2 = 2\|g\|_\pi^2 - (g, (I - P)g)_\pi \geq 0$, since $\|I - P\|_\pi \leq 2$.

9.39. This result shows that restricted to $\bigstar$, the map $\theta \mapsto F$ is the inverse of the map of Exercise 9.6(h).

9.41. Write the current as the appropriate orthogonal projection of a path from a to z and move the projection to the other side of the inner product.

9.42. Imitate the proof of Exercise 2.13 and use Exercise 9.41.

9.44. Since $\{\chi^e \; ; \; e \in \mathsf{E}_{1/2}\}$ is a basis for $\ell_-^2(\mathsf{E}, r)$ and $P_{\nabla\mathbf{HD}} = P_\Diamond^\perp - P_\bigstar$, the linear span of $\{i_\mathsf{F}^e - i_\mathsf{W}^e\}$ is dense in $\nabla\mathbf{HD}$. (This also shows that the bounded Dirichlet functions are dense in **D**. Furthermore, in combination with Exercise 2.43, it gives another proof of Corollary 9.6.)

9.45. An extension for infinite graphs H all of whose bounded harmonic functions are constant is that every bounded harmonic function on $G \mathbin{\square} H$ has the property that it does not depend on the second coordinate.

9.46. Define the random walk
$$Y_n := \begin{cases} X_n & \text{if } n \leq \tau_{\mathsf{V}\setminus\mathsf{W}} \\ X_{\tau_{\mathsf{V}\setminus\mathsf{W}}} & \text{otherwise.} \end{cases}$$
It follows from a slight extension of the ideas leading to (9.9) that for $f \in \mathbf{D}$ harmonic at all vertices in W, the sequence $\langle f(Y_n) \rangle$ is an L^2-martingale, whence $f(X_0) = \mathbf{E}[\lim_{n\to\infty} f(Y_n)]$. This is 0 if f is supported on W.

9.47. (This is an extension of Lemma 3.1 of Georgakopoulos (2010), where H_n are paths.) Follow the proof of Theorem 9.7. Choose f and e_0 as there. Let $\mathcal{P}_i$ be paths emanating from the endpoints of e_0 along which f is monotonic, increasing from the endpoint where f is larger and decreasing from the other endpoint.

9.49. Consider linearly independent elements of $\mathbf{D}/\widetilde{\mathbf{D}_0}$. This has an important refinement for Cayley graphs: see Section 10.8.

9.53. Put weights on the usual graph of $\mathbb{N}$.

9.54. If $e \in \mathsf{E}$ with $v(e^-) < \beta < v(e^+)$ and we subdivide e by a vertex x, giving the two resulting edges resistances
$$r(e^-, x) = r(e)\frac{\beta - v(e^-)}{v(e^+) - v(e^-)}$$

and

$$r(x, e^+) = r(e) \frac{v(e^+) - \beta}{v(e^+) - v(e^-)},$$

and if the corresponding random walk on the graph with e subdivided is observed only when vertices of G are visited and, further, consecutive visits to the same vertex are replaced by a single visit, then we see the original random walk on G, and $v(x)$, the probability of never visiting o, is β.

For each $k = 2, 3, \ldots$ in turn, subdivide each edge e where $v(e^-) < 1 - 1/k < v(e^+)$ as just described with a new vertex x having $v(x) = 1 - 1/k$. Let G' be the network that includes all these vertices. Let Π_k be the set of all vertices $x \in G'$ with $v(x) = 1 - 1/k$, and let τ_k be the first time the random walk on G' visits Π_k. This stopping time is finite, since $v(X_n) \to 1$ by Exercise 2.91. The limit distribution of $J(X_n)$ on the circle is the same for G as for G' and is the limit of $J(X_{\tau_k})$ since $J(X_n)$ converges a.s.

Let G'_k be the subnetwork of G' determined by all vertices x with $v(x) \le 1 - 1/k$. Because all vertices of Π_k are at the same potential in G', identifying them to a single vertex will not change the current flow in G'_k. Thus, the current flow along an edge e in G'_k incident to a vertex in Π_k is proportional to the chance that e is the edge taken when Π_k is first visited. This means that the chance that $J(X_{\tau_k})$ is an arc $J(x)$ is exactly the length of $J(x)$ if $x \in \Pi_k$. Hence the limit distribution of $J(X_{\tau_k})$ is Lebesgue measure.

Chapter 10

10.3. Let F be a finite set of edges, and compare the result on F in Wilson's algorithm on G_n to the result on F in Wilson's algorithm on G_n^W, where we root at an endpoint of F and start successive random walks at the endpoints of F.

10.4. Let $A \subset \mathsf{E}$ be a minimal set whose removal leaves at least two transient components. Show that there is a finite subset B of endpoints of the edges of A such that $\mathsf{FSF}[\exists x, y \in B \ x \leftrightarrow y] = 1 > \mathsf{WSF}[\exists x, y \in B \ x \leftrightarrow y]$. Here, $x \leftrightarrow y$ means that x and y are in the same component (tree). One can also use Exercise 9.24 with Proposition 10.14, or, alternatively, one can derive a new proof of Exercise 9.24 by using this exercise and Proposition 10.14.

10.5. This is due to Häggström (1998). The same holds if we assume merely that $\sum_n r(e_n) = \infty$ for any path $\langle e_n \rangle$ of edges in G.

10.6. **(b)** Use the bounded convergence theorem.

10.9. Use Corollary 10.5.

10.10. Use Exercise 10.4 or 10.5.

10.11. (This is due to Medolla and Soardi (1995).) Use Corollary 10.9.

10.13. The free uniform spanning forest has one tree a.s., since this joining edge is present in every finite approximation. But the wired uniform spanning forest has two a.s.; use Proposition 10.1.

10.14. Orthogonality of $\{G_x\}$ is obvious. Completeness of $\{G_x\}$ follows from the density of trigonometric polynomials in $L^2(\mathbb{T}^d)$. This proves the identity for $F \in L^2(\mathbb{T}^d)$. Furthermore, density of trigonometric polynomials in $C(\mathbb{T}^d)$ shows that f determines F uniquely (given $F \in L^1(\mathbb{T}^d)$, if $f = 0$, then $\int_{\mathbb{T}^d} F(\alpha) p(\alpha) \, d\alpha = 0$ for all trigonometric polynomials p, whence for all continuous p, so $F = 0$). Therefore, if $f \in \ell^2(\mathbb{Z}^d)$, then $F \in L^2(\mathbb{T}^d)$, so the identity also holds when $F \notin L^2(\mathbb{T}^d)$.

10.16. The assumption that $c(\cdot)$ is bounded can be replaced by the assumption that $\sum_n r(e_n) = \infty$ for any path $\langle e_n \rangle$ of edges in T.

10.17. Join $\mathbb{Z}$ and $\mathbb{Z}^3$ by an edge.

10.20. This is from BLPS (2001), Proposition 14.1.

10.21. The equation that replaces (10.41) still shows that it is expressed via a von Neumann dimension. The Hilbertable Γ-space involved does not change with the change in conductances, even though $\diamondsuit^\perp$ does change.

10.23. Show that p_c is a solution of $3 = 4p^2 + 2p^3$.

10.26. Use Proposition 6.9. In Section 14.1, we will see that the limiting speed does exist a.s. and is constant, and that when it is 0, all bounded harmonic functions are constant. This strengthens the result of this exercise in light of Exercise 9.44.

10.27. This extends to planar, transitive graphs.

10.29. Random fields converge weakly if their finite-dimensional distributions converge. The infinite sums are defined by choosing any linear ordering of $\mathsf{E}_{1/2}$. The variances of the differences of the fields at x and y are the free and wired effective resistances between x and y. Note that if jointly normal random variables Y with mean 0 have covariance matrix M, then there are independent, standard normal random variables ζ such that $Y = \sqrt{M}\zeta$. Using the notions of Gaussian Hilbert spaces (see Janson (1997)), one can give a different definition of the free and wired canonical Gaussian fields without limits.

10.31. To show that the tail σ-field is trivial, use Exercise 2.135. The covariances can be expressed via the Green function.

10.33. Use Exercise 9.45.

10.37. Use Exercise 8.47 or (10.3) on the dual graph, together with symmetry.

10.40. This analogue of Theorem 4.8 is due to Robin Pemantle after seeing Theorem 4.8.

10.41. Before Theorem 10.18 was proved, Pemantle (1991) had proved this. This property is called **strong Følner independence** and is stronger than tail triviality. To prove it, let $\langle G_n \rangle$ be an induced exhaustion of G with edge set E_n. Given n, let $K \subset \mathsf{E} \setminus \mathsf{E}_n$. Let $\mathscr{A}$ be an elementary cylinder event of the form $[D \subseteq T]$ for some set of edges $D \subseteq \mathsf{E}_n$, and let $\mathscr{B}$ be a cylinder event in $\mathscr{F}(K)$ with positive probability. By Rayleigh monotonicity, we have, for all sufficiently large $m \geq n$,

$$\mu_n^{\mathrm{W}}(\mathscr{A}) \leq \mu_m^{\mathrm{F}}(\mathscr{A} \mid \mathscr{B}) \leq \mu_n^{\mathrm{F}}(\mathscr{A}).$$

Therefore $\mu_n^{\mathrm{W}}(\mathscr{A}) \leq \mathsf{FSF}(\mathscr{A} \mid \mathscr{B}) \leq \mu_n^{\mathrm{F}}(\mathscr{A})$, and so the same is true of any $\mathscr{B} \in \mathscr{F}(K)$ of positive probability (not just cylinders $\mathscr{B}$). The hypothesis that $\mathsf{FSF} = \mathsf{WSF}$ now gives the result.

10.43. This is due to Le Gall and Rosen (1991).

10.45. This is due to BLPS (2001), Remark 9.5.

10.46. We do not know whether it is true if only one of the graphs is assumed to be transitive, but this seems likely.

10.47. Define G_0 to be two copies of $\mathbb{Z}^3$ joined by an edge. Let $G_1 := G_0 \,\square\, \mathbb{Z}$. Let G_2 be G_1 connected to a copy of $\mathbb{Z}^4$ by an edge.

(a) (This is due to BLPS (2001), Remark 9.8.) Use $G := G_1$.

(b) Use $G := G_2$.

10.50. A theorem of Gromov (1981a) and Trofimov (1984) says that all quasi-transitive graphs of at most polynomial growth satisfy the hypothesis with d an integer.

10.51. By Proposition 10.14 and Exercise 9.7, we have that $\mathsf{FSF} = \mathsf{WSF}$. Since G is nonamenable, we also know that each tree has one end. Therefore, there are finitely many points in each tree that are closest to ξ; the components can be distinguished by the degrees of these points.

10.52. This is due to BLPS (2001), Remark 9.9.

Chapter 11

11.2. The left-hand side divided by the right-hand side turns out to be 109872/109561. An outline of the calculation is given by Lyons, Peres, and Schramm (2006).

11.5. If the endpoints of e are x and y, then W is the vertex set of the component of x or the component of y in the set of edges lower than e.

11.7. Use Exercise 11.5.

11.12. Each spanning tree has all but two edges of G. Condition on the values of these two missing edges to calculate the chance that they are the missing edges. It turns out that there are three trees with probability 4/45, six with probability 7/72, and two with probability 3/40. There are three edge conductances equal to 27, two equal to 32, and one equal to 35.

11.13. There are four trees of probability 1/15 and twelve of probability 11/180.

11.15. Show that p_c is a solution of $3 = 4p^2 + 2p^3$.

11.16. (This is due to Thom (2016).) Suppose that $U(e) = p$. Then $e \notin \mathsf{FMSF}$ iff there is some simple cycle containing e all of whose other labels are $< p$. For each such simple cycle of length n, the probability that all its other labels are $< p$ is p^{n-1}. Let d be the degree of G. The number of simple cycles of length n through o is at most $(\rho d)^n$, whence

$$\mathbf{E}\left[\sum_{e \sim o} \mathbf{1}_{\{e \in \mathsf{FMSF}\}}\right] \geq \int_0^1 \left(d - \sum_{n \geq 1} (\rho d)^n p^{n-1}\right)^+ dp \,.$$

Also, the number of simple cycles of length n through o that contain e is at most $(\rho d)^{n-1}$, and the number of loops at o is at most ρd. Therefore,

$$\mathbf{E}\left[\sum_{e \sim o} \mathbf{1}_{\{e \in \mathsf{FMSF}\}}\right] \geq d(1 - \rho) \int_0^1 \left(1 - \sum_{n \geq 2} (\rho d)^{n-1} p^{n-1}\right)^+ dp \,.$$

Evaluating these integrals yields the desired bound.

If G has no loops, then the first integral, but with the sum starting at $n = 2$, yields the lower bound $1/\rho - \log(1 + 1/\rho)$, which is $\geq (1 - \log 2)/\rho$.

11.17. Whether a group as in part (a) exists was asked by Lyons, Peres, and Schramm (2006). To show that the answer is yes, use Exercise 7.25, Proposition 11.7, Corollary 7.40, and the fact that there are nonamenable groups that are not of uniformly exponential growth (see Section 3.4). Part (b) is due to Thom (2016); use Exercise 11.16 and take a power of a generating set.

11.18. The answer is $22 + 24(\log 2)^2 - 48 \log 2 = 0.26^-$. One can calculate the entire distribution of the degree from this since the expected degree equals 2.

11.19. The chance is

$$\int_0^1 \frac{1 - x^2}{1 - x^2 + x^3} \, dx = 0.72301^+ \,.$$

This is just slightly less than the chance for the uniform spanning tree.

11.20. Use Exercise 11.8.

11.21. Use Exercise 11.7. In contrast to the WUSF, the number of trees in the WMSF is not always an a.s. constant: see Example 6.2 of Lyons, Peres, and Schramm (2006).

11.22. It is a tail random variable.

11.25. This is due to Lyons, Peres, and Schramm (2006).

11.26. This is due to Lyons, Peres, and Schramm (2006).

11.28. Use Exercise 8.47.

Chapter 12

12.4. For any $\lambda > 0$, we have $\mathbf{P}[\widehat{X}_n < \lambda] = \mathbf{E}[X_n \, ; \, X_n < \lambda]/\mathbf{E}[X_n] \le \lambda \, \mathbf{P}[X_n > 0]/\mathbf{E}[X_n] \to 0$.

12.7. Use Laplace transforms.

12.9. It is not necessary to assume that A has finite mean.

12.10. We have $\mathbf{E}[A_i] \to 0$, whence $\mathbf{E}\big[\sum_{i=1}^{n} A_i/n\big] \to 0$, which implies the first result. The second result follows similarly, since $\mathbf{E}\big[\sum_{j=1}^{A_i} |C_{i,j}|\big] \le \mathbf{E}[A_i]$.

12.11. (This is due independently to J. Geiger and G. Alsmeyer, personal communications, 2000.) We need to show that for any x, we have $\mathbf{P}[A > x \mid A \ge B] \le \mathbf{P}[A > x \mid A \ge C]$. Let F be the c.d.f. of A. For any fixed x, we have

$$
\begin{aligned}
\mathbf{P}[A > x \mid A \ge B] &= \frac{\mathbf{P}[A > x, \, A \ge B]}{\mathbf{P}[A \ge B]} = \frac{\int_{y > x} \mathbf{P}[B \le y] \, dF(y)}{\int_{y \in \mathbb{R}} \mathbf{P}[B \le y] \, dF(y)} \\
&= \left(1 + \frac{\int_{y \le x} \mathbf{P}[B \le y] \, dF(y)}{\int_{y > x} \mathbf{P}[B \le y] \, dF(y)}\right)^{-1} \\
&\le \left(1 + \frac{\int_{y \le x} \mathbf{P}[C \le y] \frac{\mathbf{P}[B \le x]}{\mathbf{P}[C \le x]} \, dF(y)}{\int_{y > x} \mathbf{P}[C \le y] \frac{\mathbf{P}[B \le x]}{\mathbf{P}[C \le x]} \, dF(y)}\right)^{-1} \\
&= \left(1 + \frac{\int_{y \le x} \mathbf{P}[C \le y] \, dF(y)}{\int_{y > x} \mathbf{P}[C \le y] \, dF(y)}\right)^{-1} = \mathbf{P}[A > x \mid A \ge C].
\end{aligned}
$$

To show that the hypothesis holds for geometric random variables, we must show that for all $k \ge 1$, the function $p \mapsto (1 - p^{k+1})/(1 - p^k)$ is increasing in p. But this is clear from writing it as $(1 - p^{k+1})/(1 - p^k) = 1 + 1/\sum_{i=1}^{k} p^{-i}$.

12.12. (See Arratia and Goldstein (2010).) Apply the definition.

12.16. To deduce the Cauchy–Schwarz inequality, apply the arithmetic mean–quadratic mean inequality to the probability measure $(Y^2/\mathbf{E}[Y^2]) \, \mathbf{P}$ and the random variable X/Y.

12.19. See the proof of Theorem 12.7.

12.20. See the proof of Theorem 12.7.

12.21. (This is due to Zubkov (1975).) In the notation of the proof of Theorem 12.7, show that $i \, \mathbf{P}[X'_{n-i} > 0] \to 1$ as $i \to \infty$. See Geiger (1999) for details.

12.23. **(a)** By l'Hôpital's rule, Exercise 5.1, and Exercise 12.22, we have

$$
\begin{aligned}
\lim_{s \uparrow 1} \delta(s) &= \lim_{s \uparrow 1} \frac{f(s) - s}{(1 - s)[1 - f(s)]} = \lim_{s \uparrow 1} \frac{f'(s) - 1}{f(s) - 1 - f'(s)(1 - s)} \\
&= \lim_{s \uparrow 1} \frac{f''(s)}{2f'(s) - f''(s)(1 - s)} = \sigma^2/2 \, .
\end{aligned}
$$

(b) We have

$$
\big[1 - f^{(n)}(s)\big]^{-1} - (1 - s)^{-1} = \sum_{k=0}^{n-1} \delta\big(f^{(k)}(s)\big).
$$

Since $f^{(k)}(s_n) \to 1$ uniformly in n as $k \to \infty$, it follows that

$$
\frac{\big[1 - f^{(n)}(s_n)\big]^{-1} - (1 - s_n)^{-1}}{n} \to \sigma^2/2
$$

as $n \to \infty$. Applying the hypothesis that $n(1 - s_n) \to \alpha$ gives the result.

12.24. (i) Take $s_n := 0$ in Exercise 12.23 to get $n[1 - f^{(n)}(0)] \to 2/\sigma^2$.

(ii) By the continuity theorem for Laplace transform (Feller (1971), p. 431), it suffices to show that the Laplace transform of the law of Z_n/n conditional on $Z_n > 0$ converges to the Laplace transform of the exponential law with mean $\sigma^2/2$, that is, to $x \mapsto (1 + x\sigma^2/2)^{-1}$. Now the Laplace transform at x of the law of Z_n/n conditional on $Z_n > 0$ is

$$\mathbf{E}[e^{-xZ_n/n} \mid Z_n > 0] = \mathbf{E}[e^{-xZ_n/n} 1_{[Z_n>0]}]/\mathbf{P}[Z_n > 0] = \frac{\mathbf{E}[e^{-xZ_n/n}] - \mathbf{E}[e^{-xZ_n/n} 1_{[Z_n=0]}]}{1 - f^{(n)}(0)}$$

$$= \frac{f^{(n)}(e^{-x/n}) - f^{(n)}(0)}{1 - f^{(n)}(0)} = 1 - \frac{n[1 - f^{(n)}(e^{-x/n})]}{n[1 - f^{(n)}(0)]}.$$

Application of Exercise 12.23 with $s_n := e^{-x/n}$, together with part (i), gives the result.

Chapter 13

13.1. The expected time to exit a unary tree, once entered, is infinite; see Exercise 2.46. Alternatively, use an argument similar to that which led to (13.6) but this time with T covering an infinite graph, only finitely many of whose vertices have degree larger than 2.

13.2. See Proposition 4.2 and the solution of Exercise 4.3 in Levin and Peres (2017).

13.4. Suppose that $P^t f = \lambda^t f$ and $\max |f| = |f(x)| = 1$. Then

$$|\lambda^t| = \left| \sum_y p_t(x, y) f(y) \right| = \left| \sum_y [p_t(x, y) - \pi(y)] f(y) \right| \le \sum_y |p_t(x, y - \pi(y)| = 2\delta(t).$$

Therefore

$$t_{\mathrm{mix}}(\epsilon)\left(\frac{1}{|\lambda|} - 1\right) \ge t_{\mathrm{mix}}(\epsilon) \log\left(\frac{1}{|\lambda|}\right) \ge \log\left(\frac{1}{2\epsilon}\right).$$

13.5. By transitivity, $\widehat{D_x}^2 := \sum_y \pi(y) d^2(x, y)$ does not depend on x, so averaging over x shows that $\widehat{D_x}^2 = \widehat{D}^2$ for every x. For any three vertices x, y, z, the triangle inequality implies that $d^2(x, z) \le 2 d^2(x, y) + 2 d^2(y, z)$. Averaging over y gives $d^2(x, z) \le 4\widehat{D}^2$.

13.6. We claim that simple random walk on the hypercube satisfies

$$\mathbf{E}d(Z_t, Z_0) \ge \frac{t}{2} \qquad \forall t \le \frac{k}{4}.$$

Indeed, at each step at time $t \le k/4$, we have probability at least $\frac{3}{4}$ to increase the distance by 1 and probability at most $\frac{1}{4}$ of decreasing it by 1, so

$$\mathbf{E}d(Z_t, Z_0) \ge \frac{3}{4}t - \frac{1}{4}t = \frac{t}{2}.$$

It now follows from Jensen's inequality that $\mathbf{E}d^2(Z_t, Z_0) > t^2/4$ for $t \le k/4$, implying that the hypercubes do not have uniform Markov type 2.

13.7. Given $\epsilon > 0$, write an orthogonal decomposition $\psi = \psi_F + \chi$ where ψ_F is finitely supported and $\|\chi\| \le \epsilon$. Then $\langle P^i \psi_F, \psi_F \rangle \to 0$ by Exercise 2.106, so $\limsup_i \langle P^i \chi, 2\psi_F + \chi \rangle \le 2\epsilon$ by Cauchy–Schwarz. Adding these yields $\limsup_i \langle P^i \psi, \psi \rangle \le 2\epsilon$.

An alternative proof uses the spectral theorem. Since $\|P\| \le 1$, we may write $P = \int_{-1}^1 \lambda \, dE(\lambda)$, where $E(\cdot)$ is the spectral resolution of P. Since V is infinite, ± 1 are not eigenvalues of P. Therefore $\langle P^i \psi, \psi \rangle = \int_{(-1,1)} \lambda^i \, d\langle E(\lambda)\psi, \psi \rangle \to 0$ as $i \to \infty$.

13.9. A flow in G' gives a flow in G, while distances in G' are no smaller than in G.

13.10. Use the idea at the beginning of Section 3.1.

13.12. For the second one, use Stirling's formula.

13.13. For the upper bound, given a cut Π in T, construct a cut in $T^{[k]}$ by taking the closest ancestor in $T^{[k]}$ to each vertex in Π. For the lower bound, map flows in T to flows in $T^{[k]}$.

13.14. See Benjamini and Peres (1994b). These bounds were sharpened by Benassi (1996).

13.15. The tree T satisfies $\log(\overline{\mathrm{gr}}\,T) = \dim_{\mathrm{P}}(\partial T) = \log 2$. If the n_i increase sufficiently fast, then $\mathrm{br}\,T = 1$, as in the 1–3 tree.

13.16. (Part (b) is due to Oded Schramm, personal communication, 2007, while parts (c)–(f) are due to Mark Braverman, personal communication, 2009.) Part (b) can be modified to be a transitive, simple graph by replacing each vertex i by a set of 2^N vertices labeled from $(i, 1)$ to $(i, 2^N)$; connect all (i, j) to all $(i \pm 1, k)$ and also (i, j) to (m, j) when the distance from i to m is at least $\sqrt{N}$ (mod N). In (b), the expected distance is close to 2 for moderately large times but close to 1 for very large times. If one wants a transitive network with large diameter, one can take a Cartesian product of this network with itself many times. For (c), use reversibility and stationarity. For (d), walk $m - k$ steps from X_k to a new X'_{m-k} and use the triangle inequality $d(X_0, X_n) \leq d(X_0, X'_{m-k}) + d(X'_{m-k}, X_n)$. For (e), subdivide and re-weight the edges of the original network. For (f), use the proof of (d), not the result, for the chain Y. The same results hold for random walk on an infinite, transitive network started at any vertex. Likewise, suppose G is a quasi-transitive network with orbit representatives $o_1, \ldots, o_L$. Consider the quotient Markov chain on the set $\{o_i\}$. If this chain is reversible with stationary probability distribution v_i and $\mathbf{P}[X_0 = o_i] = v_i$, then the same results hold. In particular, if G is a unimodular, quasi-transitive network started at a π-biased normalized root, then the results hold by Exercise 8.33.

13.17. This follows from Theorem 13.1 by subtracting the conditional expectations given the past at every stage (the elementary Doob decomposition).

13.18. **(a)** This refinement of the principle of Cauchy condensation is due to Dvoretzky (1949). Choose b_n to be increasing and tending to infinity such that $\sum_n b_n a_n / n < \infty$. Define $m_1 := 1$ and then recursively $m_{k+1} := m_k + \lceil m_k / b_{m_k} \rceil$. Define $n_k \in [m_k, m_{k+1})$ so that $a_{n_k} = \min\{a_n ; n \in [m_k, m_{k+1})\}$.

(b) A special case appeared in Dvoretzky (1949). The general result is essentially due to Davenport, Erdős, and LeVeque (1963); see also Lyons (1988). Prove it first along the subsequence n_k given by (a), then for all n by imitating the proof of Theorem 13.1.

13.19. These are due to Lyons (1988).

(a) Define $n_1 := 1$ and then recursively n_{k+1} as the smallest $n > n_k$ so that $a_n / n < (n - n_k) / n_k^2$.

(b) Prove it first along the subsequence n_k given by (a), then for all n by imitating the proof of Theorem 13.1.

13.21. For (b), note that $\langle S_{n \wedge \tau} \rangle$ is also a supermartingale, where $\tau := \inf\{k ; S_k \geq L\}$.

13.22. We do the case where G is simple; for other G, consider the directed line graph of G. For positive integers L, let $N(L)$ be the number of paths in G of length L. Then $\log b(G) = \lim_{L \to \infty} \log N(L)/L$. Let A denote the directed adjacency matrix of G. Consider the Markov chain with transition probabilities $p(x, y) := A(x, y)/d(x)$. Our hypothesis on G guarantees that a stationary measure is $\sigma(x) := d(x)/D$, where $D := \sum_x d(x)$. Write $n := |V|$. The entropy of this chain is, by convexity of the function $t \mapsto t \log t$,

$$\sum_x [d(x)/D] \log d(x) = \sum_x [d(x)/D] \log[d(x)/D] + \log D \geq nz \log z + \log D = \log(D/n),$$

where $z := (1/n) \sum_x d(x)/D = 1/n$. The method of proof of (13.7) now shows the result.

13.23. This transition matrix is called the ***Shannon–Parry measure***. Let λ be the Perron eigenvalue of the directed adjacency matrix A of G with left eigenvector L and right eigenvector R. Define $\pi(x) := L(x)R(x)$, where we assume R is normalized so that this is a probability vector. Define the transition probabilities $p(x, y) := [\lambda R(x)]^{-1} A(x, y) R(y)$. If G is not simple, then one can make a similar Markov chain using the directed line graph of G.

13.24. It is $\left(3\alpha(\alpha + 1) + (1 - \alpha)\sqrt{16\alpha + 9\alpha^2}\right) / (2(2 + \alpha))$ a.s.

13.25. Let d_u denote the number of children of a vertex u in a tree. Use the SLLN for bounded martingale differences (Theorem 13.1) to compare the speed to

$$\liminf_{n \to \infty} \frac{1}{n} \sum_{k \le n} \frac{d_{X_k} - 1}{d_{X_k} + 1} \,.$$

Show that the frequency of visits to vertices with at least two children is at least $1/N$ by using the SLLN for L^2-martingale differences applied to the times between successive visits to vertices with at least two children. Note that if x has only one child, then x has a descendant at distance less than N with at least two children and also an ancestor with the same property (unless x is too close to the root).

An alternative solution goes as follows. Let $v(u)$ be the ancestor of u closest to u in the set $\{v \,;\, d_v > 2\} \cup \{o\}$ (we allow $v(u) = u$), and let $w(u)$ be the descendant of u of degree > 2 chosen such that $L(u) = |w(u)| - |u| > 0$ is minimal. Write $k(u) = |u| - |v(u)| \ge 0$, so that $k(u) + L(u) < N$ always. Check that

$$Y_t := |X_t| - [t + k(X_t) L(X_t)] / (3N)$$

is a submartingale with bounded increments by considering separately the cases where (1) $d(X_t) > 2$ or X_t is the root and (2) the remaining cases. Apply Exercise 13.17 to finish.

13.26. This is due to Lyons, Pemantle, and Peres (1997), Example 2.1.

13.27. See Example 2.2 of Lyons, Pemantle, and Peres (1997).

13.28. See Example 2.3 of Lyons, Pemantle, and Peres (1997). For more general calculations of speed on directed covers, see Takacs (1997, 1998).

13.29. It is $(3\sqrt{2} + 1)/17 = 0.31^-$.

13.31. (Compare Barlow and Perkins (1989).) Use Theorem 13.4 and summation by parts in estimating $\mathbf{P}[|X_n| \ge \sqrt{d + 2 + \epsilon} \cdot \sqrt{n \log n}]$.

Barlow and Perkins (1989) also show that there is a tree that is a subgraph of $\mathbb{Z}^2$ on which simple random walk satisfies $\limsup_{n \to \infty} \dfrac{|X_n|}{\sqrt{n \log n}} > 0$ a.s. In contrast, simple random walk on all of $\mathbb{Z}^d$ satisfies $\limsup_{n \to \infty} \dfrac{|X_n|}{\sqrt{n \log \log n}} < \infty$ a.s. by the law of iterated logarithm.

13.32. Compare Pittet and Saloff-Coste (2001).

13.34. Let $t := t_{\text{mix}}(1/4)$. By Lemma 13.10, we may write $p^t(x, y) = \frac{1}{4}\pi(y) + \frac{3}{4}Q(x, y)$ for some transition matrix Q. Construct τ as t times a geometric random variable of parameter $1/4$.

13.37. (This is due to Naor, Peres, Schramm, and Sheffield (2006).) Adapt the proof from the case of trees, Theorem 13.14.

13.38. Show first that the expected number of vertices visited by simple random walk on $\mathbb{Z}_n^3$ by time n^3 is of order n^3.

13.39. The actual asymptotic for the nonreturn probability is $\pi/\log k$, due to Dvoretzky and Erdős (1951); see Lemma 20.1 of Révész (2005).

13.40. Let $\langle A_n \rangle$ be Følner sets in G. By censoring moves of the simple random walk on G from A_n to A_n^c, we obtain a reversible Markov chain on A_n, to which we can apply the definition of Markov type 2. Fixing t and letting $n \to \infty$ yields the desired inequality, since the simple random walk on G, started at a uniformly chosen point in A_n, stays in A_n for t steps with probability that tends to 1 as $n \to \infty$. A regular tree shows that the amenability assumption is needed.

13.41. This result is due to Linial, Magen, and Naor (2002). It is open whether one can replace $\sqrt{g}$ by g: this question was asked by Linial, London, and Rabinovich (1995).

13.42. (This is Lemma 2 of Jolissaint and Valette (2014).) Define a bipartite graph on two copies of V and use Kőnig's theorem (Exercise 3.17(a)).

13.43. (This can be improved by a factor of 2 to $\log(|V|/2)/\log(k-1)$; see Jolissaint and Valette (2014).) Our proof is an example of the probabilistic method, where instead of constructing a permutation explicitly, we find it by looking at random permutations. However, we have to be a little more careful than simply using the uniform measure on permutations. Write $n := |V|$.

First, as noted before, the ball of radius r about x has at most $\left(k(k-1)^r - 2\right)/(k-2)$ vertices, which is $< \sqrt{n/2}$ when $r \le r_0 := \log(n/2)/(2\log(k-1))$. Therefore, if σ is a uniform random permutation of V, then $\mathbf{P}\big[d(x, \sigma(x)) \le r_0\big] < 1/\sqrt{2n}$. Let $B := \{x\,;\, d(x, \sigma(x)) \le r_0\}$. We have $\mathbf{E}\big[|B|\big] < \sqrt{n/2}$.

We would be done if B were empty with positive probability, but this estimate is not enough to prove that such happens. We do know that $|B| < \sqrt{n/2}$ with positive probability. Thus, we take one more step. Namely, choose a random injective mapping $\tau : B \to V$, where τ is independent of σ given B. Define

$$\tau'(x) := \begin{cases} \tau(x) & \text{if } x \in B, \\ \tau^{-1}(x) & \text{if } x \in \tau(B), \\ x & \text{otherwise.} \end{cases}$$

Consider $\tau' \circ \sigma$. Now $\mathbf{P}\big[d(x, \tau' \circ \sigma(x)) \le r_0 \mid x \in B\big] < 1/\sqrt{2n}$. Likewise, $\mathbf{P}\big[d(x, \tau' \circ \sigma(x)) \le r_0 \mid x \in \tau(B)\big] < 1/\sqrt{2n}$. Finally, $\mathbf{P}\big[d(x, \tau' \circ \sigma(x)) \le r_0 \mid x \notin B \cup \tau(B)\big] = 0$. Therefore, $\mathbf{P}\big[d(x, \tau' \circ \sigma(x)) \le r_0\big] < 1/n$, whence $\mathbf{E}\Big[\big|\{x\,;\, d(x, \tau' \circ \sigma(x)) \le r_0\}\big|\Big] < 1$. Thus there exists some permutation $\tau' \circ \sigma$ such that $d(x, \tau' \circ \sigma(x)) > r_0$ for all x.

13.44. We have written (13.38) so that instead of sums, we compare averages. But canceling common factors shows that we must prove the following inequality:

$$\frac{\lambda_1(G, c)}{4} \sum_{x \in V} \big|f(x) - f(\sigma(x))\big|^2 \le \sum_{e \in E_{1/2}} c(e)\,\|f(e^+) - f(e^-)\|^2.$$

Now by translating f, we may assume that

$$\sum_{x \in V} f(x) = 0. \tag{18.8}$$

Define the unitary operator U on $\ell^2(V, \mathscr{H})$ by $U(f) := f \circ \sigma$. If I denotes the identity map, then the triangle inequality gives us that $\|I - U\| \le 2$, whence

$$\sum_{x \in V} \big|f(x) - f(\sigma(x))\big|^2 \le 4 \sum_{x \in V} \|f(x)\|^2.$$

To complete the proof, it suffices to show that

$$\lambda_1(G, c) \sum_{x \in V} \|f(x)\|^2 \le \sum_{e \in E_{1/2}} c(e)\,\|f(e^+) - f(e^-)\|^2.$$

By considering the coordinates with respect to an orthonormal basis of $\mathscr{H}$, we see that it suffices to prove this inequality when $\mathscr{H} = \mathbb{R}$. In that case, our assumption (18.8) is that $f \perp \mathbf{1}$ and the inequality to be proved amounts to

$$\lambda_1(G, c)\,\|f\|^2 \le \left(\Delta_G f, f\right).$$

This is a consequence of the spectral theorem (as in Exercise 6.14).

13.45. Average (13.38) over all permutations.

13.46. The maximal displacement is at least $\log\big(|V|/2\big)/\big(2\log(k-1)\big)$ when G has maximum degree $k \geq 3$.

13.49. The eigenvectors of the graph Laplacian for the Cayley graph of an abelian group are the characters of the group. For the $2n$-cycle, the planar embedding as a regular polygon gives an upper bound for the distortion.

13.51. Note that Exercise 13.50 does not apply.

13.52. Let ϕ_i be embeddings of G_i. Consider embeddings of the product graph that have the form $(x_1, x_2) \mapsto \big(a\phi_1(x_1), \phi_2(x_2)\big)$ for some a.

13.54. Think about simulating large conductances via multiple edges.

13.55. Add multiple edges between vertices of a tree, where the endpoints of the new edges are at the same distance from the root.

Chapter 14

14.2. Use Exercise 2.49 or 10.16.

14.4. The difference

$$H(X, Y \mid \mathscr{F}) - H(X \mid \mathscr{F}) = \mathbf{E}\Big[\sum_{x,y} \mathbf{P}[X = x, Y = y \mid \mathscr{F}] \cdot \log \frac{\mathbf{P}[X = x, Y = y \mid \mathscr{F}]}{\mathbf{P}[X = x \mid \mathscr{F}]}\Big]$$

is an expected Kullback–Leibler divergence.

14.5. (b) If $b < d$, then simple random walk on the Cartesian product $\mathbb{T}_{b+1} \square \mathbb{T}_{d+1}$ has volume growth $v = \log d$, speed $l = (b + d - 2)/(b + d + 2)$, and entropy

$$h = \frac{(b-1)\log b + (d-1)\log d}{b + d + 2}.$$

To see the latter, condition on the number of steps taken in each coordinate.

14.6. Count states when they are visited for the last time: $\mathbf{E}_o[R_n] = \sum_{k=0}^{n-1} \mathbf{P}_o[\forall j \in (k, n-1]\ X_j \neq X_k]$. Now use transitivity. Almost sure convergence of R_n/n follows from the subadditive ergodic theorem. The limit is a.s. constant by Lemma 14.11. Compare with Exercise 2.119.

14.7. By irreducibility, $\lim_n u(X_n) = c$ almost surely-$\mathbf{P}_x$. Apply $\mathbf{E}_x$ and recall that $u(X_n)$ is a bounded martingale.

14.8. *(i)* $\Rightarrow$ *(ii):* If $f_1 \in L^\infty(M_1, \mathscr{F}_1, Q_1)$ is simple (takes finitely many values), the existence of f_2 with $f_2 \circ \Phi = f_1$ a.e. is immediate from (a). Every measurable f_1 is an a.e. limit of simple functions.

14.9. Use (a) to deduce (b) and (c).

When Γ is uncountable, the same holds in (c): see Exercise 14.33.

14.10. Write $I_n := \mathbf{1}_{A_n}$. Given $A \in \mathscr{F}$, find integers $n(k)$ such that $I_{n(k)} \to \mathbf{1}_A$ in $L^1(\Omega, \mathscr{F}, \mathbf{P})$. Passing to a subsequence if necessary, we may assume that $I_{n(k)} \to \mathbf{1}_A$ a.s.-$\mathbf{P}$. Thus, A differs from $\bigcap_{j \geq 1} \bigcup_{k \geq j} A_{n(k)}$ by a set of $\mathbf{P}$-measure 0.

14.11. Consider the analogous problem on a finite tree and take a limit. Alternatively, consider the events A_j of the form $[\mathbf{b}(X_0, X_1, \ldots)_j = \mathbf{b}(\mathbf{z})_j]$.

14.12. For $A \in \mathcal{J}$, the Markov property yields

$$\int_A \mathbf{1}_{[X_n = y, X_{n+m} = z]}\, d\mathbf{P}_x = p_n(x, y)p_m(y, z)Q_z(A) = \int_A p_n(x, y)p_m(y, z) \cdot \frac{dQ_z}{dQ_x}\, d\mathbf{P}_x.$$

Thus, the integrand on the right-hand side is a version of the conditional expectation $\mathbf{E}_x[\mathbf{1}_{[X_n=y, X_{n+m}=z]} \mid \mathcal{J}]$. The rightmost equality in (14.22) follows from (14.21) and the chain rule

$$\frac{dQ_z}{dQ_x} = \frac{dQ_y}{dQ_x} \cdot \frac{dQ_z}{dQ_y}.$$

14.13. This result is due to Sava (2010a).

(a) Extend ξ to a bi-infinite path and observe that the chain $\langle X_n \rangle$ induced on this path is a biased nearest-neighbor random walk.

(c) Let ξ_n denote the vertex of $\mathbb{T}_3$ which is n steps from o toward ξ, and let $\mathbf{W}_n$ consist of the singleton $\{(\psi, \xi_n)\}$, where $\psi(x) = 0$ if x is separated from o by ξ_n and $\psi(x) = \Psi_\infty(x)$ otherwise. Then $\mathbf{P}_{(\overline{0}, o)}[\exists m \geq n \ X_m \in \mathbf{W}_n] \geq 1/2$.

14.14. For the second part, consider $\mathbf{E}[g \circ f - g \circ f \circ T]$ for a suitable g.

14.15. (a) Use Kolmogorov's zero-one law.

14.16. Use the same proof, but note that $\bar{A}$ is $\mathcal{I}$-measurable. Also, replace all expectations by expectations conditional on $\mathcal{I}$.

14.17. Use the same proof, but note that β is $\mathcal{I}$-measurable. Also, replace all expectations by expectations conditional on $\mathcal{I}$.

14.18. Show that $\lim_n \mathbf{P}_o[\|X_{2n}\|_1 \text{ is even}] > 1/2$.

14.19. Use Exercise 14.28 and transience of $\mathbb{Z}^3$ to show that G is Liouville. In the perturbed network (G, c), the walk started from the root of T has positive probability to eventually stay in $\mathbb{Z}^4$ and positive probability to never reach $\mathbb{Z}^4$.

14.20. It is not hard to reduce to the aperiodic case. Let $\widehat{P}$ denote the reversed chain. We may assume irreducibility for the symmetrized chain $(P + \widehat{P})/2$. By considering some power P^j, we may also reduce to the case where $c := \min\{p(x, y) + \widehat{p}(x, y); \ x \sim y\} > 0$. Transitivity and Exercise 6.16 show that $|\partial_E S|_P = |\partial_E S|_{\widehat{P}}$ for all finite $S \subset V$. Since $|\partial_E S|_P + |\partial_E S|_{\widehat{P}} \geq c|\partial_E S|/2$, it follows that $|\partial_E S|_P \geq c|\partial_E S|/4$. Now the result follows from Lemma 10.46 and Corollary 6.32(i).

14.22. (This is Theorem 4.1 of Kaimanovich and Vershik (1983).) Note that $\mathbf{P}_o[X_1 = y \mid X_n = z] = p(o, y)p_{n-1}(y, z)/p_n(o, z)$.

14.23. Let $p(\cdot)$ be the law of X_1, and let $v(x) := c_\epsilon(1 + r)^{-1-\epsilon}$ for $x \in S_r$, where c_ϵ is a normalizing constant. Use the inequality $D_{\mathrm{KL}}(p(\cdot) \parallel v) \geq 0$ to show that $H(X_1) < \infty$. In addition, $\infty > \mathbf{E}_x[\log(1 + d(x, X_1))] = \int_0^\infty \mathbf{P}_x[\log(1 + d(x, X_1)) \geq s] \, ds = \int_0^\infty \mathbf{P}_x[d(x, X_1) \geq e^s - 1] \, ds$, whence for every $\epsilon > 0$, we have $\sum_{k \geq 0} \mathbf{E}_o \mathbf{P}_{X_k}[d(X_k, X_{k+1}) \geq e^{\epsilon k}] < \infty$. By the Borel–Cantelli lemma, it follows that a.s., $d(X_k, X_{k+1}) < e^{\epsilon k}$ for all large k, whence a.s., for all large n, we have $d(o, X_n) \leq ce^{\epsilon n}$ for some $c = c(\epsilon)$. Now use Theorem 14.10(ii) and polynomial growth to deduce that $\boldsymbol{h} \leq d\epsilon$, where d is a bound on the degree of growth.

In fact, $\boldsymbol{h} = 0$ as long as $H(X_1) < \infty$ and, moreover, the Markov chain is Liouville without even assuming $H(X_1) < \infty$; see Theorem 14.49.

14.24. (This is due to Kesten, Spitzer, and Whitman; see Spitzer (1964) and Whitman (1964).) $\mathbf{E}_o[R_n(K)] = \sum_{i=0}^{n-1} \sum_{\gamma \in K} \mathbf{P}_o[\forall j \in (i, n-1] \ X_i \gamma \notin X_j K]$.

14.25. (Parts (i)–(iii) are due to Kaimanovich (2000).) Use Theorem 14.10(ii). For (ii) and (iii), see the more general statements in Corollary 14.36 and Exercise 14.35; for (iv), compare with Proposition 14.42.

14.26. (a) Use Lemma 14.11.

(b) Let c be large. Show that $d(X_n, X_{n+1}) > cn$ infinitely often a.s.

(c) Discretize a Cauchy random walk, for example, by taking $V = \mathbb{Z}$ and letting (X_n, n) be the location of the first visit to the line $y = n$ by simple random walk in $\mathbb{Z}^2$.

(d) Given $\epsilon > 0$, let $r < (\boldsymbol{h} - \epsilon)/(\boldsymbol{v} + \epsilon)$ and $A_n := \{x \in V \,;\, p_n(o, x) \leq \exp(n(\epsilon - \boldsymbol{h}))\}$. Then, for large k,

$$\mathbf{P}_o\big[\exists n \geq k \ \ |X_n| \leq rn\big] \leq \mathbf{P}_o\big[\exists n \geq k \ \ X_n \notin A_n\big] + \sum_{n \geq k} \mathbf{P}_o\big[X_n \in A_n \cap B(o, rn)\big]$$

$$\leq \mathbf{P}_o\big[\exists n \geq k \ \ X_n \notin A_n\big] + \sum_{n \geq k} e^{rn(\boldsymbol{v}+\epsilon)} e^{-n(\boldsymbol{h}-\epsilon)},$$

which tends to 0 as $k \to \infty$.

14.27. **(b)** Write $\mathbb{F}_n$ for the free group on n letters. Let $\Gamma_1 := \mathbb{Z}_k \wr \mathbb{F}_2$, a generalized lamplighter group where the "lamps" take k values, with a symmetric set S_1 of $k + 3$ generators (four from $\mathbb{F}_2$ and $k - 1$ from the cyclic group). Then S_1 supports a measure with Avez entropy of order $\log k$, yet any symmetric measure μ_1 on S_1 has $\boldsymbol{l}(\mu_1) \leq 5/6$, as can be seen by bounding $\mathbf{E}\big[d(o, X_2)\big]/2$. Let $\Gamma := \Gamma_1 \times \mathbb{F}_{100}$ with generating set $S = S_1 \cup S_2$, where S_2 consists of the 200 generators of $\mathbb{F}_{100}$. If k is large enough, then symmetric measures on S maximizing $\boldsymbol{h}$ will be supported on S_1, whereas measures maximizing $\boldsymbol{l}$ will be supported on S_2.

14.28. **(a)** Let f be a tail function. The coupling implies that $\mathbf{E}_o\big[f(X_0, X_1, \ldots) \mid X_1, \ldots, X_n\big]$ does not depend on $X_1, \ldots, X_n$. Now use the Lévy zero-one law.

(b) For invariant f, the shift coupling implies that $\mathbf{E}_o\big[f(X_0, X_1, \ldots) \mid X_1, \ldots, X_n\big]$ does not depend on $X_1, \ldots, X_n$.

Both parts have converses with couplings that are not necessarily Markovian. For more on coupling and shift coupling, see Thorisson (2000).

14.30. For (b), see Lindvall and Rogers (1996), pp. 869–870.

14.31. Let G be recurrent. It suffices to show that for any two configurations (ψ, x) and (φ, y) in $G^\odot$, two lazy simple random walks started from (ψ, x) and (φ, y), respectively, can be coupled to eventually coincide. First couple the marker locations, and then the lamps, using recurrence and laziness. Appeal to Exercise 14.28.

14.32. As noted by Furstenberg (1971b), the space Ξ can be too large to provide a compactification boundary.

14.33. **(a)** Choose an element in each basic open set of Γ and take the group these elements generate.

(b) First verify this with L^1 convergence, when f is an indicator of a cylinder event. Then extend to all $f \in L^1$, and finally obtain a.e. convergence by passing to a subsequence.

(c) Let $\langle I_n \rangle$ be the invariant indicators that define the boundary map $\mathbf{b}$ in Theorem 14.31. Given $\gamma \in \Gamma$, there is a set $\Omega_\gamma \subseteq V^\mathbb{N}$ such that $\mathbf{P}_o(\Omega_\gamma) = 1$ and such that for every $k \geq 1$, there is a sequence $\langle \gamma_j \rangle$ such that $I_k \circ \gamma_j(\mathbf{x}) \to I_k \circ \gamma(\mathbf{x})$ for all $\mathbf{x} \in \Omega_\gamma$. Therefore, if $\mathbf{x}, \mathbf{y} \in \Omega_\gamma$ satisfy $I_k(\mathbf{x}) = I_k(\mathbf{y})$ for all k, then also $I_k(\gamma\mathbf{x}) = I_k(\gamma\mathbf{y})$ for all k. Thus we may define the action of $\gamma \in \Gamma$ on $\mathbf{b}(\Omega_\gamma)$ by $\gamma(\mathbf{b}(\mathbf{x})) := \mathbf{b}(\gamma\mathbf{x})$.

14.35. (This is due to Kaimanovich (2000).) Given $k \geq 1$, define $W_n(k) := \{x \in V \,;\, p_n^J(o, x) \geq e^{-n/k}\}$. By (14.26), for $n \geq N(k)$, we have $\mathbf{P}_o\big[X_n \in W_n(k)\big] > 1 - 1/k$. Write $k_n := \max\{k \leq n \,;\, N(k) \leq n\}$ and observe that $k_n \to \infty$ as $n \to \infty$. Therefore, choosing $W_n := W_n(k_n)$, we obtain that $|W_n| \leq e^{n/k_n}$.

14.36. (This is due to Furstenberg (1973), Section 9. Part (g) was first proved by Cartwright and Soardi (1989).) For (c), γ can be taken in the support of μ, since the set of γ that leaves a measure invariant is a group. A group satisfies the condition in (c) for some K iff the group is nonamenable; see Paterson (1988). Part (d) can be proved via the solution of Exercise 8.16. For part (e), observe that if ν was atomic, the finite set of its largest atoms would be $\mathbb{Z} * \mathbb{Z}$-invariant, so the technique of Exercise 8.16 can be used to rule this out. For part (g), suppose that the convergence in (b) holds, and a subsequence $X_{n_k} \to \xi \in \partial \mathbb{T}_4$ as $k \to \infty$. Then use (b) and (e) to conclude that ξ is the same for all subsequences. The assumption in (d) about the support of μ can be relaxed: it suffices that this support generates a non-abelian subgroup of $\mathbb{Z} * \mathbb{Z}$.

14.37. Verify that $\mathbf{P}(A_r) \geq c_d/r$ and $\mathbf{P}(A_r \cap A_{r+j}) \leq C_d/(rj)$ for some constants $c_d, C_d > 0$. Then use the second-moment method. See James and Peres (1996) for details.

14.38. For $\epsilon > 0$, define $\mathbf{W}_n := \mathbf{W}_{n,\epsilon}(\Psi_\infty)$ to be the set of (ψ, x) such that $n \leq \|x\|_\infty \leq n^2$ and

$$\psi(z) = \begin{cases} \Psi_\infty(z) & \text{if } \|z\|_\infty < \|x\|_\infty \\ 0 & \text{if } \|z\|_\infty > \|x\|_\infty. \end{cases}$$

By Exercise 14.37, $\limsup_{n\to\infty} \mathbf{P}_0[\exists m \geq n \ X_m \in \mathbf{W}_n] > 0$; since $|\mathbf{W}_n|^{1/n} \to 1$, Proposition 14.42 implies that $\sigma(\Psi_\infty)$ coincides with the tail $\mathcal{T}$ mod $\mathbf{P}_0$.

14.39. (This is due to Benjamini and Peres (1994b), proof of Proposition 6.2.) The subadditive ergodic theorem ensures existence of the limit a.s. and in L^1. Use L^1-convergence to deduce convergence of the expectations. Use the hint and (6.39).

Chapter 15

15.1. Show that if $\alpha_1 < \alpha_2$ and $\mathcal{H}_{\alpha_2}(E) > 0$, then $\mathcal{H}_{\alpha_1}(E) = +\infty$. (In fact, $\alpha_0 \leq d$.)

15.5. Since $\bigcap E_n \subseteq E_m$, we have $\dim \bigcap E_n \leq \dim E_m$ for each m.

15.11. It is usually, but not always, $e^{-|x|}$.

15.12. Take $E_n \subseteq \partial T$ with $\dim E_n$ decreasing to the infimum. Set $E := \bigcap E_n$.

15.16. Use Exercise 3.9 on subadditivity. We get that $\dim \sup \partial T = \inf_n \max_v \frac{1}{n} \log M_n(v)$ and $\dim \inf \partial T = \sup_n \min_v \frac{1}{n} \log M_n(v)$.

15.17. Use Furstenberg's theorem (Theorem 3.8) and Theorem 3.9.

15.19. The Fibonacci tree codes E in base 2.

15.21. Use Exercises 3.30 and 3.35.

15.22. This is due to Broman, Camia, Joosten, and Meester (2013).

15.23. Let α be the minimum appearing in Theorem 15.10. If $\gamma < 1$, then $\alpha < 1$, and so two applications of Hölder's inequality yield

$$1 = \mathbf{E}\left[\sum_1^L A_i^\alpha\right] = \mathbf{E}\left[\sum_1^L A_i^\alpha 1^{1-\alpha}\right] \leq \mathbf{E}\left[\left(\sum_1^L A_i\right)^\alpha \left(\sum_1^L 1\right)^{1-\alpha}\right]$$

$$\leq \left(\mathbf{E}\left[\sum_1^L A_i\right]\right)^\alpha \mathbf{E}[L]^{1-\alpha} = \gamma^\alpha m^{1-\alpha}.$$

If $\gamma > 1$, then $\alpha > 1$ and the inequalities are reversed. Of course, if $\gamma = 1$, then $\alpha = 1$ always.

15.25. This is due to Mauldin and Williams (1986).

15.26. (The statement of this exercise was proved in increasing generality by Hawkes (1981), Graf (1987), and Falconer (1987), Lemma 4.4(b).) Use Theorem 5.35.

15.27. These are essentially due to Hawkes (1981) and Lyons (1990), and are stated explicitly in these forms by Peres (1996).

15.28. (This is due to Peres (1996).) Use Exercise 15.27. For (a), consider inverse images under L of points, while for (b), consider inverse images of random sets $Q_{k,b}(b^{-\beta})$ with $\beta < d - \gamma$.

15.29. This is due to Frostman (1935).

(a) Take the largest b-adic cube containing the center of the ball. A bounded number of b-adic cubes of the same size cover the ball.

(b) Use the result of Exercise 3.27.

15.30. Use the law of large numbers.

15.31. Use the law of large numbers for Markov chains.

15.35. Spherically symmetric examples exist. So do subperiodic examples.

Chapter 16

16.1. Use convexity of the energy (from, say, (2.16)) to deduce that energy is minimized by the spherically symmetric flow.

16.4. The only significant change is the replacement of (16.3) and (16.4). For the former, note that diagonals of cubes are longer than sides when $d > 1$. For the latter, notice that if $|x - y| \leq b^{-n}$ and x is in a certain b-adic cube, then y must be in either the same b-adic cube or a neighboring one. See Pemantle and Peres (1995b), Theorem 3.1, for the details; they are missing the 2^{-d} factor.

16.6. (e) From the explicit form of the transition density, we get

$$G_2^*(x, y) = \int_0^\infty \frac{1}{2\pi t} \exp\left\{ -\frac{|x - y|^2}{2t} - t \right\} dt \, .$$

Thus,

$$G_2^*(x, y) = \frac{1}{2\pi} \int_0^\infty \frac{e^{-t}}{t} \int_{|x-y|^2/(2t)}^\infty e^{-s} \, ds \, dt = \frac{1}{2\pi} \int_0^\infty e^{-s} \int_{|x-y|^2/(2s)}^\infty \frac{e^{-t}}{t} \, dt \, ds \, .$$

For an upper bound, we use that

$$\int_{|x-y|^2/(2s)}^\infty \frac{e^{-t}}{t} \, dt \leq \begin{cases} \log \frac{2s}{|x-y|^2} + 1 & \text{if } |x - y|^2 \leq 2s \\ 1 & \text{if } |x - y|^2 > 2s. \end{cases}$$

For $|x - y| \leq 1$, this gives, with $\widetilde{\gamma} := \int_1^\infty e^{-s} \log s \, ds < \infty$, an upper bound of

$$G_2^*(x, y) \leq \frac{1}{2\pi} \left(1 + \log 2 + \widetilde{\gamma} - 2 \log |x - y| \right),$$

which is asymptotically equal to $-\frac{1}{\pi} \log |x - y|$. For a lower bound, we use that $e^{-t} \geq 1 - t$ for $t \in [0, 1]$, whence

$$\int_{|x-y|^2/(2s)}^\infty \frac{e^{-t}}{t} \, dt \geq \log \frac{2s}{|x - y|^2} - 1,$$

and thus with $0 < \gamma := -\int_0^\infty e^{-s} \log s \, ds$ denoting Euler's constant,

$$G_2^*(x, y) \geq \frac{1}{2\pi} \left(-1 + \log 2 - \gamma - 2 \log |x - y| \right),$$

and again this is asymptotically equal to $-\frac{1}{\pi} \log |x - y|$.

16.7. The intersection of m independent Brownian traces, stopped at independent exponential times, is intersection equivalent in the cube to the random set $Q_{2,2}(p_n^m)$, where $p_n = n/(n + 1)$ for $n \geq 1$. It is easy to see that for any m, percolation on a binary tree with edge probabilities $p_e = p_{|e|}^m$ survives with positive probability. Hence the m-wise intersection is nonempty with positive probability. We get the almost sure result the same way as in part (ii).

16.8. Follow the method of proof of Proposition 5.27.

16.10. This is due to Benjamini, Pemantle, and Peres (1995). One can also deduce Theorem 5.24 from this result: It suffices to show (5.21). Consider the Markov chain on $\partial_L T$ that moves from left to right in a planar embedding of T by simply hopping from one leaf to the next that is connected to the root. This turns out to give a kernel that differs slightly along the diagonal from the one in (5.21). Details are in Benjamini, Pemantle, and Peres (1995), which also has more applications.

16.12. Bound the potential of μ at each point x by integrating over $B_{2^{-n}}(x) \setminus B_{2^{-n-1}}(x)$, $n \geq 1$.

16.13. For the lower bound, just use the probability that $\tau \geq 1$. For the upper bound, we follow the hint. Clearly $\int_0^a h(r) f_2(r) \, dr \leq \psi(a) \int_0^a h(r) f_1(r) \, dr$. Write $T_a = \int_a^\infty f_1(r) \, dr$. We have

$$
\begin{aligned}
\int_a^\infty h(r) f_2(r) \, dr &= T_a \int_a^\infty h(r) \psi(r) \frac{f_1(r)}{T_a} \, dr \\
&\leq T_a \int_a^\infty h(r) \frac{f_1(r)}{T_a} \, dr \int_a^\infty \psi(r) \frac{f_1(r)}{T_a} \, dr \\
&= \frac{1}{T_a} \int_a^\infty h(r) f_1(r) \, dr \int_a^\infty f_2(r) \, dr
\end{aligned}
$$

by Chebyshev's inequality. Combining these two inequalities proves the lemma. Now apply the lemma with f_j the density of B_{t_j} and

$$
h(r) := \int_{|y|=r} \mathbf{P}_y\big[B(0, s) \cap A \neq \varnothing\big] \, d\sigma_r(y),
$$

where σ_r is the normalized surface area measure on the sphere $\{|y| = r\}$ in $\mathbb{R}^d$. This gives an upper bound of

$$
\frac{\mathbf{P}_0\big[B(t_2, t_2 + s) \cap A \neq \varnothing\big]}{\mathbf{P}_0\big[B(t_1, t_1 + s) \cap A \neq \varnothing\big]} \leq \frac{f_2(a)}{f_1(a)} + \frac{1}{\mathbf{P}_0\big[|B_{t_1}| > a\big]} \leq e^{|a|^2/(2t_1)} + \frac{1}{\mathbf{P}_0\big[|B_{t_1}| > a\big]} \, .
$$

Finally, let $H(I) := \mathbf{P}_0\big[B(I) \cap A \neq \varnothing\big]$, where I is an interval. Then H satisfies

$$
H(t, t + \tfrac{1}{2}) \leq C_{a,d} H(\tfrac{1}{2}, 1) \text{ for } t \geq \frac{1}{2},
$$

where $C_{a,d} = e^{|a|^2} + 1/\mathbf{P}_0\big[|B_{1/2}| > a\big]$. Hence,

$$
\begin{aligned}
\mathbf{P}_0\big[B(0, \tau) \cap A \neq \varnothing\big] = \mathbf{E}H(0, \tau) &\leq H(0, 1) + \sum_{j=2}^\infty e^{-j/2} H\left(\frac{j}{2}, \frac{j+1}{2}\right) \\
&\leq C_{a,d} \sum_{j=0}^\infty e^{-j/2} H(0, 1) = \frac{C_{a,d}}{1 - e^{-1/2}} \mathbf{P}_0\big[B(0, 1) \cap A \neq \varnothing\big].
\end{aligned}
$$

16.16. This is due to Hawkes (1970/71), but a proof that uses Exercise 15.27 and Theorem 16.11 is due to Peres (1996).

16.21. Let $X := \{0, 1\}$ and $K(x, y) := \infty \mathbf{1}_{\{x \neq y\}}$.

Chapter 17

17.4. (This generalizes Häggström (1997). It was proved in a more elementary fashion in Proposition 8.33.) If there are at least three isolated ends in some tree, then replace this tree by the tree spanned by the isolated ends. The new tree has a denumerable number of ends and still gives a translation-invariant random forest, so contradicts Corollary 17.8. If there is exactly one isolated end, go from it to the first vertex encountered of degree ≥ 3 in the reduced tree and choose two rays independently by visibility measure from there; if there are exactly two isolated ends, choose one of them at random and then do the same as when there is only one isolated end. In either case, we obtain a translation-invariant random forest with a tree containing exactly three ends, again contradicting Corollary 17.8.

17.5. Use (local) reversibility of simple random walk.

17.6. To prove this intuitively clear fact, note that the **AGW**-law of $T \setminus T^{x_{-1}}$ is **GW**, since x_{-1} is uniformly chosen from the neighbors of the root of T. Let A be the event that the walk remains in $T \setminus T^{x_{-1}}$:

$$A := \left\{ (\overleftrightarrow{x}, T) \in \mathsf{PathsInTrees}\,;\ \forall n > 0\ x_n \in T \setminus T^{x_{-1}} \right\}$$
$$= \left\{ (\overleftrightarrow{x}, T) \in \mathsf{PathsInTrees}\,;\ \vec{x} \subset T \setminus T^{x_{-1}} \right\}$$

and B_k be the event that the walk returns to the root of T exactly k times:

$$B_k := \left\{ (\overleftrightarrow{x}, T) \in \mathsf{PathsInTrees}\,;\ |\{i \geq 1\,;\ x_i = x_0\}| = k \right\}.$$

Then the $(\text{SRW} \times \text{AGW} \mid A, B_k)$-law of $(\vec{x}, T \setminus T^{x_{-1}})$ is equal to the $(\text{SRW} \times \text{GW} \mid B_k)$-law of $(\vec{x}, T)$, whence the $(\text{SRW} \times \text{AGW} \mid A)$-law of $(\vec{x}, T \setminus T^{x_{-1}})$ is equivalent to the $(\text{SRW} \times \text{GW})$-law of $(\vec{x}, T)$. By Theorem 17.13, this implies that the speed of the latter is almost surely $\mathbf{E}\left[(Z_1 - 1)/(Z_1 + 1)\right]$.

17.7. For the numerator, calculate the probability of extinction by calculating the probability that each child of the root has only finitely many descendants; whereas for the denominator, calculate the probability of extinction by regarding **AGW** as the result of joining two **GW** trees by an edge, so that extinction occurs when each of the two **GW** trees is finite.

17.8. Use Proposition 5.6.

17.10. Use concavity of log or see (6.39).

17.11. Use the Kac lemma, Exercise 2.30. Recall that the system (PathsInTrees, SRW $\times$ **AGW**, S) was proved to be ergodic in Section 17.3.

17.12. Given nonextinction, the subtree of a Galton–Watson tree consisting of those individuals with an infinite line of descent has the law of another Galton–Watson process still with mean m (Proposition 5.28). Theorem 17.27 applies to this subtree, while harmonic measure on the whole tree is equal to harmonic measure on the subtree.

17.14. From Proposition 17.31, $\mu_{\mathsf{HARM}} \leq l^{-1}\mathbf{GW}$. Therefore, wiring the first generation gives that this integral is at most $l^{-1} \sum_k p_k \log(k + 1)$.

17.15. (This is Lyons, Pemantle, and Peres (1995b), Lemma 9.1.) For a flow θ on T, define

$$\mathscr{E}_n(\theta) := \sum_{1 \leq |x| \leq n} \theta(x)^2 ,$$

so that its energy for unit conductances is $\mathscr{E}(\theta) = \lim_{n \to \infty} \mathscr{E}_n(\theta)$. Consider the sequence of numbers $a_n := \int \mathscr{E}_n(\mathsf{VIS}_T)\,d\mathbf{GW}(T)$. We have $a_0 = 0$ and

$$a_{n+1} = \int \left\{ \sum_{|x|=1} \frac{1}{Z_1^2}\left(1 + \mathscr{E}_n(\mathsf{VIS}_{T^x})\right) \right\} d\mathbf{GW}(T).$$

Conditioning on Z_1 gives

$$a_{n+1} = \sum_{k \geq 1} p_k \frac{1}{k^2} \sum_{i=1}^{k} \int \left(1 + \mathscr{E}_n(\text{VIS}_{T^{(i)}})\right) d\mathbf{GW}(T^{(i)})$$

$$= \sum_{k \geq 1} p_k \frac{1}{k^2} k (1 + a_n) = \mathbf{E}[1/Z_1](1 + a_n).$$

Therefore, by the monotone convergence theorem,

$$\int \mathscr{E}(\text{VIS}_T) \, d\mathbf{GW}(T) = \lim_{n \to \infty} a_n = \sum_{k=1}^{\infty} \mathbf{E}[1/Z_1]^k = \frac{\mathbf{E}[1/Z_1]}{1 - \mathbf{E}[1/Z_1]}.$$

17.16. Consider spherically symmetric trees.

17.17. Consider the proof of Theorem 17.4 and the number of returns to $T^{\diamond}$ until the walk moves along $T^{\diamond}$.

17.19. These observations are due to Lyons and Peres.

(a) This is an immediate consequence of the ergodic theorem.

(b) In each direction of time, the number of visits to the root is a geometric random variable. To establish that Γ_k decreases with k, first prove the elementary inequality

$$a_i > 0 \quad (1 \leq i \leq k + 1) \quad \Longrightarrow \quad \frac{1}{k+1} \sum_{i=1}^{k+1} \frac{k}{\sum_{j \neq i} a_j} \geq \frac{k+1}{\sum_{i=1}^{k+1} a_i}.$$

We have $\lim_{k \to \infty} \Gamma_k = 1/\int \gamma \, d\mathbf{GW}$; use the harmonic mean-arithmetic mean inequality in order to show uniform integrability.

(c) Let

$$N_k(\overleftrightarrow{x}, T) := \lim_{n \to \infty} \frac{\sum_{j \in D_k(\overleftrightarrow{x}, T), \, |j| \leq n} N(S^j \overleftrightarrow{x}, T)}{|D_k(\overleftrightarrow{x}, T) \cap [-n, n]|}$$

when the limit exists; this is a weighted average number of visits to vertices of degree $k + 1$. By the ergodic theorem and part (b), $N_k(\overleftrightarrow{x}, T) = 2\Gamma_k - 1$ SRW $\times$ **AGW**-a.s. Let $D'_k(\overleftrightarrow{x}, T) := \{j \in \mathbb{Z}; \deg x_j = k + 1, \ S^j(\overleftrightarrow{x}, T) \in \text{Fresh}\}$. Then

$$N_k(\overleftrightarrow{x}, T) = \lim_{n \to \infty} \frac{\sum_{j \in D'_k(\overleftrightarrow{x}, T), \, |j| \leq n} N(S^j \overleftrightarrow{x}, T)^2}{|D'_k(\overleftrightarrow{x}, T) \cap [-n, n]|}$$

when the limit exists. Thus, N_k measures the second moment of the number of visits to fresh vertices, not the first, which indeed does not depend on k (see part (d)). The fact that this decreases with k is consistent with the idea that the variance of the number of visits to a fresh vertex of degree $k + 1$ decreases in k, since a larger degree gives behavior closer to the mean.

(d) It is $1/\int \gamma \, d\mathbf{GW}$. We give two proofs.

First proof. Let this number be a_k. Then the proportion of time spent at vertices of degree $k + 1$ is proportional to $a_k p_k$, since the chance of being at a vertex of degree $k + 1$ given that the walk is at a fresh vertex is p_k. Since, however, we know that this proportion is simply p_k, it follows that a_k does not depend on k. Since it doesn't depend on k, we can simply calculate the expected number of visits to a fresh vertex. This expected number is the long-term ratio of time to number of fresh points, which we can partition into blocks between regeneration points. All visits to a fresh point occur within such a block. Thus, we want the ratio of fresh-point frequency to regeneration frequency. This is the ratio of probability of being at a fresh point to probability of being at a regeneration point, which is the same as the reciprocal of the probability of being at a regeneration point given being at a fresh point. And this is $1/\int \gamma \, d\mathbf{GW}$.

Second proof. A fresh epoch is an epoch of last visit for the reversed process. For $\text{SRW} \times \mathbf{AGW}$, if $x_n \neq x_0$ for $n > 0$, then the descendant tree T^{x_1} has an escape probability $\widehat{\gamma}$ with the size-biased distribution of the $\mathbf{GW}$-law of $\gamma(T)$. By reversibility, then, so does the tree $T^{x_{-1}}$ when $(\overleftrightarrow{x}, T) \in \mathsf{Fresh}$. Assume now that $(\overleftrightarrow{x}, T) \in \mathsf{Fresh}$ and $\deg x_0 = k + 1$. Let $y_1, \ldots, y_k$ be the neighbors of x_0 other than x_{-1}. Then $\gamma(T^{y_i})$ are i.i.d. with the distribution of the $\mathbf{GW}$-law of $\gamma(T)$ and $\widehat{\gamma}, \gamma(T^{y_1}), \ldots, \gamma(T^{y_k})$ are independent. Hence the expected number of visits to x_0 is $\mathbf{E}\big[(k+1)/(\widehat{\gamma} + \gamma(T^{y_1}) + \cdots + \gamma(T^{y_k}))\big]$. Now use Exercise 12.15.

17.20. First show the claim for automorphisms that do fix the root. For example, if $p_k > 0$, let $Z_n(k)$ denote the number of vertices in generation n that have k children; show that if $Z'_n(k)$ is independent of $Z_n(k)$ and has the same distribution, then $\lim_{n \to \infty} \mathbf{P}[Z_n(k) = Z'_n(k)] = 0$. For another approach, show that $\widetilde{W}$ has a continuous distribution by using (17.8).

17.23. See Lyons, Pemantle, and Peres (1997).

17.24. It is $(1 - \sqrt{2}) \log(\sqrt{2} - 1) + (\sqrt{2} - 2) \log(2 - \sqrt{2})$.

17.32. This exercise is relevant to the proof of Lemma 6 in Furstenberg (1970), which is incorrect. If the definition of dimension of a measure as given in Section 15.4 is used instead of Furstenberg's definition, thus implicitly revising his Lemma 6, then the present exercise together with Billingsley's Theorem 15.17 give a proof of this revision.

17.34. A similar calculation for VIS_T was made by Lyons, Pemantle, and Peres (1995b), Lemma 9.1, but it does not work for all $\lambda < m$; see Exercise 17.15. See also Pemantle and Peres (1995b), Lemma 2.2, for a related statement. See Exercise 5.45 for an upper bound on the expected effective conductance.

17.36. $S^{-1}(\mathsf{Exit})$ has the same measure as Exit and for $(\overleftrightarrow{x}, T) \in S^{-1}(\mathsf{Exit})$, the ray $x_{-\infty}$ is a path of a loop-erased simple random walk while $\overrightarrow{x}$ is a disjoint path of simple random walk.

17.37. This is due to Lyons, Pemantle, and Peres (1996a), who also show that the speed is a positive constant a.s. when $f'(q) < \lambda < m$.

Bibliography

ABÉRT, M., GLASNER, Y., and VIRÁG, B.
  (2016) The measurable Kesten theorem. *Ann. Probab.*, **44**(3), 1601–1646. MR: 3502591

ABRAMS, A. and KENYON, R.
  (2017) Fixed-energy harmonic functions. *Discrete Anal.* Paper No. 18, 21 pp. MR: 3734203

ACHLIOPTAS, D. and MCSHERRY, F.
  (2001) Fast computation of low rank matrix approximations. In *Proceedings of the Thirty-Third Annual ACM Symposium on Theory of Computing*, pages 611–618 (electronic). ACM, New York. Held in Hersonissos, 2001, available electronically at http://portal.acm.org/toc.cfm?id=380752. MR: 2120364

  (2007) Fast computation of low-rank matrix approximations. *J. ACM*, **54**(2), Art. 9. MR: 2295993

ADAMS, S.
  (1988) Indecomposability of treed equivalence relations. *Israel J. Math.*, **64**(3), 362–380. MR: 995576

  (1990) Trees and amenable equivalence relations. *Ergodic Theory Dynam. Systems*, **10**(1), 1–14. MR: 91d:28041

ADAMS, S. and LYONS, R.
  (1991) Amenability, Kazhdan's property and percolation for trees, groups and equivalence relations. *Israel J. Math.*, **75**(2–3), 341–370. MR: 93j:43001

ADAMS, S. and SPATZIER, R.J.
  (1990) Kazhdan groups, cocycles and trees. *Amer. J. Math.*, **112**(2), 271–287. MR: 91c:22011

ADLER, J. and LEV, U.
  (2003) Bootstrap percolation: Visualizations and applications. *Brazilian J. Phys.*, **33**(3), 641–644. http://dx.doi.org/10.1590/S0103-97332003000300031.

AÏDÉKON, E.
  (2008) Transient random walks in random environment on a Galton-Watson tree. *Probab. Theory Related Fields*, **142**(3–4), 525–559. MR: 2438700

  (2010) Large deviations for transient random walks in random environment on a Galton-Watson tree. *Ann. Inst. Henri Poincaré Probab. Stat.*, **46**(1), 159–189. MR: 2641775

  (2013) Note on the monotonicity of the speed of the biased random walk on a Galton-Watson tree. http://www.proba.jussieu.fr/dw/lib/exe/fetch.php?media=users:aidekon:noteaidekon.pdf.

  (2014) Speed of the biased random walk on a Galton-Watson tree. *Probab. Theory Related Fields*, **159**(3–4), 597–617. MR: 3230003

  (2020) Hausdorff dimension of the uniform measure of Galton–Watson trees without the XlogX condition. *Ann. Inst. Henri Poincaré Probab. Stat.*, **56**(4), 2301–2306. MR: 4164837

AIZENMAN, M.
  (1985) The intersection of Brownian paths as a case study of a renormalization group method for quantum field theory. *Comm. Math. Phys.*, **97**(1–2), 91–110. MR: 782960

AIZENMAN, M. and BARSKY, D.J.
  (1987) Sharpness of the phase transition in percolation models. *Comm. Math. Phys.*, **108**(3), 489–526. MR: 88c:82026

AIZENMAN, M., BURCHARD, A., NEWMAN, C.M., and WILSON, D.B.
  (1999) Scaling limits for minimal and random spanning trees in two dimensions. *Random Structures Algorithms*, **15**(3–4), 319–367. MR: 2001c:60151

AIZENMAN, M., CHAYES, J.T., CHAYES, L., and NEWMAN, C.M.
    (1988) Discontinuity of the magnetization in one-dimensional $1/|x - y|^2$ Ising and Potts models. *J. Statist. Phys.*, **50**(1–2), 1–40. MR: 89f:82072

AIZENMAN, M., KESTEN, H., and NEWMAN, C.M.
    (1987) Uniqueness of the infinite cluster and continuity of connectivity functions for short and long range percolation. *Comm. Math. Phys.*, **111**(4), 505–531. MR: 89b:82060

AIZENMAN, M. and LEBOWITZ, J.L.
    (1988) Metastability effects in bootstrap percolation. *J. Phys. A*, **21**(19), 3801–3813. MR: 968311

ALDOUS, D.J.
    (1987) On the Markov chain simulation method for uniform combinatorial distributions and simulated annealing. *Probab. Eng. Inform. Sc.*, **1**, 33–46. http://dx.doi.org/10.1017/S0269964800000267.

    (1990) The random walk construction of uniform spanning trees and uniform labelled trees. *SIAM J. Discrete Math.*, **3**(4), 450–465. MR: 91h:60013

    (1991) Random walk covering of some special trees. *J. Math. Anal. Appl.*, **157**(1), 271–283. MR: 1109456

ALDOUS, D.J. and FILL, J.A.
    (2002) *Reversible Markov Chains and Random Walks on Graphs.* Unfinished monograph, recompiled 2014 version available at http://www.stat.berkeley.edu/~aldous/RWG/book.html.

ALDOUS, D.J. and LYONS, R.
    (2007) Processes on unimodular random networks. *Electron. J. Probab.*, **12**, paper no. 54, 1454–1508 (electronic). Errata, *Electron. J. Probab.*, **22** (2017), Paper No. 51, 4 pp. and *Electron. J. Probab.* **24** (2019), Paper No. 25, 2 pp. MR: 2354165

ALELIUNAS, R., KARP, R.M., LIPTON, R.J., LOVÁSZ, L., and RACKOFF, C.
    (1979) Random walks, universal traversal sequences, and the complexity of maze problems. In *20th Annual Symposium on Foundations of Computer Science*, pages 218–223. IEEE, New York. Held in San Juan, Puerto Rico, October 29–31, 1979. MR: 598110

ALEXANDER, K.S.
    (1995a) Percolation and minimal spanning forests in infinite graphs. *Ann. Probab.*, **23**(1), 87–104. MR: 96c:60114

    (1995b) Simultaneous uniqueness of infinite clusters in stationary random labeled graphs. *Comm. Math. Phys.*, **168**(1), 39–55. Erratum: *Comm. Math. Phys.* **172**, (1995), 221. MR: 96e:60166a

ALEXANDER, K.S. and MOLCHANOV, S.A.
    (1994) Percolation of level sets for two-dimensional random fields with lattice symmetry. *J. Statist. Phys.*, **77**(3–4), 627–643. MR: 95i:82052

ALON, N.
    (1986) Eigenvalues and expanders. *Combinatorica*, **6**(2), 83–96. MR: 88e:05077

ALON, N., BENJAMINI, I., LUBETZKY, E., and SODIN, S.
    (2007) Non-backtracking random walks mix faster. *Commun. Contemp. Math.*, **9**(4), 585–603. MR: 2348845

ALON, N., BENJAMINI, I., and STACEY, A.
    (2004) Percolation on finite graphs and isoperimetric inequalities. *Ann. Probab.*, **32**(3A), 1727–1745. MR: 2073175

ALON, N., HOORY, S., and LINIAL, N.
    (2002) The Moore bound for irregular graphs. *Graphs Combin.*, **18**(1), 53–57. MR: 1892433

ALON, N. and MILMAN, V.D.
    (1985) λ_1, isoperimetric inequalities for graphs, and superconcentrators. *J. Combin. Theory Ser. B*, **38**(1), 73–88. MR: 782626

AMIR, G. and VIRÁG, B.
    (2017) Speed exponents of random walks on groups. *Int. Math. Res. Not. IMRN*, **2017**(9), 2567–2598. MR: 3658209

ANANTHARAM, V. and TSOUCAS, P.
    (1989) A proof of the Markov chain tree theorem. *Statist. Probab. Lett.*, **8**(2), 189–192. MR: 1017890

ANCONA, A.
    (1988) Positive harmonic functions and hyperbolicity. In Král, J., Lukeš, J., Netuka, I., and Veselý, J., editors, *Potential Theory—Surveys and Problems (Prague, 1987)*, pages 1–23. Springer, Berlin. MR: 973 878

ANCONA, A., LYONS, R., and PERES, Y.
(1999) Crossing estimates and convergence of Dirichlet functions along random walk and diffusion paths. *Ann. Probab.*, **27**(2), 970–989. MR: 1698991

ANGEL, O., BARLOW, M.T., GUREL-GUREVICH, O., and NACHMIAS, A.
(2016) Boundaries of planar graphs, via circle packings. *Ann. Probab.*, **44**(3), 1956–1984. MR: 3502598

ANGEL, O. and BENJAMINI, I.
(2007) A phase transition for the metric distortion of percolation on the hypercube. *Combinatorica*, **27**(6), 645–658. MR: 2384409

ANGEL, O., BENJAMINI, I., BERGER, N., and PERES, Y.
(2006) Transience of percolation clusters on wedges. *Electron. J. Probab.*, **11**, paper no. 25, 655–669 (electronic). MR: 2242658

ANGEL, O., CRAWFORD, N., and KOZMA, G.
(2014) Localization for linearly edge reinforced random walks. *Duke Math. J.*, **163**(5), 889–921. MR: 3189433

ANGEL, O., FRIEDMAN, J., and HOORY, S.
(2015) The non-backtracking spectrum of the universal cover of a graph. *Trans. Amer. Math. Soc.*, **367**(6), 4287–4318. MR: 3324928

ANGEL, O., GOODMAN, J., DEN HOLLANDER, F., and SLADE, G.
(2008) Invasion percolation on regular trees. *Ann. Probab.*, **36**(2), 420–466. MR: 2393988

ANGEL, O., GOODMAN, J., and MERLE, M.
(2013) Scaling limit of the invasion percolation cluster on a regular tree. *Ann. Probab.*, **41**(1), 229–261. MR: 3059198

ANGEL, O. and SZEGEDY, B.
(2016) Recurrence of weak limits of excluded minor graphs. In preparation.

ARRATIA, R. and GOLDSTEIN, L.
(2010) Size bias, sampling, the waiting time paradox, and infinite divisibility: When is the increment independent? Preprint, http://www.arxiv.org/abs/1007.3910.

ASADPOUR, A., GOEMANS, M.X., MĄDRY, A., OVEIS GHARAN, S., and SABERI, A.
(2010) An $O(\log n/\log \log n)$-approximation algorithm for the asymmetric traveling salesman problem. In *Proceedings of the Twenty-First Annual ACM-SIAM Symposium on Discrete Algorithms*, pages 379–389. SIAM, Philadelphia. Held in Austin, TX, January 17–19, 2010. MR: 2809683

ASMUSSEN, S. and HERING, H.
(1983) *Branching Processes*. Birkhäuser, Boston. MR: 85b:60076

ATHREYA, K.B.
(1971) A note on a functional equation arising in Galton-Watson branching processes. *J. Appl. Probability*, **8**, 589–598. MR: 45:1271

ATHREYA, K.B. and NEY, P.E.
(1972) *Branching Processes*. Vol. 196 of *Die Grundlehren der mathematischen Wissenschaften*. Springer-Verlag, New York. MR: 51:9242

ATIYAH, M.F.
(1976) Elliptic operators, discrete groups and von Neumann algebras. In *Colloque "Analyse et Topologie" en l'Honneur de Henri Cartan*, pages 43–72. Astérisque, **32–33**. Soc. Math. France, Paris. Tenu le 17–20 juin 1974 à Orsay. MR: 0420729

AVENA, L. and GAUDILLIÈRE, A.
(2018) A proof of the transfer-current theorem in absence of reversibility. *Statist. Probab. Lett.*, **142**, 17–22. MR: 3842610

AVEZ, A.
(1972) Entropie des groupes de type fini. *C. R. Acad. Sci. Paris Sér. A-B*, **275**, A1363–A1366. MR: 0324741

(1974) Théorème de Choquet-Deny pour les groupes à croissance non exponentielle. *C. R. Acad. Sci. Paris Sér. A*, **279**, 25–28. MR: 0353405

(1976) Croissance des groupes de type fini et fonctions harmoniques. In *Théorie Ergodique*, Lecture Notes in Mathematics, Vol. 532, pages 35–49. Springer, Berlin. Actes des Journées Ergodiques, Rennes, 1973/1974, Edité par J.-P. Conze et M. S. Keane. MR: 0482911

AZUMA, K.
(1967) Weighted sums of certain dependent random variables. *Tôhoku Math. J. (2)*, **19**, 357–367. MR: 0221571

BABAI, L.
(1991) Local expansion of vertex-transitive graphs and random generation in finite groups. In *STOC '91: Proceedings of the Twenty-Third Annual ACM Symposium on Theory of Computing*, pages 164–167. ACM, New York. http://dx.doi.org/10.1145/103418.103440.

(1997) The growth rate of vertex-transitive planar graphs. In *Proceedings of the Eighth Annual ACM-SIAM Symposium on Discrete Algorithms*, pages 564–573. ACM, New York. Held in New Orleans, LA, January 5–7, 1997. MR: 1447704

BABSON, E. and BENJAMINI, I.
(1999) Cut sets and normed cohomology with applications to percolation. *Proc. Amer. Math. Soc.*, **127**(2), 589–597. MR: 99g:05119

BAHAR, I., ATILGAN, A.R., and ERMAN, B.
(1997) Direct evaluation of thermal fluctuations in protein using a single parameter harmonic potential. *Folding & Design*, **2**, 173–181. http://dx.doi.org/10.1016/S1359-0278(97)00024-2.

BAKIROV, N.K., RIZZO, M.L., and SZÉKELY, G.J.
(2006) A multivariate nonparametric test of independence. *J. Multivariate Anal.*, **97**(8), 1742–1756. MR: 2298886

BALÁZS, M. and FOLLY, A.
(2016) An electric network for nonreversible Markov chains. *Amer. Math. Monthly*, **123**(7), 657–682. MR: 3539852

BALL, K.
(1992) Markov chains, Riesz transforms and Lipschitz maps. *Geom. Funct. Anal.*, **2**(2), 137–172. MR: 1159828

BALLMANN, W. and LEDRAPPIER, F.
(1994) The Poisson boundary for rank one manifolds and their cocompact lattices. *Forum Math.*, **6**(3), 301–313. MR: 1269841

BALOGH, J., BOLLOBÁS, B., DUMINIL-COPIN, H., and MORRIS, R.
(2012) The sharp threshold for bootstrap percolation in all dimensions. *Trans. Amer. Math. Soc.*, **364**(5), 2667–2701. MR: 2888224

BALOGH, J., BOLLOBÁS, B., and MORRIS, R.
(2009) Bootstrap percolation in three dimensions. *Ann. Probab.*, **37**(4), 1329–1380. MR: 2546747

BALOGH, J., PERES, Y., and PETE, G.
(2006) Bootstrap percolation on infinite trees and non-amenable groups. *Combin. Probab. Comput.*, **15**(5), 715–730. MR: 2248323

BAPAT, R.B.
(1992) Mixed discriminants and spanning trees. *Sankhyā Ser. A*, **54**(Special Issue), 49–55. Combinatorial mathematics and applications (Calcutta, 1988). MR: 1234678

BARLOW, M.T.
(2016) Loop erased walks and uniform spanning trees. In *Discrete Geometric Analysis*, vol. 34 of *MSJ Mem.*, pages 1–32. Math. Soc. Japan, Tokyo. MR: 3525847

BARLOW, M.T. and MASSON, R.
(2010) Exponential tail bounds for loop-erased random walk in two dimensions. *Ann. Probab.*, **38**(6), 2379–2417. MR: 2683633

BARLOW, M.T. and PERKINS, E.A.
(1989) Symmetric Markov chains in $\mathbf{Z}^d$: How fast can they move? *Probab. Theory Related Fields*, **82**(1), 95–108. MR: 997432

BARREIRA, L.
(2008) *Dimension and Recurrence in Hyperbolic Dynamics*. Vol. 272 of *Progress in Mathematics*. Birkhäuser, Basel. MR: 2434246

BARSKY, D.J., GRIMMETT, G.R., and NEWMAN, C.M.
(1991) Percolation in half-spaces: Equality of critical densities and continuity of the percolation probability. *Probab. Theory Related Fields*, **90**(1), 111–148. MR: 92m:60086

BARTHOLDI, L.
(1999) Counting paths in graphs. *Enseign. Math. (2)*, **45**(1–2), 83–131. MR: 1703364

(2003) A Wilson group of non-uniformly exponential growth. *C. R. Math. Acad. Sci. Paris*, **336**(7), 549–554. MR: 1981466

BARTHOLDI, L., KAIMANOVICH, V.A., and NEKRASHEVYCH, V.V.
(2010) On amenability of automata groups. *Duke Math. J.*, **154**(3), 575–598. MR: 2730578

BARTHOLDI, L. and VIRÁG, B.
(2005) Amenability via random walks. *Duke Math. J.*, **130**(1), 39–56. MR: 2176547

BASS, R.F.
(1995) *Probabilistic Techniques in Analysis*. Probability and Its Applications. Springer-Verlag, New York. MR: 96e:60001

BATEMAN, M. and KATZ, N.H.
(2008) Kakeya sets in Cantor directions. *Math. Res. Lett.*, **15**(1), 73–81. MR: 2367175

BEARDON, A.F. and STEPHENSON, K.
(1990) The uniformization theorem for circle packings. *Indiana Univ. Math. J.*, **39**(4), 1383–1425. MR: 1087197

BEKKA, M.E.B. and VALETTE, A.
(1997) Group cohomology, harmonic functions and the first L^2-Betti number. *Potential Anal.*, **6**(4), 313–326. MR: 98e:20056

BEN AROUS, G., FRIBERGH, A., GANTERT, N., and HAMMOND, A.
(2012) Biased random walks on Galton-Watson trees with leaves. *Ann. Probab.*, **40**(1), 280–338. MR: 2917774

BEN AROUS, G., FRIBERGH, A., and SIDORAVICIUS, V.
(2014) Lyons-Pemantle-Peres monotonicity problem for high biases. *Comm. Pure Appl. Math.*, **67**(4), 519–530. MR: 3168120

BEN AROUS, G., HU, Y., OLLA, S., and ZEITOUNI, O.
(2013) Einstein relation for biased random walk on Galton-Watson trees. *Ann. Inst. Henri Poincaré Probab. Stat.*, **49**(3), 698–721. MR: 3112431

BENASSI, A.
(1996) Arbres et grandes déviations. In Chauvin, B., Cohen, S., and Rouault, A., editors, *Trees*, vol. 40 of *Progr. Probab.*, pages 135–140. Birkhäuser, Basel. Proceedings of the Workshop held in Versailles, June 14–16, 1995. MR: 1439977

BENEDETTI, R. and PETRONIO, C.
(1992) *Lectures on Hyperbolic Geometry*. Universitext. Springer-Verlag, Berlin. MR: 94e:57015

BENJAMINI, I.
(1991) Instability of the Liouville property for quasi-isometric graphs and manifolds of polynomial volume growth. *J. Theoret. Probab.*, **4**(3), 631–637. MR: 1115166

BENJAMINI, I. and CURIEN, N.
(2012) Ergodic theory on stationary random graphs. *Electron. J. Probab.*, **17**, paper no. 93, 20 pp. MR: 2994841

BENJAMINI, I., DUMINIL-COPIN, H., KOZMA, G., and YADIN, A.
(2015a) Disorder, entropy and harmonic functions. *Ann. Probab.*, **43**(5), 2332–2373. MR: 3395463

(2015b) Minimal harmonic functions I, upper bounds. In preparation.

BENJAMINI, I., GUREL-GUREVICH, O., and LYONS, R.
(2007) Recurrence of random walk traces. *Ann. Probab.*, **35**(2), 732–738. MR: 2308594

BENJAMINI, I., KESTEN, H., PERES, Y., and SCHRAMM, O.
(2004) Geometry of the uniform spanning forest: Transitions in dimensions 4, 8, 12, ... *Ann. of Math. (2)*, **160**(2), 465–491. MR: 2123930

BENJAMINI, I. and KOZMA, G.
(2005) A resistance bound via an isoperimetric inequality. *Combinatorica*, **25**(6), 645–650. MR: 2199429

BENJAMINI, I., KOZMA, G., and SCHAPIRA, B.
(2011) A balanced excited random walk. *C. R. Math. Acad. Sci. Paris*, **349**(7–8), 459–462. MR: 2788390

BENJAMINI, I., LYONS, R., PERES, Y., and SCHRAMM, O.
(1999a) Critical percolation on any nonamenable group has no infinite clusters. *Ann. Probab.*, **27**(3), 1347–1356. MR: 1733 151

(1999b) Group-invariant percolation on graphs. *Geom. Funct. Anal.*, **9**(1), 29–66. MR: 99m:60149

(2001) Uniform spanning forests. *Ann. Probab.*, **29**(1), 1–65. MR: 1825 141

BENJAMINI, I., LYONS, R., and SCHRAMM, O.
    (1999) Percolation perturbations in potential theory and random walks. In Picardello, M. and Woess, W., editors, *Random Walks and Discrete Potential Theory*, Sympos. Math., XXXIX, pages 56–84. Cambridge University Press, Cambridge. Proceedings of the conference held in Cortona, June 1997. MR: 1802426

BENJAMINI, I. and MÜLLER, S.
    (2012) On the trace of branching random walks. *Groups Geom. Dyn.*, **6**(2), 231–247. MR: 2914859

BENJAMINI, I., NACHMIAS, A., and PERES, Y.
    (2011) Is the critical percolation probability local? *Probab. Theory Related Fields*, **149**(1–2), 261–269. MR: 2773031

BENJAMINI, I., PEMANTLE, R., and PERES, Y.
    (1995) Martin capacity for Markov chains. *Ann. Probab.*, **23**(3), 1332–1346. MR: 96g:60098

    (1996) Random walks in varying dimensions. *J. Theoret. Probab.*, **9**(1), 231–244. MR: 97a:60092

    (1998) Unpredictable paths and percolation. *Ann. Probab.*, **26**(3), 1198–1211. MR: 99g:60183

BENJAMINI, I. and PERES, Y.
    (1992) Random walks on a tree and capacity in the interval. *Ann. Inst. H. Poincaré Probab. Statist.*, **28**(4), 557–592. MR: 94f:60089

    (1994a) Markov chains indexed by trees. *Ann. Probab.*, **22**(1), 219–243. MR: 1258875

    (1994b) Tree-indexed random walks on groups and first passage percolation. *Probab. Theory Related Fields*, **98**(1), 91–112. MR: 94m:60141

BENJAMINI, I. and SCHRAMM, O.
    (1996a) Harmonic functions on planar and almost planar graphs and manifolds, via circle packings. *Invent. Math.*, **126**(3), 565–587. MR: 97k:31009

    (1996b) Percolation beyond $\mathbf{Z}^d$, many questions and a few answers. *Electron. Comm. Probab.*, **1**, paper no. 8, 71–82 (electronic). MR: 97j:60179

    (1996c) Random walks and harmonic functions on infinite planar graphs using square tilings. *Ann. Probab.*, **24**(3), 1219–1238. MR: 98d:60134

    (1997) Every graph with a positive Cheeger constant contains a tree with a positive Cheeger constant. *Geom. Funct. Anal.*, **7**(3), 403–419. MR: 99b:05032

    (2001a) Percolation in the hyperbolic plane. *J. Amer. Math. Soc.*, **14**(2), 487–507. MR: 1815220

    (2001b) Recurrence of distributional limits of finite planar graphs. *Electron. J. Probab.*, **6**, paper no. 23, 13 pp. (electronic). MR: 1873 300

    (2004) Pinched exponential volume growth implies an infinite dimensional isoperimetric inequality. In Milman, V.D. and Schechtman, G., editors, *Geometric Aspects of Functional Analysis*, vol. 1850 of *Lecture Notes in Math.*, pages 73–76. Springer, Berlin. Papers from the Israel Seminar (GAFA) held 2002–2003. MR: 2087152

BENNIES, J. and KERSTING, G.
    (2000) A random walk approach to Galton-Watson trees. *J. Theoret. Probab.*, **13**(3), 777–803. MR: 1785529

BERESTYCKI, N., LASLIER, B., and RAY, G.
    (2020) Dimers and imaginary geometry. *Ann. Probab.*, **48**(1), 1–52. MR: 4079430

BERESTYCKI, N., LUBETZKY, E., PERES, Y., and SLY, A.
    (2018) Random walks on the random graph. *Ann. Probab.*, **46**(1), 456–490. MR: 3758735

VAN DEN BERG, J. and KEANE, M.
    (1984) On the continuity of the percolation probability function. In Beals, R., Beck, A., Bellow, A., and Hajian, A., editors, *Conference in Modern Analysis and Probability (New Haven, Conn., 1982)*, pages 61–65. Amer. Math. Soc., Providence, RI. MR: 85g:60100

VAN DEN BERG, J. and KESTEN, H.
    (1985) Inequalities with applications to percolation and reliability. *J. Appl. Probab.*, **22**(3), 556–569. MR: 87b:60027

VAN DEN BERG, J. and MEESTER, R.W.J.
    (1991) Stability properties of a flow process in graphs. *Random Structures Algorithms*, **2**(3), 335–341. MR: 92d:90027

BIGGINS, J.D.
   (1977) Chernoff's theorem in the branching random walk. *J. Appl. Probability*, **14**(3), 630–636. MR: 0464415

BIGGS, N.L., MOHAR, B., and SHAWE-TAYLOR, J.
   (1988) The spectral radius of infinite graphs. *Bull. London Math. Soc.*, **20**(2), 116–120. MR: 89a:05103

BILLINGSLEY, P.
   (1965) *Ergodic Theory and Information*. John Wiley, New York. MR: 33:254

   (1995) *Probability and Measure*, 3rd ed. Wiley Series in Probability and Mathematical Statistics. John Wiley, New York. MR: 1324786

BJÖRKLUND, M.
   (2014) Five remarks about random walks on groups. Preprint, http://www.arxiv.org/abs/1406.0763.

BLACHÈRE, S., HAÏSSINSKY, P., and MATHIEU, P.
   (2008) Asymptotic entropy and Green speed for random walks on countable groups. *Ann. Probab.*, **36**(3), 1134–1152. MR: 2408585

BLACKWELL, D.
   (1955) On transient Markov processes with a countable number of states and stationary transition probabilities. *Ann. Math. Statist.*, **26**, 654–658. MR: 17,754d

BOLEY, D., RANJAN, G., and ZHANG, Z.L.
   (2011) Commute times for a directed graph using an asymmetric Laplacian. *Linear Algebra Appl.*, **435**(2), 224–242. MR: 2782776

BOLLOBÁS, B.
   (1998) *Modern Graph Theory*. Vol. 184 of *Graduate Texts in Mathematics*. Springer-Verlag, New York. MR: 99h:05001

BORCEA, J., BRÄNDÉN, P., and LIGGETT, T.M.
   (2009) Negative dependence and the geometry of polynomials. *J. Amer. Math. Soc.*, **22**(2), 521–567. MR: 2476782

BORCHARDT, C.W.
   (1860) Ueber eine der interpolation entsprechende darstellung der eliminations-resultante. *J. Reine Angew. Math.*, **1860**(57), 111–121.

BORRE, K. and MEISSL, P.
   (1974) Strength analysis of leveling-type networks. An application of random walk theory. *Geodaet. Inst. Medd.*, **50**, 80. MR: 0475698

BORWEIN, J.M. and ZUCKER, I.J.
   (1992) Fast evaluation of the gamma function for small rational fractions using complete elliptic integrals of the first kind. *IMA J. Numer. Anal.*, **12**(4), 519–526. MR: 1186733

BOURGAIN, J.
   (1985) On Lipschitz embedding of finite metric spaces in Hilbert space. *Israel J. Math.*, **52**(1–2), 46–52. MR: 815600

BOWDITCH, A. and TOKUSHIGE, Y.
   (2020) The speed of a biased random walk on a Galton-Watson tree is analytic. *Electron. Commun. Probab.*, **25**, Paper No. 65, 11 pp. MR: 4151882

BOWEN, L.
   (2004) Couplings of uniform spanning forests. *Proc. Amer. Math. Soc.*, **132**(7), 2151–2158 (electronic). MR: 2053 989

BRIEUSSEL, J.
   (2009) Amenability and non-uniform growth of some directed automorphism groups of a rooted tree. *Math. Z.*, **263**(2), 265–293. MR: 2534118

   (2013) Behaviors of entropy on finitely generated groups. *Ann. Probab.*, **41**(6), 4116–4161. MR: 3161471

BRIEUSSEL, J. and ZHENG, T.
   (2021) Speed of random walks, isoperimetry and compression of finitely generated groups. *Ann. of Math. (2)*, **193**(1), 1–105. MR: 4199729

BROADBENT, S.R. and HAMMERSLEY, J.M.
   (1957) Percolation processes. I. Crystals and mazes. *Proc. Cambridge Philos. Soc.*, **53**, 629–641. MR: 0091567

BRODER, A.
  (1989) Generating random spanning trees. In *30th Annual Symposium on Foundations of Computer Science (Research Triangle Park, North Carolina)*, pages 442–447. IEEE, New York. http://dx.doi.org/10.1109/SFCS.1989.63516.

BRODER, A.Z. and KARLIN, A.R.
  (1989) Bounds on the cover time. *J. Theoret. Probab.*, **2**(1), 101–120. MR: 981768

BROFFERIO, S. and SCHAPIRA, B.
  (2011) Poisson boundary of $GL_d(\mathbb{Q})$. *Israel J. Math.*, **185**, 125–140. MR: 2837130

BROMAN, E.I., CAMIA, F., JOOSTEN, M., and MEESTER, R.
  (2013) Dimension (in)equalities and Hölder continuous curves in fractal percolation. *J. Theoret. Probab.*, **26**(3), 836–854. MR: 3090553

BROOKS, R.L., SMITH, C.A.B., STONE, A.H., and TUTTE, W.T.
  (1940) The dissection of rectangles into squares. *Duke Math. J.*, **7**, 312–340. MR: 2,153d

BURTON, R.M. and KEANE, M.
  (1989) Density and uniqueness in percolation. *Comm. Math. Phys.*, **121**(3), 501–505. MR: 90g:60090

BURTON, R.M. and PEMANTLE, R.
  (1993) Local characteristics, entropy and limit theorems for spanning trees and domino tilings via transfer-impedances. *Ann. Probab.*, **21**(3), 1329–1371. MR: 94m:60019

BUSER, P.
  (1982) A note on the isoperimetric constant. *Ann. Sci. École Norm. Sup. (4)*, **15**(2), 213–230. MR: 683635

CAMPANINO, M.
  (1985) Inequalities for critical probabilities in percolation. In Durrett, R., editor, *Particle Systems, Random Media and Large Deviations*, vol. 41 of *Contemp. Math.*, pages 1–9. Amer. Math. Soc., Providence, RI. Proceedings of the AMS-IMS-SIAM joint summer research conference in the mathematical sciences on mathematics of phase transitions held at Bowdoin College, Brunswick, Maine, June 24–30, 1984. MR: 814699

CAMPBELL, G.A.
  (1911) Cisoidal oscillations. *Trans. Amer. Inst. Electrical Engineers*, **30**, 873–909. http://dx.doi.org/10.1109/T-AIEE.1911.4768303.

CANDELLERO, E. and TEIXEIRA, A.
  (2018) Percolation and isoperimetry on roughly transitive graphs. *Ann. Inst. Henri Poincaré Probab. Stat.*, **54**(4), 1819–1847. MR: 3865659

CANNON, J.W., FLOYD, W.J., KENYON, R.W., and PARRY, W.R.
  (1997) Hyperbolic geometry. In Levy, S., editor, *Flavors of Geometry*, vol. 31 of *Mathematical Sciences Research Institute Publications*, pages 59–115. Cambridge University Press, Cambridge. MR: 1491098

CANNON, J.W., FLOYD, W.J., and PARRY, W.R.
  (1994) Squaring rectangles: The finite Riemann mapping theorem. In Abikoff, W., Birman, J.S., and Kuiken, K., editors, *The Mathematical Legacy of Wilhelm Magnus: Groups, Geometry and Special Functions*, vol. 169 of *Contemp. Math.*, pages 133–212. Amer. Math. Soc., Providence, RI. Proceedings of the conference held at Polytechnic University, Brooklyn, New York, May 1–3, 1992. MR: 1292901

CAPUTO, P., LIGGETT, T.M., and RICHTHAMMER, T.
  (2010) Proof of Aldous' spectral gap conjecture. *J. Amer. Math. Soc.*, **23**(3), 831–851. MR: 2629990

CARLESON, L.
  (1967) *Selected Problems on Exceptional Sets*. Vol. 13 of *Van Nostrand Mathematical Studies*. D. Van Nostrand, Princeton, NJ. MR: 0225986

CARLITZ, L., WILANSKY, A., MILNOR, J., STRUBLE, R.A., FELSINGER, N., SIMOES, J.M.S., POWER, E.A., SHAFER, R.E., and MAAS, R.E.
  (1968) Problems and Solutions: Advanced Problems: 5600–5609. *Amer. Math. Monthly*, **75**(6), 685–687. MR: 1534960

CARMESIN, J.
  (2012) A characterization of the locally finite networks admitting non-constant harmonic functions of finite energy. *Potential Anal.*, **37**(3), 229–245. MR: 2969301

CARMESIN, J. and GEORGAKOPOULOS, A.
  (2020) Every planar graph with the Liouville property is amenable. *Random Structures Algorithms*, **57**(3), 706–729. MR: 4144081

CARNE, T.K.
  (1985) A transmutation formula for Markov chains. *Bull. Sci. Math. (2)*, **109**(4), 399–405. MR: 87m:60142

CARTWRIGHT, D.I., KAIMANOVICH, V.A., and WOESS, W.
  (1994) Random walks on the affine group of local fields and of homogeneous trees. *Ann. Inst. Fourier (Grenoble)*, **44**(4), 1243–1288. MR: 1306556

CARTWRIGHT, D.I. and SOARDI, P.M.
  (1989) Convergence to ends for random walks on the automorphism group of a tree. *Proc. Amer. Math. Soc.*, **107**(3), 817–823. MR: 90f:60137

CAYLEY, A.
  (1889) A theorem on trees. *Quart. J. Pure Appl. Math.*, **23**, 376–378. http://dx.doi.org/10.1017/CBO9780511703799.010.

ČERNÝ, J. and TEIXEIRA, A.Q.
  (2012) *From Random Walk Trajectories to Random Interlacements*. Vol. 23 of *Ensaios Matemáticos [Mathematical Surveys]*. Sociedade Brasileira de Matemática, Rio de Janeiro. MR: 3014964

CHABOUD, T. and KENYON, C.
  (1996) Planar Cayley graphs with regular dual. *Internat. J. Algebra Comput.*, **6**(5), 553–561. MR: 98a:05077

CHALUPA, J., REICH, G.R., and LEATH, P.L.
  (1979) Bootstrap percolation on a Bethe lattice. *J. Phys. C*, **12**(1), L31–L35. http://dx.doi.org/10.1088/0022-3719/12/1/008.

CHANDRA, A.K., RAGHAVAN, P., RUZZO, W.L., SMOLENSKY, R., and TIWARI, P.
  (1996/1997) The electrical resistance of a graph captures its commute and cover times. *Comput. Complexity*, **6**(4), 312–340. MR: 99h:60140

CHAUVIN, B., ROUAULT, A., and WAKOLBINGER, A.
  (1991) Growing conditioned trees. *Stochastic Process. Appl.*, **39**(1), 117–130. MR: 1135089

CHAYES, J.T., CHAYES, L., and DURRETT, R.
  (1988) Connectivity properties of Mandelbrot's percolation process. *Probab. Theory Related Fields*, **77**(3), 307–324. MR: 931500

CHAYES, J.T., CHAYES, L., and NEWMAN, C.M.
  (1985) The stochastic geometry of invasion percolation. *Comm. Math. Phys.*, **101**(3), 383–407. MR: 87i:82072

CHAYES, L.
  (1995a) Aspects of the fractal percolation process. In Bandt, C., Graf, S., and Zähle, M., editors, *Fractal Geometry and Stochastics*, vol. 37 of *Progr. Probab.*, pages 113–143. Birkhäuser, Basel. Papers from the conference held in Finsterbergen, June 12–18, 1994. MR: 1391973

  (1995b) On the absence of directed fractal percolation. *J. Phys. A.: Math. Gen.*, **28**, L295–L301. http://dx.doi.org/10.1088/0305-4470/28/10/003.

CHAYES, L., PEMANTLE, R., and PERES, Y.
  (1997) No directed fractal percolation in zero area. *J. Statist. Phys.*, **88**(5–6), 1353–1362. MR: 1478072

CHEEGER, J.
  (1970) A lower bound for the smallest eigenvalue of the Laplacian. In Gunning, R.C., editor, *Problems in Analysis*, pages 195–199. Princeton University Press, Princeton, NJ. A symposium in honor of Salomon Bochner, Princeton University, Princeton, NJ, 1–3 April 1969. MR: 53:6645

CHEEGER, J. and GROMOV, M.
  (1986) L_2-cohomology and group cohomology. *Topology*, **25**(2), 189–215. MR: 87i:58161

CHEN, D.
  (1997) Average properties of random walks on Galton-Watson trees. *Ann. Inst. H. Poincaré Probab. Statist.*, **33**(3), 359–369. MR: 1457056

CHEN, D. and PERES, Y.
  (2004) Anchored expansion, percolation and speed. *Ann. Probab.*, **32**(4), 2978–2995. With an appendix by Gábor Pete. MR: 2094436

CHEN, D. and ZHANG, F.X.
  (2007) On the monotonicity of the speed of random walks on a percolation cluster of trees. *Acta Math. Sin. (Engl. Ser.)*, **23**(11), 1949–1954. MR: 2359112

CHOQUET, G. and DENY, J.
  (1960) Sur l'équation de convolution $\mu = \mu * \sigma$. *C. R. Acad. Sci. Paris*, **250**, 799–801. MR: 22:9808

CHUNG, F.R.K., GRAHAM, R.L., FRANKL, P., and SHEARER, J.B.
  (1986) Some intersection theorems for ordered sets and graphs. *J. Combin. Theory Ser. A*, **43**(1), 23–37. MR: 859293

CHUNG, F.R.K. and TETALI, P.
  (1998) Isoperimetric inequalities for Cartesian products of graphs. *Combin. Probab. Comput.*, **7**(2), 141–148. MR: 2000c:05085

CIUCU, M.
  (1997) Enumeration of perfect matchings in graphs with reflective symmetry. *J. Combin. Theory Ser. A*, **77**(1), 67–97. MR: 1426739

COLBOURN, C.J., PROVAN, J.S., and VERTIGAN, D.
  (1995) A new approach to solving three combinatorial enumeration problems on planar graphs. *Discrete Appl. Math.*, **60**(1–3), 119–129. MR: 96e:05154

COLLEVECCHIO, A.
  (2006) On the transience of processes defined on Galton-Watson trees. *Ann. Probab.*, **34**(3), 870–878. MR: 2243872

COMBES, R.
  (2015) An extension of McDiarmid's inequality. Unpublished, http://www.arxiv.org/abs/1511.05240.

CONSTANTINE, G.M.
  (2003) Graphs, networks, and linear unbiased estimates. *Discrete Appl. Math.*, **130**(3), 381–393. MR: 1999697

COPPERSMITH, D., DOYLE, P., RAGHAVAN, P., and SNIR, M.
  (1993) Random walks on weighted graphs and applications to on-line algorithms. *J. Assoc. Comput. Mach.*, **40**(3), 421–453. MR: 1370357

COPPERSMITH, D., FEIGE, U., and SHEARER, J.
  (1996) Random walks on regular and irregular graphs. *SIAM J. Discrete Math.*, **9**(2), 301–308. MR: 1386885

COPPERSMITH, D., TETALI, P., and WINKLER, P.
  (1993) Collisions among random walks on a graph. *SIAM J. Discrete Math.*, **6**(3), 363–374. MR: 1229691

COULHON, T.
  (1996) Ultracontractivity and Nash type inequalities. *J. Funct. Anal.*, **141**(2), 510–539. MR: 1418518

COULHON, T., GRIGOR'YAN, A., and PITTET, C.
  (2001) A geometric approach to on-diagonal heat kernel lower bounds on groups. *Ann. Inst. Fourier (Grenoble)*, **51**(6), 1763–1827. MR: 1871289

COULHON, T. and SALOFF-COSTE, L.
  (1993) Isopérimétrie pour les groupes et les variétés. *Rev. Mat. Iberoamericana*, **9**(2), 293–314. MR: 94g:58263

COURNOT, A.A.
  (1847) *De l'origine et des limites de la correspondance entre l'algèbre et la géométrie.* Hachette, Paris. Available at http://gallica.bnf.fr/ark:/12148/bpt6k6563896n.

COX, J.T. and DURRETT, R.
  (1983) Oriented percolation in dimensions $d \geq 4$: Bounds and asymptotic formulas. *Math. Proc. Cambridge Philos. Soc.*, **93**(1), 151–162. MR: 84e:60150

CROYDON, D.
  (2008) Convergence of simple random walks on random discrete trees to Brownian motion on the continuum random tree. *Ann. Inst. Henri Poincaré Probab. Stat.*, **44**(6), 987–1019. MR: 2469332

CROYDON, D. and KUMAGAI, T.
  (2008) Random walks on Galton-Watson trees with infinite variance offspring distribution conditioned to survive. *Electron. J. Probab.*, **13**, paper no. 51, 1419–1441. MR: 2438812

CURIEN, N. and LE GALL, J.F.
  (2011) Random recursive triangulations of the disk via fragmentation theory. *Ann. Probab.*, **39**(6), 2224–2270. MR: 2932668

(2017) The harmonic measure of balls in random trees. *Ann. Probab.*, **45**(1), 147–209. MR: 3601648

CURIEN, N. and PERES, Y.
(2011) Random laminations and multitype branching processes. *Electron. Commun. Probab.*, **16**, 435–446. MR: 2831082

DAI, J.J.
(2005) A once edge-reinforced random walk on a Galton-Watson tree is transient. *Statist. Probab. Lett.*, **73**(2), 115–124. MR: 2159246

DAVENPORT, H., ERDŐS, P., and LEVEQUE, W.J.
(1963) On Weyl's criterion for uniform distribution. *Michigan Math. J.*, **10**, 311–314. MR: 0153656

DEKKING, F.M. and MEESTER, R.W.J.
(1990) On the structure of Mandelbrot's percolation process and other random Cantor sets. *J. Statist. Phys.*, **58**(5–6), 1109–1126. MR: 1049059

DELLACHERIE, C. and MEYER, P.A.
(1978) *Probabilities and Potential.* Vol. 29 of *North-Holland Mathematics Studies.* North-Holland, Amsterdam. MR: 521810

DEMBO, A.
(2005) Favorite points, cover times and fractals. In *Lectures on Probability Theory and Statistics*, vol. 1869 of *Lecture Notes in Math.*, pages 1–101. Springer, Berlin. Lectures from the 33rd Probability Summer School held in Saint-Flour, July 6–23, 2003, edited by Jean Picard. MR: 2228383

DEMBO, A., GANTERT, N., PERES, Y., and ZEITOUNI, O.
(2002) Large deviations for random walks on Galton-Watson trees: Averaging and uncertainty. *Probab. Theory Related Fields*, **122**(2), 241–288. MR: 1894069

DEMBO, A., GANTERT, N., and ZEITOUNI, O.
(2004) Large deviations for random walk in random environment with holding times. *Ann. Probab.*, **32**(1B), 996–1029. MR: 2044672

DEMBO, A., PERES, Y., ROSEN, J., and ZEITOUNI, O.
(2001) Thick points for planar Brownian motion and the Erdős-Taylor conjecture on random walk. *Acta Math.*, **186**(2), 239–270. MR: 1846031

DEMBO, A. and ZEITOUNI, O.
(1998) *Large Deviations Techniques and Applications*, 2nd ed. Vol. 38 of *Applications of Mathematics.* Springer-Verlag, New York. MR: 1619036

DERRIENNIC, Y.
(1976) Lois "zéro ou deux" pour les processus de Markov. Applications aux marches aléatoires. *Ann. Inst. H. Poincaré Sect. B (N.S.)*, **12**(2), 111–129. MR: 0423532

(1980) Quelques applications du théorème ergodique sous-additif. In *Journées sur les Marches Aléatoires*, vol. 74 of *Astérisque*, pages 183–201, 4. Soc. Math. France, Paris. Held at Kleebach, March 5–10, 1979. MR: 588163

DEZA, M.M. and LAURENT, M.
(1997) *Geometry of Cuts and Metrics.* Vol. 15 of *Algorithms and Combinatorics.* Springer-Verlag, Berlin. MR: 1460488

DIACONIS, P. and EVANS, S.N.
(2002) A different construction of Gaussian fields from Markov chains: Dirichlet covariances. *Ann. Inst. H. Poincaré Probab. Statist.*, **38**(6), 863–878. MR: 1955341

DIACONIS, P. and FILL, J.A.
(1990) Strong stationary times via a new form of duality. *Ann. Probab.*, **18**(4), 1483–1522. MR: 1071805

DIACONIS, P. and SALOFF-COSTE, L.
(1994) Moderate growth and random walk on finite groups. *Geom. Funct. Anal.*, **4**(1), 1–36. MR: 1254308

DIESTEL, R. and LEADER, I.
(2001) A conjecture concerning a limit of non-Cayley graphs. *J. Algebraic Combin.*, **14**(1), 17–25. MR: 2002h:05082

DING, J., LEE, J.R., and PERES, Y.
(2012) Cover times, blanket times, and majorizing measures. *Ann. of Math. (2)*, **175**(3), 1409–1471. MR: 2912708

(2013) Markov type and threshold embeddings. *Geom. Funct. Anal.*, **23**(4), 1207–1229. MR: 3077911

DISERTORI, M., SABOT, C., and TARRÈS, P.
(2015) Transience of edge-reinforced random walk. *Comm. Math. Phys.*, **339**(1), 121–148. MR: 3366053

DODZIUK, J.
(1977) de Rham-Hodge theory for L^2-cohomology of infinite coverings. *Topology*, **16**(2), 157–165. MR: 0445560

(1979) L^2 harmonic forms on rotationally symmetric Riemannian manifolds. *Proc. Amer. Math. Soc.*, **77**(3), 395–400. MR: 81e:58004

(1984) Difference equations, isoperimetric inequality and transience of certain random walks. *Trans. Amer. Math. Soc.*, **284**(2), 787–794. MR: 85m:58185

DODZIUK, J. and KENDALL, W.S.
(1986) Combinatorial Laplacians and isoperimetric inequality. In Elworthy, K.D., editor, *From Local Times to Global Geometry, Control and Physics*, pages 68–74. Longman Sci. Tech., Harlow. Papers from the Warwick symposium on stochastic differential equations and applications, held at the University of Warwick, Coventry, 1984/85. MR: 88h:58118

DOOB, J.L.
(1984) *Classical Potential Theory and Its Probabilistic Counterpart*. Springer-Verlag, New York. MR: 85k:31001

DOYLE, P.G.
(1988) Electric currents in infinite networks. Unpublished, http://www.arxiv.org/abs/math/0703899.

(2009) The Kemeny constant of a Markov chain. Unpublished, http://www.arxiv.org/abs/0909.2636.

DOYLE, P.G. and SNELL, J.L.
(1984) *Random Walks and Electric Networks*. Mathematical Association of America, Washington, DC. Also available at http://arxiv.org/abs/math/0001057. MR: 89a:94023

DOYLE, P.G. and STEINER, J.
(2008) Commuting time geometry of ergodic Markov chains. Unpublished, http://www.arxiv.org/abs/1107.2612.

DREWITZ, A., RÁTH, B., and SAPOZHNIKOV, A.
(2014) *An Introduction to Random Interlacements*. Springer Briefs in Mathematics. Springer, Cham. MR: 3308116

DUBINS, L.E. and FREEDMAN, D.A.
(1967) Random distribution functions. In *Proc. Fifth Berkeley Sympos. Math. Statist. and Probability (Berkeley, Calif., 1965/66)*, pages Vol. II: Contributions to Probability Theory, Part 1, pp. 183–214. University of California Press, Berkeley. MR: 0214109

DUFFIN, R.J.
(1962) The extremal length of a network. *J. Math. Anal. Appl.*, **5**, 200–215. MR: 0143468

DUMINIL-COPIN, H., GOSWAMI, S., RAOUFI, A., SEVERO, F., and YADIN, A.
(2020) Existence of phase transition for percolation using the Gaussian free field. *Duke Math. J.*, **169**(18), 3539–3563. MR: 4181032

DUMINIL-COPIN, H., SIDORAVICIUS, V., and TASSION, V.
(2016) Absence of infinite cluster for critical Bernoulli percolation on slabs. *Comm. Pure Appl. Math.*, **69**(7), 1397–1411. MR: 3503025

DUMINIL-COPIN, H. and SMIRNOV, S.
(2012) The connective constant of the honeycomb lattice equals $\sqrt{2 + \sqrt{2}}$. *Ann. of Math. (2)*, **175**(3), 1653–1665. MR: 2912714

DUMINIL-COPIN, H. and TASSION, V.
(2016a) A new proof of the sharpness of the phase transition for Bernoulli percolation and the Ising model. *Comm. Math. Phys.*, **343**(2), 725–745. Correction, *Commun. Math. Phys.* **359**(2), 821–822 (2018). MR: 3477351

(2016b) A new proof of the sharpness of the phase transition for Bernoulli percolation on $\mathbb{Z}^d$. *Enseign. Math.*, **62**(1–2), 199–206. MR: 3605816

DUPLANTIER, B.
(1992) Loop-erased self-avoiding walks in $2D$. *Physica A*, **191**, 516–522. http://dx.doi.org/10.1016/0378-4371(92)90575-B.

Duplantier, B. and David, F.

(1988) Exact partition functions and correlation functions of multiple Hamiltonian walks on the Manhattan lattice. *J. Statist. Phys.*, **51**(3–4), 327–434. MR: 952941

Duquesne, T. and Le Gall, J.F.

(2002) Random trees, Lévy processes and spatial branching processes. *Astérisque*, **281**, vi+147. MR: 1954248

Durrett, R.

(1986) Reversible diffusion processes. In Chao, J.A. and Woyczyński, W.A., editors, *Probability Theory and Harmonic Analysis*, pages 67–89. Dekker, New York. Papers from the conference held in Cleveland, Ohio, May 10–12, 1983. MR: 88b:60175

(2019) *Probability—Theory and Examples*, 5th ed. Vol. 49 of *Cambridge Series in Statistical and Probabilistic Mathematics*. Cambridge University Press, Cambridge. MR: 3930614

Durrett, R., Kesten, H., and Lawler, G.

(1991) Making money from fair games. In Durrett, R. and Kesten, H., editors, *Random Walks, Brownian Motion, and Interacting Particle Systems*, vol. 28 of *Progr. Probab.*, pages 255–267. Birkhäuser, Boston. A Festschrift in honor of Frank Spitzer. MR: 1146451

Dvoretzky, A.

(1949) On the strong stability of a sequence of events. *Ann. Math. Statistics*, **20**, 296–299. MR: 0031675

Dvoretzky, A. and Erdös, P.

(1951) Some problems on random walk in space. In *Proc. Second Berkeley Symposium on Math. Statist. and Probability, 1950*, pages 353–367. University of California Press, Berkeley. MR: 0047272

Dvoretzky, A., Erdös, P., and Kakutani, S.

(1950) Double points of paths of Brownian motion in *n*-space. *Acta Sci. Math. Szeged*, **12**(Leopoldo Fejer et Frederico Riesz LXX annos natis dedicatus, Pars B), 75–81. MR: 0034972

Dvoretzky, A., Erdős, P., Kakutani, S., and Taylor, S.J.

(1957) Triple points of Brownian paths in 3-space. *Proc. Cambridge Philos. Soc.*, **53**, 856–862. MR: 0094855

Dwass, M.

(1969) The total progeny in a branching process and a related random walk. *J. Appl. Probability*, **6**, 682–686. MR: 0253433

Dymarz, T.

(2010) Bilipschitz equivalence is not equivalent to quasi-isometric equivalence for finitely generated groups. *Duke Math. J.*, **154**(3), 509–526. MR: 2730576

Dynkin, E.B.

(1969) The boundary theory of Markov processes (discrete case). *Uspehi Mat. Nauk*, **24**(2), 3–42. English translation: *Russ. Math. Surv.* **24** (1969), no. 2, 1–42; http://dx.doi.org/10.1070/RM1969v024n02ABEH001341. MR: 0245096

(1980) Markov processes and random fields. *Bull. Amer. Math. Soc. (N.S.)*, **3**(3), 975–999. MR: 585179

Dynkin, E.B. and Maljutov, M.B.

(1961) Random walk on groups with a finite number of generators. *Dokl. Akad. Nauk SSSR*, **137**, 1042–1045. English translation: *Soviet Math. Dokl.* (1961) **2**, 399–402. MR: 24:A1751

Eckmann, B.

(2000) Introduction to l_2-methods in topology: Reduced l_2-homology, harmonic chains, l_2-Betti numbers. *Israel J. Math.*, **117**, 183–219. Notes prepared by Guido Mislin. MR: 1760592

Edgar, G.A.

(1990) *Measure, Topology, and Fractal Geometry*. Undergraduate Texts in Mathematics. Springer-Verlag, New York. MR: 1065392

Edmonds, J.

(1971) Matroids and the greedy algorithm. *Math. Programming*, **1**, 127–136. MR: 0297357

Enflo, P.

(1969) On the nonexistence of uniform homeomorphisms between L_p-spaces. *Ark. Mat.*, **8**, 103–105 (1969). MR: 0271719

Epstein, I. and Hjorth, G.

(2009) Rigidity and equivalence relations with infinitely many ends. Unpublished manuscript available at http://citeseerx.ist.psu.edu/viewdoc/summary?doi=10.1.1.156.8077.

ERSCHLER, A.
   (2010) Poisson-Furstenberg boundaries, large-scale geometry and growth of groups. In *Proceedings of the International Congress of Mathematicians. Volume II*, pages 681–704. Hindustan Book Agency, New Delhi. MR: 2827814

   (2011) Poisson-Furstenberg boundary of random walks on wreath products and free metabelian groups. *Comment. Math. Helv.*, **86**(1), 113–143. MR: 2745278

ERSCHLER, A. and KARLSSON, A.
   (2010) Homomorphisms to ℝ constructed from random walks. *Ann. Inst. Fourier (Grenoble)*, **60**(6), 2095–2113. MR: 2791651

ESKIN, A., FISHER, D., and WHYTE, K.
   (2012) Coarse differentiation of quasi-isometries I: Spaces not quasi-isometric to Cayley graphs. *Ann. of Math. (2)*, **176**(1), 221–260. MR: 2925383

ETHIER, S.N. and LEE, J.
   (2009) Limit theorems for Parrondo's paradox. *Electron. J. Probab.*, **14**, paper no. 62, 1827–1862. MR: 2540850

EVANS, S.N.
   (1987) Multiple points in the sample paths of a Lévy process. *Probab. Theory Related Fields*, **76**(3), 359–367. MR: 912660

EVANS, W., KENYON, C., PERES, Y., and SCHULMAN, L.J.
   (2000) Broadcasting on trees and the Ising model. *Ann. Appl. Probab.*, **10**(2), 410–433. MR: 2001g:60243

FALCONER, K.J.
   (1986) Random fractals. *Math. Proc. Cambridge Philos. Soc.*, **100**(3), 559–582. MR: 88e:28005

   (1987) Cut-set sums and tree processes. *Proc. Amer. Math. Soc.*, **101**(2), 337–346. MR: 88m:90052

   (1990) *Fractal Geometry*. Mathematical foundations and applications. John Wiley, Chichester. MR: 1102677

FAN, A.H.
   (1989) *Décompositions de Mesures et Recouvrements Aléatoires*. Ph.D. thesis, Université de Paris-Sud, Département de Mathématique, Orsay. MR: 91e:60009

   (1990) Sur quelques processus de naissance et de mort. *C. R. Acad. Sci. Paris Sér. I Math.*, **310**(6), 441–444. MR: 91d:60103

FARAUD, G.
   (2011) A central limit theorem for random walk in a random environment on marked Galton-Watson trees. *Electron. J. Probab.*, **16**, paper no. 6, 174–215. MR: 2754802

FARAUD, G., HU, Y., and SHI, Z.
   (2012) Almost sure convergence for stochastically biased random walks on trees. *Probab. Theory Related Fields*, **154**(3–4), 621–660. MR: 3000557

FEDER, T. and MIHAIL, M.
   (1992) Balanced matroids. In *Proceedings of the Twenty-Fourth Annual ACM Symposium on Theory of Computing*, pages 26–38. Association for Computing Machinery (ACM), New York. http://dx.doi.org/10.1145/129712.129716.

FELLER, W.
   (1971) *An Introduction to Probability Theory and Its Applications. Vol. II.*, 2nd ed. John Wiley, New York. MR: 42:5292

FISHER, M.E.
   (1961) Critical probabilities for cluster size and percolation problems. *J. Math. Phys.*, **2**, 620–627. MR: 0126306

FITZNER, R. and VAN DER HOFSTAD, R.
   (2015) Nearest-neighbor percolation function is continuous for $d > 10$. Preprint, http://www.arxiv.org/abs/1506.07977.

FITZSIMMONS, P.J. and SALISBURY, T.S.
   (1989) Capacity and energy for multiparameter Markov processes. *Ann. Inst. H. Poincaré Probab. Statist.*, **25**(3), 325–350. MR: 1023955

FLAJOLET, P. and SEDGEWICK, R.
   (2009) *Analytic Combinatorics*. Cambridge University Press, Cambridge. MR: 2483235

FLANDERS, H.

(1971) Infinite networks. I: Resistive networks. *IEEE Trans. Circuit Theory*, **CT–18**, 326–331. MR: 0275998

(1974) A new proof of R. Foster's averaging formula in networks. *Linear Algebra and Appl.*, **8**, 35–37. MR: 0329772

FLEISCHMANN, K. and SIEGMUND-SCHULTZE, R.

(1977) The structure of reduced critical Galton-Watson processes. *Math. Nachr.*, **79**, 233–241. MR: 0461689

FOGUEL, S.R.

(1971) On the "zero-two" law. *Israel J. Math.*, **10**, 275–280. MR: 0298759

(1976) More on the "zero-two" law. *Proc. Amer. Math. Soc.*, **61**(2), 262–264 (1977). MR: 0428076

FØLNER, E.

(1955) On groups with full Banach mean value. *Math. Scand.*, **3**, 243–254. MR: 18,51f

FONTES, L.R., SCHONMANN, R.H., and SIDORAVICIUS, V.

(2002) Stretched exponential fixation in stochastic Ising models at zero temperature. *Comm. Math. Phys.*, **228**(3), 495–518. MR: 1918786

FORD, L.R., JR. and FULKERSON, D.R.

(1962) *Flows in Networks*. Princeton University Press, Princeton, NJ. MR: 28:2917

FORTUIN, C.M.

(1972a) On the random-cluster model. II. The percolation model. *Physica*, **58**, 393–418. MR: 51:14826

(1972b) On the random-cluster model. III. The simple random-cluster model. *Physica*, **59**, 545–570. MR: 55:5127

FORTUIN, C.M. and KASTELEYN, P.W.

(1972) On the random-cluster model. I. Introduction and relation to other models. *Physica*, **57**, 536–564. MR: 0359655

FORTUIN, C.M., KASTELEYN, P.W., and GINIBRE, J.

(1971) Correlation inequalities on some partially ordered sets. *Comm. Math. Phys.*, **22**, 89–103. MR: 46:8607

FOSTER, R.M.

(1948) The average impedance of an electrical network. In *Reissner Anniversary Volume, Contributions to Applied Mechanics*, pages 333–340. J. W. Edwards, Ann Arbor, Michigan. Edited by the Staff of the Department of Aeronautical Engineering and Applied Mechanics of the Polytechnic Institute of Brooklyn. MR: 10,662a

FRISCH, J., HARTMAN, Y., TAMUZ, O., and VAHIDI FERDOWSI, P.

(2019) Choquet-Deny groups and the infinite conjugacy class property. *Ann. of Math. (2)*, **190**(1), 307–320. MR: 3990605

FRISCH, J. and TAMUZ, O.

(2016) Transitive graphs uniquely determined by their local structure. *Proc. Amer. Math. Soc.*, **144**(5), 1913–1918. MR: 3460154

FROSTMAN, O.

(1935) Potentiel d'équilibre et capacité des ensembles avec quelques applications à la théorie des fonctions. *Meddel. Lunds Univ. Mat. Sem.*, **3**, 1–118.

FUKUSHIMA, M.

(1980) *Dirichlet Forms and Markov Processes*. North-Holland, Amsterdam. MR: 81f:60105

(1985) Energy forms and diffusion processes. In Streit, L., editor, *Mathematics + Physics. Vol. 1*, pages 65–97. World Scientific, Singapore. Lectures on recent results. MR: 87m:60176

FURSTENBERG, H.

(1963) A Poisson formula for semi-simple Lie groups. *Ann. of Math. (2)*, **77**, 335–386. MR: 0146298

(1967) Disjointness in ergodic theory, minimal sets, and a problem in Diophantine approximation. *Math. Systems Theory*, **1**, 1–49. MR: 35:4369

(1970) Intersections of Cantor sets and transversality of semigroups. In Gunning, R.C., editor, *Problems in Analysis*, pages 41–59. Princeton University Press, Princeton, NJ. A symposium in honor of Salomon Bochner, Princeton University, Princeton, NJ, 1–3 April 1969. MR: 50:7040

(1971a) Boundaries of Lie groups and discrete subgroups. In *Actes du Congrès International des Mathématiciens (Nice, 1970), Tome 2*, pages 301–306. Gauthier-Villars, Paris. MR: 0430160

(1971b) Random walks and discrete subgroups of Lie groups. In *Advances in Probability and Related Topics, Vol. 1*, pages 1–63. Dekker, New York. MR: 0284569

(1973) Boundary theory and stochastic processes on homogeneous spaces. In Moore, C.C., editor, *Harmonic Analysis on Homogeneous Spaces*, pages 193–229. Amer. Math. Soc., Providence, R.I. Harmonic analysis on homogeneous spaces (Proc. Sympos. Pure Math., Vol. XXVI, Williams Coll., Williamstown, Mass., 1972). MR: 0352328

(2008) Ergodic fractal measures and dimension conservation. *Ergodic Theory Dynam. Systems*, **28**(2), 405–422. MR: 2408385

FURSTENBERG, H. and WEISS, B.
(2003) Markov processes and Ramsey theory for trees. *Combin. Probab. Comput.*, **12**(5–6), 547–563. MR: 2037069

GABORIAU, D.
(1998) Mercuriale de groupes et de relations. *C. R. Acad. Sci. Paris Sér. I Math.*, **326**(2), 219–222. MR: 99h:28034

(2000) Coût des relations d'équivalence et des groupes. *Invent. Math.*, **139**(1), 41–98. MR: 1728 876

(2002) Invariants l^2 de relations d'équivalence et de groupes. *Publ. Math. Inst. Hautes Études Sci.*, **95**, 93–150. MR: 1953 191

(2005) Invariant percolation and harmonic Dirichlet functions. *Geom. Funct. Anal.*, **15**(5), 1004–1051. MR: 2221157

GABORIAU, D. and LYONS, R.
(2009) A measurable-group-theoretic solution to von Neumann's problem. *Invent. Math.*, **177**(3), 533–540. MR: 2534099

GABORIAU, D. and TUCKER-DROB, R.
(2016) Approximations of standard equivalence relations and Bernoulli percolation at p_u. *C. R. Math. Acad. Sci. Paris*, **354**(11), 1114–1118. MR: 3566513

GANDOLFI, A., KEANE, M.S., and NEWMAN, C.M.
(1992) Uniqueness of the infinite component in a random graph with applications to percolation and spin glasses. *Probab. Theory Related Fields*, **92**(4), 511–527. MR: 93f:60149

GANTERT, N., MÜLLER, S., POPOV, S., and VACHKOVSKAIA, M.
(2012) Random walks on Galton-Watson trees with random conductances. *Stochastic Process. Appl.*, **122**(4), 1652–1671. MR: 2914767

GARBAN, C., PETE, G., and SCHRAMM, O.
(2018) The scaling limits of the Minimal Spanning Tree and Invasion Percolation in the plane. *Ann. Probab.*, **46**(6), 3501–3557. MR: 3857861

GARNETT, J.B. and MARSHALL, D.E.
(2005) *Harmonic Measure*. Vol. 2 of *New Mathematical Monographs*. Cambridge University Press, Cambridge. MR: 2150803

GAUDILLIÈRE, A. and LANDIM, C.
(2014) A Dirichlet principle for non reversible Markov chains and some recurrence theorems. *Probab. Theory Related Fields*, **158**(1–2), 55–89. MR: 3152780

GAUTERO, F. and MATHÉUS, F.
(2012) Poisson boundary of groups acting on ℝ-trees. *Israel J. Math.*, **191**(2), 585–646. MR: 3011489

GEIGER, J.
(1995) Contour processes of random trees. In Etheridge, A., editor, *Stochastic Partial Differential Equations*, vol. 216 of *London Math. Soc. Lecture Note Ser.*, pages 72–96. Cambridge University Press, Cambridge. Papers from the workshop held at the University of Edinburgh, Edinburgh, March 1994. MR: 1352736

(1999) Elementary new proofs of classical limit theorems for Galton-Watson processes. *J. Appl. Probab.*, **36**(2), 301–309. MR: 1724 856

GEL'FAND, I.M., GORESKY, R.M., MACPHERSON, R.D., and SERGANOVA, V.V.
(1987) Combinatorial geometries, convex polyhedra, and Schubert cells. *Adv. in Math.*, **63**(3), 301–316. MR: 877789

GEORGAKOPOULOS, A.
(2010) Lamplighter graphs do not admit harmonic functions of finite energy. *Proc. Amer. Math. Soc.*, **138**(9), 3057–3061. MR: 2653930

(2016) The boundary of a square tiling of a graph coincides with the Poisson boundary. *Invent. Math.*, **203**(3), 773–821. MR: 3461366

GEORGAKOPOULOS, A. and WINKLER, P.
(2014) New bounds for edge-cover by random walk. *Combin. Probab. Comput.*, **23**(4), 571–584. MR: 3217361

GERL, P.
(1988) Random walks on graphs with a strong isoperimetric property. *J. Theoret. Probab.*, **1**(2), 171–187. MR: 89g:60216

GILCH, L.A.
(2007) Rate of escape of random walks on free products. *J. Aust. Math. Soc.*, **83**(1), 31–54. MR: 2378433

(2011) Asymptotic entropy of random walks on free products. *Electron. J. Probab.*, **16**, paper no. 3, 76–105. MR: 2749773

GILCH, L.A. and MÜLLER, S.
(2011) Random walks on directed covers of graphs. *J. Theoret. Probab.*, **24**(1), 118–149. MR: 2782713

GLASNER, E. and WEISS, B.
(1997) Kazhdan's property T and the geometry of the collection of invariant measures. *Geom. Funct. Anal.*, **7**(5), 917–935. MR: 99f:28029

GLASSER, M.L. and ZUCKER, I.J.
(1977) Extended Watson integrals for the cubic lattices. *Proc. Nat. Acad. Sci. U.S.A.*, **74**(5), 1800–1801. MR: 0442300

GNEITING, T. and RAFTERY, A.E.
(2007) Strictly proper scoring rules, prediction, and estimation. *J. Amer. Statist. Assoc.*, **102**(477), 359–378. MR: 2345548

GODSIL, C.D., IMRICH, W., SEIFTER, N., WATKINS, M.E., and WOESS, W.
(1989) A note on bounded automorphisms of infinite graphs. *Graphs Combin.*, **5**(4), 333–338. MR: 1032384

GOLDMAN, J.R. and KAUFFMAN, L.H.
(1993) Knots, tangles, and electrical networks. *Adv. in Appl. Math.*, **14**(3), 267–306. MR: 94m:57013

GOUËZEL, S., MATHÉUS, F., and MAUCOURANT, F.
(2015) Sharp lower bounds for the asymptotic entropy of symmetric random walks. *Groups Geom. Dyn.*, **9**(3), 711–735. MR: 3420541

GRABOWSKI, L.
(2014) On Turing dynamical systems and the Atiyah problem. *Invent. Math.*, **198**(1), 27–69. MR: 3260857

GRAF, S.
(1987) Statistically self-similar fractals. *Probab. Theory Related Fields*, **74**(3), 357–392. MR: 88c:60038

GRAF, S., MAULDIN, R.D., and WILLIAMS, S.C.
(1988) The exact Hausdorff dimension in random recursive constructions. *Mem. Amer. Math. Soc.*, **71**(381), x+121. MR: 88k:28010

GRAVNER, J. and HOLROYD, A.E.
(2009) Local bootstrap percolation. *Electron. J. Probab.*, **14**, paper no. 14, 385–399. MR: 2480546

GREY, D.R.
(1980) A new look at convergence of branching processes. *Ann. Probab.*, **8**(2), 377–380. MR: 81e:60091

GRIFFEATH, D. and LIGGETT, T.M.
(1982) Critical phenomena for Spitzer's reversible nearest particle systems. *Ann. Probab.*, **10**(4), 881–895. MR: 84f:60140

GRIGORCHUK, R.I.
(1980) Symmetrical random walks on discrete groups. In Dobrushin, R.L., Sinaĭ, Ya.G., and Griffeath, D., editors, *Multicomponent Random Systems*, vol. 6 of *Adv. Probab. Related Topics*, pages 285–325. Dekker, New York. Translated from the Russian. MR: 599539

(1983) On the Milnor problem of group growth. *Dokl. Akad. Nauk SSSR*, **271**(1), 30–33. MR: 85g:20042

GRIGOR'YAN, A.A.
(1985) The existence of positive fundamental solutions of the Laplace equation on Riemannian manifolds. *Mat. Sb. (N.S.)*, **128(170)**(3), 354–363, 446. English translation: *Math. USSR-Sb.* **56** (1987), no. 2, 349–358. MR: 87d:58140

Grimmett, G.

(1999) *Percolation*, 2nd ed. Springer-Verlag, Berlin. MR: 1707 339

(2006) *The Random-Cluster Model*. Vol. 333 of *Grundlehren der Mathematischen Wissenschaften [Fundamental Principles of Mathematical Sciences]*. Springer-Verlag, Berlin. MR: 2243761

Grimmett, G. and Kesten, H.

(1983) Random electrical networks on complete graphs II: Proofs. Unpublished manuscript, available at http://www.arxiv.org/abs/math.PR/0107068.

Grimmett, G.R., Kesten, H., and Zhang, Y.

(1993) Random walk on the infinite cluster of the percolation model. *Probab. Theory Related Fields*, **96**(1), 33–44. MR: 94i:60078

Grimmett, G.R. and Marstrand, J.M.

(1990) The supercritical phase of percolation is well behaved. *Proc. Roy. Soc. London Ser. A*, **430**(1879), 439–457. MR: 91m:60186

Grimmett, G.R. and Newman, C.M.

(1990) Percolation in $\infty + 1$ dimensions. In Grimmett, G.R. and Welsh, D.J.A., editors, *Disorder in Physical Systems*, pages 167–190. Oxford University Press, New York. A volume in honour of John M. Hammersley on the occasion of his 70th birthday. MR: 92a:60207

Grimmett, G.R. and Stacey, A.M.

(1998) Critical probabilities for site and bond percolation models. *Ann. Probab.*, **26**(4), 1788–1812. MR: 1675079

Grimmett, G.R. and Stirzaker, D.R.

(2001) *Probability and Random Processes*, 3rd ed. Oxford University Press, New York. MR: 2059709

Gromov, M.

(1981a) Groups of polynomial growth and expanding maps. *Inst. Hautes Études Sci. Publ. Math.*, **53**, 53–73. MR: 83b:53041

(1981b) *Structures métriques pour les variétés riemanniennes*. Vol. 1 of *Textes Mathématiques [Mathematical Texts]*. CEDIC, Paris. Edited by J. Lafontaine and P. Pansu. MR: 682063

(1983) Filling Riemannian manifolds. *J. Differential Geom.*, **18**(1), 1–147. MR: 697984

(1999) *Metric Structures for Riemannian and Non-Riemannian Spaces*. Birkhäuser, Boston. Based on the 1981 French original [MR 85e:53051], with appendices by M. Katz, P. Pansu, and S. Semmes, translated from the French by Sean Michael Bates. MR: 2000d:53065

Guillotin-Plantard, N.

(2005) Gillis's random walks on graphs. *J. Appl. Probab.*, **42**(1), 295–301. MR: 2144913

Guivarc'h, Y.

(1980) Sur la loi des grands nombres et le rayon spectral d'une marche aléatoire. In *Journées sur les Marches Aléatoires*, vol. 74 of *Astérisque*, pages 47–98, 3. Soc. Math. France, Paris, Paris. Held at Kleebach, March 5–10, 1979. MR: 588157

Guttmann, A.J. and Bursill, R.J.

(1990) Critical exponent for the loop erased self-avoiding walk by Monte Carlo methods. *J. Stat. Phys.*, **59**(1–2), 1–9. http://dx.doi.org/10.1007/BF01015560.

Guttorp, P.

(1991) *Statistical Inference for Branching Processes*. Wiley Series in Probability and Mathematical Statistics. John Wiley, New York. MR: 1254434

Häggström, O.

(1994) *Aspects of Spatial Random Processes*. Ph.D. thesis, Göteborg, Sweden.

(1995) Random-cluster measures and uniform spanning trees. *Stochastic Process. Appl.*, **59**(2), 267–275. MR: 97b:60170

(1997) Infinite clusters in dependent automorphism invariant percolation on trees. *Ann. Probab.*, **25**(3), 1423–1436. MR: 98f:60207

(1998) Uniform and minimal essential spanning forests on trees. *Random Structures Algorithms*, **12**(1), 27–50. MR: 99i:05186

(2013) Two badly behaved percolation processes on a nonunimodular graph. *J. Theoret. Probab.*, **26**(4), 1165–1180. MR: 3119989

HÄGGSTRÖM, O., JONASSON, J., and LYONS, R.
    (2002) Explicit isoperimetric constants and phase transitions in the random-cluster model. *Ann. Probab.*, **30**(1), 443–473. MR: 2003e:60220

HÄGGSTRÖM, O. and PERES, Y.
    (1999) Monotonicity of uniqueness for percolation on Cayley graphs: All infinite clusters are born simultaneously. *Probab. Theory Related Fields*, **113**(2), 273–285. MR: 1676835

HÄGGSTRÖM, O., PERES, Y., and SCHONMANN, R.H.
    (1999) Percolation on transitive graphs as a coalescent process: Relentless merging followed by simultaneous uniqueness. In Bramson, M. and Durrett, R., editors, *Perplexing Problems in Probability*, pages 69–90. Birkhäuser, Boston. Festschrift in honor of Harry Kesten. MR: 2000b:60003

HÄGGSTRÖM, O., SCHONMANN, R.H., and STEIF, J.E.
    (2000) The Ising model on diluted graphs and strong amenability. *Ann. Probab.*, **28**(3), 1111–1137. MR: 2001i:60169

HAMMERSLEY, J.M.
    (1959) Bornes supérieures de la probabilité critique dans un processus de filtration. In *Le Calcul des Probabilités et ses Applications. Paris, 15–20 juillet 1958*, Colloques Internationaux du Centre National de la Recherche Scientifique, LXXXVII, pages 17–37. Centre National de la Recherche Scientifique, Paris. MR: 0105751

    (1961a) Comparison of atom and bond percolation processes. *J. Math. Phys.*, **2**, 728–733. MR: 0130722

    (1961b) The number of polygons on a lattice. *Proc. Cambridge Philos. Soc.*, **57**, 516–523. MR: 23:A814

HAN, T.S.
    (1978) Nonnegative entropy measures of multivariate symmetric correlations. *Information and Control*, **36**(2), 133–156. MR: 0464499

HARA, T. and SLADE, G.
    (1990) Mean-field critical behaviour for percolation in high dimensions. *Comm. Math. Phys.*, **128**(2), 333–391. MR: 91a:82037

    (1992) The lace expansion for self-avoiding walk in five or more dimensions. *Rev. Math. Phys.*, **4**(2), 235–327. MR: 93j:82033

    (1994) Mean-field behaviour and the lace expansion. In Grimmett, G., editor, *Probability and Phase Transition*, pages 87–122. Kluwer Academic, Dordrecht. Proceedings of the NATO Advanced Study Institute on Probability Theory of Spatial Disorder and Phase Transition held at the University of Cambridge, Cambridge, July 4–16, 1993. MR: 95d:82033

HARMER, G.P. and ABBOTT, D.
    (1999) Parrondo's paradox. *Statist. Sci.*, **14**(2), 206–213. MR: 1722065

DE LA HARPE, P. and VALETTE, A.
    (1989) La propriété (*T*) de Kazhdan pour les groupes localement compacts (avec un appendice de Marc Burger). *Astérisque*, **175**, 158. With an appendix by M. Burger. MR: 1023471

HARRIS, T.E.
    (1952) First passage and recurrence distributions. *Trans. Amer. Math. Soc.*, **73**, 471–486. MR: 0052057

    (1960) A lower bound for the critical probability in a certain percolation process. *Proc. Cambridge Philos. Soc.*, **56**, 13–20. MR: 22:6023

HASLEGRAVE, J. and PANAGIOTIS, C.
    (2021) Site percolation and isoperimetric inequalities for plane graphs. *Random Structures Algorithms*, **58**(1), 150–163. MR: 4180256

HAWKES, J.
    (1970/71) Some dimension theorems for the sample functions of stable processes. *Indiana Univ. Math. J.*, **20**, 733–738. MR: 45:1251

    (1981) Trees generated by a simple branching process. *J. London Math. Soc. (2)*, **24**(2), 373–384. MR: 83b:60072

HE, Z.X. and SCHRAMM, O.
    (1995) Hyperbolic and parabolic packings. *Discrete Comput. Geom.*, **14**(2), 123–149. MR: 1331923

HEATHCOTE, C.R.
    (1966) Corrections and comments on the paper "A branching process allowing immigration". *J. Roy. Statist. Soc. Ser. B*, **28**, 213–217. MR: 33:1896b

HEATHCOTE, C.R., SENETA, E., and VERE-JONES, D.
(1967) A refinement of two theorems in the theory of branching processes. *Teor. Verojatnost. i Primenen.*, **12**, 341–346. MR: 36:978

HEBISCH, W. and SALOFF-COSTE, L.
(1993) Gaussian estimates for Markov chains and random walks on groups. *Ann. Probab.*, **21**(2), 673–709. MR: 1217561

HERMON, J. and HUTCHCROFT, T.
(2019a) No percolation at criticality on certain groups of intermediate growth. *Int. Math. Res. Not. IMRN.* rnz265.

(2019b) Supercritical percolation on nonamenable graphs: Isoperimetry, analyticity, and exponential decay of the cluster size distribution. Preprint, http://www.arxiv.org/abs/1904.10448.

HEYDE, C.C.
(1970) Extension of a result of Seneta for the super-critical Galton-Watson process. *Ann. Math. Statist.*, **41**, 739–742. MR: 40:8136

HEYDE, C.C. and SENETA, E.
(1977) *I. J. Bienaymé. Statistical Theory Anticipated.* Vol. 3 of *Studies in the History of Mathematics and Physical Sciences.* Springer-Verlag, New York. MR: 57:2855

HIGUCHI, Y. and SHIRAI, T.
(2003) Isoperimetric constants of (d, f)-regular planar graphs. *Interdiscip. Inform. Sci.*, **9**(2), 221–228. MR: 2038013

HOEFFDING, W.
(1963) Probability inequalities for sums of bounded random variables. *J. Amer. Statist. Assoc.*, **58**, 13–30. MR: 0144363

HOFFMAN, C., HOLROYD, A.E., and PERES, Y.
(2006) A stable marriage of Poisson and Lebesgue. *Ann. Probab.*, **34**(4), 1241–1272. MR: 2257646

HOLOPAINEN, I. and SOARDI, P.M.
(1997) p-Harmonic functions on graphs and manifolds. *Manuscripta Math.*, **94**(1), 95–110. MR: 99c:31017

HOLROYD, A.E.
(2003) Sharp metastability threshold for two-dimensional bootstrap percolation. *Probab. Theory Related Fields*, **125**(2), 195–224. MR: 1961342

HOLROYD, A.E., LEVINE, L., MÉSZÁROS, K., PERES, Y., PROPP, J., and WILSON, D.B.
(2008) Chip-firing and rotor-routing on directed graphs. In Sidoravicius, V. and Vares, M.E., editors, *In and Out of Equilibrium. 2*, vol. 60 of *Progr. Probab.*, pages 331–364. Birkhäuser, Basel. Papers from the 10th Brazilian School of Probability (EBP) held in Rio de Janeiro, July 30–August 4, 2006. MR: 2477390

HOORY, S., LINIAL, N., and WIGDERSON, A.
(2006) Expander graphs and their applications. *Bull. Amer. Math. Soc. (N.S.)*, **43**(4), 439–561 (electronic). MR: 2247919

HORN, R.A. and JOHNSON, C.R.
(2013) *Matrix Analysis*, 2nd ed. Cambridge University Press, Cambridge. MR: 2978290

HOUDAYER, C.
(2012) Invariant percolation and measured theory of nonamenable groups [after Gaboriau-Lyons, Ioana, Epstein]. *Astérisque*, **348**, Exp. No. 1039, ix, 339–374. Séminaire Bourbaki: Vol. 2010/2011. Exposés 1027–1042. MR: 3051202

HOUGH, J.B., KRISHNAPUR, M., PERES, Y., and VIRÁG, B.
(2009) *Zeros of Gaussian Analytic Functions and Determinantal Point Processes.* Vol. 51 of *University Lecture Series.* American Mathematical Society, Providence, RI. MR: 2552864

HOWARD, C.D.
(2000) Zero-temperature Ising spin dynamics on the homogeneous tree of degree three. *J. Appl. Probab.*, **37**(3), 736–747. MR: 1782449

HUTCHCROFT, T.
(2016a) Critical percolation on any quasi-transitive graph of exponential growth has no infinite clusters. *C. R. Math. Acad. Sci. Paris*, **354**(9), 944–947. MR: 3535351

(2016b) Wired cycle-breaking dynamics for uniform spanning forests. *Ann. Probab.*, **44**(6), 3879–3892. MR: 3572326

(2018) Interlacements and the wired uniform spanning forest. *Ann. Probab.*, **46**(2), 1170–1200. MR: 3773383

(2019a) Harmonic Dirichlet functions on planar graphs. *Discrete Comput. Geom.*, **61**(3), 479–506. MR: 3918545

(2019b) Percolation on hyperbolic graphs. *Geom. Funct. Anal.*, **29**(3), 766–810. MR: 3962879

(2020a) The L^2 boundedness condition in nonamenable percolation. *Electron. J. Probab.*, **25**, Paper No. 127, 27 pp. MR: 4162843

(2020b) Locality of the critical probability for transitive graphs of exponential growth. *Ann. Probab.*, **48**(3), 1352–1371. MR: 4112717

(2020c) Nonuniqueness and mean-field criticality for percolation on nonunimodular transitive graphs. *J. Amer. Math. Soc.*, **33**(4), 1101–1165. MR: 4155221

(2020d) Universality of high-dimensional spanning forests and sandpiles. *Probab. Theory Related Fields*, **176**(1–2), 533–597. MR: 4055195

HUTCHCROFT, T. and NACHMIAS, A.

(2017) Indistinguishability of trees in uniform spanning forests. *Probab. Theory Related Fields*, **168**(1–2), 113–152. MR: 3651050

(2019) Uniform spanning forests of planar graphs. *Forum Math. Sigma*, **7**, e29, 55 pp. MR: 4010561

HUTCHCROFT, T. and PERES, Y.

(2017) Boundaries of planar graphs: a unified approach. *Electron. J. Probab.*, **22**, Paper No. 100, 20 pp. MR: 3733658

HUTCHCROFT, T. and PETE, G.

(2020) Kazhdan groups have cost 1. *Invent. Math.*, **221**(3), 873–891. MR: 4132958

ICHIHARA, K.

(1978) Some global properties of symmetric diffusion processes. *Publ. Res. Inst. Math. Sci.*, **14**(2), 441–486. MR: 80d:60099

IMRICH, W.

(1975) On Whitney's theorem on the unique embeddability of 3-connected planar graphs. In Fiedler, M., editor, *Recent Advances in Graph Theory (Proc. Second Czechoslovak Sympos., Prague, 1974)*, pages 303–306. (Loose errata). Academia, Prague. MR: 52:5462

INDYK, P. and MOTWANI, R.

(1999) Approximate nearest neighbors: Towards removing the curse of dimensionality. In *Proceedings of the 30th Annual ACM Symposium on Theory of Computing held in Dallas, TX, May 23–26, 1998*, pages 604–613. ACM, New York. MR: 1715608

JACKSON, T.S. and READ, N.

(2010a) Theory of minimum spanning trees. I. Mean-field theory and strongly disordered spin-glass model. *Phys. Rev. E*, **81**(2), 021130. http://dx.doi.org/10.1103/PhysRevE.81.021130.

(2010b) Theory of minimum spanning trees. II. Exact graphical methods and perturbation expansion at the percolation threshold. *Phys. Rev. E*, **81**(2), 021131. http://dx.doi.org/10.1103/PhysRevE.81.021131.

JAIN, N.C. and TAYLOR, S.J.

(1973) Local asymptotic laws for Brownian motion. *Ann. Probab.*, **1**, 527–549. MR: 0365732

JAMES, N. and PERES, Y.

(1996) Cutpoints and exchangeable events for random walks. *Teor. Veroyatnost. i Primenen.*, **41**(4), 854–868. MR: 1687097

JANSON, S.

(1997) *Gaussian Hilbert Spaces.* Vol. 129 of *Cambridge Tracts in Mathematics*. Cambridge University Press, Cambridge. MR: 1474726

JÁRAI, A.A. and REDIG, F.

(2008) Infinite volume limit of the abelian sandpile model in dimensions $d \geq 3$. *Probab. Theory Related Fields*, **141**(1–2), 181–212. MR: 2372969

JÁRAI, A.A. and WERNING, N.

(2014) Minimal configurations and sandpile measures. *J. Theoret. Probab.*, **27**(1), 153–167. MR: 3174221

JERRUM, M. and SINCLAIR, A.

(1989) Approximating the permanent. *SIAM J. Comput.*, **18**(6), 1149–1178. MR: 1025467

JOAG-DEV, K., PERLMAN, M.D., and PITT, L.D.
(1983) Association of normal random variables and Slepian's inequality. *Ann. Probab.*, **11**(2), 451–455. MR: 690142

JOAG-DEV, K. and PROSCHAN, F.
(1983) Negative association of random variables, with applications. *Ann. Statist.*, **11**(1), 286–295. MR: 85d:62058

JOFFE, A. and WAUGH, W.A.O'N.
(1982) Exact distributions of kin numbers in a Galton-Watson process. *J. Appl. Probab.*, **19**(4), 767–775. MR: 84a:60104

JOHNSON, W.B. and LINDENSTRAUSS, J.
(1984) Extensions of Lipschitz mappings into a Hilbert space. In Beals, R., Beck, A., Bellow, A., and Hajian, A., editors, *Conference in Modern Analysis and Probability*, vol. 26 of *Contemp. Math.*, pages 189–206. Amer. Math. Soc., Providence, RI. Held at Yale University, New Haven, Conn., June 8–11, 1982, Held in honor of Professor Shizuo Kakutani. MR: 737400

JOLISSAINT, P.N. and VALETTE, A.
(2014) L^p-distortion and p-spectral gap of finite graphs. *Bull. Lond. Math. Soc.*, **46**(2), 329–341. MR: 3194751

JORGENSEN, P.E.T. and PEARSE, E.P.J.
(2008) *Operator Theory of Electrical Resistance Networks*. Preprint, `http://www.arxiv.org/abs/0806.3881`.

JOYAL, A.
(1981) Une théorie combinatoire des séries formelles. *Adv. in Math.*, **42**(1), 1–82. MR: 633783

KAHN, J.
(2003) Inequality of two critical probabilities for percolation. *Electron. Comm. Probab.*, **8**, 184–187 (electronic). MR: 2042 758

KAHN, J., KIM, J.H., LOVÁSZ, L., and VU, V.H.
(2000) The cover time, the blanket time, and the Matthews bound. In *41st Annual Symposium on Foundations of Computer Science (Redondo Beach, CA, 2000)*, pages 467–475. IEEE Comput. Soc. Press, Los Alamitos, CA. MR: 1931843

KAHN, J.D., LINIAL, N., NISAN, N., and SAKS, M.E.
(1989) On the cover time of random walks on graphs. *J. Theoret. Probab.*, **2**(1), 121–128. MR: 981769

KAIMANOVICH, V.A.
(1985) An entropy criterion of maximality for the boundary of random walks on discrete groups. *Dokl. Akad. Nauk SSSR*, **280**(5), 1051–1054. MR: 780288

(1990) Boundary and entropy of random walks in random environment. In Grigelionis, B., Sazonov, V.V., and Statulevičius, V., editors, *Probability Theory and Mathematical Statistics. Vol. I*, pages 573–579. "Mokslas," Vilnius. Proceedings of the Fifth Conference held in Vilnius, June 25–July 1, 1989. MR: 1153846

(1992) Dirichlet norms, capacities and generalized isoperimetric inequalities for Markov operators. *Potential Anal.*, **1**(1), 61–82. MR: 94i:31012

(1994) The Poisson boundary of hyperbolic groups. *C. R. Acad. Sci. Paris Sér. I Math.*, **318**(1), 59–64. MR: 1260536

(1996) Boundaries of invariant Markov operators: The identification problem. In Pollicott, M. and Schmidt, K., editors, *Ergodic Theory of $\mathbf{Z}^d$ Actions*, vol. 228 of *London Math. Soc. Lecture Note Ser.*, pages 127–176. Cambridge University Press, Cambridge. Proceedings of the symposium held in Warwick, 1993–1994. MR: 1411218

(2000) The Poisson formula for groups with hyperbolic properties. *Ann. of Math. (2)*, **152**(3), 659–692. MR: 1815698

(2001) Poisson boundary of discrete groups. Preprint, `http://citeseerx.ist.psu.edu/viewdoc/summary?doi=10.1.1.6.6675`.

(2005) "Münchhausen trick" and amenability of self-similar groups. *Internat. J. Algebra Comput.*, **15**(5–6), 907–937. MR: 2197814

KAIMANOVICH, V.A. and MASUR, H.
(1996) The Poisson boundary of the mapping class group. *Invent. Math.*, **125**(2), 221–264. MR: 1395719

(1998) The Poisson boundary of Teichmüller space. *J. Funct. Anal.*, **156**(2), 301–332. MR: 1636940

KAIMANOVICH, V.A. and VERSHIK, A.M.

(1983) Random walks on discrete groups: Boundary and entropy. *Ann. Probab.*, **11**(3), 457–490. MR: 85d:60024

KAIMANOVICH, V.A. and WOESS, W.

(2002) Boundary and entropy of space homogeneous Markov chains. *Ann. Probab.*, **30**(1), 323–363. MR: 1894110

KAKUTANI, S.

(1944) Two-dimensional Brownian motion and harmonic functions. *Proc. Imp. Acad. Tokyo*, **20**, 706–714. MR: 7,315b

KAMAE, T.

(1982) A simple proof of the ergodic theorem using nonstandard analysis. *Israel J. Math.*, **42**(4), 284–290. MR: 682311

KANAI, M.

(1985) Rough isometries, and combinatorial approximations of geometries of noncompact Riemannian manifolds. *J. Math. Soc. Japan*, **37**(3), 391–413. MR: 87d:53082

(1986) Rough isometries and the parabolicity of Riemannian manifolds. *J. Math. Soc. Japan*, **38**(2), 227–238. MR: 87e:53066

KARGAPOLOV, M.I. and MERZLJAKOV, J.I.

(1979) *Fundamentals of the Theory of Groups*. Springer-Verlag, New York. Translated from the second Russian edition by Robert G. Burns. MR: 80k:20002

KARLSSON, A.

(2003) Boundaries and random walks on finitely generated infinite groups. *Ark. Mat.*, **41**(2), 295–306. MR: 2011923

KARLSSON, A. and LEDRAPPIER, F.

(2007) Linear drift and Poisson boundary for random walks. *Pure Appl. Math. Q.*, **3**(4, Special Issue: In honor of Grigory Margulis. Part 1), 1027–1036. MR: 2402595

KARLSSON, A. and WOESS, W.

(2007) The Poisson boundary of lamplighter random walks on trees. *Geom. Dedicata*, **124**, 95–107. MR: 2318539

KASSEL, A. and WILSON, D.B.

(2016) The looping rate and sandpile density of planar graphs. *Amer. Math. Monthly*, **123**(1), 19–39. MR: 3453533

KASTELEYN, P.

(1961) The statistics of dimers on a lattice I. The number of dimer arrangements on a quadratic lattice. *Physica*, **27**, 1209–1225. http://dx.doi.org/10.1007/978-0-8176-4842-8_20.

KATZNELSON, Y. and WEISS, B.

(1982) A simple proof of some ergodic theorems. *Israel J. Math.*, **42**(4), 291–296. MR: 682312

KAZAMI, T. and UCHIYAMA, K.

(2008) Random walks on periodic graphs. *Trans. Amer. Math. Soc.*, **360**(11), 6065–6087. MR: 2425703

KEANE, M.

(1995) The essence of the law of large numbers. In Takahashi, Y., editor, *Algorithms, Fractals, and Dynamics*, pages 125–129. Plenum, New York. Papers from the Hayashibara Forum '92 International Symposium on New Bases for Engineering Science, Algorithms, Dynamics and Fractals held in Okayama, November 23–28, 1992, and the Symposium on Algorithms, Fractals and Dynamics held at Kyoto University, Kyoto, November 30–December 2, 1992. MR: 1402486

KELLY, J.B.

(1970) Metric inequalities and symmetric differences. In *Inequalities, II (Proc. Second Sympos., U.S. Air Force Acad., Colo., 1967)*, pages 193–212. Academic Press, New York. MR: 0264600

KEMENY, J.G. and SNELL, J.L.

(1960) *Finite Markov Chains*. The University Series in Undergraduate Mathematics. D. Van Nostrand, Princeton, NJ. MR: 0115196

KENDALL, D.G.

(1951) Some problems in the theory of queues. *J. Roy. Statist. Soc. Ser. B.*, **13**, 151–173; discussion: 173–185. MR: 0047944

KENNELLY, A.E.

(1899) Equivalence of triangles and stars in conducting networks. *Electrical World and Engineer*, **34**, 413–414.

KENYON, R.W.

(1997) Local statistics of lattice dimers. *Ann. Inst. H. Poincaré Probab. Statist.*, **33**(5), 591–618. MR: 1473567

(1998) Tilings and discrete Dirichlet problems. *Israel J. Math.*, **105**, 61–84. MR: 99m:52026

(2000a) The asymptotic determinant of the discrete Laplacian. *Acta Math.*, **185**(2), 239–286. MR: 2002g:82019

(2000b) Long-range properties of spanning trees. *J. Math. Phys.*, **41**(3), 1338–1363. MR: 1757 962

(2001) Dominos and the Gaussian free field. *Ann. Probab.*, **29**(3), 1128–1137. MR: 1872739

(2008) Height fluctuations in the honeycomb dimer model. *Comm. Math. Phys.*, **281**(3), 675–709. MR: 2415464

KENYON, R.W., PROPP, J.G., and WILSON, D.B.

(2000) Trees and matchings. *Electron. J. Combin.*, **7**(1), Research Paper 25, 34 pp. (electronic). MR: 2001a:05123

KENYON, R.W. and WILSON, D.B.

(2015) Spanning trees of graphs on surfaces and the intensity of loop-erased random walk on planar graphs. *J. Amer. Math. Soc.*, **28**(4), 985–1030. MR: 3369907

KESTEN, H.

(1959a) Full Banach mean values on countable groups. *Math. Scand.*, **7**, 146–156. MR: 22:2911

(1959b) Symmetric random walks on groups. *Trans. Amer. Math. Soc.*, **92**, 336–354. MR: 22:253

(1967) The Martin boundary of recurrent random walks on countable groups. In *Proc. Fifth Berkeley Sympos. Math. Statist. and Probability (Berkeley, Calif., 1965/66), Vol. II: Contributions to Probability Theory, Part 2*, pages 51–74. University of California Press, Berkeley. MR: 0214137

(1980) The critical probability of bond percolation on the square lattice equals $\frac{1}{2}$. *Comm. Math. Phys.*, **74**(1), 41–59. MR: 82c:60179

(1982) *Percolation Theory for Mathematicians*. Birkhäuser, Boston. MR: 84i:60145

(1986) Subdiffusive behavior of random walk on a random cluster. *Ann. Inst. H. Poincaré Probab. Statist.*, **22**(4), 425–487. MR: 88b:60232

KESTEN, H., NEY, P., and SPITZER, F.

(1966) The Galton-Watson process with mean one and finite variance. *Teor. Verojatnost. i Primenen.*, **11**, 579–611. MR: 34:6868

KESTEN, H. and STIGUM, B.P.

(1966) A limit theorem for multidimensional Galton-Watson processes. *Ann. Math. Statist.*, **37**, 1211–1223. MR: 33:6707

KHOSHNEVISAN, D., PERES, Y., and XIAO, Y.

(2000) Limsup random fractals. *Electron. J. Probab.*, **5**, paper no. 5, 24 pp. MR: 1743726

KIFER, Y.

(1986) *Ergodic Theory of Random Transformations*. Birkhäuser, Boston. MR: 89c:58069

KINGMAN, J.F.C.

(1968) The ergodic theory of subadditive stochastic processes. *J. Roy. Statist. Soc. Ser. B*, **30**, 499–510. MR: 0254907

KIRCHHOFF, G.

(1847) Ueber die Auflösung der Gleichungen, auf welche man bei der Untersuchung der linearen Vertheilung galvanischer Ströme geführt wird. *Ann. Phys. Chem.*, **72**(12), 497–508. http://dx.doi.org/10.1002/andp.18471481202.

KLEINER, B.

(2010) A new proof of Gromov's theorem on groups of polynomial growth. *J. Amer. Math. Soc.*, **23**(3), 815–829. MR: 2629989

KOLMOGOROV, A.

(1938) On the solution of a problem in biology. *Izv. NII Matem. Mekh. Tomskogo Univ.*, **2**, 7–12.

KOLMOGOROV, A.N. and BARZDIN', Y.M.

(1967) On the realization of nets in 3-dimensional space. *Probl. Cybernet*, **19**, 261–268. In Russian. See also *Selected Works of A.N. Kolmogorov*, Vol. III, pp. 194–202 (and a remark on p. 245), Kluwer Academic, 1993. http://dx.doi.org/10.1007/978-94-017-2973-4_11.

KOREVAAR, N.J. and SCHOEN, R.M.

(1997) Global existence theorems for harmonic maps to non-locally compact spaces. *Comm. Anal. Geom.*, **5**(2), 333–387. MR: 1483983

KOTANI, M. and SUNADA, T.

(2000) Zeta functions of finite graphs. *J. Math. Sci. Univ. Tokyo*, **7**(1), 7–25. MR: 1749978

KOZÁKOVÁ, I.

(2008) Critical percolation of free product of groups. *Internat. J. Algebra Comput.*, **18**(4), 683–704. MR: 2428151

KOZDRON, M.J., RICHARDS, L.M., and STROOCK, D.W.

(2013) Determinants, their applications to Markov processes, and a random walk proof of Kirchhoff's matrix tree theorem. Preprint, http://www.arxiv.org/abs/1306.2059.

KOZMA, G.

(2011) Percolation on a product of two trees. *Ann. Probab.*, **39**(5), 1864–1895. MR: 2884876

KOZMA, G. and SCHREIBER, E.

(2004) An asymptotic expansion for the discrete harmonic potential. *Electron. J. Probab.*, **9**, paper no. 1, 1–17 (electronic). MR: 2041826

KRIKUN, M.

(2007) Connected allocation to Poisson points in $\mathbb{R}^2$. *Electron. Comm. Probab.*, **12**, 140–145 (electronic). MR: 2318161

KUIPERS, L. and NIEDERREITER, H.

(1974) *Uniform Distribution of Sequences*. Pure and Applied Mathematics. Wiley-Interscience, New York. MR: 0419394

KURATOWSKI, K.

(1966) *Topology. Vol. I*, revised and augmented ed. Academic Press, New York. Translated from the French by J. Jaworowski. MR: 36:840

LALLEY, S.P.

(1998) Percolation on Fuchsian groups. *Ann. Inst. H. Poincaré Probab. Statist.*, **34**(2), 151–177. MR: 1614583

LAMPERTI, J.

(1967) The limit of a sequence of branching processes. *Z. Wahrscheinlichkeitstheorie Verw. Gebiete*, **7**, 271–288. MR: 0217893

LARSEN, K.G. and NELSON, J.

(2016) The Johnson-Lindenstrauss lemma is optimal for linear dimensionality reduction. *43rd International Colloquium on Automata, Languages, and Programming (ICALP 2016)*, pages 82:1–82:11. http://dx.doi.org/10.4230/LIPIcs.ICALP.2016.82.

(2017) Optimality of the Johnson-Lindenstrauss lemma. In *58th Annual IEEE Symposium on Foundations of Computer Science—FOCS 2017*, pages 633–638. IEEE Computer Soc., Los Alamitos, CA. MR: 3734267

LAWLER, G., SUN, X., and WU, W.

(2019) Four-dimensional loop-erased random walk. *Ann. Probab.*, **47**(6), 3866–3910. MR: 4038044

LAWLER, G.F.

(1980) A self-avoiding random walk. *Duke Math. J.*, **47**(3), 655–693. MR: 81j:60081

(1983) A connective constant for loop-erased self-avoiding random walk. *J. Appl. Probab.*, **20**(2), 264–276. MR: 84g:60113

(1986) Gaussian behavior of loop-erased self-avoiding random walk in four dimensions. *Duke Math. J.*, **53**(1), 249–269. MR: 87i:60078

(1988) Loop-erased self-avoiding random walk in two and three dimensions. *J. Statist. Phys.*, **50**(1–2), 91–108. MR: 89f:82053

(1991) *Intersections of Random Walks*. Birkhäuser, Boston. MR: 92f:60122

(2014) The probability that planar loop-erased random walk uses a given edge. *Electron. Commun. Probab.*, **19**, paper no. 51, 13 pp. MR: 3246970

(2020) The infinite two-sided loop-erased random walk. *Electron. J. Probab.*, **25**, paper no. 87, 42 pp.

LAWLER, G.F., SCHRAMM, O., and WERNER, W.
(2004a) Conformal invariance of planar loop-erased random walks and uniform spanning trees. *Ann. Probab.*, **32**(1B), 939–995. MR: 2044 671

(2004b) On the scaling limit of planar self-avoiding walk. In Lapidus, M.L. and van Frankenhuijsen, M., editors, *Fractal Geometry and Applications: A Jubilee of Benoît Mandelbrot. Part 2*, vol. 72 of *Proc. Sympos. Pure Math.*, pages 339–364. Amer. Math. Soc., Providence, RI. Proceedings of a Special Session of the Annual Meeting of the American Mathematical Society held in San Diego, CA, January 2002. MR: 2112127

LAWLER, G.F. and SOKAL, A.D.
(1988) Bounds on the L^2 spectrum for Markov chains and Markov processes: A generalization of Cheeger's inequality. *Trans. Amer. Math. Soc.*, **309**(2), 557–580. MR: 89h:60105

LE GALL, J.F.
(1987) Le comportement du mouvement brownien entre les deux instants où il passe par un point double. *J. Funct. Anal.*, **71**(2), 246–262. MR: 880979

LE GALL, J.F. and LE JAN, Y.
(1998) Branching processes in Lévy processes: The exploration process. *Ann. Probab.*, **26**(1), 213–252. MR: 1617047

LE GALL, J.F. and ROSEN, J.
(1991) The range of stable random walks. *Ann. Probab.*, **19**(2), 650–705. MR: 1106281

LE JAN, Y.
(2011) *Markov Paths, Loops and Fields*. Vol. 2026 of *Lecture Notes in Mathematics*. Springer, Heidelberg. Lectures from the 38th Probability Summer School held in Saint-Flour, 2008, École d'Été de Probabilités de Saint-Flour. [Saint-Flour Probability Summer School]. MR: 2815763

LEDRAPPIER, F.
(1984) Frontière de Poisson pour les groupes discrets de matrices. *C. R. Acad. Sci. Paris Sér. I Math.*, **298**(16), 393–396. MR: 748930

(1985) Poisson boundaries of discrete groups of matrices. *Israel J. Math.*, **50**(4), 319–336. MR: 800190

(1992) Sharp estimates for the entropy. In Picardello, M.A., editor, *Proceedings of the International Meeting held in Frascati, July 1–10, 1991*, pages 281–288. Plenum, New York. MR: 1222466

LEE, J.R. and PERES, Y.
(2013) Harmonic maps on amenable groups and a diffusive lower bound for random walks. *Ann. Probab.*, **41**(5), 3392–3419. MR: 3127886

LEE, J.R., PERES, Y., and SMART, C.K.
(2016) A Gaussian upper bound for martingale small-ball probabilities. *Ann. Probab.*, **44**(6), 4184–4197. MR: 3572334

LEE, J.R. and QIN, T.
(2012) A note on mixing times of planar random walks. Unpublished manuscript, http://www.arxiv.org/abs/1205.3980.

LEHNER, F.
(2019) Firefighting on trees and Cayley graphs. *Australas. J. Combin.*, **75**, 66–72. MR: 3997950

LEVIN, D.A. and PERES, Y.
(2010) Pólya's theorem on random walks via Pólya's urn. *Amer. Math. Monthly*, **117**(3), 220–231. MR: 2640849

(2017) *Markov Chains and Mixing Times*. American Mathematical Society, Providence, RI. Second edition, with contributions by Elizabeth L. Wilmer and a chapter on "Coupling from the past" by James G. Propp and David B. Wilson. MR: 3726904

LEVITT, G.
(1995) On the cost of generating an equivalence relation. *Ergodic Theory Dynam. Systems*, **15**(6), 1173–1181. MR: 96i:58091

LIGGETT, T.M.
(1985) An improved subadditive ergodic theorem. *Ann. Probab.*, **13**(4), 1279–1285. MR: 806224

(1987) Reversible growth models on symmetric sets. In Itō, K. and Ikeda, N., editors, *Probabilistic Methods in Mathematical Physics*, pages 275–301. Academic Press, Boston. Proceedings of the Taniguchi International Symposium held in Katata, June 20–26, 1985, and at Kyoto University, Kyoto, June 27–29, 1985. MR: 933828

LIGGETT, T.M., SCHONMANN, R.H., and STACEY, A.M.
  (1997) Domination by product measures. *Ann. Probab.*, **25**(1), 71–95. MR: 98f:60095

LIN, S.
  (2019) Harmonic measure for biased random walk in a supercritical Galton–Watson tree. *Bernoulli*, **25**(4B), 3652–3672. MR: 4010968

LINDVALL, T. and ROGERS, L.C.G.
  (1996) On coupling of random walks and renewal processes. *J. Appl. Probab.*, **33**(1), 122–126. MR: 1371959

LINIAL, N., LONDON, E., and RABINOVICH, Y.
  (1995) The geometry of graphs and some of its algorithmic applications. *Combinatorica*, **15**(2), 215–245. MR: 1337355

LINIAL, N., MAGEN, A., and NAOR, A.
  (2002) Girth and Euclidean distortion. *Geom. Funct. Anal.*, **12**(2), 380–394. MR: 1911665

LOOMIS, L.H. and WHITNEY, H.
  (1949) An inequality related to the isoperimetric inequality. *Bull. Amer. Math. Soc*, **55**, 961–962. MR: 0031538

LOUDER, L. and SOUTO, J.
  (2012) Diameter and spectral gap for planar graphs. Preprint, http://www.arxiv.org/abs/1204.4435.

LOVÁSZ, L. and KANNAN, R.
  (1999) Faster mixing via average conductance. In *Annual ACM Symposium on Theory of Computing (Atlanta, GA, 1999)*, pages 282–287. ACM, New York. MR: 1798047

LUBETZKY, E. and PERES, Y.
  (2016) Cutoff on all Ramanujan graphs. *Geom. Funct. Anal.*, **26**(4), 1190–1216. MR: 3558308

LUBOTZKY, A., PHILLIPS, R., and SARNAK, P.
  (1988) Ramanujan graphs. *Combinatorica*, **8**(3), 261–277. MR: 963118

LÜCK, W.
  (2009) L^2-invariants from the algebraic point of view. In Bridson, M.R., Kropholler, P.H., and Leary, I.J., editors, *Geometric and Cohomological Methods in Group Theory*, vol. 358 of *London Math. Soc. Lecture Note Ser.*, pages 63–161. Cambridge University Press, Cambridge. Papers from the London Mathematical Society Symposium on Geometry and Cohomology in Group Theory held in Durham, July 2003. MR: 2605176

LYONS, R.
  (1988) Strong laws of large numbers for weakly correlated random variables. *Michigan Math. J.*, **35**(3), 353–359. MR: 90d:60038

  (1989) The Ising model and percolation on trees and tree-like graphs. *Comm. Math. Phys.*, **125**(2), 337–353. MR: 90h:82046

  (1990) Random walks and percolation on trees. *Ann. Probab.*, **18**(3), 931–958. MR: 91i:60179

  (1992) Random walks, capacity and percolation on trees. *Ann. Probab.*, **20**(4), 2043–2088. MR: 93k:60175

  (1995) Random walks and the growth of groups. *C. R. Acad. Sci. Paris Sér. I Math.*, **320**(11), 1361–1366. MR: 96e:60015

  (1996) Diffusions and random shadows in negatively curved manifolds. *J. Funct. Anal.*, **138**(2), 426–448. MR: 97d:58205

  (2000) Phase transitions on nonamenable graphs. *J. Math. Phys.*, **41**(3), 1099–1126. MR: 2001c:82028

  (2003) Determinantal probability measures. *Publ. Math. Inst. Hautes Études Sci.*, **98**(1), 167–212. MR: 2031202

  (2005) Asymptotic enumeration of spanning trees. *Combin. Probab. Comput.*, **14**(4), 491–522. MR: 2160416

  (2009) Random complexes and l^2-Betti numbers. *J. Topol. Anal.*, **1**(2), 153–175. MR: 2541759

  (2010) Identities and inequalities for tree entropy. *Combin. Probab. Comput.*, **19**(2), 303–313. MR: 2593624

  (2013a) Distance covariance in metric spaces. *Ann. Probab.*, **41**(5), 3284–3305. Errata, *Ann. Probab.* **46** (2018), no. 4, 2400–2405. MR: 3127883

  (2013b) Fixed price of groups and percolation. *Ergodic Theory Dynam. Systems*, **33**(1), 183–185. MR: 3009109

Lyons, R., Morris, B.J., and Schramm, O.
  (2008) Ends in uniform spanning forests. *Electron. J. Probab.*, **13**, paper no. 58, 1702–1725. MR: 2448128

Lyons, R. and Nazarov, F.
  (2011) Perfect matchings as IID factors on non-amenable groups. *European J. Combin.*, **32**(7), 1115–1125. MR: 2825538

Lyons, R. and Pemantle, R.
  (1992) Random walk in a random environment and first-passage percolation on trees. *Ann. Probab.*, **20**(1), 125–136. MR: 93c:60103

  (2003) Correction: "Random walk in a random environment and first-passage percolation on trees" [Ann. Probab. **20** (1992) no. 1, 125–136; MR 93c:60103]. *Ann. Probab.*, **31**(1), 528–529. MR: 1959 801

Lyons, R., Pemantle, R., and Peres, Y.
  (1995a) Conceptual proofs of $L \log L$ criteria for mean behavior of branching processes. *Ann. Probab.*, **23**(3), 1125–1138. MR: 96m:60194

  (1995b) Ergodic theory on Galton-Watson trees: Speed of random walk and dimension of harmonic measure. *Ergodic Theory Dynam. Systems*, **15**(3), 593–619. MR: 96e:60125

  (1996a) Biased random walks on Galton-Watson trees. *Probab. Theory Related Fields*, **106**(2), 249–264. MR: 97h:60094

  (1996b) Random walks on the lamplighter group. *Ann. Probab.*, **24**(4), 1993–2006. MR: 97j:60014

  (1997) Unsolved problems concerning random walks on trees. In Athreya, K.B. and Jagers, P., editors, *Classical and Modern Branching Processes*, pages 223–237. Springer, New York. Papers from the IMA Workshop held at the University of Minnesota, Minneapolis, MN, June 13–17, 1994. MR: 98j:60098

Lyons, R. and Peres, Y.
  (2015) Cycle density in infinite Ramanujan graphs. *Ann. Probab.*, **43**(6), 3337–3358. MR: 3433583

  (2021) Poisson boundaries of lamplighter groups: proof of the Kaimanovich–Vershik conjecture. *J. Eur. Math. Soc. (JEMS)*, **23**(4), 1133–1160. MR: 4228277

Lyons, R., Peres, Y., and Schramm, O.
  (2003) Markov chain intersections and the loop-erased walk. *Ann. Inst. H. Poincaré Probab. Statist.*, **39**(5), 779–791. MR: 1997 212

  (2006) Minimal spanning forests. *Ann. Probab.*, **34**(5), 1665–1692. MR: 2271476

Lyons, R., Pichot, M., and Vassout, S.
  (2008) Uniform non-amenability, cost, and the first l^2-Betti number. *Groups Geom. Dyn.*, **2**(4), 595–617. MR: 2442947

Lyons, R. and Schramm, O.
  (1999) Indistinguishability of percolation clusters. *Ann. Probab.*, **27**(4), 1809–1836. MR: 1742 889

Lyons, R. and Steif, J.E.
  (2003) Stationary determinantal processes: Phase multiplicity, Bernoullicity, entropy, and domination. *Duke Math. J.*, **120**(3), 515–575. MR: 2030095

Lyons, R. and Thom, A.
  (2016) Invariant coupling of determinantal measures on sofic groups. *Ergodic Theory Dynam. Systems*, **36**(2), 574–607. MR: 3503036

Lyons, T.
  (1983) A simple criterion for transience of a reversible Markov chain. *Ann. Probab.*, **11**(2), 393–402. MR: 84e:60102

  (1987) Instability of the Liouville property for quasi-isometric Riemannian manifolds and reversible Markov chains. *J. Differential Geom.*, **26**(1), 33–66. MR: 892030

Lyons, T.J. and Zhang, T.S.
  (1994) Decomposition of Dirichlet processes and its application. *Ann. Probab.*, **22**(1), 494–524. MR: 1258888

Lyons, T.J. and Zheng, W.A.
  (1988) A crossing estimate for the canonical process on a Dirichlet space and a tightness result. *Astérisque*, **157–158**, 249–271. Papers from the colloquium held in Palaiseau, June 22–26, 1987. MR: 976222

Maak, W.
  (1935) Eine neue Definition der fastperiodischen Funktionen. *Abh. Math. Semin. Hamb. Univ.*, **11**, 240–244. http://dx.doi.org/10.1007/BF02940727.

MACPHERSON, H.D.
(1982) Infinite distance transitive graphs of finite valency. *Combinatorica*, **2**(1), 63–69. MR: 671146

MADER, W.
(1970) Über den Zusammenhang symmetrischer Graphen. *Arch. Math. (Basel)*, **21**, 331–336. MR: 44:6534

MADRAS, N. and SLADE, G.
(1993) *The Self-Avoiding Walk.* Birkhäuser, Boston. MR: 94f:82002

MAHER, J. and TIOZZO, G.
(2018) Random walks on weakly hyperbolic groups. *J. Reine Angew. Math.*, **742**, 187–239. MR: 3849626

MAIRESSE, J. and MATHÉUS, F.
(2007a) Random walks on free products of cyclic groups. *J. Lond. Math. Soc. (2)*, **75**(1), 47–66. MR: 2302729

(2007b) Randomly growing braid on three strands and the manta ray. *Ann. Appl. Probab.*, **17**(2), 502–536. MR: 2308334

MAJUMDAR, S.N.
(1992) Exact fractal dimension of the loop-erased self-avoiding random walk in two dimensions. *Phys. Rev. Lett.*, **68**, 2329–2331. http://dx.doi.org/10.1103/PhysRevLett.68.2329.

MALYUTIN, A., NAGNIBEDA, T., and SERBIN, D.
(2017) Boundaries of $\mathbb{Z}^n$-free groups. In Ceccherini-Silberstein, T., Salvatori, M., and Sava-Huss, E., editors, *Groups, Graphs and Random Walks*, vol. 436 of *London Math. Soc. Lecture Note Ser.*, pages 355–390. Cambridge Univ. Press, Cambridge. Selected papers from the workshop held in Cortona, June 2–6, 2014. MR: 3644015

MALYUTIN, A.V.
(2003) The Poisson-Furstenberg boundary of a locally free group. *Zap. Nauchn. Sem. S.-Peterburg. Otdel. Mat. Inst. Steklov. (POMI)*, **301**(Teor. Predst. Din. Sist. Komb. i Algoritm. Metody. 9), 195–211, 245. MR: 2032055

MALYUTIN, A.V. and SVETLOV, P.
(2014) Poisson-Furstenberg boundaries of fundamental groups of closed 3-manifolds. Preprint, http://www.arxiv.org/abs/1403.2135.

MANDELBROT, B.B.
(1982) *The Fractal Geometry of Nature.* W. H. Freeman, San Francisco. Schriftenreihe für den Referenten [Series for the Referee]. MR: 665254

MANN, A.
(2012) *How Groups Grow.* Vol. 395 of *London Mathematical Society Lecture Note Series.* Cambridge University Press, Cambridge. MR: 2894945

MARCHAL, P.
(1998) The best bounds in a theorem of Russell Lyons. *Electron. Comm. Probab.*, **3**, paper no. 11, 91–94 (electronic). MR: 1650563

(2000) Loop-erased random walks, spanning trees and Hamiltonian cycles. *Electron. Comm. Probab.*, **5**, 39–50 (electronic). MR: 1736723

MARCKERT, J.F.
(2008) The lineage process in Galton-Watson trees and globally centered discrete snakes. *Ann. Appl. Probab.*, **18**(1), 209–244. MR: 2380897

MARCKERT, J.F. and MOKKADEM, A.
(2003) Ladder variables, internal structure of Galton-Watson trees and finite branching random walks. *J. Appl. Probab.*, **40**(3), 671–689. MR: 1993260

MARGULIS, G.A.
(1988) Explicit group-theoretic constructions of combinatorial schemes and their applications in the construction of expanders and concentrators. *Problemy Peredachi Informatsii*, **24**(1), 51–60. MR: 939574

MARTINEAU, S. and SEVERO, F.
(2019) Strict monotonicity of percolation thresholds under covering maps. *Ann. Probab.*, **47**(6), 4116–4136. MR: 4038050

MARTINEAU, S. and TASSION, V.
(2017) Locality of percolation for Abelian Cayley graphs. *Ann. Probab.*, **45**(2), 1247–1277. MR: 3630298

MASSEY, W.S.
(1991) *A Basic Course in Algebraic Topology.* Springer-Verlag, New York. MR: 92c:55001

MASSON, R.
(2009) The growth exponent for planar loop-erased random walk. *Electron. J. Probab.*, **14**, paper no. 36, 1012–1073. MR: 2506124

MATOUŠEK, J.
(2002) *Lectures on Discrete Geometry.* Vol. 212 of *Graduate Texts in Mathematics.* Springer-Verlag, New York. MR: 1899299

(2008) On variants of the Johnson-Lindenstrauss lemma. *Random Structures Algorithms*, **33**(2), 142–156. MR: 2436844

MATSUZAKI, K. and TANIGUCHI, M.
(1998) *Hyperbolic Manifolds and Kleinian Groups.* Oxford Mathematical Monographs. The Clarendon Press, Oxford University Press, New York. MR: 1638795

MATTHEWS, P.
(1988) Covering problems for Brownian motion on spheres. *Ann. Probab.*, **16**(1), 189–199. MR: 920264

MATTILA, P.
(1995) *Geometry of Sets and Measures in Euclidean Spaces: Fractals and Rectifiability.* Vol. 44 of *Cambridge Studies in Advanced Mathematics.* Cambridge University Press, Cambridge. MR: 1333890

MAULDIN, R.D. and WILLIAMS, S.C.
(1986) Random recursive constructions: Asymptotic geometric and topological properties. *Trans. Amer. Math. Soc.*, **295**(1), 325–346. MR: 87j:60027

McCREA, W.H. and WHIPPLE, F.J.W.
(1940) Random paths in two and three dimensions. *Proc. Roy. Soc. Edinburgh*, **60**, 281–298. MR: 2,107f

McDIARMID, C.
(1989) On the method of bounded differences. In Siemons, J., editor, *Surveys in Combinatorics, 1989*, vol. 141 of *London Math. Soc. Lecture Note Ser.*, pages 148–188. Cambridge University Press, Cambridge. Papers from the Twelfth British Combinatorial Conference held at the University of East Anglia, Norwich, 1989. MR: 1036755

McDONOUGH, T.P. and MAVRON, V.C., EDITORS
(1974) *Combinatorics.* Cambridge University Press, London. London Mathematical Society Lecture Note Series, No. 13. MR: 0345829

McGUINNESS, S.
(1989) *Random Walks on Graphs and Directed Graphs.* Ph.D. thesis, University of Waterloo.

MEDOLLA, G. and SOARDI, P.M.
(1995) Extension of Foster's averaging formula to infinite networks with moderate growth. *Math. Z.*, **219**(2), 171–185. MR: 96g:94031

MEIR, A. and MOON, J.W.
(1970) The distance between points in random trees. *J. Combinatorial Theory*, **8**, 99–103. MR: 0263685

MERKL, F. and ROLLES, S.W.W.
(2009) Recurrence of edge-reinforced random walk on a two-dimensional graph. *Ann. Probab.*, **37**(5), 1679–1714. MR: 2561431

MESTER, P.
(2013) Invariant monotone coupling need not exist. *Ann. Probab.*, **41**(3A), 1180–1190. MR: 3098675

MINC, H.
(1988) *Nonnegative Matrices.* Wiley-Interscience Series in Discrete Mathematics and Optimization. John Wiley, New York. MR: 89i:15001

MOHAR, B.
(1988) Isoperimetric inequalities, growth, and the spectrum of graphs. *Linear Algebra Appl.*, **103**, 119–131. MR: 89k:05071

MOK, N.
(1995) Harmonic forms with values in locally constant Hilbert bundles. *J. Fourier Anal. Appl.*, **Special Issue**, 433–453. Proceedings of the Conference in Honor of Jean-Pierre Kahane (Orsay, 1993). MR: 1364901

MONTROLL, E.
(1964) Lattice statistics. In Beckenbach, E., editor, *Applied Combinatorial Mathematics*, pages 96–143. John Wiley, New York. University of California Engineering and Physical Sciences Extension Series. MR: 30:4687

MOON, J.W.
(1967) Various proofs of Cayley's formula for counting trees. In Harary, F. and Beineke, L., editors, *A Seminar on Graph Theory*, pages 70–78. Holt, Rinehart and Winston, New York. MR: 35:5365

MORI, A.
(1954) A note on unramified abelian covering surfaces of a closed Riemann surface. *J. Math. Soc. Japan*, **6**, 162–176. MR: 0066468

MORRIS, B.
(2003) The components of the wired spanning forest are recurrent. *Probab. Theory Related Fields*, **125**(2), 259–265. MR: 1961 344

MORRIS, B. and PERES, Y.
(2005) Evolving sets, mixing and heat kernel bounds. *Probab. Theory Related Fields*, **133**(2), 245–266. MR: 2198701

MÖRTERS, P. and PERES, Y.
(2010) *Brownian Motion*. Cambridge Series in Statistical and Probabilistic Mathematics. Cambridge University Press, Cambridge. With an appendix by Oded Schramm and Wendelin Werner. MR: 2604525

MUCHNIK, R. and PAK, I.
(2001) Percolation on Grigorchuk groups. *Comm. Algebra*, **29**(2), 661–671. MR: 2002e:82033

MURTY, M.R.
(2003) Ramanujan graphs. *J. Ramanujan Math. Soc.*, **18**(1), 33–52. MR: 1966527

NACHBIN, L.
(1965) *The Haar Integral*. D. Van Nostrand, Princeton, NJ. MR: 0175995

NAGNIBEDA, T. and WOESS, W.
(2002) Random walks on trees with finitely many cone types. *J. Theoret. Probab.*, **15**(2), 383–422. MR: 1898814

NAOR, A.
(2010) L_1 embeddings of the Heisenberg group and fast estimation of graph isoperimetry. In Bhatia, R., Pal, A., Rangarajan, G., Srinivas, V., and Vanninathan, M., editors, *Proceedings of the International Congress of Mathematicians. Volume III*, pages 1549–1575. Hindustan Book Agency, New Delhi. MR: 2827855

NAOR, A. and PERES, Y.
(2008) Embeddings of discrete groups and the speed of random walks. *Int. Math. Res. Not. IMRN*, **2008**, Art. rnn 076, 34. MR: 2439557

NAOR, A., PERES, Y., SCHRAMM, O., and SHEFFIELD, S.
(2006) Markov chains in smooth Banach spaces and Gromov-hyperbolic metric spaces. *Duke Math. J.*, **134**(1), 165–197. MR: 2239346

NASH-WILLIAMS, C.ST.J.A.
(1959) Random walk and electric currents in networks. *Proc. Cambridge Philos. Soc.*, **55**, 181–194. MR: 23:A2239

(1961) Edge-disjoint spanning trees of finite graphs. *J. London Math. Soc.*, **36**, 445–450. MR: 0133253

NEUMANN, M. and SZE, N.S.
(2011) On the inverse mean first passage matrix problem and the inverse M-matrix problem. *Linear Algebra Appl.*, **434**(7), 1620–1630. MR: 2775741

NEVEU, J.
(1986) Arbres et processus de Galton-Watson. *Ann. Inst. H. Poincaré Probab. Statist.*, **22**(2), 199–207. MR: 88a:60150

NEVO, A. and SAGEEV, M.
(2013) The Poisson boundary of CAT(0) cube complex groups. *Groups Geom. Dyn.*, **7**(3), 653–695. MR: 3095714

NEWMAN, C.M.
(1997) *Topics in Disordered Systems*. Birkhäuser, Basel. MR: 99e:82052

NEWMAN, C.M. and SCHULMAN, L.S.

(1981) Infinite clusters in percolation models. *J. Statist. Phys.*, **26**(3), 613–628. MR: 83e:82038

NEWMAN, C.M. and STEIN, D.L.

(1996) Ground-state structure in a highly disordered spin-glass model. *J. Statist. Phys.*, **82**(3–4), 1113–1132. MR: 97a:82054

(2006) Short-range spin glasses: Selected open problems. In Bovier, A., Dunlop, F., van Enter, A., den Hollander, F., and Dalibard, J., editors, *Mathematical Statistical Physics*, pages 273–293. Elsevier, Amsterdam. Papers from the 83rd Session of the Summer School in Physics held in Les Houches, July 4–29, 2005. MR: 2581887

NIENHUIS, B.

(1982) Exact critical point and critical exponents of $O(n)$ models in two dimensions. *Phys. Rev. Lett.*, **49**(15), 1062–1065. http://dx.doi.org/10.1103/PhysRevLett.49.1062.

NILLI, A.

(1991) On the second eigenvalue of a graph. *Discrete Math.*, **91**(2), 207–210. MR: 1124768

(2004) Tight estimates for eigenvalues of regular graphs. *Electron. J. Combin.*, **11**(1), Note 9, 4 pp. (electronic). MR: 2056091

NORRIS, J., PERES, Y., and ZHAI, A.

(2017) Surprise probabilities in Markov chains. *Combin. Probab. Comput.*, **26**(4), 603–627. MR: 3656344

NORTHSHIELD, S.

(1992) Cogrowth of regular graphs. *Proc. Amer. Math. Soc.*, **116**(1), 203–205. MR: 1120509

OH, B.G.

(2020) Sharp isoperimetric inequalities for infinite plane graphs with bounded vertex and face degrees. Preprint, http://www.arxiv.org/abs/2009.04394.

ORNSTEIN, D.S. and WEISS, B.

(1987) Entropy and isomorphism theorems for actions of amenable groups. *J. Analyse Math.*, **48**, 1–141. MR: 88j:28014

ORSTEIN, D.S. and SUCHESTON, L.

(1970) An operator theorem on L_1 convergence to zero with applications to Markov kernels. *Ann. Math. Statist.*, **41**, 1631–1639. MR: 0272057

OZAWA, N.

(2018) A functional analysis proof of Gromov's polynomial growth theorem. *Ann. Sci. Éc. Norm. Supér. (4)*, **51**(3), 549–556. MR: 3831031

PAK, I. and SMIRNOVA-NAGNIBEDA, T.

(2000) On non-uniqueness of percolation on nonamenable Cayley graphs. *C. R. Acad. Sci. Paris Sér. I Math.*, **330**(6), 495–500. MR: 1756 965

PAKES, A.G. and DEKKING, F.M.

(1991) On family trees and subtrees of simple branching processes. *J. Theoret. Probab.*, **4**(2), 353–369. MR: 92f:60145

PAKES, A.G. and KHATTREE, R.

(1992) Length-biasing, characterizations of laws and the moment problem. *Austral. J. Statist.*, **34**(2), 307–322. MR: 94a:60018

PALEY, R.E.A.C. and ZYGMUND, A.

(1932) A note on analytic functions in the unit circle. *Proc. Camb. Phil. Soc.*, **28**, 266–272. http://dx.doi.org/10.1017/S0305004100010112.

PARRONDO, J.

(1996) How to cheat a bad mathematician. *EEC HC&M Network on Complexity and Chaos.* #ERBCHRX-CT940546. ISI, Torino, Italy. Unpublished. Available at http://seneca.fis.ucm.es/parr/GAMES/cheat.pdf.

PATERSON, A.L.T.

(1988) *Amenability.* American Mathematical Society, Providence, RI. MR: 90e:43001

PAUL, A. and PIPPENGER, N.

(2011) A census of vertices by generations in regular tessellations of the plane. *Electron. J. Combin.*, **18**(1), Research Paper 87, 13 pp. MR: 2795768

Peierls, R.
(1936) On Ising's model of ferromagnetism. *Math. Proc. Cambridge Philos. Soc.*, **32**, 477–481. `http://dx.doi.org/10.1017/S0305004100019174`.

Peköz, E.A. and Röllin, A.
(2011) New rates for exponential approximation and the theorems of Rényi and Yaglom. *Ann. Probab.*, **39**(2), 587–608. `MR: 2789507`

Pemantle, R.
(1988) Phase transition in reinforced random walk and RWRE on trees. *Ann. Probab.*, **16**(3), 1229–1241. `MR: 89g:60220`

(1991) Choosing a spanning tree for the integer lattice uniformly. *Ann. Probab.*, **19**(4), 1559–1574. `MR: 92g:60014`

(2000) Towards a theory of negative dependence. *J. Math. Phys.*, **41**(3), 1371–1390. `MR: 2001g:62039`

Pemantle, R. and Peres, Y.
(1995a) Critical random walk in random environment on trees. *Ann. Probab.*, **23**(1), 105–140. `MR: 96f:60123`

(1995b) Galton-Watson trees with the same mean have the same polar sets. *Ann. Probab.*, **23**(3), 1102–1124. `MR: 96i:60093`

(1996) On which graphs are all random walks in random environments transient? In Aldous, D. and Pemantle, R., editors, *Random Discrete Structures*, pages 207–211. Springer, New York. Papers from the workshop held in Minneapolis, Minnesota, November 15–19, 1993. `MR: 97f:60212`

(2000) Nonamenable products are not treeable. *Israel J. Math.*, **118**, 147–155. `MR: 2001j:43002`

(2010) The critical Ising model on trees, concave recursions and nonlinear capacity. *Ann. Probab.*, **38**(1), 184–206. `MR: 2599197`

Pemantle, R. and Stacey, A.M.
(2001) The branching random walk and contact process on Galton-Watson and nonhomogeneous trees. *Ann. Probab.*, **29**(4), 1563–1590. `MR: 1880232`

Pemantle, R. and Steif, J.E.
(1999) Robust phase transitions for Heisenberg and other models on general trees. *Ann. Probab.*, **27**(2), 876–912. `MR: 1698979`

Peres, Y.
(1996) Intersection-equivalence of Brownian paths and certain branching processes. *Comm. Math. Phys.*, **177**(2), 417–434. `MR: 98k:60143`

(1999) Probability on trees: An introductory climb. In Bernard, P., editor, *Lectures on Probability Theory and Statistics*, vol. 1717 of *Lecture Notes in Math.*, pages 193–280. Springer, Berlin. Lectures from the 27th Summer School on Probability Theory held in Saint-Flour, July 7–23, 1997. `MR: 1746302`

(2000) Percolation on nonamenable products at the uniqueness threshold. *Ann. Inst. H. Poincaré Probab. Statist.*, **36**(3), 395–406. `MR: 2001f:60114`

Peres, Y., Pete, G., and Scolnicov, A.
(2006) Critical percolation on certain nonunimodular graphs. *New York J. Math.*, **12**, 1–18 (electronic). `MR: 2217160`

Peres, Y., Schapira, B., and Sousi, P.
(2016) Martingale defocusing and transience of a self-interacting random walk. *Ann. Inst. Henri Poincaré Probab. Stat.*, **52**(3), 1009–1022. `MR: 3531697`

Peres, Y., Stauffer, A., and Steif, J.E.
(2015) Random walks on dynamical percolation: mixing times, mean squared displacement and hitting times. *Probab. Theory Related Fields*, **162**(3–4), 487–530. `MR: 3383336`

Peres, Y. and Steif, J.E.
(1998) The number of infinite clusters in dynamical percolation. *Probab. Theory Related Fields*, **111**(1), 141–165. `MR: 99e:60217`

Peres, Y. and Zeitouni, O.
(2008) A central limit theorem for biased random walks on Galton-Watson trees. *Probab. Theory Related Fields*, **140**(3–4), 595–629. `MR: 2365486`

PERLIN, A.
  (2001) *Probability Theory on Galton-Watson Trees.* Ph.D. thesis, Massachusetts Institute of Technology. Available at http://hdl.handle.net/1721.1/8673.

PESIN, Y.B.
  (1997) *Dimension Theory in Dynamical Systems.* Chicago Lectures in Mathematics. University of Chicago Press, Chicago. MR: 1489237

PETE, G.
  (2008) A note on percolation on $\mathbb{Z}^d$: Isoperimetric profile via exponential cluster repulsion. *Electron. Commun. Probab.*, **13**, 377–392. MR: 2415145

PETERSEN, K.
  (1983) *Ergodic Theory.* Cambridge University Press, Cambridge. MR: 87i:28002

PEYRE, R.
  (2008) A probabilistic approach to Carne's bound. *Potential Anal.*, **29**(1), 17–36. MR: 2421492

PIAU, D.
  (1996) Functional limit theorems for the simple random walk on a supercritical Galton-Watson tree. In Chauvin, B., Cohen, S., and Rouault, A., editors, *Trees*, vol. 40 of *Progr. Probab.*, pages 95–106. Birkhäuser, Basel. Proceedings of the Workshop held in Versailles, June 14–16, 1995. MR: 1439974

  (1998) Théorème central limite fonctionnel pour une marche au hasard en environnement aléatoire. *Ann. Probab.*, **26**(3), 1016–1040. MR: 1634413

  (2002) Scaling exponents of random walks in random sceneries. *Stochastic Process. Appl.*, **100**, 3–25. MR: 1919605

PINSKER, M.S.
  (1973) On the complexity of a concentrator. In *Proceedings of the Seventh International Teletraffic Congress (Stockholm, 1973)*, pages 318/1–318/4.

PITMAN, J.
  (1993) *Probability.* Springer, New York. http://dx.doi.org/10.1007/978-1-4612-4374-8.

  (1998) Enumerations of trees and forests related to branching processes and random walks. In Aldous, D. and Propp, J., editors, *Microsurveys in Discrete Probability*, vol. 41 of *DIMACS Ser. Discrete Math. Theoret. Comput. Sci.*, pages 163–180. Amer. Math. Soc., Providence, RI. Papers from the workshop held as part of the Dimacs Special Year on Discrete Probability in Princeton, NJ, June 2–6, 1997. MR: 1630413

PITT, L.D.
  (1982) Positively correlated normal variables are associated. *Ann. Probab.*, **10**(2), 496–499. MR: 665603

PITTET, C. and SALOFF-COSTE, L.
  (1999) Amenable groups, isoperimetric profiles and random walks. In *Geometric Group Theory Down Under (Canberra, 1996)*, pages 293–316. de Gruyter, Berlin. MR: 1714851

  (2001) A survey on the relationships between volume growth, isoperimetry, and the behavior of simple random walk on Cayley graphs, with examples. In preparation. Preliminary version available at http://www.math.cornell.edu/~lsc/surv.ps.gz.

POGHOSYAN, V.S., PRIEZZHEV, V.B., and RUELLE, P.
  (2011) Return probability for the loop-erased random walk and mean height in the Abelian sandpile model: A proof. *J. Stat. Mech.: Theory Experiment*, **2011**(10), P10004. http://iopscience.iop.org/1742-5468/2011/10/P10004.

PÓLYA, G.
  (1921) Über eine Aufgabe der Wahrscheinlichkeitsrechnung betreffend die Irrfahrt im Straßennetz. *Math. Ann.*, **84**(1–2), 149–160. MR: 1512028

PONZIO, S.
  (1998) The combinatorics of effective resistances and resistive inverses. *Inform. and Comput.*, **147**(2), 209–223. MR: 1662276

PORT, S.C. and STONE, C.J.
  (1978) *Brownian Motion and Classical Potential Theory.* Probability and Mathematical Statistics. Academic Press [Harcourt Brace Jovanovich Publishers], New York. MR: 0492329

Propp, J.G. and Wilson, D.B.
(1998) How to get a perfectly random sample from a generic Markov chain and generate a random spanning tree of a directed graph. *J. Algorithms*, **27**(2), 170–217. 7th Annual ACM-SIAM Symposium on Discrete Algorithms (Atlanta, GA, 1996). MR: 99g:60116

Puder, D.
(2015) Expansion of random graphs: New proofs, new results. *Invent. Math.*, **201**(3), 845–908. MR: 3385636

Pyke, R.
(2003) On random walks and diffusions related to Parrondo's games. In Moore, M., Froda, S., and Léger, C., editors, *Mathematical Statistics and Applications: Festschrift for Constance van Eeden*, vol. 42 of *IMS Lecture Notes Monogr. Ser.*, pages 185–216. Inst. Math. Statist., Beachwood, OH. MR: 2138293

Raoufi, A. and Yadin, A.
(2017) Indicable groups and $p_c < 1$. *Electron. Commun. Probab.*, **22**, Paper No. 13, 10 pp. MR: 3607808

Ratcliffe, J.G.
(2006) *Foundations of Hyperbolic Manifolds*, 2nd ed. Vol. 149 of *Graduate Texts in Mathematics*. Springer, New York. MR: 2249478

Raugi, A.
(2004) A general Choquet-Deny theorem for nilpotent groups. *Ann. Inst. H. Poincaré Probab. Statist.*, **40**(6), 677–683. MR: 2096214

Reimer, D.
(2000) Proof of the van den Berg-Kesten conjecture. *Combin. Probab. Comput.*, **9**(1), 27–32. MR: 1751301

Revelle, D.
(2001) Biased random walk on lamplighter groups and graphs. *J. Theoret. Probab.*, **14**(2), 379–391. MR: 1838 734

Révész, P.
(2005) *Random Walk in Random and Non-Random Environments*, 2nd ed. World Scientific, Hackensack, NJ. MR: 2168855

Rider, B. and Virág, B.
(2007) The noise in the circular law and the Gaussian free field. *Int. Math. Res. Not. IMRN*, **2007**, Art. rnm006, 33 pp. MR: 2361453

Rokhlin, V.A.
(1949) On the fundamental ideas of measure theory (Russian). *Mat. Sbornik N.S.*, **25(67)**, 107–150. English translation: *Amer. Math. Soc. Translation* **1952** (1952), no. 71, 54 pp. MR: 0030584

Rosenblatt, J.
(1981) Ergodic and mixing random walks on locally compact groups. *Math. Ann.*, **257**(1), 31–42. MR: 630645

Rosenblatt, M.
(1971) *Markov Processes. Structure and Asymptotic Behavior*. Vol. 184 of *Die Grundlehren der mathematischen Wissenschaften*. Springer-Verlag, New York. MR: 48:7379

Ross, S.M.
(1996) *Stochastic Processes*, 2nd ed. John Wiley, New York. MR: 97a:60002

Ross, S.M. and Peköz, E.A.
(2007) *A Second Course in Probability*. ProbabilityBookstore.com, Boston.

Rousselin, P.
(2018) Invariant measures, Hausdorff dimension and dimension drop of some harmonic measures on Galton-Watson trees. *Electron. J. Probab.*, **23**, Paper No. 46, 31 pp. MR: 3814240

Royden, H.L.
(1952) Harmonic functions on open Riemann surfaces. *Trans. Amer. Math. Soc.*, **73**, 40–94. MR: 0049396

(1988) *Real Analysis*, 3rd ed. Macmillan, New York. MR: 90g:00004

Rudin, W.
(1987) *Real and Complex Analysis*, 3rd ed. McGraw-Hill, New York. MR: 88k:00002

(1991) *Functional Analysis*, 2nd ed. International Series in Pure and Applied Mathematics. McGraw-Hill, New York. MR: 92k:46001

Sabidussi, G.
(1964) Vertex-transitive graphs. *Monatsh. Math.*, **68**, 426–438. MR: 0175815

SABOT, C.
  (2011) Random walks in random Dirichlet environment are transient in dimension $d \geq 3$. *Probab. Theory Related Fields*, **151**(1–2), 297–317. MR: 2834720

SABOT, C. and TARRÈS, P.
  (2015) Edge-reinforced random walk, vertex-reinforced jump process and the supersymmetric hyperbolic sigma model. *J. Eur. Math. Soc. (JEMS)*, **17**(9), 2353–2378. MR: 3420510

SABOT, C., TARRÈS, P., and ZENG, X.
  (2017) The vertex reinforced jump process and a random Schrödinger operator on finite graphs. *Ann. Probab.*, **45**(6A), 3967–3986. MR: 3729620

SABOT, C. and ZENG, X.
  (2019) A random Schrödinger operator associated with the vertex reinforced jump process on infinite graphs. *J. Amer. Math. Soc.*, **32**(2), 311–349. MR: 3904155

SALOFF-COSTE, L.
  (1995) Isoperimetric inequalities and decay of iterated kernels for almost-transitive Markov chains. *Combin. Probab. Comput.*, **4**(4), 419–442. MR: 1377559

SALVATORI, M.
  (1992) On the norms of group-invariant transition operators on graphs. *J. Theoret. Probab.*, **5**(3), 563–576. MR: 93h:60113

SAVA, E.
  (2010a) *Lamplighter Random Walks and Entropy-Sensitivity of Languages.* Ph.D. thesis, Technische Universität Graz. Available at http://www.arxiv.org/abs/1012.2757.

  (2010b) A note on the Poisson boundary of lamplighter random walks. *Monatsh. Math.*, **159**(4), 379–396. MR: 2600904

SAWYER, S. and STEGER, T.
  (1987) The rate of escape for anisotropic random walks in a tree. *Probab. Theory Related Fields*, **76**(2), 207–230. MR: 906775

SAWYER, S.A.
  (1997) Martin boundaries and random walks. In Korányi, A., editor, *Harmonic Functions on Trees and Buildings*, vol. 206 of *Contemp. Math.*, pages 17–44. Amer. Math. Soc., Providence, RI. Proceedings of the Workshop on Harmonic Functions on Graphs held at City University of New York, New York, October 30–November 3, 1995. MR: 1463727

SCHLICHTING, G.
  (1979) Polynomidentitäten und Permutationsdarstellungen lokalkompakter Gruppen. *Invent. Math.*, **55**(2), 97–106. MR: 81d:22006

SCHOENBERG, I.J.
  (1937) On certain metric spaces arising from Euclidean spaces by a change of metric and their imbedding in Hilbert space. *Ann. of Math. (2)*, **38**(4), 787–793. MR: 1503370

  (1938) Metric spaces and positive definite functions. *Trans. Amer. Math. Soc.*, **44**(3), 522–536. MR: 1501980

SCHONMANN, R.H.
  (1992) On the behavior of some cellular automata related to bootstrap percolation. *Ann. Probab.*, **20**(1), 174–193. MR: 1143417

  (1999a) Percolation in $\infty + 1$ dimensions at the uniqueness threshold. In Bramson, M. and Durrett, R., editors, *Perplexing Problems in Probability*, pages 53–67. Birkhäuser, Boston. Festschrift in honor of Harry Kesten. MR: 1703 124

  (1999b) Stability of infinite clusters in supercritical percolation. *Probab. Theory Related Fields*, **113**(2), 287–300. MR: 1676831

  (2001) Multiplicity of phase transitions and mean-field criticality on highly non-amenable graphs. *Comm. Math. Phys.*, **219**(2), 271–322. MR: 2002h:82036

SCHRAMM, O.
  (2000) Scaling limits of loop-erased random walks and uniform spanning trees. *Israel J. Math.*, **118**, 221–288. MR: 1776 084

SCHRIJVER, A.
  (2003) *Combinatorial Optimization. Polyhedra and Efficiency. Vol. B.* Vol. 24 of *Algorithms and Combinatorics.* Springer-Verlag, Berlin. Matroids, trees, stable sets, Chapters 39–69. MR: 1956925

SENETA, E.

(1968) On recent theorems concerning the supercritical Galton-Watson process. *Ann. Math. Statist.*, **39**, 2098–2102. MR: 38:2847

(1970) On the supercritical Galton-Watson process with immigration. *Math. Biosci.*, **7**, 9–14. MR: 42:5348

SHEFFIELD, S.

(2007) Gaussian free fields for mathematicians. *Probab. Theory Related Fields*, **139**(3–4), 521–541. MR: 2322706

SHEPP, L.A.

(1972) Covering the circle with random arcs. *Israel J. Math.*, **11**, 328–345. MR: 45:4468

SHIELDS, P.C.

(1987) The ergodic and entropy theorems revisited. *IEEE Trans. Inform. Theory*, **33**(2), 263–266. MR: 880168

SHIRAI, T. and TAKAHASHI, Y.

(2003a) Random point fields associated with certain Fredholm determinants. I. Fermion, Poisson and boson point processes. *J. Funct. Anal.*, **205**(2), 414–463. MR: 2018 415

(2003b) Random point fields associated with certain Fredholm determinants. II. Fermion shifts and their ergodic and Gibbs properties. *Ann. Probab.*, **31**(3), 1533–1564. MR: 1989442

SHROCK, R. and WU, F.Y.

(2000) Spanning trees on graphs and lattices in *d* dimensions. *J. Phys. A*, **33**(21), 3881–3902. MR: 2001b:05111

SINGH, M. and VISHNOI, N.K.

(2014) Entropy, optimization and counting. In *Proceedings of the 46th Annual ACM Symposium on Theory of Computing*, STOC '14, pages 50–59. ACM, New York. http://dx.doi.org/10.1145/2591796.2591803.

SLADE, G.

(2011) The self-avoiding walk: A brief survey. In *Surveys in Stochastic Processes*, EMS Ser. Congr. Rep., pages 181–199. Eur. Math. Soc., Zürich. MR: 2883859

SMIRNOV, S.

(2010) Conformal invariance in random cluster models. I. Holomorphic fermions in the Ising model. *Ann. of Math. (2)*, **172**(2), 1435–1467. MR: 2680496

SOARDI, P.M.

(1993) Rough isometries and Dirichlet finite harmonic functions on graphs. *Proc. Amer. Math. Soc.*, **119**(4), 1239–1248. MR: 94a:31004

SOARDI, P.M. and WOESS, W.

(1990) Amenability, unimodularity, and the spectral radius of random walks on infinite graphs. *Math. Z.*, **205**(3), 471–486. MR: 91m:43002

SOLOMON, F.

(1975) Random walks in a random environment. *Ann. Probab.*, **3**, 1–31. MR: 50:14943

ŠPAKULOVÁ, I.

(2009) Critical percolation of virtually free groups and other tree-like graphs. *Ann. Probab.*, **37**(6), 2262–2296. MR: 2573558

SPIELMAN, D.A. and TENG, S.H.

(2007) Spectral partitioning works: Planar graphs and finite element meshes. *Linear Algebra Appl.*, **421**(2–3), 284–305. MR: 2294342

SPITZER, F.

(1956) A combinatorial lemma and its application to probability theory. *Trans. Amer. Math. Soc.*, **82**, 323–339. MR: 0079851

(1962) Hitting probabilities. *J. Math. Mech.*, **11**, 593–614. MR: 0139219

(1964) Electrostatic capacity, heat flow, and Brownian motion. *Z. Wahrscheinlichkeitstheorie Verw. Gebiete*, **3**, 110–121. MR: 0172343

(1976) *Principles of Random Walk*, 2nd ed. Vol. 34 of *Graduate Texts in Mathematics*. Springer-Verlag, New York. MR: 52:9383

STACEY, A.M.

(1996) The existence of an intermediate phase for the contact process on trees. *Ann. Probab.*, **24**(4), 1711–1726. MR: 97j:60191

STARR, N.
  (1966) Operator limit theorems. *Trans. Amer. Math. Soc.*, **121**, 90–115. MR: 0190757

STEELE, J.M.
  (1997) *Probability Theory and Combinatorial Optimization*. Vol. 69 of *CBMS-NSF Regional Conference Series in Applied Mathematics*. Society for Industrial and Applied Mathematics (SIAM), Philadelphia. MR: 1422018

STRASSEN, V.
  (1965) The existence of probability measures with given marginals. *Ann. Math. Statist.*, **36**, 423–439. MR: 31:1693

SWART, J.M.
  (2009) The contact process seen from a typical infected site. *J. Theoret. Probab.*, **22**(3), 711–740. MR: 2530110

SZÉKELY, G.J. and RIZZO, M.L.
  (2005a) Hierarchical clustering via joint between-within distances: Extending Ward's minimum variance method. *J. Classification*, **22**(2), 151–183. MR: 2231170

  (2005b) A new test for multivariate normality. *J. Multivariate Anal.*, **93**(1), 58–80. MR: 2119764

SZÉKELY, G.J., RIZZO, M.L., and BAKIROV, N.K.
  (2007) Measuring and testing dependence by correlation of distances. *Ann. Statist.*, **35**(6), 2769–2794. MR: 2382665

SZNITMAN, A.S.
  (1998) *Brownian Motion, Obstacles and Random Media*. Springer Monographs in Mathematics. Springer-Verlag, Berlin. MR: 1717054

  (2004) Topics in random walks in random environment. In Lawler, G.F., editor, *School and Conference on Probability Theory*, ICTP Lect. Notes, XVII, pages 203–266 (electronic). Abdus Salam Int. Cent. Theoret. Phys., Trieste. Expanded lecture notes from the school and conference held in Trieste, May 2002. MR: 2198849

TAKACS, C.
  (1997) Random walk on periodic trees. *Electron. J. Probab.*, **2**, paper no. 1, 1–16 (electronic). MR: 1436761

  (1998) Biased random walks on directed trees. *Probab. Theory Related Fields*, **111**(1), 123–139. MR: 1626778

TANG, P.
  (2019) Heavy Bernoulli-percolation clusters are indistinguishable. *Ann. Probab.*, **47**(6), 4077–4115. MR: 4038049

TANNY, D.
  (1988) A necessary and sufficient condition for a branching process in a random environment to grow like the product of its means. *Stochastic Process. Appl.*, **28**(1), 123–139. MR: 90e:60105

TANUSHEV, M. and ARRATIA, R.
  (1997) A note on distributional equality in the cyclic tour property for Markov chains. *Combin. Probab. Comput.*, **6**(4), 493–496. MR: 1483432

TAYLOR, A.E. and LAY, D.C.
  (1980) *Introduction to Functional Analysis*, 2nd ed. John Wiley, New York. MR: 564653

TAYLOR, S.J.
  (1955) The α-dimensional measure of the graph and set of zeros of a Brownian path. *Proc. Cambridge Philos. Soc.*, **51**, 265–274. MR: 17,595b

TEMPERLEY, H.N.V.
  (1974) Enumeration of graphs on a large periodic lattice. In McDonough, T.P. and Mavron, V.C., editors, *Combinatorics (Proc. British Combinatorial Conf., Univ. Coll. Wales, Aberystwyth, 1973)*, vol. 13 of *London Math. Soc. Lecture Note Ser.*, pages 155–159. Cambridge University Press, London. MR: 0347616

TETALI, P.
  (1991) Random walks and the effective resistance of networks. *J. Theoret. Probab.*, **4**(1), 101–109. MR: 92c:60097

  (1994a) Design of on-line algorithms using hitting times. In *Proceedings of the Fifth Annual ACM-SIAM Symposium on Discrete Algorithms*, pages 402–411. ACM, New York. Held in Arlington, Virginia, January 23–25, 1994. MR: 1285184

(1994b) An extension of Foster's network theorem. *Combin. Probab. Comput.*, **3**(3), 421–427. MR: 1300977

THOM, A.

(2015) A remark about the spectral radius. *Int. Math. Res. Not. IMRN*, **2015**(10), 2856–2864. MR: 3352259

(2016) The expected degree of minimal spanning forests. *Combinatorica*, **36**(5), 591–600. MR: 3572426

THOMASSEN, C.

(1989) Transient random walks, harmonic functions, and electrical currents in infinite electrical networks. Technical Report Mat-Report n. 1989-07, Technical University of Denmark.

(1990) Resistances and currents in infinite electrical networks. *J. Combin. Theory Ser. B*, **49**(1), 87–102. MR: 91d:94029

(1992) Isoperimetric inequalities and transient random walks on graphs. *Ann. Probab.*, **20**(3), 1592–1600. MR: 1175279

THORISSON, H.

(2000) *Coupling, Stationarity, and Regeneration*. Probability and Its Applications. Springer-Verlag, New York. MR: 1741181

TIMÁR, Á.

(2006a) Ends in free minimal spanning forests. *Ann. Probab.*, **34**(3), 865–869. MR: 2243871

(2006b) Neighboring clusters in Bernoulli percolation. *Ann. Probab.*, **34**(6), 2332–2343. MR: 2294984

(2006c) Percolation on nonunimodular transitive graphs. *Ann. Probab.*, **34**(6), 2344–2364. MR: 2294985

(2007) Cutsets in infinite graphs. *Combin. Probab. Comput.*, **16**(1), 159–166. MR: 2286517

(2018) Indistinguishability of the components of random spanning forests. *Ann. Probab.*, **46**(4), 2221–2242. MR: 3813990

TJUR, T.

(1991) Block designs and electrical networks. *Ann. Statist.*, **19**(2), 1010–1027. MR: 1105858

TONGRING, N.

(1988) Which sets contain multiple points of Brownian motion? *Math. Proc. Cambridge Philos. Soc.*, **103**(1), 181–187. MR: 913461

TRICOT, C., JR.

(1982) Two definitions of fractional dimension. *Math. Proc. Cambridge Philos. Soc.*, **91**(1), 57–74. MR: 633256

TROFIMOV, V.I.

(1984) Graphs with polynomial growth. *Mat. Sb. (N.S.)*, **123(165)**(3), 407–421. English translation: *Math. USSR-Sb.* **51** (1985), no. 2, 405–417. MR: 735714

(1985) Groups of automorphisms of graphs as topological groups. *Mat. Zametki*, **38**(3), 378–385, 476. English translation: *Math. Notes* **38** (1985), no. 3-4, 717–720. MR: 87d:05091

TRUEMPER, K.

(1989) On the delta-wye reduction for planar graphs. *J. Graph Theory*, **13**(2), 141–148. MR: 90c:05078

TUTTE, W.T.

(1961) On the problem of decomposing a graph into n connected factors. *J. London Math. Soc.*, **36**, 221–230. MR: 0140438

UCHIYAMA, K.

(1998) Green's functions for random walks on $\mathbf{Z}^N$. *Proc. London Math. Soc. (3)*, **77**(1), 215–240. MR: 1625467

VAROPOULOS, N.TH.

(1985a) Isoperimetric inequalities and Markov chains. *J. Funct. Anal.*, **63**(2), 215–239. MR: 87e:60124

(1985b) Long range estimates for Markov chains. *Bull. Sci. Math. (2)*, **109**(3), 225–252. MR: 87j:60100

(1986) Théorie du potentiel sur des groupes et des variétés. *C. R. Acad. Sci. Paris Sér. I Math.*, **302**(6), 203–205. MR: 832044

VATUTIN, V.A. and ZUBKOV, A.M.

(1985) Branching processes. I. In *Teoriya Veroyatnostei. Matematicheskaya Statistika. Teoreticheskaya Kibernetika. Tom 23*, Itogi Nauki i Tekhniki, pages 3–67, 154. Akad. Nauk SSSR Vsesoyuz. Inst. Nauchn. i Tekhn. Inform., Moscow. English translation in *J. Soviet Math.* **39**, no. 1, pp. 2431–2475. MR: 810404

(1993) Branching processes. II. *J. Soviet Math.*, **67**(6), 3407–3485. MR: 1260986

VERSHIK, A.M. and KAIMANOVICH, V.A.
  (1979) Random walks on groups: Boundary, entropy, uniform distribution. *Dokl. Akad. Nauk SSSR*, **249**(1), 15–18. MR: 553972

VIRÁG, B.
  (2000a) Anchored expansion and random walk. *Geom. Funct. Anal.*, **10**(6), 1588–1605. MR: 1810 755

  (2000b) On the speed of random walks on graphs. *Ann. Probab.*, **28**(1), 379–394. MR: 2001g:60173

  (2002) Fast graphs for the random walker. *Probab. Theory Related Fields*, **124**(1), 50–72. MR: 1929811

VOLKOV, S.
  (2001) Vertex-reinforced random walk on arbitrary graphs. *Ann. Probab.*, **29**(1), 66–91. MR: 1825142

WALTERS, P.
  (1982) *An Introduction to Ergodic Theory*. Springer-Verlag, New York. MR: 84e:28017

WATKINS, M.E.
  (1970) Connectivity of transitive graphs. *J. Combinatorial Theory*, **8**, 23–29. MR: 0266804

  (1986) Infinite paths that contain only shortest paths. *J. Combin. Theory Ser. B*, **41**(3), 341–355. MR: 864581

WAYMIRE, E.C. and WILLIAMS, S.C.
  (1996) A cascade decomposition theory with applications to Markov and exchangeable cascades. *Trans. Amer. Math. Soc.*, **348**(2), 585–632. MR: 1322959

WHITMAN, W.W.
  (1964) *Some Strong Laws for Random Walks and Brownian Motion*. Ph.D. thesis, Cornell University. MR: 2614450

WHITNEY, H.
  (1932) Congruent graphs and the connectivity of graphs. *Amer. J. Math.*, **54**(1), 150–168. MR: 1506881

WHYTE, K.
  (1999) Amenability, bi-Lipschitz equivalence, and the von Neumann conjecture. *Duke Math. J.*, **99**(1), 93–112. MR: 1700742

WIELAND, B. and WILSON, D.B.
  (2003) Winding angle variance of Fortuin-Kasteleyn contours. *Phys. Rev. E*, **68**, 056101. http://dx.doi.org/10.1103/PhysRevE.68.056101.

WILKINSON, D. and WILLEMSEN, J.F.
  (1983) Invasion percolation: A new form of percolation theory. *J. Phys. A*, **16**(14), 3365–3376. MR: 725616

WILSON, D.B.
  (1996) Generating random spanning trees more quickly than the cover time. In *Proceedings of the Twenty-Eighth Annual ACM Symposium on the Theory of Computing*, pages 296–303. ACM, New York. Held in Philadelphia, PA, May 22–24, 1996. MR: 1427525

  (1997) Determinant algorithms for random planar structures. In *Proceedings of the Eighth Annual ACM-SIAM Symposium on Discrete Algorithms (New Orleans, LA, 1997)*, pages 258–267. ACM, New York. Held in New Orleans, LA, January 5–7, 1997. MR: 1447 672

  (2004a) Red-green-blue model. *Phys. Rev. E (3)*, **69**, 037105. http://dx.doi.org/10.1103/PhysRevE.69.037105.

WILSON, J.S.
  (2004b) On exponential growth and uniformly exponential growth for groups. *Invent. Math.*, **155**(2), 287–303. MR: 2031429

  (2004c) Further groups that do not have uniformly exponential growth. *J. Algebra*, **279**(1), 292–301. MR: 2078400

WINKLER, R.L.
  (1969) Scoring rules and the evaluation of probability assessors. *J. Amer. Stat. Assoc.*, **64**(327), 1073–1078. http://dx.doi.org/10.1007/BF02562681.

WOESS, W.
  (1991) Topological groups and infinite graphs. *Discrete Math.*, **95**(1–3), 373–384. MR: 93i:22004

  (2000) *Random Walks on Infinite Graphs and Groups*. Vol. 138 of *Cambridge Tracts in Mathematics*. Cambridge University Press, Cambridge. MR: 2001k:60006

WOLF, J.A.
  (1968) Growth of finitely generated solvable groups and curvature of Riemanniann manifolds. *J. Differential Geometry*, **2**, 421–446. MR: 0248688

YAGLOM, A.M.
  (1947) Certain limit theorems of the theory of branching random processes. *Doklady Akad. Nauk SSSR (N.S.)*, **56**, 795–798. MR: 9,149e

YAMAMOTO, K.
  (2017) An upper bound for the critical probability on the Cartesian product graph of a regular tree and a line. Preprint, http://www.arxiv.org/abs/1705.06873.

ZEITOUNI, O.
  (2004) Random walks in random environment. In *Lectures on Probability Theory and Statistics*, vol. 1837 of *Lecture Notes in Math.*, pages 189–312. Springer, Berlin. Lectures from the 31st Summer School on Probability Theory held in Saint-Flour, July 8–25, 2001, edited by Jean Picard. MR: 2071631

ZEMANIAN, A.H.
  (1976) Infinite electrical networks. *Proc. IEEE*, **64**(1), 6–17. Recent trends in system theory. MR: 0453371

ZUBKOV, A.M.
  (1975) Limit distributions of the distance to the nearest common ancestor. *Teor. Verojatnost. i Primenen.*, **20**(3), 614–623. English translation: *Theor. Probab. Appl.* **20**, 602–612. MR: 53:1770

ZUCKER, I.J.
  (2011) 70+ years of the Watson integrals. *J. Stat. Phys.*, **145**(3), 591–612. MR: 2862945

ŻUK, A.
  (1996) La propriété (T) de Kazhdan pour les groupes agissant sur les polyèdres. *C. R. Acad. Sci. Paris Sér. I Math.*, **323**(5), 453–458. MR: 1408975

Glossary of Notation

L .... offspring random variable, 134
W .... limit of martingale Z_n/m^n, 9
Z_n .... size of nth generation, 9
$\mathscr{L}$ .... offspring network random variable, 164
$\widehat{L}$ .... size-biased random variable, 414
$\overline{q}$ .... $1 - q$, 156
$f(s)$ .... offspring p.g.f., 134
m .... mean number of offspring, 8
p_k .... probability of k children in a branching process, 8
q .... probability of extinction, 134

$\mathbf{E}[X\,;\,A]$ .... expectation of X on A, xiv
$\mathcal{L}_d$ .... d-dimensional Lebesgue measure, 516
$\langle\cdots\rangle$ .... sequence, xiv
$\asymp$ .... equal up to bounded factors, 180, 545
$\upharpoonright$ .... restriction, xiv
$|\bullet|$ .... cardinality, xiv

G/e .... contraction of e in G, 2, 36, 107
$G\backslash e$ .... deletion of e in G, 107
$\deg x$ .... degree of x, 20
$\mathsf{E}(G)$ .... edge set of G, 1
$G_1 \,\square\, G_2$ .... Cartesian product graph, 1, 228
$\mathsf{E}_{1/2}$ .... one oriented edge for each unoriented edge, 32
$\mathsf{V}(G)$ .... vertex set of G, 1
G^{W} .... G with its boundary wired, 26
$\mathrm{dist}(x, y)$ .... graph distance between x and y, 2
$d(x, y)$ .... graph distance between x and y, 2
e^+ .... head of e, 2
e^- .... tail of e, 2
$\langle x, y \rangle$ .... oriented edge with endpoints x and y, 1
$-e$ .... reverse of e, 2
$[x, y]$ .... unoriented edge with endpoints x and y, 1
$x \sim y$ .... x and y are adjacent, 1
G/K .... G with K identified, 2
$\partial_{\mathsf{E}} K$ .... edge boundary of K, 112, 177
$\partial_{\mathsf{V}} K$ .... outer vertex boundary of K, 180
$\partial_{\mathsf{V}}^{\mathrm{in}} K$ .... inner vertex boundary of K, 207

Index

References to definitions have page numbers in roman font, whereas others (such as to theorems, examples, exercises) are in *slant* font. Of course, some results appear on the same page as the definition.

Ich bin dein Baum

by Friedrich Rückert

Ich bin dein Baum: o Gärtner, dessen Treue
Mich hält in Liebespfleg' und süßer Zucht,
Komm, daß ich in den Schoß dir dankbar streue
Die reife dir allein gewachs'ne Frucht.

Ich bin dein Gärtner, o du Baum der Treue!
Auf and'res Glück fühl' ich nicht Eifersucht:
Die holden Äste find' ich stets aufs neue
Geschmückt mit Frucht, wo ich gepflückt die Frucht.

I am your tree: O gardener, whose fidelity
Keeps me in loving care and sweet nurture,
Come, that in your lap I may gratefully strew
The ripened fruit, grown only for you.

I am your gardener, O you faithful tree!
I am not envious of any other happiness:
Your graceful boughs I find all newly
Adorned with fruit, even where I have plucked the fruit.

Transl. by Jonathan C. Lee and Russell D. Lyons
Set to music by Robert Schumann